Splitting Methods in Communication, Imaging, Science, and Engineering

Scientific Computation

More information about this series at http://www.springer.com/series/718

Roland Glowinski • Stanley J. Osher • Wotao Yin
Editors

Splitting Methods in Communication, Imaging, Science, and Engineering

Springer

Editors

Roland Glowinski
Department of Mathematics
University of Houston
Houston, TX, USA

Stanley J. Osher
Department of Mathematics
UCLA
Los Angeles, CA, USA

Wotao Yin
Department of Mathematics
UCLA
Los Angeles, CA, USA

ISSN 1434-8322 ISSN 2198-2589 (electronic)
Scientific Computation
ISBN 978-3-319-41587-1 ISBN 978-3-319-41589-5 (eBook)
DOI 10.1007/978-3-319-41589-5

Library of Congress Control Number: 2016951957

Mathematics Subject Classification (2010): 49-02, 65-06, 90-06, 68U10, 47N10

Printed on acid-free paper

This Springer imprint is published by Springer Nature
The registered company is Springer International Publishing AG
The registered company address is: Gewerbestrasse 11, 6330 Cham, Switzerland

Preface

Operator-splitting methods have been around for more than a century, starting with their common ancestor, the Lie scheme, introduced by Sophus Lie in the mid-1870s. It seems however that one had to wait after WW2 to see these methods joining the computational and applied mathematics mainstream, the driving force being their applicability to the solution of challenging problems from science and engineering modeled by partial differential equations. The main actors of this renewed interest in operator-splitting methods were mainly Douglas, Peaceman, Rachford, and Wachpress in the USA with the alternating direction implicit (ADI) methods and Dyakonov, Marchuk, and Yanenko in the USSR with the fractional step methods. These basic methodologies have known many variants and improvements and generated a quite important literature consisting of many articles and few books, of theoretical and applied natures, with computational mechanics and physics being the main sources of applications. In the mid-1970s, tight relationships between the augmented Lagrangian methods of Hestenes and Powell and ADI methods were identified, leading to the alternating direction methods of multipliers (ADMM). Albeit originating from problems from continuum mechanics modeled by partial differential equations and inequalities, it was quickly realized that the ADMM methodology applies to problems outside the realm of partial differential equations and inequalities, in information sciences in particular, an area where ADMM has enjoyed a very fast-growing popularity. The main reason of this popularity is that most often large-scale optimization problems have decomposition properties that ADMM can take advantage of, leading to modular algorithms, well suited to parallelization. Another factor explaining ADMM's growing popularity during the last decade was the discovery around 2007 of its many commonalities with the split-Bregman algorithm widely used first in image processing and then in compressed sensing, among other applications.

Late 2012, the three editors of this book were participating in a conference in Hong Kong, the main conference topics being scientific computing, image processing, and optimization. Since most lectures at the conference had some relations with operator splitting, ADMM, and split-Bregman algorithms, the idea of a book dedicated to these topics was explored, considering the following facts:

(i) The practitioners of the above methods have become quite specialized, forming subcommunities with very few interactions between them. (ii) New applications of operator-splitting and related algorithms appear on an almost daily basis. (iii) The diversification of the algorithms and their applications has become so large that a volume containing the contributions of a relatively large number of experts is necessary in order to interest a large audience; indeed, the last review publications on the above topics being quite specialized (as shown in Chapter 1), the editors did their very best to produce a large spectrum volume.

Following a Springer agreement to publish a book on operator splitting, ADMM, split-Bregman, and related algorithms, covering both theory and applications, experts were approached to contribute to this volume. We are pleased to say that most of them enthusiastically agreed to be part of the project.

This book is divided in chapters covering the history, foundations, applications, as well as recent developments of operator splitting, ADMM, split Bregman, and related algorithms. Due to size and time constraints, many relevant information could not be included in the book: the editors apologize to those authors whose contributions have been partially or totally overlooked.

Many thanks are in order:

- First, to the organizers of the December 2012 Hong Kong conference on *Advances in Scientific Computing, Imaging Sciences and Optimization*. Indeed, the inception of this project took place during this meeting.
- To Springer for accepting to publish this volume. The editors acknowledge in particular the assistance provided by Achi Dosanjh; she was involved with the project from day one and never lost her faith in it (and in the editors), despite the many (unavoidable) delays encountered during its completion.
- To the authors of the various chapters and to those colleagues who accepted to review them. They are really the ones who brought this book into existence.
- To Hengda Wen and Tsorng-Whay Pan for their assistance on many issues associated with the preparation of the "manuscript" (some of them LaTeXrelated). Both of them saved the day more than once.
- To the various institutions supporting the authors, the editors, and the reviewers.
- To Indhumathi at SPi Global for her leadership in transforming a complicated manuscript into a book

We would like to thank also all the scientists who contributed in their own way to operator-splitting and related methods; they made this book possible. Among them, we would like to give a special tribute to *Ernie Esser* and *Michèle Schatzman*; their untimely departure was a shock to their friends and colleagues. Both of them had outstanding contributions to various topics addressed in this book, for which we thank them and dedicate this book to their memory.

Houston, TX, USA — Roland Glowinski
Los Angeles, CA, USA — Stanley J. Osher
Los Angeles, CA, USA — Wotao Yin
February 2016

Contents

Chapter 1
Introduction

Roland Glowinski, Stanley J. Osher, and Wotao Yin

Abstract The main goal of this chapter is to present a brief overview of operator splitting methods and algorithms when applied to the solution of initial value problems and optimization problems, topics to be addressed with many more details in the following chapters of this book. The various splitting algorithms, methods, and schemes to be considered and discussed include: the Lie scheme, the Strang symmetrized scheme, the Douglas-Rachford and Peaceman-Rachford alternating direction methods, the alternating direction method of multipliers (ADMM), and the split Bregman method. This chapter also contains a brief description of (most of) the following chapters of this book.

1 Motivation and Background

In December 2012, the three editors of this volume were together in Hong-Kong, attending a conference honoring one of them (SO). A most striking fact during this event was the large number of lectures involving, if not fully dedicated to, operator-splitting methods, split Bregman and ADMM (for Alternating Direction Methods of Multipliers) algorithms, and their applications. This was not surprising considering that the title of the conference was *Advances in Scientific Computing, Imaging Science and Optimization.* Indeed, for many years, operator-splitting has provided efficient methods for the numerical solution of a large variety of problems from

R. Glowinski (✉)
Department of Mathematics, University of Houston, Houston, TX 77204, USA
e-mail: roland@math.uh.edu

S.J. Osher • W. Yin
Department of Mathematics, UCLA, Los Angeles, CA 90095, USA
e-mail: sjo@math.ucla.edu; wotaoyin@math.ucla.edu

R. Glowinski et al. (eds.), *Splitting Methods in Communication, Imaging, Science, and Engineering*, Scientific Computation, DOI 10.1007/978-3-319-41589-5_1

Mechanics, Physics, Finance, etc. modeled by linear and nonlinear partial differential equations and inequalities, with new applications appearing almost on a daily basis. Actually, there are situations where the only working solution methods are based on operator-splitting (a dramatic example being provided by the numerical simulation of super-novae explosions). Similarly, split Bregman and ADMM algorithms enjoy now a very high popularity as efficient tools for the fast solution of problems from the information sciences. Finally, ADMM algorithms have found applications to the solution of large-scale optimization problems, due to their ability at taking advantage of those special structures and associated decomposition properties, which are common in large-scale optimization.

What the three editors observed also was the lack of communications between the communities and cultures, associated with the 'vertices' of the 'triangle' *PDE oriented scientific computing - information sciences - optimization* (some bridges do exist fortunately, but it is our opinion that more are needed).

From the various facts above, the three editors came to the conclusion that time has come to have a multi-author book, with chapters dedicated to the theory, practice, and applications of operator-splitting, ADMM, and Bregman algorithms and methods, one of the many goals of this book being also to show the existing commonalities between these methods. Another justification of the present volume is that the number of publications on the above topics with a review flavor are scarce, the most visible ones being, by chronological order:

1. The 1990 article by G.I. Marchuk in the Volume I of the Handbook of Numerical Analysis [23]. This book size article (266 pages) is dedicated, mostly, to the numerical solution of linear and nonlinear time dependent partial differential equations from Mechanics and Physics, by operator-splitting and alternating direction methods, making it thus quite focused.
2. The (94 pages) article by R.I. McLachlan, G. Reinout W. Quispel in Acta Numerica 2002 [24]. Despite the broad appeal of its title, namely Splitting methods, this long article is mostly focused on the time-discretization of dynamical systems, modeled by ordinary differential equations, by (symplectic) operator-splitting schemes. It ignores for example the contributions of Douglas, Marchuk, Peaceman, Rachford (and of many other contributors to splitting methods).
3. Distributed optimization and statistical learning via the alternating direction method of multipliers [1]. This large article (122 pages) appeared in Foundations and Trends' in Machine Learning; since its publication in 2011 it has been the most cited article on ADMM methods (2, 678 citations as of November 26, 2015), an evidence of both the popularity of ADMM and of the importance of this publication. Actually, this article is mostly concerned with finite dimensional convex problems and ignores applications involving differential equations and inequalities.

It seemed to us, at the time of the above Hong-Kong conference (and also now), that a book (necessarily multi-authors), less focused than the above publications, and blending as much as possible the methods and points of view of the various splitting sub-communities, will be a useful and welcome tool for the splitting community at large. It will be useful also for those in search of efficient, modular, and relatively easy to implement solution methods for complex problems. We are glad that Springer shared this point of view and that outstanding contributors to this type of methods accepted contributing to this volume.

In the following sections of this chapter we will give a relatively short commented description of operator-splitting methods and related algorithms, and provide the historical facts we are aware of. As expected, references will be made to the following chapters and to many related publications.

2 Lie's Schemes

According to [5] it is Sophus Lie himself who introduced (in 1875) the first operator-splitting scheme recorded in history ([21], a reprint of the 1888 edition), in order to solve the following initial value problem (flow in the Dynamical System community):

$$\begin{cases} \frac{dX}{dt} + (A+B)X = 0, \text{ in } (0,T), \\ X(0) = X_0 \end{cases} \tag{1.1}$$

where in (1.1): A and B are two $d \times d$ time-independent matrices, $X_0 \in \mathbb{R}^d$ (or $\mathbb{C}^d$) and $0 < T \leq +\infty$. The solution of problem (1.1) reads as:

$$X(t) = e^{-(A+B)t} X_0. \tag{1.2}$$

From the relation $\lim_{n \to +\infty} \left(e^{-B\frac{t}{n}} e^{-A\frac{t}{n}}\right)^n = e^{-(A+B)t}$, Lie suggested the following scheme for the approximate solution of problem (1.1) (with $\Delta t > 0, t^n = n\Delta t$, and X^n an approximation of $X(t^n)$):

$$\begin{cases} X^0 = X_0, \\ X^{n+1} = e^{-B\Delta t} e^{-A\Delta t} X^n, \ \forall n \geq 0. \end{cases} \tag{1.3}$$

Since computing the exponential of a matrix is a nontrivial operation, particularly for large values of d, splitting practitioners prefer to use the following (matrix exponential free) equivalent formulation of scheme (1.3):

$$X^0 = X_0. \tag{1.4}$$

For $n \geq 0$, $X^n \to X^{n+\frac{1}{2}} \to X^{n+1}$ via:

- Solve

$$\begin{cases} \frac{dX_1}{dt} + AX_1 = 0, \text{ in } (t^n, t^{n+1}), \\ X_1(t^n) = X^n, \end{cases} \tag{1.5}$$

- and set

$$X^{n+\frac{1}{2}} = X_1(t^{n+1}). \tag{1.6}$$

- Similarly, solve

$$\begin{cases} \frac{dX_2}{dt} + BX_2 = 0, \text{ in } (t^n, t^{n+1}), \\ X_2(t^n) = X^{n+\frac{1}{2}}, \end{cases} \tag{1.7}$$

- and set

$$X^{n+1} = X_2(t^{n+1}). \tag{1.8}$$

From its equivalent formulation (1.4)–(1.8), one understands better now why the *Lie scheme* (1.3) is known as a *fractional-step* time discretization scheme. It is in fact the common ancestor to all the fractional-step and operator-splitting schemes. One can easily show that the Lie scheme (1.3), (1.4)–(1.8) is generically *first order accurate*, that is, $\|X^n - X(t^n)\| = O(\Delta t)$. If A and B *commute* the above scheme is obviously *exact*. Its generalization to splitting with more than two operators is discussed in Chapter 2. The generalization to *nonlinear* and/or *non-autonomous* initial value problems is very simple (formally at least). Indeed, let us consider the following initial value problem:

$$\begin{cases} \frac{dX}{dt} + A(X,t) = 0, \ \text{ in } (0,T), \\ X(0) = X_0, \end{cases} \tag{1.9}$$

with $A(X,t) = A_1(X,t) + A_2(X,t)$, A_1 and A_2 being possibly nonlinear. Introducing α_1, α_2 such that $0 \leq \alpha_1, \alpha_2 \leq 1, \alpha_1 + \alpha_2 = 1$, we can generalize scheme (1.4)–(1.8) as follows:

$$X^0 = X_0. \tag{1.10}$$

For $n \geq 0$, $X^n \to X^{n+\frac{1}{2}} \to X^{n+1}$ via:
For $j = 1, 2$, solve:

$$\begin{cases} \frac{dX_j}{dt} + A_j(X_j, t^n + (\sum_{k=1}^{j-1} \alpha_k)\Delta t + \alpha_j(t - t^n)) = 0, \ \text{ in } (t^n, t^{n+1}), \\ X_j(t^n) = X^{n+\frac{j-1}{2}}, \end{cases} \tag{1.11}$$

and set

$$X^{n+\frac{j}{2}} = X_j(t^{n+1}). \tag{1.12}$$

Assuming that the operator A_j are smooth enough, the generalized Lie scheme (1.10)–(1.12) is generically *first order accurate* at best. As shown in the following chapters, problem (1.9) is, despite its simplicity, a model for a large number of important applications: For example, it models the flow (in the dynamical system sense) associated with the optimality conditions of an optimization problem ($= 0$ being possibly replaced by $\ni 0$).

3 On the Strang Symmetrized Operator-Splitting Scheme

Motivated by the accurate solution of hyperbolic problems, G. Strang introduced in [30] a second order variant of the Lie scheme, based on a symmetrization principle. Let $\Delta t (> 0)$ be a time-discretization step. We can easily show that

$$e^{-(A+B)\Delta t} - e^{-A\frac{\Delta t}{2}} e^{-B\Delta t} e^{-A\frac{\Delta t}{2}} = O(\Delta t^3),$$

implying that the (symmetrized) scheme

$$\begin{cases} X^0 = X_0, \\ X^{n+1} = e^{-A\frac{\Delta t}{2}} e^{-B\Delta t} e^{-A\frac{\Delta t}{2}} X^n, \forall n \geq 0, \end{cases} \tag{1.13}$$

is *second order accurate* (and *exact* if $AB = BA$). Scheme (1.13) can be generalized as follows in order to solve the initial value problem (1.9) (with $t^{n+\frac{1}{2}} = (n+\frac{1}{2})\Delta t$):

$$X^0 = X_0. \tag{1.14}$$

For $n \geq 0, X^n \to X^{n+\frac{1}{2}} \to \hat{X}^{n+\frac{1}{2}} \to X^{n+1}$ via

- Solve

$$\begin{cases} \frac{dX_1}{dt} + A_1(X_1, t) = 0, \text{ in } (t^n, t^{n+\frac{1}{2}}), \\ X_1(t^n) = X^n, \end{cases} \tag{1.15}$$

- and set

$$X^{n+\frac{1}{2}} = X_1(t^{n+\frac{1}{2}}). \tag{1.16}$$

- Similarly, solve

$$\begin{cases} \frac{dX_2}{dt} + A_2(X_2, t^{n+\frac{1}{2}}) = 0, \text{ in } (0, \Delta t), \\ X_2(0) = X^{n+\frac{1}{2}}, \end{cases} \tag{1.17}$$

- and set

$$\hat{X}^{n+\frac{1}{2}} = X_2(\Delta t). \tag{1.18}$$

- Finally, solve

$$\begin{cases} \frac{dX_1}{dt} + A_1(X_1, t) = 0, \text{ in } (t^{n+\frac{1}{2}}, t^{n+1}), \\ X_1(t^{n+\frac{1}{2}}) = \hat{X}^{n+\frac{1}{2}}, \end{cases} \tag{1.19}$$

- and set

$$X^{n+1} = X_1(t^{n+1}). \tag{1.20}$$

Generalizing scheme (1.14)–(1.20) to decompositions of A involving more than two operators is easy: Indeed, suppose that $A = A_1 + A_2 + A_3$; we can trivially return to two operator situations by observing that we have the choice between $A = A_1 + (A_2 + A_3), A = (A_1 + A_2) + A_3$ and $A = (A_1 + \frac{1}{2}A_2) + (\frac{1}{2}A_2 + A_3)$, other decompositions being possible. The same idea applies for more than three operators. If the operators in (1.14)–(1.20) have the right regularity and monotonicity properties, the above symmetrized scheme is *second order accurate* and *unconditionally stable* (it is even A - stable), making it very popular in the computational partial differential equations community. For those situations requiring an order of accuracy higher than two, several options do exist, the best known being:

1. The 4th order *Strang-Richardson* scheme discussed in [8, 6, 7].
2. The *exponential operator-splitting schemes*. Actually, the Lie and Strang splitting schemes belong to this family of time discretization methods, whose origin (concerning schemes of order higher than two) is not easy to track back, early significant publications being [28, 29] (see also the references therein and those in [31], and Google Scholar). Arbitrary high accuracy can be obtained with these methods, the price to pay being their reduced stability and robustness (compared to the Strang scheme, for example).

We will return to these higher order schemes in Chapter 2.

4 On the Solution of the Sub-initial Value Problems

The various splitting schemes we encountered so far are generically known (for obvious reasons) as *multiplicative splitting schemes*. Actually, these schemes are only semi-constructive, in the sense that one still has to specify how to solve in practice the various sub-initial value problems they produce. For a low order scheme like the Lie scheme a very popular way to proceed is to solve the sub-initial value problems using just one step of the backward Euler scheme; applying this strategy to problem (1.9), with $A(X,t) = \sum_{j=1}^{J} A_j(X,t)$, one obtains the following scheme

$$X^0 = X_0. \tag{1.21}$$

For $n \geq 0, X^n \to X^{n+\frac{1}{J}} \to \ldots \to X^{n+\frac{j}{J}} \to \ldots \to X^{n+\frac{J-1}{J}} \to X^{n+1}$ via: $\forall j = 1, \ldots, J$, solve

$$\frac{X^{n+\frac{j}{J}} - X^{n+\frac{j-1}{J}}}{\Delta t} + A_j(X^{n+\frac{j}{J}}, t^{n+1}) = 0. \tag{1.22}$$

Scheme (1.21)–(1.22) (known by some practitioners as the *Marchuk-Yanenko* operator-splitting scheme) is first order accurate, at most, however, its robustness and simplicity make it popular for the solution of complicated problems with poorly differentiable solutions involving a large number of operators. Actually, scheme (1.21)–(1.22) can accommodate easily non-smooth, possibly multi-valued, operators.

The practical implementation of the Strang symmetrized scheme will be discussed in Chapter 2.

5 Further Comments on Multiplicative Operator-Splitting Schemes

We will conclude these generalities on *multiplicative operator-splitting methods* by mentioning their following drawback: Due to splitting errors, these methods are *asymptotically inconsistent*. What we mean by that is that when applied to the computation of steady state solutions (assuming that such solutions exist), they converge to limits, which are not steady state solutions, but just approximations of them. Typically, if the operator-splitting scheme one employs is k-order accurate, the distance of the computed steady state solutions to the exact ones is $O(|\Delta t|^k)$. One can reduce this splitting-error, via an appropriate averaging for example. Actually, in Chapter 3 of this book, *S. MacNamara* and *G. Strang* show how to modify the *Strang symmetrized operator-splitting scheme* in order to recover exact steady state solutions. Another way to combine exact steady state solutions and the operator-splitting paradigm is to use schemes such as *Peaceman-Rachford's* and *Douglas-Rachford's*, that is *ADI* (for *Alternating Direction Implicit*) type methods. These schemes will be discussed in Section 6, below.

6 On ADI Methods

To the best of our knowledge, *J. Douglas, D. Peaceman* and *H. Rachford* introduced *ADI* methods about 60 years ago (to know more about their inception, see the 2006 *SIAM News* article by *A. Usadi* and *C. Dawson* [33] on the *Celebration at Rice University of 50 Years of ADI Methods*). The founding publications, namely [27] and [9], concern the numerical solution of *elliptic* and *parabolic* equations such as the time dependent and stationary *heat equations*. However, one quickly realized that these schemes apply also to situations that are more general. Taking advantage of these generalizations, our starting point will be the following initial value problem:

$$\begin{cases} \frac{dX}{dt} + A_1(X,t) + A_2(X,t) = 0, \text{ in } (0,T), \\ X(0) = X_0, \end{cases} \tag{1.23}$$

where A_1 and A_2 operate, possibly, on an infinite dimensional space, and $0 < T \leq +\infty$. With $\Delta t (> 0)$ a time-discretization step, we denote $(n+\alpha)\Delta t$ by $t^{n+\alpha}$, and by $X^{n+\alpha}$ an approximation of $X(t^{n+\alpha})$. The idea behind the *Peaceman-Rachford* scheme is quite simple: An approximation X^n of $X(t^n)$ being known, one computes

X^{n+1} using a scheme of the *backward* (resp., *forward*) *Euler* type with respect to A_1(resp., A_2) on the time interval $[t^n, t^{n+\frac{1}{2}}]$. Then, one switches the roles of A_1 and A_2 on the time interval $[t^{n+1/2}, t^{n+1}]$. The following scheme realizes this program:

$$X^0 = X_0. \tag{1.24}$$

For $n \geq 0, X^n \to X^{n+\frac{1}{2}} \to X^{n+1}$ as follows:
Solve

$$\frac{X^{n+\frac{1}{2}} - X^n}{\Delta t/2} + A_1(X^{n+\frac{1}{2}}, t^{n+\frac{1}{2}}) + A_2(X^n, t^n) = 0, \tag{1.25}$$

and

$$\frac{X^{n+1} - X^{n+\frac{1}{2}}}{\Delta t/2} + A_1(X^{n+\frac{1}{2}}, t^{n+\frac{1}{2}}) + A_2(X^{n+1}, t^{n+1}) = 0. \tag{1.26}$$

In the particular case where A_1 and A_2 are linear and time independent linear operators which *commute*, the *Peaceman-Rachford scheme* (1.24)–(1.26) is second order accurate; it is at best *first order accurate in general.* The convergence of the above scheme has been proved in [22] and [18] under quite general *monotonicity* hypotheses concerning the operators A_1 and A_2 (see also [10, 11], and [20]); indeed, A_1 and A_2 can be nonlinear, unbounded, and even multivalued (if this is the case $\ni 0$ has to replace $= 0$ in (1.25) and/or (1.26)).

Remark 1. For those fairly common situations where A_2 is a smooth univalued operator, but operator A_1 is a 'nasty' one (discontinuous and/or multivalued, etc.), one should use the equivalent formulation of the Peaceman-Rachford scheme obtained by replacing (1.26) by

$$\frac{X^{n+1} - 2X^{n+\frac{1}{2}} + X^n}{\Delta t/2} + A_2(X^{n+1}, t^{n+1}) = A_2(X^n, t^n). \tag{1.27}$$

Further comments and remarks concerning scheme (1.24)–(1.26) may be found in [14, 16] (see also the references therein and various chapters of this book, Chapter 2 in particular). Actually, several of these comments concern the *fractional θ-scheme*, a scheme introduced in the mid-eighties for the numerical solution of the *Navier-Stokes equations* modeling *incompressible viscous flow* [13, 14]. This scheme, which is a *three sub-interval variant* of the Peaceman-Rachford scheme (1.24)–(1.26) will be discussed in Chapter 2.

A classical alternative to the *Peaceman-Rachford* scheme (1.24)–(1.26) is the *Douglas-Rachford* scheme introduced in [9]. Applied to the solution of the initial value problem (1.23), the Douglas-Rachford scheme reads as follows:

$$X^0 = X_0. \tag{1.28}$$

For $n \geq 0, X^n \to \hat{X}^{n+1} \to X^{n+1}$ as follows:
Solve

$$\frac{\hat{X}^{n+1} - X^n}{\Delta t} + A_1(\hat{X}^{n+1}, t^{n+1}) + A_2(X^n, t^n) = 0, \tag{1.29}$$

and

$$\frac{X^{n+1} - X^n}{\Delta t} + A_1(\hat{X}^{n+1}, t^{n+1}) + A_2(X^{n+1}, t^{n+1}) = 0. \tag{1.30}$$

The convergence of the above scheme has been proved in [22] and [18] under quite general *monotonicity* hypotheses concerning the operators A_1 and A_2 (see also [10, 11], and [20]); indeed, A_1 and A_2 can be nonlinear, unbounded, and even multivalued (if this is the case $\ni 0$ has to replace $= 0$ in (1.28) and/or (1.30)). As shown in, e.g., Chapter 2 the Douglas-Rachford scheme (1.28)–(1.30) is *generically first order accurate* at best, a prediction supported by the results of various numerical experiments. In order to improve the accuracy of the Douglas-Rachford scheme (1.28)–(1.30), *J. Douglas & S. Kim* introduced in the late 90s–early 2000s [33], a Crank-Nicolson based variant of the above scheme; we will briefly discuss the Douglas-Kim scheme in Chapter 2 (the price to pay for the improved accuracy is a loss of robustness).

Remark 2. For obvious reasons the *Douglas-Rachford scheme* (1.28)–(1.30) is known as an *additive operator-splitting scheme*. The same terminology applies also to the *Peaceman-Rachford scheme* (1.24)–(1.26).

Remark 3. Unlike the Peaceman-Rachford scheme (1.24)–(1.26), the Douglas-Rachford scheme (1.28)–(1.30) can be generalized to decompositions $A = \sum_{j=1}^{J} A_j$ involving *more than* two operators (see Section 2.5 of Chapter 2 for details and related references). Actually, the above observation applies also to the Douglas-Kim scheme.

Remark 4. This is the Douglas-Rachford analogue of Remark 1: For those situations where A_1 is a 'bad' operator (in the sense of Remark 1), we should use (assuming that A_2 is univalued) the equivalent formulation of the Douglas-Rachford scheme obtained by replacing (1.26) by

$$\frac{X^{n+1} - \hat{X}^{n+1}}{\Delta t} + A_2(X^{n+1}, t^{n+1}) = A_2(X^n, t^n). \tag{1.31}$$

Remark 5. At those wondering how to choose between Peaceman-Rachford and Douglas-Rachford schemes we will say that on the basis of many numerical experiments, it seems that the second scheme is more robust and faster for those situations where one of the operators is non-smooth (multivalued or singular, for example), particularly if one is interested at capturing steady state solutions. We will give a (kind of) justification in Chapter 2, based on the inspection of some simple particular cases.

Remark 6. Optimization algorithms and *ADI methods* did not interact that much for many years. The situation started changing when in the mid-1970s unexpected relationships between some *augmented Lagrangian algorithms* and the *Douglas-Rachford scheme* (1.28)–(1.30) were identified (they have been reported in Chapter 2, Section 3, and Chapter 8, Section 1, of this volume). This discovery lead to what is called now the *Alternating Direction Methods of Multipliers* (ADMM), and was known as *ALG2* at the time. Although the problems leading to ADMM were *partial differential equations* or *inequalities* related, this family of algorithms has found applications outside the realm of differential operators, as shown by [1] and several chapters of this book. Further details on the 'birth' of ADMM can be found in [15] and the Chapter 4 of [16].

The above remark is a natural introduction to the more detailed discussion, below, on the role of *operator-splitting methods* in Optimization.

7 Operator Splitting in Optimization

Several chapters of this book study operator splitting methods for solving optimization problems in (non-PDE related) signal processing, imaging, statistical and machine learning, as well as those defined on a network that require decentralized computing. They cover a large variety of problems and applications. The optimization problems considered in these chapters include both constrained and unconstrained problems, smooth and nonsmooth functionals, as well as convex and nonconvex functionals.

Operator splitting methods are remarkably powerful since they break numerically inconvenient combinations, such as smooth + nonsmooth functionals, smooth functionals + constraints, functionals of local variables + functionals of the shared global variable, and convex + nonconvex functionals, in a problem and place them in separate subproblems. It also breaks a sum of functionals that involve different parts of the input data into subproblems so that they can be solved with less amounts of memory. These features are especially important to the modern applications that routinely process a large amount of data.

Operator splitting methods appear in optimization in a variety of different special forms, and thus under different names, such as gradient-projection, proximal-gradient, alternating-direction, split Bregman [19], and primal-dual splitting methods. All of them have strong connections (often as special cases) of the forward-backward, Douglas-Rachford, and Peaceman-Rachford splitting methods. Recently, their applications in optimization have significantly increased because of the emerging need to analyze massive amounts of data in a fast, distributed, and even streaming manner.

A common technique in data analysis is *sparse optimization*, a new subfield of optimization. *Sparse* here means simple structures in the solutions — a generalization from its literal meaning of having very few nonzero elements. Owing much to statistical estimation with prior information and compressive sensing, sparse optimization has been recognized as a computational tool that plays a central role in many data processing problems. A typical sparse optimization problem has the form

$$\underset{y}{\text{minimize}}\,[r(y)+f(y)], \tag{1.32}$$

where minimizing $r(y)$ imposes a simple structure on the solution x, such as sparsity and smoothness, and minimizing $f(y)$ matches the solution x to the observations. Since f and r have distinct roles, they often possess different properties, limiting your choices of classic (non-splitting) numerical methods. Therefore, splitting methods have been the favorite numerical solution to the model (1.32).

Another frequently used model is the monotropic program

$$\begin{aligned} &\underset{x,y}{\text{minimize}}\ [p(x)+q(y)] \\ &\text{subject to}\ Ax+By=b, \end{aligned} \tag{1.33}$$

which applies to many problems in modern data sciences. Here, the functionals p,q can take the extended value $+\infty$, so the model can incorporate constraints as indicator functionals. Connecting the two functions in the objective through linear constraints in the model (1.33) significantly increases its modeling power over the model (1.32). (It is easy to see that (1.32) is a special case of (1.33) with $p=r$, $q=f$, $A=-B=I$, and $b=0$.) The model (1.33) routinely appears in the following classes of problems:

- **network-wide problems**: the linear constraints of (1.33) relate the variables defined on the network nodes and edges;
- **total variation and analysis-sparse problems**: they minimize the functional $q(Ux)$ where U represents discrete finite difference, a frame, or other linear operators; one often separates q and U by introducing the constraints $Ux-y=0$;
- **dictionary learning and dictionary-based reconstruction**;
- as well as a long list of other classes of problems ...

Operator splitting methods such as Tseng's algorithm [32] and the alternating direction method of multipliers (ADMM) [17, 12] break the functions p and q into different subproblems while classical methods such as the interior-point method and the method of multipliers cannot offer such convenience.

In some problems, the objective functions in (1.32) and (1.33) are sums of multiple subfunctionals. Operator-splitting methods can let them updated in separate subproblems and give rise to parallel or distributed implementations; therefore, they become a strong candidate for dealing with data of extreme scale (e.g., web-scale data mining) and those collected in a distributed manner (e.g., sensor networks).

This book features several chapters on operator splitting methods for optimization: Chapter 9 by Goldstein and Zhang covers ℓ_1 and compressive sensing/imaging problems, Chapter 10 by Burger, Sawatzky, and Steidl covers the operator-splitting methods for non-smooth variational problems in image processing, Chapter 12 by Darbon and Osher applies ADMM to solve previously intractable high dimensional Hamilton-Jacobi equations, Chapter 13 by Hu, Chi, and Allen introduces an approach to generate a regularization path (a sequence of solutions corresponding to different strengths of regularization) by running ADMM just once, and Chapter 14 by Giannakis, Ling, Mateos, Schizas, and Zhu covers decentralized learning problems for wireless communications and networking. In addition, Chapter 4 by Davis and Yin studies convergence rates, Chapter 5 by Yan and Yin discovers primal-dual equivalence results, and Chapter 6 by He, Liu, Lu, and Yuan introduces a Peaceman-Rachford based splitting method to the multi-block extension of the model (1.33).

Although they all have different approaches, several chapters review the important concepts of *proximal mapping* and *duality*, which is a cornerstone of many operator splitting methods for convex optimization. For a closed (that is lower semi-continuous) proper function f, its proximal mapping maps the point y to the solution of

$$\underset{x}{\text{minimize}}\,[f(x)+\frac{1}{2}\|x-y\|^2]. \tag{1.34}$$

This mapping is the *resolvent* of the subdifferential operator ∂f, a fundamental concept in the monotone operator theory. The proximal mapping is often used to deal with structured nonsmooth functions and constraints in optimization problems, giving rise to subproblems with closed-form solutions. Chapters 9 and 10 provide a list of such *proximable functions*.

Duality has always been a powerful tool in optimization. It can significantly enhance other optimization methodology including operator splitting. Traditionally, duality provides an alternative perspective to optimization problems, which gives rise to lower bounds, certificates for optimality or infeasibility, as well as alternative formulations of the same problem and thus alternative algorithms. In the context of operator splitting, duality helps split multiple functionals in the objective, as well as the different parts of a constraint. For example, the ADMM algorithm for the model (1.33) places the matrices A and B in different subproblems, and ADMM is a dual algorithm. (ADMM is equivalent to the Douglas-Rachford splitting method applied to the dual of (1.33). In fact, ADMM is also *self-dual*: it is equivalent to itself applied to the dual of (1.33), as established in Chapter 5.) In addition, in primal-dual splitting methods, duality helps split the compositions like $q(Ux) = q \circ U(x)$ so that one can apply the proximal mapping of q (often in a closed form) instead of that of $q \circ U$ (usually difficult). See Chapters 4 and 5 for primal-dual splitting methods such as BOS [37], FTVd [34], PDHG [38], Chambolle-Pock [4], and proximal/linearized-ADMM. In order to obtain parallel and distributed algorithms for problems with multiple functions in the objective and/or multiple components in the linear constraints, duality is indispensable — it helps split those functions and components into different subproblems. Chapter 14 provides a bag of such interesting examples for important applications. Often seen in a dual or primal-dual algorithm are Lagrange multipliers and convex conjugate functions; their appearance lets us easily identify the role that duality is playing.

Besides just providing faster and more scalable algorithms, there has been a novel use of operator splitting algorithm — generating a sequence of solutions for a regularization model that correspond to varying strengths of regularization, known as the *regularization path*. Throughout sparse optimization and statistical estimation models, there is a parameter that explicitly controls how simple the solution needs to be. The best choice of this parameter is practically unknown since it depends on both the model and the actual data. Solving the model for all possible values of the parameter is intractable, and even doing that for a large set of values is impractical. Homotopy techniques have been applied so that we can sequentially obtain the solutions corresponding to different parameter values much faster than repeatedly applying the same algorithm for each value from scratch. Chapter 13 by Hu, Chi, and Allen provides a much faster and simpler solution. They obtain the points that approximate the regularization path directly as the iterates of their ADMM-based algorithm. This approach leads to a substantial save in computing time for problems such as sparse linear regression, reduced-rank multi-task learning, and convex clustering. (Related to their method are the non-splitting methods: Bregman iterative regularization [25, 36], the inverse scale space method [2, 3], and differential inclusion regularization path [26].)

Last but not least, operator splitting algorithms are theoretically attractive because they converge under very few assumptions, typically, only convexity, solution existence, as well as certain constraint qualifications, imposed on the underlying problem. Convergence results have been recently developed for nonconvex problems as well. Chapter 4 by Davis and Yin covers not only the convergence but also the rates of convergence for the generic Krasnosel'skiĭ-Mann (KM) iteration of nonexpansive operators, as well as those for the forward-backward, Douglas-Rachford, Peaceman-Rachford, and ADMM splitting algorithms. All the rates given in their chapter are tight, achieved by examples in the chapter. Of particular interest is the phenomenon that, in terms of the worst-case convergence rate, the last three algorithms are as fast as the proximal-point algorithm in the ergodic sense (measuring the quality of running averages of iterates) yet also as slow as the subgradient algorithm in the non-ergodic sense (measuring the quality of last iterates).

In fact, operator splitting algorithms typically generate a sequence of iterates that have non-monotonic values of objective and constraint violations, making it very tricky to chose an algorithm parameter (in particular, line search is difficult to apply). Empirical studies strongly suggest that best performance is observed when the algorithm parameter is chosen so that primal optimality and dual optimality (which correspond to the different parts of the KKT conditions) are improving at the same pace.

8 Bregman Methods and Operator Splitting

Bregman methods are related to operator splitting through the Bregman distance, which generalizes the Euclidean distance. Let ϕ be a proper closed convex functional and $\partial\phi$ be its subdifferential. Then the Bregman distance between two points x and y is defined as

$$D^p_\phi(y,x) := \phi(y) - \phi(x) - \langle p, y-x\rangle,$$

which depends on the choice $p \in \partial\phi(x)$. In the special case of $\phi(x) = \frac{1}{2}\|x\|^2$, we have $\partial\phi(x) = \{x\}$ and thus $p = x$; therefore, we get $D^p_\phi(y,x) = \frac{1}{2}\|y-x\|^2$. In general, while $D^p_\phi(y,x)$ is not a mathematical distance, it behaves similarly to a distance. Since ϕ is convex, $D^p_\phi(y,x) \geq 0$ for any points x, y in the interior of the domain of ϕ. In addition, if a point z is further away from x than y in the sense $z = y + \alpha(y-x)$ for some $\alpha \geq 0$, then $D^p_\phi(z,x) \geq D^p_\phi(y,x)$. Therefore, minimizing over y tends to keep y close to x.

The Bregman distance is typically used in place of the Euclidean distance, for example, generalizing the proximal subproblem (1.34) to

$$\underset{x}{\text{minimize}}[f(x) + D^p_\phi(x,y)], \tag{1.35}$$

which maps the input y to the minimizer x.

Traditionally, one uses a strongly convex functional ϕ to induce the Bregman distance. In Bregman iteration regularization [2] and the Bregman algorithm [36], however, (weak) convex functionals such as the ℓ_1 norm and total variation are used to generate the sequence of points $(x^k, p^k)_{k\in\mathbb{N}}$ based on applying (1.34) recursively:

$$x^k \in \underset{x}{\text{argmin}}[f(x) + D^{p^{k-1}}_\phi(x, x^{k-1})] \tag{1.36}$$

(We use "$\in$" instead of "$=$" because the solution is not unique. The new p^k can be naturally obtained from the optimality condition of (1.36). The reader is referred to Chapters 9 and 10 for more details.) It turns out that the sequence approximates a regularization path with ϕ being the regularization function and f being the data fidelity functional. Compared to directly minimizing $f + \lambda\phi$, the points are less biased in statistical estimation and are more faithful images in image reconstruction. In addition, it was discovered that for $f(x) = \frac{1}{2}\|Ax-b\|^2$ and ϕ being a piecewise linear functional like the ℓ_1 norm, the algorithm is both fast and robust to error [35]. Although the algorithm becomes equivalent to the method of multipliers (a.k.a., augmented Lagrangian method) for this choice of f, the above results for regularization path and error robustness are both new and important to compressed sensing and statistical estimation.

The split Bregman method [19] carries the same iteration (1.36) for $\phi(x) = \phi_1(x_1) + \phi_2(x_2)$, where $x = \{x_1, x_2\}$. It is numerically observed in [19] that (1.36) can be replaced by the sequential updates:

$$x_1^k \in [\underset{x_1}{\operatorname{argmin}} f(x_1, x_2^{k-1}) + D_{\phi_1}^{p_1^{k-1}}(x_1, x_1^{k-1})] \tag{1.37a}$$

$$x_2^k \in [\underset{x_2}{\operatorname{argmin}} f(x_1^k, x_2) + D_{\phi_2}^{p_2^{k-1}}(x_2, x_2^{k-1})], \tag{1.37b}$$

while the algorithm is still running. (From the optimality conditions, one obtain $p_i^k \in \partial \phi_i(x_i^k)$ for $i = 1, 2$.) Therefore, ϕ_1 and ϕ_2 can be split into different subproblems and minimized separately. The imaging community soon recognized the advantage of such splitting and started to find good performance of this method on many challenging problems. It was then discovered that when $f(x_1, x_2) = \frac{1}{2}\|A_1x_1 + A_2x_2 - b\|^2$, algorithm (1.37) is equivalent to an ADMM algorithm.

Acknowledgements

All the chapters in this book have been peer-reviewed. We greatly appreciate the voluntary work and experted reviews by the anonymous reviewers. We want to express our deep and sincere gratitude to all the authors, who have made tremendous contributions and offered generous support to this book.

References

1. Boyd, S., Parikh, N., Chu, E., Peleato, B., Eckstein, J.: Distributed optimization and statistical learning via the alternating direction method of multipliers. Foundations and Trends® in Machine Learning **3**(1), 1–122 (2011)
2. Burger, M., Gilboa, G., Osher, S., Xu, J.: Nonlinear inverse scale space methods. Communications in Mathematical Sciences **4**(1), 179–212 (2006)
3. Burger, M., Möller, M., Benning, M., Osher, S.: An adaptive inverse scale space method for compressed sensing. Mathematics of Computation **82**(281), 269–299 (2012)
4. Chambolle, A., Pock, T.: A first-order primal-dual algorithm for convex problems with applications to imaging. Journal of Mathematical Imaging and Vision **40**(1), 120–145 (2011)
5. Chorin, A.J., Hughes, T.J., McCracken, M.F., Marsden, J.E.: Product formulas and numerical algorithms. Communications on Pure and Applied Mathematics **31**(2), 205–256 (1978)
6. Descombes, S.: Convergence of a splitting method of high order for reaction-diffusion systems. Mathematics of Computation **70**(236), 1481–1501 (2001)
7. Descombes, S., Schatzman, M.: Directions alternées d'ordre élevé en réaction-diffusion. C. R. Acad. Sci. Paris Sér. I Math. **321**(11), 1521–1524 (1995)
8. Descombes, S., Schatzman, M.: On Richardson extrapolation of Strang formula for reaction-diffusion equations. diffusion equations. In : Equations aux Dérivées Partielles et Applications : Articles dédiés á J.L. Lions pp. 429–452 (1998)
9. Douglas, J., Rachford, H.H.: On the numerical solution of heat conduction problems in two and three space variables. Transactions of the American Mathematical Society pp. 421–439 (1956)

10. Gabay, D.: Application de la méthode des multiplicateurs aux inéquations variationnelles. In: M. Fortin, R. Glowinski (eds.) Lagrangiens Augmentés: Application à la Résolution Numérique des Problèmes aux Limites pp. 279–307. Dunod, Paris (1982)
11. Gabay, D.: Applications of the method of multipliers to variational inequalities. In: M. Fortin, R. Glowinski (eds.) Augmented Lagrangians: Application to the Numerical Solution of Boundary Value Problems pp. 299–331. North–Holland, Amsterdam (1983)
12. Gabay, D., Mercier, B.: A dual algorithm for the solution of nonlinear variational problems via finite element approximation. Computers & Mathematics with Applications **2**(1), 17–40 (1976)
13. Glowinski, R.: Splitting methods for the numerical solution of the incompressible Navier-Stokes equations. In: J. A. V. Balakrishnan A. A. Dorodtnitsyn, L. Lions (eds.) Vistas in Applied Mathematics, pp. 57–95. Optimization Software (1986)
14. Glowinski, R.: Finite element methods for incompressible viscous flow. In: Ciarlet, P.G., Lions, J.L. (eds.) Handbook of Numerical Analysis **Vol. IX**, pp. 3–1176. North–Holland, Amsterdam (2003)
15. Glowinski, R.: On alternating direction methods of multipliers: A historical perspective. In: Fitzgibbon, W., Kuznetsov, Y.A., Neittaanmäki, P., Pironneau, O. (eds.) Modeling, Simulation and Optimization for Science and Technology **Vol. 34**, pp. 59–82. Springer, Dordrecht (2014)
16. Glowinski, R.: Variational Methods for the Numerical Solution of Nonlinear Elliptic Problems SIAM, Philadelphia, PA (2015)
17. Glowinski, R., Marroco, A.: On the approximation by finite elements of order one, and solution by penalisation-duality of a class of nonlinear Dirichlet problems. ESAIM: Mathematical Modelling and Numerical Analysis - Mathematical Modelling and Numerical Analysis **9**(R2), 41–76 (1975)
18. Godlewsky, E.: Méthodes à pas Multiples et de Directions Alternées pour la Discrétisation d'Equations d'Evolution. Doctoral Dissertation, Université P. & M. Curie, Paris (1980)
19. Goldstein, T., Osher, S.: The split Bregman method for L1-regularized problems. SIAM Journal on Imaging Sciences **2**(2), 323–343 (2009)
20. Layton, W.J., Maubach, J.M., Rabier, P.J.: Parallel algorithms for maximal monotone operators of local type. Numerische Mathematik **71**(1), 29–58 (1995)
21. Lie, S., Engel, F.: Theorie der transformationsgruppen (Vol. 1). American Soc., Providence, RI (1970)
22. Lions, P.L., Mercier, B.: Splitting algorithms for the sum of two nonlinear operators. SIAM Journal on Numerical Analysis **16**(6), 964–979 (1979)
23. Marchuk, G.I.: Splitting and alternating direction methods. In: Ciarlet, P.G., Lions, J.L.(eds.) Handbook of Numerical Analysis **Vol. I**, pp. 197–462. North–Holland, Amsterdam (1990)
24. McLachlan, R.I., Reinout, G., Quispel, W.: Splitting methods. Acta Numerica **11**, 341–434 (2002)
25. Osher, S., Burger, M., Goldfarb, D., Xu, J., Yin, W.: An iterative regularization method for total variation-based image restoration. Multiscale Modeling & Simulation **4**(2), 460–489 (2005)
26. Osher, S., Ruan, F., Yao, Y., Yin, W., Xiong, J.: Sparse recovery via differential inclusions. UCLA CAM Report 14–16 (2014)
27. Peaceman, D.W., Rachford Jr, H.H.: The numerical solution of parabolic and elliptic differential equations. Journal of the Society for Industrial and Applied Mathematics **3**(1), 28–41 (1955)
28. Sheng, Q.: Solving linear partial differential equations by exponential splitting. IMA Journal of Numerical Analysis **9**(2), 199–212 (1989)
29. Sheng, Q.: Global error estimates for exponential splitting. IMA Journal of Numerical Analysis **14**(1), 27–56 (1994)
30. Strang, G.: On the construction and comparison of difference schemes. SIAM Journal on Numerical Analysis **5**(3), 506–517 (1968)
31. Thalhammer, M.: High-order exponential operator splitting methods for time-dependent Schrödinger equations. SIAM Journal on Numerical Analysis **46**(4), 2022–2038 (2008)
32. Tseng, P.: Applications of a splitting algorithm to decomposition in convex programming and variational inequalities. SIAM Journal on Control and Optimization **29**(1), 119–138 (1991)

33. Usadi, A., Dawson, C.: 50 Years of ADI Methods: Celebrating the Contributions of Jim Douglas, Don Peaceman and Henry Rachford. SIAM News **39**(2) (2006)
34. Wang, Y., Yang, J., Yin, W., Zhang, Y.: A New Alternating Minimization Algorithm for Total Variation Image Reconstruction. SIAM Journal on Imaging Sciences **1**(3), 248–272 (2008)
35. Yin, W., Osher, S.: Error forgetting of Bregman iteration. Journal of Scientific Computing **54**(2–3), 684–695 (2013)
36. Yin, W., Osher, S., Goldfarb, D., Darbon, J.: Bregman iterative algorithms for ℓ_1-minimization with applications to compressed sensing. SIAM Journal on Imaging Sciences **1**(1), 143–168 (2008)
37. Zhang, X., Burger, M., Bresson, X., Osher, S.: Bregmanized nonlocal regularization for deconvolution and sparse reconstruction. SIAM Journal on Imaging Sciences **3**(3), 253–276 (2010)
38. Zhu, M., Chan, T.: An efficient primal-dual hybrid gradient algorithm for total variation image restoration. UCLA CAM Report pp. 08–34 (2008)

Chapter 2
Some Facts About Operator-Splitting and Alternating Direction Methods

Roland Glowinski, Tsorng-Whay Pan, and Xue-Cheng Tai

Abstract The main goal of this chapter is to give the reader a (relatively) brief overview of operator-splitting, augmented Lagrangian and ADMM methods and algorithms. Following a general introduction to these methods, we will give several applications in Computational Fluid Dynamics, Computational Physics, and Imaging. These applications will show the flexibility, modularity, robustness, and versatility of these methods. Some of these applications will be illustrated by the results of numerical experiments; they will confirm the capabilities of operator-splitting methods concerning the solution of problems still considered complicated by today standards.

1 Introduction

In 2004, the first author of this chapter was awarded the SIAM Von Kármán Prize for his various contributions to Computational Fluid Dynamics, the direct numerical simulation of particulate flow in particular. Consequently, he was asked by some people at SIAM to contribute an article to SIAM Review, related to the Von Kármán lecture he gave at the 2004 SIAM meeting in Portland, Oregon. Since *operator-splitting* was playing a most crucial role in the results presented during his Portland lecture, he decided to write, jointly with several collaborators (including the second author), a review article on operator-splitting methods, illustrated by several selected

R. Glowinski (✉) • T.-W. Pan
Department of Mathematics, University of Houston, Houston, TX 77204, USA
e-mail: roland@math.uh.edu; pan@math.uh.edu

X.-C. Tai
Department of Mathematics, University of Bergen, Bergen, Norway
e-mail: tai@mi.uib.no

R. Glowinski et al. (eds.), *Splitting Methods in Communication, Imaging, Science, and Engineering*, Scientific Computation, DOI 10.1007/978-3-319-41589-5_2

applications. One of the main reasons for that review article was that, to the best of our knowledge at the time, the last comprehensive publication on the subject was [121], a book-size article (266 pages) published in 1990, in the Volume I of the Handbook of Numerical Analysis. Our article was rejected, on the grounds that it was untimely. What is ironical is that the very day (of August 2005) we received the rejection e-mail message, we were having a meeting with computational scientists at Los Alamos National Laboratory (LANL) telling us that one of their *main priorities* was further investigating the various properties of operator-splitting methods, considering that these methods were (and still are) applied at LANL to solve a large variety of challenging, mostly multi-physics, problems. Another event emphasizing the importance of operator-splitting methods was the December 2005 conference, at Rice University in Houston, commemorating "50 Years of Alternating-Direction Methods" and honoring *J. Douglas, D. Peaceman* and *H. Rachford*, the inventors of those particular operator-splitting methods bearing their name. Actually, it was striking to observe during this conference that, at the time, most members of the *Partial Differential Equations* and *Optimization* communities were ignoring that most alternating-direction methods for initial value-problems are closely related to primal-dual algorithms such as *ADMM* (Alternating Direction Methods of Multipliers). In order to create a bridge between these two communities, we updated the failed SIAM Review paper and submitted it elsewhere, leading to [73] (clearly, a publication in an SIAM journal would have had more impact, worldwide). Our goal in this chapter is to present a (kind of) updated variant of [73], less CFD (resp., more ADMM) oriented. It will contain in particular applications to *Imaging*, a topic barely mentioned in reference [73]. The content of this chapter is as follows:

In Section 2, we will discuss the numerical solution of *initial value problems* by *operator-splitting* time-discretization schemes such as Peaceman-Rachford's, Douglas-Rachford's, Lie's, Strang's, Marchuk-Yanenko's, and by the fractional θ-scheme, a three-stage variation, introduced in [67] and [68], of Peaceman Rachford's scheme. We will conclude this section by some remarks on the parallelization of operator-splitting schemes.

Section 3 will be dedicated to *augmented Lagrangian* and *ADMM algorithms*. We will show in particular that some augmented Lagrangian and ADMM algorithms are nothing but disguised operator-splitting methods (justifying thus the ADMM terminology).

Following [73], we will discuss in Section 4 the operator-splitting based *direct numerical simulation of particulate flow*, in the particular case of mixtures of incompressible viscous fluids and rigid solid particles.

In Section 5, we will discuss the application of operator-splitting methods to the solution of two problems from Physics, namely the *Gross-Pitaevskii* equation, a *nonlinear Schrödinger equation* modeling *Bose-Einstein condensates*, and the *Zakharov* system, a model for the *propagation of Langmuir waves in ionized plasma*.

Next, in Section 6, we will discuss applications of augmented Lagrangian and ADMM algorithms to the solution of problems from *Imaging*, a highly popular topic nowadays (actually, the renewed interest in ADMM type algorithms that we observe currently can be largely explained by their application to Image Processing; see [156, 170]).

Finally, in Section 7, we will return to various issues that we left behind in the preceding sections of this chapter: these include augmentation parameter selection, an analysis of the asymptotic behavior of the Peaceman-Rachford and Douglas-Rachford schemes, and various comments concerning high order accurate operator-splitting schemes. Also, owing to the fact that one of the success stories of operator-splitting methods has been the numerical solution of the Navier-Stokes equations modeling viscous flow, we will conclude this section (and the chapter) by providing a (non-exhaustive) list of related references.

In addition to all the other chapters of this volume, material related to operator-splitting, augmented Lagrangian and ADMM algorithms can be found in [72] (see also the references therein). More references will be given in the following sections.

2 Operator-Splitting Schemes for the Time Discretization of Initial Value Problems

2.1 Generalities

Let us consider the following autonomous initial value problem:

$$\begin{cases} \dfrac{d\phi}{dt} + A(\phi) = 0 \text{ on } (0,T) \text{ (with } 0 < T \leq +\infty), \\ \phi(0) = \phi_0. \end{cases} \tag{2.1}$$

Operator A maps the vector space V into itself and we suppose that $\phi_0 \in V$. We suppose also that A has a *nontrivial decomposition* such as

$$A = \sum_{j=1}^{J} A_j, \tag{2.2}$$

with $J \geq 2$ (by *nontrivial* we mean that the operators A_j are individually simpler than A).

A question which arises naturally is clearly:

Can we take advantage of decomposition (2.2) for the solution of (2.1)?

It has been known for many years (see for example [36]) that the answer to the above question is definitely *yes*.

Many schemes have been designed to take advantage of the decomposition (2.2) when solving (2.1); several of them will be briefly discussed in the following paragraphs.

2.2 Time-Discretization of (2.1) by Lie's Scheme

Let $\triangle t(>0)$ be a time-discretization step (for simplicity, we suppose $\triangle t$ fixed); we denote $n\triangle t$ by t^n. With ϕ^n denoting an approximation of $\phi(t^n)$, *Lie's scheme* reads as follows (for its derivation see, e.g., [70] (Chapter 6) and Chapter 1, Section 2, of this book):

$$\phi^0 = \phi_0; \tag{2.3}$$

then, for $n \geq 0$, $\phi^n \to \phi^{n+1}$ via

$$\begin{cases} \dfrac{d\phi_j}{dt} + A_j(\phi_j) = 0 \text{ on } (t^n, t^{n+1}), \\ \phi_j(t^n) = \phi^{n+(j-1)/J}; \phi^{n+j/J} = \phi_j(t^{n+1}), \end{cases} \tag{2.4}$$

for $j = 1, \ldots, J$.

If (2.1) is taking place in a finite dimensional space and if the operators A_j are smooth enough, then $\|\phi(t^n) - \phi^n\| = O(\triangle t)$, function ϕ being the solution of (2.1).

Remark 1. The above scheme applies also for *multivalued* operators (such as the *subdifferentials* of proper lower semi-continuous convex functionals), but in such a case first order accuracy is not guaranteed anymore. A related application will be given in Section 2.7.

Remark 2. The above scheme is easy to generalize to *non-autonomous* problems by observing that

$$\begin{cases} \dfrac{d\phi}{dt} + A(\phi, t) = 0, \\ \phi(0) = \phi_0 \end{cases} \Leftrightarrow \begin{cases} \dfrac{d\phi}{dt} + A(\phi, \theta) = 0, \\ \dfrac{d\theta}{dt} - 1 = 0, \\ \phi(0) = \phi_0, \theta(0) = 0. \end{cases}$$

Remark 3. Scheme (2.3)–(2.4) is semi-constructive in the sense that we still have to solve the initial value sub-problems in (2.4) for each j. Suppose that we discretize these sub-problems using *just one step of the backward Euler scheme*. The resulting scheme reads as follows:

$$\phi^0 = \phi_0; \tag{2.5}$$

then, for $n \geq 0$, $\phi^n \to \phi^{n+1}$ via the solution of

$$\frac{\phi^{n+j/J} - \phi^{n+(j-1)/J}}{\triangle t} + A_j(\phi^{n+j/J}) = 0, \tag{2.6}$$

for $j = 1, \ldots, J$.

Scheme (2.5)–(2.6) is known as the *Marchuk-Yanenko scheme* (see, e.g., refs. [121] and [70] (Chapter 6)) for more details). Several chapters of this volume are making use of the Marchuk-Yanenko scheme.

2.3 Time-Discretization of (2.1) by Strang's Symmetrized Scheme

In order to improve the accuracy of Lie's scheme, G. Strang suggested a *symmetrized* variant of scheme (2.3)–(2.4) (ref. [153]). When applied to non-autonomous problems, in the case where $J = 2$, we obtain (with $t^{n+1/2} = (n+1/2)\triangle t$):

$$\phi^0 = \phi_0; \tag{2.7}$$

then, for $n \geq 0$, $\phi^n \to \phi^{n+1/2} \to \widehat{\phi}^{n+1/2} \to \phi^{n+1}$ via

$$\begin{cases} \dfrac{d\phi_1}{dt} + A_1(\phi_1, t) = 0 \text{ on } (t^n, t^{n+1/2}), \\ \phi_1(t^n) = \phi^n; \phi^{n+1/2} = \phi_1(t^{n+1/2}), \end{cases} \tag{2.8}$$

$$\begin{cases} \dfrac{d\phi_2}{dt} + A_2(\phi_2, t^{n+1/2}) = 0 \text{ on } (0, \triangle t), \\ \phi_2(0) = \phi^{n+1/2}; \widehat{\phi}^{n+1/2} = \phi_2(\triangle t), \end{cases} \tag{2.9}$$

$$\begin{cases} \dfrac{d\phi_1}{dt} + A_1(\phi_1, t) = 0 \text{ on } (t^{n+1/2}, t^{n+1}), \\ \phi_1(t^{n+1/2}) = \widehat{\phi}^{n+1/2}; \phi^{n+1} = \phi_1(t^{n+1}). \end{cases} \tag{2.10}$$

If (2.1) is taking place in a finite dimensional space and if operators A_1 and A_2 are smooth enough, then $\|\phi(t^n) - \phi^n\| = O(|\triangle t|^2)$, function ϕ being the solution of (2.1).

Remark 4. In order to preserve the second order accuracy of scheme (2.7)–(2.10) (assuming it takes place) we have to solve the initial value problems in (2.8), (2.9) and (2.10) by schemes which are themselves second order accurate (at least); these schemes are highly dependent of the properties of A_1 and A_2. The sub-problems (2.8), (2.9) and (2.10) are all particular cases of

$$\begin{cases} \dfrac{d\phi}{dt} + B(\phi, t) = 0 \text{ on } (t_0, t_f), \\ \phi(t_0) = \phi_0. \end{cases} \tag{2.11}$$

Suppose now that B is a (positively) monotone operator; following [70] (Chapter 6), we advocate using for the numerical integration of (2.11) the *second order implicit Runge-Kutta scheme* below:

$$\begin{cases} \phi^0 = \phi_0; \\ \text{for } q = 0,\ldots,Q-1,\ \phi^q \to \phi^{q+\theta} \to \phi^{q+1-\theta} \to \phi^{q+1} \text{ via} \\ \begin{cases} \dfrac{\phi^{q+\theta} - \phi^q}{\theta\tau} + B(\phi^{q+\theta}, t^{q+\theta}) = 0, \\ \phi^{q+1-\theta} = \dfrac{1-\theta}{\theta}\phi^{q+\theta} + \dfrac{2\theta - 1}{\theta}\phi^q, \\ \dfrac{\phi^{q+1} - \phi^{q+1-\theta}}{\theta\tau} + B(\phi^{q+1}, t^{q+1}) = 0, \end{cases} \end{cases} \tag{2.12}$$

where in (2.12):

- $Q(\geq 1)$ is an integer and $\tau = \dfrac{t_f - t_0}{Q}$.
- $\phi^{q+\alpha}$ is an approximation of $\phi(t^{q+\alpha})$, with $t^{q+\alpha} = t_0 + (q+\alpha)\tau$.
- $\theta = 1 - \dfrac{1}{\sqrt{2}}$.

It is shown in [70] (Chapter 2) that the implicit Runge-Kutta scheme (2.12) is *stiff A-stable* and "nearly" third-order accurate. It has been used, in particular, in [70] and [162] for the numerical simulation of incompressible viscous flow.

Remark 5. The main (if not the unique) drawback of Strang's symmetrized scheme (2.7)–(2.10) concerns its ability at capturing the *steady state solutions* of (2.1) (when $T = +\infty$), assuming that such solutions do exist. Indeed, the *splitting error* associated with scheme (2.7)–(2.10) prevents using large values of $\triangle t$ when integrating (2.1) from $t = 0$ to $t = +\infty$; if the sequence $\{\phi^n\}_{n\geq 0}$ converges to a limit, this limit is not, in general, a steady state solution of (2.1), albeit being close to one for small values of $\triangle t$ (a similar comment applies also to the sequences $\{\phi^{n+1/2}\}_{n\geq 0}$ and $\{\widehat{\phi}^{n+1/2}\}_{n\geq 0}$). A simple way to-partly-overcome this difficulty is to use variable time discretization steps: for example, in (2.8), (2.9) and (2.10), one can replace $\triangle t$ by τ_n (the sequence $\{\tau_n\}_{n\geq 0}$ verifying $\tau_n > 0$, $\lim_{n\to\infty} \tau_n = 0$ and $\sum_{n=0}^{\infty} \tau_n = +\infty$), and then define t^{n+1} and $t^{n+1/2}$ by $t^{n+1} = t^n + \tau_n\ \forall n \geq 0$, $t^0 = 0$, and $t^{n+1/2} = t^n + \tau^n/2$, respectively. A more sophisticated way to fix the asymptotic behavior of scheme (2.7)–(2.10) is to proceed as in the chapter by *McNamara* and *Strang* in this book (Chapter 3).

Remark 6. More comments on scheme (2.7)–(2.10) can be found in, e.g., [70] (Chapter 6), [72] (Chapter 3) and various chapters of this volume, Chapter 3 in particular. Among these comments, the generalization of the above scheme to those situations where $J \geq 3$ in (2.2) has been discussed. Conceptually, the case $J \geq 3$ is no more complicated than $J = 2$. Focusing on $J = 3$, we can return (in a nonunique way) to the case $J = 2$ by observing that

$$\begin{aligned} A &= A_1 + A_2 + A_3 = A_1 + (A_2 + A_3) = (A_1 + A_2) + A_3 \\ &= (A_1 + \frac{1}{2}A_2) + (\frac{1}{2}A_2 + A_3). \end{aligned} \tag{2.13}$$

The first (resp., second and third) arrangement in (2.13) leads to 5 (resp., 7 and 9) initial value sub-problems per time step. Scheme (2.7)–(2.10), combined with the first arrangement in (2.13), has been applied in [81] to the computation of the periodic solution of a nonlinear integro-differential equation from Electrical Engineering.

2.4 Time-Discretization of (2.1) by Peaceman-Rachford's Alternating Direction Method

Another candidate for the numerical solution of the initial value problem (2.1), or of its non-autonomous variant

$$\begin{cases} \frac{d\phi}{dt} + A(\phi,t) = 0 \text{ on } (0,T), \\ \phi(0) = \phi_0. \end{cases} \tag{2.14}$$

is provided, if $J = 2$ in (2.2), by the *Peaceman-Rachford scheme* (introduced in [139]). The idea behind the Peaceman-Rachford scheme is quite simple: the notation being like in Sections 2.1, 2.2 and 2.3, one divides the time interval $[t^n, t^{n+1}]$ into two sub-intervals of length $\Delta t/2$ using the mid-point $t^{n+1/2}$. Then assuming that the approximate solution ϕ^n is known at t^n one computes first $\phi^{n+1/2}$ using over $[t^n, t^{n+1/2}]$ a scheme of the *backward Euler* type with respect to A_1 and of the *forward Euler* type with respect to A_2; one proceeds similarly over $[t^{n+1/2}, t^{n+1}]$, switching the roles of A_1 and A_2. The following scheme, due to *Peaceman and Rachford* (see [139]), realizes precisely this program when applied to the solution of the initial value problem (2.14):

$$\begin{cases} \phi^0 = \phi_0; \\ \text{for } n \geq 0,\ \phi^n \to \phi^{n+1/2} \to \phi^{n+1} \text{ via the solution of} \\ \frac{\phi^{n+1/2} - \phi^n}{\Delta t/2} + A_1(\phi^{n+1/2}, t^{n+1/2}) + A_2(\phi^n, t^n) = 0, \\ \frac{\phi^{n+1} - \phi^{n+1/2}}{\Delta t/2} + A_1(\phi^{n+1/2}, t^{n+1/2}) + A_2(\phi^{n+1}, t^{n+1}) = 0. \end{cases} \tag{2.15}$$

The *convergence* of the *Peaceman-Rachford scheme* (2.15) has been proved in [118] and [84] under quite general *monotonicity* assumptions concerning the operators A_1 and A_2 (see also [64, 65] and [110]); indeed, A_1 and/or A_2 can be nonlinear, unbounded and even multi-valued. In general, scheme (2.15) is *first order accurate* at best; however, if the operators A_1 and A_2 are linear, time independent, and *commute* then scheme (2.15) is *second order accurate* (that is $\|\phi^n - \phi(t^n)\| = O(|\Delta t|^2)$), ϕ being the solution of problem (2.1)). Further properties of scheme (2.15) can be found in, e.g., [121, 70] (Chapter 2) and [72] (Chapter 3), including its *stability*, and its *asymptotic behavior* if $T = +\infty$; concerning this last issue, a sensible advice is to use another scheme to compute steady state solutions, scheme (2.15) *not being stiff A-stable*.

Remark 7. Scheme (2.15) belongs to the *alternating direction method* family. The reason of that terminology is well known: one of the very first applications of scheme (2.15) was the numerical solution of the *heat equation*

$$\frac{\partial \phi}{\partial t} - \frac{\partial^2 \phi}{\partial x^2} - \frac{\partial^2 \phi}{\partial y^2} = f,$$

completed by initial and boundary conditions. After finite difference discretization, the roles of A_1 and A_2 were played by the square matrices approximating the operators $-\frac{\partial^2}{\partial x^2}$ and $-\frac{\partial^2}{\partial y^2}$, respectively, explaining the terminology.

Remark 8. We observe that operators A_1 and A_2 play essentially *symmetrical* roles in scheme (2.15).

Remark 9. For those fairly common situations where operator A_2 is *uni-valued*, but operator A_1 is "nasty" (discontinuous and/or multi-valued, etc.), we should use the following equivalent formulation of the Peaceman-Rachford scheme (2.15):

$$\begin{cases} \phi^0 = \phi_0; \\ \text{for } n \geq 0,\ \phi^n \to \phi^{n+1/2} \to \phi^{n+1} \text{ via the solution of} \\ \dfrac{\phi^{n+1/2} - \phi^n}{\Delta t/2} + A_1(\phi^{n+1/2}, t^{n+1/2}) + A_2(\phi^n, t^n) = 0, \\ \dfrac{\phi^{n+1} - 2\phi^{n+1/2} + \phi^n}{\Delta t/2} + A_2(\phi^{n+1}, t^{n+1}) = A_2(\phi^n, t^n). \end{cases} \tag{2.16}$$

2.5 *Time-Discretization of (2.1) by Douglas-Rachford's Alternating Direction Method*

We assume that $J = 2$ in (2.2).

The *Douglas-Rachford* scheme (introduced in [57]) is a variant of the *Peaceman-Rachford* scheme (2.15); when applied to the numerical solution of the initial value problem (2.14) (the non-autonomous generalization of (2.1)), it takes the following form:

$$\begin{cases} \phi^0 = \phi_0; \\ \text{for } n \geq 0,\ \phi^n \to \widehat{\phi}^{n+1} \to \phi^{n+1} \text{ via the solution of} \\ \dfrac{\widehat{\phi}^{n+1} - \phi^n}{\Delta t} + A_1(\widehat{\phi}^{n+1}, t^{n+1}) + A_2(\phi^n, t^n) = 0, \\ \dfrac{\phi^{n+1} - \phi^n}{\Delta t} + A_1(\widehat{\phi}^{n+1}, t^{n+1}) + A_2(\phi^{n+1}, t^{n+1}) = 0. \end{cases} \tag{2.17}$$

The Douglas-Rachford scheme (2.17) has clearly a *predictor-corrector* flavor.

The convergence of the *Douglas-Rachford scheme* (2.17) has been proved in [118] and [84] under quite general *monotonicity* assumptions concerning the operators A_1

and A_2 (see also [64, 65] and [110]); indeed, A_1 and/or A_2 can be nonlinear, unbounded, and even multi-valued. In general, scheme (2.17) is first order accurate at best (even if the operators A_1 and A_2 are linear, time independent and commute, assumptions implying second order accuracy for the Peaceman-Rachford scheme). Further properties of scheme (2.17) can be found in, e.g., [121, 70] (Chapter 2) and [72] (Chapter 3), including its *stability*, and its *asymptotic behavior* if $T = +\infty$. Concerning this last issue, a sensible advice is to use another scheme to compute steady state solutions, scheme (2.17) *not being stiff A-stable*, a property it shares with the Peaceman-Rachford scheme (2.15).

Remark 10. Unlike the Peaceman-Rachford scheme (2.15), we observe that the roles played by operators A_1 and A_2 are *non-symmetrical* in scheme (2.17); actually, numerical experiments confirm that fact: for example, for the same $\triangle t$ the speed of convergence to a steady state solution may depend of the choice one makes for A_1 and A_2. As a rule of thumb, we advocate taking for A_2 the operator with the best continuity and monotonicity properties (see, for example, [62] (Chapter 3), [63] (Chapter 3) and [74] (Chapter 3) for more details).

Remark 11. Unlike scheme (2.15), scheme (2.17) is easy to generalize to operator decompositions involving *more than two operators*. Consider thus the numerical integration of (2.14) when $J \geq 3$ in (2.2). Following *J. Douglas* in [54] and [55] we generalize scheme (2.17) by

$$\phi^0 = \phi_0; \tag{2.18}$$

then for $n \geq 0$, ϕ^n being known, compute $\phi^{n+1/J}, \ldots, \phi^{n+j/J}, \ldots, \phi^{n+1}$ via the solution of

$$\begin{cases} \dfrac{\phi^{n+1/J}-\phi^n}{\triangle t} + \dfrac{1}{J-1}A_1(\phi^{n+1/J},t^{n+1}) + \dfrac{J-2}{J-1}A_1(\phi^n,t^n) \\ \qquad + \displaystyle\sum_{i=2}^{J} A_i(\phi^n,t^n) = 0, \end{cases} \tag{19.1}$$

$$\begin{cases} \dfrac{\phi^{n+j/J}-\phi^n}{\triangle t} + \displaystyle\sum_{i=1}^{j-1}\left[\dfrac{1}{J-1}A_i(\phi^{n+i/J},t^{n+1}) + \dfrac{J-2}{J-1}A_i(\phi^n,t^n)\right] \\ \qquad + \dfrac{1}{J-1}A_j(\phi^{n+j/J},t^{n+1}) + \dfrac{J-2}{J-1}A_j(\phi^n,t^n) \\ \qquad + \displaystyle\sum_{i=j+1}^{J} A_i(\phi^n,t^n) = 0, \end{cases} \tag{19.j}$$

$$\begin{cases} \dfrac{\phi^{n+1}-\phi^n}{\triangle t} + \displaystyle\sum_{i=1}^{J-1}\left[\dfrac{1}{J-1}A_i(\phi^{n+i/J},t^{n+1}) + \dfrac{J-2}{J-1}A_i(\phi^n,t^n)\right] \\ \qquad + \dfrac{1}{J-1}A_J(\phi^{n+1},t^{n+1}) + \dfrac{J-2}{J-1}A_J(\phi^n,t^n) = 0, \end{cases} \tag{19.J}$$

Above, $\phi^{n+i/J}$ and $\phi^{n+j/J}$ denote approximate solutions at steps i and j of the computational process; *they do not denote* approximations of $\phi(t^{n+i/J})$ and $\phi(t^{n+j/J})$ (unless $i = j = J$).

Remark 12. This is the Douglas-Rachford analog of Remark 9: for those situations where A_1 is a "bad" operator (in the sense of Remark 9), we should use (assuming that A_2 is uni-valued) the following *equivalent* formulation of the Douglas-Rachford scheme (2.17):

$$\begin{cases} \phi^0 = \phi_0; \\ \text{for } n \geq 0,\ \phi^n \to \widehat{\phi}^{n+1} \to \phi^{n+1} \text{ via the solution of} \\ \dfrac{\widehat{\phi}^{n+1} - \phi^n}{\triangle t} + A_1(\widehat{\phi}^{n+1}, t^{n+1}) + A_2(\phi^n, t^n) = 0, \\ \dfrac{\phi^{n+1} - \widehat{\phi}^{n+1}}{\triangle t} + A_2(\phi^{n+1}, t^{n+1}) = A_2(\phi^n, t^n). \end{cases} \tag{2.20}$$

Remark 13. To those wondering how to choose between the Peaceman-Rachford and Douglas-Rachford schemes, we will say that, on the basis of many numerical experiments, it seems that the second scheme is more robust and faster for those situations where one of the operators is *non-smooth* (multi-valued or singular, for example), particularly if one is interested by capturing steady state solutions. Actually, a better advice could be: consider using the *fractional θ-scheme* to be discussed in Section 2.6, below. Indeed, we have encountered situations where this *θ-scheme* outperforms both the Peaceman-Rachford and Douglas-Rachford schemes, for steady state computations in particular; such an example is provided by the *anisotropic Eikonal equation*, a nonlinear hyperbolic problem to be briefly discussed in Section 2.7. We will return to the Peaceman-Rachford vs Douglas-Rachford issue in Section 7.

2.6 Time-Discretization of (2.1) by a Fractional θ-Scheme

This scheme (introduced in [67, 68] for the solution of the *Navier-Stokes equations*) is a variant of the *Peaceman-Rachford* scheme (2.15). Let θ belong to the open interval $(0, 1/2)$ (in practice, $\theta \in [1/4, 1/3]$); the fractional θ -scheme, applied to the solution of the initial value problem (2.14) (the non-autonomous generalization of (2.1)), reads as follows if $A = A_1 + A_2$:

$$\begin{cases} \phi^0 = \phi_0; \\ \text{for } n \geq 0,\ \phi^n \to \phi^{n+\theta} \to \phi^{n+1-\theta} \to \phi^{n+1} \text{ via the solution of} \\ \dfrac{\phi^{n+\theta} - \phi^n}{\theta \triangle t} + A_1(\phi^{n+\theta}, t^{n+\theta}) + A_2(\phi^n, t^n) = 0, \\ \dfrac{\phi^{n+1-\theta} - \phi^{n+\theta}}{(1-2\theta)\triangle t} + A_1(\phi^{n+\theta}, t^{n+\theta}) + A_2(\phi^{n+1-\theta}, t^{n+1-\theta}) = 0, \\ \dfrac{\phi^{n+1} - \phi^{n+1-\theta}}{\theta \triangle t} + A_1(\phi^{n+1}, t^{n+1}) + A_2(\phi^{n+1-\theta}, t^{n+1-\theta}) = 0. \end{cases} \tag{2.21}$$

Remark 14. One should avoid confusion between scheme (2.21) and the following solution method for the initial value problem (2.14) (with $0 \leq \theta \leq 1$)

$$\begin{cases} \phi^0 = \phi_0; \\ \text{for } n \geq 0,\ \phi^n \rightarrow \phi^{n+1} \text{ via the solution of} \\ \dfrac{\phi^{n+1} - \phi^n}{\triangle t} + \theta A(\phi^{n+1}, t^{n+1}) + (1-\theta)A(\phi^n, t^n) = 0, \end{cases} \tag{2.22}$$

which is also known as a θ-scheme. We observe that if $\theta = 1$ (resp., $\theta = 0$, θ =1/2) scheme (2.22) reduces to *backward Euler's* scheme (resp., *forward Euler's* scheme, a *Crank-Nicolson's* type scheme). Another "interesting" value is $\theta = 2/3$ (for reasons detailed in, e.g., [70] (Chapter 2) and [72] (Chapter 3)). By the way, it is to avoid confusion between schemes (2.21) and (2.22) that some practitioners (*S. Turek*, in particular) call the first one a fractional θ-scheme. □

The stability and convergence properties of scheme (2.21) have been discussed in [70] (Chapter 2) and [72] (Chapter 3) for very simple finite dimensional situations where A_1 and A_2 are both positive multiples of the same symmetric positive definite matrix. Numerical experiments have shown that the good properties verified by scheme (2.21) for those simple linear situations, in particular its *stiff A-stability* for θ well chosen, still hold for more complicated problems, such as the numerical simulation of *unsteady incompressible viscous flow* modeled by the *Navier-Stokes equations* (as shown in, e.g., [23, 41, 69] and [70]).

Remark 15. We observe that operators A_1 and A_2 play *non-symmetrical* roles in scheme (2.21). Since, at each time step, one has to solve two problems (resp., one problem) associated with operator A_1 (resp., A_2) a natural choice is to take for A_1 the operator leading to the sub-problems which are the easiest to solve (that is, whose solution is the less time consuming). Less naive criteria may be used to choose A_1 and A_2, such as the *regularity* (or lack of regularity) of these operators.

Remark 16. If one takes $A_1 = A$ and $A_2 = 0$ in (2.21), the above scheme reduces to the Runge-Kutta scheme (2.12), with A replacing B.

Remark 17. The fractional θ-scheme (2.21) is a *symmetrized scheme*. From that point of view, it has some analogies with *Strang's symmetrized scheme* (2.7)–(2.10), discussed in Section 2.3.

Remark 18. This is the fractional θ-scheme analog of Remarks 9 and 12. For those situations where A_1 is a "bad" operator (in the sense of Remark 9), we advocate using the following *equivalent* formulation of the θ-scheme (2.21):

$$\begin{cases} \phi^0 = \phi_0; \\ \text{for } n \geq 0,\ \phi^n \rightarrow \phi^{n+\theta} \rightarrow \phi^{n+1-\theta} \rightarrow \phi^{n+1} \text{ via the solution of} \\ \dfrac{\phi^{n+\theta} - \phi^n}{\theta \triangle t} + A_1(\phi^{n+\theta}, t^{n+\theta}) + A_2(\phi^n, t^n) = 0, \\ \dfrac{\theta\phi^{n+1-\theta} - (1-\theta)\phi^{n+\theta} + (1-2\theta)\phi^n}{\theta(1-2\theta)\triangle t} + A_2(\phi^{n+1-\theta}, t^{n+1-\theta}) = A_2(\phi^n, t^n), \\ \dfrac{\phi^{n+1} - \phi^{n+1-\theta}}{\theta \triangle t} + A_1(\phi^{n+1}, t^{n+1}) + A_2(\phi^{n+1-\theta}, t^{n+1-\theta}) = 0. \end{cases} \tag{2.23}$$

2.7 Two Applications: Smallest Eigenvalue Computation and Solution of an Anisotropic Eikonal Equation

2.7.1 Synopsis

It is not an exaggeration to say that applications of operator-splitting methods are everywhere, new ones occurring "almost" every day; indeed, some well-known methods and algorithms are *disguised* operator-splitting schemes as we will show in Section 2.7.2, concerning the computation of the *smallest eigenvalue of a real symmetric matrix*. In Section 2.7.3, we will apply the fractional θ-scheme (2.21) to the solution of an *Eikonal equation* modeling *wave propagation in anisotropic media*. More applications will be discussed in Sections 4 and 5.

2.7.2 Application to Some Eigenvalue Computation

Suppose that $\mathbf{A}$ is a real $d \times d$ *symmetric* matrix. Ordering the eigenvalues of $\mathbf{A}$ as follows: $\lambda_1 \leq \lambda_2 \leq \cdots \leq \lambda_d$, our goal is to compute λ_1. We have (with obvious notation)

$$\lambda_1 = \min_{\mathbf{v} \in S} \mathbf{v}^t \mathbf{A}\mathbf{v}, \quad with\ S = \{\mathbf{v} | \mathbf{v} \in I\!R^d, \|\mathbf{v}\| = 1\}, \tag{2.24}$$

the norm in (2.24) being the canonical Euclidean one. The *constrained minimization* problem in (2.24) is equivalent to

$$\min_{\mathbf{v} \in I\!R^d} \left[\frac{1}{2} \mathbf{v}^t \mathbf{A}\mathbf{v} + I_S(\mathbf{v}) \right], \tag{2.25}$$

where, in (2.25), the functional $I_S : I\!R^d \to I\!R \cup \{+\infty\}$ is defined as follows

$$I_S(\mathbf{v}) = \begin{cases} 0\ if\ \mathbf{v} \in S, \\ +\infty\ otherwise, \end{cases}$$

implying that I_S is the *indicator functional* of the sphere S. Suppose that $\mathbf{u}$ is a solution of (2.25) (that is a minimizer of the functional in (2.25)); we have then

$$\mathbf{A}\mathbf{u} + \partial I_S(\mathbf{u}) \ni \mathbf{0}, \tag{2.26}$$

$\partial I_S(\mathbf{u})$ in (2.26) being a (kind of) generalized gradient of functional I_S at $\mathbf{u}$ (∂I_S is a multivalued operator). Next, we associate with the (necessary) optimality system (2.26) the following initial value problem (*flow* in the *Dynamical System* terminology):

$$\begin{cases} \dfrac{d\mathbf{u}}{dt} + \mathbf{A}\mathbf{u} + \partial I_S(\mathbf{u}) \ni \mathbf{0} \text{ in } (0, +\infty), \\ \mathbf{u}(0) = \mathbf{u}_0. \end{cases} \tag{2.27}$$

If one applies the Marchuk-Yanenko scheme (2.5)–(2.6) to the solution of problem (2.27), one obtains (with $\tau = \triangle t$):

$$\begin{cases} \mathbf{u}^0 = \mathbf{u}_0, \\ \text{for } n \geq 0,\ \mathbf{u}^n \to \mathbf{u}^{n+1/2} \to \mathbf{u}^{n+1} \text{ via the solution of} \\ \dfrac{\mathbf{u}^{n+1/2} - \mathbf{u}^n}{\tau} + \mathbf{A}\mathbf{u}^{n+1/2} = \mathbf{0}, \\ \dfrac{\mathbf{u}^{n+1} - \mathbf{u}^{n+1/2}}{\tau} + \partial I_S(\mathbf{u}^{n+1}) \ni \mathbf{0}. \end{cases} \tag{2.28}$$

The *first* finite difference equation in (2.28) implies

$$\mathbf{u}^{n+1/2} = (\mathbf{I} + \tau\mathbf{A})^{-1}\mathbf{u}^n. \tag{2.29}$$

On the other hand, the *second* finite difference equation in (2.28) can be interpreted as a necessary optimality condition for the following minimization problem

$$\min_{\mathbf{v}\in S} \left[\frac{1}{2}\|\mathbf{v}\|^2 - \mathbf{v}^t\mathbf{u}^{n+1/2}\right]. \tag{2.30}$$

Since $\|\mathbf{v}\| = 1$ over S, the solution of problem (2.30) is given by

$$\mathbf{u}^{n+1} = \frac{\mathbf{u}^{n+1/2}}{\|\mathbf{u}^{n+1/2}\|}. \tag{2.31}$$

It follows from (2.29) and (2.31) that algorithm (2.28) is nothing but the *inverse power method with shift*, a well-known algorithm from *Numerical Linear Algebra*. Indeed, if

$$0 < \tau < \frac{1}{\max(0_+, -\lambda_1)},$$

and if the projection of $\mathbf{u}_0$ on the vector space spanned by the eigenvectors of $\mathbf{A}$ associated with λ_1 is different from $\mathbf{0}$, we can easily prove that the sequence $\{\mathbf{u}^n\}_{n\geq 0}$ converges to an eigenvector of $\mathbf{A}$ associated with λ_1 and also that

$$\lim_{n\to+\infty} (\mathbf{u}^n)^t\mathbf{A}\mathbf{u}^n = \lambda_1.$$

Clearly, numerical analysts have not been waiting for operator-splitting to compute matrix eigenvalues and eigenvectors; on the other hand, operator-splitting has provided efficient algorithms for the solution of complicated problems from Differential Geometry, Mechanics, Physics, Physico-Chemistry, Finance, etc., including some nonlinear eigenvalue problems, as shown in, e.g., [72] (Chapter 7).

2.7.3 Application to the Solution of an Anisotropic Eikonal Equation from Acoustics

The next application of operator-splitting, that we are going to (briefly) consider in this chapter, was brought to our attention recently (December 2014) by our colleagues *S. Leung* and *J. Qian*. It concerns the numerical solution of the following *nonlinear hyperbolic* partial differential equation

$$|\nabla\tau| - \frac{|1 - \mathbf{V}\cdot\nabla\tau|}{c} = 0 \ \ in\ \Omega, \tag{2.32}$$

encountered in *Acoustics* and known as the *anisotropic Eikonal equation*. In (2.32), we have (see [40] for more details):

- $\Omega \subset \mathbb{R}^d$, with $d \geq 2$.
- $\tau(x)$ is the time of *1*st *arrival of the wave front* at $x \in \Omega$.
- $c > 0$ is the *wave propagation speed* in the medium filling Ω, assuming that this medium is at rest (the so-called *background medium*).
- Assuming that the ambient medium is moving, $\mathbf{V}$ is its *moving velocity*; we assume that $\mathbf{V} \in (L^\infty(\Omega))^d$.

Fast-sweeping methods have been developed for the efficient numerical solution of the classical Eikonal equation

$$|\nabla\tau| = \frac{1}{c} \ \ in\ \Omega, \tag{2.33}$$

(see, e.g., [104] and [181]); these methods provide automatically *viscosity solutions* in the sense of *Crandall* and *Lions* (see [38] for this notion). Unfortunately, as shown in [40], the fast sweeping methods developed for the solution of (2.33) cannot handle (2.32), unless one modifies them significantly, as done in [40]. Actually, there exists an alternative, simpler to implement, to the method developed in [40]: it relies on the operator-splitting methods discussed in Sections 2.3, 2.4, 2.5 and 2.6, and takes advantage of the fact that the fast-sweeping methods developed for the solution of (2.33) can be easily modified in order to handle equations such as

$$\alpha\tau - \beta\nabla^2\tau + |\nabla\tau| = f \tag{2.34}$$

and

$$\alpha\tau - \beta\nabla^2\tau - \frac{|1 - \mathbf{V}\cdot\nabla\tau|}{c} = f, \tag{2.35}$$

with $\alpha > 0$ and $\beta \geq 0$. Therefore, in order to solve problem (2.32), we associate with it the following initial value problem:

$$\begin{cases} (I - \varepsilon\nabla^2)\dfrac{\partial\tau}{\partial t} + |\nabla\tau| - \dfrac{|1 - \mathbf{V}\cdot\nabla\tau|}{c} = 0 \ \ in\ \Omega\times(0, +\infty), \\ \tau(0) = \tau_0, \end{cases} \tag{2.36}$$

whose steady state solutions are also solutions of (2.32). In (2.36), ε is a non-negative parameter (a regularizing one if $\varepsilon > 0$) and $\tau(t)$ denotes the function $t \to \tau(x,t)$. Actually, additional conditions are required to have solution uniqueness, typical ones being τ specified on a subset of $\overline{\Omega}(=\Omega \cup \partial\Omega)$, possibly reduced to just one point (a point source for the wave). A typical choice for τ_0 is the corresponding solution of problem (2.33).

The results reported in [75] show that, with $\theta = 1/3$, the fractional θ-scheme discussed in Section 2.6 outperforms the Strang's, Peaceman-Rachford's, and Douglas-Rachford's schemes when applied to the computation of the steady state solutions of (2.36). The resulting algorithm reads as follows:

$$\begin{cases} \tau^0 = \tau_0; \\ \text{for } n \geq 0,\ \tau^n \to \tau^{n+\theta} \to \tau^{n+1-\theta} \to \tau^{n+1} \text{ via the solution of} \\ (I - \varepsilon\nabla^2)\dfrac{\tau^{n+\theta} - \tau^n}{\theta \triangle t} + |\nabla \tau^{n+\theta}| - \dfrac{|1 - \mathbf{V}\cdot\nabla\tau^n|}{c} = 0, \\ (I - \varepsilon\nabla^2)\dfrac{\tau^{n+1-\theta} - \tau^{n+\theta}}{(1-2\theta)\triangle t} + |\nabla \tau^{n+\theta}| - \dfrac{|1 - \mathbf{V}\cdot\nabla\tau^{n+1-\theta}|}{c} = 0, \\ (I - \varepsilon\nabla^2)\dfrac{\tau^{n+1} - \tau^{n+1-\theta}}{\theta \triangle t} + |\nabla \tau^{n+1}| - \dfrac{|1 - \mathbf{V}\cdot\nabla\tau^{n+1-\theta}|}{c} = 0. \end{cases} \tag{2.37}$$

The three problems in (2.37) being particular cases of (2.34) and (2.35), their finite difference analogues can be solved by fast-sweeping algorithms. Physical considerations suggest that $\triangle t$ has to be of the order of the space discretization step h. Actually, the numerical results reported in [75] show that, unlike the other schemes discussed in Sections 2.2 to 2.5, scheme (2.37), with $\theta = 1/3$, has very good convergence properties, even for large values of the ratio $\dfrac{\triangle t}{h}$ (100, typically). If $\varepsilon = 0$ (resp., h^2), these numerical experiments suggest that the number of iterations (time steps), necessary to achieve convergence to a steady state solution, varies (roughly) like $h^{-1/2}$ (resp., $h^{-1/3}$), for two- and three-dimensional test problems (see [75] for further results and more details). Clearly, preconditioning does pay here (a well-known fact, in general).

Remark 19. Some readers may wonder why the authors of [75] gave the role of A_1 (resp., A_2) to the operator $\tau \to |\nabla\tau|$ (resp., $\tau \to -\dfrac{1}{c}|1 - \mathbf{V}\cdot\nabla\tau|$), and not the other way around. Let us say to these readers that the main reason behind that choice was preliminary numerical experiments showing that, for the same values of α and β, problem (2.34) is cheaper to solve that problem (2.35).

2.8 Time-Discretization of (2.1) by a Parallel Splitting Scheme

The splitting schemes presented so far have a sequential nature, i.e. the sub-problems associated with the decomposed operators are solved in a sequential manner. Actually, it is also possible to solve the sub-problems in parallel, as shown just below, using the following variant of Marchuk-Yanenko's scheme:

$$\begin{cases} \phi^0 = \phi_0; \\ \text{for } n \geq 0, \text{ we obtain } \phi^{n+1} \text{ from } \phi^n \text{ by solving first} \\ \dfrac{\phi^{n+j/2J} - \phi^n}{J\triangle t} + A_j(\phi^{n+j/2J}, t^{n+1}) = 0, \text{ for } j = 1, \ldots, J, \\ \phi^{n+1} \text{ being then obtained by averaging as follows} \\ \phi^{n+1} = \dfrac{1}{J}\sum_{j=1}^{J} \phi^{n+j/2J}. \end{cases} \tag{2.38}$$

Scheme (2.38) is nothing but Algorithm 5.1 in [119]. Under suitable conditions, it has been proved in the above reference that scheme (2.38) is first order accurate, that is $\|\phi^n - \phi(t^n)\| = O(\triangle t)$. A parallelizable algorithm with second order accuracy is presented also in [119]. The main advantage of the above schemes is that the sub-problems can be solved in parallel. Clearly, this parallel splitting idea can be used for computing the steady state solutions of (2.1). As observed in [155], the sub-problems (or at least some of them) can also be solved in parallel if the corresponding operator A_j has the right decomposition properties.

3 Augmented Lagrangian Algorithms and Alternating Direction Methods of Multipliers

3.1 Introduction

It is our opinion that a review chapter like this one has to include some material about *augmented Lagrangian* algorithms, including of course their relationships with *alternating direction methods*. On the other hand, since augmented Lagrangian algorithms and alternating direction methods of multipliers, and their last known developments, are discussed, with many details, in other chapters of this book, we will not say much about these methods in this section. However, we will give enough information so that the reader may follow Section 6 (dedicated to *Image Processing*) without spending too much time consulting the other chapters (or other references).

In Section 3.2 we will introduce several *augmented Lagrangian algorithms*, and show in section 3.3 how these algorithms relate to the *alternating direction methods* discussed in Sections 2.4 (*Peaceman-Rachford's*) and 2.5 (*Douglas-Rachford's*).

This section is largely inspired by Chapter 4 of [72].

3.2 Decomposition-Coordination Methods by Augmented Lagrangians

3.2.1 Abstract Problem Formulation. Some Examples

A large number of problems in *Mathematics, Physics, Engineering, Economics, Data Processing, Imaging, etc.* can be formulated as

$$u = \arg\min_{v\in V}[F(Bv)+G(v)], \tag{2.39}$$

where: (i) V and H are Banach spaces. (ii) $B \in L(V,H)$. (iii) $F : H \to I\!R \cup \{+\infty\}$ and $G : V \to I\!R \cup \{+\infty\}$ are *proper, lower semi-continuous*, and *convex* functionals verifying $\mathrm{dom}(F \circ B) \cap \mathrm{dom}(G) \neq \emptyset$, implying that problem (2.39) may have solutions.

Example 1. This first example concerns the following variational problem:

$$u = \arg \min_{v\in H_0^1(\Omega)} \left[\frac{\mu}{2}\int_\Omega |\nabla v|^2\,dx + \tau_y \int_\Omega |\nabla v|\,dx - \varpi \int_\Omega v dx\right], \tag{2.40}$$

where: (i) Ω is a bounded domain (that is a bounded open connected subset) of $I\!R^2$; we denote by Γ the boundary of Ω. (ii) $dx = dx_1 dx_2$. (iii) μ and τ_y are two positive constants. (iv) $|\nabla v|^2 = \left|\frac{\partial v}{\partial x_1}\right|^2 + \left|\frac{\partial v}{\partial x_2}\right|^2$ (v) The space $H_0^1(\Omega)$ (a Sobolev space) is defined by

$$H_0^1(\Omega) = \{v | v \in L^2(\Omega), \partial v/\partial x_i \in L^2(\Omega), \forall i = 1,2, v|_\Gamma = 0\}, \tag{2.41}$$

the two derivatives in (2.41) being in the *sense of distributions* (see, e.g., [148, 157] for this notion). Since Ω is bounded, $H_0^1(\Omega)$ is a Hilbert space for the inner product $\{v,w\} \to \int_\Omega \nabla v \cdot \nabla w dx$, and the associated norm. Problem (2.40) is a well-known problem from *non-Newtonian fluid mechanics*; it models the *flow of an incompressible visco-plastic fluid* (of the *Bingham* type) in an infinitely long cylinder of cross-section Ω, ϖ being the *pressure drop per unit length* and u the flow *axial velocity*. In (2.40), μ denotes the fluid *viscosity* and τ_y its *plasticity yield* (see, e.g., [59] and [83] for further information on visco-plastic fluid flows; see also the references therein). It follows from, e.g., [66] and [72], that the variational problem (2.40) has a unique solution.

Problem (2.40) is a particular case of (2.39) with $V = H_0^1(\Omega)$, $H = (L^2(\Omega))^2$, $B = \nabla$, $F(\mathbf{q}) = \tau_y \int_\Omega |\mathbf{q}|\,dx$, and $G(v) = \frac{\mu}{2}\int_\Omega |\nabla v|^2\,dx - \varpi \int_\Omega v dx$; other decompositions are possible.

Close variants of problem (2.40) are encountered in *imaging*, as shown in Section 6 (and other chapters of this volume).

Example 2. It concerns the following variant of problem (2.40):

$$u = \arg\min_{v \in K} \left[\frac{\mu}{2} \int_\Omega |\nabla v|^2 dx - C \int_\Omega v dx \right], \tag{2.42}$$

where Ω is a bounded domain of $\mathbb{R}^2$, μ is a *positive constant* and

$$K = \{v | v \in H_0^1(\Omega), |\nabla v| \leq 1 \text{ a.e. in } \Omega\}.$$

It is a classical result (see, e.g., [59]) that (2.42) models, in an appropriate system of mechanical units, the *torsion* of an infinitely long cylinder of cross-section Ω, made of an *elastic-plastic* material, C being the *torsion angle per unit length* and u a *stress potential*. It follows from, e.g., [66] and [72], that the variational problem (2.42) has a unique solution.

Problem (2.42) is a particular case of problem (2.39) with $V = H_0^1(\Omega)$, $H = (L^2(\Omega))^2$, $B = \nabla$, $G(v) = \frac{\mu}{2} \int_\Omega |\nabla v|^2 dx - C \int_\Omega v dx$, and $F(\mathbf{q}) = I_{\mathcal{K}}(\mathbf{q})$, $I_{\mathcal{K}}(\cdot)$ being the *indicator functional* of the *closed convex nonempty* subset $\mathcal{K}$ of H defined by

$$\mathcal{K} = \{\mathbf{q} | \mathbf{q} \in H, |\mathbf{q}| \leq 1 \text{ a.e. in } \Omega\}.$$

Other decompositions are possible.

Remark 20. We recall that, we have, (from the definition of *indicator functionals*)

$$I_{\mathcal{K}}(\mathbf{q}) = \begin{cases} 0 \text{ if } \mathbf{q} \in \mathcal{K}, \\ +\infty \text{ otherwise,} \end{cases}$$

implying, from the properties of $\mathcal{K}$, that $I_{\mathcal{K}}: H \to \mathbb{R} \cup \{+\infty\}$ is *convex, proper* and *lower semi-continuous.* □

Numerical methods for the solution of problem (2.42) can be found in, e.g., [66] and [76].

3.2.2 Primal-Dual Methods for the Solution of Problem (2.39): ADMM Algorithms

In order to solve problem (2.39), we are going to use a strategy introduced in [77] and [78] (to the best of our knowledge). The starting point is the obvious *equivalence* between (2.39) and the following *linearly constrained* optimization problem:

$$\{u, Bu\} = \arg \min_{\{v,q\} \in W} j(v,q), \tag{2.43}$$

where

$$j(v,q) = F(q) + G(v),$$

and

$$W = \{\{v,q\} | v \in V, q \in H, Bv - q = 0\}.$$

From now on, we will assume that V and H are (real) *Hilbert spaces*, the *H-norm* being denoted by $|\cdot|$ and the associated inner-product by $(\cdot,\cdot)$. The next step is quite natural: we associate with the minimization problem (2.43) a *Lagrangian* functional L defined by

$$L(v,q;\mu) = j(v,q) + (\mu, Bv - q),$$

and an *augmented Lagrangian* functional L_r defined (with $r > 0$) by

$$L_r(v,q;\mu) = L(v,q;\mu) + \frac{r}{2}|Bv - q|^2. \tag{2.44}$$

One can easily prove that the functionals L and L_r share the same saddle-points over $(V \times H) \times H$, and also that, if $\{\{u,p\},\lambda\}$ is such a saddle-point, then u is a solution of problem (2.39) and $p = Bu$. A classical algorithm to compute saddle-points is the so-called *Uzawa algorithm*, popularized by [3] (a book dedicated to the study of *Economics equilibria*), and further discussed in, e.g., [76]. Applying a close variant of the Uzawa algorithm to the computation of the saddle-points of L_r over $(V \times H) \times H$, we obtain

$$\begin{cases} \lambda^0 \text{ is given in } H; \\ \text{for } n \geq 0,\ \lambda^n \to \{u^n, p^n\} \to \lambda^{n+1} \text{ via} \\ \{u^n, p^n\} = \arg \min\limits_{\{v,q\} \in V \times H} L_r(v,q;\lambda^n), \\ \lambda^{n+1} = \lambda^n + \rho(Bu^n - p^n), \end{cases} \tag{2.45}$$

an algorithm called *ALG*1 by some practitioners, following a terminology introduced in [78] (an alternative name could have been *augmented Lagrangian Uzawa algorithm* which summarizes quite well what algorithm (2.45) is all about).

Concerning the *convergence* of *ALG*1 it has been proved in, e.g., [62, 63, 66] and [74] (see also [78]), that if:

(i) L has a saddle-point $\{\{u,p\},\lambda\}$ over $(V \times H) \times H$.
(ii) B is an injection and $R(B)$ is closed in H.
(iii) $\lim\limits_{|q| \to +\infty} \frac{F(q)}{|q|} = +\infty$.
(iv) $F = F_0 + F_1$ with F_0 and F_1 proper, lower semi-continuous and convex, with F_0 Gateaux-differentiable, and uniformly convex on the bounded sets of H

(the above properties imply that problem (2.39) has a unique solution), then we have, $\forall\, r > 0$ and if

$$0 < \rho < 2r,$$

the following convergence result

$$\lim_{n \to +\infty} \{u^n, p^n\} = \{u, Bu\} \text{ in } V \times H, \tag{2.46}$$

where u is the solution of problem (2.39); moreover, the convergence result (2.46) holds $\forall\, \lambda^0 \in H$. The convergence of the multiplier sequence $\{\lambda^n\}_{n\geq 0}$ is no better than *weak* in general, implying that the criterion used to stop $ALG1$ has to been chosen carefully. Of course, in *finite dimension*, the properties of B, F and G implying convergence are *less demanding* than in *infinite dimension*; for example, the *existence* of a solution to problem (2.39) is *sufficient* to imply the *existence* of a *saddle-point*.

The main difficulty with the Uzawa algorithm (2.45) is clearly the solution of the minimization problem it contains. An obvious choice to solve this problem is to use a *relaxation* method (as advocated in [77, 78]). Suppose that, as advocated in the two above references (which show that, indeed, for the nonlinear elliptic problem discussed there the number of relaxation iterations reduces quickly to two), we limit the number of relaxation iterations to one when solving the minimization problem in (2.45): we obtain then the following primal-dual algorithm (called $ALG2$ by some practitioners):

$$\{u^{-1}, \lambda^0\} \text{ is given in } V \times H; \tag{2.47}$$

for $n \geq 0$, $\{u^{n-1}, \lambda^n\} \to p^n \to u^n \to \lambda^{n+1}$ via

$$p^n = \arg\min_{q \in H} L_r(u^{n-1}, q; \lambda^n), \tag{2.48}$$

$$u^n = \arg\min_{v \in V} L_r(v, p^n; \lambda^n), \tag{2.49}$$

$$\lambda^{n+1} = \lambda^n + \rho(Bu^n - p^n). \tag{2.50}$$

Assuming that

$$0 < \rho < \frac{1+\sqrt{5}}{2} r,$$

with the other assumptions implying the convergence of $ALG1$ still holding, we have

$$\lim_{n \to +\infty} \{u^n, p^n\} = \{u, Bu\} \text{ in } V \times H,$$

where u is the solution of problem (2.39). Convergence proofs can be found in [62, 63, 66] and [74].

A simple variant (called $ALG3$) of algorithm (2.47)–(2.50) is obtained by updating the multiplier a first time immediately after (2.48); we obtain then

$$\{u^{-1}, \lambda^0\} \text{ is given in } V \times H, \tag{2.51}$$

for $n \geq 0$, $\{u^{n-1}, \lambda^n\} \to p^n \to \lambda^{n+1/2} \to u^n \to \lambda^{n+1}$ via

$$p^n = \arg\min_{q \in H} L_r(u^{n-1}, q; \lambda^n), \tag{2.52}$$

$$\lambda^{n+1/2} = \lambda^n + \rho(Bu^{n-1} - p^n), \tag{2.53}$$

$$u^n = \arg\min_{v \in V} L_r(v, p^n; \lambda^{n+1/2}), \tag{2.54}$$

$$\lambda^{n+1} = \lambda^{n+1/2} + \rho(Bu^n - p^n). \tag{2.55}$$

Most practitioners prefer *ALG*2 to *ALG*3, the main reason being that *ALG*2 is more robust than *ALG*3, in general.

Remark 21. If one takes $\rho = r$ in (2.47)–(2.50) and (2.51)–(2.55), the algorithms we obtain belong to the *Alternating Direction Methods of Multipliers* (*ADMM*) family (a terminology we will justify in Section 3.3). The convergence of *ADMM* related algorithms is rather well established in the *convex* case (see, for example, [18, 61, 95]; see also the references therein and other chapters of this book, the one by *M. Burger, A. Sawatzky & G. Steidl* in particular). On the other hand, one is still lacking a general theory for the convergence of algorithms such as *ALG*1, *ALG*2, and *ALG*3 when applied to the solution of *non-convex variational problems*. Nevertheless, the above algorithms have been successfully applied to the solution of non-convex problems as shown, for example, in [42, 72] (Chapter 4), [74], and other chapters of this book, Chapters 7 and 8, in particular.

Remark 22. An important issue with the above primal-dual algorithms is how to vary r and ρ *dynamically* in order to improve the speed of convergence of these algorithms. This issue has been addressed in, e.g., [18, 34, 45, 46] (see also the references therein).

Remark 23. An overlooked ([34] being a notable exception) property of primal-dual algorithms such as *ALG*1, *ALG*2 and *ALG*3 is that they may be constructive still, in those not so uncommon situations where in (2.39) one has $\mathrm{dom}(F \circ B) \cap \mathrm{dom}(G) = \emptyset$, implying that problem (2.39) has no solutions, strictly speaking. On the basis of the numerical results reported in [42] (see also [72] (Chapter 4) and Chapter 8 of this volume), we conjecture that if the parameters ρ and r are properly chosen, the sequence $\{\{u^n, p^n\}\}_{n \geq 0}$ converges to a pair $\{u, p\}$ minimizing the functional

$$\{v, q\} \to G(v) + F(q)$$

over the set

$$\{\{v, q\}\} | \{v, q\} \in \mathrm{dom}(G) \times \mathrm{dom}(F), |Bv - q| = \min_{\{w, \varpi\} \in \mathrm{dom}(G) \times \mathrm{dom}(F)} |Bw - \varpi|\},$$

while the sequence $\{\lambda^n\}_{n \geq 0}$ diverges *arithmetically* (that is, $|\lambda^n| \to +\infty$ like n multiplied by a positive constant, that is slowly). If the above convergence/divergence result holds true (which seems to be the case for the non-convex problem discussed in [42]), it implies that the above primal-dual algorithms solve problem (2.39) in a *least-squares sense*, a most remarkable property indeed, testifying of the robustness of these algorithms. The above results look natural, but the optimization experts we consulted had trouble to give us a precise reference (or a proof).

Remark 24. We encountered situations (in *incompressible finite elasticity* in particular; see, e.g., [74] for details) where a safe way to proceed with the above primal-dual algorithms is as follows: Employ *ALG*1 with a well-balanced (that is neither too small nor too large) stopping criterion for the relaxation algorithm used to solve the minimization problem in (2.45); it has been observed quite often that the number of relaxation iterations necessary to compute $\{u^n, p^n\}$ from λ^n goes down quickly to one or two (an observation at the origin of *ALG*2), implying that starting with *ALG*1, the algorithm switches automatically to *ALG*2. It is not uncommon that this implementation of *ALG*1 produces an algorithm faster (CPU-wise) than *ALG*2 and *ALG*3, when solving "hard" problems. □

Further information on the convergence of *Lagrange multiplier based iterative methods* can be found in other chapters of this volume, and in, e.g., [45, 60, 62, 63], [66, 74] and [100] (see also the many references therein).

3.3 On the Relationship Between Alternating Direction Methods and ALG2, ALG3

As reported in [71] and [72] (Chapter 4) some previously unknown relationships between *alternating direction methods* and *augmented Lagrangian algorithms* were identified in 1975 by T.F. Chan and the first author of this chapter, while investigating the numerical solution of some simple nonlinear elliptic problems by various iterative methods (see [30] for details). Indeed, let us consider the particular case of problem (2.39) where $V = H$, $B = I$, and F and G are both differentiable over V; then, assuming that $\rho = r$, *ALG*2 (that is algorithm (2.47)–(2.50)) takes the following form:

$$\{u^{-1}, \lambda^0\} \text{ is given in } V \times H; \tag{2.56}$$

for $n \geq 0$, $\{u^{n-1}, \lambda^n\} \to p^n \to u^n \to \lambda^{n+1}$ via

$$r(p^n - u^{n-1}) + DF(p^n) - \lambda^n = 0, \tag{2.57}$$

$$r(u^n - p^n) + DG(u^n) + \lambda^n = 0, \tag{2.58}$$

$$\lambda^{n+1} = \lambda^n + r(u^n - p^n), \tag{2.59}$$

where DF (resp., DG) denotes the differential of F (resp., G). By elimination of λ^n and λ^{n+1} in (2.57)–(2.59), we obtain

$$r(p^n - u^{n-1}) + DF(p^n) + DG(u^{n-1}) = 0,$$

$$r(u^n - u^{n-1}) + DF(p^n) + DG(u^n) = 0,$$

which imply in turn (after changing $n-1$ in n):

$$r(p^{n+1}-u^n)+DF(p^{n+1})+DG(u^n)=0, \tag{2.60}$$

$$r(u^{n+1}-u^n)+DF(p^{n+1})+DG(u^{n+1})=0. \tag{2.61}$$

Comparing to (2.17) shows that in this particular case, $ALG2$ is a disguised form of the *Douglas-Rachford* scheme discussed in Section 2.5, with $r=1/\triangle t$ and DF (resp., DG) playing the role of A_1 (resp., A_2). A similar interpretation holds for $ALG3$: indeed, if we assume again that $V=H$, $B=I$ and F and G are differentiable, then, if $\rho=r$, algorithm (2.51)–(2.55) reduces to the *Peaceman-Rachford* scheme (2.15) discussed in Section 2.4. The above equivalence result can be generalized to situations where F and/or G are not differentiable.

The reasons for which $ALG2$ and $ALG3$ are called *Alternating Direction Methods of Multipliers* (*ADMM*) by many practitioners should be clear now. For further information and details on these primal-dual equivalences, see the discussion by M. Yan and W. Yin in Chapter 5 of this book.

4 Operator-Splitting Methods for the Direct Numerical Simulation of Particulate Flow

4.1 Generalities. Problem Formulation

It is the (necessarily biased) opinion of the authors of this chapter that the *direct numerical simulation of particulate flow* has been one of the success stories of operator-splitting methods, justifying thus a dedicated section in this chapter, despite the fact that this story has been told in several publications (see, e.g., [70] (Chapters 8 & 9), [73] and [79], and the references therein). For simplicity, we will discuss only the one-particle case (however, the results of numerical experiments involving more than one particle will be presented).

Let Ω be a bounded, connected, and open region of $\mathbb{R}^d$ ($d=2$ or 3 in applications); the boundary of Ω is denoted by Γ. We suppose that Ω contains:

(i) A Newtonian incompressible viscous fluid of density ρ_f and viscosity μ_f ; ρ_f and μ_f are both positive constants.
(ii) A rigid body B of boundary ∂B, mass M, center of mass G, and inertia $\mathbf{I}$ at the center of mass (see Figure 2.1, for additional details).

The fluid occupies the region $\Omega\setminus\overline{B}$ and we suppose that *distance* $(\partial B(0),\Gamma)>0$. From now on, $\mathbf{x}=\{x_i\}_{i=1}^d$ will denote the generic point of $\mathbb{R}^d$, $d\mathbf{x}=dx_1\dots dx_d$, while $\phi(t)$ will denote the function $\mathbf{x}\to\phi(\mathbf{x},t)$. Assuming that the only external force is *gravity*, the *fluid flow-rigid body motion* coupling is modeled by

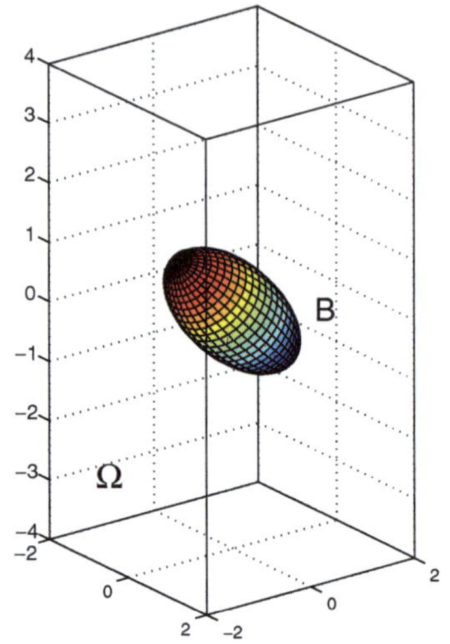

Fig. 2.1 Visualization of the rigid body and of a part of the flow region

$$\rho_f\left(\frac{\partial\mathbf{u}}{\partial t}+(\mathbf{u}\cdot\nabla)\mathbf{u}\right)-\mu_f\nabla^2\mathbf{u}+\nabla p=\rho_f\mathbf{g}\ \ in\ \Omega\setminus\overline{B(t)},\ \text{a.e. } t\in(0,T), \quad (2.62)$$

$$\nabla\cdot\mathbf{u}(t)=0\ in\ \Omega\setminus\overline{B(t)},\ \text{a.e. } t\in(0,T), \quad (2.63)$$

$$\mathbf{u}(t)=\mathbf{u}_\Gamma(t)\ on\ \Gamma,\ \text{a.e. } t\in(0,T),\ with\ \int_\Gamma\mathbf{u}_\Gamma(t)\cdot\mathbf{n}\,d\Gamma=0, \quad (2.64)$$

$$\mathbf{u}(0)=\mathbf{u}_0\ in\ \Omega\setminus\overline{B(0)}\ with\ \nabla\cdot\mathbf{u}_0=0, \quad (2.65)$$

and

$$\frac{dG}{dt}=\mathbf{V}, \quad (2.66)$$

$$M\frac{d\mathbf{V}}{dt}=M\mathbf{g}+\mathbf{R}_H, \quad (2.67)$$

$$\frac{d(\mathbf{I}\omega)}{dt}=\mathbf{T}_H, \quad (2.68)$$

$$G(0)=G_0,\mathbf{V}(0)=\mathbf{V}_0,\omega(0)=\omega_0,B(0)=B_0. \quad (2.69)$$

In relations (2.62)–(2.69):

- Vector $\mathbf{u}=\{u_i\}_{i=1}^d$ is the *fluid flow velocity* and p is the *pressure*.
- $\mathbf{u}_0$ and $\mathbf{u}_\Gamma$ are given vector-valued functions.
- $\mathbf{V}$ is the *velocity of the center of mass* of body B, while ω is the angular velocity.
- $\mathbf{R}_H$ and $\mathbf{T}_H$ denote, respectively, the *resultant* and the *torque* of the *hydrodynamical forces*, namely the forces that the fluid exerts on B; we have, actually,

$$\mathbf{R}_H=\int_{\partial B}\sigma\mathbf{n}\,d\gamma\ \ and\ \ \mathbf{T}_H=\int_{\partial B}\overrightarrow{G\mathbf{x}}\times\sigma\mathbf{n}\,d\gamma. \quad (2.70)$$

In (2.70) the *stress-tensor* $\sigma$is defined by $\sigma = 2\mu_f \mathbf{D}(\mathbf{u}) - p\mathbf{I}_d$, with $\mathbf{D}(\mathbf{v}) = \frac{1}{2}(\nabla \mathbf{v} + (\nabla \mathbf{v})^t)$, while $\mathbf{n}$ is a unit normal vector at ∂B and $\mathbf{I}_d$ is the *identity tensor*.

Concerning the compatibility conditions on ∂B we have: (i) the forces exerted by the fluid on the solid body *balance* those exerted by the solid body on the fluid, and we shall assume that: (ii) the *no-slip boundary condition* holds, namely

$$\mathbf{u}(\mathbf{x},t) = \mathbf{V}(t) + \omega(t) \times \overrightarrow{G(t)\mathbf{x}}, \ \forall \mathbf{x} \in \partial B(t). \tag{2.71}$$

Remark 25. System (2.62)–(2.65) (resp., (2.66)–(2.69)) is of the *incompressible Navier-Stokes* (resp., *Euler-Newton*) type. Also, the above model can be generalized to multiple-particles situations and/or non-Newtonian incompressible viscous fluids. □

The (local in time) *existence of weak solutions* for problems such as (2.62)–(2.69) has been proved in [52], assuming that, at $t = 0$, the particles do not touch Γ and each other (see also [87] and [145]). Concerning the numerical solution of (2.62)–(2.69), (2.71) several approaches are encountered in the literature, among them: (i) The *Arbitrary Lagrange-Euler* (*ALE*) methods; these methods, which rely on *moving meshes*, are discussed in, e.g., [98, 103] and [127]. (ii) The *fictitious boundary* method discussed in, e.g., [165], and (iii) the non-boundary fitted *fictitious domain* methods discussed in, e.g., [70, 79] and [140, 141] (and in Section 4.2, hereafter). Among other things, the methods in (ii) and (iii) have in common that the meshes used for the flow computations do not have to match the boundary of the particles.

Remark 26. Even if theory suggests that collisions may never take place in finite time (if we assume that the particles have smooth shapes and that the flow is still modeled by the Navier-Stokes equations as long as the particles do not touch each other, or Γ), near collisions take place, and after discretization particles may collide. These phenomena can be handled by introducing (as done in, e.g., [70] (Chapter 8) and [79]) well-chosen short range repulsion potentials reminiscent of those encountered in *Molecular Dynamics*, or by using *Kuhn Tucker multipliers* to authorize particle motions with *contact* but *no overlapping* (as done in, e.g., [128] and [129]). More information on the numerical treatment of particles in flow can be found in, e.g., [152] (and the references therein), and of course in Google.

4.2 A Fictitious Domain Formulation

Considering the fluid-rigid body mixture as a unique (heterogeneous) medium we are going to derive a *fictitious domain* based variational formulation to model its motion. The principle of this derivation is pretty simple: it relies on the following steps (see, e.g., [70] and [79] for more details), where in *Step a* we denote by $\mathbf{S} : \mathbf{T}$ the *Fröbenius* inner product of the tensors $\mathbf{S}$ and $\mathbf{T}$, that is (with obvious notation) $\mathbf{S} : \mathbf{T} = \sum_{1 \le i,j \le d} s_{ij} t_{ij}$:

Step a. Start from the following *global weak* formulation (of the *virtual power* type):

$$\begin{cases} \rho_f \int_{\Omega\setminus\overline{B(t)}} \left[\frac{\partial \mathbf{u}}{\partial t} + (\mathbf{u}\cdot\nabla)\mathbf{u}\right]\cdot \mathbf{v}\,d\mathbf{x} + 2\mu_f \int_{\Omega\setminus\overline{B(t)}} \mathbf{D}(\mathbf{u}) : \mathbf{D}(\mathbf{v})\,d\mathbf{x} \\ -\int_{\Omega\setminus\overline{B(t)}} p\nabla\cdot\mathbf{v}\,d\mathbf{x} + M\frac{d\mathbf{V}}{dt}\cdot\mathbf{Y} + \frac{d(\mathbf{I}\omega)}{dt}\cdot\theta \\ = \rho_f \int_{\Omega\setminus\overline{B(t)}} \mathbf{g}\cdot\mathbf{v}\,d\mathbf{x} + M\mathbf{g}\cdot\mathbf{Y}, \\ \forall\{\mathbf{v},\mathbf{Y},\theta\} \in (H^1(\Omega\setminus\overline{B(t)}))^d \times I\!R^d \times \Theta \;\; and \;\; verifying \\ \\ \mathbf{v} = 0 \; on\, \Gamma,\; \mathbf{v}(\mathbf{x}) = \mathbf{Y} + \theta \times \overrightarrow{G(t)\mathbf{x}}, \forall \mathbf{x} \in \partial B(t),\; t\in(0,T), \\ with\; \Theta = I\!R^3 \; if\; d=3,\; \Theta = \{(0,0,\theta) \mid \theta \in I\!R\} \; if \; d=2, \end{cases} \tag{2.72}$$

$$\int_{\Omega\setminus\overline{B(t)}} q\nabla\cdot\mathbf{u}(t)\,d\mathbf{x} = 0, \forall q \in L^2(\Omega\setminus\overline{B(t)}),\; t\in(0,T), \tag{2.73}$$

$$\mathbf{u}(t) = \mathbf{u}_\Gamma(t) \; on\, \Gamma,\; t\in(0,T), \tag{2.74}$$

$$\mathbf{u}(\mathbf{x},t) = \mathbf{V}(t) + \omega(t) \times \overrightarrow{G(t)\mathbf{x}}, \forall\mathbf{x}\in\partial B(t),\; t\in(0,T), \tag{2.75}$$

$$\frac{dG}{dt} = \mathbf{V}, \tag{2.76}$$

$$\mathbf{u}(\mathbf{x},0) = \mathbf{u}_0(\mathbf{x}), \forall\mathbf{x}\in\Omega\setminus\overline{B(0)}, \tag{2.77}$$

$$G(0) = G_0,\; \mathbf{V}(0) = \mathbf{V}_0,\; \omega(0) = \omega_0,\; B(0) = B_0. \tag{2.78}$$

Step b. Fill B with the surrounding fluid and impose a rigid body motion to the fluid inside B.

Step c. Modify the global weak formulation (2.72)–(2.78) accordingly, taking advantage of the fact that if $\mathbf{v}$ is a rigid body motion velocity field, then $\nabla\cdot\mathbf{v} = 0$ and $\mathbf{D}(\mathbf{v}) = \mathbf{0}$.

Step d. Use a *Lagrange multiplier* defined over B to force the rigid body motion inside B.

Assuming that B is made of a homogeneous material of density ρ_s, the above program leads to:

$$\begin{cases} \rho_f \int_{\Omega} \left[\frac{\partial \mathbf{u}}{\partial t} + (\mathbf{u}\cdot\nabla)\mathbf{u}\right]\cdot\mathbf{v}\,d\mathbf{x} + 2\mu_f\int_\Omega \mathbf{D}(\mathbf{u}):\mathbf{D}(\mathbf{v})\,d\mathbf{x} - \int_\Omega p\nabla\cdot\mathbf{v}\,d\mathbf{x} \\ +(1-\rho_f/\rho_s)\left[M\frac{d\mathbf{V}}{dt}\cdot\mathbf{Y} + \frac{d(\mathbf{I}\omega)}{dt}\cdot\theta\right] + <\lambda, \mathbf{v} - \mathbf{Y} - \theta\times\overrightarrow{G(t)\mathbf{x}}>_{B(t)} \\ = \rho_f\int_\Omega \mathbf{g}\cdot\mathbf{v}\,d\mathbf{x} + (1-\rho_f/\rho_s)M\mathbf{g}\cdot\mathbf{Y},\; \forall\{\mathbf{v},\mathbf{Y},\theta\}\in(H^1(\Omega))^d\times I\!R^d\times\Theta, \\ t\in(0,T),\;\; with\; \Theta = I\!R^3\; if\; d=3,\; \Theta=\{(0,0,\theta)\mid\theta\in I\!R\}\; if\; d=2, \end{cases} \tag{2.79}$$

$$\int_{\Omega} q\nabla \cdot \mathbf{u}(t)\, d\mathbf{x} = 0, \forall q \in L^2(\Omega),\ t \in (0,T), \tag{2.80}$$

$$\mathbf{u}(t) = \mathbf{u}_{\Gamma}(t) \text{ on } \Gamma,\ t \in (0,T), \tag{2.81}$$

$$\begin{cases} < \mu, \mathbf{u}(\mathbf{x},t) - \mathbf{V}(t) - \omega(t) \times \overrightarrow{G(t)\mathbf{x}} >_{B(t)} = 0, \\ \forall \mu \in \Lambda(t)\ (= (H^1(B(t)))^d),\ t \in (0,T), \end{cases} \tag{2.82}$$

$$\frac{dG}{dt} = \mathbf{V}, \tag{2.83}$$

$$\begin{cases} G(0) = G_0,\ \mathbf{V}(0) = \mathbf{V}_0,\ \omega(0) = \omega_0,\ B(0) = B_0, \\ \mathbf{u}(\mathbf{x},0) = \mathbf{u}_0(\mathbf{x}), \forall \mathbf{x} \in \Omega \setminus \bar{B}_0,\ \mathbf{u}(x,0) = \mathbf{V}_0 + \omega_0 \times \overrightarrow{G_0\mathbf{x}},\ \forall \mathbf{x} \in \bar{B}_0. \end{cases} \tag{2.84}$$

From a theoretical point of view, a natural choice for $< \cdot, \cdot >_{B(t)}$ is provided by, e.g.,

$$< \mu, \mathbf{v} >_{B(t)} = \int_{B(t)} [\mu \cdot \mathbf{v} + l^2 \mathbf{D}(\mu) : \mathbf{D}(\mathbf{v})]\, d\mathbf{x}; \tag{2.85}$$

in (2.85), l is a characteristic length, the diameter of B, for example. In practice, following [70] (Chapter 8) and [79], one makes things much simpler by approximating $\Lambda(t)$ by

$$\Lambda_h(t) = \{\mu \mid \mu = \sum_{j=1}^{N(t)} \mu_j \delta(\mathbf{x} - \mathbf{x}_j),\ \text{with } \mu_j \in I\!R^d,\ \forall j = 1, \ldots, N(t)\}, \tag{2.86}$$

and the pairing in (2.85) by

$$< \mu, \mathbf{v} >_{(B(t),h)} = \sum_{j=1}^{N(t)} \mu_j \cdot \mathbf{v}(\mathbf{x}_j). \tag{2.87}$$

In (2.86), (2.87), $\mathbf{x} \to \delta(\mathbf{x} - \mathbf{x}_j)$ is the Dirac measure at $\mathbf{x}_j$, and the set $\{\mathbf{x}_j\}_{j=1}^{N(t)}$ is the union of two subsets, namely: (i) The set of the points of the velocity grid contained in $B(t)$ and whose distance at $\partial B(t)$ is $\geq ch$, h being a space discretization step and c a constant ≈ 1.(ii) A set of control points located on $\partial B(t)$ and forming a mesh whose step size is of the order of h. It is clear that, using the approach above, one forces the rigid body motion inside the particle by *collocation*.

A variant of the above fictitious domain approach is discussed in [140] and [141]; after an appropriate elimination, it does not make use of Lagrange multipliers to force the rigid body motion of the particles, but uses instead projections on velocity subspaces where the rigid body motion velocity property is verified over the particles (see [140] and [141] for details).

4.3 Solving Problem (2.79)–(2.84) by Operator-Splitting

We do not consider *collisions*; after (formal) elimination of p and λ, problem (2.79)–(2.84) reduces to an initial value problem of the following form

$$\frac{d\mathbf{X}}{dt}+\sum_{j=1}^{J}A_j(\mathbf{X},t)=\mathbf{0}\ on\ (0,T),\ \ \mathbf{X}(0)=\mathbf{X}_0, \tag{2.88}$$

where $\mathbf{X}=\{\mathbf{u},\mathbf{V},\omega,G\}$ (or $\{\mathbf{u},\mathbf{V},\mathbf{I}\omega,G\}$). A typical situation will be the one where, with $J=4$, operator A_1 will be associated with *incompressibility*, A_2 with *advection*, A_3 with *diffusion*, A_4 with the *fictitious domain* treatment of the *rigid body motion*; other decompositions are possible as shown in, e.g., [70] (Chapter 8) and [79] (both references include a *collision operator*). The *Lie's scheme* (2.3), (2.4) applies "beautifully" to the solution of problem (2.79)–(2.84). The resulting method is quite modular implying that different space and time approximations can be used to treat the various sub-problems encountered at each time step; the only constraint is that two successive steps have to communicate (by projection in general) to provide the initial condition required by each initial value sub-problem.

4.4 Numerical Experiments

4.4.1 Generalities

The methodology we described (briefly) in the above paragraphs has been validated by numerous experiments (see, in particular, [70] (Chapters 8 & 9), [73, 79], [97, 137] and the related publications reported in http://www.math.uh.edu/~pan/). In this chapter, we will consider two test problems (borrowed from [73] (Section 3.4)): The first test problem involves three particles, while the second one concerns a channel flow with 300 particles. The fictitious domain/operator-splitting approach has made the solution of these problems (almost) routine nowadays. All the flow computations have been done using the *Bercovier-Pironneau finite element approximation*; namely (see [70] (Chapters 5, 8 and 9) for details), we used a globally continuous piecewise affine approximation of the velocity (resp., the pressure) associated with a triangulation (in 2-D) or tetrahedral partition (in 3-D) $\mathscr{T}_h$ (resp., $\mathscr{T}_{2h}$) of Ω, h being a space discretization step. The pressure mesh is thus twice *coarser* than the velocity one. The calculations have been done using uniform partitions $\mathscr{T}_h$ and $\mathscr{T}_{2h}$.

4.4.2 First Test Problem: Settling of Three Balls in a Vertical Narrow Tube

Our goal in this subsection is to discuss the interaction of three identical balls settling in a narrow tube of rectangular cross-section, containing an incompressible

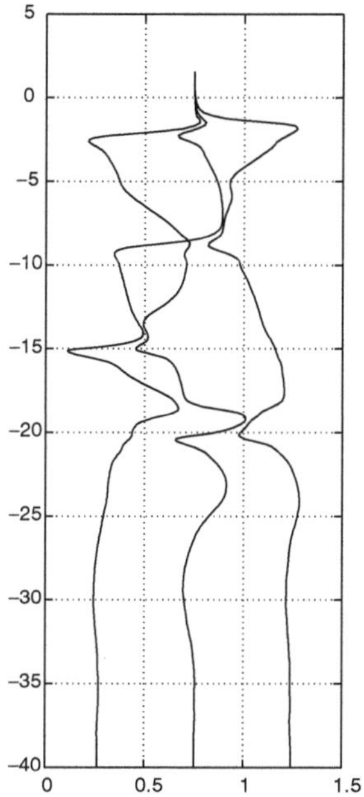

Fig. 2.2 Projections on the x_1x_3-plane of the trajectories of the mass centers of the three particles

Newtonian viscous fluid. Theoretically, the tube should be infinitely long, but for practicality we first consider the settling of the balls in a cylinder of length 6 whose cross-section is the rectangle $\Omega = (0,1.5) \times (0,0.25)$; this cylinder is moving with the balls in such a way that the center of the lower ball is in the horizontal symmetry plane (a possible, but less satisfying, alternative would be to specify periodicity in the vertical direction). At time $t = 0$, we suppose that the truncated cylinder coincides with the "box" $(0,1.5) \times (0,0.25) \times (0,6)$, and the centers of the balls are on the vertical axis of the cylinder at the points $x_1 = 0.75$, $x_2 = 0.125$, $x_3 = 1$, 1.3 and 1.6. The parameters for this test case are $\rho_s = 1.1$, $\rho_f = 1$, $\mu_f = 1$, the diameter of the balls being $d = 0.2$. The mesh size used to compute the velocity field (resp., the pressure) is $h_v = h = 1/96$ (resp., $h_p = 2h = 1/48$), while we took 1/1000 for the time-discretization step; the initial velocity of the flow is $\mathbf{0}$, while the three balls are released from rest. The velocity on the cylinder wall is $\mathbf{0}$. On the time interval $[0,15]$ the drafting, kissing and tumbling phenomenon (a terminology introduced by *D.D. Joseph*) has been observed several time before a stable quasi-horizontal configuration takes place, as shown in Figures 2.2, 2.3 and 2.4. The averaged vertical velocity of the balls is 2.4653 on the time interval $[13,15]$, while the *averaged particle Reynolds number* is 49.304 on the same time interval, a clear evidence that inertia has to be taken into account.

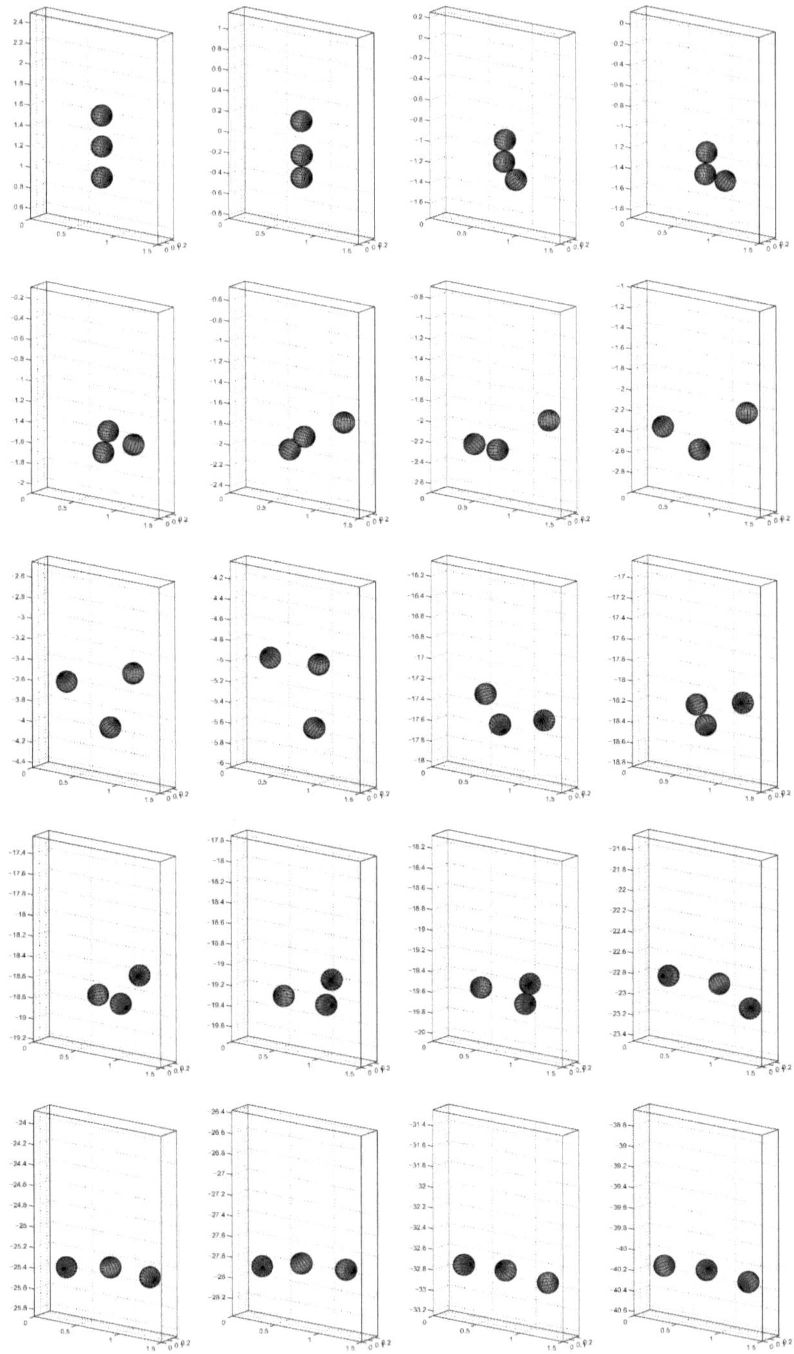

Fig. 2.3 Relative positions of the three balls at $t = 0, 0.4, 0.6, 0.65, 0.7, 0.8, 0.9, 1, 1.5, 2, 6, 6.25, 6.4, 6.6, 6.7, 8, 9, 10, 12$, and 15 (from left to right and from top to bottom)

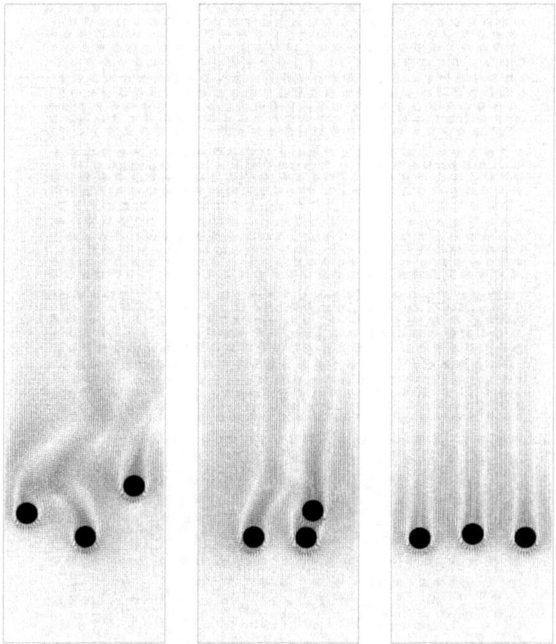

Fig. 2.4 Visualization of the flow and of the particles at $t = 1.1$, 6.6, and 15.

4.4.3 Motion of 300 Neutrally Buoyant Disks in a Two-Dimensional Horizontal Channel

This second test problem involving 300 particles and a *solid volume/fluid volume* of the order of 0.38, collisions (or near-collisions) have to be accounted for in the simulations; to do so, we have used the methods discussed in [70] (Chapter 8) and [79]. Another peculiarity of this test problem is that $\rho_s = \rho_f$ for all the particles (a neutrally buoyant situation). Indeed, neutrally buoyant models are more delicate to handle than those in the general case since $1 - \rho_f/\rho_s = 0$ in (2.79); however this difficulty can be overcome as shown in [136]. For this test problem, we have: (a) $\Omega = (0,42) \times (0,12)$. (b) Ω contains the mixture of a Newtonian incompressible viscous fluid of density $\rho_f = 1$ and viscosity $\mu_f = 1$, with 300 identical rigid solid disks of density $\rho_f = 1$ and diameter 0.9. (c) At $t = 0$, fluid and particles are at rest, the particle centers being located at the points of a regular lattice. (d) The mixture is put into motion by a uniform pressure drop of 10/9 per unit length (without the particles the corresponding steady flow would have been of the Poiseuille type with 20 as maximal flow speed). (e) The boundary conditions are given by $\mathbf{u}(x_1,x_2,t) = \mathbf{0}$ if $0 \leq x_1 \leq 42$, $x_2 = 0$ and 12, and $0 \leq t \leq 400$ (no-slip boundary condition on the horizontal parts of the boundary), and then $\mathbf{u}(0,x_2,t) = \mathbf{u}(42,x_2,t)$, $0 < x_2 < 12$, $0 \leq t \leq 400$ (space-periodic in the Ox_1 direction). (f) $h_\mathbf{v} = h = 1/10$, $h_p = 2h = 1/5$, the time-discretization step being 1/1000.

Fig. 2.5 Positions of the 300 particles at t = 100, 107.8, 114, 200, and 400 (from top to bottom).

The particle distribution at t = 100, 107.8, 114, 200, and 400 has been visualized on Figures 2.5. These figures show that, initially, we have the sliding motion of horizontal particle layers, then after some critical time a chaotic flow-motion takes place in very few time units, the highest particle concentration being along the channel axis (actually, a careful inspection of the results shows that the transition to chaos takes place just after t =107.8). The maximal speed at t =400 is 7.9, implying that the corresponding particle Reynolds number is very close to 7.1. On Figure 2.6 we show the averaged solid fraction as a function of x_2, the averaging space-time set being $\{\{x_1,t\}|0 \leq x_1 \leq 42, 380 \leq t \leq 400\}$; the particle aggregation along the channel horizontal symmetry axis appears very clearly from this figure since the solid fraction is close to 0.58 at $x_2 = 6$ while the global solid fraction is 0.38 (vertical line in the figure). Finally, we have visualized on Figure 2.7 the x_1-averaged horizontal component of the mixture velocity at t = 400, as a function of x_2. The dashed line corresponds to a horizontal velocity distribution of the steady flow of the same fluid, with no particle in the channel, for the same pressure drop; the corresponding velocity profile is (of course) of the Poiseuille type and shows that the mixture behaves like a viscous fluid whose viscosity is (approximately) 2.5 larger

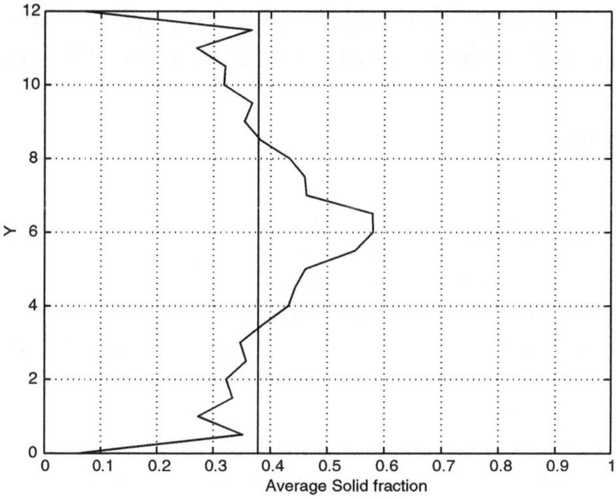

Fig. 2.6 Averaged solid fraction distribution.

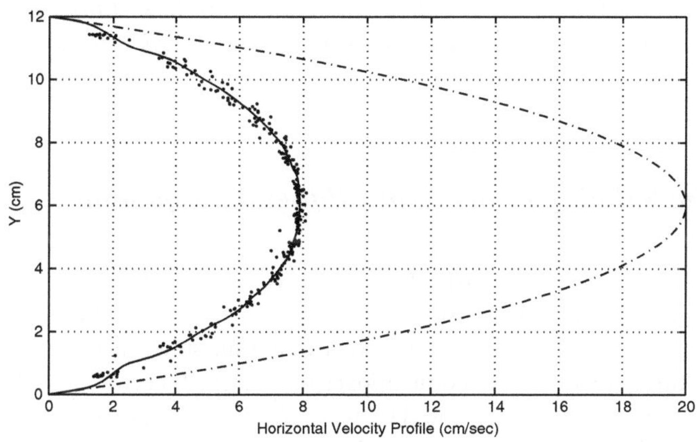

Fig. 2.7 Horizontal velocity distribution at $t = 400$.

than μ_f. Actually, a closer inspection (see [136] for details) shows that the mixture behaves like a non-Newtonian incompressible viscous fluid of the power law type, for an exponent $s = 1.7093$ ($s = 2$ corresponding to a Newtonian fluid and $s = 1$ to a perfectly plastic material). Figures 2.5, 2.6, and 2.7 show also that, as well known, some *order may be found in chaos*.

For more details and further results and comments on pressure driven neutrally buoyant particulate flows in two-dimensional channels (including simulations with much larger numbers of particles, the largest one being 1,200) see [70] (Chapter 9) and [136].

5 Operator-Splitting Methods for the Numerical Solution of Nonlinear Problems from Condensate and Plasma Physics

5.1 Introduction

Operator-splitting methods have been quite successful at solving problems in *Computational Physics*, beside those from *Computational Mechanics* (*CFD* in particular). Among these successful applications let us mention those involving *nonlinear Schrödinger equations*, as shown, for example, by [9, 10, 44] and [102]. On the basis of some very inspiring articles (see, e.g., [9, 10] and [102]) he wrote on the above topic, the editors asked their colleague *Peter Markowich* to contribute a related chapter for this book; unfortunately, Professor Markowich being busy elsewhere had to say no. Considering the importance of nonlinear Schrödinger related problems, it was decided to (briefly) discuss in this chapter the solution of some of them by operator-splitting methods (see also Chapter 18 on the propagation of laser pulses along optical fibers). In Section 5.2, we will discuss the operator-splitting solution of the celebrated *Gross-Pitaevskii* equation for *Bose-Einstein condensates*, then, in Section 5.3, we will discuss the solution of the *Zakharov system* modeling the *propagation of Langmuir waves in ionized plasma.*

5.2 On the Solution of the Gross-Pitaevskii Equation

A *Bose-Einstein condensate* (BEC) is a state of matter of a dilute gas of bosons cooled to temperatures very close to absolute zero. Under such conditions, a large fraction of the bosons occupies the lowest quantum state, at which point macroscopic quantum phenomena become apparent. The existence of Bose-Einstein condensates was predicted in the mid-1920s by *S. N. Bose* and *A. Einstein.* If dilute enough, the time evolution of a BEC is described by the following *Gross-Pitaevskii equation* (definitely of the *nonlinear Schrödinger* type and given here in a-dimensional form (following [9])):

$$i\varepsilon\frac{\partial \psi}{\partial t} = -\frac{\varepsilon^2}{2}\nabla^2\psi + V_d(x)\psi + K_d|\psi|^2\psi \text{ in } \Omega\times(0,T), \tag{2.89}$$

where, in (2.89), ψ is a complex-valued function of x and t, $i=\sqrt{-1}$, Ω is an open connected subset of $I\!R^d$ (with $d = 1$, 2 or 3), the real-valued function V_d denotes an external potential, and the real-valued parameter K_d is representative of the particles interactions. Equation (2.89) has to be completed by boundary and initial conditions. Equation (2.89) has motivated a very large literature from both physical and mathematical points of view. Let us mention among many others [1, 9, 125] and [126] (see also the many references therein). To solve equation (2.89) numerically we need to complete it by *boundary* and *initial conditions*: from now on, we will assume that

$$\psi(x,0) = \psi_0(x),\ x \in \Omega, \tag{2.90}$$

and (denoting by Γ the boundary of Ω)

$$\psi = 0 \text{ on } \Gamma \times (0,T). \tag{2.91}$$

The boundary conditions in (2.91) have been chosen for their simplicity, and also to provide an alternative to the *periodic boundary conditions* considered in [9]. An important (and very easy to prove) property of the solution of the initial boundary value problem (2.89)–(2.91) reads as:

$$\frac{d}{dt}\int_\Omega |\psi(x,t)|^2\,dx = 0 \text{ on } (0,T],$$

implying that

$$\int_\Omega |\psi(x,t)|^2\,dx = \int_\Omega |\psi_0(x)|^2\,dx \text{ on } [0,T]. \tag{2.92}$$

As done before, we denote by $\psi(t)$ the function $x \to \psi(x,t)$. Let $\triangle t(>0)$ be a time discretization step and denote $(n+\alpha)\triangle t$ by $t^{n+\alpha}$; applying to problem (2.89)–(2.91) the Strang's symmetrized scheme (2.7)–(2.10) of Section 2.3, we obtain:

$$\psi^0 = \psi_0; \tag{2.93}$$

for $n \geq 0$, $\psi^n \to \psi^{n+1/2} \to \widehat{\psi}^{n+1/2} \to \psi^{n+1}$ as follows

$$\begin{cases} i\dfrac{\partial \psi}{\partial t} + \dfrac{\varepsilon}{2}\nabla^2\psi = 0 \ \text{ in } \Omega \times (t^n, t^{n+1/2}), \\ \psi = 0 \ \text{ on } \Gamma \times (t^n, t^{n+1/2}), \\ \psi(t^n) = \psi^n;\ \psi^{n+1/2} = \psi(t^{n+1/2}), \end{cases} \tag{2.94}$$

$$\begin{cases} i\varepsilon\dfrac{\partial \psi}{\partial t} = V_d(x)\psi + K_d|\psi|^2\psi \ \text{ in } \Omega \times (0,\triangle t), \\ \psi = 0 \ \text{ on } \Gamma \times (0,\triangle t), \\ \psi(0) = \psi^{n+1/2};\ \widehat{\psi}^{n+1/2} = \psi(\triangle t), \end{cases} \tag{2.95}$$

$$\begin{cases} i\dfrac{\partial \psi}{\partial t} + \dfrac{\varepsilon}{2}\nabla^2\psi = 0 \ \text{ in } \Omega \times (t^{n+1/2}, t^{n+1}), \\ \psi = 0 \ \text{ on } \Gamma \times (t^{n+1/2}, t^{n+1}), \\ \psi(t^{n+1/2}) = \widehat{\psi}^{n+1/2};\ \psi^{n+1} = \psi(t^{n+1}). \end{cases} \tag{2.96}$$

On the solution of (2.95): Let us denote by ψ_1 (resp., ψ_2) the real (resp., imaginary) part of ψ; from (2.95), we have

$$\begin{cases} \varepsilon\dfrac{\partial \psi_1}{\partial t} = V_d(x)\psi_2 + K_d|\psi|^2\psi_2 \ \text{ in } \Omega \times (0,\triangle t), \\ \varepsilon\dfrac{\partial \psi_2}{\partial t} = -V_d(x)\psi_1 - K_d|\psi|^2\psi_1 \ \text{ in } \Omega \times (0,\triangle t), \end{cases} \tag{2.97}$$

Multiplying by ψ_1 (resp., ψ_2) the 1st (resp., the 2nd) equation in (2.97), we obtain by addition

$$\frac{\partial}{\partial t}|\psi(x,t)|^2 = 0 \text{ on}(0,\triangle t), \text{ a.e. } x \in \Omega,$$

which implies in turn that

$$|\psi(x,t)| = |\psi(x,0)| = |\psi^{n+1/2}| \text{ on}(0,\triangle t), \text{ a.e. } x \in \Omega. \tag{2.98}$$

It follows then from (2.95) and (2.98) that

$$\begin{cases} i\varepsilon\dfrac{\partial\psi}{\partial t} = V_d(x)\psi + K_d|\psi^{n+1/2}|^2\psi \text{ in } \Omega\times(0,\triangle t),\\ \psi = 0 \text{ on } \Gamma\times(0,\triangle t),\\ \psi(0) = \psi^{n+1/2};\ \widehat{\psi}^{n+1/2} = \psi(\triangle t), \end{cases}$$

which implies for $\widehat{\psi}^{n+1/2}$ the following closed-form solution

$$\widehat{\psi}^{n+1/2} = e^{-i\frac{\triangle t}{\varepsilon}\left(V_d + K_d|\psi^{n+1/2}|^2\right)}\psi^{n+1/2}. \tag{2.99}$$

On the solution of (2.94) and (2.96): The initial boundary value problems in (2.94) and (2.96) are particular cases of

$$\begin{cases} i\dfrac{\partial\phi}{\partial t} + \dfrac{\varepsilon}{2}\nabla^2\phi = 0 \text{ in } \Omega\times(t_0,t_f),\\ \phi = 0 \text{ on } \Gamma\times(t_0,t_f),\\ \phi(t_0) = \phi_0. \end{cases} \tag{2.100}$$

The above *linear Schrödinger problem* is a very classical one. Its solution is obviously given by

$$\phi(t) = e^{i\frac{\varepsilon}{2}(t-t_0)\nabla^2}\phi_0,\ \forall t\in[t_0,t_f]. \tag{2.101}$$

Suppose that $\Omega = (0,a)\times(0,b)\times(0,c)$ with $0<a<+\infty$, $0<b<+\infty$, and $0<c<+\infty$; since the eigenvalues, and related eigenfunctions, of the negative Laplace operator $-\nabla^2$, associated with the homogeneous Dirichlet boundary conditions are known explicitly, and given, for p, q and r positive integers, by

$$\begin{cases} \lambda_{pqr} = \pi^2\left(\dfrac{p^2}{a^2}+\dfrac{q^2}{b^2}+\dfrac{r^2}{c^2}\right),\\ w_{pqr}(x_1,x_2,x_3) = 2\sqrt{\dfrac{2}{abc}}\sin\left(p\pi\dfrac{x_1}{a}\right)\sin\left(q\pi\dfrac{x_2}{b}\right)\sin\left(r\pi\dfrac{x_3}{c}\right) \end{cases} \tag{2.102}$$

(we have then $\int_\Omega |w_{pqr}(x)|^2\,dx = 1$) it follows from (2.101) that

$$\phi(x,t) = \sum_{1\le p,q,r<+\infty}\phi^0_{pqr}e^{-i\frac{\varepsilon}{2}\lambda_{pqr}(t-t_0)}w_{pqr}(x), \text{ with } \phi^0_{pqr} = \int_\Omega w_{pqr}(y)\phi_0(y)\,dy. \tag{2.103}$$

In practice, one takes $1 \leq p \leq P$, $1 \leq q \leq Q$, $1 \leq r \leq R$, and uses the *Fast Fourier Transform* (FFT) to compute the coefficients ϕ^0_{pqr} and then $\phi(x,t)$.

For those more general situations where the solutions of the following *linear eigenvalue problem*

$$\begin{cases} \{w,\lambda\} \in H_0^1(\Omega) \times I\!R, \int_\Omega |w(x)|^2\,dx = 1,\ \lambda > 0, \\ \int_\Omega \nabla w \cdot \nabla v dx = \lambda \int_\Omega w v dx,\ \forall v \in H_0^1(\Omega), \end{cases} \tag{2.104}$$

are not known explicitly, one still has several options to solve (2.100), an obvious one being:

Approximate (2.104) by

$$\begin{cases} \{w,\lambda\} \in V_h \times I\!R, \int_\Omega |w(x)|^2\,dx = 1,\ \lambda > 0, \\ \int_\Omega \nabla w \cdot \nabla v dx = \lambda \int_\Omega w v dx,\ \forall v \in V_h, \end{cases} \tag{2.105}$$

where V_h is a finite dimensional sub-space of $H_0^1(\Omega)$. Then, as in, e.g., [17, 82] use an eigensolver (like the one discussed in [113]) to compute the first $Q(\leq N = \dim V_h)$ eigen-pairs solutions of (2.105), such that (with obvious notation) $\int_\Omega w_p w_q\,dx = 0$ $\forall p,q,\ 1 \leq p,q \leq Q,\ p \neq q$, and denote by V_Q the finite dimensional space span by the basis $\{w_q\}_{q=1}^Q$. Next, proceeding as in the continuous case we approximate the solution of problem (2.100) by ϕ_Q defined by

$$\phi_Q(x,t) = \sum_{q=1}^{Q} \phi_q^0 e^{-i\frac{\varepsilon}{2}\lambda_q(t-t_0)} w_q(x), \text{ with } \phi_q^0 = \int_\Omega w_q(y)\phi_0(y)\,dy. \tag{2.106}$$

For the space V_h in (2.105), we can use these *finite element* approximations of $H_0^1(\Omega)$ discussed for example in [37, 66] (Appendix 1) and [72] (Chapter 1) (see also the references therein).

Another approach, less obvious but still natural, is to observe that if ϕ, the unique solution of (2.100) is smooth enough, it is also the unique solution of

$$\begin{cases} \dfrac{\partial^2\phi}{\partial t^2} + \dfrac{\varepsilon^2}{4}\nabla^4\phi = 0 \text{ in } \Omega \times (t_0,t_f), \\ \phi = 0 \text{ and } \nabla^2\phi = 0 \text{ on } \Gamma \times (t_0,t_f), \\ \phi(t_0) = \phi_0,\ \dfrac{\partial\phi}{\partial t}(t_0) = i\dfrac{\varepsilon}{2}\nabla^2\phi_0 (= \phi_1), \end{cases} \tag{2.107}$$

a well-known model in *elasto-dynamics* (vibrations of simply supported plates). From Q, a positive integer, we define a time discretization step τ by $\tau = \dfrac{t_f - t_0}{Q}$. The initial-boundary value problem (2.107) is clearly equivalent to

$$\begin{cases} \dfrac{\partial \phi}{\partial t} = \mathbf{v} \ \text{ in } \Omega \times (t_0, t_f), \\ \dfrac{\partial \mathbf{v}}{\partial t} + \dfrac{\varepsilon^2}{4} \nabla^4 \phi = 0 \ \text{ in } \Omega \times (t_0, t_f), \\ \phi = 0 \text{ and } \nabla^2 \phi = 0 \ \text{ on } \Gamma \times (t_0, t_f), \\ \phi(t_0) = \phi_0, \ \mathbf{v}(t_0) = i\dfrac{\varepsilon}{2} \nabla^2 \phi_0 (= \mathbf{v}_0). \end{cases} \tag{2.108}$$

A time-discretization scheme for (2.107) (via (2.108)), combining good accuracy, stability, and energy conservation properties (see, e.g., [14]) reads as follows (with $\{\phi^q, \mathbf{v}^q\}$ an approximation of $\{\phi, \mathbf{v}\}$ at $t^q = t_0 + q\tau$):

$$\begin{cases} \phi^0 = \phi_0, \ \mathbf{v}^0 = \mathbf{v}_0; \\ \text{for } q = 0, \cdots, Q-1, \text{ compute } \{\phi^{q+1}, \mathbf{v}^{q+1}\} \text{ from } \{\phi^q, \mathbf{v}^q\} \text{ via the solution of} \\ \begin{cases} \dfrac{\phi^{q+1} - \phi^q}{\tau} = \dfrac{1}{2}(\mathbf{v}^{q+1} + \mathbf{v}^q), \\ \dfrac{\mathbf{v}^{q+1} - \mathbf{v}^q}{\tau} + \dfrac{\varepsilon^2}{8} \nabla^4 (\phi^{q+1} + \phi^q) = 0 \ \text{ in } \Omega, \\ \phi^{q+1} = 0 \text{ and } \nabla^2 \phi^{q+1} = 0 \ \text{ on } \Gamma. \end{cases} \end{cases} \tag{2.109}$$

By elimination of $\mathbf{v}^{q+1}$ it follows from (2.109) that ϕ^{q+1} is solution of

$$\begin{cases} \phi^{q+1} + \dfrac{(\tau\varepsilon)^2}{8} \nabla^4 \phi^{q+1} = \phi^q + \tau \mathbf{v}^q - \dfrac{(\tau\varepsilon)^2}{8} \nabla^4 \phi^q \ \text{ in } \Omega, \\ \phi^{q+1} = 0 \text{ and } \nabla^2 \phi^{q+1} = 0 \ \text{ on } \Gamma, \end{cases} \tag{2.110}$$

a bi-harmonic problem which is well posed in $H_0^1(\Omega) \cap H^2(\Omega)$. Next, one obtains easily $\mathbf{v}^{q+1}$ from

$$\mathbf{v}^{q+1} = \frac{2}{\tau}(\phi^{q+1} - \phi^q) - \mathbf{v}^q.$$

For the solution of the bi-harmonic problem (2.110) we advocate those mixed finite element approximations and conjugate gradient algorithms used in various chapters of [72] (see also the references therein).

5.3 On the Solution of Zakharov Systems

In 1972, *V.E. Zakharov* introduced a mathematical model describing the *propagation of Langmuir waves* in *ionized plasma* (ref. [179]). This model reads as follows (after rescaling):

$$\begin{cases} i\dfrac{\partial u}{\partial t} + \nabla^2 u = un, \\ \dfrac{\partial^2 n}{\partial t^2} - \nabla^2 n + \nabla^2(|u|^2) = 0, \end{cases} \tag{2.111}$$

where the complex-valued function u is associated with a highly oscillating *electric field*, while the real-valued function n denotes the *fluctuation of the plasma-ion density* from its equilibrium state. In this section, following [102], we will apply the *symmetrized Strang operator-splitting scheme* (previously discussed in Section 2.3 of this chapter) to the following generalization of the above equations:

$$\begin{cases} i\dfrac{\partial u}{\partial t} + \nabla^2 u + 2\lambda |u|^2 u + 2un = 0, \\ \dfrac{1}{c^2}\dfrac{\partial^2 n}{\partial t^2} - \nabla^2 n + \mu \nabla^2(|u|^2) = 0, \end{cases} \tag{2.112}$$

where λ and μ are real numbers and $c(>0)$ is the wave propagation speed. Following again [102], we will assume, for simplicity, that the physical phenomenon modeled by (2.112) takes place on the bounded interval $(0,L)$, with u, n, $\partial u/\partial x$ and $\partial n/\partial x$ *space-periodic*, during the time interval $[0,T]$. Thus, (2.112), completed by initial conditions, reduces to:

$$\begin{cases} i\dfrac{\partial u}{\partial t} + \dfrac{\partial^2 u}{\partial x^2} + 2\lambda |u|^2 u + 2un = 0 \text{ in } (0,L)\times(0,T), \\ \dfrac{1}{c^2}\dfrac{\partial^2 n}{\partial t^2} - \dfrac{\partial^2 n}{\partial x^2} + \mu \dfrac{\partial^2}{\partial x^2}(|u|^2) = 0 \text{ in } (0,L)\times(0,T), \\ u(0,t) = u(L,t),\ \dfrac{\partial u}{\partial x}(0,t) = \dfrac{\partial u}{\partial x}(L,t) \text{ on } (0,T), \\ n(0,t) = n(L,t),\ \dfrac{\partial n}{\partial x}(0,t) = \dfrac{\partial n}{\partial x}(L,t) \text{ on } (0,T), \\ u(0) = u_0,\ n(0) = n_0,\ \dfrac{\partial n}{\partial t}(0) = n_1. \end{cases} \tag{2.113}$$

As done previously in this chapter, we denote by $\phi(t)$ the function $x \to \phi(x,t)$.

Remark 27. Albeit considered by some as too simple from a physical point of view, *space-periodic boundary conditions* are common in plasma physics. They have been used for example in [131], a most celebrated article dedicated to the mathematical analysis of the behavior of plasma entropy (see also [163] which relates a discussion that *C. Villani* had with *E. Lieb* concerning precisely the use of space-periodic boundary conditions in plasma physics). □

From the rich structure of the Zakharov's system (2.113) it is not surprising that a variety of *operator-splitting schemes* can be applied to its numerical solution, several of these schemes being described in [102] (see also the references therein concerning splitting schemes not described in [102]). A first step to the application of operator-splitting scheme to the time-discretization of problem (2.113) is to introduce the function $p = \dfrac{\partial n}{\partial t}$ and to rewrite (2.113) as:

$$
\begin{cases}
i\dfrac{\partial u}{\partial t}+\dfrac{\partial^2 u}{\partial x^2}+2\lambda|u|^2u+2un=0 \text{ in } (0,L)\times(0,T),\\
\dfrac{\partial n}{\partial t}-p=0 \text{ in } (0,L)\times(0,T),\\
\dfrac{1}{c^2}\dfrac{\partial p}{\partial t}-\dfrac{\partial^2 n}{\partial x^2}+\mu\dfrac{\partial^2}{\partial x^2}(|u|^2)=0 \text{ in } (0,L)\times(0,T),\\
u(0,t)=u(L,t),\ \dfrac{\partial u}{\partial x}(0,t)=\dfrac{\partial u}{\partial x}(L,t) \text{ on } (0,T),\\
n(0,t)=n(L,t),\ \dfrac{\partial n}{\partial x}(0,t)=\dfrac{\partial n}{\partial x}(L,t),\ p(0,t)=p(L,t) \text{ on } (0,T),\\
u(0)=u_0,\ n(0)=n_0,\ p(0)=n_1.
\end{cases}
\tag{2.114}
$$

Applying the Strang's symmetrized scheme to the time-discretization of problem (2.114), one obtains (among other possible schemes, and with $t^{q+\alpha}=(q+\alpha)\triangle t$):

$$
\{u^0,n^0,p^0\}=\{u_0,n_0,n_1\}. \tag{2.115}
$$

For $q\geq 0$, $\{u^q,n^q,p^q\}\to\{u^{q+1/2},n^{q+1/2},p^{q+1/2}\}\to\{\hat{u}^{q+1/2},\hat{n}^{q+1/2},\hat{p}^{q+1/2}\}\to\{u^{q+1},n^{q+1},p^{q+1}\}$ via

$$
\begin{cases}
i\dfrac{\partial u}{\partial t}+\dfrac{\partial^2 u}{\partial x^2}=0 \text{ in } (0,L)\times(t^q,t^{q+1/2}),\\
\dfrac{\partial n}{\partial t}-\dfrac{p}{2}=0 \text{ in } (0,L)\times(t^q,t^{q+1/2}),\\
\dfrac{1}{c^2}\dfrac{\partial p}{\partial t}-\dfrac{\partial^2 n}{\partial x^2}=0 \text{ in } (0,L)\times(t^q,t^{q+1/2}),\\
u(0,t)=u(L,t),\ \dfrac{\partial u}{\partial x}(0,t)=\dfrac{\partial u}{\partial x}(L,t) \text{ on } (t^q,t^{q+1/2}),\\
n(0,t)=n(L,t),\ \dfrac{\partial n}{\partial x}(0,t)=\dfrac{\partial n}{\partial x}(L,t),\ p(0,t)=p(L,t) \text{ on } (t^q,t^{q+1/2}),\\
u(t^q)=u^q,\ n(t^q)=n_q,\ p(t^q)=p^q;\\
u^{q+1/2}=u(t^{q+1/2}),n^{q+1/2}=n(t^{q+1/2}),p^{q+1/2}=p(t^{q+1/2}),
\end{cases}
\tag{2.116}
$$

$$
\begin{cases}
i\dfrac{\partial u}{\partial t}+2\lambda|u|^2u+2un=0 \text{ in } (0,L)\times(0,\triangle t),\\
\dfrac{\partial n}{\partial t}-\dfrac{p}{2}=0 \text{ in } (0,L)\times(0,\triangle t),\\
\dfrac{1}{c^2}\dfrac{\partial p}{\partial t}+\mu\dfrac{\partial^2}{\partial x^2}(|u|^2)=0 \text{ in } (0,L)\times(0,\triangle t),\\
u(0,t)=u(L,t),\ \dfrac{\partial u}{\partial x}(0,t)=\dfrac{\partial u}{\partial x}(L,t) \text{ on } (0,\triangle t),\\
n(0,t)=n(L,t),\ \dfrac{\partial n}{\partial x}(0,t)=\dfrac{\partial n}{\partial x}(L,t),\ p(0,t)=p(L,t) \text{ on } (0,\triangle t),\\
u(0)=u^{q+1/2},\ n(0)=n^{q+1/2},\ p(0)=p^{q+1/2};\\
\hat{u}^{q+1/2}=u(\triangle t),\hat{n}^{q+1/2}=n(\triangle t),\hat{p}^{q+1/2}=p(\triangle t),
\end{cases}
\tag{2.117}
$$

$$\begin{cases} i\dfrac{\partial u}{\partial t}+\dfrac{\partial^2 u}{\partial x^2}=0 \text{ in } (0,L)\times(t^{q+1/2},t^{q+1}),\\ \dfrac{\partial n}{\partial t}-\dfrac{p}{2}=0 \text{ in } (0,L)\times(t^{q+1/2},t^{q+1}),\\ \dfrac{1}{c^2}\dfrac{\partial p}{\partial t}-\dfrac{\partial^2 n}{\partial x^2}=0 \text{ in } (0,L)\times(t^{q+1/2},t^{q+1}),\\ u(0,t)=u(L,t),\ \dfrac{\partial u}{\partial x}(0,t)=\dfrac{\partial u}{\partial x}(L,t) \text{ on } (t^{q+1/2},t^{q+1}),\\ n(0,t)=n(L,t),\ \dfrac{\partial n}{\partial x}(0,t)=\dfrac{\partial n}{\partial x}(L,t),\ p(0,t)=p(L,t) \text{ on } (t^{q+1/2},t^{q+1}),\\ u(t^{q+1/2})=\hat{u}^{q+1/2},\ n(t^{q+1/2})=\hat{n}^{q+1/2},\ p(t^{q+1/2})=\hat{p}^{q+1/2};\\ u^{q+1}=u(t^{q+1}),n^{q+1}=n(t^{q+1}),p^{q+1}=p(t^{q+1}). \end{cases} \tag{2.118}$$

Scheme (2.115)–(2.118) is clearly equivalent to

$$\{u^0,n^0,p^0\}=\{u_0,n_0,n_1\}. \tag{2.119}$$

For $q\geq 0$, $\{u^q,n^q,p^q\}\to\{u^{q+1/2},n^{q+1/2},p^{q+1/2}\}\to\{\hat{u}^{q+1/2},\hat{n}^{q+1/2},\hat{p}^{q+1/2}\}\to\{u^{q+1},n^{q+1},p^{q+1}\}$ via

$$\begin{cases} i\dfrac{\partial u}{\partial t}+\dfrac{\partial^2 u}{\partial x^2}=0 \text{ in } (0,L)\times(t^q,t^{q+1/2}),\\ \dfrac{2}{c^2}\dfrac{\partial^2 n}{\partial t^2}-\dfrac{\partial^2 n}{\partial x^2}=0 \text{ in } (0,L)\times(t^q,t^{q+1/2}),\\ u(0,t)=u(L,t),\ \dfrac{\partial u}{\partial x}(0,t)=\dfrac{\partial u}{\partial x}(L,t) \text{ on } (t^q,t^{q+1/2}),\\ n(0,t)=n(L,t),\ \dfrac{\partial n}{\partial x}(0,t)=\dfrac{\partial n}{\partial x}(L,t) \text{ on } (t^q,t^{q+1/2}),\\ u(t^q)=u^q,\ n(t^q)=n_q,\ \dfrac{\partial n}{\partial t}(t^q)=p^q/2;\\ u^{q+1/2}=u(t^{q+1/2}),n^{q+1/2}=n(t^{q+1/2}),p^{q+1/2}=2\dfrac{\partial n}{\partial t}(t^{q+1/2}), \end{cases} \tag{2.120}$$

$$\begin{cases} i\dfrac{\partial u}{\partial t}+2\lambda|u|^2u+2un=0 \text{ in } (0,L)\times(0,\triangle t),\\ \dfrac{2}{c^2}\dfrac{\partial^2 n}{\partial t^2}+\mu\dfrac{\partial^2}{\partial x^2}(|u|^2)=0 \text{ in } (0,L)\times(0,\triangle t),\\ u(0,t)=u(L,t),\ \dfrac{\partial u}{\partial x}(0,t)=\dfrac{\partial u}{\partial x}(L,t) \text{ on } (0,\triangle t),\\ n(0,t)=n(L,t),\ \dfrac{\partial n}{\partial x}(0,t)=\dfrac{\partial n}{\partial x}(L,t) \text{ on } (0,\triangle t),\\ u(0)=u^{q+1/2},\ n(0)=n^{q+1/2},\ \dfrac{\partial n}{\partial t}(0)=p^{q+1/2}/2;\\ \hat{u}^{q+1/2}=u(\triangle t),\hat{n}^{q+1/2}=n(\triangle t),\hat{p}^{q+1/2}=2\dfrac{\partial n}{\partial t}(\triangle t), \end{cases} \tag{2.121}$$

$$\begin{cases} i\dfrac{\partial u}{\partial t}+\dfrac{\partial^2 u}{\partial x^2}=0 \text{ in } (0,L)\times(t^{q+1/2},t^{q+1}),\\ \dfrac{2}{c^2}\dfrac{\partial^2 n}{\partial t^2}-\dfrac{\partial^2 n}{\partial x^2}=0 \text{ in } (0,L)\times(t^{q+1/2},t^{q+1}),\\ u(0,t)=u(L,t),\ \dfrac{\partial u}{\partial x}(0,t)=\dfrac{\partial u}{\partial x}(L,t) \text{ on } (t^{q+1/2},t^{q+1}),\\ n(0,t)=n(L,t),\ \dfrac{\partial n}{\partial x}(0,t)=\dfrac{\partial n}{\partial x}(L,t) \text{ on } (t^{q+1/2},t^{q+1}),\\ u(t^{q+1/2})=\hat{u}^{q+1/2},\ n(t^{q+1/2})=\hat{n}^{q+1/2},\ \dfrac{\partial n}{\partial t}(t^{q+1/2})=\hat{p}^{q+1/2}/2;\\ u^{q+1}=u(t^{q+1}), n^{q+1}=n(t^{q+1}), p^{q+1}=2\dfrac{\partial n}{\partial t}(t^{q+1}). \end{cases} \tag{2.122}$$

The *linear Schrödinger* and *wave equations* in (2.120) and (2.122) are uncoupled, implying that they can be solved by a variety of classical *spectral* or *finite difference* methods taking advantage of the space-periodic boundary conditions. On the other hand, the nonlinear system (2.121) can be solved *pointwise*: Indeed, since u and n are *real-valued* functions, it follows from the first and fifth equations in (2.121) that

$$|u(x,t)|=|u^{q+1/2}(x)|,\ \forall t\in[0,\triangle t],\ x\in[0,L]. \tag{2.123}$$

It follows then from (2.121) and (2.123) that the solution n in (2.121) is also a solution of the following *linear* problem

$$\begin{cases} \dfrac{\partial^2 n}{\partial t^2}=-\dfrac{\mu}{2}c^2\dfrac{\partial^2}{\partial x^2}(|u^{q+1/2}|^2) \text{ in } (0,L)\times(0,\triangle t),\\ n(0,t)=n(L,t) \text{ on } (0,\triangle t),\\ n(0)=n^{q+1/2},\ \dfrac{\partial n}{\partial t}(0)=p^{q+1/2}/2. \end{cases} \tag{2.124}$$

The closed form solution of (2.124) is given by

$$n(x,t)=n^{q+1/2}(x)+\frac{1}{2}p^{q+1/2}(x)t-\frac{\mu}{4}c^2\frac{\partial^2}{\partial x^2}(|u^{q+1/2}|^2)t^2 \text{ on } (0,L)\times(0,\triangle t), \tag{2.125}$$

implying, in particular, that

$$\hat{n}^{q+1/2}=n^{q+1/2}+\frac{\triangle t}{2}p^{q+1/2}-\frac{\mu}{4}(c\triangle t)^2\frac{\partial^2}{\partial x^2}(|u^{q+1/2}|^2).$$

Finally, to obtain the u solution of system (2.121), we observe that (n being known from (2.125)) it is the *unique* solution of the following *non-autonomous linear* initial value problem

$$\begin{cases} i\dfrac{\partial u}{\partial t}+2(\lambda|u^{q+1/2}|^2+n)u=0 \text{ in } (0,L)\times(0,\triangle t),\\ u(0,t)=u(L,t) \text{ on } (0,\triangle t),\\ u(0)=u^{q+1/2}, \end{cases} \tag{2.126}$$

a particular case of

$$\begin{cases} i\dfrac{\partial \phi}{\partial t} + 2(\lambda|\psi|^2 + v)\phi = 0 \ \text{ in } (0,L) \times (t_0,t_f), \\ \phi(0,t) = \phi(L,t) \ \text{ on } (t_0,t_f), \\ \phi(t_0) = \phi_0, \end{cases} \tag{2.127}$$

ψ (resp., v) being a given complex (resp., real)-valued function of x (resp., of $\{x,t\}$). With $M \geq 1$ an integer, let us define τ, a time-discretization step, by $\tau = \dfrac{t_f - t_0}{M}$, and $t^m = t_0 + m\tau$. To solve (2.127) we advocate the following time-discretization scheme of the *Crank-Nicolson* type:

$$\phi^0 = \phi_0. \tag{2.128}$$

For $m = 0, \cdots, M-1$, $\phi^m \to \phi^{m+1}$ via the solution of

$$\begin{cases} i\dfrac{\phi^{m+1} - \phi^m}{\tau} + \left[\lambda|\psi|^2 + \dfrac{v(t^{m+1}) + v(t^m)}{2}\right](\phi^{m+1} + \phi^m) = 0 \ \text{ in } (0,L), \\ \phi^{m+1}(0) = \phi^{m+1}(L). \end{cases} \tag{2.129}$$

Problem (2.129), can be solved point-wise (in practice at the grid-points of a finite difference one- dimensional "grid"). Scheme (2.128)–(2.129) is *second-order accurate* and *modulus preserving* (that is, verifies $|\phi^{m+1}| = |\phi^m|$, $\forall m = 0,\ldots,M-1$). On $[0,L]$, $\phi^{m+1}(x)$ is obtained via the solution of a 2×2 linear system (for those who prefer to use real arithmetic).

Remark 28. In [102], one advocates using instead of n the function $n - \mu|u|^2$. The numerical results reported in the above publication clearly show that operator-splitting provides efficient methods for the numerical solution of the Zakharov's system (2.112).

6 Applications of Augmented Lagrangian and ADMM Algorithms to the Solution of Problems from Imaging

6.1 Variational Models for Image Processing

6.1.1 Generalities

Usually, *image processing* refers to the processing and analysis of digital images. Variational models have become an essential part of image processing, such models relying on the minimization of a well-chosen energy functional, the minimization problem reading typically as

$$u = \arg\min_{v \in V} \left[E_{fitting}(v) + E_{regularizing}(v)\right]. \tag{2.130}$$

As shown above, the energy functional has two parts, namely a *fitting* part and a *regularizing* one. In the following we will present various variational image processing models and show that the operator-splitting and ADMM methodology provides efficient methods for the numerical solution of the related minimization problems. We will start our discussion with the well-known *Rudin-Osher-Fatemi* (ROF) model, and then follow with the presentation of some higher order models. Before going into more details, some remarks are in order, namely:

Remark 29. Most of the models we are going to consider below are not fully understood yet from a mathematical point of view, two of the main issues being, in (2.130), the choice of the space V and the weak-continuity properties of the energy functional. This will not prevent us to use these continuous models, for the simplicity of their formalism which facilitates the derivation of algorithms whose discrete analogues have provable convergence properties.

Remark 30. For image processing problems, the computational domain is always a rectangle, the image pixels providing a natural mesh for space discretization. This particularity makes easy, in general, the finite difference discretization of problem (2.130) and the implementation of iterative solution algorithms. The methodology we are going to discuss is not restricted to rectangular domains, however for domains with curved boundaries using finite-difference discretization may become complicated near the boundary; an elegant way to overcome this difficulty is to employ finite element approximations, as done in, e.g., [133].

Remark 31. A very detailed analysis of ADMM algorithms for the solution of image processing problems can be found in the chapter of this book by *M. Burger, A. Sawatzky & G. Steidl* (Chapter 10).

6.1.2 Total Variation and the ROF Model

One of the most popular variational models for image processing was proposed by *Rudin*, *Osher*, and *Fatemi* in their seminal work (ROF model) [144]. In [144], a denoised image is obtained by minimizing the following energy functional

$$E(v) = \frac{1}{2}\int_\Omega |f - v|^2\, dx + \eta \int_\Omega |\nabla v|\, dx, \tag{2.131}$$

where: $dx = dx_1 dx_2$, $f : \Omega \to I\!R$ is a given noisy image defined on Ω, $\int_\Omega |\nabla v|\, dx$ stands for the total variation of the trial function v (see [157] and [169] for a definition of the notion of total variation), and $\eta > 0$ is a positive tuning parameter controlling how much noise will be removed. The remarkable feature of the ROF model lies in its effectiveness in preserving object edges while removing noise. In fact, the total variation regularizer has been widely employed to accomplish other image processing tasks such as deblurring, segmentation, and registration.

In order to incorporate more geometrical information into the regularizer, a number of higher order regularization models have been proposed and used for image processing and computer vision problems. The ROF model has several unfavorable features. The main caveat is the stair-case effect, that is, the resulting cleaned image would present blocks even though the desired image may be smooth. Other undesirable properties include corner smearing and loss of image contrast. To remedy these drawbacks, a very rich list of results exists in the literature, see [2, 31, 120, 182, 185]. Despite the effectiveness of these models in removing the staircase effect, it is often a challenging issue to minimize the corresponding functionals. Note that if the functional E contains second-order derivatives of v, the related Euler-Lagrange equation is a fourth-order linear or nonlinear partial differential equation.

6.1.3 Regularization Using $TV2$

In [120], *Lysaker et al.* directly incorporated second order derivative information into the image denoising process, by proposing to minimize the following energy functional

$$E(v) = \frac{1}{2}\int_\Omega |f - v|^2\,dx + \eta \int_\Omega \sqrt{(v_{x_1x_1})^2 + 2(v_{x_1x_2})^2 + (v_{x_2x_2})^2}\,dx \tag{2.132}$$

This higher order energy functional is much simpler than the Elastica regularizer that we shall introduce later. Numerically, this regularizer shows rather good performance with noise suppression and edge preservation. In the literature, there exists quite a number of related models, see [20, 24, 25, 26, 28, 39, 53, 58, 89, 99, 101, 134, 138, 147, 171, 146, 180]. The well posedness of the variational problem associated with the energy functional in (2.132), and its gradient flow equation, have been studied in [88, 130]. High order models, such as the one associated with the energy in (2.132), have been discussed in, e.g., [15, 24, 32, 149, 176].

6.1.4 Regularization Using the Euler's Elastica Energy

In order to 'clean' a given function $f : \Omega \to \mathbb{R}$, the *Euler's Elastica* model relies on the minimization of the following energy functional

$$E(v) = \frac{1}{2}\int_\Omega |f - v|^2\,dx + \int_\Omega \left[a + b \left| \nabla \cdot \frac{\nabla v}{|\nabla v|} \right|^2 \right] |\nabla v|\,dx. \tag{2.133}$$

In (2.133), a and b are non-negative with $a + b > 0$. These two constants have to be chosen properly, depending of the application under consideration. The image processing model associated with the above energy functional comes from the Euler's Elastica energy for curves (see [31, 124] for the derivation of this energy): indeed, for a given curve $\Gamma \subset \mathbb{R}^2$ with curvature κ, the Euler's Elastica energy is defined (with obvious notation) by $\int_\Gamma (a + b\kappa^2)\,ds$. For a function v, the curvature of the level curve $\Gamma_c := \{x | v(x) = c\}$ is $\kappa = \nabla \cdot \dfrac{\nabla v}{|\nabla v|}$ (if $\nabla v \neq \mathbf{0}$). Thus, the Euler's Elastica energy for the level curve Γ_c is given by

$$l(c) = \int_{\Gamma_c} \left[a + b \left| \nabla \cdot \frac{\nabla v}{|\nabla v|} \right|^2 \right] ds.$$

Summing up (integrating) the Euler's Elastica energy over all the level curves Γ_c, it follows from the co-area formula (see [168]) that the total Euler's Elastica energy is given by

$$\int_{-\infty}^{\infty} l(c)\, dc = \int_{-\infty}^{\infty} \int_{\Gamma_c} \left[a + b \left| \nabla \cdot \frac{\nabla v}{|\nabla v|} \right|^2 \right] ds dc = \int_{\Omega} \left[a + b \left| \nabla \cdot \frac{\nabla v}{|\nabla v|} \right|^2 \right] |\nabla v|\, dx.$$

6.1.5 Regularization Using the Image Graph Mean Curvature

In [182], the authors proposed a variational image processing model making use of the mean curvature of the graph of function f, that is of the surface $\{x, y, z = f(x,y)\}$, to remove the noise. More specifically, the model considered in [182] employs the L^1 norm of the mean curvature of the above graph as a regularizer, the associated energy functional being defined by

$$E(v) = \frac{1}{2} \int_{\Omega} |f - v|^2 dx + \eta \int_{\Omega} \left| \nabla \cdot \frac{\nabla v}{\sqrt{1 + |\nabla v|^2}} \right| dx. \tag{2.134}$$

Above, $\eta (> 0)$ is a tuning parameter and the term $\dfrac{\nabla v}{\sqrt{1 + |\nabla v|^2}}$ is the mean curvature of the surface $\phi(x,y,z) = 0$ with $\phi(x,y,z) = u(x,y) - z$. Clearly, the model tries to fit the given noisy image surface $\{x, y, z = f(x,y)\}$ with a surface $\{x, y, z = u(x,y)\}$, u being a minimizer of the L^1-mean curvature energy functional (2.134). This idea goes back to much earlier publications, [108] for example. The model can sweep noise while keeping object edges, and it also avoids the staircase effect. More importantly, as discussed in [185], the model is also capable of preserving image contrasts as well as object corners.

6.1.6 Interface Problems: Chan-Vese Segmentation Model, Labeling Techniques, Min-Cut, and Continuous Max-Flow

In image processing, computer vision, etc., one encounters operations more complicated than denoising, *segmentation* being one of them. These applications require mathematical models more complicated (in some sense) than those considered in Sections 6.1.2 to 6.1.5, one of them being the *Chan-Vese* model introduced in [33]. Actually (as obvious from [33]), the *snake and active contour* model (ref. [106]) and the *Mumford-Shah* model (ref. [132]) can be viewed as ancestors of the Chan-Vese model. Using the notation of [33], the Chan-Vese segmentation model relies on the minimization of the following energy functional:

$$E_{CV}(\phi, d_1, d_2) = \lambda_1 \int_\Omega |f - d_1|^2 H(\phi)\, dx + \lambda_2 \int_\Omega |f - d_2|^2 [1 - H(\phi)]\, dx \quad (2.135)$$
$$+ \mu \int_\Omega |\nabla H(\phi)|\, dx + \nu \int_\Omega H(\phi)\, dx,$$

where in (2.135): (i) ϕ is a level set function whose zero level curves set represents the segmentation boundary. (ii) $H(\cdot)$ is the Heaviside function. (iii) d_1 and d_2 are two real numbers. (iv) λ_1, λ_2 and μ (resp., ν) are positive (resp., non-negative) tuning parameters (in many applications, one takes $\lambda_1 = \lambda_2 = 1$). The Euler-Lagrange equation associated with the minimization of the functional in (2.135) has been derived in [33]. In the above reference the associated gradient flow has been time-discretized by an explicit scheme to compute the solution of the above minimization problem (after an appropriate finite difference space discretization). Operator- splitting and ADMM can be used to develop algorithms with much faster convergence properties than the above explicit schemes; we will return on this issue in Section 6.2. Let us denote $H(\phi)$ by v; there is clearly equivalence between minimizing the functional defined by (2.135) and

$$\begin{cases} \inf_{\{v, d_1, d_2\} \in V \times \mathbb{R} \times \mathbb{R}} [\lambda_1 \int_\Omega |f - d_1|^2 v dx + \lambda_2 \int_\Omega |f - d_2|^2 [1 - v] dx \\ \qquad + \mu \int_\Omega |\nabla v| dx + \nu \int_\Omega v dx], \end{cases} \quad (2.136)$$

where $V = \{v | v \in L^\infty(\Omega), v(x) \in \{0, 1\}, \text{a.e. in } \Omega, \nabla v \in L^1(\Omega)\}$. The model associated with (2.136) was proposed in [117] and referred as a *binary level set* based model. More generally, we can consider the minimization, over the above set V, of energy functionals such as E_{potts} defined by

$$E_{potts}(v) = \int_\Omega f_1 v dx + \int_\Omega f_2 [1 - v] dx + \int_\Omega g |\nabla v| dx, \quad (2.137)$$

where f_1 and f_2 are given functions indicating the possibility that a point belongs to phase 0 or to phase 1, and where g is a non-negative function, possibly constant; if d_1 and d_2 are fixed in (2.136), the Chan-Vase model becomes a particular case of the model associated with the functional E_{potts} defined by (2.137). It was recently observed (see [173, 175]) that minimizing E_{potts} over the above V is a (kind of) continuous *min-cut problem*, itself equivalent (by duality) to a *max-flow problem*. Indeed, let us consider the following continuous max-flow problem

$$\begin{cases} \sup_{q_s, q_f, \mathbf{v}} \int_\Omega q_s dx \text{ subject to} \\ q_s \leq f_1, q_t \leq f_2, |\mathbf{v}| \leq g, \\ \nabla \cdot \mathbf{v} = q_s - q_t \text{ in } \Omega, \mathbf{v} \cdot \mathbf{n} = 0 \text{ on } \Gamma (= \partial\Omega), \end{cases} \quad (2.138)$$

where in (2.138)): (i) $\mathbf{v} = \{v_1, v_2\}$ and $|\mathbf{v}| = \sqrt{v_1^2 + v_2^2}$, $\mathbf{v}$ being the flow inside Ω. (ii) $\mathbf{n}$ is the unit outward vector normal at Γ. (iii) q_s (resp., q_t) represents a flow *from a source* (resp., *to a sink*). (iv) f_1 and f_2 are as in (2.137). We can also define $|\mathbf{v}|$ by $|\mathbf{v}| := |v_1| + |v_2|$; if we do so, the discretized max-flow problem can be solved by traditional graph cut methods. It follows from [175] that a dual of the max flow problem (2.138) reads as:

$$\inf_{\mu \in \Lambda} \left[\int_\Omega f_1(1-\mu)\,dx + \int_\Omega f_2 \mu\,dx + \int_\Omega g|\nabla \mu|\,dx \right], \tag{2.139}$$

where $\Lambda = \{\mu | \mu \in L^\infty(\Omega), 0 \le \mu(x) \le 1, \text{a.e. in } \Omega\} \cap W^{1,1}(\Omega)$. We have recovered thus the functional E_{potts} from (2.137) and shown a link between the Chan-Vese model and the max-flow problem. The dual problem (2.139) is known as a (continuous) *min-cut problem*. Actually, *Chan, Esedoglu* and *Nikolova* have shown in [29] that there is equivalence between (2.139) and minimizing over $V = \{v | v \in L^\infty(\Omega), v(x) \in \{0,1\}, \text{ a.e. in } \Omega, \nabla v \in L^1(\Omega)\}$ the functional E_{potts} defined by (2.137), a most remarkable result indeed since problem (2.139) is a convex variational problem whose discrete variants can be solved by ADMM type algorithms (see [5, 6, 7, 8, 114, 173, 174, 175, 178] for more details and generalizations).

Remark 32. In (2.136), (2.138) and (2.139), it is on purpose that we used inf (resp., sup) instead of min (resp., max) since we have no guarantee that the minimizing sequences of the functionals under consideration will converge weakly in the space or set where the minimization takes place.

Remark 33. Suppose that in (2.138) we replace the constraint $|\mathbf{v}| \le g$ by $|v_1| \le g_1$ and $|v_2| \le g_2$, everything else being the same; then, the dual problem of the associated variant of (2.138) reads (with Λ as in (2.139)) as

$$\inf_{\mu \in \Lambda} \left[\int_\Omega f_1(1-\mu)\,dx + \int_\Omega f_2 \mu\,dx + \int_\Omega \left(g_1 \left| \frac{\partial \mu}{\partial x_1} \right| + g_2 \left| \frac{\partial \mu}{\partial x_2} \right| \right) dx \right],$$

clearly a close variant of (2.139). Similarly, if we replace in (2.138) the constraint $|\mathbf{v}| \le g$ by $|v_1| + |v_2| \le g$, we obtain (as expected) the following dual problem

$$\inf_{\mu \in \Lambda} \left[\int_\Omega f_1(1-\mu)\,dx + \int_\Omega f_2 \mu\,dx + \int_\Omega g \sup\left(\left| \frac{\partial \mu}{\partial x_1} \right|, \left| \frac{\partial \mu}{\partial x_2} \right| \right) dx \right],$$

the set Λ being as above.

6.1.7 Segmentation Models with Higher Order Regularization

As could have been expected, first order segmentation models have limitations (discussed in [132]). To give an example let us consider the situation depicted in Figure 2.8(a) where some parts of the four letters have been erased: albeit one can easily recognize the four letters, first order segmentation models such as Chan-Vese's, might often capture the existing boundary instead of restoring the missing

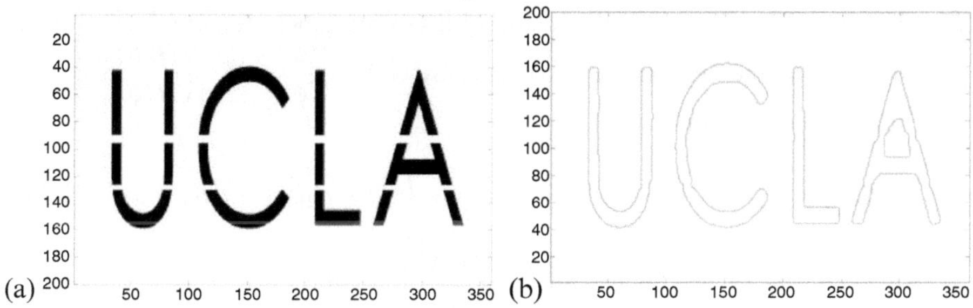

Fig. 2.8 Broken letters "UCLA" and its connected segmentation.

ones, as illustrated in Figure 2.8(b). In *inpainting* problems (see [31, 124]), missing image information is also recovered, but within given regions assigned in advance. In contrast, one would like to have a segmentation model that can interpolate the missing boundaries automatically without specifying the region of interest. To this end, one may employ the Euler's Elastica functional as a novel regularization term in the Chan-Vese's model (2.135), in order to replace the weighted TV term. Doing so we obtain the following energy functional (we assume $\nu = 0$, here):

$$E_{CVE}(\phi,d_1,d_2) = \lambda_1 \int_\Omega |f-d_1|^2 H(\phi)\,dx + \lambda_2 \int_\Omega |f-d_2|^2[1-H(\phi)]\,dx \quad (2.140)$$
$$+ \int_\Omega \left[a + b\left(\nabla \cdot \frac{\nabla\phi}{|\nabla\phi|}\right)^2\right] |\nabla H(\phi)|\,dx$$

where λ_1, λ_2, a and b are positive parameters. If ϕ is the signed distance level set function, it can be proved that the last term in (2.140) is equal to the Euler's elastica energy of the segmentation curve. This regularization was originally proposed and used in the celebrated paper on *segmentation with depth* by Nitzberg, Mumford, and Shiota (ref. [135]). Actually, it has also been used in [31] (resp., [183, 184]) for the solution of the in-painting (resp., illusory contour) problem. In [146], linear programming was used to minimize (after space discretization) curvature dependent functionals, the functional defined by (2.140) being one of those considered in this article.

Remark 34. Observe that since (formally at least, but this can be justified using a well-chosen regularization of the Heaviside function, such as $\xi \to \frac{1}{2}\left[1 + \frac{\xi}{\sqrt{\varepsilon^2+\xi^2}}\right]$) $\frac{\nabla\phi}{|\nabla\phi|} = \frac{\nabla H(\phi)}{|\nabla H(\phi)|}$, only the sign $H(\phi)$ of the function ϕ is needed when solving the segmentation problem via the functional in (2.140). This property suggests, as done in [117], to use a binary level set representation via the introduction of the function $v = H(\phi)$. Such a change of function was also used in [29] for finding the global minimizer associated with the Chan-Vese's model. More general binary level set representations with global minimization techniques have been developed (see, e.g.,

[7, 173, 174, 175, 177]) using the relationships existing between graph cuts, binary labeling and continuous max flow problems. Since $\nabla \cdot \frac{\nabla \phi}{|\nabla \phi|} = \nabla \cdot \frac{\nabla H(\phi)}{|\nabla H(\phi)|}$, one can rewrite the functional in (2.140) as

$$E(v, d_1, d_2) = \lambda_1 \int_\Omega |f - d_1|^2 v dx + \lambda_2 \int_\Omega |f - d_2|^2 [1 - v] dx + \int_\Omega \left[a + b \left(\nabla \cdot \frac{\nabla v}{|\nabla v|} \right)^2 \right] |\nabla v| dx \tag{2.141}$$

with the values taken by v being either 0 or 1. Strictly speaking the mean curvature of the graph makes sense for "smooth" functions only; to fix this issue, one relaxes the above binary restriction by replacing it by $0 \leq v \leq 1$, a less constraining condition indeed.

6.2 Fast Numerical Algorithms for Variational Image Processing Models Based on Operator- Splitting and Augmented Lagrangian Methods (ALM)

In this section, we will present operator-splitting and ALM based fast numerical algorithms, for the numerical treatment of variational image processing models.

6.2.1 Parallel Splitting Schemes for the ROF Model

The first model that we are going to consider is the ROF model discussed in Section 6.1.2. The formal *Euler-Lagrange equation* associated with the minimization of the (strictly convex) functional in (2.131) reads as

$$-\eta \nabla \cdot \frac{\nabla u}{|\nabla u|} + u = f \text{ in } \Omega, \quad \frac{\nabla u}{|\nabla u|} \cdot \mathbf{n} = 0 \text{ on } \partial\Omega, \tag{2.142}$$

with $\mathbf{n}$ the outward unit vector normal at $\partial\Omega$. In order to solve the nonlinear non-smooth elliptic equation (2.142) we associate with it an initial value problem and look for steady-state solutions. We consider thus

$$\begin{cases} \frac{\partial u}{\partial t} - \eta \nabla \cdot \frac{\nabla u}{|\nabla u|} + u = f \text{ in } \Omega \times (0, +\infty), \\ \frac{\nabla u}{|\nabla u|} \cdot \mathbf{n} = 0 \text{ on } \partial\Omega \times (0, +\infty), \\ u(0) = u_0, \end{cases} \tag{2.143}$$

an obvious choice for u_0 in (2.143) being $u_0 = f$. Actually to overcome the difficulty associated with the non-smoothness of the elliptic operator in (2.142) and (2.143), we consider the following regularized variant of (2.143):

$$\begin{cases} \dfrac{\partial u}{\partial t} - \eta \nabla \cdot \dfrac{\nabla u}{\sqrt{|\nabla u|^2 + \varepsilon^2}} + u = f \text{ in } \Omega \times (0, +\infty), \\ \dfrac{\partial u}{\partial n} = 0 \text{ on } \partial\Omega \times (0, +\infty), \\ u(0) = u_0, \end{cases} \tag{2.144}$$

with ε a small positive number. The simplest time-stepping scheme we can think about to capture the steady state solution of (2.144) is clearly the *forward-Euler scheme*. Let $\triangle t (> 0)$ be a time-discretization step; applied to the solution of (2.144) the forward Euler scheme produces the following algorithm:

$$u^0 = u_0. \tag{2.145}$$

For $n \geq 0$, $u^n \to u^{n+1}$ via

$$\begin{cases} \dfrac{u^{n+1} - u^n}{\triangle t} - \eta \nabla \cdot \dfrac{\nabla u^n}{\sqrt{|\nabla u^n|^2 + \varepsilon^2}} + u^n = f \text{ in } \Omega, \\ \dfrac{\partial u^{n+1}}{\partial n} = 0 \text{ on } \partial\Omega. \end{cases} \tag{2.146}$$

In practice, scheme (2.145)–(2.146) is applied to a discrete variant of (2.144) obtained by finite difference or finite element space discretization. Scheme (2.145)–(2.146) being explicit and the condition number of the operator in (2.146) rather large, its conditional stability requires small time steps leading to a slow convergence to a steady state solution. Suppose that Ω is the rectangle $(0, a) \times (0, b)$; in order to improve the speed of convergence to a steady state solution, we are going to apply to the solution of (2.144) the parallelizable operator-splitting scheme discussed in Section 2.8, taking advantage of the following decomposition of the operator in (2.144)

$$-\eta \nabla \cdot \frac{\nabla u}{\sqrt{|\nabla u|^2 + \varepsilon^2}} + u - f = A_1(u) + A_2(u), \tag{2.147}$$

with

$$\begin{cases} A_1(u) = -\eta \dfrac{\partial}{\partial x_1} \left(\dfrac{\dfrac{\partial u}{\partial x_1}}{\sqrt{|\nabla u|^2 + \varepsilon^2}} \right) + \dfrac{1}{2}(u - f), \\ A_2(u) = -\eta \dfrac{\partial}{\partial x_2} \left(\dfrac{\dfrac{\partial u}{\partial x_2}}{\sqrt{|\nabla u|^2 + \varepsilon^2}} \right) + \dfrac{1}{2}(u - f). \end{cases} \tag{2.148}$$

Combining the scheme we mentioned just above with a semi-explicit time discretization of the nonlinear terms we obtain

$$u^0 = u_0. \tag{2.149}$$

For $n \geq 0$, $u^n \to \{u^{n+1/4}, u^{n+2/4}\} \to u^{n+1}$ via

$$\left\{\begin{array}{l} \dfrac{u^{n+1/4}-u^n}{2\triangle t} - \eta\dfrac{\partial}{\partial x_1}\left(\dfrac{\dfrac{\partial u^{n+1/4}}{\partial x_1}}{\sqrt{|\nabla u^n|^2+\varepsilon^2}}\right) + \dfrac{u^{n+1/4}}{2} = \dfrac{f}{2} \text{ in } \Omega, \\ \dfrac{\partial u^{n+1/4}}{\partial x_1}(0,x_2) = \dfrac{\partial u^{n+1/4}}{\partial x_1}(a,x_2) = 0 \; \forall x_2 \in (0,b), \end{array}\right. \tag{150.1}$$

$$\left\{\begin{array}{l} \dfrac{u^{n+2/4}-u^n}{2\triangle t} - \eta\dfrac{\partial}{\partial x_2}\left(\dfrac{\dfrac{\partial u^{n+2/4}}{\partial x_2}}{\sqrt{|\nabla u^n|^2+\varepsilon^2}}\right) + \dfrac{u^{n+2/4}}{2} = \dfrac{f}{2} \text{ in } \Omega, \\ \dfrac{\partial u^{n+2/4}}{\partial x_2}(x_1,0) = \dfrac{\partial u^{n+2/4}}{\partial x_2}(x_1,b) = 0 \; \forall x_1 \in (0,a), \end{array}\right. \tag{150.2}$$

$$u^{n+1} = \frac{1}{2}(u^{n+1/4} + u^{n+2/4}). \tag{2.151}$$

Scheme (2.149)–(2.151) can accommodate large time steps implying a fast convergence to steady state solutions. It preserves also the symmetry of the images. Moreover since in most applications Ω is a rectangle with the image pixels uniformly distributed on it, it makes sense to use a finite difference discretization on a uniform Cartesian grid to approximate (150.1) and (150.2). For Dirichlet or Neumann boundary conditions, the finite difference discretization of (150.1) and (150.2) will produce two families of uncoupled tri-diagonal linear systems easily solvable (the good parallelization properties of the above scheme are quite obvious). The above operator-splitting scheme can be generalized to the numerical treatment of other variational models (such as Chan-Vese's, and to models involving derivatives of order higher than one, as shown in, e.g., [89]). A closely related scheme is discussed in [167].

6.2.2 A Split-Bregman Method and Related ADMM Algorithm for the ROF Model

In ref. [86], *T. Goldstein* and *S. Osher* proposed and tested a fast converging iterative method for the ROF model: this algorithm, of the *split-Bregman* type, is certainly one of the fastest numerical methods for the ROF model. It was quickly realized (see [156, 170, 172]) that the Bregman algorithm discussed in [86] is equivalent to an *ADMM* one. Here, we will explain the ideas in an informal way using the continuous model whose formalism is much simpler. As stated in Remark 29, to make our discussion more rigorous mathematically, the functional spaces for which the continuous model makes sense have to be specified (here, they are of the *bounded variation* type). This difficulty is one of the reasons explaining why some authors (as in [170]) consider discrete models, directly.

Let us denote ∇u by $\mathbf{p}$; then, it is easy to see that (from (2.131)) the ROF model is equivalent to the following linearly constrained minimization problem:

$$\{u,\mathbf{p}\} = \arg\min_{\substack{\{v,\mathbf{q}\}\\ \nabla v-\mathbf{q}=0}} \left[\eta \int_\Omega |\mathbf{q}|\,dx + \frac{1}{2}\int_\Omega |v-f|^2\,dx\right]. \tag{2.152}$$

Clearly, problem (2.152) belongs to the family of variational problems discussed in Section 3.2, the associated augmented Lagrangian being defined (with $r > 0$) by (see, e.g., [72] (Chapter 4)):

$$\begin{aligned} L_{rof}(v,\mathbf{q};\mu) = {} & \eta \int_\Omega |\mathbf{q}|\,dx + \frac{1}{2}\int_\Omega |v-f|^2\,dx \\ & + \frac{r}{2}\int_\Omega |\nabla v-\mathbf{q}|^2\,dx + \int_\Omega \mu\cdot(\nabla v-\mathbf{q})\,dx. \end{aligned} \tag{2.153}$$

Above, $u : \Omega \to I\!R$ denotes the restored image we are looking for, $\mathbf{p} = \nabla u$, μ is a Lagrange multiplier. Due to the strict convexity of the second term, the discrete analogues of the minimization problem (2.152) have a unique solution. Applying algorithm *ALG*2 of Section 3.2.2 to the solution of (2.152) we obtain the following

Algorithm 6.1: *An augmented Lagrangian method for the ROF model*

0. Initialization: $\lambda^0 = \mathbf{0}$, $u^0 = f$.

For $k = 0, 1, \cdots$, until convergence:

1. Compute $\mathbf{p}^{k+1}$ from

$$\mathbf{p}^{k+1} = \arg\min_{\mathbf{q}\in(L^2(\Omega))^2} L_{rof}(u^k,\mathbf{q};\lambda^k). \tag{2.154}$$

2. Compute u^{k+1} from

$$u^{k+1} = \arg\min_{v\in H^1(\Omega)} L_{rof}(v,\mathbf{p}^{k+1};\lambda^k). \tag{2.155}$$

3. Update λ^k by

$$\lambda^{k+1} = \lambda^k + r(\nabla u^{k+1} - \mathbf{p}^{k+1}). \tag{2.156}$$

It was observed in [156, 170] that this augmented Lagrangian algorithm is equivalent to the split-Bregman algorithm discussed in [86]. This equivalence is also explained in [172] for compressive sensing models. The minimization sub-problems (2.154) have closed form solutions which can be computed point-wise; solving them is thus quite easy. The minimization sub-problems (2.155) (in fact their discrete analogues) reduce to discrete well-posed linear Neumann problems; the associated matrix being symmetric, positive definite and sparse, these discrete elliptic problems can be solved by a large variety of direct and iterative methods (among them: sparse Cholesky, multi-level, Gauss-Seidel, conjugate gradient, FFT, etc.; see [170, 172] for more details). The convergence of algorithm (2.154)–(2.156) is discussed in [170].

Remark 35. As described above, Algorithm 6.1 is largely formal since it operates in the space $\mathbf{W} = [H^1(\Omega) \times (L^2(\Omega))^2] \times (L^2(\Omega))^2$, although the solution u of problem (2.131) may not have enough regularity to belong to $H^1(\Omega)$. However, Algorithm 6.1 makes sense for the discrete analogues of problem (2.131) and space $\mathbf{W}$ obtained by finite difference or finite element approximation; for finite element approximations in particular, the formalisms of Algorithm 6.1 and of its discrete counterparts are nearly identical. The above observation applies to most of the ADMM algorithms described below (see Remark 36, for example).

Remark 36. As shown in, e.g., [109] (for image denoising applications), Algorithm 6.1 is easy to modify in order to handle those situations where the functional $\int_\Omega |\nabla \mathbf{v}|\,dx$ is replaced by $\frac{1}{s}\int_\Omega |\nabla \mathbf{v}|^s\,dx$ with $0 < s < 1$, or by other non-convex functionals of $|\nabla \mathbf{v}|$; once discretized, these modifications of Algorithm 6.1 perform very well as shown in [109].

Remark 37. It is easy to extend algorithm (2.154)–(2.156) to the solution of the min-cut problem (2.139), since the additional constraint encountered in this last problem, namely $0 \leq \mu(x) \leq 1$, a.e. in Ω, is (relatively) easy to treat; actually, this extension has been done in [22] (see also [4, 27], and Section 6.2.3, below, for a number of related new approaches). As shown in [170] (page 320), and [142, 143], it is also easy to extend algorithm (2.154)–(2.156) to those situations where one uses vector-*TV* regularization in order to process vector- valued data.

6.2.3 An Augmented Lagrangian Method for the Continuous Min-Cut and Max-Flow Problems

The continuous max-flow problems (2.138) and (2.139) are dual to each other in the sense that if the function λ is solution of (2.139), it is a *Lagrange multiplier* for the flow conservation equation in (2.138). We can solve both problems simultaneously using a primal-dual method à la *ALG*2 relying on the following *augmented Lagrangian* functional

$$L_c(q_s, q_t, \mathbf{v}; \mu) = -\int_\Omega q_s\,dx - \int_\Omega \mu(\nabla\cdot\mathbf{v} - q_s + q_t)\,dx + \frac{r}{2}\int_\Omega (\nabla\cdot\mathbf{v} - q_s + q_t)^2\,dx, \tag{2.157}$$

where in (2.157): $r > 0$, and q_s, q_t and $\mathbf{v}$ verify, a.e. in Ω, $q_s \leq f_1$, $q_t \leq f_2$, $|\mathbf{v}| \leq g$; here $|\mathbf{v}| = \sqrt{v_1^2 + v_2^2}$, $\forall \mathbf{v} = \{v_1, v_2\}$. Applying *ALG*2 to the computation of the saddle-points of L_c over the set $(Q_1 \times Q_2 \times K) \times L^2(\Omega)$, where $Q_1 = \{q | q \in L^2(\Omega), q \leq f_1\}$, $Q_2 = \{q | q \in L^2(\Omega), q \leq f_2\}$, and $K = \{\mathbf{v} | \mathbf{v} \in (L^2(\Omega))^2, \nabla \cdot \mathbf{v} \in L^2(\Omega), \mathbf{v} \cdot \mathbf{n} = 0 \text{ on } \Gamma, |\mathbf{v}| \leq g\}$, we obtain

Algorithm 6.2: *An augmented Lagrangian method for the continuous max-flow problem*

0. Initialization: $\lambda^0 = 0$, $p_s^0 = f_1$, $p_t^0 = f_2$.

For $k = 0, 1, \cdots$, until convergence:

1. Compute $\mathbf{u}^{k+1}$ from

$$\mathbf{u}^{k+1} = \arg\min_{\mathbf{v} \in K} L_c(p_s^k, p_t^k, \mathbf{v}; \lambda^k). \tag{2.158}$$

2. Compute $\{p_s^{k+1}, p_t^{k+1}\}$ from

$$\{p_s^{k+1}, p_t^{k+1}\} = \arg\min_{\{q_s, q_t\} \in Q_1 \times Q_2} L_c(q_s, q_t, \mathbf{u}^{k+1}; \lambda^k). \tag{2.159}$$

3. Update λ^k by

$$\lambda^{k+1} = \lambda^k - r(\nabla \cdot \mathbf{u}^{k+1} - p_s^{k+1} + p_t^{k+1}). \tag{2.160}$$

We observe that (2.159) has a closed form solution (and that p_s^{k+1} and p_t^{k+1} can be computed point-wise independently of each other). The sub-problem (2.158) is a simple variant of the dual of the ROF problem (that is, the unconstrained minimization of the functional in (2.131)). We just need to solve this problem approximately; indeed, in our implementations we just used few steps of a descent algorithm, followed by a projection on the convex set $\{\mathbf{v} | \mathbf{v} \in (L^2(\Omega))^2, |\mathbf{v}| \leq g\}$ (see [173, 175] for more details on the solution of these sub-problems). The discrete variant of algorithm (2.158)–(2.160) that we implemented (via a finite difference discretization) proved being very robust with respect to initialization and to the value of the augmentation parameter r; it is also very efficient computationally.

Remark 38. As written, algorithm (2.158)–(2.160) is applicable only to the solution of two-phase flow problems. There are several ways to generalize this algorithm to models involving more than two phases, as shown in, e.g., [5, 6, 7, 8, 173, 177]. Also, we would like to emphasize the fact that the discrete analogue of algorithm (2.158)–(2.160) we implemented has good convergence properties no matter which of the following two norms we used for the flow constraint in (2.138) (see Remark 33 for the dual formulation associated with (2.162)):

$$|\mathbf{v}|_2 = \sqrt{v_1^2 + v_2^2} \tag{2.161}$$

or

$$|\mathbf{v}|_1 = |v_1| + |v_2|. \tag{2.162}$$

If one uses the meshes classically used in digital imaging, traditional graph cut methods (like those discussed in [19]) can be used to solve the discrete min-cut and max-flow problems if one uses the norm defined by (2.162) to bound $\mathbf{v}$. On the other hand, the above-mentioned graph cut methods cannot handle the norm defined by (2.161). It is also known that the solutions of the discrete min-cut and max-flow problems suffer from the matrication error if the norm in (2.162) is used. Compared to graph cut methods, ADMM algorithms such as (2.158)–(2.160) can handle both norms without particular difficulty. Moreover, these augmented Lagrangian algorithms are easy to parallelize and to implement on GPUs; also, they use much less memory than traditional graph cut methods; this enables using these algorithms for high dimensional and large size images or data.

6.2.4 A Split-Bregman Method and Related ADMM Algorithm for a Second Order Total Variation Model

Here, we will discuss the application of *ALG*2 (that is ADMM) to the solution of those image processing problems associated with the functional defined by (2.132) (also known as the *TV*2 model). The presentation follows [42, 73, 170], where the main ideas are: (i) transfer the burden of nonlinearity from the Hessian

$$\mathbf{D}^2 u \left(= \begin{pmatrix} \partial^2 u/\partial x_1^2 & \partial^2 u/\partial x_1 \partial x_2 \\ \partial^2 u/\partial x_1 \partial x_2 & \partial^2 u/\partial x_2^2 \end{pmatrix} \right)$$

to an additional unknown $\mathbf{p}$, via the relation

$$\mathbf{p} = \mathbf{D}^2 u, \tag{2.163}$$

and (ii) use a well-chosen augmented Lagrangian functional, associated with the linear relation (2.163). A similar idea has been (successfully) used in [42] for the augmented Lagrangian solution of the Dirichlet problem for the Monge-Ampère equation $\det \mathbf{D}^2 u = f$ (see also Chapter 8 of this book).

Back to the *TV*2 model (2.132), let us recall that the related minimization problem reads as

$$u = \arg\min_{v \in V} \left[\frac{1}{2} \int_\Omega |v - f|^2 dx + \eta \int_\Omega |\mathbf{D}^2 v| dx \right], \tag{2.164}$$

with $V = \{v | v \in L^2(\Omega), \mathbf{D}^2 v \in (L^1(\Omega))^{2\times 2}\}$ and $|\mathbf{M}| = \sqrt{\sum_{1 \le i,j \le 2} m_{ij}^2}$ denoting the Fröbenius norm of matrix $\mathbf{M}$. Proceeding as in Section 3.2.2, we observe the equivalence between (2.164) and

$$\{u, \mathbf{D}^2 u\} = \arg\min_{\{v,\mathbf{q}\} \in W} \left[\frac{1}{2} \int_\Omega |v - f|^2 dx + \eta \int_\Omega |\mathbf{q}| dx \right], \tag{2.165}$$

where

$$W = \{\{v,\mathbf{q}\} | v \in V, \mathbf{q} \in (L^1(\Omega))^{d\times d}, \mathbf{D}^2 v - \mathbf{q} = \mathbf{0}\},$$

an observation leading us to introduce the following augmented Lagrangian functional

$$\begin{aligned} L_{TV2}(v,\mathbf{q};\mu) = & \frac{1}{2}\int_\Omega |v-f|^2\,dx + \eta\int_\Omega |\mathbf{q}|\,dx \\ & + \frac{r}{2}\int_\Omega |\mathbf{D}^2 v - \mathbf{q}|^2\,dx + \int_\Omega \mu : (\mathbf{D}^2 v - \mathbf{q})\,dx, \end{aligned} \tag{2.166}$$

where, in (2.166), $r > 0$, and (with obvious notation) $\mathbf{S} : \mathbf{T} = \sum_{1\le i,j\le 2} s_{ij}t_{ij}$. Applying the methods discussed in Section 3.2.2 to the solution of the minimization problem (2.164) we obtain the following

Algorithm 6.3: *An augmented Lagrangian method for the TV2 model*

0. Initialization: $\lambda^0 = \mathbf{0}$, $u^0 = f$.

For $k = 0, 1, \cdots$, until convergence:

1. Compute $\mathbf{p}^{k+1}$ from

$$\mathbf{p}^{k+1} = \arg\min_{\mathbf{q}\in(L^2(\Omega))^{2\times 2}} L_{TV2}(u^k,\mathbf{q};\lambda^k). \tag{2.167}$$

2. Compute u^{k+1} from

$$u^{k+1} = \arg\min_{v\in H^2(\Omega)} L_{TV2}(v,\mathbf{p}^{k+1};\lambda^k). \tag{2.168}$$

3. Update λ^k by

$$\lambda^{k+1} = \lambda^k + r(\mathbf{D}^2 u^{k+1} - \mathbf{p}^{k+1}). \tag{2.169}$$

As with Algorithm 6.1 (that is (2.154)–(2.156)), the sub-problems (2.167) have closed-form solutions which can be computed point-wise. On the other hand, the sub-problems (2.168) reduce to linear bi-harmonic problems for the elliptic operator $I + r\nabla^4$; if properly discretized on a uniform grid (typically by finite differences), the discrete analogues of these bi-harmonic problems can be solved by FFT or by iterative methods (see [170] (page 324) for details).

Remark 39. Obviously, Remark 35 applies also to Algorithm 6.3, with $H^2(\Omega)$ playing here the role of $H^1(\Omega)$ there.

6.2.5 An Augmented Lagrangian Method for the Euler's Elastica Model

The energy functional defined by (2.133), namely

$$E(v) = \frac{1}{2}\int_\Omega |f-v|^2\,dx + \int_\Omega \left[a + b\left|\nabla\cdot\frac{\nabla v}{|\nabla v|}\right|^2\right]|\nabla v|\,dx,$$

makes no sense on the subset of Ω where ∇v vanishes. Following an approach very common in *visco-plasticity* (see, e.g., [66, 83]) one make things more rigorous by defining (following [154]) the energy functional by

$$E(v,\mathbf{m}) = \frac{1}{2}\int_\Omega |f - v|^2\, dx + \int_\Omega \left[a + b|\nabla\cdot\mathbf{m}|^2\right] |\nabla v|\, dx \tag{2.170}$$

the functions v and $\mathbf{m}$ in (2.170) verifying

$$|\nabla v| = \mathbf{m}\cdot\nabla v,\ |\mathbf{m}| \leq 1. \tag{2.171}$$

The related minimization problem reads as

$$\begin{cases} \{u,\mathbf{n}\} = \arg\min_{\{v,\mathbf{m}\}} E(v,\mathbf{m}), \\ \text{with } \{v,\mathbf{m}\} \text{ verifying (2.171).} \end{cases} \tag{2.172}$$

Introducing the vector-valued function $\mathbf{p}$ verifying $\mathbf{p} = \nabla u$, we clearly have equivalence between (2.172) and

$$\begin{cases} \{u,\mathbf{p},\mathbf{n}\} = \arg\min_{\{v,\mathbf{q},\mathbf{m}\}} \left[\frac{1}{2}\int_\Omega |f - v|^2\, dx + \int_\Omega \left[a + b|\nabla\cdot\mathbf{m}|^2\right] |\mathbf{q}|\, dx\right], \\ \text{with } \{v,\mathbf{q},\mathbf{m}\} \text{ verifying } \mathbf{q} = \nabla v,\ |\mathbf{q}| = \mathbf{m}\cdot\mathbf{q},\ |\mathbf{m}| \leq 1. \end{cases} \tag{2.173}$$

Following [154], we associate with the minimization problem (2.173) the following augmented Lagrangian functional

$$\begin{aligned} L_{elas}\{v,\mathbf{q},\mathbf{m};\mu_1,\mu_2) = {} & \frac{1}{2}\int_\Omega |v - f|^2\, dx + \int_\Omega \left[a + b|\nabla\cdot\mathbf{m}|^2\right] |\mathbf{q}|\, dx \\ & + \frac{r_1}{2}\int_\Omega |\nabla v - \mathbf{q}|^2\, dx + r_2\int_\Omega (|\mathbf{q}| - \mathbf{q}\cdot\mathbf{m})\, dx \\ & + \int_\Omega \mu_1\cdot(\nabla v - \mathbf{q})\, dx + \int_\Omega \mu_2(|\mathbf{q}| - \mathbf{q}\cdot\mathbf{m})\, dx, \end{aligned} \tag{2.174}$$

with r_1 and r_2 both positive. Suppose that in (2.174) the vector-valued function $\mathbf{m}$ belongs to $\mathbf{M}$, the closed convex set of $(L^2(\Omega))^2$ defined by

$$\mathbf{M} = \{\mathbf{m} | \mathbf{m} \in (L^2(\Omega))^2, |\mathbf{m}(x)| \leq 1, \text{a.e. in } \Omega\};$$

we have then $|\mathbf{q}| - \mathbf{q}\cdot\mathbf{m} \geq 0$, implying (since $|\mathbf{q}| - \mathbf{q}\cdot\mathbf{m} = ||\mathbf{q}| - \mathbf{q}\cdot\mathbf{m}|$)) that the variant of *ALG*2 described just below will force the condition $|\mathbf{q}| - \mathbf{q}\cdot\mathbf{m} = 0$ in the sense of $L^1(\Omega)$. This variant of *ALG*2 reads as follows when applied to the solution of problem (2.172) (below, $H(\Omega;\text{div}) = \{\mathbf{v} | \mathbf{v} \in (L^2(\Omega))^2, \nabla\cdot\mathbf{v} \in L^2(\Omega)\}$):

Algorithm 6.4: *An augmented Lagrangian method for the Euler's Elastica model*

0. Initialization: $\lambda_1^0 = \mathbf{0}$, $\lambda_2^0 = 0$, $u^0 = f$, $\mathbf{n}^0 = \mathbf{0}$.

For $k = 0, 1, \cdots$, until convergence:

1. Compute $\mathbf{p}^{k+1}$ from

$$\mathbf{p}^{k+1} = \arg\min_{\mathbf{q}\in(L^2(\Omega))^2} L_{elas}(u^k, \mathbf{q}, \mathbf{n}^k; \lambda_1^k, \lambda_2^k). \tag{2.175}$$

2. Compute $\mathbf{n}^{k+1}$ from

$$\mathbf{n}^{k+1} = \arg\min_{\mathbf{m}\in H(\Omega;\mathrm{div})\cap\mathbf{M}} L_{elas}(u^k, \mathbf{p}^{k+1}, \mathbf{m}; \lambda_1^k, \lambda_2^k). \tag{2.176}$$

3. Compute u^{k+1} from

$$u^{k+1} = \arg\min_{v\in H^1(\Omega)} L_{elas}(v, \mathbf{p}^{k+1}, \mathbf{n}^{k+1}; \lambda_1^k, \lambda_2^k). \tag{2.177}$$

4. Update $\{\lambda_1^k, \lambda_2^k\}$ by

$$\begin{cases} \lambda_1^{k+1} = \lambda_1^k + r_1(\nabla u^{k+1} - \mathbf{p}^{k+1}), \\ \lambda_2^{k+1} = \lambda_2^k + r_2(|\mathbf{p}^{k+1}| - \mathbf{p}^{k+1}\cdot\mathbf{n}^{k+1}). \end{cases} \tag{2.178}$$

Below, we will give some details and comments about the solution of the sub-problems encountered when applying algorithm (2.175)–(2.178); implementation issues will be also addressed. Further information is provided in [154].

- The minimization sub-problem (2.175) has a unique closed-form solution which can be computed point-wise.
- The minimization sub-problem (2.176) is equivalent to the following *elliptic variational inequality*

$$\begin{cases} \mathbf{n}^{k+1} \in H(\Omega;\mathrm{div})\cap\mathbf{M}, \\ b\int_\Omega |\mathbf{p}^{k+1}|\,\nabla\cdot\mathbf{n}^{k+1}\,\nabla\cdot(\mathbf{m}-\mathbf{n}^{k+1})\,dx \geq \int_\Omega (r_2+\lambda_2^k)\mathbf{p}^{k+1}\cdot(\mathbf{m}-\mathbf{n}^{k+1})\,dx, \\ \forall\mathbf{m}\in H(\Omega;\mathrm{div})\cap\mathbf{M}. \end{cases} \tag{2.179}$$

  We observe that the bilinear functional in the left-hand side of (2.179) is symmetric and positive semi-definite (indeed, $\int_\Omega |\mathbf{p}^{k+1}|(\nabla\cdot\mathbf{m})^2\,dx = 0$ if $\mathbf{m} = \nabla\times z$. However, the boundedness of $\mathbf{M}$ implies that the variational problem (2.176), (2.179) has solutions. For the solution of the discrete analogues of the above problem we advocate using few iterations of those *relaxation methods with projection* discussed in, e.g., [66, 76] (other methods are possible as shown in [154]).
- The minimization sub-problem (2.177) has a unique solution characterized by

$$\begin{cases} u^{k+1} \in H^1(\Omega), \\ \int_\Omega u^{k+1} v\,dx + r_1\int_\Omega \nabla u^{k+1}\cdot\nabla v\,dx = \int_\Omega f v\,dx + \int_\Omega (r_1\mathbf{p}^{k+1} - \lambda_1^k)\cdot\nabla v\,dx, \\ \forall v\in H^1(\Omega). \end{cases} \tag{2.180}$$

  Actually, (2.180) is nothing but a variational formulation of the following Neumann problem

$$\begin{cases} u^{k+1} - r_1\nabla^2 u^{k+1} = f - \nabla\cdot(r_1\mathbf{p}^{k+1} - \lambda_1^k) \ \text{ in } \Omega, \\ r_1\dfrac{\partial u^{k+1}}{\partial\nu} = (r_1\mathbf{p}^{k+1} - \lambda_1^k)\cdot\nu \ \text{ on } \partial\Omega, \end{cases} \tag{2.181}$$

where, in (2.181), ν denotes the outward unit vector normal at the boundary $\partial\Omega$ of Ω. The numerical solution of linear elliptic problems such as (2.181) is routine nowadays; after an appropriate space discretization it can be achieved by a large variety of direct and iterative methods (sparse Cholesky, FFT, relaxation, multilevel, etc.).

- Since the energy functional associated with the Euler's Elastica is *non-convex* (see (2.170)) the augmentation parameters r_1 and r_2 have to be chosen large enough to guarantee the convergence of algorithm (2.175)–(2.179). Actually, the tuning of r_1 and r_2 is a delicate issue in itself and we can expect (as shown for example in [133], for a problem involving three augmentation parameters) the optimal values of these parameters to be of different orders of magnitude with respect to the space discretization h.
- Another solution method for the Euler's Elastica is discussed in [21]. It relies on tractable convex relaxation in higher dimension.

Remark 40. In [154], an alternative method for the solution of the Euler's Elastica problem (2.172) is also considered. It relies on the equivalence between (2.172) and

$$\begin{cases} \{u,\mathbf{p},\mathbf{n}^1,\mathbf{n}^2\} = \underset{\{v,\mathbf{q},\mathbf{m}_1,\mathbf{m}_2\}}{\arg\min} \left[\frac{1}{2}\int_\Omega |f-v|^2\,dx + \int_\Omega \left[a+b|\nabla\cdot\mathbf{m}_1|^2\right]|\mathbf{q}|\,dx\right], \\ \text{with } \{v,\mathbf{q},\mathbf{m}_1,\mathbf{m}_2\} \text{ verifying } \mathbf{q}=\nabla v,\ \mathbf{m}_1=\mathbf{m}_2,\ |\mathbf{q}|=\mathbf{m}_2\cdot\mathbf{q},\ |\mathbf{m}_2|\leq 1. \end{cases} \tag{2.182}$$

An augmented Lagrangian associated with (2.182) is clearly the one defined by

$$\begin{aligned} L_{elas}\{v,\mathbf{q},\mathbf{m}_1,\mathbf{m}_2;\mu_1,\mu_2,\mu_3) &= \frac{1}{2}\int_\Omega |v-f|^2\,dx + \int_\Omega \left[a+b|\nabla\cdot\mathbf{m}_1|^2\right]|\mathbf{q}|\,dx \\ &+\frac{r_1}{2}\int_\Omega |\nabla v-\mathbf{q}|^2\,dx + r_2\int_\Omega (|\mathbf{q}|-\mathbf{q}\cdot\mathbf{m}_2)\,dx + r_3\int_\Omega |\mathbf{m}_1-\mathbf{m}_2|^2\,dx \\ &+\int_\Omega \mu_1\cdot(\nabla v-\mathbf{q})\,dx + \int_\Omega \mu_2(|\mathbf{q}|-\mathbf{q}\cdot\mathbf{m}_2)\,dx + \int_\Omega \mu_3\cdot(\mathbf{m}_1-\mathbf{m}_2)\,dx, \end{aligned} \tag{2.183}$$

with r_1, r_2 and r_3 all positive. From (2.184), one can easily derive a variant of algorithm (2.175)–(2.178) for the solution of the minimization problem (2.172); such an algorithm is discussed in [154]. Actually the above reference discusses also the solution by a similar methodology of the variant of problem (2.172) obtained by replacing the fidelity term $\frac{1}{2}\int_\Omega |f-v|^2\,dx$ by $\frac{1}{s}\int_\Omega |f-v|^s\,dx$ with $s\in[1,+\infty)$. Typically, one takes $s=1$ (resp., $s=2$) for salt-and-pepper noise (resp., Gaussian noise). Further details and generalizations are given in [154].

Remark 41. As shown in [186], the methodology we employed to solve the minimization problem (2.172) can be easily modified in order to handle the *Chan-Vese Elastica* model.

6.2.6 An Augmented Lagrangian Method for the L^1-Mean Curvature Model

In this section, we follow closely the presentation used in [185]. The rational of the L^1-mean curvature model has been given in Section 6.1.5, leading one to consider the following minimization problem

$$u = \arg\min_{v \in V} \left[\frac{1}{2} \int_\Omega |v - f|^2 \, dx + \eta \int_\Omega \left| \nabla \cdot \frac{\nabla v}{\sqrt{1 + |\nabla v|^2}} \right| dx \right], \tag{2.184}$$

where $\nabla = \{\partial/\partial x_i\}_{i=1}^2$. In (2.184), the choice of V is a delicate theoretical issue; indeed the safest way to proceed would be to take $V = H^2(\Omega)$ in (2.184), and to replace min by inf (a (kind) of justification for this approach can be found in [133]). Let us observe (as in [185], where a slightly different notation is used) that

$$\nabla \cdot \frac{\nabla v}{\sqrt{1 + |\nabla v|^2}} = \nabla_3 \cdot \frac{\{\nabla v, -1\}}{|\{\nabla v, -1\}|}, \tag{2.185}$$

where, in (2.185), $\nabla_3 = \{\partial/\partial x_1, \partial/\partial x_2, 0\}$,and where $\{\nabla v, -1\}$ denotes the 3-dimensional vector-valued function $\{\partial v/\partial x_1, \partial v/\partial x_2, -1\}$. In order to simplify (in some sense) the nonlinear structure of the minimization problem (2.184), we associate new unknown functions with its solution u, namely $\mathbf{p}$, $\mathbf{n}$ and ψ verifying

$$\begin{cases} \mathbf{p} = \{\nabla u, -1\}, \\ \mathbf{n} = \dfrac{\mathbf{p}}{|\mathbf{p}|}, \quad \text{or equivalently here } |\mathbf{p}| - \mathbf{p} \cdot \mathbf{n} = 0, \ |\mathbf{n}| \leq 1, \\ \psi = \nabla_3 \cdot \mathbf{n}. \end{cases} \tag{2.186}$$

From (2.185) and (2.186), there is clearly equivalence between (2.184) and

$$\begin{cases} \{u, \psi, \mathbf{p}, \mathbf{n}\} = \arg\min_{\{v, \varphi, \mathbf{q}, \mathbf{m}\}} \left[\dfrac{1}{2} \displaystyle\int_\Omega |v - f|^2 \, dx + \eta \int_\Omega |\varphi| \, dx \right], \\ \text{with } \{v, \varphi, \mathbf{q}, \mathbf{m}\} \text{ verifying } \mathbf{q} = \{\nabla v, -1\}, |\mathbf{q}| - \mathbf{q} \cdot \mathbf{m} = 0, |\mathbf{m}| \leq 1, \nabla_3 \cdot \mathbf{m} = \varphi. \end{cases} \tag{2.187}$$

In order to solve the minimization problem (2.184), taking advantage of its equivalence with (2.187), we introduce the following augmented Lagrangian functional

$$\begin{aligned} L_{MC}(v, \varphi, \mathbf{q}, \mathbf{z}, \mathbf{m}; \mu_1, \mu_2, \mu_3, \mu_4) &= \frac{1}{2} \int_\Omega |v - f|^2 \, dx + \eta \int_\Omega |\varphi| \, dx \\ &+ \frac{r_1}{2} \int_\Omega (|\mathbf{q}| - \mathbf{q} \cdot \mathbf{z}) \, dx + \int_\Omega \mu_1 (|\mathbf{q}| - \mathbf{q} \cdot \mathbf{z}) \, dx \\ &+ \frac{r_2}{2} \int_\Omega |\{\nabla v, -1\} - \mathbf{q}|^2 \, dx + \int_\Omega \mu_2 \cdot (\{\nabla v, -1\} - \mathbf{q}) \, dx \\ &+ \frac{r_3}{2} \int_\Omega \left| \varphi - \left(\frac{\partial m_1}{\partial x_1} + \frac{\partial m_2}{\partial x_2} \right) \right|^2 dx + \int_\Omega \mu_3 \left(\varphi - \left(\frac{\partial m_1}{\partial x_1} + \frac{\partial m_2}{\partial x_2} \right) \right) dx, \\ &+ \frac{r_4}{2} \int_\Omega |\mathbf{z} - \mathbf{m}|^2 \, dx + \int_\Omega \mu_4 \cdot (\mathbf{z} - \mathbf{m}) \, dx. \end{aligned} \tag{2.188}$$

The additional vector-valued function $\mathbf{z}$ has been introduced in order to decouple $\nabla_3 \cdot \mathbf{m}$ from the nonlinear relations verified by $\mathbf{m}$ in (2.187). Following [185], and taking (2.187) and (2.188) into account, we advocate the following algorithm for the solution of problem (2.184):

Algorithm 6.5: *An augmented Lagrangian method for the L^1-mean curvature model*

0. Initialization: $\lambda_1^0 = 0$, $\boldsymbol{\lambda}_2^0 = \mathbf{0}$, $\lambda_3^0 = 0$, $\boldsymbol{\lambda}_4^0 = \mathbf{0}$, $u^0 = f$, $\mathbf{p}^0 = \{\nabla u^0, -1\}$, $\mathbf{n}^0 = \mathbf{y}^0 = \dfrac{\mathbf{p}^0}{|\mathbf{p}^0|}$, $\psi^0 = \nabla_3 \cdot \mathbf{n}^0$.

For $k = 0, 1, \cdots$, until convergence:

1. Compute u^{k+1} from

$$u^{k+1} = \arg\min_{v \in H^1(\Omega)} L_{MC}(v, \psi^k, \mathbf{p}^k, \mathbf{y}^k, \mathbf{n}^k; \lambda_1^k, \boldsymbol{\lambda}_2^k, \lambda_3^k, \boldsymbol{\lambda}_4^k). \tag{2.189}$$

2. Compute ψ^{k+1} from

$$\psi^{k+1} = \arg\min_{\varphi \in L^2(\Omega)} L_{MC}(u^{k+1}, \varphi, \mathbf{p}^k, \mathbf{y}^k, \mathbf{n}^k; \lambda_1^k, \boldsymbol{\lambda}_2^k, \lambda_3^k, \boldsymbol{\lambda}_4^k). \tag{2.190}$$

3. Compute $\mathbf{p}^{k+1}$ from

$$\mathbf{p}^{k+1} = \arg\min_{\mathbf{q} \in (L^2(\Omega))^3} L_{MC}(u^{k+1}, \psi^{k+1}, \mathbf{q}, \mathbf{y}^k, \mathbf{n}^k; \lambda_1^k, \boldsymbol{\lambda}_2^k, \lambda_3^k, \boldsymbol{\lambda}_4^k). \tag{2.191}$$

4. Compute $\mathbf{y}^{k+1}$ from

$$\mathbf{y}^{k+1} = \arg\min_{\mathbf{z} \in \mathbf{Z}} L_{MC}(u^{k+1}, \psi^{k+1}, \mathbf{p}^{k+1}, \mathbf{z}, \mathbf{n}^k; \lambda_1^k, \boldsymbol{\lambda}_2^k, \lambda_3^k, \boldsymbol{\lambda}_4^k). \tag{2.192}$$

5. Compute $\mathbf{n}^{k+1}$ from

$$\mathbf{n}^{k+1} = \arg\min_{\mathbf{m} \in \mathbf{M}} L_{MC}(u^{k+1}, \psi^{k+1}, \mathbf{p}^{k+1}, \mathbf{y}^{k+1}, \mathbf{m}; \lambda_1^k, \boldsymbol{\lambda}_2^k, \lambda_3^k, \boldsymbol{\lambda}_4^k). \tag{2.193}$$

6. Update $\{\lambda_1^k, \boldsymbol{\lambda}_2^k, \lambda_3^k, \boldsymbol{\lambda}_4^k\}$ by

$$\begin{cases} \lambda_1^{k+1} = \lambda_1^k + r_1(|\mathbf{p}^{k+1}| - \mathbf{p}^{k+1} \cdot \mathbf{y}^{k+1}), \\ \boldsymbol{\lambda}_2^{k+1} = \boldsymbol{\lambda}_2^k + r_2(\{\nabla u^{k+1}, -1\} - \mathbf{p}^{k+1}), \\ \lambda_3^{k+1} = \lambda_3^k + r_3 \left(\psi^{k+1} - \left(\dfrac{\partial n_1^{k+1}}{\partial x_1} + \dfrac{\partial n_2^{k+1}}{\partial x_2} \right) \right), \\ \boldsymbol{\lambda}_4^{k+1} = \boldsymbol{\lambda}_4^k + r_4(\mathbf{y}^{k+1} - \mathbf{n}^{k+1}). \end{cases} \tag{2.194}$$

In (2.189)–(2.194), the sets $\mathbf{Z}$ and $\mathbf{M}$ are defined by

$$\mathbf{Z} = \{\mathbf{z} | \mathbf{z} \in (L^2(\Omega))^3, |\mathbf{z}(x)| \leq 1, \text{ a.e. in } \Omega\},$$

and

$$\mathbf{M} = \{\mathbf{m} | \mathbf{m} \in (L^2(\Omega))^3, \frac{\partial m_1}{\partial x_1} + \frac{\partial m_2}{\partial x_2} \in L^2(\Omega)\},$$

respectively.

We observe that the minimization sub-problems (2.190), (2.191), and (2.192) have closed form solutions which can be computed point-wise. On the other hand, the Euler-Lagrange equations of the sub-problems (2.189) and (2.193) are well-posed linear elliptic equations with constant coefficients; fast solvers exist for the solution of the discrete analogues of these elliptic problems (see [185] for details and the results of numerical experiments validating the above algorithm). An important issue is the tuning of the augmentation parameters r_1, r_2, r_3, and r_4; the comments we did in Section 6.2.5, concerning the adjustment of r_1 and r_2 in algorithm (2.176)–(2.178), still apply here.

Remark 42. Another augmented Lagrangian based solution method for the L^1-mean curvature problem (2.184) is discussed and numerically tested in ref. [133]. The related ADMM algorithm involves only three Lagrange multipliers and three augmentation parameters. Moreover, the various vector-valued functions encountered in the approach discussed in [133] map Ω into $\mathbb{R}^2$ (instead of $\mathbb{R}^3$, as it is the case for algorithm (2.189)–(2.194)).

7 Further Comments and Complements

There is much more to say about *operator-splitting* and *ADMM* algorithms; fortunately, many of these issues and topics we left behind, or said very little about, are developed in the other chapters of this book. There are however some issues we would like to-briefly-comment to conclude this chapter, namely:

(i) The convergence of operator-splitting methods and *ADMM* algorithms, when applied to the solution of problems involving *non-monotone operators* and/or *non-convex functionals*.
(ii) The choice of the *augmentation parameters* and their dynamical adjustment when applying *ADMM* algorithms.
(iii) The derivation of operator-splitting schemes of *high* (or *higher*) *orders* of accuracy.
(iv) Trying to understand why the *Douglas-Rachford* scheme is more robust than the *Peaceman-Rachford* one, using simple model problems to clarify this issue.
(v) Very few problems have generated as many operator-splitting based solution methods than the *Navier-Stokes equations* modeling viscous fluid flows. From this fact, providing the reader with a significant number of related references is a must in a book like this one. These references will conclude this chapter.

Concerning the *first issue*, to the best of our knowledge, there is no general theory concerning the convergence of operator-splitting methods and *ADMM* algorithms when the problem under consideration involves at least one non-monotone operator and/or a non-convex functional. Actually, one can find in the literature convergence results for some problems lacking monotonicity and/or convexity, but, most often, the proofs of these results are very specific of the problem under consideration, and therefore are not easy to generalize to other situations. However, some recent results obtained by *R. Luke* [96, 115] and *W. Yin* [166], and collaborators, suggest that a fairly general theory is not out of reach. However, we think that there always will be situations where one not will be able to prove the convergence of operator-splitting methods and *ADMM* algorithms. This is not surprising since these methods and algorithms have been quite successful at solving problems for which the existence of solutions has not been proved.

The *second issue* concerning the choice and the dynamical adaptation of the augmentation parameters is another complicated one, particularly for those non-convex and non-monotone situations involving more than one of such parameters. Indeed, numerical experiments have shown that the optimal values of these parameters may have several orders of magnitude (as shown in, e.g., [80] and [133]), and, from the possible existence of multiple solutions, that bifurcations can take place depending also of the values of these parameters (and of the algorithm initialization). However, for particular problems, heuristics have been found, significantly improving the speed of convergence of these *ADMM* algorithms (see, e.g., [46]).

In order to address the *high* (or *higher*) *orders* of accuracy issue (our *third issue*) we return to Section 2.3 of this chapter (the one dedicated to the Strang symmetrized operator-splitting scheme), and consider the following initial value problem

$$\begin{cases} \dfrac{dX}{dt} + (A+B)X = 0 \text{ on } (0,T), \\ X(0) = X_0, \end{cases} \tag{2.195}$$

where A and B are *linear* operators *independent* of t. When applied to the solution of the initial value problem (2.195), the Strang symmetrized scheme (2.7)–(2.10) can be written in the following more compact form

$$\begin{cases} X^0 = X_0, \\ X^{n+1} = e^{-A\triangle t/2}e^{-B\triangle t}e^{-A\triangle t/2}X^n, \ \forall n \geq 0. \end{cases} \tag{2.196}$$

The relation

$$e^{-(A+B)\triangle t} - e^{-A\triangle t/2}e^{-B\triangle t}e^{-A\triangle t/2} = O(\triangle t^3),$$

shows that scheme (2.196) is second order accurate (and exact if $AB = BA$). For those situations requiring an order of accuracy *higher than two*, several options do exist, the best known being:

(a) The 4th order *Strang-Richardson* scheme discussed in [49, 50, 48] where it is applied (among other problems) to the numerical solution of real-valued or vector-valued *reaction-diffusion* equations such as

$$\frac{\partial \mathbf{u}}{\partial t} - \mathbf{M}\nabla^2\mathbf{u} + \mathbf{F}(\mathbf{u}) = \mathbf{0},$$

where $\mathbf{u}(x,t) \in \mathbb{R}^d$, ∇^2 denotes the Laplace operator, $\mathbf{M}$ is a $d \times d$ symmetric positive definite matrix, and $\mathbf{F}$ is a smooth mapping from $\mathbb{R}^d$ into $\mathbb{R}^d$.

(b) The *exponential operator-splitting schemes*. Actually, the Lie and Strang splitting schemes belong to this family of time discretization methods, whose origin (concerning schemes of order higher than two) is not easy to track back, early significant publications being [150, 151] (see also the references therein and those in [161], and in Google Scholar). Arbitrary high accuracy can be obtained with these methods, the price to pay being their reduced stability (compared to the Strang scheme, for example).

The best way to introduce the *Strang-Richardson* scheme is to start, one more time, from the simple initial value problem (2.195). Applied to the solution of (2.195), the Strang-Richardson scheme reads as

$$\begin{cases} X^0 = X_0, \\ X^{n+1} = \frac{1}{3}\left[4e^{-A\Delta t/4}e^{-B\Delta t/2}e^{-A\Delta t/2}e^{-B\Delta t/2}e^{-A\Delta t/4}\right. \\ \qquad\qquad \left. -e^{-A\Delta t/2}e^{-B\Delta t}e^{-A\Delta t/2}\right]X^n, \ \forall n \geq 0. \end{cases} \tag{2.197}$$

A more practical equivalent formulation of the symmetrized scheme (2.197) can be found in the Chapter 6 of [70]; it avoids the use of matrix exponentials and can be generalized easily to nonlinear problems (it requires the solution of eight sub-initial value problems per time step). Scheme (2.197) is *fourth order accurate* but not as stable as the original Strang scheme (scheme (2.196)). Also, its application to decompositions involving more than two operators becomes a bit complicated to say the least (higher order methods of the same type are discussed in [85]).

In a similar fashion, we consider again the initial value problem (2.195) to introduce *exponential splitting* methods. Applied to the solution of (2.195) the typical *exponential operator-splitting scheme* reads as follows:

$$\begin{cases} X^0 = X_0, \\ X^{n+1} = \left(\prod_{j=1}^{J} e^{-b_j B\Delta t}e^{-a_j A\Delta t}\right)X^n, \ \forall n \geq 0, \end{cases} \tag{2.198}$$

where $a_j, b_j \in \mathbb{R}$, for $1 \leq j \leq J$. The Strang symmetrized scheme (2.196) is a particular case of (2.198) (corresponding to $J = 2$, $b_1 = 0$, $a_1 = 1/2$, $b_2 = 1$, $a_2 = 1/2$). By an appropriate choice of J, and of the coefficients a_j and b_j, scheme (2.198) can be made of order higher than two (as shown in, e.g., [16]), the price to pay being that some of the coefficients a_j, b_j are *negative* making the scheme inappropriate to those situations where some of the operators are dissipative. On the other hand, these higher order schemes produce spectacular results when applied to reversible systems, like those associated with some linear and nonlinear Schrödinger operators, as shown in, e.g.,[51, 161]. Their generalization to those (fairly common) situations involving more than two operators is rather complicated, although theoretically doable.

Concerning the *fourth issue*, the *Peaceman-Rachford* and *Douglas-Rachford* schemes have been briefly discussed in Sections 2.4 and 2.5, respectively. In order to have a better idea of their *accuracy* and *stability* properties, we will consider the particular situation where, in problem (2.14), $\phi_0 \in I\!R^d$, $T = +\infty$, and where A_1 (resp., A_2) is given by $A_1 = \alpha A$ (resp., $A_2 = \beta A$), A being a real symmetric positive definite $d \times d$ matrix, and α, β verifying $0 \le \alpha, \beta \le 1$ and $\alpha + \beta = 1$. The exact solution of the associated problem (2.14) reads as

$$\phi(t) = e^{-At}\phi_0, \ \forall t \ge 0,$$

which implies (by projection on an orthonormal basis of eigenvectors of matrix A, and with obvious notation)

$$\phi_i(t) = e^{-\lambda_i t}\phi_{0i}, \ \forall t \ge 0, \ \forall i = 1, \ldots, d, \tag{2.199}$$

where $0 < \lambda_1 \le \cdots \le \lambda_i \le \cdots \le \lambda_d$, the λ_i's being the eigenvalues of matrix A. Applying the *Peaceman-Rachford* scheme (2.15) to the particular problem (2.14) defined above, we obtain the following discrete analogue of (2.199):

$$\phi_i^n = (R_1(\lambda_i \triangle t))^n \phi_{0i}, \ \forall n \ge 0, \ \forall i = 1, \ldots, d, \tag{2.200}$$

R_1 being the rational function defined by

$$R_1(\xi) = \frac{\left(1 - \frac{\alpha}{2}\xi\right)\left(1 - \frac{\beta}{2}\xi\right)}{\left(1 + \frac{\alpha}{2}\xi\right)\left(1 + \frac{\beta}{2}\xi\right)}. \tag{2.201}$$

Since $|R_1(\xi)| < 1$, $\forall \xi > 0$, the Peaceman-Rachford scheme (2.15) is *unconditionally stable* in the particular case considered here. However, the property $\lim_{\xi \to +\infty} R_1(\xi) = 1$ shows that the above scheme is *not stiff A-stable*, making it not a first choice scheme to capture steady state solutions or to simulate fast transient phenomena. Actually, the stability drawback we just mentioned is not specific to the particular case we are considering, but seems to hold in general for scheme (2.15). Incidentally, the relation

$$R_1(\xi) - e^{-\xi} = O(\xi^3) \text{ in the neighborhood of } \xi = 0$$

implies that in the particular case under consideration (where A_1 and A_2 commute) scheme (2.15) is *second order accurate*. Applying now the *Douglas-Rachford* scheme (2.17) to the same particular case of problem (2.14), we obtain

$$\phi^{n+1} = (I + \alpha \triangle t A)^{-1}(I + \beta \triangle t A)^{-1}(I + \alpha\beta(\triangle t)^2 A^2)\phi^n, \ \forall n \ge 0,$$

which implies

$$\phi^n = (I + \alpha \triangle t A)^{-n}(I + \beta \triangle t A)^{-n}(I + \alpha\beta(\triangle t)^2 A^2)^n \phi_0, \ \forall n \ge 0. \tag{2.202}$$

By projection of (2.202) on an orthonormal basis of $I\!R^d$ consisting of eigenvectors of A, we obtain the following variant of (2.200):

$$\phi_i^n = (R_2(\lambda_i \triangle t))^n \phi_{0i}, \; \forall n \geq 0, \; \forall i = 1, \ldots, d, \tag{2.203}$$

R_2 being the rational function defined by

$$R_2(\xi) = \frac{1 + \alpha\beta\xi^2}{(1 + \alpha\xi)(1 + \beta\xi)}. \tag{2.204}$$

Since $0 < R_2(\xi) < 1$, $\forall \xi > 0$, the Douglas-Rachford scheme (2.17) is *unconditionally stable* in the particular case considered here. However, the property $\lim_{\xi \to +\infty} R_2(\xi) = 1$ shows that the above scheme is *not stiff A-stable*, making it not a first choice scheme to capture steady state solutions or to simulate fast transient phenomena. Actually, the stability drawback we just mentioned is not specific to the particular case we are considering, but seems to hold in general for scheme (2.17). Concerning the accuracy of scheme (2.17), we observe that in the neighborhood of $\xi = 0$, we have

$$R_2(\xi) = 1 - \xi + \xi^2 + O(\xi^3),$$

which implies, by comparison with $e^{-\xi} = 1 - \xi + \frac{\xi^2}{2} + O(\xi^3)$, that scheme (2.17) is no better than first order accurate in the particular case we are considering. Since this particular case is the most favorable one can think about, one expects the Douglas-Rachford scheme (2.17) to be *generically first order accurate*, a prediction supported by the results of various numerical experiments. It is worth mentioning that in order to improve the accuracy of the Douglas-Rachford scheme (2.17), *J. Douglas & S. Kim* introduced in the late 90s–early 2000s [56], the following variant of the above scheme

$$\phi^0 = \phi_0. \tag{2.205}$$

For $n \geq 0$, $\phi^n \to \hat{\phi}^{n+1} \to \phi^{n+1}$ as follows:

Solve

$$\frac{\hat{\phi}^{n+1} - \phi^n}{\triangle t} + A_1\left(\frac{\hat{\phi}^{n+1} + \phi^n}{2}, t^{n+1/2}\right) + A_2(\phi^n, t^n) = 0, \tag{2.206}$$

and

$$\frac{\phi^{n+1} - \phi^n}{\triangle t} + A_1\left(\frac{\hat{\phi}^{n+1} + \phi^n}{2}, t^{n+1/2}\right) + A_2\left(\frac{\phi^{n+1} + \phi^n}{2}, t^{n+1/2}\right) = 0. \tag{2.207}$$

The *Douglas-Kim* scheme (2.205)–(2.207) is clearly inspired from the *Crank-Nicolson* scheme. Scheme (2.205)–(2.207) is *second order accurate* if the operators A_1 and A_2 are sufficiently smooth, the price to pay for this accuracy enhancement being a *reduction of stability and robustness* compared to the original Douglas-Rachford scheme (2.17).

At those wondering how to choose between *Peaceman-Rachford* and *Douglas-Rachford* schemes we will say that on the basis of many numerical experiments, it seems that the second scheme is more robust and faster for those situations where one of the operators is *non-smooth* (multivalued or singular, for example), particularly if one is interested at capturing steady state solutions. Actually, this behavior is consistent with the fact that the rational function R_1 associated with the Peaceman-Rachford scheme (the one defined by (2.201)) may change sign when ξ varies on $(0,+\infty)$, unlike the rational function R_2 defined by (2.204) (the one associated with the Douglas- Rachford scheme) which stays positive on the above interval. These sign changes suggest a more oscillatory behavior for the associated scheme if fast transients take place, or if one tries to capture steady state solutions starting far away from these solutions.

As a final comment on *ADI* methods we have to mention that one of their main contributors (if not the main one), beyond their founders (*J. Douglas*, *H. Rachford*, and *D. Peaceman*), is definitely *E. Wachpress*: His wonderful book *The ADI Model Problem* [164] is an invaluable source of information and references on the Peaceman-Rachford and Douglas-Rachford methods, from the theoretical and practical points of view.

As a conclusion, let us observe that the *Navier-Stokes equations* modeling the flow of viscous fluids have been mentioned quite a few times in this chapter (Section 4 in particular), and in other chapters of this book. There is no doubt that very few partial differential equation problems have motivated such a large number of operator-splitting based solution methods. Focusing on those publications with which we have some familiarity, let us mention: [11, 12, 13, 23, 35, 43, 47, 70, 72, 73, 90, 91, 92, 93, 94, 105, 107, 111, 112, 116, 122, 123, 158, 159, 160] (see also the references therein, Google Scholar, and Chapters 21, 22 and 23 of this book).

References

1. Aftalion, A.: Vortices in Bose-Einstein Condensates, Birkhäuser, Boston, MA (2006)
2. Ambrosio, L., Masnou, S.: A direct variational approach to a problem arising in image reconstruction. Interfaces and Free Boundaries, **5**, 63–82 (2003)
3. Arrow, K., Hurwicz, L., Uzawa, H.: Studies in Linear and Nonlinear Programing. Stanford University Press, Stanford, CA (1958)
4. Aujol, J.F.: Some first-order algorithms for total variation based image restoration. Journal of Mathematical Imaging and Vision, **34**, 307–327 (2009)
5. Bae, E., Lellmann, J., Tai, X.C.: Convex relaxations for a generalized Chan-Vese model. In: Heyden, A., Kahl, F., Olsson, C., Oskarsson, M., Tai, X.C. (eds) Energy Minimization Methods in Computer Vision and Pattern Recognition, pp. 223–236. Springer, Berlin (2013)
6. Bae, E., Tai, X.C.: Efficient global minimization methods for image segmentation models with four regions. Journal of Mathematical Imaging and Vision, **51**, 71–97 (2015)
7. Bae, E., Yuan, J., Tai, X.C.: Global minimization for continuous multiphase partitioning problems using a dual approach. International Journal of Computer Vision, **92**, 112–129 (2011)
8. Bae, E., Yuan, J., Tai, X.C., Boykov, Y.: A fast continuous max-flow approach to non-convex multilabeling problems. In: Bruhn, A., Pock, T., Tai, X.C. (eds.) Efficient Algorithms for Global Optimization Methods in Computer Vision, pp. 134–154. Springer, Berlin (2014)

9. Bao, W., Jaksch, D., Markowich, P.A.: Numerical solution of the Gross-Pitaevskii equation for Bose-Einstein condensation. J. Comp. Phys., **187**, 318–342 (2003)
10. Bao, W., Jin, S., Markowich, P.A.: On time-splitting spectral approximations for the Schrödinger equation in the semi-classical regime. J. Comp. Phys, **175**, 487–524 (2002)
11. Beale, J.T., Greengard, C.: Convergence of Euler-Stokes splitting of the Navier-Stokes equations. Communications on Pure and Applied Mathematics, **47** (8),1083–115 (1994)
12. Beale, J.T., Greengard, C., Thomann, E.: Operator splitting for Navier-Stokes and Chorin-Marsden product formula. In: Vortex Flows and Related Numerical Methods, NATO ASI Series, Vol. 395, pp. 27–38. Springer-Netherlands (1993)
13. Beale, J.T., Majda, A.: Rates of convergence for viscous splitting of the Navier-Stokes equations. Mathematics of Computation, **37** (156), 243–259 (1981)
14. Belytschko, T., Hughes, T.J.R.(editors): Computational Methods for Transient Analysis. North-Holland, Amsterdam (1983)
15. Bertozzi, A.L., Greer, J.B.: Low curvature image simplifiers: global regularity of smooth solutions and Laplacian limiting schemes. Comm. Pure Appl. Math., **57**, 764–790 (2004)
16. Blanes, S., Moan, P.C.: Practical symplectic partitioned Runge-Kutta and Runge-Kutta-Nyström methods. J. Comp. Appl. Math., **142** (2), 313–330 (2002)
17. Bonito, A., Glowinski, R.: On the nodal set of the eigenfunctions of the Laplace-Beltrami operator for bounded surfaces in $\mathbf{R}^3$: A computational approach. Commun. Pure Appl. Analysis, **13**, 2115–2126 (2014)
18. Boyd, S., Parikh, N., Chu, E., Peleato, B., Eckstein, J.: Distributed optimization and statistical learning via the alternating direction method of multipliers. Foundations and Trends in Machine Learning, **3**, 1–122 (2011)
19. Boykov, Y., Kolmogorov, V.: An experimental comparison of min-cut/max-flow algorithms for energy minimization in vision. IEEE Transactions on Pattern Analysis and Machine Intelligence, **26**, 359–374 (2001)
20. Bredies, K.: Recovering piecewise smooth multichannel images by minimization of convex functionals with total generalized variation penalty. In: Bruhn, A., Pock, T., Tai, X.C. (eds.) Efficient Algorithms for Global Optimization Methods in Computer Vision, pp. 44–77. Springer, Berlin (2014)
21. Bredies, K., Pock, T., Wirth, B.: Convex relaxation of a class of vertex penalizing functionals. Journal of Mathematical Imaging and Vision, **47**, 278–302 (2013)
22. Bresson, X., Esedoglu, S., Vandergheynst, P., Thiran, J.P., Osher, S.: Fast global minimization of the active contour/snake model. Journal of Mathematical Imaging and Vision, **28**, 151–167 (2007)
23. Bristeau, M.O., Glowinski, R., Périaux, J.: Numerical methods for the Navier-Stokes equations. Application to the simulation of compressible and incompressible viscous flow. Computer Physics Reports, **6**, 73–187 (1987)
24. Brito-Loeza, C., Chen, K.: On high-order denoising models and fast algorithms for vector-valued images. IEEE Transactions on Image Processing, **19**, 1518–1527 (2010)
25. Calder, J., Mansouri, A.,Yezzi, A.: Image sharpening via Sobolev gradient flows. SIAM Journal on Imaging Sciences, **3**, 981–1014 (2010)
26. Chambolle, A., Lions, P.-L.: Image recovery via total variation minimization and related problems. Numerische Mathematik, **76**, 167–188, (1997)
27. Chambolle, A., Pock, T.: A first-order primal-dual algorithm for convex problems with applications to imaging. Journal of Mathematical Imaging and Vision, **40**, 120–145 (2011)
28. Chan, R.H., Lanza, A., Morigi, S., Sgallari, F.: An adaptive strategy for the restoration of textured images using fractional order regularization. Numerical Mathematics: Theory, Methods & Applications, **6**, 276–296 (2013)
29. Chan, T., Esedoglu, S., Nikolova, M.: Algorithms for finding global minimizers of image segmentation and denoising models. SIAM J. Appl. Math., **66**, 1632–1648 (electronic) (2006)
30. Chan, T.F., Glowinski, R.: Finite Element Approximation and Iterative Solution of a Class of Mildly Nonlinear Elliptic Equations. Stanford report STAN-CS-78-674, Computer Science Department, Stanford University, Palo Alto, CA (1978)

31. Chan, T., Kang, S.H., Shen, J.: Euler's elastica and curvature-based inpainting. SIAM Journal on Applied Mathematics, **62**, 564–592 (2002)
32. Chan, T. F., Marquina, A., Mulet, P.: High-order total variation-based image restoration. SIAM J. Sci. Comput., **22** (2), 503–516 (2000).
33. Chan, T., Vese, L.A.: Active contours without edges. IEEE Trans Image Proc., **10**, 266–277 (2001)
34. Chiche, A., Gilbert, J.C.: How the augmented Lagrangian algorithm can deal with an infeasible convex quadratic optimization problem. Journal of Convex Analysis, **22**, 30 (2015)
35. Chorin, A.J.: Numerical study of slightly viscous flow. Journal of Fluid Mechanics, **57** (4), 785–796 (1973)
36. Chorin, A.J., Hughes, T.J.R., McCracken, M.F., Marsden, J.E.: Product formulas and numerical algorithms. Com. Pure Appl. Math., **31**, 205–256 (1978)
37. Ciarlet, P.G.: The Finite Element Method for Elliptic Problems. SIAM, Philadelphia, PA (2002)
38. Crandall, M.G., Lions, P.L.: Viscosity solutions of Hamilton-Jacobi equations. Transactions of the American Mathematical Society, **277**, 1–42 (1983)
39. Cuesta, E., Kirane, M., Malik, S.A.: Image structure preserving denoising using generalized fractional time integrals. Signal Processing, **92**, 553–563 (2012)
40. Dahiya, D., Baskar, S., Coulouvrat, F.: Characteristic fast marching method for monotonically propagating fronts in a moving medium. SIAM J. Scient. Comp., **35**, A1880–A1902 (2013)
41. Dean, E.J., Glowinski, R.: On some finite element methods for the numerical simulation of incompressible viscous flow In: Gunzburger, M.D., Nicolaides, R.A. (eds.) Incompressible Computational Fluid Dynamics, pp. 109–150. Cambridge University Press, New York, NY (1993)
42. Dean, E.J., Glowinski, R.: An augmented Lagrangian approach to the numerical solution of the Dirichlet problem for the Monge-Ampère equation in two dimensions. Electronic Transactions on Numerical Analysis, **22**, 71–96 (2006)
43. Dean, E.J., Glowinski, R., Pan, T.W.: A wave equation approach to the numerical simulation of incompressible viscous fluid flow modeled by the Navier-Stokes equations. In: J.A. de Santo (ed.) Mathematical and Numerical Aspects of Wave Propagation, pp. 65–74. SIAM, Philadelphia, PA (1998)
44. Deiterding, R., Glowinski, R., Olivier, H., Poole, S.: A reliable split-step Fourier method for the propagation equation of ultra-fast pulses in single-mode optical fibers. J. Lightwave Technology, **31**, 2008–2017 (2013)
45. Delbos, F., Gilbert, J.C.: Global linear convergence of an augmented Lagrangian algorithm for solving convex quadratic optimization problems. Journal of Convex Analysis, **12**, 45–69 (2005)
46. Delbos, F., Gilbert, J.C., Glowinski, R., Sinoquet, D.: Constrained optimization in seismic reflection tomography: A Gauss-Newton augmented Lagrangian approach. Geophys. J. Internat., **164**, 670–684 (2006)
47. Demkowicz, L., Oden, J.T., Rachowicz, W.: A new finite element method for solving compressible Navier-Stokes equations based on an operator splitting method and *h-p* adaptivity. Comp. Meth. Appl. Mech. Eng., **84** (3), 275–326 (1990)
48. Descombes, S.: Convergence of splitting methods of high order for reaction-diffusion systems. Math. Comp., **70** (236), 1481–1501 (2001)
49. Descombes, S., Schatzman, M.: Directions alternées d'ordre élevé en réaction-diffusion. C.R. Acad. Sci. Paris, Sér. I, Math., **321** (11), 1521–1524 (1995)
50. Descombes, S., Schatzman, M.: On Richardson extrapolation of Strang's formula for reaction-diffusion equations. In : Equations aux Dérivées Partielles et Applications : Articles dédiés à J.L. Lions, Gauthier-Villars-Elsevier, Paris, pp. 429–452 (1998)
51. Descombes, S., Thalhammer, M.: The Lie-Trotter splitting for nonlinear evolutionary problems with critical parameters: a compact local error representation and application to nonlinear Schrödinger equations in the semiclassical regime. IMA J. Num. Anal., **33** (2), 722–745 (2013)

52. Desjardin, B., Esteban. M.: On weak solution for fluid-rigid structure interaction: compressible and incompressible models. Archives Rat. Mech. Anal., **146**, 59–71 (1999)
53. Didas, S., Weickert, J., Burgeth, B.: Properties of higher order nonlinear diffusion filtering. Journal of Mathematical Imaging and Vision, **35**, 208–226 (2009)
54. Douglas, J.: Alternating direction methods in three space variables. Numer. Math., **4**, 41–63 (1962)
55. Douglas, J.: Alternating direction methods for parabolic systems in m-space variables. J. ACM, **9**, 42–65 (1962)
56. Douglas, J., Kim, S.: Improved accuracy for locally one-dimensional methods for parabolic equations. Math. Models Meth. Appl. Sciences, **11** (9), 1563–1579 (2001)
57. Douglas, J., Rachford, H.H.: On the solution of the heat conduction problem in 2 and 3 space variables. Trans. Amer. Math. Soc., **82**, 421–439 (1956)
58. Duan, Y., Huang, W.: A fixed-point augmented Lagrangian method for total variation minimization problems. Journal of Visual Communication and Image Representation, **24**, 1168–1181 (2013)
59. Duvaut, G., Lions, J.L.: Inequalities in Mechanics and Physics. Springer, Berlin (1976)
60. Eckstein, J., Bertsekas, D.P.: On the Douglas-Rachford splitting method and the proximal point algorithm for maximal monotone operators. Math. Progam., **55**, 293–318 (1992)
61. Esser, E.: Applications of Lagrangian-based alternating direction methods and connections to split Bregman. CAM report, 9(31), Department of Mathematics, UCLA, Los Angeles, CA (2009)
62. Fortin, M., Glowinski, R.: Lagrangiens Augmentés: Application à la Résolution Numérique des Problèmes aux Limites. Dunod, Paris (1982)
63. Fortin, M., Glowinski, R.: Augmented Lagrangians: Application to the Numerical Solution of Boundary Value Problems. North-Holland, Amsterdam (1983)
64. Gabay, D.: Application de la méthode des multiplicateurs aux inéquations variationnelles. In: Fortin, M., Glowinski, R. (eds.) Lagrangiens Augmentés: Application à la Résolution Numérique des Problèmes aux Limites, pp. 279–307. Dunod, Paris (1982)
65. Gabay, D.: Application of the methods of multipliers to variational inequalities In: Fortin, M., Glowinski, R. (eds.) Augmented Lagrangians: Application to the Numerical Solution of Boundary Value Problems, pp. 299–331. North-Holland, Amsterdam (1983)
66. Glowinski, R.: Numerical Methods for Nonlinear Variational Problems. Springer, New York, NY (1984, 2nd printing: 2008)
67. Glowinski, R.: Viscous flow simulation by finite element methods and related numerical techniques. In: Murman, E.M., Abarbanel, S.S. (eds.) Progress and Supercomputing in Computational Fluid Dynamics, pp. 173–210. Birkhäuser, Boston, MA (1985)
68. Glowinski, R.: Splitting methods for the numerical solution of the incompressible Navier-Stokes equations. In: Balakrishnan, A.V., Dorodnitsyn, A.A., Lions, J.L. (eds.) Vistas in Applied Mathematics, pp. 57–95. Optimization Software, New York, NY (1986)
69. Glowinski, R.: Finite element methods for the numerical simulation of incompressible viscous flow. Application to the control of the Navier-Stokes equations. In: Anderson, C.R., Greengard, C. (eds.) Vortex Dynamics and Vortex Methods, pp. 219–301. American Mathematical Society, Providence, RI (1991)
70. Glowinski, R.: Finite element methods for incompressible viscous flow In: Ciarlet, P.G., Lions, J.L. (eds.) Handbook of Numerical Analysis, Vol. IX, pp. 3–1176. North-Holland, Amsterdam (2003)
71. Glowinski, R.: On alternating direction methods of multipliers: A historical perspective. In: Fitzgibbon, W., Kuznetsov, Y.A., Neittaanmäki, P., Pironneau, O. (eds.) Modeling, Simulation and Optimization for Science and Technology, Vol. 34, pp. 59–82. Springer, Dordrecht (2014)
72. Glowinski, R.: Variational Methods for the Numerical Solution of Nonlinear Elliptic Problems. SIAM, Philadelphia, PA (2015)
73. Glowinski, R., Dean, E.J., Guidoboni, G., Juarez, H.L., Pan, T.-W.: Applications of operator-splitting methods to the direct numerical simulation of particulate and free-surface flows and to the numerical solution of the two-dimensional elliptic Monge-Ampère equation. Japan J. Ind. Appl. Math., **25**, 1–63 (2008)

74. Glowinski, R., Le Tallec, P.: Augmented Lagrangian and Operator-Splitting Methods in Nonlinear Mechanics. SIAM, Philadelphia, PA (1989)
75. Glowinski, R., Leung, Y., Qian, J.: Operator-splitting based fast sweeping methods for isotropic wave propagation in a moving fluid. SIAM J. Scient. Comp., **38**(2), A1195–A1223 (2016)
76. Glowinski, R., Lions, J.L., Trémolières, R.: Numerical Analysis of Variational Inequalities. North-Holland, Amsterdam (1981).
77. Glowinski, R., Marrocco, A.: Sur l'approximation par éléments finis d'ordre un et la résolution par pénalisation-dualité d'une classe de problèmes de Dirichlet non-linéaires. C. R. Acad. Sci. Paris, **278A**, 1649–1652 (1974)
78. Glowinski, R., Marrocco, A.: Sur l'approximation par éléments finis d'ordre un et la résolution par pénalisation-dualité d'une classe de problèmes de Dirichlet non-linéaires. ESAIM : Math. Model. Num. Anal., **9**(R2), 41–76 (1975)
79. Glowinski, R., Pan, T.-W., Hesla, T.I., Joseph, D.D., Périaux, J.: A fictitious domain approach to the direct numerical simulation of incompressible viscous fluid flow past moving rigid bodies: application to particulate flow. J. Comp. Phys., **169**, 363–426 (2001)
80. Glowinski, R., Quaini, A.: On an inequality of C. Sundberg: A computational investigation via nonlinear programming. Journal of Optimization Theory and Applications, **158** (3), 739–772 (2013)
81. Glowinski, R., Shiau, L., Sheppard, M.: Numerical methods for a class of nonlinear integro-differential equations. Calcolo, **50**, 17–33 (2013)
82. Glowinski, R., Sorensen, D.C.: Computing the eigenvalues of the Laplace-Beltrami operator on the surface of a torus: A numerical approach. In: Glowinski, R., Neittaanmäki, P. (eds.) Partial Differential Equations: Modeling and Numerical Solution, pp. 225–232. Springer, Dordrecht (2008)
83. Glowinski, R., Wachs, A.: On the numerical simulation of visco-plastic fluid flow. In: Ciarlet, P.G., Glowinski, R., Xu, J. (eds.) Handbook of Numerical Analysis, Vol. XVI, pp. 483–717. North-Holland, Amsterdam (2011)
84. Godlewsky, E.: Méthodes à Pas Multiples et de Directions Alternées pour la Discrétisation d'Equations d'Evolution. Doctoral Dissertation, Department of Mathematics, University P. & M. Curie, Paris, France (1980)
85. Goldman, D., Kaper, T.J.: N th-order operator-splitting schemes and non-reversible systems. SIAM J. Num. Anal., **33** (1), 349–367 (1996)
86. Goldstein, T., Osher, S.: The split-Bregman method for $L1$-regularized problems. SIAM Journal on Imaging Sciences, **2**, 323–343 (2009)
87. Grandmont, C., Maday, Y.: Existence for an unsteady fluid-structure interaction problem. Math. Model. Num. Anal., **34**, 609–636 (2000)
88. Greer, J.B., Bertozzi, A.L.: Traveling wave solutions of fourth order PDEs for image processing. SIAM Journal on Mathematical Analysis, **36**, 38–68 (2004)
89. Guidotti, P., Longo, K.: Two enhanced fourth order diffusion models for image denoising. Journal of Mathematical Imaging and Vision, **40**, 188–198 (2011)
90. Guermond, J.L.: Some implementations of projection methods for Navier-Stokes equations. RAIRO-Model. Math. Anal. Num., **30** (5), 637–667 (1996)
91. Guermond, J.L., Minev, P., Shen, J.: An overview of projection methods for incompressible flows. Comp. Meth. Appl. Mech. Eng., **195** (44), 6011–6045 (2006)
92. Guermond, J.L., Quartapelle, L.: Calculation of incompressible viscous flows by an unconditionally stable projection FEM. J. Comp. Phys., **132** (1),12–33(1997)
93. Guermond, J.L., Quartapelle, L.: On the approximation of the unsteady Navier-Stokes equations by finite element projection methods. Numer. Math., **80** (2), 207–238 (1998)
94. Guermond, J.L., Shen, J.: A new class of truly consistent splitting schemes for incompressible flows. J. Comp. Phys., **192** (1), 262–276 (2003)
95. He, B., Yuan, X.: On the O(1/n) convergence rate of the Douglas-Rachford alternating direction method. SIAM J. Numer. Anal., **50**, 700–709 (2012)
96. Hesse, R, Luke, D.R.: Nonconvex notions of regularity and convergence of fundamental algorithms for feasibility problems. SIAM Journal on Optimization, **23** (4), 2397–2419 (2013)

97. Hou, S., T.-W. Pan, Glowinski, R.: Circular band formation for incompressible viscous fluid-rigid-particle mixtures in a rotating cylinder. Physical Review E, **89**, 023013 (2014)
98. Hu, H.H., Patankar, N.A., Zhu, M.Y.: Direct numerical simulation of fluid-solid systems using arbitrary Lagrangian-Eulerian techniques. J. Comp. Phys., **169**, 427–462 (2001)
99. Hu, L., Chen, D., Wei, G.W.: High-order fractional partial differential equation transform for molecular surface construction. Molecular Based Mathematical Biology, **1**, 1–25 (2013)
100. Ito, K., Kunisch, K.: Lagrange Multiplier Approach to Variational Problems. SIAM, Philadelphia, PA (2008)
101. Jidesh, P., George, S.: Fourth–order variational model with local–constraints for denoising images with textures. International Journal of Computational Vision and Robotics, **2**, 330–340 (2011)
102. Jin, S., Markowich, P.A., Zheng, C.: Numerical simulation of a generalized Zakharov system. J. Comp. Phys., **201**, 376–395 (2004)
103. Johnson, A.A., Tezduyar, T.E.: 3-D simulations of fluid-particle interactions with the number of particles reaching 100. Comp. Meth. Appl. Mech. Engrg., **145**, 301–321 (1997)
104. Kao, C.Y., Osher, S.J., Qian, J.: Lax-Friedrichs sweeping schemes for static Hamilton-Jacobi equations. J. Comput. Phys., **196**, 367–391 (2004)
105. Karniadakis, G.E., Israeli, M., Orszag, S.A.: High-order splitting methods for the incompressible Navier-Stokes equations. J. Comp. Phys., **97** (2), 414–443 (1991)
106. Kass, M., Witkin, A., Terzopoulos, D.: Snakes: Active contour models. International Journal of Computer Vision, **1**, 321–331 (1988)
107. Kim, J., Moin, P.: Application of a fractional-step method to incompressible Navier-Stokes equations. Journal of Computational Physics, **59** (2), 308–323 (1985)
108. Kimmel, R., Malladi, R., Sochen, N.: Images as embedded maps and minimal surfaces: movies, color, texture, and volumetric medical images. International Journal of Computer Vision, **39**, 111–129 (2000)
109. Lanza, A., Morigi, S., Sgallari, F.: Convex image denoising via non-convex regularization. In: Aujol, J.F., Nikolova, M., Papadakis, N. (eds.) Scale Space and Variational Methods in Computer Vision, pp. 666–677, Proceedings, LNCS 9087, Springer International Publishing (2015).
110. Layton, W.J., Maubach, J.M., Rabier, P.J.: Parallel algorithms for maximal monotone operators of local type. Numer. Math., **71**, 29–58 (1995)
111. Le, H., Moin, P.: An improvement of fractional step methods for the incompressible Navier-Stokes equations. Journal of Computational Physics, **92** (2), 369–379 (1991)
112. Lee, M.J., Do Oh, B., Kim, Y.B.: Canonical fractional-step methods and consistent boundary conditions for the incompressible Navier-Stokes equations. J. Comp. Phys., **168** (1), 73–100 (2001)
113. Lehoucq, R.B., Sorensen, D. C., Yang, C.: ARPACK Users' Guide: Solution of Large-Scale Eigenvalue Problems with Implicitly Restarted Arnoldi Methods. SIAM, Philadelphia, PA (1998)
114. Lellmann, J., Schnörr, C.: Continuous multiclass labeling approaches and algorithms. SIAM J. Imaging Sci., **4**, 1049–1096 (2011)
115. Lewis, A.S., Luke, D.R., Malick, J.: Local linear convergence for alternating and averaged nonconvex projections. Foundations of Computational Mathematics, **9** (4), 485–513 (2009)
116. Li, C.H., Glowinski, R.: Modeling and numerical simulation of low-Mach number compressible flows. Int. J. Numer. Meth. Fluids, **23** (2), 77–103 (1996)
117. Lie, J., Lysaker, M., Tai, X.-C.: A binary level set model and some applications to Mumford-Shah image segmentation. IEEE Transactions on Image Processing, **15**, 1171–1181 (2006)
118. Lions, P.L., Mercier, B.: Splitting algorithms for the sum of two nonlinear operators. SIAM J. Num. Anal., **16**, 964–979 (1979)
119. Lu, T., Neittaanmäki, P., Tai, X.-C.: A parallel splitting up method for partial differential equations and its application to Navier-Stokes equations. RAIRO Math. Model. and Numer. Anal., **26**, 673–708 (1992)
120. Lysaker, M., Lundervold, A., Tai, X.C.: Noise removal using fourth-order partial differential equation with applications to medical magnetic resonance images in space and time. Image Processing, IEEE Transactions on, **12**, 1579–1590 (2003)

121. Marchuk, G.I.: Splitting and alternating direction methods. In: Ciarlet, P.G., Lions, J.L. (eds.) Handbook of Numerical Analysis, Vol. I, pp. 197–462. North-Holland, Amsterdam (1990)
122. Marion, M., Temam, R.: Navier-Stokes equations: Theory and approximation. In: Ciarlet, P.G., Lions, J.L. (eds.) Handbook of Numerical Analysis, Vol. VI, pp. 503–689. North-Holland, Amsterdam (1998)
123. Marsden, J.: A formula for the solution of the Navier-Stokes equation based on a method of Chorin. Bulletin of the American Mathematical Society, **80** (1),154–158 (1974)
124. Masnou, S., Morel, J.M.: Level lines based disocclusions. In: Proceedings IEEE International Conferenceăon Image Processing, Chicago, IL, pp. 259–263 (1998).
125. Mason, P., Aftalion, A.: Classification of the ground states and topological defects in a rotating two-component Bose-Einstein condensate. Physical Review A, **84**, 033611 (2011)
126. Mason, P., Aftalion, A.: Vortex-peak interaction and lattice shape in rotating two-component Bose-Einstein condensates. Physical Review A, **85**, 033614 (2012)
127. Maury, B.: Direct simulation of 2-D fluid-particle flows in bi-periodic domains. J. Comp. Phys., **156**, 325–351 (1999)
128. Maury, B.: A time-stepping scheme for inelastic collisions. Numer. Math., **102**, 649–679 (2006)
129. Maury, B., Venel, J.: Handling of contacts in crowd motion simulations. In: Appert-Rolland, C., Chevoir, F., Gondret, P., Lassarre, S., Lebacque, J.P., Schreckenberg, M. (eds.) Traffic and Granular Flow'07, pp. 171–180. Springer, Berlin Heidelberg (2009)
130. Min, L., Yang, X., Gui, C.: Entropy estimates and large-time behavior of solutions to a fourth-order nonlinear degenerate equation. Communications in Contemporary Mathematics, **15**, (2013)
131. Mouhot, C., Villani, C.: On Landau damping, Acta Mathematica, **207**, 29–201 (2011)
132. Mumford, D., Shah, J.: Optimal approximation by piecewise smooth functions and associated variational problems. Comm. Pure Appl. Math., **42**, 577–685 (1989)
133. Myllykoski, M., Glowinski, R., Kärkkäinen, T., Rossi, T.: A new augmented Lagrangian approach for L^1-mean curvature image denoising. SIAM Journal on Imaging Sciences, **8**, 95–125 (2015)
134. Nadernejad, E., Forchhammer, S.: Wavelet-based image enhancement using fourth order PDE. In Intelligent Signal Processing (WISP), 2011 IEEE 7th International Symposium on, pp. 1–6. IEEE (2011)
135. Nitzberg, M., Mumford, D., Shiota, T.: Filtering, segmentation and depth. Lecture Notes in Computer Science, 662 (1993)
136. Pan, T.-W., Glowinski, R.: Direct numerical simulation of the motion of neutrally buoyant circular cylinders in plane Poiseuille flow. J. Comp. Phys., **181**, 260–279 (2002)
137. Pan, T.-W., Glowinski, R., Hou, S.: Direct numerical simulation of pattern formation in a rotating suspension of non-Brownian settling particles in a fully filled cylinder. Computers & Structures, **85**, 955–969 (2007)
138. Papafitsoros, K., Schönlieb, C.B.: A combined first and second order variational approach for image reconstruction. Journal of Mathematical Imaging and Vision, **48**, 308–338 (2014)
139. Peaceman, D.H., Rachford, H.H.: The numerical solution of parabolic and elliptic differential equations. J. Soc. Ind. Appl. Math., **3**, 28–41 (1955)
140. Prignitz, R., Bänsch, E.: Numerical simulation of suspension induced rheology. Kybernetika, **46**, 281–293 (2010)
141. Prignitz, R., Bänsch, E.: Particulate flows with the subspace projection method. J. Comp. Phys., **260**, 249–272 (2014)
142. Rosman, G., Bronstein, A.M., Bronstein, M.M., Tai, X.C., Kimmel, R.: Group-valued regularization for analysis of articulated motion. In: Fusiello, A., Murino, V., Cucchiara, R. (eds.) Computer Vision–ECCV 2012. Workshops and Demonstrations, pp. 52–62. Springer, Berlin (2012)
143. Rosman, G., Wang, Y., Tai, X.C., Kimmel, R., Bruckstein, A.M.: Fast regularization of matrix-valued images. In: Bruhn, A., Pock, T., Tai, X.C. (eds.) Efficient Algorithms for Global Optimization Methods in Computer Vision, pp. 19–43. Springer, Berlin (2014)

144. Rudin, L., Osher, S., Fatemi, E.: Nonlinear total variation based noise removal algorithms. Physica D, **60**, 259–268 (1992)
145. San Martin, J.A., Starovoitov, V., Tucksnak, M.: Global weak convergence for the two-dimensional motion of several rigid bodies in an incompressible viscous fluid. Archives Rat. Mech. Anal., **161**, 113–147 (2002)
146. Schoenemann, T., Kahl, F., Cremers, D.: Curvature regularity for region-based image segmentation and inpainting: A linear programming relaxation. In Computer Vision, 2009 IEEE 12th International Conference on, pp. 17–23. IEEE (2009)
147. Schönlieb, C.B., Bertozzi, A.: Unconditionally stable schemes for higher order inpainting. Communications in Mathematical Sciences, **9**, 413–457 (2011)
148. Schwartz, L.: Théorie des Distributions. Hermann, Paris (1966)
149. Setzer, S., Steidl, G.: Variational methods with higher order derivatives in image processing. Approximation, **12**, 360–386 (2008).
150. Sheng, Q.: Solving linear partial differential equations by exponential splitting. IMA J. Num. Anal., **9** (2), 199–212 (1989)
151. Sheng, Q.: Global error estimates for exponential splitting. IMA J. Num. Anal., **14** (1), 27–56 (1994)
152. Sigurgeirson, H., Stuart, A.M., Wan, J.: Collision detection for particles in flow. J. Comp. Phys., **172**, 766–807 (2001)
153. Strang, G.: On the construction and comparison of difference schemes. SIAM J. Num. Anal., **5**, 506–517 (1968)
154. Tai, X.C., Hahn, J., Chung, G.J.: A fast algorithm for Euler's elastica model using augmented Lagrangian method. SIAM Journal on Imaging Sciences, **4**, 313 (2011)
155. Tai, X.C., Neittaanmäki, P.: Parallel finite element splitting–up method for parabolic problems. Numerical Methods for Partial Differential Equations, **7**, 209–225 (1991)
156. Tai, X.C., Wu, C.: Augmented Lagrangian method, dual methods and split Bregman iteration for ROF model. In: Tai, X.C., Morken, K., Lysaker, M., Lie, K.-A. (eds.) Scale Space and Variational Methods in Computer Vision, pp. 502–513. Springer, Berlin (2009)
157. Tartar, L.: An Introduction to Sobolev Spaces and Interpolation Spaces. Springer, Berlin (2007)
158. Temam, R.: Sur l'approximation de la solution des équations de Navier-Stokes par la méthode des pas fractionnaires (I). Archive for Rational Mechanics and Analysis, **32** (2), 135–153 (1969)
159. Temam, R.: Sur l'approximation de la solution des équations de Navier-Stokes par la méthode des pas fractionnaires (II). Archive for Rational Mechanics and Analysis, **33** (5), 377–385 (1969)
160. Temam, R.: Navier-Stokes Equations: Theory and Numerical Analysis. American Mathematical Society, Providence, RI (2001)
161. Thalhammer, M.: High-order exponential operator-splitting methods for time-dependent Schrödinger equations. SIAM J. Num. Anal., **46** (4), 2022–2038 (2008)
162. Turek, S., Rivkind, L., Hron, J., Glowinski, R.: Numerical study of a modified time-stepping θ-scheme for incompressible flow simulations. J. Scient. Comp., **28**, 533–547 (2006)
163. Villani, C.: Birth of a Theorem: A Mathematical Adventure. Ferrar, Strauss & Giroux, New York, NY. (2015)
164. Wachpress, E.L.: The ADI model problem. Springer, New York, NY (2013)
165. Wan, D., Turek, S.: Fictitious boundary and moving mesh methods for the numerical simulation of rigid particulate flows. J. Comp. Phys., **222**, 28–56 (2007)
166. Wang, Y., Yin, W., Zeng, J.: Global convergence of ADMM in nonconvex nonsmooth optimization. arXiv preprint arXiv:1511.06324 (2015)
167. Weickert, J., ter Haar Romeny, B.M., Viergever, M.A.: Efficient and reliable schemes for nonlinear diffusion filtering. Image Processing, IEEE Transactions on, **7**, 398–410 (1998)
168. Wekipedia. Co-area formula (http://en.wikipedia.org/wiki/coarea_formula) (2013)
169. Wikipedia. Total variation (http://en.wikipedia.org/wiki/total_variation) (2014)

170. Wu, C., Tai, X.C.: Augmented Lagrangian method, dual methods, and split Bregman iteration for ROF, vectorial TV, and high order models. SIAM Journal on Imaging Sciences, **3**, 300–339 (2010)
171. Yang, F., Chen, K., Yu, B.: Efficient homotopy solution and a convex combination of ROF and LLT models for image restoration. International Journal of Numerical Analysis & Modeling, **9**, 907–927 (2012)
172. Yin, W., Osher, S., Goldfarb, D., Darbon, J.: Bregman iterative algorithms for l_1-minimization with applications to compressed sensing. SIAM Journal on Imaging Sciences, **1**, 143–168 (2008)
173. Yuan, J., Bae, E., Tai, X.C.: A study on continuous max-flow and min-cut approaches. In Computer Vision and Pattern Recognition (CVPR), 2010 IEEE Conference on, pp. 2217–2224. IEEE (2010)
174. Yuan, J., Bae, E., Tai, X.C., Boykov, Y.: A continuous max-flow approach to Potts model. In: Daniilidis, K., Maragos, P., Paragios, N. (eds.) Computer Vision–ECCV 2010, pp. 379–392. Springer, Berlin (2010)
175. Yuan, J., Bae, E., Tai, X.C., Boykov, Y.: A spatially continuous max-flow and min-cut framework for binary labeling problems. Numerische Mathematik, **126**, 559–587 (2014)
176. Yuan, J., Schnorr, C., Steidl, G.: Simultaneous higher-order optical flow estimation and decomposition. SIAM Journal on Scientific Computing, **29**, 2283–2304 (2007)
177. Yuan, J., Shi, J., Tai, X.C.: A convex and exact approach to discrete constrained $TV - L^1$ image approximation. East Asian Journal on Applied Mathematics, **1**, 172–186 (2011)
178. Zach, C., Gallup, D., Frahm, J.-M., Niethammer, M.: Fast global labeling for real-time stereo using multiple plane sweeps. In Vision, Modeling and Visualization Workshop (VMV) pp. 243–252 (2008)
179. Zakharov, V.E.: Collapse of Langmuir waves. Soviet Journal of Experimental and Theoretical Physics, **35**, 908–914 (1972)
180. Zeng, W., Lu, X., Tan, X.: Nonlinear fourth-order telegraph-diffusion equation for noise removal. IET Image Processing, **7**, 335–342 (2013)
181. Zhao, H. K.: Fast sweeping method for Eikonal equations. Math. Comp., **74**, 603–627 (2005)
182. Zhu, W., Chan, T.: Image denoising using mean curvature of image surface. SIAM Journal on Imaging Sciences, **5**, 1–32 (2012)
183. Zhu, W., Chan, T.: A variational model for capturing illusory contours using curvature. Journal of Mathematical Imaging and Vision, **27**, 29–40 (2007)
184. Zhu, W., Chan, T., Esedoglu, S.: Segmentation with depth: A level set approach. SIAM Journal on Scientific Computing, **28**, 1957–1973 (2006)
185. Zhu, W., Tai, X.-C., Chan, T.: Augmented Lagrangian method for a mean curvature based image denoising model. Inverse Problems and Imaging, **7**, 1409–1432 (2013)
186. Zhu, W., Tai, X.C., Chan, T.F.: Image segmentation using Euler's elastica as the regularization. Journal of Scientific Computing, **57** (2), 414–438 (2013).

Chapter 3
Operator Splitting

Shev MacNamara and Gilbert Strang

Abstract Operator splitting is a numerical method of computing the solution to a differential equation. The splitting method separates the original equation into two parts over a time step, separately computes the solution to each part, and then combines the two separate solutions to form a solution to the original equation. A canonical example is splitting of diffusion terms and convection terms in a convection-diffusion partial differential equation. Related applications of splitting for reaction-diffusion partial differential equations in chemistry and in biology are emphasized here. The splitting idea generalizes in a natural way to equations with more than two operators. In all cases, the computational advantage is that it is faster to compute the solution of the split terms separately, than to compute the solution directly when they are treated together. However, this comes at the cost of an error introduced by the splitting, so strategies have been devised to control this error. This chapter introduces splitting methods and surveys recent developments in the area. An interesting perspective on absorbing boundary conditions in wave equations comes via Toeplitz-plus-Hankel splitting. One recent development, balanced splitting, deserves and receives special mention: it is a new splitting method that correctly captures steady state behavior.

S. MacNamara (✉)
School of Mathematics, University of New South Wales, Sydney, NSW 2052, Australia
e-mail: s.macnamara@unsw.edu.au

G. Strang
Department of Mathematics, MIT, Cambridge, MA 02139, USA
e-mail: gs@math.mit.edu

R. Glowinski et al. (eds.), *Splitting Methods in Communication, Imaging, Science, and Engineering*, Scientific Computation, DOI 10.1007/978-3-319-41589-5_3

1 Introduction

It has been said that there are only ten big ideas in numerical analysis; all the rest are merely variations on those themes. One example of those big ideas is multi-scale computational approaches. A multi-scale motif reappears in numerous places including: multigrid for solving linear systems [51], wavelets for image processing [11], and in Multi-level Monte Carlo for the solution of stochastic differential equations [20]. Another of those big ideas could surely be *splitting* [42, 4, 57]: start with a complicated problem, split it into simpler constituent parts that can each be solved separately, and combine those separate solutions in a controlled way to solve the original overall problem. Often we solve the separate parts sequentially. The output of the first subproblem is the input to the next subproblem (within the time step).

Like all great ideas, splitting is a theme that continues to resurface in many places. Splitting principles have taken a number of generic forms:

- Split linear from nonlinear.
- Split x-direction from y-direction (dimensional splitting).
- Split terms corresponding to different physical processes. For example, split convection from diffusion in ODEs or in PDEs.
- Split a large domain into smaller pieces. For example, domain decomposition helps to solve large PDEs in parallel.
- Split objective functions in optimization.
- Split resolvents when solving linear systems: Instead of working directly with $(\lambda I-(A+B))^{-1}$, we iterate between working separately with each of $(\lambda I-A)^{-1}$ and $(\lambda I-B)^{-1}$.

Not surprisingly, a principle as fundamental as splitting finds applications in many areas. Here is a non-exhaustive list:

- A recent application of splitting is to low-rank approximation [36].
- Balanced splitting has been developed to preserve the steady state [53].
- Splitting of reaction terms from diffusion terms in reaction-diffusion PDEs is a common application of splitting in biology. Now splitting is also finding applications in stochastic, particle-based methods, such as for master equations [19, 32, 38, 37, 18, 26, 25], including analysis of sample path approaches [16].
- Splitting stochastic differential equations, by applying the component operators in a random sequence determined by coin flipping is, on average, accurate. This has applications in finance [45]. Though in a different sense, splitting also finds application in Monte Carlo estimation of expectations [2].
- Maxwell's equations for electromagnetic waves can be solved on staggered grids via Yee's method, which is closely related to splitting [55, 34].
- Motivated in part by the need for accurate oil reservoir simulations, Alternating Direction Implicit methods and Douglas-Rachford splittings have by now found wide applications [62, 47, 14].
- Split-Bregman methods are a success for compressed sensing and for image processing [21].

- Navier-Stokes equations in fluid mechanics are often approximated numerically by splitting the equations into three parts: (i) a nonlinear convection term, $u \cdot (\nabla u)$, is treated explicitly, (ii) diffusion, Δu, is treated implicitly, and (iii) continuity is imposed via Poisson's equation, $\mathrm{div} u = 0$. Chorin's splitting method is a well-known example of this approach [7, 55].
- Split the problem of finding a point in the intersection of two or more sets into alternating projections onto the individual sets [4].

This chapter places emphasis on applications to partial differential equations (PDEs) involving reaction, diffusion, and convection. *Balanced splitting*, which has found application in models of combustion, receives special attention. Computer simulation of *combustion* is important to understand how efficiently or how cleanly fuels burn. It is common to use *operator splitting* to solve the model equations. However, in practice this was observed to lead to an unacceptable error at steady state. The new method of balanced splitting was developed to correct this [53]. This balanced method might be more widely applicable because operator splitting is used in many areas. Often the steady state is important. In reaction-diffusion models, in biology for example, it is very common to split reaction terms from diffusion terms, and the steady state is almost always of interest. As we will see in the next section, the most obvious splitting scheme is only first order accurate but a symmetrized version achieves second order accuracy. Do such schemes yield the correct steady state? The answer is no, not usually. Balanced splitting corrects this.

Outline: The rest of this chapter is organized as follows. We begin with the simplest possible example of splitting. First order accurate and second order accurate splitting methods come naturally. Higher order splitting methods, and reasons why they are not always adopted, are then discussed. Next, we observe that splitting does not capture the correct *steady state*. This motivates the introduction of balanced splitting: a new splitting method that does preserve the steady state. All these ideas are illustrated by examples drawn from reaction-diffusion PDEs such as arise in mathematical biology, and from convection-diffusion-reaction PDEs such as in models of combustion. We aim especially to bring out some recent developments in these areas [59, 53]. Finally, we investigate a very special Toeplitz-plus-Hankel splitting that sheds light on the reflections at the boundary in a wave equation.

2 Splitting for Ordinary Differential Equations

The best example to start with is the linear ordinary differential equation (ODE)

$$\frac{\mathrm{d}u}{\mathrm{d}t} = (A+B)u. \tag{3.1}$$

The solution is well known to students of undergraduate differential equation courses [58]:

$$u(h) = e^{h(A+B)}u(0),$$

at time h. We are interested in splitting methods that will compute that solution for us, at least approximately. If we could simply directly compute $e^{h(A+B)}$, then we would have solved our ODE (3.1), and we would have no need for a splitting approximation. However, in applications it often happens that $e^{h(A+B)}$ is relatively difficult to compute directly, whilst there are readily available methods to compute each of e^{hA} and e^{hB} separately. For example, (3.1) may arise as a method-of-lines approximation to a PDE in which A is a finite difference approximation to diffusion, and B is a finite difference approximation to convection. (This example of a convection-diffusion PDE, together with explicit matrices, is coming next.) In that case it is natural to make the approximation

First order splitting $$e^{h(A+B)} \approx e^{hA}e^{hB}. \quad (3.2)$$

We call this approximate method of computing the solution *splitting*. This chapter stays with examples where A and B are matrices, although the same ideas apply in more general settings where A and B are operators; hence the common terminology *operator splitting*.

To begin thinking about a splitting method we need to see the matrix that appears in the ODE as the *sum of two matrices*. That sum is immediately obvious in (3.1) by the way it was deliberately written. However, had we instead been given the ODE $\mathrm{d}u/\mathrm{d}t = Mu$, then before we could apply a splitting method, we would first need to identify A and B that add up to M. Given M, identifying a good choice for A and thus for $B = M - A$ is not trivial, and the choice is critical for good splitting approximations.

When the matrices *commute*, the approximation is exact.[1] That is, if the commutator $[A,B] \equiv AB - BA = 0$, then $e^{h(A+B)} = e^{hA}e^{hB}$. Otherwise the approximation of (3.2), $e^{h(A+B)} \approx e^{hA}e^{hB}$, is only first order accurate: the Taylor series of $e^{hA}e^{hB}$ agrees with the Taylor series of $e^{h(A+B)}$ up to first order, but the second order terms differ. The Taylor series is

$$e^{h(A+B)} = I + h(A+B) + \frac{1}{2}h^2(A+B)^2 + \ldots$$

If we expand $(A+B)^2 = A^2 + AB + BA + B^2$ then we see the reason we are limited to first order accuracy. In all terms in the Taylor series for our simple splitting approximation $e^{hA}e^{hB}$, A always comes before B, so we cannot match the term BA appearing in the correct series for $e^{h(A+B)}$.

The previous observation suggests that symmetry might help and indeed it does. The symmetric Strang splitting [54, 57]

[1] An exercise in Golub and Van Loan [22] shows that $[A,B] = 0$ if and only if $e^{h(A+B)} = e^{hA}e^{hB}$ for all h.

$$\textit{Second order splitting} \qquad e^{h(A+B)} \approx e^{\frac{1}{2}hA} e^{hB} e^{\frac{1}{2}hA} \tag{3.3}$$

agrees with this Taylor series up to second order so it is a more accurate approximation than (3.2). Splitting methods have grown to have a rich history with many wonderful contributors. Marchuk is another of the important pioneers in the field. He found independently the second order accurate splitting that we develop and extend in this chapter [40, 41].

When we numerically solve the ODE (3.1) on a time interval $[0\ T]$, we usually do not compute $u(T) = e^{T(A+B)}u(0)$ in one big step time step T, nor do we approximate it by $e^{\frac{1}{2}TA}e^{TB}e^{\frac{1}{2}TA}u(0)$. Our approximations are accurate for small time steps $h > 0$ but not for large times T. Therefore, instead, we take very many small steps h that add up to T. The first step is $v_1 = e^{\frac{1}{2}hA}e^{hB}e^{\frac{1}{2}hA}u(0)$, which is our approximation to the solution of (3.1) at time h. The next step is computed from the previous step, recursively, by $v_{i+1} = e^{\frac{1}{2}hA}e^{hB}e^{\frac{1}{2}hA}v_i$, so that v_i is our approximation to the exact solution $u(ih)$ at time ih for $i = 1, 2, \ldots$. After N steps, so that $Nh = T$, we arrive at the desired approximation $v_N \approx u(T)$.

Notice that a few Strang steps in succession

$$(e^{\frac{1}{2}hA}e^{hB}e^{\frac{1}{2}hA})\,(e^{\frac{1}{2}hA}e^{hB}e^{\frac{1}{2}hA}) = (e^{\frac{1}{2}hA}e^{hB})\quad \underbrace{e^{hA}e^{hB}}_{\text{first order step}}\quad (e^{\frac{1}{2}hA})$$

is the same as a first order step in the middle, with a special half-step at the start and a different half-step at the end. This observation helps to reduce the overall work required to achieve second order accuracy when taking many steps in a row. This was noticed in the original paper [54] but perhaps it is not exploited as often as it could be.

Here is a more explicit example of first order and of second order accuracy. Notice the difference between the *local error* over one small time step h, and the *global error* over the whole time interval (those local errors grow or decay). We are interested in how fast the error decays as the time step h becomes smaller. The power of h is the key number. In general, the local error is one power of h more accurate than the global error from $1/h$ steps. Comparing Taylor series shows that

$$\text{local error} \qquad e^{\frac{1}{2}hA}e^{hB}e^{\frac{1}{2}hA} - e^{h(A+B)} = Ch^3 + \mathcal{O}(h^4).$$

where the constant is $C = \frac{1}{24}([[A,B],A] + 2[[A,B],B])$. That is, for a single small time step h, the symmetric Strang splitting (3.3) has a local error that decays like h^3. As usual, the error gets smaller as we reduce the time step h: if we reduce the time step h from 1 to 0.1, then we expect the error to reduce by a factor of $0.1^3 = 0.001$. However, to compute the solution at the final time T, we take T/h steps, thus more steps with smaller h, and the global error is (number of steps) $\times$ (error at each step) $= (1/h)h^3 = h^2$, so we say the method is second order accurate.

We have examined accuracy by directly comparing the Taylor series of the exact solution and of the approximation. Another approach is via the Baker-Campbell-Hausdorff (BCH) formula:

$$C(hA,hB) = hA + hB + \frac{1}{2}[hA,hB] + \cdots,$$

which, given A and B, is an infinite series for the matrix C such that $e^C = e^{hA}e^{hB}$. In nonlinear problems we want approximations that are *symplectic*. Then area in phase space is conserved, and approximate solutions to nearby problems remain close. The beautiful book of Hairer, Lubich, and Wanner [24] discusses the BCH formula and its connection to splitting, and when splitting methods are symplectic for nonlinear equations. Strang splitting is symplectic.

2.1 Gaining an Order of Accuracy by Taking an Average

The nonsymmetric splitting $e^{h(A+B)} \approx e^{hA}e^{hB}$ is only first order accurate. Of course, applying the operations the other way around, as in $e^{hB}e^{hA}$, is still only first order accurate. However, taking the *average* of these two first order approximations recovers a certain satisfying symmetry

$$e^{h(A+B)} \approx \frac{e^{hA}e^{hB} + e^{hB}e^{hA}}{2}.$$

Symmetry is often associated with higher order methods. Indeed this symmetric average is *second order* accurate. That is, we gain one order of accuracy by taking an average. Whilst this observation is for averages in a very simple setting, we conjecture that it is closely related to the good experience reported in the setting of finance, where a stochastic differential equation is solved with good accuracy in the weak sense even if the order of operations is randomly determined by 'coin flipping' [45].

2.2 Higher Order Methods

Naturally, we wonder about achieving higher accuracy with splitting methods. Perhaps third order splitting schemes or even higher order splitting schemes are possible. Indeed they are, at least in theory. However, they are more complicated to implement: third order or higher order splitting schemes require either substeps that go backwards in time or forward in 'complex time' [3, 65, 24, 34, 13, 12]. For diffusion equations, going backwards in time raises serious issues of numerical stability. For reaction-diffusion equations, second order splitting is still the most popular.

More generally, it is a meta-theorem of numerical analysis that second order methods often achieve the right balance between accuracy and complexity. First order methods are not accurate enough. Third order and higher order methods are accurate, but they have their own problems associated with stability or with being too complicated to implement. Dahlquist and Henrici were amongst the pioneers to uncover these themes [5, 9, 10, 31].

2.3 Convection and Diffusion

Until now, the discussion has been concerned with ODEs: time but no space. However, a big application of splitting is to PDEs: space and time.

An example is a PDE in one space dimension that models convection and diffusion. The continuous, exact solution $u(x,t)$ is to be approximated by finite differences. We compute a discrete approximation on a regular grid in space $x = \ldots, -2\Delta x, -\Delta x, 0, \Delta x, 2\Delta x, \ldots$ One part of our PDE is convection $\mathrm{d}u/\mathrm{d}t = \mathrm{d}u/\mathrm{d}x$. Convection is often represented by a one-sided finite difference matrix. For example, the finite difference approximation $\mathrm{d}u/\mathrm{d}x \approx (u(x+\Delta x) - u(x))/\Delta x$ comes via the matrix

$$P = \frac{1}{\Delta x}\begin{bmatrix} -1 & 1 & & \\ & -1 & 1 & \\ & & \ddots & \ddots \end{bmatrix}.$$

Or we can approximate convection by a centered difference matrix:

$$Q = \frac{1}{2\Delta x}\begin{bmatrix} 0 & 1 & & \\ -1 & 0 & 1 & \\ & \ddots & \ddots & \ddots \end{bmatrix}.$$

Here we think of $\mathrm{d}u/\mathrm{d}x \approx (u(x+\Delta x) - u(x-\Delta x))/2\Delta x \approx Qu$. Another part of our PDE is diffusion: $\mathrm{d}u/\mathrm{d}t = \mathrm{d}^2u/\mathrm{d}x^2$. The second spatial difference is often represented by the matrix

$$D = \frac{1}{\Delta x^2}\begin{bmatrix} -2 & 1 & & \\ 1 & -2 & 1 & \\ & \ddots & \ddots & \ddots \end{bmatrix}.$$

We will come back to this matrix at the end of the chapter in (3.8), where we change signs to $K = -D$ so that K is positive definite and K models $-\mathrm{d}^2/\mathrm{d}x^2$. In solving a simple linear PDE with convection and diffusion terms

$$\frac{\partial u}{\partial t} = \underbrace{\frac{\partial u}{\partial x}}_{\text{convection}} + \underbrace{\frac{\partial^2 u}{\partial x^2}}_{\text{diffusion}}$$

with finite differences we may thus arrive at the ODE

$$\frac{\mathrm{d}u}{\mathrm{d}t} = (P+D)u. \tag{3.4}$$

Here in the ODE we think of $u(t)$ as a column vector, with components storing the values of the solution on the spatial grid $[\ldots,u(-\Delta x),u(0),u(\Delta x),\ldots]^T$. (We are abusing notation slightly by using the same u in the PDE and in its ODE approximation.) This is a discrete-in-space and continuous-in-time, or semi-discrete approximation. We recognize it as the same ODE that we introduced at the very beginning (3.1), here with the particular choice $A = D$ and $B = P$. The solution to this semi-discrete approximation is the same $u(t) = e^{h(P+D)}u(0)$, and it is natural to consider approximating this solution by splitting into convection P and diffusion terms D.

In applying the splitting method (3.2), we somehow compute the approximation $e^{hP}e^{hD}u(0)$. Conceivably, we might choose to compute each of the matrix exponentials, e^{hP} and e^{hD}, in full, and then multiply these full matrices by the vector $u(0)$. In practice that is usually not a good idea. One reason is that the matrices are often sparse, as in the examples of D and P here, and we want to take advantage of that sparsity, whereas the matrix exponential is typically full. Moreover, computing the matrix exponential is a classical problem of numerical analysis with many challenges [43].

Usually we only want the solution vector $u(t)$, not a whole matrix. For this purpose, ODE-solvers, such as Runge-Kutta methods or Adams-Bashforth methods, are a good choice [5, 31]. The point of this chapter is merely to observe that we can still apply splitting methods. We proceed in two stages. First stage: starting from $u(0)$, solve $\mathrm{d}u/\mathrm{d}t = Du$ from time $t = 0$ to $t = h$ for the solution $w_{1/2}$. Second stage: starting from this $w_{1/2}$, solve $\mathrm{d}u/\mathrm{d}t = Pu$ from time $t = 0$ to $t = h$ for the solution $w_{2/2}$. Thus we have carried out a first order splitting: if our ODE-solvers were exact at each of the two stages, then $w_{2/2} = e^{hA}e^{hB}u(0)$. Often, we treat the diffusion term implicitly [1]. Hundsdorfer and Verwer discuss the numerical solution of convection-diffusion-reaction problems, noting additional issues when applying splitting methods to boundary value problems [30].

2.4 A Reaction-Diffusion PDE: Splitting Linear from Nonlinear

A typical example in mathematical biology is a reaction-diffusion PDE in the form

$$\frac{\partial}{\partial t}\begin{bmatrix} u \\ v \end{bmatrix} = \underbrace{\left(\nabla\cdot\begin{bmatrix} D_u & 0 \\ 0 & D_v \end{bmatrix}\nabla\right)\begin{bmatrix} u \\ v \end{bmatrix}}_{\text{diffusion}} + \underbrace{\begin{bmatrix} f(u,v) \\ g(u,v) \end{bmatrix}}_{\text{reaction}}. \tag{3.5}$$

Here D_u and D_v are positive diffusion constants for the concentrations of the two species u and v, respectively. Commonly these are modeled as genuinely constant – not spatially varying – and in the special case that $D_u = D_v = 1$, then our diffusion operator simplifies to $\nabla \cdot \nabla$, which many authors would denote by ∇^2, or in one space dimension by $\partial^2/\partial x^2$. The reactions are modeled by nonlinear functions f and g. For example, in a Gierer-Meinhardt model, $f(u,v) = a - bu + u^2/v^2$, and $g(u,v) = u^2 - v$ [44, 39].

Diffusion is linear. Reactions are nonlinear. We split these two terms and solve them separately. We solve the linear diffusion implicitly. The nonlinear reactions are solved explicitly. By analysis of our linear test problem (3.1) we found the accuracy of splitting approximations to be second order accurate, in the case of symmetric Strang splitting. However, the questions of accuracy and of stability concerning splitting approximations to the more general form

$$\frac{\mathrm{d}u}{\mathrm{d}t} = Au + f(u)$$

where A is still linear but now f is nonlinear, such as arise in reaction-diffusion PDEs, has no such simple answer [52, 1].

2.5 Stability of Splitting Methods

We have seen that splitting can be accurate. Now we wonder about stability. Together, *stability and consistency imply convergence*. That is both a real theorem for many fundamental examples, and also a meta-theorem of numerical analysis [33]. Indeed it is sometimes suggested (together with its converse) as The Fundamental Theorem of Numerical Analysis [60, 63, 8, 55].

In this context those three keywords have special meanings. *Accuracy* means that the computed approximation is close to the exact solution. A method is *stable* if over many steps, the local errors only grow slowly in a controlled way and do not come to dominate the solution. (A mathematical problem is well conditioned if small perturbations in the input only result in correspondingly small perturbations in the output. Stability for numerical algorithms is analogous to the idea of conditioning for mathematical problems.) The method is *consistent* if, over a single time step h, the numerical approximation is more and more accurate as h becomes smaller. For instance, we saw that the local error of a single step with symmetric Strang splitting (3.3), scales like h^3 as $h \to 0$, so that method is consistent. For a consistent method, we hope that the global error, after many time steps, also goes to zero as $h \to 0$: if this happens then the numerical approximation is converging to the true solution. Usually, finding a direct proof of convergence is a formidable task, whereas showing consistency and showing stability separately is more attainable. Then the theorem provides the desired assurance of convergence.

When we start thinking about the question of stability of splitting methods, typically we assume that the eigenvalues of A and of B all lie in the left half, $\mathrm{Re}(\lambda) \leq 0$, of the complex plane. Separately, each system is assumed stable.

It is natural to wonder if the eigenvalues of $(A+B)$ also lie in the left half plane. This is not true in general. Turing patterns[2] in mathematical biology are a famous instance of this [44, 39]. Typically we linearize at a steady state. For example, if J is the Jacobian matrix of the reaction terms in (3.5) at the steady state, then we study the linear equation $d\mathbf{u}/dt = (J+D)\mathbf{u}$, where $\mathbf{u} = [u\, v]^T$. Separately, the diffusion operator D, and the Jacobian J each have eigenvalues with negative real part. Analysis of Turing instability begins by identifying conditions under which an eigenvalue of $(J+D)$ can still have a positive real part.

With the assumption that all eigenvalues of M have negative real part, e^{tM} is stable for large times t. If the matrix M is real symmetric then the matrix exponential e^{hM} is also well behaved for small times. Otherwise, whilst eigenvalues do govern the long-time behavior of e^{tM}, the transient behavior can be different if there is a significant *pseudospectrum* [61]. This can happen when the eigenvectors of the matrix are not orthogonal, and the convection-diffusion operator is an important example [49].

Even if the matrix exponentials $e^{t(A+B)}$, e^{tA}, and e^{tB} are stable separately, we don't yet know about the stability of their multiplicative combination, as in say, first order splitting, $e^{hA}e^{hB}$. A *sufficient condition* for stability is that the symmetric parts

$$\text{symmetric part} \qquad A_{sym} \equiv \frac{A+A^T}{2}$$

of A and of B are negative definite. In that case we have stability of both ordinary splitting and symmetric Strang splitting because separately $||e^{hA}|| \leq 1$ and $||e^{hB}|| \leq 1$. This result was proved by Dahlquist and the idea is closely related to the log norm of a matrix. One way to show this is to observe that the derivative of $||e^{hA}u||^2$ is $((A+A^T)e^{hA}u, e^{hA}u) \leq 0$. Another way is to let $A = A_{sym} + A_{anti}$, and observe that e^{hA} is the limit of $e^{hA_{sym}/n}e^{hA_{anti}/n}e^{hA_{sym}/n}\dots e^{hA_{anti}/n}$. Each $||e^{hA_{sym}/n}|| \leq 1$ since eigenvalues of A_{sym} are negative and the matrix is symmetric. Each $||e^{hA_{anti}/n}|| \leq 1$ since eigenvalues of $e^{hA_{anti}/n}$ are purely imaginary and the matrix is orthogonal.

Returning to the convection-diffusion example (3.4), we can now see that the splitting method is stable. In that case, note that $P = Q + hD$, where Q is the anti-symmetric part and hD is the symmetric part. Observing that the diffusion matrix D is symmetric negative definite, we see that such a splitting is strongly stable with symmetric Strang splitting.

Having established that splitting is stable and accurate over finite times, we have now investigated many of the concerns of the original paper on Strang splitting:

```
Surprisingly, there seem to be no recognized rules for the
comparison of alternative difference schemes. Clearly there
are  three fundamental criteria -- accuracy, simplicity, and
stability -- and we shall evaluate each of the competing
schemes in these terms.
```

[2] See, for example, Rauch's notes on Turing instability [48].

– " On the Construction and Comparison of Difference Schemes"
Gilbert Strang, *SIAM J. Numer. Anal.*, 1968.

Perhaps another criterion could have been added – how well the method transitions from an early transient behavior to late stage steady state behavior. It turns out that most splitting schemes do not exactly capture the all-important steady state. We review next a new "balanced splitting" scheme that corrects this error.

2.6 Ordinary Splitting Does NOT Preserve the Steady State

Suppose u_∞ is a steady state of (3.1). By definition

$$(A+B)u_\infty = 0 \qquad \text{and} \qquad e^{A+B}u_\infty = u_\infty.$$

In special cases, such as when $Au_\infty = Bu_\infty = 0$, both first order splitting (3.2) and second order splitting (3.3) preserve steady states of the original ODE. However, in general, standard splitting approximations do *not* preserve the steady state u_∞: $e^{hA}e^{hB}u_\infty \neq u_\infty$ and $e^{\frac{1}{2}hA}e^{hB}e^{\frac{1}{2}hA}u_\infty \neq u_\infty$.

3 Balanced Splitting: A Symmetric Strang Splitting That Preserves the Steady State

We again consider our linear ODE $\mathrm{d}v/\mathrm{d}t = (A+B)v$. In **balanced splitting** [53] a constant vector, c, is computed at the beginning of each step. Then c is added to Av and subtracted from Bv in the substages of the splitting approximation; the parts still add to $(A+B)v$.

A first idea (simple balancing) is to choose c so that $Av+c = Bv-c$. Then the first stage solves

$$\mathrm{d}v/\mathrm{d}t = Av + c, \qquad c = \frac{1}{2}(B-A)v_0, \qquad v_0 = u_0, \tag{3.6}$$

for the solution[3] $v^+ = e^{hA}v_0 + (e^{hA}-I)A^{-1}c$, at time h. Now the second stage solves

$$\mathrm{d}v/\mathrm{d}t = Bv - c, \qquad v_0 = v^+,$$

for the solution $e^{hB}v^+ - (e^{hB}-I)B^{-1}c$ at time h. We call this method 'nonsymmetric balanced splitting'. By adding and subtracting a constant vector, we see this as

[3] Here we assume A and B are invertible. The non-invertible case is treated by the variation-of-parameters formula [58].

a modification of first order splitting (3.2), but the modified version has an advantage near steady state. Actually this choice of $c = \frac{1}{2}(B-A)v$ frequently leads to instability [53].

Of course there is also a simple modification of the second order splitting (3.3) approximation, where we add and subtract a constant at each stage. In symmetric balanced splitting, we solve the 'A stage' for a time step $\frac{1}{2}h$, then the 'B stage' for a time step h, and finally the 'A stage' again for a time step $\frac{1}{2}h$. That is, in the 'symmetric balanced splitting method', the first stage $\{ \mathrm{d}v/\mathrm{d}t = Av + c,\ v_0 = u_0 \}$, is the same as before except that we solve for the solution over a smaller interval $h/2$. Then $v^+ = e^{\frac{1}{2}hA}v_0 + (e^{\frac{1}{2}hA} - I)A^{-1}c$, is the initial condition for the second stage. We solve for v^{++} over the time interval h. The third stage is $\{ \mathrm{d}v/\mathrm{d}t = Av + c,\ v_0 = v^{++} \}$ over the remaining half step $h/2$. The output, $v(h) = Rv(0)$, is the approximation at the end of the whole step, where

$$
\begin{aligned}
R = \frac{1}{2}\Big(I - A^{-1}B + e^{\frac{1}{2}hA}e^{hB}e^{\frac{1}{2}hA}(I + A^{-1}B) \\
+ e^{\frac{1}{2}hA}(e^{hB} - I)(B^{-1}A - A^{-1}B)\Big). \qquad (3.7)
\end{aligned}
$$

In the special case that $A = B$, the formula simplifies to $R = e^{\frac{1}{2}hA}e^{hB}e^{\frac{1}{2}hA}$ so symmetric balanced splitting is identical with symmetric Strang splitting in this case. This is what we expect because in this case $c = 0$. To improve stability we may choose different balancing constants, thereby moving from simple balanced splitting to *rebalanced splitting* [53]. One good choice [53, equation 7.7] is $c_{n+1} = (-v_{n+1} + 2v_n^{++} - 2v_n^{+} + v_n)/2h + c_n$, which involves all values from the previous step.

3.1 Balanced Splitting Preserves the Steady State

Having introduced the method of balanced splitting, we now confirm its most important property. Recall that we are at a steady state if and only if the derivative is zero. Hence we may check that a steady state, u_∞, of the original system (3.1) is also a steady state of the new balanced splitting approximation by direct substitution and evaluation of the derivative. Suppose that we start at steady state, i.e., $v_0 = u(0) = u_\infty$. The first stage of balanced splitting is

$$
\mathrm{d}v/\mathrm{d}t = Au_\infty + c = Au_\infty + \frac{1}{2}(B-A)u_\infty = \frac{1}{2}(A+B)u_\infty = 0,
$$

where we have used the defining property of the steady state, i.e., $(A+B)u_\infty = 0$. Similarly for the second stage $\mathrm{d}v/\mathrm{d}t = 0$, so $v(h) = u(0) = u_\infty$. Thus a steady state of the original ODE is also a steady state of the balanced splitting method. The same observation shows that other variations of the balanced splitting method (such as symmetric balanced splitting) also preserve the steady state. This also gives the intuition behind the particular choice of the constant c – it is chosen in just the right way to 'balance' each substep.

Two special cases for which ordinary splitting may be preferable to balanced splitting are:

- In the special case that A and B commute, ordinary splitting is exact. However, balanced splitting does not share this property.
- In the special case that $Au_\infty = Bu_\infty = 0$, ordinary splitting preserves the steady state, so balanced splitting does not offer an advantage in this case.

The eigenvalues of R in (3.7) will tell us about the stability of simple balanced splitting – that remains an area of active interest [53], as does stability of operator splitting more generally [50]. The main message is that balanced splitting has applications to important problems where ordinary splitting approximations fail to capture the steady state [53].

3.2 Splitting Fast from Slow

Splitting fast processes from slow processes is very common in applied mathematics. After averaging away the fast processes, a simplified model is reached, which is sometimes known as a quasi-steady-state approximation. The principles go further than splitting, but splitting is the first step. Potentially, time-scale separation provides another application for balanced splitting: quasi-steady state approximations are not always guaranteed to preserve the steady state of the original model. We wonder if a balanced splitting can be extended to efficient simulation of stochastic processes with fast and slow time-scales [6, 15, 46].

4 A Very Special Toeplitz-Plus-Hankel Splitting

We now describe a very special splitting: a Toeplitz-plus-Hankel splitting [59]. Unlike the previous examples, where exponentials of separate terms were computed separately (e.g., in first order splitting (3.2)) as a computationally efficient approximation, in the coming example (3.11) the exact solution is split into two parts, merely to gain a novel perspective through the lens of splitting. We see solutions to the wave equation as the sum of a Toeplitz solution and a Hankel solution. It transpires that reflections at the boundary come from the Hankel part of the operator (Figure 3.1).

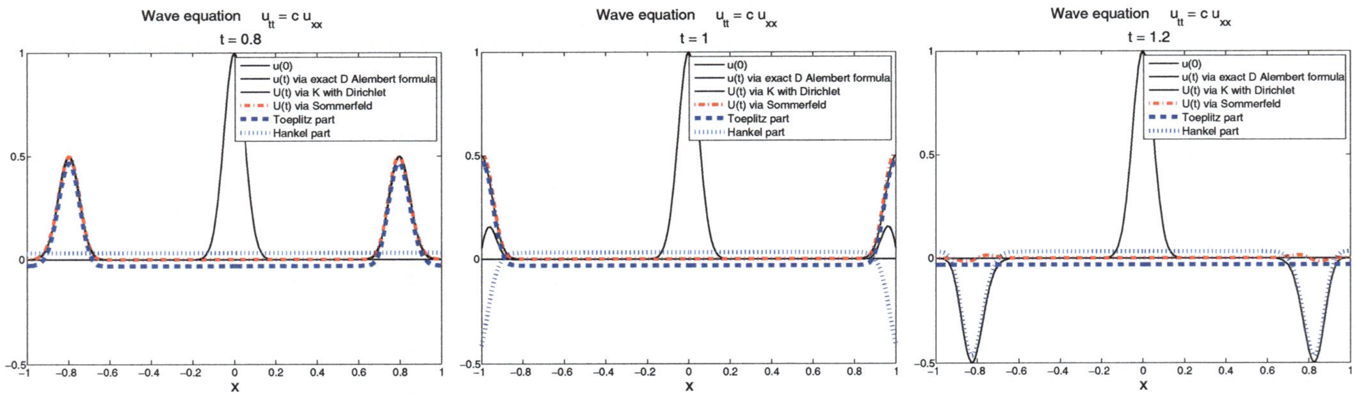

Fig. 3.1 Solutions of the wave equation $u_{tt} = u_{xx}$ at three snapshots in time: before ($t = 0.8$), during ($t = 1.0$), and after ($t = 1.2$) the wave reaches the boundary of the computational domain. The exact solution (d'Alembert's formula) and various finite difference approximations, with $\Delta x = 2/(N-1)$, $N = 201$, are shown. With the Sommerfeld radiation condition, $u_x = \pm u_t$, at the ends of our finite computational domain, we hope our computational solution is in perfect agreement with the solution to the wave equation on the whole line (with no boundary). However, a finite difference approximation of the Sommerfeld condition is not exact, and the small reflection visible at $t = 1.2$ is a manifestation of the error. With Dirichlet boundary conditions, in K, **the solution splits exactly into the sum of a Toeplitz solution and a Hankel solution.** Before reaching the boundary, the solution is purely Toeplitz, while the Hankel part is absent. After reaching the boundary, it is the other way around: the solution is purely Hankel, while the Toeplitz part is absent. Hankel parts come from the boundary. The Toeplitz part is reflectionless.

4.1 All Matrix Functions $f(K)$ Are Toeplitz-Plus-Hankel

We begin with the $N \times N$ tridiagonal, symmetric positive definite Toeplitz matrix [64, 29]:

$$K = \begin{bmatrix} 2 & -1 & & & \\ -1 & 2 & -1 & & \\ & \ddots & \ddots & \ddots & \\ & & -1 & 2 & -1 \\ & & & -1 & 2 \end{bmatrix} \qquad h = \frac{1}{N+1}. \tag{3.8}$$

Perhaps this is the most studied matrix in all of computational mathematics [55, 56]. Its eigenvalues and eigenvectors are known:

Eigenvalues of K	$\lambda_k = 2 - 2\cos(k\pi h), \qquad k = 1,\ldots,N$
Eigenvectors of K	$v_k = \sqrt{\frac{2}{N+1}}\Big(\sin(k\pi h), \sin(2k\pi h), \ldots, \sin(Nk\pi h)\Big)^T$
Function of K	$f(K)_{m,n} = \frac{2}{N+1}\sum_{k=1}^{N} f(\lambda_k)\sin(mk\pi h)\sin(nk\pi h)$

They produce the spectral decomposition

$$\textbf{Spectral theorem} \qquad K = V\Lambda V^T = \sum_1^N \lambda_k v_k v_k^T \tag{3.9}$$

where the matrix K is constructed from its eigenvalues in the diagonal matrix Λ and its eigenvectors in the columns of V. This *diagonalization* separates K into a sum of rank one symmetric matrices $\lambda_k v_k v_k^T$. Now *any matrix function* [28] comes easily via this diagonalization: $f(K) = V f(\Lambda) V^T = \sum_1^N f(\lambda_k) v_k v_k^T$.

Entries of the rank one matrices $v_k v_k^T$ are products of sines. By rewriting those products $\sin(m\theta)\sin(n\theta)$ in terms of $\cos((m-n)\theta)$ (which leads to a Toeplitz part) and $\cos((m+n)\theta)$ (which leads to a Hankel part), we learn that the rank one matrix

$$v_k v_k^T = T_k + H_k \tag{3.10}$$

is Toeplitz-plus-Hankel, for all k [59]. Explicitly,

$$\begin{aligned} \text{Toeplitz} \qquad \left(T_k\right)_{mn} &= \frac{1}{N+1}\cos\left((m-n)k\pi h\right) \\ \text{Hankel} \qquad \left(H_k\right)_{mn} &= -\frac{1}{N+1}\cos\left((m+n)k\pi h\right). \end{aligned}$$

This shows that K has the strong Toeplitz-plus-Hankel property: the rank one matrices $v_k v_k^T$ coming from the eigenvectors can be written as a sum of a Toeplitz matrix and a Hankel matrix.

We quickly recall that *Toeplitz matrices* are those with constant diagonals (entries depend on $m-n$). *Hankel matrices* have constant antidiagonals (entries depend on $m+n$). By applying a Toeplitz matrix and a Hankel matrix to the same input, you see shifts in opposite directions. Toeplitz shifts the output forwards, while Hankel shifts the output backwards:

$$\textbf{Toeplitz} \quad \begin{bmatrix} b & a & & \\ c & b & a & \\ & c & b & a \\ & & c & b \end{bmatrix} \begin{bmatrix} 0 & 0 \\ 1 & 0 \\ 0 & 1 \\ 0 & 0 \end{bmatrix} = \begin{bmatrix} a & \\ b & a \\ c & b \\ & c \end{bmatrix} \quad \text{forward shift}$$

$$\textbf{Hankel} \quad \begin{bmatrix} & & a & b \\ & a & b & c \\ a & b & c & \\ b & c & & \end{bmatrix} \begin{bmatrix} 0 & 0 \\ 1 & 0 \\ 0 & 1 \\ 0 & 0 \end{bmatrix} = \begin{bmatrix} & a \\ a & b \\ b & c \\ c & \end{bmatrix} \quad \text{backward shift}$$

Combining (3.10) with (3.9), we now see $f(K)$ as the sum of a Toeplitz matrix and a Hankel matrix:

$$\textbf{Matrix function} \quad f(K) = Vf(\Lambda)V^T = \sum_1^N f(\lambda_k)(T_k + H_k)$$
$$= \underbrace{\sum_1^N f(\lambda_k)T_k}_{\text{Toeplitz}} + \underbrace{\sum_1^N f(\lambda_k)H_k}_{\text{Hankel}} \tag{3.11}$$

If we choose $f(z) = z^{-1}$, then we split the inverse matrix into Toeplitz and Hankel parts: $K^{-1} = T + H$. With $N = 3$, this T and H are

$$K^{-1} = \frac{1}{4}\begin{bmatrix} 3 & 2 & 1 \\ 2 & 4 & 2 \\ 1 & 2 & 3 \end{bmatrix} = \frac{1}{8}\begin{bmatrix} 5 & 2 & -1 \\ 2 & 5 & 2 \\ -1 & 2 & 5 \end{bmatrix} + \frac{1}{8}\begin{bmatrix} 1 & 2 & 3 \\ 2 & 3 & 2 \\ 3 & 2 & 1 \end{bmatrix}.$$

In summary, the matrix K, and all functions of that matrix, are Toeplitz-plus-Hankel [59]. In the sequel, we make the particular choice $f(z) = \exp(\pm it\sqrt{z}/\Delta x)$ or its real part, $f(z) = \cos(t\sqrt{z}/\Delta x)$. Then $f(K)$ solves a wave equation.

Before we proceed to the wave equation, we make one small observation about the example of the convection-diffusion operator $(P+D)$ in (3.4). It is an important instance of a nonsymmetric matrix where the *pseudospectra* plays a role in the analysis [49]. That nonsymmetric matrix certainly does not have the strong Toeplitz-plus-Hankel property. However, with the help of a simple diagonal matrix Z the similar matrix $S \equiv Z(P+D)Z^{-1}$ is symmetric, and S *does* have the strong Toeplitz-plus-Hankel property. The *ith* diagonal entry $z_i = Z_{i,i}$ is found by setting $z_1 = 1$, and

$z_{i+1} = z_i\sqrt{b_i/a_i}$, where $a_i = M_{i+1,i}$, and $b_i = M_{i,i+1}$, and $M = (P+D)$. We hope to explore Toeplitz-plus-Hankel properties for convection-diffusion operators in the future.

4.2 The Wave Equation Is Toeplitz-Plus-Hankel

Our model problem is on an interval $-1 \le x \le 1$ with zero Dirichlet boundary conditions $u(-1,t) = u(1,t) = 0$. The second derivative u_{xx} is replaced by second differences at the mesh points $x = -1,\dots,-2\Delta x,-\Delta x,0,\Delta x,2\Delta x,\dots,1$, where $\Delta x = 2/(N-1)$. The familiar wave equation can be approximated with the help of the second difference matrix K:

$$\text{Wave equation} \quad \frac{\partial^2}{\partial t^2}u = \frac{\partial^2}{\partial x^2}u \qquad \text{becomes} \qquad \frac{\mathrm{d}^2}{\mathrm{d}t^2}u = -\frac{K}{\Delta x^2}u. \tag{3.12}$$

Time remains continuous in this finite difference, semi-discrete approximation. One solution to the semi-discrete approximation in (3.12) involves exponentials or cosines of matrices:

$$\text{Solution} \qquad u(t) = f(K)u(0) = \cos\left(t\sqrt{K}/\Delta x\right)u(0).$$

Our purpose here is to apply the Toeplitz-plus-Hankel splitting, so we again set $T = \sum_k f(\lambda_k)T_k$ and $H = \sum_k f(\lambda_k)H_k$, with T_k and H_k as in (3.10). Now with $f(z) = \cos(t\sqrt{z}/\Delta x)$ in (3.11) we see this same solution as the sum of two parts:

$$u(t) = f(K)u(0) = \underbrace{Tu(0)}_{\text{Toeplitz}} + \underbrace{Hu(0)}_{\text{Hankel}}.$$

Unlike the approximate splitting into products of exponentials discussed in the previous sections of this chapter, here we see an exact splitting into a sum. Thus we have split the wave equation into a Toeplitz part and a Hankel part. Now we can separately investigate the behavior of the solutions coming from each part.

Figure 3.1 shows the exact solution to the wave equation via d'Alembert's formula, as if the equation was on the whole real line with no boundaries. We use this as a reference to compare to the solution of the same problem with Dirichlet boundary conditions. We see the consequences of the boundary in the differences between these solutions. Figure 3.1 also shows, separately, the solutions coming from the Toeplitz part ($Tu(0)$) and the Hankel part ($Hu(0)$). Their sum solves the Dirichlet problem exactly. The most interesting behavior happens at the boundary. Before reaching the boundary, the solution is essentially Toeplitz. After reaching the boundary, the solution is essentially Hankel. The reflection at the boundary comes from the Hankel part of the operator.

The Toeplitz-plus-Hankel splitting described here is very special, but in this example the splitting does show reflections at the boundary in a new light: the reflections come from the Hankel part of the operator. The design of absorbing boundary conditions, or perfectly matched layers, is a big subject in computational science, that we will leave untouched [17, 27, 35]. We conjecture that Figure 3.1 can be understood by the *method of images* [59, 23]. That would involve identifying the solution of the Dirichlet boundary condition version of the problem here with the solution of a closely related problem having *periodic* boundary conditions. Periodic behavior is far from ideal when designing absorbing boundary conditions – we don't want the wave to come back to the domain later. Nevertheless, whilst the approximation of the Sommerfeld boundary condition results in a small reflection here, it is intriguing that the Toeplitz part of the solution is seemingly reflectionless in Figure 3.1.

Outlook

This chapter introduced the basic ideas behind operator splitting methods. We focused on the application of splitting methods to solve differential equations. Historically, that has been their greatest application, though by now the splitting idea has found wide applications such as in optimization. We reviewed some of the main applications in biology especially, such as splitting reaction terms from diffusion terms in reaction-diffusion PDEs. Operator splitting is an old idea of numerical analysis, so it is pleasing that new ideas and new applications keep appearing even today. Perhaps one of the biggest contemporary applications of splitting involves coupling models across scales, such as appropriate coupling of mesoscopic reaction diffusion master equation models to finer, microscopic models [19, 18, 26, 25]. On that front, we are sure that more great work on splitting methods is still to come.

References

1. Ascher, U., Ruuth, S., Wetton, B.: Implicit-explicit methods for time-dependent partial differential equations. SIAM Journal on Numerical Analysis (1995)
2. Asmussen, S., Glynn, P.W.: Stochastic Simulation: Algorithms and Analysis. Springer (2007)
3. Blanes, S., Casas, F., Chartier, P., Murua, A.: Optimized high-order splitting methods for some classes of parabolic equations. Math. Comput. (2012)
4. Boyd, S., Parikh, N., Chu, E., Peleato, B., Eckstein, J.: Distributed Optimization and Statistical Learning via the Alternating Direction Method of Multipliers. Foundations and Trends in Machine Learning pp. 1–122 (2011)
5. Butcher, J.: Numerical Methods for Ordinary Differential Equations. Wiley (2003)
6. Cao, Y., Gillespie, D.T., Petzold, L.R.: The slow-scale stochastic simulation algorithm. J. Chem. Phys. **122**, 014,116–1–18 (2005)
7. Chorin, A.J.: Numerical solution of the Navier-Stokes equations. Math. Comp. pp. 745–762 (1968)

8. Dahlquist, G.: Convergence and stability in the numerical integration of ordinary differential equations. Math. Scand. pp. 33–53 (1956)
9. Dahlquist, G.: A special stability problem for linear multistep methods. BIT Numerical Mathematics pp. 27–43 (1963)
10. Dahlquist, G., Björk, A.: Numerical Methods. Prentice-Hall (1974)
11. Daubechies, I.: Ten Lectures on Wavelets. SIAM (1992)
12. Descombes, S.: Convergence of a splitting method of high order for reaction-diffusion systems. Mathematics of Computation **70**(236), 1481–1501 (2001)
13. Descombes, S., Schatzman, M.: Directions alternées d'ordre élevé en réaction-diffusion. Comptes Rendus de l'Académie des Sciences. Série 1, Mathématique **321**(11), 1521–1524 (1995)
14. Douglas, J., Rachford, H.H.: On the numerical solution of heat conduction problems in two and three space variables. Trans. Amer. Math. Soc. pp. 421–439 (1956)
15. E, W., Liu, D., Vanden-Eijnden, E.: Nested stochastic simulation algorithm for chemical kinetic systems with disparate rates. J. Chem. Phys. **123**(19), 194,107 (2005)
16. Engblom, S.: Strong convergence for split-step methods in stochastic jump kinetics (2014). http://arxiv.org/abs/1412.6292
17. Engquist, B., Majda, A.: Absorbing boundary conditions for numerical simulation of waves. Proc Natl Acad Sci U S A **74**, 765–1766 (1977)
18. Ferm, L., Lötstedt, P.: Numerical method for coupling the macro and meso scales in stochastic chemical kinetics. BIT Numerical Mathematics **47**(4), 735–762 (2007)
19. Ferm, L., Lötstedt, P., Hellander, A.: A hierarchy of approximations of the master equation scaled by a size parameter. J. Sci. Comput. **34**, 127–151 (2008)
20. Giles, M.: Multi-level Monte Carlo path simulation. Operations Research **56**, 607–617 (2008)
21. Goldstein, T., Osher, S.: The Split Bregman Method for L1-Regularized Problems. SIAM Journal on Imaging Sciences **2**(2), 323–343 (2009)
22. Golub, G.H., Van Loan, C.F.: Matrix Computations. Johns Hopkins (1996)
23. Haberman, R.: Applied Partial Differential Equations. Prentice Hall (2013)
24. Hairer, E., Lubich, C., Wanner, G.: Geometric numerical integration: Structure-preserving algorithms for ordinary differential equations. Springer (2006)
25. Hellander, A., Hellander, S., Lötstedt, P.: Coupled mesoscopic and microscopic simulation of stochastic reaction-diffusion processes in mixed dimensions. Multiscale Model. Simul. pp. 585–611 (2012)
26. Hellander, A., Lawson, M., Drawert, B., Petzold, L.: Local error estimates for adaptive simulation of the reaction-diffusion master equation via operator splitting. J. Comput. Phys (2014)
27. Higdon, R.: Numerical Absorbing Boundary Conditions for the Wave Equation. Mathematics of Computation **49**, 65–90 (1987)
28. Higham, N.J.: Functions of Matrices. SIAM (2008)
29. Horn, R., Johnson, C.: Matrix Analysis. Cambridge University Press (2013)
30. Hundsdorfer, W., Verwer, J.: Numerical Solution of Time-Dependent Advection-Diffusion-Reaction Equations. Springer (2003)
31. Iserles, A.: A First Course in the Numerical Analysis of Differential Equations. Cambridge University Press (1996)
32. Jahnke, T., Altintan, D.: Efficient simulation of discrete stochastic reaction systems with a splitting method. BIT **50**, 797–822 (2010)
33. Lax, P.D., Richtmyer, R.D.: Survey of the stability of linear finite difference equations. Comm. Pure Appl. Math pp. 267–293 (1956)
34. Lee, J., Fornberg, B.: A split step approach for the 3-D Maxwell's equations. J. Comput. Appl. Math. **158**, 485–505 (2003)
35. Loh, P.R., Oskooi, A.F., Ibanescu, M., Skorobogatiy, M., Johnson, S.G.: Fundamental relation between phase and group velocity, and application to the failure of perfectly matched layers in backward-wave structures. Phys. Rev. E **79**, 065,601 (2009). DOI 10.1103/PhysRevE.79.065601. URL http://link.aps.org/doi/10.1103/PhysRevE.79.065601
36. Lubich, C., Oseledets, I.: A projector-splitting integrator for dynamical low-rank approximation. BIT Numerical Mathematics **54**, 171–188 (2014)

37. MacNamara, S., Burrage, K., Sidje, R.: Application of the Strang Splitting to the chemical master equation for simulating biochemical kinetics. The International Journal of Computational Science **2**, 402–421 (2008)
38. MacNamara, S., Burrage, K., Sidje, R.: Multiscale Modeling of Chemical Kinetics via the Master Equation. SIAM Multiscale Model. & Sim. **6**(4), 1146–1168 (2008)
39. Maini, P.K., Baker, R.E., Chuong, C.M.: The Turing model comes of molecular age. Science **314**, 1397–1398 (2006)
40. Marchuk, G.I.: Some application of splitting-up methods to the solution of mathematical physics problems. Aplikace Matematiky pp. 103–132 (1968)
41. Marchuk, G.I.: Splitting and alternating direction methods. In: P. Ciarlet, J. Lions (eds.) Handbook of Numerical Analysis, vol. 1, pp. 197–462. North-Holland, Amsterdam (1990)
42. McLachlan, R., Reinout, G., Quispel, W.: Splitting methods. Acta Numer. **11**, 341–434 (2002)
43. Moler, C., Van Loan, C.: Nineteen Dubious Ways to Compute the Exponential of a Matrix, 25 Years Later. SIAM Review **45(1)**, 3–49 (2003)
44. Murray, J.: Mathematical Biology: An Introduction. New York : Springer (2002)
45. Ninomiya, S., Victoir, N.: Weak approximation of stochastic differential equations and application to derivative pricing. Applied Mathematical Finance **15** (2008)
46. Pavliotis, G., Stuart, A.: Multiscale Methods: Averaging and Homogenization. Springer (2008)
47. Peaceman, D.W., Rachford Jr., H.H.: The numerical solution of parabolic and elliptic differential equations. Journal of the Society for Industrial and Applied Mathematics (1955)
48. Rauch, J.: The Turing Instability. URL: http://www.math.lsa.umich.edu/~rauch/
49. Reddy, S.C., Trefethen, L.N.: Pseudospectra of the convection-diffusion operator. SIAM J. Appl. Math (1994)
50. Ropp, D., Shadid, J.: Stability of operator splitting methods for systems with indefinite operators: Reaction-diffusion systems. Journal of Computational Physics **203**, 449–466 (2005)
51. Saad, Y.: Iterative Methods for Sparse Linear Systems. SIAM, Society for Industrial and Applied Mathematics, Philadelphia (2003)
52. Schatzman, M.: Toward non commutative numerical analysis: High order integration in time. Journal of Scientific Computing **17**, 99–116 (2002)
53. Speth, R., Green, W., MacNamara, S., Strang, G.: Balanced splitting and rebalanced splitting. SIAM Journal of Numerical Analysis **51**(6), 3084–3105 (2013)
54. Strang, G.: On the construction and comparison of difference schemes. SIAM J. Numer. Anal. **5**(2), 506–517 (1968)
55. Strang, G.: Computational Science and Engineering. Wellesley-Cambridge Press (2007)
56. Strang, G.: Introduction to Linear Algebra. Wellesley-Cambridge Press (2009)
57. Strang, G.: Essays in Linear Algebra. Wellesley-Cambridge Press (2012)
58. Strang, G.: Differential Equations and Linear Algebra. Wellesley-Cambridge Press (2014)
59. Strang, G., MacNamara, S.: Functions of difference matrices are Toeplitz plus Hankel. SIAM Review **56**(3), 525–546 (2014)
60. Trefethen, L.: Numerical analysis. In: T. Gowers, J. Barrow-Green, I. Leader (eds.) Princeton Companion to Mathematics, pp. 604–615. Princeton University Press (2008)
61. Trefethen, L., Embree, M.: Spectra and Pseudospectra: The Behavior of Nonnormal Matrices and Operators. Princeton University Press (2005)
62. Usadi, A., Dawson, C.: 50 years of ADI methods: Celebrating the contributions of Jim Douglas, Don Peaceman, and Henry Rachford. SIAM News **39** (2006)
63. Wanner, G.: Dahlquist's classical papers on stability theory. BIT Numerical Mathematics **46**, 671–683 (2006)
64. Widom, H.: Toeplitz matrices. In: I.I. Hirschman (ed.) Studies in Real and Complex Analysis. Prentice-Hall (1965)
65. Yoshida, H.: Construction of higher order symplectic integrators. Phys. Lett. A. **150**, 262–268 (1990)

Chapter 4
Convergence Rate Analysis of Several Splitting Schemes*

Damek Davis and Wotao Yin

Abstract Operator-splitting schemes are iterative algorithms for solving many types of numerical problems. A lot is known about these methods: they converge, and in many cases we know how quickly they converge. But when they are applied to optimization problems, there is a gap in our understanding: The theoretical speed of operator-splitting schemes is nearly always measured in the ergodic sense, but ergodic operator-splitting schemes are rarely used in practice. In this chapter, we tackle the discrepancy between theory and practice and uncover fundamental limits of a class of operator-splitting schemes. Our surprising conclusion is that the relaxed Peaceman-Rachford splitting algorithm, a version of the Alternating Direction Method of Multipliers (ADMM), is nearly as fast as the proximal point algorithm in the ergodic sense and nearly as slow as the subgradient method in the nonergodic sense. A large class of operator-splitting schemes extend from the relaxed Peaceman-Rachford splitting algorithm. Our results show that this class of operator-splitting schemes is also nearly as slow as the subgradient method. The tools we create in this chapter can also be used to prove nonergodic convergence rates of more general splitting schemes, so they are interesting in their own right.

* This work is supported in part by NSF grants DMS-1317602 and ECCS-1462398.

D. Davis (✉) • W. Yin
Department of Mathematics, University of California, Los Angeles, CA 90025, USA
e-mail: damek@math.ucla.edu; wotaoyin@math.ucla.edu

R. Glowinski et al. (eds.), *Splitting Methods in Communication, Imaging, Science, and Engineering*, Scientific Computation, DOI 10.1007/978-3-319-41589-5_4

1 Introduction

Operator-splitting schemes are iterative algorithms for solving optimization problems (and more generally PDE) [39, 52, 32, 47, 27]. These algorithms are useful for solving medium to large-scale problems[1] in signal processing and machine learning (see [10, 51, 54]). Operator-splitting schemes converge, and in many cases, we know how quickly they converge [4, 54, 9, 16, 50, 35, 22, 38, 41, 42, 43, 25, 26]. On the surface, we seem to have a complete understanding of these algorithms. However, there is a missing piece hidden beneath the contributions in the literature: The theoretical speed of operator-splitting schemes is nearly always measured in the ergodic sense, but ergodic operator-splitting schemes are rarely used in practice[2].

In this chapter, we tackle the discrepancy between theory and practice and uncover fundamental limits of a class of operator-splitting schemes. Many of the most powerful operator-splitting schemes extend from a single algorithm called the relaxed *Peaceman-Rachford splitting algorithm* (PRS). For most of the chapter, we will only study relaxed PRS, but along the way, we develop tools that will help us analyze other algorithms.[3] These tools are key to analyzing the speed of operator-splitting schemes in theory and in practice. We also determine exactly how fast the relaxed PRS algorithm can be; these results uncover fundamental limits on the speed of the large class operator-splitting schemes that extend relaxed PRS [21, 23, 53, 14, 7, 8, 19, 6, 20, 48, 16].[4]

Relaxed PRS is an iterative algorithm for solving

$$\underset{x\in\mathcal{H}}{\text{minimize}}\ f(x)+g(x) \tag{4.1}$$

where $\mathcal{H}$ is a Hilbert space (e.g., $\mathbf{R}^n$ for some $n\in\mathbf{N}$), and $f,g:\mathcal{H}\to(-\infty,\infty]$ are closed (i.e., lower semi-continuous), proper, and convex functions. The algorithm is easy to state: given $\gamma>0$, $z^0\in\mathcal{H}$, and $(\lambda_j)_{j\geq 0}\in[0,2]$, define

$$\text{for all } k\in\mathbf{N}\quad \begin{cases} x_g^k = \arg\min_{x\in\mathcal{H}}\left\{g(x)+\frac{1}{2\gamma}\|x-z^k\|^2\right\}; \\ x_f^k = \arg\min_{x\in\mathcal{H}}\left\{f(x)+\frac{1}{2\gamma}\|x-(2x_g^k-z^k)\|^2\right\}; \\ z^{k+1} = z^k+\lambda_k(x_f^k-x_g^k). \end{cases} \tag{4.2}$$

Although there is no precise mathematical definition of the term *operator-splitting scheme*, the iteration in (4.2) is a classic example of such an algorithm because it splits the difficult problem (1) into the sequence of simpler x_g^k and x_f^k updates.

---

[1] E.g., with gigabytes to terabytes of data.

[2] By ergodic, we mean the final approximate solution returned by the algorithm is an average over the history of all approximate solutions formed throughout the algorithm.

[3] After the initial release of this chapter, we used these tools to study several more general algorithms [25, 26, 27]

[4] This list is not exhaustive. See the comments after Theorem 8 for more details.

A large class of algorithms [21, 23, 53, 14, 7, 8, 19, 6, 20, 48, 16] extends relaxed PRS, and we can demonstrate how the theoretical analysis of such algorithm differs from how we use them in practice by measuring the convergence speed with the objective error[5] $(f+g)(x^k)-(f+g)(x^*)$ evaluated at a sequence $(x^j)_{j\geq 0}\subseteq \mathscr{H}$. Because relaxed PRS is an iterative algorithm, it outputs the final approximate solution x^k after $k\in \mathbf{N}$ iterations of the algorithm. In practice, it is common to output the *nonergodic* iterate $x^k\in\{x_g^k,x_f^k\}$ as the final approximate solution. But most theoretical analysis of splitting schemes assumes that the *ergodic* iterate $x^k\in\{(1/\sum_{i=0}^k\lambda_i)\sum_{i=0}^k x_g^i,(1/\sum_{i=0}^k\lambda_i)\sum_{i=0}^k x_f^i\}$ is the final approximate solution [16, 9, 41, 42, 43]. In many applications, the nonergodic iterates are qualitatively and quantitatively better approximate solutions to (1) than the ergodic iterates (see [27, Figure 5.3(c)] for an example). The difference in performance is particularly large in sparse optimization problems because, unlike the solution to (1), the ergodic iterate is often a dense vector. In this chapter, we create some tools to analyze the nonergodic convergence rate of the objective error in relaxed PRS. These tools can also be used to prove nonergodic convergence rates of more general splitting schemes [25, 26], so they are interesting in their own right.

Though practical experience suggests that the nonergodic iterates are better approximate solutions than the ergodic iterates, our theoretical analysis actually predicts that they are quantitatively worse. The difference between the theoretical speed of the two iterates is large: we prove that the ergodic iterates have a convergence rate of $(f+g)(x^k)-(f+g)(x^*)=O(1/(k+1))$, while the best convergence rate that we can prove for the nonergodic iterates is $(f+g)(x^k)-(f+g)(x^*)=o(1/\sqrt{k+1})$. Moreover, we provide examples of functions f and g which show that these two rates are tight. This result proves that there are fundamental limits on how quickly the ergodic and nonergodic iterates of operator-splitting schemes converge, at least for the large class of algorithms that extend relaxed PRS. These results complement, but do not follow from, the well-developed lower bounds for (sub)gradient algorithms [44, 45], which do not address the PRS algorithm.

Our analysis also applies to the Alternating Direction Method of Multipliers (ADMM) algorithm, which is an iterative algorithm for solving:

$$\begin{aligned}&\underset{x\in\mathscr{H}_1,\ y\in\mathscr{H}_2}{\text{minimize}}\ f(x)+g(y)\\&\text{subject to}\ Ax+By=b\end{aligned}\tag{4.3}$$

where $\mathscr{H}_1,\mathscr{H}_2$, and $\mathscr{G}$ are Hilbert spaces, $b\in\mathscr{G}$, and $A:\mathscr{H}_1\to\mathscr{G}$ and $B:\mathscr{H}_2\to\mathscr{G}$ are bounded linear operators. We present these results at the end of the chapter.

We prove several other theoretical results that are of independent interest: we prove the fixed-point residual of the Krasnosel'skiĭ-Mann algorithm (KM) with summable errors has convergence rate $o(1/\sqrt{k+1})$, and we show that the rate is tight; we prove that relaxed PRS can converge arbitrarily slowly; we prove convergence rates and lower bounds for the proximal point algorithm and the forward-backward splitting algorithm (see Appendix A); and we give several examples of our results on concrete applications (see Appendix D).

[5] $x^*\in\mathscr{H}$ is a minimizer of (1).

1.1 Notation

The symbols $\mathscr{H}, \mathscr{H}_1, \mathscr{H}_2, \mathscr{G}$ denote (possibly infinite dimensional) Hilbert spaces. Sequences $(\lambda_j)_{j\geq 0} \subset \mathbf{R}_+$ denote relaxation parameters, and

$$\Lambda_k := \sum_{i=0}^{k} \lambda_i$$

denote kth partial sums. The reader may assume that $\lambda_k \equiv (1/2)$ and $\Lambda_k = (k+1)/2$ in the DRS algorithm or that $\lambda_k \equiv 1$ and $\Lambda_k = (k+1)$ in the PRS algorithm. Given a sequence $(x^j)_{j\geq 0} \subset \mathscr{H}$, we let $\overline{x}^k = (1/\Lambda_k)\sum_{i=0}^{k} \lambda_i x^i$ denote its kth average with respect to the sequence $(\lambda_j)_{j\geq 0}$.

Given a closed, proper, convex function $f : \mathscr{H} \to (-\infty, \infty]$, the set $\partial f(x)$ denotes its subdifferential at x, and we let

$$\widetilde{\nabla} f(x) \in \partial f(x), \tag{4.4}$$

denotes an arbitrary subgradient; the actual choice of the subgradient $\widetilde{\nabla} f(x)$ will always be clear from the context. (This notation was used in [5, Eq. (1.10)].) The convex conjugate of a proper, closed, and convex function f is $f^*(y) := \sup_{x\in\mathscr{H}}\{\langle y, x\rangle - f(x)\}$. Let $I_{\mathscr{H}}$ denote the identity map. For any $x \in \mathscr{H}$ and scalar $\gamma \in \mathbf{R}_{++}$, we let

$$\mathbf{prox}_{\gamma f}(x) := \operatorname*{arg\,min}_{y\in\mathscr{H}} \left\{ f(y) + \frac{1}{2\gamma}\|y - x\|^2 \right\} \quad \text{and} \quad \mathbf{refl}_{\gamma f} := 2\mathbf{prox}_{\gamma f} - I_{\mathscr{H}},$$

be the *proximal* and *reflection* operators, and we define the PRS operator:

$$T_{\mathrm{PRS}} := \mathbf{refl}_{\gamma f} \circ \mathbf{refl}_{\gamma g}. \tag{4.5}$$

1.2 Assumptions

Assumption 1. Every function we consider is closed, proper, and convex.

Unless otherwise stated, a function is not necessarily differentiable.

Assumption 2 (Differentiability). Every differentiable function we consider is Fréchet differentiable [2, Definition 2.45].

Assumption 3 (Solution Existence). Functions $f, g : \mathscr{H} \to (-\infty, \infty]$ satisfy $\operatorname{zer}(\partial f + \partial g) \neq \emptyset$.

Note that this last assumption is slightly stronger than the existence of a minimizer, because $\operatorname{zer}(\partial f + \partial g) \neq \operatorname{zer}(\partial(f+g))$, in general [2, Remark 16.7]. Nevertheless, this assumption is standard.

1.3 The Algorithms

In this chapter we study the *relaxed PRS* algorithm:

Algorithm 1: Relaxed Peaceman-Rachford splitting (relaxed PRS)

input : $z^0 \in \mathcal{H}$, $\gamma > 0$, $(\lambda_j)_{j\geq 0} \subseteq (0,1]$
for $k = 0, 1, \ldots$ **do**
  $z^{k+1} = (1-\lambda_k)z^k + \lambda_k \mathbf{refl}_{\gamma f} \circ \mathbf{refl}_{\gamma g}(z^k)$;

The special cases $\lambda_k \equiv 1/2$ and $\lambda_k \equiv 1$ are called the DRS and PRS algorithms, respectively. The relaxed PRS algorithm can be applied to problem (4.3). To this end, we define the Lagrangian:

$$\mathcal{L}_\gamma(x,y;w) := f(x) + g(y) - \langle w, Ax + By - b\rangle + \frac{\gamma}{2}\|Ax + By - b\|^2.$$

Section 8 presents Algorithm 1 applied to the Lagrange dual of (4.3), which reduces to the following algorithm:

Algorithm 2: Relaxed alternating direction method of multipliers (relaxed ADMM)

input : $w^{-1} \in \mathcal{H}, x^{-1} = 0, y^{-1} = 0, \lambda_{-1} = 1/2, \gamma > 0, (\lambda_j)_{j\geq 0} \subseteq (0,1]$
for $k = -1, 0, \ldots$ **do**
  $y^{k+1} = \arg\min_y \mathcal{L}_\gamma(x^k, y; w^k) + \gamma(2\lambda_k - 1)\langle By, (Ax^k + By^k - b)\rangle$;
  $w^{k+1} = w^k - \gamma(Ax^k + By^{k+1} - b) - \gamma(2\lambda_k - 1)(Ax^k + By^k - b)$;
  $x^{k+1} = \arg\min_x \mathcal{L}_\gamma(x, y^{k+1}; w^{k+1})$;

If $\lambda_k \equiv 1/2$, Algorithm 2 recovers the standard ADMM.

Each of the above algorithms is a special case of the Krasnosel'skiĭ-Mann (KM) iteration [37, 40, 29]. An *averaged operator* is the average of a nonexpansive operator $T : \mathcal{H} \to \mathcal{H}$ and the identity mapping $I_{\mathcal{H}}$. That is, for all $\lambda \in (0,1)$, the operator

$$T_\lambda := (1-\lambda)I_{\mathcal{H}} + \lambda T \tag{4.6}$$

is called λ-*averaged* and every λ-averaged operator is exactly of the form T_λ for some nonexpansive map T.

Given a nonexpansive map T, we define the fixed-point iteration of the map T_λ:

Algorithm 3: Krasnosel'skiĭ-Mann (KM)

input : $z^0 \in \mathcal{H}, (\lambda_j)_{j\geq 0} \subseteq (0,1]$
for $k = 0, 1, \ldots$ **do**
  $z^{k+1} = T_{\lambda_k}(z^k)$;

1.4 Basic Properties of Averaged Operators

The following properties are included in textbooks such as [2].

Proposition 1. *Let $\mathcal{H}$ be a Hilbert space; let $f,g : \mathcal{H} \to (-\infty,\infty]$ be closed, proper, and convex functions; and let $T : \mathcal{H} \to \mathcal{H}$ be a nonexpansive operator*

1. *Let $x \in \mathcal{H}$. Then $x^+ = \mathbf{prox}_{\gamma f}(x)$ if, and only if, $(1/\gamma)(x - x^+) \in \partial f(x^+)$.*
2. *The operator $\mathbf{refl}_{\gamma f} : \mathcal{H} \to \mathcal{H}$ is nonexpansive. Therefore,*

$$T_{\mathrm{PRS}} := \mathbf{refl}_{\gamma f} \circ \mathbf{refl}_{\gamma g} \tag{4.7}$$

3. *For all $\lambda \in (0,1]$ and $(x,y) \in \mathcal{H} \times \mathcal{H}$, the operator T_λ (see (4.6)) satisfies*

$$\|T_\lambda x - T_\lambda y\|^2 \leq \|x-y\|^2 - \frac{1-\lambda}{\lambda}\|(I_{\mathcal{H}} - T_\lambda)x - (I_{\mathcal{H}} - T_\lambda)y\|^2.$$

4. *The operator $\mathbf{prox}_{\gamma f} : \mathcal{H} \to \mathcal{H}$ is $\frac{1}{2}$-averaged.*
5. *T_λ and T have the same set of fixed points.*

2 Summable Sequence Lemma

Summable sequences of positive numbers always converge to 0. We can even determine how quickly they converge:

Lemma 1 (Summable Sequence Convergence Rates). *Let nonnegative scalar sequences $(\lambda_j)_{j\geq 0}$ and $(a_j)_{j\geq 0}$ satisfy $\sum_{i=0}^{\infty} \lambda_i a_i < \infty$. Let $\Lambda_k := \sum_{i=0}^{k} \lambda_i$ for $k \geq 0$.*

1. **Monotonicity:** *If $(a_j)_{j\geq 0}$ is* monotonically nonincreasing, *then*

$$a_k \leq \frac{1}{\Lambda_k}\left(\sum_{i=0}^{\infty} \lambda_i a_i\right) \quad \text{and} \quad a_k = o\left(\frac{1}{\Lambda_k - \Lambda_{\lceil k/2 \rceil}}\right). \tag{4.8}$$

In particular,

(a) *if $(\lambda_j)_{j\geq 0}$ is bounded away from 0, then $a_k = o(1/(k+1))$;*
(b) *if $\lambda_k = (k+1)^p$ for some $p \geq 0$ and all $k \geq 1$, then $a_k = o(1/(k+1)^{p+1})$;*
(c) *as a special case, if $\lambda_k = (k+1)$ for all $k \geq 0$, then $a_k = o(1/(k+1)^2)$.*

2. **Monotonicity up to errors:** *Let $(e_j)_{j\geq 0}$ be a sequence of scalars. Suppose that $a_{k+1} \leq a_k + e_k$ for all $k \geq 0$ and that $\sum_{i=0}^{\infty} \Lambda_i e_i < \infty$. Then*

$$a_k \leq \frac{1}{\Lambda_k}\left(\sum_{i=0}^{\infty} \lambda_i a_i + \sum_{i=0}^{\infty} \Lambda_i e_i\right) \quad \text{and} \quad a_k = o\left(\frac{1}{\Lambda_k - \Lambda_{\lceil k/2 \rceil}}\right). \tag{4.9}$$

The rates of a_k in Parts 1(a), 1(b), and 1(c) continue to hold as long as $\sum_{i=0}^{\infty} \Lambda_i e_i < \infty$. In particular, the rates hold if $e_k = O(1/(k+1)^q)$ for some $q > 2$, $q > p+2$, and $q > 3$, in parts 1(a), 1(b), and 1(c), respectively.

3. **Faster rates:** *Suppose that $(b_j)_{j\geq 0}$ and $(e_j)_{j\geq 0}$ are nonnegative scalar sequences, that $\sum_{i=0}^{\infty} b_j < \infty$, and that $\sum_{i=0}^{\infty}(i+1)e_i < \infty$, and that for all $k \geq 0$ we have $\lambda_k a_k \leq b_k - b_{k+1} + e_k$. Then the following sum is finite:*

$$\sum_{i=0}^{\infty}(i+1)\lambda_i a_i \leq \sum_{i=0}^{\infty} b_i + \sum_{i=0}^{\infty}(i+1)e_i < \infty.$$

In particular,

(a) if $(\lambda_j)_{j\geq 0}$ is bounded away from 0, then $a_k = o(1/(k+1)^2)$;
(b) if $\lambda_k = (k+1)^p$ for some $p \geq 0$ and all $k \geq 1$, then $a_k = o(1/(k+1)^{p+2})$.

4. **No monotonicity:** *For all $k \geq 0$, define the sequence of indices*

$$k_{\text{best}} := \arg\min_i \{a_i | i = 0, \cdots, k\}.$$

Then $(a_{j_{\text{best}}})_{j\geq 0}$ is monotonically nonincreasing and the above bounds continue to hold when a_k is replaced with $a_{k_{\text{best}}}$.

Proof. Fix $k \geq 0$.

Part 1. For all $i \leq k$, we have $a_k \leq a_i$ and $\lambda_i a_i \geq 0$ and hence, $\Lambda_k a_k \leq \sum_{i=0}^{k} \lambda_i a_i \leq \sum_{i=0}^{\infty} \lambda_i a_i$. This shows the left part of (4.8). To prove the right part of (4.8), observe that

$$(\Lambda_k - \Lambda_{\lceil k/2 \rceil})a_k = \sum_{i=\lceil k/2 \rceil+1}^{k} \lambda_i a_k \leq \sum_{i=\lceil k/2 \rceil+1}^{k} \lambda_i a_i \overset{k\to\infty}{\to} 0.$$

Part 1(a). Let $\underline{\lambda} := \inf_{j\geq 0} \lambda_j > 0$. For every integer $k \geq 2$, we have $\lceil k/2 \rceil \leq (k+1)/2$. Thus, $\Lambda_k - \Lambda_{\lceil k/2 \rceil} \geq \underline{\lambda}(k - \lceil k/2 \rceil) \geq \underline{\lambda}(k-1)/2 \geq \underline{\lambda}(k+1)/6$. Hence, $a_k = o(1/(\Lambda_k - \Lambda_{\lceil k/2 \rceil})) = o(1/(k+1))$ follows from (4.8).

Part 1(b). For every integer $k \geq 3$, we have $\lceil k/2 \rceil + 1 \leq (k+3)/2 \leq 3(k+1)/4$ and $\Lambda_k - \Lambda_{\lceil k/2 \rceil} = \sum_{i=\lceil k/2 \rceil+1}^{k} \lambda_i = \sum_{i=\lceil k/2 \rceil+1}^{k}(i+1)^p \geq \int_{\lceil k/2 \rceil}^{k}(t+1)^p dt = (p+1)^{-1}((k+1)^{p+1} - (\lceil k/2 \rceil + 1)^{p+1}) \geq (p+1)^{-1}(1-(3/4)^{p+1})(k+1)^{p+1}$. Therefore, $a_k = o(1/(\Lambda_k - \Lambda_{\lceil k/2 \rceil})) = o(1/(k+1)^{p+1})$ follows from (4.8).

Part 1(c) directly follows from Part 1(b).

Part 2. For every integer $0 \leq i \leq k$, we have $a_k \leq a_i + \sum_{j=i}^{k-1} e_j$. Thus, $\Lambda_k a_k = \sum_{i=0}^{k} \lambda_i a_k \leq \sum_{i=0}^{k} \lambda_i a_i + \sum_{i=0}^{k} \lambda_i (\sum_{j=i}^{k-1} e_j) = \sum_{i=0}^{k} \lambda_i a_i + \sum_{i=0}^{k-1} e_i (\sum_{j=0}^{i} \lambda_j) = \sum_{i=0}^{k} \lambda_i a_i + \sum_{i=0}^{k-1} \Lambda_i e_i \leq \sum_{i=0}^{\infty} \lambda_i a_i + \sum_{i=0}^{\infty} \Lambda_i e_i$, from which the left part of (4.9) follows. The proof for the right part of (4.9) is similar to Part 1. The condition $e_k = O(1/(k+1)^q)$ for appropriate q is used to ensure that $\sum_{i=0}^{\infty} \Lambda_i e_i < \infty$ for each setting of λ_k in the previous Parts 1(a), 1(b), and 1(c).

Part 3. Note that

$$\begin{aligned}\lambda_k(k+1)a_k &\le (k+1)b_k-(k+1)b_{k+1}+(k+1)e_k\\ &= b_{k+1}+((k+1)b_k-(k+2)b_{k+1})+(k+1)e_k.\end{aligned}$$

Thus, because the upper bound on $(k+1)\lambda_k a_k$ is the sum of a telescoping term and a summable term, we have $\sum_{i=0}^{\infty}(i+1)\lambda_i a_i \le \sum_{i=0}^{\infty} b_i+\sum_{i=0}^{\infty}(i+1)e_i<\infty$. Parts 3(a) and 3(b) are similar to Part 1(b).

Part 4 is straightforward, so we omit its proof. □

Part 1 of Lemma 1 generalizes [36, Theorem 3.3.1] and [28, Lemma 1.2].

3 Iterative Fixed-Point Residual Analysis

In this section we determine how quickly the fixed *fixed-point residual* (FPR) $\|Tz^k - z^k\|^2$ converges to 0 in Algorithm 3.

Algorithm 3 always converges weakly to a fixed-point of T under mild conditions on $(\lambda_j)_{j\ge 0}$ [18, Theorem 3.1] (also see [22, 38]). Because strong convergence of Algorithm 3 may fail (when $\mathcal{H}$ is infinite dimensional), the quantity $\|z^k - z^*\|$, where z^* is a fixed point of T, may not converge to zero. However, the property $\lim_{k\to\infty}\|Tz^k - z^k\| = 0$, known as *asymptotic regularity* [15], holds when a fixed point of T exists. Thus, we can always measure the convergence rate of the FPR.

We measure $\|Tz^k - z^k\|^2$ when we could just as well measure $\|Tz^k - z^k\|$. We choose to measure the squared norm because it naturally appears in our analysis. In addition, it is summable and monotonic, which is analyzable by Lemma 1.

In first-order optimization algorithms, the FPR typically relates to the size of objective gradient. For example, in the unit-step gradient descent algorithm, defined for all $k\ge 0$ by the recursion $z^{k+1} = z^k - \nabla f(z^k)$, the FPR is given by $\|\nabla f(z^k)\|^2$. In the proximal point algorithm, defined for all $k \ge 0$ by the recursion $z^{k+1} = \mathbf{prox}_f(z^k)$, the FPR is given by $\|\widetilde{\nabla} f(z^{k+1})\|^2$ where $\widetilde{\nabla} f(z^{k+1}) := (z^k - z^{k+1}) \in \partial f(z^{k+1})$ (see Part 1 of Proposition 1). When the objective is the sum of multiple functions, the FPR is a combination of the (sub)gradients of those functions in the objective. In this chapter we will use the subgradient inequality and bounds on the FPR to measure how quickly $f(z^k) - f(x^*)$ converges to 0 for the minimizers x^* of f.

3.1 $o(1/(k+1))$ *FPR of Averaged Operators*

In the following theorem, the little-o convergence rates in Equation (4.12) and Part 5 are new; the rest of the results can be found in [2, Proof of Proposition 5.14], [22, Proposition 11], and [38].

Theorem 1 (Convergence Rate of Averaged Operators). *Let $T : \mathcal{H} \to \mathcal{H}$ be a nonexpansive operator, let z^* be a fixed point of T, let $(\lambda_j)_{j\geq 0} \subseteq (0,1]$ be a sequence of positive numbers, let $\tau_k := \lambda_k(1-\lambda_k)$, and let $z^0 \in \mathcal{H}$. Suppose that $(z^j)_{j\geq 0} \subseteq \mathcal{H}$ is generated by Algorithm 3: for all $k \geq 0$, let*

$$z^{k+1} = T_{\lambda_k}(z^k), \tag{4.10}$$

where T_λ is defined in (4.6). *Then, the following results hold*

1. $\|z^k - z^*\|^2$ *is monotonically nonincreasing;*
2. $\|Tz^k - z^k\|^2$ *is monotonically nonincreasing;*
3. $\tau_k\|Tz^k - z^k\|^2$ *is summable:*

$$\sum_{i=0}^{\infty} \tau_i \|Tz^i - z^i\|^2 \leq \|z^0 - z^*\|^2; \tag{4.11}$$

4. *if $\tau_k > 0$ for all $k \geq 0$, then the convergence rates hold:*

$$\|Tz^k - z^k\|^2 \leq \frac{\|z^0 - z^*\|^2}{\sum_{i=0}^{k} \tau_i} \tag{4.12}$$

$$\text{and} \quad \|Tz^k - z^k\|^2 = o\left(\frac{1}{\sum_{i=\lceil \frac{k}{2} \rceil + 1}^{k} \tau_i}\right).$$

   In particular, if $(\tau_j)_{j\geq 0} \subseteq (\varepsilon, \infty)$ for some $\varepsilon > 0$, then $\|Tz^k - z^k\|^2 = o(1/(k+1))$.

5. *Instead of Iteration* (4.10), *for all $k \geq 0$, let*

$$z^{k+1} := T_{\lambda_k}(z^k) + \lambda_k e^k \tag{4.13}$$

   for an error sequence $(e^j)_{j\geq 0} \subseteq \mathcal{H}$ that satisfies $\sum_{i=0}^{k} \lambda_i \|e^i\| < \infty$ and $\sum_{i=0}^{\infty} (i+1)\lambda_i^2 \|e^i\|^2 < \infty$. (Note that these bounds hold, for example, when for all $k \geq 0$ $\lambda_k\|e^k\| \leq \omega_k$ for a sequence $(\omega_j)_{j\geq 0}$ that is nonnegative, summable, and monotonically nonincreasing.) Then if $(\tau_j)_{j\geq 0} \subseteq (\varepsilon, \infty)$ for some $\varepsilon > 0$, we continue to have $\|Tz^k - z^k\|^2 = o(1/(k+1))$.

Proof. As noted before the Theorem, for Parts 1 through 4, we only need to prove the little-o convergence rate. This follows from the monotonicity of $(\|Tz^j - z^j\|^2)_{j\geq 0}$, Equation (4.11), and Part 1 of Lemma 1.

Part 5: We first show that the condition involving the sequence $(\omega_j)_{j\geq 0}$ is sufficient to guarantee the error bounds. We have $\sum_{i=0}^{\infty} \lambda_i \|e^i\| \leq \sum_{i=0}^{\infty} \omega_i < \infty$ and $\sum_{i=0}^{\infty} (i+1)\lambda_i^2 \|e^i\|^2 \leq \sum_{i=0}^{\infty} (i+1)\omega_i^2 < \infty$, where the last inequality is shown as follows. By Part 1 of Lemma 1, we have $\omega_k = o(1/(k+1))$. Therefore, there exists a finite K such that $(k+1)\omega_k < 1$ for $k > K$. Therefore, $\sum_{i=0}^{\infty} (i+1)\omega_i^2 < \sum_{i=0}^{K} (i+1)\omega_i^2 + \sum_{i=K+1}^{\infty} \omega_i < \infty$.

Fix $k \geq 0$. For simplicity, introduce $p^k := Tz^k - z^k$, $p^{k+1} := Tz^{k+1} - z^{k+1}$, and $r^k := z^{k+1} - z^k$. Then from (4.6) and (4.13), we have $p^k = \frac{1}{\lambda_k}(r^k - \lambda_k e^k)$. Also introduce $q^k := Tz^{k+1} - Tz^k$. Then, $p^{k+1} - p^k = q^k - r^k$.

We will show: (i) $\|p^{k+1}\|^2 \leq \|p^k\|^2 + \frac{\lambda_k^2}{\tau_k}\|e^k\|^2$ and (ii) $\sum_{i=0}^{\infty} \tau_i \|p^i\|^2 < \infty$. Then, applying Part 2 of Lemma 1 (with $a_k = \|p^k\|^2$, $e_k = \frac{\lambda_k^2}{\tau_k}\|e^k\|^2$, and $\lambda_k = 1$ for which we have $\Lambda_k = \sum_{i=0}^{k} \lambda_i = (k+1)$) and noticing that $\tau_k \geq \varepsilon$ uniformly, we obtain the rate $\|Tz^k - z^k\|^2 = o(1/(k+1))$.

Part (i): We have

$$\begin{aligned}\|p^{k+1}\|^2 &= \|p^k\|^2 + \|p^{k+1} - p^k\|^2 + 2\langle p^{k+1} - p^k, p^k\rangle \\ &= \|p^k\|^2 + \|q^k - r^k\|^2 + \frac{2}{\lambda_k}\langle q^k - r^k, r^k - \lambda_k e^k\rangle.\end{aligned}$$

By the nonexpansiveness of T, we have $\|q^k\|^2 \leq \|r^k\|^2$ and thus

$$2\langle q^k - r^k, r^k\rangle = \|q^k\|^2 - \|r^k\|^2 - \|q^k - r^k\|^2 \leq -\|q^k - r^k\|^2.$$

Therefore,

$$\begin{aligned}\|p^{k+1}\|^2 &\leq \|p^k\|^2 - \frac{1-\lambda_k}{\lambda_k}\|q^k - r^k\|^2 - 2\langle q^k - r^k, e^k\rangle. \\ &= \|p^k\|^2 - \frac{1-\lambda_k}{\lambda_k}\left\|q^k - r^k + \frac{\lambda^k}{1-\lambda_k}e^k\right\|^2 + \frac{\lambda^k}{1-\lambda_k}\|e^k\|^2 \\ &\leq \|p^k\|^2 + \frac{\lambda_k^2}{\tau_k}\|e^k\|^2.\end{aligned}$$

Part (ii): First, $\|z^k - z^*\|$ is uniformly bounded because $\|z^{k+1} - z^*\| \leq (1 - \lambda_k)\|z^k - z^*\| + \lambda_k\|Tz^k - z^*\| + \lambda_k\|e^k\| \leq \|z^k - z^*\| + \lambda_k\|e^k\|$ by the triangle inequality and the nonexpansiveness of T. From [2, Corollary 2.14], we have

$$\begin{aligned}\|z^{k+1} - z^*\|^2 &= \|(1-\lambda_k)(z^k - z^*) + \lambda_k(Tz^k - z^* + e^k)\|^2 \\ &= (1-\lambda_k)\|z^k - z^*\|^2 + \lambda_k\|Tz^k - z^* + e^k\|^2 - \lambda_k(1-\lambda_k)\|p^k + e^k\|^2 \\ &= (1-\lambda_k)\|z^k - z^*\|^2 + \lambda_k\left(\|Tz^k - z^*\|^2 + 2\lambda_k\langle Tz^k - z^*, e^k\rangle + \lambda_k\|e^k\|^2\right) \\ &\quad - \lambda_k(1-\lambda_k)\left(\|p^k\|^2 + 2\langle p^k, e^k\rangle + \|e^k\|^2\right) \\ &\leq \|z^k - z^*\|^2 - \tau_k\|p^k\|^2 + \underbrace{\lambda_k^2\|e^k\|^2 + 2\lambda_k\|Tz^k - z^*\|\|e^k\| + 2\tau_k\|p^k\|\|e^k\|}_{=:\xi_k}.\end{aligned}$$

Because we have shown (a) $\|Tz^k - z^*\|$ and $\|p^k\|$ are bounded, (b) $\sum_{i=0}^{\infty} \tau_i\|e^i\| \leq \sum_{i=0}^{\infty} \lambda_i\|e^i\| < \infty$, and (c) $\sum_{i=0}^{\infty} \lambda_i^2\|e^i\|^2 < \infty$, we have $\sum_{i=0}^{\infty} \xi_k < \infty$ and thus $\sum_{i=0}^{\infty} \tau_i\|Tz^i - z^i\|^2 \leq \|z^0 - z^*\|^2 + \sum_{i=0}^{\infty} \xi_k < \infty$. □

3.1.1 Notes on Theorem 1

The FPR, $\|Tz^k - z^k\|^2$, is a normalization of the successive iterate difference $z^{k+1} - z^k = \lambda_k(Tz^k - z^k)$. Thus, the convergence rates of $\|Tz^k - z^k\|^2$ naturally imply convergence rates of $\|z^{k+1} - z^k\|^2$.

Note that $o(1/(k+1))$ is the optimal convergence rate for the class of nonexpansive operators [12, Remarque 4]. In the special case that $T = \mathbf{prox}_{\gamma f}$ for some closed, proper, and convex function f, the rate of $\|Tz^k - z^k\|^2$ improves to $O(1/(k+1)^2)$ [12, Théorème 9]. See Section 6 for more optimality results. Also, the (error free) little-o convergence rate of the fixed-point residual associated with the resolvent of a maximal monotone linear operator was shown in [12, Proposition 4]. Finally, we mention the parallel work [24], which proves a similar little-o convergence rate for the fixed-point residual of the relaxed proximal point algorithm.

In general, it is possible that the nonexpansive operator, $T : \mathcal{H} \to \mathcal{H}$, is already averaged, i.e., there exists a nonexpansive operator $N : \mathcal{H} \to \mathcal{H}$ and a positive constant $\alpha \in (0,1]$ such that $T = (1-\alpha)I_{\mathcal{H}} + \alpha N$, where T and N share the same fixed point set. Thus, we can apply Theorem 1 to $N = (1-(1/\alpha))I_{\mathcal{H}} + (1/\alpha)T$. Furthermore, $N_\lambda = (1-\lambda/\alpha)I_{\mathcal{H}} + (\lambda/\alpha)T$. Thus, when we translate this back to an iteration on T, it enlarges the region of relaxation parameters to $\lambda_k \in (0, 1/\alpha)$ and modifies τ_k accordingly to $\tau_k = \lambda_k(1-\alpha\lambda_k)/\alpha$. The same convergence results continue to hold.

To the best of our knowledge, the little-o rates produced in Theorem 1 have never been established for the KM iteration. See [22, 38] for similar big-O results. Note that our rate in Part 5 is strictly better than the one shown in [38], and it is given under a much weaker condition on the error because [38] proves an $O(1/(k+1))$ convergence rate only when $\sum_{i=0}^{\infty}(i+1)\|e^i\| < \infty$, which implies that $\min_{i=0,\cdots,k}\{\|e^i\|\} = o(1/(k+1)^2)$ by Lemma 1. In contrast, any error sequence of the form $\|e^k\| = O(1/(k+1)^\alpha)$ with $\alpha > 1$ will satisfy Part 5 of our Theorem 1. Finally, for Banach spaces, we cannot improve big-O rates to little-o [22, Section 2.4].

3.2 $o(1/(k+1))$ *FPR of Relaxed PRS*

In this section, we apply Theorem 1 to the T_{PRS} operator defined in (4.5). For the special case of DRS ((1/2)-averaged PRS), it is straightforward to establish the rate of the FPR

$$\|(T_{\mathrm{PRS}})_{1/2}z^k - z^k\|^2 = O\left(\frac{1}{k+1}\right)$$

from two existing results: (i) the DRS iteration is a proximal iteration applied to a certain monotone operator [29, Section 4]; (ii) the convergence rate of the FPR for proximal iterations is $O(1/(k+1))$ [12, Proposition 8] whenever a fixed point exists. We improve this rate to $o(1/(k+1))$ for general relaxed PRS operators.

Corollary 1 (Convergence Rate of Relaxed PRS). *Let z^* be a fixed point of T_{PRS}, let $(\lambda_j)_{j\geq 0} \subseteq (0,1]$ be a sequence of positive numbers, let $\tau_k := \lambda_k(1-\lambda_k)$ for all $k \geq 0$, and let $z^0 \in \mathcal{H}$. Suppose that $(z^j)_{j\geq 0} \subseteq \mathcal{H}$ is generated by Algorithm 1. Then the sequence $\|z^k - z^*\|^2$ is monotonically nonincreasing,and if $\underline{\tau} := \inf_{j\geq 0} \tau_j > 0$, then the following convergence rates hold:*

$$\|T_{\mathrm{PRS}} z^k - z^k\|^2 \leq \frac{\|z^0 - z^*\|^2}{\underline{\tau}(k+1)} \quad \text{and} \quad \|T_{\mathrm{PRS}} z^k - z^k\|^2 = o\left(\frac{1}{k+1}\right). \tag{4.14}$$

3.3 $O(1/\Lambda_k^2)$ Ergodic FPR of Fejér Monotone Sequences

The following definition is often used in the analysis of optimization algorithms [17].

Definition 1. A sequence $(z^j)_{j\geq 0} \subseteq \mathcal{H}$ is *Fejér monotone* with respect to a nonempty set $C \subseteq \mathcal{H}$ if for all $z \in C$, we have $\|z^{k+1} - z\|^2 \leq \|z^k - z\|^2$.

The following fact is trivial, but useful.

Theorem 2. *Let $(z^j)_{j\geq 0}$ be a Fejér monotone sequence with respect to a nonempty set $C \subseteq \mathcal{H}$. Suppose that for all $k \geq 0$, $z^{k+1} - z^k = \lambda_k(x^k - y^k)$ for a sequence $((x^j, y^j))_{j\geq 0} \subseteq \mathcal{H}^2$ and a sequence of positive real numbers $(\lambda_j)_{j\geq 0}$. For all $k \geq 0$, let $\overline{z}^k := (1/\Lambda_k)\sum_{i=0}^k \lambda_i z^i$, let $\overline{x}^k := (1/\Lambda_k)\sum_{i=0}^k \lambda_i x^i$, and let $\overline{y}^k := (1/\Lambda_k)\sum_{i=0}^k \lambda_i y^i$. Then the following bound holds for all $z \in C$:*

$$\|\overline{x}^k - \overline{y}^k\|^2 \leq \frac{4\|z^0 - z\|^2}{\Lambda_k^2}.$$

Proof. $\Lambda_k \|\overline{x}^k - \overline{y}^k\| = \left\|\sum_{i=0}^k \left(z^{k+1} - z^k\right)\right\| = \left\|z^{k+1} - z^0\right\| \leq 2\left\|z^0 - z\right\|.$ □

Part 1 of Theorem 1 shows that any sequence $(z^j)_{j\geq 0}$ generated by Algorithm 3 is Fejér monotone with respect to the set of fixed-points of T. Therefore, Theorem 2 applies to iteration Equation (4.10) with the choice $x^k = Tz^k$ and $y^k = z^k$ for all $k \geq 0$.

The interested reader can proceed to Section 6 for several examples that show the optimality of the rates predicted in this section.

4 Subgradients and Fundamental Inequalities

This section establishes fundamental inequalities that connect the FPR in Section 3 to the *objective error* of the relaxed PRS algorithm.

In first-order optimization algorithms, we only have access to (sub)gradients and function values. Consequently, the FPR is usually a linear combination of (sub)gradients. In simple first-order algorithms, like the (sub)gradient method, a (sub)gradient is drawn from a single point at each iteration. In splitting algorithms for minimizing sums of convex functions, each function draws at subgradient at a different point. There is no natural point at which we can evaluate the entire objective function; this complicates the analysis of the relaxed PRS algorithm.

In the relaxed PRS algorithm, the two operators $\mathbf{refl}_{\gamma f}$ and $\mathbf{refl}_{\gamma g}$ are calculated one after another at different points, neither of which equals z^k or z^{k+1}. Consequently, the expression $z^k - z^{k+1}$ is more complicated, and the analysis for the standard (sub)gradient iteration does not carry through.

We **let x_f and x_g be the points where subgradients of f and g are drawn, respectively**, and use Figure 4.1 to prove algebraic relations among points z, x_f, and x_g. We use these relations many times. Propositions 2 and 3 use these algebraic relations to bound the objective error by the FPR. In these bounds, the objective errors of f and g are measured at two points x_f and x_g such that $x_f \neq x_g$. Later we will assume that one of the objectives is Lipschitz continuous and evaluate both functions at the same point (See Corollaries 2 and 3).

We conclude this introduction by combining the subgradient notation in Equation (4.4) and Part 1 of Proposition 1 to arrive at the expressions

$$\begin{aligned} &\mathbf{prox}_{\gamma f}(x) = x - \gamma \widetilde{\nabla} f(\mathbf{prox}_{\gamma f}(x)) \\ \text{and} \quad &\mathbf{refl}_{\gamma f}(x) = x - 2\gamma \widetilde{\nabla} f(\mathbf{prox}_{\gamma f}(x)). \end{aligned} \tag{4.15}$$

4.1 A Subgradient Representation of Relaxed PRS

In this section we write the relaxed PRS algorithm in terms of subgradients. Lemma 2, Table 4.1, and Figure 4.1 summarize a single iteration of relaxed PRS.

Lemma 2. *Let $z \in \mathscr{H}$. Define points $x_g := \mathbf{prox}_{\gamma g}(z)$ and $x_f := \mathbf{prox}_{\gamma f}(\mathbf{refl}_{\gamma g}(z))$. Then the identities hold:*

$$x_g = z - \gamma \widetilde{\nabla} g(x_g) \quad \text{and} \quad x_f = x_g - \gamma \widetilde{\nabla} g(x_g) - \gamma \widetilde{\nabla} f(x_f).$$

where $\widetilde{\nabla} g(x_g) := (1/\gamma)(z - x_g) \in \partial g(x_g)$ and $\widetilde{\nabla} f(x_f) := (1/\gamma)(2x_g - z - x_f) \in \partial f(x_f)$. In addition, each relaxed PRS step has the following representation:

$$(T_{\mathrm{PRS}})_\lambda(z) - z = 2\lambda(x_f - x_g) = -2\lambda\gamma(\widetilde{\nabla} g(x_g) + \widetilde{\nabla} f(x_f)). \tag{4.16}$$

Proof. Figure 4.1 provides an illustration. Equation (2) follows from $\mathbf{refl}_{\gamma g}(z) = 2x_g - z = x_g - \gamma \widetilde{\nabla} g(x_g)$ and Equation (4.15). Now, we can compute' $T_{\mathrm{PRS}}(z) - z$:

$$\begin{aligned} T_{\mathrm{PRS}}(z) - z &\overset{(4.7)}{=} \mathbf{refl}_{\gamma f}(\mathbf{refl}_{\gamma g}(z)) - z = 2x_f - \mathbf{refl}_{\gamma g}(z) - z \\ &= 2x_f - (2x_g - z) - z = 2(x_f - x_g). \end{aligned}$$

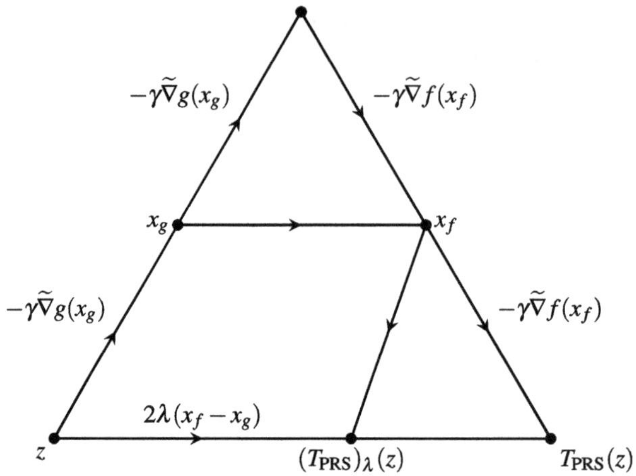

Fig. 4.1 A single relaxed PRS iteration, from z to $(T_{\mathrm{PRS}})_\lambda(z)$.

The subgradient identity in (4.16) follows from (2). Finally, Equation (4.16) follows from $(T_{\mathrm{PRS}})_\lambda(z) - z = (1-\lambda)z + \lambda T_{\mathrm{PRS}}(z) - z = \lambda(T_{\mathrm{PRS}}(z) - z)$. □

Point	Operator identity	Subgradient identity
x_g^s	$= \mathbf{prox}_{\gamma g}(z^s)$	$= z^s - \gamma\widetilde{\nabla}g(x_g^s)$
x_f^s	$= \mathbf{prox}_{\gamma f}(\mathbf{refl}_{\gamma g}(z^s))$	$= x_g^s - \gamma(\widetilde{\nabla}g(x_g^s) + \widetilde{\nabla}f(x_f^s))$
$(T_{\mathrm{PRS}})_\lambda(z^s)$	$= (1-\lambda)z^s + \lambda T_{\mathrm{PRS}}(z^s)$	$= z^s - 2\gamma\lambda(\widetilde{\nabla}g(x_g^s) + \widetilde{\nabla}f(x_f^s))$

Table 4.1 Overview of the main identities used throughout the chapter. The letter s denotes a superscript (e.g., $s = k$ or $s = *$). See Lemma 2 for a proof.

4.2 Optimality Conditions of Relaxed PRS

The following lemma characterizes the zeros of $\partial f + \partial g$ in terms of the fixed points of the PRS operator.

Lemma 3 (Optimality Conditions of T_{PRS}). *The following identity holds:*

$$\mathrm{zer}(\partial f + \partial g) = \{\mathbf{prox}_{\gamma g}(z) \mid z \in \mathcal{H}, T_{\mathrm{PRS}}z = z\}. \tag{4.17}$$

That is, if z^* is a fixed point of T_{PRS}, then $x^* = x_g^* = x_f^*$ solves Problem 1 and

$$\widetilde{\nabla}g(x^*) := \frac{1}{\gamma}(z^* - x^*) \in \partial g(x^*). \tag{4.18}$$

Proof. See [2, Proposition 25.1] for the proof of Equation (4.17). Equation (4.18) follows because $x^* = \mathbf{prox}_{\gamma g}(z^*)$ if, and only if, $z^* - x^* \in \gamma\partial g(x^*)$. □

4.3 Fundamental Inequalities

We now compute upper and lower bounds of the quantities $f(x_f^k)+g(x_g^k)-g(x^*)-f(x^*)$. Note that x_f^k and x_g^k are not necessarily equal, so this quantity can be negative.

The most important properties of the inequalities we establish below are:

1. The upper fundamental inequality has a telescoping structure in z^k and z^{k+1}.
2. They can be bounded in terms of $\|z^{k+1}-z^k\|^2$.

Properties 1 and 2 will be used to deduce ergodic and nonergodic rates, respectively.

Proposition 2 (Upper Fundamental Inequality). *Let $z \in \mathcal{H}$, let $z^+ := (T_{\mathrm{PRS}})_\lambda(z)$, and let x_f and x_g be defined as in Lemma 2. Then for all $x \in \mathrm{dom}(f) \cap \mathrm{dom}(g)$*

$$\begin{aligned} 4\gamma\lambda(f(x_f)&+g(x_g)-f(x)-g(x)) \\ &\leq \|z-x\|^2-\|z^+-x\|^2+\left(1-\frac{1}{\lambda}\right)\|z^+-z\|^2. \end{aligned}$$

Proof. We use the subgradient inequality and (4.16) in the following derivation:

$$\begin{aligned} 4\gamma\lambda(f(x_f)+g(x_g)&-f(x)-g(x)) \\ &\leq 4\lambda\gamma\left(\langle x_f-x, \widetilde{\nabla} f(x_f)\rangle + \langle x_g-x, \widetilde{\nabla} g(x_g)\rangle\right) \\ &= 4\lambda\gamma\left(\langle x_f-x_g, \widetilde{\nabla} f(x_f)\rangle + \langle x_g-x, \widetilde{\nabla} f(x_f)+\widetilde{\nabla} g(x_g)\rangle\right) \\ &= 2\left(\langle z^+-z, \gamma\widetilde{\nabla} f(x_f)\rangle + \langle z^+-z, x-x_g\rangle\right) \\ (\because -x_g = \gamma\widetilde{\nabla} g(x_g)-z) \quad &= 2\langle z^+-z, x+\gamma(\widetilde{\nabla} g(x_g)+\widetilde{\nabla} f(x_f))-z\rangle \\ &= 2\langle z^+-z, x-\frac{1}{2\lambda}(z^+-z)-z\rangle \\ &= \|z-x\|^2-\|z^+-x\|^2+\left(1-\frac{1}{\lambda}\right)\|z^+-z\|^2. \end{aligned}$$

Proposition 3 (Lower Fundamental Inequality). *Let z^* be a fixed point of T_{PRS} and $x^* := \mathbf{prox}_{\gamma g}(z^*)$. For all $x_f \in \mathrm{dom}(f)$ and $x_g \in \mathrm{dom}(g)$, the following bound holds:*

$$f(x_f)+g(x_g)-f(x^*)-g(x^*) \geq \frac{1}{\gamma}\langle x_g-x_f, z^*-x^*\rangle. \tag{4.19}$$

Proof. Let $\widetilde{\nabla} g(x^*) = (z^*-x^*)/\gamma \in \partial g(x^*)$ and let $\widetilde{\nabla} f(x^*) = -\widetilde{\nabla} g(x^*) \in \partial f(x^*)$. Then the result follows by adding $f(x_f)-f(x^*) \geq \langle x_f-x^*, \widetilde{\nabla} f(x^*)\rangle$ and $g(x_g)-g(x^*) \geq \langle x_g-x_f, \widetilde{\nabla} g(x^*)\rangle + \langle x_f-x^*, \widetilde{\nabla} g(x^*)\rangle$. □

5 Objective Convergence Rates

In this section we will prove ergodic and nonergodic convergence rates of relaxed PRS when f and g are closed, proper, and convex functions that are possibly nonsmooth. For concise notation, we let

$$\boxed{h(x,y) := f(x) + g(y) - f(x^*) - g(x^*).} \tag{4.20}$$

To ease notational memory, the reader may assume that $\lambda_k = (1/2)$ for all $k \geq 0$, which implies that $\Lambda_k = (1/2)(k+1)$, and $\tau_k = \lambda_k(1-\lambda_k) = (1/4)$ for all $k \geq 0$.

Throughout this section the point z^* denotes an arbitrary fixed point of T_{PRS}, and we define a minimizer of $f+g$ by the formula (Lemma 3):

$$x^* = \mathbf{prox}_{\gamma g}(z^*).$$

The constant $(1/\gamma)\|z^* - x^*\|$ appears in the bounds of this section. This term is independent of γ: For any fixed point z^* of T_{PRS}, the point $x^* = \mathbf{prox}_{\gamma g}(z^*)$ is a minimizer and $z^* - \mathbf{prox}_{\gamma g}(z^*) = \gamma\widetilde{\nabla} g(x^*) \in \gamma\partial g(x^*)$. Conversely, if $x^* \in \operatorname{zer}(\partial f + \partial g)$ and $\widetilde{\nabla} g(x^*) \in (-\partial f(x^*)) \cap \partial g(x^*)$, then $z^* = x^* + \gamma\widetilde{\nabla} g(x^*)$ is a fixed point. Note that in all of our bounds, we can always replace $(1/\gamma)\|z^* - x^*\| = \|\widetilde{\nabla} g(x^*)\|$ by the infimum $\inf_{z^* \in \mathrm{Fix}(T_{\mathrm{PRS}})}(1/\gamma)\|z^* - \mathbf{prox}_{\gamma g}(z^*)\|$ (the infimum might not be attained).

5.1 Ergodic Convergence Rates

In this section, we analyze the ergodic convergence of relaxed PRS. The proof follows the telescoping property of the upper and lower fundamental inequalities and an application of Jensen's inequality.

Theorem 3 (Ergodic Convergence of Relaxed PRS). *For all $k \geq 0$, let $\lambda_k \in (0,1]$. Then we have the following convergence rate*

$$-\frac{2}{\gamma\Lambda_k}\|z^0 - z^*\|\|z^* - x^*\| \leq h(\overline{x}_f^k, \overline{x}_g^k) \leq \frac{1}{4\gamma\Lambda_k}\|z^0 - x^*\|^2.$$

In addition, the following feasibility bound holds:

$$\|\overline{x}_g^k - \overline{x}_f^k\| \leq \frac{2}{\Lambda_k}\|z^0 - z^*\|. \tag{4.21}$$

Proof. Equation (4.21) follows directly from Theorem 2 because $(z^j)_{j\geq 0}$ is Fejér monotone with respect to $\mathrm{Fix}(T)$ and for all $k \geq 0$, we have $z^{k+1} - z^k = \lambda_k(x_f^k - x_g^k)$.

Recall the upper fundamental inequality from Proposition 2:

$$4\gamma\lambda_k h(x_f^k, x_g^k) \leq \|z^k - x^*\|^2 - \|z^{k+1} - x^*\|^2 + \left(1 - \frac{1}{\lambda_k}\right)\|z^{k+1} - z^k\|^2.$$

Because $\lambda_k \leq 1$, it follows that $(1-(1/\lambda_k)) \leq 0$. Thus, we sum Equation (5.1) from $i=0$ to k, divide by Λ_k, and apply Jensen's inequality to get

$$\frac{1}{4\gamma\Lambda_k}(\|z^0 - x^*\|^2 - \|z^{k+1} - x^*\|^2) \geq \frac{1}{\Lambda_k}\sum_{i=0}^{k}\lambda_i h(x_f^i, x_g^i) \geq h(\overline{x}_f^k, \overline{x}_g^k).$$

The lower bound is a consequence of the fundamental lower inequality and (4.21)

$$h(\overline{x}_f^k, \overline{x}_g^k) \overset{(4.19)}{\geq} \frac{1}{\gamma}\langle \overline{x}_g^k - \overline{x}_f^k, z^* - x^*\rangle \geq -\frac{2}{\gamma\Lambda_k}\|z^0 - z^*\|\|z^* - x^*\|. \qquad \square$$

In general, $x_f^k \notin \text{dom}(g)$ and $x_g^k \notin \text{dom}(f)$, so we cannot evaluate g at x_f^k or f at x_g^k. But the conclusion of Theorem 3 is improved if f or g is Lipschitz continuous. The following proposition is a sufficient condition for Lipschitz continuity on a ball:

Proposition 4 (Lipschitz Continuity on a Ball). *Suppose that $f : \mathcal{H} \to (-\infty, \infty]$ is proper and convex. Let $\rho > 0$ and let $x_0 \in \mathcal{H}$. If $\delta = \sup_{x,y \in B(x_0, 2\rho)} |f(x) - f(y)| < \infty$, then f is (δ/ρ)-Lipschitz on $B(x_0, \rho)$.*

Proof. See [2, Proposition 8.28]. $\square$

To use this fact, we need to show that the sequences $(x_f^j)_{j\geq 0}$, and $(x_g^j)_{j\geq 0}$ are bounded. Recall that $x_g^s = \mathbf{prox}_{\gamma g}(z^s)$ and $x_f^s = \mathbf{prox}_{\gamma f}(\mathbf{refl}_{\gamma g}(z^s))$, for $s \in \{*, k\}$ (recall that $*$ is used for quantities associated with a fixed point). Proximal and reflection maps are nonexpansive, so we have the following bound:

$$\max\{\|x_f^k - x^*\|, \|x_g^k - x^*\|\} \leq \|z^k - z^*\| \leq \|z^0 - z^*\|.$$

Thus, $(x_f^j)_{j\geq 0}, (x_g^j)_{j\geq 0} \subseteq \overline{B(x^*, \|z^0 - z^*\|)}$. By the convexity of the closed ball, we also have $(\overline{x}_f^j)_{j\geq 0}, (\overline{x}_g^j)_{j\geq 0} \subseteq \overline{B(x^*, \|z^0 - z^*\|)}$.

Corollary 2 (Ergodic Convergence with Single Lipschitz Function). *Let the notation be as in Theorem 3. Suppose that f (respectively g) is L-Lipschitz continuous on $\overline{B(x^*, \|z^0 - z^*\|)}$, and let $x^k = x_g^k$ (respectively $x^k = x_f^k$). Then the following convergence rate holds*

$$0 \leq h(\overline{x}^k, \overline{x}^k) \leq \frac{1}{4\gamma\Lambda_k}\|z^0 - x^*\|^2 + \frac{2L}{\Lambda_k}\|z^0 - z^*\|.$$

Proof. From Equation (4.21), we have $\|\overline{x}_g^k - \overline{x}_f^k\| \leq (2/\Lambda_k)\|z^0 - z^*\|$. In addition, $(x_f^j)_{j\geq 0}, (x_g^j)_{j\geq 0} \subseteq \overline{B(x^*, \|z^0 - z^*\|)}$. Thus, it follows that

$$0 \leq h(\overline{x}^k, \overline{x}^k) \leq h(\overline{x}_f^k, \overline{x}_g^k) + L\|\overline{x}_f^k - \overline{x}_g^k\| \overset{(4.21)}{\leq} h(\overline{x}_f^k, \overline{x}_g^k) + \frac{2L}{\Lambda_k}\|z^0 - z^*\|.$$

The upper bound follows from this equation and Theorem 3. $\square$

5.2 Nonergodic Convergence Rates

In this section, we prove the nonergodic convergence rate of Algorithm 1 whenever $\underline{\tau} := \inf_{j\geq 0} \tau_j > 0$. The proof uses Theorem 1 to bound the fundamental inequalities in Propositions 2 and 3.

Theorem 4 (Nonergodic Convergence of Relaxed PRS). *For all $k \geq 0$, let $\lambda_k \in (0,1)$. Suppose that $\underline{\tau} := \inf_{j\geq 0} \lambda_k(1-\lambda_k) > 0$. Recall that the function h is defined in* (4.20)*. Then we have the convergence rates:*

1. In general, we have the bounds:

$$-\frac{\|z^0 - z^*\|\|z^* - x^*\|}{2\gamma\sqrt{\underline{\tau}(k+1)}} \leq h(x_f^k, x_g^k) \leq \frac{(\|z^0 - z^*\| + \|z^* - x^*\|)\|z^0 - z^*\|}{2\gamma\sqrt{\underline{\tau}(k+1)}}$$

and $|h(x_f^k, x_g^k)| = o\left(1/\sqrt{k+1}\right)$.

2. If $\mathscr{H} = \mathbf{R}$ and $\lambda_k \equiv 1/2$, then for all $k \geq 0$,

$$\frac{\|z^0 - z^*\|\|z^* - x^*\|}{\sqrt{2}\gamma(k+1)} \leq h(x_f^{k+1}, x_g^{k+1}) \leq \frac{(\|z^0 - z^*\| + \|z^* - x^*\|)\|z^0 - z^*\|}{\sqrt{2}\gamma(k+1)}$$

and $|h(x_f^{k+1}, x_g^{k+1})| = o\left(1/(k+1)\right)$.

Proof. We prove Part 1 first. For all $\lambda \in [0,1]$, let $z_\lambda = (T_{\mathrm{PRS}})_\lambda(z^k)$. Evaluate the upper inequality in Equation (2) at $x = x^*$ to get

$$4\gamma\lambda h(x_f^k, x_g^k) \leq \|z^k - x^*\|^2 - \|z_\lambda - x^*\|^2 + \left(1 - \frac{1}{\lambda}\right)\|z_\lambda - z^k\|^2.$$

Recall the following identity:

$$\|z^k - x^*\|^2 - \|z_\lambda - x^*\|^2 - \|z_\lambda - z^k\|^2 = 2\langle z_\lambda - x^*, z^k - z_\lambda\rangle.$$

By the triangle inequality, because $\|z_\lambda - z^*\| \leq \|z^k - z^*\|$, and because $(\|z^j - z^*\|)_{j\geq 0}$ is monotonically nonincreasing (Corollary 1), it follows that

$$\|z_\lambda - x^*\| \leq \|z_\lambda - z^*\| + \|z^* - x^*\| \leq \|z^0 - z^*\| + \|z^* - x^*\|. \tag{4.22}$$

Thus, we have the bound:

$$\begin{aligned}
h(x_f^k, x_g^k) &\leq \inf_{\lambda\in[0,1]} \frac{1}{4\gamma\lambda}\left(2\langle z_\lambda - x^*, z^k - z_\lambda\rangle + 2\left(1 - \frac{1}{2\lambda}\right)\|z_\lambda - z^k\|^2\right) \\
&\leq \frac{1}{\gamma}\|z_{1/2} - x^*\|\|z^k - z_{1/2}\| \\
&\overset{(4.22)}{\leq} \frac{1}{\gamma}\left(\|z^0 - z^*\| + \|z^* - x^*\|\right)\|z^k - z_{1/2}\| \\
&\overset{(4.14)}{\leq} \frac{(\|z^0 - z^*\| + \|z^* - x^*\|)\|z^0 - z^*\|}{2\gamma\sqrt{\underline{\tau}(k+1)}}.
\end{aligned}$$

The lower bound follows from the identity $x_g^k - x_f^k = (1/2\lambda_k)(z^k - z^{k+1})$ and the fundamental lower inequality in Equation (4.19):

$$h(x_f^k, x_g^k) \geq \frac{1}{2\gamma\lambda_k}\langle z^k - z^{k+1}, z^* - x^*\rangle \geq -\frac{\|z^{k+1} - z^k\|\|z^* - x^*\|}{2\gamma\lambda_k}$$
$$\overset{(4.14)}{\geq} -\frac{\|z^0 - z^*\|\|z^* - x^*\|}{2\gamma\sqrt{\underline{\tau}(k+1)}}.$$

Finally, the $o(1/\sqrt{k+1})$ convergence rate follows from Equations (5.2) and (5.2) combined with Corollary 1 because each upper bound is of the form (bounded quantity)$\times\sqrt{\mathrm{FPR}}$, and $\sqrt{\mathrm{FPR}}$ has rate $o(1/\sqrt{k+1})$.

Part 2 follows by the same analysis but uses Theorem 12 in Appendix (Page 150) to estimate the FPR convergence rate. □

Whenever f or g is Lipschitz, we can compute the convergence rate of $f+g$ evaluated at the same point. The next theorem is similar to Corollary 2 in the ergodic case. The proof is a combination of the nonergodic convergence rate in Theorem 4 and the convergence rate of $\|x_f^k - x_g^k\| = (1/\lambda_k)\|z^{k+1} - z^k\|$ shown in Corollary 1.

Corollary 3 (Nonergodic Convergence with Lipschitz Assumption). *Let the notation be as in Theorem 4. Suppose that f (respectively g) is L-Lipschitz continuous on $\overline{B(x^*, \|z^0 - z^*\|)}$, and let $x^k = x_g^k$ (respectively $x^k = x_f^k$). Recall that the objective-error function h is defined in* (4.20)*. Then we have the convergence rates of the nonnegative term:*

1. *In general, we have the bounds:*

$$0 \leq h(x^k, x^k) \leq \frac{\left(\|z^0 - z^*\| + \|z^* - x^*\| + \gamma L\right)\|z^0 - z^*\|}{2\gamma\sqrt{\underline{\tau}(k+1)}}$$

and $h(x^k, x^k) = o\left(1/\sqrt{k+1}\right)$.

2. *If $\mathcal{H} = \mathbf{R}$ and $\lambda_k \equiv 1/2$, then for all $k \geq 0$,*

$$0 \leq h(x^{k+1}, x^{k+1}) \leq \frac{\left(\|z^0 - z^*\| + \|z^* - x^*\| + \gamma L\right)\|z^0 - z^*\|}{\sqrt{2}\gamma(k+1)}$$

and $h(x^{k+1}, x^{k+1}) = o\left(1/(k+1)\right)$.

Proof. We prove Part 1 first. Recall that $\|x_g^k - x_f^k\| = (1/(2\lambda_k))\|z^{k+1} - z^k\|$. In addition, $(x_f^j)_{j\geq 0}, (x_g^j)_{j\geq 0} \subseteq \overline{B(x^*, \|z^0 - z^*\|)}$ (See Section 5.1). Thus, it follows that

$$h(x^k, x^k) \leq h(x_f^k, x_g^k) + L\|x_f^k - x_g^k\| = h(x_f^k, x_g^k) + \frac{L\|z^{k+1} - z^k\|}{2\lambda_k}$$
$$\overset{(4.14)}{\leq} h(x_f^k, x_g^k) + \frac{L\|z^0 - z^*\|}{2\sqrt{\underline{\tau}(k+1)}}.$$

Therefore, the upper bound follows from Theorem 4 and Equation (5.2). In addition, the $o(1/\sqrt{k+1})$ bound follows from Theorem 4 combined with Equation (5.2) and Corollary 1 because each upper bound is of the form (bounded quantity)$\times\sqrt{\text{FPR}}$, and $\sqrt{\text{FPR}}$ has rate $o(1/\sqrt{k+1})$.

Part 2 follows by the same analysis, but uses Theorem 12 in Appendix (Page 150) to estimate the FPR convergence rate. □

6 Optimal FPR Rate and Arbitrarily Slow Convergence

In this section, we provide two examples in which the DRS algorithm converges slowly. Both examples are special cases of the following example, which originally appeared in [1, Section 7].

Example 1 (DRS Applied to Two Subspaces). Let $\mathcal{H} = \ell_2^2(\mathbf{N}) = \{(z_j)_{j\geq 0} \mid \forall j \in \mathbf{N}, z_j \in \mathbf{R}^2, \sum_{i=0}^{\infty} \|z_j\|_{\mathbf{R}^2}^2 < \infty\}$. Let R_θ denote counterclockwise rotation in $\mathbf{R}^2$ by θ radians. Let $e_0 := (1,0)$ denote the standard unit vector, and let $e_\theta := R_\theta e_0$. Suppose that $(\theta_j)_{j\geq 0}$ is a sequence in $(0,\pi/2]$ and $\theta_i \to 0$ as $i \to \infty$. We define two subspaces:

$$U := \bigoplus_{i=0}^{\infty} \mathbf{R}e_0 \quad \text{and} \quad V := \bigoplus_{i=0}^{\infty} \mathbf{R}e_{\theta_i}$$

where $\mathbf{R}e_0 = \{\alpha e_0 : \alpha \in \mathbf{R}\}$ and $\mathbf{R}e_{\theta_i} = \{\alpha e_{\theta_i} : \alpha \in \mathbf{R}\}$. Let $T := (T_{\mathrm{PRS}})_{1/2}$ be applied to $f = \iota_V$ and $g = \iota_U$. The next identities and properties were shown in [1, Section 7]:

$$(P_U)_i = \begin{bmatrix} 1 & 0 \\ 0 & 0 \end{bmatrix} \quad \text{and} \quad (P_V)_i = \begin{bmatrix} \cos^2(\theta_i) & \sin(\theta_i)\cos(\theta_i) \\ \sin(\theta_i)\cos(\theta_i) & \sin^2(\theta_i) \end{bmatrix};$$

$$T = c_0 R_{\theta_0} \oplus c_1 R_{\theta_1} \oplus \cdots;$$

and $(z^j)_{j\geq 0}$, recursively defined for all k by $z^{k+1} = Tz^k$, converges in norm to $z^* = 0$ for any initial point z^0. □

6.1 Optimal FPR Rates

The following theorem shows that the FPR estimates derived in Corollary 1 are tight.

Theorem 5 (Lower FPR Complexity of DRS). *There is a Hilbert space $\mathcal{H}$ and two closed subspaces U and V with zero intersection, $U \cap V = \{0\}$, such that for* *every $\alpha > 1/2$, there exists $z^0 \in \mathcal{H}$ such that if $(z^j)_{j\geq 0}$ is generated by $T = (T_{\mathrm{PRS}})_{1/2}$ applied to $f = \iota_V$ and $g = \iota_U$, then for all $k \geq 1$, we have the bound:*

$$\|Tz^k - z^k\|^2 \geq \frac{1}{(k+1)^{2\alpha}}.$$

Proof. We assume the setting of Example 1. For all $i \geq 0$ set $c_i = (i/(i+1))^{1/2}$. Then for all $i \geq 0$,

$$I_{\mathbf{R}^2} - \cos(\theta_i)R_{\theta_i} = \begin{bmatrix} \sin^2(\theta_i) & \sin(\theta_i)\cos(\theta_i) \\ -\sin(\theta_i)\cos(\theta_i) & \sin^2(\theta_i) \end{bmatrix} = \begin{bmatrix} \frac{1}{i+1} & \frac{\sqrt{i}}{i+1} \\ -\frac{\sqrt{i}}{i+1} & \frac{1}{i+1} \end{bmatrix}. \quad (4.23)$$

Therefore, the point $z^0 = (\sqrt{2\alpha e}((1/(j+1)^\alpha, 0))_{j\geq 0} \in \mathcal{H}$ has image

$$w^0 = (I-T)z^0 = \left(\sqrt{2\alpha e}\left(\frac{1}{(j+1)^{\alpha+1}}, \frac{-\sqrt{j}}{(j+1)^{\alpha+1}}\right)\right)_{j\geq 0}.$$

and for all $i \geq 1$, we have $\|w_i^0\|_{\mathbf{R}^2} = \sqrt{2\alpha e}(i+1)^{-(1+2\alpha)/2}$. Thus, for all $k \geq 1$,

$$\|Tz^k - z^k\|^2 = \|T^k w^0\|^2 = \sum_{i=0}^{\infty} c_i^{2k}\|w_i^0\|_{\mathbf{R}^2}^2 \geq \sum_{i=k}^{\infty} \frac{i^k}{(i+1)^k}\frac{2\alpha e}{(i+1)^{1+2\alpha}} \geq \frac{1}{(k+1)^{2\alpha}}. \quad \square$$

Remark 1. In the proof of Theorem 5, if $\alpha = 1/2$, then $\|z^0\| = \infty$.

6.1.1 Notes on Theorem 5

With this new optimality result in hand, we can make the following list of optimal FPR rates, not to be confused with optimal rates in objective error, for a few standard splitting schemes:

Proximal point algorithm (PPA): For the class of monotone operators, the counterexample in [12, Remarque 4] shows that there is a maximal monotone operator A such that when iteration (4.10) is applied to the resolvent $J_{\gamma A}$, the rate $o(1/(k+1))$ is tight. In addition, if $A = \partial f$ for some closed, proper, and convex function f, then the FPR rate improves to $O(1/(k+1)^2)$ [12, Théorème 9]. We improve this result to $o(1/(k+1)^2)$ in Theorem 12 in Appendix (Page 150). This result is new and is optimal by [12, Remarque 6].

Forward backward splitting (FBS): The FBS method reduces to the proximal point algorithm when the differentiable (or single valued operator) term is trivial. Thus, for the class of monotone operators, the $o(1/(k+1))$ FPR rate is optimal by [12, Remarque 4]. We improve this rate to $o(1/(k+1)^2)$ in Theorem 12 in Appendix (Page 150). This result is new, and is optimal by [12, Remarque 6].

Douglas-Rachford splitting/ADMM: Theorem 5 shows that the optimal FPR rate is $o(1/(k+1))$. Because the DRS iteration is equivalent to a proximal point algorithm (PPA) applied to a special monotone operator [29, Section 4], Theorem 5 provides an alternative counterexample to [12, Remarque 4]. In particular, Theorem 5 shows that, in general, there is no closed, proper, convex function f such that $(T_{\mathrm{PRS}})_{1/2} = \mathbf{prox}_{\gamma f}$. In the one-dimensional case, we improve the FPR to $o(1/(k+1)^2)$ in Theorem 13 in Appendix (Page 151).

Miscellaneous methods: By similar arguments we deduce the tight FPR iteration complexity for the following methods, each of which has rate at least $o(1/(k+1))$ by Theorem 1: *Standard Gradient descent* $o(1/(k+1)^2)$: (the rate follows from Theorem 12 in Appendix (Page 150). Optimality follows from the fact that PPA is equivalent to gradient descent on Moreau envelope [2, Proposition 12.29] and [12, Remarque 4]); *Forward-Douglas Rachford splitting* [13]: $o(1/(k+1))$ (choose the zero cocoercive operator and use Theorem 5); *Chambolle and Pock's primal-dual algorithm* [16] $o(1/(k+1))$: (reduce to DRS ($\sigma = \tau = 1$) [16, Section 4.2] and apply Theorem 5 using the transformation $z^k = \mathrm{primal}_k + \mathrm{dual}_k$ [16, Equation (24)] and the lower bound

$$\|z^{k+1} - z^k\|^2 \leq 2\|\mathrm{primal}_{k+1} - \mathrm{primal}_k\|^2 + 2\|\mathrm{dual}_{k+1} - \mathrm{dual}_k\|^2;$$

Vũ/Condat's primal-dual algorithm [53, 23] $o(1/(k+1))$: (extends from Chambolle and Pock's method [16]).

Note that the rate established in Theorem 1 has broad applicability, and this list is hardly extensive. For PPA, FBS, and standard gradient descent, the FPR always has rate that is the square of the objective value convergence rate. We will see that the same is true for DRS in Theorem 8.

6.2 Arbitrarily Slow Convergence

In [1, Section 7], DRS applied to Example 1 is shown to converge in norm, but not linearly. We improve this result and show that a proper choice of parameters yields arbitrarily slow convergence in norm.

The following technical lemma will help us construct a sequence that converges arbitrarily slowly. The proof idea follows from the proof of [30, Theorem 4.2], which shows that the method of alternating projections can converge arbitrarily slowly.

Lemma 4. *Suppose that $h : \mathbf{R}_+ \to (0,1)$ is a function that is strictly decreasing to zero such that $\{1/(j+1) \mid j \in \mathbf{N}\backslash\{0\}\} \subseteq \mathrm{range}(h)$. Then there exists a monotonic sequence $(c_j)_{j\geq 0} \subseteq (0,1)$ such that $c_k \to 1^-$ as $k \to \infty$ and an increasing sequence of integers $(n_j)_{j\geq 0} \subseteq \mathbf{N} \cup \{0\}$ such that for all $k \geq 0$,*

$$\frac{c_{n_k}^{k+1}}{n_k + 1} > h(k+1)e^{-1}.$$

Proof. Let h_2 be the inverse of the strictly increasing function $(1/h) - 1$, let $[x]$ denote the integer part of x, and for all $k \geq 0$, let

$$c_k = \frac{h_2(k+1)}{1 + h_2(k+1)}.$$

Note that because $\{1/(j+1) \mid j \in \mathbf{N}\backslash\{0\}\} \subseteq \text{range}(h)$, c_k is well defined. Indeed, $k+1 \in \text{dom}(h_2) \cap \mathbf{N}$ if, and only if, there is a $y \in \mathbf{R}_+$ such that $(1/h(y)) - 1 = k+1 \iff h(y) = 1/(k+2)$. It follows that $(c_j)_{j \geq 0}$ is monotonic and $c_k \to 1^-$.

For all $x \geq 0$, we have $h_2^{-1}(x) = 1/h(x) - 1 \leq [1/h(x)]$, thus, $x \leq h_2([1/h(x)])$. To complete the proof, choose $n_k \geq 0$ such that $n_k + 1 = [1/h(k+1)]$ and note that

$$\frac{c_{n_k}^{k+1}}{n_k+1} \geq h(k+1)\left(\frac{k+1}{1+(k+1)}\right)^{k+1} \geq h(k+1)e^{-1}. \quad \square$$

Theorem 6 (Arbitrarily Slow Convergence of DRS). *There is a point $z_0 \in \ell_2^2(\mathbf{N})$, such that for every function $h : \mathbf{R}_+ \to (0,1)$ that strictly decreases to zero and satisfies $\{1/(j+1) \mid j \in \mathbf{N}\backslash\{0\}\} \subseteq \text{range}(h)$, there are two closed subspaces U and V with zero intersection, $U \cap V = \{0\}$, such that the relaxed PRS sequence $(z^j)_{j\geq 0}$ generated with the functions $f = \iota_V$ and $g = \iota_U$ and relaxation parameters $\lambda_k \equiv 1/2$ converges in norm but satisfies the bound*

$$\|z^k - z^*\| \geq e^{-1}h(k).$$

Proof. We assume the setting of Example 1. Suppose that $z^0 = (z_j^0)_{j\geq 0}$, where for all $k \geq 0$, $z_k^0 \in \mathbf{R}^2$, and $\|z_k^0\|_{\mathbf{R}^2} = 1/(k+1)$. Then it follows that $\|z^0\|^2 = \sum_{i=0}^{\infty} 1/(k+1)^2 < \infty$ and so $z^0 \in \mathcal{H}$. Thus, for all $k,n \geq 0$,

$$\|T^{k+1}z^0\| \geq c_n^{k+1}\|z_n^0\|_{\mathbf{R}^2} = \frac{1}{n+1}c_n^{k+1}.$$

Therefore, we can achieve arbitrarily slow convergence by picking $(c_j)_{j\geq 0}$, and a subsequence $(n_j)_{j\geq 0} \subseteq \mathbf{N}$ using Lemma 4. $\square$

7 Optimal Objective Rates

In this section we construct four examples that show the nonergodic and ergodic convergence rates in Corollary 3 and Theorem 3 are optimal up to constant factors.

7.1 Ergodic Convergence of Minimization Problems

In this section, we will construct an example where the ergodic rates of convergence in Section 5.1 are optimal up to constant factors. Our example only converges in the ergodic sense and diverges otherwise. Throughout this section, we let $\gamma = 1$ and $\lambda_k \equiv 1$, we work in the Hilbert space $\mathcal{H} = \mathbf{R}$, and we use the following objective functions: for all $x \in \mathbf{R}$, let

$$g(x) = 0 \quad \text{and} \quad f(x) = |x|.$$

Recall that for all $x \in \mathbf{R}$

$$\mathbf{prox}_g(x) = x \quad \text{and} \quad \mathbf{prox}_f(x) = \max(|x| - 1, 0)\,\mathrm{sign}(x).$$

The proof of the following lemma is simple, so we omit it.

Lemma 5. *The unique minimizer of $f + g$ is equal to $0 \in \mathbf{R}$. Furthermore, 0 is the unique fixed point of T_{PRS}.*

Because of Lemma 5, we will use the notation:

$$z^* = 0 \quad \text{and} \quad x^* = 0.$$

We are ready to prove our main optimality result.

Proposition 5 (Optimality of Ergodic Convergence Rates). *Suppose that $z^0 = 2 - \varepsilon$ for some $\varepsilon \in (0,1)$. Then the PRS algorithm applied to f and g with initial point z^0 does not converge. As ε goes to 0, the ergodic objective convergence rate in Theorem 3 is tight, the ergodic objective convergence rate in Corollary 2 is tight up to a factor of $5/2$, and the feasibility convergence rate of Theorem 3 is tight up to a factor of 4.*

Proof. We compute the sequences $(z^j)_{j\geq 0}$, $(x_g^j)_{j\geq 0}$, and $(x_f^j)_{j\geq 0}$ by induction: First $x_g^0 = \mathbf{prox}_{\gamma g}(z^0) = z^0$ and $x_f^0 = \mathbf{prox}_{\gamma f}(2x_g^0 - z^0) = \max(|z^0| - 1, 0)\,\mathrm{sign}(z^0) = 1 - \varepsilon$. Thus, it follows that $z^1 = z^0 + 2(x_f^0 - x_g^0) = 2 - \varepsilon + 2(1 - \varepsilon - (2 - \varepsilon)) = z^0 = -\varepsilon$. Similarly, $x_g^1 = z^1 = -\varepsilon$. Finally, $x_f^1 = \max(\varepsilon - 1, 0)\,\mathrm{sign}(-\varepsilon) = 0$ and $z^2 = z^1 + 2(x_f^1 - x_g^1) = z^1 + 2\varepsilon = \varepsilon$. We only examined the base case, but it is clear that by induction we have the following identities:

$$z^k = (-1)^k \varepsilon, \qquad x_g^k = (-1)^k \varepsilon, \qquad x_f^k = 0, \qquad k = 1, 2, \ldots.$$

The sequences $(z^j)_{j\geq 0}$ and $(x_g^j)_{j\geq 0}$ do not converge; they oscillate around $0 \in \mathrm{Fix}(T)$.

We will now compute the ergodic iterates:

$$\overline{x}_g^k = \frac{1}{k+1}\sum_{i=0}^{k} x_g^i \overset{(7.1)}{=} \begin{cases} \frac{2-\varepsilon}{k+1} & \text{if } k \text{ is even;} \\ \frac{2-2\varepsilon}{k+1} & \text{otherwise.} \end{cases}$$

$$\overline{x}_f^k = \frac{1}{k+1}\sum_{i=0}^{k} x_f^i \overset{(7.1)}{=} \frac{1-\varepsilon}{k+1}.$$

Let us use these formulas to compute the objective values:

$$f(\overline{x}_f^k) + g(\overline{x}_f^k) - f(0) - g(0) \overset{(7.1)}{=} \frac{1-\varepsilon}{k+1}$$

$$f(\overline{x}_g^k) + g(\overline{x}_g^k) - f(0) - g(0) \overset{(7.1)}{=} \begin{cases} \frac{2-\varepsilon}{k+1} & \text{if } k \text{ is even;} \\ \frac{2-2\varepsilon}{k+1} & \text{otherwise.} \end{cases}$$

Theorem 3 upper bounds the objective error at $\overline{x}_f^k$ by

$$\frac{|z^0 - x^*|^2}{4(k+1)} = \frac{4-4\varepsilon}{4(k+1)} + \frac{\varepsilon^2}{4(k+1)} = \frac{1-\varepsilon}{k+1} + \frac{\varepsilon^2}{4(k+1)}.$$

By taking ε to 0, we see that this bound is tight. Because f is 1-Lipschitz continuous, Corollary 2 bounds the objective error at $\overline{x}_g^k$ with

$$\frac{|z^0 - x^*|^2}{4(k+1)} + \frac{2|z^0 - z^*|}{(k+1)} \stackrel{(7.1)}{=} \frac{1-\varepsilon}{k+1} + \frac{\varepsilon^2}{4(k+1)} + 2\frac{2-\varepsilon}{k+1} = \frac{5-3\varepsilon}{k+1} + \frac{\varepsilon^2}{4(k+1)}.$$

As we take ε to 0, we see that this bound it tight up to a factor of $5/2$. Finally, consider the feasibility convergence rate:

$$|\overline{x}_g^k - \overline{x}_f^k| \stackrel{(7.1)}{=} \begin{cases} \frac{1}{k+1} & \text{if } k \text{ is even;} \\ \frac{1-\varepsilon}{k+1} & \text{otherwise.} \end{cases}.$$

Theorem 3 predicts the following upper bound for Equation (7.1):

$$\frac{2|z^0 - z^*|}{k+1} = 2\frac{2-\varepsilon}{k+1} = \frac{4-2\varepsilon}{k+1}.$$

By taking ε to 0, we see that this bound is tight up to a factor of 4. □

7.2 *Optimal Nonergodic Objective Rates*

Our aim in this section is to show that if $\lambda_k \equiv 1/2$, then the non-ergodic convergence rate of $o(1/\sqrt{k+1})$ in Corollary 3 is essentially tight. In particular, for every $\alpha > 1/2$, we provide examples of f and g such that f is 1-Lipschitz and

$$h(x_g^k, x_g^k) = \Omega\left(\frac{1}{(k+1)^\alpha}\right),$$

where h is defined in (4.20) and Ω gives a lower bound. Our example uses point-to-set distance functions.

Proposition 6. *Let C be a closed, convex subset of $\mathcal{H}$ and let $d_C : \mathcal{H} \to \mathcal{H}$ be defined by $d_C(\cdot) := \min_{y \in C} \|\cdot - y\|$. Then $d_C(x)$ is 1-Lipschitz and for all $x \in \mathcal{H}$,*

$$\mathbf{prox}_{\gamma d_C}(x) = \theta P_C(x) + (1-\theta)x \quad \textit{where } \theta = \begin{cases} \frac{\gamma}{d_C(x)} & \textit{if } \gamma \le d_C(x); \\ 1 & \textit{otherwise.} \end{cases}$$

Proof. Follows from the formula for the subgradient of d_C [2, Example 16.49]. □

Proposition 6 says that $\mathbf{prox}_{\gamma d_C}(x)$ reduces to a projection map whenever x is close enough to C. Proposition 7 constructs a family of examples such that if γ is chosen large enough, then DRS does not distinguish between indicator functions and distance functions.

Proposition 7. *Suppose that V and U are linear subspaces of $\mathcal{H}$, $U \cap V = \{0\}$, and $z^0 \in \mathcal{H}$. If $\gamma \geq \|z^0\|$ and $\lambda_k = 1/2$ for all $k \geq 0$, then Algorithm 1 applied to the either pair of objective functions $(f = \iota_V, g = \iota_U)$ and $(f = d_V, g = \iota_U)$ produces the same sequence $(z^j)_{j\geq 0}$.*

Proof. Let $(z_1^j)_{j\geq 0}$, $(x_{g,1}^j)_{j\geq 0}$, and $(x_{f,1}^j)_{j\geq 0}$ be sequences generated by the function pair $(f = \iota_V, g = \iota_U)$, and let $(z_2^j)_{j\geq 0}$, $(x_{g,2}^j)_{j\geq 0}$, and $(x_{f,2}^j)_{j\geq 0}$ be sequences generated by the function pair $(f = d_V, g = \iota_U)$. Define operators $T_{\mathrm{PRS},1}$ and $T_{\mathrm{PRS},2}$ likewise. Observe that $x^* := 0$ is a minimizer of both functions pairs and $z^* := 0$ is a fixed point of $(T_{\mathrm{PRS},1})_{1/2}$. To show that $z_1^k = z_2^k$ for all $k \geq 0$, we just need to show that $\mathbf{prox}_{\gamma d_V}(\mathbf{refl}_g(z^k)) = x_{f,1}^k = x_{f,2}^k = P_V(\mathbf{refl}_g(z^k))$ for all $k \geq 0$. In view of Proposition 6, the identity will follow if

$$\gamma \geq d_V(\mathbf{refl}_g(z^k)) = \|\mathbf{refl}_g(z^k) - P_V(\mathbf{refl}_g(z^k))\|.$$

However, this is always the case because $(\mathbf{refl}_g(z^*) - P_V(\mathbf{refl}_g(z^*))) = 0$ and

$$\begin{aligned}
&\|\mathbf{refl}_g(z^k) - P_V(\mathbf{refl}_g(z^k)) - (\mathbf{refl}_g(z^*) - P_V(\mathbf{refl}_g(z^*)))\|^2 \\
&\quad + \|P_V(\mathbf{refl}_g(z^k)) - P_V(\mathbf{refl}_g(z^*))\|^2 \\
&\leq \|\mathbf{refl}_g(z^k) - \mathbf{refl}_g(z^*)\|^2 \leq \|z^k - z^*\|^2 \leq \|z^0 - z^*\|^2 = \|z^0\|^2 \leq \gamma^2
\end{aligned}$$

because P_V is $\frac{1}{2}$-averaged. □

For the rest of this section, we define for all $i \geq 0$,

$$\theta_i := \cos^{-1}\left(\sqrt{\frac{i}{i+1}}\right) \quad \text{and} \quad c_i := \cos(\theta_i) = \sqrt{\frac{i}{i+1}}.$$

Theorem 7. *Assume the notation of Theorem 5. Then for all $\alpha > 1/2$, there exists a point $z^0 \in \mathcal{H}$ such that if $\gamma \geq \|z^0\|$ and $(z^j)_{j\geq 0}$ is generated by DRS applied to the functions $(f = d_V, g = \iota_U)$, then $d_V(x^*) = 0$ and*

$$d_V(x_g^k) = \Omega\left(\frac{1}{(k+1)^\alpha}\right).$$

Proof. Fix $k \geq 0$. Let $z^0 = ((1/(j+1)^\alpha, 0))_{j\geq 0} \in \mathcal{H}$. Now, choose $\gamma \geq \|z^0\| = \left(\sum_{i=0}^\infty 1/(i+1)^{2\alpha}\right)^{1/2}$. Define $w^0 \in \mathcal{H}$ using Equation (4.23):

$$w^0 = (I - T)z^0 = \left(\frac{1}{(j+1)^\alpha}\left(\frac{1}{j+1}, \frac{-\sqrt{j}}{j+1}\right)\right)_{j\geq 0}.$$

Then $\|w_i^0\| = 1/(1+i)^{(1+2\alpha)/2}$.

Now we will calculate $d_V(x_g^k) = \|P_V x_g^k - x_g^k\|$. First, recall that $T^k = \bigoplus_{i=0}^\infty c_i^k R_{k\theta_i}$, where for all $\theta \in \mathbf{R}$,

$$R_\theta = \begin{bmatrix} \cos(\theta) & -\sin(\theta) \\ \sin(\theta) & \cos(\theta) \end{bmatrix}.$$

Thus,

$$
\begin{aligned}
x_g^k := P_U(z^k) &= \left(\begin{bmatrix} 1 & 0 \\ 0 & 0 \end{bmatrix} c_j^k R_{k\theta} \left(\frac{1}{(j+1)^\alpha}, 0 \right) \right)_{j \geq 0} \\
&= \left(\begin{bmatrix} 1 & 0 \\ 0 & 0 \end{bmatrix} c_j^k \frac{1}{(j+1)^\alpha} (\cos(k\theta_j), \sin(k\theta_j)) \right)_{j \geq 0} \\
&= \left(c_j^k \frac{\cos(k\theta_j)}{(j+1)^\alpha} (1,0) \right)_{j \geq 0}.
\end{aligned}
$$

Furthermore, from the identity

$$
(P_V)_i = \begin{bmatrix} \cos^2(\theta_i) & \sin(\theta_i)\cos(\theta_i) \\ \sin(\theta_i)\cos(\theta_i) & \sin^2(\theta_i) \end{bmatrix} = \begin{bmatrix} \frac{i}{i+1} & \frac{\sqrt{i}}{i+1} \\ \frac{\sqrt{i}}{i+1} & \frac{1}{i+1} \end{bmatrix},
$$

we have

$$
P_V x_g^k = \left(c_j^k \frac{\cos(k\theta_j)}{(j+1)^\alpha} \left(\frac{j}{j+1}, \frac{\sqrt{j}}{j+1} \right) \right)_{j \geq 0}.
$$

Thus, the difference has the following form:

$$
x_g^k - P_V x_g^k = \left(c_j^k \frac{\cos(k\theta_j)}{(j+1)^\alpha} \left(\frac{1}{j+1}, \frac{-\sqrt{j}}{j+1} \right) \right)_{j \geq 0}.
$$

Now we derive the lower bound:

$$
\begin{aligned}
d_V^2(x_g^k) = \|x_g^k - P_V x_g^k\|^2 &= \sum_{i=0}^{\infty} c_i^{2k} \frac{\cos^2(k\theta_i)}{(i+1)^{2\alpha+1}} \\
&= \sum_{i=0}^{\infty} c_i^{2k} \frac{\cos^2\left(k \cos^{-1}\left(\sqrt{\frac{i}{i+1}}\right)\right)}{(i+1)^{2\alpha+1}} \\
&\geq \frac{1}{e} \sum_{i=k}^{\infty} \frac{\cos^2\left(k \cos^{-1}\left(\sqrt{\frac{i}{i+1}}\right)\right)}{(i+1)^{2\alpha+1}}.
\end{aligned} \tag{4.24}
$$

The next three lemmas will focus on estimating the order of the sum in Equation (4.24). After which, Theorem 7 will follow from Equation (4.24) and Lemma 8, below. □

Lemma 6. *Let $h : \mathbf{R}_+ \to \mathbf{R}_+$ be a continuously differentiable function such that $h \in L_1(\mathbf{R}_+)$ and $\sum_{i=1}^{\infty} h(i) < \infty$. Then for all positive integers k,*

$$
\left| \int_k^\infty h(y)dy - \sum_{i=k}^{\infty} h(i) \right| \leq \sum_{i=k}^{\infty} \max_{y \in [i,i+1]} |h'(y)|.
$$

Proof. We just apply the Mean Value Theorem:

$$\left|\int_k^\infty h(y)dy - \sum_{i=k}^\infty h(i)\right| \leq \left|\sum_{i=k}^\infty \int_i^{i+1} (h(y)-h(i))dy\right| \leq \sum_{i=k}^\infty \int_i^{i+1} |h(y)-h(i)|dy$$
$$\leq \sum_{i=k}^\infty \max_{y\in[i,i+1]} |h'(y)|. \qquad \square$$

The following lemma will quantify the deviation of integral from the sum.

Lemma 7. *The following bound holds:*

$$\left|\sum_{i=k}^\infty \frac{\cos^2\left(k\cos^{-1}\left(\sqrt{\frac{i}{i+1}}\right)\right)}{(i+1)^{2\alpha+1}} - \int_k^\infty \frac{\cos^2\left(k\cos^{-1}\left(\sqrt{\frac{y}{y+1}}\right)\right)}{(y+1)^{2\alpha+1}}dy\right| = O\left(\frac{1}{(k+1)^{2\alpha+\frac{1}{2}}}\right). \tag{4.25}$$

Proof. We will use Lemma 6 with

$$h(y) = \frac{\cos^2\left(k\cos^{-1}\left(\sqrt{\frac{y}{y+1}}\right)\right)}{(y+1)^{2\alpha+1}}.$$

to deduce an upper bound on the absolute value. Indeed,

$$|h'(y)| = \left|\frac{k\sin\left(k\cos^{-1}\left(\sqrt{\frac{y}{y+1}}\right)\right)\cos\left(k\cos^{-1}\left(\sqrt{\frac{y}{y+1}}\right)\right)}{\sqrt{y}(y+1)(y+1)^{2\alpha+1}} - \frac{\cos^2\left(k\cos^{-1}\left(\sqrt{\frac{y}{y+1}}\right)\right)}{(y+1)^{2\alpha+2}}\right|$$
$$= O\left(\frac{k}{(y+1)^{2\alpha+1+3/2}} + \frac{1}{(y+1)^{2\alpha+2}}\right).$$

Therefore, we can bound Equation (4.25) by the following sum:

$$\sum_{i=k}^\infty \max_{y\in[i,i+1]} |h'(y)| = O\left(\frac{k}{(k+1)^{2\alpha+3/2}} + \frac{1}{(k+1)^{2\alpha+1}}\right) = O\left(\frac{1}{(k+1)^{2\alpha+1/2}}\right). \qquad \square$$

In the following lemma, we estimate the order of the oscillatory integral approximation to the sum in Equation (4.24). The proof follows by a change of variables and an integration by parts.

Lemma 8. *The following bound holds:*

$$\sum_{i=k}^\infty \frac{\cos^2\left(k\cos^{-1}\left(\sqrt{\frac{i}{i+1}}\right)\right)}{(i+1)^{2\alpha+1}}dy = \Omega\left(\frac{1}{(k+1)^{2\alpha}}\right). \tag{4.26}$$

Proof. Fix $k \geq 1$. We first perform a change of variables $u = \cos^{-1}(\sqrt{y/(y+1)})$ on the integral approximation of the sum:

$$\int_k^\infty \frac{\cos^2\left(k\cos^{-1}\left(\sqrt{\frac{y}{y+1}}\right)\right)}{(y+1)^{2\alpha+1}}dy$$
$$= 2\int_0^{\cos^{-1}\left(\sqrt{k/(k+1)}\right)} \cos^2(ku)\cos(u)\sin^{4\alpha-1}(u)du. \tag{4.27}$$

We will show that the right-hand side of Equation (4.27) is of order $\Omega\left(1/(k+1)^{2\alpha}\right)$. Then Equation (4.26) will follow by Lemma 7.

Let $\rho := \cos^{-1}(\sqrt{k/(k+1)})$. We have

$$2\int_0^\rho \cos^2(ku)\cos(u)\sin^{4\alpha-1}(u)du = \int_0^\rho (1+\cos(2ku))\cos(u)\sin^{4\alpha-1}(u)du$$
$$= p_1 + p_2 + p_3$$

where

$$p_1 = \int_0^\rho 1\cdot\cos(u)\sin^{4\alpha-1}(u)du = \frac{1}{4\alpha}\sin^{4\alpha}(\rho);$$
$$p_2 = \frac{1}{2k}\sin(2k\rho)\cos(\rho)\sin^{4\alpha-1}(\rho);$$
$$p_3 = -\frac{1}{2k}\int_0^\rho \sin(2ku)d(\cos(u)\sin^{4\alpha-1}(u));$$

and we have applied integration by parts for $\int_0^\rho \cos(2ku)\cos(u)\sin^{4\alpha-1}(u)du = p_2 + p_3$.

Because $\sin(\cos^{-1}(x)) = \sqrt{1-x^2}$, for all $\eta > 0$, we get

$$\sin^\eta(\rho) = \sin^\eta\cos^{-1}\left(\sqrt{k/(k+1)}\right) = \frac{1}{(k+1)^{\eta/2}}.$$

In addition, we have $\cos(\rho) = \cos\cos^{-1}\left(\sqrt{k/(k+1)}\right) = \sqrt{k/(k+1)}$ and the trivial bounds $|\sin(2k\rho)| \leq 1$ and $|\sin(2ku)| \leq 1$.

Therefore, the following bounds hold:

$$p_1 = \frac{1}{4\alpha(k+1)^{2\alpha}} \quad \text{and} \quad |p_2| \leq \frac{\sqrt{k/(k+1)}}{2k(k+1)^{2\alpha-1/2}} = O\left(\frac{1}{(k+1)^{2\alpha+1/2}}\right).$$

In addition, for p_3, we have $d(\cos(u)\sin^{4\alpha-1}(u)) = \sin^{4\alpha-2}(u)((4\alpha-1)\cos^2(u) - \sin^2(u))du$. Furthermore, for $u \in [0,\rho]$ and $\alpha > 1/2$, we have $\sin^{4\alpha-2}(u) \in [0, 1/(k+1)^{2\alpha-1}]$ and the following lower bound: $(4\alpha-1)\cos^2(u) - \sin^2(u) \geq (4\alpha-1)\cos^2(\rho) - \sin^2(\rho) = (4\alpha-1)(k/(k+1)) - 1/(k+1) > 0$ as long as $k \geq 1$. Therefore, we have $\sin^{4\alpha-2}(u)((4\alpha-1)\cos^2(u) - \sin^2(u)) \geq 0$ for all $u \in [0,\rho]$, and thus,

$$|p_3| \leq \frac{1}{2k}\cos(\rho)\sin^{4\alpha-1}(\rho) = \frac{\sqrt{k/(k+1)}}{2k(k+1)^{2\alpha-1/2}} = O\left(\frac{1}{(k+1)^{2\alpha+1/2}}\right).$$

Therefore, $p_1+p_2+p_3 \geq p_1 - |p_2| - |p_3| = \Omega\left((k+1)^{-2\alpha}\right)$. □

We deduce the following theorem from the sum estimation in Lemma 8:

Theorem 8 (Lower Complexity of DRS). *There exists closed, proper, and convex functions $f, g : \mathcal{H} \to (-\infty, \infty]$ such that f is 1-Lipschitz and for every $\alpha > 1/2$, there is a point $z^0 \in \mathcal{H}$ and $\gamma \in \mathbf{R}_{++}$ such that if $(z^j)_{j\geq 0}$ is generated by Algorithm 1 with $\lambda_k = 1/2$ for all $k \geq 0$, then*

$$h(x_g^k, x_g^k) = \Omega\left(\frac{1}{(k+1)^{\alpha}}\right),$$

where the objective-error function h is defined in (4.20).

Proof. Assume the setting of Theorem 7. Then $f = d_V$ and $g = \iota_U$, and by Lemma 8, we have $h(x_g^k, x_g^k) = d_V(x_g^k) = \Omega\left(1/(k+1)^{\alpha}\right)$. □

Theorem 8 shows that the DRS algorithm *is nearly as slow* as the subgradient method. We use the word nearly because the subgradient method has complexity $O(1/\sqrt{k+1})$, while DRS has complexity $o(1/\sqrt{k+1})$. To the best of our knowledge, this is the first *lower complexity* result for DRS algorithm. Note that Theorem 8 implies the same lower bound for the Forward-Douglas-Rachford splitting algorithm [13] and the many primal-dual operator-splitting schemes [21, 23, 53, 14, 7, 8, 19, 6, 20, 16] (this list is not exhaustive) that contain Chambolle and Pock's algorithm [16] as a special case because the algorithm is known to contain Douglas-Rachford splitting as a special case; see the comments in Section 6.1.1.

8 From Relaxed PRS to Relaxed ADMM

It is well known that ADMM is equivalent to DRS applied to the Lagrange dual of Problem (4.3) [31].[6] Thus, if we let $d_f(w) := f^*(A^*w)$ and $d_g(w) := g^*(B^*w) - \langle w, b\rangle$, then relaxed ADMM is equivalent to relaxed PRS applied to the following problem:

$$\underset{w\in\mathcal{G}}{\text{minimize}}\ d_f(w) + d_g(w).$$

We make two assumptions regarding d_f and d_g:

Assumption 4 (Solution Existence). Functions $f, g : \mathcal{H} \to (-\infty, \infty]$ satisfy

$$\text{zer}(\partial d_f + \partial d_g) \neq \emptyset.$$

This is a restatement of Assumption 3, which is used in our analysis of the primal case.

[6] A recent result reported in Chapter 5 [55] of this volume shows the direct (non-duality) equivalence between ADMM and DRS when they are both applied to Problem (4.3).

Assumption 5. The following differentiation rule holds:

$$\partial d_f(x) = A^* \circ (\partial f^*) \circ A \quad \text{and} \quad \partial d_g(x) = B^* \circ (\partial g^*) \circ B - b.$$

See [2, Theorem 16.37] for conditions that imply this identity, of which the weakest are $0 \in \text{sri}(\text{range}(A^*) - \text{dom}(f^*))$ and $0 \in \text{sri}(\text{range}(B^*) - \text{dom}(g^*))$, where sri is the strong relative interior of a convex set. This assumption may seem strong, but it is standard in the analysis of ADMM. because it implies the dual proximal operator identities in (4.33) on Page 154 below.

8.1 Primal Objective Convergence Rates in ADMM

With a little effort (see Appendix 3.3), we can show the following convergence rates for ADMM.

Theorem 9 (Ergodic Primal Convergence of ADMM). *Define the ergodic primal iterates by the formulas:* $\overline{x}^k = (1/\Lambda_k)\sum_{i=0}^k \lambda_i x^i$ *and* $\overline{y}^k = (1/\Lambda_k)\sum_{i=0}^k \lambda_i y^i$. *Then*

$$-\frac{2\|w^*\|\|z^0 - z^*\|}{\gamma\Lambda_k} \leq h(\overline{x}^k, \overline{y}^k) \leq \frac{\|z^0 - (z^* - w^*)\|^2}{4\gamma\Lambda_k},$$

where h is defined in (4.20).

The ergodic rate presented here is stronger and easier to interpret than the one in [34] for the ADMM algorithm ($\lambda_k \equiv 1/2$). Indeed, the rate presented in [34, Theorem 4.1] shows the following bound: for all $k \geq 1$ and for any bounded set $\mathscr{D} \subseteq \text{dom}(f) \times \text{dom}(g) \times \mathscr{G}$, we have the following variational inequality bound

$$\sup_{(x,y,w)\in\mathscr{D}} \left(h(\overline{x}^{k-1}, \overline{y}^k) + \langle \overline{w}_{d_g}^k, Ax + By - b\rangle - \langle A\overline{x}^{k-1} + B\overline{y}^k - b, w\rangle \right)$$
$$\leq \frac{\sup_{(x,y,w)\in\mathscr{D}} \|(x,y,w) - (x^0, y^0, w_{d_g}^0)\|^2}{2(k+1)}.$$

If $(x^*, y^*, w^*) \in \mathscr{D}$, then the supremum is positive and bounds the deviation of the primal objective from the lower fundamental inequality.

Theorem 10 (Nonergodic Primal Convergence of ADMM). *For all* $k \geq 0$, *let* $\tau_k = \lambda_k(1 - \lambda_k)$. *In addition, suppose that* $\underline{\tau} = \inf_{j\geq 0} \tau_j > 0$. *Recall that the objective-error function h is defined in* (4.20). *Then*

1. *In general, we have the bounds:*

$$\frac{-\|z^0 - z^*\|\|w^*\|}{2\sqrt{\underline{\tau}(k+1)}} \leq h(x^k, y^k) \leq \frac{\|z^0 - z^*\|(\|z^0 - z^*\| + \|w^*\|)}{2\gamma\sqrt{\underline{\tau}(k+1)}}$$

   and $|h(x^k, y^k)| = o(1/\sqrt{k+1})$.

2. *If* $\mathscr{G} = \mathbf{R}$ *and* $\lambda_k \equiv 1/2$, *then for all* $k \geq 0$,

$$\frac{-\|z^0 - z^*\|\|w^*\|}{\sqrt{2}(k+1)} \leq h(x^{k+1}, y^{k+1}) \leq \frac{\|z^0 - z^*\|(\|z^0 - z^*\| + \|w^*\|)}{\sqrt{2}\gamma(k+1)}$$

and $|h(x^{k+1}, y^{k+1})| = o(1/(k+1))$.

The rates presented in Theorem 10 are new and, to the best of our knowledge, they are the first nonergodic rate results for ADMM primal objective error.

9 Conclusion

In this chapter, we provided a comprehensive convergence rate analysis of the FPR and objective error of several splitting algorithms under general convexity assumptions. We showed that the convergence rates are essentially optimal in all cases. All results follow from some combination of a lemma that deduces convergence rates of summable monotonic sequences (Lemma 1), a simple diagram (Figure 4.1), and fundamental inequalities (Propositions 2 and 3) that relate the FPR to the objective error of the relaxed PRS algorithm. The most important open question is whether and how the rates we derived will improve when we enforce stronger assumptions, such as Lipschitz differentiability and/or strong convexity, on f and g. This will be the subject of future work.

References

1. Bauschke, H.H., Bello Cruz, J.Y., Nghia, T.T.A., Phan, H.M., Wang, X.: The rate of linear convergence of the Douglas-Rachford algorithm for subspaces is the cosine of the Friedrichs angle. Journal of Approximation Theory **185**(0), 63–79 (2014)
2. Bauschke, H.H., Combettes, P.L.: Convex Analysis and Monotone Operator Theory in Hilbert Spaces. Springer (2011)
3. Bauschke, H.H., Deutsch, F., Hundal, H.: Characterizing arbitrarily slow convergence in the method of alternating projections. International Transactions in Operational Research **16**(4), 413–425 (2009)
4. Beck, A., Teboulle, M.: A fast iterative shrinkage-thresholding algorithm for linear inverse problems. SIAM Journal on Imaging Sciences **2**(1), 183–202 (2009)
5. Bertsekas, D.P.: Incremental gradient, subgradient, and proximal methods for convex optimization: A survey. Optimization for Machine Learning pp. 85–120 (2011)
6. Boţ, R.I., Csetnek, E.R.: On the convergence rate of a forward-backward type primal-dual splitting algorithm for convex optimization problems. Optimization **64**(1), 5–23 (2015)
7. Boţ, R.I., Hendrich, C.: A Douglas–Rachford type primal-dual method for solving inclusions with mixtures of composite and parallel-sum type monotone operators. SIAM Journal on Optimization **23**(4), 2541–2565 (2013)
8. Boţ, R.I., Hendrich, C.: Solving monotone inclusions involving parallel sums of linearly composed maximally monotone operators. arXiv:1306.3191 [math] (2013)

9. Boţ, R.I., Hendrich, C.: Convergence analysis for a primal-dual monotone+ skew splitting algorithm with applications to total variation minimization. Journal of Mathematical Imaging and Vision pp. 1–18 (2014)
10. Boyd, S., Parikh, N., Chu, E., Peleato, B., Eckstein, J.: Distributed optimization and statistical learning via the alternating direction method of multipliers. Foundations and Trends in Machine Learning **3**(1), 1–122 (2011)
11. Bredies, K.: A forward–backward splitting algorithm for the minimization of non-smooth convex functionals in Banach space. Inverse Problems **25**(1), 015,005 (2009)
12. Brézis, H., Lions, P.L.: Produits infinis de résolvantes. Israel Journal of Mathematics **29**(4), 329–345 (1978)
13. Briceño-Arias, L.M.: Forward-Douglas-Rachford splitting and forward-partial inverse method for solving monotone inclusions. Optimization **64**(5), 1239–1261 (2015)
14. Briceno-Arias, L.M., Combettes, P.L.: A monotone+ skew splitting model for composite monotone inclusions in duality. SIAM Journal on Optimization **21**(4), 1230–1250 (2011)
15. Browder, F.E., Petryshyn, W.V.: The solution by iteration of nonlinear functional equations in Banach spaces. Bulletin of the American Mathematical Society **72**(3), 571–575 (1966)
16. Chambolle, A., Pock, T.: A first-order primal-dual algorithm for convex problems with applications to imaging. Journal of Mathematical Imaging and Vision **40**(1), 120–145 (2011)
17. Combettes, P.L.: Quasi-Fejérian analysis of some optimization algorithms. Studies in Computational Mathematics **8**, 115–152 (2001)
18. Combettes, P.L.: Solving monotone inclusions via compositions of nonexpansive averaged operators. Optimization **53**(5–6), 475–504 (2004)
19. Combettes, P.L.: Systems of structured monotone inclusions: duality, algorithms, and applications. SIAM Journal on Optimization **23**(4), 2420–2447 (2013)
20. Combettes, P.L., Condat, L., Pesquet, J.C., Vu, B.C.: A forward-backward view of some primal-dual optimization methods in image recovery. In: 2014 IEEE International Conference on Image Processing (ICIP), pp. 4141–4145 (2014)
21. Combettes, P.L., Pesquet, J.C.: Primal-dual splitting algorithm for solving inclusions with mixtures of composite, Lipschitzian, and parallel-sum type monotone operators. Set-Valued and variational analysis **20**(2), 307–330 (2012)
22. Cominetti, R., Soto, J.A., Vaisman, J.: On the rate of convergence of Krasnosel'skiĭ-Mann iterations and their connection with sums of Bernoullis. Israel Journal of Mathematics pp. 1–16 (2014)
23. Condat, L.: A primal–dual splitting method for convex optimization involving Lipschitzian, proximable and linear composite terms. Journal of Optimization Theory and Applications **158**(2), 460–479 (2013)
24. Corman, E., Yuan, X.: A generalized proximal point algorithm and its convergence rate. SIAM Journal on Optimization **24**(4), 1614–1638 (2014)
25. Davis, D.: Convergence Rate Analysis of Primal-Dual Splitting Schemes. SIAM Journal on Optimization **25**(3), 1912–1943 (2015)
26. Davis, D.: Convergence Rate Analysis of the Forward-Douglas-Rachford Splitting Scheme. SIAM Journal on Optimization **25**(3), 1760–1786 (2015)
27. Davis, D., Yin, W.: A three-operator splitting scheme and its optimization applications. arXiv preprint arXiv:1504.01032v1 (2015)
28. Deng, W., Lai, M.J., Yin, W.: On the $o(1/k)$ convergence and parallelization of the alternating direction method of multipliers. arXiv preprint arXiv:1312.3040 (2013)
29. Eckstein, J., Bertsekas, D.P.: On the Douglas-Rachford splitting method and the proximal point algorithm for maximal monotone operators. Mathematical Programming **55**(1–3), 293–318 (1992)
30. Franchetti, C., Light, W.: On the von Neumann alternating algorithm in Hilbert space. Journal of mathematical analysis and applications **114**(2), 305–314 (1986)
31. Gabay, D.: Application of the methods of multipliers to variational inequalities. In: M. Fortin, R. Glowinski (eds.) Augmented Lagrangians: Application to the Numerical Solution of Boundary Value Problems, pp. 299–331. North-Holland, Amsterdam (1983)

32. Glowinski, R., Marrocco, A.: Sur l'approximation, par éléments finis d'ordre un, et la résolution, par pénalisation-dualité d'une classe de problèmes de Dirichlet nonlinéaires. Rev. Francaise dAut. Inf. Rech. Oper **R-2**, 41–76 (1975)
33. Güler, O.: On the Convergence of the Proximal Point Algorithm for Convex Minimization. SIAM Journal on Control and Optimization **29**(2), 403–419 (1991)
34. He, B., Yuan, X.: On the $O(1/n)$ convergence rate of the Douglas-Rachford alternating direction method. SIAM Journal on Numerical Analysis **50**(2), 700–709 (2012)
35. He, B., Yuan, X.: On non-ergodic convergence rate of Douglas–Rachford alternating direction method of multipliers. Numerische Mathematik **130**(3), 567–577 (2015)
36. Knopp, K.: Infinite Sequences and Series. Courier Dover Publications (1956)
37. Krasnosel'skiĭ, M.A.: Two remarks on the method of successive approximations. Uspekhi Matematicheskikh Nauk **10**(1), 123–127 (1955)
38. Liang, J., Fadili, J., Peyré, G.: Convergence rates with inexact nonexpansive operators. arXiv preprint arXiv:1404.4837. Mathematical Programming 159:403 (2016)
39. Lions, P.L., Mercier, B.: Splitting algorithms for the sum of two nonlinear operators. SIAM Journal on Numerical Analysis **16**(6), 964–979 (1979)
40. Mann, W.R.: Mean value methods in iteration. Proceedings of the American Mathematical Society **4**(3), 506–510 (1953)
41. Monteiro, R.D., Svaiter, B.F.: Iteration-complexity of block-decomposition algorithms and the alternating direction method of multipliers. SIAM Journal on Optimization **23**(1), 475–507 (2013)
42. Monteiro, R.D.C., Svaiter, B.F.: On the complexity of the hybrid proximal extragradient method for the iterates and the ergodic mean. SIAM Journal on Optimization **20**(6), 2755–2787 (2010)
43. Monteiro, R.D.C., Svaiter, B.F.: Complexity of Variants of Tseng's Modified F-B Splitting and Korpelevich's Methods for Hemivariational Inequalities with Applications to Saddle-point and Convex Optimization Problems. SIAM Journal on Optimization **21**(4), 1688–1720 (2011)
44. Nemirovsky, A.S., Yudin, D.B.: Problem Complexity and Method Efficiency in Optimization. Wiley, New York (1983)
45. Nesterov, Y.: Introductory Lectures on Convex Optimization, vol. 87. Springer US, Boston, MA (2004)
46. Ogura, N., Yamada, I.: Non-strictly convex minimization over the fixed point set of an asymptotically shrinking nonexpansive mapping. Numerical Functional Analysis and Optimization **23**(1–2), 113–137 (2002)
47. Passty, G.B.: Ergodic convergence to a zero of the sum of monotone operators in Hilbert space. Journal of Mathematical Analysis and Applications **72**(2), 383 – 390 (1979)
48. Pock, T., Cremers, D., Bischof, H., Chambolle, A.: An algorithm for minimizing the Mumford-Shah functional. In: Computer Vision, 2009 IEEE 12th International Conference on, pp. 1133–1140. IEEE (2009)
49. Schizas, I.D., Ribeiro, A., Giannakis, G.B.: Consensus in ad hoc WSNs with noisy links. Part I: Distributed estimation of deterministic signals. Signal Processing, IEEE Transactions on **56**(1), 350–364 (2008)
50. Shefi, R., Teboulle, M.: Rate of convergence analysis of decomposition methods based on the proximal method of multipliers for convex minimization. SIAM Journal on Optimization **24**(1), 269–297 (2014)
51. Shi, W., Ling, Q., Yuan, K., Wu, G., Yin, W.: On the linear convergence of the ADMM in decentralized consensus optimization. IEEE Transactions on Signal Processing **62**(7), 1750–1761 (2014)
52. Tseng, P.: A modified forward-backward splitting method for maximal monotone mappings. SIAM Journal on Control and Optimization **38**(2), 431–446 (2000)
53. Vũ, B.C.: A splitting algorithm for dual monotone inclusions involving cocoercive operators. Advances in Computational Mathematics **38**(3), 667–681 (2013)
54. Wei, E., Ozdaglar, A.: Distributed alternating direction method of multipliers. In: Decision and Control (CDC), 2012 IEEE 51st Annual Conference on, pp. 5445–5450. IEEE (2012)

55. Yan, M., Yin, W.: Self equivalence of the alternating direction method of multipliers. In: R. Glowinski, S. Osher, W. Yin (eds.) Splitting Methods in Communication and Imaging, Science and Engineering. Springer (2016)

A Further Applications of the Results of Section 3

1.1 $o(1/(k+1)^2)$ *FPR of FBS and PPA*

In problem (1), let g be a C^1 function with Lipschitz derivative. The *forward-backward splitting* (FBS) algorithm is the iteration:

$$z^{k+1} = \mathbf{prox}_{\gamma f}(z^k - \gamma \nabla g(z^k)), \quad k = 0, 1, \ldots. \tag{4.28}$$

The FBS algorithm generalizes and has the following subgradient representation:

$$z^{k+1} = z^k - \gamma \widetilde{\nabla} f(z^{k+1}) - \gamma \nabla g(z^k) \tag{4.29}$$

where $\widetilde{\nabla} f(z^{k+1}) := (1/\gamma)(z^k - z^{k+1} - \gamma \nabla g(z^k)) \in \partial f(z^{k+1})$, and z^{k+1} and $\widetilde{\nabla} f(z^{k+1})$ are unique given z^k and $\gamma > 0$.

In this section, we analyze the convergence rate of the FBS algorithm given in Equations (4.28) and (4.29). If $g = 0$, FBS reduces to the proximal point algorithm (PPA) and $\beta = \infty$. If $f = 0$, FBS reduces to gradient descent. The FBS algorithm can be written in the following operator form:

$$T_{\mathrm{FBS}} := \mathbf{prox}_{\gamma f} \circ (I - \gamma \nabla g).$$

Because $\mathbf{prox}_{\gamma f}$ is $(1/2)$-averaged and $I - \gamma \nabla g$ is $\gamma/(2\beta)$-averaged [46, Theorem 3(b)], it follows that T_{FBS} is α_{FBS}-averaged for

$$\alpha_{\mathrm{FBS}} := \frac{2\beta}{4\beta - \gamma} \in (1/2, 1)$$

whenever $\gamma < 2\beta$ [2, Proposition 4.32]. Thus, we have $T_{\mathrm{FBS}} = (1 - \alpha_{\mathrm{FBS}})I + \alpha_{\mathrm{FBS}} T$ for a certain nonexpansive operator T, and $T_{\mathrm{FBS}}(z^k) - z^k = \alpha_{\mathrm{FBS}}(Tz^k - z^k)$. In particular, for all $\gamma < 2\beta$ the following sum is finite:

$$\sum_{i=0}^{\infty} \|T_{\mathrm{FBS}}(z^k) - z^k\|^2 \overset{(4.11)}{\leq} \frac{\alpha_{\mathrm{FBS}} \|z^0 - z^*\|^2}{(1 - \alpha_{\mathrm{FBS}})}.$$

To analyze the FBS algorithm we need to derive a joint subgradient inequality for $f + g$. First, we recall the following sufficient descent property for Lipschitz differentiable functions.

Theorem 11 (Descent Theorem [2, Theorem 18.15(iii)]). *If g is differentiable and ∇g is $(1/\beta)$-Lipschitz, then for all $x, y \in \mathcal{H}$ we have the upper bound*

$$g(x) \leq g(y) + \langle x - y, \nabla g(y) \rangle + \frac{1}{2\beta}\|x - y\|^2.$$

Corollary 4 (Joint Descent Theorem). *If g is differentiable and ∇g is $(1/\beta)$-Lipschitz, then for all points $x, y \in \mathrm{dom}(f)$ and $z \in \mathcal{H}$, and subgradients $\widetilde{\nabla} f(x) \in \partial f(x)$, we have*

$$f(x) + g(x) \leq f(y) + g(y) + \langle x - y, \nabla g(z) + \widetilde{\nabla} f(x) \rangle + \frac{1}{2\beta}\|z - x\|^2. \quad (4.30)$$

Proof. Inequality (4.30) follows from adding the upper bound

$$g(x) - g(y) \leq g(z) - g(y) + \langle x - z, \nabla g(z) \rangle + \frac{1}{2\beta}\|z - x\|^2 \leq \langle x - y, \nabla g(z) \rangle + \frac{1}{2\beta}\|z - x\|^2$$

with the subgradient inequality: $f(x) \leq f(y) + \langle x - y, \widetilde{\nabla} f(x) \rangle$. □

We now improve the $O(1/(k+1)^2)$ FPR rate for PPA in [12, Théorème 9] by showing that the FPR rate of FBS is actually $o(1/(k+1)^2)$.

Theorem 12 (Objective and FPR Convergence of FBS). *Let $z^0 \in \mathrm{dom}(f) \cap \mathrm{dom}(g)$ and let x^* be a minimizer of $f + g$. Suppose that $(z^j)_{j \geq 0}$ is generated by FBS (iteration (4.28)) where ∇g is $(1/\beta)$-Lipschitz and $\gamma < 2\beta$. Then for all $k \geq 0$,*

$$h(z^{k+1}, z^{k+1}) \leq \frac{\|z^0 - x^*\|^2}{k+1} \times \begin{cases} \frac{1}{2\gamma} & \text{if } \gamma \leq \beta; \\ \left(\frac{1}{2\gamma} + \left(\frac{1}{2\beta} - \frac{1}{2\gamma}\right)\frac{\alpha_{\mathrm{FBS}}}{(1-\alpha_{\mathrm{FBS}})}\right) & \text{otherwise,} \end{cases}$$

and

$$h(z^{k+1}, z^{k+1}) = o(1/(k+1)),$$

where the objective-error function h is defined in (4.20). *In addition, for all $k \geq 0$, we have $\|T_{\mathrm{FBS}} z^{k+1} - z^{k+1}\|^2 = o(1/(k+1)^2)$ and*

$$\|T_{\mathrm{FBS}} z^{k+1} - z^{k+1}\|^2 \leq \frac{\|z^0 - x^*\|^2}{\left(\frac{1}{\gamma} - \frac{1}{2\beta}\right)(k+1)^2} \times \begin{cases} \frac{1}{2\gamma} & \text{if } \gamma \leq \beta; \\ \left(\frac{1}{2\gamma} + \left(\frac{1}{2\beta} - \frac{1}{2\gamma}\right)\frac{\alpha_{\mathrm{FBS}}}{(1-\alpha_{\mathrm{FBS}})}\right) & \text{otherwise.} \end{cases}$$

Proof. Recall that $z^k - z^{k+1} = \gamma \widetilde{\nabla} f(z^{k+1}) + \gamma \nabla g(z^k) \in \gamma \partial f(z^{k+1}) + \gamma \nabla g(z^k)$ for all $k \geq 0$. Thus, the joint descent theorem shows that for all $x \in \mathrm{dom}(f)$, we have

$$\begin{aligned} f(z^{k+1}) + g(z^{k+1}) - f(x) - g(x) &\overset{(4.30)}{\leq} \frac{1}{\gamma}\langle z^{k+1} - x, z^k - z^{k+1} \rangle + \frac{1}{2\beta}\|z^k - z^{k+1}\|^2 \\ &= \frac{1}{2\gamma}\left(\|z^k - x\|^2 - \|z^{k+1} - x\|^2\right) + \left(\frac{1}{2\beta} - \frac{1}{2\gamma}\right)\|z^{k+1} - z^k\|^2. \end{aligned} \quad (4.31)$$

If we set $x = x^*$ in Equation (4.31), we see that $(h(z^{j+1}, z^{j+1}))_{j\geq 0}$ is positive, summable, and

$$\sum_{i=0}^{\infty} h(z^{j+1}, z^{j+1}) \leq \begin{cases} \frac{1}{2\gamma}\|z^0 - x^*\|^2 & \text{if } \gamma \leq \beta; \\ \left(\frac{1}{2\gamma} + \left(\frac{1}{2\beta} - \frac{1}{2\gamma}\right)\frac{\alpha_{\mathrm{FBS}}}{(1-\alpha_{\mathrm{FBS}})}\right)\|z^0 - x^*\|^2 & \text{otherwise.} \end{cases} \tag{4.32}$$

In addition, if we set $x = z^k$ in Equation (4.31), then we see that $(h(z^{j+1}, z^{j+1}))_{j\geq 0}$ is decreasing:

$$\left(\frac{1}{\gamma} - \frac{1}{2\beta}\right)\|z^{k+1} - z^k\|^2 \leq h(z^k, z^k) - h(z^{k+1}, z^{k+1}).$$

Therefore, the rates for $h(z^{k+1}, z^{k+1})$ follow by Lemma 1 Part (a), with $a_k = h(z^{k+1}, z^{k+1})$ and $\lambda_k \equiv 1$.

Now we prove the rates for $\|T_{\mathrm{FBS}} z^{k+1} - z^{k+1}\|^2$. We apply Part 3 of Lemma 1 with $a_k = (1/\gamma - 1/(2\beta))\|z^{k+2} - z^{k+1}\|^2$, $\lambda_k \equiv 1$, $e_k = 0$, and $b_k = h(z^{k+1}) - h(x^*)$ for all $k \geq 0$, to show that $\sum_{i=0}^{\infty}(i+1)a_i$ is less than the sum in Equation (4.32). Part 2 of Theorem 1 shows that $(a_j)_{j\geq 0}$ is monotonically nonincreasing. Therefore, the convergence rate of $(a_j)_{j\geq 0}$ follows from Part (b) of Lemma 1. □

When $f = 0$, the objective error upper bound in Theorem 12 is strictly better than the bound provided in [45, Corollary 2.1.2]. In FBS, the objective error rate is the same as the one derived in [4, Theorem 3.1], when $\gamma \in (0, \beta]$, and is the same as the one given in [11] in the case that $\gamma \in (0, 2\beta)$. The little-o FPR rate is new in all cases except for the special case of PPA ($g \equiv 0$) under the condition that the sequence $(z^j)_{j\geq 0}$ strongly converges to a minimizer [33].

1.2 $o(1/(k+1)^2)$ FPR of One Dimensional DRS

Whenever the operator $(T_{\mathrm{PRS}})_{1/2}$ is applied in $\mathbf{R}$, the convergence rate of the FPR improves to $o(1/(k+1)^2)$.

Theorem 13. *Suppose that $\mathcal{H} = \mathbf{R}$, and suppose that $(z^j)_{j\geq 0}$ is generated by the DRS algorithm, i.e., Algorithm 1 with $\lambda_k \equiv 1/2$. Then for all $k \geq 0$,*

$$|(T_{\mathrm{PRS}})_{1/2} z^{k+1} - z^{k+1}|^2 = \frac{|z^0 - z^*|^2}{2(k+1)^2} \quad \text{and} \quad |(T_{\mathrm{PRS}})_{1/2} z^{k+1} - z^{k+1}|^2 = o\left(\frac{1}{(k+1)^2}\right).$$

Proof. Note that $(T_{\mathrm{PRS}})_{1/2}$ is $(1/2)$-averaged, and, hence, it is the resolvent of some maximal monotone operator on $\mathbf{R}$ [2, Corollary 23.8]. Furthermore, every maximal monotone operator on $\mathbf{R}$ is the subdifferential operator of a closed, proper, and convex function [2, Corollary 22.19]. Therefore, DRS is equivalent to the proximal point algorithm applied to a certain convex function on $\mathbf{R}$. Thus, the result follows by Theorem 12 applied to this function. □

B Further Lower Complexity Results

2.1 Ergodic Convergence of Feasibility Problems

Proposition 8. *The ergodic feasibility convergence rate in Equation* (4.21) *is optimal up to a factor of two.*

Proof. Algorithm 1 with $\lambda_k = 1$ for all $k \geq 0$ (i.e., PRS) is applied to the functions $f = \iota_{\{(x_1,x_2)\in\mathbf{R}^2 | x_1=0\}}$ and $g = \iota_{\{(x_1,x_2)\in\mathbf{R}^2 | x_2=0\}}$ with the initial iterate $z^0=(1,1) \in \mathbf{R}^2$. Because $T_{\mathrm{PRS}} = -I_{\mathcal{H}}$, it is easy to see that the only fixed point of T_{PRS} is $z^* = (0,0)$. In addition, the following identities are satisfied:

$$x_g^k = \begin{cases} (1,0) & \text{even } k; \\ (-1,0) & \text{odd } k. \end{cases} \quad z^k = \begin{cases} (1,1) & \text{even } k; \\ (-1,-1) & \text{odd } k. \end{cases} \quad x_f^k = \begin{cases} (0,-1) & \text{even } k; \\ (0,1) & \text{odd } k. \end{cases}$$

Thus, the PRS algorithm oscillates around the solution $x^* = (0,0)$. However, note that the averaged iterates satisfy:

$$\overline{x}_g^k = \begin{cases} (\frac{1}{k+1},0) & \text{even } k; \\ (0,0) & \text{odd } k. \end{cases} \quad \text{and} \quad \overline{x}_f^k = \begin{cases} (0,\frac{-1}{k+1}) & \text{even } k; \\ (0,0) & \text{odd } k. \end{cases}$$

It follows that $\|\overline{x}_g^k - \overline{x}_f^k\| = (1/(k+1))\|(1,-1)\| = (1/(k+1))\|z^0 - z^*\|$, $\forall k \geq 0$. □

2.2 Optimal Objective and FPR Rates with Lipschitz Derivative

The following examples show that the objective and FPR rates derived in Theorem 12 are essentially optimal. The setup of the following counterexample already appeared in [12, Remarque 6] but the objective function lower bounds were not shown.

Theorem 14 (Lower Complexity of PPA). *There exists a Hilbert space $\mathcal{H}$, and a closed, proper, and convex function f such that for all $\alpha > 1/2$, there exists $z^0 \in \mathcal{H}$ such that if $(z^j)_{j\geq 0}$ is generated by PPA, then*

$$\|\mathbf{prox}_{\gamma f}(z^k) - z^k\|^2 \geq \frac{\gamma^2}{(1+2\alpha)e^{2\gamma}(k+\gamma)^{1+2\alpha}},$$
$$f(z^{k+1}) - f(x^*) \geq \frac{1}{4\alpha e^{2\gamma}(k+1+\gamma)^{2\alpha}}.$$

Proof. Let $\mathcal{H} = \ell_2(\mathbf{R})$, and define a linear map $A : \mathcal{H} \to \mathcal{H}$:

$$A(z_1, z_2, \cdots, z_n, \cdots) = \left(z_1, \frac{z_2}{2}, \cdots, \frac{z_n}{n}, \cdots\right).$$

For all $z \in \mathscr{H}$, define $f(x) = (1/2)\langle Az, z\rangle$. Thus, we have the proximal identity for f and

$$\mathbf{prox}_{\gamma f}(z) = (I+\gamma A)^{-1}(z) = \left(\frac{j}{j+\gamma}z_j\right)_{j\geq 1} \quad \text{and} \quad (I-\mathbf{prox}_{\gamma f})(z) = \left(\frac{\gamma}{j+\gamma}z_j\right)_{j\geq 1}.$$

Now let $z^0 = (1/(j+\gamma)^{\alpha})_{j\geq 1} \subset \mathscr{H}$, and set $T = \mathbf{prox}_{\gamma f}$. Then we get the following FPR lower bound:

$$\begin{aligned}\|z^{k+1}-z^k\|^2 = \|T^k(T-I)z^0\|^2 &= \sum_{i=1}^{\infty}\left(\frac{i}{i+\gamma}\right)^{2k}\frac{\gamma^2}{(i+\gamma)^{2+2\alpha}}\\ &\geq \sum_{i=k}^{\infty}\left(\frac{i}{i+\gamma}\right)^{2k}\frac{\gamma^2}{(i+\gamma)^{2+2\alpha}}\\ &\geq \frac{\gamma^2}{(1+2\alpha)e^{2\gamma}(k+\gamma)^{1+2\alpha}}.\end{aligned}$$

Furthermore, the objective lower bound holds

$$\begin{aligned}f(z^{k+1})-f(x^*) = \frac{1}{2}\langle Az^{k+1}, z^{k+1}\rangle &= \frac{1}{2}\sum_{i=1}^{\infty}\frac{1}{i}\left(\frac{i}{i+\gamma}\right)^{2(k+1)}\frac{1}{(i+\gamma)^{2\alpha}}\\ &\geq \frac{1}{2}\sum_{i=k+1}^{\infty}\left(\frac{i}{i+\gamma}\right)^{2(k+1)}\frac{1}{(i+\gamma)^{1+2\alpha}}\\ &\geq \frac{1}{4\alpha e^{2\gamma}(k+1+\gamma)^{2\alpha}}.\end{aligned}$$

□

C ADMM Convergence Rate Proofs

Given an initial vector $z^0 \in \mathscr{G}$, Lemma 2 shows that at each iteration relaxed PRS performs the following computations:

$$\begin{cases} w_{d_g}^k = \mathbf{prox}_{\gamma d_g}(z^k);\\ w_{d_f}^k = \mathbf{prox}_{\gamma d_f}(2w_{d_g}^k - z^k);\\ z^{k+1} = z^k + 2\lambda_k(w_{d_f}^k - w_{d_g}^k).\end{cases}$$

In order to apply the relaxed PRS algorithm, we need to compute the proximal operators of the dual functions d_f and d_g.

Lemma 9 (Proximity Operators on the Dual). *Let $w, v \in \mathscr{G}$. Then the update formulas $w^+ = \mathbf{prox}_{\gamma d_f}(w)$ and $v^+ = \mathbf{prox}_{\gamma d_g}(v)$ are equivalent to the following computations*

$$\begin{cases} x^+ = \arg\min_{x\in\mathscr{H}_1} f(x) - \langle w, Ax\rangle + \frac{\gamma}{2}\|Ax\|^2; \\ w^+ = w - \gamma Ax^+; \end{cases}$$

$$\begin{cases} y^+ = \arg\min_{y\in\mathscr{H}_2} g(y) - \langle v, By-b\rangle + \frac{\gamma}{2}\|By-b\|^2; \\ v^+ = v - \gamma(By^+ - b); \end{cases} \tag{4.33}$$

respectively. In addition, the subgradient inclusions hold: $A^*w^+ \in \partial f(x^+)$ and $B^*v^+ \in \partial g(y^+)$. Finally, w^+ and v^+ are independent of the choice of x^+ and y^+, respectively, even if they are not unique solutions to the minimization subproblems.

We can use Lemma 9 to derive the relaxed form of ADMM in Algorithm 2. Note that this form of ADMM eliminates the "hidden variable" sequence $(z^j)_{j\geq 0}$ in Equation (C). This following derivation is not new, but is included for the sake of completeness. See [31] for the original derivation.

Proposition 9 (Relaxed ADMM). *Let $z^0 \in \mathscr{G}$, and let $(z^j)_{j\geq 0}$ be generated by the relaxed PRS algorithm applied to the dual formulation in Equation (8). Choose initial points $w_{d_g}^{-1} = z^0, x^{-1} = 0$ and $y^{-1} = 0$ and initial relaxation $\lambda_{-1} = 1/2$. Then we have the following identities starting from $k = -1$:*

$$\begin{aligned} y^{k+1} &= \arg\min_{y\in\mathscr{H}_2} g(y) - \langle w_{d_g}^k, Ax^k + By - b\rangle + \\ &\qquad \frac{\gamma}{2}\|Ax^k + By - b + (2\lambda_k - 1)(Ax^k + By^k - b)\|^2 \\ w_{d_g}^{k+1} &= w_{d_g}^k - \gamma(Ax^k + By^{k+1} - b) - \gamma(2\lambda_k - 1)(Ax^k + By^k - b) \\ x^{k+1} &= \arg\min_{x\in\mathscr{H}_1} f(x) - \langle w_{d_g}^{k+1}, Ax + By^{k+1} - b\rangle + \frac{\gamma}{2}\|Ax + By^{k+1} - b\|^2 \\ w_{d_f}^{k+1} &= w_{d_g}^{k+1} - \gamma(Ax^{k+1} + By^{k+1} - b) \end{aligned}$$

Proof. By Equation (C) and Lemma 9, we get the following formulation for the k-th iteration: Given $z^0 \in \mathscr{H}$

$$\begin{cases} y^k &= \arg\min_{y\in\mathscr{H}_2} g(y) - \langle z^k, By - b\rangle + \frac{\gamma}{2}\|By-b\|^2; \\ w_{d_g}^k &= z^k - \gamma(By^k - b); \\ x^k &= \arg\min_{x\in\mathscr{H}_1} f(x) - \langle 2w_{d_g}^k - z^k, Ax\rangle + \frac{\gamma}{2}\|Ax\|^2; \\ w_{d_f}^k &= 2w_{d_g}^k - z^k - \gamma Ax^k; \\ z^{k+1} &= z^k + 2\lambda_k(w_{d_f}^k - w_{d_g}^k). \end{cases}$$

We will use this form to get to the claimed iteration. First,

$$2w_{d_g}^k - z^k = w_{d_g}^k - \gamma(By^k - b) \quad \text{and} \quad w_{d_f}^k = w_{d_g}^k - \gamma(Ax^k + By^k - b). \tag{4.34}$$

Furthermore, we can simplify the definition of x^k:

$$\begin{aligned} x^k &= \underset{x\in\mathscr{H}_1}{\arg\min} f(x) - \langle 2w_{d_g}^k - z^k, Ax\rangle + \frac{\gamma}{2}\|Ax\|^2 \\ &\overset{(4.34)}{=} \underset{x\in\mathscr{H}_1}{\arg\min} f(x) - \langle w_{d_g}^k - \gamma(By^k - b), Ax\rangle + \frac{\gamma}{2}\|Ax\|^2 \\ &= \underset{x\in\mathscr{H}_1}{\arg\min} f(x) - \langle w_{d_g}^k, Ax + By^k - b\rangle + \frac{\gamma}{2}\|Ax + By^k - b\|^2. \end{aligned}$$

Note that the last two lines of Equation (C) differ by terms independent of x.

We now eliminate the z^k variable from the y^k subproblem: because $w_{d_f}^k + z^k = 2w_{d_g}^k - \gamma Ax^k$, we have

$$\begin{aligned} z^{k+1} &= z^k + 2\lambda_k(w_{d_f}^k - w_{d_g}^k) \\ &\overset{(4.34)}{=} z^k + w_{d_f}^k - w_{d_g}^k + \gamma(2\lambda_k - 1)(Ax^k + By^k - b) \\ &= w_{d_g}^k - \gamma Ax^k - \gamma(2\lambda_k - 1)(Ax^k + By^k - b). \end{aligned}$$

We can simplify the definition of y^{k+1} by applying the identity in Equation (C):

$$\begin{aligned} y^{k+1} &= \underset{y\in\mathscr{H}_2}{\arg\min}\, g(y) - \langle z^{k+1}, By - b\rangle + \frac{\gamma}{2}\|By - b\|^2 \\ &\overset{(C)}{=} \underset{y\in\mathscr{H}_2}{\arg\min}\, g(y) - \langle w_{d_g}^k - \gamma Ax^k - \gamma(2\lambda_k - 1)(Ax^k + By^k - b), By - b\rangle + \frac{\gamma}{2}\|By - b\|^2 \\ &= \underset{y\in\mathscr{H}_2}{\arg\min}\, g(y) - \langle w_{d_g}^k, Ax^k + By - b\rangle + \frac{\gamma}{2}\|Ax^k + By - b + (2\lambda_k - 1)(Ax^k + By^k - b)\|^2. \end{aligned}$$

The result then follows from Equations (C), (4.34), (C), and (C), combined with the initial conditions listed in the statement of the proposition. In particular, note that the updates of x, y, w_{d_f}, and w_{d_g} do not explicitly depend on z. □

Remark 2. Proposition 9 proves that $w_{d_f}^{k+1} = w_{d_g}^{k+1} - \gamma(Ax^{k+1} + By^{k+1} - b)$. Recall that by Equation (C), $z^{k+1} - z^k = 2\lambda_k(w_{d_f}^k - w_{d_g}^k)$. Therefore, it follows that

$$z^{k+1} - z^k = -2\gamma\lambda_k(Ax^k + By^k - b). \tag{4.35}$$

3.1 Dual Feasibility Convergence Rates

We can apply the results of Section 5 to deduce convergence rates for the dual objective functions. Instead of restating those theorems, we just list the following bounds on the feasibility of the primal iterates.

Theorem 15. *Suppose that $(z^j)_{j\geq 0}$ is generated by Algorithm 2, and let $(\lambda_j)_{j\geq 0} \subseteq (0,1]$. Then the following convergence rates hold:*

1. **Ergodic convergence:** *The feasibility convergence rate holds:*

$$\|A\overline{x}^k + B\overline{y}^k - b\|^2 = \frac{4\|z^0 - z^*\|^2}{\gamma\Lambda_k^2}.$$

2. **Nonergodic convergence:** *Suppose that $\underline{\tau} = \inf_{j\geq 0} \lambda_j(1-\lambda_j) > 0$. Then*

$$\|Ax^k + By^k - b\|^2 \leq \frac{\|z^0 - z^*\|^2}{4\gamma^2\underline{\tau}(k+1)} \quad \text{and} \quad \|Ax^k + By^k - b\|^2 = o\left(\frac{1}{k+1}\right).$$

Proof. Parts 1 and 2 are straightforward applications of Corollary 1. and the FPR identity: $z^k - z^{k+1} \overset{(4.35)}{=} 2\gamma\lambda_k(Ax^k + By^k - b)$. □

3.2 Converting Dual Inequalities to Primal Inequalities

The ADMM algorithm generates the following five sequences of iterates:

$$(z^j)_{j\geq 0}, (w_{d_f}^j)_{j\geq 0}, \text{ and } (w_{d_g}^j)_{j\geq 0} \subseteq \mathcal{G} \quad \text{and} \quad (x^j)_{j\geq 0} \in \mathcal{H}_1, (y^j)_{j\geq 0} \in \mathcal{H}_2.$$

The dual variables do not necessarily have a meaningful interpretation, so it is desirable to derive convergence rates involving the primal variables. In this section we will apply the Fenchel-Young inequality [2, Proposition 16.9] to convert the dual objective into a primal expression.

The following proposition will help us derive primal fundamental inequalities akin to Proposition 2 and 3.

Proposition 10. *Suppose that $(z^j)_{j\geq 0}$ is generated by Algorithm 2. Let z^* be a fixed point of T_{PRS} and let $w^* = \mathbf{prox}_{\gamma d_f}(z^*)$. Then the following identity holds:*

$$\begin{aligned} 4\gamma\lambda_k(h(x^k, y^k)) &= -4\gamma\lambda_k(d_f(w_{d_f}^k) + d_g(w_{d_g}^k) - d_f(w^*) - d_g(w^*)) \\ &\quad + \left(2\left(1 - \frac{1}{2\lambda_k}\right)\|z^k - z^{k+1}\|^2 + 2\langle z^k - z^{k+1}, z^{k+1}\rangle\right). \end{aligned} \tag{4.36}$$

Proof. We have the following subgradient inclusions from Proposition 9: $A^*w_{d_f}^k \in \partial f(x^k)$ and $B^*w_{d_g}^k \in \partial g(y^k)$. From the Fenchel-Young inequality [2, Proposition 16.9] we have the expression for f and g:

$$d_f(w_{d_f}^k) = \langle A^*w_{d_f}^k, x^k\rangle - f(x^k) \quad \text{and} \quad d_f(w_{d_g}^k) = \langle B^*w_{d_g}^k, y^k\rangle - g(y^k) - \langle w_{d_g}^k, b\rangle.$$

Therefore,

$$-d_f(w_{d_f}^k) - d_g(w_{d_g}^k) = f(x^k) + g(y^k) - \langle Ax^k + By^k - b, w_{d_f}^k\rangle - \langle w_{d_g}^k - w_{d_f}^k, By^k - b\rangle.$$

Let us simplify this bound with an identity from Proposition 9: from $w_{d_f}^k - w_{d_g}^k = -\gamma(Ax^k + By^k - b)$, it follows that

$$-d_f(w_{d_f}^k) - d_g(w_{d_g}^k) = f(x^k) + g(y^k) + \frac{1}{\gamma}\langle w_{d_f}^k - w_{d_g}^k, w_{d_f}^k + \gamma(By^k - b)\rangle. \quad (4.37)$$

Recall that $\gamma(By^k - b) = z^k - w_{d_g}^k$. Therefore

$$w_{d_f}^k + \gamma(By^k - b) = z^k + (w_{d_f}^k - w_{d_g}^k) = z^k + \frac{1}{2\lambda_k}(z^{k+1} - z^k) = \frac{1}{2\lambda_k}(2\lambda_k - 1)(z^k - z^{k+1}) + z^{k+1},$$

and the inner product term can be simplified as follows:

$$\begin{aligned}\frac{1}{\gamma}\langle w_{d_f}^k - w_{d_g}^k, w_{d_f}^k + \gamma(By^k - b)\rangle &= \frac{1}{\gamma}\langle \frac{1}{2\lambda_k}(z^{k+1} - z^k), \frac{1}{2\lambda_k}(2\lambda_k - 1)(z^k - z^{k+1})\rangle \\ &\quad + \frac{1}{\gamma}\langle \frac{1}{2\lambda_k}(z^{k+1} - z^k), z^{k+1}\rangle \\ &= -\frac{1}{2\gamma\lambda_k}\left(1 - \frac{1}{2\lambda_k}\right)\|z^{k+1} - z^k\|^2 \\ &\quad - \frac{1}{2\gamma\lambda_k}\langle z^k - z^{k+1}, z^{k+1}\rangle. \end{aligned} \quad (4.38)$$

Now we derive an expression for the dual objective at a dual optimal w^*. First, if z^* is a fixed point of T_{PRS}, then $0 = T_{\mathrm{PRS}}(z^*) - z^* = 2(w_{d_g}^* - w_{d_f}^*) = -2\gamma(Ax^* + By^* - b)$. Thus, from Equation (4.37) with k replaced by $*$, we get

$$-d_f(w^*) - d_g(w^*) = f(x^*) + g(y^*) + \langle Ax^* + Bx^* - b, w^*\rangle = f(x^*) + g(y^*). \quad (4.39)$$

Therefore, Equation (4.36) follows by subtracting (4.39) from Equation (4.37), rearranging and using the identity in Equation (4.38). □

The following two propositions prove two fundamental inequalities that bound the primal objective.

Proposition 11 (ADMM Primal Upper Fundamental Inequality). *Let z^* be a fixed point of T_{PRS} and let $w^* = \mathbf{prox}_{\gamma d_g}(z^*)$. Then for all $k \geq 0$, we have the bound:*

$$4\gamma\lambda_k h(x^k, y^k) \leq \|z^k - (z^* - w^*)\|^2 - \|z^{k+1} - (z^* - w^*)\|^2 + \left(1 - \frac{1}{\lambda_k}\right)\|z^k - z^{k+1}\|^2, \quad (4.40)$$

where the objective-error function h is defined in (4.20).

Proof. The lower inequality in Proposition 3 applied to $d_f + d_g$ shows that

$$-4\gamma\lambda_k(d_f(w_{d_f}^k) + d_g(w_{d_g}^k) - d_f(w^*) - d_g(w^*)) \leq 2\langle z^{k+1} - z^k, z^* - w^*\rangle.$$

The proof then follows from Proposition 10, and the simplification:

$$2\langle z^k - z^{k+1}, z^{k+1} - (z^* - w^*)\rangle + 2\left(1 - \frac{1}{2\lambda_k}\right)\|z^k - z^{k+1}\|^2$$
$$= \|z^k - (z^* - w^*)\|^2 - \|z^{k+1} - (z^* - w^*)\|^2 + \left(1 - \frac{1}{\lambda_k}\right)\|z^k - z^{k+1}\|^2. \qquad \square$$

Remark 3. Note that Equation (4.40) is nearly identical to the upper inequality in Proposition 2, except that $z^* - w^*$ appears in the former where x^* appears in the latter.

Proposition 12 (ADMM Primal Lower Fundamental Inequality). *Let z^* be a fixed point of T_{PRS} and let $w^* = \mathbf{prox}_{\gamma d_g}(z^*)$. Then for all $x \in \mathcal{H}_1$ and $y \in \mathcal{H}_2$ we have the bound:*

$$h(x,y) \geq \langle Ax + By - b, w^*\rangle, \tag{4.41}$$

where the objective-error function h is defined in (4.20).

Proof. The lower bound follows from the subgradient inequalities:

$$f(x) - f(x^*) \geq \langle x - x^*, A^*w^*\rangle \quad \text{and} \quad g(y) - g(y^*) \geq \langle y - y^*, B^*w^*\rangle.$$

We sum these inequalities and use $Ax^* + By^* = b$ to get Equation (4.41). $\square$

Remark 4. We use Inequality (4.41) in two special cases:

$$h(x^k, y^k) \geq \frac{1}{\gamma}\langle w_{d_g}^k - w_{d_f}^k, w^*\rangle$$
$$h(\overline{x}^k, \overline{y}^k) \geq \frac{1}{\gamma}\langle \overline{w}_{d_g}^k - \overline{w}_{d_f}^k, w^*\rangle.$$

These bounds are nearly identical to the fundamental lower inequality in Proposition 3, except that w^* appears in the former where $z^* - x^*$ appeared in the latter.

3.3 Converting Dual Convergence Rates to Primal Convergence Rates

We can use the inequalities deduced in Section 3.2 to derive convergence rates for the primal objective values. The structure of the proofs of Theorems 9 and 10 are exactly the same as in the primal convergence case in Section 5, except that we use the upper and lower inequalities derived in the Section 3.2 instead of the fundamental upper and lower inequalities in Propositions 2 and 3. This amounts to replacing the term $z^* - x^*$ and x^* by w^* and $z^* - w^*$, respectively, in all of the inequalities from Section 5. Thus, we omit the proofs.

D Examples

In this section, we apply relaxed PRS and relaxed ADMM to concrete problems and explicitly bound the associated objectives and FPR terms with the convergence rates we derived in the previous sections.

4.1 Feasibility Problems

Suppose that C_f and C_g are closed convex subsets of $\mathcal{H}$, with nonempty intersection. The goal of the feasibility problem is to find a point in the intersection of C_f and C_g. In this section, we present one way to model this problem using convex optimization and apply the relaxed PRS algorithm to reach the minimum.

In general, we cannot expect linear convergence of relaxed PRS algorithm for the feasibility problem. We showed this in Theorem 6 by constructing an example for which the DRS iteration converges in norm but does so *arbitrarily slow*. A similar result holds for the alternating projection (AP) algorithm [3]. Thus, in this section we focus on the convergence rate of the FPR.

Let ι_{C_f} and ι_{C_g} be the indicator functions of C_f and C_g. Then $x \in C_f \cap C_g$, if, and only if, $\iota_{C_f}(x) + \iota_{C_g}(x) = 0$, and the sum is infinite otherwise. Thus, a point is in the intersection of C_f and C_g if, and only if, it is the minimizer of the following problem:

$$\underset{x \in \mathcal{H}}{\text{minimize}}\, \iota_{C_f}(x) + \iota_{C_g}(x).$$

The relaxed PRS algorithm applied to this problem, with $f = \iota_{C_f}$ and $g = \iota_{C_g}$, has the following form: Given $z^0 \in \mathcal{H}$, for all $k \geq 0$, let

$$\begin{cases} x_g^k = P_{C_g}(z^k); \\ x_f^k = P_{C_f}(2x_g^k - z^k); \\ z^{k+1} = z^k + 2\lambda_k(x_f^k - x_g^k). \end{cases}$$

Because $f = \iota_{C_f}$ and $g = \iota_{C_g}$ only take on the values 0 and ∞, the objective value convergence rates derived earlier do not provide meaningful information, other than $x_f^k \in C_f$ and $x_g^k \in C_g$. However, from the FPR identity $x_f^k - x_g^k = 1/(2\lambda_k)(z^{k+1} - z^k)$, we find that after k iterations, Corollary 1 produces the bound

$$\max\{d_{C_g}^2(x_f^k), d_{C_f}^2(x_g^k)\} \leq \|x_f^k - x_g^k\|^2 = o\left(\frac{1}{k+1}\right)$$

whenever $(\lambda_j)_{j\geq 0}$ is bounded away from 0 and 1. Theorem 5 showed that this rate is optimal. Furthermore, if we average the iterates over all k, Theorem 3 gives the improved bound

$$\max\{d_{C_g}^2(\overline{x}_f^k), d_{C_f}^2(\overline{x}_g^k)\} \leq \|\overline{x}_f^k - \overline{x}_g^k\|^2 = O\left(\frac{1}{\Lambda_k^2}\right),$$

which is optimal by Proposition 8. Note that the averaged iterates satisfy $\overline{x}_f^k = (1/\Lambda_k)\sum_{i=0}^k \lambda_i x_f^i \in C_f$ and $\overline{x}_g^k = (1/\Lambda_k)\sum_{i=0}^k \lambda_i x_g^i \in C_g$, because C_f and C_g are convex. Thus, we can state the following proposition:

Proposition 13. *After k iterations the relaxed PRS algorithm produces a point in each set with distance of order $O(1/\Lambda_k)$ from each other.*

4.2 Parallelized Model Fitting and Classification

The following general scenario appears in [10, Chapter 8]. Consider the following general convex model fitting problem: Let $M : \mathbf{R}^n \to \mathbf{R}^m$ be a *feature matrix*, let $b \in \mathbf{R}^m$ be the *output* vector, let $l : \mathbf{R}^m \to (-\infty, \infty]$ be a *loss function* and let $r : \mathbf{R}^n \to (-\infty, \infty]$ be a *regularization function*. The *model fitting problem* is formulated as the following minimization:

$$\underset{x\in\mathbf{R}^n}{\text{minimize}}\ l(Mx - b) + r(x). \tag{4.42}$$

The function l is used to enforce the constraint $Mx = b + \nu$ up to some noise ν in the measurement, while r enforces the *regularity* of x by incorporating *prior knowledge* of the form of the solution. The function r can also be used to enforce the uniqueness of the solution of $Mx = b$ in ill-posed problems.

We can solve Equation (4.42) by a direct application of relaxed PRS and obtain $O(1/\Lambda_k)$ ergodic convergence and $o\left(1/\sqrt{k+1}\right)$ nonergodic convergence rates. Note that these rates do not require differentiability of f or g. In contrast, the FBS algorithm requires differentiability of one of the objective functions and a knowledge of the Lipschitz constant of its gradient. The advantage of FBS is the $o(1/(k+1))$ convergence rate shown in Theorem 12. However, we do not necessarily assume that l is differentiable, so we may need to compute $\mathbf{prox}_{\gamma l(M(\cdot)-b)}$, which can be significantly more difficult than computing $\mathbf{prox}_{\gamma l}$. Thus, in this section we separate M from l by rephrasing Equation (4.42) in the form of Problem (4.3).

In this section, we present several different ways to split Equation (4.42). Each splitting gives rise to a different algorithm and can be applied to general convex functions l and r. Our results predict convergence rates that hold for primal objectives, dual objectives, and the primal feasibility. Note that in parallelized model fitting, it is not always desirable to take the time average of all of the iterates. Indeed, when r enforces sparsity, averaging the current r-iterate with old iterates, all of which are sparse, can produce a non-sparse iterate. This will slow down vector additions and prolong convergence.

4.2.1 Auxiliary Variable

We can split Equation (4.42) by defining an auxiliary variable for $My-b$:

$$\begin{aligned} &\underset{x\in\mathbf{R}^m, y\in\mathbf{R}^n}{\text{minimize}}\ l(x)+r(y) \\ &\text{subject to } My-x=b. \end{aligned} \tag{4.43}$$

The constraint in Equation (4.43) reduces to $Ax+By=b$ where $B=M$ and $A=-I_{\mathbf{R}^m}$. If we set $f=l$ and $g=r$ and apply ADMM, the analysis of Section 3.3 shows that

$$|l(x^k)+r(y^k)-l(My^*-b)-r(y^*)| = o\left(\frac{1}{\sqrt{k+1}}\right)$$
$$\|My^k-b-x^k\|^2 = o\left(\frac{1}{k+1}\right).$$

In particular, if l is Lipschitz, $|l(x^k)-l(My^k-b)| = o\left(1/\sqrt{k+1}\right)$. Thus, we have

$$|l(My^k-b)+r(y^k)-l(My^*-b)-r(y^*)| = o\left(\frac{1}{\sqrt{k+1}}\right).$$

A similar analysis shows that

$$|l(M\overline{y}^k-b)+r(\overline{y}^k)-l(My^*-b)-r(y^*)| = O\left(\frac{1}{\Lambda_k}\right)$$
$$\|M\overline{y}^k-b-\overline{x}^k\|^2 = O\left(\frac{1}{\Lambda_k^2}\right).$$

In the next two splittings, we leave the derivation of convergence rates to the reader.

4.2.2 Splitting Across Examples

We assume that l is block separable: we have $l(Mx-b)=\sum_{i=1}^R l_i(M_ix-b_i)$ where

$$M=\begin{bmatrix} M_1 \\ \vdots \\ M_R \end{bmatrix} \quad \text{and} \quad b=\begin{bmatrix} b_1 \\ \vdots \\ b_R \end{bmatrix}.$$

Each $M_i\in\mathbf{R}^{m_i\times n}$ is a submatrix of M, each $b_i\in\mathbf{R}^{m_i}$ is a subvector of b, and $\sum_{i=1}^R m_i = m$. Therefore, an equivalent form of Equation (4.42) is given by

$$\begin{aligned} &\underset{x_1,\cdots,x_R,y\in\mathbf{R}^n}{\text{minimize}}\ \sum_{i=1}^R l_i(M_ix_i-b_i)+r(y) \\ &\text{subject to } x_r-y=0, \quad r=1,\cdots,R. \end{aligned} \tag{4.44}$$

We say that Equation (4.44) is *split across examples*. Thus, to apply ADMM to this problem, we simply stack the vectors x_i, $i = 1, \cdots, R$ into a vector $x = (x_1, \cdots, x_R)^T \in \mathbf{R}^{nR}$. Then the constraints in Equation (4.44) reduce to $Ax + By = 0$ where $A = I_{\mathbf{R}^{nR}}$ and $By = (-y, \cdots, -y)^T$.

4.2.3 Splitting Across Features

We can also split Equation (4.42) *across features*, whenever r is block separable in x, in the sense that there exists $C > 0$, such that $r = \sum_{i=1}^{C} r_i(x_i)$, and $x_i \in \mathbf{R}^{n_i}$ where $\sum_{i=1}^{C} n_i = n$. This splitting corresponds to partitioning the columns of M, i.e., $M = [M_1, \cdots, M_C]$, and $M_i \in \mathbf{R}^{m \times n_i}$, for all $i = 1, \cdots, C$. For all $y \in \mathbf{R}^n$, $My = \sum_{i=1}^{C} M_i y_i$. With this notation, we can derive an equivalent form of Equation (4.42) given by

$$\begin{aligned} &\underset{x,y \in \mathbf{R}^n}{\text{minimize}}\; l\left(\sum_{i=1}^{C} x_i - b\right) + \sum_{i=1}^{C} r_i(y_i) \\ &\text{subject to } \; x_i - M_i y_i = 0, \quad i = 1, \cdots, C. \end{aligned} \tag{4.45}$$

The constraint in Equation (4.45) reduces to $Ax + By = 0$ where $A = I_{\mathbf{R}^{mC}}$ and $By = -(M_1 y_1, \cdots, M_C y_C)^T \in \mathbf{R}^{nC}$.

4.3 Distributed ADMM

In this section our goal is to use Algorithm 2 for

$$\underset{x \in \mathcal{H}}{\text{minimize}} \sum_{i=1}^{m} f_i(x)$$

by using the splitting in [49]. Note that we could minimize this function by reformulating it in the product space $\mathcal{H}^m$ as follows:

$$\underset{\mathbf{x} \in \mathcal{H}^m}{\text{minimize}} \sum_{i=1}^{m} f_i(x_i) + \iota_D(\mathbf{x}),$$

where $D = \{(x, \cdots, x) \in \mathcal{H}^m \mid x \in \mathcal{H}\}$ is the diagonal set. Applying relaxed PRS to this problem results in a parallel algorithm where each function performs a local minimization step and then communicates its local variable to a *central processor*. In this section, we assign each function a local variable but we never communicate it to a central processor. Instead, each function only communicates with *neighbors*.

Formally, we assume there is a simple, connected, undirected graph $G = (V, E)$ on $|V| = m$ vertices with edges E that describe a connection among the different functions. We introduce a variable $x_i \in \mathcal{H}$ for each function f_i, and, hence, we set $\mathcal{H}_1 = \mathcal{H}^m$, (see Section 8). We can encode the constraint that each node communicates with neighbors by introducing an auxiliary variable for each edge in the graph:

$$\begin{aligned}&\underset{\mathbf{x}\in\mathcal{H}^m,\mathbf{y}\in\mathcal{H}^{|E|}}{\text{minimize}} \sum_{i=1}^{m} f_i(x_i)\\ &\text{subject to } x_i = y_{ij}, x_j = y_{ij}, \text{ for all } (i,j)\in E.\end{aligned} \tag{4.46}$$

The linear constraints in Equation (4.46) can be written in the form of $A\mathbf{x}+B\mathbf{y}=0$ for proper matrices A and B. Thus, we reformulate Equation (4.46) as

$$\begin{aligned}&\underset{\mathbf{x}\in\mathcal{H}^m,\mathbf{y}\in\mathcal{H}^{|E|}}{\text{minimize}} \sum_{i=1}^{m} f_i(x_i)+g(\mathbf{y})\\ &\text{subject to } A\mathbf{x}+B\mathbf{y}=0,\end{aligned} \tag{4.47}$$

where $g : \mathcal{H}^{|E|} \to \mathbf{R}$ is the zero map.

Because we only care about finding the value of the variable $\mathbf{x}\in\mathcal{H}^m$, the following simplification can be made to the sequences generated by ADMM applied to Equation (4.47) with $\lambda_k = 1/2$ for all $k \geq 1$ [51]: Let $\mathcal{N}_i$ denote the set of neighbors of $i \in V$ and set $x_i^0 = \alpha_i^0 = 0$ for all $i \in V$. Then for all $k \geq 0$,

$$\begin{cases} x_i^{k+1} = \arg\min_{x_i\in\mathcal{H}} f_i(x) + \frac{\gamma|\mathcal{N}_i|}{2}\|x_i - x_i^k - \frac{1}{|\mathcal{N}_i|}\sum_{j\in\mathcal{N}_i} x_j^k + \frac{1}{\gamma|\mathcal{N}_i|}\alpha_i\|^2 + \frac{\gamma|\mathcal{N}_i|}{2}\|x_i\|^2 \\ \alpha_i^{k+1} = \alpha_i^k + \gamma\left(|\mathcal{N}_i| x_i^{k+1} - \sum_{j\in\mathcal{N}_i} x_j^{k+1}\right). \end{cases}$$

The above iteration is truly distributed because each node $i\in V$ only requires information from its local neighbors at each iteration.

In [51], linear convergence is shown for this algorithm provided that f_i are strongly convex and ∇f_i are Lipschitz. For general convex functions, we can deduce the nonergodic rates from Theorem 10

$$\left|\sum_{i=1}^{m} f_i(x_i^k) - f(x^*)\right| = o\left(\frac{1}{\sqrt{k+1}}\right)$$

$$\sum_{\substack{i\in V\\ j\in N_i}} \|x_i^k - z_{ij}^k\|^2 + \sum_{\substack{i\in V\\ i\in N_j}} \|x_j^k - z_{ij}^k\|^2 = o\left(\frac{1}{k+1}\right),$$

and the ergodic rates from Theorem 9

$$\left|\sum_{i=1}^{m} f_i(\overline{x}_i^k) - f(x^*)\right| = O\left(\frac{1}{k+1}\right)$$

$$\sum_{\substack{i\in V\\ j\in N_i}} \|\overline{x}_i^k - \overline{z}_{ij}^k\|^2 + \sum_{\substack{i\in V\\ i\in N_j}} \|\overline{x}_j^k - \overline{z}_{ij}^k\|^2 = O\left(\frac{1}{(k+1)^2}\right).$$

These convergence rates are new and complement the linear convergence results in [51]. In addition, they complement the similar ergodic rate derived in [54] for a different distributed splitting.

Chapter 5
Self Equivalence of the Alternating Direction Method of Multipliers

Ming Yan and Wotao Yin

Abstract The alternating direction method of multipliers (ADM or ADMM) breaks a complex optimization problem into much simpler subproblems. The ADM algorithms are typically short and easy to implement yet exhibit (nearly) state-of-the-art performance for large-scale optimization problems.

To apply ADM, we first formulate a given problem into the "ADM-ready" form, so the final algorithm depends on the formulation. A problem like $\text{minimize}_{\mathbf{x}}\, u(\mathbf{x}) + v(\mathbf{C}\mathbf{x})$ has six different "ADM-ready" formulations. They can be in the primal or dual forms, and they differ by how dummy variables are introduced. To each "ADM-ready" formulation, ADM can be applied in two different orders depending on how the primal variables are updated. Finally, we get twelve different ADM algorithms! How do they compare to each other? Which algorithm should one choose? In this chapter, we show that many of the different ways of applying ADM are equivalent. Specifically, we show that ADM applied to a primal formulation is equivalent to ADM applied to its Lagrange dual; ADM is equivalent to a primal-dual algorithm applied to the saddle-point formulation of the same problem. These results are surprising since the primal and dual variables in ADM are seemingly treated very differently, and some previous work exhibit preferences in one over the other on specific problems.

M. Yan (✉)
Department of Computational Mathematics, Science and Engineering, Michigan State University, East Lansing, MI 48824, USA

Department of Mathematics, Michigan State University, East Lansing, MI 48824, USA
e-mail: yanm@math.msu.edu

W. Yin
Department of Mathematics, University of California, Los Angeles, CA 90095, USA
e-mail: wotaoyin@math.ucla.edu

R. Glowinski et al. (eds.), *Splitting Methods in Communication, Imaging, Science, and Engineering*, Scientific Computation, DOI 10.1007/978-3-319-41589-5_5

In addition, when one of the two objective functions is quadratic, possibly subject to an affine constraint, we show that swapping the update order of the two primal variables in ADM gives the same algorithm. These results identify the few truly different ADM algorithms for a problem, which generally have different forms of subproblems from which it is easy to pick one with the most computationally friendly subproblems.

1 Introduction

The alternating direction method of multipliers (ADM or ADMM) is a very popular algorithm with a wide range of applications in signal and image processing, machine learning, statistics, compressive sensing, and operations research. Combined with problem reformulation tricks, the method can reduce a complicated problem into much simpler subproblems.

The vanilla ADM applies to a linearly constrained problem with a separable convex objective function in the following "ADM-ready" form:

$$\begin{cases} \underset{\mathbf{x},\mathbf{y}}{\text{minimize}} \ f(\mathbf{x})+g(\mathbf{y}) \\ \text{subject to } \mathbf{Ax}+\mathbf{By}=\mathbf{b}, \end{cases} \tag{P1}$$

where functions f,g are proper, closed (i.e., lower semi-continuous), convex but not necessarily differentiable. ADM reduces (P1) into two simpler subproblems and then iteratively updates $\mathbf{x}$, $\mathbf{y}$, as well as a multiplier (dual) variable $\mathbf{z}$. Given $(\mathbf{x}^k,\mathbf{y}^k,\mathbf{z}^k)$, ADM generates $(\mathbf{x}^{k+1},\mathbf{y}^{k+1},\mathbf{z}^{k+1})$ as follows

1. $\mathbf{x}^{k+1} \in \arg\min_{\mathbf{x}} f(\mathbf{x}) + (\lambda/2)\|\mathbf{Ax}+\mathbf{By}^k-\mathbf{b}+\lambda^{-1}\mathbf{z}^k\|_2^2,$
2. $\mathbf{y}^{k+1} \in \arg\min_{\mathbf{y}} g(\mathbf{y}) + (\lambda/2)\|\mathbf{Ax}^{k+1}+\mathbf{By}-\mathbf{b}+\lambda^{-1}\mathbf{z}^k\|_2^2,$
3. $\mathbf{z}^{k+1} = \mathbf{z}^k + \lambda(\mathbf{Ax}^{k+1}+\mathbf{By}^{k+1}-\mathbf{b}),$

where $\lambda > 0$ is a fixed parameter. We use "$\in$" since the subproblems do not necessarily have unique solutions.

Since $\{f,\mathbf{A},\mathbf{x}\}$ and $\{g,\mathbf{B},\mathbf{y}\}$ are in symmetric positions in (P1), swapping them does not change the problem. This corresponds to switching the order that $\mathbf{x}$ and $\mathbf{y}$ are updated in each iteration. But, since the variable updated first is used in the updating of the other variable, this swap leads to a different sequence of variables and thus a different algorithm.

Note that the order switch does not change the per-iteration cost of ADM. Also note that one, however, cannot mix the two update orders at different iterations because it will generally cause divergence, even when the primal-dual solution to (P1) is unique. For example, let us apply ADMM with mixed update orders of x and y and parameter $\lambda = 1$ to the problem

$$\underset{x,y}{\text{minimize}} \ 2|x-10|+|y| \quad \text{subject to} \quad x-y=0,$$

which has the unique primal-dual solution $(x^*,y^*,z^*) = (10,10,1)$. Set initial values $(x^0,y^0,z^0) = (3,2,2)$. At odd iterations, we apply the update order: x, y, and z; at even iterations, we apply the update order: y, x, and z. Then we obtain $(x^k,y^k,z^k) = (2,3,1)$ for odd k and $(x^k,y^k,z^k) = (3,2,2)$ for even k.

1.1 ADM Works in Many Different Ways

In spite of its popularity and vast literature, there are still simple unanswered questions about ADM: how many ways can ADM be applied? and which ways work better? Before answering these questions, let us examine the following problem, to which we can find **twelve different ways to apply ADM:**

$$\underset{\mathbf{x}}{\text{minimize}}\ u(\mathbf{x}) + v(\mathbf{C}\mathbf{x}), \tag{5.1}$$

where u and v are proper, closed, convex functions and $\mathbf{C}$ is a linear mapping. Problem (5.1) generalizes a large number of signal and image processing problems, inverse problems, and machine learning models.

We shall reformulate (5.1) into the form of (P1). By introducing dummy variables in two different ways, we obtain two ADM-ready formulations of problem (5.1):

$$\begin{cases} \underset{\mathbf{x},\mathbf{y}}{\text{minimize}}\ u(\mathbf{x}) + v(\mathbf{y}) \\ \text{subject to } \mathbf{C}\mathbf{x} - \mathbf{y} = 0 \end{cases} \quad \text{and} \quad \begin{cases} \underset{\mathbf{x},\bar{\mathbf{y}}}{\text{minimize}}\ u(\mathbf{x}) + v(\mathbf{C}\bar{\mathbf{y}}) \\ \text{subject to } \mathbf{x} - \bar{\mathbf{y}} = 0. \end{cases} \tag{5.2}$$

If $\mathbf{C} = \mathbf{I}$, these two formulations are exactly the same. In addition, we can derive the dual problem of (5.1):

$$\underset{\mathbf{v}}{\text{minimize}}\ u^*(-\mathbf{C}^*\mathbf{v}) + v^*(\mathbf{v}), \tag{5.3}$$

where u^*, v^* are the convex conjugates (i.e., Legendre transforms) of functions u, v, respectively, $\mathbf{C}^*$ is the adjoint of $\mathbf{C}$, and $\mathbf{v}$ is the dual variable. (The steps to derive (5.3) from (5.1) are standard and thus omitted.) Then, we also reformulate (5.3) into two ADM-ready forms, which use different dummy variables:

$$\begin{cases} \underset{\mathbf{u},\mathbf{v}}{\text{minimize}}\ u^*(\mathbf{u}) + v^*(\mathbf{v}) \\ \text{subject to } \mathbf{u} + \mathbf{C}^*\mathbf{v} = 0 \end{cases} \quad \text{and} \quad \begin{cases} \underset{\bar{\mathbf{u}},\mathbf{v}}{\text{minimize}}\ u^*(\mathbf{C}^*\bar{\mathbf{u}}) + v^*(\mathbf{v}) \\ \text{subject to } \bar{\mathbf{u}} + \mathbf{v} = 0. \end{cases} \tag{5.4}$$

Clearly, ADM can be applied to all of the four formulations in (5.2) and (5.4), and including the update order swaps, there are *eight different ways* to apply ADM.

Under some technical conditions such as the existence of saddle-point solutions, all the eight ADM will converge to a saddle-point solution or solution for problem (5.1). In short, they all work.

It is worth noting that by the Moreau identity, the subproblems involving u^* and v^* can be easily reduced to subproblems involving u and v, respectively. No significant computing is required.

The two formulations in (5.2), however, lead to significantly different ADM subproblems. In the ADM applied to the left formulation, u and $\mathbf{C}$ will appear in one subproblem and v in the other subproblem. To the right formulation, u will be alone while v and $\mathbf{C}$ will appear in the same subproblem. This difference applies to the two formulations in (5.4) as well. It depends on the structures of $u, v, \mathbf{C}$ to determine the better choices. Therefore, out of the eight, four will have (more) difficult subproblems than the rest.

There are *another four ways* to apply ADM to problem (5.1). Every one of them will have three subproblems that separately involve $u, v, \mathbf{C}$, so they are all different from the above eight. To get the first two, let us take the left formulation in (5.2) and introduce a dummy variable $\mathbf{s}$, obtaining a new equivalent formulation

$$\begin{cases} \underset{\mathbf{x},\mathbf{y},\mathbf{s}}{\text{minimize}} \ u(\mathbf{s}) + v(\mathbf{y}) \\ \text{subject to } \mathbf{C}\mathbf{x} - \mathbf{y} = 0, \\ \qquad\qquad\quad \mathbf{x} - \mathbf{s} = 0. \end{cases} \tag{5.5}$$

It turns out that the same "dummy variable" trick applied to the right formulation in (5.2) also gives (5.5), up to a change of variable names. Although there are three variables, we can group $(\mathbf{y}, \mathbf{s})$ and treat $\mathbf{x}$ and $(\mathbf{y}, \mathbf{s})$ as the two variables. Then problem (5.5) has the form (P1). Hence, we have two ways to apply ADM to (5.5) with two different update orders. Note that $\mathbf{y}$ and $\mathbf{s}$ do not appear together in any equation or function, so the ADM subproblem that updates $(\mathbf{y}, \mathbf{s})$ will further decouple to two separable subproblems of $\mathbf{y}$ and $\mathbf{s}$; in other words, the resulting ADM has three subproblems involving $\{\mathbf{x}, \mathbf{C}\}$, $\{\mathbf{y}, v\}$, $\{\mathbf{s}, u\}$ separately. The other two ways are results of the same "dummy variable" trick applied to either formulation in (5.4). Again, since now $\mathbf{C}$ has its own subproblem, these four ways are distinct from the previous eight ways.

As demonstrated through an example, there are quite many ways to formulate the same optimization problem into "ADM-ready" forms and obtain different ADM algorithms. While most ADM users choose just one way without paying much attention to the other choices, some show preferences toward a specific formulation. For example, some prefer (5.5) over those in (5.2) and (5.4) since $\mathbf{C}$, u, v all end up in separate subproblems. When applying ADM to certain ℓ_1 minimization problems, the authors of [24, 25] emphasize on the dual formulations, and later the authors of [23] show a preference over the primal formulations. When ADM was proposed to solve a traffic equilibrium problem, it was first applied to the dual formulation in [13] and, years later, to the primal formulation in [12]. Regarding which one of the two variables should be updated first in ADM, neither a rule nor an equivalence claim is found in the literature. Other than giving preferences to ADM with simpler subproblems, there is no results that compare the different formulations.

1.2 Contributions

This chapter shows that, applied to certain pairs of different formulations of the same problem, ADM will generate equivalent sequences of variables that can be mapped exactly from one to another at every iteration. Specifically, between the sequence of an ADM algorithm on a primal formulation and that on the corresponding dual formulation, such maps exist. For a special class of problems, this mapping is provided in [9].

We also show that whenever at least one of f and g is a quadratic function (including affine function as a special case), possibly subject to an affine constraint, the sequence of an ADM algorithm can be mapped to that of the ADM algorithm using the opposite order for updating their variables.

Abusing the word "equivalence", we say that ADM has "primal-dual equivalence" and "update-order equivalence (with a quadratic objective function)." Equivalent ADM algorithms take the same number of iterations to reach the same accuracy. (However, it is possible that one algorithm is slightly better than the other in terms of numerical stability, for example, against round-off errors.)

Equipped with these equivalence results, the first eight ways to apply ADM to problem (5.1) that were discussed in Section 1.1 are reduced to four ways in light of primal-dual equivalence, and the four will further reduce to two whenever u or v, or both, is a quadratic function.

The last four ways to apply ADM on problem (5.1) discussed in Section 1.1, which yield three subproblems that separately involve u, v, and $\mathbf{C}$, are all equivalent and reduce to just one due to primal-dual equivalence and one variable in them is associated with 0 objective (for example, variable $\mathbf{x}$ has 0 objective in problem (5.5)).

Take the ℓ_p-regularization problem, $p \in [1,\infty]$,

$$\underset{\mathbf{x}}{\text{minimize}} \ \|\mathbf{x}\|_p + f(\mathbf{C}\mathbf{x}) \tag{5.6}$$

as an example, which is a special case of problem (5.1) with a quadratic function u when $p = 2$. We list its three different formulations, whose ADM algorithms are truly different, as follows. When $p \neq 2$ *and* f is non-quadratic, each of the first two formulations leads to a pair of different ADM algorithms with different orders of variable update; otherwise, each pair of algorithms is equivalent.

1. Left formulation of (5.2):

$$\begin{cases} \underset{\mathbf{x},\mathbf{y}}{\text{minimize}} \ \|\mathbf{x}\|_p + f(\mathbf{y}) \\ \text{subject to } \mathbf{C}\mathbf{x} - \mathbf{y} = 0. \end{cases}$$

   The subproblem for $\mathbf{x}$ involves ℓ_p-norm and $\mathbf{C}$. The other one for $\mathbf{y}$ involves f.
2. Right formulation of (5.2):

$$\begin{cases} \underset{\mathbf{x},\mathbf{y}}{\text{minimize}} \ \|\mathbf{x}\|_p + f(\mathbf{C}\mathbf{y}) \\ \text{subject to } \mathbf{x} - \mathbf{y} = 0. \end{cases}$$

The subproblem for $\mathbf{x}$ involves ℓ_p-norm and, for $p = 1$ and 2, has a closed-form solution. The other subproblem for $\mathbf{y}$ involves $f(\mathbf{C}\cdot)$.

3. Formulation (5.5): for any $\mu > 0$,

$$\begin{cases} \underset{\mathbf{x},\mathbf{y},\mathbf{s}}{\text{minimize}} \ \|\mathbf{s}\|_p + f(\mathbf{y}) \\ \text{subject to } \mathbf{C}\mathbf{x} - \mathbf{y} = 0, \\ \qquad\qquad \mu(\mathbf{x} - \mathbf{s}) = 0. \end{cases}$$

The subproblem for $\mathbf{x}$ is a quadratic program involving $\mathbf{C}^*\mathbf{C} + \mu\mathbf{I}$. The subproblem for $\mathbf{s}$ involves ℓ_p-norm. The subproblem for $\mathbf{y}$ involves f. The subproblems for $\mathbf{s}$ and $\mathbf{y}$ are independent.

The best choice depends on which has the simplest subproblems.

The result of ADM's primal-dual equivalence is surprising for three reasons. Firstly, ADM iteration updates *two* primal variable, $\mathbf{x}^k$ and $\mathbf{y}^k$ in (P1) and *one* dual variable, all in different manners. The updates to the primal variables are done in a Gauss-Seidel manner and involve minimizing functions f and g, but the update to the dual variable is explicit and linear. Surprisingly, ADM actually treats one of the two primal variables and the dual variable equally as we will later show. Secondly, most literature describes ADM as an inexact version of the augmented Lagrangian method (ALM) [17], which updates $(\mathbf{x}, \mathbf{y})$ together rather than one after another. Although ALM maintains the primal variables, under the hood ALM is the dual-only proximal-point algorithm that iterates the dual variable. It is commonly believed that ADM is an inexact dual algorithm. Thirdly, primal and dual problems typically have different sizes and regularity properties, causing the same algorithm, even if it is applicable to both, to exhibit different performance. For example, the primal and dual variables may have different dimensions. If the primal function f is Lipschitz differentiable, the dual function f^* is strongly convex but can be non-differentiable, and vice versa. Such primal-dual differences often mean that it is numerically advantageous to solve one rather than the other, yet our result means that there is no such primal-dual difference on ADM.

Our maps between equivalent ADM sequences have very simple forms, as the reader will see below. Besides the technical proofs that establish the maps, it is interesting to mention the operator-theoretic perspective of our results. It is shown in [13] that the dual-variable sequence of ADM coincides with a sequence of the Douglas-Rachford splitting (DRS) algorithm [7, 18]. Our ADM's primal-dual equivalence can be obtained through the above ADM–DRS relation and the Moreau identity: $\mathbf{prox}_h + \mathbf{prox}_{h^*} = \mathbf{I}$, applied to the proximal maps of f and f^* and those of g and g^*. The details are omitted in this chapter. Here, $\mathbf{prox}_h(x) := \arg\min_s h(s) + \frac{1}{2}\|s - x\|^2$.

Our results of primal-dual and update-order equivalence for ADM extends to the Peaceman-Rachford splitting (PRS) algorithm. Let the PRS operator [19] be denoted as $\mathbf{T}_{\text{PRS}} = (2\mathbf{prox}_f - \mathbf{I}) \circ (2\mathbf{prox}_g - \mathbf{I})$. The DRS operator is the average of the identity map and the PRS operator: $\mathbf{T}_{\text{DRS}} = \frac{1}{2}\mathbf{I} + \frac{1}{2}\mathbf{T}_{\text{PRS}}$, and the relaxed PRS (RPRS) operator is a weighted-average: $\mathbf{T}_{\text{RPRS}} = (1 - \alpha)\mathbf{I} + \alpha\mathbf{T}_{\text{PRS}}$, where $\alpha \in (0, 1]$. The DRS and PRS algorithms that iteratively apply their operators to find a fixed point were originally proposed for evolving PDEs with two spatial dimensions in the 1950s and then extended to finding a root of the sum of two maximal monotone (set-valued) mappings by Lions and Mercier [18]. Eckstein showed, in [8, Chapter 3.5], that DRS/PRS applied to the primal problem (5.1) is equivalent to DRS/PRS applied to the dual problem (5.4) when $\mathbf{C} = \mathbf{I}$. We will show that RPRS applied to (5.1) is equivalent to RPRS applied to (5.3) for all $\mathbf{C}$.

In addition to the aforementioned primal-dual and update-order equivalence, we obtain a primal-dual algorithm for the saddle-point formulation of (P1) that is also equivalent to the ADM. This primal-dual algorithm is generally *different* from the primal-dual algorithm proposed by Chambolle and Pock [3], while they become the same in a special case. The connection between these two algorithms will be explained.

Even when using the same number of dummy variables, truly different ADM algorithms can have different iteration complexities (do not confuse them with the difficulties of their subproblems). The convergence analysis of ADM, such as conditions for sublinear or linear convergence, involves many different scenarios [6, 5, 4]. The discussion of convergence rates of ADM algorithms is beyond the scope of this chapter. Our focus is on the equivalence.

1.3 Organization

This chapter is organized as follows. Section 2 specifies our notation, definitions, and basic assumptions. The three equivalence results for ADM are shown in Sections 4, 5, and 6: The primal-dual equivalence of ADM is discussed in Section 4; ADM is shown to be equivalent to a primal-dual algorithm applied to the saddle-point formulation in Section 5; In Section 6, we show the update-order equivalence of ADM if f or g is a quadratic function, possibly subject to an affine constraint. Sections 4 to 6 do not require any knowledge of monotone operators. The primal-dual and update-order equivalence of RPRS is shown in Section 7 based on monotone operator properties. We conclude this chapter with the application of our results on total variation image denoising in Section 8.

2 Notation, Definitions, and Assumptions

Let $\mathscr{H}_1$, $\mathscr{H}_2$, and $\mathscr{G}$ be (possibly infinite dimensional) Hilbert spaces. Bold lowercase letters such as $\mathbf{x}$, $\mathbf{y}$, $\mathbf{u}$, and $\mathbf{v}$ are used for points in the Hilbert spaces. In the example of (P1), we have $\mathbf{x} \in \mathscr{H}_1$, $\mathbf{y} \in \mathscr{H}_2$, and $\mathbf{b} \in \mathscr{G}$. When the Hilbert space a point belongs to is clear from the context, we do not specify it for the sake of simplicity. The inner product between points $\mathbf{x}$ and $\mathbf{y}$ is denoted by $\langle \mathbf{x}, \mathbf{y} \rangle$, and $\|\mathbf{x}\|_2 := \sqrt{\langle \mathbf{x}, \mathbf{x} \rangle}$ is the corresponding norm; $\|\cdot\|_1$ and $\|\cdot\|_\infty$ denote the ℓ_1 and ℓ_∞ norms, respectively. Bold uppercase letters such as $\mathbf{A}$ and $\mathbf{B}$ are used for both continuous linear mappings and matrices. $\mathbf{A}^*$ denotes the adjoint of $\mathbf{A}$. $\mathbf{I}$ denotes the identity mapping.

If $\mathscr{C}$ is a convex and nonempty set, the indicator function $\iota_\mathscr{C}$ is defined as follows:

$$\iota_{\mathscr{C}}(\mathbf{x}) = \begin{cases} 0, & \text{if } \mathbf{x} \in \mathscr{C}, \\ \infty, & \text{if } \mathbf{x} \notin \mathscr{C}. \end{cases}$$

Both lower and uppercase letters such as f, g, F, and G are used for functions. Let $\partial f(\mathbf{x})$ be the subdifferential of function f at $\mathbf{x}$. The proximal operator $\mathbf{prox}_{f(\cdot)}$ of function f is defined as

$$\mathbf{prox}_{f(\cdot)}(\mathbf{x}) = \arg\min_{\mathbf{y}} f(\mathbf{y}) + \frac{1}{2}\|\mathbf{y}-\mathbf{x}\|_2^2,$$

where the minimization problem has a unique solution. The convex conjugate f^* of function f is defined as

$$f^*(\mathbf{v}) = \sup_{\mathbf{x}}\{\langle \mathbf{v}, \mathbf{x}\rangle - f(\mathbf{x})\}.$$

Let $\mathbf{L} : \mathscr{H} \to \mathscr{G}$, the *infimal postcomposition* [1, Def. 12.33] of $f : \mathscr{H} \to (-\infty, +\infty]$ by $\mathbf{L}$ is given by

$$\mathbf{L} \triangleright f : \mathbf{s} \mapsto \inf f(\mathbf{L}^{-1}(\mathbf{s})) = \inf_{\mathbf{x}:\mathbf{L}\mathbf{x}=\mathbf{s}} f(\mathbf{x}),$$

with $\mathrm{dom}(\mathbf{L} \triangleright f) = \mathbf{L}(\mathrm{dom}(f))$.

Lemma 1. *If f is convex and $\mathbf{L}$ is affine and expressed as $\mathbf{L}(\cdot) = \mathbf{A}\cdot + \mathbf{b}$, then $\mathbf{L} \triangleright f$ is convex and the convex conjugate of $\mathbf{L} \triangleright f$ can be found as follows:*

$$(\mathbf{L} \triangleright f)^*(\cdot) = f^*(\mathbf{A}^*\cdot) + \langle \cdot, \mathbf{b}\rangle.$$

Proof. Following from the definitions of convex conjugate and infimal postcomposition, we have

$$\begin{aligned}(\mathbf{L} \triangleright f)^*(\mathbf{v}) &= \sup_{\mathbf{y}} \langle \mathbf{v}, \mathbf{y}\rangle - \mathbf{L} \triangleright f(\mathbf{y}) = \sup_{\mathbf{x}} \langle \mathbf{v}, \mathbf{A}\mathbf{x} + \mathbf{b}\rangle - f(\mathbf{x}) \\ &= \sup_{\mathbf{x}} \langle \mathbf{A}^*\mathbf{v}, \mathbf{x}\rangle - f(\mathbf{x}) + \langle \mathbf{v}, \mathbf{b}\rangle = f^*(\mathbf{A}^*\mathbf{v}) + \langle \mathbf{v}, \mathbf{b}\rangle.\end{aligned}$$

Definition 1. We say that an algorithm **I** applied to a problem is *equivalent to* an algorithm **II** applied to either the same or an equivalent problem if, given the set of parameters and a sequence of iterates $\{\xi_2^k\}_{k\geq 0}$ of algorithm **II**, i.e., $\xi_2^{k+1} = A_2(\xi_2^k, \xi_2^{k-1}, \cdots, \xi_2^{k-\Delta_1})$ with $\Delta_1 \geq 0$, there exist a set of parameters and a sequence of iterates $\{\xi_1^k\}_{k\geq 0}$ of algorithm **I** such that $\xi_1^k = T(\xi_2^k, \xi_2^{k-1}, \cdots, \xi_2^{k-\Delta})$ for some transformation T and $\Delta \geq 0$.

Definition 2. An optimization algorithm is called *primal-dual equivalent* if this algorithm applied to the primal formulation is equivalent to the same algorithm applied to its Lagrange dual.

It is important to note that most algorithms are not primal-dual equivalent. ALM applied to the primal problem is equivalent to proximal point method applied to the dual problem [20], but both algorithms are not primal-dual equivalent. In this chapter, we will show that ADM and RPRS are primal-dual equivalent.

We make the following assumptions throughout the chapter:

Assumption 1. *All the functions in this chapter are assumed to be proper, closed, and convex.*

Assumption 2. *The saddle-point solutions to all the optimization problems in this chapter are assumed to exist.*

3 Equivalent Problems

A *primal formulation* equivalent to (P1) is

$$\begin{cases} \underset{\mathbf{s},\mathbf{t}}{\text{minimize}}\ F(\mathbf{s})+G(\mathbf{t}) \\ \text{subject to } \mathbf{s}+\mathbf{t}=\mathbf{0}, \end{cases} \tag{P2}$$

where $\mathbf{s},\mathbf{t}\in\mathscr{G}$ and

$$F(\mathbf{s}) := \min_{\mathbf{x}} f(\mathbf{x}) + \iota_{\{\mathbf{x}:\mathbf{Ax}=\mathbf{s}\}}(\mathbf{x}), \tag{5.7a}$$

$$G(\mathbf{t}) := \min_{\mathbf{y}} g(\mathbf{y}) + \iota_{\{\mathbf{y}:\mathbf{By}-\mathbf{b}=\mathbf{t}\}}(\mathbf{y}). \tag{5.7b}$$

Remark 1. If we define $\mathbf{L}_f$ and $\mathbf{L}_g$ as $\mathbf{L}_f(\mathbf{x})=\mathbf{Ax}$ and $\mathbf{L}_g(\mathbf{y})=\mathbf{By}-\mathbf{b}$, respectively, then

$$F=\mathbf{L}_f\triangleright f, \qquad G=\mathbf{L}_g\triangleright g.$$

The Lagrange dual of (P1) is

$$\underset{\mathbf{v}}{\text{minimize}}\quad f^*(-\mathbf{A}^*\mathbf{v})+g^*(-\mathbf{B}^*\mathbf{v})+\langle\mathbf{v},\mathbf{b}\rangle, \tag{5.8}$$

which can be derived from $\underset{\mathbf{v}}{\text{minimize}}\left(-\min_{\mathbf{x},\mathbf{y}} L(\mathbf{x},\mathbf{y},\mathbf{v})\right)$ with the Lagrangian defined as follows:

$$L(\mathbf{x},\mathbf{y},\mathbf{v}) = f(\mathbf{x})+g(\mathbf{y})+\langle\mathbf{v},\mathbf{Ax}+\mathbf{By}-\mathbf{b}\rangle.$$

An *ADM-ready* formulation of (5.8) is

$$\begin{cases} \underset{\mathbf{u},\mathbf{v}}{\text{minimize}}\ f^*(-\mathbf{A}^*\mathbf{u})+g^*(-\mathbf{B}^*\mathbf{v})+\langle\mathbf{v},\mathbf{b}\rangle \\ \text{subject to } \mathbf{u}-\mathbf{v}=\mathbf{0}. \end{cases} \tag{D1}$$

When ADM is applied to an ADM-ready formulation of the Lagrange dual problem, we call it *Dual ADM*. The original ADM is called *Primal ADM*.

Following similar steps, the ADM ready formulation of the Lagrange dual of (P2) is

$$\begin{cases} \underset{\mathbf{u},\mathbf{v}}{\text{minimize}}\ F^*(-\mathbf{u}) + G^*(-\mathbf{v}) \\ \text{subject to } \mathbf{u} - \mathbf{v} = \mathbf{0}. \end{cases} \tag{D2}$$

The equivalence between (D1) and (D2) is trivial since

$$\begin{aligned} F^*(\mathbf{u}) &= f^*(\mathbf{A}^*\mathbf{u}), \\ G^*(\mathbf{v}) &= g^*(\mathbf{B}^*\mathbf{v}) - \langle \mathbf{v}, \mathbf{b} \rangle, \end{aligned}$$

which follows from Lemma 1.

Although there can be multiple equivalent formulations of the same problem (e.g., (P1), (P2), (5.8), and (D1)/(D2) are equivalent), an algorithm may or may not be applicable to some of them. Even when they are, on different formulations, their behaviors such as convergence and speed of convergence are different. In particular, most algorithms have different behaviors on primal and dual formulations of the same problem. An algorithm applied to a primal formulation does not dictate the behaviors of the same algorithm applied to the related dual formulation. The simplex method in linear programming has different performance when applied to both the primal and dual problems, i.e., the primal simplex method starts with a primal basic feasible solution (dual infeasible) until the dual feasibility conditions are satisfied, while the dual simplex method starts with a dual basic feasible solution (primal infeasible) until the primal feasibility conditions are satisfied. The ALM also has different performance when applied to the primal and dual problems, i.e., ALM applied to the primal problem is equivalent to proximal point method applied to the related dual problem, and proximal point method is, in general, different from ALM on the same problem.

4 Primal-Dual Equivalence of ADM

In this section we show the primal-dual equivalence of ADM. Algorithms 1–3 describe how ADM is applied to (P1), (P2), and (D1)/(D2) [14, 15].

Algorithm 1 ADM on (P1)

initialize $\mathbf{x}_1^0$, $\mathbf{z}_1^0$, $\lambda > 0$
for $k = 0, 1, \cdots$ **do**
  $\mathbf{y}_1^{k+1} \in \arg\min_{\mathbf{y}} g(\mathbf{y}) + (2\lambda)^{-1}\|\mathbf{A}\mathbf{x}_1^k + \mathbf{B}\mathbf{y} - \mathbf{b} + \lambda\mathbf{z}_1^k\|_2^2$
  $\mathbf{x}_1^{k+1} \in \arg\min_{\mathbf{x}} f(\mathbf{x}) + (2\lambda)^{-1}\|\mathbf{A}\mathbf{x} + \mathbf{B}\mathbf{y}_1^{k+1} - \mathbf{b} + \lambda\mathbf{z}_1^k\|_2^2$
  $\mathbf{z}_1^{k+1} = \mathbf{z}_1^k + \lambda^{-1}(\mathbf{A}\mathbf{x}_1^{k+1} + \mathbf{B}\mathbf{y}_1^{k+1} - \mathbf{b})$
end for

Algorithm 2 ADM on (P2)

initialize $\mathbf{s}_2^0$, $\mathbf{z}_2^0$, $\lambda > 0$
for $k = 0, 1, \cdots$ **do**
$\mathbf{t}_2^{k+1} = \arg\min_{\mathbf{t}} G(\mathbf{t}) + (2\lambda)^{-1}\|\mathbf{s}_2^k + \mathbf{t} + \lambda\mathbf{z}_2^k\|_2^2$
$\mathbf{s}_2^{k+1} = \arg\min_{\mathbf{s}} F(\mathbf{s}) + (2\lambda)^{-1}\|\mathbf{s} + \mathbf{t}_2^{k+1} + \lambda\mathbf{z}_2^k\|_2^2$
$\mathbf{z}_2^{k+1} = \mathbf{z}_2^k + \lambda^{-1}(\mathbf{s}_2^{k+1} + \mathbf{t}_2^{k+1})$
end for

Algorithm 3 ADM on (D1)/(D2)

initialize $\mathbf{u}_3^0$, $\mathbf{z}_3^0$, $\lambda > 0$
for $k = 0, 1, \cdots$ **do**
$\mathbf{v}_3^{k+1} = \arg\min_{\mathbf{v}} G^*(-\mathbf{v}) + \frac{\lambda}{2}\|\mathbf{u}_3^k - \mathbf{v} + \lambda^{-1}\mathbf{z}_3^k\|_2^2$
$\mathbf{u}_3^{k+1} = \arg\min_{\mathbf{u}} F^*(-\mathbf{u}) + \frac{\lambda}{2}\|\mathbf{u} - \mathbf{v}_3^{k+1} + \lambda^{-1}\mathbf{z}_3^k\|_2^2$
$\mathbf{z}_3^{k+1} = \mathbf{z}_3^k + \lambda(\mathbf{u}_3^{k+1} - \mathbf{v}_3^{k+1})$
end for

The $\mathbf{y}_1^k$ and $\mathbf{x}_1^k$ in Algorithm 1 may not be unique because of the matrices $\mathbf{A}$ and $\mathbf{B}$, while $\mathbf{A}\mathbf{x}_1^k$ and $\mathbf{B}\mathbf{y}_1^k$ are unique. In addition, $\mathbf{A}\mathbf{x}_1^k$ and $\mathbf{B}\mathbf{y}_1^k$ are calculated for twice and thus stored in the implementation of Algorithm 1 to save the second calculation. Following the equivalence of Algorithms 1 and 2 in Part 1 of the following Theorem 1, we can view problem (P2) as the *master problem* of (P1). We can say that ADM is essentially an algorithm applied only to the master problem (P2), which is Algorithm 2; this fact has been obscured by the often-seen Algorithm 1, which integrates ADM on the master problem with the independent subproblems in (5.7).

Theorem 1 (Equivalence of Algorithms 1–3). *Suppose* $\mathbf{A}\mathbf{x}_1^0 = \mathbf{s}_2^0 = \mathbf{z}_3^0$ *and* $\mathbf{z}_1^0 = \mathbf{z}_2^0 = \mathbf{u}_3^0$ *and that the same parameter* λ *is used in Algorithms 1–3. Then, their equivalence can be established as follows.*

1. From $\mathbf{x}_1^k$, $\mathbf{y}_1^k$, $\mathbf{z}_1^k$ *of Algorithm 1, we obtain* $\mathbf{t}_2^k$, $\mathbf{s}_2^k$, $\mathbf{z}_2^k$ *of Algorithm 2 through:*

$$\mathbf{t}_2^k = \mathbf{B}\mathbf{y}_1^k - \mathbf{b}, \tag{5.9a}$$

$$\mathbf{s}_2^k = \mathbf{A}\mathbf{x}_1^k, \tag{5.9b}$$

$$\mathbf{z}_2^k = \mathbf{z}_1^k. \tag{5.9c}$$

From $\mathbf{t}_2^k$, $\mathbf{s}_2^k$, $\mathbf{z}_2^k$ *of Algorithm 2, we obtain* $\mathbf{y}_1^k$, $\mathbf{x}_1^k$, $\mathbf{z}_1^k$ *of Algorithm 1 through:*

$$\mathbf{y}_1^k = \arg\min_{\mathbf{y}}\{g(\mathbf{y}) : \mathbf{B}\mathbf{y} - \mathbf{b} = \mathbf{t}_2^k\}, \tag{5.10a}$$

$$\mathbf{x}_1^k = \arg\min_{\mathbf{x}}\{f(\mathbf{x}) : \mathbf{A}\mathbf{x} = \mathbf{s}_2^k\}, \tag{5.10b}$$

$$\mathbf{z}_1^k = \mathbf{z}_2^k. \tag{5.10c}$$

2. *We can recover the iterates of Algorithms 2 and 3 from each other through*

$$\mathbf{u}_3^k = \mathbf{z}_2^k, \qquad \mathbf{z}_3^k = \mathbf{s}_2^k. \tag{5.11}$$

Proof. Part 1. Proof by induction.

We argue that under (5.9b) and (5.9c), Algorithms 1 and 2 have *essentially identical* subproblems in their *first* steps at the kth iteration. Consider the following problem, which is obtained by plugging the definition of $G(\cdot)$ into the $\mathbf{t}_2^{k+1}$-subproblem of Algorithm 2:

$$(\mathbf{y}_1^{k+1}, \mathbf{t}_2^{k+1}) = \arg\min_{\mathbf{y},\mathbf{t}} g(\mathbf{y}) + \iota_{\{(\mathbf{y},\mathbf{t}):\mathbf{B}\mathbf{y}-\mathbf{b}=\mathbf{t}\}}(\mathbf{y},\mathbf{t}) + (2\lambda)^{-1}\|\mathbf{s}_2^k + \mathbf{t} + \lambda \mathbf{z}_2^k\|_2^2. \tag{5.12}$$

If one minimizes over $\mathbf{y}$ first while keeping $\mathbf{t}$ as a variable, one eliminates $\mathbf{y}$ and recovers the $\mathbf{t}_2^{k+1}$-subproblem of Algorithm 2. If one minimizes over $\mathbf{t}$ first while keeping $\mathbf{y}$ as a variable, then after plugging in (5.9b) and (5.9c), problem (5.12) reduces to the $\mathbf{y}_1^{k+1}$-subproblem of Algorithm 1. In addition, $(\mathbf{y}_1^{k+1}, \mathbf{t}_2^{k+1})$ obeys

$$\mathbf{t}_2^{k+1} = \mathbf{B}\mathbf{y}_1^{k+1} - \mathbf{b}, \tag{5.13}$$

which is (5.9a) at $k+1$. Plugging $\mathbf{t} = \mathbf{t}_2^{k+1}$ into (5.12) yields problem (5.10a) for $\mathbf{y}_1^{k+1}$, which must be equivalent to the $\mathbf{y}_1^{k+1}$-subproblem of Algorithm 2. Therefore, the $\mathbf{y}_1^{k+1}$-subproblem of Algorithm 1 and the $\mathbf{t}_2^{k+1}$-subproblem of Algorithm 2 are equivalent through (5.9a) and (5.10a) at $k+1$, respectively.

Similarly, under (5.13) and (5.9c), we can show that the $\mathbf{x}_1^{k+1}$-subproblem of Algorithm 1 and the $\mathbf{s}_2^{k+1}$-subproblem of Algorithm 2 are equivalent through the formulas for (5.9b) and (5.10b) at $k+1$, respectively.

Finally, under (5.9a) and (5.9b) at $k+1$ and $\mathbf{z}_2^k = \mathbf{z}_1^k$, the formulas for $\mathbf{z}_1^{k+1}$ and $\mathbf{z}_2^{k+1}$ in Algorithms 1 and 2 are identical, and they return $\mathbf{z}_1^{k+1} = \mathbf{z}_2^{k+1}$, which is (5.9c) and (5.10c) at $k+1$.

Part 2. Proof by induction. Suppose that (5.11) holds. We shall show that (5.11) holds at $k+1$. Starting from the optimality condition of the $\mathbf{t}_2^{k+1}$-subproblem of Algorithm 2, we derive

$$\begin{aligned}
&\mathbf{0} \in \partial G(\mathbf{t}_2^{k+1}) + \lambda^{-1}(\mathbf{s}_2^k + \mathbf{t}_2^{k+1} + \lambda \mathbf{z}_2^k)\\
\iff& \mathbf{t}_2^{k+1} \in \partial G^*(-\lambda^{-1}(\mathbf{s}_2^k + \mathbf{t}_2^{k+1} + \lambda \mathbf{z}_2^k))\\
\iff& \lambda\left[\lambda^{-1}(\mathbf{s}_2^k + \mathbf{t}_2^{k+1} + \lambda \mathbf{z}_2^k)\right] - (\lambda \mathbf{z}_2^k + \mathbf{s}_2^k) \in \partial G^*(-\lambda^{-1}(\mathbf{s}_2^k + \mathbf{t}_2^{k+1} + \lambda \mathbf{z}_2^k))\\
\iff& -\lambda\left[\lambda^{-1}(\mathbf{s}_2^k + \mathbf{t}_2^{k+1} + \lambda \mathbf{z}_2^k)\right] + (\lambda \mathbf{u}_3^k + \mathbf{z}_3^k) \in -\partial G^*(-\lambda^{-1}(\mathbf{s}_2^k + \mathbf{t}_2^{k+1} + \lambda \mathbf{z}_2^k))\\
\iff& \mathbf{0} \in -\partial G^*(-\lambda^{-1}(\mathbf{s}_2^k + \mathbf{t}_2^{k+1} + \lambda \mathbf{z}_2^k)) - \lambda\left[\mathbf{u}_3^k - \lambda^{-1}(\mathbf{s}_2^k + \mathbf{t}_2^{k+1} + \lambda \mathbf{z}_2^k) + \lambda^{-1}\mathbf{z}_3^k\right]\\
\iff& \mathbf{v}_3^{k+1} = \lambda^{-1}(\mathbf{s}_2^k + \mathbf{t}_2^{k+1} + \lambda \mathbf{z}_2^k) = \lambda^{-1}(\mathbf{z}_3^k + \mathbf{t}_2^{k+1} + \lambda \mathbf{z}_2^k),
\end{aligned}$$

where the last equivalence follows from the optimality condition for the $\mathbf{v}_3^{k+1}$-subproblem of Algorithm 3.

Starting from the optimality condition of the $\mathbf{s}_2^{k+1}$-subproblem of Algorithm 2, and applying the update, $\mathbf{z}_2^{k+1} = \mathbf{z}_2^k + \lambda^{-1}(\mathbf{s}_2^{k+1} + \mathbf{t}_2^{k+1})$, in Algorithm 2 and the identity of $\mathbf{t}_2^{k+1}$ obtained above, we derive

$$\begin{aligned}
&\mathbf{0} \in \partial F(\mathbf{s}_2^{k+1}) + \lambda^{-1}(\mathbf{s}_2^{k+1} + \mathbf{t}_2^{k+1} + \lambda \mathbf{z}_2^k) \\
\Longleftrightarrow\ &\mathbf{0} \in \partial F(\mathbf{s}_2^{k+1}) + \mathbf{z}_2^{k+1} \\
\Longleftrightarrow\ &\mathbf{0} \in \mathbf{s}_2^{k+1} - \partial F^*(-\mathbf{z}_2^{k+1}) \\
\Longleftrightarrow\ &\mathbf{0} \in \lambda(\mathbf{z}_2^{k+1} - \mathbf{z}_2^k) - \mathbf{t}_2^{k+1} - \partial F^*(-\mathbf{z}_2^{k+1}) \\
\Longleftrightarrow\ &\mathbf{0} \in \lambda(\mathbf{z}_2^{k+1} - \mathbf{z}_2^k) + \mathbf{z}_3^k + \lambda(\mathbf{z}_2^k - \mathbf{v}_3^{k+1}) - \partial F^*(-\mathbf{z}_2^{k+1}) \\
\Longleftrightarrow\ &\mathbf{0} \in -\partial F^*(-\mathbf{z}_2^{k+1}) + \lambda(\mathbf{z}_2^{k+1} - \mathbf{v}_3^{k+1} + \lambda^{-1}\mathbf{z}_3^k) \\
\Longleftrightarrow\ &\mathbf{z}_2^{k+1} = \mathbf{u}_3^{k+1}.
\end{aligned}$$

where the last equivalence follows from the optimality condition for the $\mathbf{u}_3^{k+1}$-subproblem of Algorithm 3. Finally, combining the update formulas of $\mathbf{z}_2^{k+1}$ and $\mathbf{z}_3^{k+1}$ in Algorithms 2 and 3, respectively, as well as the identities for $\mathbf{u}_3^{k+1}$ and $\mathbf{v}_3^{k+1}$ obtained above, we obtain

$$\begin{aligned}
\mathbf{z}_3^{k+1} &= \mathbf{z}_3^k + \lambda(\mathbf{u}_3^{k+1} - \mathbf{v}_3^{k+1}) = \mathbf{s}^k + \lambda(\mathbf{z}_2^{k+1} - \mathbf{z}_2^k - \lambda^{-1}(\mathbf{s}_2^k + \mathbf{t}_2^{k+1})) \\
&= \lambda(\mathbf{z}_2^{k+1} - \mathbf{z}_2^k) - \mathbf{t}_2^{k+1} = \mathbf{s}_2^{k+1}.
\end{aligned}$$

□

Remark 2. Part 2 of the theorem (ADM's primal-dual equivalence) can also be derived by combining the following two equivalence results: (i) the equivalence between ADM on the primal problem and the Douglas-Rachford splitting (DRS) algorithm [7, 18] on the dual problem [13], and (ii) the equivalence result between DRS algorithms applied to the master problem (P2) and its dual problem (cf. [8, Chapter 3.5] [9]). In this chapter, however, we provide an elementary algebraic proof in order to derive the formulas in Theorem 1 that recover the iterates of one algorithm from another.

Part 2 of the theorem shows that ADM is a symmetric primal-dual algorithm. The reciprocal positions of parameter λ indicates its function to "balance" the primal and dual progresses.

Part 2 of the theorem also shows that Algorithms 2 and 3 have no difference, in terms of per-iteration complexity and the number of iterations needed to reach an accuracy. However, Algorithms 1 and 2 have difference in terms of per-iteration complexity. In fact, Algorithm 2 is implemented for Algorithm 1 because Algorithm 2 has smaller complexity than Algorithm 1. See the examples in Sections 4.2 and 4.3.

4.1 Primal-Dual Equivalence of ADM on (5.1) with Three Subproblems

In Section 1.1, we introduced four different ways to apply ADM on (5.1) with three subproblems. The ADM-ready formulation for the primal problem is (5.5), and the ADM applied to this formulation is

$$\mathbf{x}^{k+1} = \arg\min_{\mathbf{x}} \|\mathbf{x} - \mathbf{s}^k + \lambda \mathbf{z}_{\mathbf{s}}^k\|_2^2 + \|\mathbf{C}\mathbf{x} - \mathbf{y}^k + \lambda \mathbf{z}_{\mathbf{y}}^k\|_2^2, \tag{5.14a}$$

$$\mathbf{s}^{k+1} = \arg\min_{\mathbf{s}} u(\mathbf{s}) + (2\lambda)^{-1}\|\mathbf{x}^{k+1} - \mathbf{s} + \lambda \mathbf{z}_{\mathbf{s}}^k\|_2^2, \tag{5.14b}$$

$$\mathbf{y}^{k+1} = \arg\min_{\mathbf{y}} v(\mathbf{y}) + (2\lambda)^{-1}\|\mathbf{C}\mathbf{x}^{k+1} - \mathbf{y} + \lambda \mathbf{z}_{\mathbf{y}}^k\|_2^2, \tag{5.14c}$$

$$\mathbf{z}_{\mathbf{s}}^{k+1} = \mathbf{z}_{\mathbf{s}}^k + \lambda^{-1}(\mathbf{x}^{k+1} - \mathbf{s}^{k+1}), \tag{5.14d}$$

$$\mathbf{z}_{\mathbf{y}}^{k+1} = \mathbf{z}_{\mathbf{y}}^k + \lambda^{-1}(\mathbf{C}\mathbf{x}^{k+1} - \mathbf{y}^{k+1}). \tag{5.14e}$$

Similarly, we can introduce a dummy variable $\mathbf{t}$ into the left formulation in (5.4) and obtain a new equivalent formulation

$$\begin{cases} \underset{\mathbf{u},\mathbf{v},\mathbf{t}}{\text{minimize}} \; u^*(\mathbf{u}) + v^*(\mathbf{t}) \\ \text{subject to } \mathbf{C}^*\mathbf{v} + \mathbf{u} = 0, \; \mathbf{v} - \mathbf{t} = 0. \end{cases} \tag{5.15}$$

The ADM applied to (5.15) is

$$\mathbf{v}^{k+1} = \arg\min_{\mathbf{v}} \|\mathbf{C}^*\mathbf{v} + \mathbf{u}^k + \lambda^{-1}\mathbf{z}_{\mathbf{u}}^k\|_2^2 + \|\mathbf{v} - \mathbf{t}^k + \lambda^{-1}\mathbf{z}_{\mathbf{t}}^k\|_2^2, \tag{5.16a}$$

$$\mathbf{u}^{k+1} = \arg\min_{\mathbf{u}} u^*(\mathbf{u}) + \frac{\lambda}{2}\|\mathbf{C}^*\mathbf{v}^{k+1} + \mathbf{u} + \lambda^{-1}\mathbf{z}_{\mathbf{u}}^k\|_2^2, \tag{5.16b}$$

$$\mathbf{t}^{k+1} = \arg\min_{\mathbf{t}} v^*(\mathbf{t}) + \frac{\lambda}{2}\|\mathbf{v}^{k+1} - \mathbf{t} + \lambda^{-1}\mathbf{z}_{\mathbf{t}}^k\|_2^2, \tag{5.16c}$$

$$\mathbf{z}_{\mathbf{u}}^{k+1} = \mathbf{z}_{\mathbf{u}}^k + \lambda(\mathbf{C}^*\mathbf{v}^{k+1} + \mathbf{u}^{k+1}), \tag{5.16d}$$

$$\mathbf{z}_{\mathbf{t}}^{k+1} = \mathbf{z}_{\mathbf{t}}^k + \lambda(\mathbf{v}^{k+1} - \mathbf{t}^{k+1}). \tag{5.16e}$$

Interestingly, as shown in the following theorem, ADM algorithms (5.14) and (5.16) applied to (5.5) and (5.15) are equivalent.

Theorem 2. *If the initialization for algorithms* (5.14) *and* (5.16) *satisfies* $\mathbf{z}_{\mathbf{y}}^0 = \mathbf{t}^0$, $\mathbf{z}_{\mathbf{s}}^0 = \mathbf{u}^0$, $\mathbf{s}^0 = -\mathbf{z}_{\mathbf{u}}^0$, *and* $\mathbf{y}^0 = \mathbf{z}_{\mathbf{t}}^0$. *Then for* $k \geq 1$, *we have the following equivalence results between the iterations of the two algorithms:*

$$\mathbf{z}_{\mathbf{y}}^k = \mathbf{t}^k, \quad \mathbf{z}_{\mathbf{s}}^k = \mathbf{u}^k, \quad \mathbf{s}^k = -\mathbf{z}_{\mathbf{u}}^k, \quad \mathbf{y}^k = \mathbf{z}_{\mathbf{t}}^k.$$

The proof is similar to the proof of Theorem 1 and is omitted here.

4.2 Example: Basis Pursuit

The basis pursuit problem seeks for the minimal ℓ_1 solution to a set of linear equations:

$$\operatorname*{minimize}_{\mathbf{u}} \|\mathbf{u}\|_1 \quad \text{subject to } \mathbf{A}\mathbf{u} = \mathbf{b}. \tag{5.17}$$

Its Lagrange dual is

$$\operatorname*{minimize}_{\mathbf{x}} -\mathbf{b}^T\mathbf{x} \quad \text{subject to } \|\mathbf{A}^*\mathbf{x}\|_\infty \leq 1. \tag{5.18}$$

The YALL1 algorithms [24] implement ADMs on a set of primal and dual formulations for basis pursuit and LASSO, yet ADM for (5.17) is not given (however, a linearized ADM is given for (5.17)). Although seemingly awkward, problem (5.17) can be turned equivalently into the ADM-ready form

$$\operatorname*{minimize}_{\mathbf{u},\mathbf{v}} \|\mathbf{v}\|_1 + \iota_{\{\mathbf{u}:\mathbf{A}\mathbf{u}=\mathbf{b}\}}(\mathbf{u}) \quad \text{subject to } \mathbf{u} - \mathbf{v} = \mathbf{0}. \tag{5.19}$$

Similarly, problem (5.18) can be turned equivalently into the ADM-ready form

$$\operatorname*{minimize}_{\mathbf{x},\mathbf{y}} -\mathbf{b}^T\mathbf{x} + \iota_{B_1^\infty}(\mathbf{y}) \quad \text{subject to } \mathbf{A}^*\mathbf{x} - \mathbf{y} = \mathbf{0}, \tag{5.20}$$

where $B_1^\infty = \{\mathbf{y} : \|\mathbf{y}\|_\infty \leq 1\}$.

For simplicity, let us suppose that $\mathbf{A}$ has full row rank so the inverse of $\mathbf{A}\mathbf{A}^*$ exists. (Otherwise, $\mathbf{A}\mathbf{u} = \mathbf{b}$ are redundant whenever they are consistent; and $(\mathbf{A}\mathbf{A}^*)^{-1}$ shall be replaced by the pseudo-inverse below.) ADM for problem (5.19) can be simplified to the iteration:

$$\mathbf{v}_3^{k+1} = \arg\min_{\mathbf{v}} \|\mathbf{v}\|_1 + \frac{\lambda}{2}\|\mathbf{u}_3^k - \mathbf{v} + \frac{1}{\lambda}\mathbf{z}_3^k\|_2^2, \tag{5.21a}$$

$$\mathbf{u}_3^{k+1} = \mathbf{v}_3^{k+1} - \frac{1}{\lambda}\mathbf{z}_3^k - \mathbf{A}^*(\mathbf{A}\mathbf{A}^*)^{-1}(\mathbf{A}(\mathbf{v}_3^{k+1} - \frac{1}{\lambda}\mathbf{z}_3^k) - \mathbf{b}), \tag{5.21b}$$

$$\mathbf{z}_3^{k+1} = \mathbf{z}_3^k + \lambda(\mathbf{u}_3^{k+1} - \mathbf{v}_3^{k+1}). \tag{5.21c}$$

And ADM for problem (5.20) can be simplified to the iteration:

$$\mathbf{y}_1^{k+1} = \mathscr{P}_{B_1^\infty}(\mathbf{A}^*\mathbf{x}_1^k + \lambda\mathbf{z}_1^k), \tag{5.22a}$$

$$\mathbf{x}_1^{k+1} = (\mathbf{A}\mathbf{A}^*)^{-1}(\mathbf{A}\mathbf{y}_1^{k+1} - \lambda(\mathbf{A}\mathbf{z}_1^k - \mathbf{b})), \tag{5.22b}$$

$$\mathbf{z}_1^{k+1} = \mathbf{z}_1^k + \lambda^{-1}(\mathbf{A}^*\mathbf{x}_1^{k+1} - \mathbf{y}_1^{k+1}), \tag{5.22c}$$

where $\mathscr{P}_{B_1^\infty}$ is the projection onto B_1^∞. Looking into the iteration in (5.22), we can find that $\mathbf{A}^*\mathbf{x}_1^k$ is used in both the kth and $k+1$st iterations. To save the computation, we can store $\mathbf{A}^*\mathbf{x}_1^k$ as $\mathbf{s}_2^k$. In addition, let $\mathbf{t}_2^k = \mathbf{y}_1^k$ and $\mathbf{z}_2^k = \mathbf{z}_1^k$, we have

$$\mathbf{t}_2^{k+1} = \mathscr{P}_{B_1^\infty}(\mathbf{s}_2^k + \lambda \mathbf{z}_2^k), \tag{5.23a}$$

$$\mathbf{s}_2^{k+1} = \mathbf{A}^*(\mathbf{A}\mathbf{A}^*)^{-1}(\mathbf{A}(\mathbf{t}_2^{k+1} - \lambda \mathbf{A}\mathbf{z}_2^k) + \lambda \mathbf{b})), \tag{5.23b}$$

$$\mathbf{z}_2^{k+1} = \mathbf{z}_2^k + \lambda^{-1}(\mathbf{s}_2^{k+1} - \mathbf{t}_2^{k+1}), \tag{5.23c}$$

which is exactly Algorithm 2 for (5.20). Thus, Algorithm 2 has smaller complexity than Algorithm 1, i.e., one matrix vector multiplication $\mathbf{A}^*\mathbf{x}_1^k$ is saved from Algorithm 2.

The corollary below follows directly from Theorem 1 by associating (5.20) and (5.19) as (P1) and (D2), and (5.22) and (5.21) with the iterations of Algorithms 1 and 3, respectively.

Corollary 1. *Suppose that* $\mathbf{A}\mathbf{u} = \mathbf{b}$ *are consistent. Consider ADM iterations* (5.21) *and* (5.22)*. Let* $\mathbf{u}_3^0 = \mathbf{z}_1^0$ *and* $\mathbf{z}_3^0 = \mathbf{A}^*\mathbf{x}_1^0$*. Then, for* $k \geq 1$*, iterations* (5.21) *and* (5.22) *are equivalent. In particular,*

- *From* $\mathbf{x}_1^k$, $\mathbf{z}_1^k$ *in* (5.22)*, we obtain* $\mathbf{u}_3^k$, $\mathbf{z}_3^k$ *in* (5.21) *through:*

$$\mathbf{u}_3^k = \mathbf{z}_1^k, \qquad \mathbf{z}_3^k = \mathbf{A}^*\mathbf{x}_1^k.$$

- *From* $\mathbf{u}_3^k$, $\mathbf{z}_3^k$ *in* (5.21)*, we obtain* $\mathbf{x}_1^k$, $\mathbf{z}_1^k$ *in* (5.22) *through:*

$$\mathbf{x}_1^k = (\mathbf{A}\mathbf{A}^*)^{-1}\mathbf{A}\mathbf{z}_3^k, \qquad \mathbf{z}_1^k = \mathbf{u}_3^k.$$

4.3 Example: Basis Pursuit Denoising

The basis pursuit denoising problem is

$$\underset{\mathbf{u}}{\text{minimize}}\, \|\mathbf{u}\|_1 + \frac{1}{2\alpha}\|\mathbf{A}\mathbf{u} - \mathbf{b}\|_2^2 \tag{5.24}$$

and its Lagrange dual, in the ADM-ready form, is

$$\underset{\mathbf{x},\mathbf{y}}{\text{minimize}} - \langle \mathbf{b}, \mathbf{x}\rangle + \frac{\alpha}{2}\|\mathbf{x}\|_2^2 + \iota_{B_1^\infty}(\mathbf{y}) \quad \text{subject to } \mathbf{A}^*\mathbf{x} - \mathbf{y} = \mathbf{0}. \tag{5.25}$$

The iteration of ADM for (5.25) is

$$\mathbf{y}_1^{k+1} = \mathscr{P}_{B_1^\infty}(\mathbf{A}^*\mathbf{x}_1^k + \lambda \mathbf{z}_1^k), \tag{5.26a}$$

$$\mathbf{x}_1^{k+1} = (\mathbf{A}\mathbf{A}^* + \alpha\lambda \mathbf{I})^{-1}(\mathbf{A}\mathbf{y}_1^{k+1} - \lambda(\mathbf{A}\mathbf{z}_1^k - \mathbf{b})), \tag{5.26b}$$

$$\mathbf{z}_1^{k+1} = \mathbf{z}_1^k + \lambda^{-1}(\mathbf{A}^*\mathbf{x}_1^{k+1} - \mathbf{y}_1^{k+1}). \tag{5.26c}$$

Looking into the iteration in (5.26), we can find that $\mathbf{A}^*\mathbf{x}_1^k$ is used in both the kth and $k+1$st iterations. To save the computation, we can store $\mathbf{A}^*\mathbf{x}_1^k$ as $\mathbf{s}_2^k$. In addition, let $\mathbf{t}_2^k = \mathbf{y}_1^k$ and $\mathbf{z}_2^k = \mathbf{z}_1^k$, we have

$$\mathbf{t}_2^{k+1} = \mathscr{P}_{B_1^\infty}(\mathbf{s}_2^k + \lambda \mathbf{z}_2^k), \tag{5.27a}$$
$$\mathbf{s}_2^{k+1} = \mathbf{A}^*(\mathbf{A}\mathbf{A}^* + \alpha\lambda \mathbf{I})^{-1}(\mathbf{A}(\mathbf{t}_2^{k+1} - \lambda \mathbf{z}_2^k) + \lambda \mathbf{b})), \tag{5.27b}$$
$$\mathbf{z}_2^{k+1} = \mathbf{z}_2^k + \lambda^{-1}(\mathbf{s}_2^{k+1} - \mathbf{t}_2^{k+1}), \tag{5.27c}$$

which is exactly Algorithm 2 for (5.25). Thus, Algorithm 2 has a lower per iteration complexity than Algorithm 1, i.e., one matrix vector multiplication $\mathbf{A}^*\mathbf{x}_1^k$ is saved from Algorithm 2. In addition, if $\mathbf{A}^*\mathbf{A} = \mathbf{I}$, (5.27b) becomes

$$\mathbf{s}_2^{k+1} = (\alpha\lambda + 1)^{-1}(\mathbf{t}_2^{k+1} - \lambda \mathbf{z}_2^k + \lambda \mathbf{A}^*\mathbf{b}), \tag{5.28}$$

and no matrix vector multiplications is needed during the iteration because $\lambda \mathbf{A}^*\mathbf{b}$ can be precalculated.

The ADM-ready form of the original problem (5.24) is

$$\underset{\mathbf{u},\mathbf{v}}{\text{minimize}}\, \|\mathbf{v}\|_1 + \frac{1}{2\alpha}\|\mathbf{A}\mathbf{u} - \mathbf{b}\|_2^2 \quad \text{subject to } \mathbf{u} - \mathbf{v} = \mathbf{0}, \tag{5.29}$$

whose ADM iteration is

$$\mathbf{v}_3^{k+1} = \arg\min_{\mathbf{v}} \|\mathbf{v}\|_1 + \frac{\lambda}{2}\|\mathbf{u}_3^k - \mathbf{v} + \frac{1}{\lambda}\mathbf{z}_3^k\|_2^2, \tag{5.30a}$$
$$\mathbf{u}_3^{k+1} = (\mathbf{A}^*\mathbf{A} + \alpha\lambda \mathbf{I})^{-1}(\mathbf{A}^*\mathbf{b} + \alpha\lambda \mathbf{v}_3^{k+1} - \alpha \mathbf{z}_3^k), \tag{5.30b}$$
$$\mathbf{z}_3^{k+1} = \mathbf{z}_3^k + \lambda(\mathbf{u}_3^{k+1} - \mathbf{v}_3^{k+1}). \tag{5.30c}$$

The corollary below follows directly from Theorem 1.

Corollary 2. *Consider ADM iterations* (5.26) *and* (5.30)*. Let* $\mathbf{u}_3^0 = \mathbf{z}_1^0$ *and* $\mathbf{z}_3^0 = \mathbf{A}^*\mathbf{x}_1^0$*. For* $k \geq 1$*, ADM on the dual and primal problems* (5.26) *and* (5.30) *are equivalent in the following way:*

- *From* $\mathbf{x}_1^k$, $\mathbf{z}_1^k$ *in* (5.26)*, we recover* $\mathbf{u}_3^k$, $\mathbf{z}_3^k$ *in* (5.30) *through:*

$$\mathbf{u}_3^k = \mathbf{z}_1^k, \qquad \mathbf{z}_3^k = \mathbf{A}^*\mathbf{x}_1^k.$$

- *From* $\mathbf{u}_3^k$, $\mathbf{z}_3^k$ *in* (5.30)*, we recover* $\mathbf{x}_1^k$, $\mathbf{z}_1^k$ *in* (5.26) *through:*

$$\mathbf{x}_1^k = -(\mathbf{A}\mathbf{u}_3^k - \mathbf{b})/\alpha, \qquad \mathbf{z}_1^k = \mathbf{u}_3^k.$$

Remark 3. Iteration (5.30) is different from that of ADM for another ADM-ready form of (5.24)

$$\underset{\mathbf{u},\mathbf{v}}{\text{minimize}}\, \|\mathbf{u}\|_1 + \frac{1}{2\alpha}\|\mathbf{v}\|_2^2 \quad \text{subject to } \mathbf{A}\mathbf{u} - \mathbf{v} = \mathbf{b}, \tag{5.31}$$

which is used in [24]. In general, there are different ADM-ready forms and their ADM algorithms yield different iterates. ADM on one ADM-ready form is equivalent to it on the corresponding dual ADM-ready form.

5 ADM as a Primal-Dual Algorithm on the Saddle-Point Problem

As shown in Section 4, ADM on a pair of convex primal and dual problems are equivalent, and there is a connection between $\mathbf{z}_1^k$ in Algorithm 1 and dual variable $\mathbf{u}_3^k$ in Algorithm 3. This primal-dual equivalence naturally suggests that ADM is also equivalent to a primal-dual algorithm involving both primal and dual variables.

We derive problem (P1) into an equivalent primal-dual saddle-point problem (5.33) as follows:

$$\begin{aligned} &\min_{\mathbf{y},\mathbf{x}} g(\mathbf{y}) + f(\mathbf{x}) + \iota_{\{(\mathbf{x},\mathbf{y}):\mathbf{A}\mathbf{x}=\mathbf{b}-\mathbf{B}\mathbf{y}\}}(\mathbf{x},\mathbf{y}) \\ =&\min_{\mathbf{y}} g(\mathbf{y}) + F(\mathbf{b}-\mathbf{B}\mathbf{y}) \\ =&\min_{\mathbf{y}}\max_{\mathbf{u}} g(\mathbf{y}) + \langle -\mathbf{u}, \mathbf{b}-\mathbf{B}\mathbf{y}\rangle - F^*(-\mathbf{u}) \qquad (5.32) \\ =&\min_{\mathbf{y}}\max_{\mathbf{u}} g(\mathbf{y}) + \langle \mathbf{u}, \mathbf{B}\mathbf{y}-\mathbf{b}\rangle - f^*(-\mathbf{A}^*\mathbf{u}). \qquad (5.33) \end{aligned}$$

A primal-dual algorithm for solving (5.33) is described in Algorithm 4. Theorem 3 establishes the equivalence between Algorithms 1 and 4.

Algorithm 4 Primal-dual formulation of ADM on problem (5.33)

initialize $\mathbf{u}_4^0$, $\mathbf{u}_4^{-1}$, $\mathbf{y}_4^0$, $\lambda > 0$
for $k = 0, 1, \cdots$ **do**
  $\bar{\mathbf{u}}_4^k = 2\mathbf{u}_4^k - \mathbf{u}_4^{k-1}$
  $\mathbf{y}_4^{k+1} = \arg\min_{\mathbf{y}} g(\mathbf{y}) + (2\lambda)^{-1}\|\mathbf{B}\mathbf{y} - \mathbf{B}\mathbf{y}_4^k + \lambda\bar{\mathbf{u}}_4^k\|_2^2$
  $\mathbf{u}_4^{k+1} = \arg\min_{\mathbf{u}} f^*(-\mathbf{A}^*\mathbf{u}) - \langle \mathbf{u}, \mathbf{B}\mathbf{y}_4^{k+1} - \mathbf{b}\rangle + \lambda/2\|\mathbf{u} - \mathbf{u}_4^k\|_2^2$
end for

Remark 4. Publication [3] proposed a primal-dual algorithm for (5.32) and obtained its connection to ADM [10]: When $\mathbf{B} = \mathbf{I}$, ADM is equivalent to the primal-dual algorithm in [3]; When $\mathbf{B} \neq \mathbf{I}$, the primal-dual algorithm is a preconditioned ADM as an additional proximal term $\delta/2\|\mathbf{y} - \mathbf{y}_4^k\|_2^2 - (2\lambda)^{-1}\|\mathbf{B}\mathbf{y} - \mathbf{B}\mathbf{y}_4^k\|_2^2$ is added to the subproblem for $\mathbf{y}_4^{k+1}$. This is also a special case of inexact ADM in [6]. Our Algorithm 4 is a primal-dual algorithm that is equivalent to ADM in the general case.

Theorem 3 (Equivalence between Algorithms 1 and 4). *Suppose that* $\mathbf{A}\mathbf{x}_1^0 = \lambda(\mathbf{u}_4^0 - \mathbf{u}_4^{-1}) + \mathbf{b} - \mathbf{B}\mathbf{y}_4^0$ *and* $\mathbf{z}_1^0 = \mathbf{u}_4^0$. *Then, Algorithms 1 and 4 are equivalent with the identities:*

$$\mathbf{A}\mathbf{x}_1^k = \lambda(\mathbf{u}_4^k - \mathbf{u}_4^{k-1}) + \mathbf{b} - \mathbf{B}\mathbf{y}_4^k, \qquad \mathbf{z}_1^k = \mathbf{u}_4^k, \tag{5.34}$$

for all $k > 0$.

Proof. By assumption, (5.34) holds at iteration $k = 0$.
Proof by induction. Suppose that (5.34) holds at iteration $k \geq 0$. We shall establish (5.34) at iteration $k+1$. From the first step of Algorithm 1, we have

$$\begin{aligned}
\mathbf{y}_1^{k+1} &= \arg\min_{\mathbf{y}} g(\mathbf{y}) + (2\lambda)^{-1}\|\mathbf{A}\mathbf{x}_1^k + \mathbf{B}\mathbf{y} - \mathbf{b} + \lambda\mathbf{z}_1^k\|_2^2 \\
&= \arg\min_{\mathbf{y}} g(\mathbf{y}) + (2\lambda)^{-1}\|\lambda(\mathbf{u}_4^k - \mathbf{u}_4^{k-1}) + \mathbf{B}\mathbf{y} - \mathbf{B}\mathbf{y}_4^k + \lambda\mathbf{u}_4^k\|_2^2,
\end{aligned}$$

which is the same as the first step in Algorithm 4. Thus we have $\mathbf{y}_1^{k+1} = \mathbf{y}_4^{k+1}$.

Combing the second and third steps of Algorithm 1, we have

$$\mathbf{0} \in \partial f(\mathbf{x}_1^{k+1}) + \lambda^{-1}\mathbf{A}^*(\mathbf{A}\mathbf{x}_1^{k+1} + \mathbf{B}\mathbf{y}_1^{k+1} - \mathbf{b} + \lambda\mathbf{z}_1^k) = \partial f(\mathbf{x}_1^{k+1}) + \mathbf{A}^*\mathbf{z}_1^{k+1}.$$

Therefore,

$$\begin{aligned}
& \mathbf{x}_1^{k+1} \in \partial f^*(-\mathbf{A}^*\mathbf{z}_1^{k+1}) \\
\Longrightarrow\ & \mathbf{A}\mathbf{x}_1^{k+1} \in \partial F^*(-\mathbf{z}_1^{k+1}) \\
\Longleftrightarrow\ & \lambda(\mathbf{z}_1^{k+1} - \mathbf{z}_1^k) + \mathbf{b} - \mathbf{B}\mathbf{y}_1^{k+1} \in \partial F^*(-\mathbf{z}_1^{k+1}) \\
\Longleftrightarrow\ & \mathbf{z}_1^{k+1} = \arg\min_{\mathbf{z}} F^*(-\mathbf{z}) - \langle \mathbf{z}, \mathbf{B}\mathbf{y}_1^{k+1} - \mathbf{b}\rangle + \lambda/2\|\mathbf{z} - \mathbf{z}_1^k\|_2^2 \\
\Longleftrightarrow\ & \mathbf{z}_1^{k+1} = \arg\min_{\mathbf{z}} f^*(-\mathbf{A}^*\mathbf{z}) - \langle \mathbf{z}, \mathbf{B}\mathbf{y}_4^{k+1} - \mathbf{b}\rangle + \lambda/2\|\mathbf{z} - \mathbf{u}_4^k\|_2^2,
\end{aligned}$$

where the last line is the second step of Algorithm 4. Therefore, we have $\mathbf{z}_1^{k+1} = \mathbf{u}_4^{k+1}$ and $\mathbf{A}\mathbf{x}_1^{k+1} = \lambda(\mathbf{z}_1^{k+1} - \mathbf{z}_1^k) + \mathbf{b} - \mathbf{B}\mathbf{y}_1^{k+1} = \lambda(\mathbf{u}_4^{k+1} - \mathbf{u}_4^k) + \mathbf{b} - \mathbf{B}\mathbf{y}_4^{k+1}$. □

6 Equivalence of ADM for Different Orders

In both problem (P1) and Algorithm 1, we can swap **x** and **y** and obtain Algorithm 5, which is still an ADM algorithm. In general, the two algorithms are different. In this section, we show that for a certain type of functions f (or g), Algorithms 1 and 5 become equivalent.

Algorithm 5 ADM2 on (P1)

initialize $\mathbf{y}_5^0$, $\mathbf{z}_5^0$, $\lambda > 0$
for $k = 0, 1, \cdots$ **do**
$\mathbf{x}_5^{k+1} = \arg\min_{\mathbf{x}} f(\mathbf{x}) + (2\lambda)^{-1}\|\mathbf{Ax} + \mathbf{By}_5^k - \mathbf{b} + \lambda\mathbf{z}_5^k\|_2^2$
$\mathbf{y}_5^{k+1} = \arg\min_{\mathbf{y}} g(\mathbf{y}) + (2\lambda)^{-1}\|\mathbf{Ax}_5^{k+1} + \mathbf{By} - \mathbf{b} + \lambda\mathbf{z}_5^k\|_2^2$
$\mathbf{z}_5^{k+1} = \mathbf{z}_5^k + \lambda^{-1}(\mathbf{Ax}_5^{k+1} + \mathbf{By}_5^{k+1} - \mathbf{b})$
end for

The assumption that we need is that either $\mathbf{prox}_{F(\cdot)}$ or $\mathbf{prox}_{G(\cdot)}$ is affine (cf. (5.7) for the definitions of F and G). The definition of affine mapping is given in Definition 3.

Definition 3. A mapping T is affine if $T(\mathbf{r}) - T(\mathbf{0})$ is linear in $\mathbf{r}$, i.e.,

$$T(\alpha\mathbf{r}_1 + \beta\mathbf{r}_2) - T(\mathbf{0}) = \alpha[T(\mathbf{r}_1) - T(\mathbf{0})] + \beta[T(\mathbf{r}_2) - T(\mathbf{0})], \quad \forall\alpha, \beta \in \mathbf{R}. \tag{5.35}$$

A mapping T is affine if and only if it can be written as a linear mapping plus a constant, and the following proposition provides several equivalent statements for $\mathbf{prox}_{G(\cdot)}$ being affine.

Proposition 1. *Let $\lambda > 0$. The following statements are equivalent:*

1. $\mathbf{prox}_{G(\cdot)}$ *is affine;*
2. $\mathbf{prox}_{\lambda G(\cdot)}$ *is affine;*
3. $a\mathbf{prox}_{G(\cdot)} \circ b\mathbf{I} + c\mathbf{I}$ *is affine for any scalars a, b and c;*
4. $\mathbf{prox}_{G^*(\cdot)}$ *is affine;*
5. *G is convex quadratic (or, affine or constant) and its domain* $\mathrm{dom}(G)$ *is either* $\mathscr{G}$ *or the intersection of hyperplanes in* $\mathscr{G}$.

In addition, if function g is convex quadratic and its domain is the intersection of hyperplanes, then function G defined in (5.7b) *satisfies Part 5 above.*

Proposition 2. *If* $\mathbf{prox}_{G(\cdot)}$ *is affine, then the following holds for any* $\mathbf{r}_1$ *and* $\mathbf{r}_2$*:*

$$\mathbf{prox}_{G(\cdot)}(2\mathbf{r}_1 - \mathbf{r}_2) = 2\mathbf{prox}_{G(\cdot)}\mathbf{r}_1 - \mathbf{prox}_{G(\cdot)}\mathbf{r}_2. \tag{5.36}$$

Proof. Equation (5.36) is obtained by letting $\alpha = 2$ and $\beta = -1$ in (5.35). □

Theorem 4 (Equivalence of Algorithms 1 and 5).

1. *Assume that* $\mathbf{prox}_{\lambda G(\cdot)}$ *is affine. Given the sequences* $\mathbf{y}_5^k$, $\mathbf{z}_5^k$, *and* $\mathbf{x}_5^k$ *of Algorithm 5, if* $\mathbf{y}_5^0$ *and* $\mathbf{z}_5^0$ *satisfy* $-\mathbf{z}_5^0 \in \partial G(\mathbf{By}_5^0 - \mathbf{b})$, *then we can initialize Algorithm 1 with* $\mathbf{x}_1^0 = \mathbf{x}_5^1$ *and* $\mathbf{z}_1^0 = \mathbf{z}_5^0 + \lambda^{-1}(\mathbf{Ax}_5^1 + \mathbf{By}_5^0 - \mathbf{b})$, *and recover the sequences* $\mathbf{x}_1^k$ *and* $\mathbf{z}_1^k$ *of Algorithm 1 through*

$$\mathbf{x}_1^k = \mathbf{x}_5^{k+1}, \tag{5.37a}$$
$$\mathbf{z}_1^k = \mathbf{z}_5^k + \lambda^{-1}(\mathbf{Ax}_5^{k+1} + \mathbf{By}_5^k - \mathbf{b}). \tag{5.37b}$$

2. *Assume that* $\mathbf{prox}_{\lambda F(\cdot)}$ *is affine. Given the sequences* $\mathbf{x}_1^k$, $\mathbf{z}_1^k$, *and* $\mathbf{y}_1^k$ *of Algorithm 1, if* $\mathbf{x}_1^0$ *and* $\mathbf{z}_1^0$ *satisfy* $-\mathbf{z}_1^0 \in \partial F(\mathbf{A}\mathbf{x}_1^0)$, *then we can initialize Algorithm 5 with* $\mathbf{y}_5^0 = \mathbf{y}_1^1$ *and* $\mathbf{z}_5^0 = \mathbf{z}_1^0 + \lambda^{-1}(\mathbf{A}\mathbf{x}_1^0 + \mathbf{B}\mathbf{y}_1^1 - \mathbf{b})$, *and recover the sequences* $\mathbf{y}_5^k$ *and* $\mathbf{z}_5^k$ *of Algorithm 5 through*

$$\mathbf{y}_5^k = \mathbf{y}_1^{k+1}, \tag{5.38a}$$
$$\mathbf{z}_5^k = \mathbf{z}_1^k + \lambda^{-1}(\mathbf{A}\mathbf{x}_1^k + \mathbf{B}\mathbf{y}_1^{k+1} - \mathbf{b}). \tag{5.38b}$$

Proof. We prove Part 1 only by induction. (The proof for the other part is similar.) The initialization of Algorithm 1 clearly follows (5.37) at $k = 0$. Suppose that (5.37) holds at $k \geq 0$. We shall show that (5.37) holds at $k+1$. We first show from the affine property of $\mathbf{prox}_{\lambda G(\cdot)}$ that

$$\mathbf{B}\mathbf{y}_1^{k+1} = 2\mathbf{B}\mathbf{y}_5^{k+1} - \mathbf{B}\mathbf{y}_5^k. \tag{5.39}$$

The optimization subproblems for $\mathbf{y}_1$ and $\mathbf{y}_5$ in Algorithms 1 and 5, respectively, are as follows:

$$\mathbf{y}_1^{k+1} = \operatorname*{arg\,min}_{\mathbf{y}} g(\mathbf{y}) + (2\lambda)^{-1}\|\mathbf{A}\mathbf{x}_1^k + \mathbf{B}\mathbf{y} - \mathbf{b} + \lambda\mathbf{z}_1^k\|_2^2,$$
$$\mathbf{y}_5^{k+1} = \operatorname*{arg\,min}_{\mathbf{y}} g(\mathbf{y}) + (2\lambda)^{-1}\|\mathbf{A}\mathbf{x}_5^{k+1} + \mathbf{B}\mathbf{y} - \mathbf{b} + \lambda\mathbf{z}_5^k\|_2^2.$$

Following the definition of G in (5.7), we have

$$\mathbf{B}\mathbf{y}_1^{k+1} - \mathbf{b} = \mathbf{prox}_{\lambda G(\cdot)}(-\mathbf{A}\mathbf{x}_1^k - \lambda\mathbf{z}_1^k), \tag{5.40a}$$
$$\mathbf{B}\mathbf{y}_5^{k+1} - \mathbf{b} = \mathbf{prox}_{\lambda G(\cdot)}(-\mathbf{A}\mathbf{x}_5^{k+1} - \lambda\mathbf{z}_5^k), \tag{5.40b}$$
$$\mathbf{B}\mathbf{y}_5^k \quad - \mathbf{b} = \mathbf{prox}_{\lambda G(\cdot)}(-\mathbf{A}\mathbf{x}_5^k - \lambda\mathbf{z}_5^{k-1}). \tag{5.40c}$$

The third step of Algorithm 5 is

$$\mathbf{z}_5^k = \mathbf{z}_5^{k-1} + \lambda^{-1}(\mathbf{A}\mathbf{x}_5^k + \mathbf{B}\mathbf{y}_5^k - \mathbf{b}). \tag{5.41}$$

(Note that for $k = 0$, the assumption $-\mathbf{z}_5^0 \in \partial G(\mathbf{B}\mathbf{y}_5^0 - \mathbf{b})$ ensures the existence of $\mathbf{z}_5^{-1}$ in (5.40c) and (5.41).) Then, (5.37) and (5.41) give us

$$\begin{aligned}
\mathbf{A}\mathbf{x}_1^k + \lambda\mathbf{z}_1^k &\overset{(5.37)}{=} \mathbf{A}\mathbf{x}_5^{k+1} + \lambda\mathbf{z}_5^k + \mathbf{A}\mathbf{x}_5^{k+1} + \mathbf{B}\mathbf{y}_5^k - \mathbf{b} \\
&= 2(\mathbf{A}\mathbf{x}_5^{k+1} + \lambda\mathbf{z}_5^k) - (\lambda\mathbf{z}_5^k - \mathbf{B}\mathbf{y}_5^k + \mathbf{b}) \\
&\overset{(5.41)}{=} 2(\mathbf{A}\mathbf{x}_5^{k+1} + \lambda\mathbf{z}_5^k) - (\mathbf{A}\mathbf{x}_5^k + \lambda\mathbf{z}_5^{k-1}).
\end{aligned}$$

Since $\mathbf{prox}_{\lambda G(\cdot)}$ is affine, we have (5.36). Once we plug in (5.36): $\mathbf{r}_1 = -\mathbf{A}\mathbf{x}_5^{k+1} - \lambda\mathbf{z}_5^k$, $\mathbf{r}_2 = -\mathbf{A}\mathbf{x}_5^k - \lambda\mathbf{z}_5^{k-1}$, and $2\mathbf{r}_1 - \mathbf{r}_2 = -\mathbf{A}\mathbf{x}_1^k - \lambda\mathbf{z}_1^k$ and then apply (5.40), we obtain (5.39).

Next, the third step of Algorithm 5 and (5.39) give us

$$\begin{aligned}\mathbf{B}\mathbf{y}_1^{k+1}-\mathbf{b}+\lambda\mathbf{z}_1^k &\overset{(5.39)}{=} 2(\mathbf{B}\mathbf{y}_5^{k+1}-\mathbf{b})-(\mathbf{B}\mathbf{y}_5^k-\mathbf{b})+\lambda\mathbf{z}_5^k+(\mathbf{A}\mathbf{x}_5^{k+1}+\mathbf{B}\mathbf{y}_5^k-\mathbf{b})\\ &= (\mathbf{B}\mathbf{y}_5^{k+1}-\mathbf{b})+\lambda\mathbf{z}_5^k+(\mathbf{A}\mathbf{x}_5^{k+1}+\mathbf{B}\mathbf{y}_5^{k+1}-\mathbf{b})\\ &= (\mathbf{B}\mathbf{y}_5^{k+1}-\mathbf{b})+\lambda\mathbf{z}_5^{k+1}.\end{aligned}$$

This identity shows that the updates of $\mathbf{x}_1^{k+1}$ and $\mathbf{x}_5^{k+2}$ in Algorithms 1 and 5, respectively, have identical data, and therefore, we recover $\mathbf{x}_1^{k+1}=\mathbf{x}_5^{k+2}$.

Lastly, from the third step of Algorithm 1 and the identities above, it follows that

$$\begin{aligned}\mathbf{z}_1^{k+1} &= \mathbf{z}_1^k+\lambda^{-1}(\mathbf{A}\mathbf{x}_1^{k+1}+\mathbf{B}\mathbf{y}_1^{k+1}-\mathbf{b})\\ &= \mathbf{z}_1^k+\lambda^{-1}\left(\mathbf{A}\mathbf{x}_5^{k+2}+(\mathbf{B}\mathbf{y}_5^{k+1}-\mathbf{b}+\lambda\mathbf{z}_5^{k+1}-\lambda\mathbf{z}_1^k)\right)\\ &= \mathbf{z}_5^{k+1}+\lambda^{-1}(\mathbf{A}\mathbf{x}_5^{k+2}+\mathbf{B}\mathbf{y}_5^{k+1}-\mathbf{b}).\end{aligned}$$

Therefore, we obtain (5.37) at $k+1$. □

Remark 5. We can avoid the technical condition $-\mathbf{z}_5^0\in\partial G(\mathbf{B}\mathbf{y}_5^0-\mathbf{b})$ on Algorithm 5 in Part 1 of Theorem 4. When it does not hold, we can use the always-true relation $-\mathbf{z}_5^1\in\partial G(\mathbf{B}\mathbf{y}_5^1-\mathbf{b})$ instead; correspondingly, we shall add one iteration to the iterates of Algorithm 5, namely, initialize Algorithm 1 with $\mathbf{x}_1^0=\mathbf{x}_5^2$ and $\mathbf{z}_1^0=\mathbf{z}_5^1+\lambda^{-1}(\mathbf{A}\mathbf{x}_5^2+\mathbf{B}\mathbf{y}_5^1-\mathbf{b})$ and recover the sequences $\mathbf{x}_1^k$ and $\mathbf{z}_1^k$ of Algorithm 1 through

$$\mathbf{x}_1^k=\mathbf{x}_5^{k+2}, \tag{5.42a}$$

$$\mathbf{z}_1^k=\mathbf{z}_5^{k+1}+\lambda^{-1}(\mathbf{A}\mathbf{x}_5^{k+2}+\mathbf{B}\mathbf{y}_5^{k+1}-\mathbf{b}). \tag{5.42b}$$

Similar arguments apply to the other part of Theorem 4.

7 Equivalence Results of Relaxed PRS

In this section, we consider the following convex problem:

$$\underset{\mathbf{x}}{\text{minimize}} \quad f(\mathbf{x})+g(\mathbf{A}\mathbf{x}), \tag{P3}$$

and its corresponding Lagrangian dual

$$\underset{\mathbf{v}}{\text{minimize}} \quad f^*(\mathbf{A}^*\mathbf{v})+g^*(-\mathbf{v}). \tag{D3}$$

In addition, we introduce another primal-dual pair equivalent to (P3)–(D3):

$$\underset{\mathbf{y}}{\text{minimize}} \quad (f^*\circ\mathbf{A}^*)^*(\mathbf{y})+g(\mathbf{y}), \tag{P4}$$

$$\underset{\mathbf{u}}{\text{minimize}} \quad f^*(\mathbf{u})+(g\circ\mathbf{A})^*(-\mathbf{u}). \tag{D4}$$

Here (P4) is obtained as the dual of (D3) by reformulating (D3) as

$$\underset{\mathbf{v},\bar{\mathbf{v}}}{\text{minimize}} \quad f^*(\mathbf{A}^*\mathbf{v}) + g^*(-\bar{\mathbf{v}}) \text{ subject to } \mathbf{v} = \bar{\mathbf{v}},$$

and (D4) is obtained as the dual of (P3) in a similar way. Lemma 2 below will establish the equivalence between the two primal-dual pairs.

Remark 6. When $\mathbf{A} = \mathbf{I}$, we have $(f^* \circ \mathbf{A}^*)^* = f$, and problem (P3) is exactly the same as problem (P4). Similarly, problem (D3) is exactly the same as problem (D4).

Lemma 2. *Problems* (P3) *and* (P4) *are equivalent in the following sense:*

- *Given any solution* $\mathbf{x}^*$ *to* (P3), $\mathbf{y}^* = \mathbf{A}\mathbf{x}^*$ *is a solution to* (P4),
- *Given any solution* $\mathbf{y}^*$ *to* (P4), $\mathbf{x}^* \in \underset{\mathbf{x}:\mathbf{A}\mathbf{x}=\mathbf{y}^*}{\arg\min} f(\mathbf{x})$ *is a solution to* (P3).

The equivalence between problems (D3) *and* (D4) *is similar:*

- *Given any solution* $\mathbf{v}^*$ *to* (D3), $\mathbf{A}^*\mathbf{v}^*$ *is a solution to* (D4),
- *Given any solution* $\mathbf{u}^*$ *to* (D4), $\mathbf{v}^* \in \underset{\mathbf{v}:\mathbf{A}^*\mathbf{v}=\mathbf{u}^*}{\arg\min} g^*(-\mathbf{v})$ *is a solution to* (D3).

Proof. We prove only the equivalence of (P3) and (P4), the proof for the equivalence of (D3) and (D4) is similar.

Part 1: If $\mathbf{x}^*$ is a solution to (P3), we have $\mathbf{0} \in \partial f(\mathbf{x}^*) + \mathbf{A}^*\partial g(\mathbf{A}\mathbf{x}^*)$. Assume that there exists $\mathbf{q}$ such that $-\mathbf{q} \in \partial g(\mathbf{A}\mathbf{x}^*)$ and $\mathbf{A}^*\mathbf{q} \in \partial f(\mathbf{x}^*)$. Then we have

$$\begin{aligned}\mathbf{A}^*\mathbf{q} \in \partial f(\mathbf{x}^*) &\Longleftrightarrow \mathbf{x}^* \in \partial f^*(\mathbf{A}^*\mathbf{q})\\ &\Longrightarrow \mathbf{A}\mathbf{x}^* \in \mathbf{A}\partial f^*(\mathbf{A}^*\mathbf{q}) = \partial(f^* \circ \mathbf{A}^*)(\mathbf{q})\\ &\Longleftrightarrow \mathbf{q} \in \partial(f^* \circ \mathbf{A}^*)^*(\mathbf{A}\mathbf{x}^*).\end{aligned}$$

Therefore,

$$\mathbf{0} \in \partial(f^* \circ \mathbf{A}^*)^*(\mathbf{A}\mathbf{x}^*) + \partial g(\mathbf{A}\mathbf{x}^*)$$

and $\mathbf{A}\mathbf{x}^*$ is a solution to (P4).

Part 2: If $\mathbf{y}^*$ is a solution to (P4), the optimality condition gives us

$$\mathbf{0} \in \partial(f^* \circ \mathbf{A}^*)^*(\mathbf{y}^*) + \partial g(\mathbf{y}^*).$$

Assume that there exists $\mathbf{q}$ such that $-\mathbf{q} \in \partial g(\mathbf{y}^*)$ and $\mathbf{q} \in \partial(f^* \circ \mathbf{A}^*)^*(\mathbf{y}^*)$. Then we have

$$\mathbf{q} \in \partial(f^* \circ \mathbf{A}^*)^*(\mathbf{y}^*) \Longleftrightarrow \mathbf{y}^* \in \partial(f^* \circ \mathbf{A}^*)(\mathbf{q}). \tag{5.43}$$

Consider the following optimization problem for finding $\mathbf{x}^*$ from $\mathbf{y}^*$

$$\underset{\mathbf{x}}{\text{minimize}}\, f(\mathbf{x}) \qquad \text{subject to } \mathbf{A}\mathbf{x} = \mathbf{y}^*,$$

and the corresponding dual problem

$$\underset{\mathbf{v}}{\text{maximize}} -f^*(\mathbf{A}^*\mathbf{v}) + \langle \mathbf{v}, \mathbf{y}^* \rangle.$$

It is easy to obtain from (5.43) that $\mathbf{q}$ is a solution of the dual problem. The optimal duality gap is zero and the strong duality gives us

$$f(\mathbf{x}^*) = f(\mathbf{x}^*) - \langle \mathbf{q}, \mathbf{A}\mathbf{x}^* - \mathbf{y}^* \rangle = -f^*(\mathbf{A}^*\mathbf{q}) + \langle \mathbf{q}, \mathbf{y}^* \rangle. \tag{5.44}$$

Thus $\mathbf{x}^*$ is a solution of $\underset{\mathbf{x}}{\text{minimize}} f(\mathbf{x}) - \langle \mathbf{A}^*\mathbf{q}, \mathbf{x} \rangle$ and

$$\mathbf{A}^*\mathbf{q} \in \partial f(\mathbf{x}^*) \Longleftrightarrow \mathbf{0} \in \partial f(\mathbf{x}^*) - \mathbf{A}^*\mathbf{q}. \tag{5.45}$$

Because $-\mathbf{q} \in \partial g(\mathbf{y}^*) = \partial g(\mathbf{A}\mathbf{x}^*)$,

$$\mathbf{0} \in \partial f(\mathbf{x}^*) + \mathbf{A}^*\partial \mathbf{g}(\mathbf{A}\mathbf{x}^*) = \partial f(\mathbf{x}^*) + \partial (\mathbf{g} \circ \mathbf{A})(\mathbf{x}^*) \tag{5.46}$$

Therefore $\mathbf{x}^*$ is a solution of (P3). □

Next we will show the equivalence between the RPRS to the primal and dual problems:

$$\boxed{\text{RPRS on (P3)}} \Longleftrightarrow \boxed{\text{RPRS on (D4)}}$$
$$\boxed{\text{RPRS on (P4)}} \Longleftrightarrow \boxed{\text{RPRS on (D3)}}$$

We describe the RPRS on (P3) in Algorithm 6, and the RPRS on other problems can be obtained in the same way.

Algorithm 6 RPRS on (P3)

initialize $\mathbf{w}^0$, $\lambda > 0$, $0 < \alpha \leq 1$.
for $k = 0, 1, \cdots$ **do**
  $\mathbf{x}^{k+1} = \mathbf{prox}_{\lambda f(\cdot)}\mathbf{w}^k$
  $\mathbf{w}^{k+1} = (1-\alpha)\mathbf{w}^k + \alpha(2\mathbf{prox}_{\lambda g \circ \mathbf{A}(\cdot)} - \mathbf{I})(2\mathbf{x}^{k+1} - \mathbf{w}^k)$
end for

Theorem 5 (Primal-dual equivalence of RPRS). *RPRS on* (P3) *is equivalent to RPRS on* (D4). *RPRS on* (P4) *is equivalent to RPRS on* (D3).

Before proving this theorem, we introduce a lemma, which was also given in [8, Proposition 3.34]. Here, we prove it in a different way using the generalized Moreau decomposition.

Lemma 3. *For* $\lambda > 0$, *we have*

$$\begin{aligned} \lambda^{-1}(2\mathbf{prox}_{\lambda F(\cdot)} - \mathbf{I})\mathbf{w} &= (\mathbf{I} - 2\mathbf{prox}_{\lambda^{-1}F^*(\cdot)})(\mathbf{w}/\lambda) \\ &= (2\mathbf{prox}_{\lambda^{-1}F^*(-\cdot)} - \mathbf{I})(-\mathbf{w}/\lambda). \end{aligned} \tag{5.47}$$

Proof. We prove it using the generalized Moreau decomposition [11, Theorem 2.3.1]

$$\mathbf{w} = \mathbf{prox}_{\lambda F(\cdot)}(\mathbf{w}) + \lambda \mathbf{prox}_{\lambda^{-1}F^*(\cdot)}(\mathbf{w}/\lambda). \tag{5.48}$$

Using the generalized Moreau decomposition, we have

$$\begin{aligned}\lambda^{-1}(2\mathbf{prox}_{\lambda F(\cdot)} - \mathbf{I})\mathbf{w} &= 2\lambda^{-1}\mathbf{prox}_{\lambda F(\cdot)}(\mathbf{w}) - \mathbf{w}/\lambda \\ &\overset{(5.48)}{=} 2\lambda^{-1}(\mathbf{w} - \lambda \mathbf{prox}_{\lambda^{-1}F^*(\cdot)}(\mathbf{w}/\lambda)) - \mathbf{w}/\lambda \\ &= \mathbf{w}/\lambda - 2\mathbf{prox}_{\lambda^{-1}F^*(\cdot)}(\mathbf{w}/\lambda) \\ &= (\mathbf{I} - 2\mathbf{prox}_{\lambda^{-1}F^*(\cdot)})(\mathbf{w}/\lambda).\end{aligned}$$

The last equality of (5.47) comes from

$$\mathbf{prox}_{\lambda^{-1}F^*(-\cdot)}(-\mathbf{w}/\lambda) = -\mathbf{prox}_{\lambda^{-1}F^*(\cdot)}(\mathbf{w}/\lambda).$$

□

Proof (Proof of Theorem 5). We will prove only the equivalence of RPRS on (P3) and (D4). The proof for the other equivalence is the same. The RPRS on (P3) and (D4) can be formulated as

$$\mathbf{w}_1^{k+1} = (1-\alpha)\mathbf{w}_1^k + \alpha(2\mathbf{prox}_{\lambda g\circ\mathbf{A}(\cdot)} - \mathbf{I})(2\mathbf{prox}_{\lambda f(\cdot)} - \mathbf{I})\mathbf{w}_1^k, \tag{5.49}$$

and

$$\mathbf{w}_2^{k+1} = (1-\alpha)\mathbf{w}_2^k + \alpha(2\mathbf{prox}_{\lambda^{-1}(g\circ\mathbf{A})^*(-\cdot)} - \mathbf{I})(2\mathbf{prox}_{\lambda^{-1}f^*(\cdot)} - \mathbf{I})\mathbf{w}_2^k, \tag{5.50}$$

respectively. In addition, we can recover the variables $\mathbf{x}^k$ (or $\mathbf{v}^k$) from $\mathbf{w}_1^k$ (or $\mathbf{w}_2^k$) using the following:

$$\mathbf{x}^{k+1} = \mathbf{prox}_{\lambda f(\cdot)}\mathbf{w}_1^k, \tag{5.51}$$

$$\mathbf{v}^{k+1} = \mathbf{prox}_{\lambda^{-1}f^*(\cdot)}\mathbf{w}_2^k. \tag{5.52}$$

Proof by induction. Suppose $\mathbf{w}_2^k = \mathbf{w}_1^k/\lambda$ holds. We next show that $\mathbf{w}_2^{k+1} = \mathbf{w}_1^{k+1}/\lambda$.

$$\begin{aligned}\mathbf{w}_2^{k+1} &= (1-\alpha)\mathbf{w}_1^k/\lambda + \alpha(2\mathbf{prox}_{\lambda^{-1}(g\circ\mathbf{A})^*(-\cdot)} - \mathbf{I})(2\mathbf{prox}_{\lambda^{-1}f^*(\cdot)} - \mathbf{I})(\mathbf{w}_1^k/\lambda) \\ &\overset{(5.47)}{=} (1-\alpha)\mathbf{w}_1^k/\lambda + \alpha(2\mathbf{prox}_{\lambda^{-1}(g\circ\mathbf{A})^*(-\cdot)} - \mathbf{I})(-\lambda^{-1}(2\mathbf{prox}_{\lambda f(\cdot)} - \mathbf{I})\mathbf{w}_1^k) \\ &\overset{(5.47)}{=} (1-\alpha)\mathbf{w}_1^k/\lambda + \alpha\lambda^{-1}(2\mathbf{prox}_{\lambda(g\circ\mathbf{A})(-\cdot)} - \mathbf{I})(2\mathbf{prox}_{\lambda f(\cdot)} - \mathbf{I})\mathbf{w}_1^k \\ &= \lambda^{-1}[(1-\alpha)\mathbf{w}_1^k + \alpha(2\mathbf{prox}_{\lambda(g\circ\mathbf{A})(\cdot)} - \mathbf{I})(2\mathbf{prox}_{\lambda f(\cdot)} - \mathbf{I})\mathbf{w}_1^k] \\ &= \mathbf{w}_1^{k+1}/\lambda.\end{aligned}$$

In addition we have

$$\begin{aligned}\mathbf{x}^{k+1}+\lambda\mathbf{v}^{k+1} &= \mathbf{prox}_{\lambda f(\cdot)}\mathbf{w}_1^k+\lambda\mathbf{prox}_{\lambda^{-1}f^*(\cdot)}\mathbf{w}_2^k\\ &= \mathbf{prox}_{\lambda f(\cdot)}\mathbf{w}_1^k+\lambda\mathbf{prox}_{\lambda^{-1}f^*(\cdot)}(\mathbf{w}_1^k/\lambda)=\mathbf{w}_1^k.\end{aligned}$$

□

Remark 7. Eckstein showed in [8, Chapter 3.5] that DRS/PRS on (P3) is equivalent to DRS/PRS on (D3) when $\mathbf{A}=\mathbf{I}$. This special case can be obtained from this theorem immediately because when $\mathbf{A}=\mathbf{I}$, (D3) is exactly the same as (D4) and we have

$$\boxed{\text{DRS/PRS on (P3)}} \Longleftrightarrow \boxed{\text{DRS/PRS on (D4)}}$$
$$\Longleftrightarrow \boxed{\text{DRS/PRS on (D3)}} \Longleftrightarrow \boxed{\text{DRS/PRS on (P4)}}.$$

Remark 8. In order to make sure that RPRS on the primal and dual problems are equivalent, the initial conditions and parameters have to satisfy conditions described in the proof of Theorem 5. We need the initial condition to satisfy $\mathbf{w}_2^0=\mathbf{w}_1^0/\lambda$ and the parameter for RPRS on the dual problem has to be chosen as λ^{-1}, see the differences in (5.49) and (5.50).

Similar to the ADM, we can swap f and $g\circ A$ and obtain a new RPRS. The iteration in Algorithm 6 can be written as

$$\mathbf{w}_1^{k+1}=(1-\alpha)\mathbf{w}_1^k+\alpha(2\mathbf{prox}_{\lambda g\circ\mathbf{A}(\cdot)}-\mathbf{I})(2\mathbf{prox}_{\lambda f(\cdot)}-\mathbf{I})\mathbf{w}_1^k, \tag{5.53}$$

and the RPRS after the swapping is

$$\mathbf{w}_3^{k+1}=(1-\alpha)\mathbf{w}_3^k+\alpha(2\mathbf{prox}_{\lambda f(\cdot)}-\mathbf{I})(2\mathbf{prox}_{\lambda g\circ\mathbf{A}(\cdot)}-\mathbf{I})\mathbf{w}_3^k. \tag{5.54}$$

We show below that for a certain type of function f (or g), (5.53) and (5.54) are equivalent.

Theorem 6. *1. Assume that* $\mathbf{prox}_{\lambda f(\cdot)}$ *is affine. If* (5.53) *and* (5.54) *initially satisfy* $\mathbf{w}_3^0=(2\mathbf{prox}_{\lambda f(\cdot)}-\mathbf{I})\mathbf{w}_1^0$, *then* $\mathbf{w}_3^k=(2\mathbf{prox}_{\lambda f(\cdot)}-\mathbf{I})\mathbf{w}_1^k$ *for all k.*
2. Assume that $\mathbf{prox}_{\lambda g\circ\mathbf{A}(\cdot)}$ *is affine. If* (5.53) *and* (5.54) *initially satisfy* $\mathbf{w}_1^0=(2\mathbf{prox}_{\lambda g\circ\mathbf{A}(\cdot)}-\mathbf{I})\mathbf{w}_3^0$, *then* $\mathbf{w}_1^k=(2\mathbf{prox}_{\lambda g\circ\mathbf{A}(\cdot)}-\mathbf{I})\mathbf{w}_3^k$ *for all k.*

Proof. We only prove the first statement, as the second one can be proved in a similar way. We apply proof by induction. Suppose $\mathbf{w}_3^k=(2\mathbf{prox}_{\lambda f(\cdot)}-\mathbf{I})\mathbf{w}_1^k$ holds. From (5.54), we have

$$\begin{aligned}\mathbf{w}_3^{k+1} &= (1-\alpha)(2\mathbf{prox}_{\lambda f(\cdot)}-\mathbf{I})\mathbf{w}_1^k\\ &\quad +\alpha(2\mathbf{prox}_{\lambda f(\cdot)}-\mathbf{I})(2\mathbf{prox}_{\lambda g\circ\mathbf{A}(\cdot)}-\mathbf{I})(2\mathbf{prox}_{\lambda f(\cdot)}-\mathbf{I})\mathbf{w}_1^k\\ &= (2\mathbf{prox}_{\lambda f(\cdot)}-\mathbf{I})\left[(1-\alpha)\mathbf{w}_1^k+\alpha(2\mathbf{prox}_{\lambda g\circ\mathbf{A}(\cdot)}-\mathbf{I})(2\mathbf{prox}_{\lambda f(\cdot)}-\mathbf{I})\mathbf{w}_1^k\right]\\ &= (2\mathbf{prox}_{\lambda f(\cdot)}-\mathbf{I})\mathbf{w}_1^{k+1}.\end{aligned}$$

The first equality holds because $2\mathbf{prox}_{\lambda f(\cdot)} - \mathbf{I}$ is affine, which comes from the assumption that $\mathbf{prox}_{\lambda f(\cdot)}$ is affine and Lemma 2. The second equality comes from (5.53). □

8 Application: Total Variation Image Denoising

ADM (or split Bregman [16]) has been applied on many image processing applications, and we apply the previous equivalence results of ADM to derive several equivalent algorithms for total variation denoising.

The total variation (ROF model [21]) applied on image denoising is

$$\underset{x \in BV(\Omega)}{\text{minimize}} \int_\Omega |\nabla x| + \frac{\alpha}{2}\|x-b\|_2^2$$

where x stands for an image, and $BV(\Omega)$ is the set of all bounded variation functions on Ω. The first term is known as the total variation of x, minimizing which tends to yield a piece-wise constant solution. The discrete version is as follows:

$$\underset{\mathbf{x}}{\text{minimize}} \|\nabla \mathbf{x}\|_{2,1} + \frac{\alpha}{2}\|\mathbf{x}-\mathbf{b}\|_2^2,$$

where $\nabla \mathbf{x}$ is a finite difference approximation of the gradient, which can be expressed as a linear operator. Without loss of generality, we consider the two-dimensional image $\mathbf{x}$, and the discrete total variation $\|\nabla \mathbf{x}\|_{2,1}$ of image $\mathbf{x}$ is defined as

$$\|\nabla \mathbf{x}\|_{2,1} = \sum_{ij} |(\nabla \mathbf{x})_{ij}|,$$

where $|\cdot|$ is the 2-norm of a vector. The equivalent ADM-ready form [16, Equation (3.1)] is

$$\underset{\mathbf{x},\mathbf{y}}{\text{minimize}} \ \|\mathbf{y}\|_{2,1} + \frac{\alpha}{2}\|\mathbf{x}-\mathbf{b}\|_2^2 \qquad \text{subject to } \mathbf{y} - \nabla \mathbf{x} = \mathbf{0}, \tag{5.55}$$

and its dual problem in ADM-ready form [2, Equation (8)] is

$$\underset{\mathbf{v},\mathbf{u}}{\text{minimize}} \ \frac{1}{2\alpha}\|\text{div}\, \mathbf{u} + \alpha \mathbf{b}\|_2^2 + \iota_{\{\mathbf{v}:\|\mathbf{v}\|_{2,\infty}\le 1\}}(\mathbf{v}) \qquad \text{subject to } \mathbf{u} - \mathbf{v} = \mathbf{0}, \tag{5.56}$$

where $\|\mathbf{v}\|_{2,\infty} = \max_{ij} |(\mathbf{v})_{ij}|$ and div $\mathbf{u}$ is the finite difference approximation of divergence that satisfies $\langle \mathbf{x}, \text{div}\, \mathbf{u}\rangle = -\langle \nabla \mathbf{x}, \mathbf{u}\rangle$ for any $\mathbf{x}$ and $\mathbf{u}$. In addition, the equivalent saddle-point problem is

$$\underset{\mathbf{x}}{\text{minimize}} \, \underset{\mathbf{v}}{\text{maximize}} \quad \frac{1}{2\alpha}\|\mathbf{x}-\mathbf{b}\|_2^2 + \langle \mathbf{v}, \nabla \mathbf{x}\rangle - \iota_{\{\mathbf{v}:\|\mathbf{v}\|_{2,\infty}\le 1\}}(\mathbf{v}). \tag{5.57}$$

We list the following equivalent algorithms for solving the total variation image denoising problem. The equivalence result stated in Corollary 3 can be obtained from Theorems 1–4.

1. Algorithm 1 (primal ADM) on (5.55) is

$$\mathbf{x}_1^{k+1} = \arg\min_{\mathbf{x}} \frac{\alpha}{2}\|\mathbf{x}-\mathbf{b}\|_2^2 + (2\lambda)^{-1}\|\nabla\mathbf{x}-\mathbf{y}_1^k+\lambda\mathbf{z}_1^k\|_2^2, \tag{5.58a}$$

$$\mathbf{y}_1^{k+1} = \arg\min_{\mathbf{y}} \|\mathbf{y}\|_{2,1} + (2\lambda)^{-1}\|\nabla\mathbf{x}_1^{k+1}-\mathbf{y}+\lambda\mathbf{z}_1^k\|_2^2, \tag{5.58b}$$

$$\mathbf{z}_1^{k+1} = \mathbf{z}_1^k + \lambda^{-1}(\nabla\mathbf{x}_1^{k+1}-\mathbf{y}_1^{k+1}). \tag{5.58c}$$

2. Algorithm 3 (dual ADM) on (5.56) is

$$\mathbf{u}_3^{k+1} = \arg\min_{\mathbf{u}} \frac{1}{2\alpha}\|\mathrm{div}\,\mathbf{u}+\alpha\mathbf{b}\|_2^2 + \frac{\lambda}{2}\|\mathbf{v}_3^k-\mathbf{u}+\lambda^{-1}\mathbf{z}_3^k\|_2^2, \tag{5.59a}$$

$$\mathbf{v}_3^{k+1} = \arg\min_{\mathbf{v}} \iota_{\{\mathbf{v}:\|\mathbf{v}\|_{2,\infty}\le 1\}}(\mathbf{v}) + \frac{\lambda}{2}\|\mathbf{v}-\mathbf{u}_3^{k+1}+\lambda^{-1}\mathbf{z}_3^k\|_2^2, \tag{5.59b}$$

$$\mathbf{z}_3^{k+1} = \mathbf{z}_3^k + \lambda(\mathbf{v}_3^{k+1}-\mathbf{u}_3^{k+1}). \tag{5.59c}$$

3. Algorithm 4 (primal-dual) on (5.57) is

$$\bar{\mathbf{v}}_4^k = 2\mathbf{v}_4^k - \mathbf{v}_4^{k-1}, \tag{5.60a}$$

$$\mathbf{x}_4^{k+1} = \arg\min_{\mathbf{x}} \frac{\alpha}{2}\|\mathbf{x}-\mathbf{b}\|_2^2 + (2\lambda)^{-1}\|\nabla\mathbf{x}-\nabla\mathbf{x}_4^k+\lambda\bar{\mathbf{v}}_4^k\|_2^2, \tag{5.60b}$$

$$\mathbf{v}_4^{k+1} = \arg\min_{\mathbf{v}} \iota_{\{\mathbf{v}:\|\mathbf{v}\|_{2,\infty}\le 1\}}(\mathbf{v}) - \langle\mathbf{v},\nabla\mathbf{x}_4^{k+1}\rangle + \frac{\lambda}{2}\|\mathbf{v}-\mathbf{v}^k\|_2^2. \tag{5.60c}$$

4. Algorithm 5 (primal ADM with order swapped) on (5.55) is

$$\mathbf{y}_5^{k+1} = \arg\min_{\mathbf{y}} \|\mathbf{y}\|_{2,1} + (2\lambda)^{-1}\|\nabla\mathbf{x}_5^k-\mathbf{y}+\lambda\mathbf{z}_5^k\|_2^2, \tag{5.61a}$$

$$\mathbf{x}_5^{k+1} = \arg\min_{\mathbf{x}} \frac{\alpha}{2}\|\mathbf{x}-\mathbf{b}\|_2^2 + (2\lambda)^{-1}\|\nabla\mathbf{x}-\mathbf{y}_5^{k+1}+\lambda\mathbf{z}_5^k\|_2^2, \tag{5.61b}$$

$$\mathbf{z}_5^{k+1} = \mathbf{z}_5^k + \lambda^{-1}(\nabla\mathbf{x}_5^{k+1}-\mathbf{y}_5^{k+1}). \tag{5.61c}$$

Corollary 3. *Let* $\mathbf{x}_5^0 = \mathbf{b} + \alpha^{-1}\mathrm{div}\,\mathbf{z}_5^0$. *If the initialization for all algorithms* (5.58)-(5.61) *satisfies* $\mathbf{y}_1^0 = -\mathbf{z}_3^0 = \nabla\mathbf{x}_4^0 - \lambda(\mathbf{v}_4^0 - \mathbf{v}_4^{-1}) = \mathbf{y}_5^1$ *and* $\mathbf{z}_1^0 = \mathbf{v}_3^0 = \mathbf{v}_4^0 = \mathbf{z}_5^0 + \lambda^{-1}(\nabla\mathbf{x}_5^0 - \mathbf{y}_5^1)$. *Then for* $k \ge 1$, *we have the following equivalence results between the iterations of the four algorithms:*

$$\begin{aligned}
\mathbf{y}_1^k &= -\mathbf{z}_3^k = \nabla\mathbf{x}_4^k - \lambda(\mathbf{v}_4^k - \mathbf{v}_4^{k-1}) = \mathbf{y}_5^{k+1},\\
\mathbf{z}_1^k &= \mathbf{v}_3^k \;\;= \mathbf{v}_4^k \qquad\qquad\qquad\qquad\; = \mathbf{z}_5^k + \lambda^{-1}(\nabla\mathbf{x}_5^k - \mathbf{y}_5^{k+1}).
\end{aligned}$$

Remark 9. In any of the four algorithms, the ∇ or div operator is separated in a different subproblem from the term $\|\cdot\|_{2,1}$ or its dual norm $\|\cdot\|_{2,\infty}$. The ∇ or div operator is translation invariant, so their subproblems can be solved by a diagonalization trick [22]. The subproblems involving the term $\|\cdot\|_{2,1}$ or the indicator function $\iota_{\{\mathbf{v}:\|\mathbf{v}\|_{2,\infty}\leq 1\}}$ have closed-form solutions. Therefore, in addition to the equivalence results, all the four algorithms have essentially the same per-iteration costs.

Acknowledgments

This work is supported by NSF Grants DMS-1349855 and DMS-1317602 and ARO MURI Grant W911NF-09-1-0383. We thank Jonathan Eckstein for bringing his early work [8, Chapter 3.5] and [9] to our attention and anonymous reviewers for their helpful comments and suggestions.

References

1. Bauschke, H.H., Combettes, P.L.: Convex Analysis and Monotone Operator Theory in Hilbert Spaces. Springer (2011)
2. Chambolle, A.: An algorithm for total variation minimization and applications. Journal of Mathematical Imaging and Vision **20**(1-2), 89–97 (2004)
3. Chambolle, A., Pock, T.: A first-order primal-dual algorithm for convex problems with applications to imaging. Journal of Mathematical Imaging and Vision **40**(1), 120–145 (2011)
4. Davis, D., Yin, W.: Faster convergence rates of relaxed Peaceman-Rachford and ADMM under regularity assumptions. arXiv preprint arXiv:1407.5210 (2014)
5. Davis, D., Yin, W.: Convergence rate analysis of several splitting schemes. In: R. Glowinski, S. Osher, W. Yin (eds.) Splitting Methods in Communication and Imaging, Science and Engineering, Chapter 4. Springer (2016)
6. Deng, W., Yin, W.: On the global and linear convergence of the generalized alternating direction method of multipliers. Journal of Scientific Computing **66**(3), 889–916 (2015)
7. Douglas, J., Rachford, H.H.: On the numerical solution of heat conduction problems in two and three space variables. Transactions of the American Mathematical Society **82**(2), 421–439 (1956)
8. Eckstein, J.: Splitting methods for monotone operators with applications to parallel optimization. Ph.D. thesis, Massachusetts Institute of Technology (1989)
9. Eckstein, J., Fukushima, M.: Some reformulations and applications of the alternating direction method of multipliers. In: Large Scale Optimization, pp. 115–134. Springer (1994)
10. Esser, E., Zhang, X., Chan, T.: A general framework for a class of first order primal-dual algorithms for convex optimization in imaging science. SIAM Journal on Imaging Sciences **3**(4), 1015–1046 (2010)
11. Esser, J.: Primal dual algorithms for convex models and applications to image restoration, registration and nonlocal inpainting. Ph.D. thesis, University of California, Los Angeles (2010)
12. Fukushima, M.: The primal Douglas-Rachford splitting algorithm for a class of monotone mappings with application to the traffic equilibrium problem. Mathematical Programming **72**(1), 1–15 (1996)

13. Gabay, D.: Applications of the method of multipliers to variational inequalities. In: M. Fortin, R. Glowinski (eds.) Augmented Lagrangian Methods: Applications to the Solution of Boundary-Value Problems. North-Holland: Amsterdam, Amsterdam (1983)
14. Gabay, D., Mercier, B.: A dual algorithm for the solution of nonlinear variational problems via finite element approximation. Computers & Mathematics with Applications **2**(1), 17–40 (1976)
15. Glowinski, R., Marroco, A.: Sur l'approximation, par éléments finis d'ordre un, et la résolution, par pénalisation-dualité d'une classe de problèmes de dirichlet non linéaires. Rev. Française d'Automat. Inf. Recherche Opérationelle **9**(2), 41–76 (1975)
16. Goldstein, T., Osher, S.: The split Bregman method for l1-regularized problems. SIAM Journal on Imaging Sciences **2**(2), 323–343 (2009)
17. Hestenes, M.: Multiplier and gradient methods. Journal of Optimization Theory and Applications **4**(5), 303–320 (1969)
18. Lions, P., Mercier, B.: Splitting algorithms for the sum of two nonlinear operators. SIAM Journal on Numerical Analysis **16**(6), 964–979 (1979)
19. Peaceman, D.W., Rachford, H.H.: The numerical solution of parabolic and elliptic differential equations. Journal of the Society for Industrial and Applied Mathematics **3**(1), 28–41 (1955)
20. Rockafellar, R.T.: A dual approach to solving nonlinear programming problems by unconstrained optimization. Mathematical Programming **5**(1), 354–373 (1973)
21. Rudin, L.I., Osher, S., Fatemi, E.: Nonlinear total variation based noise removal algorithms. Physica D: Nonlinear Phenomena **60**(1–4), 259–268 (1992)
22. Wang, Y., Yang, J., Yin, W., Zhang, Y.: A new alternating minimization algorithm for total variation image reconstruction. SIAM Journal on Imaging Sciences **1**(3), 248–272 (2008)
23. Xiao, Y., Zhu, H., Wu, S.Y.: Primal and dual alternating direction algorithms for ℓ_1-ℓ_1-norm minimization problems in compressive sensing. Computational Optimization and Applications **54**(2), 441–459 (2013)
24. Yang, J., Zhang, Y.: Alternating direction algorithms for ℓ_1-problems in compressive sensing. SIAM Journal on Scientific Computing **33**(1), 250–278 (2011)
25. Yang, Y., Möller, M., Osher, S.: A dual split Bregman method for fast ℓ^1 minimization. Mathematics of Computation **82**(284), 2061–2085 (2013)

Chapter 6
Application of the Strictly Contractive Peaceman-Rachford Splitting Method to Multi-Block Separable Convex Programming

Bingsheng He, Han Liu, Juwei Lu, and Xiaoming Yuan

Abstract Recently, a strictly contractive Peaceman-Rachford splitting method (SC-PRSM) was proposed to solve a convex minimization model with linear constraints and a separable objective function which is the sum of two functionals without coupled variables. We show by an example that the SC-PRSM cannot be directly extended to the case where the objective function is the sum of three or more functionals. To solve such a multi-block model, if we treat its variables and functions as two groups and directly apply the SC-PRSM, then at least one of SC-PRSM subproblems involves more than one function and variable which might not be easy to solve. One way to improve the solvability for this direct application of the SC-PRSM is to further decompose such a subproblem so as to generate easier decomposed subproblems which could potentially be simple enough to have closed-form solutions for some specific applications. The curse accompanying this improvement in solvability is that the SC-PRSM with further decomposed subproblems is not necessarily convergent, either. We will show its divergence by the same example. Our main goal in this chapter is to show that the convergence can be guaranteed if the further decomposed subproblems of the direct application of the SC-PRSM are regularized by the proximal regularization. As a result,

B. He
International Centre of Management Science and Engineering, School of Management and Engineering, Nanjing University, Nanjing, China

Department of Mathematics, Nanjing University, Nanjing 200093, China
e-mail: hebma@nju.edu.cn

H. Liu • J. Lu
Department of Operations Research and Financial Engineering, Princeton University, Princeton, NJ 08544, USA
e-mail: hanliu@princeton.edu; junweil@princeton.edu

X. Yuan (✉)
Department of Mathematics, Hong Kong Baptist University, Kowloon Tong, Hong Kong
e-mail: xmyuan@hkbu.edu.hk

R. Glowinski et al. (eds.), *Splitting Methods in Communication, Imaging, Science, and Engineering*, Scientific Computation, DOI 10.1007/978-3-319-41589-5_6

an SC-PRSM-based splitting algorithm with provable convergence and easy implementability is proposed for multi-block convex minimization models. We analyze the convergence for the derived algorithm, including proving its global convergence and establishing its worst-case convergence rate measured by the iteration complexity. The efficiency of the new algorithm is illustrated by testing some applications arising in image processing and statistical learning.

1 Introduction

We first consider a convex minimization model with linear constraints and an objective function in form of the sum of two functions without coupled variables:

$$\min\{\theta_1(x)+\theta_2(y) \mid Ax+By=b,\, x\in\mathscr{X}, y\in\mathscr{Y}\}, \tag{6.1}$$

where $A\in\Re^{m\times n_1}$, $B\in\Re^{m\times n_2}$, $\mathscr{X}\subset\Re^{n_1}$ and $\mathscr{Y}\subset\Re^{n_2}$ are closed convex sets, θ_1 and θ_2 are convex but not necessarily smooth functions. A typical application of (6.1) is that θ_1 refers to a data-fidelity term and θ_2 denotes a regularization term. Concrete applications of the model (6.1) arise frequently in many areas such as image processing, statistical learning, computer vision, etc., where θ_1 and θ_2 could be further specified by particular physical or industrial elaboration for a given scenario.

A benchmark solver for (6.1) is the alternating direction method of multipliers (ADMM) originally proposed in [23] (see also [6, 19]). Applying to the solution of problem (6.1), ADMM reads as

$$\begin{cases} x^{k+1}=\arg\min\{\theta_1(x)-(\lambda^k)^T(Ax+By^k-b)+\frac{\beta}{2}\|Ax+By^k-b\|^2 \mid x\in\mathscr{X}\},\\ y^{k+1}=\arg\min\{\theta_2(y)-(\lambda^k)^T(Ax^{k+1}+By-b)+\frac{\beta}{2}\|Ax^{k+1}+By-b\|^2 \mid y\in\mathscr{Y}\},\\ \lambda^{k+1}=\lambda^k-\beta(Ax^{k+1}+By^{k+1}-b), \end{cases} \tag{6.2}$$

where $\lambda^k,\lambda^{k+1}\in\Re^m$ are Lagrange multipliers and $\beta>0$ is a penalty parameter. Throughout we assume that the penalty parameter β is fixed. As analyzed intensively in [18, 22] (and observed for the first time in [6]), the scheme (6.2) can be regarded as an application of the Douglas-Rachford splitting method (DRSM), which is well known in the PDE literature (see [12, 34]). The ADMM algorithm can also be regarded as a splitting version of the augmented Lagrangian method (ALM), introduced in [31, 41]; and it outperforms the direct application of the ALM in that the functions θ_1 and θ_2 are treated individually and thus the splitted subproblems in (9.48) could be much easier than the original ALM subproblems. Recently, this feature has found impressive applications in a variety of areas, and it has inspired a "renaissance" of ADMM in the literature. We refer to [4, 13, 20] for some review papers of the ADMM.

In addition to DRSM, some authors (see, e.g., [24, 33]) have also investigated how to apply the Peacemen-Rachford splitting method (PRSM) (another well-known operator-splitting method, introduced in [39] and further discussed in, e.g.,

[34]) to the separable convex minimization model (6.1). This motivation was further enhanced by the observation that "very often PRSM is faster than DRSM whenever it converges", as pointed out in [2, 21, 22]. More specifically, PRSM applied to the solution of (6.1) leads to

$$\begin{cases} x^{k+1} = \arg\min\{\theta_1(x) - (\lambda^k)^T(Ax+By^k-b) + \frac{\beta}{2}\|Ax+By^k-b\|^2 \mid x\in\mathscr{X}\}, \\ \lambda^{k+\frac{1}{2}} = \lambda^k - \beta(Ax^{k+1}+By^k-b), \\ y^{k+1} = \arg\min\{\theta_2(y) - (\lambda^{k+\frac{1}{2}})^T(Ax^{k+1}+By-b) + \frac{\beta}{2}\|Ax^{k+1}+By-b\|^2 \mid y\in\mathscr{Y}\}, \\ \lambda^{k+1} = \lambda^{k+\frac{1}{2}} - \beta(Ax^{k+1}+By^{k+1}-b), \end{cases} \tag{6.3}$$

which differs from the DRSM scheme (6.2) in that it updates the Lagrange multipliers twice at each iteration. The PRSM algorithm has the disadvantage of being "less stable than DRSM", although it does outperform it in most situations where it is convergent (see [34, 21]). In [26], this disadvantage was explained by the fact that the sequence generated by PRSM is not necessarily strictly contractive with respect to the solution set of (6.1) (assuming that this set is nonempty), while the sequence generated by DRSM is strictly contractive. Note that we follow the definition of a strictly contractive sequence given in [3]. In [10], a counterexample showing that the sequence generated by PRSM could maintain a constant distance to the solution set was constructed. Thus, the PRSM algorithm (6.3) is not necessarily convergent. To reinforce the robustness of the PRSM algorithm (6.3), with provable convergence property, the following strictly contractive variant of (6.3) (denoted by SC-PRSM) was proposed in [26]:

$$\begin{cases} x^{k+1} = \arg\min\{\theta_1(x) - (\lambda^k)^T(Ax+By^k-b) + \frac{\beta}{2}\|Ax+By^k-b\|^2 \mid x\in\mathscr{X}\}, \\ \lambda^{k+\frac{1}{2}} = \lambda^k - \alpha\beta(Ax^{k+1}+By^k-b), \\ y^{k+1} = \arg\min\{\theta_2(y) - (\lambda^{k+\frac{1}{2}})^T(Ax^{k+1}+By-b) + \frac{\beta}{2}\|Ax^{k+1}+By-b\|^2 \mid y\in\mathscr{Y}\}, \\ \lambda^{k+1} = \lambda^{k+\frac{1}{2}} - \alpha\beta(Ax^{k+1}+By^{k+1}-b), \end{cases} \tag{6.4}$$

where $\alpha \in (0,1)$. It was shown in [26] that the parameter α plays the role of enforcing the sequence generated by (6.4) to be strictly contractive with respect to the solution set of (6.1). Hence, the convergence of the SC-PRSM algorithm (6.4) can be proved by standard techniques in the literature (e.g., [3]). In [26], the efficiency of the SC-PRSM algorithm (6.4) was also verified numerically.

In addition to the model (6.1), we encounter numerous applications where the objective function has a higher degree of separability such that it can be expressed as the sum of more than two functionals without coupled variables. To expose our main idea with easier notation, let us only focus on the case with three functionals in the objective

$$\min\{\theta_1(x)+\theta_2(y)+\theta_3(z) \mid Ax+By+Cz=b, x\in\mathscr{X}, y\in\mathscr{Y}, z\in\mathscr{Z}\}, \tag{6.5}$$

where $A\in\Re^{m\times n_1}$, $B\in\Re^{m\times n_2}$, $C\in\Re^{m\times n_3}$, $b\in\Re^m$, $\mathscr{X}\subset\Re^{n_1}$ and $\mathscr{Y}\subset\Re^{n_2}$, $\mathscr{Z}\subset\Re^{n_3}$ are closed convex sets, θ_i $(i=1,2,3)$ are convex functions. Throughout, the solution set of (6.5) is assumed to be nonempty. Some typical applications in the form of (6.5) include the robust principal component analysis model with noisy and incomplete data in [44], the latent variable Gaussian graphical model selection in [7], the robust alignment model for linearly correlated images in [40], the quadratic discriminant analysis model in [36], and many others.

To solve (6.5), one natural idea is to directly extend the ADMM algorithm (6.2). The resulting scheme is

$$\begin{cases} x^{k+1} = \arg\min\{\mathscr{L}_\beta(x,y^k,z^k,\lambda^k) \mid x \in \mathscr{X}\}, \\ y^{k+1} = \arg\min\{\mathscr{L}_\beta(x^{k+1},y,z^k,\lambda^k) \mid y \in \mathscr{Y}\}, \\ z^{k+1} = \arg\min\{\mathscr{L}_\beta(x^{k+1},y^{k+1},z,\lambda^k) \mid z \in \mathscr{Z}\}, \\ \lambda^{k+1} = \lambda^k - \beta(Ax^{k+1}+By^{k+1}+Cz^{k+1}-b), \end{cases} \tag{6.6}$$

where $\mathscr{L}_\beta(x,y,z,\lambda)$ is the augmented Lagrangian function of (6.5) defined as

$$\mathscr{L}_\beta(x,y,z,\lambda) := \theta_1(x)+\theta_2(y)+\theta_3(z)-\lambda^T(Ax+By+Cz-b)+\frac{\beta}{2}\|Ax+By+Cz-b\|^2$$

and where $\lambda^k, \lambda^{k+1} \in \Re^m$ are Lagrange multipliers, $\beta > 0$ is a penalty parameter. Algorithm (6.6) is a direct extension of the ADMM algorithm (6.2); from now on we will denote (6.6) by E-ADMM; E-ADMM perfectly inherits the advantage of the ADMM algorithm (6.2), and it can be obtained by simply decomposing the ALM subproblem into 3 subproblems in Gauss-Seidel manner at each iteration. Empirically, it often works very well, see, e.g., [40, 44] for some applications. However, it was shown in [9] that the E-ADMM (6.6) is not necessarily convergent. We refer to [27, 28] for some methods whose main common idea is to ensure the convergence via correcting the output of (6.6) appropriately.

Similarly, for solving the multi-block convex minimization model (6.5), we may wish to consider directly extending the SC-PRSM scheme (6.4) as

$$\begin{cases} x^{k+1} = \arg\min\{\theta_1(x)-(\lambda^k)^T(Ax+By^k+Cz^k-b)+\frac{\beta}{2}\|Ax+By^k+Cz^k-b\|^2 \mid x\in\mathscr{X}\}, \\ \lambda^{k+\frac{1}{3}} = \lambda^k-\alpha\beta(Ax^{k+1}+By^k+Cz^k-b), \\ y^{k+1} = \arg\min\{\theta_2(y)-(\lambda^{k+\frac{1}{3}})^T(Ax^{k+1}+By+Cz^k-b)+\frac{\beta}{2}\|Ax^{k+1}+By+Cz^k-b\|^2 \mid y\in\mathscr{Y}\}, \\ \lambda^{k+\frac{2}{3}} = \lambda^{k+\frac{1}{3}}-\alpha\beta(Ax^{k+1}+By^{k+1}+Cz^k-b), \\ z^{k+1} = \arg\min\{\theta_3(z)-(\lambda^{k+\frac{2}{3}})^T(Ax^{k+1}+By^{k+1}+Cz-b)+\frac{\beta}{2}\|Ax^{k+1}+By^{k+1}+Cz-b\|^2 \mid z\in\mathscr{Z}\}, \\ \lambda^{k+1} = \lambda^{k+\frac{2}{3}}-\alpha\beta(Ax^{k+1}+By^{k+1}+Cz^{k+1}-b). \end{cases} \tag{6.7}$$

Hereafter, we denote algorithm (6.7) by E-SC-PRSM. Our first purpose is to show that E-SC-PRSM is not necessarily convergent, as shown in Section 5.2, the method to prove this property being similar to the one used in [9]. From this possible divergence property, E-SC-PRSM (6.7) cannot be used directly to solve (6.5).

Alternatively, one may wish to use the original SC-PRSM algorithm (6.4) directly by regarding $\theta_2(y)+\theta_3(z)$ as the second functional in (6.1) and regrouping (y,z) and (B,C) as the second variable and coefficient matrix in (6.1), respectively. The direct application of the SC-PRSM algorithm (6.4) to the solution of problem (6.5) leads to the following iterative method:

$$\begin{cases} x^{k+1} = \arg\min\{\theta_1(x) - (\lambda^k)^T(Ax+By^k+Cz^k-b) + \frac{\beta}{2}\|Ax+By^k+Cz^k-b\|^2 \mid x\in\mathscr{X}\},\\ \lambda^{k+\frac{1}{2}} = \lambda^k - \alpha\beta(Ax^{k+1}+By^k+Cz^k-b),\\ (y^{k+1},z^{k+1}) = \arg\min\left\{\begin{array}{l}\theta_2(y)+\theta_3(z) - (\lambda^{k+\frac{1}{2}})^T(Ax^{k+1}+By+Cz-b)\\ +\frac{\beta}{2}\|Ax^{k+1}+By+Cz-b\|^2 \mid y\in\mathscr{Y},\, z\in\mathscr{Z}\end{array}\right\},\\ \lambda^{k+1} = \lambda^{k+\frac{1}{2}} - \alpha\beta(Ax^{k+1}+By^{k+1}+Cz^{k+1}-b).\end{cases} \tag{6.8}$$

Provided that the two minimization subproblems in (6.8) are solved exactly, this direct application of SC-PRSM is successful, its convergence being guaranteed automatically. This is the blessing of applying SC-PRSM directly to the solution of (6.5). For many concrete applications of (6.5), such as the ones mentioned above, it is not wise, however, to do so because the (y,z)-subproblem in (6.8) must treat θ_2 and θ_3 aggregately even though both could be very simple. This is the curse accompanying algorithm (6.8). Under the assumption that each function θ_i in (6.5) is well structured or has some special properties in the sense that treating a minimization problem involving only one of them is easy (e.g., when the resolvent operator of $\partial\theta_i$ has a closed-form solution, as it is the case if θ_i is a l_1-norm term), a natural idea to overcome the curse associated with (6.8) is to further decompose the (y,z)-subproblem in (6.8) in a Jacobian style. Thus, we solve approximately the (y,z)-subproblem in (6.8) by replacing it with

$$\begin{cases} y^{k+1} = \arg\min\{\theta_2(y) - (\lambda^{k+\frac{1}{2}})^T(Ax^{k+1}+By+Cz^k-b) + \frac{\beta}{2}\|Ax^{k+1}+By+Cz^k-b\|^2 \mid y\in\mathscr{Y}\},\\ z^{k+1} = \arg\min\{\theta_3(z) - (\lambda^{k+\frac{1}{2}})^T(Ax^{k+1}+By^k+Cz-b) + \frac{\beta}{2}\|Ax^{k+1}+By^k+Cz-b\|^2 \mid z\in\mathscr{Z}\}.\end{cases} \tag{6.9}$$

The two subproblems in (6.9) are in general easier to solve than the (y,z)-subproblem in (6.8), since each of them only involves one θ_i in its objective function. Another reason for implementing this Jacobian style decomposition is that the two subproblems in (6.9) are well suited for parallel computation. This decomposition makes some particular sense for large scale cases of the model (6.5) arising from high dimension statistical learning problems or some image processing applications. With the further decomposition (6.9) for the (y,z)-subproblem in (6.8), the direct application of the SC-PRSM (6.4) to the model (6.5) becomes

$$\begin{cases} x^{k+1} = \arg\min\{\theta_1(x) - (\lambda^k)^T(Ax+By^k+Cz^k-b) + \frac{\beta}{2}\|Ax+By^k+Cz^k-b\|^2 \mid x\in\mathscr{X}\},\\ \lambda^{k+\frac{1}{2}} = \lambda^k - \alpha\beta(Ax^{k+1}+By^k+Cz^k-b),\\ y^{k+1} = \arg\min\{\theta_2(y) - (\lambda^{k+\frac{1}{2}})^T(Ax^{k+1}+By+Cz^k-b) + \frac{\beta}{2}\|Ax^{k+1}+By+Cz^k-b\|^2 \mid y\in\mathscr{Y}\},\\ z^{k+1} = \arg\min\{\theta_3(z) - (\lambda^{k+\frac{1}{2}})^T(Ax^{k+1}+By^k+Cz-b) + \frac{\beta}{2}\|Ax^{k+1}+By^k+Cz-b\|^2 \mid z\in\mathscr{Z}\}.\\ \lambda^{k+1} = \lambda^{k+\frac{1}{2}} - \alpha\beta(Ax^{k+1}+By^{k+1}+Cz^{k+1}-b).\end{cases} \tag{6.10}$$

Compared with (6.8), algorithm (6.10) is much easier to implement because its subproblems are much simpler. The properties of θ_i's, if any, can thus be fully exploited by algorithm (6.10). However, it is easy to understand that despite of the guaranteed convergence of (6.8), the convergence of (6.10) might not hold because the original (y,z)-subproblem in (6.8) is solved only approximately via (6.9). In Section 5.1, we will use the same example showing the divergence of the E-SC-PRSM algorithm (6.7) to prove the divergence of algorithm (6.10). Thus, it may be not reasonable to use either the E-SC-PRSM algorithm (6.7) or to apply directly the SC-PRSM algorithm (6.10), with its decomposed subproblems, to the solution of the multi-block convex minimization problem (6.5).

Our second goal is to show that the convergence of (6.10) can be guaranteed if the decomposed y- and z-subproblems in (6.10) are further regularized by quadratic proximal terms. This idea inspires us to propose the following SC-PRSM algorithm with proximal regularization (SC-PRSM-PR)

$$\begin{cases} x^{k+1} = \arg\min\{\theta_1(x) - (\lambda^k)^T(Ax+By^k+Cz^k-b) + \frac{\beta}{2}\|Ax+By^k+Cz^k-b\|^2 \mid x \in \mathscr{X}\}, \\ \lambda^{k+\frac{1}{2}} = \lambda^k - \alpha\beta(Ax^{k+1}+By^k+Cz^k-b), \\ y^{k+1} = \arg\min\left\{ \begin{array}{l} \theta_2(y) - (\lambda^{k+\frac{1}{2}})^T(Ax^{k+1}+By+Cz^k-b)+ \\ \frac{\beta}{2}\|Ax^{k+1}+By+Cz^k-b\|^2 + \frac{\mu\beta}{2}\|B(y-y^k)\|^2 \end{array} \middle| y \in \mathscr{Y} \right\}, \\ z^{k+1} = \arg\min\left\{ \begin{array}{l} \theta_3(z) - (\lambda^{k+\frac{1}{2}})^T(Ax^{k+1}+By^k+Cz-b)+ \\ \frac{\beta}{2}\|Ax^{k+1}+By^k+Cz-b\|^2 + \frac{\mu\beta}{2}\|C(z-z^k)\|^2 \end{array} \middle| z \in \mathscr{Z} \right\}, \\ \lambda^{k+1} = \lambda^{k+\frac{1}{2}} - \alpha\beta(Ax^{k+1}+By^{k+1}+Cz^{k+1}-b), \end{cases} \tag{6.11}$$

where $\alpha \in (0,1)$ and $\mu > \alpha$. Note that the added quadratic proximal terms $\frac{\mu\beta}{2}$ $\|B(y-y^k)\|^2$ and $\frac{\mu\beta}{2}\|C(z-z^k)\|^2$ enjoy the same explanation than the original proximal point algorithm which has been well studied in the literature, see e.g. [10, 35, 42]. An intuitive illustration is that since the objective functions in (6.9) are only approximation to the objective function in the (y,z)-subproblem in (6.8), we use the quadratic terms to control the proximity of the new iterate to the previous iterate. The requirement $\mu \geq \alpha$ is in certain sense to control such proximity. Note that the subproblems in (6.11) are of the same difficulty as those in (6.10); while the convergence of (6.11) can be rigorously proved (see Section 3).

As a customized application of the original SC-PRSM algorithm (6.4) to the specific multi-block convex minimization problem (6.5), the SC-PRSM-PR algorithm (6.11) is equally implementable as (6.4) in the sense that their subproblems are of the same level of difficulty. Moreover, we will show that the SC-PRSM-PR algorithm (6.11) is also globally convergent and its worst-case convergence rate measured by the iteration complexity in both the ergodic and a nonergodic senses can be established. Thus, besides its implementability, the SC-PRSM-PR algorithm (6.11) also fully inherits the theoretical properties of the original SC-PRSM algorithm (6.4) established in [26]. This is the main goal of Sections 3 and 4. In Section 5, as mentioned, we will construct an example to show the divergence of the E-SC-PRSM algorithm (6.7), and of the SC-PRSM algorithm (6.10) if applied directly. As mentioned already, the SC-PRSM-PR (6.11) is motivated by some practical applications, particular cases of the abstract model problem (6.5). Indeed, the efficiency of algorithm (6.11) will be tested in Section 6 via its application to the solution of useful practical problems arising in Image Processing and Statistical Learning, with the results of numerical experiments being reported. Finally, some concluding remarks will be made in Section 7.

2 Preliminaries

In this section, we summarize some known results in the literature which are useful for our analysis later; we will also define some auxiliary variables which can simplify the notation of our analysis.

2.1 The Variational Inequality Reformulation of (6.5)

We first reformulate the multi-block convex minimization problem (6.5) as a variational inequality (VI): Find $w^* = (x^*, y^*, z^*, \lambda^*)$ such that

$$\text{VI}(\Omega, F, \theta): \qquad w^* \in \Omega, \quad \theta(u) - \theta(u^*) + (w - w^*)^T F(w^*) \geq 0, \quad \forall w \in \Omega, \tag{6.12a}$$

where

$$u = \begin{pmatrix} x \\ y \\ z \end{pmatrix}, \qquad w = \begin{pmatrix} x \\ y \\ z \\ \lambda \end{pmatrix}, \qquad \theta(u) = \theta_1(x) + \theta_2(y) + \theta_3(z),$$

$$F(w) = \begin{pmatrix} -A^T\lambda \\ -B^T\lambda \\ -C^T\lambda \\ Ax + By + Cz - b \end{pmatrix} \qquad \text{and} \qquad \Omega := \mathscr{X} \times \mathscr{Y} \times \mathscr{Z} \times \Re^m. \tag{6.12b}$$

Obviously, the mapping $F(w)$ defined in (6.12b) is affine with a skew-symmetric matrix; it is thus monotone. We denote by Ω^* the solution set of $\text{VI}(\Omega, F, \theta)$, and it is not empty under the nonempty assumption of the solution set of (6.5).

Note that x^k is not required to generate the new $(k+1)$-th iteration in all the DRSM- or PRSM-based algorithms mentioned previously, see (6.2), (6.3), (6.4), (6.6), (6.7), (6.8), (6.10), and (6.11). That is, such an algorithm only requires (y^k, z^k, λ^k) to generate the next new iterate. Thus, as mentioned in [4], x is an **intermediate** variable in all the mentioned DRSM- or PRSM-based schemes. For this reason, in the following analysis, we use the notations $v^k = (y^k, z^k, \lambda^k)$ and $\mathscr{V} = \mathscr{Y} \times \mathscr{Z} \times \Re^m$, and we let

$$\mathscr{V}^* := \{v^* = (y^*, z^*, \lambda^*) \,|\, w^* = (x^*, y^*, z^*, \lambda^*) \in \Omega^*\}.$$

2.2 Some Notation

We define some auxiliary variables in this subsection which will help us alleviate the notation in the convergence analysis and improve the presentation.

First of all, we introduce a new sequence $\{\tilde{w}^k\}$ by

$$\tilde{w}^k = \begin{pmatrix} \tilde{x}^k \\ \tilde{y}^k \\ \tilde{z}^k \\ \tilde{\lambda}^k \end{pmatrix} = \begin{pmatrix} x^{k+1} \\ y^{k+1} \\ z^{k+1} \\ \lambda^k - \beta(Ax^{k+1} + By^k + Cz^k - b) \end{pmatrix}, \tag{6.13}$$

where $(x^{k+1}, y^{k+1}, z^{k+1})$ is generated by the scheme (6.11) from (y^k, z^k, λ^k). Accordingly, we have

$$\tilde{v}^k = \begin{pmatrix} \tilde{y}^k \\ \tilde{z}^k \\ \tilde{\lambda}^k \end{pmatrix}, \tag{6.14}$$

where $(\tilde{y}^k, \tilde{z}^k, \tilde{\lambda}^k)$ is defined in (6.13).

In fact, using the definition of $\lambda^{k+\frac{1}{2}}$ in (6.11), we have

$$\lambda^{k+1} = \lambda^k - 2\alpha\beta\big(Ax^{k+1} + \frac{1}{2}B(y^k + y^{k+1}) + \frac{1}{2}C(z^k + z^{k+1}) - b\big).$$

According to (6.13), because

$$x^{k+1} = \tilde{x}^k, \qquad y^{k+1} = \tilde{y}^k \qquad z^{k+1} = \tilde{z}^k,$$

we have

$$\tilde{\lambda}^k = \lambda^k - \beta(A\tilde{x}^k + By^k + Cz^k - b) \qquad \text{and} \qquad \lambda^{k+\frac{1}{2}} = \lambda^k - \alpha(\lambda^k - \tilde{\lambda}^k). \tag{6.15}$$

By a manipulation, the updated form of λ^{k+1} (6.11) can be represented as

$$\begin{aligned} \lambda^{k+1} &= \lambda^k - \alpha(\lambda^k - \tilde{\lambda}^k) - \alpha\beta(A\tilde{x}^k + B\tilde{y}^k + C\tilde{z}^k - b) \\ &= \lambda^k - \alpha(\lambda^k - \tilde{\lambda}^k) - \alpha\beta\left[(A\tilde{x}^k + By^k + Cz^k - b) - B(y^k - \tilde{y}^k) - C(z^k - \tilde{z}^k)\right] \\ &= \lambda^k - \alpha(\lambda^k - \tilde{\lambda}^k) - \alpha\left[(\lambda^k - \tilde{\lambda}^k) - \beta B(y^k - \tilde{y}^k) - \beta C(z^k - \tilde{z}^k)\right] \\ &= \lambda^k - \left[2\alpha(\lambda^k - \tilde{\lambda}^k) - \alpha\beta B(y^k - \tilde{y}^k) - \alpha\beta C(z^k - \tilde{z}^k)\right]. \end{aligned} \tag{6.16}$$

In the following lemma, we establish the relationship between the iterates v^k and v^{k+1} generated by the SC-PRSM-PR algorithm (6.11) and the auxiliary variable $\tilde{v}^k$ defined in (6.13).

Lemma 1. *Let v^{k+1} be generated by the SC-PRSM-PR algorithm (6.11) with v^k given and $\tilde{v}^k$ defined by (6.14). Then, we have*

$$v^{k+1} = v^k - M(v^k - \tilde{v}^k), \tag{6.17}$$

where

$$M = \begin{pmatrix} I & 0 & 0 \\ 0 & I & 0 \\ -\alpha\beta B & -\alpha\beta C & 2\alpha I \end{pmatrix}. \tag{6.18}$$

Proof. Together with $y^{k+1} = \tilde{y}^k$ and $z^{k+1} = \tilde{z}^k$ and using (6.16), we have the following relationship:

$$\begin{pmatrix} y^{k+1} \\ z^{k+1} \\ \lambda^{k+1} \end{pmatrix} = \begin{pmatrix} y^k \\ z^k \\ \lambda^k \end{pmatrix} - \begin{pmatrix} I & 0 & 0 \\ 0 & I & 0 \\ -\alpha\beta B & -\alpha\beta C & 2\alpha I \end{pmatrix} \begin{pmatrix} y^k - \tilde{y}^k \\ z^k - \tilde{z}^k \\ \lambda^k - \tilde{\lambda}^k \end{pmatrix}.$$

This can be rewritten as a compact form of (6.17), where M is defined in (6.18).

3 Global Convergence

In this section, we show that the sequence generated by the SC-PRSM-PR algorithm (6.11) globally converges to a solution point of VI(Ω, F, θ). We first prove some inequalities which are crucial for establishing the strict contraction for the sequence generated by algorithm (6.11). We summarize them in several lemmas.

Lemma 2. *Let $\{w^k = (x^k, y^k, z^k, \lambda^k)\}$ be the sequence generated by the SC-PRSM-PR algorithm (6.11) and denote (y^k, z^k, λ^k) by v^k; The sequences $\{\tilde{w}^k\}$ and $\{\tilde{v}^k\}$ being still defined by (6.13) and (6.14), respectively, then we have*

$$\theta(u) - \theta(\tilde{u}^k) + (w - \tilde{w}^k)^T F(\tilde{w}^k) \geq (v - \tilde{v}^k)^T Q(v^k - \tilde{v}^k), \quad \forall w \in \Omega, \tag{6.19}$$

where

$$Q = \begin{pmatrix} (1+\mu)\beta B^T B & 0 & -\alpha B^T \\ 0 & (1+\mu)\beta C^T C & -\alpha C^T \\ -B & -C & \frac{1}{\beta} I \end{pmatrix}. \tag{6.20}$$

Proof. Since $\tilde{x}^k = x^{k+1}$, deriving the first-order optimality condition of the x-subproblem in (6.11), we have

$$\theta_1(x) - \theta_1(\tilde{x}^k) + (x - \tilde{x}^k)^T \{A^T[\beta(A\tilde{x}^k + By^k + Cz^k - b) - \lambda^k]\} \geq 0, \forall x \in \mathscr{X}.$$

Substituting $\tilde{\lambda}^k = \lambda^k - \beta(A\tilde{x}^k + By^k + Cz^k - b)$ (see (6.15)) into the above inequality, we obtain

$$\theta_1(x) - \theta_1(\tilde{x}^k) + (x - \tilde{x}^k)^T \{-A^T \tilde{\lambda}^k\} \geq 0, \quad \forall x \in \mathscr{X}. \tag{6.21}$$

Using $\lambda^{k+\frac{1}{2}} = \lambda^k - \alpha(\lambda^k - \tilde{\lambda}^k)$ (also see (6.15)), the y-minimization problem in (6.11) can be written as

$$\tilde{y}^k = \arg\min \left\{ \begin{array}{l} \theta_2(y) - [\lambda^k - \alpha(\lambda^k - \tilde{\lambda}^k)]^T (A\tilde{x}^k + By + Cz^k - b) \\ + \frac{\beta}{2} \|A\tilde{x}^k + By + Cz^k - b\|^2 + \frac{\mu\beta}{2} \|B(y - y^k)\|^2 \end{array} \middle| \, y \in \mathscr{Y} \right\},$$

and its first-order optimality condition gives us

$$\theta_2(y)-\theta_2(\tilde{y}^k)+(y-\tilde{y}^k)^T\{-B^T[\lambda^k-\alpha(\lambda^k-\tilde{\lambda}^k)]$$
$$+\beta B^T(A\tilde{x}^k+B\tilde{y}^k+Cz^k-b)+\mu\beta B^TB(\tilde{y}^k-y^k)\}\geq 0,\ \forall y\in\mathcal{Y}. \quad (6.22)$$

Again, using (6.15), we have

$$\begin{aligned}
&-[\lambda^k-\alpha(\lambda^k-\tilde{\lambda}^k)]+\beta(A\tilde{x}^k+B\tilde{y}^k+Cz^k-b)+\mu\beta B(\tilde{y}^k-y^k)\\
&=-[\lambda^k-\alpha(\lambda^k-\tilde{\lambda}^k)]+\beta(A\tilde{x}^k+By^k+Cz^k-b)+(1+\mu)\beta B(\tilde{y}^k-y^k)\\
&=-[\lambda^k-\alpha(\lambda^k-\tilde{\lambda}^k)]+(\lambda^k-\tilde{\lambda}^k)+(1+\mu)\beta B(\tilde{y}^k-y^k)\\
&=-\tilde{\lambda}^k+(1+\mu)\beta B(\tilde{y}^k-y^k)-\alpha(\tilde{\lambda}^k-\lambda^k).
\end{aligned}$$

Consequently, it follows from (6.22) that

$$\theta_2(y)-\theta_2(\tilde{y}^k)+(y-\tilde{y}^k)^T\{-B^T\tilde{\lambda}^k+(1+\mu)\beta B^TB(\tilde{y}^k-y^k)-\alpha B^T(\tilde{\lambda}^k-\lambda^k)\}\geq 0,\ \forall y\in\mathcal{Y}. \quad (6.23)$$

Analogously, from the z-minimization problem in (6.11), we get

$$\theta_3(z)-\theta_3(\tilde{z}^k)+(z-\tilde{z}^k)^T\{-C^T\tilde{\lambda}^k+(1+\mu)\beta C^TC(\tilde{z}^k-z^k)-\alpha C^T(\tilde{\lambda}^k-\lambda^k)\}\geq 0,\ \forall z\in\mathcal{Z}. \quad (6.24)$$

In addition, it follows from the last equation in (6.13) that

$$(A\tilde{x}^k+B\tilde{y}^k+C\tilde{z}^k-b)-B(\tilde{y}^k-y^k)-C(\tilde{z}^k-z^k)+\frac{1}{\beta}(\tilde{\lambda}^k-\lambda^k)=0,$$

which can be rewritten as

$$\tilde{\lambda}^k\in\Re^m,\ (\lambda-\tilde{\lambda}^k)^T\{(A\tilde{x}^k+B\tilde{y}^k+C\tilde{z}^k-b)-B(\tilde{y}^k-y^k)-C(\tilde{z}^k-z^k)+\frac{1}{\beta}(\tilde{\lambda}^k-\lambda^k)\}\geq 0,\ \forall\lambda\in\Re^m. \quad (6.25)$$

Combining (6.21), (6.23), (6.24), and (6.25) together, we get

$$\tilde{w}^k\in\Omega,\quad \theta(u)-\theta(\tilde{u}^k)+\begin{pmatrix} x-\tilde{x}^k\\ y-\tilde{y}^k\\ z-\tilde{z}^k\\ \lambda-\tilde{\lambda}^k\end{pmatrix}^T\left\{\begin{pmatrix} -A^T\tilde{\lambda}^k\\ -B^T\tilde{\lambda}^k\\ -C^T\tilde{\lambda}^k\\ A\tilde{x}^k+B\tilde{y}^k+C\tilde{z}^k-b\end{pmatrix}\right.$$
$$\left.+\begin{pmatrix} 0\\ (1+\mu)\beta B^TB(\tilde{y}^k-y^k)-\alpha B^T(\tilde{\lambda}^k-\lambda^k)\\ (1+\mu)\beta C^TC(\tilde{z}^k-z^k)-\alpha C^T(\tilde{\lambda}^k-\lambda^k)\\ -B(\tilde{y}^k-y^k)-C(\tilde{z}^k-z^k)+\frac{1}{\beta}(\tilde{\lambda}^k-\lambda^k)\end{pmatrix}\right\}\geq 0,\qquad \forall w\in\Omega.$$

From the definition of $F(w)$ (see (6.12)) and of the matrix Q (see (6.20)), the assertion (6.19) follows directly from the last inequality above, completing the proof of the lemma.

Before we proceed the proof, recall we have defined the matrix M by (6.18). Then, together with the matrix Q defined in (6.20), let us define a new matrix H as

$$H = QM^{-1}. \tag{6.26}$$

Some useful properties of H are summarized in the following.

Proposition 1. *The matrix H defined in (6.26) is symmetric and it can be written as*

$$H = \begin{pmatrix} (1+\mu-\frac{1}{2}\alpha)\beta B^T B & -\frac{1}{2}\alpha\beta B^T C & -\frac{1}{2}B^T \\ -\frac{1}{2}\alpha\beta C^T B & (1+\mu-\frac{1}{2}\alpha)\beta C^T C & -\frac{1}{2}C^T \\ -\frac{1}{2}B & -\frac{1}{2}C & \frac{1}{2\alpha\beta}I \end{pmatrix}. \tag{6.27}$$

Moreover, for any fixed $\alpha \in (0,1)$ and $\mu \geq \alpha$, H is positive definite.

Proof. The proof requires some linear algebra knowledge. First, note that $H = QM^{-1}$. For the matrix M defined in (6.18), we have

$$M^{-1} = \begin{pmatrix} I & 0 & 0 \\ 0 & I & 0 \\ \frac{\beta}{2}B & \frac{\beta}{2}C & \frac{1}{2\alpha}I \end{pmatrix}.$$

Then, by a manipulation, we obtain

$$H = \begin{pmatrix} (1+\mu)\beta B^T B & 0 & -\alpha B^T \\ 0 & (1+\mu)\beta C^T C & -\alpha C^T \\ -B & -C & \frac{1}{\beta}I \end{pmatrix} \begin{pmatrix} I & 0 & 0 \\ 0 & I & 0 \\ \frac{\beta}{2}B & \frac{\beta}{2}C & \frac{1}{2\alpha}I \end{pmatrix}$$
$$= \begin{pmatrix} (1+\mu-\frac{1}{2}\alpha)\beta B^T B & -\frac{1}{2}\alpha\beta B^T C & -\frac{1}{2}B^T \\ -\frac{1}{2}\alpha\beta C^T B & (1+\mu-\frac{1}{2}\alpha)\beta C^T C & -\frac{1}{2}C^T \\ -\frac{1}{2}B & -\frac{1}{2}C & \frac{1}{2\alpha\beta}I \end{pmatrix}.$$

This is just the form of (6.27) and H is symmetric. The first part is proved.

To show the positive definiteness of matrix H, we need only to inspect the following 3×3 matrix

$$\begin{pmatrix} (1+\mu-\frac{1}{2}\alpha) & -\frac{1}{2}\alpha & -\frac{1}{2} \\ -\frac{1}{2}\alpha & (1+\mu-\frac{1}{2}\alpha) & -\frac{1}{2} \\ -\frac{1}{2} & -\frac{1}{2} & \frac{1}{2\alpha} \end{pmatrix}.$$

Since $\alpha \in (0,1)$ and $\mu \geq \alpha$, we have

$$1+\mu-\frac{1}{2}\alpha > 0 \qquad \text{and} \qquad \begin{vmatrix} (1+\mu-\frac{1}{2}\alpha) & -\frac{1}{2}\alpha \\ -\frac{1}{2}\alpha & (1+\mu-\frac{1}{2}\alpha) \end{vmatrix} > 0.$$

Note that

$$\begin{vmatrix} (1+\mu-\frac{1}{2}\alpha) & -\frac{1}{2}\alpha & -\frac{1}{2} \\ -\frac{1}{2}\alpha & (1+\mu-\frac{1}{2}\alpha) & -\frac{1}{2} \\ -\frac{1}{2} & -\frac{1}{2} & \frac{1}{2\alpha} \end{vmatrix}$$

$$= -\frac{1}{2}\begin{vmatrix} -\frac{1}{2}\alpha & -\frac{1}{2} \\ (1+\mu-\frac{1}{2}\alpha) & -\frac{1}{2} \end{vmatrix} + \frac{1}{2}\begin{vmatrix} (1+\mu-\frac{1}{2}\alpha) & -\frac{1}{2} \\ -\frac{1}{2}\alpha & -\frac{1}{2} \end{vmatrix} + \frac{1}{2\alpha}\begin{vmatrix} (1+\mu-\frac{1}{2}\alpha) & -\frac{1}{2}\alpha \\ -\frac{1}{2}\alpha & (1+\mu-\frac{1}{2}\alpha) \end{vmatrix}$$

$$= \begin{vmatrix} (1+\mu-\frac{1}{2}\alpha) & -\frac{1}{2} \\ -\frac{1}{2}\alpha & -\frac{1}{2} \end{vmatrix} + \frac{1}{2\alpha}\big((1+\mu-\frac{1}{2}\alpha)^2 - (\frac{1}{2}\alpha)^2\big)$$

$$= \begin{vmatrix} (1+\mu) & 0 \\ -\frac{1}{2}\alpha & -\frac{1}{2} \end{vmatrix} + \frac{1}{2\alpha}(1+\mu)(1+\mu-\alpha)$$

$$= \frac{1}{2\alpha}(1+\mu)(1+\mu-2\alpha).$$

Since $\alpha \in (0,1)$ and $\mu \geq \alpha$, H is positive definite.

The assertion in Proposition 1 helps us present the convergence analysis in succinct notation. Now, with the defined matrices M, Q and H, we can further analyze the conclusion proved in Lemma 2. More specifically, let us inspect first the right-hand side of the inequality (6.19) and rewrite it as the sum of some quadratic terms under certain matrix norms. This is done in the following lemma.

Lemma 3. *Let $\{w^k = (x^k, y^k, z^k, \lambda^k)\}$ be the sequence generated by the SC-PRSM-PR algorithm (6.11) and denote (y^k, z^k, λ^k) by v^k. The sequences $\{\tilde{w}^k\}$ and $\{\tilde{v}^k\}$ being still defined by (6.13) and (6.14), respectively, we then have*

$$(v-\tilde{v}^k)^T Q(v^k-\tilde{v}^k) = \frac{1}{2}\big(\|v-v^{k+1}\|_H^2 - \|v-v^k\|_H^2\big) + \frac{1}{2}\|v^k-\tilde{v}^k\|_G^2, \tag{6.28}$$

with

$$G = Q^T + Q - M^T HM, \tag{6.29}$$

where the matrices M, Q, and H are defined in (6.18), (6.20), and (6.27), respectively.

Proof. The proof only requires some elementary manipulations. More specifically, using the fact that $Q = HM$ (see (6.26)) and the relation (6.17), the right-hand side of (6.19) can be written as

$$(v-\tilde{v}^k)^T Q(v^k-\tilde{v}^k) = (v-\tilde{v}^k)^T H(v^k - v^{k+1}). \tag{6.30}$$

Applying the identity

$$(a-b)^T H(c-d) = \frac{1}{2}\{\|a-d\|_H^2 - \|a-c\|_H^2\} + \frac{1}{2}\{\|c-b\|_H^2 - \|d-b\|_H^2\},$$

to the right-hand side of (6.30) with

$$a = v, \quad b = \tilde{v}^k, \quad c = v^k, \quad \text{and} \quad d = v^{k+1},$$

we thus obtain

$$(v-\tilde{v}^k)^T Q(v^k-\tilde{v}^k) = \frac{1}{2}(\|v-v^k\|_H^2 - \|v-v^{k+1}\|_H^2) + \frac{1}{2}(\|v^k-\tilde{v}^k\|_H^2 - \|v^{k+1}-\tilde{v}^k\|_H^2). \tag{6.31}$$

For the last term in (6.31), we have

$$\begin{aligned}
&\|v^k-\tilde{v}^k\|_H^2 - \|v^{k+1}-\tilde{v}^k\|_H^2 \\
&= \|v^k-\tilde{v}^k\|_H^2 - \|(v^k-\tilde{v}^k)-(v^k-v^{k+1})\|_H^2 \\
&\overset{(6.17)}{=} \|v^k-\tilde{v}^k\|_H^2 - \|(v^k-\tilde{v}^k)-M(v^k-\tilde{v}^k)\|_H^2 \\
&= 2(v^k-\tilde{v}^k)HM(v^k-\tilde{v}^k) - (v^k-\tilde{v}^k)M^T HM(v^k-\tilde{v}^k) \\
&\overset{(6.26)}{=} (v^k-\tilde{v}^k)^T(Q^T+Q-M^T HM)(v^k-\tilde{v}^k).
\end{aligned} \tag{6.32}$$

By using (6.31), (6.32), and (6.29), the assertion of Lemma 3 is proved.

In Lemma 3, a new matrix G is introduced in order to improve the inequality (6.19) in Lemma 2. Let us hold on the proof temporarily and take a closer look at the matrix G just defined in (6.29). Some properties of this matrix are summarized in the following.

Proposition 2. *The symmetric matrix G defined in (6.29) can be rewritten as*

$$G = \begin{pmatrix} (1+\mu-\alpha)\beta B^T B & -\alpha\beta B^T C & -(1-\alpha)B^T \\ -\alpha\beta C^T B & (1+\mu-\alpha)\beta C^T C & -(1-\alpha)C^T \\ -(1-\alpha)B & -(1-\alpha)C & \frac{2-2\alpha}{\beta} I \end{pmatrix}. \tag{6.33}$$

Moreover, for a fixed $\alpha \in (0,1)$ and any $\mu \geq \alpha$ (resp. $\mu > \alpha$), G is positive semi-definite (resp., positive definite).

Proof. For the matrix G defined in (6.29), since $Q = HM$ (see (6.26)), we have

$$G = Q^T + Q - M^T HM = Q^T + Q - M^T Q.$$

Using the matrices M and Q (see (6.18) and (6.20)), we obtain

$$G = (Q^T + Q) - \begin{pmatrix} I & 0 & -\alpha\beta B^T \\ 0 & I & -\alpha\beta C^T \\ 0 & 0 & 2\alpha I \end{pmatrix} \begin{pmatrix} (1+\mu)\beta B^T B & 0 & -\alpha B^T \\ 0 & (1+\mu)\beta C^T C & -\alpha C^T \\ -B & -C & \frac{1}{\beta} I \end{pmatrix}$$

$$= \begin{pmatrix} (2+2\mu)\beta B^T B & 0 & -(1+\alpha)B^T \\ 0 & (2+2\mu)\beta C^T C & -(1+\alpha)C^T \\ -(1+\alpha)B & -(1+\alpha)C & \frac{2}{\beta} I \end{pmatrix} - \begin{pmatrix} (1+\mu+\alpha)\beta B^T B & \alpha\beta B^T C & -2\alpha B^T \\ -\alpha\beta C^T B & (1+\mu+\alpha)\beta C^T C & -2\alpha C^T \\ -2\alpha B & -2\alpha C & \frac{2\alpha}{\beta} I \end{pmatrix}$$

$$= \begin{pmatrix} (1+\mu-\alpha)\beta B^T B & -\alpha\beta B^T C & -(1-\alpha)B^T \\ \alpha\beta C^T B & (1+\mu-\alpha)\beta C^T C & -(1-\alpha)C^T \\ -(1-\alpha)B & -(1-\alpha)C & \frac{2-2\alpha}{\beta} I \end{pmatrix}.$$

This is just the matrix G we announced in (6.33), which completes the first part of the proof of the lemma.

To show the positive semi-definiteness (resp., positive definiteness) of G, we need only to inspect the following 3×3 matrix

$$\begin{pmatrix} (1+\mu-\alpha) & -\alpha & -(1-\alpha) \\ -\alpha & (1+\mu-\alpha) & -(1-\alpha) \\ -(1-\alpha) & -(1-\alpha) & 2(1-\alpha) \end{pmatrix}.$$

Since $\alpha \in (0,1)$ and $\mu \geq \alpha$, we have

$$1+\mu-\alpha > 0 \qquad \text{and} \qquad \begin{vmatrix} (1+\mu-\alpha) & -\alpha \\ -\alpha & (1+\mu-\alpha) \end{vmatrix} > 0.$$

Note that

$$\begin{vmatrix} (1+\mu-\alpha) & -\alpha & -(1-\alpha) \\ -\alpha & (1+\mu-\alpha) & -(1-\alpha) \\ -(1-\alpha) & -(1-\alpha) & 2(1-\alpha) \end{vmatrix}$$

$$= -(1-\alpha) \begin{vmatrix} -\alpha & -(1-\alpha) \\ (1+\mu-\alpha) & -(1-\alpha) \end{vmatrix} + (1-\alpha) \begin{vmatrix} (1+\mu-\alpha) & -(1-\alpha) \\ -\alpha & -(1-\alpha) \end{vmatrix}$$

$$+2(1-\alpha) \begin{vmatrix} (1+\mu-\alpha) & -\alpha \\ -\alpha & (1+\mu-\alpha) \end{vmatrix}$$

$$= 2(1-\alpha)\begin{vmatrix} (1+\mu-\alpha) & -(1-\alpha) \\ -\alpha & -(1-\alpha) \end{vmatrix} + 2(1-\alpha)\begin{vmatrix} (1+\mu-\alpha) & -\alpha \\ -\alpha & (1+\mu-\alpha) \end{vmatrix}$$

$$= 2(1-\alpha)\begin{vmatrix} (1+\mu) & 0 \\ -\alpha & -(1-\alpha) \end{vmatrix} + 2(1-\alpha)((1+\mu-\alpha)^2-\alpha^2)$$

$$= -2(1-\alpha)^2(1+\mu)+2(1-\alpha)(1+\mu)(1+\mu-2\alpha)$$

$$= 2(1-\alpha)(1+\mu)(\mu-\alpha).$$

Thus, for a fixed $\alpha \in (0,1)$ and any $\mu \geq \alpha$ (resp. $\mu > \alpha$), G is positive semi-definite (resp. positive definite). The proof is complete.

Now, with the proved propositions and lemmas, the inequality (6.19) in Lemma 2 can be significantly polished. We summarize it as a theorem and will use it later.

Theorem 1. *Let $\{w^k = (x^k, y^k, z^k, \lambda^k)\}$ be the sequence generated by the SC-PRSM-PR algorithm (6.11) and denote (y^k, z^k, λ^k) by v^k; The sequences $\{\tilde{w}^k\}$ and $\{\tilde{v}^k\}$ being still defined by (6.13) and (6.14), respectively, we then have*

$$\theta(u)-\theta(\tilde{u}^k)+(w-\tilde{w}^k)^T F(\tilde{w}^k) \geq \frac{1}{2}\left(\|v-v^{k+1}\|_H^2 - \|v-v^k\|_H^2\right)+\frac{1}{2}\|v^k-\tilde{v}^k\|_G^2,\ \forall w \in \Omega, \quad (6.34)$$

where H and G are defined by (6.27) and (6.29), respectively.

Proof. It is trivial by combining the assertions (6.19) and (6.28).

Now we are ready to show that the sequence $\{v^k\}$ generated by the SC-PRSM-PR algorithm (6.11) with $\alpha \in (0,1)$ and $\mu > \alpha$ is strictly contractive with respect to the solution of the VI (6.12a). Note that for this case the matrix G defined in (6.29) is positive definite as proved in Proposition 2.

Theorem 2. *Let $\{w^k = (x^k, y^k, z^k, \lambda^k)\}$ be the sequence generated by the SC-PRSM-PR algorithm (6.11); and denote (y^k, z^k, λ^k) by v^k. The sequences $\{\tilde{w}^k\}$ and $\{\tilde{v}^k\}$ being still defined by (6.13) and (6.14), respectively, we then have*

$$\|v^{k+1}-v^*\|_H^2 \leq \|v^k-v^*\|_H^2 - \|v^k-\tilde{v}^k\|_G^2, \quad \forall v^* \in \mathcal{V}^*, \quad (6.35)$$

where H and G are defined by (6.27) and (6.29), respectively.

Proof. Setting $w = w^*$ in (6.34), we get

$$\|v^k-v^*\|_H^2 - \|v^{k+1}-v^*\|_H^2 \geq \|v^k-\tilde{v}^k\|_G^2 + 2\{\theta(\tilde{u}^k)-\theta(u^*)+(\tilde{w}^k-w^*)^T F(\tilde{w}^k)\}. \quad (6.36)$$

By using the optimality of w^* and the monotonicity of F, we have

$$\theta(\tilde{u}^k)-\theta(u^*)+(\tilde{w}^k-w^*)^T F(\tilde{w}^k) \geq \theta(\tilde{u}^k)-\theta(u^*)+(\tilde{w}^k-w^*)^T F(w^*) \geq 0$$

and thus

$$\|v^k - v^*\|_H^2 - \|v^{k+1} - v^*\|_H^2 \ge \|v^k - \tilde{v}^k\|_G^2. \tag{6.37}$$

The assertion (6.35) follows directly.

The assertion (6.35) thus implies the strict contraction of the sequence $\{v^k\}$ generated by the SC-PRSM-PR algorithm (6.11). We can thus easily prove the convergence based on this assertion, as stated in the following theorem. This assertion is also the basis for establishing the worst-case convergence rate in a nonergodic sense in Section 4.1.

Theorem 3. *Let $\{w^k = (x^k, y^k, z^k, \lambda^k)\}$ be the sequence generated by the SC-PRSM-PR algorithm (6.11). The sequence $\{v^k = (y^k, z^k, \lambda^k)\}$ converges to some v^∞ which belongs to $\mathcal{V}^*$.*

Proof. According to (6.35), the sequence $\{v^k\}$ is bounded and

$$\lim_{k\to\infty} \|v^k - \tilde{v}^k\| = 0. \tag{6.38}$$

So, $\{\tilde{v}^k\}$ is also bounded. Let v^∞ be a cluster point of $\{\tilde{v}^k\}$ and $\{\tilde{v}^{k_j}\}$ a subsequence which converges to v^∞. Let $\{\tilde{w}^k\}$ and $\{\tilde{w}^{k_j}\}$ be the sequences induced by $\{\tilde{v}^k\}$ and $\{\tilde{v}^{k_j}\}$, respectively. It follows from (6.19) that

$$\tilde{w}^{k_j} \in \Omega, \quad \theta(u) - \theta(u^\infty) + (w - \tilde{w}^{k_j})^T F(\tilde{w}^{k_j}) \ge (v - v^{k_j})^T Q(v^{k_j} - \tilde{v}^{k_j}), \quad \forall\, w \in \Omega.$$

Since the matrix Q is not singular, it follows from the continuity of θ and F that

$$w^\infty \in \Omega, \quad \theta(u) - \theta(u^\infty) + (w - w^\infty)^T F(w^\infty) \ge 0, \quad \forall\, w \in \Omega.$$

The above variational inequality indicates that w^∞ is a solution of VI(Ω, F). By using (6.38) and $\lim_{j\to\infty} v^{k_j} = v^\infty$, the subsequence $\{v^{k_j}\}$ converges to v^∞. Due to (6.35), we have

$$\|v^{k+1} - v^\infty\|_H \le \|v^k - v^\infty\|_H$$

and thus $\{v^k\}$ converges to v^∞. The proof is complete.

4 Worst-Case Convergence Rate

In this section, we establish the worst-case $O(1/t)$ convergence rate measured by the iteration complexity in both the ergodic and a nonergodic senses for the SC-PRSM-PR algorithm (6.11), where t is the iteration counter. Note that as in the publications [37, 38], and many others, a worst-case $O(1/t)$ convergence rate measured by the iteration complexity means that an approximate solution with an accuracy of $O(1/t)$ can be found based on t iterations of an iterative scheme; or equivalently, it requires at most $O(1/\varepsilon)$ iterations to find an approximate solution with an accuracy of ε.

4.1 Worse-Case Convergence Rate in a Nonergodic Sense

We first establish the worst-case $O(1/t)$ convergence rate in a nonergodic sense for the SC-PRSM-PR algorithm (6.11). The starting point for the analysis is the assertion (6.35) in Theorem 2, and the analytic framework follows from the work in [30] for the ADMM scheme (6.2).

First, recall the matrices M, H, and G defined respectively by (6.18), (6.27), and (6.29). Since both matrices G and M^THM are positive definite, there exists a constant $c > 0$ such that

$$\|M(v^k - \tilde{v}^k)\|_H^2 \le c\|v^k - \tilde{v}^k\|_G^2.$$

Substituting it into (6.35) and using (6.17), it follows that

$$\|v^{k+1} - v^*\|_H^2 \le \|v^k - v^*\|_H^2 - \frac{1}{c}\|v^k - v^{k+1}\|_H^2, \quad \forall v^* \in \mathcal{V}^*. \tag{6.39}$$

In the following, we will show that the sequence $\{\|v^k - v^{k+1}\|_H^2\}$ is monotonically non-increasing. That is, we have

$$\|v^{k+1} - v^{k+2}\|_H^2 \le \|v^k - v^{k+1}\|_H^2, \quad \forall k \ge 0.$$

The following lemma establishes an important inequality for this purpose.

Lemma 4. *Let $\{w^k = (x^k, y^k, z^k, \lambda^k)\}$ be the sequence generated by the SC-PRSM-PR algorithm (6.11), and denote (y^k, z^k, λ^k) by v^k. The sequences $\{\tilde{w}^k\}$ and $\{\tilde{v}^k\}$ being still defined by (6.13) and (6.14), respectively, we then have*

$$(\tilde{v}^k - \tilde{v}^{k+1})^T Q\{(v^k - v^{k+1}) - (\tilde{v}^k - \tilde{v}^{k+1})\} \ge 0, \tag{6.40}$$

where the matrix Q is defined by (6.20).

Proof. Setting $w = \tilde{w}^{k+1}$ in (6.19), we have

$$\theta(\tilde{u}^{k+1}) - \theta(\tilde{u}^k) + (\tilde{w}^{k+1} - \tilde{w}^k)^T F(\tilde{w}^k) \ge (\tilde{v}^{k+1} - \tilde{v}^k)^T Q(v^k - \tilde{v}^k). \tag{6.41}$$

Note that (6.19) is also true for $k := k+1$ and thus

$$\theta(u) - \theta(\tilde{u}^{k+1}) + (w - \tilde{w}^{k+1})^T F(\tilde{w}^{k+1}) \ge (v - \tilde{v}^{k+1})^T Q(v^{k+1} - \tilde{v}^{k+1}), \quad \forall w \in \Omega.$$

Set $w = \tilde{w}^k$ in the above inequality, we obtain

$$\theta(\tilde{u}^k) - \theta(\tilde{u}^{k+1}) + (\tilde{w}^k - \tilde{w}^{k+1})^T F(\tilde{w}^{k+1}) \ge (\tilde{v}^k - \tilde{v}^{k+1})^T Q(v^{k+1} - \tilde{v}^{k+1}). \tag{6.42}$$

Adding (6.41) and (6.42) and using the monotonicity of F, we get (6.40) immediately.

One more inequality is needed; we summarize it in the following lemma.

Lemma 5. *Let $\{w^k=(x^k,y^k,z^k,\lambda^k)\}$ be the sequence generated by the SC-PRSM-PR algorithm (6.11), and denote (y^k,z^k,λ^k) by v^k. The sequences $\{\tilde{w}^k\}$ and $\{\tilde{v}^k\}$ being still defined by (6.13) and (6.14), respectively, we then have*

$$(v^k-\tilde{v}^k)^T M^T HM\{(v^k-\tilde{v}^k)-(v^{k+1}-\tilde{v}^{k+1})\} \geq \frac{1}{2}\|(v^k-\tilde{v}^k)-(v^{k+1}-\tilde{v}^{k+1})\|^2_{(Q^T+Q)}. \tag{6.43}$$

where the matrices M, H, and Q are defined by (6.18), (6.27), and (6.20), *respectively.*

Proof. Adding the equation

$$\{(v^k-v^{k+1})-(\tilde{v}^k-\tilde{v}^{k+1})\}^T Q\{(v^k-v^{k+1})-(\tilde{v}^k-\tilde{v}^{k+1})\} = \frac{1}{2}\|(v^k-\tilde{v}^k)-(v^{k+1}-\tilde{v}^{k+1})\|^2_{(Q^T+Q)}$$

to the both sides of (6.40), we get

$$(v^k-v^{k+1})^T Q\{(v^k-v^{k+1})-(\tilde{v}^k-\tilde{v}^{k+1})\} \geq \frac{1}{2}\|(v^k-\tilde{v}^k)-(v^{k+1}-\tilde{v}^{k+1})\|^2_{(Q^T+Q)}. \tag{6.44}$$

By using (see (6.17) and (6.26))

$$v^k-v^{k+1}=M(v^k-\tilde{v}^k) \qquad \text{and} \qquad Q=HM,$$

to the term v^k-v^{k+1} in the left-hand side of (6.44), we obtain

$$(v^k-\tilde{v}^k)^T M^T HM\{(v^k-v^{k+1})-(\tilde{v}^k-\tilde{v}^{k+1})\} \geq \frac{1}{2}\|(v^k-\tilde{v}^k)-(v^{k+1}-\tilde{v}^{k+1})\|^2_{(Q^T+Q)}.$$

and the lemma is proved.

Now, we are ready to show that the sequence $\{\|M(v^k-\tilde{v}^k)\|^2_H\}$ is non-increasing.

Theorem 4. *Let $\{w^k=(x^k,y^k,z^k,\lambda^k)\}$ be the sequence generated by the SC-PRSM-PR algorithm (6.11), and denote (y^k,z^k,λ^k) by v^k. The sequences $\{\tilde{w}^k\}$ and $\{\tilde{v}^k\}$ being still defined by (6.13) and (6.14), respectively, we then have*

$$\|M(v^{k+1}-\tilde{v}^{k+1})\|^2_H \leq \|M(v^k-\tilde{v}^k)\|^2_H, \tag{6.45}$$

where the matrices M and H defined by (6.18) and (6.27), respectively.

Proof. Setting $a=M(v^k-\tilde{v}^k)$ and $b=M(v^{k+1}-\tilde{v}^{k+1})$ in the identity

$$\|a\|^2_H-\|b\|^2_H=2a^T H(a-b)-\|a-b\|^2_H,$$

we obtain

$$\begin{aligned}&\|M(v^k-\tilde{v}^k)\|^2_H-\|M(v^{k+1}-\tilde{v}^{k+1})\|^2_H\\ &=2(v^k-\tilde{v}^k)^T M^T HM\{(v^k-\tilde{v}^k)-(v^{k+1}-\tilde{v}^{k+1})\}-\|M(v^k-\tilde{v}^k)-M(v^{k+1}-\tilde{v}^{k+1})\|^2_H.\end{aligned}$$

Inserting (6.43) into the first term of the right-hand side of the last equality, we obtain

$$\|M(v^k - \tilde{v}^k)\|_H^2 - \|M(v^{k+1} - \tilde{v}^{k+1})\|_H^2 \geq \|(v^k - \tilde{v}^k) - (v^{k+1} - \tilde{v}^{k+1})\|^2_{(Q^T+Q-M^THM)}.$$

The assertion (6.45) follows from the above inequality and Lemma 2 immediately.

Now, according to (6.45) and (6.17), we have

$$\|v^{k+1} - v^{k+2}\|_H^2 \leq \|v^k - v^{k+1}\|_H^2. \tag{6.46}$$

That is, the monotonicity of the sequence $\{\|v^k - v^{k+1}\|_H^2\}$ is proved. Then, with (6.39) and (6.46), we can prove a worst-case $O(1/t)$ convergence rate in a nonergodic sense for the SC-PRSM-PR algorithm (6.11) with $\alpha \in [0,1)$. We summarize the result in the following.

Theorem 5. *Let $\{w^k = (x^k, y^k, z^k, \lambda^k)\}$ be the sequence generated by the SC-PRSM-PR algorithm (6.11) and $\{v^k = (y^k, z^k, \lambda^k)\}$. Then we have*

$$\|v^k - v^{k+1}\|_H^2 \leq \frac{c}{(k+1)}\|v^0 - v^*\|_H^2, \quad \forall v^* \in \mathcal{V}^*, \tag{6.47}$$

where the matrix H is defined by (6.27).

Proof. First, it follows from (6.35) that

$$\frac{1}{c}\sum_{t=0}^{\infty}\|v^t - v^{t+1}\|_H^2 \leq \|v^0 - v^*\|_H^2, \quad \forall v^* \in \mathcal{V}^*. \tag{6.48}$$

According to Theorem 4, the sequence $\{\|v^t - v^{t+1}\|_H^2\}$ is monotonically non-increasing. Therefore, we have

$$(k+1)\|v^k - v^{k+1}\|_H^2 < \sum_{i=0}^{k}\|v^i - v^{i+1}\|_H^2. \tag{6.49}$$

The assertion (6.47) follows from (6.48) and (6.49) immediately.

Notice that $\mathcal{V}^*$ is convex and closed. Let $d := \inf\{\|v^0 - v^*\|_H \,|\, v^* \in \Omega^*\}$. Then, for any given $\varepsilon > 0$, Theorem 5 shows that the scheme (6.11) needs at most $\lfloor d^2/\varepsilon \rfloor$ iterations to ensure that $\|v^k - v^{k+1}\|_H^2 \leq \varepsilon$. Recall that w^{k+1} is a solution of VI(Ω, F, θ) if $\|v^k - v^{k+1}\|_H^2 = 0$. A worst-case $O(1/t)$ convergence rate in a nonergodic sense is thus established for the SC-PRSM-PR algorithm (6.11) in Theorem 5.

4.2 Worse-Case Convergence Rate in the Ergodic Sense

In this subsection, we establish a worst-case $O(1/t)$ convergence rate in the ergodic sense for the SC-PRSM-PR algorithm (6.11). For this purpose, we only need the positive semi-definiteness of the matrix G defined in (6.29). Thus, as asserted in

Proposition 2, we just choose $\alpha \in (0,1)$ and $\mu \geq \alpha$ for the SC-PRSM-PR algorithm (6.11). For the analysis, the starting point is Theorem 1, and its follows the work in [29] for the ADMM algorithm (6.2).

Let us first recall Theorem 2.3.5 in [14], which provides us an insightful characterization for the solution set of a generic VI. In the following theorem, we specific the above theorem from [14] for the particular VI(Ω, F, θ) under consideration.

Theorem 6. *The solution set of VI(Ω, F, θ) is convex and it can be characterized as*

$$\Omega^* = \bigcap_{w \in \Omega} \{\tilde{w} \in \Omega : (\theta(u) - \theta(\tilde{u})) + (w - \tilde{w})^T F(w) \geq 0\}. \tag{6.50}$$

Proof. The proof is an incremental extension of Theorem 2.3.5 in [14], or, alternatively, see the proof of Theorem 2.1 in [29].

Theorem 6 thus implies that $\tilde{w} \in \Omega$ is an approximate solution of VI(Ω, F, θ) with the accuracy $\varepsilon > 0$ if it satisfies

$$\theta(u) - \theta(\tilde{u}) + F(w)^T (w - \tilde{w}) \geq -\varepsilon, \ \forall w \in \Omega \cap \mathscr{D}(\tilde{u}),$$

where

$$\mathscr{D}(\tilde{u}) = \{u \,|\, \|u - \tilde{u}\| \leq 1\}.$$

In the remainder, our purpose is to show that based on t iterations of the SC-PRSM-PR algorithm (6.11), we can find $\tilde{w} \in \Omega$ such that

$$\theta(\tilde{u}) - \theta(u) + (\tilde{w} - w)^T F(w) \leq \varepsilon, \ \forall w \in \Omega \cap \mathscr{D}(\tilde{u}), \tag{6.51}$$

with $\varepsilon = O(1/t)$. That is, an approximate solution of VI(Ω, F, θ) with an accuracy of $O(1/t)$ can be found based on t iterations of the SC-PRSM-PR (6.11).

For the coming analysis, let us slightly refine the assertion (6.34) in Theorem 1. Using the fact (see the definition of F in (6.12b))

$$(w - \tilde{w}^k)^T F(w) = (w - \tilde{w}^k)^T F(\tilde{w}^k),$$

then it follows from (6.34) that

$$\theta(u) - \theta(\tilde{u}^k) + (w - \tilde{w}^k)^T F(w) + \frac{1}{2}\|v - v^k\|_H^2 \geq \frac{1}{2}\|v - v^{k+1}\|_H^2, \ \forall w \in \Omega. \tag{6.52}$$

Then, we summarize the worst-case convergence $O(1/t)$ convergence rate in the ergodic sense for the SC-PRSM-PR algorithm (6.11) in the following.

Theorem 7. *Let $\{w^k\}$ be the sequence generated by the SC-PRSM-PR algorithm (6.11), $\{\tilde{w}^k\}$ be defined by (6.13), and t be an integer. Let $\tilde{w}_t$ be defined as the average of $\tilde{w}^k$ for $k = 1, 2, \cdots t$, i.e.,*

$$\tilde{w}_t = \frac{1}{t+1} \sum_{k=0}^{t} \tilde{w}^k. \tag{6.53}$$

Then, we have $\tilde{w}_t \in \Omega$ *and*

$$\theta(\tilde{u}_t) - \theta(u) + (\tilde{w}_t - w)^T F(w) \le \frac{1}{2(t+1)} \|v - v^0\|_H^2, \quad \forall w \in \Omega, \tag{6.54}$$

where H is defined by (6.27).

Proof. First, from (6.13), it holds that $\tilde{w}^k \in \Omega$ for all $k \ge 0$. Together with the con vexity of $\mathscr{X}$ and $\mathscr{Y}$, (6.53) implies that $\tilde{w}_t \in \Omega$. Summing the inequality (6.52) over $k = 0, 1, \ldots, t$, we obtain

$$(t+1)\theta(u) - \sum_{k=0}^{t} \theta(\tilde{u}^k) + \Big((t+1)w - \sum_{k=0}^{t} \tilde{w}^k\Big)^T F(w) + \frac{1}{2}\|v - v^0\|_H^2 \ge 0, \quad \forall w \in \Omega.$$

Use the definition of $\tilde{w}_t$, the above inequality can be written as

$$\frac{1}{t+1} \sum_{k=0}^{t} \theta(\tilde{u}^k) - \theta(u) + (\tilde{w}_t - w)^T F(w) \le \frac{1}{2(t+1)} \|v - v^0\|_H^2, \quad \forall w \in \Omega. \tag{6.55}$$

Since θ is convex and

$$\tilde{u}_t = \frac{1}{t+1} \sum_{k=0}^{t} \tilde{u}^k,$$

we have that

$$\theta(\tilde{u}_t) \le \frac{1}{t+1} \sum_{k=0}^{t} \theta(\tilde{u}^k).$$

Substituting it in (6.55), the assertion of this theorem follows directly.

Remembering (6.51), relation (6.54) shows us that based on t iteration of the SC-PRSM-PR algorithm (6.11), we can find $\tilde{w}_t$, defined by (6.53), which is an approximate solution of (6.5) with an accuracy of $O(1/t)$. That is, a worst-case $O(1/t)$ convergence rate in the ergodic sense is established for the SC-PRSM-PR algorithm (6.11) in Theorem 7.

5 A Divergence Example

It has been shown in the last section that the SC-PRSM-PR algorithm (6.11) is convergent when $\alpha \in (0,1)$ and $\mu > \alpha$. In this section, we give an example showing that both the direct extension of the SC-PRSM algorithm (6.7) and the direct application of the SC-PRSM algorithm (6.10) are not necessarily convergent. Thus, the motivation of considering the SC-PRSM-PR algorithm (6.11) for the multi-block convex minimization model (6.5) is justified.

The example is inspired by the counter-example in [9], showing the divergence of the E-ADMM algorithm (6.6). More specifically, we consider the following system of linear equations:

$$Ax + By + Cz = 0, \tag{6.56}$$

where $A, B, C \in \Re^4$ are linearly independent such that the matrix (A, B, C) is full rank and x, y, z are all in $\Re$. This is a special case of the model (6.5) with $\theta_1 = \theta_2 = \theta_3 = 0$, $m = 4$, $n_1 = n_2 = n_3 = 1$, $\mathscr{X} = \mathscr{Y} = \mathscr{Z} = \Re$; and the coefficients matrices are A, B, and C, respectively. Obviously, the system of linear equation (6.56) has the unique solution $x^* = y^* = z^* = 0$. In particular, we consider

$$(A, B, C) = \begin{pmatrix} 1 & 1 & 1 \\ 1 & 1 & 2 \\ 1 & 2 & 2 \\ 1 & 2 & 2 \end{pmatrix}. \tag{6.57}$$

5.1 Divergence of the Direct Application of the SC-PRSM Algorithm (6.10)

First, we show that the direct application of the SC-PRSM (6.10) is not necessarily convergent.

Applying the scheme (6.10) to the homogeneous system of linear equation (6.56), the resulting scheme can be written as

$$\begin{cases} A^T[\beta(Ax^{k+1} + By^k + Cz^k) - \lambda^k] = 0 \\ \alpha\beta(Ax^{k+1} + By^k + Cz^k) + \lambda^{k+\frac{1}{2}} - \lambda^k = 0 \\ B^T[\beta(Ax^{k+1} + By^{k+1} + Cz^k) - \lambda^{k+\frac{1}{2}}] = 0 \\ C^T[\beta(Ax^{k+1} + By^k + Cz^{k+1}) - \lambda^{k+\frac{1}{2}}] = 0 \\ \alpha\beta(Ax^{k+1} + By^{k+1} + Cz^{k+1}) + \lambda^{k+1} - \lambda^{k+\frac{1}{2}} = 0. \end{cases} \tag{6.58}$$

It follows from the first equation in (6.58) that

$$x^{k+1} = \frac{1}{A^T A}(-A^T By^k - A^T Cz + A^T \lambda^k/\beta). \tag{6.59}$$

For simplicity, let us denote λ^k/β by μ^k. Then, plugging (6.59) into the other equations in (6.58), we obtain

$$\begin{pmatrix} B^T B & 0 & 0 \\ 0 & C^T C & 0 \\ \alpha B & \alpha C & I \end{pmatrix} \begin{pmatrix} y^{k+1} \\ z^{k+1} \\ \mu^{k+1} \end{pmatrix}$$

$$= \left[\begin{pmatrix} -\alpha B^T B & -(\alpha+1)B^T C & B^T \\ -(\alpha+1)C^T B & -\alpha C^T C & C^T \\ -\alpha B & -\alpha C & I \end{pmatrix} - \frac{1}{A^T A} \begin{pmatrix} (\alpha+1)B^T A \\ (\alpha+1)C^T A \\ 2\alpha A \end{pmatrix} (-A^T B, -A^T C, A^T) \right] \begin{pmatrix} y^k \\ z^k \\ \mu^k \end{pmatrix}.$$

Let

$$L_1 = \begin{pmatrix} B^T B & 0 & 0 \\ 0 & C^T C & 0 \\ \alpha B & \alpha C & I \end{pmatrix},$$

$$R_1 = \begin{pmatrix} -\alpha B^T B & -(\alpha+1)B^T C & B^T \\ -(\alpha+1)C^T B & -\alpha C^T C & C^T \\ -\alpha B & -\alpha C & I \end{pmatrix} - \frac{1}{A^T A} \begin{pmatrix} (\alpha+1)B^T A \\ (\alpha+1)C^T A \\ 2\alpha A \end{pmatrix} (-A^T B, -A^T C, A^T),$$

and denote

$$M_1 = L_1^{-1} R_1.$$

Then, the scheme (6.58) can be written compactly as

$$\begin{pmatrix} y^k \\ z^k \\ \mu^k \end{pmatrix} = M_1^k \begin{pmatrix} y^0 \\ z^0 \\ \mu^0 \end{pmatrix}.$$

Obviously, if the spectral radius of M_1, denoted by $\rho(M_1) := |\lambda_{\max}(M_1)|$ (the largest eigenvalue of M_1), is not smaller than 1, then the sequence generated by the scheme above is not possible to converge to the solution point $(x^*, y^*, z^*) = (0,0,0)$ of the system (6.56) for any starting point.

Consider the example where (A,B,C) in (6.56) is given by (6.57). Then, with trivial manipulation, we know that

$$L_1 = \begin{pmatrix} 10 & 0 & 0 & 0 & 0 & 0 \\ 0 & 13 & 0 & 0 & 0 & 0 \\ \alpha & \alpha & 1 & 0 & 0 & 0 \\ \alpha & 2\alpha & 0 & 1 & 0 & 0 \\ 2\alpha & 2\alpha & 0 & 0 & 1 & 0 \\ 2\alpha & 2\alpha & 0 & 0 & 0 & 1 \end{pmatrix}$$

and

$$R_1 = \frac{1}{4} \begin{pmatrix} 36-4\alpha & -2\alpha-2 & -6\alpha-2 & -6\alpha-2 & 2-6\alpha & 2-6\alpha \\ -2\alpha-2 & 49-3\alpha & -7\alpha-3 & 1-7\alpha & 1-7\alpha & 1-7\alpha \\ 8\alpha & 10\alpha & 4-2\alpha & -2\alpha & -2\alpha & -2\alpha \\ 8\alpha & 6\alpha & -2\alpha & 4-2\alpha & -2\alpha & -2\alpha \\ 4\alpha & 6\alpha & -2\alpha & -2\alpha & 4-2\alpha & -2\alpha \\ 4\alpha & 6\alpha & -2\alpha & -2\alpha & -2\alpha & 4-2\alpha \end{pmatrix}$$

In Figure 6.1, we plot the values of $\rho(M_1)$ for different values of α varying from 0 to 1 with a 0.02 increment. It is obvious that $\rho(M_1) \geq 1$ for all tested cases, and it is monotonically increasing with respect to $\alpha \in (0,1)$. Therefore, the sequence generated by the above scheme is not convergent to the solution point of the system (6.56) for any starting point. It illustrates thus that the direct application of the SC-PRSM algorithm (6.10) is not always convergent.

5.2 Divergence of the E-SC-PRSM Algorithm (6.7)

Now we turn our attention to the divergence of the E-SC-PRSM algorithm (6.7) when it is applied to the solution of the same example involving (6.56). When algorithm (6.7) is applied to the solution of the above particular problem, it can be written as

$$\begin{cases} A^T[\beta(Ax^{k+1}+By^k+Cz^k)-\lambda^k]=0 \\ \alpha\beta(Ax^{k+1}+By^k+Cz^k)+\lambda^{k+\frac{1}{3}}-\lambda^k=0 \\ B^T[\beta(Ax^{k+1}+By^{k+1}+Cz^k)-\lambda^{k+\frac{1}{3}}]=0 \\ \alpha\beta(Ax^{k+1}+By^{k+1}+Cz^k)+\lambda^{k+\frac{2}{3}}-\lambda^{k+\frac{1}{3}}=0 \\ C^T[\beta(Ax^{k+1}+By^{k+1}+Cz^{k+1})-\lambda^{k+\frac{2}{3}}]=0 \\ \alpha\beta(Ax^{k+1}+By^{k+1}+Cz^{k+1})+\lambda^{k+1}-\lambda^{k+\frac{2}{3}}=0. \end{cases} \tag{6.60}$$

Similarly as in the last subsection, we can solve x^{k+1} first based on the first equation in (6.60) and then substitute it into the other equations. This leads to the following equation:

$$\begin{pmatrix} B^TB & 0 & 0 \\ (1+\alpha)C^TB & C^TC & 0 \\ 2\alpha B & \alpha C & I \end{pmatrix}\begin{pmatrix} y^{k+1} \\ z^{k+1} \\ \mu^{k+1} \end{pmatrix}$$
$$= \left[\begin{pmatrix} -\alpha B^TB & -(\alpha+1)B^TC & B^T \\ -\alpha C^TB & -2\alpha C^TC & C^T \\ -\alpha B & -2\alpha C & I \end{pmatrix} - \frac{1}{A^TA}\begin{pmatrix} (\alpha+1)B^TA \\ (2\alpha+1)C^TA \\ 3\alpha A \end{pmatrix}(-A^TB,-A^TC,A^T)\right]\begin{pmatrix} y^k \\ z^k \\ \mu^k \end{pmatrix}.$$

Then, we denote

$$L_2 = \begin{pmatrix} B^TB & 0 & 0 \\ (1+\alpha)C^TB & C^TC & 0 \\ 2\alpha B & \alpha C & I \end{pmatrix},$$

$$R_2 = \begin{pmatrix} -\alpha B^TB & -(\alpha+1)B^TC & B^T \\ -\alpha C^TB & -2\alpha C^TC & C^T \\ -\alpha B & -2\alpha C & I \end{pmatrix} - \frac{1}{A^TA}\begin{pmatrix} (\alpha+1)B^TA \\ (2\alpha+1)C^TA \\ 3\alpha A \end{pmatrix}\left(-A^TB,-A^TC,A^T\right).$$

and

$$M_2 = L_2^{-1}R_2.$$

With the specific definitions of (A,B,C) in (6.57), we can easily show that

$$L_2 = \begin{pmatrix} 10 & 0 & 0&0&0&0 \\ 11\alpha+11 & 13 & 0&0&0&0 \\ 2\alpha & \alpha & 1&0&0&0 \\ 2\alpha & 2\alpha & 0&1&0&0 \\ 4\alpha & 2\alpha & 0&0&1&0 \\ 4\alpha & 2\alpha & 0&0&0&1 \end{pmatrix}$$

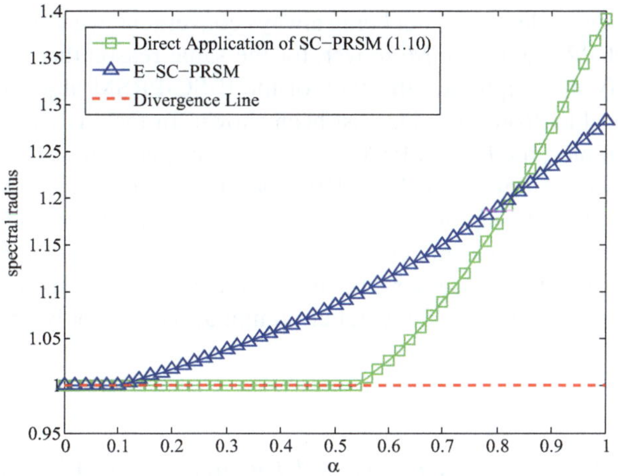

Fig. 6.1 The magnitude of the spectral radius for M_1 and M_2 with respect to $\alpha \in (0,1)$

and

$$R_2 = \frac{1}{4}\begin{pmatrix} 36-4\alpha & -2\alpha-2 & -6\alpha-2 & -6\alpha-2 & 2-6\alpha & 2-6\alpha \\ 40\alpha+42 & 49-6\alpha & -14\alpha-3 & 1-14\alpha & 1-14\alpha & 1-14\alpha \\ 14\alpha & 13\alpha & 4-3\alpha & -3\alpha & -3\alpha & -3\alpha \\ 14\alpha & 5\alpha & -3\alpha & 4-3\alpha & -3\alpha & -3\alpha \\ 10\alpha & 5\alpha & -3\alpha & -3\alpha & 4-3\alpha & -3\alpha \\ 10\alpha & 5\alpha & -3\alpha & -3\alpha & -3\alpha & 4-3\alpha \end{pmatrix}.$$

Hence, to check the spectral radius of $\rho(M_2) := \lambda_{\max}(M_2)$ (the largest eigenvalue of M_2), we take different values of α varying from 0 to 1 with a 0.02 increment and plot the values of $\rho(M_2)$. In Figure 6.1, we show that $\rho(M_2) \geq 1$ for all the tested cases, and it is monotonically increasing with respect to $\alpha \in (0,1)$. Therefore, the sequence generated by the scheme above is not convergent to the solution point of the system (6.56). It thus illustrates that the E-SC-PRSM algorithm (6.7) does not always converge.

6 Numerical Results

In this section, we test the proposed SC-PRSM-PR algorithm (6.11) for some applications arising from image processing and statistical learning domains, and report the numerical results. Since the SC-PRSM-PR algorithm (6.11) can be regarded as a customized application of the original SC-PRSM algorithm (6.4) to the specific multi-block convex minimization problem (6.5) and that it is an operator-splitting algorithm, we will mainly compare it with some methods of the same kind. In particular, as well justified in the literature (e.g., [40, 44]), the E-ADMM algorithm (6.6) often performs well despite of not being always convergent; we thus compare it with the SC-PRSM-PR algorithm. Moreover, for the same reason than (6.6), it is interesting to verify the empirical efficiency of the E-SC-PRSM algorithm (6.7), even though its possible divergence has just been shown. In fact, as we shall report, for the tested examples, the E-SC-PRSM algorithm does perform well too.

Our code was written in Matlab 2010b and all the numerical experiments were performed on a Dell(R) laptop computer with 1.5GHz AMD(TM) A8 processor and a 4GB memory. Since our numerical experiments were conducted on an ordinary laptop without parallel processors, for the y- and z-subproblems in (6.11) at each iteration (which are eligible for parallel computation), we only count the longer time.

6.1 Image Restoration with Mixed Impulsive and Gaussian Noises

Let $u \in \Re^n$ represent a digital image with $n = n_1 \times n_2$. Note that a two-dimensional image can be represented by vectorizing it as a one-dimensional vector in certain order, e.g., the lexicographic order. Suppose that the clean image u is corrupted by both blur and noise. We consider the case where the noise is the mixture of an additive Gaussian white noise and an impulse noise. The corrupted (also observed) image is denoted by u^0. Image restoration is to recover the clean image u from the observed image u^0.

We consider the following image restoration model for mixed noise removal which was proposed in [32]:

$$\min_{u,f} \left\{\tau\|u\|_{\mathrm{TV}} + \frac{\rho}{2}\|u-f\|^2 + \|P_{\mathscr{A}}(Hf-u^0)\|_1\right\}. \tag{6.61}$$

In (6.61), $\|\cdot\|$ and $\cdot\|_1$ denote the l_2- and l_1 norms, respectively; $\|\cdot\|_{\mathrm{TV}}$ is the discrete total variation defined by

$$\|u\|_{\mathrm{TV}} = \sum_{1\le i,j\le n} \sqrt{\left|(\nabla_1 u)_{j,k}\right|^2 + \left|(\nabla_2 u)_{j,k}\right|^2},$$

where $\nabla_1 : \Re^n \to \Re^n$ and $\nabla_2 : \Re^n \to \Re^n$ are the discrete horizontal and vertical partial derivatives, respectively; and we denote $\nabla = (\nabla_1, \nabla_2)$, see [43]; thus $\|u\|_{\mathrm{TV}}$ can be written as $\|\nabla u\|_1$; H is the matrix representation (convolution operator) of a spatially invariant blur; $\mathscr{A}$ represents the set of pixels which are corrupted by the impulsive noise (all the pixels outside $\mathscr{A}$ are corrupted by the Gaussian noise); $P_{\mathscr{A}}$ is the characteristic function of the set $\mathscr{A}$, i.e., $P_{\mathscr{A}}(u)$ has the value 1 for any pixel within $\mathscr{A}$ and 0 for any pixel outside $\mathscr{A}$; u^0 is the corrupted image with blurry and mixed noise; and τ and ρ are positive constants. To identify the set $\mathscr{A}$, it was suggested in [32] to apply the adaptive median filter (AMF) first to remove most of the impulsive noise within $\mathscr{A}$.

We first show that the minimization problem (6.61) can be reformulated as a special case of (6.5). In fact, by introducing the auxiliary variables w, v, and z, we can reformulate (6.61) as

$$\begin{aligned} \min\ & \tau\|w\|_1 + \tfrac{\rho}{2}\|v\|^2 + \|P_{\mathscr{A}}(z)\|_1 \\ \text{s.t. } & w = \nabla u \\ & v = u - f \\ & z = Hf - u^0, \end{aligned} \tag{6.62}$$

which is a special case of the abstract problem (6.5) with the following specifications:

- $x := u$, $y := f$ and $z := (w, v, z)$; $\mathscr{X}$, $\mathscr{Y}$ and $\mathscr{Z}$ are all real spaces with appropriate dimensionality;
- $\theta_1(x) := 0$, $\theta_2(y) := 0$ and $\theta_3(z) := \theta_3(w, v, z) = \tau\|w\|_1 + \frac{\rho}{2}\|v\|^2 + \|P_{\mathscr{A}}(z)\|_1$;
- and

$$A := \begin{bmatrix} \nabla \\ I \\ 0 \end{bmatrix},\ B := \begin{bmatrix} 0 \\ -I \\ H \end{bmatrix},\ C := \begin{bmatrix} -I & 0 & 0 \\ 0 & -I & 0 \\ 0 & 0 & -I \end{bmatrix},\ b := \begin{bmatrix} 0 \\ 0 \\ u^0 \end{bmatrix}. \tag{6.63}$$

Thus, the methods "E-ADMM", "E-SC-PRSM", and "SC-PRSM-PR" are all applicable to the minimization problem (6.62). Below we elaborate on the minimization subproblems arising in the SC-PRSM-PR algorithm (6.11) and show that they all have closed-form solutions. We skip the elaboration on the sub-problems of the E-ADMM and E-SC-PRSM algorithms which can be easily found in the literature or similar to those of the SC-PRSM-PR algorithm.

When the SC-PRSM-PR algorithm (6.11) is applied to the minimization problem (6.62), the first sub-problem (i.e., the u-subproblem) is

$$u^{k+1} = \arg\min_{u \in \Re^{n\times n}} \left\{ \|\nabla u - w^k - \frac{\lambda_1^k}{\beta}\|^2 + \|u - f^k - v^k - \frac{\lambda_2^k}{\beta}\|^2 \right\},$$

whose solution is given by

$$\beta(\nabla^T\nabla + I)u^{k+1} = \lambda_2^k + \beta(f^k + v^k) + \nabla^T(\lambda_1^k + \beta w^k),$$

which can be solved efficiently by the fast Fourier transform (FFT) or the discrete cosine transform (DCT) (see, e.g., [25] for details). In fact, applying the FFT to diagonalize ∇ such that $\nabla = \mathscr{F}^{-1}D\mathscr{F}$, where $\mathscr{F}$ is the Fourier transformation matrix and D is a diagonal matrix, we can rewrite the equation above as

$$\beta(D^TD + I)\mathscr{F}u^{k+1} = \mathscr{F}\lambda_2^k + \beta(\mathscr{F}f^k + \mathscr{F}v^k) + D^T(\mathscr{F}\lambda_1^k + \beta\mathscr{F}w^k),$$

where $\mathscr{F}u^{k+1}$ can be obtained by FFT and then u^{k+1} is recovered by inverse FFT.

After updating the Lagrange multiplier $\lambda_j^{k+1/2}$ for $j = 1, 2, 3$ according to (6.11), the second subproblem (i.e., the f-subproblem) reads as

$$f^{k+1} = \arg\min_{f \in \Re^{n\times n}} \left\{ \|u^k - f - v^k - \lambda_2^{k+1/2}/\beta\|^2 + \|Hf - z^k - u^0 - \lambda_3^{k+1/2}/\beta\|^2 + \mu\|f - f^k\|^2 \right\},$$

whose solution is given by

$$\beta(H^T H+(\mu+1))f^{k+1} = H^T(\lambda_3^{k+1/2}+\beta(z^k+u^0))+\beta(u^k-v^k)-\lambda_2^{k+1/2}+\beta\mu f^k.$$

For this system of equations, since the matrix H is a spatially convolution operator which can be diagonalized by FFT, we can compute f^{k+1} efficiently, using a strategy similar to the one we used to compute u^{k+1}.

The third sub-problem (i.e., the (w,v,z)-subproblem) in the SC-PRSM-PR algorithm (6.11) can be split into three smaller subproblems as follows:

$$\begin{aligned}
w^{k+1} &= \arg\min_w \left\{ \tau\|w\|_1 + \tfrac{\beta}{2}\|\nabla u^{k+1} - w - \tfrac{\lambda_1^{k+1/2}}{\beta}\|^2 + \tfrac{\mu\beta}{2}\|w-w^k\|^2 \right\} \\
&= \mathrm{Shrink}_{\frac{\tau}{\beta(1+\mu)}}\left(\tfrac{1}{\mu+1}(\nabla u^{k+1} - \lambda_1^{k+1/2}/\beta + \mu w^k)\right); \\
v^{k+1} &= \arg\min_v \left\{ \rho\|v\|^2 + \beta\|u^{k+1} - f^k - v - \lambda_2^{k+1/2}/\beta\|^2 + \mu\beta\|v-v^k\|^2 \right\} \\
&= (\beta(u^{k+1}-f^k) - \lambda_2^{k+1/2} + \mu\beta v^k)/(\rho+(1+\mu)\beta) \\
z^{k+1} &= \arg\min_z \left\{ \|P_{\mathscr{A}}(z)\|_1 + \tfrac{\beta}{2}\|Hf^k - u^0 - \lambda_2^k/\beta\|^2 + \tfrac{\mu\beta}{2}\|z-z^k\|^2 \right\},
\end{aligned}$$

with

$$(z^{k+1})_i = \begin{cases} \mathrm{Shrink}_{1/((1+\mu)\beta)}\left(\frac{1}{\mu+1}(Hf^k - u^0 - \lambda_3^{k+1/2}/\beta + \mu z^k)\right), & \text{if } i \in \mathscr{A}; \\ \frac{1}{\mu+1}(Hf^k - u^0 - \lambda_3^{k+1/2}/\beta + \mu z^k), & \text{otherwise.} \end{cases}$$

and $\mathrm{Shrink}_\sigma(\cdot)$ denotes the shrinkage operator (see e.g. [11]). That is:

$$\mathrm{Shrink}_\sigma(a) = a/|a| \circ \max\{|a|-\sigma, 0\}, \forall a \in \mathscr{R}^n,$$

with $\sigma > 0$, $|\cdot|$ is the Euclidian norm, and the operator "$\circ$" stands for the componentwise scalar multiplication.

Finally, we update λ^{k+1} based on $\lambda^{k+1/2}$ and the just-computed u^{k+1}, f^{k+1} and $(w^{k+1}, v^{k+1}, z^{k+1})$.

For problem (6.62), we tested two images, namely: Cameraman.png and House.png. Both are of size 256×256. These two images were first convoluted by a blurring kernel with radius 3 and then corrupted by impulsive noise with intensity 0.7 and Gaussian white noise with variance 0.01. We first applied the AMF (see, e.g., [32]) with window size 19 to identify the corrupted index set $\mathscr{A}$ and remove the impulsive noise and get the filtered images. The original, degraded, and filtered images are shown in Figure 6.2.

We now numerically compare SC-PRSM-PR with E-ADMM and E-SC-PRSM. We took $\tau = 0.02$ and $\rho = 1$ in (6.62). For a fair comparison, we chose the same values for the parameters common to various methods, that is $\beta = 6$ for all of them and $\alpha = 0.15$ for SC-PRSM-PR and E-SC-PRSM. For the additional parameter μ of SC-PRSM-PR, we took $\mu = 0.16$. We used as stopping criterion the one to be defined by (6.80), with $\varepsilon = 4 \times 10^{-5}$ and the maximum iteration number was set to be 1000. The initial iterates for all methods were the degraded images. We used the signal-to-noise ratio (SNR) in the dB unit as the measure of the performance of the restored images from all methods. The SNR is defined by

$$\mathrm{SNR} = 10\log_{10}\frac{\|u\|^2}{\|\hat{u}-u\|^2},$$

where u is the original image and $\hat{u}$ is the restored image. In the experiment, we use a stopping criterion, which is popularly adopted in the literature of image processing, namely:

$$\mathrm{Tol} := \frac{\|f^{k+1}-f^k\|_F}{1+\|f^k\|_F} < 4 \times 10^{-5}. \tag{6.64}$$

The images restored by SC-PRSM-PR are shown in Figure 6.2. In Figure 6.3, we plotted the evolution curves of the SNR values with respect to the computing time in seconds for the tested images. Table 6.1 reports some statistics of the comparison between these methods, including the number of iterations ("Iter."), computing time in second ("CPU(s)") and the SNR value in dB ("SNR(dB)") of the restored image. This set of experiment shows that: 1) the E-SC-PRSM algorithm (6.7), as we have expected, does work well for the tested examples empirically even though its lack of convergence has been demonstrated in Section 5.2; and 2) the proposed SC-PRSM-PR algorithm with proved convergence is very competitive with, and sometimes is even faster than, E-ADMM and E-SC-PRSM whose convergence is unproven. Note that the SC-PRSM-PR algorithm (6.11) usually requires more iterations; but the two sub-problems it contains can be solved in parallel at each iteration. This helps saving computing time.

Table 6.1 Numerical Comparison for the image restoration problem (6.61).

	Cameraman.png			House.png		
Algorithm	Iter.	CPU (s)	SNR (dB)	Iter.	CPU (s)	SNR (dB)
E-ADMM	901	31.73	19.20	814	28.49	23.84
E-SC-PRSM	767	31.47	19.19	695	27.81	23.84
SC-PRSM-PR	**935**	**29.76**	**19.20**	**845**	**25.77**	**23.85**

6.2 *Robust Principal Component Analysis with Missing and Noisy Data*

It is worth mentioning that the problem (6.5) and the previous analysis are for vector spaces. But without any difficulty, they can be trivially extended to the case where matrix variables are considered and some linear mappings rather than matrices are

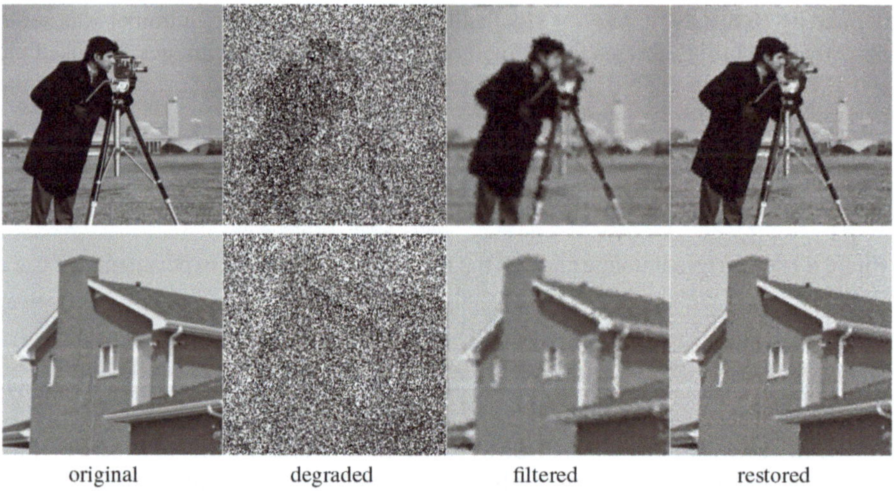

Fig. 6.2 The original images (first column), the degraded images (second column), the filtered images by AMF (third column), and the restored images by SC-PRSM-PR.

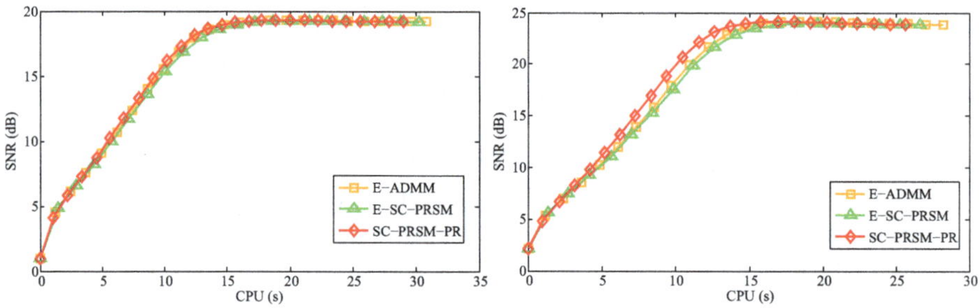

Fig. 6.3 Image restoration from the mixture of noises (left: Cameraman.png and right: House.png on the right): evolutions of the SNR (dB) with respect to the CPU time.

accompanying the variables in the constraints. In the upcoming subsections, we will test two particular cases of problem (6.5) with matrix variables, both arising in statistical learning.

We tested first the model of robust principal component analysis (RPCA) with missing and noisy data. This model aims at decomposing a matrix $M \in \Re^{m\times n}$ as the sum of a low-rank matrix $R \in \Re^{m\times n}$ and a sparse matrix $S \in \Re^{m\times n}$; but not all the entries of M are known and M is corrupted by a noisy matrix. More specifically, we focus on the unconstrained problem studied in [44], namely:

$$\min \; ||R||_* + \gamma ||S||_1 + \tfrac{\nu}{2} ||P_\Omega(M - R - S)||_F^2, \tag{6.65}$$

where $||\cdot||_*$ is the nuclear norm defined as the sum of all singular values of a matrix, $||\cdot||_1$ is the sum of the absolute values of all the entries of a matrix, $||\cdot||_F$ is the Frobenius norm which is the square root of the sum of the squares of all the entries of a matrix; $\gamma > 0$ is a constant balancing the low-rank and sparsity and $\nu > 0$ is a constant reflecting the Gaussian noise level; Ω is a subset of the index set of the entries $\{1,2,\cdots,m\} \times \{1,2,\cdots,n\}$, and we assume that only those entries $\{C_{ij},(i,j) \in \Omega\}$ can be observed; the incomplete observation information is summarized by the operator $P_\Omega : \Re^{m\times n} \to \Re^{m\times n}$, which is the orthogonal projection onto the span of matrices vanishing outside of Ω so that the ij-th entry of $P_\Omega(X)$ is X_{ij} if $(i,j) \in \Omega$ and zero otherwise. Note that problem (6.65) is a generalization of the matrix decomposition problem in [8] and of the robust principal component analysis problem in [5].

Introducing an auxiliary variable $Z \in \Re^{m\times n}$, we can reformulate (6.65) as

$$\begin{array}{ll} \min & \gamma\|S\|_1 + \|R\|_* + \frac{\nu}{2}\|P_\Omega(Z)\|_F^2, \\ s.t. & S+R+Z = M. \end{array} \tag{6.66}$$

which is a special case of (6.5) with $x = S$, $y = R$, $z = Z$; A, B, and C are all identity mappings; $b = M$; $\theta_1(x) = \gamma\|S\|_1$, $\theta_2(y) = \|R\|_*$, $\theta_3(z) = \frac{\nu}{2}\|P_\Omega(Z)\|_F^2$, $\mathscr{X} = \mathscr{Y} = \mathscr{Z} = \Re^{m\times n}$. Then, applying the SC-PRSM-PR algorithm (6.11) to the solution of problem (6.66), and omitting some details, we can see that all the resulting subproblems have closed-form solutions. More specifically, the resulting SC-PRSM-PR algorithm reads as follows

$$\begin{cases} S^{k+1} &= \mathscr{S}_{\frac{1}{\beta}}(-R^k - Z^k + M + \frac{1}{\beta}\lambda^k) \\ \lambda^{k+\frac{1}{2}} &= \lambda^k - \alpha\beta(R^{k+1} + S^k + Z^k - M) \\ R^{k+1} &= \mathscr{D}_{\frac{\gamma}{\beta(\mu+1)}}\left(\frac{\mu}{\mu+1}S^k - \frac{1}{\mu+1}(S^{k+1} + Z^k - M - \frac{1}{\beta}\lambda^{k+\frac{1}{2}})\right) \\ Z^{k+1} &= \tilde{Z}^k \\ \lambda^{k+1} &= \lambda^{k+\frac{1}{2}} - \alpha\beta(R^{k+1} + S^{k+1} + Z^{k+1} - M), \end{cases} \tag{6.67}$$

where $\tilde{Z}^k$ is given by

$$\tilde{Z}_{ij}^k = \begin{cases} \frac{\mu\beta}{\nu+(\mu+1)\beta}Z^k - \frac{\beta}{\nu+(\mu+1)\beta}(S^{k+1} + R^{k+1} - M - \frac{1}{\beta}\lambda^{k+\frac{1}{2}}), & \text{if } (i,j) \subset \Omega; \\ \frac{\mu}{\mu+1}Z^k - \frac{1}{\mu+1}(S^{k+1} + R^{k+1} - M - \frac{1}{\beta}\lambda^{k+\frac{1}{2}}), & \text{otherwise.} \end{cases}$$

Note that in (6.67), $\mathscr{S}_{\frac{1}{\beta}}$ is the matrix version of the shrinkage operator defined before, that is

$$(S_{\frac{1}{\beta}})_{ij} = (1 - \frac{1}{\beta}/|X_{ij}|)_+ \cdot X_{ij}, \quad 1 \le i \le m, 1 \le j \le n, \tag{6.68}$$

for $\beta > 0$. In addition, $\mathscr{D}_\tau(X)$ with $\tau > 0$ is the singular value soft-thresholding operator defined as follows. If a matrix X with rank r has the singular value decomposition (SVD)

$$X = U\Lambda V^*, \quad \Lambda = \text{diag}(\{\sigma_i\}_{1\le i\le r}), \tag{6.69}$$

then we define

$$\mathscr{D}_\tau(X) = U\mathscr{D}_\tau(\Lambda)V^*, \quad \mathscr{D}_\tau(\Lambda) = \text{diag}(\{(\sigma_i - \tau)_+\}_{1\le i\le r}). \tag{6.70}$$

As tested in [5, 40, 44], one application of the RPCA model (6.66) is to extract the background from a surveillance video. For this application, the low-rank and sparse components, R and S, represent the background and foreground of the video M, respectively. We test the surveillance video at the hall of an airport, which is available at http://perception.i2r.a-star.edu.sg/bk_model/bk_index.html. The video consists of 200 frames of size of 114×176. Thus, the video data can be realigned as a matrix $M \in \Re^{m\times n}$ with $m = 25,344, n = 200$. We obtain the observed subset Ω by randomly generating 80% of the entries of D and add the Gaussian noise with mean zero and variance 10^{-3} to D to obtain the observed matrix M. According to [5], we took for regularization parameters $\gamma = 1/\sqrt{m}$ and $\nu = 100$. The 10th, 100th, and 200th frames of the original and the corrupted video are displayed at the first and second rows in Figure 6.4, respectively.

To implement E-ADMM, E-SC-PRSM, and SC-PRSM-PR, we set

$$\beta = 0.005|\Omega|/\|M\|_1$$

for all these methods, choosing $\alpha = 0.25$ for E-SC-PRSM and SC-PRSM-PR. For the additional parameter μ of SC-PRSM-PR we took $\mu = 0.26$. The initial iterates all start at zero matrices and the stopping criteria for all these methods were taken as

$$\max\left\{\frac{\|R^{k+1}-R^k\|_F}{1+\|R^k\|_F}, \frac{\|S^{k+1}-S^k\|_F}{1+\|S^k\|_F}\right\} < 10^{-2}.$$

Some frames of the foreground recovered by SC-PRSM-PR are shown in the third row of Figure 6.4. In Figure 6.5, we plotted the respective evolutions of the primal and dual residuals for all the methods under comparison with respect to both the computing time and number of iterations. Table 6.2 reports some quantitative comparisons among these methods, including the number of iterations ("Iter"), computing time in seconds ("CPU(s)"), and rank of the recovery video foreground ("rank($\hat{R}$)") and the number of nonzero entries of the video background ("$|\text{supp}(\hat{S})|$"). The statistics in Table 6.2 and curves in Figure 6.5 demonstrate that SC-PRSM-PR outperforms the other two methods. This set of experiments further verifies the efficiency of the proposed SC-PRSM-PR algorithm (6.11).

6.3 Quadratic Discriminant Analysis

Then we tested a quadratic discriminant analysis (QDA) problem. A rather new and challenging problem in the statistical learning area, the QDA aims at classifying two sets of normal distribution data (denoted by X_1 and X_2) with different but close covariance matrices generalized from the linear discriminant analysis, see, e.g., [15, 17, 36] for details. An assumption for the QDA problem is that the data vector $X_1 \in \Re^{n_1\times d}$ is generated from $\mathcal{N}(\mu_1, \Sigma_1)$, while the data vector $X_2 \in \Re^{n_2\times d}$ is generated from $\mathcal{N}(\mu_2, \Sigma_2)$, where d is the data dimension, n_1 and n_2 are the sample

Fig. 6.4 The 10th, 100th, and 200th frames of the original surveillance video (first row), the corrupted video (second row), and the sparse video recovered from SC-PRSM-PR (third row).

Table 6.2 Numerical Comparison for the RPCA problem (6.66)

Algorithm	Iter.	CPU(s)	rank($\hat{R}$)	$\lvert\mathrm{supp}(\hat{S})\rvert$
E-ADMM	44	48.73	7	243,991
E-SC-PRSM	46	79.28	14	579,658
SC-PRSM-PR	**33**	**40.97**	**9**	**301,387**

size, respectively. We denote by $\Sigma = \Sigma_1^{-1} - \Sigma_2^{-1}$ the difference between the inverse covariance matrices Σ_1^{-1} and Σ_1^{-1}; estimating Σ is important for classification in high dimensional statistics. In order to estimate a high-dimensional covariance, a QDA model usually assumes that Σ has some special features, one of these features being sparsity, see, e.g., [15].

In this subsection, we propose a new model under the assumption that the matrix Σ, the difference between the inverse covariance matrices Σ_1^{-1} and Σ_1^{-1}, is represented by $\Sigma = S + R$ where S is a sparse matrix and R is a low-rank matrix. Considering the sparsity and low-rank features simultaneously is indeed important especially for some high-dimensional statistical learning problems, see, e.g., [1, 16]. Let the sample covariance matrices of X_1 and X_2 be $\hat{\Sigma}_1 = X_1^T X_1/n_1$ and $\hat{\Sigma}_2 = X_2^T X_2/n_2$,

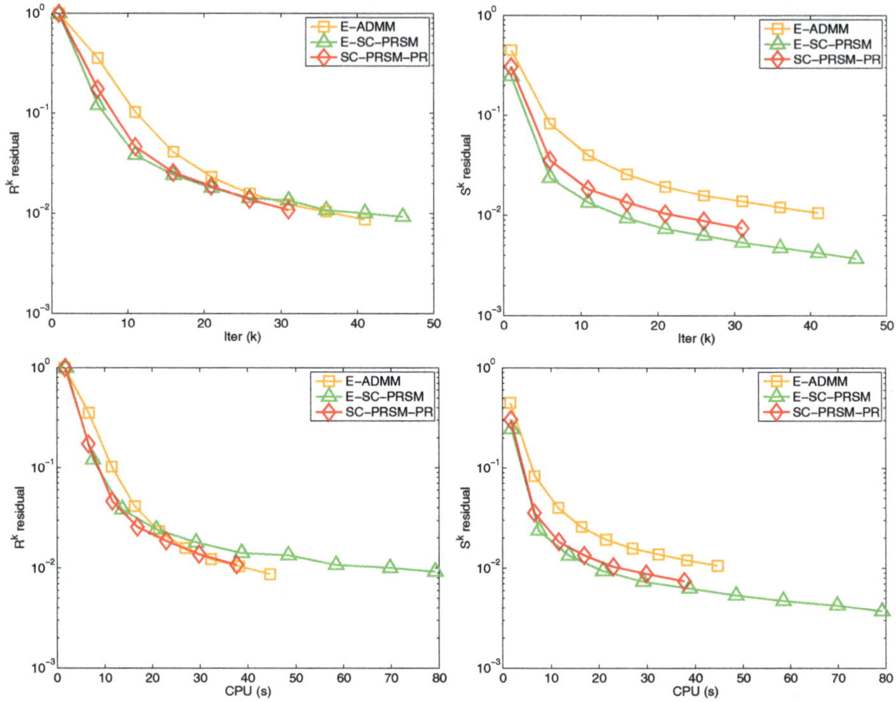

Fig. 6.5 RPCA problem (6.66): evolution of the primal and dual residuals for E-ADMM, E-SC-PRSM, and SC-PRSM-PR *w.r.t.* the number of iterations (first row) and computing time (second row).

respectively. Obviously, we have $\Sigma_1 \Omega \Sigma_2 = \Sigma_2 - \Sigma_1$. Assuming the sparsity and low-rank features of Σ simultaneously and considering the scenarios with noise on the data sets, we propose the following novel formulation to estimate $\Sigma = S + R$:

$$\begin{aligned} &\min\ \gamma ||S||_1 + ||R||_* \\ &s.t.\ ||\hat{\Sigma}_1 (S+R) \hat{\Sigma}_2 - (\hat{\Sigma}_2 - \hat{\Sigma}_1)||_\infty \le r, \end{aligned} \tag{6.71}$$

where again $\gamma > 0$ is a trade-off constant between the sparsity and low-rank features, $r > 0$ is a tolerance reflecting the noise level, $\|\cdot\|_*$ and $\|\cdot\|_1$ are defined as before in (6.65), and $\|\cdot\|_\infty := \max_{i,j} |U_{ij}|$ denotes the entry-wise maximum norm of a matrix.

Introducing an auxiliary variable U, we can reformulate (6.71) as

$$\begin{aligned} &\min\ ||R||_* + \gamma ||S||_1 \\ &s.t.\ U - \hat{\Sigma}_1 (R+S) \hat{\Sigma}_2 = \hat{\Sigma}_1 - \hat{\Sigma}_2, \\ &\quad\ \ ||U||_\infty \le r, \end{aligned} \tag{6.72}$$

which is a special case of (6.5) with $x = U$, $y = S$, $z = R$, A is the identity mapping, B and C are the mappings defined by $\hat{\Sigma}_1$ and $\hat{\Sigma}_2$, $b = \hat{\Sigma}_1 - \hat{\Sigma}_2$, $\mathscr{X} := \{U \in \Re^{d \times d},\ |\ ||U||_\infty \le r\}$, $\mathscr{Y} = \mathscr{Z} = \Re^{d \times d}$. To further see why problem (6.72) can be casted into (6.5), we can vectorize the matrices R and S, and then write the constraint in (6.72) as

$$\mathbf{vec}(U) - (\hat{\Sigma}_2^T \otimes \hat{\Sigma}_1)[\mathbf{vec}(R) + \mathbf{vec}(S)] = \mathbf{vec}(\hat{\Sigma}_1 - \hat{\Sigma}_2), \tag{6.73}$$

where $\mathbf{vec}(X)$ denotes the vectorization of the matrix $X \in \Re^{n\times n}$ by stacking the columns of X into a single column vector in $\Re^{n^2}$, and $\otimes$ is a Kronecker product.

Applying the SC-PRSM-PR algorithm (6.11) to the solution of problem (6.72) we obtain

$$\begin{cases} U^{k+1} &= \arg\min_{||U||_\infty<r}\{\frac{\beta}{2}||U - \hat{\Sigma}_1(R^k+S^k)\hat{\Sigma}_2 - (\hat{\Sigma}_1 - \hat{\Sigma}_2) - \frac{1}{\beta}\lambda^k||_F^2\} \\ \lambda^{k+\frac{1}{2}} &= \lambda^k - \alpha\beta(U^{k+1} - \hat{\Sigma}_1(R^k+S^k)\hat{\Sigma}_2 - (\hat{\Sigma}_1 - \hat{\Sigma}_2)) \\ S^{k+1} &= \arg\min\{\gamma||S||_1 + \frac{\beta}{2}||U^{k+1} - \hat{\Sigma}_1(R^k+S)\hat{\Sigma}_2 - (\hat{\Sigma}_1 - \hat{\Sigma}_2) - \frac{1}{\beta}\lambda^{k+1/2}||_F^2 + \frac{\mu\beta}{2}||\hat{\Sigma}_1(S-S^k)\hat{\Sigma}_2||^2\} \\ R^{k+1} &= \arg\min\{||R||_* + \frac{\beta}{2}||U^{k+1} - \hat{\Sigma}_1(R+S^k)\hat{\Sigma}_2 - (\hat{\Sigma}_1 - \hat{\Sigma}_2) - \frac{1}{\beta}\lambda^{k+1/2}||_F^2 + \frac{\mu\beta}{2}||\hat{\Sigma}_1(R-R^k)\hat{\Sigma}_2||^2\} \\ \lambda^{k+1} &= \lambda^{k+\frac{1}{2}} - \alpha\beta(U^{k+1} - \hat{\Sigma}_1(R^{k+1}+S^{k+1})\hat{\Sigma}_2 - (\hat{\Sigma}_1 - \hat{\Sigma}_2)), \end{cases} \tag{6.74}$$

Now, let us elaborate on the sub-problems in (6.74). First, the U sub-problem in (6.74) has a closed-form solution given by

$$U^{k+1} = \mathscr{T}_r\left(\hat{\Sigma}_1(R^k+S^k)\hat{\Sigma}_2 + (\hat{\Sigma}_1 - \hat{\Sigma}_2) + \frac{1}{\beta}\lambda^k\right), \tag{6.75}$$

where $(\mathscr{T}_r(A))_{ij}$ is defined as

$$(\mathscr{T}_r(A))_{ij} = \mathrm{sign}(A_{ij}) \cdot \max(|A_{ij}|, r).$$

The S- and R-sub-problems in (6.74) do not have closed-form solutions and must be solved iteratively. Again, we just used the ADMM algorithm (6.2) to solve them. For example, the S-sub-problem can be reformulated as

$$\begin{aligned} &\min\ \gamma||S||_1 + \frac{\beta}{2}||U^{k+1} - \hat{\Sigma}_1(R^k+A)\hat{\Sigma}_2 - (\hat{\Sigma}_1 - \hat{\Sigma}_2) - \frac{1}{\beta}\lambda^{k+1/2}||_F^2 + \frac{\mu\beta}{2}||\hat{\Sigma}_1(A-S^k)\hat{\Sigma}_2||^2 \\ &s.t.\ A - S = 0, \end{aligned} \tag{6.76}$$

where A is an auxiliary variable. Applying the ADMM algorithm (6.2), with $\beta = 1$, to the solution of problem (6.76), we obtain

$$\begin{cases} \mathbf{vec}(A^{k+1}) = \left([\hat{\Sigma}_2^T\hat{\Sigma}_2] \otimes [(\mu+1)\hat{\Sigma}_1^T\hat{\Sigma}_1] + I_{d^2}\right)^{-1}\mathbf{vec}\big(\hat{\Sigma}_1^T(U^{k+1} - (\hat{\Sigma}_1 - \hat{\Sigma}_2) - \frac{1}{\beta}\lambda^k)\hat{\Sigma}_2^T \\ \qquad\qquad + \mu\hat{\Sigma}_1^T\hat{\Sigma}_1R^k\hat{\Sigma}_2\hat{\Sigma}_2^T + A^k + \frac{1}{\beta}\lambda_S^k\big), \\ S^{k+1} = \mathscr{S}_{\frac{\gamma}{\beta}}(A^{k+1} - \frac{1}{\beta}\lambda_S^k), \\ \lambda_S^{k+1} = \lambda_S^k - (A^{k+1} - S^{k+1}). \end{cases} \tag{6.77}$$

Similarly, we can reformulate the R-subproblem in (6.74) as

$$\begin{aligned} &\min\ ||R||_* + \frac{\beta}{2}||U^{k+1} - \hat{\Sigma}_1(A+S^k)\hat{\Sigma}_2 - (\hat{\Sigma}_1 - \hat{\Sigma}_2) - \frac{1}{\beta}\lambda^{k+1/2}||_F^2 + \frac{\mu\beta}{2}||\hat{\Sigma}_1(A-R^k)\hat{\Sigma}_2||^2 \\ &s.t.\ A - R = 0, \end{aligned} \tag{6.78}$$

where A is an auxiliary variable; and apply the ADMM algorithm (6.2), to the solution of problem (6.78). The resulting algorithm reads as

$$\begin{cases} \mathbf{vec}(A^{k+1}) = \left([\hat{\Sigma}_2^T\hat{\Sigma}_2]\otimes[(\mu+1)\hat{\Sigma}_1^T\hat{\Sigma}_1]+I_{d^2}\right)^{-1}\mathbf{vec}\big(\hat{\Sigma}_1^T(U^{k+1}-(\hat{\Sigma}_1-\hat{\Sigma}_2)-\frac{1}{\beta}\lambda^k)\hat{\Sigma}_2^T \\ \qquad\qquad +\mu\hat{\Sigma}_1^T\hat{\Sigma}_1 S^k\hat{\Sigma}_2\hat{\Sigma}_2^T + A^k + \frac{1}{\beta}\lambda_R^k\big), \\ R^{k+1} = \mathscr{D}_{\frac{1}{\beta}}(A^{k+1}-\frac{1}{\beta}\lambda_R^k), \\ \lambda_R^{k+1} = \lambda_R^k - (A^{k+1}-R^{k+1}). \end{cases} \tag{6.79}$$

Note that the operators $\mathscr{S}_{\frac{\gamma}{\beta}}$ in (6.77) and $\mathscr{D}_{\frac{1}{\beta}}$ in (6.79) are defined by (6.68) and (6.70), respectively.

For generating the two d-dimensional normal distributions $\mathscr{N}(\mu_1,\Sigma_1)$ and $\mathscr{N}(\mu_2,\Sigma_2)$, we set $d=20$ and the mean vector as $\mu_1=\mu_2=(0,0,\ldots,0)^T$. We first generated a random matrix $U\in\Re^{20\times 20}$ whose entries are i.i.d. $\mathscr{N}(0,1)$ and a random diagonal matrix $D\in\Re^{20\times 20}$ whose diagonal elements are i.i.d uniform distribution on $[1,2]$, then let $\Sigma_1=U^TDU$. To obtain a low rank semi-positive definite matrix, we generated a random matrix $R_1\in\Re^{20\times 10}$ whose entries are i.i.d. $\mathscr{N}(0,1)$ and let $R=R_1R_1^T$. Therefore, we have $\text{rank}(R)=10$. To obtain a sparse positive definite matrix, we first generated a sparse symmetric matrix $S_1\in\Re^{20\times 20}$ with 50 nonzero entries and each nonzero entries are i.i.d. $\mathscr{N}(0,1)$. In order to guarantee the positive definiteness of S, we let $S=S_1+2|\lambda_{\min}(S_1)|I_{20}$, where $\lambda_{\min}(S_1)$ is the smallest eigenvalue of S_1. Let $\Sigma_2=(\Sigma_1^{-1}+S+R)^{-1}$; we then have $\Omega=\Sigma_1^{-1}-\Sigma_2^{-1}=S+R$. We generated $n=5000$ data $X_1\in\Re^{n\times d}\sim\mathscr{N}(\mu_1,\Sigma_1)$ and $X_2\in\Re^{n\times d}\sim\mathscr{N}(\mu_2,\Sigma_2)$ and obtained the sample covariance matrix $\hat{\Sigma}_1=X_1^TX_1/d$ and $\hat{\Sigma}_2=X_2^TX_2/d$. We set the regularization parameters $\gamma=0.5/\sqrt{d}$ and $r=2\sqrt{d}\approx 8.944$ in the problem (6.71).

To compare SC-PRSM-PR with the E-ADMM and E-SC-PRSM, we set $\beta=2$ for all these algorithms; took $\alpha=0.1$ for SC-PRSM and SC-PRSM-PR; and $\mu=0.11$ for SC-PRSM-PR. All the initial values were zeros matrices, and the same ADMM algorithm (6.2) with $\beta=1$ was used for solving the sub-problems in all three schemes. The algorithms used for the solution of the sub-problems are similar to those in (6.77) and (6.79). As stopping criterion we used

$$\text{Tol} := \max\{\beta\|R^{k+1}-R^k\|,\beta\|S^{k+1}-S^k\|,\frac{1}{\beta}\|\lambda^{k+1}-\lambda^k\|\}\le d\varepsilon, \tag{6.80}$$

with $\varepsilon=1\times 10^{-2}$ and the maximum iteration number was set as 200. As in [4], the quantities $\beta\|R^{k+1}-R^k\|$ and $\beta\|S^{k+1}-S^k\|$ measure the primal residual and $\frac{1}{\beta}\|\lambda^{k+1}-\lambda^k\|$ measures the dual residual of the optimality of an iterate generated by scheme (6.74).

In Figure 6.6, we plotted the evolution curves of the objective function values, the primal and dual residuals with respect to the iteration number and computing time. These curves show that E-ADMM and E-SC-PRSM are stuck in reducing the primal and dual residuals (see the figures in the third and fourth columns in Figure 6.6), and the stopping criterion (6.80) is not fulfilled after running out of the maximal num-

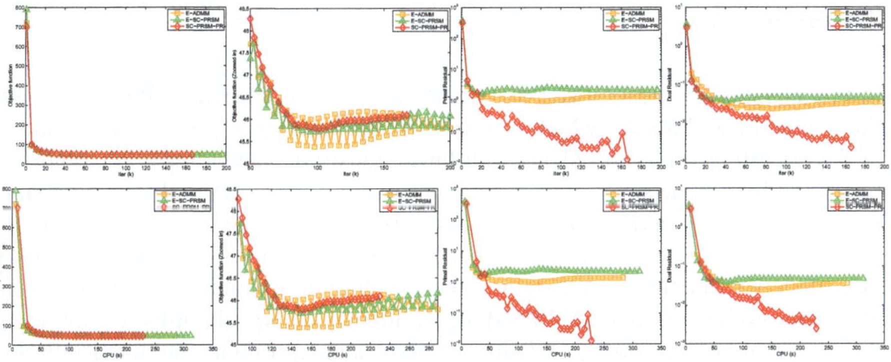

Fig. 6.6 Quadratic discriminant analysis problem (6.72): evolution of (from left to right) the objective function value, the zoomed objective value after the 50th iteration, primal residual and dual residual for E-ADMM, E-SC-PRSM, and SC-PRSM-PR *w.r.t* the number of iterations (first row) and computing time (second row).

ber of iterations set beforehand. Moreover, from the zoomed figures (see the second column in Figure 6.6), we see that the objective function values associated with the iterations of E-ADMM and E-SC-PRSM are oscillating, which corresponds to the curves showing that the reductions of the primal and dual residuals get stuck. In fact, the curves in Figure 6.6 also further show the divergence of E-ADMM and E-SC-PRSM. On the contrary, the proposed SC-RPSM-PR algorithm (6.11) shows good convergence properties for reducing at once the objective function values, and the primal and dual residuals. In Table 6.3, we reported some statistics on the comparison of these three methods, when applied to the solution of problem (6.71); They include the number of iterations ("Iter"), computing time ("CPU(s)"), the rank of R ("rank($\hat{R}$)"), the number of nonzero entries of S ("$|\mathrm{supp}(\hat{S})|$"), and, finally, the number of violation of the constraints $\|U\|_\infty$. Recall that our simulated data set requires that $\|U\|_\infty \leq r \approx 8.944$. Thus, according to Figure 6.6 and Table 6.3, although these three methods perform almost similarly at recovering the rank of R and the sparsity of S, E-SC-PRSM clearly outperforms the other algorithms at reducing the objective function values and the constraint violations (which are exactly the measurement of optimality) for the problem (6.71). These better convergence properties clearly show the superiority of the proposed E-SC-PRSM algorithm.

Table 6.3 Numerical comparison for the QDA problem (6.71).

Algorithm	Iter.	CPU(s)	rank($\hat{R}$)	$\lvert\mathrm{supp}(\hat{S})\rvert$	$\|U\|_\infty$
E-ADMM	200	288.732	17	67	9.29
E-SC-PRSM	200	318.59	17	69	9.30
SC-PRSM-PR	**168**	**230.73**	**17**	**64**	**8.96**

7 Conclusions

In this chapter, we generalized the strictly contractive Peaceman-Rachford splitting method (SC-PRSM), which was recently proposed (see [26]) for the solution of convex minimization problems with linear constraints and a separable objective function which is the sum of two functionals without coupled variables. Our goal in this chapter was to address the multi-block solution of convex minimization problems with a higher degree of separability, where the objective function is the sum of more than two functionals. We showed, via a well-chosen example, that natural generalizations of the original SC-PRSM algorithm may diverge. In order to solve these more general minimization problems, we advocated regrouping the functionals and variables of the original multi-block problem as two blocks, then apply the original SC-PRSM algorithm, the resulting dub-problems being split into smaller ones, easier to solve in principle, and finally to regularize these sub-problems by proximal regularization in order to insure convergence. The resulting algorithm, called SC-PRSM with proximal regularization (SC-PRSM-PR), preserves the implementation simplicity of operator-splitting type methods, with easy to solve sub-problems, and provable convergence (a big theoretical plus). We discussed also the worst-case convergence rate measured by the iteration complexity. The efficiency of the SC-PRSM-PR algorithm was verified by the numerical results obtained by applying this novel algorithm to the solution of problems from Image Processing and Statistical Learning.

This chapter illustrates the fact that a given operator-splitting method, with proved convergence properties when applied to the solution of convex minimization problems with a two-block separable structure, cannot always be applied directly to the solution of similar problems having a higher degree of separability. To insure convergence, some tricky treatment of the sub-problems may be necessary, such as our strategy of proximally regularizing the decomposed sub-problems. This chapter can also be viewed as an example on how to design an algorithm for the solution of some separable convex programming problems, starting from an algorithm with proved convergence property applicable to the solution of simpler separable convex problems. The approach we took in this chapter may help designing customized algorithms in other contexts. For example, we can apply directly the original ADMM algorithm (6.2) to the solution of the multi-block convex minimization problem (6.5). Then, just as we did for SC-PRSM-PR, we can further decompose the resulting (y,z)-sub-problems and regularize the sub-sub-problems by proximal terms. We can also consider employing an alternating decomposition for the sub-problems in (6.8). The algorithmic design approach we follow in this chapter, and the corresponding analytic framework, can be used to construct such a variant and prove its convergence rigorously. To simplify our presentation, we focused on minimization problems with a three-block separable structure. However, the approach we took in this chapter can be generalized to separable problems with an arbitrary number of blocks; of course, we expect the convergence analysis to be more complicated.

Acknowledgements This author was supported by the NSFC grants 11471156 and 91530115. The author "H. Liu" was supported by NSF Grant III-1116730. This author was supported by the General Research Fund from Research Grants Council of Hong Kong: 12302514.

References

1. Agarwal, A., Negahban, S., Wainwright, M.J.: Noisy matrix decomposition via convex relaxation: Optimal rates in high dimensions. The Annals of Statistics **40**(2), 1171–1197 (2012)
2. Bertsekas, D.P.: Constrained Optimization and Lagrange Multiplier Methods. Athena Scientific (1996)
3. Blum, E., Oettli, W.: Mathematische Optimierung: Grundlagen und Verfahren. Springer, Berlin (2012)
4. Boyd, S., Parikh, N., Chu, E., Peleato, B., Eckstein, J.: Distributed optimization and statistical learning via the alternating direction method of multipliers. Foundations and Trends in Machine Learning **3**(1), 1–122 (2011)
5. Candès, E.J., Li, X., Ma, Y., Wright, J.: Robust principal component analysis? Journal of the ACM **58**(3), 1–37 (2011)
6. Chan, T.F., Glowinski, R.: Finite element approximation and iterative solution of a class of mildly non-linear elliptic equations. Report STAN-CS-78-674, Computer Science Department, Stanford University (1978)
7. Chandrasekaran, V., Parrilo, P.A., Willsky, A.S.: Latent variable graphical model selection via convex optimization. In: Proceedings of the 48th Annual Allerton Conference on Communication, Control, and Computing, pp. 1610–1613 (2010)
8. Chandrasekaran, V., Sanghavi, S., Parrilo, P.A., Willsky, A.S.: Rank-sparsity incoherence for matrix decomposition. SIAM Journal on Optimization **21**(2), 572–596 (2011)
9. Chen, C., He, B., Ye, Y., Yuan, X.: The direct extension of ADMM for multi-block convex minimization problems is not necessarily convergent. Mathematical Programming **155**(1-2), 57–79 (2016)
10. Corman, E., Yuan, X.: A generalized proximal point algorithm and its convergence rate. SIAM Journal on Optimization **24**(4), 1614–1638 (2014)
11. Donoho, D.: Compressed sensing. IEEE Transactions on Information Theory **52**(4), 1289–1306 (2006)
12. Douglas, J., Rachford, H.H.: On the numerical solution of heat conduction problems in two and three space variables. Transactions of the American Mathematical Society **82**(2), 421–421 (1956)
13. Eckstein, J., Yao, W.: Augmented Lagrangian and alternating direction methods for convex optimization: A tutorial and some illustrative computational results. RUTCOR Research Reports **32** (2012)
14. Facchinei, F., Pang, J.S.: Finite-Dimensional Variational Inequalities and Complementarity Problems. Springer, New York, NY (2004)
15. Fan, J., Ke, Z.T., Liu, H., Xia, L.: QUADRO: A supervised dimension reduction method via Rayleigh quotient optimization. The Annals of Statistics **43**(4), 1498–1534 (2015)
16. Fan, J., Liao, Y., Mincheva, M.: Large covariance estimation by thresholding principal orthogonal complements. Journal of the Royal Statistical Society: Series B (Statistical Methodology) **75**(4), 603–680 (2013)
17. Friedman, J.H.: Regularized discriminant analysis. Journal of the American Statistical Association **84**(405), 165–175 (1989)
18. Gabay, D.: Applications of the method of multipliers to variational inequalities. In: M. Fortin, R. Glowinski (eds.) Augmented Lagrange Methods: Applications to the Solution of Boundary-valued Problems, pp. 299–331. North-Holland, Amsterdam (1983)

19. Gabay, D., Mercier, B.: A dual algorithm for the solution of nonlinear variational problems via finite element approximation. Computers & Mathematics with Applications **2**(1), 17–40 (1976)
20. Glowinski, R.: On alternating direction methods of multipliers: A historical perspective. In: W. Fitzgibbon, Y.A. Kuznetsov, P. Neittaanmäki, O. Pironneau (eds.) Modeling, Simulation and Optimization for Science and Technology, vol. 34, pp. 59–82. Springer Netherlands, Dordrecht (2014)
21. Glowinski, R., Karkkainen, T., Majava, K.: On the convergence of operator-splitting methods. In: E. Heikkola, Y.A. Kuznetsov, P. Neittaanmaki, O. Pironneau (eds.) Numerical Methods for Scientific Computing, Variational Problems and Applications, pp. 67–79. CIMNE, Barcelona (2003)
22. Glowinski, R., Le Tallec, P.: Augmented Lagrangian and Operator-Splitting Methods in Nonlinear Mechanics. Society for Industrial and Applied Mathematics (1989)
23. Glowinski, R., Marroco, A.: Sur l'approximation, par éléments finis d'ordre un, et la résolution, par pénalisation-dualité d'une classe de problèmes de Dirichlet non linéaires. ESAIM: Mathematical Modelling and Numerical Analysis – Modélisation Mathématique et Analyse Numérique **9**(R2), 41–76 (1975)
24. Han, D., Yuan, X.: Convergence analysis of the Peaceman-Rachford splitting method for nonsmooth convex optimization. Optimization Online (2012)
25. Hansen, P.C., Nagy, J.G., O'Leary, D.P.: Deblurring Images: Matrices, Spectra, and Filtering. SIAM, Philadelphia (2006)
26. He, B., Liu, H., Wang, Z., Yuan, X.: A strictly contractive Peaceman–Rachford splitting method for convex programming. SIAM Journal on Optimization **24**(3), 1011–1040 (2014)
27. He, B., Tao, M., Yuan, X.: Alternating direction method with Gaussian back substitution for separable convex programming. SIAM Journal on Optimization **22**(2), 313–340 (2012)
28. He, B., Tao, M., Yuan, X.: Convergence rate analysis for the alternating direction method of multipliers with a substitution procedure for separable convex programming. Mathematics of Operations Research, to appear
29. He, B., Yuan, X.: On the $O(1/n)$ convergence rate of the Douglas–Rachford alternating direction method. SIAM Journal on Numerical Analysis **50**(2), 700–709 (2012)
30. He, B., Yuan, X.: On non-ergodic convergence rate of Douglas–Rachford alternating direction method of multipliers. Numerische Mathematik **130**(3), 567–577 (2015)
31. Hestenes, M.R.: Multiplier and gradient methods. Journal of Optimization Theory and Applications **4**(5), 303–320 (1969)
32. Huang, Y.M., Ng, M.K., Wen, Y.W.: Fast image restoration methods for impulse and Gaussian noises removal. IEEE Signal Processing Letters **16**(6), 457–460 (2009)
33. Kiwiel, K.C., Rosa, C.H., Ruszczynski, A.: Proximal decomposition via alternating linearization. SIAM Journal on Optimization **9**(3), 668–689 (1999)
34. Lions, P.L., Mercier, B.: Splitting algorithms for the sum of two nonlinear operators. SIAM Journal on Numerical Analysis **16**(6), 964–979 (1979)
35. Martinet, B.: Régularisation d'inéquations variationelles par approximations successives. Revue Française Info. Rech. Opér. **R4**(3), 154–158 (1970)
36. McLachlan, G.J.: Discriminant Analysis and Statistical Pattern Recognition. Wiley, Hoboken, NJ (2004)
37. Nemirovski, A., Yudin, D.B.: Problem Complexity and Method Efficiency in Optimization. Wiley, New York (1983)
38. Nesterov, Y.: A method of solving a convex programming problem with convergence rate $O(1/k^2)$. Soviet Mathematics Doklady **27**, 372–376 (1983)
39. Peaceman, D.W., Rachford, H.H.: The numerical solution of parabolic and elliptic differential equations. Journal of the Society for Industrial and Applied Mathematics **3**(1), 28–41 (1955)
40. Peng, Y., Ganesh, A., Wright, J., Xu, W., Ma, Y.: RASL: Robust alignment by sparse and low-rank decomposition for linearly correlated images. In: Proceedings of the IEEE Conference on Computer Vision and Pattern Recognition (CVPR), pp. 763–770 (2010)
41. Powell, M.J.D.: A method for nonlinear constraints in minimization problems. In: R. Fletcher (ed.) Optimization, pp. 283–298. Academic Press, London (1969)

42. Rockafellar, R.T.: Monotone operators and the proximal point algorithm. SIAM Journal on Control and Optimization **14**(5), 877–898 (1976)
43. Rudin, L.I., Osher, S., Fatemi, E.: Nonlinear total variation based noise removal algorithms. Physica D: Nonlinear Phenomena **60**(1-4), 259–268 (1992)
44. Tao, M., Yuan, X.: Recovering low-rank and sparse components of matrices from incomplete and noisy observations. SIAM Journal on Optimization **21**(1), 57–81 (2011)

Chapter 7
Nonconvex Sparse Regularization and Splitting Algorithms

Rick Chartrand and Wotao Yin

Abstract Nonconvex regularization functions such as the ℓ^p quasinorm $(0 < p < 1)$ can recover sparser solutions from fewer measurements than the convex ℓ^1 regularization function. They have been widely used for compressive sensing and signal processing. This chapter briefly reviews the development of algorithms for nonconvex regularization. Because nonconvex regularization usually has different regularity properties from other functions in a problem, we often apply operator splitting (forward-backward splitting) to develop algorithms that treat them separately. The treatment on nonconvex regularization is via the proximal mapping.

We also review another class of coordinate descent algorithms that work for both convex and nonconvex functions. They split variables into small, possibly parallel, subproblems, each of which updates a variable while fixing others. Their theory and applications have been recently extended to cover nonconvex regularization functions, which we review in this chapter.

Finally, we also briefly mention an ADMM-based algorithm for nonconvex regularization, as well as the recent algorithms for the so-called nonconvex sort ℓ^1 and $\ell^1 - \ell^2$ minimization.

1 Early History of Nonconvex Regularization for Sparsity

The attempt to compute a sparse solution of a problem (such as a linear system of equations) by minimizing a nonconvex penalty function can be traced back at least to Leahy and Jeffs [31], who used a simplex algorithm (essentially a nonlinear version

R. Chartrand (✉)
Descartes Labs, Los Alamos, NM 87544, USA
e-mail: rick@descarteslabs.com

W. Yin
Department of Mathematics, University of California, Los Angeles, CA 90095, USA
e-mail: wotaoyin@math.ucla.edu

R. Glowinski et al. (eds.), *Splitting Methods in Communication, Imaging, Science, and Engineering*, Scientific Computation, DOI 10.1007/978-3-319-41589-5_7

of linear programming) to minimize the ℓ^p norm[1] subject to a linear constraint. They describe the algorithm as similar to one of Barrodale and Roberts [3], where ℓ^p norm minimization is considered in a different context.

The next algorithmic development came from Gorodnitsky and Rao, named FOCUSS (for FOCal Underdetermined System Solver) [25]. In fact, the approach was a much older method, iteratively reweighted least squares (IRLS) [30], applied to ℓ^0 norm minimization. This was extended to general ℓ^p minimization by Rao and Kreutz-Delgado [45]. Global convergence was erroneously claimed in [26], based on Zangwill's Global Convergence Theorem [65], which only provides that subsequential limits are local minima.

Attention to nonconvex regularization for sparsity was next spurred by the development of compressive sensing [12, 23], which mostly featured ℓ^1 minimization. Generalization to ℓ^p minimization with $p < 1$ was carried out by Chartrand, initially with a projected gradient algorithm [15], followed by an IRLS approach with Yin [18]. A crucial difference between this work and the earlier FOCUSS work was the use of iterative mollification, where $|x|^p$ was replaced by $(x^2 + \varepsilon_n)^{p/2}$ for a sequence (ε_n) converging geometrically to zero. This approach, reminiscent of the graduated nonconvexity approach of Blake and Zisserman [7] resulted in far better signal reconstruction results, seemingly due to much better avoidance of local minima. A similar approach was developed independently by Mohimani et al. [39], except with iterative mollification of the ℓ^0 norm. In addition, Candès et al. [13] developed a reweighted ℓ^1 algorithm, using a fixed mollifying ε. If the same iterative mollification approach is used, empirical evidence suggests that reweighted ℓ^1 and IRLS are equally effective.

2 Forward-Backward Splitting and Thresholdings

First, let us examine the proximal mapping of the ℓ^0 norm:

$$\mathrm{prox}_{\lambda\|\cdot\|_0}(x) \stackrel{\text{def}}{=} \arg\min_{w} \|w\|_0 + \tfrac{1}{2\lambda}\|w-x\|_2^2, \tag{7.1}$$

where $\lambda > 0$. The optimization problem in (7.1) is clearly separable. For an input component x_i, since $\|w\|_0$ only depends on whether w_i is nonzero, the minimizing w_i is clearly either 0 or x_i. If $w_i = 0$, the ith term of the objective function is $x_i^2/(2\lambda)$, while if $w_i = x_i \neq 0$, the value is 1. We thus obtain

$$\left[\mathrm{prox}_{\lambda\|\cdot\|_0}(x)\right]_i = \begin{cases} 0 & \text{if } |x_i| < \sqrt{2\lambda}; \\ \{0, x_i\} & \text{if } |x_i| = \sqrt{2\lambda}; \\ x_i & \text{if } |x_i| > \sqrt{2\lambda}. \end{cases} \tag{7.2}$$

[1] We use "norm" loosely, to refer to such things as the ℓ^p quasinorm, or the ℓ^0 penalty function (which has no correct norm-like name).

This motivates the use of the mapping known as *hard thresholding*, defined componentwise as follows:

$$H_t(x)_i = \begin{cases} 0 & \text{if } |x_i| \leq t; \\ x_i & \text{if } |x_i| > t. \end{cases} \tag{7.3}$$

(Our choice of value at the discontinuity is arbitrary.) Thus $H_{\sqrt{2\lambda}}(x)$ gives a (global) minimizer of the problem in (7.1).

Now consider the following optimization problem:

$$\min_x \|x\|_0 + \tfrac{1}{2\lambda}\|Ax-b\|_2^2. \tag{7.4}$$

Such a problem computes a sparse, approximate solution of the linear system $Ax = b$, and is an ℓ^0 analog of basis pursuit denoising (BPDN) [20]. This is one of the standard problems considered in compressive sensing, where A would be the product of a measurement matrix and a sparse representation matrix (or *dictionary*), and b the (possibly noisy) measurement data. If A is simply a dictionary, then (7.4) is an example of sparse coding.

If we apply forward-backward splitting (FBS) to (7.4), we essentially obtain the Iterative Hard Thresholding algorithm (IHT) of Blumensath and Davies [8, 9]:

$$x^{n+1} = H_t(x^n - \mu A^T(Ax^n - b)). \tag{7.5}$$

The one difference between FBS and IHT is that in IHT, the threshold value t is adaptive, chosen so that the outcome of the thresholding has at most K nonzero components, for some positive integer K.

The discontinuities of the ℓ^0 norm and hard thresholding are obstacles to good algorithmic performance. For example, substituting hard thresholding for soft thresholding in ADMM gives an algorithm that oscillates fiercely, though Dong and Zhang [22] managed to tame the beast by using "double augmented Lagrangian" [48] and using the mean of the iterates as the solution.

Using the ℓ^p norm, with $p \in (0,1)$, provides a better-behaved alternative:

$$\min_x \|x\|_p^p + \tfrac{1}{2\lambda}\|Ax-b\|_2^2. \tag{7.6}$$

Note that we use the pth power of the ℓ^p norm, because it is easier to compute with and satisfies the triangle inequality.

We can try to compute the proximal mapping:

$$\text{prox}_\lambda \|\cdot\|_p^p(x) = \arg\min_w \|w\|_p^p + \tfrac{1}{2\lambda}\|w-x\|_2^2. \tag{7.7}$$

As before, this is a separable problem; for simplicity, assume w and x are scalars, and without loss of generality, assume $x > 0$. If we seek a nonzero minimizer, we need to solve $pw^{p-1} + (w-x)/2 = 0$. Eliminating the negative exponent gives $w^{2-p} - xw^{1-p} + 2p = 0$. This is not analytically solvable for general p. The cases where it can be solved are $p = 1/2$, where we have a cubic equation in $\sqrt{w}$, and $p = 2/3$, where we have a quartic equation in $\sqrt[3]{w}$. While this may seem restrictive, there is anecdotal, empirical evidence suggesting $p = 1/2$ is a good choice in at least a broad range of circumstances [18, 49, 63].

The first paper to make use of these special cases is by Krishnan and Fergus [29]. They consider a nonconvex generalization of TV-regularized deblurring, with (anisotropic) TV replaced by the ℓ^p norm of the gradient, for $p \in \{1/2, 2/3\}$. Their approach generalizes the splitting approach of FTVd [54], with soft thresholding replaced by the appropriate proximal mapping. They consider solving the cubic or quartic equation by radicals, and by means of a lookup table, and report the latter to be faster while giving the same quality.

Xu et al. [62] conduct further analysis of the $\ell^{1/2}$ proximal mapping, considering the solution of the cubic equation in terms of the cosine and arccosine functions. Similarly, Cao et al. [14] develop a closed-form formula for the $\ell^{2/3}$ proximal mapping. They show that the proximal mapping can be considered as a thresholding mapping, in the sense of each component of the output being the same scalar function of the corresponding component of the input, and that the scalar function maps all inputs of magnitude below some threshold to zero. They use the thresholding mapping within an application of forward-backward splitting to the $\ell^{1/2}$ generalization of BPDN, as well as part of alternative approach to the algorithm of Krishnan and Fergus.

Given the difficulty of computing proximal mappings for a wide range of nonconvex penalty functions, it is reasonable to consider reversing the process: specifying a thresholding mapping, and then determining whether it is a proximal mapping of a penalty function. Antoniadis considers a general class of thresholding functions, and then shows that these are proximal mappings of penalty functions [1]:

Theorem 1 (Antoniadis). *Let $\delta_\lambda : \mathbb{R} \to \mathbb{R}$ be a thresholding function that is increasing and antisymmetric such that $0 \le \delta_\lambda(x) \le x$ for $x \ge 0$ and $\delta_\lambda(x) \to \infty$ as $x \to \infty$. Then there exists a continuous positive penalty function p_λ, with $p_\lambda(x) \le p_\lambda(y)$ whenever $|x| \le |y|$, such that $\delta_\lambda(z)$ is the unique solution of the minimization problem $\min_\theta (z\theta)^2 + 2p_\lambda(|\theta|)$ for every z at which δ_λ is continuous.*

Such thresholding functions are considered in the context of thresholding of wavelet coefficients, such as for denoising, and not for use with splitting algorithms.

Chartrand [16, 17] proves a similar theorem, but by making use of tools from convex analysis in the proof, is able to add first-order information about the penalty function to the conclusions:

Theorem 2 (Chartrand). *Suppose $s = s^\lambda : \mathbb{R}_+ \to \mathbb{R}_+$ is continuous, satisfies $x \le \lambda \Rightarrow s(x) = 0$ for some $\lambda > 0$, is strictly increasing on $[\lambda, \infty)$, and $s(x) \le x$. Define $S = S^\lambda$ on $\mathbb{R}^n$ by $S(\mathbf{x})_i = s(|x_i|)\operatorname{sign}(x_i)$ for each i. Then S is the proximal mapping of a penalty function $G(\mathbf{x}) = \sum_i g(x_i)$ where g is even, strictly increasing* *and continuous on* $[0,\infty)$*, differentiable on* $(0,\infty)$*, and nondifferentiable at 0 with* $\partial g(0) = [-1,1]$*. If also* $x - s(x)$ *is nonincreasing on* $[\lambda,\infty)$*, then* g *is concave on* $[0,\infty)$ *and* G *satisfies the triangle inequality.*

The proximal mappings are considered for use in generalizations of ADMM. A new thresholding function, a C^∞ approximation of hard thresholding, is used to give a state of the art image reconstruction result. Subsequently, however, the older and simpler *firm thresholding* function of Gao and Bruce [24] was found to perform at least as well.

It can be shown (cf. [58]) that when all of the hypotheses of Thm. 2 are satisfied, the resulting mapping G is *weakly convex* (also known as *semiconvex*. This is the property that $G(x)+\rho\|x\|_2^2$ is convex for sufficiently large ρ. Bayram has shown [4] that the generalization of FBS for weakly convex penalty functions converges to a global minimizer of the analog of (7.6):

$$\min_x G(x) + \tfrac{1}{2\lambda}\|Ax-b\|_2^2. \tag{7.8}$$

Theorem 3 (Bayram). *Let* A *in* (7.8) *have smallest and largest singular values* σ_m *and* σ_M*, respectively. Suppose* G *is proper, lower semi-continuous, and such that* $G(x)+\rho\|x\|_2^2$ *is convex for* $\rho \geq 2\sigma_m^2$*. Suppose also that the set of minimizers of* (7.8) *is nonempty. Then if* $\lambda < 1/\sigma_M^2$*, the sequence generated by FBS converges to a global minimizer of* (7.8) *and monotonically decreases the cost.*

3 Coordinate Descent Methods

Coordinate descent has been applied for problems with nonconvex regularization functions. This section reviews the coordinate methods, as well as the work that adapts them to nonconvex problems.

Coordinate descent is a class of method that solves a large problem by updating one, or a block of, coordinates at each step. The method works with *smooth* and *nonsmooth-separable* functions, as well as convex and nonconvex functions. We call a function *separable* if it has the form $f(\mathbf{x}) = f_1(x_1)+\cdots+f_n(x_n)$. (Coordinate descent methods generally cannot handle functions that are simultaneously nonsmooth and nonseparable because they can cause convergence to a non-critical point [57].) There has been very active research on coordinate descent methods because their steps have an *small memory footprint*, making them suitable for solving large-sized problems. Certain coordinate descent methods can be parallelized. In addition, even on small- and mid-sized problems, coordinate descent methods can run much faster than the traditional first-order methods because each coordinate descent step can use a much larger step size and the selection of updating coordinates can be chosen in favor of the problem structure and data.

Let us consider the problem of

$$\underset{\mathbf{x}}{\text{minimize}}\, f(x_1,\ldots,x_n).$$

At each step of coordinate descent, all but some ith coordinate of x are fixed. Let

$$x_{-i} := (x_1,\ldots,x_{i-1},x_{i+1},\ldots,x_n),$$

collect all components of $\mathbf{x}$ but x_i. The general framework of coordinate descent is: For $k = 0,1,\ldots,$ perform:

1. Select a coordinate i_k
2. Do

$$\begin{cases} \text{update } x_{i_k}^{k+1}, \\ x_j^{k+1} = x_j^k, \qquad \text{for all } j \neq i_k, \end{cases}$$

3. Check stopping criteria.

In the algorithm, each update to x_i^{k+1} can take one of the following forms:

$$x_i^{k+1} \leftarrow \underset{x_i}{\arg\min}\, f(x_i, x_{-i}^k), \tag{7.9a}$$

$$x_i^{k+1} \leftarrow \underset{x_i}{\arg\min}\, f(x_i, x_{-i}^k) + \frac{1}{2\eta_k}\|x_i - x_i^k\|^2, \tag{7.9b}$$

$$x_i^{k+1} \leftarrow \underset{x_i}{\arg\min}\, \langle \nabla_i f(x^k), x_i \rangle + \frac{1}{2\eta_k}\|x_i - x_i^k\|^2, \tag{7.9c}$$

$$x_i^{k+1} \leftarrow \underset{x_i}{\arg\min}\, \langle \nabla_i f^{\text{diff}}(x^k), x_i \rangle + f_i^{\text{prox}}(x_i) + \frac{1}{2\eta_k}\|x_i - x_i^k\|^2, \tag{7.9d}$$

which are called *direct* update (first appeared in [57]), *proximal* update (first appeared in [2]), *gradient* update, and *prox-gradient* update, respectively. The *gradient* update (7.9c) is equivalent to the x_i-directional gradient descent with step size η_k:

$$x_i^{k+1} \leftarrow \left(x^k - \eta_k \nabla f(x^k)\right)_i.$$

The *prox-gradient* update (7.9d) is also known as the *prox-linear* update and *forward-backward* update. It is applied to the objective function of the following form

$$f(x_1,\ldots,x_n) = f^{\text{diff}}(x_1,\ldots,x_n) + \sum_{i=1}^{n} f_i^{\text{prox}}(x_i),$$

where f^{diff} is a differentiable function and each f_i^{prox} is a proximal function. The problem (7.6) is an example of this form. In addition to the four forms of updates in (7.9), there are recent other forms of coordinate descent: a gradient $\nabla_i f(x^k)$ can be replaced by its *stochastic approximation* and/or be evaluated at a point $\hat{x}^k$ extrapolating x^k and x^{k-1}.

Different forms of coordinate updates can be mixed up in a coordinate descent algorithm, but each coordinate x_i is usually updated by one chosen form throughout the iterations. The best choice is based on the problem structure and required convergence properties.

As a very nice feature, coordinate descent can select updating coordinates i_k at different orders. Such flexibilities lead to numerical advantages. The most common choice is the *cyclic* order. Recently, *random* and *shuffled cyclic* orders are found to be very efficient. On nonconvex problems, the stochasticity of these two orders helps avoid low-quality local minimizers.

Although less common, the *greedy* order has started to gain popularity quickly (e.g., [5, 32, 59, 21, 43, 40]). It chooses next coordinate that minimizes the objective function most. The Gauss-Southwell selection rule, a commonly used greedy order with very long history, selects the coordinate with the largest component-wise derivative. In general, greedy coordinate descent is only applicable when certain scores can be quickly computed for all the coordinates and then used to choose the next i_k. The greedy order is more expensive at each iteration but also reduces the total number of iterations. The tradeoff depends on the problem. For problems with sparse solutions, the greedy order has been found *highly effective* because some coordinates are kept always at zero. Other orders, however, will waste lots of computation on updating these coordinates yet eventually setting them back to nearly zero.

Parallel coordinate descent has been recently developed for multiple computing nodes to perform coordinate descent simultaneously [10, 50, 43, 47, 36]. The selection of coordinates that update in parallel can also be deterministic (all coordinates are updated), stochastic, or greedy. These parallel algorithms partition computation into smaller subproblems, which are solved simultaneously and then synchronized to ensure up-to-date information at all computing nodes.

The recent development [34, 33, 42], *asynchronous parallel* coordinate descent or coordinate update, brings significant further improvement because every computing node can perform a new update without waiting for other nodes to finish their running updates. Asynchrony appears originally in solving linear equations [19] and later introduced to optimization [6, 51]. On large-scale problems, asynchrony greatly reduces processor idling, especially when the different computing nodes take different amounts of time to finish their coordinate updates. Furthermore, asynchrony spreads out communication over time and thus avoids slowdown due to communication congestion. On the downside, asynchrony causes computing nodes to perform updates with possibly out-of-date information, thus potentially increasing the total number of updates for convergence. Despite this, the tradeoff still appears to vastly favor asynchrony over synchrony [34, 33, 42].

One must be cautious at applying coordinate descent to *nonconvex functions*, for which the most natural form of update (7.9a) may not converge. Powell [44] illustrated this through an example with a differentiable nonconvex objective:

$$f(x_1,x_2,x_3) = -(x_1x_2+x_2x_3+x_1x_3)+\sum_{i=1}^{3}\max\{|x_i|-1,0\}^2.$$

The minimizers are $(x_1,x_2,x_3) = \pm(1,1,1)$. However, the subproblem (7.9a) generates a sequence of points that approximately cycle near the other six points in the set $\{(x_1,x_2,x_3) : x_i = \pm 1, i = 1,2,3\}$. Nonetheless, each of other forms of updates in (7.9) has been shown to converge (under various conditions) for differentiable nonconvex functions, which bring us to the next issue of *nonsmooth* functions. Warga's example [57],

$$f(x_1,x_2) = \begin{cases} x_1 - 2x_2, & 0 \le x_2 \le x_1 \le 1, \\ x_2 - 2x_1, & 0 \le x_1 \le x_2 \le 1, \end{cases}$$

is a convex, nonsmooth function, which has its minimizer at $(x_1,x_2) = (1,1)$. Starting from an arbitrary point (x_1,x_2) where $x_1 \neq 1$ and $x_2 \neq 1$, all the existing coordinate descent algorithms will fail at a non-optimal point $x_1 = x_2$. Although this example is simple, such a phenomenon generally occurs with many nonsmooth functions that couple multiple variables. To deal with this problem, we need to apply objective smoothing, variable splitting, or primal-dual splitting, which we do not review here.

Regularity assumptions on the nonconvex objective function are required to show convergence of coordinate descent algorithms. Grippo and Sciandrone [27] require that f is component-wise strictly quasiconvex with respect to all but 2 blocks, or that f is pseudoconvex and has bounded level sets. Tseng [52] requires that either f is pseudoconvex in every coordinate-pair of among $n-1$ coordinates or f has at most one minimizer for each of $(n-2)$ coordinates, along with other assumptions. Tseng and Yun [53] studied coordinate descent based on (7.9d) under an assumed local error bound that holds if f^{diff} is nonconvex quadratic or equals $g(Ax)$ for a strongly convex function g and matrix A. However, their local error bound requires the regularization function to be convex and polyhedral, too strong for ℓ^p-norm and other nonconvex regularization functions to hold.

The paper [60] considers multiconvex[2] smooth functions plus separable proximal functions in the objective. It analyzes the mixed use of different update forms, extrapolation, as well as the functions satisfying the Kurdyka-Łojasiewicz (KL) property, which improves subsequential convergence to global convergence. The BSUM framework [46] aims at analyzing different forms of coordinate descent in a unified approach. Their analysis has allowed the differentiable part of the objective to be nonconvex, along with other conditions. Checking their conditions, piecewise linear regularization functions are permitted. However, the assumptions in the above papers exclude other major nonconvex regularization functions.

The papers [38, 11] are among the first to apply coordinate descent to regularized statistical regression problems with nonconvex regularization functions including the minimax concave penalty (MCP) and the smoothly clipped absolute deviation (SCAD). They require properties on the fitting terms such that the sum of fitting and regularization terms is convex and the results from [53] can apply.

[2] The objective is convex in each of coordinates while the other coordinates are fixed.

Coordinate descent for nonconvex ℓ^p-norm minimization recently appeared in [37], where the algorithm uses the update form (7.9d) and the cyclic order. Although the proximal map of ℓ^p is set-valued in general, a single-valued selection is used. The work establishes subsequential convergence and, under a further assumption of scalable restricted isometry property, full sequential convergence to a local minimizer. The algorithm in [35] analyzes a random prox-gradient coordinate descent algorithm. It accepts an objective of the smooth+proximal form where both functions can be nonconvex. More recently, the work [61] extends the analysis of the previous work [60] by including nonconvex regularization functions such as the minimax concave penalty (MCP), the smoothly clipped absolute deviation (SCAD), and others. The work also analyzes both deterministic and randomly shuffled orders and reports better performance using the latter order. Another recent work [66] establishes that the support set converges in finitely many iterations and provides new conditions to ensure convergence to a local minimizer. The papers [61, 66] also apply the KL property for full sequential convergence.

Last but not least, to have strong performance, coordinate descent algorithms rely on solving smaller, simpler subproblems that have low complexities and low memory requirements. Not all problem structures are amenable for coordinate descent. Hence, identifying coordinate friendly structures in a problem is crucial to implementing coordinate descent algorithms effectively. See the recent work [41].

4 Other Methods

This section briefly mentions some other algorithms for problems with nonconvex regularization functions.

The work [56] analyzes the convergence of the alternating direction method of multipliers (ADMM) for minimizing a nonconvex and possibly nonsmooth objective function, $f(x_1,\ldots,x_p,y)$, subject to linear constraints. Unlike the majority of coordinate descent algorithms, ADMM handles constraints by involving dual variables. The proposed ADMM in [56] sequentially updates the primal variables in the order $x_1,\ldots,x_p,y$, followed by updating the dual variable. For convergence to a critical point, the objective function on the y-block must be smooth (possibly nonconvex) and the objective function of each x_i-block can be smooth plus nonsmooth. A variety of nonconvex functions such as piecewise linear functions, ℓ^p norm, Schatten-p norm ($0 < p < 1$), SCAD, as well as the indicator functions of compact smooth manifolds (e.g., spherical, Stiefel, and Grassman manifolds), can be applied to the x_i variables.

The sorted ℓ^1 function [28] is a *nonconvex* regularization function. It is a weighted ℓ^1 function where the weights have fixed values but are dynamically assigned to the components under the principle that components with larger magnitudes get smaller weights. This is the same principle in reweighted ℓ^2 and ℓ^1 algorithms, where the assignments are fixed and the weights are dynamic. Sorted ℓ^1 regularization is related to iterative support detection [55] and iterative hard thresholding [8].

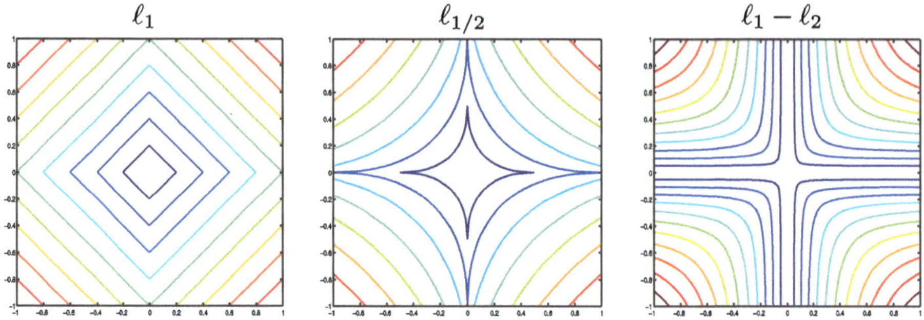

The difference of ℓ^1 and ℓ^2, $\|x\|_1 - \|x\|_2$, favors sparse vectors [64], so it can also serve as a nonconvex regularization function for generating sparse solutions. The picture is courtesy of [64]. The authors analyzed the global solution and proposed an iterative algorithm based on the *difference of convex functions* algorithm. Their simulation suggests stronger performance when the sampling matrix in compressed sensing is ill-conditioned (the restricted isometry property is not satisfied) than other compared algorithms.

References

1. Antoniadis, A.: Wavelet methods in statistics: Some recent developments and their applications. Statistics Surveys **1**, 16–55 (2007)
2. Auslender, A.: Asymptotic properties of the Fenchel dual functional and applications to decomposition problems. Journal of Optimization Theory and Applications **73**(3), 427–449 (1992)
3. Barrodale, I., Roberts, F.D.K.: Applications of mathematical programming to ℓ_p approximation. In: J.B. Rosen, O.L. Mangasarian, K. Ritter (eds.) Nonlinear Programming, Madison, Wisconsin, May 4–6, 1970, pp. 447–464. Academic Press, New York (1970)
4. Bayram, I.: On the convergence of the iterative shrinkage/thresholding algorithm with a weakly convex penalty. IEEE Transactions on Signal Processing **64**(6), 1597–1608 (2016)
5. Bertsekas, D.P.: Nonlinear Programming, 2nd edition edn. Athena Scientific (1999)
6. Bertsekas, D.P., Tsitsiklis, J.N.: Parallel and Distributed Computation: Numerical Methods. Prentice Hall, Englewood Cliffs, N.J (1989)
7. Blake, A., Zisserman, A.: Visual Reconstruction. MIT Press, Cambridge, MA (1987)
8. Blumensath, T., Davies, M.E.: Iterative hard thresholding for compressed sensing. Applied and Computational Harmonic Analysis **27**(3), 265–274 (2009)
9. Blumensath, T., Davies, M.E.: Normalized iterative hard thresholding: Guaranteed stability and performance. IEEE Journal of Selected Topics in Signal Processing **4**(2), 298–309 (2010)
10. Bradley, J.K., Kyrola, A., Bickson, D., Guestrin, C.: Parallel coordinate descent for l1-regularized loss minimization. arXiv preprint arXiv:1105.5379 (2011)
11. Breheny, P., Huang, J.: Coordinate descent algorithms for nonconvex penalized regression, with applications to biological feature selection. The Annals of Applied Statistics **5**(1), 232–253 (2011)
12. Candès, E.J., Romberg, J., Tao, T.: Robust uncertainty principles: Exact signal reconstruction from highly incomplete frequency information. IEEE Transactions on Information Theory **52**(2), 489–509 (2006)

13. Candès, E.J., Wakin, M.B., Boyd, S.P.: Enhancing sparsity by reweighted ℓ_1 minimization. Journal of Fourier Analysis and Applications **14**(5–6), 877–905 (2008)
14. Cao, W., Sun, J., Xu, Z.: Fast image deconvolution using closed-form thresholding formulas of $l_q(q = \frac{1}{2}, \frac{2}{3})$ regularization. Journal of Visual Communication and Image Representation **24**(1), 31–41 (2013)
15. Chartrand, R.: Exact reconstructions of sparse signals via nonconvex minimization. IEEE Signal Processing Letters **14**(10), 707–710 (2007)
16. Chartrand, R.: Generalized shrinkage and penalty functions. In: IEEE Global Conference on Signal and Information Processing, p. 616. Austin, TX (2013)
17. Chartrand, R.: Shrinkage mappings and their induced penalty functions. In: IEEE International Conference on Acoustics, Speech, and Signal Processing. Florence, Italy (2014)
18. Chartrand, R., Yin, W.: Iteratively reweighted algorithms for compressive sensing. In: IEEE International Conference on Acoustics, Speech, and Signal Processing, pp. 3869–3872. Las Vegas, NV (2008)
19. Chazan, D., Miranker, W.: Chaotic relaxation. Linear Algebra and its Applications **2**(2), 199–222 (1969)
20. Chen, S.S., Donoho, D.L., Saunders, M.A.: Atomic decomposition by basis pursuit. SIAM Journal on Scientific Computing **20**, 33–61 (1998)
21. Dhillon, I.S., Ravikumar, P.K., Tewari, A.: Nearest Neighbor based Greedy Coordinate Descent. In: J. Shawe-Taylor, R.S. Zemel, P.L. Bartlett, F. Pereira, K.Q. Weinberger (eds.) Advances in Neural Information Processing Systems 24, pp. 2160–2168. Curran Associates, Inc. (2011)
22. Dong, B., Zhang, Y.: An efficient algorithm for ℓ_0 minimization in wavelet frame based image restoration. Journal of Scientific Computing **54**(2–3), 350–368 (2013)
23. Donoho, D.L.: Compressed sensing. IEEE Transactions on Information Theory **52**(4), 1289–1306 (2006)
24. Gao, H.Y., Bruce, A.G.: Waveshrink with firm shrinkage. Statistica Sinica **7**(4), 855–874 (1997)
25. Gorodnitsky, I.F., Rao, B.D.: A new iterative weighted norm minimization algorithm and its applications. In: IEEE Sixth SP Workshop on Statistical Signal and Array Processing, pp. 412–415 (1992)
26. Gorodnitsky, I.F., Rao, B.D.: Convergence analysis of a class of adaptive weighted norm extrapolation algorithms. In: Asilomar Conference on Signals, Systems, and Computers, vol. 1, pp. 339–343. Pacific Grove, CA (1993)
27. Grippo, L., Sciandrone, M.: On the convergence of the block nonlinear Gauss–Seidel method under convex constraints. Operations Research Letters **26**(3), 127–136 (2000)
28. Huang, X.L., Shi, L., Yan, M.: Nonconvex sorted ℓ_1 minimization for sparse approximation. Journal of the Operations Research Society of China **3**(2), 207–229 (2015)
29. Krisnan, D., Fergus, R.: Fast image deconvolution using hyper-Laplacian priors. In: Y. Bengio, D. Schuurmans, J. Lafferty, C. Williams, A. Culotta (eds.) Advances in Neural Information Processing Systems, vol. 22, pp. 1033–1041. Vancouver, BC (2009)
30. Lawson, C.L.: Contributions to the theory of linear least maximum approximation. Ph.D. thesis, University of California, Los Angeles (1961)
31. Leahy, R.M., Jeffs, B.D.: On the design of maximally sparse beamforming arrays. IEEE Transactions on Antennas and Propagation **39**(8), 1178–1187 (1991)
32. Li, Y., Osher, S.: Coordinate descent optimization for ℓ_1 minimization with application to compressed sensing; a greedy algorithm. Inverse Problems and Imaging **3**(3), 487–503 (2009)
33. Liu, J., Wright, S.J.: Asynchronous Stochastic Coordinate Descent: Parallelism and Convergence Properties. SIAM Journal on Optimization **25**, 351–376 (2015)
34. Liu, J., Wright, S.J., Ré, C., Bittorf, V., Sridhar, S.: An Asynchronous Parallel Stochastic Coordinate Descent Algorithm. Journal of Machine Learning Research **16**(1), 285–322 (2015)
35. Lu, Z., Xiao, L.: Randomized block coordinate non-monotone gradient method for a class of nonlinear programming. arXiv preprint arXiv:1306.5918 (2013)

36. Mareček, J., Richtárik, P., Takáč, M.: Distributed Block Coordinate Descent for Minimizing Partially Separable Functions. In: M. Al-Baali, L. Grandinetti, A. Purnama (eds.) Numerical Analysis and Optimization, no. 134 in Springer Proceedings in Mathematics and Statistics, pp. 261–288. Springer International Publishing (2015)
37. Marjanovic, G., Solo, V.: Sparsity Penalized Linear Regression With Cyclic Descent. IEEE Transactions on Signal Processing **62**(6), 1464–1475 (2014)
38. Mazumder, R., Friedman, J.H., Hastie, T.: SparseNet: Coordinate Descent With Nonconvex Penalties. Journal of the American Statistical Association **106**(495), 1125–1138 (2011)
39. Mohimani, H., Babaie-Zadeh, M., Jutten, C.: A fast approach for overcomplete sparse decomposition based on smoothed L0 norm. IEEE Transactions on Signal Processing **57**(1), 289–301 (2009)
40. Nutini, J., Schmidt, M., Laradji, I.H., Friedlander, M., Koepke, H.: Coordinate descent converges faster with the Gauss-Southwell rule than random selection. In Proceedings of the 32nd International Conference on Machine Learning, Lille, France, (2015)
41. Peng, Z., Wu, T., Xu, Y., Yan, M., Yin, W.: Coordinate friendly structures, algorithms and applications. Annals of Mathematical Sciences and Applications **1** (2016)
42. Peng, Z., Xu, Y., Yan, M., Yin, W.: ARock: an Algorithmic Framework for Asynchronous Parallel Coordinate Updates. SIAM Journal on Scientific Computing **38**(5), A2851-A2879, (2015)
43. Peng, Z., Yan, M., Yin, W.: Parallel and distributed sparse optimization. In: Signals, Systems and Computers, 2013 Asilomar Conference on, pp. 659–646. IEEE (2013)
44. Powell, M.J.D.: On search directions for minimization algorithms. Mathematical Programming **4**(1), 193–201 (1973)
45. Rao, B.D., Kreutz-Delgado, K.: An affine scaling methodology for best basis selection. IEEE Transactions on Signal Processing **47**(1), 187–200 (1999)
46. Razaviyayn, M., Hong, M., Luo, Z.: A Unified Convergence Analysis of Block Successive Minimization Methods for Nonsmooth Optimization. SIAM Journal on Optimization **23**(2), 1126–1153 (2013)
47. Richtárik, P., Takáč, M.: Parallel coordinate descent methods for big data optimization. Mathematical Programming **156**(1–2), 433–484 (2015)
48. Rockafellar, R.T.: Augmented Lagrangians and applications of the proximal point algorithm in convex programming. Mathematics of Operations Research **1**(2), 97–116 (1976)
49. Saab, R., Chartrand, R., Özgür Yilmaz: Stable sparse approximations via nonconvex optimization. In: IEEE International Conference on Acoustics, Speech, and Signal Processing (2008)
50. Scherrer, C., Halappanavar, M., Tewari, A., Haglin, D.: Scaling up coordinate descent algorithms for large ℓ_1 regularization problems. arXiv preprint arXiv:1206.6409 (2012)
51. Tseng, P.: On the rate of convergence of a partially asynchronous gradient projection algorithm. SIAM Journal on Optimization **1**(4), 603–619 (1991)
52. Tseng, P.: Convergence of a block coordinate descent method for nondifferentiable minimization. Journal of Optimization Theory and Applications **109**(3), 475–494 (2001)
53. Tseng, P., Yun, S.: A coordinate gradient descent method for nonsmooth separable minimization. Mathematical Programming **117**(1–2), 387–423 (2009)
54. Wang, Y., Yang, J., Yin, W., Zhang, Y.: A new alternating minimization algorithm for total variation image reconstruction. SIAM Journal on Imaging Science **1**(3), 248–272 (2008)
55. Wang, Y., Yin, W.: Sparse Signal Reconstruction via Iterative Support Detection. SIAM Journal on Imaging Sciences **3**(3), 462–491 (2010)
56. Wang, Y., Yin, W., Zeng, J.: Global Convergence of ADMM in Nonconvex Nonsmooth Optimization. arXiv:1511.06324 [cs, math] (2015)
57. Warga, J.: Minimizing certain convex functions. Journal of the Society for Industrial & Applied Mathematics **11**(3), 588–593 (1963)
58. Woodworth, J., Chartrand, R.: Compressed sensing recovery via nonconvex shrinkage penalties. http://arxiv.org/abs/1504.02923
59. Wu, T.T., Lange, K.: Coordinate descent algorithms for lasso penalized regression. The Annals of Applied Statistics **2**(1), 224–244 (2008)

60. Xu, Y., Yin, W.: A Block Coordinate Descent Method for Regularized Multiconvex Optimization with Applications to Nonnegative Tensor Factorization and Completion. SIAM Journal on Imaging Sciences **6**(3), 1758–1789 (2013)
61. Xu, Y., Yin, W.: A globally convergent algorithm for nonconvex optimization based on block coordinate update. arXiv:1410.1386 [math] (2014)
62. Xu, Z., Chang, X., Xu, F., Zhang, H.: $L_{1/2}$ regularization: A thresholding representation theory and a fast solver. IEEE Transactions on Neural Networks and Learning Systems **23**(7), 1013–1027 (2012)
63. Xu, Z.B., Guo, H.L., Wang, Y., Zhang, H.: Representative of $L_{1/2}$ regularization among L_q $(0 < q \leq 1)$ regularizations: an experimental study based on phase diagram. Acta Automatica Sinica **38**(7), 1225–1228 (2012)
64. Yin, P., Lou, Y., He, Q., Xin, J.: Minimization of ℓ_{1-2} for compressed sensing. SIAM Journal on Scientific Computing **37**(1), A536–A563 (2015)
65. Zangwill, W.I.: Nonlinear Programming: A Unified Approach. Prentice-Hall, Englewood Cliffs, NJ (1969)
66. Zeng, J., Peng, Z., Lin, S.: A Gauss-Seidel Iterative Thresholding Algorithm for l_q Regularized Least Squares Regression. arXiv:1507.03173 [cs] (2015)

Chapter 8
ADMM and Non-convex Variational Problems

Roland Glowinski

Abstract Our main goal in this chapter is to discuss the application of Alternating Direction Methods of Multipliers (ADMM) to the numerical solution of non-convex (and possibly non-smooth) variational problems. After giving a relatively detailed history of the ADMM methodology, we will discuss its application to the solution of problems from nonlinear Continuum Mechanics, nonlinear Elasticity, in particular. The ADMM solution of the two-dimensional Dirichlet problem for the Monge-Ampère equation will be discussed also. The results of numerical experiments will be reported, in order to illustrate the capabilities of the methodology under consideration

1 Introduction and Synopsis

To the best of our knowledge, *ADMM* was discovered *accidentally* in the mid-seventies when R. Glowinski and A. Marrocco were investigating the numerical solution of the following *nonlinear* (if $s \neq 2$) *Poisson problem*:

$$\begin{cases} -\nabla \cdot (|\nabla u|^{s-2} \nabla u) = f \text{ in } \Omega, \\ u = 0 \text{ on } \Gamma (= \partial \Omega), \end{cases} \tag{8.1}$$

with $s \in (1, +\infty)$ in (8.1). If the function f is a constant and Ω is a bounded domain of $\mathbb{R}^2$, system (8.1) models, in an appropriate system of units, the steady flow of a power-law incompressible viscous fluid in an infinitely long cylinder of cross-section Ω, f being the pressure drop per unit length and u the axial velocity.

R. Glowinski (✉)
Department of Mathematics, University of Houston, Houston, TX 77204, USA
e-mail: roland@math.uh.edu

R. Glowinski et al. (eds.), *Splitting Methods in Communication, Imaging, Science, and Engineering*, Scientific Computation, DOI 10.1007/978-3-319-41589-5_8

Actually, (8.1) is the *Euler-Lagrange equation* of the following problem from *Calculus of Variations*

$$\begin{cases} u \in W_0^{1,s}(\Omega), \\ J(u) \leq J(v),\ \forall v \in W_0^{1,s}(\Omega), \end{cases} \tag{8.2}$$

where (assuming that $\Omega \subset \mathbb{R}^d$) the *Sobolev space* $W_0^{1,s}(\Omega)$ is defined by

$$W_0^{1,s}(\Omega) = \{v | v \in L^s(\Omega), \frac{\partial v}{\partial x_i} \in L^s(\Omega), \forall i = 1,\ldots,d, v = 0 \text{ on } \Gamma\}, \tag{8.3}$$

and

$$J(v) = \frac{1}{s}\int_\Omega |\nabla v|^s\, d\mathbf{x} - < f, v >; \tag{8.4}$$

in (8.4), we have $d\mathbf{x} = dx_1 \ldots dx_d$, $|\mathbf{z}| = \sqrt{\sum_{i=1}^d z_i^2}$, $\forall \mathbf{z} = \{z_i\}_{i=1}^d \in \mathbb{R}^d$, and $< \cdot, \cdot >$ denotes the duality pairing between the dual space $W^{-1,\frac{s}{s-1}}(\Omega)$ of $W_0^{1,s}(\Omega)$ and $W_0^{1,s}(\Omega)$ which coincides with the canonical inner-product of $L^2(\Omega)$ if the first argument is smooth enough. The derivatives in (8.1), (8.3), and (8.4) are in the sense of *distributions* (see, e.g., [59] and [18] for this important notion).

Problem (8.2) is a well-posed (strictly) *convex* variational problem. In order to solve it, Glowinski and Marrocco advocated an *augmented Lagrangian* approach relying on the equivalence between (8.2) and

$$\begin{cases} \{u, \mathbf{p}\} \in \mathbf{W}, \\ j(u, \mathbf{p}) \leq j(v, \mathbf{q}),\ \forall \{v, \mathbf{q}\} \in \mathbf{W}, \end{cases} \tag{8.5}$$

where

$$\mathbf{W} = \{\{v, \mathbf{q}\} | \{v, \mathbf{q}\} \in W_0^{1,s}(\Omega) \times (L^s(\Omega))^d,\ \nabla v - \mathbf{q} = \mathbf{0}\}, \tag{8.6}$$

and

$$j(v, \mathbf{q}) = \frac{1}{s}\int_\Omega |\mathbf{q}|^s\, d\mathbf{x} - < f, v > . \tag{8.7}$$

Let us denote $\frac{s}{s-1}$ by s'; following Glowinski & Marrocco (see, e.g., [41] and [42]), we associate with the minimization problem (8.5) the following *augmented Lagrangian functional* (where $r > 0$):

$$\mathscr{L}_r(v, \mathbf{q}; \mu) = j(v, \mathbf{q}) + \frac{r}{2}\int_\Omega |\nabla v - \mathbf{q}|^2\, d\mathbf{x} + \int_\Omega (\nabla v - \mathbf{q}) \cdot \mu\, d\mathbf{x}, \tag{8.8}$$

where in (8.8): $\mathbf{a} \cdot \mathbf{b} = \sum_{i=1}^d a_i b_i$, $\mathbf{a} = \{a_i\}_{i=1}^d$, $\mathbf{b} = \{b_i\}_{i=1}^d \in \mathbb{R}^d$. Suppose that $\{\{u, \mathbf{p}\}, \lambda\}$ is a *saddle-point* of $\mathscr{L}_r$ over $W_0^{1,s}(\Omega) \times (L^s(\Omega))^d \times (L^{s'}(\Omega))^d$, that is

$$\begin{cases} \{\{u, \mathbf{p}\}, \lambda\} \in (W_0^{1,s}(\Omega) \times (L^s(\Omega))^d) \times (L^{s'}(\Omega))^d, \\ \mathscr{L}_r(u, \mathbf{p}; \mu) \leq \mathscr{L}_r(u, \mathbf{p}; \lambda) \leq \mathscr{L}_r(v, \mathbf{q}; \lambda), \\ \forall \{\{v, \mathbf{q}\}, \mu\} \in (W_0^{1,s}(\Omega), \times (L^s(\Omega))^d) \times (L^{s'}(\Omega))^d, \end{cases} \tag{8.9}$$

then, one can prove easily that u is the solution of problem (8.2) and that $\mathbf{p} = \nabla u$. Natural candidates to capture a saddle-point solution of problem (8.9) are provided by those *Uzawa's algorithms* whose convergence is discussed in, e.g., [40, 29, 30, 31, 39], and [32]. Applied to the solution of problem (8.9), such an Uzawa's algorithm reads as follows (for simplicity, we have denoted $W_0^{1,s}(\Omega)$ by V, $(L^s(\Omega))^d$ by $\mathbf{Q}$, and $(L^{s'}(\Omega))^d$ by Λ):

$$\boldsymbol{\lambda}^0 \text{ is given in } \Lambda (\boldsymbol{\lambda}^0 = \mathbf{0}, \text{ for example}). \tag{8.10}$$

For $n \geq 0$, $\boldsymbol{\lambda}^n \to \{u^n, \mathbf{p}^n\} \to \boldsymbol{\lambda}^{n+1}$ as follows:

$$\{u^n, \mathbf{p}^n\} = \arg\min_{\{v,\mathbf{q}\} \in V \times \mathbf{Q}} \mathscr{L}_r(v, \mathbf{q}; \boldsymbol{\lambda}^n), \tag{8.11}$$

$$\boldsymbol{\lambda}^{n+1} = \boldsymbol{\lambda}^n + \rho(\nabla u^n - \mathbf{p}^n), \tag{8.12}$$

where $\rho > 0$ ($\rho = r$ being a safe choice, in general). Actually, taking advantage of the *convexity* and *differentiability* properties of the functional $j(\cdot,\cdot)$, one can prove, as shown in, e.g., [42], that, $\forall \boldsymbol{\lambda}^0$, the sequence $\{\{u^n, \mathbf{p}^n\}\}_n$ converges to $\{u, \nabla u\}$, if $0 < \rho < 2r$. Of course, the main difficulty associated with algorithm (8.10)–(8.12) (called *ALG*1 in the above reference, and in subsequent ones) is the solution of problem (8.11). Following [13], an obvious choice to solve (8.11) was to use a (kind of) *block relaxation method*, namely

$$u^{n,0} = u^{n-1}. \tag{8.13}$$

For $n \geq 0$, $u^{n,k-1} \to \mathbf{p}^{n,k} \to u^{n,k}$ via the solution of

$$\mathbf{p}^{n,k} = \arg\min_{\mathbf{q} \in \mathbf{Q}} \mathscr{L}_r(u^{n,k-1}, \mathbf{q}; \boldsymbol{\lambda}^n), \tag{8.14}$$

and

$$u^{n,k} = \arg\min_{v \in V} \mathscr{L}_r(v, \mathbf{p}^{n,k}; \boldsymbol{\lambda}^n). \tag{8.15}$$

It follows from (8.8) that more explicit (and practical) formulations of problems (8.14) and (8.15) are given by

$$|\mathbf{p}^{n,k}|^{s-2}\mathbf{p}^{n,k} + r\mathbf{p}^{n,k} = r\nabla u^{n,k-1} + \boldsymbol{\lambda}^n, \tag{8.16}$$

and

$$\begin{cases} -r\nabla^2 u^{n,k} = f - \nabla \cdot (r\mathbf{p}^{n,k} - \boldsymbol{\lambda}^n) \text{ in } \Omega, \\ u^{n,k} = 0 \text{ on } \Gamma, \end{cases} \tag{8.17}$$

respectively. Problem (8.16) can be solved point-wise, while problem (8.17) is nothing but a 'nice' linear Poisson problem.

When applying, algorithm (8.10)–(8.12), (8.13)–(8.15), with $\rho = r$, to the solution of problem (8.1), for various values of s, Glowinski and Marrocco observed that the number of relaxation iterations was quickly converging to one, even for a rather demanding tolerance in the stopping criterion of algorithm (8.13)–(8.15).

This observation suggested limiting to one, from the start, the number of relaxation iterations in (8.13)–(8.15), leading thus to the following variant (called *ALG*2, at the time) of algorithm (8.10)–(8.12):

$$\{u^{-1}, \lambda^0\} \text{ is given in } V \times \Lambda. \tag{8.18}$$

For $n \geq 0$, $\{u^{n-1}, \lambda^n\} \to \mathbf{p}^n \to u^n \to \lambda^{n+1}$ via

$$\mathbf{p}^n = \arg\min_{\mathbf{q}\in\mathbf{Q}} \mathscr{L}_r(u^{n-1}, \mathbf{q}; \lambda^n), \tag{8.19}$$

and

$$u^n = \arg\min_{v\in V} \mathscr{L}_r(v, \mathbf{p}^n; \lambda^n), \tag{8.20}$$

and

$$\lambda^{n+1} = \lambda^n + \rho(\nabla u^n - \mathbf{p}^n). \tag{8.21}$$

A nice property of algorithm (8.18)–(8.21) is clearly the *decoupling* that ALG2 realizes between the nonlinearity and the differential operators.

As far as we know, what we just described is the way *ADMM* was discovered, indeed largely by accident. Actually, the authors of [41, 42] realized immediately the applicability of algorithms such as *ALG*1 and *ALG*2 to variational problems such as

$$\begin{cases} u \in V, \\ J(u) \leq J(v), \ \forall v \in V, \end{cases} \tag{8.22}$$

where in (8.22):

(i) V is a real Hilbert space for the inner product $(\cdot,\cdot)$ and the associated norm $\|\cdot\|$ (defined by $\|v\| = \sqrt{(v, v)}$.

(ii) $J(v) = F(Bv) + G(v)$, with $B \in \mathscr{L}(V, H)$, H being also a Hilbert space, $F : H \to I\!R \cup \{+\infty\}$ (resp., $G : V \to I\!R \cup \{+\infty\}$) being convex, proper and lower semi-continuous (l.s.c.) with

$$\mathrm{dom}(F \circ B) \cap \mathrm{dom}(G) \neq \emptyset.$$

(iii) $\lim_{\|v\| \to +\infty} J(v) = +\infty$

If the above assumptions hold, the minimization problem (8.22) has a solution; if J is strictly convex this solution is unique (as shown in, e.g., [50] and [27]). The two following examples were provided in [42]:

Example 1. Let us consider the following constrained minimization problem

$$\begin{cases} u \in K, \\ J(u) \leq J(v), \ \forall v \in K, \end{cases} \tag{8.23}$$

where, in (8.23),

$$K = \{v | v \in H_0^1(\Omega), |\nabla v(x)| \leq 1 \ a.e. \text{ in } \Omega\}, \tag{8.24}$$

and

$$J(v) = \frac{\mu}{2}\int_{\Omega} |\nabla v|^2 \, d\mathbf{x} - C\int_{\Omega} v d\mathbf{x}, \tag{8.25}$$

with $H_0^1(\Omega) = W_0^{1,2}(\Omega)$ and

- μ is a positive constant.
- Ω is a bounded domain of $\mathbb{R}^d$ (with $d \geq 1$).
- C is a constant.

The set K is clearly a closed convex nonempty subset of $H_0^1(\Omega)$. Problem (8.23) is a particular case of (8.22), corresponding to

$$V = H_0^1(\Omega), \text{ and } H = (L^2(\Omega))^d, \tag{8.26}$$

$$G(v) = \frac{\mu}{2}\int_{\Omega} |\nabla v|^2 \, d\mathbf{x} - C\int_{\Omega} v d\mathbf{x}, \tag{8.27}$$

and

$$F(\mathbf{q}) = I_{\mathscr{K}}(\mathbf{q}) \tag{8.28}$$

where, in (8.28), the set $\mathscr{K}$ is defined by

$$\mathscr{K} = \{\mathbf{q} | \mathbf{q} \in (L^2(\Omega))^d, |\mathbf{q}(x)| \leq 1 \; a.e. \text{ in } \Omega\}, \tag{8.29}$$

and $I_{\mathscr{K}}$ is the *indicator functional* of $\mathscr{K}$, that is, $I_{\mathscr{K}}$ is defined by

$$I_{\mathscr{K}}(\mathbf{q}) = \begin{cases} 0, \text{ if } \mathbf{q} \in \mathscr{K}, \\ +\infty, \text{ if } \mathbf{q} \in (L^2(\Omega))^d \setminus \mathscr{K}, \end{cases} \tag{8.30}$$

implying that the functional $I_{\mathscr{K}}$ is proper, convex and lower semi-continuous. Since Ω is bounded, the semi-norm $v \to \sqrt{\int_{\Omega} |\nabla v|^2 \, d\mathbf{x}}$ defines a norm over $H_0^1(\Omega)$ which is equivalent to the canonical $H^1(\Omega)$-norm (see, e.g., the Appendix 1 of [31] for a proof); it follows from this norm equivalence that problem (8.23) is well posed and can be solved by the variant of algorithm (8.18)–(8.21) associated with the augmented Lagrangian functional $\mathscr{L}_r$ defined by

$$\mathscr{L}_r(v, \mathbf{q}; \mu) = \frac{\mu}{2}\int_{\Omega} |\nabla v|^2 \, d\mathbf{x} - C\int_{\Omega} v \, d\mathbf{x} + I_{\mathscr{K}}(\mathbf{q}) + \frac{r}{2}\int_{\Omega} |\nabla v - \mathbf{q}|^2 \, d\mathbf{x} + \int_{\Omega} \mu \cdot (\nabla v - \mathbf{q}) \, d\mathbf{x}.$$

Other decompositions are possible; we can define, for example, the functional F (resp., G) by

$$F(\mathbf{q}) = \frac{\mu}{2}\int_{\Omega} |\mathbf{q}|^2 \, d\mathbf{x} + I_{\mathscr{K}}(\mathbf{q}) \;\; (\text{resp., } G(v) = -C\int_{\Omega} v \, d\mathbf{x}).$$

The algorithms of type *ALG*1 and *ALG*2 associated with the above two decompositions of problem (8.23) behave similarly.

Remark 1. If Ω is a simply connected bounded domain of $\mathbb{R}^2$, (8.23) models (in an appropriate system of units) the torsion of an infinitely long cylinder of cross-section Ω, made of an *elastic-plastic* material, μ being characteristic of the elasticity properties of the material; in this model, C is the torsion angle per unit length, and u a stress potential.

Example 2. It is the variant of problem (8.23) defined by

$$\begin{cases} u \in V, \\ J(u) \leq J(v), \ \forall v \in V, \end{cases} \tag{8.31}$$

where, in (8.31),

$$V = H_0^1(\Omega), \tag{8.32}$$

and

$$J(v) = \frac{\mu}{2}\int_\Omega |\nabla v|^2 \, d\mathbf{x} + \tau_y \int_\Omega |\nabla v| \, d\mathbf{x} - C\int_\Omega v \, d\mathbf{x}, \tag{8.33}$$

with

- Ω is a bounded domain of $\mathbb{R}^2$.
- μ and τ_y are positive constants and $C \in \mathbb{R}$.

It follows from, e.g., [23, 26] and [44], that (8.31) models the flow of a *Bingham* incompressible *visco-plastic* fluid in an infinitely long cylindrical duct (pipe) of cross section Ω, μ, τ_y and C being the *fluid viscosity*, the *plasticity yield* and the *pressure drop per unit length*, respectively. The unique solution u of problem (8.31) is the *flow axial velocity*.

Problem (8.31) is a particular case of (8.22), corresponding to

$$V = H_0^1(\Omega), \text{ and } H = (L^2(\Omega))^2, \tag{8.34}$$

$$G(v) = \frac{\mu}{2}\int_\Omega |\nabla v|^2 \, d\mathbf{x} - C\int_\Omega v d\mathbf{x}, \tag{8.35}$$

and

$$F(\mathbf{q}) = \tau_y \int_\Omega |\mathbf{q}| \, d\mathbf{x}, \tag{8.36}$$

as with Example 1, other decompositions are possible.

Problem (8.31) is well posed and can be solved by the variant of algorithm (8.18)–(8.21) associated with the augmented Lagrangian functional $\mathcal{L}_r$ defined by

$$\begin{aligned} \mathcal{L}_r(v, \mathbf{q}; \mu) = &\frac{\mu}{2}\int_\Omega |\nabla v|^2 \, d\mathbf{x} - C\int_\Omega v \, d\mathbf{x} + \tau_y \int_\Omega |\mathbf{q}| \, d\mathbf{x} + \\ &\frac{r}{2}\int_\Omega |\nabla v - \mathbf{q}|^2 \, d\mathbf{x} + \int_\Omega \mu \cdot (\nabla v - \mathbf{q}) \, d\mathbf{x}. \end{aligned}$$

This algorithm, and other methods for the solution of problem (8.31), are discussed in [23] and [44] (see also [46]).

Back to problem (8.22), we associate with it the augmented Lagrangian functional $\mathscr{L}_r$ defined by

$$\mathscr{L}_r(v,q;\mu) = F(q) + G(v) + \frac{r}{2}|Bv - q|^2 + [\mu, Bv - q], \tag{8.37}$$

where $[\cdot,\cdot]$ (resp.,$|\cdot|$) denotes the inner-product over H (resp., its associated norm). Assuming that $\mathscr{L}_r$ has a saddle-point over $(V \times H) \times H$, it follows from the discussion concerning the solution of problem (8.2) that algorithms (8.10)–(8.12) and (8.18)–(8.21) can be easily generalized in order to solve problem (8.22); we obtain then:

$$\lambda^0 \text{ is given in } H \ (\lambda^0 = 0, \text{ for example}). \tag{8.38}$$

For $n \geq 0$, $\lambda^n \to \{u^n, p^n\} \to \lambda^{n+1}$ as follows:

$$\{u^n, p^n\} = \arg\min_{\{v,q\} \in V \times H} \mathscr{L}_r(v,q;\lambda^n), \tag{8.39}$$

$$\lambda^{n+1} = \lambda^n + \rho(Bu^n - p^n), \tag{8.40}$$

and

$$\{u^{-1}, \lambda^0\} \text{ is given in } V \times H. \tag{8.41}$$

For $n \geq 0$, $\{u^{n-1}, \lambda^n\} \to p^n \to u^n \to \lambda^{n+1}$ via

$$p^n = \arg\min_{q \in H} \mathscr{L}_r(u^{n-1}, q; \lambda^n), \tag{8.42}$$

and

$$u^n = \arg\min_{v \in V} \mathscr{L}_r(v, p^n; \lambda^n), \tag{8.43}$$

and

$$\lambda^{n+1} = \lambda^n + \rho(Bu^n - p^n). \tag{8.44}$$

For simplicity we still call (8.38)–(8.40) (resp., (8.41)–(8.44)) *ALG*1 (resp., *ALG*2). More explicit formulations of the two above algorithms read, respectively, as

*Explicit formulation of ALG*1:

$$\lambda^0 \text{ is given in } H \ (\lambda^0 = 0, \text{ for example}). \tag{8.45}$$

For $n \geq 0$, $\lambda^n \to \{u^n, p^n\} \to \lambda^{n+1}$ as follows:

$$\begin{cases} \{u^n, p^n\} \in V \times H, \\ F(q) - F(p^n) + G(u^n) - G(v) + r[Bu^n - p^n, B(v - u^n) - (q - p^n)] + \\ [\lambda^n, B(v - u^n) - (q - p^n)] \geq 0, \ \forall \{v,q\} \in V \times H, \end{cases} \tag{8.46}$$

then

$$\lambda^{n+1} = \lambda^n + \rho(Bu^n - p^n), \tag{8.47}$$

and

Explicit formulation of ALG2:

$$\{u^{-1}, \lambda^0\} \text{ is given in } V \times H. \tag{8.48}$$

For $n \geq 0$, $\lambda^n \to p^n \to u^n \to \lambda^{n+1}$ via

$$\begin{cases} p^n \in H, \\ F(q) - F(p^n) + r[p^n, q - p^n] \geq [\lambda^n + rBu^{n-1}, q - p^n], \ \forall q \in H, \end{cases} \tag{8.49}$$

and

$$\begin{cases} u^n \in V, \\ G(v) - G(u^n) + r[Bu^n, B(v - u^n)] \geq [r\mathbf{p}^n - \lambda^n, B(v - u^n)], \ \forall v \in V, \end{cases} \tag{8.50}$$

then

$$\lambda^{n+1} = \lambda^n + \rho(Bu^n - p^n). \tag{8.51}$$

The variational problems (8.46), (8.49), and (8.50) are particular cases of those elliptic variational inequality problems discussed in, e.g., [40, 31] and [34].

Concerning the convergence of algorithm (8.45)–(8.47) (resp., (8.48)–(8.51)) it has been proved in, e.g., [29, 30, 31] and [39] (see also [47]) that under mild assumptions on operator B, and on the convex functionals F and G, then the following convergence result holds:

$$\lim_{n \to +\infty} \{u^n, p^n\} = \{u, Bu\} \text{ in } V \times H, \ \forall \lambda^0 \in H, \text{ if } \ 0 < \rho < 2r \tag{8.52}$$

(resp.,

$$\begin{cases} \lim_{n \to +\infty} \{u^n, p^n\} = \{u, Bu\} \text{ in } V \times H, \ \forall \{u^{-1}, \lambda^0\} \in V \times H, \text{ if} \\ 0 < \rho < \dfrac{1 + \sqrt{5}}{2} r). \end{cases} \tag{8.53}$$

Proving the convergence result (8.52) (essentially by an energy method) is relatively easy as shown in, e.g., [29, 30, 31] and [39]. Proving (8.53) (still by an energy method) is a bit more complicated, but the most complicated part of the convergence analysis is to prove that if the assumptions on B, F, G, and ρ, implying (8.52) and (8.53), hold then

$$\lim_{n \to +\infty} \lambda^n = \lambda \text{ weakly in } H, \tag{8.54}$$

$\{\{u, Bu\}, \lambda\}$ being a saddle-point of $\mathscr{L}_r$ over $(V \times H) \times H$. Indeed, to prove (8.54), one can use the results on the convergence of multiplier sequences available in, e.g., the Appendix 2 of [40] and the Chapter 4 of [32].

Remark 2. The convergence results (8.52), (8.53), and (8.54) apply to the solution, via *ALG*1 and *ALG*2, of the variational problems considered in Examples 1 and 2. Focusing on Example 2, applying *ALG*2 to the solution of problem (8.31) we obtain the following algorithm:

$$\{u^{-1}, \lambda^0\} \text{ is given in } H_0^1(\Omega) \times \Lambda, \tag{8.55}$$

with $\Lambda = (L^2(\Omega))^2$.

For $n \geq 0$, $\{u^{n-1}, \lambda^n\} \to \mathbf{p}^n \to u^n \to \lambda^{n+1}$ via

$$\mathbf{p}^n(x) = \frac{\mathbf{X}^n(x)}{r}\left[1 - \frac{\tau_y}{|\mathbf{X}^n(x)|}\right]^+, \; a.e. \text{ in } \Omega, \tag{8.56}$$

with $\mathbf{X}^n = \lambda^n + r\nabla u^{n-1}$ and $\xi^+ = \max(0, \xi)$, $\forall \xi \in I\!R$, followed by

$$\begin{cases} u^n \in H_0^1(\Omega), \\ (r+\mu)\int_\Omega \nabla u^n \cdot \nabla v \, d\mathbf{x} = C\int_\Omega v \, d\mathbf{x} + \int_\Omega (r\mathbf{p}^n - \lambda^n) \cdot \nabla v \, d\mathbf{x}, \\ \forall v \in H_0^1(\Omega), \end{cases} \tag{8.57}$$

$$\lambda^{n+1} = \lambda^n + \rho(\nabla u^n - \mathbf{p}^n). \tag{8.58}$$

Problem (8.57) is a *linear Poisson-Dirichlet* problem written in *variational form*. On the other hand, it follows from (8.56) that $\mathbf{p}^n$ is obtained from $\mathbf{X}^n$ by application of what those scientists of the *Image Restoration community* call a *shrinking operator*.

The augmented Lagrangian solution of problems from visco-plasticity, more complicated than (8.31), is discussed in [44] (see also the references therein).

Remark 3. A natural variant of *ALG*2 reads as follows:

$$\{u^{-1}, \lambda^0\} \text{ is given in } V \times H. \tag{8.59}$$

For $n \geq 0$, $\{u^{n-1}, \lambda^n\} \to p^n \to \lambda^{n+1/2} \to u^n \to \lambda^{n+1}$ via

$$p^n = \text{arg min}_{q \in H} \mathscr{L}_r(u^{n-1}, q; \lambda^n), \tag{8.60}$$

$$\lambda^{n+1/2} = \lambda^n + \rho(Bu^{n-1} - p^n), \tag{8.61}$$

$$u^n = \text{arg min}_{v \in V} \mathscr{L}_r(v, p^n; \lambda^{n+1/2}), \tag{8.62}$$

and

$$\lambda^{n+1} = \lambda^{n+1/2} + \rho(Bu^n - p^n). \tag{8.63}$$

Algorithm (8.59)–(8.63) was called *ALG*3 in [29, 30] and [39]. A more explicit formulation of *ALG*3 is given by:

*Explicit formulation of ALG*3:

$$\{u^{-1}, \lambda^0\} \text{ is given in } V \times H. \tag{8.64}$$

For $n \geq 0$, $\lambda^n \to p^n \to \lambda^{n+1/2} \to u^n \to \lambda^{n+1}$ via

$$\begin{cases} p^n \in H, \\ F(q) - F(p^n) + r[p^n, q-p] \geq [\lambda^n + rBu^{n-1}, q - p^n], \ \forall q \in H, \end{cases} \tag{8.65}$$

$$\lambda^{n+1/2} = \lambda^n + \rho(Bu^{n-1} - p^n), \tag{8.66}$$

$$\begin{cases} u^n \in V, \\ G(v) - G(u^n) + r[Bu^n, B(v-u^n)] \geq [rp^n - \lambda^{n+1/2}, B(v-u^n)], \ \forall v \in V, \end{cases} \tag{8.67}$$

then

$$\lambda^{n+1} = \lambda^{n+1/2} + \rho(Bu^n - p^n). \tag{8.68}$$

Remark 4. Let us consider the particular case of problem (8.22) where $H = V, B = I$, and where F and G are both differentiable with $A_1 = DF$ and $A_2 = DG$. Assuming that $\rho = r$, it follows from (8.48) to (8.51) that $ALG2$ reduces to:

$$\{u^{-1}, \lambda^0\} \text{ is given in } V \times H. \tag{8.69}$$

For $n \geq 0$, $\{u^{n-1}, \lambda^n\} \to p^n \to u^n \to \lambda^{n+1}$ via

$$rp^n + A_1(p^n) = ru^{n-1} + \lambda^n, \tag{8.70}$$

$$ru^n + A_2(u^n) = rp^n - \lambda^n, \tag{8.71}$$

$$\lambda^{n+1} = \lambda^n + \rho(u^n - p^n). \tag{8.72}$$

By elimination of λ^n and λ^{n+1}, it follows from (8.70) to (8.72) that

$$r(p^{n+1} - u^n) + A_1(p^{n+1}) + A_2(u^n) = 0, \tag{8.73}$$

$$r(u^{n+1} - u^n) + A_1(p^{n+1}) + A_2(u^{n+1}) = 0. \tag{8.74}$$

Denote p^{n+1} by $u^{n+1/2}$, it follows then from (8.73) and (8.74) that

$$r(u^{n+1/2} - u^n) + A_1(u^{n+1/2}) + A_2(u^n) = 0, \tag{8.75}$$

$$r(u^{n+1} - u^n) + A_1(u^{n+1/2}) + A_2(u^{n+1}) = 0. \tag{8.76}$$

It follows from (8.75) and (8.76) that in the particular case considered here, $ALG2$ coincides with the celebrated *Douglas-Rachford alternating direction method* (introduced in [25]).

Similarly, it follows from (8.64) to (8.68) that, if $\rho = r$, $ALG3$ reduces to

$$\{u^{-1}, \lambda^0\} \text{ is given in } V \times H. \tag{8.77}$$

For $n \geq 0$, $\{u^{n-1}, \lambda^n\} \to p^n \to \lambda^{n+1/2} \to u^n \to \lambda^{n+1}$ via

$$rp^n + A_1(p^n) = ru^{n-1} + \lambda^n, \tag{8.78}$$
$$\lambda^{n+1/2} = \lambda^n + r(u^{n-1} - p^n). \tag{8.79}$$
$$ru^n + A_2(u^n) = rp^n - \lambda^{n+1/2}, \tag{8.80}$$
$$\lambda^{n+1} = \lambda^{n+1/2} + r(u^n - p^n). \tag{8.81}$$

By elimination of λ^n, $\lambda^{n+1/2}$ and λ^{n+1}, it follows from (8.78)–(8.81) that

$$r(p^{n+1} - u^n) + A_1(p^{n+1}) + A_2(u^n) = 0, \tag{8.82}$$
$$r(u^{n+1} - p^{n+1}) + A_1(p^{n+1}) + A_2(u^{n+1}) = 0. \tag{8.83}$$

As above, denote p^{n+1} by $u^{n+1/2}$, it follows then from (8.82) and (8.83) that

$$r(u^{n+1/2} - u^n) + A_1(u^{n+1/2}) + A_2(u^n) = 0, \tag{8.84}$$
$$r(u^{n+1} - u^{n+1/2}) + A_1(u^{n+1/2}) + A_2(u^{n+1}) = 0, \tag{8.85}$$

showing that in the particular case considered here, *ALG*3 coincides with another celebrated algorithm, namely the *Peaceman-Rachford alternating direction method* (introduced in [56]).

The above observations justify the terminology *Alternating Direction Methods of Multipliers* used to denote nowadays algorithms such as *ALG*2 and *ALG*3. To the best of our knowledge this equivalence between some alternating direction methods and augmented Lagrangian algorithms was discovered, in 1975, by Chan and Glowinski when solving numerically *mildly nonlinear elliptic equations* such as

$$\begin{cases} -\nabla^2 u + \phi(u) = f \ \text{ in } \Omega, \\ u = g \ \text{ on } \Gamma, \end{cases} \tag{8.86}$$

where, in (8.86), ϕ is a non-decreasing continuous function from $I\!R$ into $I\!R$. The circumstances of this discovery are reported with more details in [14, 33] and [34] (see also [31]).

All the problems considered so far have been *convex variational problems*. Actually, albeit ADMM can be applied to the solution of *non-convex* problems, as one will see shortly, one is still lacking a general convergence theory. Indeed, there are only four pages dedicated to non-convex problems in [5], a very popular review article on ADMM. To the best of our knowledge, the first significant *non-convex* application of ADMM took place in the late seventies/early eighties when COFLEXIP, a French company specialized in the manufacturing of flexible pipelines used in offshore Oil & Gas operations, asked our assistance for the numerical simulation of the static and dynamic behaviors of its products. Considering these pipelines as *inextensible elastic beams* (a reasonable assumption in practice), a very simple (but typical) related problem reads as follows:

$$\begin{cases} \mathbf{x} \in \mathbf{E}, \\ J(\mathbf{x}) \leq J(\mathbf{y}), \ \forall \mathbf{y} \in \mathbf{E}, \end{cases} \tag{8.87}$$

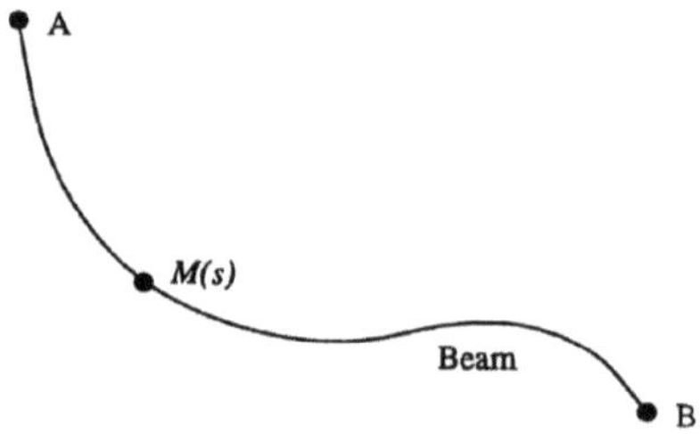

Fig. 8.1 Beam visualization and notation

where:

$$J(\mathbf{y}) = \frac{EI}{2}\int_0^L |\mathbf{y}''|^2\, ds - \rho_b \int_0^L \mathbf{g}\cdot\mathbf{y}\, ds \tag{8.88}$$

and

$$\begin{aligned}\mathbf{E} = \{\ & \mathbf{y}|\mathbf{y} \in (H^2(0,L))^3, |\mathbf{y}'(s)| = 1, \forall s \in [0,L], \mathbf{y}(0) = \mathbf{A}, \mathbf{y}(L) = \mathbf{B},\\ & \mathbf{y}'(0) = \alpha, \mathbf{y}'(L) = \beta\}.\end{aligned} \tag{8.89}$$

In (8.87)–(8.89):

- $\mathbf{x}$ (resp., $\mathbf{y}$) denotes the equilibrium (resp., an admissible) displacement of the beam.
- L (resp., EI and ρ_b) denotes the length (resp., the flexural stiffness and the linear density) of the beam.
- $\mathbf{g} = \{0,0,-g\}$ denotes the gravity field.
- $\mathbf{a}\cdot\mathbf{b} = \sum_{i=1}^3 a_i b_i,\ \mathbf{a} = \{a_i\}_{i=1}^3, \mathbf{b} = \{b_i\}_{i=1}^3 \in I\!R^3$.
- s is a curvilinear abscissa (here, the arc-length originating from $\mathbf{A}$, as shown in Figure 8.1).
- $\mathbf{y}' = \dfrac{d\mathbf{y}}{ds}, \mathbf{y}'' = \dfrac{d^2\mathbf{y}}{ds^2}$.
- The relation $|\mathbf{y}'(s)| \left(= \sqrt{\sum_{i=1}^3 |y_i'(s)|^2}\right) = 1$, $\forall s \in [0,L]$, models the inextensibility condition.
- $\alpha, \beta \in I\!R^3$, $|\alpha| = |\beta| = 1$.
- $H^2(0,L) = \{\phi|\phi \in C^1[0,L], \phi'' \in L^2(0,L)\}$.

Above, we have assumed that the beam is *sufficiently flexible* so that the *strain-stress* relation is *linear*, and the *torsional effects* have been *neglected*. Problem (8.87) is related to the celebrated *Euler's Elastica* problem.

If $|\overrightarrow{\mathbf{AB}}| < L$, the set $\mathbf{E}$ is non-empty and weakly closed in $(H^2(0,L))^3$, implying the existence of solutions to problem (8.87).

From a *computational point of view*, the main difficulty is the *inextensibility condition*

$$|\mathbf{y}'(s)| = 1, \forall s \in [0,L] \Leftrightarrow |\mathbf{y}'(s)|^2 \left(= \sum_{i=1}^3 |y_i'(s)|^2\right) = 1, \forall s \in [0,L], \tag{8.90}$$

a *point-wise nonlinear equality constraint*. In order to handle the constraint (8.90) several approaches are available: (i) One can treat (8.90) by *penalty* (as done for example in [36]), or (ii) Introduce a *Lagrange multiplier* associated with (8.90) and solve the resulting nonlinear Kuhn-Tucker system by a Newton's or quasi-Newton's method. Actually, our method of choice was derived from the equivalence between (8.87) and the following minimization problem:

$$\begin{cases} \{\mathbf{x},\mathbf{p}\} \in \mathbf{W}, \\ J(\mathbf{x},\mathbf{p}) \leq J(\mathbf{y},\mathbf{q}),\ \forall \{\mathbf{y},\mathbf{q}\} \in \mathbf{W}, \end{cases} \tag{8.91}$$

with

$$J(\mathbf{y},\mathbf{q}) = \frac{EI}{2}\int_0^L |\mathbf{y}''|^2\,ds - \rho_b \int_0^L \mathbf{g}\cdot\mathbf{y}\,ds \tag{8.92}$$

and

$$\begin{aligned} \mathbf{W} = \{\{\mathbf{y},\mathbf{q}\} \mid \mathbf{y} \in (H^2(0,L))^3, \mathbf{y}(0) = \mathbf{A}, \mathbf{y}(L) = \mathbf{B}, \mathbf{y}'(0) = \alpha, \mathbf{y}'(L) = \beta, \\ \mathbf{q} \in (L^2(0,L))^3, |\mathbf{q}(s)| = 1 \text{ a.e. on } (0,L), \mathbf{y}' - \mathbf{q} = \mathbf{0}\}. \end{aligned} \tag{8.93}$$

We associate with (8.91)–(8.93) the sets

$$\mathbf{Y}_{ad} = \{\mathbf{y} | \mathbf{y} \in (H^2(0,L))^3, \mathbf{y}(0) = \mathbf{A}, \mathbf{y}(L) = \mathbf{B}, \mathbf{y}'(0) = \alpha, \mathbf{y}'(L) = \beta\}, \tag{8.94}$$

$$\mathbf{S} = \{\mathbf{q} | \mathbf{q} \in (L^2(0,L))^3, |\mathbf{q}(s)| = 1 \text{ a.e. on } (0,L)\}, \tag{8.95}$$

and the *augmented Lagrangian functional*

$$\mathscr{L}_r : (H^2(0,L) \times L^2(0,L))^3 \times (L^2(0,L))^3 \to I\!R$$

defined (with $r > 0$) by

$$\begin{aligned} \mathscr{L}_r(\mathbf{y},\mathbf{q};\mu) = & \frac{EI}{2}\int_0^L |\mathbf{y}''|^2\,ds - \rho_b \int_0^L \mathbf{g}\cdot\mathbf{y}\,ds + \\ & \frac{r}{2}\int_0^L |\mathbf{y}' - \mathbf{q}|^2\,ds + \int_0^L \mu\cdot(\mathbf{y}' - \mathbf{q})\,ds. \end{aligned} \tag{8.96}$$

One can easily show that if $\{\mathbf{x},\mathbf{p};\lambda\}$ is a saddle-point of $\mathscr{L}_r$ over $(\mathbf{Y}_{ad} \times \mathbf{S}) \times (L^2(0,L))^3$, that is

$$\begin{cases} \{\{\mathbf{x},\mathbf{p}\},\lambda\} \in (\mathbf{Y}_{ad} \times \mathbf{S}) \times (L^2(0,L))^3, \\ \mathscr{L}_r(\mathbf{x},\mathbf{p};\mu) \leq \mathscr{L}_r(\mathbf{x},\mathbf{p};\lambda) \leq \mathscr{L}_r(\mathbf{y},\mathbf{q};\lambda), \\ \forall \{\{\mathbf{y},\mathbf{q}\},\mu\} \in (\mathbf{Y}_{ad} \times \mathbf{S}) \times (L^2(0,L))^3, \end{cases} \tag{8.97}$$

then $\mathbf{x}$ is a solution of problem (8.87) and $\mathbf{p} = \mathbf{x}'$. This result, which is very easy to prove, suggests using *ALG*1, *ALG*2, or *ALG*3 to solve the non-convex variational problem (8.87). Focusing on *ALG*2, we obtain the following algorithm:

$$\{\mathbf{x}^{-1},\lambda^0\} \text{ is given in } \mathbf{Y}_{ad} \times (L^2(0,L))^3. \tag{8.98}$$

For $n \geq 0$, $\{\mathbf{x}^{n-1}, \lambda^n\} \to \mathbf{p}^n \to \mathbf{x}^n \to \lambda^{n+1}$ via

$$\mathbf{p}^n = \arg\min_{\mathbf{q}\in\mathbf{S}} \mathscr{L}_r(\mathbf{x}^{n-1}, \mathbf{q}; \lambda^n), \tag{8.99}$$

and

$$\mathbf{x}^n = \arg\min_{\mathbf{y}\in\mathbf{Y}_{ad}} \mathscr{L}_r(\mathbf{y}, \mathbf{p}^n; \lambda^n), \tag{8.100}$$

and

$$\lambda^{n+1} = \lambda^n + \rho(\frac{d}{ds}\mathbf{x}^n - \mathbf{p}^n). \tag{8.101}$$

It follows from (8.96) that (8.99) reduces to

$$\mathbf{p}^n = \arg\min_{\mathbf{q}\in\mathbf{S}} \left[\frac{r}{2}\int_0^L |\mathbf{q}|^2\,ds - \int_0^L (r\frac{d}{ds}\mathbf{x}^{n-1} + \lambda^n)\cdot\mathbf{q}\,ds \right],$$

that is, since $\int_0^L |\mathbf{q}|^2\,ds = L$,

$$\mathbf{p}^n = \arg\max_{\mathbf{q}\in\mathbf{S}} \int_0^L (r\frac{d}{ds}\mathbf{x}^{n-1} + \lambda^n)\cdot\mathbf{q}\,ds. \tag{8.102}$$

Let us denote by $\mathbf{X}^n$ the vector-valued function $r\frac{d}{ds}\mathbf{x}^{n-1} + \lambda^n$; it follows then from (8.102) that

$$\mathbf{p}^n(s) = \frac{\mathbf{X}^n(s)}{|\mathbf{X}^n(s)|} \text{ if } \mathbf{X}^n(s) \neq \mathbf{0},\ a.e. \text{ on } (0,L). \tag{8.103}$$

It is worth mentioning that our many numerical experiments, with the discrete analogues of algorithm (8.98)–(8.101), never encountered the situation $\mathbf{X}^n(s) = \mathbf{0}$ when taking $\rho = r$ in (8.101) and r sufficiently large. The *normalization operator* associated with (8.103) plays, for problem (8.87), the role played by the *shrinking operator* in (8.56) for problem (8.31).

Concerning problem (8.100), one can easily show that $\mathbf{x}^n$ is also the unique solution of the following well-posed linear variational problem:

$$\begin{cases} \mathbf{x} \in \mathbf{Y}_{ad}, \\ EI\int_0^L \frac{d^2\mathbf{x}^n}{ds^2}\cdot\frac{d^2\mathbf{y}}{ds^2}\,ds + r\int_0^L \frac{d\mathbf{x}^n}{ds}\cdot\frac{d\mathbf{y}}{ds}\,ds = \\ \int_0^L \frac{d\mathbf{x}^n}{ds}(r\mathbf{p}^n - \lambda^n)\cdot\frac{d\mathbf{y}}{ds}\,ds + \rho_b\int_0^L \mathbf{g}\cdot\mathbf{y}\,ds, \forall\mathbf{y} \in (H_0^2(0,L))^3. \end{cases} \tag{8.104}$$

Here $H_0^2(0,L) = \{\phi | \phi \in H^2(0,L), \phi(0) = \phi(L) = 0, \phi'(0) = \phi'(L) = 0\}$. Those scientists with some knowledge in *elasticity* will recognize immediately that the linear variational problem (8.104) is a one-dimensional 4th order elliptic problem of the *Euler-Bernoulli* type (see, e.g., [48] for the modeling and control of elastic beams and plates). For the numerical implementation of algorithm (8.98)–(8.101), and of its variants associated with other boundary conditions and external forces, we systematically employed Hermite cubic based finite element approximations (see [4, 29, 30, 39] and [36] for details).

The numerical experiments reported in the five above references show that the ADMM algorithm (8.98)–(8.101) is easy to implement, due its modularity, and has good convergence properties if one takes $\rho = r$ and r sufficiently large.

The generalization to time dependent variants of problem (8.87) is discussed in the five above references, and in [33] and [34].

It is worth noticing that productions codes, used in Oil & Gas Industry, were developed in the eighties (and still used in the nineties), relying on algorithm (8.98)–(8.101), and on its generalizations associated with more complicated models (including, for example, *acceleration terms*, *interaction with water*, *obstacles*, and *torsional effects*; see [4, 39], and the references therein for details).

As far as we know, the *second encounter* between *ADMM* and *Nonlinear Elasticity* took place in 1979 during a visit of the author at UT Austin, where his former (French) Master thesis student Patrick Le Tallec was working on a PhD thesis in Aerospace and Mechanical Engineering, under the supervision of J.T. Oden. The main topics of Le Tallec thesis were the analysis of *Finite Elasticity* models of the *Mooney-Rivlin* type (a particular attention being given to *incompressible materials*), and the computation of the solutions of the related *equilibrium* problems. It was quickly realized that the ADMM based methodology used to solve (8.87) (another Nonlinear Elasticity problem) could be easily adapted to the solution of problems in incompressible Finite Elasticity, the *incompressibility condition* playing for these problems the role played by the inextensibility condition $|y'| = 1$ in problem (8.87). Taking advantage of the modularity of the ADMM methodology we were able to develop, in just few days, a finite element code able to solve two-dimensional equilibrium problems for Mooney-Rivlin incompressible elastic materials, including nontrivial situations with cracks (developing a three-dimensional code took more time, as expected). We will return in Section 2 to the ADMM based solution of Finite Elasticity equilibrium problems for incompressible Mooney-Rivlin materials. In Section 3, we will discuss the ADMM solution of the Dirichlet problem for the two-dimensional *Monge-Ampère equation*. Finally, in Section 4, we will apply ADMM to the solution of a *non-smooth nonlinear eigenvalue problem from visco-plasticity*.

2 On the ADMM Based Solution of Equilibrium Problems in Incompressible Finite Elasticity

2.1 Introduction. Problem Formulation

Section 2 is dedicated to what we consider to be one of the most dramatic applications of ADMM ever, namely the computation of the solutions of equilibrium Finite Elasticity problems, for *Mooney-Rivlin* incompressible materials. However, in order to avoid lengthy developments we will focus mostly on two-dimensional problems, more general situations being discussed in, e.g., [38, 39] and [49].

Following [37], we consider the deformation relative to a *fixed* reference configuration of an *incompressible hyper-elastic* body. The displacement of this body is characterized by the displacement field $\mathbf{u}(\mathbf{x})$, where $\mathbf{x} = \{x_i\}_{i=1}^d$ denotes the position of a material particle in the reference configuration ($d = 2$ or 3, in practice); we assume that the reference configuration of the body occupies Ω, a *bounded* domain of $\mathbb{R}^d$. The body is subjected to *body forces* of intensity $\mathbf{f}$ per unit mass in the reference configuration, and to (algebraic) *surface tractions* $\mathbf{t}$ per unit area in the reference configuration; $\mathbf{t}$ (resp., $\mathbf{u}$) is prescribed on a subset Γ_1 (resp., Γ_0) of the boundary Γ of Ω, with Γ_0 and Γ_1 verifying

$$\Gamma = \Gamma_0 \cup \Gamma_1,\ \Gamma_0 \cap \Gamma_1 \neq \emptyset,\ \text{meas.}(\Gamma_i) > 0, \forall i = 0, 1. \tag{8.105}$$

The *internal elastic energy* of the body is typically of the form

$$E_e(\mathbf{u}) = \int_\Omega \sigma(\mathbf{x}, \nabla\mathbf{u})\, d\mathbf{x}, \tag{8.106}$$

where the *stored energy function* σ is of the *Caratheodory* type, that is,

- $\forall \mathbf{v}$, the function $\mathbf{x} \to \sigma(\mathbf{x}, \nabla\mathbf{v})$ is measurable,
- $\forall \mathbf{x}$, the function $\mathbf{q} \to \sigma(\mathbf{x}, \mathbf{q})$ is measurable and differentiable in $\mathbb{R}^{d\times d}$.

The specific form of $\sigma(\mathbf{x}, \mathbf{q})$ characterizes the material of which the body is made. In the case of a *Mooney-Rivlin* material, the stored energy function σ is of the form

$$\sigma(\mathbf{x}, \mathbf{q}) = C_1[|\mathbf{I} + \mathbf{q}|^2 - d] + C_2[|\mathrm{adj}(\mathbf{I} + \mathbf{q})|^2 - d], \tag{8.107}$$

where, in (8.107), $\mathbf{I}$ is the $d \times d$ identity matrix, C_1 and C_2 are two material dependent positive constants, and where (with obvious notation)

$$|\mathbf{T}|^2 = \mathrm{trace}(\mathbf{T}^t\mathbf{T}) \left(= \sum_{1\leq i,j\leq d} t_{ij}^2 \right), \forall \mathbf{T} \in \mathbb{R}^{d\times d}. \tag{8.108}$$

In (8.107), adj$\mathbf{T}$ is the *adjugate* of $\mathbf{T}$ (that is, the transpose of its cofactor matrix).

Last (but not least), since we consider incompressible materials, the displacement field $\mathbf{u}$ has to obey the *local incompressibility condition*; this condition reads as:

$$\det(\mathbf{I} + \nabla\mathbf{u}) = 1 \text{ a.e. in } \Omega. \tag{8.109}$$

For simplicity, we will assume that $\mathbf{f}$ and $\mathbf{t}$ verify the *dead loading hypothesis*, that is, do not depend of $\mathbf{u}$. It is thus reasonable to assume that *stable equilibria* are solutions (local or global) of the following variational problem:

$$\begin{cases} \mathbf{u} \in \mathbf{E}, \\ J(\mathbf{u}) \leq J(\mathbf{v}),\ \forall \mathbf{u} \in \mathbf{E}, \end{cases} \tag{8.110}$$

where in (8.110):

- The functional J is defined by

$$J(\mathbf{v}) = \int_{\Omega} \sigma(\mathbf{x}, \nabla \mathbf{v})\, d\mathbf{x} - \int_{\Omega} \rho\, \mathbf{f} \cdot \mathbf{v}\, d\mathbf{x} - \int_{\Gamma_1} \mathbf{t} \cdot \mathbf{v}\, d\Gamma, \tag{8.111}$$

  with ρ the body density.
- The set $\mathbf{E}$ of *admissible displacements* is defined by

$$\mathbf{E} = \{\mathbf{v} | \mathbf{v} \in \mathbf{V}, \mathbf{v} = \mathbf{u}_0 \text{ on } \Gamma_0, \det(\mathbf{I} + \nabla \mathbf{v}) = 1 \text{ a.e. in } \Omega\}. \tag{8.112}$$

We have, typically, $\mathbf{V} = (W^{1,s}(\Omega))^d$ in (8.112), with s depending of the stored energy function σ; in order to have $\mathbf{E} \neq \emptyset$, we assume that $\mathbf{u}_0$ is the trace on Γ_0 of a function $\tilde{u}_0 \in \mathbf{V}$ verifying $\det(\mathbf{I} + \nabla \tilde{u}_0) = 1$. If σ is of the *Mooney-Rivlin* type (i.e., is defined by (8.107)) then a natural choice for the space $\mathbf{V}$ in (8.112) is $(W^{1,s}(\Omega))^d$, with $s = 2d - 2$. We observe that for *Mooney-Rivlin* materials J is *convex* if $d = 2$; this is a direct consequence of the relation $|\mathbf{T}| = |\text{adj}\mathbf{T}|$, $\forall \mathbf{T} \in \mathbb{R}^{2\times 2}$. On the other hand J is *non-convex* if $d = 3$. The set $\mathbf{E}$ is non-convex if $d \geq 2$. From now on, in addition to the dead loading assumption, we will assume that the body is cylindrical along Ox_3 and made of a homogeneous Mooney-Rivlin incompressible material. If $\mathbf{f}$, $\mathbf{t}$ and $\mathbf{u}$ are parallel to the plane (Ox_1, Ox_2), it makes sense to look for solutions of the form $\{u_1, u_2, 0\}$ *(plane strain* solutions). Below, we will denote by Ω the cross-section of the elastic slab under consideration, which we assume thick enough so that the two-dimensional simplification takes place. Taking into account the various assumptions and simplifications we mentioned above, problem (8.110) reduces to:

$$\mathbf{u} = \arg\min_{\mathbf{v}\in\mathbf{E}} \left[\frac{C}{2} \int_{\Omega} |\mathbf{I} + \nabla \mathbf{v}|^2 d\mathbf{x} - \rho \int_{\Omega} \mathbf{f} \cdot \mathbf{v}\, d\mathbf{x} - \int_{\Gamma_1} \mathbf{t} \cdot \mathbf{v}\, d\Gamma \right], \tag{8.113}$$

where, in (8.113), C is a material dependent positive constant, and where

$$\mathbf{E} = \{\mathbf{v} | \mathbf{v} \in (H^1(\Omega))^2, \mathbf{v} = \mathbf{u}_0 \text{ on } \Gamma_0, \det(\mathbf{I} + \nabla \mathbf{v}) = 1 \text{ a.e. in } \Omega\}. \tag{8.114}$$

From a computational point of view we will make things simpler by introducing the vector-valued function $\tilde{u}$ defined by

$$\tilde{\mathbf{u}}(\mathbf{x}) = \mathbf{x} + \mathbf{u}(\mathbf{x}), \text{ a.e. in } \Omega. \tag{8.115}$$

We have then

$$\tilde{\mathbf{u}} = \arg\min_{\mathbf{v}\in\widetilde{\mathbf{E}}} \left[\frac{C}{2} \int_{\Omega} |\nabla \mathbf{v}|^2 d\mathbf{x} - \rho \int_{\Omega} \mathbf{f} \cdot \mathbf{v}\, d\mathbf{x} - \int_{\Gamma_1} \mathbf{t} \cdot \mathbf{v}\, d\Gamma \right], \tag{8.116}$$

with

$$\widetilde{\mathbf{E}} = \{\mathbf{v} | \mathbf{v} \in (H^1(\Omega))^2, \mathbf{v} = \tilde{\mathbf{u}}_0 \text{ on } \Gamma_0, \det \nabla \mathbf{v} = 1 \text{ a.e. in } \Omega\}. \tag{8.117}$$

where $\tilde{\mathbf{u}}_0(\mathbf{x}) = \mathbf{x} + \mathbf{u}_0(\mathbf{x})$, a.e. on Γ_0. In the (relatively) simple situation under consideration, proving the *existence* of solutions to problem (8.116) is fairly easy as shown in the following subsection.

2.2 On the Existence of Solutions to Problem (8.116)

Concerning the solutions to problem (8.116) we have the following

Theorem 1. *Let assume that the linear functional* $\mathbf{v} \to \rho \int_\Omega \mathbf{f}\cdot\mathbf{v}\,d\mathbf{x} + \int_{\Gamma_1} \mathbf{t}\cdot\mathbf{v}\,d\Gamma$ *is continuous over* $(H^1(\Omega))^2$*. Then, problem (8.116) has a solution.*

Proof. Let us denote by $\widetilde{J}$ the functional in (8.116) and suppose that $\{\mathbf{u}^n\}_{n\geq 0}$ is a minimizing sequence associated with the above problem; we have then

$$\mathbf{u}^n \in \widetilde{\mathbf{E}}, \forall n \geq 0, \text{and } \lim_{n\to+\infty} \widetilde{J}(\mathbf{u}^n) = \inf_{\mathbf{v}\in\widetilde{\mathbf{E}}} \widetilde{J}(\mathbf{v}); \tag{8.118}$$

without loss of generality we can assume that

$$\widetilde{J}(\mathbf{u}^n) \leq \widetilde{J}(\mathbf{u}^{n-1}) \leq \cdots \leq \widetilde{J}(\mathbf{u}^1) \leq \widetilde{J}(\mathbf{u}^0), \forall n \geq 2. \tag{8.119}$$

Let observe that the functional $\mathbf{v} \to \sqrt{\int_\Omega |\nabla\mathbf{v}|^2\,d\mathbf{x}}$ defines over the space

$$\mathbf{V}_0 = \{\mathbf{v}|\mathbf{v} \in (H^1(\Omega))^2, \mathbf{v} = \mathbf{0} \text{ on } \Gamma_0\}$$

a norm equivalent to the one induced by $(H^1(\Omega))^2$; combining this property with (8.119), we can easily show that the sequence $\{\mathbf{u}^n\}_{n\geq 0}$ is bounded in $(H^1(\Omega))^2$. The above space being a Hilbert space, it follows from the boundedness of $\{\mathbf{u}^n\}_{n\geq 0}$ that we can extract from the above sequence a subsequence -still denoted by $\{\mathbf{u}^n\}_{n\geq 0}$- such that

$$\lim_{n\to+\infty} \mathbf{u}^n = \tilde{\mathbf{u}} \text{ weakly in } (H^1(\Omega))^2. \tag{8.120}$$

Since the functional $\widetilde{J}$ is convex and continuous over $(H^1(\Omega))^2$, it is also weakly lower semi-continuous, implying (from (8.120)) that

$$\widetilde{J}(\tilde{\mathbf{u}}) \leq \liminf_{n\to+\infty} \widetilde{J}(\mathbf{u}^n) = \inf_{\mathbf{v}\in\widetilde{\mathbf{E}}} \widetilde{J}(\mathbf{v}). \tag{8.121}$$

If we can prove that $\tilde{\mathbf{u}} \in \widetilde{\mathbf{E}}$, then the proof will be complete. Actually, it follows from Lemma 1, below, that $\widetilde{\mathbf{E}}$ is *weakly closed* in $(H^1(\Omega))^2$, which completes the proof of our theorem.

Lemma 1. *Assuming that* Ω *is bounded in* $\mathbb{R}^2$*, the set* $\widetilde{\mathbf{E}}$ *defined by (8.117) is weakly closed in* $(H^1(\Omega))^2$*.*

Proof. Suppose that $\{\mathbf{v}, \phi\} \in (C^\infty(\overline{\Omega}))^2 \times \mathscr{D}(\Omega)$, $\mathscr{D}(\Omega)$ being the space of those real-valued functions which are infinitely differentiable over $\overline{\Omega}$ and have a compact support in Ω. Integration by parts, taking advantage of the fact that ϕ vanishes in the neighborhood of Γ, give us

$$\begin{cases} \int_\Omega (\det \nabla \mathbf{v})\phi \, d\mathbf{x} = \int_\Omega \left(\frac{\partial v_1}{\partial x_1}\frac{\partial v_2}{\partial x_2} - \frac{\partial v_1}{\partial x_2}\frac{\partial v_2}{\partial x_1} \right) \phi \, d\mathbf{x} = \\ \frac{1}{2}\int_\Omega v_1 \left(\frac{\partial v_2}{\partial x_1}\frac{\partial \phi}{\partial x_2} - \frac{\partial v_2}{\partial x_2}\frac{\partial \phi}{\partial x_1} \right) d\mathbf{x} + \frac{1}{2}\int_\Omega v_2 \left(\frac{\partial v_1}{\partial x_2}\frac{\partial \phi}{\partial x_1} - \frac{\partial v_1}{\partial x_1}\frac{\partial \phi}{\partial x_2} \right) d\mathbf{x}, \\ \forall \mathbf{v} = \{v_1, v_2\} \in (C^\infty(\overline{\Omega}))^2, \ \forall \phi \in \mathscr{D}(\Omega). \end{cases} \tag{8.122}$$

The density of $C^\infty(\overline{\Omega})$ in $H^1(\Omega)$, and the continuity of the functionals in relation (8.122), imply that (8.122) still holds if $\mathbf{v}$ belongs to $(H^1(\Omega))^2$. Proving that the set $\widetilde{\mathbf{E}}$ is weakly closed is pretty easy now: indeed let us consider a sequence $\{\mathbf{w}^n\}_{n\geq 0}$ of elements of $\widetilde{\mathbf{E}}$ converging weakly to $\mathbf{w}$ in $(H^1(\Omega))^2$. Since $\mathbf{w}^n \in \widetilde{\mathbf{E}}$, $\forall n \geq 0$, it follows from (8.122) that (with obvious notation)

$$\begin{cases} \frac{1}{2}\int_\Omega w_1^n \left(\frac{\partial w_2^n}{\partial x_1}\frac{\partial \phi}{\partial x_2} - \frac{\partial w_2^n}{\partial x_2}\frac{\partial \phi}{\partial x_1} \right) d\mathbf{x} + \\ \quad \frac{1}{2}\int_\Omega w_2^n \left(\frac{\partial w_1^n}{\partial x_2}\frac{\partial \phi}{\partial x_1} - \frac{\partial w_1^n}{\partial x_1}\frac{\partial \phi}{\partial x_2} \right) d\mathbf{x} \\ = \int_\Omega (\det \nabla \mathbf{w}^n)\phi \, d\mathbf{x} = \int_\Omega \phi \, d\mathbf{x}, \forall n \geq 0, \ \forall \phi \in \mathscr{D}(\Omega). \end{cases} \tag{8.123}$$

The weak convergence of $\{\mathbf{w}^n\}_{n\geq 0}$ to $\mathbf{w}$ in $(H^1(\Omega))^2$ implies

$$\lim_{n\to+\infty} \nabla \mathbf{w}^n = \nabla \mathbf{w} \text{ weakly in } (L^2(\Omega))^{2\times 2}, \tag{8.124}$$

$$\lim_{n\to+\infty} \mathbf{w}^n = \mathbf{w} \text{ in } (L^2(\Omega))^2, \tag{8.125}$$

$$\lim_{n\to+\infty} \mathbf{w}^n|_\Gamma = \mathbf{w}|_\Gamma \text{ in } (L^2(\Gamma))^2 \tag{8.126}$$

(see, e.g., [1, 52, 53] and [62] for these results). It follows from relations (8.122)–(8.126), and from $\mathbf{w}^n|_{\Gamma_0} = \tilde{\mathbf{u}}_0$, $\forall n \geq 0$, that

$$\mathbf{w}|_{\Gamma_0} = \tilde{\mathbf{u}}_0 \tag{8.127}$$

and

$$\begin{cases} \int_\Omega (\det \nabla \mathbf{w})\phi \, d\mathbf{x} = \frac{1}{2}\int_\Omega w_1 \left(\frac{\partial w_2}{\partial x_1}\frac{\partial \phi}{\partial x_2} - \frac{\partial w_2}{\partial x_2}\frac{\partial \phi}{\partial x_1} \right) d\mathbf{x} + \\ \quad \frac{1}{2}\int_\Omega w_2 \left(\frac{\partial w_1}{\partial x_2}\frac{\partial \phi}{\partial x_1} - \frac{\partial w_1}{\partial x_1}\frac{\partial \phi}{\partial x_2} \right) d\mathbf{x} = \\ \int_\Omega \phi \, d\mathbf{x}, \ \forall \phi \in \mathscr{D}(\Omega). \end{cases} \tag{8.128}$$

Relation (8.128) implies that $\det \nabla \mathbf{w} = 1$ in the sense of distributions, that is a.e. in Ω, which combined with (8.127) implies in turn that $\mathbf{w} \in \widetilde{\mathbf{E}}$, which completes the proof of the lemma.

Remark 5. Situations where problem (8.113) has multiple solutions, due to buckling phenomena for example, are common in practice, as shown in, e.g., [37].

2.3 On the ADMM Solution of Problem (8.116)

The mathematical structure of problem (8.116) is very close to those of the convex and non-convex variational problems encountered in Section 1. As above, our starting point is the equivalence between problem (8.116) and

$$\{\tilde{\mathbf{u}},\tilde{\mathbf{p}}\} = \arg\min_{\{\mathbf{v},\mathbf{q}\}\in\mathscr{E}} \widetilde{J}(\mathbf{v},\mathbf{q}), \tag{8.129}$$

with

$$\widetilde{J}(\mathbf{v},\mathbf{q}) = \frac{C}{2}\int_\Omega |\nabla\mathbf{v}|^2\,d\mathbf{x} - \rho\int_\Omega \mathbf{f}\cdot\mathbf{v}\,d\mathbf{x} - \int_{\Gamma_1}\mathbf{t}\cdot\mathbf{v}\,d\Gamma, \tag{8.130}$$

and

$$\mathscr{E} = \{\{\mathbf{v},\mathbf{q}\} \mid \mathbf{v}\in(H^1(\Omega))^2, \mathbf{q}\in(L^2(\Omega))^{2\times 2}, \mathbf{v}=\tilde{\mathbf{u}}_0 \text{ on } \Gamma_0,$$
$$\det\mathbf{q} = 1 \text{ a.e. in } \Omega, \nabla\mathbf{v}-\mathbf{q}=\mathbf{0}\}. \tag{8.131}$$

With $r>0$, we associate with (8.130) and (8.131) the *augmented Lagrangian functional*

$$\mathscr{L}_r : ((H^1(\Omega))^2 \times (L^2(\Omega))^{2\times 2} \times (L^2(\Omega))^{2\times 2} \to I\!R$$

defined by

$$\mathscr{L}_r(\mathbf{v},\mathbf{q};\mu) = \frac{C}{2}\int_\Omega |\nabla\mathbf{v}|^2\,d\mathbf{x} - \rho\int_\Omega \mathbf{f}\cdot\mathbf{v}\,d\mathbf{x} - \int_{\Gamma_1}\mathbf{t}\cdot\mathbf{v}\,d\Gamma$$
$$+\frac{r}{2}\int_\Omega |\nabla\mathbf{v}-\mathbf{q}|^2\,d\mathbf{x} + \int_\Omega \mu:(\nabla\mathbf{v}-\mathbf{q})\,d\mathbf{x}; \tag{8.132}$$

in (8.132), we have $\mathbf{S}:\mathbf{T} = \sum_{1\le i,j\le 2} s_{ij}t_{ij}\ \forall\mathbf{S}=(s_{ij})_{1\le i,j\le 2}, \mathbf{T}=(t_{ij})_{1\le i,j\le 2}\in I\!R^{2\times 2}$.
Let us define now $\mathbf{V}_{\tilde{\mathbf{u}}_0}$ and Σ by

$$\mathbf{V}_{\tilde{\mathbf{u}}_0} = \{\mathbf{v}|\mathbf{v}\in(H^1(\Omega))^2, \mathbf{v}=\tilde{\mathbf{u}}_0 \text{ on } \Gamma_0\}, \tag{8.133}$$
$$\Sigma = \{\mathbf{q}|\mathbf{q}\in(L^2(\Omega))^{2\times 2}, \det\mathbf{q}=1 \text{ a.e. in } \Omega\}, \tag{8.134}$$

respectively. Suppose now that $\{\{\tilde{\mathbf{u}},\tilde{\mathbf{p}}\},\lambda\}$ is a saddle-point of $\mathscr{L}_r$ over $(\mathbf{V}_{\tilde{\mathbf{u}}_0}\times\Sigma)\times(L^2(\Omega))^{2\times 2}$, that is

$$\begin{cases} \{\{\tilde{\mathbf{u}},\tilde{\mathbf{p}}\},\lambda\}\in(\mathbf{V}_{\tilde{\mathbf{u}}_0}\times\Sigma)\times(L^2(\Omega))^{2\times 2}, \\ \mathscr{L}_r(\tilde{\mathbf{u}},\tilde{\mathbf{p}};\mu) \le \mathscr{L}_r(\tilde{\mathbf{u}},\tilde{\mathbf{p}};\lambda) \le \mathscr{L}_r(\mathbf{v},\mathbf{q};\lambda), \\ \forall\{\{\mathbf{v},\mathbf{q}\},\mu\}\in(\mathbf{V}_{\tilde{\mathbf{u}}_0}\times\Sigma)\times(L^2(\Omega))^{2\times 2}, \end{cases} \tag{8.135}$$

then $\tilde{\mathbf{u}}$ is a solution to problem (8.116) and $\tilde{\mathbf{p}} = \nabla\tilde{\mathbf{u}}$, implying that *ALG*1, *ALG*2 and *ALG*3 are natural candidates for the solution of problem (8.116). Focusing on *ALG*2, we obtain:

$$\{\mathbf{u}^{-1},\lambda^0\} \text{ is given in } \mathbf{V}_{\tilde{\mathbf{u}}_0}\times(L^2(\Omega))^{2\times 2}. \tag{8.136}$$

For $n \geq 0$, $\{\mathbf{u}^{n-1}, \lambda^n\} \to \mathbf{p}^n \to \mathbf{u}^n \to \lambda^{n+1}$ via

$$\mathbf{p}^n = \arg\min_{\mathbf{q} \in \Sigma} \mathscr{L}_r(\mathbf{u}^{n-1}, \mathbf{q}; \lambda^n), \tag{8.137}$$

and

$$\mathbf{u}^n = \arg\min_{\mathbf{v} \in \mathbf{V}_{\tilde{\mathbf{u}}_0}} \mathscr{L}_r(\mathbf{v}, \mathbf{p}^n; \lambda^n), \tag{8.138}$$

and

$$\lambda^{n+1} = \lambda^n + \rho(\nabla \mathbf{u}^n - \mathbf{p}^n). \tag{8.139}$$

The minimization problem (8.138) is well posed; indeed $\mathbf{u}^n$ is the unique solution (from the *Lax-Milgram theorem*; see, e.g., Appendix 1 of [31] and Chapter 1 of [34]) of the following *linear variational problem*:

$$\begin{cases} \mathbf{u}^n \in \mathbf{V}_{\tilde{\mathbf{u}}_0}, \\ (C+r)\displaystyle\int_\Omega \nabla \mathbf{u}^n : \nabla \mathbf{v}\, d\mathbf{x} = \rho \int_\Omega \mathbf{f} \cdot \mathbf{v}\, d\mathbf{x} + \int_{\Gamma_1} \mathbf{t} \cdot \mathbf{v}\, d\Gamma + \\ \quad \displaystyle\int_\Omega (r\mathbf{p}^n - \lambda^n) : \nabla \mathbf{v}\, d\mathbf{x}; \forall \mathbf{v} \in \mathbf{V}_0, \end{cases} \tag{8.140}$$

with $\mathbf{V}_0 = \{\mathbf{v} | \mathbf{v} \in (H^1(\Omega))^2, \mathbf{v} = \mathbf{0} \text{ on } \Gamma_0\}$. The elliptic problem (8.140) is a Poisson system with mixed boundary conditions (of the Dirichlet-Neumann type). The two components of $\mathbf{u}^n$ can be computed independently (and also in parallel).

The minimization problem (8.137) can be solved *point-wise*. Indeed, a.e. in Ω, we have to solve a *quadratically constrained* minimization problem in $\mathbb{R}^4$, namely

$$\mathbf{p}^n = \arg\min_{\mathbf{G} \in \Sigma_4} \left[\frac{r}{2} |\mathbf{G}|^2 - \mathbf{X}^n(\mathbf{x}) : \mathbf{G} \right], \tag{8.141}$$

where, in (8.141), the set Σ_4 and the matrix-valued function $\mathbf{X}^n$ are defined by

$$\Sigma_4 = \{\mathbf{G} | \mathbf{G} = (G_{ij})_{1 \leq i,j \leq 2} \in \mathbb{R}^{2\times 2}, G_{11}G_{22} - G_{12}G_{21} = 1\},$$

and

$$\mathbf{X}^n = r\nabla \mathbf{u}^{n-1} + \lambda^n,$$

respectively. In order to facilitate the solution of problem (8.141), we introduce the vector $\mathbf{z} = \{z_i\}_{i=1}^4 \in \mathbb{R}^4$ defined by

$$\begin{cases} \sqrt{2}\, z_1 = G_{11} + G_{22}, \sqrt{2}\, z_2 = G_{11} - G_{22}, \\ \sqrt{2}\, z_3 = G_{12} + G_{21}, \sqrt{2}\, z_4 = G_{12} - G_{21}. \end{cases} \tag{8.142}$$

We have then

$$\begin{pmatrix} p_{11}^n(\mathbf{x}) \\ p_{22}^n(\mathbf{x}) \\ p_{12}^n(\mathbf{x}) \\ p_{21}^n(\mathbf{x}) \end{pmatrix} = \mathbf{S} \arg\min_{\mathbf{z} \in \mathbf{Z}_4} \left[\frac{r}{2} |\mathbf{z}|^2 - \mathbf{b}^n(\mathbf{x}) \cdot \mathbf{z} \right], \tag{8.143}$$

where, in (8.143),

$$\mathbf{S} = \frac{1}{\sqrt{2}}\begin{pmatrix} 1 & 1 & 0 & 0 \\ 1 & -1 & 0 & 0 \\ 0 & 0 & 1 & 1 \\ 0 & 0 & 1 & -1 \end{pmatrix}, \quad \begin{pmatrix} b_1^n(\mathbf{x}) \\ b_2^n(\mathbf{x}) \\ b_3^n(\mathbf{x}) \\ b_4^n(\mathbf{x}) \end{pmatrix} = \mathbf{S}\begin{pmatrix} X_{11}^n(\mathbf{x}) \\ X_{22}^n(\mathbf{x}) \\ X_{12}^n(\mathbf{x}) \\ X_{21}^n(\mathbf{x}) \end{pmatrix}$$

and

$$\mathbf{Z}_4 = \{\mathbf{z} | \mathbf{z} = \{z_i\}_{i=1}^4 \in \mathbb{R}^4, z_1^2 - z_2^2 - z_3^2 + z_4^2 = 1\}.$$

The minimization problem in (8.143) is clearly a *generalized eigenvalue problem.* Let us denote by $\mathbf{y}$ a solution to the above minimization problem; introducing a *Lagrange multiplier* λ associated with the constraint $\frac{1}{2}(z_1^2 - z_2^2 - z_3^2 + z_4^2 - 1) = 0$, the corresponding optimality system reads as

$$\begin{cases} ry_1 = b_1 + \lambda y_1, \\ ry_2 = b_2 - \lambda y_2, \\ ry_3 = b_3 - \lambda y_3, \\ ry_4 = b_4 + \lambda y_4, \\ y_1^2 - y_2^2 - y_3^2 + y_4^2 = 1 \end{cases} \tag{8.144}$$

which implies in turn that λ is solution to

$$\frac{b_1^2 + b_4^2}{(r-\lambda)^2} = \frac{b_2^2 + b_3^2}{(r+\lambda)^2} + 1. \tag{8.145}$$

The solution of the minimization problem (8.143), via (8.144), (8.145), is thoroughly discussed in [37] for which we refer also for the results of numerical experiments showing that the ADMM algorithm (8.136)–(8.139) has good convergence properties if r lies in an appropriate interval (see also [39]).

Remark 6. As shown in, e.g., [37, 38] and [39], the ADMM based methodology, we discussed above, can be generalized in order to solve *axisymmetric* and *three-dimensional* equilibrium problems for elastic bodies made of incompressible Mooney-Rivlin materials. The *finite element* implementation of these algorithms is discussed in the above three references, and their *parallelization* in [58].

Remark 7. The numerical experiments reported in, e.g., [39] show that a robust strategy to solve equilibrium problems in Finite Elasticity is to use *ALG*1 with an appropriate stopping criterion for the relaxation iterations. Indeed, this makes the resulting algorithm less sensitive to initialization than *ALG*2 and *ALG*3. Most often, the number of relaxation iterations reduces to one or two rather quickly. The fact that *ALG*1 is more robust than *ALG*2 and *ALG*3 is not surprising: After all, *ALG*2 and *ALG*3 are just 'cheap' approximations of *ALG*1 where one limits to one the number of relaxation iterations used to solve the minimization problem (39) in algorithm (38)-(40). To be more precise, what *ALG*1 does is to solve the *dual problem*

$$\lambda = \arg\max_{\mu \in H} \min_{\{v,q\} \in V \times H} \mathscr{L}_r(v, q; \mu)$$

by a *gradient ascent method* with fixed step ρ, since the vector $Bu^n - p^n$ is nothing but the gradient at λ^n of the *dual functional*

$$\mu \to \min_{\{v,q\} \in V \times H} \mathscr{L}_r(v,q;\mu).$$

On the other hand *ALG*2 and *ALG*3 rely on ascent vectors at λ^n which have been obtained from an incomplete (to say the least) solution of problem (8.39), leading to mediocre approximations of the gradient (during the first iterations at least), and thus to less robustness. Other evidences of the superior robustness of *ALG*1 are: (i) the convergence condition (for convex problems) $\rho \in (0, 2r)$, compared to $\rho \in (0, \frac{1+\sqrt{5}}{2} r)$ for *ALG*2, and (ii) the fact that the number of outer iterations necessary to achieve convergence is a decreasing function of r (until round-off and truncation errors catch up), while, for *ALG*2 and *ALG*3, r has to be neither too small nor too large to obtain optimal speed of convergence (assuming that r is fixed). To be honest these comments about *ALG*1 being more robust than *ALG*2 and *ALG*3 apply mostly to convex problems (albeit they apply also to non-convex problems such as (8.110)). One may find in [64] examples of non-convex problems for which *ALG*1 does not converge to a solution while *ALG*2 does.

3 On the ADMM Based Solution of the Dirichlet Problem for the Elliptic Monge-Ampère Equation in Dimension Two

3.1 Introduction. Synopsis

If $f > 0$, the two-dimensional canonical real *Monge-Ampère equation*

$$\det \mathbf{D}^2 u = f \tag{8.146}$$

is certainly the simplest example of *second order fully nonlinear elliptic equations*. In (8.146), $\mathbf{D}^2 u$ denotes the *Hessian* of the real-valued unknown function u, that is

$$\mathbf{D}^2 u = \begin{pmatrix} \dfrac{\partial^2 u}{\partial x_1^2} & \dfrac{\partial^2 u}{\partial x_1 \partial x_2} \\ \dfrac{\partial^2 u}{\partial x_1 \partial x_2} & \dfrac{\partial^2 u}{\partial x_2^2} \end{pmatrix}$$

and therefore

$$\det \mathbf{D}^2 u = \frac{\partial^2 u}{\partial x_1^2} \frac{\partial^2 u}{\partial x_2^2} - \left| \frac{\partial^2 u}{\partial x_1 \partial x_2} \right|^2. \tag{8.147}$$

Actually, equation (8.146) is trickier than what it looks like; to be convinced let us consider the seemingly simple following boundary value problem

$$\begin{cases} \det \mathbf{D}^2 u = 1 \text{ in } \Omega \\ u = 0 \text{ on } \partial\Omega, \end{cases} \tag{8.148}$$

with $\Omega = (0,1)^2$; problem (8.148) is a typical *Dirichlet problem* for the Monge-Ampère equation (8.146). It is clear that problem (8.148) has no smooth solution on $\overline{\Omega}$ since, for such a solution, the condition $u = 0$ on $\partial\Omega$ implies that the product $\frac{\partial^2 u}{\partial x_1^2}\frac{\partial^2 u}{\partial x_2^2}$ and the cross derivative $\frac{\partial^2 u}{\partial x_1 \partial x_2}$ vanish at the boundary, implying in turn that $\det \mathbf{D}^2 u$ is strictly less than 1 in some neighborhood of $\partial\Omega$, which contradicts $\det \mathbf{D}^2 u = 1$ in Ω. Actually, the non-existence of classical solutions to (8.148) stems from the *non-strict convexity* of Ω, not from the fact that it has corners. In order to overcome this type of difficulties, various generalizations of the concept of solution were introduced, the most popular ones being the *generalized solutions in the sense of Alexandroff* and the *viscosity solutions*. *Generalized solutions* to Monge-Ampère and other fully nonlinear elliptic equations are discussed in, e.g., [11, 12, 45] and the Chapter 4 of [63] (see also the many references therein); the third of the above references contains a discussion of the intricate relations existing between these various notions of generalized solutions. For years, the numeric of fully nonlinear elliptic equations has been far behind their analysis, two notable exceptions being [55] and [2]. Fortunately, these past few years have been witnessing a fast increasing interest by the computational and applied mathematics community for the numerical analysis and solution of fully nonlinear elliptic equations, some recent and very recent related publications being [3, 54, 7] and [8]. Actually, from 2000 to the present days, a rather large variety of methods have been investigated to achieve the numerical solution of fully nonlinear elliptic equations (finite differences, finite elements, discontinuous Galerkin, mesh-less, viscosity solutions, vanishing moments, Newton's, multilevel, least-squares, etc.); some of these methods are discussed in [28], a review article which describes also various situations from *Mechanics* and *Physics* leading to *fully nonlinear elliptic equations*, including some from *Cosmology* (see, e.g., [6]). What about *augmented Lagrangians* and *ADMM*? As far as we know they have been used in two instances:

(i) In [2], where *ALG*2 was used to solve a variant of the *Monge-Ampère equation* associated with the *Monge-Kantorovich optimal transportation problem* (this was the *first time* that, to the best of our knowledge, an augmented Lagrangian algorithm had been applied to the solution of a Monge-Ampère type problem).

(ii) In [20, 21, 22] and [35], where *ALG*2 (combined with mixed finite element approximations) was applied to the solution of a *nonlinearly constrained minimization problem* associated with the Monge-Ampère problem under consideration (see also Section 3.1 of [28]). This approach will be discussed in Sections 3.3 and 3.4.

3.2 Problem Formulation

Let Ω be a *bounded convex* domain of $\mathbb{R}^2$; from now on, we will denote by Γ the boundary $\partial\Omega$ of Ω. The *two-dimensional Dirichlet problem* for the *Monge-Ampère equation* (8.146) reads as follows:

$$\begin{cases} \det \mathbf{D}^2 u = f \text{ in } \Omega \\ u = g \text{ on } \partial\Omega, \end{cases} \tag{8.149}$$

where $f(>0)$ and g are two given functions. Since (8.149) may have multiple solutions (*two* at most actually, as shown in [19]), we will look for *convex solutions* only. The *existence and uniqueness* of convex solutions (classical or generalized) to (8.149) is discussed in, e.g., [12, 45] and [11] (see also the references therein).

3.3 An Augmented Lagrangian Approach for the Solution of Problem (8.149)

Suppose that in (8.149) we have $f(>0) \in L^1(\Omega)$ and $g \in H^{3/2}(\Gamma)$; it makes sense then to attempt solving problem (8.149) in $H^2(\Omega)$ by considering it as a *nonlinear bi-harmonic problem*. A way to do so is to consider the following problem from *Calculus of Variations*

$$u = \arg\min_{v \in E^+_{fg}} \frac{1}{2}\int_\Omega |\nabla^2 v|^2 \, d\mathbf{x}, \tag{8.150}$$

with

$$\begin{aligned} E^+_{fg} &= \{v | v \in V_g, \det \mathbf{D}^2 v = f, \ v \, convex\} \\ &= \{v | v \in V_g, \det \mathbf{D}^2 v = f, \ \frac{\partial^2 v}{\partial x_1^2} > 0, \ \frac{\partial^2 v}{\partial x_2^2} > 0\} \end{aligned}$$

and

$$V_g = \{v | v \in H^2(\Omega), \ v = g \text{ on } \Gamma\}.$$

If problem (8.149) has a *convex* solution u in $H^2(\Omega)$, it is *unique* and is also the *unique* solution of problem (8.150). Problem (8.150) is in turn equivalent to

$$\{u, \mathbf{D}^2 u\} = \arg\min_{\{v,\mathbf{q}\} \in \mathscr{E}^+_{fg}} \frac{1}{2}\int_\Omega |\nabla^2 v|^2 \, d\mathbf{x}, \tag{8.151}$$

with

$$\mathscr{E}^+_{fg} = \{\{v, \mathbf{q}\} | v \in V_g, \mathbf{q} \in \mathbf{Q}, \mathbf{D}^2 v - \mathbf{q} = 0, \det \mathbf{q} = f, \ q_{11} > 0, q_{22} > 0\}$$

and

$$\mathbf{Q} = \{\mathbf{q} | \mathbf{q} = (q_{ij})_{1 \le i,j \le 2}, q_{12} = q_{21}, \ q_{ij} \in L^2(\Omega), \ 1 \le i, j \le 2\}.$$

Following Section 1, we associate with problem (8.151):

(i) The *augmented Lagrangian functional* $\mathscr{L}_r : (H^2(\Omega) \times \mathbf{Q}) \times \mathbf{Q} \to \mathbb{R}$ defined, with $r > 0$, by

$$\mathscr{L}_r(v, \mathbf{q}; \mu) = \frac{1}{2}\int_\Omega |\nabla^2 v|^2 \, d\mathbf{x} + \frac{r}{2}\int_\Omega |\mathbf{D}^2 v - \mathbf{q}|^2 \, d\mathbf{x} + \int_\Omega \mu : (\mathbf{D}^2 v - \mathbf{q}) \, d\mathbf{x}, \tag{8.152}$$

with $\mathbf{S} : \mathbf{T} = \sum_{1 \le i,j \le 2} s_{ij} t_{ij}$ if $\mathbf{S} = (s_{ij})_{1 \le i,j \le 2}$, $\mathbf{T} = (t_{ij})_{1 \le i,j \le 2}$, and $|\mathbf{S}| = \sqrt{\mathbf{S} : \mathbf{S}}$.

(ii) The *saddle-point* problem

$$\begin{cases} \{\{u, \mathbf{p}\}, \lambda\} \in (V_g \times \mathbf{Q}_f^+) \times \mathbf{Q}, \\ \mathscr{L}_r(u, \mathbf{p}; \mu) \le \mathscr{L}_r(u, \mathbf{p}; \lambda) \le \mathscr{L}_r(v, \mathbf{q}; \lambda), \\ \forall \{\{v, \mathbf{q}\}, \mu\} \in (V_g \times \mathbf{Q}_f^+) \times \mathbf{Q}, \end{cases} \tag{8.153}$$

with $\mathbf{Q}_f^+ = \{\mathbf{q} | \mathbf{q} \in \mathbf{Q}, \det \mathbf{q} = f, q_{11} > 0, q_{22} > 0\}$.

One can easily show that if $\{\{u, \mathbf{p}\}, \lambda\}$ is a solution of the saddle-point problem (8.153), then u is a *convex solution* of the Monge-Ampère problem (8.149), $\mathbf{p} = \mathbf{D}^2 u$, and λ is a *Lagrange multiplier* associated with the relation $\mathbf{D}^2 u - \mathbf{p} = \mathbf{0}$.

From the structure of the saddle-point problem (8.153), an obvious candidate for its iterative solution is clearly the algorithm *ALG*2 already considered in Section 1 (despite the fact that we are in a *non-convex* environment due to the *non-convexity* of the set $\mathbf{Q}_f^+$). Applying *ALG*2 to the solution of (8.153), we obtain:

$$\{u^{-1}, \lambda^0\} \text{ is given in } V_g \times \mathbf{Q}. \tag{8.154}$$

For $n \ge 0$, $\{u^{n-1}, \lambda^n\} \to \mathbf{p}^n \to u^n \to \lambda^{n+1}$ via

$$\begin{cases} \mathbf{p}^n \in \mathbf{Q}_f^+, \\ \mathscr{L}_r(u^{n-1}, \mathbf{p}^n; \lambda^n) \le \mathscr{L}_r(u^{n-1}, \mathbf{q}; \lambda^n), \forall \mathbf{q} \in \mathbf{Q}_f^+, \end{cases} \tag{8.155}$$

$$\begin{cases} u^n \in V_g, \\ \mathscr{L}_r(u^n, \mathbf{p}^n; \lambda^n) \le \mathscr{L}_r(v, \mathbf{p}^n; \lambda^n), \forall v \in V_g, \end{cases} \tag{8.156}$$

$$\lambda^{n+1} = \lambda^n + \rho(\mathbf{D}^2 u^n - \mathbf{p}^n). \tag{8.157}$$

Algorithm (8.154)–(8.157) deserves several comments, among them:

- Concerning the initialization of algorithm (8.154)–(8.157), we advocate taking $\lambda^0 = \mathbf{0}$ and u^{-1} as the solution of the linear Poisson-Dirichlet problem

$$\begin{cases} \nabla^2 u = 2\sqrt{f} \ \text{ in } \Omega, \\ u^{-1} = g \ \text{ on } \Gamma. \end{cases} \tag{8.158}$$

The rational of such a choice stems from the fact that if we denote by λ_1 and λ_2 the eigenvalues of $\mathbf{D}^2u$ we have $\lambda_1+\lambda_2=\nabla^2 u$ and $\lambda_1\lambda_2=\det\mathbf{D}^2u=f$. Suppose now that λ_1 and λ_2 are close to each other; it follows then from the identity $4\lambda_1\lambda_2\equiv(\lambda_1+\lambda_2)^2-(\lambda_1-\lambda_2)^2$ that we have $\nabla^2 u\approx 2\sqrt{\det\mathbf{D}^2u}=2\sqrt{f}$, justifying (8.158). Actually, numerical experiments have persistently shown the soundness of (8.158), even when λ_1 and λ_2 are not that close to each other.

- We always took $\rho=r$ in (8.157) when applying the above algorithm (actually its finite element analogues) to the numerical solution of the particular cases of problem (8.149) we used as test problems.
- Algorithm (8.154)–(8.157) may seem a bit abstract, however, relations (8.155), (8.156) can be reformulated, respectively, as

$$\mathbf{p}^n=\arg\min_{\mathbf{q}\in\mathbf{Q}_f^+}\left[\frac{1}{2}\int_\Omega|\mathbf{q}|^2\,d\mathbf{x}-\int_\Omega(\mathbf{D}^2u^{n-1}+\frac{1}{r}\lambda^n):\mathbf{q}\,d\mathbf{x}\right],\tag{8.159}$$

$$\begin{cases}u^n\in V_g,\\ \displaystyle\int_\Omega\nabla^2u^n\nabla^2v\,d\mathbf{x}+r\int_\Omega\mathbf{D}^2u^n:\mathbf{D}^2v\,d\mathbf{x}=\int_\Omega(r\mathbf{p}^n-\lambda^n):\mathbf{D}^2v\,d\mathbf{x},\\ \forall v\in V_0\end{cases}\tag{8.160}$$

with $V_0=H^2(\Omega)\cap H_0^1(\Omega)$. The solution of the sub-problems (8.159) and (8.160) will be discussed just below.

On the solution of (8.159): The minimization problem (8.159) can be solved *pointwise*; indeed, one has to solve for almost every $\mathbf{x}$ in Ω (in practice, for the interior vertices of a finite element triangulation of Ω) a tri-dimensional minimization problem of the following type:

$$\begin{cases}\{p_{11}^n(\mathbf{x}),p_{22}^n(\mathbf{x}),p_{12}^n(\mathbf{x})\}=\\ \arg\min_{\mathbf{z}\in\mathbf{Z}(f,\mathbf{x})}\left[\frac{1}{2}(z_1^2+z_2^2+2z_3^2)-b_1^n(\mathbf{x})z_1-b_2^n(\mathbf{x})z_2-2b_3^n(\mathbf{x})z_3\right],\end{cases}\tag{8.161}$$

where

$$\mathbf{Z}(f,\mathbf{x})=\{\mathbf{z}|\mathbf{z}=\{z_i\}_{i=1}^3\in I\!R^3,z_1,z_2>0,z_1z_2-z_3^2=f(\mathbf{x})\}$$

and

$$\begin{cases}b_1^n(\mathbf{x})=\dfrac{\partial^2u^{n-1}}{\partial x_1^2}(\mathbf{x})+\dfrac{1}{r}\lambda_{11}(\mathbf{x}),\\ b_2^n(\mathbf{x})=\dfrac{\partial^2u^{n-1}}{\partial x_2^2}(\mathbf{x})+\dfrac{1}{r}\lambda_{22}(\mathbf{x}),\\ b_3^n(\mathbf{x})=\dfrac{\partial^2u^{n-1}}{\partial x_1\partial x_2}(\mathbf{x})+\dfrac{1}{r}\lambda_{12}(\mathbf{x}).\end{cases}$$

To transform the three-dimensional *constrained minimization* problem (8.161) into an *unconstrained* two-dimensional one, we perform the following change of variables:

$$z_1=\sqrt{f(\mathbf{x})}e^\rho\cosh\theta,z_2=\sqrt{f(\mathbf{x})}e^{-\rho}\cosh\theta,\text{and } z_3=\sqrt{f(\mathbf{x})}\sinh\theta.$$

There is then equivalence between (8.161) and the following two-dimensional minimization problem:

$$\begin{cases} \{\rho^n, \theta^n\} \in \mathbb{R}^2, \\ j_n(\rho^n, \theta^n) \leq j_n(\rho, \theta), \forall \{\rho, \theta\} \in \mathbb{R}^2, \end{cases} \tag{8.162}$$

where

$$\begin{aligned} j_n(\rho, \theta) = & \frac{\sqrt{f(\mathbf{x})}}{2}(\cosh 2\rho \ \cosh 2\theta + \cosh 2\rho + \cosh 2\theta - 1) \\ & -(b_1^n(\mathbf{x})e^{\rho} + b_2^n(\mathbf{x})e^{-\rho})\cosh\theta - 2b_3^n(\mathbf{x})\sinh\theta. \end{aligned}$$

This in turn leads to the solution of

$$Dj_n(\rho^n, \theta^n) = \mathbf{0}, \tag{8.163}$$

where the vector-valued function Dj_n is the differential of the functional j_n. To solve the nonlinear system (8.163), we can use the *Newton's method*; we obtain then (after dropping the subscript and superscript n):

$$\begin{cases} \{\rho^0, \theta^0\} \text{ is given in } \mathbb{R}^2. \\ \text{For } k \geq 0, \{\rho^k, \theta^k\} \to \{\rho^{k+1}, \theta^{k+1}\} \text{ via the solution of} \\ D^2 j(\rho^k, \theta^k) \begin{pmatrix} \rho^{k+1} - \rho^k \\ \theta^{k+1} - \theta^k \end{pmatrix} = -Dj(\rho^k, \theta^k), \end{cases} \tag{8.164}$$

where $D^2 j(\rho, \theta)$ is the Hessian of j at $\{\rho, \theta\}$.

An alternative to the above Newton's method can be found in [60]; numerical experiments show that this new algorithm is both faster and more robust than the Newton's algorithm (8.164). Detailed comparisons between the two algorithms can be found in [10].

On the solution of (8.160): The sub-problems (8.160) are all members of the following family of *well-posed linear variational problems*:

$$\begin{cases} u \in V_g, \\ \int_\Omega \nabla^2 u \nabla^2 v \, d\mathbf{x} + r \int_\Omega \mathbf{D}^2 u : \mathbf{D}^2 v \, d\mathbf{x} = L(v), \forall v \in V_0, \end{cases} \tag{8.165}$$

with the functional L linear and continuous from $H^2(\Omega)$ into $\mathbb{R}$; problem (8.165) is clearly of the *bi-harmonic* type. The *conjugate gradient* solution of linear variational problems in Hilbert spaces, such as (8.165), has been discussed in the Chapter 3 of [32] and in the Chapter 2 of [34]. Following the two above references, we are going to solve (8.165) by a conjugate gradient algorithm operating in the spaces V_0 and V_g, both spaces being equipped with the inner product defined by

$\{v,w\} \to \int_\Omega \nabla^2 v \nabla^2 w \, d\mathbf{x}$ and the associated norm. When applied to the solution of problem (8.165), this conjugate gradient algorithm reads as follows:

$$u^0 \text{ is given in } V_g; \tag{8.166}$$

solve then

$$\begin{cases} g^0 \in V_0, \\ \int_\Omega \nabla^2 g^0 \nabla^2 v \, d\mathbf{x} = \int_\Omega \nabla^2 u^0 \nabla^2 v \, d\mathbf{x} + r \int_\Omega \mathbf{D}^2 u^0 : \mathbf{D}^2 v \, d\mathbf{x} - L(v), \forall v \in V_0. \end{cases} \tag{8.167}$$

If $\dfrac{\int_\Omega |\nabla^2 g^0|^2 \, d\mathbf{x}}{\int_\Omega |\nabla^2 u^0|^2 \, d\mathbf{x}} \leq tol^2$, take $u = u^0$; otherwise, set

$$w^0 = g^0. \tag{8.168}$$

Then, for $k \geq 0$, u^k, g^k, w^k being known, the last two different from 0, we compute u^{k+1}, g^{k+1}, and if necessary w^{k+1}, as follows:

Solve

$$\begin{cases} \bar{g}^k \in V_0, \\ \int_\Omega \nabla^2 \bar{g}^k \nabla^2 v \, d\mathbf{x} = \int_\Omega \nabla^2 w^k \nabla^2 v \, d\mathbf{x} + r \int_\Omega \mathbf{D}^2 w^k : \mathbf{D}^2 v \, d\mathbf{x}, \forall v \in V_0, \end{cases} \tag{8.169}$$

and compute

$$\rho_k = \frac{\int_\Omega |\nabla^2 g^k|^2 \, d\mathbf{x}}{\int_\Omega \nabla^2 \bar{g}^k \nabla^2 w^k \, d\mathbf{x}}, \tag{8.170}$$

$$u^{k+1} = u^k - \rho_k w^k, \tag{8.171}$$

$$g^{k+1} = g^k - \rho_k \bar{g}^k. \tag{8.172}$$

If $\dfrac{\int_\Omega |\nabla^2 g^{k+1}|^2 \, d\mathbf{x}}{\max\left[\int_\Omega |\nabla^2 g^0|^2 \, d\mathbf{x}, \int_\Omega |\nabla^2 u^{k+1}|^2 \, d\mathbf{x}\right]} \leq tol^2$, take $u = u^{k+1}$; else, compute

$$\gamma_k = \frac{\int_\Omega |\nabla^2 g^{k+1}|^2 \, d\mathbf{x}}{\int_\Omega |\nabla^2 g^k|^2 \, d\mathbf{x}}, \tag{8.173}$$

and

$$w^{k+1} = g^k + \gamma_k w^k. \tag{8.174}$$

Do $k+1 \to k$ and return to (8.169).

Concerning the choice of *tol*, see the two above references.

Numerical experiments have shown that algorithm (8.166)–(8.174) (in fact its discrete variants) has excellent convergence properties when applied to the solution of the Monge-Ampère problem (8.149); when combined with an appropriate *mixed finite element approximation* of (8.149), it requires the solution of two discrete Poisson problems at each iteration. For more details, see [21, 22] and also the Chapter 9 of [34] where the *least-squares/mixed finite elements* solution of some *fully nonlinear elliptic equations* (including (8.149)) is discussed.

3.4 Numerical Experiments

A *mixed finite element* implementation of the augmented Lagrangian algorithm (8.154)–(8.157) has been discussed in [21] and [22]. Here, we summarize the main numerical results and observations from the two above publications. Below, h denotes the mesh size and $\{u_h^c, \mathbf{p}_h^c\}$ the computed mixed finite element approximation of $\{u, \mathbf{p}\}$. We consider below three test problems, all associated with $\Omega = (0,1)^2$, the finite element grids used for these computations being uniform ones like the one shown in Figure 8.2 (with $h = 1/4$).

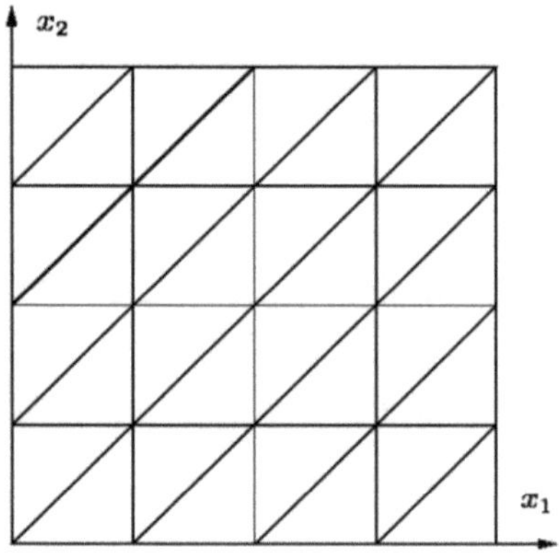

Fig. 8.2 A uniform triangulation of the unit square ($h = 1/4$)

The *first test problem* is defined as follows (with $|\mathbf{x}| = \sqrt{x_1^2 + x_2^2}$):

$$\begin{cases} \det \mathbf{D}^2 u = \dfrac{1}{|\mathbf{x}|} \text{ in } \Omega \\ u = \dfrac{(2|\mathbf{x}|)^{\frac{3}{2}}}{3} \text{ on } \Gamma. \end{cases} \tag{8.175}$$

The unique convex solution of (8.175) is given by $u(\mathbf{x}) = \frac{(2|\mathbf{x}|)^{\frac{3}{2}}}{3}$. If we set $r = 1$, and use $\|\mathbf{D}_h^2 u_h^n - \mathbf{p}_h^n\|_{0h} \leq 10^{-6}$ as as stopping criterion, then the discrete analogue of algorithm (8.154)–(8.157) converges in about 150 iterations to an approximate solution u_h^c verifying $\|u_h^c - u\|_{0h} = O(h^2)$, which is generically optimal for second order elliptic problems when using piecewise affine approximations (above, $\|\cdot\|_{0h}$ denotes the approximation of $\|\cdot\|_{L^2(\Omega)}$, associated with the trapezoidal rule). Moreover, the number of iterations required to achieve convergence increases slowly with h^{-1} (actually, much slower than $h^{-1/2}$). It is worth noticing that, the function u defined by $u(\mathbf{x}) = \frac{1}{3}(2|\mathbf{x}|)^{\frac{3}{2}}$ does not belong to $C^2(\overline{\Omega})$, but since it belongs to $W^{2,p}(\Omega)$ for all $p \in [1,4)$, it has enough regularity-in principle-to be captured by algorithm (8.154)–(8.157), which is indeed the case.

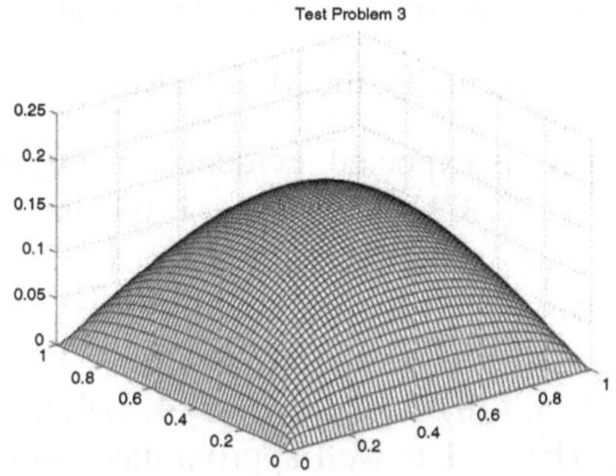

Fig. 8.3 Third test problem: Graph of $-u_h^c$ ($h = 1/64$)

The *second test problem* is

$$\begin{cases} \det \mathbf{D}^2 u = \dfrac{R^2}{(R^2 - |\mathbf{x}|^2)^2} \text{ in } \Omega \\ u = -\sqrt{R^2 - |\mathbf{x}|^2} \text{ on } \Gamma, \end{cases} \tag{8.176}$$

with $R \geq \sqrt{2}$. The *convex* solution of problem (8.176) is given by $u(\mathbf{x}) = -\sqrt{R^2 - |\mathbf{x}|^2}$. It is easy to check that $u \in C^\infty(\overline{\Omega})$ if $R > \sqrt{2}$, and $u \in W^{1,p}(\Omega)$ for $p \in [1,4)$ (which is sharp) if $R = \sqrt{2}$. We set again $r = 1$. For $R = 2$ and $R = \sqrt{2} + 0.1$, the number of iterations to achieve convergence is essentially independent of h and is close to 80 (resp., 120) for $R = 2$ (resp., $R = \sqrt{2} + 0.1$); we still have $\|u_h^c - u\|_{0h} = O(h^2)$. On the other hand, if $R = \sqrt{2} + 0.01$, the discrete analogue of algorithm (8.154)–(8.157) does not converge, although $u \in C^\infty(\overline{\Omega})$(but 'barely').

The *third test problem* is

$$\begin{cases} \det \mathbf{D}^2 u = 1 \text{ in } \Omega, \\ u = 0 \text{ on } \Gamma. \end{cases}$$

We recall that if $\Omega = (0,1)^2$ then, despite the smoothness of its data, the above problem has no smooth solution, due the non-strict convexity of Ω. After applying, for various values of h, the discrete variant of algorithm (8.154)–(8.157) discussed in [21] and [22], the following phenomena were observed (we used $\|u_h^n - u_h^{n-1}\|_{0h} \leq 10^{-6}$ as stopping criterion).

(i) For r sufficiently small, but not too small ($r = 1$, here), the sequence $\{\{u_h^n, \mathbf{p}_h^n\}\}_{n\geq 0}$ converges geometrically (albeit slowly since the number of iterations required to achieve convergence increases as $h^{-1/2}$) to a limit $\{u_h^c, \mathbf{p}_h^c\}$, which is clearly (see Figure 8.3) the approximation of a function in V_g. On the other hand, the sequence $\{\lambda_h^n\}_{n\geq 0}$ diverges *arithmetically*.

(ii) A close-up look at the graph of $-u_h^c$ with $h = 1/64$ shows (see Figure 8.3) that the curvature of the graph becomes negative near the corners, which (locally) violates the Monge-Ampère equation (we recall that the *Gaussian curvature* K of the graph of a smooth two variable function ψ is given by $K = \dfrac{\det \mathbf{D}^2\psi}{(1+|\nabla\psi|^2)^2}$). Actually, the Monge-Ampère equation is also violated along the boundary, as expected, evidences being that $\|\mathbf{D}_h^2 u_h^c - \mathbf{p}_h^c\|_{0h,\Omega} = 1.8 \times 10^{-2}$ if $h = 1/32$, 3.3×10^{-2} if $h = 1/64$, and 4.2×10^{-2} if $h = 1/128$, while $\|\mathbf{D}_h^2 u_h^c - \mathbf{p}_h^c\|_{0h,\Omega_1} = 2.7 \times 10^{-4}$ if $h = 1/32$, 4.1×10^{-4} if $h = 1/64$, and 4.9×10^{-4} if $h = 1/128$, and finally $\|\mathbf{D}_h^2 u_h^c - \mathbf{p}_h^c\|_{0h,\Omega_2} = 4.4 \times 10^{-5}$ if $h = 1/32$, 4.9×10^{-5} if $h = 1/64$, and 5.1×10^{-5} if $h = 1/128$. Above, we have $\Omega_1 = (1/8, 7/8)^2$ and $\Omega_2 = (1/4, 3/4)^2$. These results show that the condition $\det\mathbf{D}^2 u = 1$ is well approximated not too far away from the boundary.

Since u_h^c does not vary much with h, one might wonder what kind of generalized solution is captured by algorithm (8.154)–(8.157). Since, in this particular case, *ALG*2 is an *Uzawa type algorithm* applied to a problem where the constraint set is *empty*, it was conjectured in [21] (by analogy with simpler situations) that the sequence $\{\{u_h^n, \mathbf{p}_h^n\}\}_{n\geq 0}$ converges to a limit solving the Monge-Ampère problem (8.148) in a *least-squares sense*; actually, this was verified numerically, but proving it theoretically seems to be out of reach at the moment.

The *least-squares* solution of the Monge-Ampère problem (8.149) (and of the *Pucci equation*, another fully nonlinear elliptic equation) is discussed in the Chapter 9 of [34].

Remark 8. One of the main difficulties we encountered, when using *ALG*2 to solve problem (8.149), was finding the proper value of the *augmentation parameter* r. With r large, the convergence may be very slow, while with r too small, the algorithm may diverge. This behavior is typical of *non-convex* problems, and justifies the search for methods adjusting r automatically (such a method is discussed in [24] where it has been applied to the solution of an *inverse problem* from *Geophysics*).

4 Application to the Solution of a Non-smooth Eigenvalue Problem from Visco-plasticity

4.1 Formulation. Motivation

Our goal in this section is to discuss the *augmented Lagrangian* solution of a *non-smooth* and *non-convex* problem from *Calculus of Variations*, namely:

Compute

$$\gamma = \inf_{v \in \Sigma} \int_\Omega |\nabla v| \, d\mathbf{x}, \tag{8.177}$$

where Ω is a *bounded* domain of $\mathbb{R}^2$ and

$$\Sigma = \{v | v \in H_0^1(\Omega), \int_\Omega |v|^2 \, d\mathbf{x} = 1\}. \tag{8.178}$$

Problem (8.177) has, clearly, the features of an *eigenvalue problem.*

Concerning γ, it has been proved in [61] that $\gamma = 2\sqrt{\pi}$, independently of the shape and size of Ω (even if Ω is not simply connected). A question arising naturally is then

Why solve numerically a problem whose exact solution is known?

The main reasons for doing so are:

(i) Since its exact solution is known, problem (8.177) is a good problem to test and validate solution methods for non-smooth non-convex variational problems.

(ii) The functional $v \to \int_\Omega |\nabla v| \, d\mathbf{x}$ arises in a variety of problems from *image processing* and *plasticity*.

(iii) Since we expect the *minimizing sequences* associated with (8.177) to converge to a limit *outside* of Σ, it is interesting, mathematically, to investigate, via numerical methods, the limit of these sequences and what kind of convergence will take place.

Actually, our main motivation for investigating (8.177) stems from the following problem from *visco-plasticity* (*unsteady flow* of a *Bingham fluid* in an infinitely long cylindrical pipe of cross-section Ω):

Find $u \in L^2(0,T; H_0^1(\Omega)) \cap C^0([0,T]; L^2(\Omega))$ such that, a.e. on $(0,T)$,

$$\begin{cases} \rho \left\langle \dfrac{\partial u}{\partial t}(t), v - u(t) \right\rangle + \mu \displaystyle\int_\Omega \nabla u(t) \cdot \nabla(v - u(t)) \, d\mathbf{x} \\ + \tau_y [\displaystyle\int_\Omega |\nabla v| \, d\mathbf{x} - \int_\Omega |\nabla u(t)| \, d\mathbf{x}] \geq C \int_\Omega (v - u(t)) \, d\mathbf{x}, \forall v \in H_0^1(\Omega), \\ u(0) = u_0, \end{cases} \tag{8.179}$$

where: (i) Ω is a bounded domain of $I\!R^2$. (ii) $\rho(>0)$ is the fluid *density*, $\tau_y(>0)$ is the fluid *plasticity yield*, and $\mu(>0)$ is the fluid *viscosity*. (iii) $u(t)$ denotes the function $\mathbf{x} \to u(\mathbf{x},t)$, $u(\mathbf{x},t)$ being the axial velocity of the fluid at the point $\mathbf{x}$ of the cross-section and at time t. (iv) $C(t)$ is the *pressure drop per unit length* at time t. (v) $< \cdot,\cdot >$ denotes the duality pairing between $H^{-1}(\Omega)$ and $H_0^1(\Omega)$, verifying $< f, v >= \int_\Omega f v\, d\mathbf{x}$ if $f \in L^2(\Omega)$. (vi) $u_0 \in L^2(\Omega)$.

The parabolic variational inequality problem (8.179) is well posed as shown in [26].

Suppose now that $C \equiv 0$ and $T = +\infty$ in (8.179). We can easily prove (see, e.g., [31] and [23]) that

$$u(t) = 0, \forall t \geq T_c, \tag{8.180}$$

with

$$T_c = \frac{\rho}{\lambda_0 \mu} \ln\left(1 + \frac{\lambda_0 \mu}{\gamma \tau_y} \|u_0\|_{L^2(\Omega)}\right), \tag{8.181}$$

λ_0 being the *smallest eigenvalue* of $-\nabla^2$ in $H_0^1(\Omega)$. As shown in the two above references, a similar cut-off property holds if, after an appropriate *finite element approximation* we use the *backward Euler scheme* for the time discretization of (8.179), with λ_0 and γ replaced by their discrete analogues λ_{0h} and γ_h. Suppose that the space discretization is achieved via a C^0-piecewise linear finite element approximation like those discussed in, e.g., the Appendix 1 of [31]; we have then

$$|\lambda_{0h} - \lambda_0| = O(h^2),$$

as shown in, e.g., [57], but what can we say about $|\gamma_h - \gamma|$? One of the goals of this section is to find an answer to the above question.

4.2 Some Regularization Procedures

There are several ways to approximate the minimization problem (8.177)-at the continuous level- by a better posed and/or smoother variational problem. The most obvious candidate is clearly

$$\gamma_\varepsilon = \inf_{v \in \Sigma} \int_\Omega \sqrt{\varepsilon + |\nabla v|^2}\, d\mathbf{x}, \tag{8.182}$$

with $\varepsilon > 0$. Problem (8.182) involves a *regularization* procedure which has been quite popular in *Image Processing*. Assuming that the above problem has a minimizer u_ε, this minimizer verifies the following *Euler-Lagrange equation* (reminiscent of the celebrated *mean curvature* equation)

$$\begin{cases} -\nabla \cdot \left(\dfrac{\nabla u_\varepsilon}{\sqrt{\varepsilon + |\nabla u_\varepsilon|^2}} \right) = \gamma_\varepsilon u_\varepsilon \text{ in } \Omega, \\ u_\varepsilon = 0 \text{ on } \Gamma, \\ \displaystyle\int_\Omega |u_\varepsilon|^2 \, d\mathbf{x} = 1. \end{cases} \tag{8.183}$$

Problem (8.183) is clearly a *nonlinear eigenvalue problem* for a close variant of the *mean curvature operator*, the eigenvalue being γ_ε.

Another regularization, more sophisticated in some sense (since this time the regularized problem has minimizers), is provided (with $\varepsilon > 0$) by

$$\gamma_\varepsilon = \inf_{v \in \Sigma} \left[\frac{\varepsilon}{2} \int_\Omega |\nabla v|^2 \, d\mathbf{x} + \int_\Omega |\nabla v| \, d\mathbf{x} \right]. \tag{8.184}$$

An associated *Euler-Lagrange (multivalued) equation* reads as follows:

$$\begin{cases} -\varepsilon \nabla^2 u_\varepsilon + \partial j(u_\varepsilon) \ni \gamma_\varepsilon u_\varepsilon \text{ in } \Omega, \\ u_\varepsilon = 0 \text{ on } \Gamma, \\ \displaystyle\int_\Omega |u_\varepsilon|^2 \, d\mathbf{x} = 1. \end{cases} \tag{8.185}$$

where, in (8.185), $\partial j(u_\varepsilon)$ is the sub-differential at u_ε of the functional $j : H_0^1(\Omega) \to I\!R$ defined by $j(v) = \int_\Omega |\nabla v| \, d\mathbf{x}$. Problem (8.185) is clearly a *nonlinear non-smooth eigenvalue* problem with γ_ε the eigenvalue. The numerical solution of problems such as (8.185) is discussed in [51]; the method used in the above reference is based on *operator-splitting*, giving it the features of an *inverse power method.*

In order to avoid dealing simultaneously with *two small parameters*, namely ε and h, we will address the solution of problem (8.177) *without using any regularization* (unless one considers the *space approximation* of (8.177) as a kind of regularization, which is indeed the case since the *discrete analogues* of (8.177) have minimizers).

4.3 Finite Element Approximation of Problem (8.177)

In order to approximate problem (8.177), we will proceed as follows:

(i) First, we introduce a family $\{\Omega_h\}_h$ of polygonal approximations of Ω verifying

$$\lim_{h \to 0} \Omega_h = \Omega$$

(that is: (a) If $\mathscr{U}$ is an *open* set containing $\overline{\Omega}$, we have $\overline{\Omega}_h \subset \mathscr{U}$ for *h sufficiently small*; (b) if K is a *compact* set contained in Ω, we have $K \subset \Omega_h$ for *h sufficiently small*).

(ii) With each Ω_h we associate a *finite element triangulation* $\mathscr{T}_h$ verifying the usual assumptions of *compatibility* between triangles and of *regularity* (see the Appendix 1 of [31] for details).

(iii) With each $\mathscr{T}_h$ we associate the *finite dimensional* space V_{0h} defined by

$$V_{0h} = \{v | v \in C^0(\overline{\Omega}_h), v|_K \in P_1, \forall K \in \mathscr{T}_h, v = 0 \text{ on } \partial\Omega_h\}, \quad (8.186)$$

with P_1 the space of the polynomials of two variables of degree ≤ 1.

We approximate then problem (8.177) by

$$\gamma_h = \inf_{v \in \Sigma_h} \int_{\Omega_h} |\nabla v| \, d\mathbf{x}, \quad (8.187)$$

with

$$\Sigma_h = \{v | v \in V_{0h}, \int_{\Omega_h} |v|^2 \, d\mathbf{x} = 1\}. \quad (8.188)$$

In [9] one proved the following

Theorem 2. *The minimization problem in (8.187) has a solution, that is there exists* $u_h \in \Sigma_h$ *such that*

$$\gamma_h = \int_{\Omega_h} |\nabla u_h| \, d\mathbf{x}.$$

Moreover

$$\lim_{h \to 0} \gamma_h = \gamma (= 2\sqrt{\pi}). \quad (8.189)$$

The existence of u_h follows from the fact that the functional $v \to \int_{\Omega_h} |\nabla u_h| \, d\mathbf{x}$ defines a norm over V_{0h} (which implies its continuity) and from the *compactness* of Σ_h in V_{0h}. To prove (8.189), one can take advantage of the density of $\mathscr{D}(\Omega)$ in $H_0^1(\Omega)$, as done in the above reference.

From the *non-smoothness* of the problem, we do not expect $|\gamma_h - \gamma| = O(h^2)$. On the other hand, we expect $|\gamma_h - \gamma| = O(h^\alpha)$, with $0 < \alpha < 2$ and α depending of the shape of Ω.

Remark 9. There is no difficulty at expressing Σ_h as a function of the values taken by v at the vertices of $\mathscr{T}_h$ (using the two-dimensional Simpson rule on each triangle of $\mathscr{T}_h$). However, we found more convenient, from a computational point of view, to use the following approximation of Σ, derived from (8.178) by application of the *trapezoidal rule*:

$$\Sigma_h^* = \{v | v \in V_{0h}, \|v\|_{0h} = 1\}, \quad (8.190)$$

with

$$\|v\|_{0h} = \sqrt{\frac{1}{3} \sum_{i=1}^{N_{0h}} |\omega_i| |v(Q_i)|^2} \quad (8.191)$$

where: (a) $\{Q_i\}_{i=1}^{N_{0h}}$ is the set of the vertices of $\mathscr{T}_h$ interior to Ω_h; we have then $N_{0h} = \dim V_{0h}$. (b) ω_i is the polygonal union of those triangles of $\mathscr{T}_h$ which have Q_i has a common vertex, and $|\omega_i| =$ measure of ω_i.

4.4 *Applying ALG2 to the Solution of Problem (8.187)*

When addressing the numerical solution of the minimization problem in (8.187), we are going to privilege *robustness, modularity* and *programming simplicity* instead of performances measured in number of elementary operations (this is not Image Processing and/or real time).

4.4.1 An Equivalent Formulation and Its Associated Augmented Lagrangian

For formalism simplicity, we will use the *continuous problem* notation. We observe that there is *equivalence* between

$$\gamma = \inf_{v\in\Sigma}\int_\Omega |\nabla v|\, d\mathbf{x},$$

and

$$\gamma = \inf_{\{v,\mathbf{q},z\}\in\mathbf{E}}\int_\Omega |\mathbf{q}|\, d\mathbf{x}, \tag{8.192}$$

where $\mathbf{E}$ is defined by

$$\mathbf{E} = \{\{v,\mathbf{q},z\} \mid v \in H_0^1(\Omega), \mathbf{q} \in (L^2(\Omega))^2, z \in L^2(\Omega), \nabla v - \mathbf{q} = \mathbf{0},\\ v - z = 0, \|z\|_{L^2(\Omega)} = 1\}. \tag{8.193}$$

We associate with (8.192) and (8.193) the following augmented Lagrangian functional

$$\mathscr{L}_{\mathbf{r}} : (H_0^1(\Omega)\times\mathbf{Q}\times L^2(\Omega))\times(\mathbf{Q}\times L^2(\Omega)) \to I\!R$$

defined, with $\mathbf{Q} = (L^2(\Omega))^2$ and $\mathbf{r} = \{r_1, r_2\} (r_1, r_2 > 0)$, by

$$\begin{cases} \mathscr{L}_{\mathbf{r}}(v,\mathbf{q},z;\mu_1,\mu_2) = \displaystyle\int_\Omega |\mathbf{q}|\, d\mathbf{x} + \frac{r_1}{2}\int_\Omega |\nabla v - \mathbf{q}|^2\, d\mathbf{x} \\ +\displaystyle\frac{r_2}{2}\int_\Omega |v-z|^2\, d\mathbf{x} + \int_\Omega \boldsymbol{\mu}_1\cdot(\nabla v - \mathbf{q})\, d\mathbf{x} + \int_\Omega \mu_2(v-z)\, d\mathbf{x}. \end{cases} \tag{8.194}$$

Next, we consider the following *saddle-point problem*

$$\begin{cases} \{\{u,\mathbf{p},y\},\{\boldsymbol{\lambda}_1,\lambda_2\}\} \in (H_0^1(\Omega)\times\mathbf{Q}\times S)\times(\mathbf{Q}\times L^2(\Omega)) \text{ such that} \\ \mathscr{L}_{\mathbf{r}}(u,\mathbf{p},y;\boldsymbol{\mu}_1,\mu_2) \le \mathscr{L}_{\mathbf{r}}(u,\mathbf{p},y;\boldsymbol{\lambda}_1,\lambda_2) \le \mathscr{L}_{\mathbf{r}}(v,\mathbf{q},z;\boldsymbol{\lambda}_1,\lambda_2), \\ \forall\{\{v,\mathbf{q},z\},\{\boldsymbol{\mu},\mu_2\}\} \in (H_0^1(\Omega)\times\mathbf{Q}\times S)\times(\mathbf{Q}\times L^2(\Omega)), \end{cases} \tag{8.195}$$

with

$$S = \{z | z \in L^2(\Omega), \|z\|_{L^2(\Omega)} = 1\}. \tag{8.196}$$

Suppose that the above saddle-point problem has a solution, then $\mathbf{p} = \nabla u$ and $y = u$, u being a solution of the minimization problem in (8.177) (actually such a minimizer does not exist for (8.177), but exists for discrete analogues such as (8.187), or its variant obtained by replacing Σ_h by Σ_h^* defined by (190)).

4.4.2 An ADMM Type Algorithm for the Solution of (8.177)

To solve the saddle-point problem (8.195) (and also its discrete analogues), we advocate one more time *ALG2* (despite the *non-convexity* of the set *S*). We obtain then the following algorithm:

$$\{u^{-1}, \{\lambda_1^0, \lambda_2^0\}\} \text{ is given in } H_0^1(\Omega) \times (\mathbf{Q} \times L^2(\Omega)). \tag{8.197}$$

For $n \geq 0$, $\{u^{n-1}, \{\lambda_1^n, \lambda_2^n\}\} \to \{\mathbf{p}^n, y^n\} \to u^n \to \{\lambda_1^{n+1}, \lambda_2^{n+1}\}$ via

$$\{\mathbf{p}^n, y^n\} = \arg\min_{\{\mathbf{q},z\} \in \mathbf{Q} \times S} \mathscr{L}_{\mathbf{r}}(u^{n-1}, \mathbf{q}, z; \lambda_1^n, \lambda_2^n), \tag{8.198}$$

$$u^n = \arg\min_{v \in H_0^1(\Omega)} \mathscr{L}_{\mathbf{r}}(v, \mathbf{p}^n, y^n; \lambda_1^n, \lambda_2^n), \tag{8.199}$$

$$\begin{cases} \lambda_1^{n+1} = \lambda_1^n + r_1(\nabla u^n - \mathbf{p}^n), \\ \lambda_2^{n+1} = \lambda_2^n + r_2(u^n - y^n). \end{cases} \tag{8.200}$$

Problem (8.199) is *equivalent* to the following *well-posed linear variational problem*

$$\begin{cases} u^n \in H_0^1(\Omega), \\ r_1 \int_\Omega \nabla u^n \cdot \nabla v \, d\mathbf{x} + r_2 \int_\Omega u^n v \, d\mathbf{x} = \\ \int_\Omega (r_1 \mathbf{p}^n - \lambda_1^n) \cdot \nabla v \, d\mathbf{x} + \int_\Omega (r_2 y^n - \lambda_2^n) v \, d\mathbf{x}, \ \forall v \in H_0^1(\Omega), \end{cases} \tag{8.201}$$

which implies in turn that u^n is the unique solution in $H_0^1(\Omega)$ of the following Dirichlet problem

$$\begin{cases} -r_1 \nabla^2 u^n + r_2 u^n = -\nabla \cdot (r_1 \mathbf{p}^n - \lambda_1^n) + r_2 y^n - \lambda_2^n \text{ in } \Omega, \\ u^n = 0 \text{ in } \Gamma. \end{cases} \tag{8.202}$$

The solution of the discrete analogues of problem (8.201), (8.202), by direct or iterative methods, is routine nowadays.

Problem (8.198) decouples as

$$\mathbf{p}^n = \arg\min_{\mathbf{q} \in \mathbf{Q}} \left[\frac{r_1}{2} \int_\Omega |\mathbf{q}|^2 \, d\mathbf{x} + \int_\Omega |\mathbf{q}| \, d\mathbf{x} - \int_\Omega (r_1 \nabla u^{n-1} + \lambda_1^n) \cdot \mathbf{q} \, d\mathbf{x} \right], \tag{8.203}$$

$$y^n = \arg\min_{z \in S} \left[\frac{r_2}{2} \int_\Omega |z|^2 \, d\mathbf{x} - \int_\Omega (r_2 u^{n-1} + \lambda_2^n) z \, d\mathbf{x} \right]. \tag{8.204}$$

Both problems (8.203) and (8.204) have *closed form solutions*, namely

$$\mathbf{p}^n = \frac{1}{r_1}\left(1 - \frac{1}{|\mathbf{X}^n(\mathbf{x})|}\right)^+ \mathbf{X}^n(\mathbf{x}), \text{ a.e. on } \Omega \tag{8.205}$$

(with $\mathbf{X}^n = r_1 \nabla u^{n-1} + \lambda_1^n$), and

$$y^n = \frac{r_2 u^{n-1} + \lambda_2^n}{\|r_2 u^{n-1} + \lambda_2^n\|_{L^2(\Omega)}}. \tag{8.206}$$

The finite element implementation of algorithm (8.197)–(8.200) is discussed in [9]; it relies on the choice of (8.190) and (8.191) to approximate Σ.

Remark 10. One of the main issues (if not the main issue), concerning the implementation of algorithm (8.197)–(8.200), is the proper choice of the two augmentation parameters r_1 and r_2; as we already know this is a general problem concerning the use of *ADMM* type algorithms. In the particular case of algorithm (8.197)–(8.200), the inspection of (8.201), (8.202) can give some hint concerning the choice of the ratio r_2/r_1. An obvious (and standard) approach is to balance the contributions of the operators $-r_1\nabla^2$ and $r_2 I$ to the main diagonal of the *stiffness matrix* originating from the finite element approximation of the elliptic problem (8.201), (8.202): this will lead to $r_2/r_1 = Ch^2$, with C of the order of 1. This strategy was the one employed in [9], but we think that it may be too 'radical', a 'milder' choice could have been to take the geometric mean of the above Ch^2 with the smallest eigenvalue of $-\nabla^2$ in $H_0^1(\Omega)$ (denoted by λ_0 in Section 4.1); this alternative was not tested.

4.5 Numerical Experiments

4.5.1 Synopsis

In this section, we will present the results obtained when applying the *augmented Lagrangian/finite element methodology* discussed in Sections 4.3 and 4.4 to a variety of test problems. We will discuss in particular:

(i) The influence of the shape of Ω on the order of convergence to zero of $|\gamma_h - \gamma|$.
(ii) The behavior of the minimizing sequences $\{u_h\}_h$ as $h \to 0$.

Concerning (ii), we will see that, if Ω is *simply connected*, then u_h converges in $BV(\Omega)$ to a limit u which is (modulo a multiplicative constant) the characteristic function of a disk contained in Ω ($BV(\Omega)$ is the celebrated space of those functions which have a *bounded variation* over Ω).

4.5.2 Numerical Results for Disk Shaped Domains

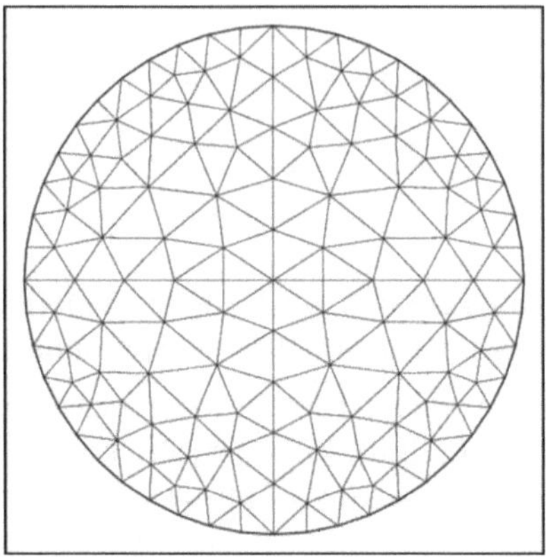

Fig. 8.4 A finite element triangulation of the unit disk

The *first test problem* that we consider concerns the case where Ω is the *unit disk* centered at the origin, that is

$$\Omega = \{\mathbf{x}|\mathbf{x} = \{x_1, x_2\} \in \mathbb{R}^2, x_1^2 + x_2^2 < 1\}.$$

In Figure 8.4 we have visualized a (relatively coarse) finite element triangulation $\mathscr{T}_h$ of Ω; this triangulation is *unstructured* but reasonably *isotropic* (as will be the other ones associated with the smaller values of h). When applying the discrete analogue of algorithm (8.197)–(8.200) to the solution of the approximate problem (8.187) we used (with obvious notation) $\{\lambda_{1h}^0, \lambda_{2h}^0\} = \{\mathbf{0}, 0\}$ and took for u_h^{-1} the function of V_{0h} taking the value $1/\sqrt{\pi}$ at the vertices of $\mathscr{T}_h$ contained in Ω. Concerning $\mathbf{r}$, we took $r_1 = 1$ and $r_2 = 1,000$

Using $\sqrt{\sum_{i=1}^{N_{0h}} |u_h^n(Q_i) - u_h^{n-1}(Q_i)|^2} \leq 10^{-5}$ as stopping criterion, around 2,000 iterations are needed to obtain convergence of the discrete analogue of algorithm (8.197)–(8.200). In Table 8.1 we have reported the values of γ_h as a function of the mesh size (here the size of the largest edge(s) of $\mathscr{T}_h$);

h	γ_h
1.2446127×10^{-1}	3.6245176
2.2604052×10^{-2}	3.5578964
1.0673766×10^{-2}	3.5528046
8.3407899×10^{-3}	3.5489181
$2\sqrt{\pi} = 3.5449077$	

Table 8.1 Unit disk test problem: Variation of γ_h versus the mesh size of the triangulation $\mathscr{T}_h$

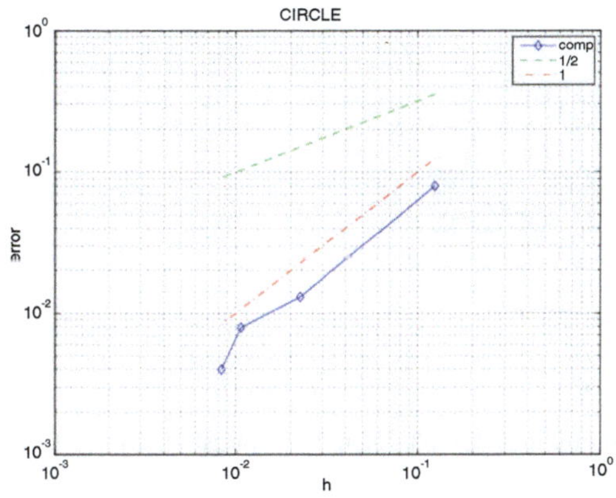

Fig. 8.5 Unit disk test problem: Variation of $\gamma_h - \gamma$ versus h (log-log scale)

the results of Table 8.1 strongly support the convergence result (8.189). In Figure 8.5, we have visualized, using a log-log scale, the variations versus h of the difference $\gamma_h - \gamma$; these results suggest $\gamma_h - \gamma = O(h)$.

In Figure 8.6 we have visualized the *contours* of the computed solutions associated with the *coarsest* (left; $h = 1.2446127 \times 10^{-1}$) and the *finest* (right; $h = 8.3407899 \times 10^{-3}$) triangulations We observe that the solution is *constant*, except in a *numerical boundary layer* whose thickness is of the order of h and where the solution jumps from 0 to a constant value converging to $1/\sqrt{\pi}(= 0.564189\ldots)$ as $h \to 0$ (for the finest mesh, this constant value is $0.567243\ldots$, which compares well with $1/\sqrt{\pi}$).

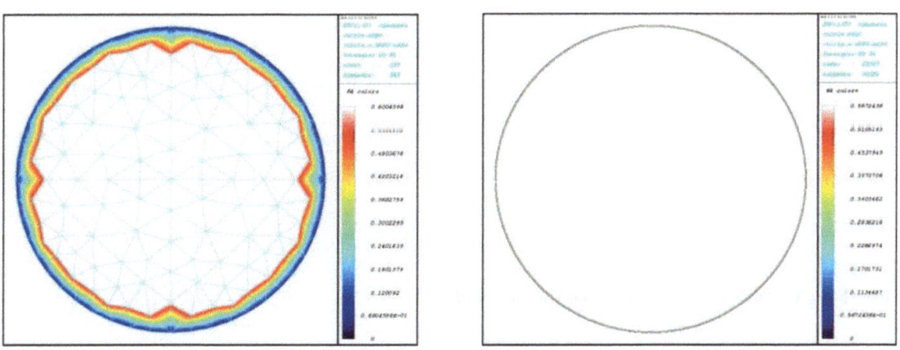

Fig. 8.6 Unit disk test problem: Contours of the computed solutions. *Left*: Coarse mesh solution ($h = 1.2446127 \times 10^{-1}$). *Right*: Fine mesh solution ($h = 8.3407899 \times 10^{-3}$)

In Figure 8.7, we have visualized the *graphs* of the computed solutions associated with the coarsest (left) and finest (right) triangulations. We note that the finite element approximation we use has no problem at handling the sharp discontinuity taking place at the boundary of the disk: no overshots or undershots are observed (actually, this property was already visible on Figure 8.6).

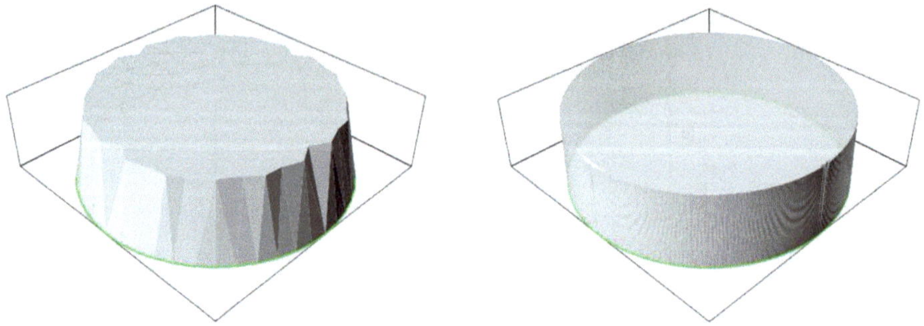

Fig. 8.7 Unit disk test problem: Graphs of the computed solutions. *Left*: Coarse mesh solution *Right*: Fine mesh solution

4.5.3 Numerical Results for Square Shaped Domains

The second test problem that we consider concerns the case where Ω is the unit square $(0,1)^2$.

In Figure 8.8 we have visualized a (relatively coarse) finite element triangulation of $\mathscr{T}_h$ of Ω; this triangulation (of the 'British flag' type) is *uniform*, the finer ones being obtained by refinement of this first one.

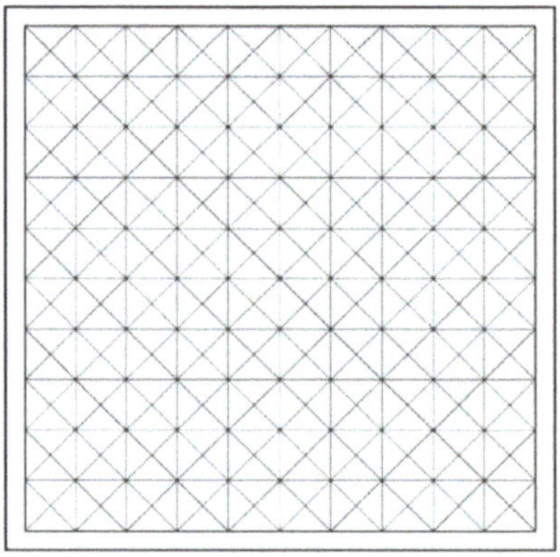

Fig. 8.8 A uniform triangulation of the unit square

When applying the discrete analogue of algorithm (8.197)–(8.200) to the solution of the approximate problem (8.187) we used (with obvious notation) $\{\lambda_{1h}^0, \lambda_{2h}^0\} = \{\mathbf{0}, 0\}$ and took for u_h^{-1} the function of V_{0h} taking the value 1 at the vertices of $\mathscr{T}_h$ contained in Ω. Concerning $\mathbf{r}$, we took $r_1 = 1$ and $r_2 = 1{,}000$, that is the same values we used in Section 4.5.2. The number of iterations necessary to obtain convergence (using again $\sqrt{\sum_{i=1}^{N_{0h}} |u_h^n(Q_i) - u_h^{n-1}(Q_i)|^2} \leq 10^{-5}$ as stopping criterion) is still about 2,000. In Table 8.2 we have reported the values of γ_h as a function of the mesh size; the results in Table 8.2 support the convergence result (8.189). In Figure 8.9, we have visualized, using a log-log scale, the variations versus h of the difference $\gamma_h - \gamma$; these results suggest $\gamma_h - \gamma = O(\sqrt{h})$.

h	γ_h
7.0710651×10^{-2}	3.6705782
1.7677653×10^{-2}	3.6123807
8.8388053×10^{-3}	3.5925411
5.8925086×10^{-3}	3.5827775
$2\sqrt{\pi} = 3.5449077$	

Table 8.2 Unit square test problem: Variation of γ_h versus the mesh size of the triangulation $\mathscr{T}_h$

In Figure 8.10 we have visualized the *contours* of the computed solutions associated with the *coarsest* (left; $h = 7.0710651 \times 10^{-2}$) and the *finest* (right; $h = 5.8925086 \times 10^{-3}$) triangulations

Finally, we have visualized in Figure 8.11, the *graphs* of the computed solutions associated with the coarsest (left) and finest (right) triangulations. Here, again, we observe the ability of our continuous piecewise affine finite element approximation at capturing a sharp discontinuity without overshots or undershots.

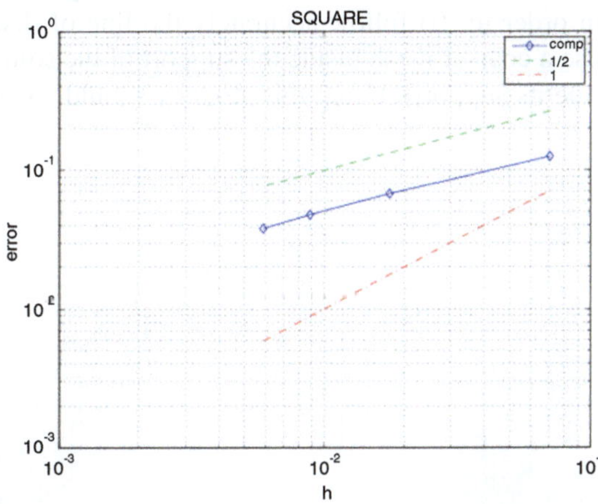

Fig. 8.9 Unit square test problem: Variation of $\gamma_h - \gamma$ versus h (log-log scale)

It is worth noting that the numerical results obtained with uniform triangulations like the one in Figure 8.2 in Section 3.4, or unstructured isotropic ones, are quite close to those reported in this subsection (obtained with uniform triangulations like the one in Figure 8.8); in particular, the convergence property $\gamma_h - \gamma = O(\sqrt{h})$ is still accurately verified.

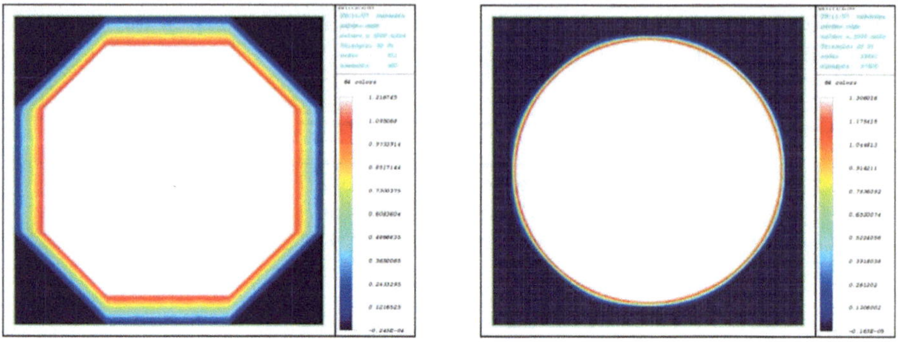

Fig. 8.10 Unit square test problem: Contours of the computed solutions. *Left*: Coarse mesh solution ($h = 7.0710651 \times 10^{-2}$). *Right*: Fine mesh solution ($h = 5.8925086 \times 10^{-3}$)

4.5.4 Further Comments and Remarks

The results of additional test problems can be found in [9]; they concern non-convex domains ω, including domains with holes and reentrant corners.

From the results presented in Sections 4.5.2 and 4.5.3, it is quite clear that the disk has the optimal shape concerning approximation accuracy; indeed, its shape matches the shape of the support of the limit of the minimizing sequence $\{u_h\}_h$ as $h \to 0$. An approach we did not follow, and even investigate, is to use *adaptive mesh refinement* in order to: (i) follow accurately the line of discontinuity of the gradient, and (ii) use a coarser mesh inside the support of the solution (the solution being constant inside its support). Using this strategy should significantly improve the convergence.

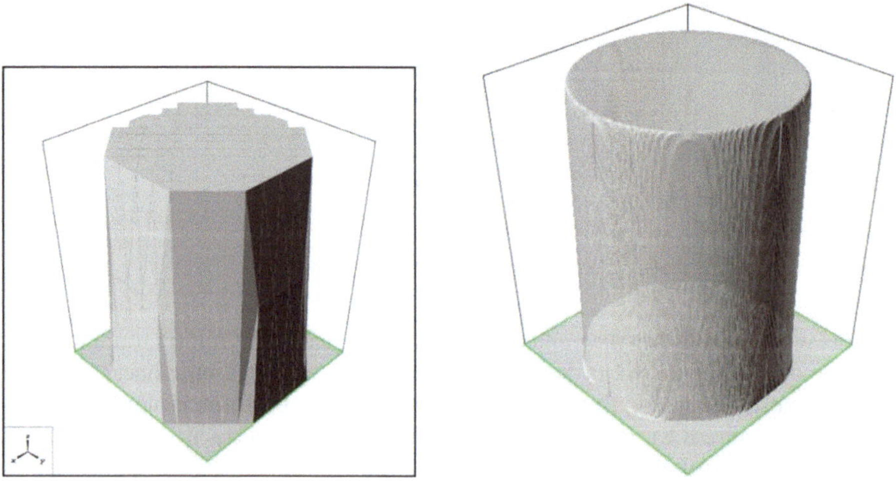

Fig. 8.11 Unit square test problem: Graphs of the computed solutions. *Left*: Coarse mesh solution *Right*: Fine mesh solution

Remark 11. The sequence $\{u_h\}_h$ being bounded in $BV(\Omega)$ is *pre-compact* in $L^1(\Omega)$; actually, our numerical results strongly support the following stronger property: $\{u_h\}_h$ converges to its limit in $L^s(\Omega)$, $\forall s \in [1, +\infty)$ and in $L^\infty(\Omega)$-weak *.

Remark 12. The results presented above required a large number of iterations (about 2,000). Several factors may explain this not so impressive behavior of the discrete analogue of algorithm (8.197)–(8.200):

(i) *ALG*1, *ALG*2 and *ALG*3 were designed to solve, in *Hilbert spaces*, variational problems with the appropriate structure. The natural functional space for problem (8.177) being $BV(\Omega)$, a close 'relative' to the space $W^{1,1}(\Omega)$, it is not surprising that one has a slow convergence when applying a discrete analogue of algorithm (8.197)–(8.200) to a finite element approximation of (8.177): indeed, for h sufficiently small, the approximate problems introduced in Section 4.3 retain many of the features of the continuous problems they come from, despite their finite dimensionality.
(ii) No serious effort was made to adjust r_1 and r_2, beyond taking r_2 substantially larger than r_1.
(iii) We did not take advantage, via a *cascadic* approach for example, of the fact that we have been solving a (finite) sequence of problems indexed by $h_1 > h_2 > h_3 > h_4$.

There is therefore a lot of room for improving the speed of convergence of algorithm (8.197)–(8.200), the basis of the methodology we used to solve problem (8.177).

However, we would like to insist that our main goals here were to investigate: (a) the order of convergence of the difference $\gamma_h - \gamma$ as $h \to 0$, and (b) the limit of the minimizing sequences, knowing in advance that they will converge outside of $H_0^1(\Omega)$. Owing to the modularity of our methodology, making it easy to implement, we had very quick answers to the questions associated with (a) and (b).

5 Further Comments

Albeit a firm convergence theory is still missing, ADMM based algorithms are increasingly applied to the solution of non-convex variational problems, some of them considered in Chapter 7 of this volume. Let us mention, among others, the contributions of R. Chartrand and his collaborators to *signal processing* and *medical imaging*, discussed in Chapter 7 of this volume and in [15, 16] and [17]. Personally, we doubt that a general theory exits for the convergence of ADMM when applied to the solution of non-convex problems. This belief stems from the few non-convex examples where convergence has been proved: each time, full advantage has been taken of the very specific properties of the problem under consideration. From the very large variety of non-convex problems encountered in applications, a general theory covering the known existing cases will have to be quite large and highly technical, but we expect it not to be large and technical enough to handle some of the new problems occurring on an almost daily basis. The likely non-existence of such a general theory for non-convex problems is not shocking in itself: after all there is no general theory for the existence of solutions of nonlinear partial differential equations, beyond the case of monotone operators (which is consistent with the fact that the differential of a non-convex functional is not a monotone operator). At any rate, one may find in Section 7 of Chapter 2 some comments and references concerning the "ADMM for non-convex problems" issue.

To conclude on a personal note, let us mention [43], an article dedicated to the numerical solution of a relatively complicated *non-convex variational problem* (suggested by C. Sundberg, UT- Knoxville), namely

$$\begin{cases} \text{Compute} \\ \gamma = \sup_{v \in E} \dfrac{\displaystyle\int_0^1 \frac{|v'|^4}{v^6}\, dx}{1 + \displaystyle\int_0^1 |v''|^2\, dx}, \\ \text{with} \\ E = \{v | v \in H^2(0,1), v(0) = v(1), v'(0) = v'(1), v \geq 1\}, \end{cases}$$

a problem with the features of an *obstacle* problem and of a *nonlinear eigenvalue* problem, that one could handle via ADMM (using three *augmentation parameters* and three *Lagrange multipliers* functions to treat the linear constraints $v - q_0 = 0$, $v' - q_1 = 0$, $v'' - q_2 = 0$).

Acknowledgments

The author wants to thank this chapter referee and his present and former colleagues and collaborators J.F. Bourgat, A. Caboussat, E.J. Dean, J.M. Dumay, P. Le Tallec, A. Quaini, T.W. Pan, V. Pons, and L. Tartar for their invaluable help and suggestions. The support of NSF grants DMS 0412267 and DMS 0913982 is also acknowledged.

References

1. Adams, R.A.: Sobolev Spaces. Academic Press, New York, NY (1975)
2. Benamou, J.D., Brenier, Y.: A computational fluid mechanics solution to the Monge-Kantorovich mass transfer problem. Numer. Math., **84**, 375–393 (2000)
3. Böhmer, K.: On finite element methods for nonlinear elliptic equations of second order. SIAM J. Numer. Anal., **46**, 1212–1249 (2008)
4. Bourgat, J.F., Dumay, J.M., Glowinski, R.: Large displacement calculations of flexible pipelines by finite element and nonlinear programming methods. SIAM J. Sci. Stat. Comput., **1** (1), 34–81 (1980)

5. Boyd, S., Parikh, N., Chu, E., Peleato, B., Eckstein, J.: Distributed optimization and statistical learning via the alternating direction method of multipliers. Foundations and Trends in Machine Learning, **3** (1), 1–122 (2011)
6. Brenier, Y., Frisch, U., Hénon, M., Loeper, G., Mattarese, S., Mohayahee, R., Sobolevskii, A.: Reconstruction of the early Universe as a convex optimization problem. Month. Notices Roy. Astron. Soc.,**346** (2), 501–524 (2003)
7. Brenner, S.C., T. Gudi, T., Neilan, M., L.Y. Sung, L.Y.: C^0 penalty methods for the fully nonlinear Monge-Ampère equation. Math. Comp., **80** (276), 1979–1995 (2011)
8. Brenner, S.C., Neilan, M.: Finite element approximations of the three-dimensional Monge-Ampère equation. ESAIM: Math. Model. Numer. Anal., **46** (5), 979–1001 (2012)
9. Caboussat, A., Glowinski, R., Pons, V.: An augmented Lagrangian approach to the numerical solution of a non-smooth eigenvalue problem. Journal of Numerical Mathematics, **17** (1), 3–26 (2009)
10. Caboussat, A., Glowinski, R., Sorensen, D.C.: A least-squares method for the numerical solution of the Dirichlet problem for the elliptic Monge-Ampère equation in dimension two. ESAIM: Control Optim. Calcul Variations, **19** (3), 780–810 (2013)
11. Cabré, X.: Topics in regularity and qualitative properties of solutions of nonlinear elliptic equations. Discr. Cont. Dyn, Syst., **8** (2), 331–360 (2002)
12. Caffarelli, L.A., Cabré, X.: Fully Nonlinear Elliptic Equations. AMS, Providence, RI (1995)
13. Cea, J., Glowinski, R.: Sur des méthodes d'optimisation par relaxation. ESAIM: Math. Model. Num. Anal., **7** (R3),5–31 (1973)
14. Chan, T.F., Glowinski, R.: Finite Element Approximation and Iterative Solution of a Class of Mildly Nonlinear Elliptic Equations. Stanford report STAN-CS-78-674, Computer Science Department, Stanford University, Palo Alto, CA (1978)
15. Chartrand, R.: Non-convex splitting for regularized low-rank + sparse decomposition. IEEE Transactions on Signal Processing, **60** (11), 5810–5819 (2012)
16. Chartrand, R., Wohlberg, B.: A non-convex ADMM algorithm for group sparsity with sparse groups. In Proceedings of the IEEE 2013 International Conference on Acoustics, Speech and Signal Processing (ICASSP), pp. 6009–6013, IEEE (2013)
17. Chartrand, R., Sidky, E.Y., Pan, X.: Non-convex compressive sensing for X-ray CT: an algorithm comparison. In Proceedings of the IEEE 2013 Asilomar Conference on Signals, Systems and Computers, pp. 665–669, IEEE (2013)
18. Ciarlet, P.G.: Linear and Nonlinear Functional Analysis with Applications. SIAM, Philadelphia, PA (2013)
19. Courant, R., Hilbert, D.: Methods of Mathematical Physics. Volume II: Partial Differential Equations. J. Wiley, New York, NY (1989)
20. Dean, E.J., Glowinski, R.: Numerical solution of the two-dimensional elliptic Monge-Ampère equation with Dirichlet boundary conditions: an augmented Lagrangian approach. C. R. Math. Acad. Sci. Paris, **336** (9), 779–784 (2003)
21. Dean, E.J., Glowinski, R.: An augmented Lagrangian approach to the numerical solution of the Dirichlet problem for the elliptic Monge-Ampère equation in two dimensions. Electron. Transact. Num. Anal.,**22**, 71–96 (2006)
22. Dean, E.J., Glowinski, R.: Numerical methods for fully nonlinear elliptic equations of the Monge-Ampère type. Comp. Meth. Appl. Mech. Eng., **195** (13), 1344–1386 (2006)
23. Dean, E.J., Glowinski, R., Guidoboni, G.: On the numerical simulation of Bingham visco-plastic flow: Old and new results. J. Non-Newtonian Fluid Mech., **142** (1–3), 36–62 (2007)
24. Delbos, F., Gilbert, J.C., Glowinski, R., Sinoquet, D.: Constrained optimization in seismic reflection tomography: a Gauss-Newton augmented Lagrangian approach. Geophys. J. International, 164(3), 670–684 (2006)
25. Douglas, J., Rachford, H.H.: On the solution of the heat conduction problem in 2 and 3 space variables. Trans. Amer. Math. Soc., **82**, 421–439 (1956)
26. Duvaut, G., Lions, J.L.: Inequalities in Mechanics and Physics. Springer, Berlin (1976)
27. Ekeland, I., Temam, R.: Convex Analysis and Variational Problems. SIAM, Philadelphia, PA (1999)

28. Feng, X., Glowinski, R., Neilan, M.: Recent developments in numerical methods for fully nonlinear partial differential equations. SIAM Rev., **55** (2), 205–267 (2013)
29. Fortin, M., Glowinski, R.: Méthodes de Lagrangiens Augmentés: Application à la Résolution Numérique des Problèmes aux Limites. Dunod, Paris (1982)
30. Fortin, M., Glowinski, R.: Augmented Lagrangian Methods: Applications to the Numerical Solution of Boundary Value Problems. North-Holland, Amsterdam (1983)
31. Glowinski, R.: Numerical Methods for Nonlinear Variational Problems. Springer, New York NY (1984) (2nd printing: 2008)
32. Glowinski, R.: Finite element methods for incompressible viscous flow. In: Ciarlet, P.G., Lions, J.L. (eds.) Handbook of Numerical Analysis, Vol. IX, pp. 3–1176. North-Holland, Amsterdam (2003)
33. Glowinski, R.: On alternating direction methods of multipliers: a historical perspective. In Fitzgibbon, W., Kuznetsov, Y.A., Neittannmäki, P., Pironneau, O. (eds.) Modeling, Simulation and Optimization for Science and Technology, pp. 52–82. Springer, Dordrecht (2014)
34. Glowinski, R.: Variational Methods for the Numerical Solution of Nonlinear Elliptic Problems. SIAM, Philadelphia, PA (2015).
35. Glowinski, R., Dean, E.J., Guidoboni, G., Juarez, H.L., Pan, T.W.: Applications of operator-splitting methods to the direct numerical simulation of particulate and free surface flows and to the numerical solution of the two-dimensional elliptic Monge-Ampère equation. Japan J. Ind. Appl. Math., **25** (1), 1–63 (2008)
36. Glowinski, R., Hölmström, M.: Constrained motion problems with applications by nonlinear programming methods. Survey on Mathematics for Industry, **5**, 75–108 (1995)
37. Glowinski, R., Le Tallec, P.: Numerical solution of problems in incompressible finite elasticity by augmented Lagrangian methods. I. Two-dimensional and axisymmetric problems. SIAM J. Appl. Math., **42** (2), 400–429 (1982)
38. Glowinski, R., Le Tallec, P.: Numerical solution of problems in incompressible finite elasticity by augmented Lagrangian methods. II. Three-dimensional problems. SIAM J. Appl. Math., **44** (4), 710–733 (1984)
39. Glowinski, R., Le Tallec, P.: Augmented Lagrangian and Operator-Splitting Methods in Nonlinear Mechanics. SIAM, Philadelphia, PA (1989)
40. Glowinski, R., Lions, J.L., Trémolières, R.: Numerical Analysis of Variational Inequalities. North-Holland, Amsterdam (1981)
41. Glowinski, R., Marrocco, A.: Sur l'approximation par éléments finis d'ordre un et la résolution par pénalisation-dualité d'une classe de problèmes de Dirichlet non linéaires. C.R. Acad. Sci. Paris, **278A**, 1649–1652 (1974)
42. Glowinski, R., Marrocco, A.: Sur l'approximation par éléments finis d'ordre un et la résolution par pénalisation-dualité d'une classe de problèmes de Dirichlet non linéaires. ESAIM: Math. Model. Num. Anal., **9** (R2), 41–76 (1975)
43. Glowinski, R., Quaini, A.: On an inequality of C. Sundberg: A computational investigation via nonlinear programming. J. Optim. Theory Appl., **158** (3), 739–772 (2013)
44. Glowinski, R., Wachs, A.: On the numerical simulation of visco-plastic fluid flow. In: Ciarlet, P.G., Glowinski, R., Xu, J. (eds.) Handbook of Numerical Analysis, Vol. XVI, North-Holland, Amsterdam, pp. 483–717 (2011)
45. Guttiérez, C.: The Monge-Ampère Equation. Birkhäuser, Boston, MA (2001)
46. He, J.W., Glowinski, R.: Steady Bingham fluid flow in cylindrical pipes: a time dependent approach to the iterative solution. Num. Linear Algebra Appl., **7** (6), 381–428 (2000)
47. Ito, K., Kunish, K.: Lagrange Multiplier Approach to Variational Problems and Applications. SIAM, Philadelphia, PA (2008)
48. Lagnese, J., Lions, J.L.: Modelling, Analysis and Control of Thin Plates. Masson, Paris (1988)
49. Le Tallec, P.: Numerical methods for nonlinear three-dimensional elasticity. In: Ciarlet, P.G., Lions, J.L., (eds.) Handbook of Numerical Analysis, Vol. 3, North-Holland, Amsterdam, pp. 465–622 (1994)
50. Lions, J.L.: Optimal Control of Systems Governed by Partial Differential Equations. Springer, Berlin (1971)

51. Majava, K., Glowinski, R., Kärkkäinen, T.: Solving a non-smooth eigenvalue problem using operator-splitting methods. Inter. J. Comp. Math., **84** (6), 825–846 (2007)
52. Nečas, J.: Les Méthodes Directes en Théorie des Equations Elliptiques. Masson, Paris (1967)
53. Nečas, J.: Direct Methods in the Theory of Elliptic Equations. Springer, Heidelberg (2011)
54. Neilan, M.: A nonconforming Morley finite element method for the fully nonlinear Monge-Ampère equation. Numer. Math., **115** (3), 371–394 (2010)
55. Oliker, V., Prussner, L.: On the numerical solution of the equation $z_{xx}z_{yy} - z_{xy} = f$ and its discretization. I, Numer. Math., **54** (3), 271–293 (1988)
56. Peaceman, D.H., Rachford, H.H.: The numerical solution of parabolic and elliptic differential equations. J. SIAM, **3**, 28–41 (1955)
57. Raviart, P.A., Thomas, J.M.: Introduction à l'Analyse Numérique des Equations aux Dérivées Partielles. Masson, Paris (1983)
58. Schäfer, M.: Parallel algorithms for the numerical solution of incompressible finite elasticity problems. SIAM J. Sci. Stat. Comput., **12**, 247–259 (1991)
59. Schwartz, L.: Théorie des Distributions. Hermann, Paris (1966)
60. Sorensen, D.C., Glowinski, R.: A quadratically constrained minimization problem arising from PDE of Monge-Ampère type. Numer. Algor.. **53** (1), 53–66 (2010)
61. Talenti, G.: Best constant in Sobolev inequality. Ann. Mat. Pura Appl., **110** (1), 353–372 (1976)
62. Tartar, L.: An Introduction to Sobolev Spaces and Interpolation Spaces. Springer, Berlin (2007)
63. Villani, C.: Topics in Optimal Transportation. AMS, Providence, RI (2003)
64. Wang, Y., Yin, W., Zeng, J.: Global convergence of ADMM in nonconvex nonsmooth optimization. arXiv:1511.06324 [cs, math] (2015)

Chapter 9
Operator Splitting Methods in Compressive Sensing and Sparse Approximation

Tom Goldstein and Xiaoqun Zhang

Abstract Compressive sensing and sparse approximation have many emerging applications, and are a relatively new driving force for the development of splitting methods in optimization. Many sparse coding problems are well described by variational models with ℓ_1-norm penalties and constraints that are designed to promote sparsity. Successful algorithms need to take advantage of the separable structure of potentially complicated objectives by "splitting" them into simpler pieces and iteratively solving a sequence of simpler convex minimization problems. In particular, isolating ℓ_1 terms from the rest of the objective leads to simple soft thresholding updates or ℓ_1 ball projections. A few basic splitting techniques can be used to design a huge variety of relevant algorithms. This chapter will focus on operator splitting strategies that are based on proximal operators, duality, and alternating direction methods. These will be explained in the context of basis pursuit variants and through compressive sensing applications.

1 Introduction

Operator splitting and alternating direction methods are extremely useful in scientific computing and image processing. Classical splitting methods were not originally developed for optimization purposes, but rather for solving other problems in numerical analysis. For example, the coordinate descent method for quadratic

T. Goldstein (✉)
Department of Computer Science, University of Maryland, College Park, MD, USA
e-mail: tomg@cs.umd.edu

X. Zhang
Shanghai Jiao Tong University, Institute of Natural Sciences and
School of Mathematical Sciences, Shanghai, China
e-mail: xqzhang@sjtu.edu.cn

R. Glowinski et al. (eds.), *Splitting Methods in Communication, Imaging, Science, and Engineering*, Scientific Computation, DOI 10.1007/978-3-319-41589-5_9

minimization is equivalent to the Gauss-Seidel method for solving positive definite linear systems of equation. A more complex example is the Douglas-Rachford method for function minimization, which has its roots in alternating direction implicit methods for partial-differential equations [52].

More recently, splitting methods have been adapted to solve *composite* optimization problems, which involve sums of convex functions composed with linear operators, by breaking them down into simple sub-problems. Such methods have found widespread use in compressive sensing, image processing, and machine learning applications. Some of the most versatile methods for minimizing large-scale non-differentiable optimization problems are variants of the proximal point method [116, 95], the method of multipliers [83, 110], and the alternating direction method of multipliers (ADMM) [71, 69]. There are close connections between splitting methods for optimization and those used in scientific computing. For example, ADMM is closely related to the Douglas-Rachford splitting [68, 57] (see also Chapter 2, Section 3.3, Chapter 8, Section 1, and the references therein, for details on these relations and their discovery) and involves alternately minimizing an augmented Lagrangian in a Gauss-Seidel fashion. Applications of ADMM can be found in [70, 54, 14, 74, 137, 16].

Recent applications have motivated many new improvements to splitting methods such as more parallelizable variants, generalized applicability, preconditioning, adaptive parameters, and acceleration to improve convergence rates [75]. This chapter will, however, focus on the splitting techniques themselves and also on how they can be used to design effective algorithms for models in compressive sensing.

Sparse Models in Compressive Sensing

The ℓ_1 norm appears in many compressive sensing models because of its utility in promoting sparsity and robustly handling outliers within a convex optimization framework. Many successful sparse approximation models are built around ℓ_1 penalties and constraints. One of the most fundamental models in compressive sensing is basis pursuit [39], which aims to find a sparse solution to an underdetermined system of linear equations by solving

$$\min_x \|x\|_1 \qquad \text{s.t.} \qquad Ax = b\,. \tag{9.1}$$

Related formulations that allow for some noise in the measurements include the unconstrained problem

$$\min_x \lambda \|x\|_1 + \frac{1}{2}\|Ax - b\|^2\,, \tag{9.2}$$

the Lasso problem [125]

$$\min_x \frac{1}{2}\|Ax - b\|^2 \qquad \text{s.t.} \qquad \|x\|_1 \leq \tau\,, \tag{9.3}$$

and the basis pursuit denoising problem [49, 27]

$$\min_x \|x\|_1 \qquad \text{s.t.} \qquad \|Ax-b\|^2 \leq \sigma\,. \tag{9.4}$$

See [9] for a discussion of connections between these formulations.

In the context of compressive sensing, the measurement matrix A is constructed so that ℓ_1 minimization is capable of stably recovering the sparse signal of interest [26, 48]. Good measurement matrices should have no sparse vectors in their nullspace, and in fact they should approximately behave like orthogonal matrices when applied to sparse vectors. These principles can be quantified in terms of a nullspace property [50, 138] and the Restricted Isometry Property (RIP) [28, 4] respectively.

Other sparse approximation models may involve penalties of the form $\|\Phi x\|_1$ or constraints of the form $\|\Phi x\|_1 \leq \varepsilon$ in any combination with other convex functions and constraints. One illustrative example is total variation (TV) denoising [118],

$$\min_u \lambda \|u\|_{TV} + \frac{1}{2}\|u-f\|^2. \tag{9.5}$$

The total variation regularizer $\|u\|_{TV} = \|\nabla u\|_1$ penalizes the ℓ_1 norm of the image gradient to find an optimal piecewise constant image $u^\star$ that well approximates the noisy image f.

All of the models mentioned so far involve the ℓ_1 norm. Alternating direction methods are well suited for such problems because they lead to iterative procedures where the ℓ_1 terms can be separated from the other convex functions. They are then able to take advantage of the simple, separable structure of the ℓ_1 norm.

Organization of This Chapter

We begin with some theoretical background on convex functions, including classification of convex functions and basic calculus of non-differentiable functions. Using these preliminaries, we introduce the forward-backward splitting (also called the proximal gradient method) for solving composite minimization problems without constraints (see Section 3). Before moving on to discuss more complex splitting methods, we first present a short review on duality in Section 4. Using duality theory, we then derive the method of multipliers for solving problems with constraints in Section 5. After discussing many different applications of formulations for the classical method of multipliers, we introduce the alternating direction method of multipliers (ADMM) in Section 6. Finally, we study example applications of ADMM in compressive sensing in Section 7.

2 Preliminaries: Convex Functions

Most of this chapter is devoted to the minimization of convex functions. Throughout this chapter, when we say *convex function* we mean a function $F : \mathbb{R}^n \to (-\infty, \infty]$ that is not only convex but additionally is closed, proper and has bounded sub-level sets. A *closed* function is lower semicontinuous, which is equivalent to the epigraph $\mathrm{Epi}\, F = \{(x,z) : x \in \mathbb{R}^n, z \in \mathbb{R}, z \geq F(x)\}$ being closed. A *proper* function is not identically infinity. We further assume the existence of a minimizer. This assumption excludes, for example, the function $F(x) = 1/x$, which is convex but has no minimizer. We refer the reader to [41] for an in-depth discussion of such technicalities.

The remainder of this section discusses some important properties of convex functions and their derivatives.

Smooth Functions: Gradients and Inequalities

The simplest strategy for minimizing a smooth function F is gradient descent, which relies on the update

$$x^{k+1} = x^k - \alpha \nabla F(x^k), \tag{9.6}$$

where α is some positive stepsize parameter and $\nabla F(x^k)$ is the gradient of F evaluated at the iterate x^k.

When F is convex and differentiable at some point x, the gradient satisfies the following important identity:

$$F(y) \geq F(x) + (y-x)^T \nabla F(x), \ \forall y. \tag{9.7}$$

In plain words, a convex function always lies above its linear approximation. One simple result of (9.7) is that any stationary point of F (i.e., a point where the gradient is zero) must be a global minimizer.

When the gradient of F is Lipschitz continuous, we can also generate an upper bound for F. Let L be a Lipschitz constant for ∇F satisfying

$$\|\nabla F(x) - \nabla F(y)\| \leq L\|x-y\|, \forall x, y.$$

We then have

$$F(y) \leq F(x) + (y-x)^T \nabla F(x) + \frac{L}{2}\|x-y\|^2, \ \forall x, y. \tag{9.8}$$

Equation (9.8) tells us that the Lipschitz constant L is an upper bound on the curvature of F; the function F always lies below the quadratic function with curvature L defined at the right-hand side of (9.8).

Non-smooth Functions: Sub-differentials and Proximal Operators

When F is non-differentiable, gradient descent methods are unavailable. In this case, we resort to methods that only require the existence of the *sub-differential* of F, which is defined as

$$\partial F(x) = \{g | \; F(y) \geq F(x) + (y-x)^T g, \forall y\}. \tag{9.9}$$

The sub-differential contains all vectors that *behave* like gradients, in that they satisfy an inequality similar to (9.7). In the one-dimensional case, the sub-differential contains the slopes of all lines that lower-bound F at x. A particular vector in the sub-differential is referred as a *sub-gradient.*

As an alternative to gradient descent, many algorithms for non-differentiable problems use the *proximal mapping* of F given by

$$\text{prox}_F(z,\alpha) = \arg\min_x F(x) + \frac{1}{2\alpha}\|x-z\|^2. \tag{9.10}$$

The proximal mapping (9.10) is sometimes called a *backward* gradient descent step. To understand this terminology, we examine Equation (9.10). Any point $x^\star$ that minimizes (9.10) must satisfy the optimality condition

$$0 = \alpha g + (x^\star - z), \tag{9.11}$$

where $g \in \partial F(x^\star)$ is some sub-gradient of F at x^*. Note that when F is differentiable we simply have $g = \nabla F(x^\star)$. Equation (9.11) rearranges to

$$x^\star = \text{prox}_F(z,\alpha) = z - \alpha g.$$

This shows that $x^\star$ is obtained from z by marching down the sub-gradient of F at x^*. For this reason, the proximal operator performs a gradient descent step. Because the sub-gradient g is evaluated at the final point $x^\star$ rather than the starting point z, this is called *backward* gradient descent.

Equation (9.11) is equivalent to the set inclusion $0 \in \alpha\partial F(x^\star) + (x^\star - z)$, which rearranges to

$$z \in \alpha\partial F(x^\star) + x^\star = (\alpha\partial F + I)x^\star.$$

For this reason, the proximal operator (9.10) is sometimes written

$$x^\star = (\alpha\partial F + I)^{-1} z = J_{\alpha\partial F}(z),$$

where $J_{\alpha\partial F} = (\alpha\partial F + I)^{-1}$ is the *resolvent operator* of $\alpha\partial F$. Regardless of notation, the behavior of these operators is simple: the proximal/resolvent operator performs a backward gradient descent step starting at z.

The use of a backward step for F is advantageous in several ways. First, it may be difficult to choose a sub-gradient of F in cases where the sub-differential ∂F has a complex form or contains many vectors. In contrast, it can be shown that problem

(9.10) always has a unique, well-defined solution, and (as we will see later) it is often possible to solve this problem in simple closed form.

A particularly important proximal operator arises when $F(x) = \|x\|_1$. In this case we have

$$S_\alpha(v) = \arg\min_x \|x\|_1 + \frac{1}{2\alpha}\|x - v\|^2 \,. \tag{9.12}$$

The solution to (9.12) is the well-known soft thresholding or shrinkage formula, which is defined component-wise as

$$S_\alpha(v)_j = \begin{cases} v_j - \alpha\,\text{sign}(v_j) & \text{if } |v_j| > \alpha \\ 0 & \text{otherwise} \,. \end{cases} \tag{9.13}$$

The orthogonal projection of a vector v onto the set $\{x : \|x\|_1 \leq \tau\}$ is defined by

$$\Pi_{\|\cdot\|_1 \leq \tau}(v) = \arg\min_x \frac{1}{2}\|x - v\|^2 \qquad \text{s.t.} \qquad \|x\|_1 \leq \tau \tag{9.14}$$

and also amounts to soft thresholding. Although in this case the appropriate threshold depends on the input, the overall projection can nonetheless be reduced to projection onto a simplex and computed in linear time [19].

A closed proper function can be minimized simply by repeated application of the proximal operator. This process is called the proximal point algorithm (PPA). This method iterates

$$x^{k+1} = \text{prox}_F(x^k, \alpha_k) \tag{9.15}$$

for some (fixed or adaptive) stepsize parameter $\alpha_k > 0$. The PPA minimizes F simply by backward gradient descent. The PPA is seldom used in its raw form, however we will see that many useful optimization schemes can be reduced to the PPA (or its generalizations) by a change of variables. This fact is extremely useful for analyzing algorithms, as the PPA converges under very general circumstances. In particular, convergence is guaranteed if the convex function F has a minimizer and if $\liminf_{k\to\infty} \alpha_k > 0$ [113].

In the following, we will introduce splitting methods that are based on the PPA, including the forward-backward splitting method, method of multipliers, and alternating direction methods for ℓ_1 based convex optimization with separable structures. Finally we will present state-of-the-art algorithms for compressive sensing and sparse approximation applications in signal and image processing.

3 Forward-Backward Splitting

The ℓ_1-ℓ_2 minimization problem in Equation (9.12) can be directly solved by the soft thresholding formula, but problem (9.2) is not so simple because the variables x are coupled by the matrix A. However, by separating A from the ℓ_1 norm, we can minimize (9.2) by computing a sequence of soft thresholding steps. This is accomplished using the proximal gradient, or forward-backward splitting (FBS) method.

In this section we present FBS in its classical form. We then explore various interpretations and generalizations of FBS by relating it to majorization-minimization (MM) algorithms from statistics as well as the classical proximal-point algorithm.

Forward-Backward Splitting Methods and IST

Forward-backward splitting (also called proximal gradient) methods solve general minimization problems of the form [93, 105, 67, 42, 131, 41]

$$\min_x F(x) + G(x) , \tag{9.16}$$

where F is a closed, proper convex function and G is differentiable with a Lipschitz continuous gradient. The general FBS iteration is

$$x^{k+1} = \arg\min_x F(x) + \frac{1}{2\alpha_k}\|x - x^k + \alpha_k \nabla G(x^k)\|^2 . \tag{9.17}$$

If L is the Lipschitz constant of ∇G, then the method can be shown to converge for step sizes $0 < \alpha_k < \frac{2}{L}$ [42].

The *iterative soft thresholding (IST)* method solves (9.2) using FBS with $G = \frac{1}{2}\|Ax - b\|^2$ and $F = \lambda\|x\|_1$. The IST method can be written succinctly as

$$\begin{aligned} x^{k+1} &= \arg\min_x \lambda\|x\|_1 + \frac{1}{2\alpha_k}\|x - x^k + \alpha_k A^T(Ax^k - b)\|^2 && (9.18)\\ &= \text{prox}_F(x^k - \alpha_k A^T(Ax^k - b), \alpha_k) && (9.19)\\ &= S_{\lambda\alpha_k}(x^k - \alpha_k A^T(Ax^k - b)) . && (9.20) \end{aligned}$$

Variants of IST exist for solving the Lasso problem (9.3) and other constrained problems. For such problems, it is sometimes helpful to work with convex functions that take on infinite values. Let χ_C denote the indicator function for a closed convex set C

$$\chi_C = \begin{cases} 0 & x \in C \\ \infty & \text{otherwise} . \end{cases}$$

We can solve the Lasso problem (9.3) by replacing $\lambda\|x\|_1$ in (9.18) with the indicator function for $C = \{x : \|x\|_1 \le \tau\}$. The resulting update is

$$x^{k+1} = \arg\min_x \frac{1}{2}\|x - x^k + \alpha_k A^T(Ax^k - b)\|^2 \qquad \text{s.t.} \qquad \|x\|_1 \le \tau , \tag{9.21}$$

which corresponds to one iteration of FBS for solving (9.3). The iteration can equivalently be written as

$$x^{k+1} = \Pi_{\|\cdot\|_1 \le \tau}(x^k - \alpha_k A^T(Ax^k - b)) ,$$

where $\Pi_{\|\cdot\|_1 \le \tau}$ denotes the orthogonal projection onto the ℓ_1 ball of radius τ. As with IST, by separating A from the ℓ_1 constraint, gradient projection can solve the Lasso problem by a sequence of ℓ_1 ball projections that are easy to compute.

The fixed point interpretation of proximal gradient methods gives us a different perspective on operator splitting. As explained in [78, 7, 103], we can derive a forward backward splitting method for (9.16) through a fixed point iteration designed to handle F implicitly and G explicitly. The optimality condition for (9.16) is $0 \in \partial(F(x) + G(x))$. Assuming strict feasibility, this can simply be written $0 \in \partial F(x) + \nabla G(x)$, which is equivalent to $0 \in x + \alpha\partial F(x) - x + \alpha\nabla G(x)$. This relation can be represented as $x = (I + \alpha\partial F)^{-1}(x - \alpha\nabla G(x))$, where the resolvent $(I + \alpha\partial F)^{-1}$ corresponds to $\text{prox}_{\alpha F}$, which is to say that given b,

$$(I + \alpha\partial F)^{-1}(b) = \arg\min_x F(x) + \frac{1}{2\alpha}\|x - b\|^2 .$$

FBS can then be seen as the fixed point iteration

$$x^{k+1} = (I + \alpha\partial F)^{-1}(x^k - \alpha\nabla G(x^k)) .$$

If F and G are both differentiable, we can think of FBS as a way of finding a steady state of

$$\frac{\mathrm{d}x(t)}{\mathrm{d}t} = -\nabla F(x(t)) - \nabla G(x(t))$$

using the semi-implicit discretization

$$\frac{x^{k+1} - x^k}{\alpha} = -\nabla F(x^{k+1}) - \nabla G(x^k)$$

that combines a backward Euler step for F with a forward Euler step for G.

There are many variants and interpretations of IST algorithms [51, 31, 61, 44, 42, 60, 58, 78, 6]. We will review some of these interpretations in the sequel.

Interpretation as Proximal Point Algorithm

Another useful interpretation of FBS is as a special case of the PPA, where Bregman distances are used in place of quadratic penalties [29, 36]. The Bregman distance [18] associated with a convex function $J(x)$ is

$$D_J^p(x,y) = J(x) - J(y) - \langle p, x - y\rangle, \qquad p \in \partial J(y) \tag{9.22}$$

where $\partial J(y)$ denotes the sub-differential of J, defined in (9.9). Although $D_J^p(x,y)$ is not in fact a distance function (because $D_J^p(x,y) \neq D_J^p(y,x)$), it still satisfies $D_J^p(x,y) \ge 0$ and $D_J^p(x,y) = 0$ if $x = y$. Moreover, if J is strictly convex then $D_J^p(x,y) = 0$ only if $x = y$. The Bregman distance can be used to define proximal penalties as an alternative to the squared ℓ_2-norm. Applying the Bregman variant of the proximal point algorithm to (9.16) yields the iterations

$$x^{k+1} = \arg\min_x F(x) + G(x) + D_J^{p^k}(x, x^k) , \tag{9.23}$$

which converges to a minimizer if one exists and if J is strictly convex and differentiable for $p^k = \nabla J(x^k)$ [36, 55]. Assuming ∇G has Lipschitz constant L, let $J(x) = \frac{1}{2\alpha_k}\|x\|^2 - G(x)$ for $\alpha_k < \frac{1}{L}$. Then J is convex and

$$D_J^{p^k}(x,x^k) = \frac{1}{2\alpha_k}\|x\|^2 - \frac{1}{2\alpha_k}\|x^k\|^2 - \langle \frac{x^k}{\alpha_k}, x - x^k\rangle - G(x) + G(x^k) + \langle \nabla G(x^k), x - x^k\rangle .$$

With this choice of J, (9.23) reduces to the FBS update (9.17).

Iterative soft thresholding can also be viewed as an instance of the PPA for minimizing a convex function [116, 103]. The matrix norm $\|x\|_M^2 = \langle x, Mx\rangle$ (for a symmetric positive definite matrix M) can be used to construct proximal point iterations. The iterations

$$x^{k+1} = \arg\min_x \lambda\|x\|_1 + \frac{1}{2}\|Ax - b\|^2 + \frac{1}{2}\|x - x^k\|_{M_k}^2$$

are equivalent to iterative soft thresholding (9.18) if we let $M_k = \frac{1}{\alpha_k}I - A^TA$ with $\alpha_k < \frac{1}{\|A^TA\|}$ to ensure positive definiteness.

Interpretation as Majorization-Minimization Algorithm

The FBS method is even further generalized by the majorization-minimization (MM) algorithm. Majorization-minimization [62, 7], sometimes also described in terms of surrogate functionals or optimization transfer [44, 90], is a particularly useful perspective. Classical algorithms such as the proximal point algorithm [116] and scaled gradient projection [12] have majorization minimization interpretations, and the same technique can also be used in many other contexts, for example to turn implicit updates into explicit ones in linearized variants of ADMM [137, 129].

MM algorithms solve the generic problem

$$\min_x F(x)$$

by iteratively minimizing simpler *majorizing* functions $G_k(x)$. A function G_k is said to majorize F at a point x^k if $G_k(x^k) = F(x^k)$ and $G_k(x) \geq F(x)$ for all x. Successive iterates are defined by

$$x^{k+1} = \arg\min_x G_k(x) . \tag{9.24}$$

It follows directly that

$$F(x^{k+1}) \leq G_k(x^{k+1}) \leq G_k(x^k) = F(x^k),$$

so that the objective is monotonically nonincreasing. This iterative approach can be advantageous when the functions G_k are easy to minimize. Often they can be chosen to be convex quadratic functions.

In the context of iterative soft thresholding, let

$$F(x) = \lambda \|x\|_1 + \frac{1}{2}\|Ax - b\|^2.$$

Now, leave the ℓ_1 term alone and replace $\|Ax - b\|^2$ with its quadratic majorizer to obtain

$$G_k(x) = \lambda \|x\|_1 + \frac{1}{2}\|Ax^k - b\|^2 + \langle x - x^k, A^T(Ax^k - b)\rangle + \frac{1}{2\alpha_k}\|x - x^k\|^2 \geq F(x)$$

for $\alpha_k < \frac{1}{\|A^T A\|}$. With this definition of $G_k(x)$, the majorization-minimization algorithm (9.24) is equivalent to the IST method (9.18). Note, however that for the MM interpretation to hold we need $\alpha_k < \frac{1}{L}$ for $L = \|A^T A\|$, while we know from the FBS analysis that the method converges for any $\alpha_k < \frac{2}{L}$.

Preconditioning

Iterative soft thresholding and gradient projection iterations can be slow when ∇G is not well conditioned. It is possible to add preconditioning from the perspective of either the MM, FBS, or fixed-point derivations. Consider the Bregman distance generalization of the proximal point method (9.23) applied to (9.16), and choose $J(x) = \frac{1}{2\alpha_k}\|x\|^2_{H_k} - G(x)$ for some symmetric positive definite matrix H_k. Then the resulting iterations would be

$$\begin{aligned} x^{k+1} &= \arg\min_x F(x) + G(x) + D_J^{p^k}(x, x^k) \\ &= \arg\min_x F(x) + \frac{1}{2\alpha_k}\|x - x^k + \alpha_k {H_k}^{-1}\nabla G(x^k)\|^2_{H_k}. \end{aligned} \tag{9.25}$$

With the Lasso problem for example, this can be interpreted as scaled gradient projection [14]. As long as the choice of H_k does not make the minimization problem too difficult to solve, it is beneficial to approximate the Hessian of G at x^k, in which case (9.25) becomes a proximal Newton method [11, 121, 91, 8]. In the case where $F(x)$ equals $\lambda\|x\|_1$ or $\chi_{\|\cdot\|_1 \leq \tau}(x)$, a diagonal preconditioner H_k would preserve the simplicity of the iterations since soft thresholding and ℓ_1 ball projections are still straightforward to compute with a weighted ℓ_1 norm.

4 Duality

Duality is a useful tool for representing convex functions in different ways and reformulating convex minimization problems so that they are easier to solve. Dual formulations of convex problems can sometimes be simplified to remove constraints. Dual formulations can also be smoother, enabling the use of gradient methods on problems that are non-differentiable in their original form. For example, it is possible to write down unconstrained differentiable dual problems for basis pursuit (9.1). The dual formulation is then easily solved using gradient descent. Dual formulations can also be better suited for problems with separable structure.

Legendre-Fenchel Transform

There is more than one way to construct dual formulations of convex problems. They are usually expressed in terms of Legendre-Fenchel transforms of the convex functions in the primal objective. The Legendre-Fenchel transform, or convex conjugate, of a closed proper convex function J is defined by

$$J^*(p) = \sup_u \langle u, p \rangle - J(u) \,. \tag{9.26}$$

When J is differentiable, $p = \nabla J(u^*)$ where the sup is attained at u^*. The conjugate J^* of a convex function is also proper, closed and convex, which can be seen by interpreting it as a supremum of affine functions [113],

$$J^*(p) = \sup_{(u,z)\in \mathrm{Epi} J} \langle u, p \rangle - z$$

where $\mathrm{Epi} J = \{(u,z) | J(u) \leq z\}$ represents the *epigraph* of J. If J is closed, proper, and convex then the biconjugate J^{**} is simply the original function J (when these conditions do not hold, the biconjugate is the convex hull of J). This can be thought of as a dual way of representing J as a pointwise supremum of affine functions, namely

$$J(u) = \sup_{(p,q)\in \mathrm{Epi} J^*} \langle u, p \rangle - q \,.$$

Legendre transforms have several uses. First, they provide an alternative way of writing optimality conditions. It follows directly from the definition (9.26) that $p \in \partial J(u)$ is equivalent to $u \in \partial J^*(p)$. Also, properties of J can often be inferred by examining J^*. For example, for finite-valued J, if J is strongly convex, then J^* has Lipschitzian gradient.

An important example is the convex conjugate of a norm. Let $J(u) = \|u\|$ be some arbitrary norm, and let $\|u\|_*$ be the corresponding dual norm. Then

$$J^*(p) = \sup_u \langle u, p\rangle - \|u\|$$
$$= \begin{cases} 0 & \text{if } \sup_{\|u\|\le 1}\langle u, p\rangle \le 1 \\ \infty & \text{otherwise} \end{cases}$$
$$= \begin{cases} 0 & \text{if } \|p\|_* \le 1 \\ \infty & \text{otherwise ,} \end{cases}$$

which is the indicator function $\chi_{\|\cdot\|_*\le 1}$ for the unit ball in the dual norm. The convex conjugates of the ℓ_1 norm and the indicator function for the ℓ_1 ball of radius τ can be used to construct dual problems for the unconstrained (9.2) and Lasso (9.3) basis pursuit variants. If $J(u) = \|u\|_1$ then $J^*(p) = \chi_{\|\cdot\|_\infty \le 1}(u)$. Conversely, if $J(u) = \chi_{\|\cdot\|_1\le\tau}(u)$, then $J^*(p) = \tau\|p\|_\infty$.

Moreau Decomposition

An extremely useful identity for converting between primal and dual problems is the Moreau decomposition [97, 42]. For a convex function J and a symmetric positive definite matrix M, under suitable constraint qualification, a vector v can be decomposed into the sum

$$v = A\left(\arg\min_p \left\{J(p) + \frac{1}{2}\|Ap - v\|_M^2\right\}\right) + M^{-1}\arg\min_u \left\{J^*(A^T u) + \frac{1}{2}\|u - Mv\|_{M^{-1}}^2\right\}. \quad (9.27)$$

This can be thought of as a generalization of decomposing v into a sum of projections onto orthogonal subspaces. For instance, if $M = I$, $A = I$ and J is the indicator function for a subspace $\mathfrak{L}$, then J^* is the indicator function for the orthogonal complement $\mathfrak{L}^\perp$ and the Moreau decomposition becomes $v = \Pi_{\mathfrak{L}}(v) + \Pi_{\mathfrak{L}^\perp}(v)$.

Applying the Moreau decomposition to the ℓ_1-penalized least squares problem (9.2) yields

$$b = A\left(\arg\min_x \left\{\lambda\|x\|_1 + \frac{1}{2}\|Ax - b\|^2\right\}\right) + \arg\min_p \left\{\chi_{\|\cdot\|_\infty\le\lambda}(A^T p) + \frac{1}{2}\|p - b\|^2\right\}.$$

We can think of

$$\min_p \chi_{\|\cdot\|_\infty\le\lambda}(A^T p) + \frac{1}{2}\|p - b\|^2 \quad (9.28)$$

as a dual problem for (9.2). The dual solution p^* is related to a primal solution x^* by $p^* = b - Ax^*$. Similarly, applying the Moreau decomposition to (9.3) implies

$$b = A\left(\arg\min_x \left\{\chi_{\|\cdot\|_1\le\tau} + \frac{1}{2}\|Ax - b\|^2\right\}\right) + \arg\min_q \left\{\tau\|A^T q\|_\infty + \frac{1}{2}\|q - b\|^2\right\},$$

where again the dual and primal solutions are related by $q^* = b - Ax^*$.

Both dual objectives are strictly convex, so p^*, q^*, and Ax^* are all uniquely determined even though x^* might not be unique. One insight that follows from comparing these dual problems is that a choice of λ that makes the problems equivalent is $\lambda = \|A^T q^*\|_\infty$, where q^* solves the Lasso dual problem

$$\min_q \tau \|A^T q\|_\infty + \frac{1}{2}\|q-b\|^2 . \tag{9.29}$$

With this choice of λ, $p = q^*$ is feasible for (9.28), but by the assumption that q^* minimizes (9.29), there can be no feasible p such that $\|A^T p\|_\infty \leq \|A^T q^*\|_\infty$ and $\|p-b\| < \|q^*-b\|$. Thus q^* also minimizes (9.28) if $\lambda = \|A^T q^*\|_\infty$.

The Moreau decomposition can also be used to derive the soft thresholding formula (9.13). We write

$$\begin{aligned} S_\alpha(v) &= \arg\min_x \alpha\|x\|_1 + \frac{1}{2}\|x-v\|^2 \\ &= v - \arg\min_p \left\{ \chi_{\|\cdot\|_\infty \leq \alpha}(p) + \frac{1}{2}\|p-v\|^2 \right\} \\ &= v - \Pi_{\|\cdot\|_\infty \leq \alpha}(v) . \end{aligned}$$

The elements of $S_\alpha(v)$ are then defined by

$$\begin{aligned} S_\alpha(v)_i &= v_i - \frac{v_i}{\max\left(\frac{|v_i|}{\alpha}, 1\right)} \\ &= \begin{cases} (|v_i| - \alpha)\frac{v_i}{|v_i|} & \text{if } |v_i| \geq \alpha \\ 0 & \text{otherwise} . \end{cases} \end{aligned}$$

The same derivation extends to mixed norms. When working with total variation regularization or the ℓ_1 norm of complex valued vectors or other models requiring group sparsity such as the group Lasso [135, 2], we replace $\|x\|_1 = \sum_j |x_j|$ with $\sum_g \|x_g\|$, where x_g are non-overlapping subsets of x indexed by g. The dual norm of $\sum_g \|x_g\|$ is $\max_g \|x_g\|$. Let

$$x^* = \arg\min_x \alpha \sum_g \|x_g\| + \frac{1}{2}\|x-v\|^2 .$$

Then for each vector x_g^*, the soft thresholding formula is

$$x_g^* = \begin{cases} (\|v_g\| - \alpha)\frac{v_g}{\|v_g\|} & \text{if } \|v_g\| \geq \alpha \\ 0 & \text{otherwise} . \end{cases} \tag{9.30}$$

This can also be seen by noting that x_g^* has to be in the direction of v_g. By replacing x_g with $\|x_g\|\frac{v_g}{\|v_g\|}$, the problem can be reduced to scalar soft thresholding.

The PPA (9.15) for a general unconstrained convex minimization problems, $\min_u J(u)$, can be interpreted as a gradient descent method on the Moreau envelope of J. The Moreau envelope of J is defined by the *infimal convolution* of J and $\frac{1}{2}\|\cdot\|_{M^{-1}}^2$,

$$E_M(y) = \min_u J(u) + \frac{1}{2}\|u - y\|_{M^{-1}}^2 ,$$

where M is a symmetric positive definite matrix. Its Legendre transform is

$$E_M^*(p) = J^*(p) + \frac{1}{2}\|p\|_M^2 ,$$

which is strongly convex, meaning that E_M is differentiable. Let $g = \nabla E_M(y)$. Then $y \in \partial E_M^*(g)$. Using the definition of E_M^*, $0 \in \partial J^*(g) + Mg - y$, which is exactly the optimality condition for

$$g = \arg\min_p J^*(p) + \frac{1}{2}\|p - M^{-1}y\|_M^2 .$$

The Moreau decomposition (9.27) then implies that

$$g = M^{-1}(y - \arg\min_u J(u) + \frac{1}{2}\|u - y\|_{M^{-1}}^2) .$$

Let $\bar{u}(y) = \arg\min_u J(u) + \frac{1}{2}\|u - y\|_{M^{-1}}^2$. It follows that

$$0 \in \partial J(\bar{u}(y)) + M^{-1}(\bar{u}(y) - y) .$$

Since $\nabla E_M(y) = M^{-1}(y - \bar{u}(y)) \in \partial J(\bar{u}(y))$, finding y such that $\nabla E_M(y) = 0$ yields $\bar{u}(y)$ that minimizes J. ∇E_M is Lipschitz continuous whereas J is not necessarily differentiable. If we apply PPA on $\min_u J(u)$, we obtain the scheme

$$u^{k+1} = \mathrm{prox}_J(u^k, \alpha_k).$$

For $\alpha_k = 1$, we obtain $0 \in \partial J(u^{k+1}) + u^{k+1} - u^k$. This can be interpreted as a gradient descent method on E_M with stepsize 1 and $M = I$, i.e., $u^{k+1} = u^k - \nabla E_M(u^k)$. Different choices of M can also lead to improved proximal point methods such as the variable metric proximal point algorithm [22], which is scaled gradient descent applied to E_M with its gradient scaled by an approximation to its inverse Hessian.

Lagrange Duality

Dual problems can also be derived from a Lagrangian perspective. Consider a generic primal problem

$$\min_u J(u) \quad \text{s.t.} \quad Au = b . \tag{P}$$

Associated with (P) is a Lagrangian

$$L(u,p) = J(u) + \langle p, b - Au \rangle \,, \tag{9.31}$$

where the dual variables p can be thought of as a vector of Lagrange multipliers for the $Au = b$ constraints. $L(u,p)$ is a convex-concave function whose saddle points characterize both primal and dual solutions. The dual function $q(p)$ associated with L is defined by

$$q(p) = \inf_u L(u,p) \tag{9.32}$$

$$= \langle p, b \rangle - \sup_u \langle A^T p, u \rangle - J(u) \tag{9.33}$$

$$= \langle p, b \rangle - J^*(A^T p) \,. \tag{9.34}$$

The dual function $q(p)$ is concave and the associated dual problem is

$$\max_p q(p) \,. \tag{D}$$

By weak duality, $q(p) \leq J(u)$. Assuming J is convex and (P) has a solution u^*, and strong duality holds (constraint qualification holds), which means a solution p^* to (D) exists and $J(u^*) = q(p^*)$ [113, 12].

When the primal and dual problems are not easily solvable, the *saddle point* characterization of the optimal point becomes useful. When strong duality holds, u^* solves (P) and p^* solves (D) if and only if (u^*, p^*) is a saddle point of $L(u,p)$,

$$L(u^*, p) \leq L(u^*, p^*) \leq L(u, p^*) \quad \text{for all } u, p \,.$$

From this we can also see that the necessary conditions for optimality are

$$\begin{cases} Au^* = b \\ A^T p^* \in \partial J(u^*) \qquad \text{or equivalently} \qquad u^* \in \partial J^*(A^T p^*) \,. \end{cases}$$

As an example, we can define the same dual problem (9.28) for (9.2) by first reformulating (9.2) to

$$\min_{x,z} \lambda \|x\|_1 + \frac{1}{2}\|z\|^2 \qquad \text{s.t.} \qquad z = b - Ax$$

and forming the Lagrangian

$$L(x,z,p) = \lambda \|x\|_1 + \frac{1}{2}\|z\|^2 + \langle p, b - Ax - z \rangle \,.$$

The dual function can then be defined by

$$\begin{aligned} q(p) &= \inf_{x,z} L(x,z,p) \\ &= \langle p,b\rangle - \{\sup_x \langle A^T p,x\rangle - \lambda\|x\|_1\} - \{\sup_z \langle p,z\rangle - \frac{1}{2}\|z\|^2\} \\ &= -\chi_{\|\cdot\|_\infty \le \lambda}(A^T p) - \frac{1}{2}\|p\|^2 + \langle p,b\rangle \,. \end{aligned}$$

A maximizer of $q(p)$ can be equivalently found by minimizing (9.28). Note however that the objective for (9.28) differs from $q(p)$ by the constant $\frac{1}{2}\|b\|^2$.

Although converting a convex minimization problem into a saddle point problem may seem to make things more difficult, it leads to useful algorithms. When the primal problem has separable structure it can be reformulated in such a way that iterative methods for finding saddle points of an associated Lagrangian also take advantage of this structure. Constraints can be decoupled from the objective and dealt with one at a time. If the objective is a sum of simpler functions, these can also be decoupled.

Uzawa's Method

One of the simplest strategies for finding a saddle point of L (9.31) is Uzawa's method [1], which iterates

$$u^{k+1} \in \arg\min_u J(u) + \langle p^k, b - Au\rangle \tag{9.35}$$

$$p^{k+1} = p^k + \delta(b - Au^{k+1}) \,. \tag{9.36}$$

The optimality condition for the u^{k+1} subproblem is $0 \in \partial J(u^{k+1}) - A^T p^k$, which is equivalent to $u^{k+1} \in \partial J^*(A^T p^k)$. This means $p^{k+1} \in p^k + \delta(b - A\partial J^*(A^T p^k))$. If J^* is differentiable, then $\partial J^* = \nabla J^*$ and $p^{k+1} = p^k + \delta \nabla q(p^k)$ is exactly a gradient ascent step for maximizing the dual objective, $\langle p,b\rangle - J^*(A^T p)$. However, we need strong convexity of J for this interpretation to hold. ∇J^* is Lipschitz continuous with constant $\frac{1}{\sigma}$ if and only if J is strongly convex with modulus σ, which means $J(u) - \frac{\sigma}{2}\|u\|^2$ is convex [117].

One way to introduce strong convexity is to change the objective slightly by adding a small strongly convex term. This slightly smooths the dual objective so that gradient ascent can be applied. The linearized Bregman method [134] for ℓ_1 minimization is simply an instance of Uzawa's method – a relationship that was explored in [23, 24, 132].

By adding $\frac{1}{2\alpha}\|u\|^2$ to $J(u)$ in (P), Uzawa's method becomes

$$\begin{cases} u^{k+1} = \arg\min J(u) + \frac{1}{2\alpha}\|u - \alpha A^T p^k\|^2 \\ p^{k+1} = p^k + \delta(b - Au^{k+1}) \end{cases} \tag{9.37}$$

for $0 < \delta < \frac{2}{\alpha\|A^T A\|}$. So in the case of basis pursuit (9.1) where $J(u) = \|u\|_1$, linearized Bregman solves a slightly perturbed problem by a sequence of soft thresholding steps. However, for α large enough the perturbed objective satisfies an exact regularization property so that its solutions are the same as solutions of the unperturbed basis pursuit problem [65, 132]. The linearized Bregman method can also deal with objectives that have a special kind of separable structure. If $J(u) = \sum_g J_g(u_g)$, where u_g denotes nonoverlapping subsets of u, then the minimization subproblem (9.35) decouples into separate minimization problems involving each J_g separately.

5 Method of Multipliers

The method of multipliers [83, 110] can be viewed as a way to solve (P) by a dual gradient ascent method without requiring strong convexity [10]. Based on the augmented Lagrangian

$$L_\delta(u,p) = J(u) + \langle p, b - Au\rangle + \frac{\delta}{2}\|Au - b\|^2 ,$$

the method of multipliers iterates

$$\begin{cases} u^{k+1} \in \arg\min_u L_\delta(u, p^k) \\ p^{k+1} = p^k + \delta(b - Au^{k+1}) . \end{cases} \tag{9.38}$$

It can be shown by examining the optimality conditions that each iteration is finding a saddle point (u^{k+1}, p^{k+1}) of

$$L(u,p) - \frac{1}{2\delta}\|p - p^k\|^2 .$$

From this, we can show that the p^{k+1} update is simply the proximal point iteration for maximizing $q(p)$ [115, 14]. The existence of a saddle point implies

$$\min_u \max_p L(u,p) - \frac{1}{2\delta}\|p - p^k\|^2 = \max_p \min_u L(u,p) - \frac{1}{2\delta}\|p - p^k\|^2 ,$$

which by the definition of the dual function is exactly the proximal iteration

$$\max_p q(p) - \frac{1}{2\delta}\|p - p^k\|^2 .$$

The proximal point algorithm interpretation for the method of multipliers means that it can be seen as a gradient ascent method on the Moreau envelope of $q(p)$ defined by

$$q_\delta(p^k) = \max_p q(p) - \frac{1}{2\delta}\|p - p^k\|^2 .$$

As explained for example in [14], $q_\delta(p)$ can be interpreted as an alternate dual function for (P) defined via the augmented Lagrangian by

$$q_\delta(p) = \inf_u L_\delta(u,p) \; .$$

The method of multipliers has many useful interpretations and advantages for ℓ_1 minimization problems. These were largely studied and developed from the perspective of Bregman iteration [99, 134], which solves (P) by iterating

$$\begin{cases} u^{k+1} \in \arg\min_u D_J^{v^k}(u,u^k) + \dfrac{\delta}{2}\|Au-b\|^2 \\ v^{k+1} = v^k + \delta A^T(b - Au^{k+1}) \in \partial J(u^{k+1}) \; , \end{cases} \tag{9.39}$$

where $D_J^{v^k}$ denotes the Bregman distance [18], and the v^{k+1} update ensures that $v^{k+1} \in \partial J(u^{k+1})$. The vector v^k is related to the Lagrange multiplier p^k in (9.38) by $v^k = A^T p^k$.

Bregman iteration (a special case of the method of multipliers) has numerous useful properties when used to solve ℓ_1 regularized problems. When applied to the equality constrained basis pursuit problem (9.1), the method of multipliers exhibits finite convergence, something that holds when the proximal point method is applied to linear programs such as the basis pursuit dual in this case [14]. Although this application of Bregman iteration to basis pursuit will eventually find a solution that satisfies $Au = b$, it is still effective for solving noisy basis pursuit problems. Often only a handful of iterations are needed. Each iteration requires solving an unconstrained problem of the form (9.2). These iterations have the intriguing interpretation of solving unconstrained denoising problems and then adding the noise back to the residual for the next iteration [99]. A stopping condition based on the norm of the residual being sufficiently small can be justified by Morozov's discrepancy principle as a regularization approach for dealing with noise [64]. Another major advantage for ℓ_1 minimization problems is an error cancellation property that holds when the unconstrained subproblems are inexactly solved [133].

In addition to Bregman iteration (9.39), the method of multipliers has been studied under various other names and settings. This method is known as iteration refinement in the field of inverse problems. The iterative refinement procedure has been generalized to a time continuous formulation, known as the inverse scale space (ISS) method [120, 20]. Many variants of the ISS method are also proposed in the context of compressive sensing and sparse approximation [21, 96, 100].

Method of Multipliers for Composite Objectives

The method of multipliers can also be an effective method for more complicated objectives of the form

$$\sum_j F_j(b_j - A_j u) + G(u) \; , \tag{P1}$$

where F_j and G are convex and G is also differentiable with Lipschitz continuous gradient. This model can include linear equality constraints by letting F_j be the indicator function of 0 (more general convex constraints are also possible). Introduce new variables $z_j = b_j - A_j u$ and define the augmented Lagrangian

$$L_\delta(u,z,p) = G(u) + \sum_j F_j(z_j) + \langle p_j, b_j - A_j u - z_j \rangle + \frac{\delta}{2} \|b_j - A_j u - z_j\|^2 .$$

By explicitly minimizing with respect to z_j, each F_j is replaced by its infimal convolution with $\frac{\delta}{2}\|\cdot\|^2$. Let

$$\tilde{F}(y) = \min_z F(z) + \frac{\delta}{2}\|z - y\|^2 .$$

Then the augmented Lagrangian becomes

$$L_\delta(u,p) = G(u) + \sum_j \tilde{F}_j\left(\frac{p_j}{\delta} + b_j - A_j u\right) - \frac{1}{2\delta}\|p_j\|^2 .$$

This is related to the penalty Lagrangian formulation in [114] and a special case of the generalized method of multipliers in [55]. The resulting iterations are

$$\begin{cases} u^{k+1} \in \arg\min_u G(u) + \sum_j \tilde{F}_j\left(\dfrac{p_j^k}{\delta} + b_j - A_j u\right) \\ p_j^{k+1} = \arg\min_p F_j^*(p) + \dfrac{1}{2\delta}\|p - (p_j^k + \delta(b_j - A_j u^{k+1}))\|^2 . \end{cases} \tag{9.40}$$

The minimization sub-problem for updating u^{k+1} has Lipschitz continuous gradient, and so gradient methods or even limited memory quasi Newton methods can be used. If F_j is the indicator function for 0, then $\tilde{F}_j(\frac{p_j}{\delta} + b_j - A_j u) = \frac{1}{2\delta}\|p_j + \delta(b_j - A_j u)\|^2$ and $p_j^{k+1} = p_j^k + \delta(b_j - A_j u^{k+1})$, which are consistent with the standard method of multipliers with linear equality constraints. Applying gradient methods to solve for u^{k+1} naturally decouples the F_j since the gradients of $\tilde{F}_j$ can be computed independently. These gradients are of the form

$$\nabla(\tilde{F}_j(\frac{p_j^k}{\delta} + b_j - A_j u))|_{u^k} = -A_j^T \left(\arg\min_p F_j^*(p) + \frac{1}{2\delta}\|p - (p_j^k + \delta(b_j - A_j u^k))\|^2\right),$$

and can thus be computed using the same proximal mappings needed to update the dual variables. As an example application to (9.4), we could consider $G = 0$, $F_1(z) = \|z\|_1$, $F_2(z) = \chi_{\|\cdot\| \le \sigma}(z)$ and solve $\min_u F_1(u) + F_2(b - Au)$.

Linearized Method of Multipliers

Sometime the $u-$subproblem in (9.38) does not have a closed form solution. This is often the case when the matrix A is more complex than a simple identity. To further split the matrix A from the constraint term, the method of multipliers can be combined with some of the linearization (i.e., majorization) techniques described in Section 3. Generalizing the proximal method of multipliers [115] and the Bregman proximal point algorithm [29], it was shown in [55] that the Bregman distance could be used to add a proximal term to the method of multipliers. Applied to (P), the quadratic penalty $\frac{\delta}{2}\|Au-b\|^2$ can be linearized by adding the Bregman distance between u and u^k of $\frac{1}{2\alpha}\|u\|^2 - \frac{\delta}{2}\|Au-b\|^2$ for $0 < \alpha < \frac{1}{\delta\|A^TA\|}$. This Bregman distance is given by

$$\frac{1}{2\alpha}\|u\|^2 - \frac{1}{2\alpha}\|u^k\|^2 - \langle \frac{u^k}{\alpha}, u-u^k\rangle - \frac{\delta}{2}\|Au-b\|^2 + \frac{\delta}{2}\|Au^k-b\|^2 + \langle \delta A^T(Au^k-b), u\rangle$$

and is equivalent to the proximal penalty

$$\frac{1}{2}\langle u-u^k, (\frac{1}{\alpha}I - \delta A^T A)(u-u^k)\rangle \ .$$

This modification results in the following linearized method of multipliers:

$$\begin{cases} u^{k+1} = \arg\min_u J(u) + \frac{1}{2\alpha}\|u-u^k+\alpha\delta A^T(Au^k-b-\frac{p^k}{\delta})\|^2 \\ p^{k+1} = p^k + \delta(b-Au^{k+1}) \ . \end{cases} \tag{9.41}$$

This algorithm was proposed as Bregman operator splitting (BOS) and split inexact Uzawa method in [136, 137] and applied to sparse reconstruction and compressive sensing problems. Every iteration is finding a saddle point (u^{k+1}, p^{k+1}) of

$$J(u) + \langle p, b-Au\rangle - \frac{1}{2\delta}\|p-p^k\|^2 + \frac{1}{2}\|u-u^k\|_D^2,$$

where $D = \frac{1}{\alpha}I - \delta A^T A$ is positive definite, so it can also be viewed as a minimax application of the proximal point method [115].

As with linearized Bregman, BOS can also take advantage of the separable structure of $J(u) = \sum_g J_g(u_g)$, where u_g are non-overlapping subsets of u. In this case the primal update decouples into separate proximal minimizations for each J_g. However, BOS can also solve much more complicated problems by a similar sequence of iterations. Consider (P1) in the case when $G=0$. This now has the form of a monotropic programming problem [112, 13], and duality can be used to decouple the F_j. Once again, introduce variables $z_j = b_j - A_j u$ and let $A = [A_1^T, \cdots, A_N^T]^T$. The reformulated problem becomes

$$\min_{u,z} \sum_j F_j(z_j) \qquad \text{s.t.} \qquad z_j = b_j - A_j u \ . \tag{9.42}$$

Adding Lagrange multipliers for the equality constraints and minimizing the resulting Lagrangian with respect to the primal variables leads to the dual problem

$$\min_p \sum_j F_j^*(p_j) - \langle p_j, b_j \rangle \qquad \text{s.t.} \qquad A^T p = 0, \tag{9.43}$$

where $p = [p_1^T \cdots p_N^T]^T$. Lagrange multipliers for the $A^T p = 0$ constraint correspond to the primal variables u. Therefore we can solve the primal problem by applying BOS to its dual, which has the kind of separable structure that allows proximal minimization steps for each F_j^* to be computed independently and in parallel. Other related strategies of adding proximal penalties to the Lagrangian can achieve similar decoupling of separable objective functions, for example the predictor-corrector proximal method [37].

More general Bregman distances can also be added to the method of multipliers to linearize smooth objectives. This can be valuable even when nothing in the original problem is smooth. Consider applying (9.40) to $\min_u G(u) + F(b - Au)$, which corresponds to basis pursuit denoising (9.4) when $G(u) = \|u\|_1$ and $F(z) = \chi_{\|\cdot\| \leq \sigma}(z)$. Note that G is no longer assumed to be differentiable. The primal update requires solving

$$u^{k+1} = \arg\min_u G(u) + \tilde{F}\left(\frac{p^k}{\delta} + b - Au\right).$$

Consider adding the Bregman distance between u and u^k of $\frac{1}{2\alpha}\|u\|^2 - \tilde{F}(\frac{p^k}{\delta} + b - Au)$ for $\delta\alpha < \frac{1}{\|A^T A\|}$ in order to linearize $\tilde{F}$, which here can be interpreted as the distance squared to the ℓ_2 ball of radius σ. Using the fact that

$$\nabla \tilde{F}\left(\frac{p^k}{\delta} + b - Au\right)|_{u^k} = -A^T\left(p^k + \delta(b - Au^k) - \Pi_{\|\cdot\| \leq \delta\sigma}(p^k + \delta(b - Au^k))\right),$$

the overall iterations simplify to

$$\begin{cases} u^{k+1} = S_\alpha(u^k + \alpha A^T \bar{p}^k) \\ p^{k+1} = p^k + \delta(b - Au^{k+1}) - \Pi_{\|\cdot\| \leq \delta\sigma}(p^k + \delta(b - Au^{k+1})) \\ \bar{p}^{k+1} = p^{k+1} + \delta(b - Au^{k+1}) - \Pi_{\|\cdot\| \leq \delta\sigma}(p^{k+1} + \delta(b - Au^{k+1})) \end{cases} \tag{9.44}$$

and amount to iterating one soft thresholding step followed by two ℓ_2 ball projections. Preconditioning can also be introduced as with (9.46) by using the Bregman distance of $\frac{1}{2\alpha}\|u\|_D^2 - \tilde{F}(\frac{p^k}{\delta} + b - Au)$ for a positive definite matrix D and possibly also using a matrix norm M when defining the infimal convolution $\tilde{F}$.

Preconditioning

Preconditioning can improve the convergence rate of Lagrangian methods when A is ill-conditioned. Introduce symmetric positive definite matrices M and D and let $D_M = \frac{1}{\delta}D^{-2} - \alpha AMA^T$ with α and δ chosen to ensure D_M is positive definite. A preconditioned application of BOS to (9.43) iteratively finds saddle points of

$$F^*(p) + \langle p, Au - b\rangle + \frac{1}{2}\|p - p^k\|_{D_M}^2 - \frac{1}{2\alpha}\|u - u^k\|_{M^{-1}}^2 , \tag{9.45}$$

leading to the iterations

$$\begin{cases} p_j^{k+1} = \arg\min_p F_j^*(p) + \frac{1}{2\delta}\|D^{-1}(p - p_j^k) + \delta D(Au^k - b + \alpha AMA^T p_j^k)\|^2 \\ u^{k+1} = u^k + \alpha MA^T p^{k+1}. \end{cases} \tag{9.46}$$

These iterations are more efficient when D and M are chosen so that $DAMA^TD$ is well conditioned. Depending on F_j, it is often the case that diagonal matrices D don't overly complicate the minimization problems. The difficulty of these proximal steps is not affected by M. Related preconditioning strategies can be found in [136, 107].

As an example application, we consider applying dual BOS to the Lasso problem (9.3) with $D = I$ and $M = I$. This yields the iterations

$$\begin{cases} p_1^{k+1} = \frac{1}{1+\delta}(p_1^k - \delta A(u^k + \alpha(A^T p_1^k + p_2^k) - b) \\ p_2^{k+1} = p_2^k - \delta(u^k + \alpha(A^T p_1^{k+1} + p_2)) - \Pi_{\|\cdot\|_1 \le \tau\delta}(p_2^k - \delta(u^k + \alpha(A^T p_1^{k+1} + p_2))) \\ u^{k+1} = u^k + \alpha(A^T p_1^{k+1} + p_2^{k+1}) . \end{cases} \tag{47}$$

Although the iterations are simple to compute, more dual variables were introduced than were really necessary to achieve this decoupling. Alternative approaches based on ADMM will be discussed in Section 6.

6 Alternating Direction Methods

The alternating direction method of multipliers (ADMM) [71, 69] is a versatile method for solving convex minimization problems of the form

$$\min_{u,z} H(u) + F(z) \qquad \text{s.t.} \qquad Au + Bz = b . \tag{P2}$$

Although the objective is a sum of only two convex functions, we can often rewrite more complicated problems in this way. For example, analogous to how (P1) was written as (9.42), it can be written as (P2) with $B = I$ and G replaced by a possibly nondifferentiable convex function H. Compared to dual BOS (9.46), ADMM's more implicit iterations can be more effective for some problems.

As with the method of multipliers, ADMM finds a saddle point of the augmented Lagrangian

$$\begin{aligned} L_\delta(u,z,p) &= H(u)+F(z)+\langle p, b-Bz-Au\rangle + \frac{\delta}{2}\|Au+Bz-b\|^2 \\ &= L(u,z,p)+\frac{\delta}{2}\|Au+Bz-b\|^2 , \end{aligned}$$

but it does so by alternately minimizing L_δ with respect to u and z in a Gauss-Seidel fashion before updating the Lagrange multipliers p. Each iteration of ADMM contains the updates

$$\begin{cases} u^{k+1} \in \arg\min_u L_\delta(u,z^k,p^k) \\ z^{k+1} \in \arg\min_z L_\delta(u^{k+1},z,p^k) \\ p^{k+1} = p^k + \delta(b - Bz^{k+1} - Au^{k+1}) . \end{cases} \tag{9.48}$$

Algorithm (9.48) is guaranteed to converge under fairly broad conditions. In particular, it is required that $H(u)$ and $F(z)$ are closed proper convex functions, there exists a saddle point of $L(u,z,p)$, all sub-problems in (9.48) have solutions, and $\delta > 0$ [38, 45]. ADMM was shown in [68, 63] to be equivalent to Douglas-Rachford splitting on the dual problem

$$\min_p H^*(A^T p) - \langle p,b\rangle + F^*(B^T p) . \tag{D2}$$

The convergence of Douglas-Rachford splitting was shown in [93] from the perspective of finding zeros of sums of two maximal monotone operators. ADMM's interpretation as a proximal point algorithm in [54, 57] allowed the convergence analysis to be generalized. More recent proximal point and gradient method interpretations have led to preconditioned and accelerated generalizations [17, 106, 73].

Convex optimization algorithms based on ADMM were proposed for a wide variety of applications in [70, 54, 14]. In the signal and image processing community there was a lot of renewed interest in ADMM after it was shown to be a good approach for general ℓ_1 minimization problems, including those involving total variation regularization. This made ADMM a perfect fit not only for emerging compressive sensing applications but also for many types of regularized inverse problems based on convex models with separable structure of some sort.

ADMM was proposed for ℓ_1 and TV minimization problems from the perspective of split Bregman [74], which had many advantages over other popular approaches at the time. For example, there is no need to smooth the objective as with lagged diffusion [127, 34] or gradient descent. For moderate accuracy requirements it outperforms the dual approach in [30] as well as primal-dual Newton method in [33], although superlinearly convergent Newton-based methods [33, 89] can ultimately be more efficient when very high accuracy is needed. Split Bregman and other fast TV minimization algorithms based on the primal-dual hybrid gradient (PDHG) [140] and quadratic penalty [130] methods all showed the benefit of operator splitting for such problems. However, PDHG had not been proved to converge at the time. Its convergence is now better understood [59, 15, 81], but it is still not as generally applicable. In addition, compared to the quadratic penalty approach in [130], split Bregman and ADMM do not require penalty parameters to tend to infinity.

As an example application of ADMM to TV denoising, the objective in (9.5) can be written in the form of (P2) by letting $F(z) = \lambda \|z\|_{1,2}$, $H(u) = \frac{1}{2}\|u-f\|^2$, $B = I$, $A = -D$ and $b = 0$. Here, the $\|\cdot\|_{1,2}$ notation denotes a mixed ℓ_1-ℓ_2 norm [123] corresponding to a sum of ℓ_2 norms. For images with pixel indices (i,j), $\|Du\|_{1,2}$ corresponds to the discretization of

$$\|u\|_{TV} = \sum_{i,j} \frac{1}{h} \left\| \begin{bmatrix} u_{i,j+1} - u_{i,j} \\ u_{i+1,j} - u_{i,j} \end{bmatrix} \right\|,$$

which equals $\|Du\|_{1,2}$ if D is a difference operator such that Du is a concatenation of all 2×1 discrete gradients of the form $\begin{bmatrix} u_{i,j+1} - u_{i,j} \\ u_{i+1,j} - u_{i,j} \end{bmatrix}$. For the augmented Lagrangian

$$L_\delta(u,z,p) = \lambda \|z\|_{1,2} + \frac{1}{2}\|u-f\|^2 + \langle p, Du - z \rangle + \frac{\delta}{2}\|Du - z\|^2 ,$$

the ADMM iterations are given by

$$\begin{cases} u^{k+1} = (I + \delta D^T D)^{-1}(f - D^T p^k + \delta D^T z^k) \\ z^{k+1} = \arg\min_z \lambda \|z\|_{1,2} + \frac{\delta}{2}\|z - Du^{k+1} - \frac{p^k}{\delta}\|^2 \\ p^{k+1} = p^k + \delta(Du^{k+1} - z^{k+1}) . \end{cases}$$

If D is defined using periodic or Neumann boundary conditions, then $(I + \delta D^T D)^{-1}$ can be efficiently applied using the fast Fourier transform or discrete cosine transform respectively (see the example applications in Section 7). The z^{k+1} update can be computed using soft thresholding (9.30).

There is a lot of flexibility in designing ADMM iterations so that the minimization subproblems are easy to solve. For example, consider applying ADMM to a TV-ℓ_1 image deblurring model

$$\min_u \lambda \|Du\|_{1,2} + \|Ku - f\|_1 ,$$

which can again be written in the form of (P2) by introducing $z_1 = Du$ and $z_2 = Ku - f$ and letting $B = I$, $A = \begin{bmatrix} -D \\ -K \end{bmatrix}$ and $b = \begin{bmatrix} 0 \\ -f \end{bmatrix}$. The resulting ADMM iterations require applying $(D^T D + K^T K)^{-1}$ which may or may not be easy depending on whether fast transforms can simultaneously diagonalize $D^T D$ and $K^T K$. If not, it may still be the case that $I+D^TD$ and $I+K^TK$ are easy to invert, in which case we could consider a reformulation of the problem where

$$B=\begin{bmatrix} I & 0 & 0 \\ 0 & I & 0 \\ 0 & 0 & I \\ 0 & 0 & I \end{bmatrix}, A=\begin{bmatrix} -D & 0 \\ 0 & -K \\ -I & 0 \\ 0 & -I \end{bmatrix}, \text{ and } b=\begin{bmatrix} 0 \\ -f \\ 0 \\ 0 \end{bmatrix}.$$

Any problem of the form (P2) can always be reformulated so that A^TA is as simple of a block diagonal matrix as desired. While ADMM iterations can be simplified in this way, adding too many extra variables and constraints can lead to less efficient algorithms. Rather than introducing additional auxiliary variables, some ADMM variants update more than two blocks in a Gauss-Seidel style. This approach sometimes converges considerably faster and uses less memory than standard ADMM applied to a formulation with many additional auxiliary variables and constraints. However, the straightforward multi-block extension of ADMM is not always guaranteed to converge [35] and requires additional assumptions to be applicable [85, 46]. ADMM algorithms can be generalized to handle three objective terms (rather than two) by adding correction steps [80].

Alternative Interpretations of ADMM

The classical ADMM is fundamentally a two block method [56]. From its equivalence to Douglas-Rachford splitting on a dual problem, ADMM can be interpreted as a method for finding a fixed point of a composition of two nonexpansive mappings [56, 54, 57]. When the dual problem (D2) is a feasibility problem of finding a point in the intersection of two convex sets, then Douglas-Rachford splitting has the interesting geometric interpretation of averaged alternating reflections (AAR) [5]. Suppose the dual objective is a sum of indicator functions $\chi_{C_1}+\chi_{C_2}$ for convex sets C_1 and C_2. Then the Douglas-Rachford iterations are given by

$$y^{k+1}=\frac{1}{2}(R_{C_1}(R_{C_2}(y^k))+y^k),$$

where R_C denotes the reflection defined by $R_C(y)=2\Pi_C(y)-y$. Using the definition of R_C and introducing $p^{k+1}=\Pi_{C_2}(y^{k+1})$, the AAR iterations can be written as

$$\begin{cases} y^{k+1}=\Pi_{C_1}(2p^k-y^k)-p^k+y^k \\ p^{k+1}=\Pi_{C_2}(y^{k+1}). \end{cases} \tag{9.49}$$

Generalizing from this feasibility problem to solving (D2), replace $\Pi_{C_1}(f)$ by

$$\arg\min_p H^*(A^Tp)-\langle p,b\rangle+\frac{1}{2}\|p-f\|^2$$

and replace $\Pi_{C_2}(f)$ by

$$\arg\min_p F^*(B^T p) + \frac{1}{2\alpha}\|p-f\|^2$$

to obtain the Douglas-Rachford method for solving (D2). The equivalence between ADMM on (P2) and Douglas-Rachford on (D2) can be derived by applying the Moreau decomposition twice to (9.48) and letting $y^k = p^k + \delta B z^k$.

There is also an alternating direction implicit (ADI) method interpretation of ADMM [70]. Suppose F^* and H^* are differentiable. Then ADMM, via its Douglas-Rachford interpretation, can be viewed as an ADI method for finding a steady state of

$$\frac{\mathrm{d}p(t)}{\mathrm{d}t} = -\nabla\left(F^*(B^T p) + H^*(A^T p) - \langle p, b\rangle\right)$$

by iterating

$$\frac{q^k - p^k}{\delta} = -A\nabla H^*(A^T q^k) + b - B\nabla F^*(B^T p^k)$$

$$\frac{p^{k+1} - p^k}{\delta} = -A\nabla H^*(A^T q^k) + b - B\nabla F^*(B^T p^{k+1}),$$

where $q^k = y^{k+1} - y^k + p^k$ with y^k and p^k having the same interpretation here as (9.49) and (9.48). For nondifferentiable F^* and H^*, ADMM can still be interpreted as an analogous ADI method for a differential inclusion [70, 57].

Variants of ADMM

Strategies for designing ADMM algorithms with simple iterations have been proposed for example in [54, 14, 70, 66] as well as in many more recent works. As with the method of multipliers, ADMM can also be combined with linearization techniques. ADMM-based methods that linearize the quadratic penalty terms include the split inexact Uzawa (SIU) method in [137], the proximal alternating direction based contraction methods in [79], and the alternating direction proximal method of multipliers in [122]. There are also extensions that include linearizations of smooth functions [43] as well as the addition of Bregman distances to ADMM minimization subproblems [129].

A particularly useful modification is obtained by "linearizing" ADMM by adding proximal penalties of the form $\frac{1}{2}\|u-u^k\|_{M_u}^2$ and $\frac{1}{2}\|z-z^k\|_{M_z}^2$ to the u and z updates in (9.48). Choosing positive definite $M_u = \frac{1}{\alpha}I - \delta A^T A$ and $M_z = \frac{1}{\alpha}I - \delta B^T B$ will effectively linearize the augmented Lagrangian quadratic penalty terms in those updates. If only the u^{k+1} update is linearized, the resulting method is equivalent to modified PDHG, so called in [59] because of its close resemblance to the PDHG method in [140]. The method was proposed for an image segmentation application in [108] as a variant of a method in [109]. The method was further generalized in [59, 32], accelerated and preconditioned in [32, 107], and shown to have a proximal point method interpretation in [82]. Adaptive variants of the method that allow for automated stepsize choices were studied in [72]. The combination of implicit and explicit steps has made this a useful method in many applications.

Consider problems of the form (P1) but with G a possibly nondifferentiable convex function. Rewriting as

$$\min_{u,z} \sum_j F_j(z_j) + G(u) \qquad \text{s.t.} \qquad z_j + A_j u = b_j$$

puts it in the form of (P2) with $B = I$ and with its dual problem given by

$$\min_p G^*(A^T p) + \sum_j F_j^*(p_j) - \langle p_j, b_j \rangle .$$

Introduce $q = A^T p$ and consider an ADMM scheme based on the augmented Lagrangian

$$L_\alpha(p,q,u) = F^*(p) + G^*(q) - \langle p, b \rangle + \langle u, A^T p - q \rangle + \frac{\alpha}{2} \|A^T p - q\|_M^2$$

that updates p, then q and then u every iteration. If we linearize the p update by adding $\frac{1}{2}\langle p - p^k, (\frac{D^{-2}}{\delta} - \alpha AMA^T)(p - p^k)\rangle$ for some positive definite matrix $\frac{D^{-2}}{\delta} - \alpha AMA^T$, then we obtain a preconditioned instance of modified PDHG that has the form

$$\begin{cases} u^{k+1} = \arg\min_u G(u) - \langle u, A^T p^k \rangle + \dfrac{1}{2\alpha} \|u - u^k\|_{M^{-1}}^2 \\ p_j^{k+1} = \arg\min_p F_j^*(p_j) + \langle A_j(2u^{k+1} - u^k) - b_j, p_j \rangle + \dfrac{1}{2\delta} \|p_j - p_j^k\|_{D^{-2}}^2 . \end{cases} \tag{9.50}$$

In this way the individual F_j functions are decoupled and their most recent proximal minimizations are used in the proximal minimization of G.

As an example, consider applying ADMM and modified PDHG to the dual of the Lasso problem. Letting $F(z) = \frac{1}{2}\|z\|^2$ and $G(u) = \chi_{\|\cdot\|_1 \le \tau}(u)$ the lasso problem (9.3) has the form of (P2). ADMM applied to (D2) can be written as

$$\begin{cases} u^{k+1} = \Pi_{\|\cdot\|_1 \le \tau}(u^k + \alpha A^T p^k) \\ p^{k+1} = p^k + (I + \alpha AA^T)^{-1}(b - A(2u^{k+1} - u^k) - p^k) . \end{cases} \tag{9.51}$$

Applying (9.50) to the Lasso dual leads to very similar iterations given by

$$\begin{cases} u^{k+1} = \Pi_{\|\cdot\|_1 \le \tau}(u^k + \alpha A^T p^k) \\ p^{k+1} = p^k + \dfrac{\delta}{\delta + 1}(b - A(2u^{k+1} - u^k) - p^k) . \end{cases} \tag{9.52}$$

The modified PDHG application replaces $(I+\alpha AA^T)^{-1}$ with $\frac{\delta}{\delta+1}$ and is therefore equivalent to ADMM if $AA^T = I$ and $\alpha = \frac{1}{\delta}$. Both methods are simpler than BOS applied to the Lasso problem (47) and tend to work better in practice. For underdetermined matrices A, $I+\alpha AA^T$ is smaller and often easier to invert in the dual ADMM method than the matrix $I+\delta A^T A$ that would appear in a primal application of ADMM to the Lasso problem. However, in either case the Woodbury matrix identity could be used in order to work with the smaller of the two.

As another example, consider applying ADMM to the basis pursuit denoising problem (9.4) based on the augmented Lagrangian

$$L_\delta(u,z,p) = \|u\|_1 + \delta_{\|\cdot\|\le\sigma}(z) + \langle p, b-z-Au\rangle + \frac{\delta}{2}\|Au+z-b\|^2 .$$

Linearizing the u update leads to the iterations

$$\begin{cases} u^{k+1} = S_\alpha(u^k - \alpha A^T(p^k + \delta(b-Au^k) - \delta z^k) \\ z^{k+1} = \Pi_{\|\cdot\|\le\sigma}(\frac{p^k}{\delta} + b - Au^{k+1}) \\ p^{k+1} = p^k + \delta(b - Au^{k+1} - z^{k+1}) . \end{cases} \tag{9.53}$$

This is closely related to the Bregman method of multipliers applied to the same problem (9.44). If the z update is repeated after the p update using p^{k+1} in place of p^k, then the methods are equivalent.

There continues to be active research into extensions and generalizations of ADMM and linearized ADMM. For example, as with accelerated gradient methods based on [98] and accelerated proximal gradient methods such as FISTA [6], there has also been work to accelerate the convergence rate of ADMM and its variants [73, 102, 106, 106]. There are extensions to nonconvex problems that can still guarantee convergence to a stationary point [126, 92]. From an operator splitting perspective, multiblock ADMM [85, 46] as well as coordinate descent and stochastic versions of ADMM [124, 84, 128, 101, 139] have a lot of potential for designing efficient methods for large scale optimization problems where the objective contains a sum of many simple functions.

7 Compressive Sensing Examples

Most compressive signal reconstruction problems are either of the *analysis* or *synthesis* type. Both problem forms assume an unknown signal u is sparse under some transform D. We thus expect Du to have small ℓ_1-norm. Suppose further that we obtain linear measurements of the form $b = Au + \eta$, where A is a sensing matrix with appropriate compressive properties and η is a vector of additive noise. If the sparsifying transform D is invertible (for example, if D is a wavelet or discrete cosine transform) then the signal can be written $u = D^{-1}v$ for some sparse v, and can be recovered by solving the synthesis problem

$$v^* = \arg\min_v \lambda \|v\|_1 + \|AD^{-1}v - b\|^2 \quad (9.54)$$

for some appropriately chosen scalar λ. However, if D is a non-invertible transform (such as the discrete gradient operator or overcomplete wavelet transform), then we must solve the analysis problem

$$u^* = \arg\min_u \lambda \|Du\|_1 + \|Au - b\|^2. \quad (9.55)$$

The synthesis problem (9.54) is very straightforward to solve using the forward-backward splitting method described in Section 3. The analysis problem (9.55) is considerably more flexible (as it allows for non-invertible D) and is often used in image processing applications. However, analysis formulations generally require more complex splitting methods than synthesis formulations. For this reason, we focus here on numerous analysis problems in compressive sensing (for a recent review of synthesis problems, see [75]).

The Single-Pixel Camera and the Stone Transform

Compressive imaging devices allow high-resolution images to be recovered from a small amount of data. Unlike conventional cameras that measure each pixel with a separate detector, compressive imaging devices acquire information about an image/scene but measuring *transform coefficients* such as Fourier/Hadamard transform modes of pixels. If the measurement transform is chosen appropriately, images can be reconstructed accurately from under-sampled measurements [25, 3]. However, the problem of reconstructing an image from under-sampled data requires solving a variational reconstruction problem. Compressive imaging exploits the tradeoff between measurement complexity and reconstruction complexity — compressive methods speed up data acquisition (by reducing the number of measurements needed for accurate reconstruction), but the resulting image reconstruction problem becomes more computationally intense. Numerous compressive imaging devices have been proposed [76, 53, 77, 111, 111, 94]. We will focus here on the spatially multiplexing Single Pixel Camera (SPC) as described in [76, 53], however the numerical methods discussed are broadly applicable.

An SPC consists of a lens, a Digital Micro-mirror Device (DMD), and a photo detector (see Figure 9.1) [53]. Each mirror on the DMD modulates an individual pixel by diverting light either towards or away from the detector. This results in a combination coefficient for that pixel of $+1$ or -1, respectively.[1] Because SPCs are built around a single photodetector rather than a large photodetector array, they are

[1] It is more correct to say light diverted towards/away from the detector results in a $0/1$ coefficient. In practice, $0/1$ measurements are converted to $+1/-1$ measurements by subtracting the average

advantageous in applications where sensor construction is extremely costly, such as imaging in the Short-Wave Infrared (SWIR) spectrum.

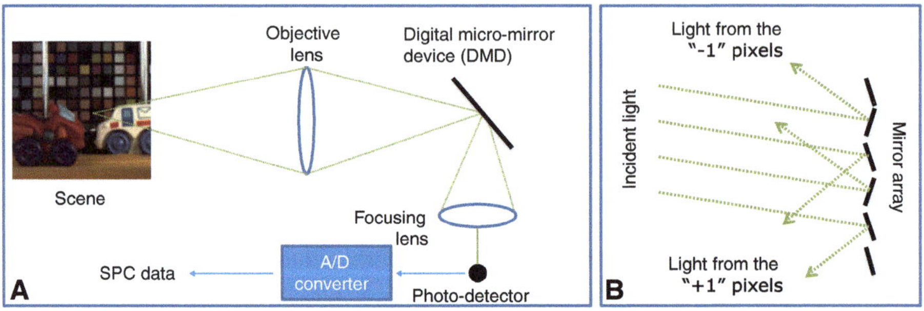

Fig. 9.1 Schematic diagram of a single-pixel camera. (A) Light from the scene is focused onto a Micro Mirror Device (MMD), and then redirected onto a single detector. An analog-to-digital converter reads out a measurement of light intensity impinging on the detector. (B) The MMD contains an array of mirrors that can be individually "tilted" to produce different measurement patterns.

An SPC obtains multiple measurements from a scene before image reconstruction. The ith measurement is an inner product $\langle A_i, u\rangle$ where u is the vectorized image, and A_i is a vector of ± 1's encoding the orientation of the mirrors. Once M measurements have been collected, the observed information can be written

$$Au = b + \eta, \tag{9.56}$$

where A is a "measurement matrix" having the vectors $\{A_i\}$ as its rows, b is the vector of measurements, and η represents noise. If u contains N pixels, then A is an $M \times N$ matrix.

The problem (9.56) is underdetermined. Therefore a prior is necessary to make solutions to this problem unique. In the context of compressive sensing, we assume the images are sparse in some transform domain. Images are then recovered by solving

$$\min \lambda \|Du\|_1 + \frac{1}{2}\|Au - b\|^2, \tag{9.57}$$

where D is some transform that sparsifies the image, and λ is a regularization parameter that controls the strength of the ℓ_1 penalty. When $D = \nabla$ is a discrete gradient operator (which generates the differences between adjacent pixels), the regularizer $\|Du\|_1$ becomes the well-known total variation (TV) semi-norm. Another common choice for D is a discrete wavelet transform.

image intensity. Measurement matrices with $+1/-1$ coefficients are more well conditioned than their $0/1$ counterparts, and are easier to handle numerically.

Choosing the Measurements Operator

There is a lot of flexibility when choosing a measurement operator A. Clearly, the sensing matrix needs to be binary, otherwise it cannot be represented on the DMD. In addition, the rows of A should be sub-sampled from an orthogonal matrix. This condition guarantees that A will satisfy an uncertainty principle (with high-probability), and thus will be an effective compressive sensing matrix [25]. Practical implementations further require the orthogonal matrix to have a fast transform algorithm, such as the Hadamard transform, to enable fast reconstruction.

Finally, some practical sensing matrices are designed to enable reconstruction via both compressive sensing and traditional Nyquist (one-pixel-per-measurement) methods. Such measurement matrices can immediately reconstruct low-resolution images using a single fast transform, or can optionally form high-resolution reconstructions from the same data using the iterative methods discussed below. This was originally accomplished with DSS matrices [119, 104] that allow reconstruction at two resolutions, and later by the Stone transform [76] which allows reconstruction at multiple resolutions and additionally has a fast transform.

Regardless of which measurement framework is chosen, all conventional measurement matrices contain rows that are sub-sampled from orthogonal matrices. This important property is exploited in our discussion of numerical methods.

Solving the Reconstruction Problem

As discussed above, most compressive imaging systems rely on subsampled orthogonal measurement matrices. In such cases, the image recovery problem (9.57) has the form

$$\min \lambda \|Du\|_1 + \frac{1}{2}\|RTu - b\|^2, \tag{9.58}$$

where T is some orthogonal transform (e.g., a Hadamard or Stone transform [76]), and R is a row-selector matrix that selects the outputs from T that are measured by the device and throws away the unmeasured entries.

Most modern compressive reconstructions rely on non-invertible sparsifying transforms such as total variation, tight frames, or over-complete dictionaries. In this case, variants of ADMM become a powerful tool. To apply ADMM, we begin with (9.58) and make the change of variables $z \leftarrow Du$ to arrive at

$$\min_{u,z} \lambda \|z\|_1 + \frac{1}{2}\|RTu - b\|^2 \quad \text{s.t.} \quad z = Du.$$

The augmented Lagrangian is then

$$\lambda \|z\|_1 + \frac{1}{2}\|RTu - b\|^2 + \langle p, z - Du\rangle + \frac{\delta}{2}\|z - Du\|^2. \tag{9.59}$$

The ADMM algorithm (9.48) requires that we minimize the augmented Lagrangian for u and then z. The u-update requires the inversion of $(T^T R^T RT + \delta D^T D)$. If D is a tight frame, then $D^T D = I$ and this operator might be easily inverted. However, in general it happens that T and D do not share common properties and this system cannot be directly inverted. For example, when T is a Hadamard or Stone transform and D is a gradient operator, this system has no closed form solution. The authors of [53] proposed solving this system using a truncated conjugate gradient method. However, one can avoid the difficulty of solving a system involving both T and D by decoupling the terms in the augmented Lagrangian (9.59) using the linearized methods discussed in Section 6.

To decouple the data and sparsity terms in (9.59), we add the proximal penalty $\frac{1}{2}\|u - u^k\|_M^2$ to the augmented Lagrangian, where $M = \frac{1}{\alpha}I - \delta D^T D$. The resulting minimization sub-problems are

$$\begin{cases} u^{k+1} \in \arg\min_u \frac{1}{2}\|RTu - b\|^2 + \langle p^k, -Du\rangle + \frac{\delta}{2}\|z^k - Du\|^2 + \frac{1}{2}\|u - u^k\|_M^2 \\ z^{k+1} = \arg\min_z \lambda\|z\|_1 + \langle p^k, z\rangle + \frac{\delta}{2}\|z - Du^{k+1}\|^2 \ \left(= S_{\lambda/\delta}(Du^{k+1} - \delta^{-1}p^k)\right) \\ p^{k+1} = p^k + \delta(z^{k+1} - Du^{k+1}). \end{cases} \tag{9.60}$$

The z- and p-updates in this method are explicit. The u-update has the optimality condition

$$(T^T R^T RT + \frac{1}{\alpha}I)u^{k+1} = T^T R^T b + D^T p^k + \frac{1}{\alpha}u^k - \delta D^T(Du^k - z^k). \tag{9.61}$$

Thus, the u-update requires us to solve $(T^T R^T RT + \frac{1}{\alpha}I)^{-1}r$, where r denotes the right-hand side of (9.61). When T is an orthogonal transform, the solution is given by

$$(T^T R^T RT + \frac{1}{\alpha}I)^{-1}r = (T^T(R^T R + \frac{1}{\alpha}I)T)^{-1}r = T^T(R^T R + \frac{1}{\alpha}I)^{-1}Tr.$$

Note the matrix $(R^T R + \frac{1}{\alpha}I)$ is a diagonal operator, and is thus easily inverted. This preconditioned ADMM strategy is advantageous because every substep of the method has a closed-form solution. The strategy described here is equivalent to the PDHG method [59] which has been applied for reconstruction of compressive video [76]. A drawback of this strategy is the need to choose two different stepsize parameters. This issue is addressed using adaptive stepsize selection in [72].

Applications Involving the Fourier Transform: Compressive Fourier Sampling and Deblurring

Many problems in image processing and compressive sensing require the reconstruction of images from under-sampled Fourier measurements. For example, in compressive magnetic resonance imaging (MRI) it is common to have a measurement matrix of the form $R\mathscr{F}$ where $\mathscr{F}$ is a matrix representing the Fourier transform operator. Reconstruction of images from such measurements is often achieved using the variational formulation

$$\min_u \lambda \|\nabla u\|_1 + \frac{1}{2}\|R\mathscr{F}u - b\|^2, \tag{9.62}$$

where λ is a regularization parameter that controls the tradeoff between the data term and the total variation penalty. This is a special case of (9.58). However we address it separately because it is possible to exploit the structure of the Fourier transform.

Image deblurring is another problem class that relies of the special problem form (9.62). In this case, we are given blurred measurements of the form $f = Kx + \eta$ where K is a linear blur operator, x is the true unknown image, and η is noise. Image recovery relies on the total variation regularized problem

$$\min_u \lambda \|\nabla u\|_1 + \frac{1}{2}\|Ku - f\|^2. \tag{9.63}$$

If we observe that convolution matrices are diagonalized by the Fourier transform, we can write $\mathscr{F}^{-1}R\mathscr{F}$ where R is a diagonal matrix. If $\mathscr{F}$ is a unitary Fourier operator, then the ℓ_2 norm of a vector is invariant under $\mathscr{F}$, and we have

$$\frac{1}{2}\|Ku - f\|^2 = \frac{1}{2}\|\mathscr{F}^{-1}R\mathscr{F}u - f\|^2 = \frac{1}{2}\|R\mathscr{F}u - \mathscr{F}f\|^2.$$

Thus, problem (9.63) is equivalent to a problem of the form (9.62) with $b = \mathscr{F}f$. This property was utilized to solve TV based image deblurring problems in [130] and [74].

Problem (9.62) is easily addressed using the ADMM updates (9.48). The resulting updates are

$$\begin{cases} u^{k+1} \in \arg\min_u \frac{1}{2}\|R\mathscr{F}u - b\|^2 + \langle p^k, -\nabla u\rangle + \frac{\delta}{2}\|z^k - \nabla u\|^2 \\ z^{k+1} = \arg\min_z \lambda\|z\|_1 + \langle p^k, z\rangle + \frac{\delta}{2}\|z - \nabla u^{k+1}\|^2 \big(= S_{\lambda/\delta}(\nabla u^{k+1} - \delta^{-1}p^k)\big) \\ p^{k+1} = p^k + \delta(z^{k+1} - \nabla u^{k+1}). \end{cases} \tag{9.64}$$

The u update in this formulation is simply a quadratic minimization with optimality condition

$$(\mathscr{F}^T R^T R\mathscr{F} + \delta\nabla^T\nabla)u^{k+1} = (\mathscr{F}^T R^T R\mathscr{F} - \delta\Delta)u^{k+1} = \mathscr{F}^T R^T b + \nabla^T p^k + \delta\nabla^T z^k,$$

where Δ represents the discrete two-dimensional Laplace operator. When circulant boundary conditions are used, the Laplacian operator is diagonalized by the Fourier transform, and can be written $\Delta = \mathscr{F}^T K\mathscr{F}$ for some (diagonal) Fourier kernel K.

The system to be inverted thus has the from $(\mathscr{F}^T R^T R\mathscr{F} - \delta\mathscr{F}^T K\mathscr{F}) = \mathscr{F}^T(R^T R - \delta K)\mathscr{F}$, and the solution is given by

$$u^{k+1} = \mathscr{F}^T(R^T R - \delta K)^{-1}\mathscr{F}(\mathscr{F}^T R^T b + \nabla^T p^k + \delta\nabla^T z^k).$$

Thus every step in (9.64) has a closed form solution, and the runtime is dominated by the two fast Fourier transforms needed to update u.

This straightforward application of ADMM was made possible by the elegant combination of total variation with the Fourier transform. When unusual boundary conditions are used or more complex formulations are needed, then the steps of (9.64) may not have closed form solutions. In this case linearized ADMM might be required to simplify the iterations. We see one such situation in the next example.

Parallel MRI Reconstruction

The data acquisition step in conventional MRI is a relatively low-speed sampling procedure. *Parallel magnetic resonance image* (pMRI) is a technique that has been widely adopted in clinical radiology to accelerate the sampling speed of conventional MRI. By surrounding the scanned objects by an array containing multiple sensing coils, pMRI is able to extract spatial information from many coils in parallel, resulting in accelerated data acquisition. Because of discrepancies between coils in the Fourier domain, some aliasing artifacts will arise in the reconstructed image. The total variation regularization for pMRI image reconstruction has been considered in [40, 88, 87] and here we will consider wavelet frame regularization [47] as an illustrative example.

The most common image-domain-based parallel imaging method, namely *Sensitivity encoding* (SENSE), is based on the following acquisition model : for $j = 1\cdots, J$

$$R\mathscr{F}S_j u = b_j + \eta, \tag{9.65}$$

where u is the unknown image, b_j is the vector of measured partial Fourier coefficients at the jth receiver, R is a diagonal sub-sampling operator, $\mathscr{F}$ is the Fourier transform, η is the Gaussian noise, and J is the total number of coils. The operator S_j is a diagonal matrix *sensitivity mapping* for the jth receiver, as is used to compensate for the decay of signal intensity with distance from each pixel. In practice, the sensitivity map S_j can be estimated in advance (see, e.g., [86]).

Denote $A_j = R\mathscr{F}S_j$. The sparse reconstruction model has the analysis formulation

$$u^* = \arg\min_u \lambda\|Wu\|_1 + \frac{1}{2}\sum_{j=1}^{J}\|A_j u - b_j\|^2. \tag{9.66}$$

Here W denotes a wavelet frame transform satisfying $W^T W = I$ and λ is a weight parameter. As the operator $\sum_{j=1}^{J} A_j^T A_j$ is not diagonalizable by the Fourier transform, we apply the linearized ADMM algorithm, i.e., split Inexact Uzawa method, to solve this problem. By introducing a positive definite matrix M, the numerical scheme is as follows:

$$\begin{cases} u^{k+1} = \arg\min_u \frac{1}{2}\sum_{j=1}^{J} \|A_j u - b_j\|^2 + \langle p^k, -Wu\rangle + \frac{\delta}{2}\|z^k - Wu\|^2 + \frac{1}{2}\|u - u^k\|_M \\ z^{k+1} = \arg\min_z \lambda\|z\|_1 + \langle p^k, z\rangle + \frac{\delta}{2}\|z - Wu^{k+1}\|^2 \\ p^{k+1} = p^k + \delta(z^{k+1} - Wu^{k+1}). \end{cases} \tag{9.67}$$

For the u subproblem, we observe that $W^T W = I$ and choose $M = \alpha - \sum_{j=1}^{J} A_j^T A_j$. We arrive at the closed form solution

$$\begin{cases} u^{k+1} = \dfrac{\alpha u^k - \sum_{j=1}^{J} A_j^T(A_j u^k - b_j) + W^T(p^k + \delta z^k)}{\delta + \alpha} \\ z^{k+1} = S_{\lambda/\delta}(Wu^{k+1} - \delta^{-1}p^k) \\ p^{k+1} = p^k + \delta(z^{k+1} - Wu^{k+1}). \end{cases} \tag{9.68}$$

Note that alternative splitting methods could also be obtained using different auxiliary variables and constraints. For example, by substituting $v_j = u$ and $F_j(v_j) = \|A_j v_j - b_j\|^2$ for the data term and using the composite objective scheme (P1), we obtain a method that allows a parallel processing on each coil data.

In our experiments, the diagonal down-sampling operator R is implemented with a downsampling ratio $r = 2,4$ along one dimension, i.e. the full data is half/quarter-sampled. The test data can be found in the online MATLAB toolbox PULSAR [86], where the data is acquired using an eight-channel head array. We refer the reader to [86] for more details of the machine configuration. Figure 9.2 shows the eight coils of the brain data by applying a direct inverse Fourier transform. The sensitivity map S_j is estimated by the built-in function in PULSAR. The wavelet tight frame transform operator W we adopt for this simulation is associated with the piecewise linear spline tight frame system (see e.g. [47]) with decomposition level simply set to 1. Results from several other popular for pMRI reconstruction, including image-domain-based and k-space-based methods like SENSE, SPACE-RIP, GRAPPA, are shown in Figure 9.3. The computation time is also listed in Table 9.1. From Figure 9.3 one may observe that, compared with other pMRI image reconstruction methods, SIU with wavelet frames can achieve the best image reconstruction quality for both tests in a reasonable computing time. We note that the scheme can be easily parallelized since each subproblem is completely decoupled.

Sampling ratio	SENSE	SPACE-RIP	GRAPPA	SIU
$r = 2$	7.08	173.78	61.82	5.32
$r = 4$	3.91	98.51	58.34	12.75

Table 9.1 Computation time (listed in *seconds*) of different method for the eight-channel brain data with subsampling ratio $r = 2,4$. The size of the image is 256×256.

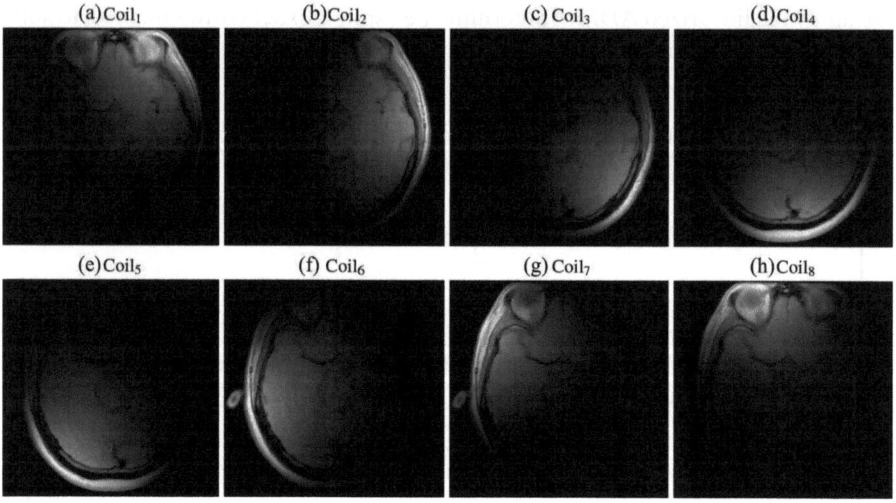

Fig. 9.2 Full data of the brain image: (a)–(h) eight-channel brain data.

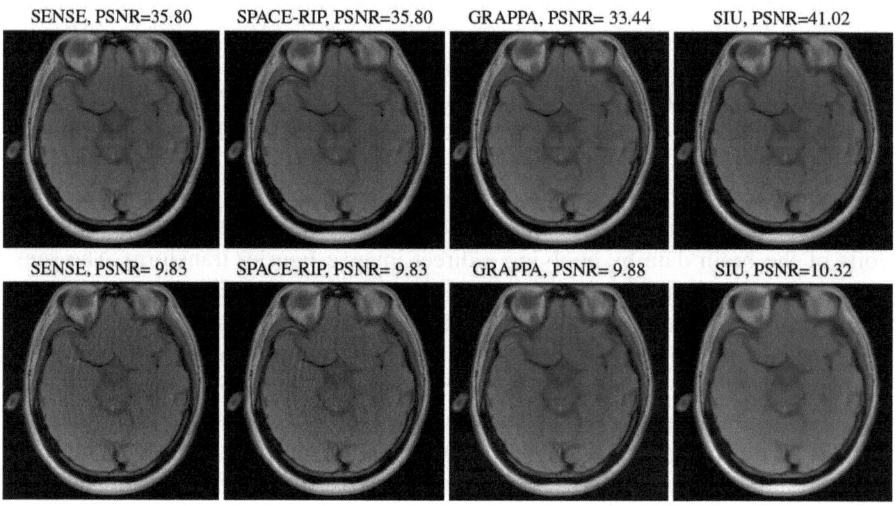

Fig. 9.3 Reconstruction results from the eight-channel brain data with subsampling ratio $r = 2$ and $r = 4$.

References

1. Arrow, K.J., Hurwicz, L., Uzawa, H.: Studies in Linear and Non-Linear Programming. Stanford University Press (1958)
2. Bach, F., Jenatton, R., Mairal, J., Obozinski, G.: Convex optimization with sparsity-inducing norms. In: S. Sra, S. Nowozin, S. Wright (eds.) Optimization for Machine Learning, pp. 19–53. The MIT Press, Cambridge, MA (2012)
3. Bajwa, W.U., Sayeed, A.M., Nowak, R.: A restricted isometry property for structurally-subsampled unitary matrices. In: Communication, Control, and Computing, 2009. Allerton 2009. 47th Annual Allerton Conference on, pp. 1005–1012. IEEE (2009)
4. Baraniuk, R., Davenport, M., DeVore, R., Wakin, M.: A simple proof of the restricted isometry property for random matrices. Constructive Approximation **28**(3), 253–263 (2008)
5. Bauschke, H.H., Combettes, P.L., Luke, D.R.: Finding best approximation pairs relative to two closed convex sets in Hilbert spaces. Journal of Approximation Theory **127**(2), 178–192 (2004)
6. Beck, A., Teboulle, M.: A fast iterative shrinkage-thresholding algorithm for linear inverse problems. SIAM Journal on Imaging Sciences **2**(1), 183–202 (2009)
7. Beck, A., Teboulle, M.: Gradient-based algorithms with applications to signal recovery. In: D. Palomar, Y. Eldar (eds.) Convex Optimization in Signal Processing and Communications, pp. 3–51. Cambridge University Press, Cambridge, UK (2010)
8. Becker, S., Fadili, J.: A quasi-Newton proximal splitting method. In: Advances in Neural Information Processing Systems (NIPS), pp. 2618–2626 (2012)
9. Van den Berg, E., Friedlander, M.: Probing the Pareto frontier for basis pursuit solutions. SIAM Journal on Scientific Computing **31**(2), 890–912 (2008)
10. Bertsekas, D.P.: Constrained Optimization and Lagrange Multiplier Methods. Academic Press, Cambridge, MA (1982)
11. Bertsekas, D.P.: Projected Newton methods for optimization problems with simple constraints. SIAM Journal on Control and Optimization **20**(2), 221–246 (1982)
12. Bertsekas, D.P.: Nonlinear Programming, 2nd edn. Athena Scientific, Nashua, NH (1999)
13. Bertsekas, D.P.: Extended monotropic programming and duality. Journal of Optimization Theory and Applications **139**(2), 209–225 (2008)
14. Bertsekas, D.P., Tsitsiklis, J.N.: Parallel and Distributed Computation: Numerical Methods, vol. 23. Prentice Hall, Englewood Cliffs, NJ (1989)
15. Bonettini, S., Ruggiero, V.: On the convergence of primal-dual hybrid gradient algorithms for total variation image restoration. Journal of Mathematical Imaging and Vision **44**(3), 236–253 (2012)
16. Boyd, S., Parikh, N., Chu, E., Peleato, B., Eckstein, J.: Distributed optimization and statistical learning via the alternating direction method of multipliers. Foundations and Trends in Machine Learning **3**(1), 1–122 (2011)
17. Bredies, K., Sun, H.: Preconditioned Douglas-Rachford splitting methods for convex-concave saddle-point problems. SIAM Journal on Numerical Analysis **53**(1), 421–444 (2015)
18. Bregman, L.M.: The relaxation method of finding the common point of convex sets and its application to the solution of problems in convex programming. USSR Computational Mathematics and Mathematical Physics **7**(3), 200–217 (1967)
19. Brucker, P.: An O(n) algorithm for quadratic knapsack problems. Operations Research Letters **3**(3), 163–166 (1984)
20. Burger, M., Gilboa, G., Osher, S., Xu, J.: Nonlinear inverse scale space methods. Communications in Mathematical Sciences **4**(1), 179–212 (2006)
21. Burger, M., Möller, M., Benning, M., Osher, S.: An adaptive inverse scale space method for compressed sensing. Mathematics of Computation **82**(281), 269–299 (2013)
22. Burke, J.V., Qian, M.: A variable metric proximal point algorithm for monotone operators. SIAM Journal on Control and Optimization **37**(2), 353–375 (1999)

23. Cai, J.F., Candès, E.J., Shen, Z.: A singular value thresholding algorithm for matrix completion. SIAM Journal on Optimization **20**(4), 1956–1982 (2010)
24. Cai, J.F., Osher, S., Shen, Z.: Linearized Bregman iterations for compressed sensing. Mathematics of Computation **78**(267), 1515–1536 (2009)
25. Candès, E., Romberg, J.: Sparsity and incoherence in compressive sampling. Inverse Problems **23**(3), 969 (2007)
26. Candès, E.J., Romberg, J., Tao, T.: Robust uncertainty principles: Exact signal reconstruction from highly incomplete frequency information. Information Theory, IEEE Transactions on **52**(2), 489–509 (2006)
27. Candès, E.J., Romberg, J.K., Tao, T.: Stable signal recovery from incomplete and inaccurate measurements. Communications on pure and applied mathematics **59**(8), 1207–1223 (2006)
28. Candès, E.J., Tao, T.: Decoding by linear programming. Information Theory, IEEE Transactions on **51**(12), 4203–4215 (2005)
29. Censor, Y., Zenios, S.A.: Proximal minimization algorithm with D-functions. Journal of Optimization Theory and Applications **73**(3), 451–464 (1992)
30. Chambolle, A.: An algorithm for total variation minimization and applications. Journal of Mathematical Imaging and Vision **20**(1–2), 89–97 (2004)
31. Chambolle, A., De Vore, R.A., Lee, N.Y., Lucier, B.J.: Nonlinear wavelet image processing: variational problems, compression, and noise removal through wavelet shrinkage. Image Processing, IEEE Transactions on **7**(3), 319–335 (1998)
32. Chambolle, A., Pock, T.: A first-order primal-dual algorithm for convex problems with applications to imaging. Journal of Mathematical Imaging and Vision **40**(1), 120–145 (2011)
33. Chan, T.F., Golub, G.H., Mulet, P.: A nonlinear primal-dual method for total variation-based image restoration. SIAM Journal on Scientific Computing **20**(6), 1964–1977 (1999)
34. Chan, T.F., Mulet, P.: On the convergence of the lagged diffusivity fixed point method in total variation image restoration. SIAM Journal on Numerical Analysis **36**(2), 354–367 (1999)
35. Chen, C., He, B., Ye, Y., Yuan, X.: The direct extension of ADMM for multi-block convex minimization problems is not necessarily convergent. Mathematical Programming **155**(1–2), 57–79 (2016)
36. Chen, G., Teboulle, M.: Convergence analysis of a proximal-like minimization algorithm using Bregman functions. SIAM Journal on Optimization **3**(3), 538–543 (1993)
37. Chen, G., Teboulle, M.: A proximal-based decomposition method for convex minimization problems. Mathematical Programming **64**(1), 81–101 (1994)
38. Chen, L., Sun, D., Toh, K.C.: A note on the convergence of ADMM for linearly constrained convex optimization problems. arXiv preprint arXiv:1507.02051 (2015)
39. Chen, S.S., Donoho, D.L., Saunders, M.A.: Atomic decomposition by basis pursuit. SIAM Review **43**(1), 129–159 (2001)
40. Chen, Y., Hager, W., Huang, F., Phan, D., Ye, X., Yin, W.: Fast algorithms for image reconstruction with application to partially parallel MR imaging. SIAM Journal on Imaging Sciences **5**(1), 90–118 (2012)
41. Combettes, P.L., Pesquet, J.C.: Proximal splitting methods in signal processing. In: Fixed-Point Algorithms for Inverse Problems in Science and Engineering, pp. 185–212. Springer (2011)
42. Combettes, P.L., Wajs, V.R.: Signal recovery by proximal forward-backward splitting. Multiscale Modeling & Simulation **4**(4), 1168–1200 (2005)
43. Condat, L.: A primal-dual splitting method for convex optimization involving Lipschitzian, proximable and linear composite terms. Journal of Optimization Theory and Applications **158**(2), 460–479 (2013)
44. Daubechies, I., Defrise, M., De Mol, C.: An iterative thresholding algorithm for linear inverse problems with a sparsity constraint. Communications on Pure and Applied Mathematics **57**(11), 1413–1457 (2004)
45. Davis, D., Yin, W.: Convergence rate analysis of several splitting schemes. In: R. Glowinski, S. Osher, W. Yin (eds.) Splitting Methods in Communication and Imaging, Science and Engineering, Chapter 4. Springer (2016)

46. Deng, W., Lai, M.J., Peng, Z., Yin, W.: Parallel multi-block ADMM with o(1/k) convergence. arXiv preprint arXiv:1312.3040 (2013)
47. Dong, B., Shen, Z., et al.: MRA based wavelet frames and applications. IAS Lecture Notes Series, Summer Program on "The Mathematics of Image Processing", Park City Mathematics Institute **19** (2010)
48. Donoho, D.: Compressed sensing. Information Theory, IEEE Transactions on **52**(4), 1289–1306 (2006)
49. Donoho, D.: For most large underdetermined systems of equations, the minimal ℓ_1-norm near solution approximates the sparsest near-solution. Communications on Pure and Applied Mathematics **59**(7), 907–934 (2006)
50. Donoho, D., Huo, X.: Uncertainty principles and ideal atomic decomposition. Information Theory, IEEE Transactions on **47**(7), 2845–2862 (2001)
51. Donoho, D.L., Johnstone, J.M.: Ideal spatial adaptation by wavelet shrinkage. Biometrika **81**(3), 425–455 (1994)
52. Douglas, J., Rachford, H.H.: On the numerical solution of heat conduction problems in two and three space variables. Transactions of the American mathematical Society **82**(2), 421–439 (1956)
53. Duarte, M., Davenport, M., Takhar, D., Laska, J., Sun, T., Kelly, K., Baraniuk, R.: Single-pixel imaging via compressive sampling: Building simpler, smaller, and less-expensive digital cameras. Signal Processing Magazine, IEEE **25**(2), 83–91 (2008)
54. Eckstein, J.: Splitting methods for monotone operators with applications to parallel optimization. Ph.D. thesis, Massachusetts Institute of Technology (1989)
55. Eckstein, J.: Nonlinear proximal point algorithms using bregman functions, with applications to convex programming. Mathematics of Operations Research **18**(1), 202–226 (1993)
56. Eckstein, J.: Augmented Lagrangian and alternating direction methods for convex optimization: A tutorial and some illustrative computational results. RUTCOR Research Reports **32** (2012)
57. Eckstein, J., Bertsekas, D.P.: On the Douglas-Rachford splitting method and the proximal point algorithm for maximal monotone operators. Mathematical Programming **55**(1–3), 293–318 (1992)
58. Elad, M.: Why simple shrinkage is still relevant for redundant representations? Information Theory, IEEE Transactions on **52**(12), 5559–5569 (2006)
59. Esser, E., Zhang, X., Chan, T.F.: A general framework for a class of first order primal-dual algorithms for convex optimization in imaging science. SIAM Journal on Imaging Sciences **3**(4), 1015–1046 (2010)
60. Fadili, M., Starck, J.: Sparse representation-based image deconvolution by iterative thresholding. Astronomical Data Analysis (ADA) '06, Marseille, France (2006)
61. Figueiredo, M., Nowak, R.D.: An EM algorithm for wavelet-based image restoration. Image Processing, IEEE Transactions on **12**(8), 906–916 (2003)
62. Figueiredo, M.A., Bioucas-Dias, J.M., Nowak, R.D.: Majorization–minimization algorithms for wavelet-based image restoration. Image Processing, IEEE Transactions on **16**(12), 2980–2991 (2007)
63. Fortin, M., Glowinski, R.: Augmented Lagrangian Methods: Applications to the Numerical Solution of Boundary-Value Problems. North-Holland, Amsterdam (1983)
64. Frick, K., Lorenz, D.A., Resmerita, E.: Morozov's principle for the augmented Lagrangian method applied to linear inverse problems. Multiscale Modeling & Simulation **9**(4), 1528–1548 (2011)
65. Friedlander, M., Tseng, P.: Exact regularization of convex programs. SIAM Journal on Optimization **18**(4), 1326–1350 (2007)
66. Fukushima, M.: Application of the alternating direction method of multipliers to separable convex programming problems. Computational Optimization and Applications **1**(1), 93–111 (1992)
67. Fukushima, M., Mine, H.: A generalized proximal point algorithm for certain non-convex minimization problems. International Journal of Systems Science **12**(8), 989–1000 (1981)

68. Gabay, D.: Applications of the method of multipliers to variational inequalities. In: M. Fortin, R. Glowinski (eds.) Augmented Lagrangian Methods: Applications to the Numerical Solution of Boundary-Value Problems. North-Holland, Amsterdam (1983)
69. Gabay, D., Mercier, B.: A dual algorithm for the solution of nonlinear variational problems via finite element approximation. Computers & Mathematics with Applications **2**(1), 17–40 (1976)
70. Glowinski, R., Le Tallec, P.: Augmented Lagrangian and Operator-Splitting Methods in Nonlinear Mechanics. SIAM, Philadelphia, PA (1989)
71. Glowinski, R., Marroco, A.: Sur l'approximation, par éléments finis d'ordre un, et la résolution, par pénalisation-dualité d'une classe de problèmes de Dirichlet non linéaires. ESAIM: Mathematical Modelling and Numerical Analysis **9**(R2), 41–76 (1975)
72. Goldstein, T., Li, M., Yuan, X.: Adaptive primal-dual splitting methods for statistical learning and image processing. In: Advances in Neural Information Processing Systems (NIPS), pp. 2080–2088 (2015)
73. Goldstein, T., O'Donoghue, B., Setzer, S., Baraniuk, R.: Fast alternating direction optimization methods. SIAM Journal on Imaging Sciences **7**(3), 1588–1623 (2014)
74. Goldstein, T., Osher, S.: The split Bregman method for L1-regularized problems. SIAM Journal on Imaging Sciences **2**(2), 323–343 (2009)
75. Goldstein, T., Studer, C., Baraniuk, R.: A field guide to forward-backward splitting with a FASTA implementation. arXiv preprint arXiv:1411.3406 (2014)
76. Goldstein, T., Xu, L., Kelly, K.F., Baraniuk, R.: The STONE transform: Multi-resolution image enhancement and compressive video. IEEE Transactions on Image Processing **24**(12), 5581–5593 (2015)
77. Gupta, M., Agrawal, A., Veeraraghavan, A., Narasimhan, S.G.: Flexible voxels for motion-aware videography. In: Proc. of the 11th European Conference on Computer Vision: Part I, ECCV'10, pp. 100–114. Springer-Verlag, Berlin, Heidelberg (2010)
78. Hale, E.T., Yin, W., Zhang, Y.: Fixed-point continuation for ℓ_1-minimization: Methodology and convergence. SIAM Journal on Optimization **19**(3), 1107–1130 (2008)
79. He, B., Peng, Z., Wang, X.: Proximal alternating direction-based contraction methods for separable linearly constrained convex optimization. Frontiers of Mathematics in China **6**(1), 79–114 (2011)
80. He, B., Tao, M., Xu, M., Yuan, X.: An alternating direction-based contraction method for linearly constrained separable convex programming problems. Optimization **62**(4), 573–596 (2013)
81. He, B., You, Y., Yuan, X.: On the convergence of primal-dual hybrid gradient algorithm. SIAM Journal on Imaging Sciences **7**(4), 2526–2537 (2014)
82. He, B., Yuan, X.: Convergence analysis of primal-dual algorithms for a saddle-point problem: from contraction perspective. SIAM Journal on Imaging Sciences **5**(1), 119–149 (2012)
83. Hestenes, M.R.: Multiplier and gradient methods. Journal of Optimization Theory and Applications **4**(5), 303–320 (1969)
84. Hong, M., Chang, T., Wang, X., Razaviyayn, M., Ma, S., Luo, Z.Q.: A block successive upper bound minimization method of multipliers for linearly constrained convex optimization. arXiv preprint arXiv:1401.7079 (2014)
85. Hong, M., Luo, Z.Q.: On the linear convergence of the alternating direction method of multipliers. arXiv preprint arXiv:1208.3922 (2012)
86. Ji, J.X., Son, J.B., Rane, S.D.: PULSAR: A Matlab toolbox for parallel magnetic resonance imaging using array coils and multiple channel receivers. Concepts in Magnetic Resonance Part B: Magnetic Resonance Engineering **31**(1), 24–36 (2007)
87. Keeling, S.L., Clason, C., Hintermüller, M., Knoll, F., Laurain, A., Von Winckel, G.: An image space approach to cartesian based parallel MR imaging with total variation regularization. Medical Image Analysis **16**(1), 189–200 (2012)
88. Knoll, F., Clason, C., Bredies, K., Uecker, M., Stollberger, R.: Parallel imaging with nonlinear reconstruction using variational penalties. Magnetic Resonance in Medicine **67**(1), 34–41 (2012)

89. Kunisch, K., Hintermüller, M.: Total bounded variation regularization as a bilaterally constrained optimization problem. SIAM Journal on Applied Mathematics **64**(4), 1311–1333 (2004)
90. Lange, K., Hunter, D., Yang, I.: Optimization transfer using surrogate objective functions. Journal of Computational and Graphical Statistics **9**(1), 1–20 (2000)
91. Lee, J., Sun, Y., Saunders, M.: Proximal Newton-type methods for convex optimization. In: Advances in Neural Information Processing Systems (NIPS), pp. 836–844 (2012)
92. Li, G., Pong, T.K.: Global convergence of splitting methods for nonconvex composite optimization. SIAM Journal on Optimization **25**(4), 2434–2460 (2015)
93. Lions, P., Mercier, B.: Splitting algorithms for the sum of two nonlinear operators. SIAM Journal on Numerical Analysis **16**(6), 964–979 (1979)
94. Marcia, R.F., Harmany, Z.T., Willett, R.M.: Compressive coded aperture imaging. In: Proc. SPIE, p. 72460 (2009)
95. Martinet, B.: Régularisation d'inéquations variationnelles par approximations successives. Revue Française d'Informatique et de Recherche Opérationnelle **4**(3), 154–158 (1970)
96. Moeller, M., Zhang, X.: Fast sparse reconstruction: Greedy inverse scale space flows. Mathematics of Computation **85**(297), 179–208 (2016)
97. Moreau, J.J.: Proximité et dualité dans un espace Hilbertien. Bulletin de la Société Mathématique de France **93**, 273–299 (1965)
98. Nesterov, Y.: A method for unconstrained convex minimization problem with the rate of convergence $O(1/k^2)$. Soviet Mathematics Doklady **269**(3), 543–547 (1983)
99. Osher, S., Burger, M., Goldfarb, D., Xu, J., Yin, W.: An iterative regularization method for total variation-based image restoration. Multiscale Modeling & Simulation **4**(2), 460–489 (2005)
100. Osher, S., Ruan, F., Xiong, J., Yao, Y., Yin, W.: Sparse recovery via differential inclusions. Applied and Computational Harmonic Analysis (2016)
101. Ouyang, H., He, N., Tran, L., Gray, A.: Stochastic alternating direction method of multipliers. In: Proceedings of the 30th International Conference on Machine Learning, vol. 28, pp. 80–88 (2013)
102. Ouyang, Y., Chen, Y., Lan, G., Eduardo Pasiliao, J.: An accelerated linearized alternating direction method of multipliers. SIAM Journal on Imaging Sciences **8**(1), 644–681 (2015)
103. Parikh, N., Boyd, S.P.: Proximal algorithms. Foundations and Trends in Optimization **1**(3), 127–239 (2014)
104. Park, J.Y., Wakin, M.B.: Multiscale algorithm for reconstructing videos from streaming compressive measurements. Journal of Electronic Imaging **22**(2), 021,001–021,001 (2013)
105. Passty, G.B.: Ergodic convergence to a zero of the sum of monotone operators in Hilbert space. Journal of Mathematical Analysis and Applications **72**(2), 383–390 (1979)
106. Patrinos, P., Stella, L., Bemporad, A.: Douglas-Rachford splitting: Complexity estimates and accelerated variants. In: Decision and Control (CDC), 2014 IEEE 53rd Annual Conference on, pp. 4234–4239 (2014)
107. Pock, T., Chambolle, A.: Diagonal preconditioning for first order primal-dual algorithms in convex optimization. In: Computer Vision (ICCV), 2011 IEEE International Conference on, pp. 1762–1769. IEEE (2011)
108. Pock, T., Cremers, D., Bischof, H., Chambolle, A.: An algorithm for minimizing the Mumford-Shah functional. In: 2009 IEEE 12th International Conference on Computer Vision, pp. 1133–1140. IEEE (2009)
109. Popov, L.: A modification of the Arrow-Hurwicz method for search of saddle points. Mathematical Notes **28**(5), 845–848 (1980)
110. Powell, M.: A method for nonlinear constraints in minimization problems. In: R. Fletcher (ed.) Optimization. Academic Press, New York, NY (1969)
111. Reddy, D., Veeraraghavan, A., Chellappa, R.: P2C2: Programmable pixel compressive camera for high speed imaging. In: IEEE Conference on Computer Vision and Pattern Recognition, CVPR '11, pp. 329–336 (2011)
112. Rockafellar, R.: Monotropic programming: descent algorithms and duality. Nonlinear Programming **4**, 327–366 (1981)

113. Rockafellar, R.T.: Convex Analysis. Princeton University Press (1970)
114. Rockafellar, R.T.: A dual approach to solving nonlinear programming problems by unconstrained optimization. Mathematical Programming **5**(1), 354–373 (1973)
115. Rockafellar, R.T.: Augmented Lagrangians and applications of the proximal point algorithm in convex programming. Mathematics of Operations Research **1**(2), 97–116 (1976)
116. Rockafellar, R.T.: Monotone operators and the proximal point algorithm. SIAM Journal on Control and Optimization **14**(5) (1976)
117. Rockafellar, R.T., Wets, R.J.B.: Variational Analysis. Springer Dordrecht (2009)
118. Rudin, L.I., Osher, S., Fatemi, E.: Nonlinear total variation based noise removal algorithms. Physica D: Nonlinear Phenomena **60**(1), 259–268 (1992)
119. Sankaranarayanan, A.C., Studer, C., Baraniuk, R.G.: CS-MUVI: Video compressive sensing for spatial-multiplexing cameras. In: IEEE International Conference on Computational Photography (ICCP), pp. 1–10. IEEE (2012)
120. Scherzer, O., Groetsch, C.: Inverse scale space theory for inverse problems. In: Scale-Space and Morphology in Computer Vision, vol. 2106, pp. 317–325. Springer, Berlin Heidelberg (2001)
121. Schmidt, M., Kim, D., Sra, S.: Projected Newton-type methods in machine learning. In: S. Sra, S. Nowozin, S. Wright (eds.) Optimization for Machine Learning, pp. 305–330. MIT Press (2011)
122. Shefi, R., Teboulle, M.: Rate of convergence analysis of decomposition methods based on the proximal method of multipliers for convex minimization. SIAM Journal on Optimization **24**(1), 269–297 (2014)
123. Sra, S.: Fast projections onto mixed-norm balls with applications. Data Mining and Knowledge Discovery **25**(2), 358–377 (2012)
124. Suzuki, T.: Stochastic dual coordinate ascent with alternating direction method of multipliers. In: Proceedings of the 31st International Conference on Machine Learning (ICML-14), pp. 736–744 (2014)
125. Tibshirani, R.: Regression shrinkage and selection via the Lasso. Journal of the Royal Statistical Society. Series B (Methodological) **58**(1), 267–288 (1996)
126. Valkonen, T.: A primal–dual hybrid gradient method for nonlinear operators with applications to MRI. Inverse Problems **30**(5), 055,012 (2014)
127. Vogel, C., Oman, M.: Iterative methods for total variation denoising. SIAM Journal on Scientific Computing **17**(1), 227–238 (1996)
128. Wang, H., Banerjee, A.: Online alternating direction method (longer version). arXiv preprint arXiv:1306.3721 (2013)
129. Wang, H., Banerjee, A.: Bregman alternating direction method of multipliers. In: Z. Ghahramani, M. Welling, C. Cortes, N. Lawrence, K. Weinberger. (eds.) Advances in Neural Information Processing Systems 27 (NIPS), pp. 2816–2824 (2014)
130. Wang, Y., Yang, J., Yin, W., Zhang, Y.: A new alternating minimization algorithm for total variation image reconstruction. SIAM Journal on Imaging Sciences **1**(3), 248–272 (2008)
131. Wright, S.J., Nowak, R.D., Figueiredo, M.A.: Sparse reconstruction by separable approximation. Signal Processing, IEEE Transactions on **57**(7), 2479–2493 (2009)
132. Yin, W.: Analysis and generalizations of the linearized Bregman method. SIAM Journal on Imaging Sciences **3**(4), 856–877 (2010)
133. Yin, W., Osher, S.: Error forgetting of Bregman iteration. Journal of Scientific Computing **54**(2–3), 684–695 (2012)
134. Yin, W., Osher, S., Goldfarb, D., Darbon, J.: Bregman iterative algorithms for ℓ_1-minimization with applications to compressed sensing. SIAM Journal on Imaging Sciences **1**(1), 143–168 (2008)
135. Yuan, M., Lin, Y.: Model selection and estimation in regression with grouped variables. Journal of the Royal Statistical Society: Series B (Statistical Methodology) **68**(1), 49–67 (2006)
136. Zhang, X., Burger, M., Bresson, X., Osher, S.: Bregmanized nonlocal regularization for deconvolution and sparse reconstruction. SIAM Journal on Imaging Sciences **3**(3), 253–276 (2010)

137. Zhang, X., Burger, M., Osher, S.: A unified primal-dual algorithm framework based on Bregman iteration. Journal of Scientific Computing **46**(1), 20–46 (2010)
138. Zhang, Y.: Theory of compressive sensing via ℓ_1-minimization: a Non-RIP analysis and extensions. Journal of the Operations Research Society of China **1**(1), 79–105 (2013)
139. Zhong, W., Kwok, J.: Fast stochastic alternating direction method of multipliers. In: Proceedings of the 31st International Conference on Machine Learning (ICML), pp. 46–54 (2014)
140. Zhu, M., Chan, T.: An efficient primal-dual hybrid gradient algorithm for total variation image restoration. CAM report 08-34, UCLA (2008)

Chapter 10
First Order Algorithms in Variational Image Processing

M. Burger, A. Sawatzky, and G. Steidl

Abstract The success of non-smooth variational models in image processing is heavily based on efficient algorithms. Taking into account the specific structure of the models as sum of different convex terms, splitting algorithms are an appropriate choice. Their strength consists in the splitting of the original problem into a sequence of smaller proximal problems which are easy and fast to compute.

Operator splitting methods were first applied to linear, single-valued operators for solving partial differential equations in the 60th of the last century. More than 20 years later these methods were generalized in the convex analysis community to the solution of inclusion problems, where the linear operators have to be replaced by nonlinear, set-valued, monotone operators. Again after more than 20 years splitting methods became popular in image processing. In particular, operator splittings in combination with (augmented) Lagrangian methods and primal-dual methods have been applied very successfully.

In this chapter we give an overview of first order algorithms recently used to solve convex non-smooth variational problems in image processing. We present computational studies providing a comparison of different methods and also illustrating their success in applications.

M. Burger (✉) • A. Sawatzky
Department of Mathematics and Computer Science, University of Münster, 48149 Münster, Germany
e-mail: martin.burger@wwu.de

G. Steidl
Department of Mathematics, University of Kaiserslautern, Paul Ehrlich Str. 31, 67653 Kaiserslautern, Germany

R. Glowinski et al. (eds.), *Splitting Methods in Communication, Imaging, Science, and Engineering*, Scientific Computation, DOI 10.1007/978-3-319-41589-5_10

1 Introduction

Variational methods in imaging are nowadays developing towards a quite universal and flexible tool, allowing for highly successful approaches on tasks like denoising, deblurring, inpainting, segmentation, super-resolution, disparity, and optical flow estimation. The overall structure of such approaches is of the form

$$\mathcal{D}(Ku) + \alpha\mathcal{R}(u) \to \min_u,$$

where the functional $\mathcal{D}$ is a data fidelity term also depending on some input data f and measuring the deviation of Ku from such and $\mathcal{R}$ is a regularization functional. Moreover K is a (often linear) forward operator modeling the dependence of data on an underlying image, and α is a positive regularization parameter. While $\mathcal{D}$ is often smooth and (strictly) convex, the current practice almost exclusively uses nonsmooth regularization functionals. The majority of successful techniques is using nonsmooth and convex functionals like the total variation and generalizations thereof, cf. [28, 31, 41], or ℓ_1-norms of coefficients arising from scalar products with some frame system, cf. [78] and references therein.

The efficient solution of such variational problems in imaging demands for appropriate algorithms. Taking into account the specific structure as a sum of very different terms to be minimized, splitting algorithms are a quite canonical choice. Consequently this field has revived the interest in techniques like operator splittings or augmented Lagrangians. In this chapter we shall provide an overview of methods currently developed and recent results as well as some computational studies providing a comparison of different methods and also illustrating their success in applications.

We start with a very general viewpoint in the first sections, discussing basic notations, properties of proximal maps, firmly non-expansive and averaging operators, which form the basis of further convergence arguments. Then we proceed to a discussion of several state-of-the-art algorithms and their (theoretical) convergence properties. In this chapter we focus on the so-called first order methods involving only subgradients of the functional, but no higher order derivatives. After a section discussing issues related to the use of analogous iterative schemes for ill-posed problems, we present some practical convergence studies in numerical examples related to PET and spectral CT reconstruction.

2 Notation

In the following we summarize the notations and definitions that will be used throughout the present chapter:

- $x_+ := \max\{x, 0\}$, $x \in \mathbb{R}^d$, whereby the maximum operation has to be interpreted componentwise.

- ι_C is the indicator function of a set $C \subseteq \mathbb{R}^d$ given by

$$\iota_C(x) := \begin{cases} 0 & \text{if } x \in C, \\ +\infty & \text{otherwise.} \end{cases}$$

- $\Gamma_0(\mathbb{R}^d)$ is a set of proper, convex, and lower semi-continuous functions mapping from $\mathbb{R}^d$ into the extended real numbers $\mathbb{R} \cup \{+\infty\}$.
- $\mathrm{dom} f := \{x \in \mathbb{R}^d : f(x) < +\infty\}$ denotes the *effective domain* of f.
- $\partial f(x_0) := \{p \in \mathbb{R}^d : f(x) - f(x_0) \geq \langle p, x - x_0 \rangle \; \forall x \in \mathbb{R}^d\}$ denotes the *subdifferential* of $f \in \Gamma_0(\mathbb{R}^d)$ at $x_0 \in \mathrm{dom} f$ and is the set consisting of the *subgradients* of f at x_0. If $f \in \Gamma_0(\mathbb{R}^d)$ is differentiable at x_0, then $\partial f(x_0) = \{\nabla f(x_0)\}$. Conversely, if $\partial f(x_0)$ contains only one element then f is differentiable at x_0 and this element is just the gradient of f at x_0. By *Fermat's rule*, $\hat{x}$ is a global minimizer of $f \in \Gamma_0(\mathbb{R}^d)$ if and only if

$$0 \in \partial f(\hat{x}).$$

- $f^*(p) := \sup_{x \in \mathbb{R}^d} \{\langle p, x \rangle - f(x)\}$ is the (Fenchel) *conjugate* of f. For proper f, we have $f^* = f$ if and only if $f(x) = \frac{1}{2}\|x\|_2^2$. If $f \in \Gamma_0(\mathbb{R}^d)$ is *positively homogeneous*, i.e., $f(\alpha x) = \alpha f(x)$ for all $\alpha > 0$, then

$$f^*(x^*) = \iota_{C_f}(x^*), \quad C_f := \{x^* \in \mathbb{R}^d : \langle x^*, x \rangle \leq f(x) \; \forall x \in \mathbb{R}^d\}.$$

In particular, the conjugate functions of ℓ_p-norms, $p \in [1, +\infty]$, are given by

$$\|\cdot\|_p^*(x^*) = \iota_{B_q(1)}(x^*) \tag{10.1}$$

where $\frac{1}{p} + \frac{1}{q} = 1$ and as usual $p = 1$ corresponds to $q = \infty$ and conversely, and $B_q(\lambda) := \{x \in \mathbb{R}^d : \|x\|_q \leq \lambda\}$ denotes the ball of radius $\lambda > 0$ with respect to the ℓ_q-norm.

3 Proximal Operator

The algorithms proposed in this chapter to solve various problems in digital image analysis and restoration have in common that they basically reduce to the evaluation of a series of proximal problems. Therefore we start with these kind of problems. For a comprehensive overview on proximal algorithms we refer to [139].

3.1 Definition and Basic Properties

For $f \in \Gamma_0(\mathbb{R}^d)$ and $\lambda > 0$, the *proximal operator* $\mathrm{prox}_{\lambda f} : \mathbb{R}^d \to \mathbb{R}^d$ of λf is defined by

$$\mathrm{prox}_{\lambda f}(x) := \underset{y \in \mathbb{R}^d}{\mathrm{argmin}} \left\{ \frac{1}{2\lambda}\|x - y\|_2^2 + f(y) \right\}. \tag{10.2}$$

It compromises between minimizing f and being near to x, where λ is the trade-off parameter between these terms. The *Moreau envelope* or *Moreau-Yoshida regularization* ${}^{\lambda}f : \mathbb{R}^d \to \mathbb{R}$ is given by

$$^{\lambda}f(x) := \min_{y \in \mathbb{R}^d} \left\{ \frac{1}{2\lambda} \|x-y\|_2^2 + f(y) \right\}.$$

A straightforward calculation shows that ${}^{\lambda}f = (f^* + \frac{\lambda}{2}\|\cdot\|_2^2)^*$. The following theorem ensures that the minimizer in (10.2) exists, is unique, and can be characterized by a variational inequality. The Moreau envelope can be considered as a smooth approximation of f. For the proof we refer to [8].

Theorem 1. *Let $f \in \Gamma_0(\mathbb{R}^d)$. Then,*

i) *For any $x \in \mathbb{R}^d$, there exists a unique minimizer $\hat{x} = \mathrm{prox}_{\lambda f}(x)$ of* (10.2).
ii) *The variational inequality*

$$\frac{1}{\lambda}\langle x-\hat{x}, y-\hat{x}\rangle + f(\hat{x}) - f(y) \le 0 \qquad \forall y \in \mathbb{R}^d. \tag{10.3}$$

is necessary and sufficient for $\hat{x}$ to be the minimizer of (10.2).
iii) *$\hat{x}$ is a minimizer of f if and only if it is a fixed point of* $\mathrm{prox}_{\lambda f}$, *i.e.,*

$$\hat{x} = \mathrm{prox}_{\lambda f}(\hat{x}).$$

iv) *The Moreau envelope ${}^{\lambda}f$ is continuously differentiable with gradient*

$$\nabla\left({}^{\lambda}f\right)(x) = \frac{1}{\lambda}\left(x - \mathrm{prox}_{\lambda f}(x)\right). \tag{10.4}$$

v) *The set of minimizers of f and ${}^{\lambda}f$ are the same.*

Rewriting iv) as $\mathrm{prox}_{\lambda f}(x) = x - \lambda\nabla\left({}^{\lambda}f\right)(x)$ *we can interpret* $\mathrm{prox}_{\lambda f}(x)$ *as a gradient descent step with step size λ for minimizing ${}^{\lambda}f$.*

Example 1. Consider the univariate function $f(y) := |y|$ and

$$\mathrm{prox}_{\lambda f}(x) = \operatorname*{argmin}_{y \in \mathbb{R}} \left\{ \frac{1}{2\lambda}(x-y)^2 + |y| \right\}.$$

Then, a straightforward computation yields that $\mathrm{prox}_{\lambda f}$ is the *soft-shrinkage* function S_λ with threshold λ (see Figure 10.1) defined by

$$S_\lambda(x) := (x-\lambda)_+ - (-x-\lambda)_+ = \begin{cases} x-\lambda & \text{for } x > \lambda, \\ 0 & \text{for } x \in [-\lambda, \lambda], \\ x+\lambda & \text{for } x < -\lambda. \end{cases}$$

Setting $\hat{x} := S_\lambda(x) = \mathrm{prox}_{\lambda f}(x)$, we get

$$ {}^\lambda f(x) = |\hat{x}| + \frac{1}{2\lambda}(x-\hat{x})^2 = \begin{cases} x - \frac{\lambda}{2} & \text{for } x > \lambda, \\ \frac{1}{2\lambda}x^2 & \text{for } x \in [-\lambda, \lambda], \\ -x - \frac{\lambda}{2} & \text{for } x < -\lambda. \end{cases} $$

This function ${}^\lambda f$ is known as *Huber function* (see Figure 10.1).

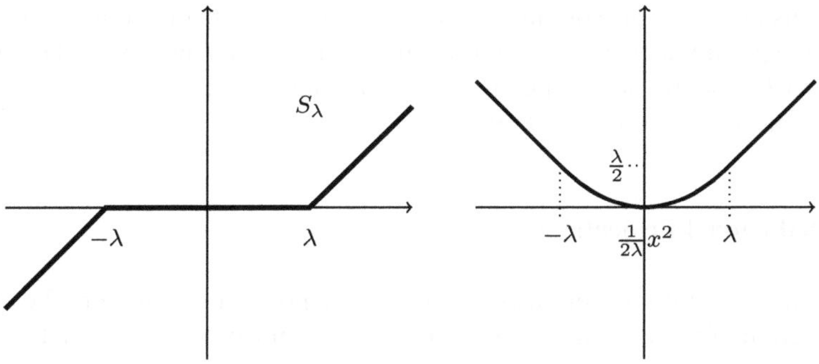

Fig. 10.1: Left: Soft-shrinkage function $\mathrm{prox}_{\lambda f} = S_\lambda$ for $f(y) = |y|$. Right: Moreau envelope ${}^\lambda f$.

Theorem 2 (Moreau Decomposition). *For $f \in \Gamma_0(\mathbb{R}^d)$ the following decomposition holds:*

$$ \mathrm{prox}_f(x) + \mathrm{prox}_{f^*}(x) = x, $$
$$ {}^1 f(x) + {}^1 f^*(x) = \frac{1}{2}\|x\|_2^2. $$

For a proof we refer to [148, Theorem 31.5].

Remark 1 (Proximal Operator and Resolvent). The subdifferential operator is a set-valued function $\partial f : \mathbb{R}^d \to 2^{\mathbb{R}^d}$. For $f \in \Gamma_0(\mathbb{R}^d)$, we have by Fermat's rule and subdifferential calculus that $\hat{x} = \mathrm{prox}_{\lambda f}(x)$ if and only if

$$ 0 \in \hat{x} - x + \lambda \partial f(\hat{x}), $$
$$ x \in (I + \lambda \partial f)(\hat{x}), $$

which implies by the uniqueness of the proximum that $\hat{x} = (I + \lambda \partial f)^{-1}(x)$. In particular, $J_{\lambda \partial f} := (I + \lambda \partial f)^{-1}$ is a single-valued operator which is called the *resolvent* of the set-valued operator $\lambda \partial f$. In summary, the proximal operator of λf coincides with the resolvent of $\lambda \partial f$, i.e.,

$$ \mathrm{prox}_{\lambda f} = J_{\lambda \partial f}. $$

The proximal operator (10.2) and the proximal algorithms described in Section 5 can be generalized by introducing a symmetric, positive definite matrix $Q \in \mathbb{R}^{d,d}$ as follows:

$$\mathrm{prox}_{Q,\lambda f} := \operatorname*{argmin}_{y\in\mathbb{R}^d}\left\{\frac{1}{2\lambda}\|x-y\|_Q^2 + f(y)\right\}, \tag{10.5}$$

where $\|x\|_Q^2 := x^{\mathrm{T}}Qx$, see, e.g., [52, 57, 190].

3.2 *Special Proximal Operators*

Algorithms involving the solution of proximal problems are only efficient if the corresponding proximal operators can be evaluated in an efficient way. In the following we collect frequently appearing proximal mappings in image processing. For epigraphical projections see [12, 50, 94].

3.2.1 Orthogonal Projections

The proximal operator generalizes the orthogonal projection operator. The orthogonal projection of $x \in \mathbb{R}^d$ onto a nonempty, closed, convex set C is given by

$$\Pi_C(x) := \operatorname*{argmin}_{y\in C}\|x-y\|_2$$

and can be rewritten for any $\lambda > 0$ as

$$\Pi_C(x) = \operatorname*{argmin}_{y\in\mathbb{R}^d}\left\{\frac{1}{2\lambda}\|x-y\|_2^2 + \iota_C(y)\right\} = \mathrm{prox}_{\lambda\iota_C}(x).$$

Some special sets C are considered next.

Affine set

$C := \{y \in \mathbb{R}^d : Ay = b\}$ with $A \in \mathbb{R}^{m,d}$, $b \in \mathbb{R}^m$.
In case of $\|x-y\|_2 \to \min_y$ subject to $Ay = b$ we substitute $z := x - y$ which leads to

$$\|z\|_2 \to \min_z \quad \text{subject to} \quad Az = r := Ax - b.$$

This can be directly solved, see [20], and leads after back-substitution to

$$\Pi_C(x) = x - A^\dagger(Ax-b),$$

where $A^\dagger$ denotes the Moore-Penrose inverse of A.

Halfspace

$C := \{y \in \mathbb{R}^d : a^\mathrm{T} y \le b\}$ with $a \in \mathbb{R}^d$, $b \in \mathbb{R}$.
A straightforward computation gives

$$\Pi_C(x) = x - \frac{(a^\mathrm{T} x - b)_+}{\|a\|_2^2} a.$$

Box and Nonnegative Orthant

$C := \{y \in \mathbb{R}^d : l \le y \le u\}$ with $l, u \in \mathbb{R}^d$.
The proximal operator can be applied componentwise and gives

$$(\Pi_C(x))_k = \begin{cases} l_k & \text{if } x_k < l_k, \\ x_k & \text{if } l_k \le x_k \le u_k, \\ u_k & \text{if } x_k > u_k. \end{cases}$$

For $l = 0$ and $u = +\infty$ we get the orthogonal projection onto the non-negative orthant

$$\Pi_C(x) = x_+.$$

Probability Simplex

$C := \{y \in \mathbb{R}^d : \mathbf{1}^\mathrm{T} y = \sum_{k=1}^d y_k = 1,\ y \ge 0\}$.
Here we have

$$\Pi_C(x) = (x - \mu \mathbf{1})_+,$$

where $\mu \in \mathbb{R}$ has to be determined such that $h(\mu) := \mathbf{1}^\mathrm{T}(x - \mu\mathbf{1})_+ = 1$. Now μ can be found, e.g., by bisection with starting interval $[\max_k x_k - 1, \max_k x_k]$ or by a method similar to those described in subSection 3.2.2 for projections onto the ℓ_1-ball. Note that h is a linear spline function with knots $x_1, \ldots, x_d$ so that μ is completely determined if we know the neighbor values x_k of μ.

3.2.2 Vector Norms

We consider the proximal operator of $f = \|\cdot\|_p$, $p \in [1, +\infty]$. By the Moreau decomposition in Theorem 2, regarding $(\lambda f)^* = \lambda f^*(\cdot/\lambda)$ and by (10.1) we obtain

$$\begin{aligned} \operatorname{prox}_{\lambda f}(x) &= x - \operatorname{prox}_{\lambda f^*(\cdot/\lambda)}(x) \\ &= x - \Pi_{B_q(\lambda)}(x), \end{aligned}$$

where $\frac{1}{p} + \frac{1}{q} = 1$. Thus the proximal operator can be simply computed by the projections onto the ℓ_q-ball. In particular, it follows for $p = 1, 2, \infty$:

$p = 1, q = \infty$:

For $k = 1, \ldots, d$,

$$\left(\Pi_{B_\infty(\lambda)}(x)\right)_k = \begin{cases} x_k \text{ if } |x_k| \le \lambda, \\ \lambda \operatorname{sgn}(x_k) \text{ if } |x_k| > \lambda, \end{cases} \quad \text{and} \quad \operatorname{prox}_{\lambda\|\cdot\|_1}(x) = S_\lambda(x),$$

where $S_\lambda(x)$, $x \in \mathbb{R}^d$, denotes the componentwise soft-shrinkage with threshold λ.

$p = q = 2$:

$$\Pi_{B_{2,\lambda}}(x) = \begin{cases} x & \text{if } \|x\|_2 \le \lambda, \\ \lambda \frac{x}{\|x\|_2} & \text{if } \|x\|_2 > \lambda, \end{cases} \quad \text{and} \quad \operatorname{prox}_{\lambda\|\cdot\|_2}(x) = \begin{cases} 0 & \text{if } \|x\|_2 \le \lambda, \\ x\left(1 - \frac{\lambda}{\|x\|_2}\right) & \text{if } \|x\|_2 > \lambda. \end{cases}$$

$p = \infty, q = 1$:

$$\Pi_{B_{1,\lambda}}(x) = \begin{cases} x & \text{if } \|x\|_1 \le \lambda, \\ S_\mu(x) & \text{if } \|x\|_1 > \lambda, \end{cases}$$

and

$$\operatorname{prox}_{\lambda\|\cdot\|_\infty}(x) = \begin{cases} 0 & \text{if } \|x\|_1 \le \lambda, \\ x - S_\mu(x) & \text{if } \|x\|_1 > \lambda, \end{cases}$$

with $\mu := \frac{|x_{\pi(1)}| + \ldots + |x_{\pi(m)}| - \lambda}{m}$, where $|x_{\pi(1)}| \ge \ldots \ge |x_{\pi(d)}| \ge 0$ are the sorted absolute values of the components of x and $m \le d$ is the largest index such that $|x_{\pi(m)}|$ is positive and $\frac{|x_{\pi(1)}| + \ldots + |x_{\pi(m)}| - \lambda}{m} \le |x_{\pi(m)}|$, see also [62, 67]. Another method follows similar lines as the projection onto the probability simplex in the previous subsection.

Further, grouped/mixed ℓ_2-ℓ_p-norms are defined for $x = (x_1, \ldots, x_n)^{\mathrm{T}} \in \mathbb{R}^{dn}$ with $x_j := (x_{jk})_{k=1}^d \in \mathbb{R}^d$, $j = 1, \ldots, n$ by

$$\|x\|_{2,p} := \|\,(\|x_j\|_2)_{j=1}^n\,\|_p.$$

For the ℓ_2-ℓ_1-norm we see that

$$\operatorname{prox}_{\lambda\|\cdot\|_{2,1}}(x) = \operatorname*{argmin}_{y \in \mathbb{R}^{dn}} \left\{ \frac{1}{2\lambda} \|x - y\|_2^2 + \|y\|_{2,1} \right\}$$

can be computed separately for each j which results by the above considerations for the ℓ_2-norm for each j in

$$\mathrm{prox}_{\lambda\|\cdot\|_2}(x_j) = \begin{cases} 0 & \text{if } \|x_j\|_2 \le \lambda, \\ x_j(1 - \frac{\lambda}{\|x_j\|_2}) & \text{if } \|x_j\|_2 > \lambda. \end{cases}$$

The procedure for evaluating $\mathrm{prox}_{\lambda\|\cdot\|_{2,1}}$ is sometimes called *coupled or grouped shrinkage*.

Finally, we provide the following rule from [56, Prop. 3.6].

Lemma 1. *Let* $f = g + \mu|\cdot|$, *where* $g \in \Gamma_0(\mathbb{R})$ *is differentiable at* 0 *with* $g'(0) = 0$. *Then* $\mathrm{prox}_{\lambda f} = \mathrm{prox}_{\lambda g} \circ S_{\lambda\mu}$.

Example 2. Consider the *elastic net* regularizer $f(x) := \frac{1}{2}\|x\|_2^2 + \mu\|x\|_1$, see [192]. Setting the gradient in the proximal operator of $g := \frac{1}{2}\|\cdot\|_2^2$ to zero we obtain

$$\mathrm{prox}_{\lambda g}(x) = \frac{1}{1+\lambda} x.$$

The whole proximal operator of f can be then evaluated componentwise and we see by Lemma 1 that

$$\mathrm{prox}_{\lambda f}(x) = \mathrm{prox}_{\lambda g}\left(S_{\lambda\mu}(x)\right) = \frac{1}{1+\lambda} S_{\mu\lambda}(x).$$

3.2.3 Matrix Norms

Next we deal with proximation problems involving matrix norms. For $X \in \mathbb{R}^{m,n}$, we are looking for

$$\mathrm{prox}_{\lambda\|\cdot\|}(X) = \operatorname*{argmin}_{Y \in \mathbb{R}^{m,n}} \left\{ \frac{1}{2\lambda}\|X - Y\|_{\mathcal{F}}^2 + \|Y\| \right\}, \tag{10.6}$$

where $\|\cdot\|_{\mathcal{F}}$ is the Frobenius norm and $\|\cdot\|$ is any unitarily invariant matrix norm, i.e., $\|X\| = \|UXV^T\|$ for all unitary matrices $U \in \mathbb{R}^{m,m}, V \in \mathbb{R}^{n,n}$. Von Neumann (1937) [176] has characterized the unitarily invariant matrix norms as those matrix norms which can be written in the form

$$\|X\| = g(\sigma(X)),$$

where $\sigma(X)$ is the vector of singular values of X and g is a symmetric *gauge function*, see [182]. Recall that $g : \mathbb{R}^d \to \mathbb{R}_+$ is a symmetric gauge function if it is a positively homogeneous convex function which vanishes at the origin and fulfills

$$g(x) = g(\varepsilon_1 x_{k_1}, \ldots, \varepsilon_k x_{k_d})$$

for all $\varepsilon_k \in \{-1,1\}$ and all permutations $k_1,\dots,k_d$ of indices. An analogous result was given by Davis [63] for symmetric matrices, where V^{T} is replaced by U^{T} and the singular values by the eigenvalues.
We are interested in the *Schatten-p norms* for $p = 1,2,\infty$ which are defined for $X \in \mathbb{R}^{m,n}$ and $t := \min\{m,n\}$ by

$$\|X\|_* := \sum_{i=1}^{t} \sigma_i(X) = g_*(\sigma(X)) = \|\sigma(X)\|_1, \qquad \text{(Nuclear norm)}$$

$$\|X\|_{\mathcal{F}} := (\sum_{i=1}^{m}\sum_{j=1}^{n} x_{ij}^2)^{\frac{1}{2}} = (\sum_{i=1}^{t} \sigma_i(X)^2)^{\frac{1}{2}} = g_{\mathcal{F}}(\sigma(X)) = \|\sigma(X)\|_2, \quad \text{(Frobenius norm)}$$

$$\|X\|_2 := \max_{i=1,\dots,t} \sigma_i(X) = g_2(\sigma(X)) = \|\sigma(X)\|_\infty, \qquad \text{(Spectral norm)}.$$

The following theorem shows that the solution of (10.6) reduces to a proximal problem for the vector norm of the singular values of X. Another proof for the special case of the nuclear norm can be found in [37].

Theorem 3. *Let $X = U\Sigma_X V^{T}$ be the singular value decomposition of X and $\|\cdot\|$ a unitarily invariant matrix norm. Then* $\mathrm{prox}_{\lambda\|\cdot\|}(X)$ *in* (10.6) *is given by* $\hat{X} = U\Sigma_{\hat{X}}V^{T}$*, where the singular values $\sigma(\hat{X})$ in $\Sigma_{\hat{X}}$ are determined by*

$$\sigma(\hat{X}) := \mathrm{prox}_{\lambda g}(\sigma(X)) = \operatorname*{argmin}_{\sigma\in\mathbb{R}^t}\{\frac{1}{2}\|\sigma(X)-\sigma\|_2^2 + \lambda g(\sigma)\} \tag{10.7}$$

with the symmetric gauge function g corresponding to $\|\cdot\|$.

Proof. By Fermat's rule we know that the solution $\hat{X}$ of (10.6) is determined by

$$0 \in \hat{X} - X + \lambda\partial\|\hat{X}\| \tag{10.8}$$

and from [182] that

$$\partial\|X\| = \mathrm{conv}\{UDV^{\mathrm{T}} : X = U\Sigma_X V^{\mathrm{T}}, D = \mathrm{diag}(d), d \in \partial g(\sigma(X))\}. \tag{10.9}$$

We now construct the unique solution $\hat{X}$ of (10.8). Let $\hat{\sigma}$ be the unique solution of (10.7). By Fermat's rule $\hat{\sigma}$ satisfies $0 \in \hat{\sigma} - \sigma(X) + \lambda\partial g(\hat{\sigma})$ and consequently there exists $d \in \partial g(\hat{\sigma})$ such that

$$0 = U\big(\mathrm{diag}(\hat{\sigma}) - \Sigma_X + \lambda\mathrm{diag}(d)\big)V_F^{\mathrm{T}} \Leftrightarrow 0 = U\,\mathrm{diag}(\hat{\sigma})\,V^{\mathrm{T}} - X + \lambda U\,\mathrm{diag}(d)\,V^{\mathrm{T}}.$$

By (10.9) we see that $\hat{X} := U\,\mathrm{diag}(\hat{\sigma})\,V^{\mathrm{T}}$ is a solution of (10.8). This completes the proof. □

For the special matrix norms considered above, we obtain by the previous subsection

$$\begin{aligned} \|\cdot\|_*: \quad & \sigma(\hat{X}) := \sigma(X) - \Pi_{B_{\infty,\lambda}}(\sigma(X)), \\ \|\cdot\|_{\mathcal{F}}: \quad & \sigma(\hat{X}) := \sigma(X) - \Pi_{B_{2,\lambda}}(\sigma(X)), \\ \|\cdot\|_2: \quad & \sigma(\hat{X}) := \sigma(X) - \Pi_{B_{1,\lambda}}(\sigma(X)). \end{aligned}$$

4 Fixed Point Algorithms and Averaged Operators

An operator $T : \mathbb{R}^d \to \mathbb{R}^d$ is *contractive* if it is Lipschitz continuous with Lipschitz constant $L < 1$, i.e., there exists a norm $\|\cdot\|$ on $\mathbb{R}^d$ such that

$$\|Tx - Ty\| \le L\|x - y\| \quad \forall x, y \in \mathbb{R}^d.$$

In case $L = 1$, the operator is called *nonexpansive*. A function $T : \mathbb{R}^d \supset \Omega \to \mathbb{R}^d$ is *firmly nonexpansive* if it fulfills for all $x, y \in \mathbb{R}^d$ one of the following equivalent conditions [12]:

$$\begin{aligned} &\|Tx - Ty\|_2^2 \le \langle x - y, Tx - Ty\rangle, \\ &\|Tx - Ty\|_2^2 \le \|x - y\|_2^2 - \|(I - T)x - (I - T)y\|_2^2. \end{aligned} \tag{10.10}$$

In particular we see that a firmly nonexpansive function is nonexpansive.

Lemma 2. *For $f \in \Gamma_0(\mathbb{R}^d)$, the proximal operator* $\mathrm{prox}_{\lambda f}$ *is firmly nonexpansive. In particular the orthogonal projection onto convex sets is firmly nonexpansive.*

Proof. By Theorem 1ii) we have that

$$\frac{1}{\lambda}\langle x - \mathrm{prox}_{\lambda f}(x), z - \mathrm{prox}_{\lambda f}(x)\rangle + f(\mathrm{prox}_{\lambda f}(x)) - f(z) \le 0 \quad \forall z \in \mathbb{R}^d.$$

With $z := \mathrm{prox}_{\lambda f}(y)$ this gives

$$\langle x - \mathrm{prox}_{\lambda f}(x), \mathrm{prox}_{\lambda f}(y) - \mathrm{prox}_{\lambda f}(x)\rangle + \lambda f(\mathrm{prox}_{\lambda f}(x)) - \lambda f(\mathrm{prox}_{\lambda f}(y)) \le 0$$

and similarly

$$\langle y - \mathrm{prox}_{\lambda f}(y), \mathrm{prox}_{\lambda f}(x) - \mathrm{prox}_{\lambda f}(y)\rangle + \lambda f(\mathrm{prox}_{\lambda f}(y)) - \lambda f(\mathrm{prox}_{\lambda f}(x)) \le 0.$$

Adding these inequalities we obtain

$$\begin{aligned} &\langle x - \mathrm{prox}_{\lambda f}(x) + \mathrm{prox}_{\lambda f}(y) - y, \mathrm{prox}_{\lambda f}(y) - \mathrm{prox}_{\lambda f}(x)\rangle \le 0, \\ &\qquad \|\mathrm{prox}_{\lambda f}(y) - \mathrm{prox}_{\lambda f}(x)\|_2^2 \le \langle y - x, \mathrm{prox}_{\lambda f}(y) - \mathrm{prox}_{\lambda f}(x)\rangle. \end{aligned}$$

□

The Banach fixed point theorem guarantees that a contraction has a unique fixed point and that the *Picard sequence*

$$x^{(r+1)} = Tx^{(r)} \tag{10.11}$$

converges to this fixed point for every initial element $x^{(0)}$. However, in many applications the contraction property is too restrictive in the sense that we often do not have a unique fixed point. Indeed, it is quite natural in many cases that the reached fixed point depends on the starting value $x^{(0)}$. Note that if T is continuous and $(T^r x^{(0)})_{r\in\mathbb{N}}$ is convergent, then it converges to a fixed point of T. In the following, we denote by $\mathrm{Fix}(T)$ the *set of fixed points* of T. Unfortunately, we do not have convergence of $(T^r x^{(0)})_{r\in\mathbb{N}}$ just for nonexpansive operators as the following example shows.

Example 3. In $\mathbb{R}^2$ we consider the reflection operator

$$R := \begin{pmatrix} 1 & 0 \\ 0 & -1 \end{pmatrix}.$$

Obviously, R is nonexpansive and we only have convergence of $(R^r x^{(0)})_{r\in\mathbb{N}}$ if $x^{(0)} \in \mathrm{Fix}(R) = \mathrm{span}\{(1,0)^{\mathrm{T}}\}$. A possibility to obtain a 'better' operator is to average R, i.e., to build

$$T := \alpha I + (1-\alpha)R = \begin{pmatrix} 1 & 0 \\ 0 & 2\alpha - 1 \end{pmatrix}, \quad \alpha \in (0,1).$$

By

$$Tx = x \Leftrightarrow \alpha x + (1-\alpha)R(x) = x \Leftrightarrow (1-\alpha)R(x) = (1-\alpha)x, \tag{10.12}$$

we see that R and T have the same fixed points. Moreover, since $2\alpha - 1 \in (-1,1)$, the sequence $(T^r x^{(0)})_{r\in\mathbb{N}}$ converges to $(x_1^{(0)},0)^{\mathrm{T}}$ for every $x^{(0)} = (x_1^{(0)},x_2^{(0)})^{\mathrm{T}} \in \mathbb{R}^2$.

An operator $T : \mathbb{R}^d \to \mathbb{R}^d$ is called *averaged* if there exists a nonexpansive mapping R and a constant $\alpha \in (0,1)$ such that

$$T = \alpha I + (1-\alpha)R.$$

Following (10.12) we see that

$$\mathrm{Fix}(R) = \mathrm{Fix}(T).$$

Historically, the concept of averaged mappings can be traced back to [112, 120, 156], where the name 'averaged' was not used yet. Results on averaged operators can also be found, e.g., in [12, 36, 55].

Lemma 3 (Averaged, (Firmly) Nonexpansive and Contractive Operators).

i) *Every averaged operator is nonexpansive.*
ii) *A contractive operator $T : \mathbb{R}^d \to \mathbb{R}^d$ with Lipschitz constant $L < 1$ is averaged with respect to all parameters $\alpha \in (0,(1-L)/2]$.*
iii) *An operator is firmly nonexpansive if and only if it is averaged with $\alpha = \frac{1}{2}$.*

Proof. i) Let $T = \alpha I + (1-\alpha)R$ be averaged. Then the first assertion follows by

$$\|T(x) - T(y)\|_2 \le \alpha\|x-y\|_2 + (1-\alpha)\|R(x) - R(y)\|_2 \le \|x-y\|_2.$$

ii) We define the operator $R := \frac{1}{1-\alpha}(T - \alpha I)$. It holds for all $x, y \in \mathbb{R}^d$ that

$$
\begin{aligned}
\|Rx - Ry\|_2 &= \frac{1}{1-\alpha}\|(T-\alpha I)x - (T-\alpha I)y\|_2, \\
&\leq \frac{1}{1-\alpha}\|Tx - Ty\|_2 + \frac{\alpha}{1-\alpha}\|x-y\|_2, \\
&\leq \frac{L+\alpha}{1-\alpha}\|x-y\|_2,
\end{aligned}
$$

so R is nonexpansive if $\alpha \leq (1-L)/2$.
iii) With $R := 2T - I = T - (I - T)$ we obtain the following equalities

$$
\begin{aligned}
\|Rx - Ry\|_2^2 &= \|Tx - Ty - ((I-T)x - (I-T)y)\|_2^2 \\
&= -\|x-y\|_2^2 + 2\|Tx - Ty\|_2^2 + 2\|(I-T)x - (I-T)y\|_2^2
\end{aligned}
$$

and therefore after reordering

$$
\begin{aligned}
&\|x-y\|_2^2 - \|Tx - Ty\|_2^2 - \|(I-T)x - (I-T)y\|_2^2 \\
&= \|Tx - Ty\|_2^2 + \|(I-T)x - (I-T)y\|_2^2 - \|Rx - Ry\|_2^2 \\
&= \frac{1}{2}(\|x-y\|_2^2 + \|Rx - Ry\|_2^2) - \|Rx - Ry\|_2^2 \\
&= \frac{1}{2}(\|x-y\|_2^2 - \|Rx - Ry\|_2^2).
\end{aligned}
$$

If R is nonexpansive, then the last expression is ≥ 0 and consequently (10.10) holds true so that T is firmly nonexpansive. Conversely, if T fulfills (10.10), then

$$
\frac{1}{2}(\|x-y\|_2^2 - \|Rx - Ry\|_2^2) \geq 0
$$

so that R is nonexpansive. This completes the proof. □

By the following lemma averaged operators are closed under composition.

Lemma 4 (Composition of Averaged Operators).

i) *Suppose that $T : \mathbb{R}^d \to \mathbb{R}^d$ is averaged with respect to $\alpha \in (0,1)$. Then, it is also averaged with respect to any other parameter $\tilde{\alpha} \in (0, \alpha]$.*
ii) *Let $T_1, T_2 : \mathbb{R}^d \to \mathbb{R}^d$ be averaged operators. Then, $T_2 \circ T_1$ is also averaged.*

Proof. i) By assumption, $T = \alpha I + (1-\alpha)R$ with R nonexpansive. We have

$$
T = \tilde{\alpha} I + ((\alpha - \tilde{\alpha})I + (1-\alpha)R) = \tilde{\alpha} I + (1-\tilde{\alpha}) \underbrace{\left(\frac{\alpha - \tilde{\alpha}}{1-\tilde{\alpha}} I + \frac{1-\alpha}{1-\tilde{\alpha}} R\right)}_{\tilde{R}}
$$

and for all $x,y \in \mathbb{R}^d$ it holds that

$$\|\tilde{R}(x) - \tilde{R}(y)\|_2 \le \frac{\alpha - \tilde{\alpha}}{1-\tilde{\alpha}}\|x-y\|_2 + \frac{1-\alpha}{1-\tilde{\alpha}}\|R(x) - R(y)\|_2 \le \|x-y\|_2.$$

So, $\tilde{R}$ is nonexpansive.
ii) By assumption there exist nonexpansive operators R_1, R_2 and $\alpha_1, \alpha_2 \in (0,1)$ such that

$$\begin{aligned} T_2(T_1(x)) &= \alpha_2 T_1(x) + (1-\alpha_2) R_2(T_1(x)) \\ &= \alpha_2(\alpha_1 x + (1-\alpha_1) R_1(x)) + (1-\alpha_2) R_2(T_1(x)) \\ &= \underbrace{\alpha_2\alpha_1}_{:=\alpha} x + (\alpha_2 - \underbrace{\alpha_2\alpha_1}_{=\alpha}) R_1(x) + (1-\alpha_2) R_2(T_1(x)) \\ &= \alpha x + (1-\alpha) \underbrace{\left(\frac{\alpha_2 - \alpha}{1-\alpha} R_1(x) + \frac{1-\alpha_2}{1-\alpha} R_2(T_1(x)) \right)}_{=:R} \end{aligned}$$

The concatenation of two nonexpansive operators is nonexpansive. Finally, the convex combination of two nonexpansive operators is nonexpansive so that R is indeed nonexpansive. □

An operator $T : \mathbb{R}^d \to \mathbb{R}^d$ is called *asymptotically regular* if it holds for all $x \in \mathbb{R}^d$ that

$$\left(T^{r+1}x - T^r x\right) \to 0 \quad \text{for} \quad r \to +\infty.$$

Note that this property does not imply convergence, even boundedness cannot be guaranteed. As an example consider the partial sums of a harmonic sequence.

Theorem 4 (Asymptotic Regularity of Averaged Operators).
Let $T : \mathbb{R}^d \to \mathbb{R}^d$ be an averaged operator with respect to the nonexpansive mapping R and the parameter $\alpha \in (0,1)$. Assume that $\mathrm{Fix}(T) \neq \emptyset$. *Then, T is asymptotically regular.*

Proof. Let $\hat{x} \in \mathrm{Fix}(T)$ and $x^{(r)} = T^r x^{(0)}$ for some starting element $x^{(0)}$. Since T is nonexpansive, i.e., $\|x^{(r+1)} - \hat{x}\|_2 \le \|x^{(r)} - \hat{x}\|_2$ we obtain

$$\lim_{r\to\infty} \|x^{(r)} - \hat{x}\|_2 = d \ge 0. \tag{10.13}$$

Using $\mathrm{Fix}(T) = \mathrm{Fix}(R)$ it follows

$$\limsup_{r\to\infty} \|R(x^{(r)}) - \hat{x}\|_2 = \limsup_{r\to\infty} \|R(x^{(r)}) - R(\hat{x})\|_2 \le \lim_{r\to\infty} \|x^{(r)} - \hat{x}\|_2 = d. \tag{10.14}$$

Assume that $\|x^{(r+1)} - x^{(r)}\|_2 \not\to 0$ for $r \to \infty$. Then, there exists a subsequence $(x^{(r_l)})_{l\in\mathbb{N}}$ such that

$$\|x^{(r_l+1)} - x^{(r_l)}\|_2 \ge \varepsilon$$

for some $\varepsilon > 0$. By (10.13) the sequence $(x^{(r_l)})_{l\in\mathbb{N}}$ is bounded. Hence there exists a convergent subsequence $(x^{(r_{l_j})})$ such that

$$\lim_{j\to\infty} x^{(r_{l_j})} = a,$$

where $a \in S(\hat{x},d) := \{x \in \mathbb{R}^d : \|x-\hat{x}\|_2 = d\}$ by (10.13). On the other hand, we have by the continuity of R and (10.14) that

$$\lim_{j\to\infty} R(x^{(r_{l_j})}) = b, \quad b \in B(\hat{x},d).$$

Since $\varepsilon \le \|x^{(r_{l_j}+1)} - x^{(r_{l_j})}\|_2 = \|(\alpha-1)x^{(r_{l_j})} + (1-\alpha)R(x^{(r_{l_j})})\|_2$ we conclude by taking the limit $j \to \infty$ that $a \neq b$. By the continuity of T and (10.13) we obtain

$$\lim_{j\to\infty} T(x^{(r_{l_j})}) = c, \quad c \in S(\hat{x},d).$$

However, by the strict convexity of $\|\cdot\|_2^2$ this yields the contradiction

$$\begin{aligned}\|c-\hat{x}\|_2^2 &= \lim_{j\to\infty} \|T(x^{(r_{l_j})}) - \hat{x}\|_2^2 = \lim_{j\to\infty} \|\alpha(x^{(r_{l_j})} - \hat{x}) + (1-\alpha)(R(x^{(r_{l_j})}) - \hat{x})\|_2^2 \\ &= \|\alpha(a-\hat{x}) + (1-\alpha)(b-\hat{x})\|_2^2 < \alpha\|a-\hat{x}\|_2^2 + (1-\alpha)\|b-\hat{x}\|_2^2 \\ &\le d^2.\end{aligned}$$

□

The following theorem was first proved for operators on Hilbert spaces by Opial [133, Theorem 1] based on the results in [29], where convergence must be replaced by weak convergence in general Hilbert spaces. A shorter proof can be found in the appendix of [61]. For finite dimensional spaces the proof simplifies as follows.

Theorem 5 (Opial's Convergence Theorem).
Let $T : \mathbb{R}^d \to \mathbb{R}^d$ fulfill the following conditions: $\mathrm{Fix}(T) \neq \emptyset$, *$T$ is nonexpansive and asymptotically regular. Then, for every $x^{(0)} \in \mathbb{R}^d$, the sequence of Picard iterates $(x^{(r)})_{r\in\mathbb{N}}$ generated by $x^{(r+1)} = Tx^{(r)}$ converges to an element of* $\mathrm{Fix}(T)$.

Proof. Since T is nonexpansive, we have for any $\hat{x} \in \mathrm{Fix}(T)$ and any $x^{(0)} \in \mathbb{R}^d$ that

$$\|T^{r+1}x^{(0)} - \hat{x}\|_2 \le \|T^r x^{(0)} - \hat{x}\|_2.$$

Hence $(T^r x^{(0)})_{r\in\mathbb{N}}$ is bounded and there exists a subsequence $(T^{r_l}x^{(0)})_{l\in\mathbb{N}} = (x^{(r_l)})_{l\in\mathbb{N}}$ which converges to some $\tilde{x}$. If we can show that $\tilde{x} \in \mathrm{Fix}(T)$ we are done because in this case

$$\|T^r x^{(0)} - \tilde{x}\|_2 \le \|T^{r_l}x^{(0)} - \tilde{x}\|_2, \quad r \ge r_l$$

and thus the whole sequence converges to $\tilde{x}$.

Since T is asymptotically regular it follows that

$$(T-I)(T^{r_l}x^{(0)}) = T^{r_l+1}x^{(0)} - T^{r_l}x^{(0)} \to 0$$

and since $(T^{r_l}x^{(0)})_{l\in\mathbb{N}}$ converges to $\tilde{x}$ and T is continuous we get that $(T-I)(\tilde{x}) = 0$, i.e., $\tilde{x} \in \mathrm{Fix}(T)$. □

Combining the above Theorems 4 and 5 we obtain the following main result.

Theorem 6 (Convergence of Averaged Operator Iterations). *Let $T : \mathbb{R}^d \to \mathbb{R}^d$ be an averaged operator such that $\mathrm{Fix}(T) \neq \emptyset$. Then, for every $x^{(0)} \in \mathbb{R}^d$, the sequence $(T^r x^{(0)})_{r\in\mathbb{N}}$ converges to a fixed point of T.*

5 Proximal Algorithms

5.1 Proximal Point Algorithm

By Theorem 1 iii) the minimizer of a function $f \in \Gamma_0(\mathbb{R}^d)$, which we suppose to exist, is characterized by the fixed point equation

$$\hat{x} = \mathrm{prox}_{\lambda f}(\hat{x}).$$

The corresponding Picard iteration gives rise to the following *proximal point algorithm* which dates back to [121, 147]. Since $\mathrm{prox}_{\lambda f}$ is firmly nonexpansive by Lemma 2 and thus averaged, the algorithm converges by Theorem 6 for any initial value $x^{(0)} \in \mathbb{R}^d$ to a minimizer of f if there exits one.

Algorithm 1 Proximal Point Algorithm (PPA)

Initialization: $x^{(0)} \in \mathbb{R}^d$, $\lambda > 0$
Iterations: For $r = 0, 1, \ldots$
$x^{(r+1)} = \mathrm{prox}_{\lambda f}(x^{(r)}) = \mathrm{argmin}_{x\in\mathbb{R}^d} \left\{ \frac{1}{2\lambda} \|x^{(r)} - x\|_2^2 + f(x) \right\}$

The PPA can be generalized for the sum $\sum_{i=1}^n f_i$ of functions $f_i \in \Gamma_0(\mathbb{R}^d)$, $i = 1, \ldots, n$. Popular generalizations are the so-called cyclic PPA [18] and the parallel PPA [53].

5.2 Proximal Gradient Algorithm

We are interested in minimizing functions of the form $f = g + h$, where $g : \mathbb{R}^d \to \mathbb{R}$ is convex, differentiable with Lipschitz continuous gradient and Lipschitz constant L, i.e.,

$$\|\nabla g(x) - \nabla g(y)\|_2 \le L\|x - y\|_2 \quad \forall x, y \in \mathbb{R}^d, \tag{10.15}$$

and $h \in \Gamma_0(\mathbb{R}^d)$. Note that the Lipschitz condition on ∇g implies

$$g(x) \le g(y) + \langle \nabla g(y), x - y \rangle + \frac{L}{2}\|x-y\|_2^2 \quad \forall x, y \in \mathbb{R}^d, \tag{10.16}$$

see, e.g., [134]. We want to solve

$$\underset{x \in \mathbb{R}^d}{\operatorname{argmin}} \{g(x) + h(x)\}. \tag{10.17}$$

By Fermat's rule and subdifferential calculus we know that $\hat{x}$ is a minimizer of (10.17) if and only if

$$\begin{aligned} 0 &\in \nabla g(\hat{x}) + \partial h(\hat{x}), \\ \hat{x} - \eta \nabla g(\hat{x}) &\in \hat{x} + \eta \partial h(\hat{x}), \\ \hat{x} &= (I + \eta \partial h)^{-1} (\hat{x} - \eta \nabla g(\hat{x})) = \operatorname{prox}_{\eta h} (\hat{x} - \eta \nabla g(\hat{x})). \end{aligned} \tag{10.18}$$

This is a fixed point equation for the minimizer $\hat{x}$ of f. The corresponding Picard iteration is known as *proximal gradient algorithm* or as *proximal forward-backward splitting*.

Algorithm 2 Proximal Gradient Algorithm (FBS)

Initialization: $x^{(0)} \in \mathbb{R}^d$, $\eta \in (0, 2/L)$
Iterations: For $r = 0, 1, \ldots$
$x^{(r+1)} = \operatorname{prox}_{\eta h}\left(x^{(r)} - \eta \nabla g(x^{(r)})\right)$

In the special case when $h := \iota_C$ is the indicator function of a nonempty, closed, convex set $C \subset \mathbb{R}^d$, the above algorithm for finding

$$\underset{x \in C}{\operatorname{argmin}}\, g(x)$$

becomes the gradient descent re-projection algorithm, also known as "gradient projection algorithm".

Algorithm 3 Gradient Descent Re-Projection Algorithm

Initialization: $x^{(0)} \in \mathbb{R}^d$, $\eta \in (0, 2/L)$
Iterations: For $r = 0, 1, \ldots$
$x^{(r+1)} = \Pi_C\left(x^{(r)} - \eta \nabla g(x^{(r)})\right)$

It is also possible to use flexible variables $\eta_r \in (0, \frac{2}{L})$ in the proximal gradient algorithm. For further details, modifications, and extensions see also [72, Chapter 12]. The convergence of the algorithm follows by the next theorem.

Theorem 7 (Convergence of Proximal Gradient Algorithm). *Let $g : \mathbb{R}^d \to \mathbb{R}$ be a convex, differentiable function on $\mathbb{R}^d$ with Lipschitz continuous gradient and Lipschitz constant L and $h \in \Gamma_0(\mathbb{R}^d)$. Suppose that a solution of* (10.17) *exists. Then, for every initial point $x^{(0)}$ and $\eta \in (0, \frac{2}{L})$, the sequence $\{x^{(r)}\}_r$ generated by the proximal gradient algorithm converges to a solution of* (10.17).

Proof. We show that $\text{prox}_{\eta h}(I - \eta \nabla g)$ is averaged. Then we are done by Theorem 6. By Lemma 2 we know that $\text{prox}_{\eta h}$ is firmly nonexpansive. By the Baillon-Haddad Theorem [12, Corollary 16.1] the function $\frac{1}{L}\nabla g$ is also firmly nonexpansive, i.e., it is averaged with parameter $\frac{1}{2}$. This means that there exists a nonexpansive mapping R such that $\frac{1}{L}\nabla g = \frac{1}{2}(I + R)$ which implies

$$I - \eta \nabla g = I - \tfrac{\eta L}{2}(I + R) = (1 - \tfrac{\eta L}{2})I + \tfrac{\eta L}{2}(-R).$$

Thus, for $\eta \in (0, \frac{2}{L})$, the operator $I - \eta \nabla g$ is averaged. Since the concatenation of two averaged operators is averaged again we obtain the assertion. □

Under the above conditions a linear convergence rate can be achieved in the sense that

$$f(x^{(r)}) - f(\hat{x}) = \mathcal{O}(1/r),$$

see, e.g., [49, 125][1].

Example 4. For solving

$$\operatorname*{argmin}_{x \in \mathbb{R}^d} \{ \underbrace{\frac{1}{2}\|Kx - b\|_2^2}_{g} + \underbrace{\lambda \|x\|_1}_{h} \}$$

we compute $\nabla g(x) = K^{\mathrm{T}}(Kx - b)$ and use that the proximal operator of the ℓ_1-norm is just the componentwise soft-shrinkage. Then the proximal gradient algorithm becomes

$$x^{(r+1)} = \text{prox}_{\lambda\eta\|\cdot\|_1}\left(x^{(r)} - \eta K^{\mathrm{T}}(Kx^{(r)} - b)\right) = S_{\eta\lambda}\left(x^{(r)} - \eta K^{\mathrm{T}}(Kx^{(r)} - b)\right).$$

This algorithm is known as *iterative soft-thresholding algorithm* (ISTA) and was developed and analyzed through various techniques by many researchers. For a general Hilbert space approach of ISTA, see, e.g., [61].

The FBS algorithm has been recently extended to the case of non-convex functions in [6, 7, 22, 52, 132]. The convergence analysis mainly rely on the assumption that the objective function $f = g + h$ satisfies the Kurdyka-Lojasiewicz inequality which is indeed fulfilled for a wide class of functions as $\log - \exp$, semi-algebraic, and subanalytic functions which are of interest in image processing.

[1] There exist different notations for the convergence rate of algorithms in the literature. Sometimes the notation of this chapter is also called "superlinear convergence" while $\|\hat{x} - x^{(r)}\| \le C\delta^r$, $\delta < 1$ is used for the linear convergence. But if $C = C(r) \to 0$ as $r \to +\infty$ in the last formula, this could be also meant by "superlinear convergence".

5.3 Accelerated Algorithms

For large scale problems as those arising in image processing a major concern is to find efficient algorithms solving the problem in a reasonable time. While each FBS step has low computational complexity, it may suffer from slow linear convergence [49]. Using a simple extrapolation idea with appropriate parameters τ_r, the convergence can often be accelerated:

$$y^{(r)} = x^{(r)} + \tau_r \left(x^{(r)} - x^{(r-1)} \right),$$
$$x^{(r+1)} = \mathrm{prox}_{\eta h} \left(y^{(r)} - \eta \nabla g(y^{(r)}) \right). \tag{10.19}$$

By the next Theorem 8 we will see that $\tau_r = \frac{r-1}{r+2}$ appears to be a good choice. Clearly, we can vary η in each step again. Choosing θ_r such that $\tau_r = \frac{\theta_r(1-\theta_{r-1})}{\theta_{r-1}}$, e.g., $\theta_r = \frac{2}{r+2}$ for the above choice of τ_r, the algorithm can be rewritten as follows:

Algorithm 4 Fast Proximal Gradient Algorithm

Initialization: $x^{(0)} = z^{(0)} \in \mathbb{R}^d$, $\eta \in (0, 1/L)$, $\theta_r = \frac{2}{r+2}$
Iterations: For $r = 0, 1, \ldots$
$y^{(r)} \quad = (1-\theta_r)x^{(r)} + \theta_r z^{(r)}$
$x^{(r+1)} = \mathrm{prox}_{\eta h} \left(y^{(r)} - \eta \nabla g(y^{(r)}) \right)$
$z^{(r+1)} = x^{(r)} + \frac{1}{\theta_r} \left(x^{(r+1)} - x^{(r)} \right)$

By the following standard theorem the extrapolation modification of the FBS algorithm ensures a quadratic convergence rate see also [125, 171].

Theorem 8. *Let $f = g + h$, where $g : \mathbb{R}^d \to \mathbb{R}$ is a convex, Lipschitz differentiable function with Lipschitz constant L and $h \in \Gamma_0(\mathbb{R}^d)$. Assume that f has a minimizer $\hat{x}$. Then the fast proximal gradient algorithm fulfills*

$$f(x^{(r)}) - f(\hat{x}) = \mathcal{O}\left(1/r^2\right).$$

Proof. First we consider the progress in one step of the algorithm. By the Lipschitz differentiability of g in (10.16) and since $\eta < \frac{1}{L}$ we know that

$$g(x^{(r+1)}) \le g(y^{(r)}) + \langle \nabla g(y^{(r)}), x^{(r+1)} - y^{(r)} \rangle + \frac{1}{2\eta} \|x^{(r+1)} - y^{(r)}\|_2^2 \tag{10.20}$$

and by the variational characterization of the proximal operator in Theorem 1ii) for all $u \in \mathbb{R}^d$ that

$$h(x^{(r+1)}) \le h(u) + \frac{1}{\eta}\langle y^{(r)} - \eta\nabla g(y^{(r)}) - x^{(r+1)}, x^{(r+1)} - u\rangle$$
$$\le h(u) - \langle \nabla g(y^{(r)}), x^{(r+1)} - u\rangle + \frac{1}{\eta}\langle y^{(r)} - x^{(r+1)}, x^{(r+1)} - u\rangle. \quad (10.21)$$

Adding the main inequalities (10.20) and (10.21) and using the convexity of g yields

$$f(x^{(r+1)}) \le f(u) \underbrace{-g(u) + g(y^{(r)}) + \langle \nabla g(y^{(r)}), u - y^{(r)}\rangle}_{\le 0}$$
$$+ \frac{1}{2\eta}\|x^{(r+1)} - y^{(r)}\|_2^2 + \frac{1}{\eta}\langle y^{(r)} - x^{(r+1)}, x^{(r+1)} - u\rangle$$
$$\le f(u) + \frac{1}{2\eta}\|x^{(r+1)} - y^{(r)}\|_2^2 + \frac{1}{\eta}\langle y^{(r)} - x^{(r+1)}, x^{(r+1)} - u\rangle.$$

Combining these inequalities for $u := \hat{x}$ and $u := x^{(r)}$ with $\theta_r \in [0,1]$ gives

$$\theta_r\left(f(x^{(r+1)}) - f(\hat{x})\right) + (1-\theta_r)\left(f(x^{(r+1)}) - f(x^{(r)})\right)$$
$$= f(x^{(r+1)}) - f(\hat{x}) + (1-\theta_r)\left(f(\hat{x}) - f(x^{(r)})\right)$$
$$\le \frac{1}{2\eta}\|x^{(r+1)} - y^{(r)}\|_2^2 + \frac{1}{\eta}\langle y^{(r)} - x^{(r+1)}, x^{(r+1)} - \theta_r\hat{x} - (1-\theta_r)x^{(r)}\rangle$$
$$= \frac{1}{2\eta}\left(\|y^{(r)} - \theta_r\hat{x} - (1-\theta_r)x^{(r)}\|_2^2 - \|x^{(r+1)} - \theta_r\hat{x} - (1-\theta_r)x^{(r)}\|_2^2\right)$$
$$= \frac{\theta_r^2}{2\eta}\left(\|z^{(r)} - \hat{x}\|_2^2 - \|z^{(r+1)} - \hat{x}\|_2^2\right).$$

Thus, we obtain for a single step

$$\frac{\eta}{\theta_r^2}\left(f(x^{(r+1)}) - f(\hat{x})\right) + \frac{1}{2}\|z^{(r+1)} - \hat{x}\|_2^2 \le \frac{\eta(1-\theta_r)}{\theta_r^2}\left(f(x^{(r)}) - f(\hat{x})\right) + \frac{1}{2}\|z^{(r)} - \hat{x}\|_2^2.$$

Using the relation recursively on the right-hand side and regarding that $\frac{(1-\theta_r)}{\theta_r^2} \le \frac{1}{\theta_{r-1}^2}$ we obtain

$$\frac{\eta}{\theta_r^2}\left(f(x^{(r+1)}) - f(\hat{x})\right) \le \frac{\eta(1-\theta_0)}{\theta_0^2}\left(f(x^{(0)}) - f(\hat{x})\right) + \frac{1}{2}\|z^{(0)} - \hat{x}\|_2^2 = \frac{1}{2}\|x^{(0)} - \hat{x}\|_2^2.$$

This yields the assertion

$$f(x^{(r+1)}) - f(\hat{x}) \le \frac{2}{\eta(r+2)^2}\|x^{(0)} - \hat{x}\|_2^2.$$

□

There exist many variants or generalizations of the above algorithm (with certain convergence rates under special assumptions):

- *Nesterov's algorithms* [126, 128], see also [60, 171]; this includes approximation algorithms for nonsmooth g [14, 129] as NESTA,
- *fast iterative shrinkage algorithms* (FISTA) by Beck and Teboulle [13],
- *variable metric strategies* [24, 33, 57, 138], where based on (10.5) step (10.19) is replaced by

$$x^{(r+1)} = \mathrm{prox}_{Q_r,\eta_r h}\left(y^{(r)} - \eta_r Q_r^{-1}\nabla g(y^{(r)})\right) \tag{10.22}$$

  with symmetric, positive definite matrices Q_r.

Line search strategies can be incorporated [89, 93, 127]. Finally we mention Barzilei-Borwein step size rules [11] based on a Quasi-Newton approach and relatives, see [79] for an overview and the cyclic proximal gradient algorithm related to the cyclic Richardson algorithm [165].

6 Primal-Dual Methods

6.1 Basic Relations

The following minimization algorithms closely rely on the primal-dual formulation of problems. We consider functions $f = g + h(A\,\cdot)$, where $g \in \Gamma_0(\mathbb{R}^d)$, $h \in \Gamma_0(\mathbb{R}^m)$, and $A \in \mathbb{R}^{m,d}$, and ask for the solution of the primal problem

$$(P) \quad \underset{x\in\mathbb{R}^d}{\mathrm{argmin}}\,\{g(x) + h(Ax)\}, \tag{10.23}$$

that can be rewritten as

$$(P) \quad \underset{x\in\mathbb{R}^d, y\in\mathbb{R}^m}{\mathrm{argmin}}\;\{g(x) + h(y) \quad \text{s.t.} \quad Ax = y\}. \tag{10.24}$$

The *Lagrangian* of (10.24) is given by

$$L(x,y,p) := g(x) + h(y) + \langle p, Ax - y\rangle \tag{10.25}$$

and the *augmented Lagrangian* by

$$\begin{aligned} L_\gamma(x,y,p) &:= g(x) + h(y) + \langle p, Ax - y\rangle + \frac{\gamma}{2}\|Ax - y\|_2^2, \quad \gamma > 0,\\ &= g(x) + h(y) + \frac{\gamma}{2}\|Ax - y + \frac{p}{\gamma}\|_2^2 - \frac{1}{2\gamma}\|p\|_2^2. \end{aligned} \tag{10.26}$$

Based on the Lagrangian (10.25), the primal and dual problem can be written as

$$(P) \quad \underset{x\in\mathbb{R}^d, y\in\mathbb{R}^m}{\mathrm{argmin}}\;\sup_{p\in\mathbb{R}^m}\{g(x) + h(y) + \langle p, Ax - y\rangle\}, \tag{10.27}$$

$$(D) \quad \underset{p\in\mathbb{R}^m}{\mathrm{argmax}}\;\inf_{x\in\mathbb{R}^d, y\in\mathbb{R}^m}\{g(x) + h(y) + \langle p, Ax - y\rangle\}. \tag{10.28}$$

Since

$$\min_{y\in\mathbb{R}^m}\{h(y)-\langle p,y\rangle\} = -\max_{y\in\mathbb{R}^m}\{\langle p,y\rangle - h(y)\} = -h^*(p)$$

and in (10.23) further

$$h(Ax) = \max_{p\in\mathbb{R}^m}\{\langle p,Ax\rangle - h^*(p)\},$$

the primal and dual problem can be rewritten as

$$\begin{aligned}(P)\quad & \operatorname*{argmin}_{x\in\mathbb{R}^d}\sup_{p\in\mathbb{R}^m}\{g(x)-h^*(p)+\langle p,Ax\rangle\},\\ (D)\quad & \operatorname*{argmax}_{p\in\mathbb{R}^m}\inf_{x\in\mathbb{R}^d}\{g(x)-h^*(p)+\langle p,Ax\rangle\}.\end{aligned}$$

If the infimum exists, the dual problem can be seen as Fenchel dual problem

$$(D)\quad \operatorname*{argmax}_{p\in\mathbb{R}^m}\{-g^*(-A^{\mathrm{T}}p)-h^*(p)\}. \tag{10.29}$$

Recall that $((\hat{x},\hat{y}),\hat{p}) \in \mathbb{R}^{dm,m}$ is a *saddle point* of the Lagrangian L in (10.25) if

$$L((\hat{x},\hat{y}),p) \le L((\hat{x},\hat{y}),\hat{p}) \le L((x,y),\hat{p}) \qquad \forall (x,y)\in\mathbb{R}^{dm},\, p\in\mathbb{R}^m.$$

If $((\hat{x},\hat{y}),\hat{p}) \in \mathbb{R}^{dm,m}$ is a saddle point of L, then $(\hat{x},\hat{y})$ is a solution of the primal problem (10.27) and $\hat{p}$ is a solution of the dual problem (10.28). The converse is also true. However the existence of a solution of the primal problem $(\hat{x},\hat{y}) \in \mathbb{R}^{dm}$ does only imply under additional qualification constraint that there exists $\hat{p}$ such that $((\hat{x},\hat{y}),\hat{p}) \in \mathbb{R}^{dm,m}$ is a saddle point of L.

6.2 Alternating Direction Method of Multipliers

Based on the Lagrangian formulation (10.27) and (10.28), a first idea to solve the optimization problem would be to alternate the minimization of the Lagrangian with respect to (x,y) and to apply a gradient ascent approach with respect to p. This is known as general Uzawa method [5]. More precisely, noting that for differentiable $v(p) := \inf_{x,y} L(x,y,p) = L(\tilde{x},\tilde{y},p)$ we have $\nabla v(p) = A\tilde{x}-\tilde{y}$, the algorithm reads

$$\begin{aligned}(x^{(r+1)},y^{(r+1)}) &\in \operatorname*{argmin}_{x\in\mathbb{R}^d,y\in\mathbb{R}^m} L(x,y,p^{(r)}),\\ p^{(r+1)} &= p^{(r)}+\gamma(Ax^{(r+1)}-y^{(r+1)}), \qquad \gamma>0.\end{aligned} \tag{10.30}$$

Linear convergence can be proved under certain conditions (strong convexity of f) [87]. The assumptions on f to ensure convergence of the algorithm can be relaxed by replacing the Lagrangian by the augmented Lagrangian L_γ (10.26) with fixed parameter γ:

$$(x^{(r+1)}, y^{(r+1)}) \in \operatorname*{argmin}_{x\in\mathbb{R}^d, y\in\mathbb{R}^m} L_\gamma(x,y,p^{(r)}), \tag{10.31}$$
$$p^{(r+1)} = p^{(r)} + \gamma(Ax^{(r+1)} - y^{(r+1)}), \qquad \gamma > 0.$$

This augmented Lagrangian method is known as *method of multipliers* [101, 141, 147]. It can be shown [35, Theorem 3.4.7], [17] that the sequence $(p^{(r)})_r$ generated by the algorithm coincides with the proximal point algorithm applied to $-v(p)$, i.e.,

$$p^{(r+1)} = \mathrm{prox}_{-\gamma v}\left(p^{(r)}\right).$$

The improved convergence properties came at a cost. While the minimization with respect to x and y can be separately computed in (10.30) using $\langle p^{(r)}, (A| -I)\begin{pmatrix}x\\y\end{pmatrix}\rangle = \langle \begin{pmatrix}A^{\mathrm{T}}\\-I\end{pmatrix} p^{(r)}, \begin{pmatrix}x\\y\end{pmatrix}\rangle$, this is no longer possible for the augmented Lagrangian. A remedy is to alternate the minimization with respect to x and y which leads to

$$x^{(r+1)} \in \operatorname*{argmin}_{x\in\mathbb{R}^d} L_\gamma(x,y^{(r)},p^{(r)}), \tag{10.32}$$
$$y^{(r+1)} = \operatorname*{argmin}_{y\in\mathbb{R}^m} L_\gamma(x^{(r+1)},y,p^{(r)}), \tag{10.33}$$
$$p^{(r+1)} = p^{(r)} + \gamma(Ax^{(r+1)} - y^{(r+1)}).$$

This is the *alternating direction method of multipliers (ADMM)* which dates back to [82, 83, 88].

Algorithm 5 Alternating Direction Method of Multipliers (ADMM)

Initialization: $y^{(0)} \in \mathbb{R}^m$, $p^{(0)} \in \mathbb{R}^m$
Iterations: For $r = 0, 1, \ldots$

$$x^{(r+1)} \in \mathrm{argmin}_{x\in\mathbb{R}^d}\left\{g(x) + \tfrac{\gamma}{2}\|\tfrac{1}{\gamma}p^{(r)} + Ax - y^{(r)}\|_2^2\right\}$$
$$y^{(r+1)} = \mathrm{argmin}_{y\in\mathbb{R}^m}\left\{h(y) + \tfrac{\gamma}{2}\|\tfrac{1}{\gamma}p^{(r)} + Ax^{(r+1)} - y\|_2^2\right\} = \mathrm{prox}_{\frac{1}{\gamma}h}(\tfrac{1}{\gamma}p^{(r)} + Ax^{(r+1)})$$
$$p^{(r+1)} = p^{(r)} + \gamma(Ax^{(r+1)} - y^{(r+1)})$$

Setting $b^{(r)} := p^{(r)}/\gamma$ we obtain the following (scaled) ADMM:
A good overview on the ADMM algorithm and its applications is given in [27], where in particular the important issue of choosing the parameter $\gamma > 0$ is addressed. Convergence of the ADMM under various assumptions was proved, e.g., in [83, 96, 116, 170]. The ADMM can be considered for more general problems

$$\operatorname*{argmin}_{x_i\in\mathbb{R}^{d_i}}\left\{\sum_{i=1}^m g_i(x) \quad \text{s.t.} \quad A_i x_i = c\right\}. \tag{10.34}$$

Algorithm 6 Alternating Direction Method of Multipliers (scaled ADMM)

Initialization: $y^{(0)} \in \mathbb{R}^m$, $b^{(0)} \in \mathbb{R}^m$
Iterations: For $r = 0, 1, \ldots$

$$x^{(r+1)} \in \mathrm{argmin}_{x\in\mathbb{R}^d} \left\{ g(x) + \tfrac{\gamma}{2}\|b^{(r)} + Ax - y^{(r)}\|_2^2 \right\}$$
$$y^{(r+1)} = \mathrm{argmin}_{y\in\mathbb{R}^m} \left\{ h(y) + \tfrac{\gamma}{2}\|b^{(r)} + Ax^{(r+1)} - y\|_2^2 \right\} = \mathrm{prox}_{\frac{1}{\gamma}h}(b^{(r)} + Ax^{(r+1)})$$
$$b^{(r+1)} = b^{(r)} + Ax^{(r+1)} - y^{(r+1)}$$

Here we refer to [47] and the references therein. We will see that for our problem (10.24) the convergence follows by the relation of the ADMM to the so-called Douglas-Rachford splitting algorithm where convergence can be shown using averaged operators. Few bounds on the global convergence rate of the algorithm can be found in [68] (linear convergence for linear programs depending on a variety of quantities), [102] (linear convergence for sufficiently small step size) and on the local behavior of a specific variation of the ADMM during the course of iteration for quadratic programs in [21]. Further global rates of the ADMM are given in the recent preprints [64, 65].

Theorem 9 (Convergence of ADMM). *Let $g \in \Gamma_0(\mathbb{R}^d)$, $h \in \Gamma_0(\mathbb{R}^m)$ and $A \in \mathbb{R}^{m,d}$. Assume that the Lagrangian* (10.25) *has a saddle point. Then, for $r \to \infty$, the sequence $\gamma\left(b^{(r)}\right)_r$ converges to a solution of the dual problem. If in addition the first step* (10.32) *in the ADMM algorithm has a unique solution, then $\left(x^{(r)}\right)_r$ converges to a solution of the primal problem.*

There exist different modifications of the ADMM algorithm presented above:

- *inexact computation* of the first step (10.32) [48, 69] such that it might be handled by an iterative method,
- *variable parameter and metric strategies* [27, 95, 96, 98, 111] where the fixed parameter γ can vary in each step, or the quadratic term $(\gamma/2)\|Ax - y\|_2^2$ within the augmented Lagrangian (10.26) is replaced by the more general proximal operator based on (10.5) such that the ADMM updates (10.32) and (10.33) receive the form

$$x^{(r+1)} \in \underset{x\in\mathbb{R}^d}{\mathrm{argmin}} \left\{ g(x) + \frac{1}{2}\|b^{(r)} + Ax - y^{(r)}\|_{Q_r}^2 \right\},$$
$$y^{(r+1)} = \underset{y\in\mathbb{R}^m}{\mathrm{argmin}} \left\{ h(y) + \frac{1}{2}\|b^{(r)} + Ax^{(r+1)} - y\|_{Q_r}^2 \right\},$$

respectively, with symmetric, positive definite matrices Q_r. The variable parameter strategies might mitigate the performance dependency on the initial chosen fixed parameter [27, 98, 111, 181] and include monotone conditions [96, 111] or more flexible non-monotone rules [27, 95, 98].

ADMM from the Perspective of Variational Inequalities

The ADMM algorithm presented above from the perspective of Lagrangian functions has been also studied extensively in the area of *variational inequalities (VIs)*, see, e.g., [82, 95, 170]. A VI problem consists of finding for a mapping $F : \mathbb{R}^l \to \mathbb{R}^l$ a vector $\hat{z} \in \mathbb{R}^l$ such that

$$\langle z - \hat{z}, F(\hat{z})\rangle \geq 0, \qquad \forall z \in \mathbb{R}^l. \tag{10.35}$$

In the following, we consider the minimization problem (10.24), i.e.,

$$\operatorname*{argmin}_{x\in\mathbb{R}^d, y\in\mathbb{R}^m} \{g(x) + h(y) \quad \text{s.t.} \quad Ax = y\},$$

for $g \in \Gamma_0(\mathbb{R}^d)$, $h \in \Gamma_0(\mathbb{R}^m)$. The discussion can be extended to the more general problem (10.34) [95, 170]. Considering the Lagrangian (10.25) and its optimality conditions, solving (10.24) is equivalent to find a triple $\hat{z} = ((\hat{x}, \hat{y}), \hat{p}) \in \mathbb{R}^{dm,m}$ such that (10.35) holds with

$$z = \begin{pmatrix} x \\ y \\ p \end{pmatrix}, \qquad F(z) = \begin{pmatrix} \partial g(x) + A^{\mathrm{T}} p \\ \partial h(y) - p \\ Ax - y \end{pmatrix},$$

where ∂g and ∂h have to be understood as any element of the corresponding subdifferential for simplicity. Note that ∂g and ∂h are maximal monotone operators [12]. A VI problem of this form can be solved by ADMM as proposed by Gabay [82] and Gabay and Mercier [83]: for a given triple $(x^{(r)}, y^{(r)}, p^{(r)})$ generate new iterates $(x^{(r+1)}, y^{(r+1)}, p^{(r+1)})$ by

i) find $x^{(r+1)}$ such that

$$\langle x - x^{(r+1)}, \partial g(x^{(r+1)}) + A^{\mathrm{T}}(p^{(r)} + \gamma(Ax^{(r+1)} - y^{(r)}))\rangle \geq 0, \quad \forall x \in \mathbb{R}^d, \tag{10.36}$$

ii) find $y^{(r+1)}$ such that

$$\langle y - y^{(r+1)}, \partial h(y^{(r+1)}) - (p^{(r)} + \gamma(Ax^{(r+1)} - y^{(r+1)}))\rangle \geq 0, \quad \forall y \in \mathbb{R}^m, \tag{10.37}$$

iii) update $p^{(r+1)}$ via

$$p^{(r+1)} = p^{(r)} + \gamma(Ax^{(r+1)} - y^{(r+1)}),$$

where $\gamma > 0$ is a fixed penalty parameter. To corroborate the equivalence of the iteration scheme above to ADMM in Algorithm 5, note that (10.35) reduces to $\langle \hat{z}, F(\hat{z})\rangle \geq 0$ for $z = 2\hat{z}$. On the other hand, (10.35) is equal to $\langle \hat{z}, F(\hat{z})\rangle \leq 0$ when $z = -\hat{z}$. The both cases transform (10.35) to find a solution $\hat{z}$ of a system of equations $F(\hat{z}) = 0$. Thus, the VI sub-problems (10.36) and (10.37) can be reduced to find a pair $(x^{(r+1)}, y^{(r+1)})$ with

$$\partial g(x^{(r+1)}) + A^{\mathrm{T}}(p^{(r)} + \gamma(Ax^{(r+1)} - y^{(r)})) = 0,$$
$$\partial h(y^{(r+1)}) - (p^{(r)} + \gamma(Ax^{(r+1)} - y^{(r+1)})) = 0.$$

The both equations correspond to optimality conditions of the minimization subproblems (10.32) and (10.33) of the ADMM algorithm, respectively. The theoretical properties of ADMM from the perspective of VI problems were studied extensively and a good reference overview can be found in [95].

Relation to Douglas-Rachford Splitting

Finally we want to point out the relation of the ADMM to the Douglas-Rachford splitting algorithm applied to the dual problem, see [44, 69, 70, 82, 86, 164]. We consider again the problem (10.17), i.e.,

$$\operatorname*{argmin}_{x \in \mathbb{R}^d} \{g(x) + h(x)\},$$

where we assume this time only $g, h \in \Gamma_0(\mathbb{R}^d)$ and that g or h is continuous at a point in $\mathrm{dom} g \cap \mathrm{dom} h$. Fermat's rule and subdifferential calculus imply that $\hat{x}$ is a minimizer if and only if

$$0 \in \partial g(\hat{x}) + \partial h(\hat{x}) \quad \Leftrightarrow \quad \exists \hat{\xi} \in \eta \partial g(\hat{x}) \text{ such that } \hat{x} = \mathrm{prox}_{\eta h}(\hat{x} - \hat{\xi}). \tag{10.38}$$

The basic idea for finding such minimizer is to set up a 'nice' operator $T : \mathbb{R}^d \to \mathbb{R}^d$ by

$$T := \mathrm{prox}_{\eta h}(2\mathrm{prox}_{\eta g} - I) - \mathrm{prox}_{\eta g} + I, \tag{10.39}$$

whose fixed points $\hat{t}$ are related to the minimizers as follows: setting $\hat{x} := \mathrm{prox}_{\eta g}(\hat{t})$, i.e., $\hat{t} \in \hat{x} + \eta \partial g(\hat{x})$ and $\hat{\xi} := \hat{t} - \hat{x} \in \eta \partial g(\hat{x})$ we see that

$$\begin{aligned} \hat{t} = T(\hat{t}) &= \mathrm{prox}_{\eta h}(2\hat{x} - \hat{t}) - \hat{x} + \hat{t}, \\ \hat{\xi} + \hat{x} &= \mathrm{prox}_{\eta h}(\hat{x} - \hat{\xi}) + \hat{\xi}, \\ \hat{x} &= \mathrm{prox}_{\eta h}(\hat{x} - \hat{\xi}), \end{aligned}$$

which coincides with (10.38). By the proof of the next theorem, the operator T is firmly nonexpansive such that by Theorem 6 a fixed point of T can be found by Picard iterations. This gives rise to the following *Douglas-Rachford splitting algorithm (DRS)*.

Algorithm 7 Douglas-Rachford Splitting Algorithm (DRS)

Initialization: $x^{(0)}, t^{(0)} \in \mathbb{R}^d$, $\eta > 0$
Iterations: For $r = 0, 1, \ldots$

$$\begin{aligned} t^{(r+1)} &= \mathrm{prox}_{\eta h}(2x^{(r)} - t^{(r)}) + t^{(r)} - x^{(r)}, \\ x^{(r+1)} &= \mathrm{prox}_{\eta g}(t^{(r+1)}). \end{aligned}$$

The following theorem verifies the convergence of the DRS algorithm. For a recent convergence result see also [97].

Theorem 10 (Convergence of Douglas-Rachford Splitting Algorithm). *Let* $g,h \in \Gamma_0(\mathbb{R}^d)$ *where one of the functions is continuous at a point in* $\operatorname{dom} g \cap \operatorname{dom} h$. *Assume that a solution of* $\operatorname{argmin}_{x\in\mathbb{R}^d}\{g(x)+h(x)\}$ *exists. Then, for any initial* $t^{(0)}, x^{(0)} \in \mathbb{R}^d$ *and any* $\eta > 0$, *the DRS sequence* $(t^{(r)})_r$ *converges to a fixed point* $\hat{t}$ *of* T *in* (10.39) *and* $(x^{(r)})_r$ *to a solution of the minimization problem.*

Proof. It remains to show that T is firmly nonexpansive. We have for $R_{\eta g} := 2\operatorname{prox}_{\eta g} - I$ and $R_{\eta h} := 2\operatorname{prox}_{\eta h} - I$ that

$$\begin{aligned} 2T &= 2\operatorname{prox}_{\eta h}(2\operatorname{prox}_{\eta g} - I) - (2\operatorname{prox}_{\eta g} - I) + I = R_{\eta h} \circ R_{\eta g} + I, \\ T &= \tfrac{1}{2} I + \tfrac{1}{2} R_{\eta h} \circ R_{\eta g}. \end{aligned}$$

The operators $R_{\eta g}, R_{\eta h}$ are nonexpansive since $\operatorname{prox}_{\eta g}$ and $\operatorname{prox}_{\eta h}$ are firmly nonexpansive. Hence $R_{\eta h} \circ R_{\eta g}$ is nonexpansive and we are done. □

For variety of (sharp) convergence rates of DRS we refer to [64, 65].
The relation of the ADMM algorithm and DRS algorithm applied to the Fenchel dual problem (10.29), i.e.,

$$\begin{aligned} t^{(r+1)} &= \operatorname{prox}_{\eta g^* \circ (-A^\mathrm{T})}(2p^{(r)} - t^{(r)}) + t^{(r)} - p^{(r)}, \\ p^{(r+1)} &= \operatorname{prox}_{\eta h^*}(t^{(r+1)}), \end{aligned} \tag{10.40}$$

is given by the following theorem, see [69, 82].

Theorem 11 (Relation Between ADMM and DRS). *The ADMM sequences* $\left(b^{(r)}\right)_r$ *and* $\left(y^{(r)}\right)_r$ *are related to the sequences* (10.40) *generated by the DRS algorithm applied to the dual problem by* $\eta = \gamma$ *and*

$$\begin{aligned} t^{(r)} &= \eta(b^{(r)} + y^{(r)}), \\ p^{(r)} &= \eta b^{(r)}. \end{aligned} \tag{10.41}$$

Proof. First, we show that

$$\hat{p} = \operatorname*{argmin}_{p\in\mathbb{R}^m} \left\{ \frac{\eta}{2} \|Ap - q\|_2^2 + g(p) \right\} \quad \Rightarrow \quad \eta(A\hat{p} - q) = \operatorname{prox}_{\eta g^* \circ (-A^\mathrm{T})}(-\eta q) \tag{10.42}$$

holds true. The left-hand side of (10.42) is equivalent to

$$0 \in \eta A^\mathrm{T}(A\hat{p} - q) + \partial g(\hat{p}) \quad \Leftrightarrow \quad \hat{p} \in \partial g^*\big(-\eta A^\mathrm{T}(A\hat{p} - q)\big).$$

Applying $-\eta A$ on both sides and using the chain rule implies

$$-\eta A\hat{p} \in -\eta A \partial g^*\big(-\eta A^\mathrm{T}(A\hat{p} - q)\big) = \eta\, \partial\big(g^* \circ (-A^\mathrm{T})\big)\big(\eta(A\hat{p} - q)\big).$$

Adding $-\eta q$ we get

$$-\eta q \in \big(I + \eta\,\partial(g^* \circ (-A^{\mathrm{T}}))\big)\big(\eta(A\hat{p} - q)\big),$$

which is equivalent to the right-hand side of (10.42) by the definition of the resolvent (see Remark 1).
Secondly, applying (10.42) to the first ADMM step with $\gamma = \eta$ and $q := y^{(r)} - b^{(r)}$ yields

$$\eta(b^{(r)} + Ax^{(r+1)} - y^{(r)}) = \mathrm{prox}_{\eta g^* \circ (-A^{\mathrm{T}})}(\eta(b^{(r)} - y^{(r)})).$$

Assume that the ADMM and DRS iterates have the identification (10.41) up to some $r \in \mathbb{N}$. Using this induction hypothesis it follows that

$$\eta(b^{(r)} + Ax^{(r+1)}) = \mathrm{prox}_{\eta g^* \circ (-A^{\mathrm{T}})}(\underbrace{\eta(b^{(r)} - y^{(r)})}_{2p^{(r)} - t^{(r)}}) + \underbrace{\eta y^{(r)}}_{t^{(r)} - p^{(r)}} \overset{(10.40)}{=} t^{(r+1)} \qquad (10.43)$$

By definition of $b^{(r+1)}$ we see that $\eta(b^{(r+1)} + y^{(r+1)}) = t^{(r+1)}$. Next we apply (10.42) in the second ADMM step where we replace g by h and A by $-I$ and use $q := -b^{(r)} - Ax^{(r+1)}$. Together with (10.43) this gives

$$\eta(-y^{(r+1)} + b^{(r)} + Ax^{(r+1)}) = \mathrm{prox}_{\eta h^*}(\underbrace{\eta(b^{(r)} + Ax^{(r+1)})}_{t^{(r+1)}}) \overset{(10.40)}{=} p^{(r+1)}. \qquad (10.44)$$

Using again the definition of $b^{(r+1)}$ we obtain $\eta b^{(r+1)} = p^{(r+1)}$ which completes the proof. □

A recent work [186] shows that the ADMM is in some sense self-dual, i.e., it is equivalent not only to the DRS applied to the dual problem, but also to the primal one.

6.3 Primal Dual Hybrid Gradient Algorithms

The first ADMM step (10.32) requires in general the solution of a linear system of equations. This can be avoided by modifying this step using the Taylor expansion at $x^{(r)}$:

$$\frac{\gamma}{2}\|\frac{1}{\gamma}p^{(r)} + Ax - y^{(r)}\|_2^2 \approx \mathrm{const} + \gamma\langle A^{\mathrm{T}}(Ax^{(r)} - y^{(r)} + \frac{1}{\gamma}p^{(r)}), x\rangle + \frac{\gamma}{2}(x - x^{(r)})^{\mathrm{T}} A^{\mathrm{T}} A (x - x^{(r)}),$$

approximating $A^{\mathrm{T}}A \approx \frac{1}{\gamma\tau}I$, setting $\gamma := \sigma$ and using $p^{(r)}/\sigma$ instead of $Ax^{(r)} - y^{(r)} + p^{(r)}/\sigma$ we obtain (note that $p^{(r+1)}/\sigma = p^{(r)}/\sigma + Ax^{(r+1)} - y^{(r+1)}$):

$$\begin{aligned}
x^{(r+1)} &= \operatorname{argmin}_{x\in\mathbb{R}^d}\left\{g(x)+\tfrac{1}{2\tau}\|x-\left(x^{(r)}-\tau A^{\mathrm{T}}p^{(r)}\right)\|_2^2\right\} = \operatorname{prox}_{\tau g}\left(x^{(r)}-\tau A^{\mathrm{T}}p^{(r)}\right),\\
y^{(r+1)} &= \operatorname{argmin}_{y\in\mathbb{R}^m}\left\{h(y)+\tfrac{\sigma}{2}\|\tfrac{1}{\sigma}p^{(r)}+Ax^{(r+1)}-y\|_2^2\right\} = \operatorname{prox}_{\frac{1}{\sigma}h}\left(\tfrac{1}{\sigma}p^{(r)}+Ax^{(r+1)}\right),\\
p^{(r+1)} &= p^{(r)}+\sigma(Ax^{(r+1)}-y^{(r+1)}).
\end{aligned} \tag{10.45}$$

The above algorithm can be deduced in another way by the Arrow-Hurwicz method: we alternate the minimization in the primal and dual problems (10.27) and (10.28) and add quadratic terms. The resulting sequences

$$\begin{aligned}
x^{(r+1)} &= \operatorname*{argmin}_{x\in\mathbb{R}^d}\left\{g(x)+\langle p^{(r)},Ax\rangle+\frac{1}{2\tau}\|x-x^{(r)}\|_2^2\right\},\\
&= \operatorname{prox}_{\tau g}(x^{(r)}-\tau A^{\mathrm{T}}p^{(r)}) && (10.46)\\
p^{(r+1)} &= \operatorname*{argmin}_{p\in\mathbb{R}^m}\left\{h^*(p)-\langle p,Ax^{(r+1)}\rangle+\frac{1}{2\sigma}\|p-p^{(r)}\|_2^2\right\},\\
&= \operatorname{prox}_{\sigma h^*}(p^{(r)}+\sigma Ax^{(r+1)}) && (10.47)
\end{aligned}$$

coincide with those in (10.45) which can be seen as follows: For $x^{(r)}$ the relation is straightforward. From the last equation we obtain

$$\begin{aligned}
p^{(r)}+\sigma Ax^{(r+1)} &\in p^{(r+1)}+\sigma\partial h^*(p^{(r+1)}),\\
\frac{1}{\sigma}(p^{(r)}-p^{(r+1)})+Ax^{(r+1)} &\in \partial h^*(p^{(r+1)}),
\end{aligned}$$

and using that $p\in\partial h(x)\Leftrightarrow x\in\partial h^*(p)$ further

$$p^{(r+1)}\in\partial h(\underbrace{\frac{1}{\sigma}(p^{(r)}-p^{(r+1)})+Ax^{(r+1)}}_{y^{(r+1)}}).$$

Setting

$$y^{(r+1)}:=\frac{1}{\sigma}(p^{(r)}-p^{(r+1)})+Ax^{(r+1)},$$

we get

$$p^{(r+1)}=p^{(r)}+\sigma(Ax^{(r+1)}-y^{(r+1)}) \tag{10.48}$$

and $p^{(r+1)}\in\partial h(y^{(r+1)})$ which can be rewritten as

$$\begin{aligned}
y^{(r+1)}+\frac{1}{\sigma}p^{(r+1)} &\in y^{(r+1)}+\frac{1}{\sigma}\partial h(y^{(r+1)}),\\
\frac{1}{\sigma}p^{(r)}+Ax^{(r+1)} &\in y^{(r+1)}+\frac{1}{\sigma}\partial h(y^{(r+1)}),\\
y^{(r+1)} &= \operatorname{prox}_{\frac{1}{\sigma}h}\left(\frac{1}{\sigma}p^{(r)}+Ax^{(r+1)}\right).
\end{aligned}$$

There are several modifications of the basic "linearized" ADMM which improve its convergence properties as

- the predictor corrector proximal multiplier method [48],
- the primal dual hybrid gradient method (PDHG) [191] with convergence proof in [23],
- primal dual hybrid gradient method with extrapolation of the primal or dual variable [43, 140], a preconditioned version [42] and a generalization [58], Douglas-Rachford-type algorithms [25, 26] for solving inclusion equations, see also [54, 177], as well an extension allowing the operator A to be nonlinear [172].

A good overview on primal-dual methods is also given in [110]. Here is the algorithm proposed by Chambolle, Cremers, and Pock [43, 140].

Algorithm 8 Primal Dual Hybrid Gradient Method with Extrapolation of Dual Variable (PDHGMp)

Initialization: $y^{(0)}, b^{(0)} = b^{(-1)} \in \mathbb{R}^m$, $\tau, \sigma > 0$ with $\tau\sigma < 1/\|A\|_2^2$ and $\theta \in (0,1]$
Iterations: For $r = 0, 1, \dots$

$$x^{(r+1)} = \text{argmin}_{x\in\mathbb{R}^d} \left\{ g(x) + \tfrac{1}{2\tau}\|x - (x^{(r)} - \tau\sigma A^{\mathrm{T}}\bar{b}^{(r)})\|_2^2 \right\}$$
$$y^{(r+1)} = \text{argmin}_{y\in\mathbb{R}^m} \left\{ h(y) + \tfrac{\sigma}{2}\|b^{(r)} + Ax^{(r+1)} - y\|_2^2 \right\}$$
$$b^{(r+1)} = b^{(r)} + Ax^{(r+1)} - y^{(r+1)}.$$
$$\bar{b}^{(r+1)} = b^{(r+1)} + \theta(b^{(r+1)} - b^{(r)})$$

Note that the new first updating step can be also deduced by applying the so-called *inexact Uzawa algorithm* to the first ADMM step (see Section 6.4). Furthermore, it can be directly seen that for A being the identity and $\theta = 1$ and $\gamma = \sigma = \frac{1}{\tau}$, the PDHGMp algorithm corresponds to the ADMM as well as the Douglas-Rachford splitting algorithm as proposed in Section 6.2. The following theorem and convergence proof are based on [43].

Theorem 12. *Let $g \in \Gamma_0(\mathbb{R}^d)$, $h \in \Gamma_0(\mathbb{R}^m)$ and $\theta \in (0,1]$. Let $\tau, \sigma > 0$ fulfill*

$$\tau\sigma < 1/\|A\|_2^2. \tag{10.49}$$

Suppose that the Lagrangian $L(x,p) := g(x) - h^*(p) + \langle Ax, p\rangle$ has a saddle point. Then the sequence $\{(x^{(r)}, p^{(r)})\}_r$ produced by PDGHMp converges to a saddle point of the Lagrangian.

Proof. We restrict the proof to the case $\theta = 1$. For arbitrary $\bar{x} \in \mathbb{R}^d, \bar{p} \in \mathbb{R}^m$ consider according to (10.46) and (10.47) the iterations

$$x^{(r+1)} = (I + \tau\partial g)^{-1}\left(x^{(r)} - \tau A^{\mathrm{T}}\bar{p}\right),$$
$$p^{(r+1)} = (I + \sigma\partial h^*)^{-1}\left(p^{(r)} + \sigma A\bar{x}\right),$$

i.e.,

$$\frac{x^{(r)} - x^{(r+1)}}{\tau} - A^{\mathrm{T}}\bar{p} \in \partial g\left(x^{(r+1)}\right), \quad \frac{p^{(r)} - p^{(r+1)}}{\sigma} + A\bar{x} \in \partial h^*\left(p^{(r+1)}\right).$$

By definition of the subdifferential we obtain for all $x \in \mathbb{R}^d$ and all $p \in \mathbb{R}^m$ that

$$\begin{aligned} g(x) &\geq g(x^{(r+1)}) + \frac{1}{\tau}\langle x^{(r)} - x^{(r+1)}, x - x^{(r+1)}\rangle - \langle A^{\mathrm{T}}\bar{p}, x - x^{(r+1)}\rangle, \\ h^*(p) &\geq h^*(p^{(r+1)}) + \frac{1}{\sigma}\langle p^{(r)} - p^{(r+1)}, p - p^{(r+1)}\rangle + \langle p - p^{(r+1)}, A\bar{x}\rangle \end{aligned}$$

and by adding the equations

$$\begin{aligned} 0 \geq\ & g(x^{(r+1)}) - h^*(p) - \left(g(x) - h^*(p^{(r+1)})\right) - \langle A^{\mathrm{T}}\bar{p}, x - x^{(r+1)}\rangle + \langle p - p^{(r+1)}, A\bar{x}\rangle \\ & + \frac{1}{\tau}\langle x^{(r)} - x^{(r+1)}, x - x^{(r+1)}\rangle + \frac{1}{\sigma}\langle p^{(r)} - p^{(r+1)}, p - p^{(r+1)}\rangle. \end{aligned}$$

By

$$\langle x^{(r)} - x^{(r+1)}, x - x^{(r+1)}\rangle = \frac{1}{2}\left(\|x^{(r)} - x^{(r+1)}\|_2^2 + \|x - x^{(r+1)}\|_2^2 - \|x - x^{(r)}\|_2^2\right)$$

this can be rewritten as

$$\begin{aligned} & \frac{1}{2\tau}\|x - x^{(r)}\|_2^2 + \frac{1}{2\sigma}\|p - p^{(r)}\|_2^2 \\ \geq\ & \frac{1}{2\tau}\|x^{(r)} - x^{(r+1)}\|_2^2 + \frac{1}{2\tau}\|x - x^{(r+1)}\|_2^2 + \frac{1}{2\sigma}\|p^{(r)} - p^{(r+1)}\|_2^2 + \frac{1}{2\sigma}\|p - p^{(r+1)}\|_2^2 \\ & + \left(g(x^{(r+1)}) - h^*(p) + \langle p, Ax^{(r+1)}\rangle\right) - \left(g(x) - h^*(p^{(r+1)}) + \langle p^{(r+1)}, Ax\rangle\right) \\ & - \langle p, Ax^{(r+1)}\rangle + \langle p^{(r+1)}, Ax\rangle - \langle \bar{p}, A(x - x^{(r+1)})\rangle + \langle p - p^{(r+1)}, A\bar{x}\rangle \\ =\ & \frac{1}{2\tau}\|x^{(r)} - x^{(r+1)}\|_2^2 + \frac{1}{2\tau}\|x - x^{(r+1)}\|_2^2 + \frac{1}{2\sigma}\|p^{(r)} - p^{(r+1)}\|_2^2 + \frac{1}{2\sigma}\|p - p^{(r+1)}\|_2^2 \\ & + \left(g(x^{(r+1)}) - h^*(p) + \langle p, Ax^{(r+1)}\rangle\right) - \left(g(x) - h^*(p^{(r+1)}) + \langle p^{(r+1)}, Ax\rangle\right) \\ & + \langle p^{(r+1)} - p, A(x^{(r+1)} - \bar{x})\rangle - \langle p^{(r+1)} - \bar{p}, A(x^{(r+1)} - x)\rangle. \end{aligned}$$

For any saddle point (x^*, p^*) we have that $L(x^*, p) \leq L(x^*, p^*) \leq L(x, p^*)$ for all x, p so that in particular $0 \leq L(x^{(r+1)}, p^*) - L(x^*, p^{(r+1)})$. Thus, using $(x, p) := (x^*, p^*)$ in the above inequality, we get

$$\begin{aligned} & \frac{1}{2\tau}\|x^* - x^{(r)}\|_2^2 + \frac{1}{2\sigma}\|p^* - p^{(r)}\|_2^2 \\ \geq\ & \frac{1}{2\tau}\|x^{(r)} - x^{(r+1)}\|_2^2 + \frac{1}{2\tau}\|x^* - x^{(r+1)}\|_2^2 + \frac{1}{2\sigma}\|p^{(r)} - p^{(r+1)}\|_2^2 + \frac{1}{2\sigma}\|p^* - p^{(r+1)}\|_2^2 \\ & + \langle p^{(r+1)} - p^*, A(x^{(r+1)} - \bar{x})\rangle - \langle p^{(r+1)} - \bar{p}, A(x^{(r+1)} - x^*)\rangle. \end{aligned}$$

In the algorithm we use $\bar{x} := x^{(r+1)}$ and $\bar{p} := 2p^{(r)} - p^{(r-1)}$. Note that $\bar{p} = p^{(r+1)}$ would be the better choice, but this is impossible if we want to keep on an explicit algorithm. For these values the above inequality further simplifies to

$$\begin{aligned}
&\frac{1}{2\tau}\|x^* - x^{(r)}\|_2^2 + \frac{1}{2\sigma}\|p^* - p^{(r)}\|_2^2 \\
\geq\ &\frac{1}{2\tau}\|x^{(r)} - x^{(r+1)}\|_2^2 + \frac{1}{2\tau}\|x^* - x^{(r+1)}\|_2^2 + \frac{1}{2\sigma}\|p^{(r)} - p^{(r+1)}\|_2^2 + \frac{1}{2\sigma}\|p^* - p^{(r+1)}\|_2^2 \\
&+ \langle p^{(r+1)} - 2p^{(r)} + p^{(r-1)}, A(x^* - x^{(r+1)})\rangle.
\end{aligned}$$

We estimate the last summand using Cauchy-Schwarz's inequality as follows:

$$\begin{aligned}
&\langle p^{(r+1)} - p^{(r)} - (p^{(r)} - p^{(r-1)}), A(x^* - x^{(r+1)})\rangle \\
=\ &\langle p^{(r+1)} - p^{(r)}, A(x^* - x^{(r+1)})\rangle - \langle p^{(r)} - p^{(r-1)}, A(x^* - x^{(r)})\rangle \\
&- \langle p^{(r)} - p^{(r-1)}, A(x^{(r)} - x^{(r+1)})\rangle \\
\geq\ &\langle p^{(r+1)} - p^{(r)}, A(x^* - x^{(r+1)})\rangle - \langle p^{(r)} - p^{(r-1)}, A(x^* - x^{(r)})\rangle \\
&- \|A\|_2 \|x^{(r+1)} - x^{(r)}\|_2 \, \|p^{(r)} - p^{(r-1)}\|_2.
\end{aligned}$$

Since

$$2uv \leq \alpha u^2 + \frac{1}{\alpha}v^2, \quad \alpha > 0, \tag{10.50}$$

we obtain

$$\begin{aligned}
\|A\|_2 \|x^{(r+1)} - x^{(r)}\|_2 \, \|p^{(r)} - p^{(r-1)}\|_2 &\leq \frac{\|A\|_2}{2}\left(\alpha\|x^{(r+1)} - x^{(r)}\|_2^2 + \frac{1}{\alpha}\|p^{(r)} - p^{(r-1)}\|_2^2\right) \\
&= \frac{\|A\|_2 \alpha\tau}{2\tau}\|x^{(r+1)} - x^{(r)}\|_2^2 + \frac{\|A\|_2\sigma}{2\alpha\sigma}\|p^{(r)} - p^{(r-1)}\|_2^2.
\end{aligned}$$

With $\alpha := \sqrt{\sigma/\tau}$ the relation

$$\|A\|_2 \alpha\tau = \frac{\|A\|_2\sigma}{\alpha} = \|A\|_2\sqrt{\sigma\tau} < 1$$

holds true. Thus, we get

$$\begin{aligned}
&\frac{1}{2\tau}\|x^* - x^{(r)}\|_2^2 + \frac{1}{2\sigma}\|p^* - p^{(r)}\|_2^2 \\
\geq\ &\frac{1}{2\tau}\|x^* - x^{(r+1)}\|_2^2 + \frac{1}{2\sigma}\|p^* - p^{(r+1)}\|_2^2 \\
&+ \frac{1}{2\tau}(1 - \|A\|_2\sqrt{\sigma\tau})\|x^{(r+1)} - x^{(r)}\|_2^2 + \frac{1}{2\sigma}\|p^{(r+1)} - p^{(r)}\|_2^2 - \frac{\|A\|_2\sqrt{\sigma\tau}}{2\sigma}\|p^{(r)} - p^{(r-1)}\|_2^2 \\
&+ \langle p^{(r+1)} - p^{(r)}, A(x^* - x^{(r+1)})\rangle - \langle p^{(r)} - p^{(r-1)}, A(x^* - x^{(r)})\rangle.
\end{aligned} \tag{10.51}$$

Summing up these inequalities from $r = 0$ to $N - 1$ and regarding that $p^{(0)} = p^{(-1)}$, we obtain

$$\begin{aligned}
&\frac{1}{2\tau}\|x^* - x^{(0)}\|_2^2 + \frac{1}{2\sigma}\|p^* - p^{(0)}\|_2^2 \\
&\geq \frac{1}{2\tau}\|x^* - x^{(N)}\|_2^2 + \frac{1}{2\sigma}\|p^* - p^{(N)}\|_2^2 \\
&\quad + (1 - \|A\|_2\sqrt{\sigma\tau})\left(\frac{1}{2\tau}\sum_{r-1}^{N}\|x^{(r)} - x^{(r-1)}\|_2^2 + \frac{1}{2\sigma}\sum_{r-1}^{N-1}\|p^{(r)} - p^{(r-1)}\|_2^2\right) \\
&\quad + \frac{1}{2\sigma}\|p^{(N)} - p^{(N-1)}\|_2^2 + \langle p^{(N)} - p^{(N-1)}, A(x^* - x^{(N)})\rangle
\end{aligned}$$

By

$$\langle p^{(N)} - p^{(N-1)}, A(x^{(N)} - x^*)\rangle \leq \frac{1}{2\sigma}\|p^{(N)} - p^{(N-1)}\|_2^2 + \frac{\|A\|_2^2\sigma\tau}{2\tau}\|x^{(N)} - x^*\|_2^2$$

this can be further estimated as

$$\begin{aligned}
&\frac{1}{2\tau}\|x^* - x^{(0)}\|_2^2 + \frac{1}{2\sigma}\|p^* - p^{(0)}\|_2^2 \\
&\geq \frac{1}{2\tau}(1 - \|A\|_2^2\sigma\tau)\|x^* - x^{(N)}\|_2^2 + \frac{1}{2\sigma}\|p^* - p^{(N)}\|_2^2 \\
&\quad + (1 - \|A\|_2\sqrt{\sigma\tau})\left(\frac{1}{2\tau}\sum_{r=1}^{N}\|x^{(r)} - x^{(r-1)}\|_2^2 + \frac{1}{2\sigma}\sum_{r=1}^{N-1}\|p^{(r)} - p^{(r-1)}\|_2^2\right). \quad (10.52)
\end{aligned}$$

By (10.52) we conclude that the sequence $\{(x^{(n)}, p^{(n)})\}_n$ is bounded. Thus, there exists a convergent subsequence $\{(x^{(n_j)}, p^{(n_j)})\}_j$ which converges to some point $(\hat{x}, \hat{p})$ as $j \to \infty$. Further, we see by (10.52) that

$$\lim_{n\to\infty}\left(x^{(n)} - x^{(n-1)}\right) = 0, \quad \lim_{n\to\infty}\left(p^{(n)} - p^{(n-1)}\right) = 0.$$

Consequently,

$$\lim_{j\to\infty}\left(x^{(n_j-1)} - \hat{x}\right) = 0, \quad \lim_{j\to\infty}\left(p^{(n_j-1)} - \hat{p}\right) = 0$$

holds true. Let T denote the iteration operator of the PDHGMp cycles, i.e., $T(x^{(r)}, p^{(r)}) = (x^{(r+1)}, p^{(r+1)})$. Since T is the concatenation of affine operators and proximation operators, it is continuous. Now we have that $T\left(x^{(n_j-1)}, p^{(n_j-1)}\right) = \left(x^{(n_j)}, p^{(n_j)}\right)$ and taking the limits for $j \to \infty$ we see that $T(\hat{x}, \hat{p}) = (\hat{x}, \hat{p})$ so that $(\hat{x}, \hat{p})$ is a fixed point of the iteration and thus a saddle point of L. Using this particular saddle point in (10.51) and summing up from $r = n_j$ to $N-1$, $N > n_j$ we obtain

$$
\begin{aligned}
&\frac{1}{2\tau}\|\hat{x}-x^{(n_j)}\|_2^2+\frac{1}{2\sigma}\|\hat{p}-p^{(n_j)}\|_2^2\\
&\geq \frac{1}{2\tau}\|\hat{x}-x^{(N)}\|_2^2+\frac{1}{2\sigma}\|\hat{p}-p^{(N)}\|_2^2\\
&+(1-\|A\|_2\sqrt{\sigma\tau})\left(\frac{1}{2\tau}\sum_{r=n_j}^{N-1}\|x^{(r+1)}-x^{(r)}\|_2^2+\frac{1}{2\sigma}\sum_{r=n_j+1}^{N-1}\|p^{(r)}-p^{(r-1)}\|^2\right)\\
&+\frac{1}{2\sigma}\|p^{(N)}-p^{(N-1)}\|^2-\frac{\|A\|_2\sqrt{\sigma\tau}}{2\sigma}\|p^{(n_j)}-p^{(n_j-1)}\|_2^2\\
&+\langle p^{(N)}-p^{(N-1)},A(\hat{x}-x^{(N)})\rangle-\langle p^{(n_j)}-p^{(n_j-1)},A(\hat{x}-x^{(n_j)})\rangle
\end{aligned}
$$

and further

$$
\begin{aligned}
&\frac{1}{2\tau}\|\hat{x}-x^{(n_j)}\|_2^2+\frac{1}{2\sigma}\|\hat{p}-p^{(n_j)}\|_2^2\\
&\geq \frac{1}{2\tau}\|\hat{x}-x^{(N)}\|_2^2+\frac{1}{2\sigma}\|\hat{p}-p^{(N)}\|_2^2\\
&+\frac{1}{2\sigma}\|p^{(N)}-p^{(N-1)}\|_2^2-\frac{\|A\|_2\sqrt{\sigma\tau}}{2\sigma}\|p^{(n_j)}-p^{(n_j-1)}\|_2^2\\
&+\langle p^{(N)}-p^{(N-1)},A(\hat{x}-x^{(N)})\rangle-\langle p^{(n_j)}-p^{(n_j-1)},A(\hat{x}-x^{(n_j)})\rangle
\end{aligned}
$$

For $j\to\infty$ this implies that $(x^{(N)},p^{(N)})$ converges also to $(\hat{x},\hat{y})$ and we are done. □

6.4 Proximal ADMM

To avoid the computation of a linear system of equations in the first ADMM step (10.32), we can more generally use the proximal ADMM algorithm [95, 190] that can be interpreted as a preconditioned variant of ADMM. In this variant the minimization step (10.32) is replaced by a proximal-like iteration based on the general proximal operator (10.5),

$$
x^{(r+1)}=\operatorname*{argmin}_{x\in\mathbb{R}^d}\{L_\gamma(x,y^{(r)},p^{(r)})+\frac{1}{2}\|x-x^{(r)}\|_R^2\} \tag{10.53}
$$

with a symmetric, positive definite matrix $R\in\mathbb{R}^{d,d}$. The introduction of R provides an additional flexibility to cancel out linear operators which might be difficult to invert. In addition the modified minimization problem is strictly convex inducing a unique minimizer. In the same manner the second ADMM step (10.33) can also be extended by a proximal term $(1/2)\|y-y^{(r)}\|_S^2$ with a symmetric, positive definite matrix $S\in\mathbb{R}^{m,m}$ [190]. The convergence analysis of the proximal ADMM was provided in [190] and the algorithm can be also classified as an inexact Uzawa method. A generalization, where the matrices R and S can non-monotonically vary in each iteration step, was analyzed in [95], additionally allowing an inexact computation of minimization problems.

In case of the PDHGMp algorithm, it was mentioned that the first updating step can be deduced by applying the inexact Uzawa algorithm to the first ADMM step. Using the proximal ADMM, it is straightforward to see that the first updating step of the PDHGMp algorithm with $\theta = 1$ corresponds to (10.53) in case of $R = \frac{1}{\tau}I - \sigma A^{\mathrm{T}}A$ with $0 < \tau < 1/\|\sigma A^{\mathrm{T}}A\|$, see [43, 71]. Further relations of the (proximal) ADMM to primal dual hybrid methods discussed above can be found in [71].

6.5 Bregman Methods

Bregman methods became very popular in image processing by a series papers of Osher and co-workers, see, e.g., [91, 135]. Many of these methods can be interpreted as ADMM methods and its linearized versions. In the following we briefly sketch these relations.

The PPA is a special case of the Bregman PPA. Let $\varphi : \mathbb{R}^d \to \mathbb{R} \cup \{+\infty\}$ be a convex function. Then the *Bregman distance* $D_\varphi^p : \mathbb{R}^d \times \mathbb{R}^d \to \mathbb{R}$ is given by

$$D_\varphi^p(x,y) = \varphi(x) - \varphi(y) - \langle p, x - y \rangle$$

with $p \in \partial\varphi(y)$, $y \in \mathrm{dom} f$. If $\partial\varphi(y)$ contains only one element, we just write D_φ. If φ is smooth, then the Bregman distance can be interpreted as subtracting the first order Taylor expansion of $\varphi(x)$ at y.

Example 5. (Special Bregman Distances)

1. The Bregman distance corresponding to $\varphi(x) := \frac{1}{2}\|x\|_2^2$ is given by $D_\varphi(x,y) = \frac{1}{2}\|x-y\|_2^2$.
2. For the negative Shannon entropy $\varphi(x) := \langle 1_d, x \log x \rangle$, $x > 0$ we obtain the (discrete) I-divergence or generalized Kullback-Leibler entropy $D_\varphi(x,y) = x \log \frac{x}{y} - x + y$.

For $f \in \Gamma_0(\mathbb{R}^d)$ we consider the generalized proximal problem

$$\operatorname*{argmin}_{y \in \mathbb{R}^d} \left\{ \frac{1}{\gamma} D_\varphi^p(x,y) + f(y) \right\}.$$

The Bregman Proximal Point Algorithm (BPPA) for solving this problem reads as follows:

Algorithm 9 Bregman Proximal Point Algorithm (BPPA)

Initialization: $x^{(0)} \in \mathbb{R}^d$, $p^{(0)} \in \partial\varphi(x^{(0)})$, $\gamma > 0$
Iterations: For $r = 0, 1, \ldots$
$x^{(r+1)} = \mathrm{argmin}_{y \in \mathbb{R}^d} \left\{ \frac{1}{\gamma} D_\varphi^{p^{(r)}}(y, x^{(r)}) + f(y) \right\}$
$p^{(r+1)} \in \partial\varphi(x^{(r+1)})$

The BPPA converges for any initialization $x^{(0)}$ to a minimizer of f if $f \in \Gamma_0(\mathbb{R}^d)$ attains its minimum and φ is finite, lower semi-continuous and strictly convex. For convergence proofs we refer, e.g., to [107, 108]. We are interested again in the problem (10.24), i.e.,

$$\underset{x\in\mathbb{R}^d, y\in\mathbb{R}^m}{\operatorname{argmin}} \; \{\Phi(x,y) \quad \text{s.t.} \quad Ax = y\}, \quad \Phi(x,y) := g(x) + h(y).$$

We consider the BPP algorithm for the objective function $f(x,y) := \frac{1}{2}\|Ax-y\|^2$ with the Bregman distance

$$D_\Phi^{(p_x^{(r)},p_y^{(r)})}\left((x,y),(x^{(r)},y^{(r)})\right) \doteq \Phi(x,y) - \Phi(x^{(r)},y^{(r)}) - \langle p_x^{(r)}, x-x^{(r)}\rangle - \langle p_y^{(r)}, y-y^{(r)}\rangle,$$

where $\left(p_x^{(r)}, p_y^{(r)}\right) \in \partial\Phi(x^{(r)},y^{(r)})$. This results in

$$(x^{(r+1)},y^{(r+1)}) = \underset{x\in\mathbb{R}^d, y\in\mathbb{R}^m}{\operatorname{argmin}} \; \left\{\frac{1}{\gamma} D_\Phi^{(p_x^{(r)},p_y^{(r)})}\left((x,y),(x^{(r)},y^{(r)})\right) + \frac{1}{2}\|Ax-y\|^2\right\},$$

$$p_x^{(r+1)} = p_x^{(r)} - \gamma A^*(Ax^{(r+1)} - y^{(r+1)}), \tag{10.54}$$

$$p_y^{(r+1)} = p_y^{(r)} + \gamma(Ax^{(r+1)} - y^{(r+1)}), \tag{10.55}$$

where we have used that the first equation implies

$$0 \in \frac{1}{\gamma}\partial\left(\Phi(x^{(r+1)},y^{(r+1)}) - (p_x^{(r)},p_y^{(r)})\right) + \left(A^*(Ax^{(r+1)} - y^{(r+1)}), -(Ax^{(r+1)} - y^{(r+1)})\right),$$

$$0 \in \partial\Phi(x^{(r+1)},y^{(r+1)}) - (p_x^{(r+1)},p_y^{(r+1)}),$$

so that $\left(p_x^{(r)}, p_y^{(r)}\right) \in \partial\Phi(x^{(r)},y^{(r)})$. From (10.54) and (10.55) we see by induction that $p_x^{(r)} = -A^* p_y^{(r)}$. Setting $p^{(r)} = p_y^{(r)}$ and regarding that

$$\begin{aligned}
&\frac{1}{\gamma} D_\Phi^{p^{(r)}}\left((x,y),(x^{(r)},y^{(r)})\right) + \frac{1}{2}\|Ax-y\|_2^2 \\
&= \text{const} + \frac{1}{\gamma}\left(\Phi(x,y) + \langle A^* p^{(r)}, x\rangle - \langle p^{(r)}, y\rangle\right) + \frac{1}{2}\|Ax-y\|_2^2 \\
&= \text{const} + \frac{1}{\gamma}\left(\Phi(x,y) + \frac{\gamma}{2}\|\frac{p^{(r)}}{\gamma} + Ax - y\|_2^2\right)
\end{aligned}$$

we obtain the following split Bregman method, see [91]:
Obviously, this is exactly the form of the augmented Lagrangian method in (10.31). Concerning the elastic net example 2 we refer to [38, 114, 187] for convergence results of the so-called linearized Bregman algorithm.

Algorithm 10 Split Bregman Algorithm

Initialization: $x^{(0)} \in \mathbb{R}^d$, $p^{(0)}$, $\gamma > 0$
Iterations: For $r = 0, 1, \ldots$

$$(x^{(r+1)}, y^{(r+1)}) = \underset{x\in\mathbb{R}^d, y\in\mathbb{R}^m}{\operatorname{argmin}} \left\{ \Phi(x,y) + \tfrac{\gamma}{2}\|\tfrac{p^{(r)}}{\gamma} + Ax - y\|_2^2 \right\}$$

$$p^{(r+1)} = p^{(r)} + \gamma(Ax^{(r+1)} - y^{(r+1)})$$

7 Iterative Regularization for Ill-Posed Problems

So far we have discussed the use of splitting methods for the numerical solution of well-posed variational problems, which arise in a discrete setting and in particular for the standard approach of Tikhonov-type regularization in inverse problems in imaging. The latter is based on minimizing a weighted sum of a data fidelity and a regularization functional, and can be more generally analyzed in Banach spaces, cf. [163]. However, such approaches have several disadvantages, in particular it has been shown that they lead to unnecessary bias in solutions, e.g., a contrast loss in the case of total variation regularization, cf. [136, 31]. A successful alternative to overcome this issue is iterative regularization, which directly applies iterative methods to solve the constrained variational problem

$$\underset{x\in\mathcal{X}}{\operatorname{argmin}} \{g(x) \quad \text{s.t.} \quad Ax = f\}. \tag{10.56}$$

Here $A : \mathcal{X} \to \mathcal{Y}$ is a bounded linear operator between Banach spaces (also nonlinear versions can be considered, cf. [9, 105]) and f are given data. In the well-posed case, (10.56) can be rephrased as the saddle-point problem

$$\min_{x\in\mathcal{X}} \sup_{q} \left(g(x) - \langle q, Ax - f\rangle\right) \tag{10.57}$$

The major issue compared to the discrete setting is that for many prominent examples the operator A does not have a closed range (and hence a discontinuous pseudo-inverse), which makes (10.56) ill-posed. From the optimization point of view, this raises two major issues:

- *Emptyness of the constraint set:* In the practically relevant case of noisy measurements one has to expect that f is not in the range of A, i.e., the constraint cannot be satisfied exactly. Reformulated in the constrained optimization view, the standard paradigm in iterative regularization is to construct an iteration slowly increasing the functional g while decreasing the error in the constraint.
- *Nonexistence of saddle points:* Even if the data or an idealized version Ax^* to be approximated are in the range of A, the existence of a saddle point (x^*, q^*) of (10.57) is not guaranteed. The optimality condition for the latter would yield

$$A^* q^* \in \partial g(x^*), \tag{10.58}$$

which is indeed an abstract smoothness condition on the subdifferential of g at x^* if A and consequently A^* are smoothing operators, it is known as source condition in the field of inverse problems, cf. [31].

Due to the above reasons the use of iterative methods for solving respectively approximating (10.56) has a different flavor than iterative methods for well-posed problems. The key idea is to employ the algorithm as an iterative regularization method, cf. [105], where appropriate stopping in dependence on the noise, i.e. a distance between Ax^* and f, needs to be performed in order to obtain a suitable approximation. The notion to be used is called semiconvergence, i.e., if $\delta > 0$ denotes a measure for the data error (noise level) and $\hat{r}(\delta)$ is the stopping index of the iteration in dependence on δ, then we look for convergence

$$x^{(\hat{r}(\delta))} \to x^* \qquad \text{as } \delta \to 0, \tag{10.59}$$

in a suitable topology. The minimal ingredient in the convergence analysis is the convergence $x^{(r)} \to x^*$, which already needs different approaches as discussed above. For iterative methods working on primal variables one can at least use the existence of (10.56) in this case, while real primal-dual iterations still suffer from the potential nonexistence of solutions of the saddle point problem (10.57).
A well-understood iterative method is the Bregman iteration

$$x^{(r+1)} \in \operatorname*{argmin}_{x \in \mathcal{X}} \left(\frac{\mu}{2} \|Ax - f\|^2 + D_g^{p^{(r)}}(x, x^{(r)}) \right), \tag{10.60}$$

with $p^{(r)} \in \partial g(x^{(r)})$, which has been analyzed as an iterative method in [136], respectively for nonlinear A in [9]. Note that again with $p^{(r)} = A^* q^{(r)}$ the Bregman iteration is equivalent to the augmented Lagrangian method for the saddle-point problem (10.57). With such iterative regularization methods superior results compared to standard variational methods can be computed for inverse and imaging problems, in particular bias can be eliminated, cf. [31].
The key properties are the *decrease of the data fidelity*

$$\|Ax^{(r+1)} - f\|^2 \leq \|Ax^{(r)} - f\|^2, \tag{10.61}$$

for all r and the *decrease of the Bregman distance to the clean solution*

$$D_g^{p^{(r+1)}}(x^*, x^{(r+1)}) \leq D_g^{p^{(r)}}(x^*, x^{(r)}) \tag{10.62}$$

for those r such that

$$\|Ax^{(r)} - f\| \geq \|Ax^* - f\| = \delta.$$

Together with a more detailed analysis of the difference between consecutive Bregman distances, this can be used to prove semiconvergence results in appropriate weak topologies, cf. [136, 163]. In [9] further variants approximating the quadratic terms, such as the linearized Bregman iteration, are analyzed, however with further restrictions on g. For all other iterative methods discussed above, a convergence

analysis in the case of ill-posed problems is completely open and appears to be a valuable task for future research. Note that in the realization of the Bregman iteration, a well-posed but complex variational problem needs to be solved in each step. By additional operator splitting in an iterative regularization method one could dramatically reduce the computational effort.
If the source condition is satisfied, i.e. if there exists a saddle-point (x^*,q^*), one can further exploit the decrease of *dual distances*

$$\|q^{(r+1)} - q^*\| \leq \|q^{(r)} - q^*\| \tag{10.63}$$

to obtain a quantitative estimate on the convergence speed, we refer to [30, 32] for a further discussion.

8 Applications

So far we have focused on technical aspects of first order algorithms whose (further) development has been heavily forced by practical applications. In this section we give a rough overview of the use of first order algorithms in practice. We start with applications from classical imaging tasks such as computer vision and image analysis and proceed to applications in natural and life sciences. From the area of biomedical imaging, we will present the *Positron Emission Tomography (PET)* and *Spectral X-ray CT* in more detail and show some results reconstructed with first order algorithms.
At the beginning it is worth to emphasize that many algorithms based on proximal operators such as proximal point algorithm, proximal forward-backward splitting, ADMM, or Douglas-Rachford splitting have been introduced in the 1970s, cf. [83, 116, 146, 147]. However these algorithms have found a broad application in the last two decades, mainly caused by the technological progress. Due to the ability of distributed convex optimization with ADMM related algorithms, these algorithms seem to be qualified for 'big data' analysis and large-scale applications in applied statistics and machine learning, e.g., in areas as artificial intelligence, internet applications, computational biology and medicine, finance, network analysis, or logistics [27, 139]. Another boost for the popularity of first order splitting algorithms was the increasing use of sparsity-promoting regularizers based on ℓ_1- or L_1-type penalties [91, 151], in particular in combination with inverse problems considering nonlinear image formation models [9, 172] and/or statistically derived (inverse) problems [31]. The latter problems lead to non-quadratic fidelity terms which result from the non-Gaussian structure of the noise model. The overview given in the following mainly concentrates on the latter mentioned applications.
The most classical application of first order splitting algorithms is image analysis such as denoising, where these methods were originally pushed by the Rudin, Osher, and Fatemi (ROF) model [151]. This model and its variants are frequently used as prototypes for total variation methods in imaging to illustrate the applicability of

proposed algorithms in case of non-smooth cost functions, cf. [43, 70, 71, 91, 136, 164, 191]. Since the standard L_2 fidelity term is not appropriate for non-Gaussian noise models, modifications of the ROF problem have been considered in the past and were solved using splitting algorithms to denoise images perturbed by non-Gaussian noise, cf. [19, 43, 76, 153, 168]. Due to the close relation of total variation techniques to image segmentation [31, 140], first order algorithms have been also applied in this field of applications (cf. [42, 43, 90, 140]). Other image analysis tasks where proximal based algorithms have been applied successfully are deblurring and zooming (cf. [23, 43, 71, 166]), inpainting [43], stereo and motion estimation [46, 43, 188], and segmentation [10, 115, 140, 106].

Due to increasing noise level in modern biomedical applications, the requirement on statistical image reconstruction methods has been risen recently and the proximal methods have found access to many applied areas of biomedical imaging. Among the enormous amount of applications from the last two decades, we only give the following selection and further links to the literature:

- *X-ray CT:* Recently statistical image reconstruction methods have received increasing attention in X-ray CT due to increasing noise level encountered in modern CT applications such as sparse/limited-view CT and low-dose imaging, cf., e.g., [173, 178, 179], or K-edge imaging where the concentrations of K-edge materials are inherently low, see, e.g., [159, 158, 160]. In particular, first order splitting methods have received strong attention due to the ability to handle non-standard noise models and sparsity-promoting regularizers efficiently. Beside the classical fan-beam and cone-beam X-ray CT (see, e.g., [4, 45, 51, 104, 130, 145, 167, 173]), the algorithms have also found applications in emerging techniques such as spectral CT, see Section 8.2 and [85, 155, 185] or phase contrast CT [59, 131, 184].
- *Magnetic resonance imaging (MRI):* Image reconstruction in MRI is mainly achieved by inverting the Fourier transform which can be performed efficiently and robustly if a sufficient number of Fourier coefficients is measured. However, this is not the case in special applications such as fast MRI protocols, cf., e.g., [117, 142], where the Fourier space is undersampled so that the Nyquist criterion is violated and Fourier reconstructions exhibit aliasing artifacts. Thus, compressed sensing theory have found the way into MRI by exploiting sparsity-promoting variational approaches, see, e.g., [15, 103, 118, 144]. Furthermore, in advanced MRI applications such as velocity-encoded MRI or diffusion MRI, the measurements can be modeled more accurately by nonlinear operators and splitting algorithms provide the ability to handle the increased reconstruction complexity efficiently [172].
- *Emission tomography:* Emission tomography techniques used in nuclear medicine such as *positron emission tomography (PET)* and *single photon emission computed tomography (SPECT)* [183] are classical examples for inverse problems in biomedical imaging where statistical modeling of the reconstruction problem is essential due to Poisson statistics of the data. In addition, in cases where short time or low tracer dose measurements are available (e.g., using cardiac and/or respiratory gating [34]) or tracer with a short radioactive half-life are used (e.g., radioactive water $H_2{}^{15}O$ [157]), the measurements

suffer from inherently high noise level and thus a variety of first order splitting algorithms has been utilized in emission tomography, see, e.g., [4, 16, 122, 123, 143, 153].

- *Optical microscopy:* In modern light microscopy techniques such as *stimulated emission depletion (STED)* or *4Pi-confocal fluorescence microscopy* [99, 100] resolutions beyond the diffraction barrier can be achieved allowing imaging at nanoscales. However, by reaching the diffraction limit of light, measurements suffer from blurring effects and Poisson noise with low photon count rates [66, 162], in particular in live imaging and in high resolution imaging at nanoscopic scales. Thus, regularized (blind) deconvolution addressing appropriate Poisson noise is quite beneficial and proximal algorithms have been applied to achieve this goal, cf., e.g., [30, 81, 153, 166].
- *Other modalities:* It is quite natural that first order splitting algorithms have found a broad usage in biomedical imaging, in particular in such applications where the measurements are highly perturbed by noise and thus regularization with probably a proper statistical modeling are essential as, e.g., in optical tomography [1, 80], medical ultrasound imaging [154], hybrid photo-/optoacoustic tomography [84, 180], or electron tomography [92].

8.1 Positron Emission Tomography (PET)

PET is a biomedical imaging technique visualizing biochemical and physiological processes such as glucose metabolism, blood flow, or receptor concentrations, see, e.g., [183]. This modality is mainly applied in nuclear medicine and the data acquisition is based on weak radioactively marked pharmaceuticals (so-called tracers), which are injected into the blood circulation. Then bindings dependent on the choice of the tracer to the molecules are studied. Since the used markers are radio-isotopes, they decay by emitting a positron which annihilates almost immediately with an electron. The resulting emission of two photons is detected and, due to the radioactive decay, the measured data can be modeled as an inhomogeneous Poisson process with a mean given by the X-ray transform of the spatial tracer distribution (cf., e.g., [124, 174]). Note that, up to notation, the X-ray transform coincides with the more popular Radon transform in the two dimensional case [124]. Thus, the underlying reconstruction problem can be modeled as

$$\sum_{m=1}^{M} \big((Ku)_m - f_m \log((Ku)_m)\big) + \alpha R(u) \to \min_{u\geq 0}, \qquad \alpha > 0, \tag{10.64}$$

where M is the number of measurements, f are the given data, and K is the system matrix which describes the full properties of the PET data acquisition.

To solve (10.64), algorithms discussed above can be applied and several of them have been already studied for PET recently. In the following, we will give a (certainly incomplete) performance discussion of different first order splitting algorithms on synthetic PET data and highlight the strengths and weaknesses of them which could be carried over to many other imaging applications. For the study below, the total variation (TV) was applied as regularization energy R in (10.64) and the following algorithms and parameter settings were used for the performance evaluation:

- ***FB-EM-TV:*** The FB-EM-TV algorithm [153] represents an instance of the proximal forward-backward (FB) splitting algorithm discussed in Section 5.2 using a variable metric strategy (10.22). The preconditioned matrices $Q^{(r)}$ in (10.22) are chosen in a way that the gradient descent step corresponds to an expectation-maximization (EM) reconstruction step. The EM algorithm is a classically applied (iterative) reconstruction method in emission tomography [124, 174]. The TV proximal problem was solved by an adopted variant of the modified Arrow-Hurwicz method proposed in [43] since it was shown to be the most efficient method for TV penalized weighted least-squares denoising problems in [152]. Furthermore, a warm starting strategy was used to initialize the dual variables within the TV proximal problem and the inner iteration sequence was stopped if the relative error of primal and dual optimality conditions was below an error tolerance δ, i.e., using the notations from [43], if

$$\max\{d^{(r)}, p^{(r)}\} \leq \delta \tag{10.65}$$

  with

$$\begin{aligned} d^{(r)} &= \|(y^{(r)} - y^{(r-1)})/\sigma_{r-1} + \nabla(x^{(r)} - x^{(r-1)})\| \,/\, \|\nabla x^{(r)}\|, \\ p^{(r)} &= \|x^{(r)} - x^{(r-1)}\| \,/\, \|x^{(r)}\|. \end{aligned}$$

  The damping parameter $\eta^{(r)}$ in (10.22) was set to $\eta^{(r)} = 1$ as indicated in [153].
- ***FB-EM-TV-Nes83:*** A modified version of FB-EM-TV described above using the acceleration strategy proposed by Nesterov in [128]. This modification can be seen as a variant of FISTA [13] with a variable metric strategy (10.22). Here, $\eta^{(r)}$ in (10.22) was chosen fixed (i.e. $\eta^{(r)} = \eta$) but has to be adopted to the predefined inner accuracy threshold δ (10.65) to guarantee the convergence of the algorithm and it was to be done manually.
- ***CP-E:*** The fully explicit variant of the Chambolle-Pock's primal-dual algorithm [43] (cf. Section 6.3) studied for PET reconstruction problems in [3] (see *CP2TV* in [3]). The dual step size σ was set manually and the primal one corresponding to [43] as $\tau\sigma(\|\nabla\|^2 + \|K\|^2) = 1$, where $\|K\|$ was pre-estimated using the Power method.
- ***Precond-CP-E:*** The CP-E algorithm described above but using the diagonal preconditioning strategy proposed in [42] with $\alpha = 1$ in [42, Lemma 2].
- ***CP-SI:*** The semi-implicit variant of the Chambolle-Pock's primal-dual algorithm [43] (cf. Section 6.3) studied for PET reconstruction problems in [3] (see *CP1TV* in [3]). The difference to CP-E is that a TV proximal problem has to be solved in each iteration step. This was performed as in case of FB-EM-TV method. Furthermore, the dual step size σ was set manually and the primal one corresponding to [43] as $\tau\sigma\|K\|^2 = 1$, where $\|K\|$ was pre-estimated using the Power method.

- ***Precond-CP-SI:*** The CP-SI algorithm described above but using the diagonal preconditioning strategy proposed in [42] with $\alpha = 1$ in [42, Lemma 2].
- ***PIDSplit+:*** An ADMM based algorithm (cf. Section 6.2) that has been discussed for Poisson deblurring problems of the form (10.64) in [166]. However, in case of PET reconstruction problems, the solution of a linear system of equations of the form
$$(I + K^{\mathrm{T}}K + \nabla^{\mathrm{T}}\nabla)u^{(r+1)} = z^{(r)} \tag{10.66}$$
has to be computed in a different way in contrast to deblurring problems. This was done by running two preconditioned conjugate gradient (PCG) iterations with warm starting and cone filter preconditioning whose effectiveness has been validated in [145] for X-ray CT reconstruction problems. The cone filter was constructed as described in [73, 74] and diagonalized by the discrete cosine transform (DCT-II) supposing Neumann boundary conditions. The PIDSplit+ algorithm described above can be accomplished by a strategy of adaptive augmented Lagrangian parameters γ in (10.26) as proposed for the PIDSplit+ algorithm in [27, 169]. The motivation behind this strategy is to mitigate the performance dependency on the initial chosen fixed parameter that may strongly influence the speed of convergence of ADMM based algorithms.

All algorithms were implemented in MATLAB and executed on a machine with 4 CPU cores, each 2.83 GHz, and 7.73 GB physical memory, running a 64 bit Linux system and MATLAB 2013b. The built-in multi-threading in MATLAB was disabled such that all computations were limited to a single thread. The algorithms were evaluated on a simple object (image size 256×256) and the synthetic 2D PET measurements were obtained via a Monte-Carlo simulation with 257 radial and 256 angular samples, using one million simulated events (see Figure 10.2). Due to sustainable image and measurement dimensions, the system matrix K was precomputed for all reconstruction runs. To evaluate the performance of algorithms described above, the following procedure was applied. First, since K is injective and thus an unique minimizer of (10.64) is guaranteed [153], we can run a well-performing method for a very long time to compute a "ground truth" solution u_α^* for a fixed α. To this end, we have run the Precond-CP-E algorithm for 100,000 iterations for the following reasons: (1) all iteration steps can be solved exactly such that the solution cannot be influenced by inexact computations (see discussion below); (2) due to preconditioning strategy, no free parameters are available those may influence the speed of convergence negatively such that u_α^* is expected to be of high accuracy after 100,000 iterations. Having u_α^*, each algorithm was applied until the relative error
$$\|u_\alpha^{(r)} - u_\alpha^*\| / \|u_\alpha^*\| \tag{10.67}$$
was below a predefined threshold ε (or a maximum number of iterations adopted for each algorithm individually was reached). The "ground truth" solutions for three different values of α are shown in Figure 10.3.

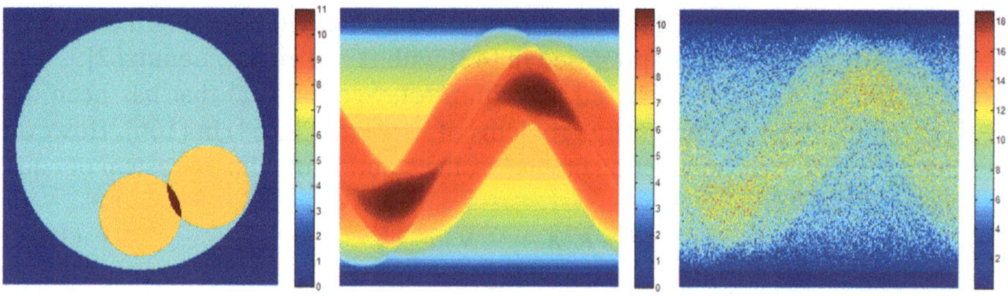

Fig. 10.2: Synthetic 2D PET data. *Left:* Exact object. *Middle:* Exact Radon data. *Right:* Simulated PET measurements via a Monte-Carlo simulation using one million events.

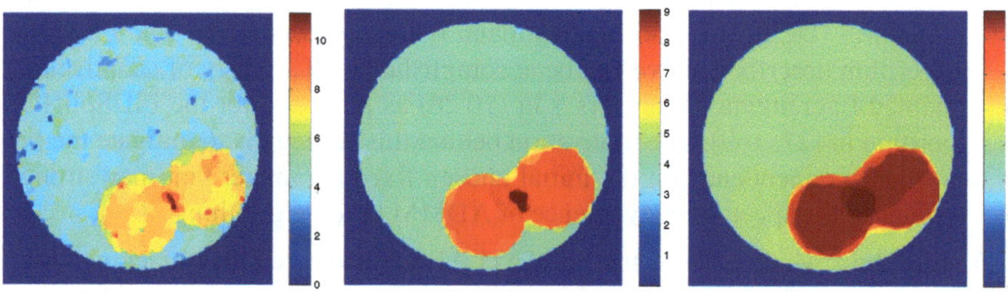

Fig. 10.3: The "ground truth" solutions for regularization parameter values $\alpha = 0.04$ (left), $\alpha = 0.08$ (middle), and $\alpha = 0.20$ (right).

Figures 10.4–10.8 show the performance evaluation of algorithms plotting the propagation of the relative error (10.67) in dependency on the number of iterations and CPU time in seconds. Since all algorithms have a specific set of unspecified parameters, different values of them are plotted to give a representative overall impression. The reason for showing the performance both in dependency on the number of iterations and CPU time is twofold: (1) in the presented case where the PET system matrix K is pre-computed and thus available explicitly, the evaluation of forward and backward projections is nearly negligible and TV relevant computations have the most contribution to the run time such that the CPU time will be a good indicator for algorithm's performance; (2) in practically relevant cases where the forward and backward projections have to be computed in each iteration step implicitly and in general are computationally consuming, the number of iterations and thus the number of projection evaluations will be the crucial factor for algorithm's efficiency. In the following, we individually discuss the behavior of algorithms observed for the regularization parameter $\alpha = 0.08$ (10.64) with the "ground truth" solution shown in Figure 10.3:

- ***FB-EM-TV(-Nes83):*** The evaluation of FB-EM-TV based algorithms is shown in Figure 10.4. The major observation for any δ in (10.65) is that the inexact computations of TV proximal problems lead to a restrictive approximation of the "ground truth" solution where the approximation accuracy stagnates after a specific number of iterations depending on δ. In addition, it can also be observed that the relative error (10.67) becomes better with more accurate TV proximal solutions (i.e., smaller δ) indicating that a decreasing sequence $\delta^{(r)}$ should be used to converge against the solution of (10.64) (see, e.g., [161, 175] for convergence analysis of inexact proximal gradient algorithms). However, as indicated in [119] and is shown in Figure 10.4, the choice of δ provides a trade-off between the approximation accuracy and computational cost such that the convergence rates proved in [161, 175] might be computationally not optimal. Another observation concerns the accelerated modification FB-EM-TV-Nes83. In Figure 10.4 we can observe that the performance of FB-EM-TV can actually be improved by FB-EM-TV-Nes83 regarding the number of iterations but only for smaller values of δ. One reason might be that using FB-EM-TV-Nes83 we have seen in our experiments that the gradient descent parameter $0 < \eta \leq 1$ [153] in (10.22) has to be chosen smaller with increased TV proximal accuracy (i.e., smaller δ). Since in such cases the effective regularization parameter value in each TV proximal problem is $\eta\alpha$, a decreasing η will result in poorer denoising properties increasing the inexactness of TV proximal operator. Recently, an (accelerated) inexact variable metric proximal gradient method was analyzed in [52] providing a theoretical view on such a type of methods.

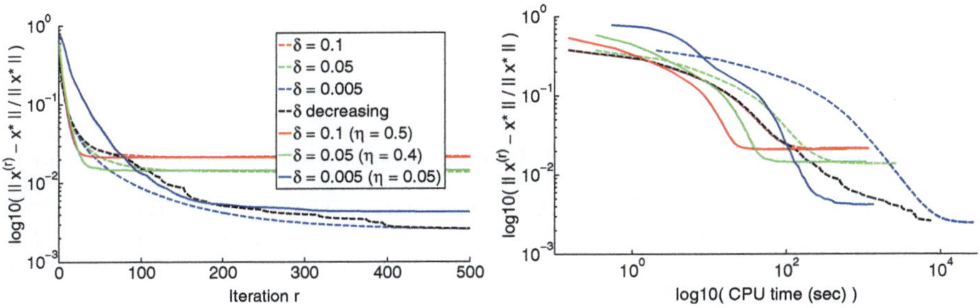

Fig. 10.4: Performance of FB-EM-TV (dashed lines) and FB-EM-TV-Nes83 (solid lines) for different accuracy thresholds δ (10.65) within the TV proximal step. Evaluation of relative error (10.67) is shown as a function of the number of iterations (left) and CPU time in seconds (right).

- ***(Precond-)CP-E:*** In Figure 10.5, the algorithms CP-E and Precond-CP-E are evaluated. In contrast to FB-EM-TV-(Nes83), the approximated solution cannot be influenced by inexact computations such that a decaying behavior of relative error can be observed. The single parameter that affects the convergence rate is the dual steplength σ and we observe in Figure 10.5(a) that some values yield a fast initial convergence (see, e.g., $\sigma = 0.05$ and $\sigma = 0.1$), but are less suited to achieve fast asymptotic convergence and vice versa (see, e.g., $\sigma = 0.3$ and $\sigma = 0.5$). However, the plots in Figure 10.5(b) indicate that $\sigma \in [0.2, 0.3]$ may provide an acceptable trade-off between initial and asymptotic convergence in terms of the number of iterations and CPU time. Regarding the latter mentioned aspect we note that in case of CP-E the more natural setting of σ would be $\sigma = \sqrt{\|\nabla\|^2 + \|K\|^2}$ what is approximately 0.29 in our experiments providing acceptable trade-off between initial and asymptotic convergence. Finally, no acceleration was observed in case of Precond-CP-E algorithm due to the regular structure of linear operators ∇ and K in our experiments such that the performance is comparable to CP-E with $\sigma = 0.5$ (see Figure 10.5(a)).

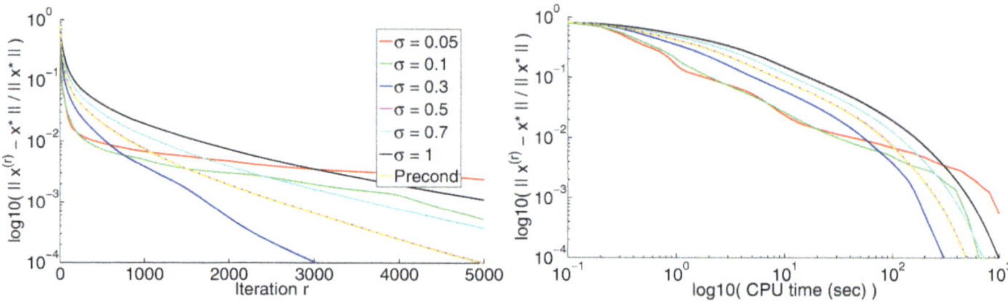

(a) Evaluation of relative error for fixed dual step sizes σ.

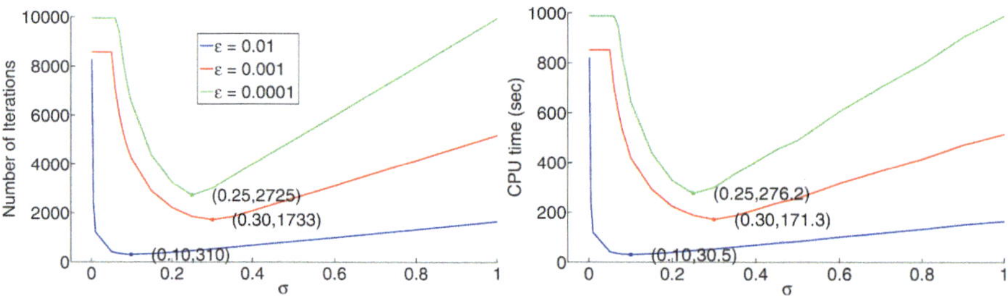

(b) Performance to get the relative error below the threshold ϵ as a function of dual step size σ.

Fig. 10.5: Performance of Precond-CP-E (dashed lines in ((a))) and CP-E (solid lines) for different dual step sizes σ. ((a)) Evaluation of relative error as a function of the number of iterations (left) and CPU time in seconds (right). ((b)) Required number of iterations (left) and CPU time (right) to get the relative error below a predefined threshold ε as a function of σ.

- ***(Precond-)CP-SI:*** In Figures 10.6 and 10.7, the evaluation of CP-SI and Precond-CP-SI is presented. Since a TV proximal operator has to be approximated in each iteration step, the same observations can be made as in case of FB-EM-TV that depending on δ the relative error stagnates after a specific number of iterations and that the choice of δ provides a trade-off between approximation accuracy and computational time (see Figure 10.6 for Precond-CP-SI). In addition, since the performance of CP-SI not only depends on δ but also on the dual steplength σ, the evaluation of CP-SI for different values of σ and two stopping values δ is shown in Figure 10.7. The main observation is that for smaller σ a better initial convergence can be achieved in terms of the number of iterations but results in less efficient performance regarding the CPU time. The reason is that the effective regularization parameter within the TV proximal problem is $\tau\alpha$ (see (10.22)) with $\tau = (\sigma\|K\|^2)^{-1}$ and a decreasing σ leads to an increasing TV denoising effort. Thus, in practically relevant cases, σ should be chosen optimally in a way balancing the required number of iterations and TV proximal computation.
- ***PIDSplit+:*** In Figure 10.8 the performance of PIDSplit+ is shown. It is well evaluated that the convergence of ADMM based algorithms is strongly depen-

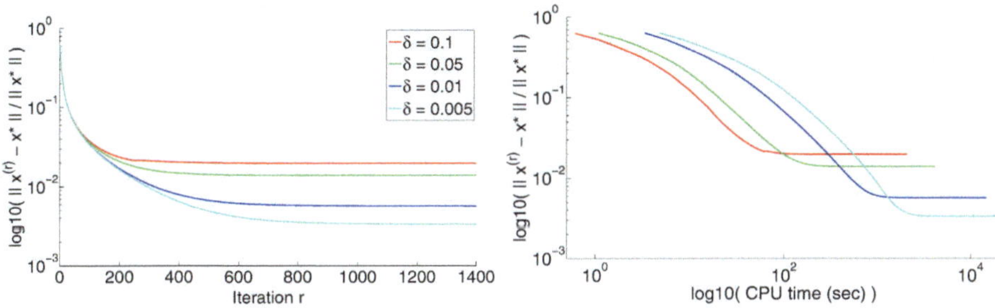

Fig. 10.6: Performance of Precond-CP-SI for different accuracy thresholds δ (10.65) within the TV proximal step. Evaluation of relative error as a function of number of iterations (left) and CPU time in seconds (right).

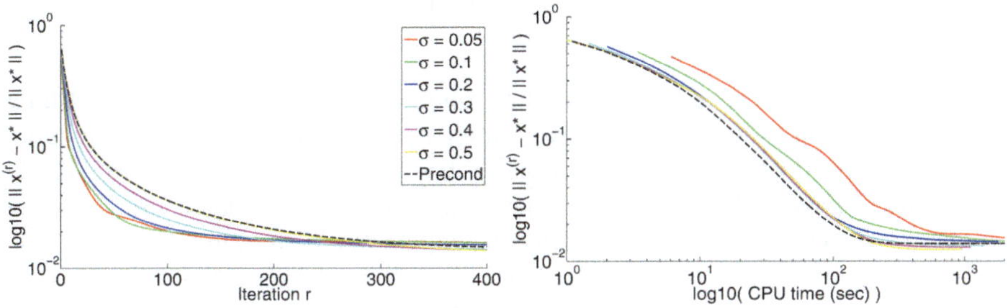

(a) Performance with TV proximal accuracy $\delta = 0.05$ (65).

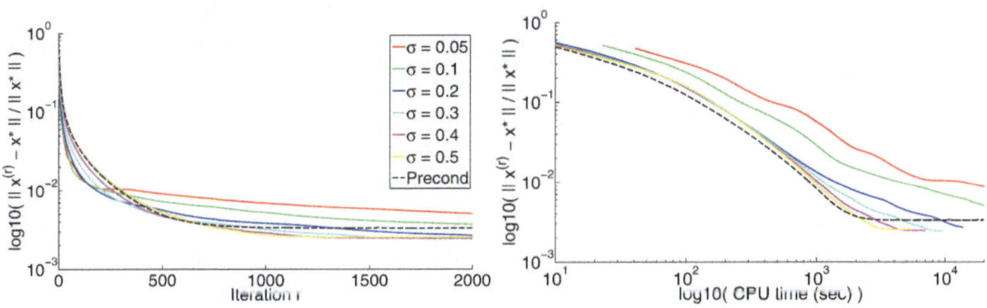

(b) Performance with TV proximal accuracy $\delta = 0.005$ (65).

Fig. 10.7: Performance of Precond-CP-SI (dashed lines) and CP-SI (solid lines) for different dual step sizes σ. Evaluation of relative error as a function of number of iterations (left) and CPU time in seconds (right) for accuracy thresholds $\delta = 0.05$ ((a)) and $\delta = 0.005$ ((b)) within the TV proximal problem.

dent on the augmented Lagrangian parameter γ (10.26) and that some values yield a fast initial convergence but are less suited to achieve a fast asymptotic convergence and vice versa. This behavior can also be observed in Figure 10.8 (see $\gamma = 30$ in upper row).

Finally, to get a feeling how the algorithms perform against each other, the required CPU time and number of projection evaluations to get the relative error (10.67)

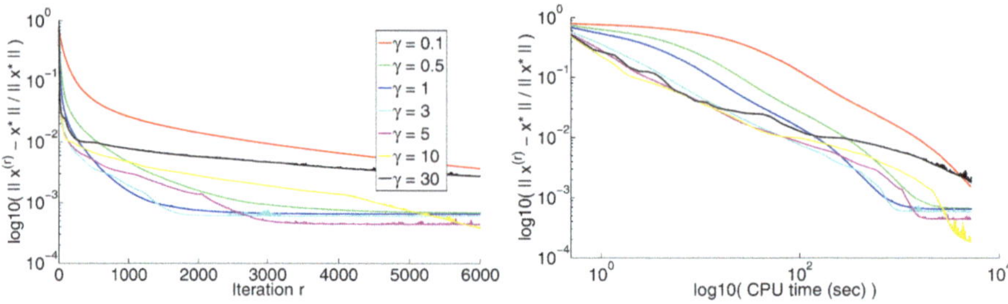

Fig. 10.8: Performance of PIDSplit+ for fixed augmented Lagrangian penalty parameters γ (10.26). Evaluation of relative error as a function of number of iterations (left) and CPU time in seconds (right).

below a predefined threshold are shown in Table 10.1 for two different values of ε. The following observations can be made:

- The FB-EM-TV based algorithms are competitive in terms of required number of projection evaluations but have a higher CPU time due to the computation of TV proximal operators, in particular the CPU time strongly grows with decreasing ε since TV proximal problems have to be approximated with increased accuracy. However, in our experiments, a fixed δ was used in each TV denoising step and thus the performance can be improved utilizing the fact that a rough accuracy is sufficient at the beginning of the iteration sequence without influencing the performance regarding the number of projector evaluations negatively (cf. Figure 10.4). Thus, a proper strategy to iteratively decrease δ in (10.65) can strongly improve the performance of FB-EM-TV based algorithms.
- The CP-E algorithm is optimal in our experiments in terms of CPU time since the TV regularization is computed by the shrinkage operator and thus is simply to evaluate. However, this algorithm needs almost the highest number of projection evaluations that will result in a slow algorithm in practically relevant cases where the projector evaluations are highly computationally expansive.
- The PIDSplit+ algorithm is slightly poorer in terms of CPU time than CP-E but required a smaller number of projector evaluations. However, we remind that this performance may probably be improved since two PCG iterations were used in our experiments and thus two forward and backward projector evaluations are required in each iteration step of PIDSplit+ method. Thus, if only one PCG step is used, the CPU time and number of projector evaluations can be decreased leading to a better performing algorithm. However, in the latter case, the total number of iteration steps might be increased since a poorer approximation of (10.66) will be performed if only one PCG step is used. Another opportunity to improve the performance of PIDSplit+ algorithm is to use the proximal ADMM strategy described in Section 6.4, namely, to remove $K^{\mathrm{T}}K$ from (10.66). That will result in only a single evaluation of forward and backward projectors in each iteration step but may lead to an increased number of total number of algorithm iterations.

Table 10.1: Performance evaluation of algorithms described above for $\alpha = 0.08$ (see u^*_α in Figure 10.3 (middle)). The table displays the CPU time in seconds and required number of forward and backward projector evaluations (K/K^{T}) to get the relative error (10.67) below the error tolerance ε. For each algorithm the best performance regarding the CPU time and K/K^{T} evaluations are shown where ıı means that the value coincides with the value directly above.

		$\varepsilon = 0.05$		$\varepsilon = 0.005$	
		K/K^{T}	CPU	K/K^{T}	CPU
FB-EM-TV	(best K/K^{T})	20	40.55	**168**	4999.71
ıı	(best CPU)	ıı	ıı	230	3415.74
FB-EM-TV-Nes83	(best K/K^{T})	**15**	14.68	231	308.74
ıı	(best CPU)	ıı	ıı	ıı	ıı
CP-E	(best K/K^{T})	48	**4.79**	696	**69.86**
ıı	(best CPU)	ıı	ıı	ıı	ıı
CP-SI	(best K/K^{T})	22	198.07	456	1427.71
ıı	(best CPU)	25	23.73	780	1284.56
PIDSplit+	(best K/K^{T})	30	7.51	698	179.77
ıı	(best CPU)	ıı	ıı	ıı	ıı

Finally, to study the algorithm's stability regarding the choice of regularization parameter α, we have run the algorithms for two additional values of α using the parameters shown the best performance in Table 10.1. The additional penalty parameters include a slightly under-smoothed and over-smoothed result respectively as shown in Figure 10.3 and the evaluation results are shown in Tables 10.2 and 10.3. In the following we describe the major observations:

- The FB-EM-TV method has the best efficiency in terms of projector evaluations, independently from the penalty parameter α, but has the disadvantage of solving a TV proximal problem in each iteration step which get harder to solve with increasing smoothing level (i.e., larger α) leading to a negative computational time. The latter observation holds also for the CP-SI algorithm. In case of a rough approximation accuracy (see Table 10.2), the FB-EM-TV-Nes83 scheme is able to improve the overall performance, respectively at least the computational time for higher accuracy in Table 10.3, but here the damping parameter η in (10.22) has to be chosen carefully to ensure the convergence (cf. Table 10.2 and 10.3 in case of $\alpha = 0.2$). Additionally based on Table 10.1, a proper choice of η is dependent not only on α but also on the inner accuracy of TV proximal problems.
- In contrast to FB-EM-TV and CP-SI, the remaining algorithms provide a superior computational time due to the solution of TV related steps by the shrinkage formula but show a strongly increased requirements on projector evaluations across all penalty parameters α. In addition, the performance of these algorithms is strongly dependent on the proper setting of free parameters (σ in case

of CP-E and γ in PIDSplit+) which unfortunately are able to achieve only a fast initial convergence or a fast asymptotic convergence. Thus different parameter settings of σ and γ were used in Tables 10.2 and 10.3.

Table 10.2: Performance evaluation for different values of α (see Figure 10.3). The table displays the CPU time in seconds and required number of forward and backward projector evaluations (K/K^{T}) to get the relative error (10.67) below the error tolerance $\varepsilon = 0.05$. For each α, the algorithms were run using the following parameters: FB-EM-TV ($\delta = 0.1$), FB-EM-TV-Nes83 ($\delta = 0.1$, $\eta = 0.5$), CP-E ($\sigma = 0.07$), CP-SI ($\delta = 0.1$, $\sigma = 0.05$), PIDSplit+ ($\gamma = 10$), which were chosen based on the "best" performance regarding K/K^{T} for $\varepsilon = 0.05$ in Table 10.1.

	$\alpha = 0.04$		$\alpha = 0.08$		$\alpha = 0.2$	
	K/K^{T}	CPU	K/K^{T}	CPU	K/K^{T}	CPU
FB-EM-TV	28	16.53	20	40.55	**19**	105.37
FB-EM-TV-Nes83	**17**	**5.26**	**15**	14.68	-	-
CP-E	61	6.02	48	**4.79**	51	**5.09**
CP-SI	-	-	25	23.73	21	133.86
PIDSplit+	32	8.08	30	7.51	38	9.7

Table 10.3: Performance evaluation for different values of α (see Figure 10.3) as in Table 10.2 but for $\varepsilon = 0.005$ and using the following parameters: FB-EM-TV ($\delta = 0.005$), FB-EM-TV-Nes83 ($\delta = 0.005$, $\eta = 0.05$), CP-E ($\sigma = 0.2$), CP-SI ($\delta = 0.005$, $\sigma = 0.3$), PIDSplit+ ($\gamma = 3$).

	$\alpha = 0.04$		$\alpha = 0.08$		$\alpha = 0.2$	
	K/K^{T}	CPU	K/K^{T}	CPU	K/K^{T}	CPU
FB-EM-TV	**276**	2452.14	**168**	4999.71	**175**	12612.7
FB-EM-TV-Nes83	512	222.98	231	308.74	-	-
CP-E	962	**94.57**	696	**69.86**	658	**65.42**
CP-SI	565	1117.12	456	1427.71	561	7470.18
PIDSplit+	932	239.35	698	179.77	610	158.94

8.2 Spectral X-Ray CT

Conventional X-ray CT is based on recording changes in the X-ray intensity due to attenuation of X-ray beams traversing the scanned object and has been applied in clinical practice for decades. However, the transmitted X-rays carry more information than just intensity changes since the attenuation of an X-ray depends strongly

on its energy [2, 109]. It is well understood that the transmitted energy spectrum contains valuable information about the structure and material composition of the imaged object and can be utilized to better distinguish different types of absorbing material, such as varying tissue types or contrast agents. But the detectors employing in traditional CT systems provide an integral measure of absorption over the transmitted energy spectrum and thus eliminate spectral information [39, 158]. Even so, spectral information can be obtained by using different input spectra [77, 193] or using the concept of dual-layer (integrating) detectors [40]. This has limited the practical usefulness of energy-resolving imaging, also referred to as spectral CT, to dual energy systems. Recent advances in detector technology towards binned photon-counting detectors have enabled a new generation of detectors that can measure and analyze incident photons individually [39] providing the availability of more than two spectral measurements. This development has led to a new imaging method named K-edge imaging [113] that can be used to selectively and quantitatively image contrast agents loaded with K-edge materials [75, 137]. For a compact overview on technical and practical aspects of spectral CT we refer to [39, 158].
Two strategies have been proposed to reconstruct material specific images from spectral CT projection data and we refer to [158] for a compact overview. Either of them is a projection-based material decomposition with a subsequent image reconstruction. This means that in the first step, estimates of material-decomposed sinograms are computed from the energy-resolved measurements, and in the second step, material images are reconstructed from the decomposed material sinograms. A possible decomposition method to estimate the material sinograms f_l, $l = 1,\ldots,L$, from the acquired data is a maximum-likelihood estimator assuming a Poisson noise distribution [150], where L is the number of materials considered. An accepted noise model for line integrals f_l is a multivariate Gaussian distribution [158, 159] leading to a penalized weighted least squares (PWLS) estimator to reconstruct material images u_l:

$$\frac{1}{2}\|f - (I_L \otimes K)u\|^2_{\Sigma^{-1}} + \alpha R(u) \to \min_u, \quad \alpha > 0, \tag{10.68}$$

where $f = (f_1^{\mathrm{T}},\ldots,f_L^{\mathrm{T}})^{\mathrm{T}}$, $u = (u_1^{\mathrm{T}},\ldots,u_L^{\mathrm{T}})^{\mathrm{T}}$, I_L denotes the $L \times L$ identity matrix, $\otimes$ represents the Kronecker product, and K is the forward projection operator. The given block matrix Σ is the covariance matrix representing the (multivariate) Gaussian distribution, where the off-diagonal block elements describe the inter-sinogram correlations, and can be estimated, e.g., using the inverse of the Fisher information matrix [149, 159]. Since f is computed from a common set of measurements, the correlation of the decomposed data is very high and thus a significant improvement can in fact be expected intuitively by exploiting the fully populated covariance matrix Σ in (10.68). In the following, we exemplary show reconstruction results on spectral CT data where (10.68) was solved by a proximal ADMM algorithm with a material independent total variation penalty function R as discussed in [155]. For a discussion why ADMM based methods are more preferable for PWLS problems in X-ray CT than, e.g., gradient descent based techniques, we refer to [145].

Figures 10.10 and 10.11 show an example for a statistical image reconstruction method applied to K-edge imaging. A numerical phantom as shown in Figure 10.9 was employed in a spectral CT simulation study assuming a photon-counting detector. Using an analytical spectral attenuation model, spectral measurements were computed. The assumed X-ray source spectrum and detector response function of the photon-counting detector were identical to those employed in a prototype spectral CT scanner described in [160]. The scan parameters were set to tube voltage 130 kVp, anode current 200 μA, detector width/height 946.38/1.14 mm, number of columns 1024, source-to-isocenter/-detector distance 570/1040 mm, views per turn 1160, time per turn 1 s, and energy thresholds to 25, 46, 61, 64, 76, and 91 keV. The spectral data were then decomposed into 'photo-electric absorption', 'Compton effect', and 'ytterbium' by performing a maximum-likelihood estimation [150]. By computing the covariance matrix Σ of the material decomposed sinograms via the Fisher information matrix [149, 159] and treating the sinograms as the mean and Σ as the variance of a Gaussian random vector, noisy material sinograms were computed. Figures 10.10 and 10.11 show material images that were then reconstructed using the traditional filtered backprojection (upper row) and proximal ADMM algorithm as described in [155] (middle and lower row). In the latter case, two strategies were performed: (1) keeping only the diagonal block elements of Σ in (10.68) and thus neglecting cross-correlations and decoupling the reconstruction of material images (middle row); (2) using the fully populated covariance matrix Σ in (10.68) such that all material images have to be reconstructed jointly (lower row). The results suggest, best visible in the K-edge images in Figure 10.11, that the iterative reconstruction method, which exploits knowledge of the inter-sinogram correlations, produces images that possess a better reconstruction quality. For comparison of iterative reconstruction strategies, the regularization parameters were chosen manually so the reconstructed images possessed approximately the same variance within the region indicated by the dotted circle in Figure 10.9. Further (preliminary) results that demonstrate advantages of exploiting inter-sinogram correlations on computer-simulated and experimental data in spectral CT can be found in [155, 189].

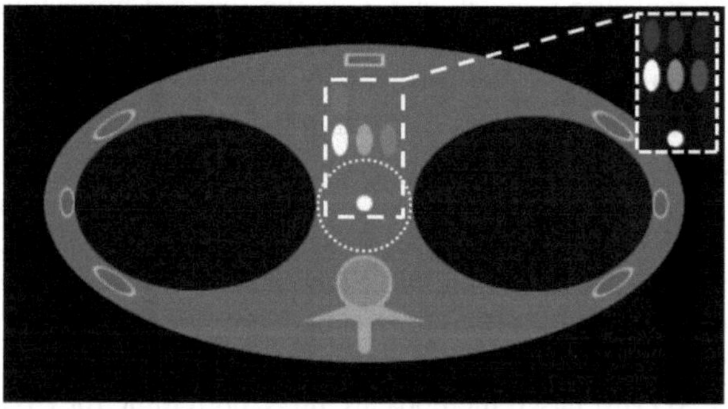

Fig. 10.9: Software thorax phantom comprising sternum, ribs, lungs, vertebrae, and one circle and six ellipsoids containing different concentrations of K-edge material ytterbium [137]. The phantom was used to simulate spectral CT measurements with a six-bin photon-counting detector.

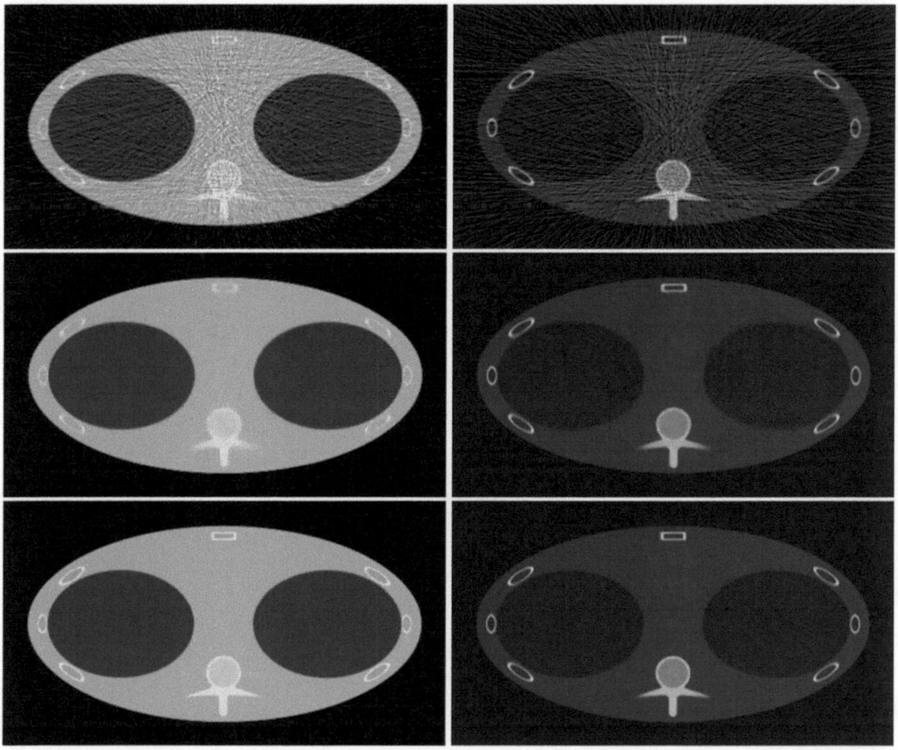

Fig. 10.10: Reconstructions based on the thorax phantom (see Figure 10.9) using the traditional filtered backprojection with Shepp-Logan filter (upper row) and a proximal ADMM algorithm as described in [155] (middle and lower row). The middle row shows results based on (10.68) neglecting cross-correlations between the material decomposed sinograms and lower row using the fully populated covariance matrix Σ. The material images show the results for the 'Compton effect' (left column) and 'photo-electric absorption' (right column). The K-edge material 'ytterbium' is shown in Figure 10.11.

Acknowledgements

The authors thank Frank Wübbeling (University of Münster, Germany) for providing the Monte-Carlo simulation for the synthetic 2D PET data. The authors also thank Thomas Koehler (Philips Technologie GmbH, Innovative Technologies, Hamburg, Germany) and Simon Setzer (G-Research, London) for comments that improved the manuscript and Eva-Maria Brinkmann (University of Münster, Germany) for the careful proofreading. The work on spectral CT was performed when A. Sawatzky was with the Computational Bioimaging Laboratory at Washington University in St. Louis, USA, and was supported in part by NIH award EB009715 and funded from Philips Research North America. G. Steidl's work was partially supported within the BMBF Project 05M13UKA.

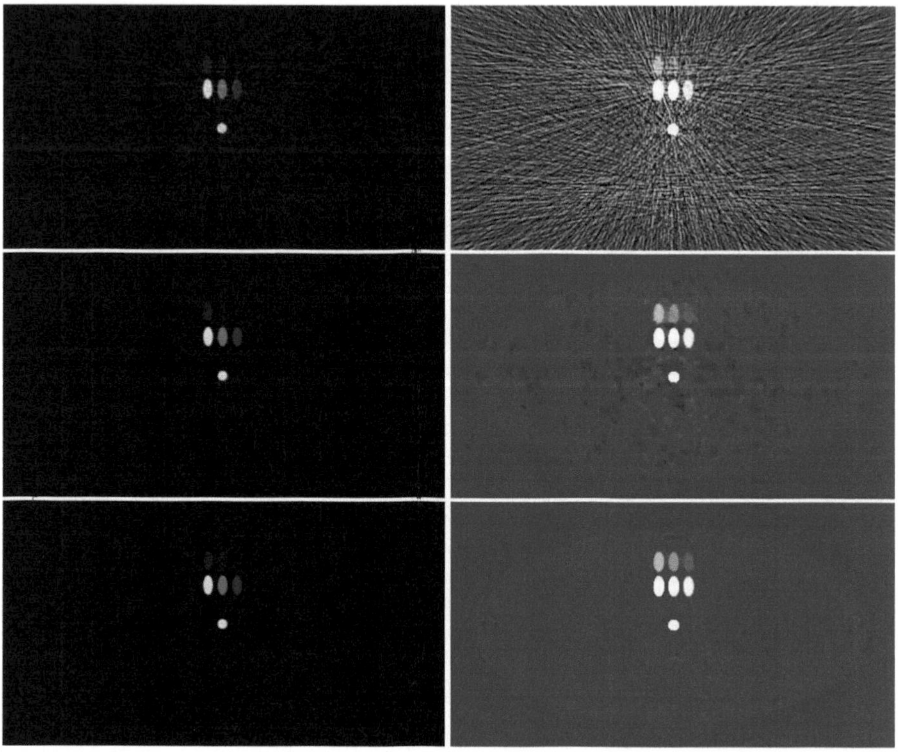

Fig. 10.11: Reconstructions of the K-edge material ‘ytterbium’ using the thorax phantom shown in Figure 10.9. For details see Figure 10.10. To recognize the differences, the maximal intensity value of original reconstructed images shown in left column was set down in the right column.

References

1. J. F. P.-J. Abascal, J. Chamorro-Servent, J. Aguirre, S. Arridge, T. Correia, J. Ripoll, J. J. Vaquero, and M. Desco. Fluorescence diffuse optical tomography using the split Bregman method. *Med. Phys.*, 38:6275, 2011.
2. R. E. Alvarez and A. Macovski. Energy-selective reconstructions in X-ray computerized tomography. *Phys. Med. Biol.*, 21(5):733–744, 1976.
3. S. Anthoine, J.-F. Aujol, Y. Boursier, and C. Mélot. On the efficiency of proximal methods in CBCT and PET. In *Proc. IEEE Int. Conf. Image Proc. (ICIP)*, 2011.
4. S. Anthoine, J.-F. Aujol, Y. Boursier, and C. Mélot. Some proximal methods for CBCT and PET. In *Proc. SPIE (Wavelets and Sparsity XIV)*, volume 8138, 2011.
5. K. J. Arrow, L. Hurwitz, and H. Uzawa. *Studies in Linear and Nonlinear Programming*. Stanford University Press, 1958.
6. H. Attouch and J. Bolte. On the convergence of the proximal algorithm for nonsmooth functions involving analytic features. *Math. Program.*, 116(1–2):5–16, 2009.
7. H. Attouch, J. Bolte, and B. F. Svaiter. Convergence of descent methods for semi-algebraic and tame problems: proximal algorithms, forward-backward splitting, and regularized Gauss-Seidel methods. *Math. Program. Series A*, 137(1–2):91–129, 2013.
8. J.-P. Aubin. *Optima and Equilibria: An Introduction to Nonlinear Analysis*. Springer, Berlin, Heidelberg, New York, 2nd edition, 2003.

9. M. Bachmayr and M. Burger. Iterative total variation schemes for nonlinear inverse problems. *Inverse Problems*, 25(10):105004, 2009.
10. E. Bae, J. Yuan, and X.-C. Tai. Global minimization for continuous multiphase partitioning problems using a dual approach. *International Journal of Computer Vision*, 92(1):112–129, 2011.
11. J. Barzilai and J. M. Borwein. Two-point step size gradient methods. *IMA J. Numer. Anal.*, 8(1):141–148, 1988.
12. H. H. Bauschke and P. L. Combettes. *Convex Analysis and Monotone Operator Theory in Hilbert Spaces*. Springer, New York, 2011.
13. A. Beck and M. Teboulle. Fast gradient-based algorithms for constrained total variation image denoising and deblurring. *SIAM J. Imag. Sci.*, 2:183–202, 2009.
14. S. Becker, J. Bobin, and E. J. Candès. NESTA: a fast and accurate first-order method for sparse recovery. *SIAM J. Imag. Sci.*, 4(1):1–39, 2011.
15. M. Benning, L. Gladden, D. Holland, C.-B. Schönlieb, and T. Valkonen. Phase reconstruction from velocity-encoded MRI measurement - A survey of sparsity-promoting variational approaches. *J. Magn. Reson.*, 238:26–43, 2014.
16. M. Benning, P. Heins, and M. Burger. A solver for dynamic PET reconstructions based on forward-backward-splitting. In *AIP Conf. Proc.*, volume 1281, pages 1967–1970, 2010.
17. D. P. Bertsekas. *Constrained Optimization and Lagrange Multiplier Methods*. Academic Press, New York, 1982.
18. D. P. Bertsekas. Incremental proximal methods for large scale convex optimization. *Math. Program., Ser. B*, 129(2):163–195, 2011.
19. J. M. Bioucas-Dias and M. A. T. Figueiredo. Multiplicative noise removal using variable splitting and constrained optimization. *IEEE Trans. Image Process.*, 19(7):1720–1730, 2010.
20. A. Björck. *Least Squares Problems*. SIAM, Philadelphia, 1996.
21. D. Boley. Local linear convergence of the alternating direction method of multipliers on quadratic or linear programs. *SIAM J. Optim.*, 2014.
22. J. Bolte, S. Sabach, and M. Teboulle. Proximal alternating linearized minimization for nonconvex and nonsmooth problems. *Math. Program., Series A*, 2013.
23. S. Bonettini and V. Ruggiero. On the convergence of primal-dual hybrid gradient algorithms for total variation image restoration. *J. Math. Imaging Vis.*, 44:236–253, 2012.
24. J. F. Bonnans, J. C. Gilbert, C. Lemaréchal, and C. A. Sagastizábal. A family of variable metric proximal methods. *Mathematical Programming*, 68:15–47, 1995.
25. R. I. Boţ and C. Hendrich. A Douglas-Rachford type primal-dual method for solving inclusions with mixtures of composite and parallel-sum type monotone operators. *SIAM Journal on Optimization*, 23(4):2541–2565, 2013.
26. R. I. Boţ and C. Hendrich. Convergence analysis for a primal-dual monotone + skew splitting algorithm with applications to total variation minimization. *Journal of Mathematical Imaging and Vision*, 49(3):551–568, 2014.
27. S. Boyd, N. Parikh, E. Chu, B. Peleato, and J. Eckstein. Distributed optimization and statistical learning via the alternating direction method of multipliers. *Foundations and Trends in Machine Learning*, 3(1):101–122, 2011.
28. K. Bredies, K. Kunisch, and T. Pock. Total generalized variation. *SIAM Journal on Imaging Sciences*, 3(3):492–526, 2010.
29. F. E. Browder and W. V. Petryshyn. The solution by iteration of nonlinear functional equations in Banach spaces. *Bulletin of the American Mathematical Society*, 72:571–575, 1966.
30. C. Brune, A. Sawatzky, and M. Burger. Primal and dual Bregman methods with application to optical nanoscopy. *International Journal of Computer Vision*, 92(2):211–229, 2010.
31. M. Burger and S. Osher. A guide to the TV zoo. In *Level Set and PDE Based Reconstruction Methods in Imaging*, pages 1–70. Springer, 2013.
32. M. Burger, E. Resmerita, and L. He. Error estimation for Bregman iterations and inverse scale space methods in image restoration. *Computing*, 81(2–3):109–135, 2007.
33. J. V. Burke and M. Qian. A variable metric proximal point algorithm for monotone operators. *SIAM Journal on Control and Optimization*, 37:353–375, 1999.

34. F. Büther, M. Dawood, L. Stegger, F. Wübbeling, M. Schäfers, O. Schober, and K. P. Schäfers. List mode-driven cardiac and respiratory gating in PET. *J. Nucl. Med.*, 50(5):674–681, 2009.
35. D. Butnariu and A. N. Iusem. *Totally convex functions for fixed points computation and infinite dimensional optimization*, volume 40 of *Applied Optimization*. Kluwer, Dordrecht, 2000.
36. C. Byrne. A unified treatment of some iterative algorithms in signal processing and image reconstruction. *Inverse Problems*, 20:103–120, 2004.
37. J.-F. Cai, E. J. Candès, and Z. Shen. A singular value thresholding algorithm for matrix completion. *SIAM Journal on Optimization*, 20(4):1956–1982, 2010.
38. J.-F. Cai, S. Osher, and Z. Shen. Convergence of the linearized Bregman iteration for ℓ_1-norm minimization. *Mathematics of Computation*, 78(268):2127–2136, 2009.
39. J. Cammin, J. S. Iwanczyk, and K. Taguchi. *Emerging Imaging Technologies in Medicine*, chapter Spectral/Photo-Counting Computed Tomography, pages 23–39. CRC Press, 2012.
40. R. Carmi, G. Naveh, and A. Altman. Material separation with dual-layer CT. In *Proc. IEEE Nucl. Sci. Symp. Conf. Rec.*, pages 1876–1878, 2005.
41. A. Chambolle, V. Caselles, D. Cremers, M. Novaga, and T. Pock. An introduction to total variation for image analysis. In *Theoretical Foundations and Numerical Methods for Sparse Recovery*, volume 9 of *Radon Series Compl. Appl. Math.*, pages 263–340. Walter de Gruyter, 2010.
42. A. Chambolle and T. Pock. Diagonal preconditioning for first order primal-dual algorithms in convex optimization. *ICCV*, pages 1762–1769, 2011.
43. A. Chambolle and T. Pock. A first-order primal-dual algorithm for convex problems with applications to imaging. *Journal of Mathematical Imaging and Vision*, 40(1):120–145, 2011.
44. T. F. Chan and R. Glowinski. Finite element approximation and iterative solution of a class of mildly non-linear elliptic equations. Technical report, STAN-CS-78-674, Stanford University, 1978.
45. R. Chartrand, E. Y. Sidky, and X. Pan. Nonconvex compressive sensing for X-ray CT: an algorithm comparison. In *Asilomar Conference on Signals, Systems, and Computers*, 2013.
46. C. Chaux, M. El-Gheche, J. Farah, J. Pesquet, and B. Popescu. A parallel proximal splitting method for disparity estimation from multicomponent images under illumination variation. *Journal of Mathematical Imaging and Vision*, 47(3):1–12, 2012.
47. C. H. Chen, B. S. He, Y. Y. Ye, and X. M. Yuan. The direct extension of ADMM for multi-block convex minimization problems is not necessarily convergent. *Mathematical Programming*, to appear.
48. G. Chen and M. Teboulle. A proximal-based decomposition method for convex minimization problems. *Mathematical Programming*, 64:81–101, 1994.
49. G. H.-G. Chen and R. T. Rockafellar. Convergence rates in forward-backward splitting. *SIAM Journal on Optimization*, 7:421–444, 1997.
50. G. Chierchia, N. Pustelnik, J.-C. Pesquet, and B. Pesquet-Popescu. Epigraphical projection and proximal tools for solving constrained convex optimization problems: Part I. arXiv preprint arXiv:1210.5844 (2014).
51. K. Choi, J. Wang, L. Zhu, T.-S. Suh, S. Boyd, and L. Xing. Compressed sensing based cone-beam computed tomography reconstruction with a first-order method. *Med. Phys.*, 37(9):5113–5125, 2010.
52. E. Chouzenoux, J.-C. Pesquet, and A. Repetti. Variable metric forward-backward algorithm for minimizing the sum of a differentiable function and a convex function. *J. Optim. Theory Appl.*, 2013.
53. P. Combettes and J.-C. Pesquet. Proximal splitting methods in signal processing. In *Fixed-Point Algorithms for Inverse Problems in Science and Engineering*, pages 185–212. Springer, 2011.
54. P. Combettes and J.-C. Pesquet. Primal-dual splitting algorithm for solving inclusions with mixture of composite, Lipschitzian, and parallel-sum type monotone operators. *Set-Valued and Variational Analysis*, 20(2):307–330, 2012.

55. P. L. Combettes. Solving monotone inclusions via compositions of nonexpansive averaged operators. *Optimization*, 53(5–6):475–504, 2004.
56. P. L. Combettes and J.-C. Pesquet. Proximal thresholding algorithm for minimization over orthonormal bases. *SIAM Journal on Optimization*, 18(4):1351–1376, 2007.
57. P. L. Combettes and B. C. Vu. Variable metric forward-backward splitting with applications to monotone inclusions in duality. *Optimization*, pages 1–30, 2012.
58. L. Condat. A primal-dual splitting method for convex optimization involving Lipschitzian, proximable and linear composite terms. *J. Optim. Theory Appl.*, 158(2):460–479, 2013.
59. W. Cong, J. Yang, and G. Wang. Differential phase-contrast interior tomography. *Phys. Med. Biol.*, 57:2905–2914, 2012.
60. J. Dahl, P. J. Hansen, S. H. Jensen, and T. L. Jensen. Algorithms and software for total variation image reconstruction via first order methods. Numerical Algorithms, 53:67, 2010.
61. I. Daubechies, M. Defrise, and C. De Mol. An iterative thresholding algorithm for linear inverse problems with a sparsity constraint. *Communications on Pure and Applied Mathematics*, 51:1413–1541, 2004.
62. I. Daubechies, M. Fornasier, and I. Loris. Accelerated projected gradient methods for linear inverse problems with sparsity constraints. *The Journal of Fourier Analysis and Applications*, 14(5–6):764–792, 2008.
63. C. Davis. All convex invariant functions of Hermitian matrices. *Archive in Mathematics*, 8:276–278, 1957.
64. D. Davis and W. Yin. Convergence rate analysis of several splitting schemes. In: R. Glowinski, S. Osher, W. Yin (eds.) Splitting Methods in Communication and Imaging, Science and Engineering. Springer, 2016.
65. D. Davis and W. Yin. Faster convergence rates of relaxed Peaceman-Rachford and ADMM under regularity assumptions. *ArXiv Preprint 1407.5210*, 2014.
66. N. Dey, L. Blanc-Feraud, C. Zimmer, P. Roux, Z. Kam, J.-C. Olivo-Marin, and J. Zerubia. Richardson-Lucy algorithm with total variation regularization for 3D confocal microscope deconvolution. *Microsc. Res. Tech.*, 69:260–266, 2006.
67. J. Duchi, S. Shalev-Shwartz, Y. Singer, and T. Chandra. Efficient projections onto the l1-ball for learning in high dimensions. In *ICML '08 Proceedings of the 25th International Conference on Machine Learning*, ACM New York, 2008.
68. J. Eckstein and D. P. Bertsekas. An alternating direction method for linear programming. *Tech. Report MIT Lab. for Info. and Dec. Sys.*, 1990.
69. J. Eckstein and D. P. Bertsekas. On the Douglas-Rachford splitting method and the proximal point algorithm for maximal monotone operators. *Mathematical Programming*, 55:293–318, 1992.
70. E. Esser. Applications of Lagrangian-based alternating direction methods and connections to split Bregman. Technical report, UCLA Computational and Applied Mathematics, March 2009.
71. E. Esser, X. Zhang, and T. F. Chan. A general framework for a class of first order primal-dual algorithms for convex optimization in imaging science. *SIAM J Imag Sci*, 3(4):1015–1046, 2010.
72. F. Facchinei and J.-S. Pang. *Finite-Dimensional Variational Inequalities and Complementarity Problems*, volume II. Springer, New York, 2003.
73. J. A. Fessler. Conjugate-gradient preconditioning methods: Numerical results. Technical Report 303, Commun. Signal Process. Lab., Dept. Elect. Eng. Comput. Sci., Univ. Michigan, Ann Arbor, MI, Jan. 1997. available from http://web.eecs.umich.edu/~fessler/.
74. J. A. Fessler and S. D. Booth. Conjugate-gradient preconditioning methods for shift-variant PET image reconstruction. *IEEE Trans. Image Process.*, 8(5):688–699, 1999.
75. S. Feuerlein, E. Roessl, R. Proksa, G. Martens, O. Klass, M. Jeltsch, V. Rasche, H.-J. Brambs, M. H. K. Hoffmann, and J.-P. Schlomka. Multienergy photon-counting K-edge imaging: Potential for improved luminal depiction in vascular imaging. *Radiology*, 249(3):1010–1016, 2008.

76. M. Figueiredo and J. Bioucas-Dias. Deconvolution of Poissonian images using variable splitting and augmented Lagrangian optimization. In *IEEE Workshop on Statistical Signal Processing*, Cardiff, 2009.
77. T. G. Flohr, C. H. McCollough, H. Bruder, M. Petersilka, K. Gruber, C. Süß, M. Grasruck, K. Stierstorfer, B. Krauss, R. Raupach, A. N. Primak, A. Küttner, S. Achenbach, C. Becker, A. Kopp, and B. M. Ohnesorge. First performance evaluation of a dual-source CT (DSCT) system. *Eur. Radiol.*, 16:256–268, 2006.
78. M. Fornasier. *Theoretical Foundations and Numerical Methods for Sparse Recovery*, volume 9. Walter de Gruyter, 2010.
79. G. Frassoldati, L. Zanni, and G. Zanghirati. New adaptive stepsize selections in gradient methods. *Journal of Industrial and Management Optimization*, 4(2):299–312, 2008.
80. M. Freiberger, C. Clason, and H. Scharfetter. Total variation regularization for nonlinear fluorescence tomography with an augmented Lagrangian splitting approach. *Appl. Opt.*, 49(19):3741–3747, 2010.
81. K. Frick, P. Marnitz, and A. Munk. Statistical multiresolution estimation for variational imaging: With an application in Poisson-biophotonics. *J. Math. Imaging Vis.*, 46:370–387, 2013.
82. D. Gabay. Applications of the method of multipliers to variational inequalities. In M. Fortin and R. Glowinski, editors, *Augmented Lagrangian Methods: Applications to the Solution of Boundary Value Problems*, chapter IX, pages 299–340. North-Holland, Amsterdam, 1983.
83. D. Gabay and B. Mercier. A dual algorithm for the solution of nonlinear variational problems via finite element approximations. *Computer and Mathematics with Applications*, 2:17–40, 1976.
84. H. Gao, S. Osher, and H. Zhao. *Mathematical Modeling in Biomedical Imaging II: Optical, Ultrasound, and Opto-Acoustic Tomographies*, chapter Quantitative Photoacoustic Tomography, pages 131–158. Springer, 2012.
85. H. Gao, H. Yu, S. Osher, and G. Wang. Multi-energy CT based on a prior rank, intensity and sparsity model (PRISM). *Inverse Problems*, 27(11):115012, 2011.
86. R. Glowinski. On alternating direction methods of multipliers: a historical perspective. In *Modeling, Simulation and Optimization for Science and Technology*, pages 59–82. Springer, 2014.
87. R. Glowinski and P. Le Tallec. *Augmented Lagrangian and Operator-Splitting Methods in Nonlinear Mechanics*, volume 9 of *SIAM Studies in Applied and Numerical Mathematics*. SIAM, Philadelphia, 1989.
88. R. Glowinski and A. Marroco. Sur l'approximation, par éléments finis d'ordre un, et la résolution, par pénalisation-dualité d'une classe de problèmes de Dirichlet non linéaires. *Revue française d'automatique, informatique, recherche opérationnelle. Analyse numérique*, 9(2):41–76, 1975.
89. D. Goldfarb and K. Scheinberg. Fast first-order methods for composite convex optimization with line search. *SIAM Journal on Imaging Sciences*, 2011.
90. T. Goldstein, X. Bresson, and S. Osher. Geometric applications of the split Bregman method: Segmentation and surface reconstruction. *J. Sci. Comput.*, 45:272–293, 2010.
91. T. Goldstein and S. Osher. The split Bregman method for L1-regularized problems. *SIAM Journal on Imaging Sciences*, 2(2):323–343, 2009.
92. B. Goris, W. Van den Broek, K. J. Batenburg, H. H. Mezerji, and S. Bals. Electron tomography based on a total variation minimization reconstruction technique. *Ultramicroscopy*, 113:120–130, 2012.
93. O. Güler. New proximal point algorithms for convex minimization. *SIAM J. Optim.*, 2(4):649–664, 1992.
94. S. Harizanov, J.-C. Pesquet, and G. Steidl. Epigraphical projection for solving least squares Anscombe transformed constrained optimization problems. In A. K. et al., editor, *Scale-Space and Variational Methods in Computer Vision. Lecture Notes in Computer Science, SSVM 2013, LNCS 7893*, pages 125–136, Berlin, 2013. Springer.
95. B. He, L.-Z. Liao, D. Han, and H. Yang. A new inexact alternating directions method for monotone variational inequalities. *Math. Program., Ser. A*, 92(1):103–118, 2002.

96. B. He and H. Yang. Some convergence properties of the method of multiplieres for linearly constrained monotone variational operators. *Operation Research Letters*, 23:151–161, 1998.
97. B. He and X. Yuan. On the $\mathcal{O}(1/n)$ convergence rate of the Douglas-Rachford alternating direction method. *SIAM Journal on Numerical Analysis*, 2:700–709, 2012.
98. B. S. He, H. Yang, and S. L. Wang. Alternating direction method with self-adaptive penalty parameters for monotone variational inequalties. *J. Optimiz. Theory App.*, 106(2):337–356, 2000.
99. S. W. Hell. Toward fluorescence nanoscopy. *Nat. Biotechnol.*, 21(11):1347–1355, 2003.
100. S. W. Hell. Far-field optical nanoscopy. *Science*, 316(5828):1153–1158, 2007.
101. M. R. Hestenes. Multiplier and gradient methods. *Journal of Optimization Theory and Applications*, 4:303–320, 1969.
102. M. Hong and Z. Q. Luo. On linear convergence of the alternating direction method of multipliers. *Arxiv preprint 1208.3922*, 2012.
103. J. Huang, S. Zhang, and D. Metaxas. Efficient MR image reconstruction for compressed MR imaging. *Med. Image Anal.*, 15:670–679, 2011.
104. X. Jia, Y. Lou, R. Li, W. Y. Song, and S. B. Jiang. GPU-based fast cone beam CT reconstruction from undersampled and noisy projection data via total variation. *Med. Phys.*, 37(4):1757–1760, 2010.
105. B. Kaltenbacher, A. Neubauer, and O. Scherzer. *Iterative regularization methods for nonlinear ill-posed problems*, volume 6. Walter de Gruyter, 2008.
106. S. H. Kang, B. Shafei, and G. Steidl. Supervised and transductive multi-class segmentation using p-Laplacians and RKHS methods. *J. Visual Communication and Image Representation*, 25(5):1136–1148, 2014.
107. K. C. Kiwiel. Free-steering relaxation methods for problems with strictly convex costs and linear constraints. *Mathematics of Operations Research*, 22(2):326–349, 1997.
108. K. C. Kiwiel. Proximal minimization methods with generalized Bregman functions. *SIAM Journal on Control and Optimization*, 35(4):1142–1168, 1997.
109. G. F. Knoll. *Radiation Detection and Measurement*. Wiley, 3rd edition, 2000.
110. N. Komodakis and J.-C. Pesquet. Playing with duality: an overview of recent primal-dual approaches for solving large-scale optimization problems. *ArXiv Preprint arXiv:1406.5429*, 2014.
111. S. Kontogiorgis and R. R. Meyer. A variable-penalty alternating directions method for convex optimization. *Math. Program.*, 83(1–3):29–53, 1998.
112. M. A. Krasnoselskii. Two observations about the method of successive approximations. *Uspekhi Matematicheskikh Nauk*, 10:123–127, 1955. In Russian.
113. R. A. Kruger, S. J. Riederer, and C. A. Mistretta. Relative properties of tomography, K-edge imaging, and K-edge tomography. *Med. Phys.*, 4(3):244–249, 1977.
114. M.-J. Lai and W. Yin. Augmented ℓ_1 and nuclear-norm models with a globally linearly convergent algorithm. *SIAM Journal on Imaging Sciences*, 6(2):1059–1091, 2013.
115. J. Lellmann, J. Kappes, J. Yuan, F. Becker, and C. Schnörr. Convex multi-class image labeling with simplex-constrained total variation. In X.-C. Tai, K. Morken, M. Lysaker, and K.-A. Lie, editors, *Scale Space and Variational Methods, volume 5567 of LNCS*, volume 5567 of *Lecture Notes in Computer Science*, pages 150–162. Springer, 2009.
116. P. L. Lions and B. Mercier. Splitting algorithms for the sum of two linear operators. *SIAM Journal on Numerical Analysis*, 16:964–976, 1979.
117. M. Lustig, D. Donoho, and J. M. Pauly. Sparse MRI: The application of compressed sensing for rapid MR imaging. *Magn. Reson. Med.*, 58:1182–1195, 2007.
118. S. Ma, W. Y. Y. Zhang, and A. Chakraborty. An efficient algorithm for compressed MR imaging using total variation and wavelets. In *Proc. IEEE Comput. Vision Pattern Recognit.*, 2008.
119. P. Machart, S. Anthoine, and L. Baldassarre. Optimal computational trade-off of inexact proximal methods. arXiv preprint arXiv:1210.5034, 2012.
120. W. R. Mann. Mean value methods in iteration. *Proceedings of the American Mathematical Society*, 16(4):506–510, 1953.

121. B. Martinet. Régularisation d'inéquations variationnelles par approximations successives. *Revue Française d'Informatique et de Recherche Operationelle*, 4(3):154–158, 1970.
122. A. Mehranian, A. Rahmim, M. R. Ay, F. Kotasidis, and H. Zaidi. An ordered-subsets proximal preconditioned gradient algorithm for edge-preserving PET image reconstruction. *Med. Phys.*, 40(5):052503, 2013.
123. J. Müller, C. Brune, A. Sawatzky, T. Kösters, F. Wübbeling, K. Schäfers, and M. Burger. Reconstruction of short time PET scans using Bregman iterations. In *Proc. IEEE Nucl. Sci. Symp. Conf. Rec.*, 2011.
124. F. Natterer and F. Wübbeling. *Mathematical Methods in Image Reconstruction*. SIAM, 2001.
125. A. S. Nemirovsky and D. B. Yudin. *Problem Complexity and Method Efficiency in Optimization*. J. Wiley & Sons, Ltd., 1983.
126. Y. Nesterov. *Introductory Lectures on Convex Optimization - A Basic Course*, volume 87 of *Applied Optimization*. Springer US, 2004.
127. Y. Nesterov. Gradient methods for minimizing composite functions. *Math. Program., Series B*, 140(1):125–161, 2013.
128. Y. E. Nesterov. A method of solving a convex programming problem with convergence rate $O(1/k^2)$. *Soviet Mathematics Doklady*, 27(2):372–376, 1983.
129. Y. E. Nesterov. Smooth minimization of non-smooth functions. *Mathematical Programming*, 103:127–152, 2005.
130. H. Nien and J. A. Fessler. Fast X-ray CT image reconstruction using the linearized augmented Lagrangian method with ordered subsets. arXiv preprint arXiv:1402.4381, 2014.
131. M. Nilchian, C. Vonesch, P. Modregger, M. Stampanoni, and M. Unser. Fast iterative reconstruction of differential phase contrast X-ray tomograms. *Optics Express*, 21(5):5511–5528, 2013.
132. P. Ochs, Y. Chen, T. Brox, and T. Pock. iPiano: Inertial proximal algorithm for nonconvex optimization. *SIAM J Imaging Sci*, 7(2):1388–1419, 2014.
133. Z. Opial. Weak convergence of a sequence of successive approximations for nonexpansive mappings. *Bulletin of the American Mathematical Society*, 73:591–597, 1967.
134. J. M. Ortega and W. C. Rheinboldt. *Iterative Solution of Nonlinear Equations in Several Variables*. SIAM, New York, 1970.
135. S. Osher, M. Burger, D. Goldfarb, J. Xu, and W. Yin. An iterative regularization method for the total variation based image restoration. *Multiscale Modeling and Simulation*, 4:460–489, 2005.
136. S. Osher, M. Burger, D. Goldfarb, J. Xu, and W. Yin. An iterative regularization method for total variation-based image restoration. *Multiscale Modeling & Simulation*, 4(2):460–489, 2005.
137. D. Pan, C. O. Schirra, A. Senpan, A. H. Schmieder, A. J. Stacy, E. Roessl, A. Thran, S. A. Wickline, R. Proksa, and G. M. Lanza. An early investigation of ytterbium nanocolloids for selective and quantitative "multicolor" spectral CT imaging. *ACS Nano*, 6(4):3364–3370, 2012.
138. L. A. Parente, P. A. Lotito, and M. V. Solodov. A class of inexact variable metric proximal point algorithms. *SIAM J. Optim.*, 19(1):240–260, 2008.
139. N. Parikh and S. Boyd. Proximal algorithms. *Foundations and Trends in Optimization*, 1(3):123–231, 2013.
140. T. Pock, A. Chambolle, D. Cremers, and H. Bischof. A convex relaxation approach for computing minimal partitions. *IEEE Conference on Computer Vision and Pattern Recognition*, pages 810–817, 2009.
141. M. J. D. Powell. A method for nonlinear constraints in minimization problems. *Optimization*, 1972.
142. K. P. Pruessmann, M. Weiger, M. B. Scheidegger, and P. Boesiger. SENSE: Sensitivity encoding for fast MRI. *Magn. Reson. Med.*, 42:952–962, 1999.
143. N. Pustelnik, C. Chaux, J.-C. Pesquet, and C. Comtat. Parallel algorithm and hybrid regularization for dynamic PET reconstruction. In *Proc. IEEE Nucl. Sci. Symp. Conf. Rec.*, 2010.

144. S. Ramani and J. A. Fessler. Parallel MR image reconstruction using augmented Lagrangian methods. *IEEE Trans. Med. Imag.*, 30(3):694–706, 2011.
145. S. Ramani and J. A. Fessler. A splitting-based iterative algorithm for accelerated statistical X-ray CT reconstruction. *IEEE Trans. Med. Imag.*, 31(3):677–688, 2012.
146. R. T. Rockafellar. Augmented Lagrangians and applications of the proximal point algorithm in convex programming. *Math. Oper. Res.*, 1(2):97–116, 1976.
147. R. T. Rockafellar. Monotone operators and the proximal point algorithm. *SIAM Journal on Control and Optimization*, 14:877–898, 1976.
148. R. T. Rockafellar. *Convex Analysis*. Princeton University Press, Princeton, 10 edition, 1997.
149. E. Roessl and C. Herrmann. Cramér-Rao lower bound of basis image noise in multiple-energy x-ray imaging. *Phys. Med. Biol.*, 54(5):1307–1318, 2009.
150. E. Roessl and R. Proksa. K-edge imaging in x-ray computed tomography using multi-bin photon counting detectors. *Phys. Med. Biol.*, 52(15):4679–4696, 2007.
151. L. Rudin, S. Osher, and E. Fatemi. Nonlinear total variation based noise removal algorithms. *Physica D*, 60:259–268, 1992.
152. A. Sawatzky. Performance of first-order algorithms for TV penalized weighted least-squares denoising problem. In *Image and Signal Processing*, volume 8509 of *Lecture Notes in Computer Science*, pages 340–349. Springer International Publishing, 2014.
153. A. Sawatzky, C. Brune, T. Kösters, F. Wübbeling, and M. Burger. EM-TV methods for inverse problems with Poisson noise. In *Level Set and PDE Based Reconstruction Methods in Imaging*, pages 71–142. Springer, 2013.
154. A. Sawatzky, D. Tenbrinck, X. Jiang, and M. Burger. A variational framework for region-based segmentation incorporating physical noise models. *J. Math. Imaging Vis.*, 47(3):179–209, 2013.
155. A. Sawatzky, Q. Xu, C. O. Schirra, and M. A. Anastasio. Proximal ADMM for multi-channel image reconstruction in spectral X-ray CT. *IEEE Trans. Med. Imag.*, 33(8):1657–1668, 2014.
156. H. Schäfer. Über die Methode sukzessiver Approximationen. *Jahresbericht der Deutschen Mathematiker-Vereinigung*, 59:131–140, 1957.
157. K. P. Schäfers, T. J. Spinks, P. G. Camici, P. M. Bloomfield, C. G. Rhodes, M. P. Law, C. S. R. Baker, and O. Rimoldi. Absolute quantification of myocardial blood flow with $H_2{}^{15}O$ and 3-dimensional PET: An experimental validation. *J. Nucl. Med.*, 43:1031–1040, 2002.
158. C. O. Schirra, B. Brendel, M. A. Anastasio, and E. Roessl. Spectral CT: a technology primer for contrast agent development. *Contrast Media Mol. Imaging*, 9(1):62–70, 2014.
159. C. O. Schirra, E. Roessl, T. Koehler, B. Brendel, A. Thran, D. Pan, M. A. Anastasio, and R. Proksa. Statistical reconstruction of material decomposed data in spectral CT. *IEEE Trans. Med. Imag.*, 32(7):1249–1257, 2013.
160. J. P. Schlomka, E. Roessl, R. Dorscheid, S. Dill, G. Martens, T. Istel, C. Baumer, C. Herrmann, R. Steadmann, G. Zeitler, A. Livne, and R. Proksa. Experimental feasibility of multi-energy photon-counting K-edge imaging in pre-clinical computed tomography. *Phys. Med. Biol.*, 53(15):4031–4047, 2008.
161. M. Schmidt, N. Le Roux, and F. Bach. Convergence rates of inexact proximal-gradient methods for convex optimization. Technical report, arXiv e-print, 2011. http://arxiv.org/abs/1109.2415.
162. M. Schrader, S. W. Hell, and H. T. M. van der Voort. Three-dimensional super-resolution with a 4Pi-confocal microscope using image restoration. *J. Appl. Phys.*, 84(8):4033–4042, 1998.
163. T. Schuster, B. Kaltenbacher, B. Hofmann, and K. S. Kazimierski. *Regularization methods in Banach spaces*, volume 10. Walter de Gruyter, 2012.
164. S. Setzer. Operator splittings, Bregman methods and frame shrinkage in image processing. *International Journal of Computer Vision*, 92(3):265–280, 2011.
165. S. Setzer, G. Steidl, and J. Morgenthaler. A cyclic projected gradient method. *Computational Optimization and Applications*, 54(2):417–440, 2013.
166. S. Setzer, G. Steidl, and T. Teuber. Deblurring Poissonian images by split Bregman techniques. *J. Vis. Commun. Image R.*, 21:193–199, 2010.

167. E. Y. Sidky, J. H. Jörgensen, and X. Pan. Convex optimization problem prototyping for image reconstruction in computed tomography with the Chambolle-Pock algorithm. *Phys. Med. Biol.*, 57(10):3065–3091, 2012.
168. G. Steidl and T. Teuber. Removing multiplicative noise by Douglas-Rachford splitting methods. *J. Math. Imaging Vis.*, 36:168–184, 2010.
169. T. Teuber. *Anisotropic Smoothing and Image Restoration Facing Non-Gaussian Noise*. PhD thesis, Technische Universität Kaiserslautern, Apr. 2012. available from https://kluedo.ub.uni-kl.de/frontdoor/index/index/docId/3219.
170. P. Tseng. Applications of a splitting algorithm to decomposition in convex programming and variational inequalities. *SIAM Journal on Control and Optimization*, 29:119–138, 1991.
171. P. Tseng. On accelerated proximal gradient methods for convex-concave optimization. Technical report, Dept. of Mathematics, University of Washington, Seattle, 2008.
172. T. Valkonen. A primal-dual hybrid gradient method for nonlinear operators with applications to MRI. *Inverse Problems*, 30(5):055012, 2014.
173. B. Vandeghinste, B. Goossens, J. D. Beenhouwer, A. Pizurica, W. Philips, S. Vandenberghe, and S. Staelens. Split-Bregman-based sparse-view CT reconstruction. *Proc. Int. Meeting Fully 3D Image Recon. Rad. Nucl. Med.*, pages 431–434, 2011.
174. Y. Vardi, L. A. Shepp, and L. Kaufman. A statistical model for positron emission tomography. *J. Am. Stat. Assoc.*, 80(389):8–20, 1985.
175. S. Villa, S. Salzo, L. Baldassarre, and A. Verri. Accelerated and inexact forward-backward algorithms. *SIAM J. Optim.*, 23(3):1607–1633, 2013.
176. J. von Neumann. Some matrix inequalities and metrization of matrix-space. In *Collected Works, Pergamon, Oxford, 1962, Volume IV, 205–218*, pages 286–300. Tomsk University Review, 1937.
177. B. C. Vũ. A splitting algorithm for dual monotone inclusions involving cocoercive operators. *Advances in Computational Mathematics*, 38(3):667–681, 2013.
178. G. Wang, H. Yu, and B. D. Man. An outlook on x-ray CT research and development. *Med. Phys.*, 35(3):1051–1064, 2008.
179. J. Wang, T. Li, H. Lu, and Z. Liang. Penalized weighted least-squares approach to sinogram noise reduction and image reconstruction for low-dose X-ray computed tomography. *IEEE Trans. Med. Imag.*, 25(10):1272–1283, 2006.
180. K. Wang, R. Su, A. A. Oraevsky, and M. A. Anastasio. Investigation of iterative image reconstruction in three-dimensional optoacoustic tomography. *Phys. Med. Biol.*, 57:5399–5423, 2012.
181. S. L. Wang and L. Z. Liao. Decomposition method with a variable parameter for a class of monotone variational inequality problems. *J. Optimiz. Theory App.*, 109(2):415–429, 2001.
182. G. A. Watson. Characterization of the subdifferential of some matrix norms. *Linear Algebra and its Applications*, 170:33–45, 1992.
183. M. N. Wernick and J. N. Aarsvold, editors. *Emission Tomography: The Fundamentals of PET and SPECT*. Elsevier Academic Press, 2004.
184. Q. Xu, A. Sawatzky, and M. A. Anastasio. A multi-channel image reconstruction method for grating-based X-ray phase-contrast computed tomography. In *Proc. SPIE 9033, Medical Imaging 2014: Physics of Medical Imaging*, 2014.
185. Q. Xu, A. Sawatzky, M. A. Anastasio, and C. O. Schirra. Sparsity-regularized image reconstruction of decomposed K-edge data in spectral CT. *Phys. Med. Biol.*, 59(10):N65, 2014.
186. M. Yan and W. Yin. Self equivalence of the alternating direction method of multipliers. In: R. Glowinski, S. Osher, W. Yin (eds.) Splitting Methods in Communication and Imaging, Science and Engineering. Springer, 2016.
187. W. Yin. Analysis and generalizations of the linearized Bregman method. *SIAM Journal on Imaging Sciences*, 3(4):856–877, 2010.
188. J. Yuan, C. Schnörr, and G. Steidl. Simultaneous higher order optical flow estimation and decomposition. *SIAM Journal on Scientific Computing*, 29(6):2283–2304, 2007.
189. R. Zhang, J.-B. Thibault, C. A. Bouman, and K. D. S. J. Hsieh. A model-based iterative algorithm for dual-energy X-ray CT reconstruction. In *Proc. Int. Conf. Image Form. in X-ray CT*, pages 439–443, 2012.

190. X. Zhang, M. Burger, and S. Osher. A unified primal-dual algorithm framework based on Bregman iteration. *J. Sci. Comput.*, 46(1):20–46, 2011.
191. M. Zhu and T. F. Chan. An efficient primal-dual hybrid gradient algorithm for total variation image restoration. *UCLA CAM Report 08-34*, 2008.
192. H. Zou and T. Hastie. Regularization and variable selection via the elastic net. *Journal of Royal Statistical Society: Series B (Statistical Methodology)*, 67:301–320, 2005.
193. Y. Zou and M. D. Silver. Analysis of fast kV-switching in dual energy CT using a pre-reconstruction decomposition technique. In *Proc. SPIE (Medical Imaging 2008)*, volume 6913, page 691313, 2008.

Chapter 11
A Parameter Free ADI-Like Method for the Numerical Solution of Large Scale Lyapunov Equations

Danny C. Sorensen

Abstract This work presents an algorithm for constructing an approximate numerical solution to a large scale Lyapunov equation in low rank factored form. The algorithm is based upon a synthesis of an approximate power method and an alternating direction implicit (ADI) method. The former is parameter free and tends to be efficient in practice but there is little theoretical understanding of its convergence properties. The ADI method has a well-understood convergence theory, but the method relies upon selection of shift parameters and a poor shift selection can lead to very slow convergence in practice. The algorithm presented here uses an approximate power method iteration to obtain a basis update and then constructs a re-weighting of this basis to provide a factorization update that satisfies ADI-like convergence properties.

1 Introduction

This chapter is concerned with approximating solutions to large scale Lyapunov equations of the form

$$\mathbf{A}P + P\mathbf{A}^T + \mathbf{B}\mathbf{B}^T = \mathbf{0} \tag{11.1}$$

where $\mathbf{A} \in \mathbb{R}^{n\times n}$ and $\mathbf{B} \in \mathbb{R}^{n\times p}$ with $\mathbf{A}$ stable (all eigenvalues in the open left half-plane) and $p \leq n$. It is well known that the solution P is symmetric and positive semidefinite and that P is definite if and only if $\text{rank}[\lambda\mathbf{I} - \mathbf{A}, \mathbf{B}] = n$ for all $\lambda \in \mathbb{C}$ (see e.g., [22]). However, even when P is positive definite, it is often the case that the

D.C. Sorensen (✉)
Department of Computational and Applied Mathematics, Rice University, Houston, TX 77251-1892, USA
e-mail: sorensen@rice.edu

R. Glowinski et al. (eds.), *Splitting Methods in Communication, Imaging, Science, and Engineering*, Scientific Computation, DOI 10.1007/978-3-319-41589-5_11

eigenvalues of P decay rapidly so that P may be well approximated by a low rank matrix [15, 3]. This decay is essential for obtaining a practical numerical solution with an iterative method.

This rapid decay of eigenvalues arises in numerous applications involving control of dynamical systems. In that setting, there is considerable interest in obtaining approximate solutions $P \approx \mathbf{L}\mathbf{L}^T$ in *low rank factored form*, meaning that $\mathbf{L} \in \mathbb{R}^{n\times k}$ is a rank k matrix factor with $k \ll n$. If such an approximation has been obtained, one can efficiently construct an approximate balanced reduction of a large scale linear time invariant system [1].

A number methods for the numerical solution of large Lyapunov equations have been proposed and investigated. These include matrix sign function methods [4, 10, 5] and Krylov Projection Methods [6, 18]. Two methods that are of particular interest are approximate power methods [9] and ADI or Smith methods [19, 15, 11, 8]. Approximate power methods can be quite effective in practice but very little is known about convergence. On the other hand, there is an elegant convergence theory for ADI methods that is complete and well understood [11, 8]. Unfortunately, the performance of ADI methods is heavily dependent upon shift parameters and a poor shift selection can easily lead to very slow convergence. There is a theory of optimal shift selection for the symmetric case and some results for spectra in restricted regions of the complex plane discussed in Ellner and Wachspress [7]. A more practical suboptimal heuristic choice was suggested by Penzl [15]. An extremely effective shift selection scheme was developed by J. Sabino in his Ph.D. thesis [17].

The purpose of this chapter is to derive an algorithm that retains ADI convergence properties without the need to select parameters. Parameter selection is done automatically through the approximate power method. This can be an advantage over explicit parameter selection schemes in the absence of detailed information about the spectrum of $\mathbf{A}$.

2 The Alternating Direction Implicit Method

This section will provide a brief derivation of the ADI or Smith method for Lyapunov Equations.

From the original equation

$$\mathbf{A}P + P\mathbf{A}^T + \mathbf{B}\mathbf{B}^T = 0$$

Apply a real shift $\mu < 0$ from left

$$P = -(\mathbf{A} + \mu\mathbf{I})^{-1}\left[P(\mathbf{A} - \mu\mathbf{I})^T + \mathbf{B}\mathbf{B}^T\right] \tag{11.2}$$

Then, apply shift μ from right (Alternate Direction)

$$P = -\left[(\mathbf{A} - \mu\mathbf{I})P + \mathbf{B}\mathbf{B}^T\right](\mathbf{A} + \mu\mathbf{I})^{-T} \tag{11.3}$$

Finally, substitute the formula for P in (11.2) into the right-hand side of (11.3) to get a Stein Equation of the form

$$P = \mathbf{A}_\mu P \mathbf{A}_\mu^T + \mathbf{B}_\mu \mathbf{B}_\mu^T \iff \mathbf{A}P + P\mathbf{A}^T + \mathbf{B}\mathbf{B}^T = 0, \tag{11.4}$$

where

$$\mathbf{A}_\mu = (\mathbf{A} - \mu\mathbf{I})(\mathbf{A} + \mu\mathbf{I})^{-1}, \ \mathbf{B}_\mu = \sqrt{2|\mu|}(\mathbf{A} + \mu\mathbf{I})^{-1}\mathbf{B}.$$

A convenient aspect of this formulation is that the analytic solution can be written as an infinite matrix series

$$P = \sum_{j=0}^{\infty} \mathbf{A}_\mu^j \mathbf{B}_\mu \mathbf{B}_\mu^T (\mathbf{A}_\mu^j)^T = \mathbf{L}\mathbf{L}^T, \tag{11.5}$$

where $\mathbf{L} = [\mathbf{B}_\mu, \ \mathbf{A}_\mu\mathbf{B}_\mu, \ \mathbf{A}_\mu^2\mathbf{B}_\mu, \ \ldots]$ and hence this expresses the solution in factored form. This formulation has two advantages. First, a solution constructed this way is automatically positive definite when $\mathbf{L}$ has full rank. Second, in practice, the successive terms usually decay rapidly and hence a low rank approximation to the solution is naturally obtained. This is the key idea (going back to Penzl [15]) that makes the approximation to solutions of large scale Lyapunov equations computationally tractable. Moreover, this formulation suggests the iteration

$$P_{j+1} = \mathbf{A}_\mu P_j \mathbf{A}_\mu^T + \mathbf{B}_\mu \mathbf{B}_\mu^T, \ j = 0, 1, 2, \ldots,$$

with $P_0 = \mathbf{0}$ as a means to construct an approximate solution.

The original formulation of ADI by Peaceman and Rachford [14] did not combine the two steps (11.2) and (11.3) into the single step indicated in Equation (11.4). In fact, the original ADI formulation was not explicitly expressed in terms of matrix equations. Following the original formulation, the ADI iteration has traditionally been expressed in "half-steps". Thus, if P_j is the approximate solution at step j then the approximate solution $P_{j+\frac{1}{2}}$ is obtained after solving the "split" Equation (11.2) in the form

$$P_{j+\frac{1}{2}} = -(\mathbf{A} + \mu\mathbf{I})^{-1} \left[P_j(\mathbf{A} - \mu\mathbf{I})^T + \mathbf{B}\mathbf{B}^T\right] \tag{11.6}$$

Then P_{j+1} is obtained by solving Equation (11.3) in the form

$$P_{j+1} = -\left[(\mathbf{A} - \mu\mathbf{I})P_{j+\frac{1}{2}} + \mathbf{B}\mathbf{B}^T\right](\mathbf{A} + \mu\mathbf{I})^{-T}. \tag{11.7}$$

This iteration has been analyzed and used extensively throughout the numerical PDE literature. However the one step formulation of Equation (11.4) exposes the insight of the factored form of the solution in Equation (11.5) and also lends itself to some fairly straightforward analysis.

For example, since $\mathbf{A}$ is asymptotically stable, it follows that the spectral radius of $\mathbf{A}_\mu$ is less than 1. Therefore, since

$$P_m = \sum_{j=0}^{m} \mathbf{A}_\mu^j \mathbf{B}_\mu \mathbf{B}_\mu^T (\mathbf{A}_\mu^j)^T$$

and

$$P_{m+1} = \mathbf{A}_\mu P_m \mathbf{A}_\mu^T + \mathbf{B}_\mu \mathbf{B}_\mu^T,$$

it is easily shown that

$$E_{m+1} = \mathbf{A}_\mu E_m \mathbf{A}_\mu^T = \mathbf{A}_\mu^{m+2} P (\mathbf{A}_\mu^{m+2})^T \rightarrow \mathbf{0}, \tag{11.8}$$

$$\text{where } \mathrm{E_m} = \mathrm{P} - \mathrm{P_m}.$$

When the shift μ is complex with $\rho = \mathrm{Real}(\mu) < 0$, essentially the same derivation and analysis will be valid with

$$\mathbf{A}_\mu = (\mathbf{A} - \bar{\mu}\mathbf{I})(\mathbf{A} + \mu\mathbf{I})^{-1}, \quad \mathbf{B}_\mu = \sqrt{2|\rho|}(\mathbf{A} + \mu\mathbf{I})^{-1}\mathbf{B},$$

and with "transpose" replaced by "conjugate transpose" in the above equations.

The Low-Rank Smith method developed by Penzl [15] repeatedly updates the factored form $P_m = \mathbf{L}_m \mathbf{L}_m^T$ via

$$\begin{aligned} \mathbf{L}_{m+1} &= [\mathbf{A}_\mu \mathbf{L}_m, \mathbf{B}_\mu] \\ &= [\mathbf{A}_\mu^{m+1} \mathbf{B}_\mu, \mathbf{L}_m]. \end{aligned}$$

The second formulation of the update may be implemented by $\mathbf{Z}_0 = \mathbf{B}_j$, $\mathbf{Z}_{m+1} \leftarrow \mathbf{A}_\mu \mathbf{Z}_m$ Since $\mathbf{L}$ typically has many more columns than $\mathbf{B}$, this is far less expensive than the first formulation but is potentially unstable when $\mathbf{A}$ is highly non-normal. Both formulations may require prohibitively large amounts of storage.

An asymptotic convergence rate is obtained from the spectral radius $\rho(\mathbf{A}_\mu)$ and one may attempt to select an optimal value of μ to minimize this spectral radius. Usually a single shift μ is not sufficient to obtain a spectral radius significantly less than one. Therefore multiple shifts are recommended. However each individual shift requires a matrix factorization which must be stored or re-computed each time a particular shift is applied. In [8] a Modified Smith Method was developed to overcome the storage difficulties and also to enable applications of multiple shifts with only one matrix factorization per shift.

A full implementation of the multi-shift method, shown in Figure 11.1, propagates a low rank SVD approximation to $\mathbf{L}_m$ and aggregates the shift application so that all instances of a particular shift are applied before the next shift is brought in. In addition, a stopping rule would be specified in place of a fixed number (`ksteps`) of shift applications. Moreover, at each shift application, the SVD of $\mathbf{L}$ is updated and truncated in a manner that limits storage but maintains sufficient accuracy of the SVD approximation to $\mathbf{L}$. The aggregated shift formulation may be derived from the Residual Equation (11.8). This modified Smith method is much faster and requires far less storage than the original method.

```
Z = B; L = [];
for j = 1 : kshifts,
  μ = −u(j);
  ρ = √(2μ);
  % A sparse factorization of A − μI_n may be computed and
  % re-used for the solve here and in the j-loop below.
  Z = ((A − μI_n)\Z);
  L = [ρZ, L];
  for j = 1 : ksteps,
    Z = (A + μI_n)((A − μI_n)\Z);
    L = [ρZ, L];  % The SVD of L may be updated and truncated here.
  end
  Z = (A + μI_n)Z;
end
```

Fig. 11.1 Modified Low Rank Smith (Multishift-Smith)

The key to performance, however, is still the selection of shifts to minimize the spectral radius of the product $\mathbf{A}_{\mu_1}\mathbf{A}_{\mu_2}\cdots\mathbf{A}_{\mu_k}$. The complete details of this implementation and its performance are available in [8].

3 The Approximate Power Method (APM)

A very different approach called the Approximate Power Method (APM) was suggested by Hodel et al. [9]. The idea is to utilize a subspace iteration technique to approximately compute the dominant invariant subspace of P. The difficulty with such an approach is that the matrix P is not available and thus approximate matrix-matrix products of the form $\mathbf{Z} = P\mathbf{U}$ must be provided indirectly. The idea is to use the structure of the Lyapunov equation and note that if P is a solution and $\mathbf{U}$ is any $n \times k$ matrix then

$$\mathbf{A}P\mathbf{U} + P\mathbf{A}^T\mathbf{U} + \mathbf{B}\mathbf{B}^T\mathbf{U} = \mathbf{0}.$$

Simply adding and subtracting the term $P\mathbf{U}\mathbf{U}^T\mathbf{A}^T\mathbf{U}$ gives

$$\mathbf{A}P\mathbf{U} + P\mathbf{U}\mathbf{U}^T\mathbf{A}^T\mathbf{U} + \mathbf{B}\mathbf{B}^T\mathbf{U} + P(\mathbf{I} - \mathbf{U}\mathbf{U}^T)\mathbf{A}^T\mathbf{U} = \mathbf{0}.$$

If we think of $P(\mathbf{I} - \mathbf{U}\mathbf{U}^T)\mathbf{A}^T\mathbf{U}$ as a small error term then

$$\mathbf{A}P\mathbf{U} + P\mathbf{U}\mathbf{H}^T + \mathbf{B}\mathbf{B}^T\mathbf{U} \approx \mathbf{0}, \quad \text{where} \quad \mathbf{H} = \mathbf{U}^T\mathbf{A}\mathbf{U}.$$

Thus, solving

$$\mathbf{A}\mathbf{Z} + \mathbf{Z}\mathbf{H}^T + \mathbf{B}\mathbf{B}^T\mathbf{U} = \mathbf{0} \tag{11.9}$$

gives an approximation

$$\mathbf{Z} \approx P\mathbf{U}$$

to the desired matrix-matrix product.

The simplest form of the Approximate Power Method iterates this calculation with

$$\mathbf{H}_j = \mathbf{U}_j^T \mathbf{A} \mathbf{U}_j, \quad [\mathbf{U}_{j+1}, \mathbf{S}_{j+1}, \mathbf{W}_{j+1}] = \text{svd}(\mathbf{Z}_j) \quad \text{for} \quad j = 1, 2, ...$$

where $\mathbf{Z}_j$ is obtained by solving (11.9). The APM iteration may be initialized by setting $\mathbf{U}_0 = \mathbf{Q}$ where the $n \times k$ orthogonal matrix $\mathbf{Q}$ is obtained from a QR-factorization of a random $n \times k$ matrix. Variants of the APM pose the results of (11.9) as an update equation (see [2]).

This iteration can often provide very good results and convergence is rapid in many cases. However, there is no assurance that the discarded error term $P(\mathbf{I} - \mathbf{U}\mathbf{U}^T)\mathbf{A}^T\mathbf{U}$ is small and when it is not, the convergence may be slow or non-existent. This difficulty has been addressed with some success in [23, 21]. However, the approach put forth in the next section will finesse this difficulty.

4 A Parameter Free ADI Method

The ultimate goal of both the APM and the ADI methods is to construct a basis $\mathbf{U}$ for the dominant invariant subspace of P corresponding its largest eigenvalues. The shortcomings of each of these can be overcome by combining the two in a way that automatically selects shifts for the ADI method and also overcomes the possible effect of the troublesome error term in APM.

Observe that whenever $\mathbf{U}$ is a basis for an invariant subspace of P then $P\mathbf{U} = \mathbf{U}\hat{P}$ (for some matrix $\hat{P}$) and

$$\mathbf{0} = \mathbf{U}^T(\mathbf{A}P\mathbf{U} + P\mathbf{A}\mathbf{U} + \mathbf{B}\mathbf{B}^T\mathbf{U}) = \mathbf{H}\hat{P} + \hat{P}\mathbf{H}^T + \hat{\mathbf{B}}\hat{\mathbf{B}}^T,$$

where $\mathbf{H} = \mathbf{U}^T\mathbf{A}\mathbf{U}$ and $\hat{\mathbf{B}} = \mathbf{U}^T\mathbf{B}$. This projected Lyapunov equation will play a key role in a synthesis of the APM and ADI methods. It is combined with the Sylvester equation of the APM.

This synthesis of the APM and ADI methods consists of a Parameter Free ADI Iteration (PFADI) that has four basic steps. At the j-th iterate of an outer iteration, an orthogonal basis $\mathbf{U}_j$ for an approximate invariant subspace of P is assumed to be available. This basis provides for an approximate factorization $P_j = \mathbf{L}_j\mathbf{L}_j^T$ with $\mathbf{L}_j = \mathbf{U}_j\mathbf{S}_j$ and $P_j(\mathbf{U}_j) = \mathbf{U}_j\mathbf{S}_j^2$. Thus, $\mathbf{U}_j$ is an orthonormal basis for the range of P_j and is at the same time the basis for the dominant invariant subspace of $\mathbf{U}_j$. A selected subspace of Range($\mathbf{U}$) is spanned by the columns of an ortho-normal matrix $\mathbf{U}_j$ typically with fewer columns than $\mathbf{U}$. Typically, we think of $\mathbf{U}_j$ as a basis for the dominant invariant subspace of P_j, but other choices are possible. The number of columns of $\mathbf{U}_j$ is unspecified and can, for example, be a small number and vary at each outer step j. At step j, the four inner iteration steps shown in Figure 11.2 are performed. To begin the iteration, initialize $\mathbf{L}_1 = \mathbf{0}$ and $\mathbf{B}_1 = \mathbf{B}$.

Step 1: Solve the reduced order Lyapunov equation for $\hat{\mathscr{P}}_j$

Solve $\mathbf{H}_j\hat{\mathscr{P}}_j + \hat{\mathscr{P}}_j\mathbf{H}_j^T + \hat{\mathbf{B}}_j\hat{\mathbf{B}}_j^T = \mathbf{0}$,
with $\mathbf{H}_j = \mathbf{U}_j^T\mathbf{A}\mathbf{U}_j$, $\hat{\mathbf{B}}_j = \mathbf{U}_j^T\mathbf{B}_j$.

Step 2: (APM step) Solve a projected Sylvester equation for $\mathbf{Z}$

$\mathbf{A}\mathbf{Z}_j + \mathbf{Z}_j\mathbf{H}_j^T + \mathbf{B}_j\hat{\mathbf{B}}_j^T = \mathbf{0}$,

Step 3: Modify $\mathbf{B}_j$

Update $\mathbf{B}_{j+1} \leftarrow (\mathbf{I} - \mathbf{Z}_j\hat{\mathscr{P}}_j^{-1}\mathbf{U}_j^T)\mathbf{B}_j$.

Step 4: (ADI step) Update factorization and basis $\mathbf{U}$

Re-scale $\mathbf{Z}_j \leftarrow \mathbf{Z}_j\hat{\mathscr{P}}_j^{-1/2}$.
Update (and truncate) $[\mathbf{U},\mathbf{S}] \leftarrow \mathrm{svd}[\mathbf{L}_j, \mathbf{Z}_j]$.
$\mathbf{L}_{j+1} = \mathbf{U}\mathbf{S}$.
$\mathbf{U}_{j+1} \leftarrow \mathbf{U}(:, 1 : k_j)$, basis for dominant subspace of dimension k_j.

Fig. 11.2 A Parameter Free Lyapunov Solver (PFADI)

The motivation for these steps resides in the following results. The re-scaling at Step 4 provides an update to the factorization that will always maintain $P \succeq P_j$. The modification to $\mathbf{B}_j$ at Step 4 provides an updated Lyapunov equation for the difference $P - P_{j+1}$. Each subsequent iteration of these four basic steps provides an incremental update to the approximate factorization such that $P \succeq P_{j+1} \succeq P_j$. Hence, the iteration will automatically be convergent.

The reduced Lyapunov equation at Step 1 might not possess a stable matrix $\mathbf{H}_j$ (although this would be assured near convergence). Thus, the additional assumption that $\mathbf{A} + \mathbf{A}^T$ is negative definite shall be made. This condition is often satisfied in practice and it assures that $\mathbf{H}_j = \mathbf{U}_j^T\mathbf{A}\mathbf{U}_j$ is always stable. To see this, just note that if $\mathbf{H}_j\mathbf{x} = \mathbf{x}\lambda$ with $\mathbf{x}^*\mathbf{x} = 1$, then

$$0 > \mathbf{y}^*(\mathbf{A} + \mathbf{A}^T)\mathbf{y} = \mathbf{y}^*\mathbf{A}\mathbf{y} + \overline{\mathbf{y}^*\mathbf{A}\mathbf{y}} = \lambda + \bar{\lambda} = 2Re(\lambda),$$

since $\lambda = \mathbf{x}^*\mathbf{x}\lambda = \mathbf{x}^*\mathbf{H}\mathbf{x} = \mathbf{y}^*\mathbf{A}\mathbf{y}$ with $\mathbf{y} = \mathbf{U}_j\mathbf{x}$. Hence $Re(\lambda) < 0$ for any eigenvalue λ of $\mathbf{H}_j$.

The following lemmas will establish the convergence properties. In the following discussion, it is assumed that $\hat{P}$ is always positive definite. In fact, it is possible to modify this algorithm to assure that cond($\hat{P}$) uniformly bounded. This amounts to selecting a subspace corresponding to the leading eigenvalues of $\hat{P}$ and projecting to this subspace. The following results are also valid for this modification which will be introduced in detail in Section 5 where implementation issues are discussed.

Lemma 1. *Let* $\mathbf{Z}_j, \hat{P}_j, \mathbf{B}_j$ *and* $\mathbf{H}_j$ *be defined as in Figure 11.2 in Steps 1–3, and let* $\mathbf{G}_j = \hat{P}_j^{-1}$. *Then*

$$\mathbf{A}\mathbf{Z}_j\mathbf{G}_j\mathbf{Z}_j^T + \mathbf{Z}_j\mathbf{G}_j\mathbf{Z}_j^T\mathbf{A}^T + \mathbf{B}_j\mathbf{B}_j^T = \mathbf{B}_{j+1}\mathbf{B}_{j+1}^T,$$

where $\mathbf{B}_{j+1} \equiv (\mathbf{I} - \mathbf{Z}_j\mathbf{G}_j\mathbf{U}_j^T)\mathbf{B}_j$.

Proof: Set $\mathbf{G}_j = \hat{P}_j^{-1}$ and note $\mathbf{G}_j = \mathbf{G}_j^T$ since $\hat{P}_j$ is symmetric. From the equations in Steps 1 and 2 of Figure 11.2, we obtain

$$\mathbf{A}\mathbf{Z}_j\mathbf{G}_j\mathbf{Z}_j^T = -[\mathbf{Z}_j\mathbf{H}_j^T + \mathbf{B}_j\mathbf{B}_j^T\mathbf{U}_j]\mathbf{G}_j\mathbf{Z}_j^T$$

after multiplication on the right by $\mathbf{G}_j\mathbf{Z}_j^T$. Thus

$$\begin{aligned}\mathbf{A}\mathbf{Z}_j\mathbf{G}_j\mathbf{Z}_j^T + \mathbf{Z}_j\mathbf{G}_j\mathbf{Z}_j^T\mathbf{A}^T &= -\mathbf{Z}_j[\mathbf{H}_j^T\mathbf{G}_j + \mathbf{G}_j\mathbf{H}_j]\mathbf{Z}_j^T - \mathbf{B}_j\mathbf{B}_j^T\mathbf{U}_j\mathbf{G}_j\mathbf{Z}_j^T - \mathbf{Z}_j\mathbf{G}_j\mathbf{U}_j^T\mathbf{B}_j\mathbf{B}_j^T \\ &= \mathbf{Z}_j[\mathbf{G}_j\mathbf{U}_j^T\mathbf{B}_j\mathbf{B}_j^T\mathbf{U}_j\mathbf{G}_j]\mathbf{Z}_j^T - \mathbf{B}_j\mathbf{B}_j^T\mathbf{U}_j\mathbf{G}_j\mathbf{Z}_j^T - \mathbf{Z}_j\mathbf{G}_j\mathbf{U}_j^T\mathbf{B}_j\mathbf{B}_j^T,\end{aligned}$$

and hence

$$\begin{aligned}&\mathbf{A}\mathbf{Z}_j\mathbf{G}_j\mathbf{Z}_j^T + \mathbf{Z}_j\mathbf{G}_j\mathbf{Z}_j^T\mathbf{A}^T + \mathbf{B}_j\mathbf{B}_j^T \\ &\quad = \mathbf{Z}_j\mathbf{G}_j\mathbf{U}_j^T\mathbf{B}_j\mathbf{B}_j^T\mathbf{U}_j\mathbf{G}_j\mathbf{Z}_j^T - \mathbf{B}_j\mathbf{B}_j^T\mathbf{U}_j\mathbf{G}_j\mathbf{Z}_j^T - \mathbf{Z}_j\mathbf{G}_j\mathbf{U}_j^T\mathbf{B}_j\mathbf{B}_j^T + \mathbf{B}_j\mathbf{B}_j^T \\ &\quad = (\mathbf{I} - \mathbf{Z}_j\mathbf{G}_j\mathbf{U}_j^T)\mathbf{B}_j\mathbf{B}_j^T(\mathbf{I} - \mathbf{U}_j\mathbf{G}_j\mathbf{Z}_j^T) \\ &\quad = \mathbf{B}_{j+1}\mathbf{B}_{j+1}^T.\end{aligned}$$

■

This result leads immediately to an equation for the difference $P - P_j$.

Lemma 2. *Let the hypothesis of Lemma 1 hold. Then*

$$\mathbf{A}P_j + P_j\mathbf{A}^T + \mathbf{B}\mathbf{B}^T = \mathbf{B}_j\mathbf{B}_j^T, \tag{11.10}$$

and

$$\mathbf{A}(P - P_j) + (P - P_j)\mathbf{A}^T + \mathbf{B}_j\mathbf{B}_j^T = \mathbf{0}. \tag{11.11}$$

As a consequence, $P \succeq P_{j+1} \succeq P_j \succeq \mathbf{0}$ *for* $j = 1, 2, \ldots$.

Proof: The proof will be an induction. Since $\mathbf{B}_1 = \mathbf{B}$ and $P_1 = \mathbf{0}$, Equations (11.10) and (11.11) are trivially satisfied when $j = 1$. Assume that Equations (11.10) and (11.11) are true for some $j \geq 1$. It follows from Lemma 1 that

$$\begin{aligned}\mathbf{A}P_j + P_j\mathbf{A}^T + \mathbf{B}\mathbf{B}^T &= \mathbf{B}_j\mathbf{B}_j^T \\ &= \mathbf{B}_{j+1}\mathbf{B}_{j+1}^T - [\mathbf{A}\mathbf{Z}_j\mathbf{G}_j\mathbf{Z}_j^T + \mathbf{Z}_j\mathbf{G}_j\mathbf{Z}_j^T\mathbf{A}^T].\end{aligned}$$

Hence,

$$\mathbf{A}(P_j + \mathbf{Z}_j\mathbf{G}_j\mathbf{Z}_j^T) + (P_j + \mathbf{Z}_j\mathbf{G}_j\mathbf{Z}_j^T)\mathbf{A}^T + \mathbf{B}\mathbf{B}^T = \mathbf{B}_{j+1}\mathbf{B}_{j+1}^T$$

to establish

$$\mathbf{A}P_{j+1} + P_{j+1}\mathbf{A}^T + \mathbf{B}\mathbf{B}^T = \mathbf{B}_{j+1}\mathbf{B}_{j+1}^T, \tag{11.12}$$

since $P_{j+1} = P_j + \mathbf{Z}_j \mathbf{G}_j \mathbf{Z}_j^T$. The update of Equation (11.11) readily follows from Equation (11.12) since P solves the original Lyapunov equation and the induction is complete.

Equation (11.11) implies $P \succeq P_j$ for all j and clearly $P_{j+1} \succeq P_j \succeq \mathbf{0}$ since $\mathbf{G}_j$ is positive definite for all j and this completes the proof. ■

Lemma 11.11 establishes that the iteration will provide a sequence of symmetric positive semidefinite matrices $P_j = \mathbf{L}_j \mathbf{L}_j^T$ which satisfy $P \succeq P_{j+1} \succeq P_j$ for all $j = 1, 2, \ldots$. The following lemma will establish that the sequence P_j is convergent. This lemma is proved in far greater generality for bounded symmetric linear operators in Hilbert space in Riesz and SZ.-Nagy [16] where the result is attributed to J.P. Vigier in his Ph.D. thesis.

Lemma 3. *Suppose P_ℓ is a sequence of symmetric matrices such that $P \succeq P_{\ell+1} \succeq P_\ell$, Where P is a fixed symmetric matrix. Then $\lim_{\ell \to \infty} P_\ell = P_o \preceq P$ exists.*

Proof: Let $\rho_{ij}^{(\ell)}$ denote the ij-th element of P_ℓ and let ρ_{ij} denote the ij-th element of P. The definition of the partial ordering $\succeq$ implies that $\mathbf{e}_j^T P \mathbf{e}_j \geq \mathbf{e}_j^T P_{\ell+1} \mathbf{e}_j \geq \mathbf{e}_j^T P_\ell \mathbf{e}_j$ and hence that $\rho_{jj} \geq \rho_{jj}^{(\ell+1)} \geq \rho_{jj}^{(\ell)}$. Thus $\{\rho_{jj}^{(\ell)}\}$ is an increasing sequence bounded above and hence convergent to a limit $\rho_{jj}^{(o)} \leq \rho_{jj}$ for each $j = 1, 2, \ldots, n$. Moreover, the fact that $\mu_{ii} + \mu_{jj} \geq 2|\mu_{ij}|$ for any symmetric positive semi-definite matrix $\mathbf{M} = (\mu_{ij})$ will imply that

$$(\rho_{ii}^{(\ell+1)} - \rho_{ii}^{(\ell)}) + (\rho_{jj}^{(\ell+1)} - \rho_{jj}^{(\ell)}) \geq 2|\rho_{ij}^{(\ell+1)} - \rho_{ij}^{(\ell)}|,$$

for all ℓ and for all (i, j). Observe that

$$\sum_{\ell=1}^{\infty} (\rho_{ii}^{(\ell+1)} - \rho_{ii}^{(\ell)}) + (\rho_{jj}^{(\ell+1)} - \rho_{jj}^{(\ell)}) \leq (\rho_{ii} - \rho_{ii}^{(1)}) + (\rho_{jj} - \rho_{jj}^{(1)})$$

since the series is telescoping. Therefore, the series

$$\sum_{\ell=1}^{\infty} |\rho_{ij}^{(\ell+1)} - \rho_{ij}^{(\ell)}| \leq \frac{1}{2}[(\rho_{ii} - \rho_{ii}^{(1)}) + (\rho_{jj} - \rho_{jj}^{(1)})]$$

is convergent. Now, since

$$|\rho_{ij}^{(m_2)} - \rho_{ij}^{(m_1)}| \leq \sum_{\ell=m_1}^{m_2-1} |\rho_{ij}^{(\ell+1)} - \rho_{ij}^{(\ell)}|$$

for any positive integers $m_1 < m_2$, it follows that the sequence $\{\rho_{ij}^{(\ell)}\}$ is Cauchy and hence convergent for each (i, j). This concludes the proof. ■

Another consequence of Lemma 1 is that the iteration is norm decreasing for $\mathbf{B}_j$ in the sense that $\|\mathbf{B}_{j+1}\|_F < \|\mathbf{B}_j\|_F$ for all j.

Lemma 4. *Let the hypothesis of Lemma 1 hold, and let* $\mathbf{B}_{j+1}$ *be defined as in Lemma 2 and let* $\gamma_{min} \leq \mathbf{v}^T(\mathbf{A}+\mathbf{A}^T)\mathbf{v} \leq \gamma_{max} < 0$ *for all* $\mathbf{v}$ *of norm one. Then* $\|\mathbf{B}_{j+1}\|_F \leq \|\mathbf{B}_j\|_F + \gamma_{max}\text{trace}(\mathbf{Z}_j\mathbf{G}_j\mathbf{Z}_j^T) < \|\mathbf{B}_j\|_F$ *for all j.*

Proof: Let $\mathbf{Z}_j\mathbf{G}_j\mathbf{Z}_j^T = \mathbf{V}_j\hat{\mathbf{S}}_j\mathbf{V}_j^T$ where $\hat{\mathbf{S}}_j$ is diagonal with positive diagonal elements $\hat{\sigma}_i$ and $\mathbf{V}_j^T\mathbf{V}_j = \mathbf{I}$. From Lemma 1,

$$\begin{aligned}\|\mathbf{B}_{j+1}\|_F^2 &= \text{trace}(\mathbf{B}_{j+1}\mathbf{B}_{j+1}^T)\\ &= \text{trace}[\mathbf{A}\mathbf{V}_j\hat{\mathbf{S}}_j\mathbf{V}_j^T + \mathbf{V}_j\hat{\mathbf{S}}_j\mathbf{V}_j^T\mathbf{A}^T] + \text{trace}(\mathbf{B}_j\mathbf{B}_j^T).\end{aligned}$$

However,

$$\begin{aligned}\text{trace}[\mathbf{A}\mathbf{V}_j\hat{\mathbf{S}}_j\mathbf{V}_j^T + \mathbf{V}_j\hat{\mathbf{S}}_j\mathbf{V}_j^T\mathbf{A}^T] &= \text{trace}[\mathbf{V}_j^T\mathbf{A}\mathbf{V}_j\hat{\mathbf{S}}_j + \hat{\mathbf{S}}_j\mathbf{V}_j^T\mathbf{A}^T\mathbf{V}_j]\\ &= \sum_{i=1}^{k_j} \mathbf{v}_i^T(\mathbf{A}+\mathbf{A}^T)\mathbf{v}_i\hat{\sigma}_i\\ &\leq \gamma_{max}\text{trace}[\hat{\mathbf{S}}_j]\\ &< 0.\end{aligned}$$

■

Lemma 5. *Assume* $\mathbf{A}\mathbf{X} = \mathbf{X}\Lambda$ *is diagonalizable and let the hypothesis of Lemma 1 hold. Suppose also that* $\mathbf{B} = \mathbf{b}$ *is a vector and that* $r = 1$ *throughout the iteration, i.e., that* $\mathbf{U}_j = \mathbf{u}_j$ *with* $P_j\mathbf{u}_j = \mathbf{u}_j\sigma_j$ *the dominant eigenvector of* P_j*. Let* $\mathbf{b}_j$ *be the rhs at step j. Then*

$$\|\mathbf{b}_{j+1}\| \leq \rho_j\|\mathbf{b}_j\|,$$

where $\rho_j = \rho((\gamma_j\mathbf{I}-\mathbf{A})(\gamma_j\mathbf{I}+A)^{-1})$ *is the spectral radius, with* $\gamma_j = \frac{1}{2}\mathbf{u}_j^T(\mathbf{A}+\mathbf{A}^T)\mathbf{u}_j < 0$*. Moreover, there is a constant* $\rho < 1$ *such that* $\rho_j \leq \rho$ *for all j sufficiently large.*

Proof: From the definition of the quantities in the iteration, we have

$$\mathbf{A}\mathbf{z}_j + \mathbf{z}_j\gamma_j + \mathbf{b}_j\beta_j = 0 \;\text{ and }\; \gamma_j\hat{\eta}_j + \hat{\eta}_j\gamma_j + \beta_j^2 = 0,$$

where $\beta_j = \mathbf{u}_j^T\mathbf{b}_j$ and $\eta_j = \hat{P}_j$ is used to emphasize that P_j is a scalar in this case. Thus

$$\mathbf{z}_j = -(\gamma_j\mathbf{I}+\mathbf{A})^{-1}\mathbf{b}_j\beta_j \;\text{ and }\; \hat{\eta}_j = -\frac{\beta_j^2}{2\gamma_j},$$

so that

$$\begin{aligned}
\mathbf{b}_{j+1} &= \mathbf{b}_j - \mathbf{z}_j(\hat{\eta}_j)^{-1}\mathbf{u}_j^T\mathbf{b}_j \\
&= \mathbf{b}_j - (\gamma_j\mathbf{I}+\mathbf{A})^{-1}\mathbf{b}_j\beta_j(\frac{2\gamma_j}{\beta_j^2})\beta_j \\
&- \mathbf{b}_j - 2\gamma_j(\gamma_j\mathbf{I}+\mathbf{A})^{-1}\mathbf{b}_j \\
&= (\gamma_j\mathbf{I}+\mathbf{A})^{-1}[(\gamma_j\mathbf{I}+\mathbf{A}) - 2\gamma_j\mathbf{I}]\mathbf{b}_j \\
&= -(\gamma_j\mathbf{I}+\mathbf{A})^{-1}(\gamma_j\mathbf{I}-\mathbf{A})\mathbf{b}_j.
\end{aligned}$$

Since $\mathbf{A}$ is stable and $\gamma_j < 0$, the eigenvalues of $\mathbf{A}$ are mapped to the interior of the unit disc under the Cayley transformation $\mathbf{C}_j = (\gamma_j\mathbf{I}-\mathbf{A})(\gamma_j\mathbf{I}+A)^{-1}$. Hence, the spectral radius satisfies $\rho_j = \rho((\gamma_j\mathbf{I}-\mathbf{A})(\gamma_j\mathbf{I}+A)^{-1}) < 1$ for all j. Since $\gamma_j \to \gamma < 0$ is convergent and since $\rho((\gamma\mathbf{I}-\mathbf{A})(\gamma\mathbf{I}+A)^{-1}) < 1$ there is a number $\rho < 1$ such that $\rho_j \leq \rho$ for all j sufficiently large. Moreover, a standard argument will show that $\|\mathbf{b}_j\| \leq \rho^{(j-m)}\|\mathbf{b}\|$ for all $j > m$, where m is a positive integer such that $\text{cond}(\mathbf{X})\prod_{i=1}^{m}\rho_i < 1$. This concludes the proof. ■

Note that a consequence of this result is that $\mathbf{b}_j \to \mathbf{0}$ and hence that $P_j \to P$ (i.e., convergence to the true P at a linear rate).

A more general result will provide a direct relationship to the ADI method. In fact, it can be shown that this parameter free method provides exactly the same update as ADI would construct with shifts given by the eigenvalues of the projected matrix $\mathbf{H}$. The advantages of the parameter free approach is that these complex shifts are automatically selected and they are applied indirectly through the solution of the Sylvester equation which may be done in real arithmetic.

It is necessary to establish a formula for the update provided by ADI with a set of shifts $\mu_j,\ \ j = 1,2,\ldots k$. First, suppose that μ is a shift with $\rho = \text{Real}(\mu) < 0$. Then

$$P = \mathbf{A}_\mu P\mathbf{A}_\mu^* + \mathbf{B}_\mu\mathbf{B}_\mu^* \ \text{ and thus } \ P = 2|\rho|\sum_{j=0}^{\infty}\mathbf{A}_\mu^j\mathbf{B}_\mu\mathbf{B}_\mu^*(\mathbf{A}_\mu^j)^*,$$

where

$$\mathbf{A}_\mu = (\mathbf{A}-\bar{\mu}\mathbf{I})(\mathbf{A}+\mu\mathbf{I})^{-1}, \ \ \mathbf{B}_\mu = (\mathbf{A}+\mu\mathbf{I})^{-1}\mathbf{B}.$$

Let $\mathbf{E} = 2|\rho|\mathbf{B}_\mu\mathbf{B}_\mu^*$. Then

$$\begin{aligned}
\mathbf{A}(P-\mathbf{E})+(P-\mathbf{E})\mathbf{A}^T &= 2|\rho|\left[\mathbf{A}\left(\sum_{j=1}^{\infty}\mathbf{A}_\mu^j\mathbf{B}_\mu\mathbf{B}_\mu^T(\mathbf{A}_\mu^j)^*\right) + \left(\sum_{j=1}^{\infty}\mathbf{A}_\mu^j\mathbf{B}_\mu\mathbf{B}_\mu^T(\mathbf{A}_\mu^j)^*\right)\mathbf{A}^T\right] \\
&= 2|\rho|\mathbf{A}_\mu\left[\mathbf{A}\left(\sum_{j=0}^{\infty}\mathbf{A}_\mu^j\mathbf{B}_\mu\mathbf{B}_\mu^T(\mathbf{A}_\mu^j)^*\right) + \left(\sum_{j=0}^{\infty}\mathbf{A}_\mu^j\mathbf{B}_\mu\mathbf{B}_\mu^T(\mathbf{A}_\mu^j)^*\right)\mathbf{A}^T\right]\mathbf{A}_\mu^* \\
&= \mathbf{A}_\mu\left[\mathbf{A}P+P\mathbf{A}^T\right]\mathbf{A}_\mu^* \\
&= -\mathbf{A}_\mu\left[\mathbf{B}\mathbf{B}^T\right]\mathbf{A}_\mu^*
\end{aligned}$$

Thus,

$$\mathbf{A}(P-\mathbf{E})+(P-\mathbf{E})\mathbf{A}^T+\mathbf{A}_\mu\mathbf{B}\mathbf{B}^T\mathbf{A}_\mu^*=\mathbf{0}.$$

Now, define

$$\mathbf{B}_{\mu_i}\equiv\left(\prod_{\ell=1}^{i}\mathbf{A}_{\mu_\ell}\right)\mathbf{B},\ \ \mathbf{B}^{(i)}\equiv\sqrt{2|\rho_i|}(\mathbf{A}+\mu_i\mathbf{I})^{-1}\mathbf{B}_{\mu_{i-1}},\ \text{ and }\ \mathbf{E}_i\equiv\mathbf{B}^{(i)}\mathbf{B}^{(i)*}\quad(11.13)$$

with $\mathbf{B}_{\mu_0}\equiv\mathbf{B}$. Then

$$\mathbf{A}(P-E_i)+(P-E_i)\mathbf{A}^T+\mathbf{B}_{\mu_i}\mathbf{B}_{\mu_i}^*=\mathbf{0},\ \text{ for } i=1,2,\ldots,k,\qquad(11.14)$$

where $E_i\equiv\mathbf{E}_1+\mathbf{E}_2+\cdots+\mathbf{E}_i$.

The discussion leading to Equations (11.13), (11.14) are valid for arbitrary $\mathbf{B}$. However, when $\mathbf{B}$ is a vector it is possible to establish a direct relationship with the ADI method.

Lemma 6. *Suppose that $\mathbf{B}=\mathbf{b}$ is a vector. Let $\mathbf{H}_j=\mathbf{U}_j^T\mathbf{A}\mathbf{U}_j\in\mathbb{R}^{k\times k}$ and $\hat{\mathbf{B}}_j=\mathbf{U}_j^T\mathbf{B}_j$ where $\mathbf{B}_j$ is the rhs at Step j. Then the update $P_{j+1}=P_j+\mathbf{Z}_j\hat{P}^{-1}\mathbf{Z}_j^T$ is precisely the same as the update that would be obtained by applying k steps of ADI with shifts $\{\mu_1,\mu_2,\ldots,\mu_k\}=\sigma(\mathbf{H}_j)$ (the spectrum of $\mathbf{H}_j$).*

Proof: From Equations (11.13), (11.14), the update obtained by applying shifts $\{\mu_1,\mu_2,\ldots,\mu_k\}=\sigma(\mathbf{H}_j)$ is of the form $\mathbf{L}_j\mathbf{L}_j^T$ where

$$\mathbf{L}_j=\left[\mathbf{B}^{(1)},\mathbf{B}^{(2)},\ldots,\mathbf{B}^{(k)}\right],\ \text{ with }\ \mathbf{B}^{(i)}\equiv\sqrt{2\rho_i}(\mathbf{A}+\mu_j\mathbf{I})^{-1}\left(\prod_{\ell=1}^{i-1}\mathbf{A}_{\mu_\ell}\right)\mathbf{B}_j,$$

and with $\rho_i=|\mathrm{Real}(\mu_i)|$.

Now, regardless of how the solution to $\mathbf{A}\mathbf{Z}+\mathbf{Z}\mathbf{H}_j^T+\mathbf{B}_j\hat{\mathbf{B}}_j^T=\mathbf{0}$ is obtained, it is mathematically equivalent to the solution obtained by the ADI method. This solution is given by $\mathbf{Z}_j=\mathbf{L}_j\hat{\mathbf{L}}_j^T$ where

$$\hat{\mathbf{B}}^{(i)}\equiv\sqrt{2\rho_i}(\hat{\mathbf{H}}_j+\mu_j\mathbf{I})^{-1}\left(\prod_{\ell=1}^{i-1}\hat{\mathbf{H}}_{\mu_i}\right)\hat{\mathbf{B}}_j\ \text{ and }\ \hat{\mathbf{L}}_j=\left[\hat{\mathbf{B}}^{(1)},\hat{\mathbf{B}}^{(2)},\ldots,\hat{\mathbf{B}}^{(k)}\right].$$

Moreover, $\hat{P}_j=\hat{\mathbf{L}}_j\hat{\mathbf{L}}_j^T$ is the solution to

$$\mathbf{H}_j\hat{P}+\hat{P}\mathbf{H}_j^T+\hat{\mathbf{B}}_j\hat{\mathbf{B}}_j=\mathbf{0}.$$

Thus

$$\mathbf{Z}_j\hat{P}^{-1}\mathbf{Z}_j^T=\mathbf{L}_j\mathbf{L}_j^T$$

which is the same as would have been obtained by the ADI method with shifts μ_i. This concludes the proof. ■

This result together with the fact that P_j is convergent can be used to establish a rate of convergence involving the spectral radius $\rho(\prod_{i=1}^{k} \mathbf{A}_{\mu_i})$, where the $\{\mu_i\}$ are the limits of the eigenvalues of $\mathbf{H}_j$. Unfortunately, the connection of the parameter free iteration with ADI is not so straightforward in the case that $\mathbf{B}$ has $m > 1$ columns and a rate of convergence has not yet been established for this case.

This automatic shift selection strategy is quite different than existing schemes proposed in the literature such as the sophisticated scheme proposed in [12]. In particular, for complex spectra, the Lu-Wachspress scheme [12] requires explicit knowledge of all of the eigenvalues of the matrix $\mathbf{A}$ for an optimal shift selection. At the very least, this theory requires that all eigenvalues of $\mathbf{A}$ lie within a certain "elliptic function domain". Such information is generally not available in the large scale setting. Moreover, if $\mathbf{A}$ is non-normal then its effective spectral properties with respect to iterative methods are determined by pseudo-spectra rather than the traditional spectrum of $\mathbf{A}$ (see [20]). The PFADI scheme proposed here doesn't require any a priori knowledge of the spectrum of $\mathbf{A}$. The shift parameters (which are implicitly applied) are determined by the small projected matrix $\mathbf{H}$ whose spectrum is automatically determined and influenced by the pseudo-spectrum of $\mathbf{A}$ when appropriate.

5 Implementation Details

There are a number of implementation details that must be specified in order to turn the basic of PFADI into a complete algorithm (Figure 11.3). These details are briefly outlined in this section.

Controlling the Condition of $\hat{P}$: Computing the update $\mathbf{ZGZ}^T$ and the modification $\mathbf{B} \leftarrow (\mathbf{I} - \mathbf{ZGU}^T)\mathbf{B}$ involves some delicate numerical issues that can introduce instabilities into the algorithm if not properly handled. This problem arises when $\hat{P}$ becomes ill-conditioned.

Recall that $\hat{P}$ is obtained from the equation $\mathbf{H}\hat{P} + \hat{P}\mathbf{H}^T + \hat{\mathbf{B}}\hat{\mathbf{B}}^T = \mathbf{0}$. The convergence of P_j may be used to show that $\mathbf{ZGU}^T\mathbf{B}$ must be well behaved in exact arithmetic even though $\mathbf{G} = \hat{P}^{-1}$ becomes arbitrarily large in norm as the iteration progresses. However, in finite precision this term can be problematic and must be handled with care.

One way to deal with this term is to control the conditioning of $\hat{P}$. To do this, one can just compute the eigensystem $\hat{P} = \mathbf{Q}\hat{\mathbf{S}}^2\mathbf{Q}^T$ with $\hat{\sigma}_1 \geq \hat{\sigma}_2 \geq \cdots \geq \hat{\sigma}_m$ the diagonal elements of $\hat{\mathbf{S}}$. The strategy is to maintain a uniformly bounded condition number for $\hat{P}$ via truncation. Namely, if $\hat{\sigma}_k \geq \tau\hat{\sigma}_1 \geq \hat{\sigma}_{k+1}$ then the search subspace is reduced to order k using the leading k columns $\mathbf{Q}_k$ of $\mathbf{Q}$ and the leading principal order k submatrix $\hat{\mathbf{S}}_k$ of $\hat{\mathbf{S}}$.

The Sylvester equation is then modified as follows:

$$1) \ \mathbf{Z}_j \leftarrow \mathbf{Z}_j\mathbf{Q}_k\hat{\mathbf{S}}_k^{-1} \quad \text{and} \quad 2) \ \mathbf{B}_{j+1} = \mathbf{B}_j - \mathbf{Z}_j\hat{\mathbf{S}}_k^{-1}\mathbf{Q}_k^T\mathbf{U}_j^T\mathbf{B}_j$$

The convergence theory will still be valid after this modification which should be carried out prior to solving for $\mathbf{Z}$.

Stopping Rules: The PFADI method has a very convenient and effective stopping rule which follows directly from the equation

$$\mathbf{A}P_j + P_j\mathbf{A}^T + \mathbf{B}\mathbf{B}^T = \mathbf{B}_j\mathbf{B}_j^T.$$

Hence, $\|\mathbf{A}P_j + P_j\mathbf{A}^T + \mathbf{B}\mathbf{B}^T\| = \|\mathbf{B}_j\|^2$ and thus with $\mathbf{B}_j$ comes an immediate calculation of the residual norm without having to explicitly compute the residual. This suggest stopping the iteration when one or both of the following conditions are satisfied:

1. $\dfrac{\|P_{j+1} - P_j\|_2}{\|P_{j+1}\|_2} = \dfrac{\|\mathbf{Z}_j \hat{P}_j^{-1/2}\|_2^2}{\|\mathbf{S}_{j+1}\|_2^2} \leq tol$
2. $\dfrac{\|\mathbf{B}_j\|_2}{\|\mathbf{B}\|_2} \leq \sqrt{tol}.$

Step 1: Solve the reduced order Lyapunov equation for $\mathscr{P}_j$

i) Solve $\mathbf{H}_j\mathscr{P}_j + \mathscr{P}_j\mathbf{H}_j^T + \hat{\mathbf{B}}_j\hat{\mathbf{B}}_j^T = \mathbf{0}$.

ii) Reduce size of subspace so that cond($\mathscr{P}_j$) $> \frac{1}{\tau}$.

Step 2: (APM step) Solve a projected Sylvester equation for $\mathbf{Z}$

$\mathbf{A}\mathbf{Z}_j + \mathbf{Z}_j\mathbf{H}_j^T + \mathbf{B}_j\hat{\mathbf{B}}_j^T = \mathbf{0}$, with $\mathbf{H}_j = \mathbf{U}_j^T\mathbf{A}\mathbf{U}_j$, $\hat{\mathbf{B}}_j = \mathbf{U}_j^T\mathbf{B}_j$.

Step 3: Modify $\mathbf{B}_j$

Update $\mathbf{B}_{j+1} \leftarrow (\mathbf{I} - \mathbf{Z}_j\mathscr{P}_j^{-1}\mathbf{U}_j^T)\mathbf{B}_j$.

Step 4: (ADI step) Update factorization and basis $\mathbf{U}$

Re-scale $\mathbf{Z}_j \leftarrow \mathbf{Z}\mathscr{P}_j^{-1/2}$.
Update (and truncate) $[\mathbf{U},\mathbf{S}] \leftarrow \mathrm{svd}[\mathbf{L}_j, \mathbf{Z}_j]$.
$\mathbf{L}_{j+1} = \mathbf{U}\mathbf{S}$.
$\mathbf{U}_{j+1} \leftarrow \mathbf{U}(:, 1:k_j)$, basis for dominant subspace of dimension k_j.

Fig. 11.3 A Parameter Free Lyapunov Solver (PFADI)

Solving the projected Sylvester equation:

In order to compute the update $\mathbf{Z}$ one must solve the projected Sylvester equation

$$\mathbf{A}\mathbf{Z} + \mathbf{Z}\mathbf{H}^T + \mathbf{B}\hat{\mathbf{B}}^T = \mathbf{0}, \text{ where } \mathbf{H} = \mathbf{U}^T\mathbf{A}\mathbf{U} \text{ and } \hat{\mathbf{B}} = \mathbf{U}^T\mathbf{B}.$$

To solve this equation, one could utilize a minor variant of the Bartels-Stewart algorithm. If $\mathbf{H}^T = \mathbf{Q}\mathbf{R}\mathbf{Q}^*$ is a Schur decomposition of $\mathbf{H}^T$, then one may solve

$$\mathbf{A}\tilde{\mathbf{Z}} + \tilde{\mathbf{Z}}\mathbf{R} + \mathbf{B}\tilde{\mathbf{B}}^T = \mathbf{0}, \text{ where } \tilde{\mathbf{Z}} = \mathbf{Z}\mathbf{Q} \text{ and } \tilde{\mathbf{B}} = \mathbf{Q}^*\hat{\mathbf{B}}.$$

Then put $\mathbf{Z} = \tilde{\mathbf{Z}}\mathbf{Q}^*$. This takes the form for j = 1:k,

Solve $(\mathbf{A} - \rho_{jj}\mathbf{I})\mathbf{z}_j = \mathbf{B}\tilde{\mathbf{B}}^T\mathbf{e}_j - \Sigma_{i=1}^{j-1}\mathbf{z}_i\rho_{i,j}$;

end

where the elements of $\mathbf{R}$ have been denoted as $\rho_{i,j}$.

This approach has the advantage that the assumption that $\mathbf{A}+\mathbf{A}^T$ is negative definite is no longer needed. The algorithm is valid regardless of the stability of $\mathbf{H}$ so long as $\sigma(\mathbf{A}) \cap \sigma(-\mathbf{H}) = \emptyset$. On the other hand, each step requires the sparse direct factorization of the matrix $\mathbf{A} - \rho_{jj}\mathbf{I}$ and this must be done in complex arithmetic whenever an eigenvalue ρ_{jj} of $\mathbf{H}$ is complex.

An alternative approach is to construct a basis $[\mathbf{X}^T, \mathbf{Y}^T]^T$ for a block upper triangular matrix:

$$\begin{bmatrix} \mathbf{A} & \mathbf{M} \\ \mathbf{0} & -\mathbf{H}^T \end{bmatrix} \begin{bmatrix} \mathbf{X} \\ \mathbf{Y} \end{bmatrix} = \begin{bmatrix} \mathbf{X} \\ \mathbf{Y} \end{bmatrix} \hat{\mathbf{R}}, \tag{11.15}$$

where $\mathbf{M} = \mathbf{B}\hat{\mathbf{B}}^T$. It is easily seen that the solution $\mathbf{Z}$ to the Sylvester equation is recovered as $\mathbf{Z} = \mathbf{X}\mathbf{Y}^{-1}$ whenever $\mathbf{Y}$ is nonsingular.

Notice that the spectrum of the block upper triangular matrix in Equation (11.15) is $\sigma(\mathbf{A}) \cup \sigma(-\mathbf{H}^T$ and it is easily seen that $\mathbf{Y}$ is nonsingular if and only if $\sigma(\hat{\mathbf{R}}) = \sigma(-\mathbf{H}^T)$. This is established formally with the following lemma.

Lemma 7. *In Equation* (11.15)*,* $\mathbf{Y}$ *is nonsingular if and only if* $\sigma(\hat{\mathbf{R}}) = \sigma(-\mathbf{H}^T)$.

Proof: The equation $-\mathbf{H}^T\mathbf{Y} = \mathbf{Y}\hat{\mathbf{R}}$ follows directly from Equation 11.15. Thus, whenever $\mathbf{Y}$ is nonsingular, $\sigma(\hat{\mathbf{R}}) = \sigma(-\mathbf{H}^T)$.

Now, suppose $\sigma(\hat{\mathbf{R}}) = \sigma(-\mathbf{H}^T)$. If $\mathbf{Y}$ is singular, let $\mathbf{N}$ be a basis for $Null(\mathbf{Y})$ and note $Rank(\mathbf{N}) \geq 1$. Observe that

$$\mathbf{0} = -\mathbf{H}^T\mathbf{Y}\mathbf{N} = \mathbf{Y}\hat{\mathbf{R}}\mathbf{N},$$

which shows each column of $\hat{\mathbf{R}}\mathbf{N}$ is in $Null(\mathbf{Y})$. Hence $\hat{\mathbf{R}}\mathbf{N} = \mathbf{N}\mathbf{K}$ which implies the spectrum $\sigma(\mathbf{K}) \subset \sigma(\hat{\mathbf{R}}) = \sigma(-\mathbf{H}^T)$. Now, it follows from Equation (11.15) that

$$\mathbf{X}\mathbf{N}\mathbf{K} = \mathbf{X}\hat{\mathbf{R}}\mathbf{N} = \mathbf{A}\mathbf{X}\mathbf{N} + \mathbf{M}\mathbf{Y}\mathbf{N} = \mathbf{A}\mathbf{X}\mathbf{N}. \tag{11.16}$$

Since $\begin{bmatrix} \mathbf{X}\mathbf{N} \\ \mathbf{0} \end{bmatrix} = \begin{bmatrix} \mathbf{X} \\ \mathbf{Y} \end{bmatrix} \mathbf{N}$ is full rank, it follows that $\mathbf{X}\mathbf{N}$ is full rank and hence Equation (11.16) implies $\sigma(\mathbf{K}) \subset \sigma(\mathbf{A})$. This is a contradiction since $\sigma(-\mathbf{H}^T) \cap \sigma(\mathbf{A}) =$ which in turn implies $\mathbf{Y}$ must be nonsingular. ■

The important consequence of this result is that the relevant invariant subspace problem is highly structured. Since $\mathbf{H}$ is relatively small, its eigenvalues may be computed directly and then used as a test and also as an aid to efficiently solve the desired invariant subspace problem.

To illustrate some possibilities, define

$$\tilde{\mathbf{A}} \equiv \begin{bmatrix} \mathbf{A} & \mathbf{M} \\ \mathbf{0} & -\mathbf{H}^T \end{bmatrix}.$$

One could just apply an iterative eigenvalue method to compute a basis for the invariant subspace of $\tilde{\mathbf{A}}$ corresponding to the eigenvalues of $-\mathbf{H}^T$. The dimension of this subspace and the eigenvalues are determined by $\mathbf{H}$. Obviously such iterative methods are accelerated by employing spectral transformations such as shift-invert or Cayley transformations.

Now, consider a real positive *shift* $\mu \in (0, \min Re(\sigma(-\mathbf{H}^T))$. Since the eigenvalues of $\mathbf{H}$ may be computed directly, placement of μ can be done with precision and the shift-invert operator $(\tilde{\mathbf{A}} - \mu\mathbf{I})^{-1}$ maps the eigenvalues of $\mathbf{A}$ into the open left half plane and the eigenvalues of $-\mathbf{H}^T$ into the open right half plane with the eigenvalues of $-\mathbf{H}^T$ closest to the imaginary axis mapped to extreme eigenvalues of the shift-invert operator. Positive real shifts taken in the interior of the spectrum of $-\mathbf{H}^T$ could be used to enhance the magnitude of the interior eigenvalues.

Another very useful possibility for utilization of the shift μ is to consider a Cayley transformation

$$\mathbf{C}_{\tilde{\mathbf{A}},\mu} \equiv (\mu\mathbf{I} - \tilde{\mathbf{A}})^{-1}(\mu\mathbf{I} + \tilde{\mathbf{A}}).$$

Under this transformation, an eigenvalue λ of $\tilde{\mathbf{A}}$ is mapped to an eigenvalue $\omega = \frac{\mu+\lambda}{\mu-\lambda}$ of $\mathbf{C}_{\tilde{\mathbf{A}},\mu}$. These eigenvalues have the same eigenvector. The eigenvalues of $\tilde{\mathbf{A}}$ consist of $\sigma(\mathbf{C}(\mathbf{A},\mu)) \cup \sigma(\mathbf{C}(-\mathbf{H}^T,\mu))$. Since both $\mathbf{A}$ and $\mathbf{H}$ are stable, the eigenvalues of $\mathbf{A}$ are mapped strictly to the interior of the unit disc while the eigenvalues of $-\mathbf{H}^T$ are mapped strictly to the exterior of the unit disc. An iterative method (such as the implicitly restarted Arnoldi method available in ARPACK) may be used to rapidly compute the k eigenvalues of largest magnitude for $\mathbf{C}_\mu$. This only requires a single sparse direct factorization which can be done in real arithmetic. The parameter μ may be chosen to enhance convergence of the iterative eigenvalue method. When a sparse direct factorization is not affordable, one can resort to an iterative approach to applying the Cayley transformation to approximate a matrix-vector product with $\mathbf{C}_{\tilde{\mathbf{A}},\mu}$ as required.

Ryan Nong investigated effective choices of μ in his Ph.D. thesis [13]. The results were quite encouraging and the reader is referred to the thesis for further detail. Extensive computational results concerning the performance of PFADI and the possible choices of a real shift μ are presented there.

Acknowledgments

I am greatly indebted to Mark Embree and also Serkan Gugercin for several discussions on this material. This work was originally presented at the GAMM-SIAM Linear Algebra Meeting, held in Dusseldorf during July 2006.

References

1. A. Antoulas, D. Sorensen, and S. Gugercin, *A survey of model reduction methods for large-scale systems*, Contemporary Mathematics, AMS Publications, 280 (2001), pp. 193–219.
2. A. C. Antoulas and D. C. Sorensen, *Approximation of large-scale dynamical systems: An overview*, International J. of Applied Mathematics and Computational Science, 11 (2001), pp. 1093–1121.
3. A. C. Antoulas, D. C. Sorensen, and Y. Zhou, *On the decay rate of Hankel singular values and related issues*, Systems and Control Lett., 46 (2002), pp. 323–342.
4. A. N. Beavers and E. D. Dennam, *A new solution method for the Lyapunov matrix equations*, SIAM J. Appl. Math., 29 (1975), pp. 416–421.
5. P. Benner and E. S. Quintana-Orti, *Solving stable generalized Lyapunov equations with the matrix sign function*, Sonderforschungsbereich, 393 (1997).
6. D. Calvetti, B. Lewis, and L. Reichel, *On the solution of large Sylvester-observer equations*, Numer. Linear Algebra Appl., 8 (2001), p. 35–451.
7. N. Ellner and E. Wachspress, *Alternating direction implicit iteration for systems with complex spectra*, SIAM J. Num. Anal., 28(3) (1991), pp. 859–870.
8. S. Gugercin, D. Sorensen, and A. Antoulas, *A modified low-rank Smith method for large-scale Lyapunov equations*, Numerical Algorithms, 32 (2003), pp. 27–55.
9. A. S. Hodel, B. Tenison, and K. R. Poolla, *Numerical solution of the Lyapunov equation by approximate power iteration*, Linear Algebra and Its Applications, 236 (1996), pp. 205–230.
10. W. D. Hoskins, D. S. Meek, and D. J. Walton, *The numerical solution of A'Q+QA=-C*, IEEE Trans. Automat. Control, AC-22 (1977), pp. 882–883.
11. J.-R. Li, *Model Reduction of Large Linear Systems via Low Rank System Gramians*, PhD thesis, Massachusetts Institute of Technology, 2000.
12. A. Lu and E. Wachspress, *Solution of Lyapunov equations by alternating direction implicit iteration*, Computers & Mathematics with Applications, 21(9) (1991), pp. 43–58.
13. H. D. Nong, *Numerical Solutions of Matrix Equations Arising in Model Reduction of Large-Scale Linear-Time-Invariant Systems*, PhD thesis, Rice University, 2009.
14. D. Peaceman and H. Rachford, *The numerical solution of elliptic and parabolic differential equations*, J. Soc. Indust. Appl. Math., 3 (1955), pp. 28–41.
15. T. Penzl, *A cyclic low-rank Smith method for large sparse Lyapunov equations*, SIAM J. Sci. Comput., 21 (2000), pp. 1401–1418.
16. F. Riesz and B. S. Nagy, *Functional Analysis*, Ungar, second ed., 1955.
17. J. Sabino, *Solution of Large-Scale Lyapunov Equations via the Block Modified Smith Method*, PhD thesis, Rice University, 2006.
18. V. Simoncini, *A new iterative method for solving large-scale Lyapunov matrix equations*, SIAM J. Sci. Comput., 29 (2007), p. 1268–1288.
19. R. Smith, *Matrix equation $XA + BX = C$*, SIAM J. Appl. Math., 16 (1968).
20. L. N. Trefethen and M. Embree, *Spectra and Pseudospectra*, Princeton University Press: Princeton, Oxford, 2005.
21. D. Vasilyev and J. White, *A more reliable reduction algorithm for behavior model extraction*, Proceedings of the IEEE/ACM International Conference on Computer-Aided Design, (2005), pp. 813–820.
22. K. Zhou, J. C. Doyle, and K. Glover, *Robust and Optimal Control*, Prentice Hall, 1996.
23. Y. Zhou and D. Sorensen, *Approximate implicit subspace iteration with alternating directions for LTI system model reduction*, Numer. Linear Algebra Appl. 2008, 15 (2008), pp. 873–886.

Chapter 12
Splitting Enables Overcoming the Curse of Dimensionality

Jérôme Darbon and Stanley J. Osher

Abstract In this chapter we briefly outline a new and remarkably fast algorithm for solving a large class of high dimensional Hamilton-Jacobi (H-J) initial value problems arising in optimal control and elsewhere [1]. This is done without the use of grids or numerical approximations. Moreover, by using the level set method [8] we can rapidly compute projections of a point in $\mathbb{R}^n$, n large to a fairly arbitrary compact set [2]. The method seems to generalize widely beyond what will we present here to some nonconvex Hamiltonians, new linear programming algorithms, differential games, and perhaps state dependent Hamiltonians.

1 Introduction

We begin with the Hamilton-Jacobi (HJ) initial value problem

$$\begin{cases} \dfrac{\partial \varphi}{\partial t}(x,t) + H(\nabla_x \varphi(x,t)) = 0 & \text{in } \mathbb{R}^n \times (0,+\infty) \\ \varphi(x,0) = J(x) & \forall x \in \mathbb{R}^n. \end{cases} \tag{12.1}$$

We assume $J : \mathbb{R}^n \to \mathbb{R}$ is convex and one coercive, i.e., $\lim_{\|x\|_2 \to +\infty} \frac{J(x)}{\|x\|_2} = +\infty$, $H : \mathbb{R}^n \to \mathbb{R}$ is convex and positively one homogeneous (we sometimes relax all these assumptions).

J. Darbon (✉)
CNRS/CMLA-Ecole Normale Supérieure de Cachan, Cachan, France
e-mail: darbon@cmla.ens-cachan.fr

S.J. Osher
Department of Mathematics, UCLA, Los Angeles, CA 90095-1555, USA
e-mail: sjo@math.ucla.edu

R. Glowinski et al. (eds.), *Splitting Methods in Communication, Imaging, Science, and Engineering*, Scientific Computation, DOI 10.1007/978-3-319-41589-5_12

A good example of this is

$$H(v) = \|v\|_2.$$

Here $\|v\|_p = \left(\Sigma_{i=1}^n |v_i|^p\right)^{\frac{1}{p}}$ for $p \geq 1$ and $\langle x, v\rangle = \Sigma_{i=1}^n x_i v_i$.

Let us consider a convex Lipschitz function J having the property that, for Ω a convex compact set of $\mathbb{R}^n$

$$\begin{cases} J(x) < 0 & \text{for any } x \in \text{ int } \Omega \\ J(x) = 0 & \text{for any } x \in (\Omega \setminus \text{ int } \Omega) \\ J(x) > 0 & \text{for any } x \in (\mathbb{R}^n \setminus \Omega). \end{cases}$$

We call this level set initial data. Then the set of points for which $\varphi(x,t) = 0,\ t > 0$ are exactly those at a distance t, from the boundary of Ω. In fact given $\bar{x} \notin \Omega$, then the closest point x_{opt} from $\bar{x}$ to $(\Omega \setminus \text{ int } \Omega)$ is exactly

$$x_{\text{opt}} = \bar{x} - t\frac{\nabla\varphi(\bar{x},t)}{\|\nabla\varphi(\bar{x},t)\|_2}. \tag{12.2}$$

To solve (12.1) we use the Hopf formula [5]

$$\varphi(x,t) = (J^* + tH)^*(x) = -\min_{v \in R^n}\{J^*(v) + tH(v) - \langle x, v\rangle\},$$

where the Fenchel-Legendre transform $f^* : \mathbb{R}^n \to R \cup (+\infty)$ of the convex function f is defined by

$$f^*(v) = \sup_{x \in R^n}\{\langle v, x\rangle - f(x)\}.$$

Moreover, for free we get that the minimizer satisfies

$$\arg\min_{v \in \mathbb{R}^n}\{J^*(v) + tH(v) - \langle x, v\rangle\} = \nabla_x\varphi(x,t). \tag{12.3}$$

whenever $\varphi(\cdot,t)$ is differentiable at x. Let us note here that our algorithm computes $\varphi(x,t)$ but also $\nabla_x\varphi(x,t)$.

Also, we can use the Hopf-Lax formula [5, 6] to solve (12.1).

$$\varphi(x,t) = \min_{z \in \mathbb{R}^n}\left\{J(z) + tH^*\left(\frac{x-z}{t}\right)\right\} \tag{12.4}$$

for convex H.

From (12.4) it is easy to show that if we have k different initial value problems $i = 1, \ldots k$

$$\begin{cases} \dfrac{\partial\varphi_i}{\partial t}(x,t) + H(\nabla_x\varphi_i(x,t)) = 0, & \text{in } \mathbb{R}^n \times (0, +\infty) \\ \varphi_i(x,0) = J_i(x) & \forall x \in \mathbb{R}^n \end{cases}$$

with the usual hypotheses, then (12.4) implies, for any $x \in \mathbb{R}^n$, $t > 0$

$$\varphi_i(x,t) = \min_{z \in R^n} \left\{ J_i(z) + tH^* \left(\frac{x-z}{t} \right) \right\}.$$

So

$$\min_{i=1,k} \varphi_i(x,t) = \min_{z \in R^n} \left\{ \min_{i=1,\dots,k} \left\{ J_i(z) + tH^* \left(\frac{x-z}{t} \right) \right\} \right\}$$

solves the initial value problem

$$\begin{cases} \dfrac{\partial \varphi}{\partial t}(x,t) + H(\nabla_x \varphi(x,t)) = 0, & \text{in } \mathbb{R}^n \times (0,+\infty) \\ \varphi(x,0) = \min\limits_{i=1,\dots,k} J_i(x) & \forall x \in \mathbb{R}^n. \end{cases} \tag{12.5}$$

This means that if $\Omega = \cup_{i=1,\dots,k} \Omega_i$, where each Ω_i is compact and convex and may overlap, then we can easily compute the set of all points at distance t from Ω which is exactly the solution to (12.5) where each J_i is a level set function for Ω_i. Moreover, at every point $\bar{x}$ outside of $\bar{\Omega}$ for which there is one i such that $\varphi_i(\bar{x},t) < \varphi_{i'}(\bar{x},t)$ for any $i \neq i'$, then the closest point x_{opt} to $\bar{x}$ and Ω is again

$$x_{\text{opt}} = \bar{x} - t \frac{\nabla_x \varphi_i(\bar{x},t)}{|\nabla_x \varphi_i(\bar{x},t)|}.$$

If there are several i for which $\varphi_i(\bar{x},t)$ is the minimum among all k of them, then $\nabla_x \varphi$ will be "multivalued", i.e., it will have jumps, but any of the x_{opt} defined above will be a closest point on Ω to $\bar{x}$.

2 Split Bregman

We solve the optimization problem (12.3) by using the split Bregman algorithm [4, 3, 9] as follows

$$v^{k+1} = \arg\min_{v \in \mathbb{R}^n} \{ J^*(v) - \langle x, v \rangle + \frac{\lambda}{2} \| d^k - v - b^k \|_2^2 \}, \tag{12.6}$$

$$d^{k+1} = \arg\min_{d \in \mathbb{R}^n} \left\{ tH(d) + \frac{\lambda}{2} \| d - v^{k+1} - b^k \|_2^2 \right\} \tag{12.7}$$

$$b^{k+1} = b^k + v^{k+1} - d^{k+1}. \tag{12.8}$$

Here the sequences $(v^k)_{k \in \mathbb{N}}, (d^k)_{k \in \mathbb{N}}$ both converge to $\nabla_x \varphi(x,t)$. Let us emphasize again that our numerical algorithm not only computes the solution $\varphi(x,t)$ but also computes $\nabla_x \varphi(x,t)$ when $\varphi(\cdot,t)$ is differentiable.

Both (12.6) and (12.7), up to change of variables, can be reformulated as finding the unique minimizer of

$$\arg\min_w \left\{ \alpha f(w) + \frac{1}{2}\|w-z\|_2^2 \right\}$$

which is the proximal map of f. Equation (12.6) can be solved if either J^* or J have easily computable proximal maps, which often occurs, especially if one of them is smooth.

Equation (12.7) can be easily solved if $H(d) = \|d\|_2$ via the shrink_2 operator defined by

$$\text{shrink}_2(z,\alpha) = \begin{cases} \frac{z}{\|z\|_2}\max(\|z\|_2-\alpha,0) & \text{if } z \neq 0 \\ 0 & \text{if } z = 0 \end{cases}$$

and we have

$$\arg\min_w \left\{ \alpha\|w\|_2 + \frac{1}{2}\|w-z\|_2^2 \right\} = \text{shrink}_2(z,\alpha)$$

If $H(d) = \|d\|_1$ we use shrink_1 operator defined as follows for any $i = 1,\dots,n$

$$(\text{shrink}_1(z,\alpha))_i = \begin{cases} z_i - \alpha & \text{if } z_i > \alpha \\ 0 & \text{if } |z_i| \leq \alpha \\ z_i + \alpha & \text{if } z_i < -\alpha \end{cases}$$

and we have

$$\arg\min_w \left\{ \alpha\|w\|_1 + \frac{1}{2}\|w-z\|_2^2 \right\} = \text{shrink}_1(z,\alpha).$$

To solve (12.7) for more general $H(d)$ convex one homogeneous or to find the proximal map for f of that type we use the fact that H^* is the characteristic function of a closed convex set $C \subset \mathbb{R}^n$

$$H^* = I_c.$$

By using the Moreau identity [7] we realize that the proximal map of H can be obtained by projecting onto C. To do this projection, we merely solve the eikonal equation with level set initial data for C via split Bregman as above in (12.6), (12.7), (12.8) with $H(d) = \|d\|_2$. This is easy using the shrink_2 operator. We then use (12.2) to obtain the projection and repeat the entire iteration.

3 Numerical Experiments

Numerical experiments on an Intel Laptop Core i5-5300U running at 2.3 GHz are now presented. We consider diagonal matrices D defined by $D_{ii} = 1 + \frac{1+i}{n}$ for $i = 1,\dots,n$. We also consider matrices A defined by $A_{ii} = 2$ for $i = 1,\dots,n$ and $A_{ij} = 1$

for $i,j = 1,\dots,n$. Table 12.1 presents the average time (in seconds) to evaluate the solution over 1,000,000 samples (x,t) uniformly drawn in $[-10,10]^n \times [0,10]$. The convergence is remarkably rapid: 10^{-6} to 10^{-8} seconds on a standard laptop, per function evaluation. Figure 12.1 depicts 2-dimensional slices at different times for the (H-J) equation with a weighted ℓ_1 Hamiltonian $H = \|D\cdot\|_1$, initial data $J = \frac{1}{2}\|\cdot\|_2^2$ and $n = 8$.

n	$\|y\|_1$	$\|y\|_2$	$\|y\|_\infty$	$\|y\|_D$	$\|y\|_A$
4	6.36e-08	1.20e-07	2.69e-07	7.00e-07	8.83e-07
8	6.98e-08	1.28e-07	4.89e-07	1.07e-06	1.57e-06
12	8.72e-08	1.56e-07	7.09e-07	1.59e-06	2.23e-06
16	9.24e-08	1.50e-07	9.92e-07	2.04e-06	2.95e-06

Table 12.1 Time results in seconds for the average time per call for evaluating the solution of the HJ-PDE with the initial data $J = \frac{1}{2}\|\cdot\|_2^2$, several Hamiltonians and various dimensions n.

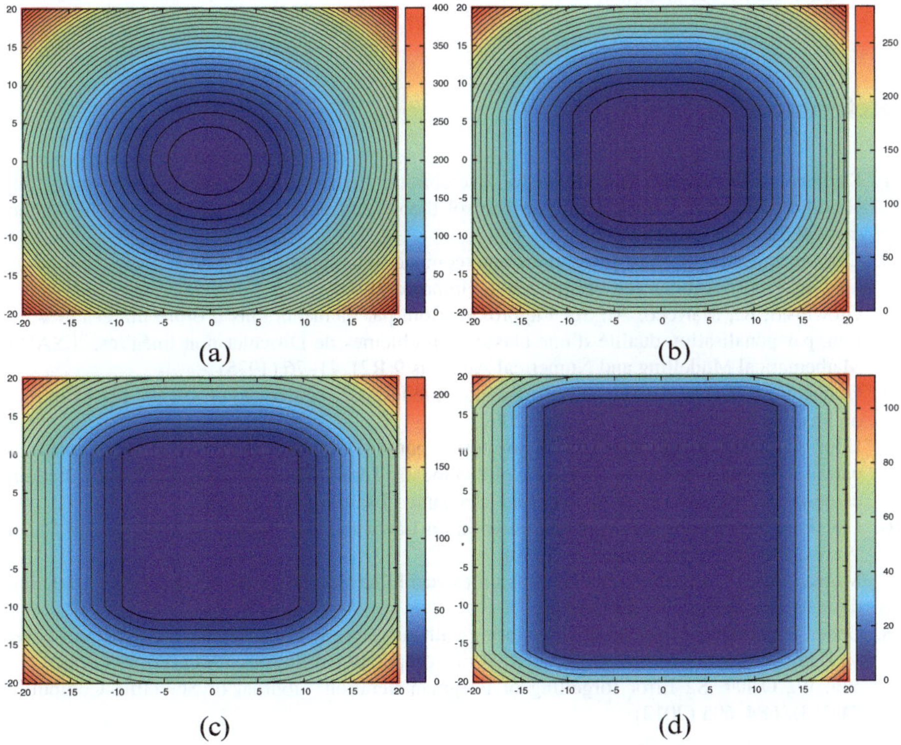

Fig. 12.1 Evaluation of the solution $\phi((x_1,x_2,0,0,0,0,0,0)^\dagger,t)$ of the HJ-PDE with initial data $J = \frac{1}{2}\|\cdot\|_2^2$ and Hamiltonian $H = \|D\cdot\|_1$ for $(x_1,x_2) \in [-20,20]^2$ for different times t. Plots for $t = 0,3,5,8$ and respectively depicted in (a), (b), (c), and (d). The level lines multiple of 10 are superimposed on the plots.

4 Summary and Future Work

We have derived a very fast and totally parallelizable method to solve a large class of high dimensional, state independent H-J initial value problems. We do this suing the Hopf formula and convex optimization via splitting, which overcomes the "curse of dimensionality". This is also done without the use of grids or numerical approximations, yielding not only the solution, but also its gradient.

We also, as a step in this procedure, very rapidly compute the projections from a point in $\mathbb{R}^n$, n large, to a fairly arbitrary compact set.

In future work, we expect to extend this set of ideas to nonconvex Hamiltonians, including some that arise in differential games, to new linear programming algorithms, to fast methods for redistancing in level set methods and, hopefully, to a wide class of state dependent Hamiltonians.

Acknowledgements Research supported by ONR grants N000141410683, N000141210838 and DOE grant DE-SC00183838.

References

1. Darbon, J., Osher, S.: Algorithms for overcoming the curse of dimensionality for certain Hamilton-Jacobi equations arising in control theory and elsewhere. Research in the Mathematical Sciences (to appear)
2. Darbon, J., Osher, S.: Fast projections onto compact sets in high dimensions using the level set method, Hopf formulas and optimization. (In preparation)
3. Glowinski, R., Marroco, A.: Sur l'approximation, par éléments finis d'ordre un, et la résolution, par pénalisation-dualité d'une classe de problèmes de Dirichlet non linéaires. ESAIM: Mathematical Modelling and Numerical Analysis **9**(R2), 41–76 (1975)
4. Goldstein, T., Osher, S.: The split Bregman method for L1-regularized problems. SIAM Journal on Imaging Sciences **2**(2), 323–343 (2009)
5. Hopf, E.: Generalized solutions of non-linear equations of first order (First order nonlinear partial differential equation discussing global locally-Lipschitzian solutions via Jacoby theorem extension). Journal of Mathematics and Mechanics **14**, 951–973 (1965)
6. Lax, P.D.: Hyperbolic Systems of Conservation Laws and the Mathematical Theory of Shock Waves. SIAM, Philadelphia, PA (1990)
7. Moreau, J.J.: Proximité et dualité dans un espace hilbertien. Bulletin de la Société Mathématique de France **93**, 273–299 (1965)
8. Osher, S., Sethian, J.A.: Fronts propagating with curvature-dependent speed: Algorithms based on Hamilton-Jacobi formulations. Journal of Computational Physics **79**(1), 12–49 (1988)
9. Yin, W., Osher, S.: Error forgetting of Bregman iteration. Journal of Scientific Computing **54**(2–3), 684–695 (2013)

Chapter 13
ADMM Algorithmic Regularization Paths for Sparse Statistical Machine Learning

Yue Hu, Eric C. Chi, and Genevera I. Allen

Abstract Optimization approaches based on operator splitting are becoming popular for solving sparsity regularized statistical machine learning models. While many have proposed fast algorithms to solve these problems for a single regularization parameter, conspicuously less attention has been given to computing regularization paths, or solving the optimization problems over the full range of regularization parameters to obtain a sequence of sparse models. In this chapter, we aim to quickly approximate the sequence of sparse models associated with regularization paths for the purposes of statistical model selection by using the building blocks from a classical operator splitting method, the Alternating Direction Method of Multipliers (ADMM). We begin by proposing an ADMM algorithm that uses warm-starts to quickly compute the regularization path. Then, by employing approximations along this warm-starting ADMM algorithm, we propose a novel concept that we term the *ADMM Algorithmic Regularization Path*. Our method can quickly outline the sequence of sparse models associated with the regularization path in computational time that is often less than that of using the ADMM algorithm to solve the problem

---

Y. Hu
Department of Statistics, Rice University, Houston, TX, USA
e-mail: yh6@rice.edu

E.C. Chi
Department of Statistics, North Carolina State University, Raleigh, NC, USA
e-mail: eric_chi@ncsu.edu

G.I. Allen (✉)
Department of Statistics and Electrical and Computer Engineering, Rice University,
Jan and Dan Duncan Neurological Research Institute, Baylor College of Medicine, Houston, TX, USA
e-mail: gallen@rice.edu

R. Glowinski et al. (eds.), *Splitting Methods in Communication, Imaging, Science, and Engineering*, Scientific Computation, DOI 10.1007/978-3-319-41589-5_13

for a single regularization parameter. We demonstrate the applicability and substantial computational savings of our approach through three popular examples, sparse linear regression, reduced-rank multi-task learning, and convex clustering.

1 Introduction

With the rise of Big Data and the subsequent explosion of statistical machine learning methods to analyze it, statisticians have become avid consumers of large-scale optimization procedures to estimate sparse models. The estimation problem is often cast as an optimization problem of the form:

$$\underset{\boldsymbol{\beta}}{\text{minimize}} \quad L(\boldsymbol{\beta};\mathbf{W}) + \lambda P(\boldsymbol{\beta}), \tag{13.1}$$

where $\boldsymbol{\beta}$ is a parameter which specifies a statistical model, $L(\boldsymbol{\beta};\mathbf{W})$ is a smooth loss function or data-fidelity term that quantifies the discrepancy between the data, $\mathbf{W}$, and the model specified by $\boldsymbol{\beta}$, and $P(\boldsymbol{\beta})$ is a nonsmooth penalty that encourages sparsity in model parameter $\boldsymbol{\beta}$ [3, 4, 15]. A regularization parameter, $\lambda \geq 0$, explicitly trades off the model fit and the model complexity.

Directly solving the optimization problem (13.1) is often challenging. Operator splitting methods, such as the Alternating Direction Method of Multipliers (ADMM), have become popular because they convert solving the problem into solving a sequence of simpler optimization problems that involve only the smooth loss or nonsmooth penalty. By breaking up the problem into smaller ones, ADMM may end up taking more iterations than directly solving (13.1), but it often runs in less total time since the subproblems are typically easy to solve. Clearly in the context of Big Data, faster algorithms are indispensable, and the numerical optimization community has devoted a great deal of effort to solving (13.1) rapidly for a fixed value of λ. This goal, however, is not necessarily aligned with the application of statistical machine learning problems to real data.

In practice, statisticians are interested in finding the best sparse model that represents the data. Achieving this typically entails a two-step procedure: (i) model selection, or selecting the best sparse model or equivalently the best subset of parameters, and (ii) model fitting, or fitting the model by minimizing the loss function over the selected parameters [15]. The first step is often the most challenging computationally as this entails searching the combinatorial space of all possible sparse models. As this combinatorial search is infeasible for large-scale problems, many consider convex relaxations through constraints or penalties as computationally feasible surrogates to help search through the space of sparse models. Consider for example, sparse linear regression, where the goal is to find the subset of variables or inputs that best predicts the response or output. Searching over all possible subsets of variables, however, is an NP hard problem. Instead, many have employed the penalty or constraint, $P(\boldsymbol{\beta}) = \|\boldsymbol{\beta}\|_1$, which is the tightest convex relaxation to performing best subset selection and whose solution can be computed in polynomial time.

The nonsmooth penalty term, $P(\boldsymbol{\beta})$, then serves to translate an infeasible computational problem into a tractable one for model selection purposes.

Suppose now that we focus on selecting the best sparse model by means of penalized statistical estimation as in (13.1). As λ varies, we trace out a continuous parametric curve $\hat{\boldsymbol{\beta}}(\lambda) \in \mathbb{R}^p$. Since this curve cannot be determined analytically in general, the curve is estimated for a finite sequence of regularization parameters. To choose the best model, statisticians inspect the sequence of sparse solutions to (13.1) over the full range of regularization parameters: $\{\hat{\boldsymbol{\beta}}(\lambda_n) : 0 \leq \lambda_1 \leq \cdots \leq \lambda_{\max}\}$, where $\lambda_{\max}$ is the value of λ at which $\hat{\boldsymbol{\beta}}(\lambda_{\max}) = \mathbf{0}$, the maximally sparse solution. This sequence of sparse solutions is often called the *regularization path* [9, 10, 14]. For model selection purposes, however, the actual parameter values, $\hat{\boldsymbol{\beta}}(\lambda)$, as λ varies in the regularization paths are less important than identifying the nonzero components of $\hat{\boldsymbol{\beta}}(\lambda)$. (Note that the parameter values for the optimal model are typically re-fit anyways in the second model fitting stage.) Instead, the support of $\hat{\boldsymbol{\beta}}(\lambda)$ or the sequence of *active sets* defined as $\mathscr{A}(\lambda) = \{j : \hat{\beta}_j(\lambda) \neq 0\}$, are the important items; these yield a good sequence of sparse models to consider that limit computationally intensive exploration of a combinatorial model space. Out of this regularization path or sequence of active sets, the optimal model can be chosen via a number of popular techniques such as minimizing the trade-off in model complexity as with the AIC and BIC, the prediction error as with cross-validation [15] or the model stability as with stability selection [23].

To apply sparse statistical learning methods to large-scale problems, we need fast algorithms not only to fit (13.1) for one value of λ, but also to estimate the entire sequence of sparse models in the model selection stage. Our objective in this chapter is to study the latter, which has received relatively little attention from the optimization community. Specifically, we seek to develop a new method to approximate the sequence of active sets associated with regularization paths that is (i) computationally fast and (ii) comprehensively explores the space of sparse models at a sufficiently fine resolution. In doing so, we will not try to closely approximate the parameter values, $\hat{\boldsymbol{\beta}}(\lambda)$, but instead try to closely approximate the sparsity of the parameters, $\mathscr{A}(\lambda)$, for the statistical learning problem (13.1).

To rapidly approximate the sequence of active sets associated with regularization paths, we turn to the ADMM optimization framework. We first introduce a procedure to estimate the regularization path by using the ADMM algorithm with warm starts over a range of regularization parameters to yield a path-like sequence of solutions. Extending this, we preform a one-step approximation along each point on this path, yielding the novel method that we term *ADMM Algorithmic Regularization Paths*. Our procedure can closely approximate active sets given by regularization paths at a fine resolution, but dramatically reduces computational time. This new approach to estimating a sequence of sparse models opens many interesting questions from both statistical and optimization perspectives. In this chapter, however, we focus on motivating our approach and demonstrating its computational advantages on several sparse statistical machine learning examples.

This chapter is organized as follows. We first review, in Section 1.1, how ADMM algorithms have been used in the statistical machine learning literature. Then, to motivate our approach, we consider, in Section 1.2, application of ADMM to the familiar example of sparse linear regression. In Section 2, we introduce our novel Algorithmic Regularization Paths for general sparse statistical machine learning procedures. We then demonstrate how to apply our methods through some popular machine learning problems in Section 3; specifically, we consider three examples – sparse linear regression (Section 3.1), reduced-rank multi-task learning (Section 3.2), and convex clustering (Section 3.3) – where our Algorithm Paths yield substantial computational benefits. We conclude with a discussion of our work and the many open questions it raises in Section 4.

1.1 ADMM in Statistical Machine Learning

The ADMM algorithm has become popular in statistical machine learning in recent years because the resulting algorithms are typically simple to code and can scale efficiently to large problems. Although ADMM has been successfully applied over a diverse spectrum of problems, there are essentially two thematic challenges among the problems that ADMM has proven adept at addressing: (i) decoupling constraints and regularizers that are straightforward to handle individually, but not in conjunction; and (ii) simplifying fusion type penalties. We make note of these two types of problems because the ADMM Algorithmic Regularization Path we introduce in this chapter can be applied to either type of problem.

An illustrative example of the first thematic challenge arises in sparse principal component analysis (PCA). In [36] Vu et al. propose estimating sparse principal subspace estimator $\hat{\mathbf{B}}$ of a symmetric input matrix $\mathbf{S}$ with the solution to the following semidefinite program:

$$\underset{\mathbf{B}}{\text{minimize}} \quad -\langle \mathbf{S},\mathbf{B}\rangle + \lambda \|\mathbf{B}\|_1 \quad \text{subject to} \quad \mathbf{B} \in \mathscr{F}^d,$$

where $\|\mathbf{B}\|_1$ is 1-norm of the vectorization of $\mathbf{B}$, the set $\mathscr{F}^d = \{\mathbf{B} : \mathbf{0} \preceq \mathbf{B} \preceq \mathbf{I}, \text{tr}(\mathbf{B}) = d\}$ is a closed and convex set called the Fantope, and $\lambda \geq 0$ is a regularization parameter. The main algorithmic challenge is the interaction between the Fantope constraint and the ℓ_1-norm penalty. If only either the penalty or constraint were present the problem would be straightforward to solve. Consider the following equivalent problem to which ADMM can be readily applied:

$$\underset{\mathbf{B},\mathbf{Z}}{\text{minimize}} \quad \delta_{\mathscr{F}^d}(\mathbf{B}) - \langle \mathbf{S},\mathbf{B}\rangle + \lambda \|\mathbf{Z}\|_1 \quad \text{subject to} \quad \mathbf{Z} - \mathbf{B} = \mathbf{0},$$

where $\delta_C(\Sigma)$ denotes the indicator function of the closed convex set C, namely the function that is 0 on C and ∞ otherwise. By minimizing an augmented Lagrangian over $\mathbf{B}$ and the copy variable $\mathbf{Z}$, and updating the scaled dual variable $\mathbf{U}$ as outlined in [3], we arrive at the following ADMM updates:

$$\begin{aligned}
\mathbf{B}^k &= \arg\min_{\mathbf{B}} \frac{1}{2}\|\mathbf{B}-(\mathbf{Z}^{k-1}-\mathbf{U}^{k-1}+\rho^{-1}\mathbf{S})\|_{\mathrm{F}}^2+\delta_{\mathscr{F}^d}(\mathbf{B}) \\
&= \mathscr{P}_{\mathscr{F}^d}(\mathbf{Z}^{k-1}-\mathbf{U}^{k-1}+\rho^{-1}\mathbf{S}) \\
\mathbf{Z}^k &= \arg\min_{\mathbf{Z}} \frac{\lambda}{\rho}\|\mathbf{Z}\|_1+\frac{1}{2}\|\mathbf{B}^k+\mathbf{U}^{k-1}-\mathbf{Z}\|_{\mathrm{F}}^2 = S_{\lambda/\rho}(\mathbf{B}^k+\mathbf{U}^{k-1}) \\
\mathbf{U}^k &= \mathbf{U}^{k-1}+\mathbf{B}^k-\mathbf{Z}^k.
\end{aligned}$$

Thus, the penalty and constraint are effectively decoupled resulting in simple updates: the update for $\mathbf{B}$ involves the projection onto the Fantope, denoted by $\mathscr{P}_{\mathscr{F}^d}$, which has a closed form solution given in [36], and the update for $\mathbf{Z}$ requires the soft-thresholding operator, $S_\mu(\mathbf{x}) = \text{sign}(\mathbf{x})(|\mathbf{x}|-\mu)_+$.

The literature abounds with many more examples of using the ADMM splitting strategy to decouple an otherwise challenging optimization problem into simpler subproblems. Boyd et al. [3] review many such applications. Other example applications include decoupling trace or nuclear norm penalties as in robust PCA [42], latent variable graphical models [21], and tensor completion [20]; decoupling different types of hierarchical constraints [2], decoupling a series of loss functions [18], decoupling joint graphical models [6], and decoupling large linear programming problems [1], among many others.

The second thematic challenge that ADMM algorithms have been used to solve involve fusion or non-separable penalties. A good illustrative example of this challenge arises in total variation (TV) denoising [32]. Consider the simple version of this problem, specifically finding a smooth estimate of a noisy one-dimensional signal $\mathbf{y} \in \Re^n$:

$$\underset{\boldsymbol{\beta}}{\text{minimize}} \quad \frac{1}{2}\|\mathbf{y}-\boldsymbol{\beta}\|_2^2+\lambda\sum_{i=1}^{n-1}|\beta_i-\beta_{i+1}|,$$

where the tuning parameter $\lambda \geq 0$ trades off the smoothness of the approximation with the goodness of fit with the data $\mathbf{y}$. What makes this problem challenging is that the fusion penalty couples the non-smooth terms so that they are non-separable. Note that this penalty can be written more compactly as $\|\mathbf{A}\mathbf{x}\|_1$ where $\mathbf{A}$ is the discrete first order differences operator matrix. More generally, this second class of problems consist of problems of the form, $L(\boldsymbol{\beta};\mathbf{W})+\lambda P(\mathbf{A}\boldsymbol{\beta})$. In the machine learning context these penalties arise because we often wish to impose structure, not on a latent variable of interest directly, but rather on a linear transformation of it. In the TV denoising example we seek sparsity in differences of adjacent time points of the latent signal.

Previously, we could break the objective into a sum of simpler objectives. The issue here is different; specifically the composition of the regularizer with a linear mapping complicates matters. ADMM can again greatly simplify this problem if we let the *ADMM copy variable* duplicate the linearly transformed parameters:

$$\underset{\boldsymbol{\beta},\mathbf{z}}{\text{minimize}} \quad L(\boldsymbol{\beta};\mathbf{W})+\lambda P(\mathbf{z}) \quad \text{subject to} \quad \mathbf{z}-\mathbf{A}\boldsymbol{\beta}=\mathbf{0}.$$

The ADMM subproblems for iteratively solving this problem then have the following simple form:

$$\begin{aligned}
\boldsymbol{\beta}^k &\in \arg\min_{\boldsymbol{\beta}} \;\; L(\boldsymbol{\beta};\mathbf{W}) + \frac{\rho}{2}\|\mathbf{A}\boldsymbol{\beta} - \mathbf{z}^{k-1} + \mathbf{u}^{k-1}\|_2^2 \\
\mathbf{z}^k &= \arg\min_{\mathbf{z}} \;\; \lambda P(\mathbf{z}) + \frac{\rho}{2}\|\mathbf{z} - \mathbf{A}\boldsymbol{\beta}^k - \mathbf{u}^{k-1}\|_2^2 \\
\mathbf{u}^k &= \mathbf{u}^{k-1} + \boldsymbol{\beta}^k - \mathbf{z}^k.
\end{aligned}$$

Note that we have eliminated having to minimize any functions containing the troublesome composition penalty. In the context of the TV denoising example, the $\boldsymbol{\beta}$ update requires solving a linear system of equations, and the $\mathbf{z}$ update involves a straightforward soft-threshold.

The ADMM algorithm has been used to decouple fusion or non-separable types of penalties in many statistical learning problems. These include more general instances of total variation [37, 12], a convex formulation of clustering [5], joint graphical model selection [24, 25], overlapping group lasso penalties [40], and more generally for structured sparsity [22].

Overall, while the ADMM algorithm is gaining more widespread application in statistics and machine learning, the algorithm is applied in the traditional sense to solve a composite optimization problem for one value of the regularization parameter. In this chapter, we seek to investigate the ADMM algorithm for a different purpose, namely to find a sequence of sparse models associated with regularization paths.

1.2 Developing an Algorithmic Regularization Path: Sparse Regression

Our goal is to quickly generate a sequence of candidate sparse solutions for model selection purposes. To achieve this, we will propose a method of approximating the sequence of active sets given by regularization paths, or the path-like sequence solutions of penalized statistical models over the full range of regularization parameters. To motivate our approach, we study the familiar example of sparse linear regression. Suppose we observe a covariate matrix $\mathbf{X} \in \Re^{n\times p}$ consisting of n independent and identically distributed (iid) observations of p variables and an associated response variable $\mathbf{y} \in \Re^n$. We are interested in fitting the linear model $\mathbf{y} = \mathbf{X}\boldsymbol{\beta} + \varepsilon$ where ε is independent noise, but assume that the linear coefficient vector $\boldsymbol{\beta}$ is sparse, $\|\boldsymbol{\beta}\|_0 \ll p$ where $\|\cdot\|_0$ is the ℓ_0 "norm" or the number of nonzero elements of $\boldsymbol{\beta}$. Minimizing a criterion subject to a constraint of the form $\|\boldsymbol{\beta}\|_0 \leq k$ for some k, becomes a combinatorially hard task. To estimate a sparse model in reasonable time, many have proposed to use the tightest convex relaxation, the ℓ_1-norm penalty, commonly called the LASSO [34] in the statistical literature:

$$\underset{\boldsymbol{\beta}}{\text{minimize}} \quad \frac{1}{2n}\|\mathbf{y}-\mathbf{X}\boldsymbol{\beta}\|_2^2+\lambda\|\boldsymbol{\beta}\|_1 \tag{13.2}$$

where $\lambda \geq 0$ is the regularization parameter controlling the sparsity of $\boldsymbol{\beta}$.

The full regularization path of solutions for the LASSO is the set of regression coefficients $\{\hat{\boldsymbol{\beta}}(\lambda) : \forall\, 0 \leq \lambda \leq \lambda_{\max}\}$ where $\lambda_{\max} = \frac{1}{n}\|\mathbf{X}^\top \mathbf{y}\|_\infty$ is the smallest amount of regularization that yields the sparse solution $\hat{\boldsymbol{\beta}} = \mathbf{0}$. The regularization paths for the LASSO have been well studied and, in particular, are continuous and piece-wise linear [28, 8, 31]. These paths also outline a sequence of active sets or sparse models that smoothly increase in sparsity levels as λ decreases from the fully sparse solution at $\lambda = \lambda_{\max}$ to the fully dense solution at $\lambda = 0$. Hence for model selection, one can limit exploration of the combinatorial space of sparse models to that of the sequence of active sets outlined by the LASSO regularization paths.

Computing the full regularization paths, however, can be a computational challenge. Several path following algorithms for the LASSO [28, 31] and closely related algorithms such as Least Angle Regression (LAR) [8] and Orthogonal Matching Pursuit (OMP) [7] have been proposed; their computational complexity, however, is $\mathcal{O}(p^3)$ which is prohibitive for large-scale problems. Consequently, many have suggested to closely approximate these paths by solving a series of optimization problems over a grid of regularization parameter values. Specifically, this is typically done for a sequence of 100 log-spaced values from $\lambda_{\max}$ to $\lambda_1 = 0$. Statisticians often employ homotopy, or warm-starts, to speed computation along the regularization path [9]. Warm-starts use the solution from the previous value of λ_j, $\hat{\boldsymbol{\beta}}(\lambda_j)$, as the initialization for the optimization algorithm to solve the problem at λ_{j+1}. As the coefficients, $\boldsymbol{\beta}$, change continuously in λ, warm-starts can dramatically reduce the number of iterations needed for convergence as $\hat{\boldsymbol{\beta}}(\lambda_j)$ is expected to be close to $\hat{\boldsymbol{\beta}}(\lambda_{j+1})$ for small changes from λ_j to λ_{j+1}. Many consider shooting methods, or coordinate descent procedures [9, 38], that use warm-starts and iterate over the active set for 100 log-spaced values of λ [10] to be the fastest approximate solvers of the LASSO regularization path.

We seek to further speed the computation of the sequence of active sets given by the regularization path by using a single path approximating algorithm instead of solving separate optimization problems over a grid of regularization parameter values. Our approach is motivated by two separate observations: (i) the evolution of the sparsity level of the iterates of the ADMM algorithm used to fit (13.2) for one value of λ, and (ii) the behavior of a new version of the ADMM algorithm that incorporates warm-starts to expedite computation of regularization paths. We study each of these motivations separately, beginning with the first.

Consider using ADMM to solve the LASSO problem. First, we split the differentiable loss function from the non-differentiable penalty term by introducing a copy $\mathbf{z}$ of the variable $\boldsymbol{\beta}$ in the penalty function, and adding an equality constraint forcing them to be equal. The LASSO problem (13.2) can then be rewritten as:

$$\underset{\boldsymbol{\beta},\mathbf{z}}{\text{minimize}} \quad \frac{1}{2n}\|\mathbf{y}-\mathbf{X}\boldsymbol{\beta}\|_2^2+\lambda\|\mathbf{z}\|_1 \quad \text{subject to} \quad \boldsymbol{\beta} = \mathbf{z},$$

with its associated augmented Lagrangian:

$$\mathscr{L}(\boldsymbol{\beta},\mathbf{z},\mathbf{u}) = \frac{1}{2n}\|\mathbf{y}-\mathbf{X}\boldsymbol{\beta}\|_2^2 + \lambda\|\mathbf{z}\|_1 + \frac{\rho}{2}\|\boldsymbol{\beta}-\mathbf{z}+\mathbf{u}\|_2^2.$$

Here, $\mathbf{u}$ is the scaled dual variable of the same dimension as $\boldsymbol{\beta}$ and ρ is the algorithm tuning parameter. The ADMM algorithm then follows three steps (subproblems) to solve the LASSO:

$$\boldsymbol{\beta}\text{-subproblem: } \boldsymbol{\beta}^k = \arg\min_{\boldsymbol{\beta}} \frac{1}{2n}\|\mathbf{y}-\mathbf{X}\boldsymbol{\beta}\|_2^2 + \frac{\rho}{2}\|\boldsymbol{\beta}-\mathbf{z}^{k-1}+\mathbf{u}^{k-1}\|_2^2$$

$$\mathbf{z}\text{-subproblem: } \mathbf{z}^k = \arg\min_{\mathbf{z}} \lambda\|\mathbf{z}\|_1 + \frac{\rho}{2}\|\mathbf{z}-\boldsymbol{\beta}^k-\mathbf{u}^{k-1}\|_2^2$$

$$\text{Dual update: } \mathbf{u}^k = \mathbf{u}^{k-1}+\boldsymbol{\beta}^k-\mathbf{z}^k.$$

The benefit of solving this reformulation is simpler iterative updates. These three steps are iterated until convergence, typically measured by the primal and dual residuals [3]. The $\boldsymbol{\beta}$-subproblem solves a linear regression with an additional quadratic ridge penalty. Solving the $\mathbf{z}$-subproblem introduces sparsity. Notice that this is the proximal operator of the ℓ_1-norm applied to $\boldsymbol{\beta}^k - \mathbf{u}^k$ which is solved analytically via soft-thresholding.Finally, the dual update ensures that $\boldsymbol{\beta}$ is squeezed towards $\mathbf{z}$ and primal feasibility as the algorithm progresses.

Consider the sparsity of the $\mathbf{z}$ iterates, $\|\mathbf{z}^k\|_0$, for the LASSO problem. Notice that as the algorithm proceeds, $\mathbf{z}^k$ becomes increasingly sparse; this is illustrated for a small simulated example in the left panel of Figure 13.1. Let us study why this occurs and its implications. Regardless of λ, the ADMM algorithm begins with a fully dense $\boldsymbol{\beta}^1$ as this is the solution to a ridge problem with parameter ρ. Soft-thresholding in the $\mathbf{z}$-subproblem then sets coefficients of small magnitude to zero. The first dual update, $\mathbf{u}^1$, has magnitude at most $|\lambda|$, meaning that the second $\boldsymbol{\beta}^2$ update is essentially shrunken towards $\boldsymbol{\beta}^1$. Smaller coefficients decrease further in magnitude and soft-thresholding in the $\mathbf{z}$-subproblem sets even more coefficients to zero. The algorithm thus proceeds until the sparsity of the $\mathbf{z}^k$ stabilizes to that of the solution, $\hat{\boldsymbol{\beta}}(\lambda)$. Hence, the support of the $\mathbf{z}^k$ has approximated the active set of the solution long before the iterates of the $\boldsymbol{\beta}$-subproblem; the latter typically does not reach the sparsity of the solution until convergence when primal feasibility is achieved. While Figure 13.1 only illustrates that $\mathbf{z}^k$ quickly converges to the correct sparsity level, we have observed empirically in all our examples that the active set outlined by $\|\mathbf{z}^k\|_0$ also quickly identifies the true nonzero elements of the solution, $\hat{\boldsymbol{\beta}}(\lambda)$.

Interestingly then, the $\mathbf{z}^k$ quickly explore a sequence of sparse models going from dense to sparse, similar in nature to the sequence of sparse models outlined by the regularization path. While from Figure 13.1 we can see that this sequence of sparse models is not desirable as it does not smoothly transition in sparsity and does not fully explore the sparse model space, we nonetheless learn two important items from this: (i) We are motivated to consider using the algorithm iterates of the $\mathbf{z}$-subproblem, as a possible means of quickly exploring the sparse model space; and (ii) we are motivated to consider a sequence of solutions going from dense to sparse as this naturally aligns with the sparsity levels observed in the ADMM algorithm iterates. Given these, we ask: Is it possible to use or modify the iterates of the ADMM algorithm to achieve a path-like smooth transition in sparsity levels similar in nature to the sparsity levels and active sets corresponding to regularization paths?

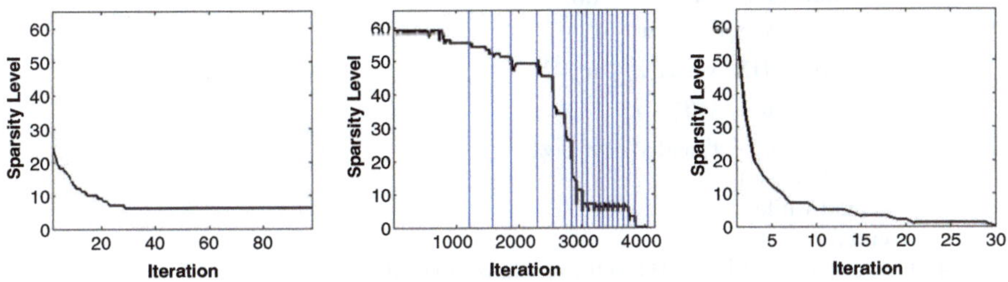

Fig. 13.1 Sparsity levels outlining a sequence of active sets for a simulated sparse linear regression example. (Left) Sparsity levels of the $\mathbf{z}$-subproblem iterates of the ADMM algorithm, $\|\mathbf{z}^k\|_0$, fit for one fixed value of λ. (Middle) Sparsity levels of the $\mathbf{z}$-subproblem over the iterates of our path approximating ADMM Algorithm with Warm Starts for a small range of $\boldsymbol{\lambda}$. Vertical lines denote the start of the algorithm for an increased value of λ. (Right) Sparsity levels of the $\mathbf{z}$-subproblem over the iterates of our novel ADMM Algorithmic Regularization Path.

One possible solution would be to employ warm-starts in the ADMM algorithm along a grid of regularization parameters similar to other popular algorithms for approximating regularization paths. Recall that warm-starts use the solution obtained at the previous value of λ as initialization for the next value of λ. We first introduce this new extension of the ADMM algorithm for approximating regularization paths in Algorithm 1 and then return to our motivation of studying the sequence of active sets outlined by this algorithm.

Our ADMM algorithm with warm starts is an alternative algorithm for fitting regularization paths. It begins with λ small corresponding to a dense model, fits the ADMM algorithm to obtain the solution, and then uses the previous solution, $\boldsymbol{\beta}(\lambda_{j-1})$, and dual variable, $\mathbf{u}(\lambda_{j-1})$, to initialize the ADMM algorithm for λ_j.

Before considering the sequence of active sets outlined by this algorithm, we pause to discuss some noteworthy features. First, notice that the ADMM tuning parameter, ρ, does not appear in this algorithm. We have omitted this as a parameter by fixing $\rho = 1$ throughout the algorithm. Fixing ρ stands in contrast with the burgeoning literature on how to dynamically update ρ for ADMM algorithms [3]. For example, adaptive procedures that change ρ to speed up convergence are introduced in [16]. Others have proposed accelerated versions of the ADMM algorithm that achieve a similar phenomenon [11]. Changing the algorithm tuning parameter, however, is not conducive to achieving a path-like algorithm using warm-starts. Consider the $\mathbf{z}$-subproblem which is solving by soft-thresholding at the level λ_j/ρ. Thus, if ρ is changed in the algorithm, the sparsity levels of $\mathbf{z}$ dramatically change, eliminating the advantages of using warm-starts to achieve smooth transitions in sparsity levels. Second, notice that we have switched the order of the sub-problems by beginning with the $\mathbf{z}$-subproblem. While technically the order of the subproblems does not

Algorithm 1 ADMM with Warm Starts: Sparse Regression

1. Initialize $\beta^0 = \mathbf{0}$, $\mathbf{u}^0 = \mathbf{0}$, and M log-spaced values, $\lambda = \{\lambda_1 < \lambda_2 < \cdots < \lambda_M\}$, for $\lambda_1 = 0$ and $\lambda_M = \lambda_{\max}$.
2. Precompute matrix inverse $\mathbf{H} = (\mathbf{X}^\top\mathbf{X}/n + \mathbf{I})^{-1}$ and $\mathbf{H}\mathbf{X}^\top\mathbf{y}$.
3. **for** $j = 1 \ldots M$ **do**
   - **while** $\|\mathbf{r}^k\| \wedge \|\mathbf{s}^k\| > \varepsilon^{\text{tol}}$ **do**
     - $\mathbf{z}_j^k = S_{\lambda_j}(\beta_j^{k-1} + \mathbf{u}_j^{k-1})$
     - $\beta_j^k = \mathbf{H}\mathbf{X}^\top\mathbf{y} + \mathbf{H}(\mathbf{z}_j^k - \mathbf{u}_j^{k-1})$
     - $\mathbf{u}_j^k = \mathbf{u}^{k-1} + \beta_j^k - \mathbf{z}_j^k$
     - $\mathbf{r}^k = \beta_j^k - \mathbf{z}_j^k$ and $\mathbf{s}^k = \mathbf{z}_j^k - \mathbf{z}_j^{k-1}$
     - $k = k+1$
   - **end while**

   end for
4. Output $\{\beta_j : j = 1, \cdots, M\}$ as the regularization path.

matter [39], we begin with the **z**-subproblem as this is where the sparsity is achieved through soft-thresholding at the value, λ; hence, the solution for **z** is what changes when λ is increased.

Next, notice that our regularization paths go from dense to sparse, or λ small to large, which is the opposite of other path-wise algorithms and algorithms that approximate regularization paths over a grid of λ values [10]. Recall that our objective is to obtain a smooth path-like transition in sparsity levels corresponding to a sequence of active sets that fully explores the space of sparse models. Our warm-start procedure naturally aligns with the sparsity levels of the iterates of the ADMM algorithm which go from dense to sparse, thus ensuring a smooth transition in the sparsity level of **z** as λ is increased. Our warm-start procedure could certainly be employed going in the reverse direction from sparse to dense, but we have observed that this introduces discontinuities in the $\mathbf{z}^k$ iterates and consequently their active sets as well, thus requiring more iterations for convergence. This behavior occurs as the solution of the $\boldsymbol{\beta}$-subproblem is always more dense than that of the **z**-subproblem, even when employing warm-starts.

Now, let us return to our motivation and consider the sparsity levels and corresponding sequence of active sets achieved by the iterates of our new path-approximating ADMM Algorithm. The sparsity of the iterates of the **z**-subproblems, $\|\mathbf{z}^k\|_0$, are plotted for 30 log-spaced values of λ for the same simulated example in the middle panel of Figure 13.1. The iterates over all values of λ are plotted on the x-axis with vertical lines denoting the increase to the next λ value. Carefully consider the sparsity levels of the **z** iterates for each fixed value of λ in our ADMM algorithm with warm starts. Notice that the sparsity levels of **z** typically stabilize to that of the solution within the first few iterations after λ is increased. The remaining iterations and a large proportion of the computational time are spent on squeezing $\boldsymbol{\beta}$ towards **z** to satisfy primal feasibility. This means that the **z**-subproblem has achieved the sparsity associated with the active set of $\hat{\boldsymbol{\beta}}(\lambda)$ within a few iterations of increasing λ. One could surmise that if the increase in λ were small enough, then the $\mathbf{z}$-subproblem could correctly approximate the active set corresponding to λ within one iteration when using this warm-start procedure. The right panel of Figure 13.1 illustrates the sparsity levels achieved by the $\mathbf{z}$-subproblem for this sequence of one-step approximations to our ADMM algorithm with warm-starts. Notice that this procedure achieves a smooth transition in sparsity levels corresponding to a sequence of active sets that fully explore the range of possible sparse models, but requires only a fraction of the total number of iterations and compute time. This, then is the motivation for our new ADMM Algorithmic Regularization Paths introduced in the next section.

2 The Algorithmic Regularization Path

Our objective is to use the ADMM splitting method as the foundation upon which to develop a new approximation to the sequence of sparse solutions outlined by regularization paths. In doing so, we are not interested in estimating parameter values by solving a statistical learning optimization problem with high precision. Instead, we are interested in quickly exploring the space of sparse model at a fine resolution for model selection purposes by approximating the sequence of active sets given by the regularization path.

Again, consider the general sparse statistical machine learning problem of the following form:

$$\underset{\boldsymbol{\beta}}{\text{minimize}} \quad L(\boldsymbol{\beta};\mathbf{W}) + \lambda P(\boldsymbol{\beta}),$$

where $\mathbf{W}$ denotes the "data" (for the sparse linear regression example, $\mathbf{W} = \{\mathbf{X}, \mathbf{y}\}$), the loss function, $L(\boldsymbol{\beta};\mathbf{W})$ is a differentiable, convex function of $\boldsymbol{\beta}$, and the regularization term, $P : \Re^p \to \Re^+$ is a convex and non-differentiable penalty function. As before, $\lambda \geq 0$ is the regularization parameter controlling the trade-off between the penalty and loss function. Following the setup of the ADMM algorithm, consider splitting the smooth loss from the nonsmooth penalty through the copy variable, $\mathbf{z}$:

$$\underset{\boldsymbol{\beta},\mathbf{z}}{\text{minimize}} \quad L(\boldsymbol{\beta};\mathbf{W}) + \lambda P(\mathbf{z}) \quad \text{subject to } \boldsymbol{\beta} = \mathbf{z}, \tag{13.3}$$

With scaled dual variable $\mathbf{u}$, the augmented Lagrangian of general problem (13.3) is

$$\mathscr{L}(\boldsymbol{\beta},\mathbf{z},\mathbf{u}) = L(\boldsymbol{\beta};\mathbf{W}) + \lambda P(\mathbf{z}) + \frac{\rho}{2}\|\boldsymbol{\beta} - \mathbf{z} + \mathbf{u}\|_2^2.$$

Now following from the motivations discussed in the previous section, there are three key ingredients that we employ in our Algorithmic Regularization Paths: (i) warm-starts to go from a dense to a sparse solution, (ii) the sparsity patterns of the $\mathbf{z}$-subproblem iterates, and (iii) one-step approximations at each regularization level. We put these ingredients together in Algorithm 2 to give our Algorithmic Regularization Paths:

Algorithm 2 Algorithmic Regularization Path for Sparse Statistical Learning

1. Initialize $\mathbf{z}^0 = \mathbf{0}$, $\mathbf{u}^0 = \mathbf{0}$, $\gamma^0 = \varepsilon$, $k = 1$, and set $t > 0$.
2. While $\|\mathbf{z}^k\| \neq 0$

   $\gamma^k = \gamma^{k-1} + t$ (or $\gamma^k = \gamma^{k-1} t$)

   $\boldsymbol{\beta}^k = \underset{\boldsymbol{\beta}}{\text{minimize}} \quad L(\boldsymbol{\beta};\mathbf{W}) + \frac{1}{2}\|\boldsymbol{\beta} - \mathbf{z}^{k-1} + \mathbf{u}^{k-1}\|_2^2$

   $\mathbf{z}^k = \underset{\mathbf{z}}{\text{minimize}} \quad \gamma^k P(\mathbf{z}) + \frac{1}{2}\|\mathbf{z} - \boldsymbol{\beta}^k - \mathbf{u}^{k-1}\|_2^2$ (Record $\mathbf{z}^k$ at each iteration.)

   $\mathbf{u}^k = \mathbf{u}^{k-1} + \boldsymbol{\beta}^k - \mathbf{z}^k$

   $k = k+1$

   end
3. Output $\{\mathbf{z}^k : k = 1, \cdots, K\}$ as algorithmic regularization path.

Our Algorithmic Regularization Path, Algorithm Path for short, outlines a sequence of sparse models going from fully dense to fully sparse. This can be used as an approximation to the sequence of active sets given by regularization paths for the purpose of model selection. Consider that the algorithm begins will the fully dense ridge solution. It then gradually increases the amount of regularization, γ, performing one full iterate of the ADMM algorithm ($\boldsymbol{\beta}$-subproblem, $\mathbf{z}$-subproblem, and dual update) for each new level of regularization. The regularization level is increased until the $\mathbf{z}$-subproblem becomes fully sparse.

One may ask why we would expect our Algorithm Path to yield a sequence of active sets that well approximate those of the regularization path? While a mathematical proof of this is beyond the scope of this chapter, we outline the intuition stemming from our three key ingredients. (Note that we also demonstrate this through specific examples in the next section).

(i) Warm-starts from dense to sparse. Beginning with a dense solution and gradually increasing the amount of regularization ensures a smooth decrease in the sparsity levels corresponding to a smooth pruning of the active set as this naturally aligns with sparsity levels of the ADMM algorithm iterates.
(ii) $\mathbf{z}$-subproblem iterates. The iterates of the $\mathbf{z}$-subproblem encode the sparsity of the active set, $\hat{\boldsymbol{\beta}}(\lambda)$, quickly as compared to the $\boldsymbol{\beta}$-subproblem which achieves sparsity only in the limit upon algorithm convergence.
(iii) One-step approximations. For a small increase in regularization when using warm-starts, the iterates of the $\mathbf{z}$-subproblem often achieve the sparsity level of the active set within one-step.

Notice that if we iterated the three subproblems of our Algorithm Path fully until convergence, then our algorithm would be equivalent to our ADMM Algorithm with warm starts; thus, the one-step approximation is the major difference between Algorithms 1 and 2. Because of this one-step approximation, we are not fully solving (13.1) and thus the parameter values, $\boldsymbol{\beta}$, will never stabilize to that of the regularization path. Instead, our Algorithm Path quickly approximates the sequence of active sets corresponding to the regularization path, as encoded in the $\mathbf{z}$-subproblem iterates.

The astute reader will notice that we have denoted the regularization parameters in Algorithm 2 as γ instead of λ as in (13.1). This was intentional since due to the one-step approximation, we are not solving (13.1) and thus the level of regularization achieved, γ, does not correspond to λ from (13.1). Also notice that we have introduced a step size, t, that increases the regularization level, γ, at each iteration. The sequence of γ's can either be linearly spaced, as with additive t, or geometrically spaced, as with multiplicative t. Again, if t is very small, then we expect the sparsity patterns of our Algorithm Paths to well approximate the active sets of regularization paths.

We will explore the behavior and benefits of our Algorithm Paths through demonstrations on popular sparse statistical learning problems in the next section. Before presenting specific examples, however, we pause to outline three important advantages that are general to sparse statistical learning problems of the form (13.1).

1. Easy to implement. Our Algorithm Path is much simpler than other algorithms to approximate regularization paths. The hardest parts, the $\boldsymbol{\beta}$ and $\mathbf{z}$ subproblems, often have analytical solutions for many popular statistical learning methods. Then, with only one loop, our algorithm can often be implemented in a few lines of code. This is in contrast to other algorithm paths which require multiple loops and much overhead to track algorithm convergence or the coordinates of active sets [10].
2. Finer resolution exploration of the model space. Our Algorithm Path has the potential to explore the space of sparse models at a much finer resolution than other fast methods for approximating regularization paths over a grid of λ values. Consider that as the later are computed over M, typically $M = 100$, λ values, these can yield an upper bound of M distinct active sets; usually these yield much less than M distinct models. In contrast, our Algorithm Path yields an upper bound of K distinct models where K is the number of iterations needed, depending on the step-size t, to fully explore the sequence of sparse models. As K will often be much greater than M, our Algorithm Path will often explore a sequence of many more active sets and at a finer resolution than comparable methods.
3. Computationally fast. Our Algorithm Path has the potential to yield a sequence of sparse solutions much faster than other methods for computing regularization paths. Consider that our algorithm takes at most K iterations. In contrast, regularization paths of a grid of M λ values require M times the number of iterations needed to fully estimate $\hat{\boldsymbol{\beta}}(\lambda_j)$; often this will be much larger than K. In each iteration of our algorithm, the $\boldsymbol{\beta}$ and $\mathbf{z}$ subproblems require the most computational time. The $\boldsymbol{\beta}$ subproblem consists of the loss function with a quadratic penalty which can be solved via an analytical form for many loss functions. The $\mathbf{z}$ subproblem has the form of the proximal operator of P [29]: $\text{prox}_{\lambda P}(\mathbf{x}) = \arg\min_{\mathbf{u}} \|\mathbf{x} - \mathbf{u}\|_2^2 + \lambda P(\mathbf{u})$. For many popular convex penalty-types such as the ℓ_1-norm, group lasso, and nuclear norm, the proximal operator has an analytical solution. Thus, for a large number of statistical machine learning problems, the iterations of our Algorithm Path are inexpensive to compute.

Overall, our Algorithmic Regularization Paths give a novel method for finding a sequence of sparse solutions by approximating the active sets of regularization paths. Our methods can be used in place of regularization paths for model selection purposes with many sparse statistical learning problems. In this chapter, instead of studying the mathematical and statistical properties of our new Algorithm Paths, which we leave for future work, we study our method through applications to several statistical learning problems in the next section.

3 Examples

To demonstrate the versatility and advantages of our ADMM Algorithmic Regularization Paths, we present several example applications to sparse statistical learning methods: sparse linear regression, reduced-rank multi-task learning, and convex clustering.

3.1 Sparse Linear Regression

As our first example, we revisit the motivating example of sparse linear regression discussed in Section 1.2. We reproduce the problem here for convenience:

$$\underset{\boldsymbol{\beta}}{\text{minimize}} \quad \frac{1}{2n}\|\mathbf{y}-\mathbf{X}\boldsymbol{\beta}\|_2^2+\lambda\|\boldsymbol{\beta}\|_1$$

And, our Algorithmic Regularization Path for this example is presented in Algorithm 3:

Algorithm 3 Algorithmic Regularization Path for Sparse Regression

1. Initialize $\mathbf{z}^0 = \mathbf{0}$, $\mathbf{u}^0 = \mathbf{0}$, $\gamma^0 = \varepsilon$, $k = 1$, and set $t > 0$.
2. Precompute matrix inverse $\mathbf{H} = (\mathbf{X}^\top \mathbf{X}/n + \mathbf{I})^{-1}$ and $\mathbf{H}\mathbf{X}^\top \mathbf{y}$.
3. While $\|\mathbf{z}^k\| \neq 0$

    $\gamma^k = \gamma^{k-1} + t$

    $\boldsymbol{\beta}^k = \mathbf{H}\mathbf{X}^\top \mathbf{y} + \mathbf{H}(\mathbf{z}^{k-1} - \mathbf{u}^{k-1})$

    $\mathbf{z}^k = S_{\gamma^k}(\boldsymbol{\beta}^k + \mathbf{u}^{k-1})$ (Record $\mathbf{z}^k$ at each iteration.)

    $\mathbf{u}^k = \mathbf{u}^{k-1} + \boldsymbol{\beta}^k - \mathbf{z}^k$

    $k = k+1$

   end
4. Output $\{\mathbf{z}^k : k = 1, \cdots, K\}$ as the algorithmic regularization path.

Let us first discuss computational aspects of our Algorithm Path for sparse linear regression. Notice that the $\boldsymbol{\beta}$-subproblem consists of solving a ridge-like regression problem. Much of the computations involved, however, can be pre-computed, specifically the matrix inversion, $(\mathbf{X}^\top \mathbf{X}/n + \mathbf{I})^{-1}$, and matrix-vector multiplication, $\mathbf{X}^\top \mathbf{y}$. In cases where $p \gg n$, inverting a $p \times p$ matrix is highly computationally intensive, requiring $\mathscr{O}(p^3)$ operations. We can reduce the computational cost to $\mathscr{O}(n^3)$, however, by invoking the Woodbury Matrix Identity [13]: $(\mathbf{X}^\top \mathbf{X}/n + \mathbf{I}_p)^{-1} = \mathbf{I}_p - \mathbf{X}^\top (n\mathbf{I}_n + \mathbf{X}\mathbf{X}^\top)^{-1}\mathbf{X}$ and caching the Cholesky decomposition of the smaller n-by-n matrix $n\mathbf{I}_n + \mathbf{X}\mathbf{X}^\top$. Thus, the iterative updates for $\boldsymbol{\beta}^k$ are reduced to $\mathscr{O}(n^2)$, the cost of solving two n-by-n triangular linear systems of equations. The $\mathbf{z}$-subproblem is solved via soft-thresholding, which requires only $\mathscr{O}(p)$ operations.

We study our Algorithmic Regularization Path for sparse linear regression through a real data example. We use the publicly available 14-cancer microarray data from [15] to form our covariate matrix. This consists of gene expression measurements for $n = 198$ subjects and 16063 genes; we randomly sample $p = 2000$ genes to use as our data matrix $\mathbf{X}$. We simulate sparse true signal $\boldsymbol{\beta}^*$ with $s = 16$ nonzero features of absolute magnitude 5–10, and with the signs of the nonzero signals assigned randomly; the 16 nonzero variables were randomly chosen from the 2000 genes. The response variable $\mathbf{y}$ is generated as $\mathbf{y} = \mathbf{X}\boldsymbol{\beta}^* + \varepsilon$, where $\varepsilon \overset{i.i.d.}{\sim} N(0,1)$. A visualization of regularization paths, stability paths, and our Algorithmic Regularization Paths is given in Figure 13.2 for this example.

First, we verify empirically that our ADMM algorithm with warm starts is equivalent to the regularization path (top left and top middle). Additionally, notice that, as expected, our Algorithm Path with a tiny step size (bottom right) also well approximates the sequence of active sets given by the regularization paths. With a larger step-size, however, our algorithm path (bottom left) yields a sequence of sparse models that differ markedly from the sparsity patterns of the regularization paths. This occurs as the change in regularization levels of each step are large enough so that the sparsity levels of the $\mathbf{z}$-subproblem after the one-step approximation are not equivalent to that of the solution to (13.2).

Despite this, Figure 13.2 suggests that our Algorithm Paths with larger step sizes may have some advantages in terms of variable selection. Notice that regularization paths select many false positives (blue and gray dashed lines) before the true positives (red lines). This is expected as we used a real microarray data set for $\mathbf{X}$ consisting of strongly correlated variables that directly violate the irrepresentable conditions under which variable selection for sparse regression is achievable [4]. Our method, however, selects several true variables before the first false positive enters the model. To understand this further, we compare our approach to the Stability Paths used for stability selection [23], a re-sampling scheme with some of the strongest theoretical guarantees for variable selection. The stability paths, however, also select several false positives. This as well as other empirical results that are omitted for space reasons suggest that our Algorithm Path with moderate or larger step sizes may perform better than convex optimization techniques in terms of variable selection. While a theoretical investigation of this is beyond the scope of this book chapter, the intuition for this is readily apparent. Our Algorithm Path starts

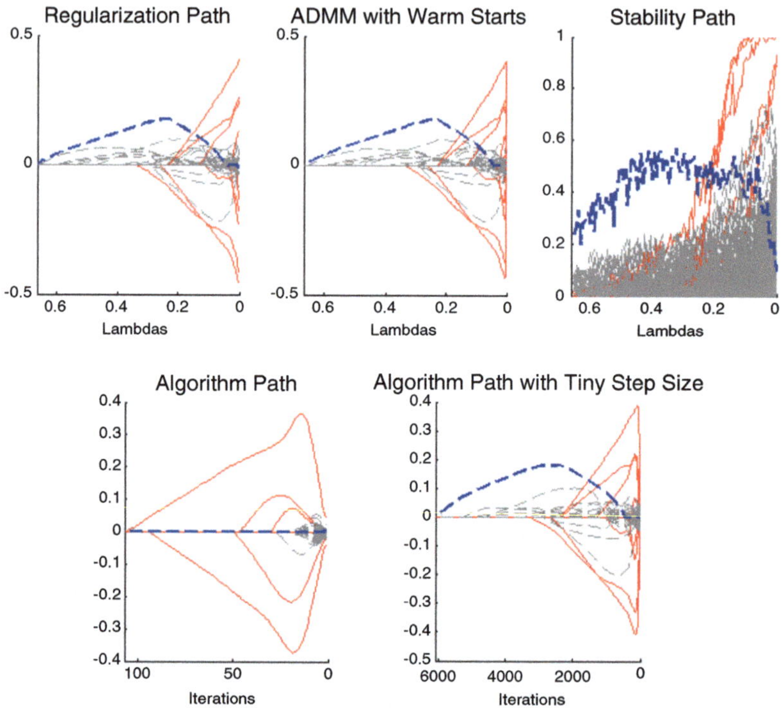

Fig. 13.2 Comparisons of Algorithmic Regularization Paths (bottom panel) to regularization paths (top left and middle) and stability paths (top right) for the sparse linear regression example. The -- lines denote false variables, — lines denote true nonzero variables, and -- lines denote some highlighted false positives. Regularization paths were computed via the popular shooting method [10] (top left) and our ADMM algorithm with warm-starts (top middle). Our Algorithmic Regularization Path with a tiny step size (bottom right) closely approximates the sparsity patterns of the regularization paths, while our method with a larger step size (bottom left) dramatically differs from the regularization paths. Notice that sparse regression in this example does a poor job of variable selection, selecting many false positives before any true features enter the model. Even the stability paths (top right) select many false positives. Our Algorithmic Regularization Path with a larger step-size, however, selects many of the true variables with much fewer false positives.

from a dense solution and uses a ridge-like penalty. Thus, coefficients of highly correlated variables are likely to be grouped and have similar magnitude coefficient values. When soft-thresholding is performed in the $\mathbf{z}$-subproblem, variables which are strongly correlated are likely to remain together in the model for at least the first several algorithm iterations. By keeping correlated variables together in the model longer and otherwise eliminating irrelevant variables, this gives our algorithm a better chance of selecting the truly nonzero variables out of a correlated set. Hence, the fact that we start with a dense solution seems to help us; this is in contrast to the LASSO, LAR, and OMP paths which are initialized with an empty active set and greedily add variables most correlated with the response [28, 8]. We plan on investigating our methods in terms of variable selection in future work.

Table 13.1 Timing comparison (averaged over 50 replications) of our ADMM Algorithmic Regularization Paths, Regularization Paths obtained from the shooting method (coordinate descent), and Stability Paths for different numbers of variables in the true model.

Time (seconds)	Algorithmic Regularization Path	Regularization Path	Stability Path
s = 20, p = 4000	0.0481	0.1322	36.6813
s = 20, p = 6000	0.0469	0.1621	43.9320

Finally, we compare our Algorithm Paths to the state-of-the-art methods for computing the sparse regression regularization paths in terms of computational time in Table 13.1. The regularization paths were computed using the `glmnet` `R` package [10] which is based on shooting (coordinate descent) routines [10]. This approach and software is widely regarded as one of the fastest solvers for sparse regression. Notice that our Algorithm Paths, coded entirely in Matlab, run in about a fifth of the time as this state-of-the-art competitor. Also, our computational time is far superior to the re-sampling schemes required to compute the stability paths.

Overall, our Algorithmic Regularization Path for sparse linear regression reveals major computational advantages for finding a sequence of sparse models that approximate the active sets of regularization paths. Additionally, empirical evidence suggests that our methods may also enjoy some important statistical advantages in terms of variable selection that we will explore in future work.

3.2 Reduced-Rank Multi-Task Learning

Our ADMM Algorithmic Regularization Path applies generally to many convex penalty types beyond the ℓ_1-norm. Here, we demonstrate our method in conjunction with a reduced-rank multi-task learning problem also called multi-response regression. This problem has been studied by [27] among many others.

Suppose we observe n iid samples measured on p covariates and for q outcomes, yielding a covariate matrix, $\mathbf{X} \in \Re^{n \times p}$, and a response matrix $\mathbf{Y} \in \Re^{n \times q}$. Then, our goal is to fit the following statistical model: $\mathbf{Y} = \mathbf{X}\mathbf{B} + \varepsilon$, where $\mathbf{B}$ is the $p \times q$ coefficient matrix which we seek to learn, and ε is independent noise. As often the number of covariates is large relative to the sample size, $pq \gg n$, many have suggested to regularize the coefficient matrix $\mathbf{B}$ by assuming it has a low-rank structure, rank($\mathbf{B}$) $< p \wedge q$. Thus, our model space of sparse solutions is given by the space of all possible reduced-rank solutions. Exploring this space is an NP hard computational problem; thus, many have employed the nuclear norm penalty, $\|\mathbf{B}\|_* = \sum_{j=1}^{p \wedge q} \sigma_j(\mathbf{B})$, which is the sum (or ℓ_1-norm) of the singular values of $\mathbf{B}$, $\sigma(\mathbf{B})$, and the tightest convex relaxation of the rank constraint. Thus, we arrive at the following optimization problem:

$$\underset{\mathbf{B}}{\text{minimize}} \quad \frac{1}{2}\|\mathbf{Y} - \mathbf{X}\mathbf{B}\|_F^2 + \lambda \|\mathbf{B}\|_* \tag{13.4}$$

Here, $\|\cdot\|_F$ is the Frobenious norm, $\lambda \geq 0$ is the regularization parameter controlling the rank of the solution and $\|\cdot\|_*$ is the nuclear norm penalty.

For model selection then, one seeks to explore the sequence of low-rank solutions obtained as λ varies. To develop our Algorithm Path for approximating this sequence of low-rank solutions, let us consider the ADMM sub-problems for solving (13.4). The augmented Lagrangian, sub-problems, dual updates are analogous to that of the sparse linear regression example, and hence we omit these here. Examining the $\mathbf{Z}$-subproblem, however, recall that this is the proximal operator for the nuclear norm penalty: $\mathbf{Z}^k = \arg\min_{\mathbf{Z}} \frac{1}{2}\|\mathbf{Z}-(\mathbf{B}^k+\mathbf{U}^k)\|_F^2 + \gamma\|\mathbf{Z}\|_*$, which can be solved by soft-thresholding the singular values: Suppose that $\mathbf{A} = \mathbf{U}\Sigma\mathbf{V}^\top$ is the SVD of $\mathbf{A}$. Then singular-value thresholding is defined as $\text{SVT}_\gamma(\mathbf{A}) = \mathbf{U}[\text{diag}((\sigma-\gamma)_+)]\mathbf{V}^\top$, where σ denotes $diag(\Sigma)$, and the solution for the $\mathbf{Z}$ sub-problem is $\mathbf{Z}^k = \text{SVT}_\gamma(\mathbf{B}^k+\mathbf{U}^k)$.

Algorithm 4 Algorithmic Regularization Path for Reduced-Rank Regression

1. Initialize: $\mathbf{Z}^0 = \mathbf{0}$, $\mathbf{U}^0 = \mathbf{0}$, $\gamma^0 = \varepsilon$, and step size $t > 0$.
2. Precompute: $\mathbf{H} = (\mathbf{X}^\top\mathbf{X}/n+\mathbf{I})^{-1}$ and $\mathbf{H}\mathbf{X}^\top\mathbf{Y}$.
3. While $\|\mathbf{Z}^k\| \neq 0$

    $\gamma^k = \gamma^{k-1}+t$ (or $\gamma^k = \gamma^{k-1}t$).

    $\mathbf{B}^k = \mathbf{H}\mathbf{X}^\top\mathbf{Y}+\mathbf{H}(\mathbf{Z}^{k-1}-\mathbf{U}^{k-1})$.

    $\mathbf{Z}^k = \text{SVT}_{\gamma^k}(\mathbf{B}^k+\mathbf{U}^{k-1})$. (Record $\mathbf{Z}^k$ at each iteration.)

    $\mathbf{U}^k = \mathbf{U}^{k-1}+\mathbf{B}^k-\mathbf{U}^k$

   end
4. Output $\{\mathbf{Z}^k : k = 1, \cdots, K\}$ as the algorithmic regularization path.

Then, following the framework of the sparse linear regression example, our ADMM Algorithmic Regularization Path for the reduced-rank multi-task learning (regression) is outlined in Algorithm 4. Notice that the algorithm has the same basic steps as in the previous example except that solving the proximal operator for the $\mathbf{Z}$ sub-problem entails singular value thresholding. This step is the most computationally time consuming aspect of the algorithm as K total SVDs must be computed to approximate the sequence of solutions. Also note that similarly to the sparse regression example, the inversion needed, $(\mathbf{X}^\top\mathbf{X}/n+\mathbf{I})^{-1}$, can be precomputed by using the matrix inversion identities as previously discussed and cached as a convenient factorization; hence, this is computationally feasible even when $p \gg n$.

To demonstrate the computational advantages of our approach, we conduct a small simulation study comparing our method to the two most commonly used algorithms for reduced-rank regression: proximal gradient descent and ADMM. First, we generate data according to the model: $\mathbf{Y} = \mathbf{X}\mathbf{B}+\varepsilon$, where $\mathbf{X}_{200\times 100}$ is generated as independent standard Gaussians, $\mathbf{B}_{100\times 100}$ is an image of the Batman symbol, and $\varepsilon_{200\times 100}$ is independent standard Gaussian noise. We set the signal in the coefficient

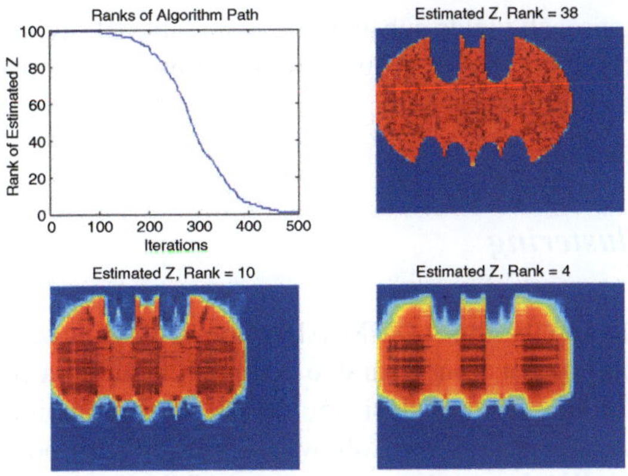

Fig. 13.3 Reduced Rank Regression simulated example. The true coefficient matrix, $\mathbf{B} \in \Re^{100\times100}$, is an image of batman that is rank 38. Our Algorithm Path provides a sequence of low-ranks solutions at a fine resolution (top left) that well approximate the low-rank signal; three such low-rank solutions (top right and bottom panel) are shown from iterates of our Algorithm Path.

	# Ranks Considered	# SVDs	Time in Seconds
Algorithm Path	90	476	2.354
Proximal Gradient	57	2519	12.424
ADMM	51	115,946	599.144

Table 13.2 Algorithm comparisons for reduced rank regression example.

matrix to be a low-rank image of the Batman symbol, $\text{rank}(\mathbf{B}) = 38$, which can be well-approximated by further reduced rank images. We applied our Algorithmic Regularization Paths to this simulated example with 500 logarithmically spaced values of γ. Results are given in Figure 13.3 and show that our Algorithm Path smoothly explores the model space of reduced rank solutions and nicely approximates the true signal as a low-rank batman image. We also conduct a timing comparison to implementations of proximal gradient descent and ADMM algorithms using warm-starts for this same example; results are given in Table 13.2. Here, we see that our approach requires much fewer SVD computations and is much faster than both algorithms, especially the ADMM algorithm. Additionally, both the ADMM and proximal gradient algorithm employed 100 logarithmically spaced values of the regularization parameter, λ. With this, however, we see that not all possible ranks of the model space are considered, with proximal gradient and ADMM considering 57 and 51 ranks out of 100 respectively. In contrast, our Algorithmic Regularization Path yields a sequence of sparse solutions at a much finer resolution, considering 90 out of the 100 possible ranks. Thus, for proximal gradient and ADMM algorithms to consider the same range of possible sparsity levels (ranks), a greater number of problems would have to be solved over a much finer grid of regularization parameters, further inflating compute times.

Overall, our approach yields substantial computational savings for computing a sequence of sparse solutions for reduced rank regression compared to other state-of-the-art methods for this problem.

3.3 Convex Clustering

Our final example applies the ADMM Algorithmic Regularization path to an example with fusion type or non-separable penalties, namely a recently introduced convex formulation of cluster analysis [5, 17, 19]. Given n points $\mathbf{y}_1, \ldots, \mathbf{y}_n$ in $\Re^p$, we pose the clustering problem as follows. Assign to each point $\mathbf{y}_i$ its own cluster center $\beta_i \in \Re^p$. We then seek an assignment of β_i that minimizes the distances between $\mathbf{y}_i$ and β_i and seeks sparsity between cluster center pairs β_i and β_j. Computing all possible cluster assignments, however, is an NP hard problem. Hence, the following relaxation poses finding the cluster assignments as a convex optimization problem:

$$\underset{\beta_1,\ldots,\beta_n}{\text{minimize}} \ \frac{1}{2}\sum_{i=1}^{n}\|\mathbf{y}_i - \beta_i\|_2^2 + \lambda \sum_{i<j} w_{ij}\|\beta_i - \beta_j\|_2, \tag{13.5}$$

where λ is a positive regularization parameter, and w_{ij} is a nonnegative weight. When $\lambda = 0$, the minimum is attained when $\beta_i = \mathbf{y}_i$, and each point occupies a unique cluster. As λ increases, the cluster centers begin to coalesce. Two points $\mathbf{y}_i$ and $\mathbf{y}_j$ with $\beta_i = \beta_j$ are said to belong to the same cluster. For sufficiently large λ all points coalesce into a single cluster at $\bar{\mathbf{y}}$, the mean of the $\mathbf{y}_i$. Because the objective in (13.5) is strictly convex and coercive, it possesses a unique minimizer for each value of λ. This is in stark contrast to other typical criteria used for clustering, which often rely on greedy algorithms that are prone to get trapped in suboptimal local minima. Because of its coalescent behavior, the resulting solution path can be considered a convex relaxation of hierarchical clustering [17].

This problem generalizes the fused LASSO [35], and as with other fused LASSO problems, penalizing affine transformations of the decision variable makes minimization challenging in general. The one exception is when a 1-norm is used instead of the 2-norm in the fusion penalty terms. In this case, the problem reduces to a weighted one-dimensional total variation denoising problem. Under other norms, including the 2-norm, the situation, is salvageable if we adopt a splitting strategy discussed earlier in Section 1.1 for dealing with fusion type or non-separable penalties. Briefly, we consider using the 2-norm in the fusion penalty to be most broadly applicable since the solutions to the convex clustering problem become invariant to rotations in the data. Consequently, clustering assignments will also be guaranteed to be rotationally invariant.

Let the variables $\mathbf{z}_{ij} \in \Re^p$ record the differences between the ith and jth points. We denote the collections of variables $\{\beta_i\}_{i=1}^n$ and $\{z_{ij}\}_{i<j}$ by β and $\mathbf{z}$ respectively. Then the original problem can be reformulated as:

$$\underset{\beta,\mathbf{z}}{\text{minimize}}\ \frac{1}{2}\sum_{i=1}^{n}\|\mathbf{y}_i-\beta_i\|_2^2+\lambda\sum_{i<j}w_{ij}\|\mathbf{z}_{ij}\|_2 \quad \text{subject to} \quad \beta_i-\beta_j-\mathbf{z}_{ij}=\mathbf{0}. \tag{13.6}$$

Consider the ADMM algorithm derived in [5] for solving (13.6). Let $\mathbf{u}_{ij}\in\Re^p$ denote the Lagrange multiplier for the ijth equality constraint. Let $\mathbf{u}$ denote the collection of variables $\{\mathbf{u}_{ij}\}_{i<j}$. The augmented Lagrangian is given by:

$$\mathscr{L}(\beta,\mathbf{z},\mathbf{u})=\frac{1}{2}\sum_{i=1}^{n}\|\mathbf{y}_i-\beta_i\|_2^2+\lambda\sum_{i<j}w_{ij}\|\mathbf{z}_{ij}\|_2+\frac{1}{2}\sum_{i<j}\|\beta_i-\beta_j-\mathbf{z}_{ij}+\mathbf{u}_{ij}\|_2^2.$$

Then, the three ADMM subproblems are given by:

$$\begin{aligned}
\beta^{k+1} &= \arg\min_{\beta}\frac{1}{2}\sum_{i=1}^{n}\|\mathbf{y}_i-\beta_i\|_2^2+\frac{1}{2}\sum_{i<j}\|\beta_i-\beta_j-\mathbf{z}_{ij}+\mathbf{u}_{ij}\|_2^2\\
\mathbf{z}^{k+1} &= \arg\min_{\mathbf{z}}\lambda\sum_{i<j}w_{ij}\|\mathbf{z}_{ij}\|_2+\frac{1}{2}\sum_{i<j}\|\beta_i-\beta_j-\mathbf{z}_{ij}+\mathbf{u}_{ij}\|_2^2\\
\mathbf{u}_{ij}^{k+1} &= \mathbf{u}_{ij}^{k}+[\mathbf{z}_{ij}^{k+1}-(\beta_i^{k+1}-\beta_j^{k+1})].
\end{aligned}$$

Splitting the variables in this manner allows us to solve a series of straightforward subproblems. Updating β involves solving a ridge regression problem. Despite the fact that the quadratic penalty term is not separable in the β, after some algebraic maneuvering, which is detailed in [5], it is possible to explicitly write down the updates for each β separately:

$$\beta_i^{k+1}=\left[\frac{1}{1+n}\mathbf{y}_i+\frac{n}{1+n}\bar{\mathbf{y}}\right]+\frac{1}{1+n}\left[\sum_{j>i}[\mathbf{u}_{ij}^k+\mathbf{z}_{ij}^k]-\sum_{j<i}[\mathbf{u}_{ji}^k+\mathbf{z}_{ji}^k]\right].$$

Updating $\mathbf{z}$ requires minimizing an objective that separates in each of the $\mathbf{z}_{ij}$,

$$\mathbf{z}_{ij}^{k+1}=\arg\min_{\mathbf{z}_{ij}}\frac{1}{2}\|\mathbf{z}_{ij}-[\beta_i^{k+1}-\beta_j^{k+1}-\mathbf{u}_{ij}^k]\|_2^2+\lambda w_{ij}\|\mathbf{z}_{ij}\|_2.$$

This step can be computed explicitly using the block-wise soft-thresholding operator, the proximal operator of the group LASSO [41], namely,

$$S(\mathbf{z},\tau)=\arg\min_{\zeta}\frac{1}{2}\|\zeta-\mathbf{z}\|_2^2+\tau\|\zeta\|_2=\left[1-\frac{\tau}{\|\mathbf{z}\|_2}\right]_+\mathbf{z},$$

where $a_+=\max(a,0)$ and $\tau\geq 0$ controls the amount of shrinkage towards zero.

For model selection purposes, one typically studies the sequence of cluster assignments given by coalescent patterns of $\boldsymbol{\beta}$, or the sparse patterns in the first differences of $\boldsymbol{\beta}$, as λ varies. We then seek to quickly approximate this sequence of active sets given by the coalescent patterns of $\boldsymbol{\beta}$ with our Algorithmic Regularization Paths, summarized in Algorithm 5.

Algorithm 5 Algorithmic Regularization Path for Convex Clustering

1. Initialize $\mathbf{z}_{ij}^0 = \mathbf{0}$, $\mathbf{u}_{ij}^0 = \mathbf{0}$, $\gamma^{(0)} = \varepsilon$, $k = 1$, and set $t > 0$.
2. While $\|\mathbf{z}^k\|_{\text{F}} > 0$:

   for all i **do**

   $\beta_i^{k+1} = \left[\frac{1}{1+n}\mathbf{Y}_i + \frac{n}{1+n}\bar{\mathbf{y}}\right] + \frac{1}{1+n}\left[\Sigma_{j>i}[\mathbf{u}_{ij}^k + \mathbf{z}_{ij}^k] - \Sigma_{j<i}[\mathbf{u}_{ji}^k + \mathbf{z}_{ji}^k]\right]$

   end for

   for all $i < j$ **do**

   $\mathbf{z}_{ij}^{k+1} = S\left(\beta_i^{k+1} - \beta_j^{k+1} - \mathbf{u}_{ij}^k, \gamma^{(k)} w_{ij}\right)$

   $\mathbf{u}_{ij}^{k+1} = \mathbf{u}_{ij}^k + [\mathbf{z}_{ij}^{k+1} - (\beta_i^{k+1} - \beta_j^{k+1})]$,

   end for

   $\gamma^{(k+1)} = t\gamma^{(k)}$
3. Output $\left\{\mathbf{z}_{ij}^k\right\}$ as the algorithm path.

As in the general case, we can use iterates of the $\mathbf{z}$-subproblem to approximate a sparse sequence of cluster assignments. Given $\mathbf{z}$, we can determine a clustering assignment in time that is linear in the number of data points n. We simply apply breadth-first search to identify the connected components of the following graph induced by the $\mathbf{z}$. The graph identifies a node with every data point and places an edge between the ith and jth node if and only if $\mathbf{z}_{ij} = \mathbf{0}$. Each connected component corresponds to a cluster.

We now illustrate on a simulated "halfmoon" data set of $n = 200$ points in $\Re^2$, that computing our Algorithm Path can lead to nontrivial computational cost savings for obtaining a sequence of clustering assignments. We first detail some preliminaries. Although we do not take the space to discuss it here, in practice the choice of weights is very important. This topic is explored in [5], and we use the sparse kernel weights which were shown to work well empirically in that paper. We created a geometric sequence of parameters $\lambda^{(k)}$ and $\gamma^{(k)}$, namely given a fixed multiplicative factor $t > 1$, we set $\lambda^{(k+1)} = t\lambda^{(k)}$. The sequence $\gamma^{(k)}$ was constructed similarly, although we study our Algorithm Paths for several multiplicative factors, $t \in \{1.1, 1.05, 1.01\}$.

In contrast to the regularization path, the Algorithm Path does not require any convergence checks since only one step is taken at each grid point. Nonetheless, we only report the number of rounds of ADMM updates taken by each approach. The Algorithm Path took $259, 1294$, and $2,536$ rounds of updates for the three step sizes considered; in contrast, the regularization path even for a very modest tolerance level, 10^{-4}, required a grand total of 30,008 rounds of updates, substantially more than our approach.

Figure 13.4 shows the ADMM Algorithmic Regularization paths and regularization path respectively for this simulated example. For each data point i we plot the sequence of the segments between consecutive estimates of its center, namely β_i^{k+1} and β_i^k. These paths begin to overlap and merge into "trunks" when center estimates for close-by data points begin to coincide as the parameters $\lambda^{(k)}$ and $\gamma^{(k)}$ becomes sufficiently large. For sufficiently small step sizes for the regularization levels the Algorithm Path and regularization path are strikingly similar, as expected and demonstrated previously in our other examples. For larger step sizes, however, the paths differ markedly, but still appear to capture the same clustering assignments. Overall, although the simulated data is relatively small, computing the whole regularization path, even for a modest stopping tolerance can, requires an order of magnitude more iterations and computational time than the Algorithm Path.

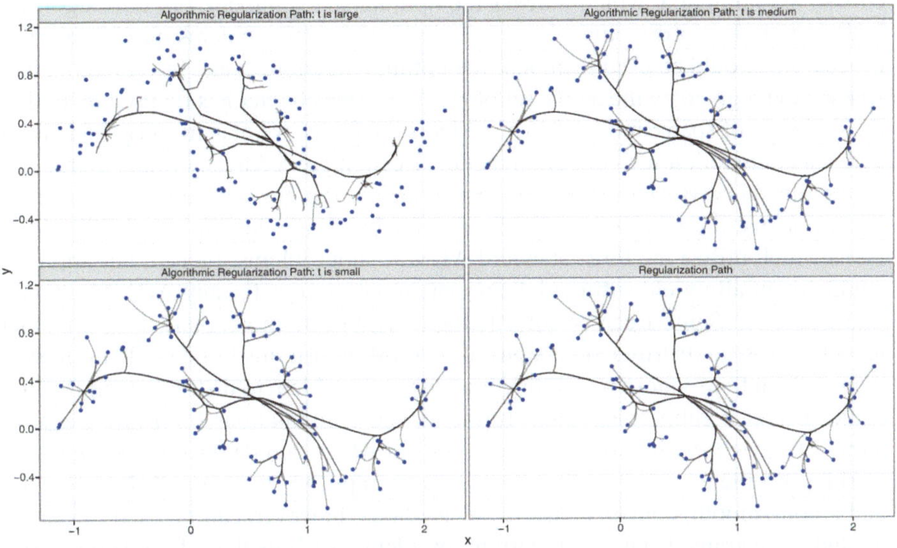

Fig. 13.4 Convex clustering on simulated data: In the first three panels (from left to right, top to bottom), lines trace the ADMM Algorithmic Regularization path of the individual cluster centers as the algorithm path parameter γ increases for $t = 1.1$ (large), 1.05 (medium), and 1.01 (small). In the panel in the lower right corner, the lines trace the regularization path of the individual cluster centers as the regularization parameter λ increases.

4 Discussion

In this chapter, we have presented a novel framework for approximating the sequence of active sets associated with regularization paths of sparse statistical learning problems. Instead of solving optimization problems over a grid of penalty parameters as in traditional regularization paths, our algorithm performs a series of one-step approximations to an ADMM algorithm employing warm-starts with the goal of estimating a good sequence of sparse models. Our approach has a number of advantages including easy implementation, exploration of the sparse model space at a fine resolution, and most importantly fast compute times; we have demonstrated these advantages through several sparse statistical learning examples.

In our demonstrations, we have focused simply on computing the full sequence of active sets corresponding to the regularization path which is the critical computationally intensive step in the process of model selection. Once the sequence of sparse models has been found, common methods for model selection such as AIC, BIC, cross-validation, and stability selection can be employed to choose the optimal model. We note that with regularization paths, model selection procedures typically choose the optimal λ which indexes the optimal sparse model. For our Algorithm Paths which do not directly solve regularized statistical problems, model selection procedures should be used to choose the optimal iteration, k, and the corresponding sparse model given by the active set of $\mathbf{z}^k$. While this chapter has focused on finding the sequence of sparse models via our Algorithm Paths, we plan to study using these paths in conjunction with common model selection procedures in future work.

As the ADMM algorithm has been widely used for sparse statistical learning problems, the mechanics are in place for broad application of our Algorithm Paths which utilize the three standard ADMM subproblems. Indeed, our approach could potentially yield substantial computational savings for any ADMM application where the $\boldsymbol{\beta}$ and $\mathbf{z}$ can be solved efficiently. Furthermore, there has been much recent interest in distributed versions of ADMM algorithms [30, 26]. Thus, there is the potential to use these in conjunction with our problem to distribute computation in the $\boldsymbol{\beta}$ and $\mathbf{z}$ subproblems and further speed computations for Big-Data problems. Also, we have focused on developing our Algorithm Path for sparse statistical learning problems that can be written as a composite of a smooth loss function and a non-smooth, convex penalty. Our methods, however, can be easily extended to study constrained statistical learning problems, such as that of the support vector machines. Finally, our framework utilizes the ADMM splitting method, but the strategies we develop could also be useful for computing a sequence of sparse models using other operator splitting algorithms.

Our work raises many questions from statistical and optimization perspectives. Further work needs to be done to characterize and study the mathematical properties of the Algorithm Paths as well as relate them to existing optimization procedures and algorithms. For example, ADMM is just one of many variants of proximal methods [29]. We suspect that other variants, such as proximal gradient descent, used to fit sparse models will also benefit from an Algorithm Path approach in expediting the model selection procedure. We leave this as future work.

In our demonstrations in Section 3, we suggested empirically that our Algorithm Paths with a tiny step size closely approximate the sequence of active sets associated with regularization paths. Further work needs to be done to verify this connection mathematically. Along these lines, a key practical question is how to choose the appropriate step size for increasing the amount of regularization as the algorithm progresses. As we have demonstrated, changing the step size yields paths with very different solutions and behaviors that warrant further investigation. For now, our recommendation is to employ a fairly small step size as these well approximate the traditional regularization paths in all of the examples we have studied. Additionally, our approach may be related to other new proposals for computing regularization paths based on partial differential equations, for example [33]; these potential connections merit further investigation.

Our work also raises a host of interesting statistical questions as well. The sparse regression example suggested that Algorithm Paths may not simply yield computational savings, but may also perform better in terms of variable selection. This raises an interesting statistical prospect that we plan to carefully study in future work.

To conclude, we have introduced a novel approach to approximating the sequence of active sets associated with regularization paths for large-scale sparse statistical learning procedures. Our methods yield substantial computational savings and raise a number of interesting open questions for future research.

Acknowledgments

The authors thank Wotao Yin for helpful discussions. Y.H. and G.A. acknowledge support from NSF DIMS 1209017, 1264058, and 1317602. E.C. acknowledges support from CIA Postdoctoral Fellowship #2012-12062800003.

References

1. Aguiar, P., Xing, E.P., Figueiredo, M., Smith, N.A., Martins, A.: An augmented Lagrangian approach to constrained MAP inference. In: Proceedings of the 28th International Conference on Machine Learning (ICML-11), pp. 169–176 (2011)
2. Bien, J., Taylor, J., Tibshirani, R.: A lasso for hierarchical interactions. The Annals of Statistics **41**(3), 1111–1141 (2013)
3. Boyd, S., Parikh, N., Chu, E., Peleato, B., Eckstein, J.: Distributed optimization and statistical learning via the alternating direction method of multipliers. Foundations and Trends® in Machine Learning **3**(1), 1–122 (2011)
4. Bühlmann, P., Van De Geer, S.: Statistics for High-Dimensional Data: Methods, Theory and Applications. Springer, Berlin Heidelberg (2011)
5. Chi, E.C., Lange, K.: Splitting methods for convex clustering. Journal of Computational and Graphical Statistics (to appear)
6. Danaher, P., Wang, P., Witten, D.M.: The joint graphical lasso for inverse covariance estimation across multiple classes. Journal of the Royal Statistical Society: Series B (Statistical Methodology) **76**(2), 373–397 (2014)
7. Donoho, D.L., Tsaig, Y., Drori, I., Starck, J.L.: Sparse solution of underdetermined systems of linear equations by stagewise orthogonal matching pursuit. Information Theory, IEEE Transactions on **58**(2), 1094–1121 (2012)
8. Efron, B., Hastie, T., Johnstone, I., Tibshirani, R.: Least angle regression. The Annals of Statistics **32**(2), 407–499 (2004)
9. Friedman, J., Hastie, T., Höfling, H., Tibshirani, R., et al.: Pathwise coordinate optimization. The Annals of Applied Statistics **1**(2), 302–332 (2007)
10. Friedman, J., Hastie, T., Tibshirani, R.: Regularization paths for generalized linear models via coordinate descent. Journal of Statistical Software **33**(1), 1–22 (2010)
11. Goldstein, T., O'Donoghue, B., Setzer, S.: Fast alternating direction optimization methods. SIAM Journal on Imaging Sciences **7**, 1588–1623 (2014)
12. Goldstein, T., Osher, S.: The split Bregman method for L1-regularized problems. SIAM J. Img. Sci. **2**(2), 323–343 (2009)
13. Hager, W.W.: Updating the inverse of a matrix. SIAM Review **31**(2), 221–239 (1989)

14. Hastie, T., Rosset, S., Tibshirani, R., Zhu, J.: The entire regularization path for the support vector machine. Journal of Machine Learning Research **5**, 1391–1415 (2004)
15. Hastie, T., Tibshirani, R., Friedman, J.: The Elements of Statistical Learning: Data Mining, Inference, and Prediction, 2 edn. Springer (2009)
16. He, B., Yang, H., Wang, S.: Alternating direction method with self-adaptive penalty parameters for monotone variational inequalities. Journal of Optimization Theory and applications **106**(2), 337–356 (2000)
17. Hocking, T., Vert, J.P., Bach, F., Joulin, A.: Clusterpath: an algorithm for clustering using convex fusion penalties. In: L. Getoor, T. Scheffer (eds.) Proceedings of the 28th International Conference on Machine Learning (ICML-11), ICML '11, pp. 745–752 (2011)
18. Hu, Y., Allen, G.I.: Local-aggregate modeling for big-data via distributed optimization: Applications to neuroimaging. Biometrics **41**(4), 905–917 (2015)
19. Lindsten, F., Ohlsson, H., Ljung, L.: Just relax and come clustering! A convexification of *k*-means clustering. Tech. rep., Linköpings Universitet (2011)
20. Liu, J., Musialski, P., Wonka, P., Ye, J.: Tensor completion for estimating missing values in visual data. In: Computer Vision, 2009 IEEE 12th International Conference on, pp. 2114–2121. IEEE (2009)
21. Ma, S., Xue, L., Zou, H.: Alternating direction methods for latent variable Gaussian graphical model selection. Neural Computation **25**(8), 2172–2198 (2013)
22. Mairal, J., Jenatton, R., Obozinski, G., Bach, F.: Convex and network flow optimization for structured sparsity. The Journal of Machine Learning Research **12**, 2681–2720 (2011)
23. Meinshausen, N., Bühlmann, P.: Stability selection. Journal of the Royal Statistical Society: Series B (Statistical Methodology) **72**(4), 417–473 (2010)
24. Mohan, K., Chung, M., Han, S., Witten, D., Lee, S.I., Fazel, M.: Structured learning of Gaussian graphical models. In: Advances in Neural Information Processing Systems, pp. 629–637 (2012)
25. Mohan, K., London, P., Fazel, M., Witten, D., Lee, S.I.: Node-based learning of multiple Gaussian graphical models. Journal of Machine Learning Research **15**, 445–488 (2014)
26. Mota, J.F., Xavier, J., Aguiar, P.M., Puschel, M.: Distributed basis pursuit. Signal Processing, IEEE Transactions on **60**(4), 1942–1956 (2012)
27. Negahban, S., Wainwright, M.J., et al.: Estimation of (near) low-rank matrices with noise and high-dimensional scaling. The Annals of Statistics **39**(2), 1069–1097 (2011)
28. Osborne, M.R., Presnell, B., Turlach, B.A.: A new approach to variable selection in least squares problems. IMA Journal of Numerical Analysis **20**(3), 389–403 (2000)
29. Parikh, N., Boyd, S.: Proximal algorithms. Foundations and Trends in Optimization **1**(3), 123–231 (2013)
30. Peng, Z., Yan, M., Yin, W.: Parallel and distributed sparse optimization. In: Signals, Systems and Computers, 2013 Asilomar Conference on, pp. 659–646. IEEE (2013)
31. Rosset, S., Ji, Z.: Piecewise linear regularized solution paths. The Annals of Statistics **35**(3), 1012–1030 (2007)
32. Rudin, L.I., Osher, S., Fatemi, E.: Nonlinear total variation based noise removal algorithms. Physica D: Nonlinear Phenomena **60**(1–4), 259–268 (1992)
33. Shi, J., Yin, W., Osher, S.: A new regularization path for logistic regression via linearized Bregman. Tech. rep., Rice CAAM Tech Report TR12-24 (2012)
34. Tibshirani, R.: Regression shrinkage and selection via the lasso. Journal of the Royal Statistical Society. Series B (Methodological) **58**(1), 267–288 (1996)
35. Tibshirani, R., Saunders, M., Rosset, S., Zhu, J., Knight, K.: Sparsity and smoothness via the fused lasso. Journal of the Royal Statistical Society: Series B (Statistical Methodology) **67**(1), 91–108 (2005)
36. Vu, V.Q., Cho, J., Lei, J., Rohe, K.: Fantope projection and selection: A near-optimal convex relaxation of sparse PCA. In: Advances in Neural Information Processing Systems 26, pp. 2670–2678 (2013)
37. Wahlberg, B., Boyd, S., Annergren, M., Wang, Y.: An ADMM algorithm for a class of total variation regularized estimation problems. System Identification **16**(1), 83–88 (2012)

38. Wu, T.T., Lange, K., et al.: Coordinate descent algorithms for lasso penalized regression. The Annals of Applied Statistics **2**(1), 224–244 (2008)
39. Yan, M., Yin, W.: Self equivalence of the alternating direction method of multipliers. In: R. Glowinski, S. Osher, W. Yin (eds.) Splitting Methods in Communication and Imaging, Science and Engineering. Springer (2016)
40. Yuan, L., Liu, J., Ye, J.: Efficient methods for overlapping group lasso. In: Advances in Neural Information Processing Systems 24 (NIPS 2011), 9, pp. 352–360 (2011)
41. Yuan, M., Lin, Y.: Model selection and estimation in regression with grouped variables. Journal of the Royal Statistical Society: Series B (Statistical Methodology) **68**(1), 49–67 (2006)
42. Yuan, X., Yang, J.: Sparse and low-rank matrix decomposition via alternating direction method. The Pacific Journal of Optimization **9**(1), 167–180 (2012)

Chapter 14
Decentralized Learning for Wireless Communications and Networking

Georgios B. Giannakis, Qing Ling, Gonzalo Mateos, Ioannis D. Schizas, and Hao Zhu

Abstract This chapter deals with decentralized learning algorithms for in-network processing of graph-valued data. A generic learning problem is formulated and recast into a separable form, which is iteratively minimized using the alternating-direction method of multipliers (ADMM) so as to gain the desired degree of parallelization. Without exchanging elements from the distributed training sets and keeping inter-node communications at affordable levels, the local (per-node) learners consent to the desired quantity inferred globally, meaning the one obtained if the entire training data set were centrally available. Impact of the decentralized learning framework to contemporary wireless communications and networking tasks is illustrated through case studies including target tracking using wireless sensor networks, unveiling Internet traffic anomalies, power system state estimation, as well as spectrum cartography for wireless cognitive radio networks.

G.B. Giannakis (✉)
University of Minnesota, 117 Pleasant St., Minneapolis, MN 55455, USA
e-mail: georgios@umn.edu

Q. Ling
University of Science and Technology of China, 443 Huangshan Road,
Hefei, Anhui 230027, China
e-mail: qingling@mail.ustc.edu.cn

G. Mateos
University of Rochester, 413 Hopeman Engineering Building, Rochester, NY 14627, USA
e-mail: gmateosb@ece.rochester.edu

I.D. Schizas
University of Texas at Arlington, 416 Yates Street, Arlington, TX 76010, USA
e-mail: schizas@uta.edu

H. Zhu
University of Illinois at Urbana-Champaign, 4058 ECE Building,
306 N. Wright Street, Urbana, IL 61801, USA
e-mail: haozhu@illinois.edu

R. Glowinski et al. (eds.), *Splitting Methods in Communication, Imaging, Science, and Engineering*, Scientific Computation, DOI 10.1007/978-3-319-41589-5_14

1 Introduction

This chapter puts forth an optimization framework for learning over networks, that entails decentralized processing of training data acquired by interconnected nodes. Such an approach is of paramount importance when communication of training data to a central processing unit is prohibited due to, e.g., communication cost or privacy reasons. The so-termed in-network processing paradigm for decentralized learning is based on successive refinements of local model parameter estimates maintained at individual network nodes. In a nutshell, each iteration of this broad class of fully decentralized algorithms comprises: (i) a communication step where nodes exchange information with their neighbors through, e.g., the shared wireless medium or Internet backbone; and (ii) an update step where each node uses this information to refine its local estimate. Devoid of hierarchy and with their decentralized in-network processing, local, e.g., estimators should eventually consent to the global estimator sought, while fully exploiting existing spatiotemporal correlations to maximize estimation performance. In most cases, consensus can formally be attained asymptotically in time. However, a finite number of iterations will suffice to obtain results that are sufficiently accurate for all practical purposes.

In this context, the approach followed here entails reformulating a generic learning task as a convex constrained optimization problem, whose structure lends itself naturally to decentralized implementation over a network graph. It is then possible to capitalize upon this favorable structure by resorting to the alternating-direction method of multipliers (ADMM), an iterative optimization method that can be traced back to [33] (see also [31]), and which is specially well suited for parallel processing [7, 9]. This way simple decentralized recursions become available to update each node's local estimate, as well as a vector of dual prices through which network-wide agreement is effected.

Problem Statement. Consider a network of n nodes in which scarcity of power and bandwidth resources encourages only single-hop inter-node communications, such that the i-th node communicates solely with nodes j in its single-hop neighborhood $\mathscr{N}_i$. Inter-node links are assumed symmetric, and the network is modeled as an undirected graph whose vertices are the nodes and its edges represent the available communication links. As it will become clear through the different application domains studied here, nodes could be wireless sensors, wireless access points (APs), electrical buses, sensing cognitive radios, or routers, to name a few examples. Node i acquires m_i measurements stacked in the vector $\mathbf{y}_i \in \mathbb{R}^{m_i}$ containing information about the unknown model parameters in $\mathbf{s} \in \mathbb{R}^p$, which the nodes need to estimate. Let $\mathbf{y} := [\mathbf{y}_1^\top, \ldots, \mathbf{y}_n^\top]^\top \in \mathbb{R}^{\sum_i m_i}$ collect measurements acquired across the entire network. Many popular centralized schemes obtain an estimate $\hat{\mathbf{s}}$ as follows

$$\hat{\mathbf{s}} \in \arg\min_{\mathbf{s}} \textstyle\sum_{i=1}^n f_i(\mathbf{s}; \mathbf{y}_i). \tag{14.1}$$

In the decentralized learning problem studied here though, the summands f_i are assumed to be local cost functions only known to each node i. Otherwise sharing this information with a centralized processor, also referred to as fusion center (FC), can be challenging in various applications of interest, or, it may be even impossible in e.g., wireless sensor networks (WSNs) operating under stringent power budget constraints. In other cases such as the Internet or collaborative healthcare studies, agents may not be willing to share their private training data $\mathbf{y}_i$ but only the learning results. Performing the optimization (14.1) in a centralized fashion raises robustness concerns as well, since the central processor represents an isolated point of failure.

In this context, the objective of this chapter is to develop a decentralized algorithmic framework for learning tasks, based on in-network processing of the locally available data. The described setup naturally suggests three characteristics that the algorithms should exhibit: c1) each node $i = 1, \ldots, n$ should obtain an estimate of $\mathbf{s}$, which coincides with the corresponding solution $\hat{\mathbf{s}}$ of the centralized estimator (14.1) that uses the entire data $\{\mathbf{y}_i\}_{i=1}^n$; c2) processing per node should be kept as simple as possible; and c3) the overhead for inter-node communications should be affordable and confined to single-hop neighborhoods. It will be argued that such an ADMM-based algorithmic framework can be useful for contemporary applications in the domain of wireless communications and networking.

Prior Art. Existing decentralized solvers of (14.1) can be classified in two categories: C1) those obtained by modifying centralized algorithms and operating in the primal domain; and C2) those handling an equivalent constrained form of (14.1) (see (14.2) in Section 2), and operating in the primal-dual domain.

Primal-domain algorithms under C1 include the (sub)gradient method and its variants [57, 62, 85, 37], the incremental gradient method [60], the proximal gradient method [16], and the dual averaging method [24, 77]. Each node in these methods averages its local iterate with those of neighbors and descends along its local negative (sub)gradient direction. However, the resultant algorithms are limited to inexact convergence when using constant stepsizes [57, 85]. If diminishing stepsizes are employed instead, the algorithms can achieve exact convergence at the price of slowing down speed [37, 60, 24]. A constant-stepsize exact first-order algorithm is also available to achieve fast and exact convergence, by correcting error terms in the distributed gradient iteration with two-step historic information [72].

Primal-dual domain algorithms under C2 solve an equivalent constrained form of (14.1), and thus drive local solutions to reach global optimality. The dual decomposition method is hence applicable because (sub)gradients of the dual function depend on local and neighboring iterates only, and can thus be computed without global cooperation [61]. ADMM modifies the dual decomposition by regularizing the constraints with a quadratic term, which improves numerical stability as well as rate of convergence, as will be demonstrated later in this chapter. Per ADMM iteration, each node solves a subproblem that can be demanding. Fortunately, these subproblems can be solved inexactly by running one-step gradient or proximal gradient descent iterations, which markedly mitigate the computation burden [43, 15]. A sequential distributed ADMM algorithm can be found in [79].

Chapter Outline. The remainder of this chapter is organized as follows. Section 2 describes a generic ADMM framework for decentralized learning over networks, which is at the heart of all algorithms described in the chapter and was pioneered in [67, 70] for in-network estimation using WSNs. Section 3 focuses on batch estimation as well as (un)supervised inference, while Section 4 deals with decentralized adaptive estimation and tracking schemes where network nodes collect data sequentially in time. Internet traffic anomaly detection and spectrum cartography for wireless CR networks serve as motivating applications for the sparsity-regularized rank minimization algorithms developed in Section 5. Fundamental results on the convergence and convergence rate of decentralized ADMM are stated in Section 6.

2 In-Network Learning with ADMM in a Nutshell

Since local summands in (14.1) are coupled through a *global* variable $\mathbf{s}$, it is not straightforward to decompose the unconstrained optimization problem in (14.1). To overcome this hurdle, the key idea is to introduce local variables $\mathscr{S} := \{\mathbf{s}_i\}_{i=1}^n$ which represent local estimates of $\mathbf{s}$ per network node i [67, 70]. Accordingly, one can formulate the *constrained* minimization problem

$$\{\hat{\mathbf{s}}_i\}_{i=1}^n \in \arg\min_{\mathscr{S}} \sum_{i=1}^n f_i(\mathbf{s}_i; \mathbf{y}_i), \quad \text{s. to} \quad \mathbf{s}_i = \mathbf{s}_j, \; j \in \mathscr{N}_i. \tag{14.2}$$

The "consensus" equality constraints in (14.2) ensure that local estimates coincide within neighborhoods. Further, if the graph is connected then consensus naturally extends to the whole network, and it turns out that problems (14.1) and (14.2) are equivalent in the sense that $\hat{\mathbf{s}} = \hat{\mathbf{s}}_1 = \ldots = \hat{\mathbf{s}}_n$ [70]. Interestingly, the formulation in (14.2) exhibits a separable structure that is amenable to decentralized minimization. To leverage this favorable structure, the alternating direction method of multipliers (ADMM), see, e.g., [7, pg. 253–261], can be employed here to minimize (14.2) in a decentralized fashion. This procedure will yield a distributed estimation algorithm whereby local iterates $\mathbf{s}_i(k)$, with k denoting iterations, provably converge to the centralized estimate $\hat{\mathbf{s}}$ in (14.1); see also Section 6.

To facilitate application of ADMM, consider the auxiliary variables $\mathscr{Z} := \{\mathbf{z}_i^j\}_{j \in \mathscr{N}_i}$, and reparameterize the constraints in (14.2) with the equivalent ones

$$\begin{aligned} &\{\hat{\mathbf{s}}_i\}_{i=1}^n \in \arg\min_{\mathscr{S}} \sum_{i=1}^n f_i(\mathbf{s}_i; \mathbf{y}_i), \\ &\text{s. to} \quad \mathbf{s}_i = \mathbf{z}_i^j \text{ and } \mathbf{s}_j = \mathbf{z}_i^j, \; i = 1, \ldots, n, \; j \in \mathscr{N}_i, \; i \neq j. \end{aligned} \tag{14.3}$$

Variables $\mathbf{z}_i^j$ are only used to derive the local recursions but will be eventually eliminated. Attaching Lagrange multipliers $\mathscr{V} := \{\{\bar{\mathbf{v}}_i^j\}_{j \in \mathscr{N}_i}, \{\tilde{\mathbf{v}}_i^j\}_{j \in \mathscr{N}_i}\}_{i=1}^n$ to the constraints (14.3), consider the augmented Lagrangian function

$$L_c[\mathscr{S},\mathscr{Z},\mathscr{V}] = \sum_{i=1}^{n} f_i(\mathbf{s}_i;\mathbf{y}_i) + \sum_{i=1}^{n}\sum_{j\in\mathscr{N}_i}\left[(\bar{\mathbf{v}}_i^j)^\top(\mathbf{s}_i-\mathbf{z}_i^j)+(\tilde{\mathbf{v}}_i^j)^\top(\mathbf{s}_j-\mathbf{z}_i^j)\right]$$
$$+\frac{c}{2}\sum_{i=1}^{n}\sum_{j\in\mathscr{N}_i}\left[\|\mathbf{s}_i-\mathbf{z}_i^j\|^2+\|\mathbf{s}_j-\mathbf{z}_i^j\|^2\right] \quad (14.4)$$

where the constant $c>0$ is a penalty coefficient. To minimize (14.2), ADMM entails an iterative procedure comprising three steps per iteration $k=1,2,\dots$

[S1] Multiplier updates:

$$\bar{\mathbf{v}}_i^j(k) = \bar{\mathbf{v}}_i^j(k-1)+c[\mathbf{s}_i(k)-\mathbf{z}_i^j(k)]$$
$$\tilde{\mathbf{v}}_i^j(k) = \tilde{\mathbf{v}}_i^j(k-1)+c[\mathbf{s}_j(k)-\mathbf{z}_i^j(k)].$$

[S2] Local estimate updates:

$$\mathscr{S}(k+1) = \arg\min_{\mathscr{S}} L_c\left[\mathscr{S},\mathscr{Z}(k),\mathscr{V}(k)\right].$$

[S3] Auxiliary variable updates:

$$\mathscr{Z}(k+1) = \arg\min_{Z} L_c\left[\mathscr{S}(k+1),\mathscr{Z},\mathscr{V}(k)\right]$$

where $i=1,\dots,n$ and $j\in\mathscr{N}_i$ in [S1]. Reformulating the generic learning problem (14.1) as (14.3) renders the augmented Lagrangian in (14.4) highly decomposable. The separability comes in two flavors, both with respect to the sets $\mathscr{S}$ and $\mathscr{Z}$ of primal variables, as well as across nodes $i=1,\dots,n$. This in turn leads to highly parallelized, simplified recursions corresponding to the aforementioned steps [S1]-[S3]. Specifically, as detailed in, e.g., [70, 68, 69, 29, 51, 48], it follows that if the multipliers are initialized to zero, the ADMM-based decentralized algorithm reduces to the following updates carried out locally at every node

In-network learning algorithm at node i, for $k=1,2,\dots$:

$$\mathbf{v}_i(k) = \mathbf{v}_i(k-1)+c\sum_{j\in\mathscr{N}_i}[\mathbf{s}_i(k)-\mathbf{s}_j(k)] \quad (14.5)$$

$$\mathbf{s}_i(k+1) = \arg\min_{\mathbf{s}_i}\left\{f_i(\mathbf{s}_i;\mathbf{y}_i)+\mathbf{v}_i^\top(k)\mathbf{s}_i+c\sum_{j\in\mathscr{N}_i}\left\|\mathbf{s}_i-\frac{\mathbf{s}_i(k)+\mathbf{s}_j(k)}{2}\right\|^2\right\} \quad (14.6)$$

where $\mathbf{v}_i(k) := 2\sum_{j\in\mathscr{N}_i}\bar{\mathbf{v}}_i^j(k)$, and all initial values are set to zero.

Recursions (14.5) and (14.6) entail local updates, which comprise the general purpose ADMM-based decentralized learning algorithm. The inherently redundant set of auxiliary variables in $\mathscr{Z}$ and corresponding multipliers have been eliminated.

Each node, say the i-th one, does not need to *separately* keep track of all its non-redundant multipliers $\{\bar{\mathbf{v}}_i^j(k)\}_{j\in\mathcal{N}_i}$, but only to update the (scaled) sum $\mathbf{v}_i(k)$. In the end, node i has to store and update only two p-dimensional vectors, namely $\{\mathbf{s}_i(k)\}$ and $\{\mathbf{v}_i(k)\}$. A unique feature of in-network processing is that nodes communicate their updated local estimates $\{\mathbf{s}_i\}$ (and not their raw data $\mathbf{y}_i$) with their neighbors, in order to carry out the tasks (14.5)–(14.6) for the next iteration.

As elaborated in Section 6, under mild assumptions on the local costs one can establish that $\lim_{k\to\infty}\mathbf{s}_i(k)=\hat{\mathbf{s}}$, for $i=1,\dots,n$. As a result, the algorithm asymptotically attains consensus and the performance of the centralized estimator [cf. (14.1)].

3 Batch In-Network Estimation and Inference

3.1 Decentralized Signal Parameter Estimation

Many workhorse estimation schemes such as maximum likelihood estimation (MLE), least-squares estimation (LSE), best linear unbiased estimation (BLUE), as well as linear minimum mean-square error estimation (LMMSE) and the maximum a posteriori (MAP) estimation, all can be formulated as a minimization task similar to (14.1); see, e.g., [38]. However, the corresponding centralized estimation algorithms fall short in settings where both the acquired measurements and computational capabilities are distributed among multiple spatially scattered sensing nodes, which is the case with WSNs. Here we outline a novel batch decentralized optimization framework building on the ideas in Section 2, that formulates the desired estimator as the solution of a separable constrained convex minimization problem tackled via ADMM; see, e.g., [7, 9, 70, 68] for further details on the algorithms outlined here.

Depending on the estimation technique utilized, the local cost functions $f_i(\cdot)$ in (14.1) should be chosen accordingly; see, e.g., [38, 70, 68]. For instance, when $\mathbf{s}$ is assumed to be an unknown deterministic vector, then:

- If $\hat{\mathbf{s}}$ corresponds to the centralized MLE then $f_i(\mathbf{s};\mathbf{y}_i) = -\ln[p_i(\mathbf{y}_i;\mathbf{s})]$ is the negative log-likelihood capturing the data probability density function (pdf), while the network-wide data $\{\mathbf{y}_i\}_{i=1}^n$ are assumed statistically independent.
- If $\hat{\mathbf{s}}$ corresponds to the BLUE (or weighted least-squares estimator) then $f_i(\mathbf{s};\mathbf{y}_i) = (1/2)\|\boldsymbol{\Sigma}_{y_i}^{-1/2}(\mathbf{y}_i - \mathbf{H}_i\mathbf{s})\|^2$, where $\boldsymbol{\Sigma}_{y_i}$ denotes the covariance of the data $\mathbf{y}_i$, and $\mathbf{H}_i$ is a known fitting matrix.

When $\mathbf{s}$ is treated as a random vector, then:

- If $\hat{\mathbf{s}}$ corresponds to the centralized MAP estimator then $f_i(\mathbf{s};\mathbf{y}_i) = -(\ln[p_i(\mathbf{y}_i|\mathbf{s})] + n^{-1}\ln[p(\mathbf{s})])$ accounts for the data pdf, and $p(\mathbf{s})$ for the prior pdf of $\mathbf{s}$, while data $\{\mathbf{y}_i\}_{i=1}^n$ are assumed conditionally independent given $\mathbf{s}$.
- If $\hat{\mathbf{s}}$ corresponds to the centralized LMMSE then $f_i(\mathbf{s};\mathbf{y}_i) = (1/2)\|\mathbf{s} - n\boldsymbol{\Sigma}_{sy_i}\mathbf{u}^i\|_2^2$, where $\boldsymbol{\Sigma}_{sy_i}$ denotes the cross-covariance of $\mathbf{s}$ with $\mathbf{y}_i$, while $\mathbf{u}^i$ stands for the i-th $m_i \times 1$ block subvector of $\mathbf{u} = \boldsymbol{\Sigma}_y^{-1}\mathbf{y}$.

Substituting in (14.6) the specific $f_i(\mathbf{s};\mathbf{y}_i)$ for each of the aforementioned estimation tasks, yields a family of batch ADMM-based decentralized estimation algorithms. The decentralized BLUE algorithm will be described in this section as an example of decentralized linear estimation.

Recent advances in cyber-physical systems have also stressed the need for decentralized nonlinear least-squares (LS) estimation. Monitoring the power grid for instance is challenged by the nonconvexity arising from the nonlinear AC power flow model; see, e.g., [82, Ch. 4], while the interconnection across local transmission systems motivates their operators to collaboratively monitor the global system state. Interestingly, this nonlinear (specifically quadratic) estimation task can be convexified to a semidefinite program (SDP) [8, pg. 168], for which a decentralized semidefinite programming (SDP) algorithm can be developed by leveraging the batch ADMM; see also [80] for an ADMM-based centralized SDP precursor.

3.1.1 Decentralized BLUE

The minimization involved in (14.6) can be performed locally at sensor i by employing numerical optimization techniques [8]. There are cases where the minimization in (14.6) yields a closed-form and easy to implement updating formula for $\mathbf{s}_i(k+1)$. If for example network nodes wish to find the BLUE estimator in a distributed fashion, the local cost is $f_i(\mathbf{s};\mathbf{y}_i) = (1/2)\|\boldsymbol{\Sigma}_{y_i}^{-1/2}(\mathbf{y}_i - \mathbf{H}_i\mathbf{s})\|^2$, and (14.6) becomes a strictly convex unconstrained quadratic program which admits the following closed-form solution (see details in [70, 54])

$$\mathbf{s}_i(k+1) = \left(\mathbf{H}_i^\top \boldsymbol{\Sigma}_{y_i}^{-1}\mathbf{H}_i + 2c|\mathscr{N}_i|\mathbf{I}_p\right)^{-1}\left[\mathbf{H}_i^\top \boldsymbol{\Sigma}_{y_i}^{-1}\mathbf{y}_i - \mathbf{v}_i(k) + c\sum_{j\in\mathscr{N}_i}(\mathbf{s}_i(k)+\mathbf{s}_j(k))\right]. \tag{14.7}$$

The pair (14.5) and (14.7) comprise the decentralized (D-) BLUE algorithm [67, 70]. For the special case where each node acquires unit-variance scalar observations y_i, there is no fitting matrix and s is scalar (i.e., $p = 1$); D-BLUE offers a decentralized algorithm to obtain the network-wide sample average $\hat{s} = (1/n)\sum_{i=1}^{n} y_i$. The update rule for the local estimate is obtained by suitably specializing (14.7) to

$$s_i(k+1) = (1+2c|\mathscr{N}_i|)^{-1}\left[y_i - v_i(k) + c\sum_{j\in\mathscr{N}_i}(s_i(k)+s_j(k))\right]. \tag{14.8}$$

Different from existing distributed averaging approaches [4, 22, 83, 84], the ADMM-based one originally proposed in [67, 70] allows the decentralized computation of general nonlinear estimators that may not be available in closed form and cannot be expressed as "averages." Further, the obtained recursions exhibit robustness in the presence of additive noise in the inter-node communication links.

3.1.2 Decentralized SDP

Consider now that each scalar y_i^ℓ in $\mathbf{y}_i$ adheres to a quadratic measurement model in $\mathbf{s}$ plus additive Gaussian noise, where the centralized MLE requires solving a nonlinear least-squares problem. To tackle the nonconvexity due to the quadratic dependence, the task of estimating the state $\mathbf{s}$ can be reformulated as that of estimating the outer-product matrix $\mathbf{S} := \mathbf{s}\mathbf{s}^\top$. In this reformulation y_i^ℓ is a linear function of $\mathbf{S}$, given by $\mathrm{Tr}(\mathbf{H}_i^\ell \mathbf{S})$ with a known matrix $\mathbf{H}_i^\ell$ [87]. Motivated by the separable structure in (14.3), the nonlinear estimation problem can be similarly formulated as

$$\{\hat{\mathbf{S}}_i\}_{i=1}^n \in \arg\min \sum_{i=1}^n \sum_\ell \left[y_i^\ell - \mathrm{Tr}(\mathbf{H}_i^\ell \mathbf{S})\right]^2,$$

$$\text{s. to} \quad \mathbf{S}_i = \mathbf{Z}_i^j \text{ and } \mathbf{S}_j = \mathbf{Z}_i^j,\ i = 1,\ldots,n,\ j \in \mathscr{N}_i,\ i \neq j$$

$$\mathbf{S}_i \succeq \mathbf{0} \text{ and } \mathrm{rank}(\mathbf{S}_i) = 1,\ i = 1,\ldots,n \tag{14.9}$$

where the positive-semidefiniteness and rank constraints ensure that each matrix $\mathbf{S}_i$ is an outer-product matrix. By dropping the non-convex rank constraints, the problem (14.9) becomes a convex semidefinite program (SDP), which can be solved in a decentralized fashion by adopting the batch ADMM iterations (14.5) and (14.6).

This decentralized SDP approach has been successfully employed for monitoring large-scale power networks [32]. To estimate the complex voltage phasor all nodes (a.k.a. power system state), measurements are collected on real/reactive power and voltage magnitude, all of which have quadratic dependence on the unknown states. Gauss-Newton iterations have been the 'workhorse' tool for this nonlinear estimation problem; see, e.g., [1, 82]. However, the iterative linearization therein could suffer from convergence issues and local optimality, especially due to the increasing variability in power grids with high penetration of renewables. With improved communication capabilities, decentralized state estimation among multiple control centers has attracted growing interest; see Figure 14.1 illustrating three interconnected areas aiming to achieve the centralized estimation collaboratively.

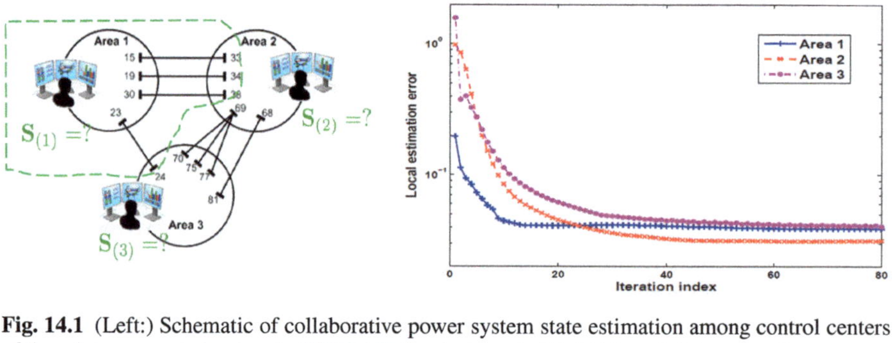

Fig. 14.1 (Left:) Schematic of collaborative power system state estimation among control centers of three interconnected networks (IEEE 118-bus test case). (Right:) Local state estimation error vs. iteration number using the decentralized SDP-based state estimation method.

A decentralized SDP-based state estimator has been developed in [87] with reduced complexity compared to (14.9). The resultant algorithm involves only internal voltages and those of next-hop neighbors in the local matrix $\mathbf{S}_{(i)}$; e.g., in Figure 14.1 $\mathbf{S}_{(1)}$ is identified by the dashed lines. Interestingly, the positive-semidefiniteness constraint for the overall $\mathbf{S}$ decouples nicely into that of all local $\{\mathbf{S}_i\}$, and the estimation error converges to the centralized performance within only a dozen iterations. The decentralized SDP framework has successfully addressed a variety of power system operational challenges, including a distributed microgrid optimal power flow solver in [18]; see also [32] for a tutorial overview of these applications.

3.2 Decentralized Inference

Along with decentralized signal parameter estimation, a variety of inference tasks become possible by relying on the collaborative sensing and computations performed by networked nodes. In the special context of resource-constrained WSNs deployed to determine the common messages broadcast by a wireless AP, the relatively limited node reception capability makes it desirable to design a decentralized *detection* scheme for all sensors to attain sufficient statistics for the *global* problem. Another exciting application of WSNs is environmental monitoring for, e.g., inferring the presence or absence of a pollutant over a geographical area. Limited by the local sensing capability, it is important to develop a decentralized *learning* framework such that all sensors can *collaboratively* approach the performance as if the network wide data had been available everywhere (or at a FC for that matter). Given the diverse inference tasks, the challenge becomes how to design the best inter-node information exchange schemes that would allow for minimal communication and computation overhead in specific applications.

3.2.1 Decentralized Detection

Message Decoding. A decentralized detection framework is introduced here for the message decoding task, which is relevant for diverse wireless communications and networking scenarios. Consider an AP broadcasting a $p \times 1$ coded block $\mathbf{s}$ to a network of sensors, all of which know the codebook $\mathscr{C}$ that $\mathbf{s}$ belongs to. For simplicity assume *binary* codewords, and that each node $i = 1, \ldots, n$ receives a same-length block of symbols $\mathbf{y}_i$ through a discrete, memoryless, symmetric channel that is conditionally independent across sensors. Sensor i knows its local channel from the AP, as characterized by the conditional pdf $p(y_{il}|s_l)$ per bit l. Due to conceivably low signal-to-noise-ratio (SNR) conditions, each low-cost sensor may be unable to reliably decode the message. Accordingly, the need arises for information exchanges among single-hop neighboring sensors to achieve the global (that is, centralized) error performance. Given $\mathbf{y}_i$ per sensor i, the assumption on memoryless and independent channels yields the centralized *maximum-likelihood* (ML) decoder as

$$\hat{\mathbf{s}}^{DEC} = \arg\max_{\mathbf{s}\in\mathscr{C}} p(\{\mathbf{y}_i\}_{i=1}^n|\mathbf{s}) = \arg\min_{\mathbf{s}\in\mathscr{C}} \Sigma_{l=1}^p \Sigma_{i=1}^n \left[-\log p(y_{il}|s_l)\right]. \tag{14.10}$$

ML decoding amounts to deciding the most likely codeword among multiple candidate ones and, in this sense, it can be viewed as a test of multiple hypotheses. In this general context, belief propagation approaches have been developed in [66], so that all nodes can cooperate to learn the centralized likelihood per hypothesis. However, even for linear binary block codes, the number of hypotheses, namely the cardinality of $\mathscr{C}$, grows exponentially with the codeword length. This introduces high communication and computation burden for the low-cost sensor designs.

The key here is to extract minimal sufficient statistics for the centralized decoding problem. For binary codes, the log-likelihood terms in (14.10) become $\log p(y_{il}|s_l) = -\gamma_{il}s_l + \log p(y_{il}|s_l = 0)$, where

$$\gamma_{il} := \log\left(\frac{p(y_{il}|s_l = 0)}{p(y_{il}|s_l = 1)}\right) \tag{14.11}$$

is the local log-likelihood ratio (LLR) for the bit s_l at sensor i. Ignoring all constant terms $\log p(y_{il}|s_l = 0)$, the ML decoding objective ends up only depending on the sum LLRs, as given by $\hat{\mathbf{s}}_{ML} = \arg\min_{\mathbf{s}\in\mathscr{C}} \Sigma_{l=1}^p (\Sigma_{i=1}^n \gamma_{il}) s_l$. Clearly, the sufficient statistic for solving (14.10) is the sum of all local LLR terms, or equivalently, the average $\bar{\gamma}_l = (1/n)\Sigma_{i=1}^n \gamma_{il}$ for each bit l. Interestingly, the average of $\{\gamma_{il}\}_{i=1}^n$ is one instance of the BLUE discussed in Section 3.1.1 when $\boldsymbol{\Sigma}_{y,i} = \mathbf{H}_j = \mathbf{I}_{p\times p}$, since

$$\bar{\gamma}_l = \arg\min_\gamma \Sigma_{i=1}^n (\gamma_{il} - \gamma)^2. \tag{14.12}$$

This way, the ADMM-based decentralized learning framework in Section 2 allows for all sensors to collaboratively attain the sufficient statistic for the decoding problem (14.10) via in-network processing. Each sensor only needs to estimate a vector of the codeword length p, which bypasses the exponential complexity under the framework of belief propagation. As shown in [89], decentralized *soft* decoding is also feasible since the *a posteriori probability (APP)* evaluator also relies on LLR averages which are sufficient statistics, where extensions to non-binary alphabet codeword constraints and random failing inter-sensor links are also considered.

The bit error rate (BER) versus SNR plot in Figure 14.2 demonstrates the performance of ADMM-based in-network decoding of a convolutional code with $p = 60$ and $|\mathscr{C}| = 40$. This numerical test involves $n = 10$ sensors and AWGN AP-sensor channels with $\sigma_i^2 = 10^{-SNR_i/10}$. Four schemes are compared: (i) the local ML decoder based on per-sensor data only (corresponds to the curve marked as $k = 0$ since it is used to initialize the decentralized iterations); (ii) the centralized benchmark ML decoder (corresponds to $k = \infty$); (iii) the in-network decoder which forms $\bar{\gamma}_l$ using "consensus-averaging" linear iterations [83]; and (iv) the ADMM-based decentralized algorithm. Indeed, the ADMM-based decoder exhibits faster convergence than its consensus-averaging counterpart; and surprisingly, only 10 iterations suffice to bring the decentralized BER very close to the centralized performance.

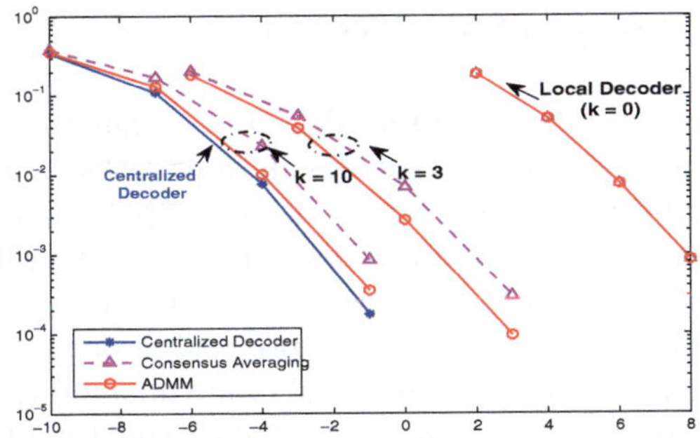

Fig. 14.2 BER vs. SNR (in dB) curves depicting the local ML decoder vs. the consensus-averaging decoder vs. the ADMM-based approach vs. the centralized ML decoder benchmark.

Message Demodulation. In a related detection scenario the common AP message $\mathbf{s}$ can be mapped to a space-time matrix, with each entry drawn from a finite alphabet $\mathscr{A}$. The received block $\mathbf{y}_i$ per sensor i typically admits a linear input/output relationship $\mathbf{y}_i = \mathbf{H}_i\,\mathbf{s} + \varepsilon_i$. Matrix $\mathbf{H}_i$ is formed from the fading AP-sensor channel, and ε_i stands for the additive white Gaussian noise of unit variance, that is assumed uncorrelated across sensors. Since low-cost sensors have very limited budget on number of antennas compared to the AP, the length of $\mathbf{y}_i$ is much shorter than $\mathbf{s}$ (i.e., $m_i < p$). Hence, the local linear demodulator using $\{\mathbf{y}_i, \mathbf{H}_i\}$ may not even be able to identify $\mathbf{s}$. Again, it is critical for each sensor i to cooperate with its neighbors to collectively form the global ML demodulator

$$\hat{\mathbf{s}}^{DEM} = \arg\max_{\mathbf{s}\in\mathscr{A}^N} -\Sigma_{i=1}^{n} \|\mathbf{y}_i - \mathbf{H}_i\mathbf{s}\|^2 = \arg\max_{\mathbf{s}\in\mathscr{A}^N} \left\{ 2\Big(\Sigma_{i=1}^{n}\mathbf{r}_i\Big)^{\top}\mathbf{s} - \mathbf{s}^{\top}\Big(\Sigma_{i=1}^{n}\mathbf{R}_i\Big)\mathbf{s} \right\} \tag{14.13}$$

where $\mathbf{r}_i := \mathbf{H}_i^{\top}\mathbf{y}_i$ and $\mathbf{R}_i := \mathbf{H}_i^{\top}\mathbf{H}_i$ are the sample (cross-)covariance terms. To solve (14.13) locally, it suffices for each sensor to acquire the network-wide average of $\{\mathbf{r}_i\}_{i=1}^{n}$, as well as that of $\{\mathbf{R}_i\}_{i=1}^{n}$, as both averages constitute the minimal sufficient statistics for the centralized demodulator. Arguments similar to decentralized decoding lead to ADMM iterations that (as with BLUE) attain locally these average terms. These iterations constitute a viable decentralized demodulation method, whose performance analysis in [88] reveals that its error diversity order can approach the centralized one within only a dozen of iterations.

As demonstrated by the decoding and demodulation tasks, the cornerstone of developing a decentralized detection scheme is to extract the minimal sufficient statistics for the centralized hypothesis testing problem. This leads to significant complexity reduction in terms of communications and computational overhead.

3.2.2 Decentralized Support Vector Machines

The merits of support vector machines (SVMs) in a centralized setting have been well documented in various supervised classification tasks including surveillance, monitoring, and segmentation, see, e.g., [71]. These applications often call for *decentralized supervised learning* solutions, when limited training data are acquired at different locations and a central processing unit is costly or even discouraged due to, e.g., scalability, communication overhead, or privacy reasons. Noteworthy examples include WSNs for environmental or structural health monitoring, as well as diagnosis of medical conditions from patient's records distributed at different hospitals.

In this in-network classification task, a labeled training set $\mathscr{T}_i := \{(\mathbf{x}_{il}, y_{il})\}$ of size m_i is available per node i, where $\mathbf{x}_{il} \in \mathbb{R}^p$ is the input data vector and $y_{il} \in \{-1, 1\}$ denotes its corresponding class label. Given all network-wide training data $\{\mathscr{T}_i\}_{i=1}^n$, the *centralized* SVM seeks a maximum-margin linear discriminant function $\hat{g}(\mathbf{x}) = \mathbf{x}^\top \hat{\mathbf{s}} + \hat{b}$, by solving the following convex optimization problem [71]

$$\begin{aligned} \{\hat{\mathbf{s}}, \hat{b}\} = \arg \min_{\mathbf{s},\, b, \{\xi_{il}\}} \quad & \frac{1}{2}\|\mathbf{s}\|^2 + C \sum_{i=1}^{n} \sum_{l=1}^{m_i} \xi_{il} \\ \text{s. to} \quad & y_{il}(\mathbf{s}^\top \mathbf{x}_{il} + b) \geq 1 - \xi_{il}, \quad i = 1, \ldots, n,\ l = 1, \ldots, m_i \\ & \xi_{il} \geq 0, \qquad\qquad\qquad\quad i = 1, \ldots, n,\ l = 1, \ldots, m_i \end{aligned} \tag{14.14}$$

where the slack variables ξ_{il} account for nonlinearly separable training sets, and C is a tunable positive scalar that allows for controlling model complexity. Nonlinear discriminant functions $g(\mathbf{x})$ can also be accommodated after mapping input vectors $\mathbf{x}_{il}$ to a higher- (possibly infinite)-dimensional space using, e.g., kernel functions, and pursuing a generalized maximum-margin linear classifier as in (14.14). Since the SVM classifier (14.14) couples the local datasets, early *distributed* designs either rely on a centralized processor so they are not decentralized [47], or, their performance is not guaranteed to reach that of the centralized SVM [56].

A fresh view of decentralized SVM classification is taken in [29], which reformulates (14.14) to estimate the parameter pair $\{\mathbf{s}, b\}$ from all local data $\mathscr{T}_i$ after eliminating slack variables ξ_{il}, namely

$$\{\hat{\mathbf{s}}, \hat{b}\} = \arg\min_{\mathbf{s},\, b} \frac{1}{2}\|\mathbf{s}\|^2 + C \sum_{i=1}^{n} \sum_{l=1}^{m_i} \max\{0, 1 - y_{il}(\mathbf{s}^\top \mathbf{x}_{il} + b)\}. \tag{14.15}$$

Notice that (14.15) has the same decomposable structure that the general decentralized learning task in (14.1), upon identifying the local cost $f_i(\bar{\mathbf{s}}; \mathbf{y}_i) = \frac{1}{2n}\|\mathbf{s}\|^2 + C\sum_{l=1}^{m_i} \max\{0, 1 - y_{il}(\mathbf{s}^\top \mathbf{x}_{il} + b)\}$, where $\bar{\mathbf{s}} := [\mathbf{s}^\top, b^\top]^\top$, and $\mathbf{y}_i := [y_{i1}, \ldots, y_{im_i}]^\top$. Accordingly, all network nodes can solve (14.15) in a decentralized fashion via iterations obtained following the ADMM-based algorithmic framework of Section 2. Such a decentralized ADMM-DSVM scheme is provably convergent to the centralized SVM classifier (14.14), and can also incorporate nonlinear discriminant functions as detailed in [29].

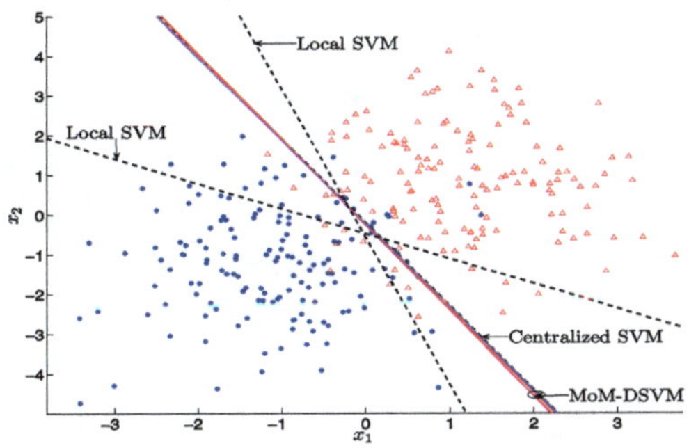

Fig. 14.3 Decision boundary comparison among ADMM-DSVM, centralized SVM, and local SVM results for synthetic data generated from two Gaussian classes, and a network of $n = 30$ nodes.

To illustrate the performance of the ADMM-DSVM algorithm in [29], consider a randomly generated network with $n = 30$ nodes. Each node acquires labeled training examples from two different classes, which are equiprobable and consist of random vectors drawn from a two-dimensional (i.e., $p = 2$) Gaussian distribution with common covariance matrix $\boldsymbol{\Sigma}_x = [1,\ 0;\ 0,\ 2]$, and mean vectors $\mu_1 = [-1,\ -1]^\top$ and $\mu_2 = [1,\ 1]^\top$, respectively. The Bayes optimal classifier for this 2-class problem is linear [25, Ch. 2]. To visualize this test case, Figure 14.3 depicts the global training set, along with the linear discriminant functions found by the centralized SVM (14.14) and the ADMM-DSVM at two different nodes after 400 iterations. Local SVM results for two different nodes are also included for comparison. It is apparent that ADMM-DSVM approaches the decision rule of its centralized counterpart, whereas local classifiers deviate since they neglect most of the training examples in the network.

3.2.3 Decentralized Clustering

Unsupervised learning using a network of wireless sensors as an exploratory infrastructure is well motivated for inferring hidden structures in distributed data collected by the sensors. Different from supervised SVM-based classification tasks, each node $i = 1, \ldots, n$ has available a set of *unlabeled* observations $\mathscr{X}_i := \{\mathbf{x}_{il},\ l = 1, \ldots, m_i\}$, drawn from a total of K classes. In this network setting, the goal is to design local clustering rules assigning each $\mathbf{x}_{il}$ to a cluster $k \in \{1, \ldots, K\}$. Again, the desiderata is a decentralized algorithm capable of attaining the performance of a benchmark clustering scheme, where all $\{\mathscr{X}_i\}_{i=1}^n$ are centrally available for joint processing.

Various criteria are available to quantify similarity among observations in a centralized setting, and a popular selection is the deterministic partitional clustering (DPC) one entailing prototypical elements (a.k.a. cluster centroids) per class in order to avoid comparisons between every pair of observations. Let μ_k denote the prototype element for class k, and ν_{ilk} the membership coefficient of $\mathbf{x}_{il}$ to class k.

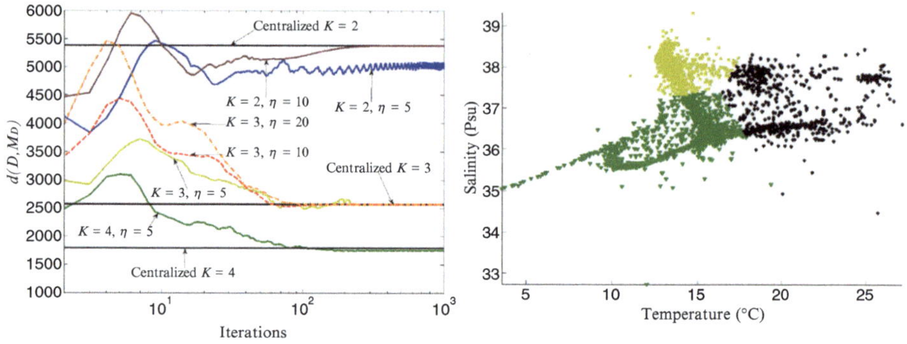

Fig. 14.4 Average performance of hard-DKM on a real data set using a WSN with $n = 20$ nodes for various values of η and K (left). Clustering with $K = 3$ and $\eta = 5$ (right) at $k = 400$ iterations.

A natural clustering problem amounts to specifying the family of K clusters with centroids $\{\mu_k\}_{k=1}^K$, such that the sum of squared-errors is minimized; that is

$$\min_{\{v_{ilk} \in \mathscr{V}\}, \{\mu_k\}} \sum_{i=1}^{n} \sum_{l=1}^{m_i} \sum_{k=1}^{K} v_{ilk}^{\rho} \|\mathbf{x}_{il} - \mu_k\|^2 \tag{14.16}$$

where $\rho \geq 1$ is a tuning parameter, and $\mathscr{V} := \{v_{ilk} : \sum_k v_{ilk}^{\rho} = 1,\ v_{ilk} \in [0,1],\ \forall i,l\}$ denotes the convex set of constraints on all membership coefficients. With $\rho = 1$ and $\{\mu_k\}$ fixed, (14.16) becomes a linear program in v_{ilk}. Consequently, (14.16) admits binary $\{0,1\}$ optimal solutions giving rise to the so-termed *hard* assignments, by choosing the cluster k for $\mathbf{x}_{il}$ whenever $v_{ilk} = 1$. Otherwise, for $\rho > 1$ the optimal coefficients generally result in *soft* membership assignments, and the optimal cluster is $k^* := \arg\max_k v_{ilk}^{\rho}$ for $\mathbf{x}_{il}$. In either case, the DPC clustering problem (14.16) is NP-hard, which motivates the (suboptimal) K-means algorithm that, on a per iteration basis, proceeds in two-steps to minimize the cost in (14.16) w.r.t.: (S1) $\mathscr{V}$ with $\{\mu_k\}$ fixed; and (S2) $\{\mu_k\}$ with $\mathscr{V}$ fixed [44]. Convergence of this two-step alternating-minimization scheme is guaranteed at least to a local minimum. Nonetheless, K-means requires central availability of global information (those variables that are fixed per step), which challenges in-network implementations. For this reason, most early attempts are either confined to specific communication network topologies, or they offer no closed-form local solutions; see, e.g., [58, 81].

To address these limitations, [30] casts (14.16) [yet another instance of (14.1)] as a decentralized estimation problem. It is thus possible to leverage ADMM iterations and solve (14.16) in a decentralized fashion through information exchanges among single-hop neighbors only. Albeit the non-convexity of (14.16), the decentralized DPC iterations in [30] provably approach a local minimum arbitrarily closely, where the asymptotic convergence holds for hard K-means with $\rho = 1$. Further extensions in [30] include a decentralized expectation-maximization algorithm for probabilistic partitional clustering, and methods to handle unknown number of classes.

Clustering of Oceanographic Data. Environmental monitoring is a typical application of WSNs. In WSNs deployed for oceanographic monitoring, the cost of computation per node is lower than the cost of accessing each node's observations [2]. This makes the option of centralized processing less attractive, thus motivating decentralized processing. Here we test the decentralized DPC schemes of [30] on real data collected by multiple underwater sensors in the Mediterranean coast of Spain [10], with the goal of identifying regions sharing common physical characteristics. A total of $5,720$ feature vectors were selected, each having entries for the temperature (°C) and salinity (psu) levels ($p = 2$). The measurements were normalized to have zero mean, unit variance, and they were grouped in $n = 20$ blocks (one per sensor) of $m_i = 286$ measurements each. The algebraic connectivity of the WSN is 0.2289 and the average degree per node is 4.9. Figure 14.4 (left) shows the performance of 25 Monte Carlo runs for the hard-DKM algorithm with different values of the parameter $c := \eta$. The best average convergence rate was obtained for $\eta = 5$, attaining the average centralized performance after 300 iterations. Tests with different values of K and η are also included in Figure 14.4 (left) for comparison. Note that for $K = 2$ and $\eta = 5$ hard-DKM hovers around a point without converging. Choosing a larger η guarantees convergence of the algorithm to a unique solution. The clustering results of hard-DKM at $k = 400$ iterations for $\eta = 5$ and $K = 3$ are depicted in Figure 14.4 (right).

4 Decentralized Adaptive Estimation

Sections 2 and 3 dealt with decentralized *batch* estimation, whereby network nodes acquire data only once and then locally exchange messages to reach consensus on the desired estimators. In many applications, however, networks are deployed to perform estimation in a constantly changing environment without having available a complete statistical description of the underlying processes of interest, e.g., with time-varying thermal or seismic sources. This motivates the development of decentralized adaptive estimation schemes, where nodes collect data sequentially in time and local estimates are recursively refined "on-the-fly". In settings where statistical state models are available, it is prudent to develop model-based tracking approaches implementing in-network Kalman or particle filters. Next, the scope of Section 2 is broadened to facilitate real-time (adaptive) processing of network data, when the local costs in (14.1) and unknown parameters are allowed to vary with time.

4.1 Decentralized Least-Mean Squares

A decentralized least-mean squares (LMS) algorithm is developed here for adaptive estimation of (possibly) nonstationary parameters, even when statistical information such as ensemble data covariances are unknown. Suppose network nodes are deployed to estimate a signal vector $\mathbf{s}(t) \in \mathbb{R}^{p\times 1}$ in a collaborative fashion subject to single-hop communication constraints, by resorting to the linear LMS criterion; see, e.g., [74, 46, 69]. Per time instant $t = 0,1,2,\ldots$, each node has available a regression vector $\mathbf{h}_i(t) \in \mathbb{R}^{p\times 1}$ and acquires a scalar observation $y_i(t)$, both assumed zero-mean without loss of generality. Introducing the global vector $\mathbf{y}(t) := [\mathbf{y}_1(t)\ldots\mathbf{y}_n(t)]^\top \in \mathbb{R}^{n\times 1}$ and matrix $\mathbf{H}(t) := [\mathbf{h}_1(t)\ldots\mathbf{h}_n(t)]^\top \in \mathbb{R}^{n\times p}$, the global time-dependent LMS estimator of interest can be written as [74, 46, 69, p. 14]

$$\hat{\mathbf{s}}(t) := \arg\min_{\mathbf{s}} \mathbb{E}\left[\|\mathbf{y}(t) - \mathbf{H}(t)\mathbf{s}\|^2\right] = \arg\min_{\mathbf{s}} \Sigma_{i=1}^n \mathbb{E}\left[(y_i(t) - \mathbf{h}_i^\top(t)\mathbf{s})^2\right]. \quad (14.17)$$

For jointly wide-sense stationary $\{\mathbf{x}(t),\mathbf{H}(t)\}$, solving (14.17) leads to the well-known Wiener filter estimate $\hat{\mathbf{s}}_W = \boldsymbol{\Sigma}_H^{-1}\boldsymbol{\Sigma}_{Hy}$, where $\boldsymbol{\Sigma}_H := \mathbb{E}[\mathbf{H}^\top(t)\mathbf{H}(t)]$ and $\boldsymbol{\Sigma}_{Hy} := \mathbb{E}[\mathbf{H}^\top(t)\mathbf{y}(t)]$; see e.g., [74, p. 15].

For the cases where the auto- and cross-covariance matrices $\boldsymbol{\Sigma}_H$ and $\boldsymbol{\Sigma}_{Hy}$ are unknown, the approach followed here to develop the decentralized (D-) LMS algorithm includes two main building blocks: (i) recast (14.17) into an equivalent form amenable to in-network processing via the ADMM framework of Section 2; and (ii) leverage stochastic approximation iterations [40] to obtain an adaptive LMS-like algorithm that can handle the unavailability/variation of statistical information. Following those algorithmic construction steps outlined in Section 2, the following updating recursions are obtained for the multipliers $\mathbf{v}_i(t)$ and the local estimates $\mathbf{s}_i(t+1)$ at time instant $t+1$ and $i = 1,\ldots,n$

$$\mathbf{v}_i(t) = \mathbf{v}_i(t-1) + c\sum_{j\in\mathcal{N}_i}[\mathbf{s}_i(t) - \mathbf{s}_j(t)] \quad (14.18)$$

$$\mathbf{s}_i(t+1) = \arg\min_{\mathbf{s}_i}\left\{\mathbb{E}\left[(y_i(t+1) - \mathbf{h}_i^\top(t+1)\mathbf{s}_i)^2\right] + \mathbf{v}_i^\top(t)\mathbf{s}_i + c\sum_{j\in\mathcal{N}_i}\left\|\mathbf{s}_i - \frac{\mathbf{s}_i(t)+\mathbf{s}_j(t)}{2}\right\|^2\right\}. \quad (14.19)$$

It is apparent that after differentiating (14.19) and setting the gradient equal to zero, $\mathbf{s}_i(t+1)$ can be obtained as the root of an equation of the form

$$\mathbb{E}[\boldsymbol{\varphi}(\mathbf{s}_i, y_i(t+1), \mathbf{h}_i(t+1))] = \mathbf{0} \quad (14.20)$$

where $\boldsymbol{\varphi}$ corresponds to the stochastic gradient of the cost in (14.19). However, the previous equation cannot be solved since the nodes do not have available any statistical information about the acquired data. Inspired by stochastic approximation techniques (such as the celebrated Robbins-Monro algorithm; see, e.g.,[40, Ch. 1]) which iteratively find the root of (14.20) given noisy observations $\{\boldsymbol{\varphi}(\mathbf{s}_i(t), y_i(t+1), \mathbf{h}_i(t+1))\}_{t=0}^\infty$, one can just drop the unknown expected value to obtain the following D-LMS (i.e., stochastic gradient) updates

$$\mathbf{s}_i(t+1) = \mathbf{s}_i(t) + \mu\left[\mathbf{h}_i(t+1)e_i(t+1) - \mathbf{v}_i(t) - c\sum_{j\in\mathcal{N}_i}[\mathbf{s}_i(t) - \mathbf{s}_j(t)]\right] \quad (14.21)$$

where μ denotes a constant step-size, and $e_i(t+1) := 2[y_i(t+1) - \mathbf{h}_i^\top(t+1)\mathbf{s}_i(t)]$ is twice the local *a priori* error.

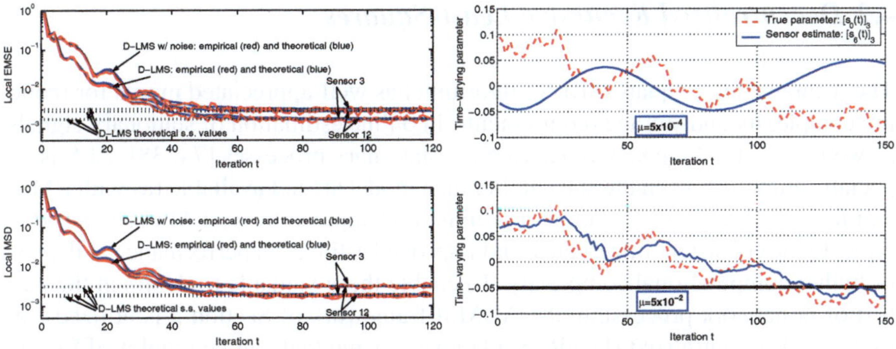

Fig. 14.5 Tracking with D-LMS. (left) Local MSE performance metrics both with and without inter-node communication noise for sensors 3 and 12; and (right) True and estimated time-varying parameters for a representative node, using slow and optimal adaptation levels.

Recursions (14.18) and (14.21) constitute the D-LMS algorithm, which can be viewed as a stochastic-gradient counterpart of D-BLUE in Section 3.1.1. D-LMS is a pioneering approach for decentralized online learning, which blends for the first time affordable (first-order) stochastic approximation steps with parallel ADMM iterations. The use of a constant step-size μ endows D-LMS with tracking capabilities. This is desirable in a constantly changing environment, within which, e.g., WSNs are envisioned to operate. The D-LMS algorithm is stable and converges even in the presence of inter-node communication noise (see details in [69, 55]). Further, closed-form expressions for the evolution and the steady-state mean-square error (MSE), as well as selection guidelines for the step-size μ can be found in [55].

Here we test the tracking performance of D-LMS with a computer simulation. For a random geometric graph with $n = 20$ nodes, network-wide observations y_i are linearly related to a large-amplitude slowly time-varying parameter vector $\mathbf{s}_0(t) \in \mathbb{R}^4$. Specifically, $\mathbf{s}_0(t) = \boldsymbol{\Theta}\mathbf{s}_0(t-1) + \boldsymbol{\zeta}(t)$, where $\boldsymbol{\Theta} = (1-10^{-4})\text{diag}(\theta_1,\ldots,\theta_p)$ with $\theta_i \sim \mathscr{U}[0,1]$. The driving noise is normally distributed with $\boldsymbol{\Sigma}_\zeta = 10^{-4}\mathbf{I}_p$. To model noisy links, additive white Gaussian noise with variance 10^{-2} is present at the receiving end. For $\mu = 5\times10^{-2}$, Figure 14.5 (left) depicts the local performance of two representative nodes through the evolution of the excess mean-square error $\text{EMSE}_i(t) = \mathbb{E}[(\mathbf{h}_i^\top(t)[\mathbf{s}_i(t-1) - \mathbf{s}_0(t-1)])^2]$ and the mean-square deviation $\text{MSD}_i(t) = \mathbb{E}[\|\mathbf{s}_i(t) - \mathbf{s}_0(t)\|^2]$ figures of merit. Both noisy and ideal links are considered, and the empirical curves closely follow the theoretical trajectories derived in [55]. Steady-state limiting values are also extremely accurate. As intuitively expected and suggested by the analysis, a performance penalty due to non-ideal links is also apparent. Figure 14.5 (right) illustrates how the adaptation level affects the resulting per-node estimates when tracking time-varying parameters with D-LMS. For $\mu = 5\times10^{-4}$ (slow adaptation) and $\mu = 5\times10^{-2}$ (near optimal adaptation), we depict the third entry of the parameter vector $[\mathbf{s}_0(t)]_3$ and the respective estimates from the randomly chosen sixth node. Under optimal adaptation the local estimate closely tracks the true variations, while – as expected – for the smaller step-size D-LMS fails to provide an accurate estimate [55, 74].

4.2 Decentralized Recursive Least-Squares

The recursive least-squares (RLS) algorithm has well-appreciated merits for reducing complexity and storage requirements, in online estimation of stationary signals, as well as for tracking slowly varying nonstationary processes [74, 38]. RLS is especially attractive when the state and/or data model are not available (as with LMS), and fast convergence rates are at a premium. Compared to the LMS scheme, RLS typically offers faster convergence and improved estimation performance at the cost of higher computational complexity. To enable these valuable tradeoffs in the context of in-network processing, the ADMM framework of Section 2 is utilized here to derive a decentralized (D-) RLS adaptive scheme that can be employed for distributed localization and power spectrum estimation (see also [54, 52] for further details on the algorithmic construction and convergence claims).

Consider the data setting and linear regression task in Section 4.1. The RLS estimator for the unknown parameter $\mathbf{s}_0(t)$ minimizes the exponentially weighted least-squares (EWLS) cost; see, e.g., [74, 38]

$$\hat{\mathbf{s}}_{\text{ewls}}(t) := \arg\min_{\mathbf{s}} \sum_{\tau=0}^{t} \sum_{i=1}^{n} \gamma^{t-\tau} \left[y_i(\tau) - \mathbf{h}_i^\top(\tau)\mathbf{s} \right]^2 + \gamma^t \mathbf{s}^T \boldsymbol{\Phi}_0 \mathbf{s} \tag{14.22}$$

where $\gamma \in (0,1]$ is a forgetting factor, while the positive definite matrix $\boldsymbol{\Phi}_0$ is included for regularization. Note that in forming the EWLS estimator at time t, the entire history of data $\{y_i(\tau), \mathbf{h}_i(\tau)\}_{\tau=0}^{t}$ for $i = 1, \ldots, n$ is incorporated in the online estimation process. Whenever $\gamma < 1$, past data are exponentially discarded thus enabling tracking of nonstationary processes.

Again to decompose the cost function in (14.22), in which summands are coupled through the global variable $\mathbf{s}$, we introduce auxiliary variables $\{\mathbf{s}_i\}_{i=1}^{n}$ that represent local estimates per node i. These local estimates are utilized to form the convex *constrained* and separable minimization problem in (14.3), which can be solved using ADMM to yield the following decentralized iterations (details in [54, 52])

$$\mathbf{v}_i(t) = \mathbf{v}_i(t-1) + c \sum_{j \in \mathcal{N}_i} [\mathbf{s}_i(t) - \mathbf{s}_j(t)] \tag{14.23}$$

$$\mathbf{s}_i(t+1) = \boldsymbol{\Phi}_i^{-1}(t+1)\boldsymbol{\psi}_i(t+1) - \frac{1}{2}\boldsymbol{\Phi}_i^{-1}(t+1)\mathbf{v}_i(t) \tag{14.24}$$

where $\boldsymbol{\Phi}_i(t+1) := \sum_{\tau=0}^{t+1} \gamma^{t+1-\tau}\mathbf{h}_i(\tau)\mathbf{h}_i^\top(\tau) + n^{-1}\gamma^{t+1}\boldsymbol{\Phi}_0$ and

$$\boldsymbol{\Phi}_i^{-1}(t+1) = \gamma^{-1}\boldsymbol{\Phi}_i^{-1}(t) - \frac{\gamma^{-1}\boldsymbol{\Phi}_i^{-1}(t)\mathbf{h}_i(t+1)\mathbf{h}_i^\top(t+1)\boldsymbol{\Phi}_i^{-1}(t)}{\gamma + \mathbf{h}_i^\top(t+1)\boldsymbol{\Phi}_i^{-1}(t)\mathbf{h}_i(t+1)} \tag{14.25}$$

$$\boldsymbol{\psi}_i(t+1) := \sum_{\tau=0}^{t+1} \gamma^{t+1-\tau}\mathbf{h}_i(\tau)y_i(\tau) = \gamma\boldsymbol{\psi}_i(t) + \mathbf{h}_i(t+1)y_i(t+1). \tag{14.26}$$

The D-RLS recursions (14.23) and (14.24) involve similar inter-node communication exchanges as in D-LMS. It is recommended to initialize the matrix recursion with $\boldsymbol{\Phi}_i^{-1}(0) = n\boldsymbol{\Phi}_0^{-1} := \delta \mathbf{I}_p$, where $\delta > 0$ is chosen sufficiently large [74]. The local estimates in D-RLS converge in the mean-sense to the true $\mathbf{s}_0$ (time-invariant case), even when information exchanges are imperfect. Closed-form expressions for the bounded estimation MSE along with numerical tests and comparisons with the incremental RLS [45] and diffusion RLS [13] algorithms can be found in [52].

Decentralized Spectrum Sensing Using WSNs. A WSN application where the need for linear regression arises is spectrum estimation for the purpose of environmental monitoring. Suppose sensors comprising a WSN deployed over some area of interest observe a narrowband source to determine its spectral peaks. These peaks can reveal hidden periodicities due to, e.g., a natural heat or seismic source. The source of interest propagates through multi-path channels and is contaminated with additive noise present at the sensors. The unknown source-sensor channels may introduce deep fades at the frequency band occupied by the source. Thus, having each sensor operating on its own may lead to faulty assessments. The available spatial diversity to effect improved spectral estimates can only be achieved via sensor collaboration as in the decentralized estimation algorithms presented in this chapter.

Let $\theta(t)$ denote the evolution of the source signal in time, and suppose that $\theta(t)$ can be modeled as an autoregressive (AR) process [76, p. 106]

$$\theta(t) = -\sum_{\tau=1}^{p} \alpha_\tau \theta(t-\tau) + w(t)$$

where p is the order of the AR process, while $\{\alpha_\tau\}$ are the AR coefficients and $w(t)$ denotes driving white noise. The source propagates to sensor i via a channel modeled as an FIR filter $C_i(z) = \sum_{l=0}^{L_i-1} c_{il} z^{-l}$, of unknown order L_i and tap coefficients $\{c_{il}\}$ and is contaminated with additive sensing noise $\bar{\varepsilon}_i(t)$ to yield the observation

$$y_i(t) = \sum_{l=0}^{L_i-1} c_{il} \theta(t-l) + \bar{\varepsilon}_i(t).$$

Since $y_i(t)$ is an autoregressive moving average (ARMA) process, then [76]

$$y_i(t) = -\sum_{\tau=1}^{p} \alpha_\tau y_i(t-\tau) + \sum_{\tau'=1}^{m} \beta_{\tau'} \tilde{\eta}_i(t-\tau') \tag{14.27}$$

where the MA coefficients $\{\beta_{\tau'}\}$ and the variance of the white noise process $\tilde{\eta}_i(t)$ depend on $\{c_{il}\}$, $\{\alpha_\tau\}$ and the variance of the noise terms $w(t)$ and $\bar{\varepsilon}_i(t)$. For the purpose of determining spectral peaks, the MA term in (14.27) can be treated as observation noise, i.e., $\varepsilon_i(t) := \sum_{\tau'=1}^{m} \beta_{\tau'} \tilde{\eta}_i(t-\tau')$. This is very important since this way sensors do not have to know the source-sensor channel coefficients as well as the noise variances. Accordingly, the spectral content of the source can be estimated provided sensors estimate the coefficients $\{\alpha_\tau\}$. To this end, let $\mathbf{s}_0 := [\alpha_1 \ldots \alpha_p]^\top$ be the unknown parameter of interest. From (14.27) the regression vectors are given as $\mathbf{h}_i(t) = [-y_i(t-1) \ldots -y_i(t-p)]^\top$, and can be acquired directly from the sensor measurements $\{y_i(t)\}$ without the need of training/estimation.

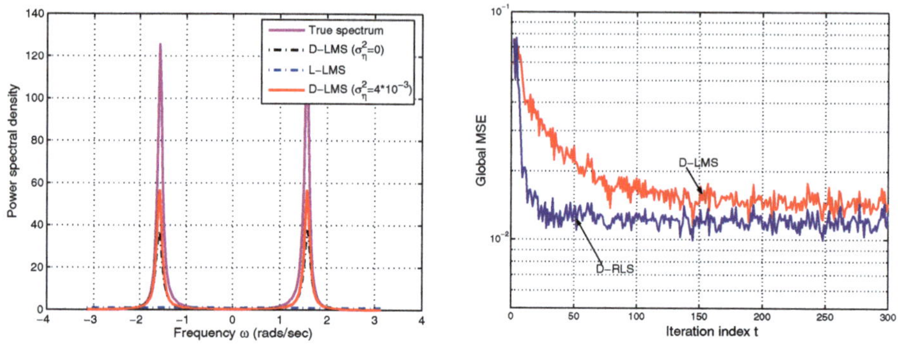

Fig. 14.6 D-LMS in a power spectrum estimation task. (left) The true narrowband spectra is compared to the estimated PSD, obtained after the WSN runs the D-LMS and (non-cooperative) L-LMS algorithms. The reconstruction results correspond to a sensor whose multipath channel from the source introduces a null at $\omega = \pi/2 = 1.57$. (right) Global MSE evolution (network learning curve) for the D-LMS and D-RLS algorithms.

Performance of the decentralized adaptive algorithms described so far is illustrated next, when applied to the aforementioned power spectrum estimation task. For the numerical experiments, an ad hoc WSN with $n = 80$ sensors is simulated as a realization of a random geometric graph. The source-sensor channels corresponding to a few of the sensors are set so that they have a null at the frequency where the AR source has a peak, namely at $\omega = \pi/2$. Figure 14.6 (left) depicts the actual power spectral density (PSD) of the source as well as the estimated PSDs for one of the sensors affected by a bad channel. To form the desired estimates in a distributed fashion, the WSN runs the local (L-) LMS and the D-LMS algorithm outlined in Section 4.1. The L-LMS is a non-cooperative scheme since each sensor, say the ith, independently runs an LMS adaptive filter fed by its local data $\{y_i(t), \mathbf{h}_i(t)\}$ only. The experiment involving D-LMS is performed under ideal and noisy inter-sensor links. Clearly, even in the presence of communication noise D-LMS exploits the spatial diversity available and allows all sensors to estimate accurately the actual spectral peak, whereas L-LMS leads the problematic sensors to misleading estimates.

For the same setup, Figure 14.6 (right) shows the global learning curve evolution $\text{MSE}(t) = (1/n)\sum_{i=1}^{n} \|y_i(t) - \mathbf{h}_i^\top(t)\mathbf{s}_i(t-1)\|^2$. The D-LMS and the D-RLS algorithms are compared under ideal communication links. It is apparent that D-RLS achieves improved performance both in terms of convergence rate and steady state MSE. As discussed in Section 4.2 this comes at the price of increased computational complexity per sensor, while the communication costs incurred are identical.

4.3 Decentralized Model-Based Tracking

The decentralized adaptive schemes in Sections 4.1 and 4.2 are suitable for tracking slowly time-varying signals in settings where no statistical models are available. In certain cases, such as target tracking, state evolution models can be derived and employed by exploiting the physics of the problem. The availability of such models paves the way for improved state tracking via Kalman filtering/smoothing techniques, e.g., see [3, 38]. Model-based decentralized Kalman filtering/smoothing as well as particle filtering schemes for multi-node networks are briefly outlined here.

Initial attempts to distribute the centralized KF recursions (see [59] and references in [68]) rely on consensus-averaging [83]. The idea is to estimate across nodes those sufficient statistics (that are expressible in terms of network-wide averages) required to form the corrected state and corresponding corrected state error covariance matrix. Clearly, there is an inherent delay in obtaining these estimates confining the operation of such schemes only to applications with slow-varying state vectors $\mathbf{s}_0(t)$, and/or fast communications needed to complete multiple consensus iterations within the time interval separating the acquisition of consecutive measurements $y_i(t)$ and $y_i(t+1)$. Other issues that may lead to instability in existing decentralized KF approaches are detailed in [68].

Instead of filtering, the delay incurred by those inner-loop consensus iterations motivated the consideration of fixed-lag decentralized Kalman smoothing (KS) in [68]. Matching consensus iterations with those time instants of data acquisition, fixed-lag smoothers allow sensors to form local MMSE optimal smoothed estimates, which take advantage of all acquired measurements within the "waiting period." The ADMM-enabled decentralized KS in [68] also overcomes the noise-related limitations of consensus-averaging algorithms [84]. In the presence of communication noise, these estimates converge in the mean sense, while their noise-induced variance remains bounded. This noise resiliency allows sensors to exchange quantized data further lowering communication cost. For a tutorial treatment of decentralized Kalman filtering approaches using WSNs (including the decentralized ADMM-based KS of [68] and strategies to reduce the communication cost of state estimation problems), the interested reader is referred to [63]. These reduced-cost strategies exploit the redundancy in information provided by individual observations collected at different sensors, different observations collected at different sensors, and different observations acquired at the same sensor.

On a related note, a collaborative algorithm is developed in [17] to estimate the channel gains of wireless links in a geographical area. Kriged Kalman filtering (KKF) [64], which is a tool with widely appreciated merits in spatial statistics and geosciences, is adopted and implemented in a decentralized fashion leveraging the ADMM framework described here. The distributed KKF algorithm requires only local message passing to track the time-variant so-termed "shadowing field" using a network of radiometers, yet it provides a global view of the radio frequency (RF) environment through consensus iterations; see also Section 5.3 for further elaboration on spectrum sensing carried out via wireless cognitive radio networks.

To wrap-up the discussion, consider a network of collaborating agents (e.g., robots) equipped with wireless sensors measuring distance and/or bearing from a target that they wish to track. Even if state models are available, the nonlinearities present in these measurements prevent sensors from employing the clairvoyant (linear) Kalman tracker discussed so far. In response to these challenges, [27] develops a set-membership constrained particle filter (PF) approach that: (i) exhibits performance comparable to the centralized PF; (ii) requires only communication of particle weights among neighboring sensors; and (iii) it can afford both consensus-based and incremental averaging implementations. Affordable inter-sensor communications are enabled through a novel distributed adaptation scheme, which considerably reduces the number of particles needed to achieve a given performance. The interested reader is referred to [36] for a recent tutorial account of decentralized PF in multi-agent networks.

5 Decentralized Sparsity-Regularized Rank Minimization

Modern network data sets typically involve a large number of attributes. This fact motivates predictive models offering a *sparse*, broadly meaning parsimonious, representation in terms of a few attributes. Such low-dimensional models facilitate interpretability and enhanced predictive performance. In this context, this section deals with ADMM-based decentralized algorithms for sparsity-regularized rank minimization. It is argued that such algorithms are key to unveiling Internet traffic anomalies given ubiquitous link-load measurements. Moreover, the notion of RF cartography is subsequently introduced to exemplify the development of a paradigm infrastructure for situational awareness at the physical layer of wireless cognitive radio (CR) networks. A (subsumed) decentralized sparse linear regression algorithm is outlined to accomplish the aforementioned cartography task.

5.1 Network Anomaly Detection via Sparsity and Low Rank

Consider a backbone IP network, whose abstraction is a graph with n nodes (routers) and L physical links. The operational goal of the network is to transport a set of F origin-destination (OD) traffic flows associated with specific OD (ingress-egress router) pairs. Let $x_{l,t}$ denote the traffic volume (in bytes or packets) passing through link $l \in \{1,\ldots,L\}$ over a fixed time interval $(t, t+\Delta t)$. Link counts across the entire network are collected in the vector $\mathbf{x}_t \in \mathbb{R}^L$, e.g., using the ubiquitous SNMP protocol. Single-path routing is adopted here, meaning a given flow's traffic is carried through multiple links connecting the corresponding source-destination pair along a single path. Accordingly, over a discrete time horizon $t \in [1,T]$ the measured link counts $\mathbf{X} := [x_{l,t}] \in \mathbb{R}^{L\times T}$ and (unobservable) OD flow traffic matrix $\mathbf{Z} := [z_{f,t}] \in \mathbb{R}^{F\times T}$, are thus related through $\mathbf{X} = \mathbf{RZ}$ [41], where the so-termed routing matrix $\mathbf{R} := [r_{l,f}] \in \{0,1\}^{L\times F}$ is such that $r_{l,f} = 1$ if link l carries the flow f, and zero otherwise. The routing matrix is 'wide,' as for backbone networks the number of OD flows is much larger than the number of physical links $(F \gg L)$. A cardinal property of the traffic matrix is noteworthy. Common temporal patterns across OD traffic flows in addition to their almost periodic behavior, render most rows (respectively columns) of the traffic matrix linearly dependent, and thus $\mathbf{Z}$ typically has *low rank*. This intuitive property has been extensively validated with real network data; see Figure 14.7 and, e.g., [41].

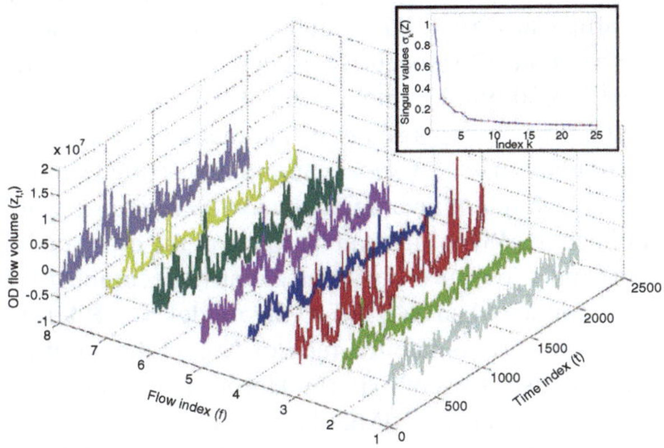

Fig. 14.7 Volumes of 6 representative (out of 121 total) OD flows, taken from the operation of Internet-2 during a seven-day period. Temporal periodicities and correlations across flows are apparent. As expected, in this case $\mathbf{Z}$ can be well approximated by a low-rank matrix, since its normalized singular values decay rapidly to zero.

It is not uncommon for some of the OD flow rates to experience unexpected abrupt changes. These so-termed *traffic volume anomalies* are typically due to (unintentional) network equipment misconfiguration or outright failure, unforeseen behaviors following routing policy modifications, or, cyber attacks (e.g., DoS attacks) which aim at compromising the services offered by the network [86, 41, 53]. Let $a_{f,t}$ denote the unknown amount of anomalous traffic in flow f at time t, which one wishes to estimate. Explicitly accounting for the presence of anomalous flows, the measured traffic carried by link l is then given by $y_{l,t} = \sum_f r_{l,f}(z_{f,t} + a_{f,t}) + \varepsilon_{l,t}$, $t = 1,\ldots,T$, where the noise variables $\varepsilon_{l,t}$ capture measurement errors and unmodeled dynamics. Traffic volume anomalies are (unsigned) sudden changes in the traffic of OD flows, and as such their effect can span multiple links in the network. A key difficulty in unveiling anomalies from link-level measurements only is that oftentimes, clearly discernible anomalous spikes in the flow traffic can be masked through "destructive interference" of the superimposed OD flows [41]. An additional challenge stems from missing link-level measurements $y_{l,t}$, an unavoidable operational reality affecting most traffic engineering tasks that rely on (indirect) measurement of traffic matrices [65]. To model missing link measurements, collect the tuples (l,t) associated with the available observations $y_{l,t}$ in the set $\Omega \subseteq [1,2,\ldots,L] \times [1,2,\ldots,T]$. Introducing the matrices $\mathbf{Y} := [y_{l,t}], \mathbf{E} := [\varepsilon_{l,t}] \in \mathbb{R}^{L\times T}$, and $\mathbf{A} := [a_{f,t}] \in \mathbb{R}^{F\times T}$, the (possibly incomplete) set of link-traffic measurements can be expressed in compact matrix form as

$$\mathscr{P}_{\Omega}(\mathbf{Y}) = \mathscr{P}_{\Omega}(\mathbf{X}+\mathbf{RA}+\mathbf{E}) \tag{14.28}$$

where the sampling operator $\mathscr{P}_{\Omega}(.)$ sets the entries of its matrix argument not in Ω to zero, and keeps the rest unchanged. Since the objective here is not to estimate the OD flow traffic matrix $\mathbf{Z}$, (14.28) is expressed in terms of the nominal (anomaly free) link-level traffic rates $\mathbf{X} := \mathbf{RZ}$, which inherits the low-rank property of $\mathbf{Z}$. Anomalies in $\mathbf{A}$ are expected to occur sporadically over time, and last for a short time relative to the (possibly long) measurement interval $[1,T]$. In addition, only a small fraction of the flows is supposed to be anomalous at a any given time instant. This renders the anomaly matrix $\mathbf{A}$ *sparse* across rows (flows) and columns (time).

Recently, a natural estimator leveraging the low rank property of $\mathbf{X}$ and the sparsity of $\mathbf{A}$ was put forth in [48], which can be found at the crossroads of compressive sampling [23] and timely low-rank plus sparse matrix decompositions [11, 14]. The idea is to fit the incomplete data $\mathscr{P}_{\Omega}(\mathbf{Y})$ to the model $\mathbf{X}+\mathbf{RA}$ [cf. (14.28)] in the LS error sense, as well as minimize the rank of $\mathbf{X}$, and the number of nonzero entries of $\mathbf{A}$ measured by its ℓ_0-(pseudo) norm. Unfortunately, albeit natural both rank and ℓ_0-norm criteria are in general NP-hard to optimize. Typically, the nuclear norm $\|\mathbf{X}\|_* := \sum_k \sigma_k(\mathbf{X})$ ($\sigma_k(\mathbf{X})$ denotes the k-th singular value of $\mathbf{X}$) and the ℓ_1-norm $\|\mathbf{A}\|_1$ are adopted as surrogates [28, 12], since they are the closest *convex* approximants to rank($\mathbf{X}$) and $\|\mathbf{A}\|_0$, respectively. Accordingly, one solves

$$\min_{\{\mathbf{X},\mathbf{A}\}} \|\mathscr{P}_{\Omega}(\mathbf{Y}-\mathbf{X}-\mathbf{RA})\|_F^2 + \lambda_*\|\mathbf{X}\|_* + \lambda_1\|\mathbf{A}\|_1 \tag{14.29}$$

where $\lambda_*, \lambda_1 \geq 0$ are rank- and sparsity-controlling parameters. While a non-smooth optimization problem, (14.29) is appealing because it is convex. An efficient accelerated proximal gradient algorithm with quantifiable iteration complexity was developed to unveil network anomalies [50]. Interestingly, (14.29) also offers a cleansed estimate of the link-level traffic $\hat{\mathbf{X}}$, that could be subsequently utilized for network tomography tasks. In addition, (14.29) *jointly* exploits the spatio-temporal correlations in link traffic as well as the sparsity of anomalies, through an optimal single-shot estimation-detection procedure that turns out to outperform the algorithms in [41] and [86] (the latter decouple the estimation and detection steps); see Figure 14.8.

5.2 In-Network Traffic Anomaly Detection

Implementing (14.29) presumes that network nodes continuously communicate their link traffic measurements to a central monitoring station, which uses their aggregation in $\mathscr{P}_{\Omega}(\mathbf{Y})$ to unveil anomalies. While for the most part this is the

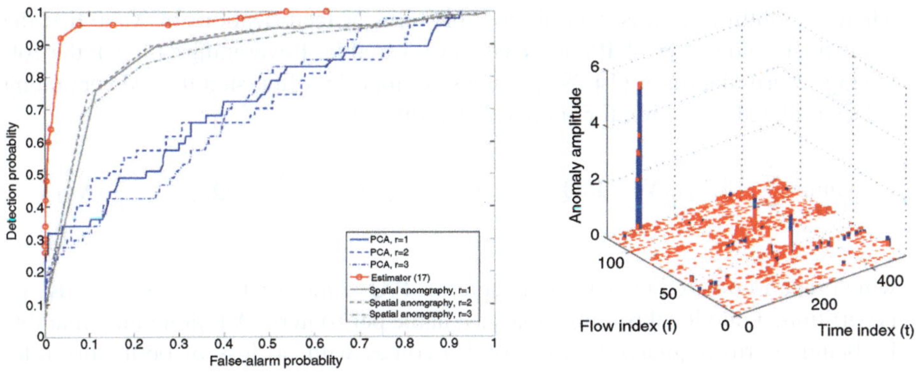

Fig. 14.8 Unveiling anomalies from Internet-2 data. (Left) ROC curve comparison between (14.29) and the PCA methods in [41, 86], for different values of the rank($\mathbf{Z}$). Leveraging sparsity and low rank jointly leads to improved performance. (Right) In red, the estimated anomaly map $\hat{\mathbf{A}}$ obtained via (14.29) superimposed to the "true" anomalies shown in blue [49].

prevailing operational paradigm adopted in current networks, it is prudent to reflect on the limitations associated with this architecture. For instance, fusing all this information may entail excessive protocol overheads. Moreover, minimizing the exchanges of raw measurements may be desirable to reduce unavoidable communication errors that translate to missing data. Solving (14.29) centrally raises robustness concerns as well, since the central monitoring station represents an isolated point of failure.

These reasons prompt one to develop *fully decentralized* iterative algorithms for unveiling traffic anomalies, and thus embed network anomaly detection functionality to the routers. As in Section 2, per iteration node i carries out simple computational tasks locally, relying on its own link count measurements (a submatrix $\mathbf{Y}_i$ within $\mathbf{Y} := [\mathbf{Y}_1^\top, \ldots, \mathbf{Y}_n^\top]^\top$ corresponding to router i's links). Subsequently, local estimates are refined after exchanging messages only with directly connected neighbors, which facilitates percolation of local information to the whole network. The end goal is for network nodes to consent on a global map of network anomalies $\hat{\mathbf{A}}$, and attain (or at least come close to) the estimation performance of the centralized counterpart (14.29) which has all data $\mathscr{P}_\Omega(\mathbf{Y})$ available.

Problem (14.29) is not amenable to distributed implementation because of the non-separable nuclear norm present in the cost function. If an upper bound rank($\hat{\mathbf{X}}$) $\leq \rho$ is a priori available [recall $\hat{\mathbf{X}}$ is the estimated link-level traffic obtained via (14.29)], the search space of (14.29) is effectively reduced, and one can factorize the decision variable as $\mathbf{X} = \mathbf{P}\mathbf{Q}^\top$, where $\mathbf{P}$ and $\mathbf{Q}$ are $L \times \rho$ and $T \times \rho$ matrices, respectively. Again, it is possible to interpret the columns of $\mathbf{X}$ (viewed as points in $\mathbb{R}^L$) as belonging to a low-rank nominal subspace, spanned by the columns of $\mathbf{P}$. The rows of $\mathbf{Q}$ are thus the projections of the columns of $\mathbf{X}$ onto the traffic subspace. Next, consider the following alternative characterization of the nuclear norm (see, e.g., [75])

$$\|\mathbf{X}\|_* := \min_{\{\mathbf{P},\mathbf{Q}\}} \quad \frac{1}{2}\left(\|\mathbf{P}\|_F^2 + \|\mathbf{Q}\|_F^2\right), \quad \text{s. to} \quad \mathbf{X} = \mathbf{P}\mathbf{Q}^\top \tag{14.30}$$

where the optimization is over all possible bilinear factorizations of $\mathbf{X}$, so that the number of columns ρ of $\mathbf{P}$ and $\mathbf{Q}$ is also a variable. Leveraging (14.30), the following reformulation of (14.29) provides an important first step towards obtaining a decentralized algorithm for anomaly identification

$$\min_{\{\mathbf{P},\mathbf{Q},\mathbf{A}\}} \sum_{i=1}^{n} \left[\|\mathscr{P}_{\Omega_i}(\mathbf{Y}_i - \mathbf{P}_i\mathbf{Q}^\top - \mathbf{R}_i\mathbf{A})\|_F^2 + \frac{\lambda_*}{2n}\left(n\|\mathbf{P}_i\|_F^2 + \|\mathbf{Q}\|_F^2\right) + \frac{\lambda_1}{n}\|\mathbf{A}\|_1 \right] \tag{14.31}$$

which is non-convex due to the bilinear terms $\mathbf{P}_i\mathbf{Q}^\top$, and where $\mathbf{R} := \left[\mathbf{R}_1^\top, \ldots, \mathbf{R}_n^\top\right]^\top$ is partitioned into local routing tables available per router i. Adopting the separable Frobenius-norm regularization in (14.31) comes with no loss of optimality relative to (14.29), provided $\text{rank}(\hat{\mathbf{X}}) \leq \rho$. By finding the global minimum of (14.31) [which could entail considerably less variables than (14.29)], one can recover the optimal solution of (14.29). But since (14.31) is non-convex, it may have stationary points which need not be globally optimum. As asserted in [48, Prop. 1], however, if a stationary point $\{\bar{\mathbf{P}}, \bar{\mathbf{Q}}, \bar{\mathbf{A}}\}$ of (14.31) satisfies $\|\mathscr{P}_{\Omega}(\mathbf{Y} - \bar{\mathbf{P}}\bar{\mathbf{Q}}^\top - \bar{\mathbf{A}})\| < \lambda_*$, then $\{\hat{\mathbf{X}} := \bar{\mathbf{P}}\bar{\mathbf{Q}}^\top, \hat{\mathbf{A}} := \bar{\mathbf{A}}\}$ is the globally optimal solution of (14.29).

To decompose the cost in (14.31), in which summands inside the square brackets are coupled through the global variables $\{\mathbf{Q}, \mathbf{A}\}$, one can proceed as in Section 2 and introduce auxiliary copies $\{\mathbf{Q}_i, \mathbf{A}_i\}_{i=1}^n$ representing local estimates of $\{\mathbf{Q}, \mathbf{A}\}$, one per node i. These local copies along with *consensus* constraints yield the decentralized estimator

$$\min_{\{\mathbf{P}_i,\mathbf{Q}_i,\mathbf{A}_i\}} \sum_{i=1}^{n} \left[\|\mathscr{P}_{\Omega_i}(\mathbf{Y}_i - \mathbf{P}_i\mathbf{Q}_i^\top - \mathbf{R}_i\mathbf{A}_i)\|_F^2 + \frac{\lambda_*}{2n}\left(n\|\mathbf{P}_i\|_F^2 + \|\mathbf{Q}_i\|_F^2\right) + \frac{\lambda_1}{n}\|\mathbf{A}_i\|_1 \right] \tag{14.32}$$

$$\text{s. to} \quad \mathbf{Q}_i = \mathbf{Q}_j, \; \mathbf{A}_i = \mathbf{A}_j, \; i = 1, \ldots, n, \; j \in \mathscr{N}_i, \; i \neq j$$

which follows the general form in (14.2), and is equivalent to (14.31) provided the network topology graph is connected. Even though consensus is a fortiori imposed within neighborhoods, it carries over to the entire (connected) network and local estimates agree on the global solution of (14.31). Exploiting the separable structure of (14.32) using the ADMM, a general framework for in-network sparsity-regularized rank minimization was put forth in [48]. In a nutshell, local tasks per iteration $k = 1, 2, \ldots$ entail solving small unconstrained quadratic programs to refine the normal subspace $\mathbf{P}_i[k]$, in addition to soft-thresholding operations to update the anomaly maps $\mathbf{A}_i[k]$ per router. Routers exchange their estimates $\{\mathbf{Q}_i[k], \mathbf{A}_i[k]\}$ only with directly connected neighbors per iteration. This way the communication overhead remains affordable, regardless of the network size n.

When employed to solve non-convex problems such as (14.32), so far ADMM offers no convergence guarantees. However, there is ample experimental evidence in the literature that supports empirical convergence of ADMM, especially when the non-convex problem at hand exhibits "favorable" structure [9]. For instance, (14.32) is a linearly constrained bi-convex problem with potentially good convergence properties – extensive numerical tests in [48] demonstrate that this is indeed the case. While establishing convergence remains an open problem, one can still prove that upon convergence the distributed iterations attain consensus and global optimality, thus offering the desirable centralized performance guarantees [48].

5.3 RF Cartography via Decentralized Sparse Linear Regression

In the domain of spectrum sensing for CR networks, RF cartography amounts to constructing in a distributed fashion: i) global power spectral density (PSD) maps capturing the distribution of radiated power across space, time, and frequency; and ii) local channel gain (CG) maps offering the propagation medium per frequency from each node to any point in space [17]. These maps enable identification of opportunistically available spectrum bands for re-use and handoff operation; as well as localization, transmit-power estimation, and tracking of primary user activities. While the focus here is on the construction of PSD maps, the interested reader is referred to [39] for a tutorial treatment on CG cartography.

A cooperative approach to RF cartography was introduced in [5], that builds on a basis expansion model of the PSD map $\Phi(\mathbf{x}, f)$ across space $\mathbf{x} \in \mathbb{R}^2$, and frequency f. Spatially distributed CRs collect smoothed periodogram samples of the received signal at given sampling frequencies, based on which the unknown expansion coefficients are determined. Introducing a virtual spatial grid of candidate source locations, the estimation task can be cast as a linear LS problem with an augmented vector of unknown parameters. Still, the problem complexity (or effective degrees of freedom) can be controlled by capitalizing on two forms of sparsity: the first one introduced by the narrow-band nature of transmit-PSDs relative to the broad swaths of usable spectrum; and the second one emerging from sparsely located active radios in the operational space (due to the grid artifact). Nonzero entries in the parameter vector sought correspond to spatial location-frequency band pairs corresponding to active transmissions. All in all, estimating the PSD map and locating the active transmitters as a byproduct boils down to a variable selection problem. This motivates well employment of the ADMM and the least-absolute shrinkage and selection operator (Lasso) for decentralized sparse linear regression [51, 48], an estimator subsumed by (14.29) when $\mathbf{X} = \mathbf{0}_{L \times T}$, $T = 1$, and matrix $\mathbf{R}$ has a specific structure that depends on the chosen bases and the path-loss propagation model.

Sparse total LS variants are also available to cope with uncertainty in the regression matrix, arising due to inaccurate channel estimation and grid-mismatch effects [39]. Nonparametric spline-based PSD map estimators [6] have been also shown effective in capturing general propagation characteristics including both shadowing and fading; see also Figure 14.9 for an actual PSD atlas spanning 14 frequency sub-bands.

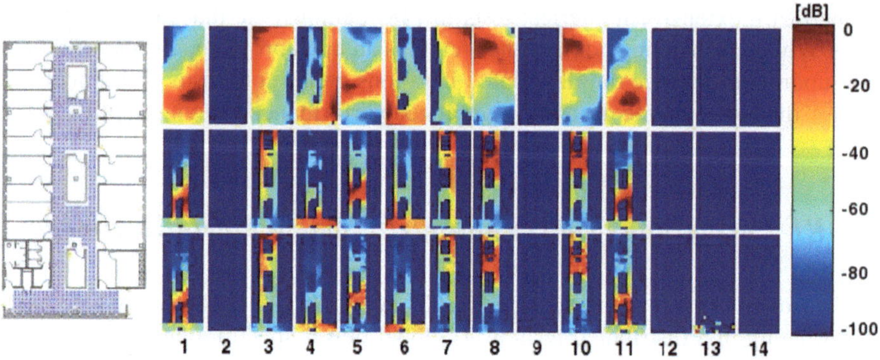

Fig. 14.9 Spline-based RF cartography from real wireless LAN data. (Left) Detailed floor plan schematic including the location of $N = 166$ sensing radios; (Right-bottom) original measurements spanning 14 frequency sub-bands; (Right-center) estimated maps over the surveyed area; and (Right-top) extrapolated maps. The proposed decentralized estimator is capable of recovering the 9 (out of 14 total) center frequencies that are being utilized for transmission. It accurately recovers the power levels in the surveyed area with a smooth extrapolation to zones were there are no measurements, and suggests possible locations for the transmitters [6].

6 Convergence Analysis

In this section we analyze the convergence and assess the rate of convergence for the decentralized ADMM algorithm outlined in Section 2. We focus on the batch learning setup, where the local cost functions are static.

6.1 Preliminaries

Network Model Revisited and Elements of Algebraic Graph Theory. Recall the network model briefly introduced in Section 1, based on a connected graph composed of a set of n nodes (agents, vertices), and a set of L edges (arcs, links). Each edge $e = (i, j)$ represents an ordered pair (i, j) indicating that node i communicates with node j. Communication is assumed bidirectional so that per edge $e = (i, j)$, the edge $e' = (j, i)$ is also present. Nodes adjacent to i are its neighbors and belong to the (neighborhood) set $\mathscr{N}_i$. The cardinality of $|\mathscr{N}_i|$ equals the degree d_i of node i. Let $\mathbf{A}_s \in \mathbb{R}^{Lp \times np}$ denote the block edge source matrix, where the block $[\mathbf{A}_s]_{e,i} = \mathbf{I}_p \in \mathbb{R}^{p \times p}$ if the edge e originates at node i, and is null otherwise. Likewise, define the block edge destination matrix $\mathbf{A}_d \in \mathbb{R}^{Lp \times np}$ where the block $[\mathbf{A}_d]_{e,j} = \mathbf{I}_p \in \mathbb{R}^{p \times p}$ if the edge e terminates at node j, and is null otherwise. The so-termed extended oriented incidence matrix can be written as $\mathbf{E}_o = \mathbf{A}_s - \mathbf{A}_d$, and the unoriented incidence matrix as $\mathbf{E}_u = \mathbf{A}_s + \mathbf{A}_d$. The extended oriented (signed) Laplacian is then given by $\mathbf{L}_o = (1/2)\mathbf{E}_o^\top \mathbf{E}_o$, the unoriented (unsigned) Laplacian by $\mathbf{L}_u = (1/2)\mathbf{E}_u^\top \mathbf{E}_u$, and the degree matrix $\mathbf{D} = \text{diag}(d_1, \ldots, d_n)$

is $\mathbf{D} = (1/2)(\mathbf{L}_o + \mathbf{L}_u)$. With Γ_u denoting the largest eigenvalue of $\mathbf{L}_u$, and γ_o the smallest nonzero eigenvalue of $\mathbf{L}_o$, basic results in algebraic graph theory establish that both Γ_u and γ_o are measures of network connectedness.

Compact Learning Problem Representation. With reference to the optimization problem (14.3), define $\mathbf{s} := [\mathbf{s}_1^\top \ldots \mathbf{s}_n^\top]^\top \in \mathbb{R}^{np}$ concatenating all local estimates $\mathbf{s}_i$, and $\mathbf{z} := [\mathbf{z}_1^\top \ldots \mathbf{z}_L^\top]^\top \in \mathbb{R}^{Lp}$ concatenating all auxiliary variables $\mathbf{z}_e = \mathbf{z}_i^j$. For notational convenience, introduce the aggregate cost function $f : \mathbb{R}^{np} \to \mathbb{R}$ as $f(\mathbf{s}) := \sum_{i=1}^n f_i(\mathbf{s}_i; \mathbf{y}_i)$. Using these definitions along with the edge source and destination matrices, (14.3) can be rewritten in compact matrix form as

$$\min_{\mathbf{s}} f(\mathbf{s}), \quad \text{s. to } \mathbf{A}_s\mathbf{s} - \mathbf{z} = \mathbf{0},\ \mathbf{A}_d\mathbf{s} - \mathbf{z} = \mathbf{0}.$$

Upon defining $\mathbf{A} := [\mathbf{A}_s^\top \mathbf{A}_d^\top]^\top \in \mathbb{R}^{2Lp \times np}$ and $\mathbf{B} := [-\mathbf{I}_{Lp}\ -\mathbf{I}_{Lp}]^\top$, (14.33) reduces to

$$\min_{\mathbf{s}} f(\mathbf{s}), \quad \text{s. to } \mathbf{As} + \mathbf{Bz} = \mathbf{0}.$$

As in Section 2, consider Lagrange multipliers $\bar{\mathbf{v}}_e = \bar{\mathbf{v}}_i^j$ associated with the constraints $\mathbf{s}_i = \mathbf{s}_i^j$, and $\tilde{\mathbf{v}}_e = \bar{\mathbf{v}}_i^j$ associated with $\mathbf{s}_j = \mathbf{s}_i^j$. Next, define the supervectors $\bar{\mathbf{v}} := [\bar{\mathbf{v}}_1^\top \ldots \bar{\mathbf{v}}_L^\top]^\top \in \mathbb{R}^{Lp}$ and $\tilde{\mathbf{v}} := [\tilde{\mathbf{v}}_1^\top \ldots \tilde{\mathbf{v}}_L^\top]^\top \in \mathbb{R}^{Lp}$, collecting those multipliers associated with the constraints $\mathbf{A}_s\mathbf{s} - \mathbf{z} = \mathbf{0}$ and $\mathbf{A}_d\mathbf{s} - \mathbf{z} = \mathbf{0}$, respectively. Finally, associate multipliers $\mathbf{v} := [\bar{\mathbf{v}}^\top\ \tilde{\mathbf{v}}^\top]^\top \in \mathbb{R}^{2Lp}$ with the constraint in (14.33), namely $\mathbf{As} + \mathbf{Bz} = \mathbf{0}$. This way, the augmented Lagrangian function of (14.33) is

$$L_c(\mathbf{s}, \mathbf{z}, \mathbf{v}) = f(\mathbf{s}) + \mathbf{v}^\top(\mathbf{As} + \mathbf{Bz}) + \frac{c}{2}\|\mathbf{As} + \mathbf{Bz}\|^2$$

where $c > 0$ is a positive constant [cf. (14.4) back in Section 2].

Assumptions and Scope of the Convergence Analysis. In the convergence analysis, we assume that (14.3) has at least a pair of primal-dual solutions. In addition, we make the following assumptions on the local cost functions f_i.

Assumption 1. The local cost functions f_i are closed, proper, and convex.

Assumption 2. The local cost functions f_i have Lipschitz gradients, meaning there exists a positive constant $M_f > 0$ such that for any node i and for any pair of points $\tilde{\mathbf{s}}_a$ and $\tilde{\mathbf{s}}_b$, it holds that $\|\nabla f_i(\tilde{\mathbf{s}}_a) - \nabla f_i(\tilde{\mathbf{s}}_b)\| \leq M_f \|\tilde{\mathbf{s}}_a - \tilde{\mathbf{s}}_b\|$.

Assumption 3. The local cost functions f_i are strongly convex; that is, there exists a positive constant $m_f > 0$ such that for any node i and for any pair of points $\tilde{\mathbf{s}}_a$ and $\tilde{\mathbf{s}}_b$, it holds that $(\tilde{\mathbf{s}}_a - \tilde{\mathbf{s}}_b)^\top(\nabla f_i(\tilde{\mathbf{s}}_a) - \nabla f_i(\tilde{\mathbf{s}}_b)) \geq m_f \|\tilde{\mathbf{s}}_a - \tilde{\mathbf{s}}_b\|^2$.

Assumption 1 implies that the aggregate function $f(\mathbf{s}) := \sum_{i=1}^n f_i(\mathbf{s}_i; \mathbf{y}_i)$ is closed, proper, and convex. Assumption 2 ensures that the aggregate cost f has Lipschitz gradients with constant M_f; thus, for any pair of points $\mathbf{s}_a$ and $\mathbf{s}_b$ it holds that

$$\|\nabla f(\mathbf{s}_a) - \nabla f(\mathbf{s}_b)\| \leq M_f \|\mathbf{s}_a - \mathbf{s}_b\|. \tag{14.33}$$

Assumption 3 guarantees that the aggregate cost f is strongly convex with constant m_f; hence, for any pair of points $\mathbf{s}_a$ and $\mathbf{s}_b$ it holds that

$$(\mathbf{s}_a - \mathbf{s}_b)^\top (\nabla f(\mathbf{s}_a) - \nabla f(\mathbf{s}_b)) \geq m_f \|\mathbf{s}_a - \mathbf{s}_b\|^2. \tag{14.34}$$

Observe that Assumptions 2 and 3 imply that the local cost functions f_i and the aggregate cost function f are differentiable. Assumption 1 is sufficient to prove global convergence of the decentralized ADMM algorithm. To establish linear rate of convergence however, one further needs Assumptions 2 and 3.

6.2 Convergence

In the sequel, we investigate convergence of the primal variables $\mathbf{s}(k)$ and $\mathbf{z}(k)$ as well as the dual variable $\mathbf{v}(k)$, to their respective optimal values. At an optimal primal solution pair $(\mathbf{s}^*, \mathbf{z}^*)$, consensus is attained and $\mathbf{s}^*$ is formed by n stacked copies of $\tilde{\mathbf{s}}^*$, while $\mathbf{z}^*$ also comprises L stacked copies of $\tilde{\mathbf{s}}^*$, where $\tilde{\mathbf{s}}^* = \hat{\mathbf{s}}$ is an optimal solution of (14.1). If the local cost functions are not strongly convex, then there may exist multiple optimal primal solutions; instead, if the local cost functions are strongly convex (i.e., Assumption 3 holds), the optimal primal solution is unique.

For an optimal primal solution pair $(\mathbf{s}^*, \mathbf{z}^*)$, there exist multiple optimal Lagrange multipliers $\mathbf{v}^* := [(\bar{\mathbf{v}}^*)^\top \, (\tilde{\mathbf{v}}^*)^\top]^\top$, where $\bar{\mathbf{v}}^* = -\tilde{\mathbf{v}}^*$ [42, 73]. In the following convergence analysis, we show that $\mathbf{v}(k)$ converges to one of such optimal dual solutions $\mathbf{v}^*$. In establishing linear rate of convergence, we require that the dual variable is initialized so that $\mathbf{v}(0)$ lies in the column space of $\mathbf{E}_o$; and consider its convergence to a unique dual solution $\mathbf{v}^* := [(\bar{\mathbf{v}}^*)^\top \, (\tilde{\mathbf{v}}^*)^\top]^\top$ in which $\bar{\mathbf{v}}^*$ and $\tilde{\mathbf{v}}^*$ also lie in the column space of $\mathbf{E}_o$. Existence and uniqueness of such a $\mathbf{v}^*$ are also proved in [42, 73].

Throughout the analysis, define

$$\mathbf{u} := \begin{bmatrix} \mathbf{s} \\ \bar{\mathbf{v}} \end{bmatrix}, \quad \mathbf{H} := \begin{bmatrix} \frac{c}{2}\mathbf{L}_u & \mathbf{0} \\ \mathbf{0} & \frac{1}{c}\mathbf{I}_{Lp} \end{bmatrix}.$$

We consider convergence of $\mathbf{u}(k)$ to its optimum $\mathbf{u}^* := [(\mathbf{s}^*)^\top \, (\bar{\mathbf{v}}^*)^\top]^\top$, where $(\mathbf{s}^*, \bar{\mathbf{v}}^*)$ is an optimal primal-dual pair. The analysis is based on several contraction inequalities, in which the distance is measured in the (pseudo) Euclidean norm with respect to the positive semi-definite matrix $\mathbf{H}$.

To situate the forthcoming results in context, notice that convergence of the *centralized* ADMM for constrained optimization problems has been proved in, e.g., [26], and its ergodic $O(1/k)$ rate of convergence is established in [34, 78]. For non-ergodic convergence, [35] proves an $O(1/k)$ rate, and [20] improves the rate to $o(1/k)$. Observe that in [35, 20] the rate refers to the speed at which the difference between two successive primal-dual iterates vanishes, different from the speed that the primal-dual optimal iterates converge to their optima. Convergence of the decentralized ADMM is presented next in the sense that the primal-dual iterates converge to their optima. The analysis proceeds in four steps:

S1. Show that $\|\mathbf{u}(k)-\mathbf{u}^*\|_{\mathbf{H}}^2$ is monotonic, namely, for all times $k \geq 0$ it holds that

$$\|\mathbf{u}(k+1)-\mathbf{u}^*\|_{\mathbf{H}}^2 \leq \|\mathbf{u}(k)-\mathbf{u}^*\|_{\mathbf{H}}^2 - \|\mathbf{u}(k+1)-\mathbf{u}(k)\|_{\mathbf{H}}^2. \tag{14.35}$$

S2. Show that $\|\mathbf{u}(k+1)-\mathbf{u}(k)\|_H^2$ is monotonically non-increasing, that is

$$\|\mathbf{u}(k+2)-\mathbf{u}(k+1)\|_{\mathbf{H}}^2 \leq \|\mathbf{u}(k+1)-\mathbf{u}(k)\|_{\mathbf{H}}^2. \tag{14.36}$$

S3. Derive an $O(1/k)$ rate in a non-ergodic sense based on (14.35) and (14.36), i.e.,

$$\|\mathbf{u}(k+1)-\mathbf{u}(k)\|_{\mathbf{H}}^2 \leq \frac{1}{k+1}\|\mathbf{u}(0)-\mathbf{u}^*\|_{\mathbf{H}}^2. \tag{14.37}$$

S4. Prove that $\mathbf{u}(k) := [\mathbf{s}(k)^\top \, \bar{\mathbf{v}}(k)^\top]^\top$ converges to a pair of optimal primal and dual solutions of (14.33).

The first three steps are similar to those discussed in [35, 20]. Proving the last step is straightforward from the KKT conditions of (14.33). Under S1-S4, the main result establishing convergence of the decentralized ADMM is as follows.

Theorem 1. *If for iterations* (14.5) *and* (14.6) *the initial multiplier* $\mathbf{v}(0) := [\bar{\mathbf{v}}(0)^\top \, \tilde{\mathbf{v}}(0)^\top]^\top$ *satisfies* $\bar{\mathbf{v}}(0) = -\tilde{\mathbf{v}}(0)$*, and* $\mathbf{z}(0)$ *is such that* $\mathbf{E}_u\mathbf{s}(0) = 2\mathbf{z}(0)$*, then with the ADMM penalty parameter* $c > 0$ *it holds under Assumption 1 that the iterates* $\mathbf{s}(k)$ *and* $\bar{\mathbf{v}}(k)$ *converge to a pair of optimal primal and dual solutions of* (14.33).

Theorem 1 asserts that under proper initialization, convergence of the decentralized ADMM only requires the local costs f_i to be closed, proper, and convex. However, it does not specify a pair of optimal primal and dual solutions of (14.33), which $(\mathbf{s}(k), \bar{\mathbf{v}}(k))$ converge to. Indeed, $\mathbf{s}(k)$ can converge to one of the optimal primal solutions $\mathbf{s}^*$, and $\bar{\mathbf{v}}(k)$ can converge to one of the corresponding optimal dual solutions $\bar{\mathbf{v}}^*$. The limit $(\mathbf{s}^*, \bar{\mathbf{v}}^*)$ is ultimately determined by the initial $\mathbf{s}(0)$ and $\bar{\mathbf{v}}(0)$. Indeed, the conditions in Theorem 1 also guarantee ergodic and non-ergodic $o(1/k)$ convergence rates in terms of objective error and successive iterate differences, as proved in the recent paper [19].

6.3 Linear Rate of Convergence

Linear rate of convergence for the *centralized* ADMM is established in [21], and for the decentralized ADMM in [73]. Similar to the convergence analysis of the last section, the proof includes the following steps:

S1'. Show that $\|\mathbf{u}(k)-\mathbf{u}^*\|_{\mathbf{H}}^2$ is contractive, namely, for all times $k \geq 0$ it holds that

$$\|\mathbf{u}(k+1)-\mathbf{u}^*\|_{\mathbf{H}}^2 \leq \frac{1}{1+\delta}\|\mathbf{u}(k)-\mathbf{u}^*\|_{\mathbf{H}}^2 \tag{14.38}$$

where $\delta > 0$ is a constant [cf. (14.40)]. Note that the contraction inequality (14.38) implies Q-linear convergence of $\|\mathbf{u}(k) - \mathbf{u}^*\|_{\mathbf{H}}^2$.

S2'. Show that $\|\mathbf{s}(k+1) - \mathbf{s}^*\|_{\mathbf{H}}^2$ is R-linearly convergent since it is upper-bounded by a Q-linear convergent sequence, meaning

$$\|\mathbf{s}(k+1) - \mathbf{s}^*\|^2 \leq \frac{1}{m_f}\|\mathbf{u}(k) - \mathbf{u}^*\|_{\mathbf{H}}^2 \tag{14.39}$$

where m_f is the strong convexity constant of the aggregate cost function f.

We now state the main result establishing linear rate of convergence for the decentralized ADMM algorithm.

Theorem 2. *If for iterations* (14.5) *and* (14.6) *the initial multiplier* $\mathbf{v}(0) := [\bar{\mathbf{v}}(0)^\top\ \tilde{\mathbf{v}}(0)^\top]^\top$ *satisfies* $\bar{\mathbf{v}}(0) = -\tilde{\mathbf{v}}(0)$*; the initial auxiliary variable* $\mathbf{z}(0)$ *is such that* $\mathbf{E}_u\mathbf{s}(0) = 2\mathbf{z}(0)$*; and the initial multiplier* $\bar{\mathbf{v}}(0)$ *lies in the column space of* $\mathbf{E}_o$*, then with the ADMM parameter* $c > 0$*, it holds under Assumptions 1–3 that the iterates* $\mathbf{s}(k)$ *and* $\bar{\mathbf{v}}(k)$ *converge R-linearly to* $(\mathbf{s}^*, \bar{\mathbf{v}}^*)$*, where* $\mathbf{s}^*$ *is the unique optimal primal solution of* (14.33)*, and* $\bar{\mathbf{v}}^*$ *is the unique optimal dual solution lying in the column space of* $\mathbf{E}_o$*.*

Theorem 2 requires the local cost functions to be closed, proper, convex, strongly convex, and have Lipschitz gradients. In addition to the initialization dictated by Theorem 1, Theorem 2 further requires the initial multiplier $\bar{\mathbf{v}}(0)$ to lie in the column space of $\mathbf{E}_o$, which guarantees that $\bar{\mathbf{v}}(k)$ converges to $\bar{\mathbf{v}}^*$, the unique optimal dual solution lying in the column space of $\mathbf{E}_o$. The primal solution $\mathbf{s}(k)$ converges to $\mathbf{s}^*$, which is unique since the original cost function in (14.1) is strongly convex.

Observe from the contraction inequality (14.38) that the speed of convergence is determined by the contraction parameter δ: A larger δ means stronger contraction and hence faster convergence. Indeed, [73] give an explicit expression of δ, that is

$$\delta = \min\left\{\frac{(\mu-1)\gamma_o}{\mu\Gamma_u}, \frac{2cm_f\gamma_o}{c^2\Gamma_u\gamma_o + \mu M_f^2}\right\} \tag{14.40}$$

where m_f is the strong convexity constant of f, M_f is the Lipschitz continuity constant of ∇f, γ_o is the smallest nonzero eigenvalue of the oriented Laplacian $\mathbf{L}_o$, Γ_u is the largest eigenvalue of the unoriented Laplacian $\mathbf{L}_u$, c is the ADMM penalty parameter, and $\mu > 1$ is an arbitrary constant.

As the current form of (14.40) does not offer insights on how the properties of the cost functions, the underlying network, and the ADMM parameter influence the speed of convergence, [42, 73] finds the largest value of δ by tuning the constant μ and the ADMM parameter c. Specifically, [42, 73] shows that

$$c = M_f\sqrt{\frac{\mu}{\Gamma_u\gamma_o}} \quad \text{and} \quad \sqrt{\frac{1}{\mu}} = \sqrt{\frac{1}{4}\frac{m_f^2}{M_f^2}\frac{\Gamma_L}{\gamma_L} + 1} - \frac{1}{2}\frac{m_f}{M_f}\sqrt{\frac{\Gamma_L}{\gamma_L}}$$

maximizes the right-hand side of (14.40), so that

$$\delta = \frac{m_f}{M_f}\left[\sqrt{\frac{1}{4}\frac{m_f^2}{M_f^2}+\frac{\gamma_o}{\Gamma_u}}-\frac{1}{2}\frac{m_f}{M_f}\right]. \tag{14.41}$$

The best contraction parameter δ is a function of the condition number M_f/m_f of the aggregate cost function f, and the condition number of the graph Γ_u/γ_o. Note that we always have $\delta < 1$, while small values of δ result when $M_f/m_f \gg 1$ or when $\Gamma_u/\gamma_o \gg 1$; that is, when either the cost function or the graph is ill conditioned. When the condition numbers are such that $\Gamma_u/\gamma_o \gg M_f^2/m_f^2$, the condition number of the graph dominates, and we obtain $\delta \approx \gamma_o/\Gamma_u$, implying that the contraction is determined by the condition number of the graph. When $M_f^2/m_f^2 \gg \Gamma_u/\gamma_o$, the condition number of the cost dominates and we have $\delta \approx (m_f/M_f)\sqrt{\gamma_o/\Gamma_u}$. In the latter case the contraction is constrained by both the condition number of the cost function and the condition number of the graph.

Acknowledgements

The authors wish to thank the following friends, colleagues, and co-authors who contributed to their joint publications that the material of this chapter was extracted from: Drs. J.A. Bazerque, A. Cano, E. Dall'Anese, S. Farahmand, N. Gatsis, P. Forero, V. Kekatos, S.-J. Kim, M. Mardani, K. Rajawat, S. Roumeliotis, A. Ribeiro, W. Shi, G. Wu, W. Yin, and K. Yuan. The lead author (and while with SPiNCOM all co-authors) were supported in part from NSF grants 1202135, 1247885 1343248, 1423316, 1442686; the MURI Grant No. AFOSR FA9550-10-1-0567; and the NIH Grant No. 1R01GM104975-01.

References

1. Abur, A., Gomez-Exposito, A.: Power System State Estimation: Theory and Implementation. Marcel Dekker, New York, NY (2004)
2. Albaladejo, C., Sanchez, P., Iborra, A., Soto, F., Lopez, J. A., Torres, R.: Wireless sensor networks for oceanographic monitoring: A systematic review. Sensors. **10**, 6948–6968 (2010)
3. Anderson, B. D., Moore, J. B.: Optimal Filtering. Prentice Hall, Englewood Cliffs, NJ (1979)
4. Barbarossa, S., Scutari, G.: Decentralized maximum likelihood estimation for sensor networks composed of nonlinearly coupled dynamical systems. IEEE Trans. Signal Process. **55** 3456–3470 (2007)
5. Bazerque, J. A., Giannakis, G. B.: Distributed spectrum sensing for cognitive radio networks by exploiting sparsity. IEEE Trans. Signal Process. **58**, 1847–1862 (2010)
6. Bazerque, J. A., Mateos, G., Giannakis, G. B.: Group Lasso on splines for spectrum cartography. IEEE Trans. Signal Process. **59**, 4648–4663 (2011)
7. Bertsekas, D. P., Tsitsiklis, J. N.: Parallel and distributed computation: Numerical methods. 2nd Edition, Athena Scientific, Boston (1997)
8. Boyd, S., Vandenberghe, L.: Convex Optimization. Cambridge Univ. Press, UK (2004)

9. Boyd, S., Parikh, N., Chu, E., Peleato, B., Eckstein, J.: Distributed optimization and statistical learning via the alternating direction method of multipliers. Found. Trends Mach. Learn. **3**, 1–122 (2011)
10. Boyer, T. P., Antonov, J. I., Garcia, H. E., Johnson, D. R., Locarnini, R. A., Mishonov, A. V., Pitcher, M. T., Baranova, O. K., Smolyar, I. V.: World Ocean Database. NOAA Atlas NESDIS. **60**, 190 (2005)
11. Candes, E. J., Li, X., Ma, Y., Wright, J.: Robust principal component analysis? Journal of the ACM. **58**, 1–37 (2011)
12. Candes, E. J., Tao, T.: Decoding by linear programming. IEEE Trans. Info. Theory. **51**, 4203–4215 (2005)
13. Cattivelli, F. S., Lopes, C. G., Sayed, A. H.: Diffusion recursive least-squares for distributed estimation over adaptive networks. IEEE Trans. Signal Process. **56**, 1865–1877 (2008)
14. Chandrasekaran, V., Sanghavi, S., Parrilo, P. R., Willsky, A. S.: Rank-sparsity incoherence for matrix decomposition. SIAM J. Optim. **21**, 572–596 (2011)
15. Chang, T., Hong, M., Wang, X.: Multiagent distributed large-scale optimization by inexact consensus alternating direction method of multipliers. In: Proceedings of IEEE International Conference on Acoustics, Speech, and Signal Processing (2014)
16. Chen, A. and Ozdaglar, A.: A fast distributed proximal-gradient method. In: Proceedings of Allerton Conference on Communication, Control, and Computing (2012)
17. Dall'Anese, E., Kim, S. J., Giannakis, G. B.: Channel gain map tracking via distributed kriging. IEEE Trans. Vehicular Tech. **60**, 1205–1211 (2011)
18. Dall'Anese, E., Zhu, H., Giannakis, G. B.: Distributed optimal power flow for smart microgrids. IEEE Trans. on Smart Grid, **4**, 1464–1475 (2013)
19. Davis, D., Yin, W.: Convergence rate analysis of several splitting schemes. In: R. Glowinski, S. Osher, W. Yin (eds.) Splitting Methods in Communication and Imaging, Science and Engineering. Springer (2016)
20. Deng, W., Lai, M., Peng, Z., Yin, W.: Parallel multi-block ADMM with $o(1/k)$ convergence. arXiv preprint arXiv:1312.3040 (2013)
21. Deng, W., Yin, W.: On the global and linear convergence of the generalized alternating direction method of multipliers. Manuscript, Journal of Scientific Computing, **66**(3):889–916 (2106)
22. Dimakis, A., Kar, S., Moura, J. M. F., Rabbat, M., Scaglione, A.: Gossip algorithms for distributed signal processing. Proc. of the IEEE. **89**, 1847–1864 (2010)
23. Donoho, D. L.: Compressed sensing. IEEE Trans. Info. Theory. **52**, 1289–1306 (2006)
24. Duchi, J., Agarwal, A., Wainwright, M.: Dual averaging for distributed optimization: Convergence analysis and network scaling. IEEE Trans. on Autom. Control, **57** 592–606 (2012)
25. Duda, R. O., Hart, P. E., Stork, D. G.: Pattern Classification. 2nd edition, Wiley, NY (2002)
26. Eckstein, J., Bertsekas, D.: On the Douglas-Rachford splitting method and the proximal point algorithm for maximal monotone operators. Mathematical Programming **55**, 293–318 (1992)
27. Farahmand, S., Roumeliotis, S. I., Giannakis, G. B.: Set-membership constrained particle filter: Distributed adaptation for sensor networks. IEEE Trans. Signal Process. **59**, 4122–4138 (2011)
28. Fazel, M.: Matrix rank minimization with applications. Ph.D. dissertation, Electrical Eng. Dept., Stanford University (2002)
29. Forero, P., Cano, A., Giannakis, G. B.: Consensus-based distributed support vector machines. Journal of Machine Learning Research. **11**, 1663–1707 (2010)
30. Forero, P., Cano, A., Giannakis, G. B.: Distributed clustering using wireless sensor networks. IEEE Journal of Selected Topics in Signal Processing. **5**, 707–724 (2011)
31. Gabay, D., Mercier, B.: A dual algorithm for the solution of nonlinear variational problems via finite-element approximations. Comp. Math. Appl. **2**, 17–40 (1976)
32. Giannakis, G. B., Kekatos, V., Gatsis, N., Kim, S.-J., Zhu, H., Wollenberg, B. F.: Monitoring and Optimization for Power Grids: A Signal Processing Perspective. IEEE Signal Processing Magazine, **30**, 107–128 (2013)

33. Glowinski, R., Marrocco, A.: Sur l'approximation, par éléments finis d'orde un, et la résolution par pénalisation-dualité d'une classe de problèmes de Dirichlet non-linéaires. Rev. Francaise d'Aut. Inf. Rech. Oper. **2**, 41–76 (1975)
34. He, B., Yuan, X.: On the $O(1/t)$ convergence rate of the alternating direction method. SIAM Journal on Numerical Analysis **50**, 700–709 (2012)
35. He, B., Yuan, X.: On non-ergodic convergence rate of Douglas-Rachford alternating direction method of multipliers. Manuscript (2012)
36. Hlinka, O., Hlawatsch, F., Djuric, P. M.: Distributed particle filtering in agent networks. IEEE Signal Process. Mag. **30**, 61–81 (2013)
37. Jakovetic, D., Xavier, J., Moura, J.: Fast distributed gradient methods. Manuscript
38. Kay, S.: Fundamentals of statistical signal processing: Estimation theory. Prentice-Hall, Englewood Cliffs (1993)
39. Kim, S.-J., Dall'Anese, E., Bazerque, J. A., Rajawat, K., Giannakis, G. B.: Advances in spectrum sensing and cross-layer design for cognitive radio networks. Elsevier E-Reference Signal Processing (2012)
40. Kushner, H. J., Yin, G. G.: Stochastic approximation and recursive algorithms and applications. 2nd Edition, Springer, Berlin, Germany (2003)
41. Lakhina, A., Crovella, M., Diot, C.: Diagnosing network-wide traffic anomalies. Proc. ACM SIGCOMM. Portland, OR (2004)
42. Ling, Q., Ribeiro, A.: Decentralized dynamic optimization through the alternating direction method of multipliers. IEEE Transactions on Signal Processing **62**, 1185–1197 (2014)
43. Ling, Q., Ribeiro, A.: Decentralized linearized alternating direction method of multipliers. In: Proceedings of IEEE Intl. Conf. on Acoustics, Speech, and Signal Processing (2014)
44. Lloyd, S. P.: Least-squares quantization in PCM. IEEE Trans. on Info. Theory. **28**, 129–137 (1982)
45. Lopes, C. G., Sayed, A. H.: Incremental adaptive strategies over distributed networks," IEEE Trans. Signal Process. **55**, 4064–4077 (2007)
46. Lopes, C. G., Sayed, A. H.: Diffusion least-mean squares over adaptive networks: Formulation and performance analysis. IEEE Trans. Signal Process. **56**, 3122–3136 (2008)
47. Lu, Y., Roychowdhury, V., Vandenberghe, L.: Distributed parallel support vector machines in strongly connected networks. IEEE Tran. on Neural Networks. **19**, 1167–1178 (2008)
48. Mardani, M., Mateos, G., Giannakis, G. B.: Decentralized sparsity-regularized rank minimization: Algorithms and applications. IEEE Trans. Signal Process. **61**, 5374–5388 (2013)
49. Mardani, M., Mateos, G., Giannakis, G. B.: Dynamic Anomalography: Tracking Network Anomalies via Sparsity and Low Rank. IEEE Journal of Selected Topics in Signal Process. **7**, 50–66 (2013)
50. Mardani, M., Mateos, G., Giannakis, G. B.: Recovery of low-rank plus compressed sparse matrices with application to unveiling traffic anomalies. IEEE Trans. Info. Theory. **59**, 5186–5205 (2013)
51. Mateos, G., Bazerque, J. A., Giannakis, G. B.: Distributed sparse linear regression. IEEE Trans. Signal Process. **58**, 5262–5276 (2010)
52. Mateos, G., Giannakis, G. B.: Distributed recursive least-squares: Stability and performance analysis. IEEE Trans. Signal Process. **60**, 3740–3754 (2012)
53. Mateos, G., Rajawat, K.: Dynamic network cartography. IEEE Signal Process. Mag. **30**, 29–143 (2013)
54. Mateos, G., Schizas, I. D., Giannakis, G. B.: Distributed recursive least-squares for consensus-based in-network adaptive estimation. IEEE Trans. Signal Process. **57**, 4583–4588 (2009)
55. Mateos, G., Schizas, I. D., Giannakis, G. B.: Performance analysis of the consensus-based distributed LMS algorithm. EURASIP J. Advances Signal Process. Article ID 981030, 1–19 (2009)
56. Navia-Vazquez, A., Gutierrez-Gonzalez, D., Parrado-Hernandez, E., Navarro-Abellan, J. J.: Distributed support vector machines. IEEE Tran. on Neural Networks. **17**, 1091–1097 (2006)
57. Nedic, A., Ozdaglar, A.: Distributed subgradient methods for multiagent optimization. IEEE Transactions on Automatic Control, **54**, 48–61 (2009)

58. Nowak, R. D.: Distributed EM algorithms for density estimation and clustering in sensor networks. IEEE Trans. on Signal Processing. **51**, 2245–2253 (2003)
59. Olfati-Saber, R.: Distributed Kalman filter with embedded consensus filters. Proc. 44th IEEE Conf. Decision and Control. Seville, Spain (2005)
60. Rabbat, M., Nowak, R.: Quantized incremental algorithms for distributed optimization. IEEE Journal on Selected Areas in Communications, **23**, 798–808 (2005)
61. Rabbat, M., Nowak, R., Bucklew, J.: Generalized consensus computation in networked systems with erasure links. In: Proceedings of IEEE International Workshop on Signal Processing Advances for Wireless Communications (2005)
62. Ram, S., Nedic, A., Veeravalli, V.: Distributed stochastic subgradient projection algorithms for convex optimization. Journal of Optimization Theory and Applications, **147**, 516–545 (2010)
63. Ribeiro, A., Schizas, I. D, Roumeliotis, S. I., Giannakis, G. B.: Kalman filtering in wireless sensor networks: Incorporating communication cost in state estimation problems. IEEE Control Syst. Mag. **30**, 66–86 (2010)
64. Ripley, B. D.: Spatial Statistics. Wiley, Hoboken, New Jersey (1981)
65. Roughan, M.: A case study of the accuracy of SNMP measurements. Journal of Electrical and Computer Engineering. Article ID 812979 (2010)
66. Saligrama, V., Alanyali, M., Savas, O.: Distributed detection in sensor networks with packet losses and finite capacity links. IEEE Trans. on Signal Processing. **54**, 4118–4132 (2006)
67. Schizas, I. D., G. B. Giannakis, G. B.: Consensus-Based Distributed Estimation of Random Signals with Wireless Sensor Networks, Proc. of 40th Asilomar Conf. on Signals, Systems, and Computers, 530–534, Pacific Grove, CA (2006)
68. Schizas, I. D., Giannakis, G. B., Roumeliotis, S. I., Ribeiro, A.: Consensus in ad hoc WSNs with noisy links - Part II: Distributed estimation and smoothing of random signals. IEEE Trans. Signal Process. **56**, 1650–1666 (2008)
69. Schizas, I. D., Mateos, G., Giannakis, G. B.: Distributed LMS for consensus-based in-network adaptive processing. IEEE Trans. Signal Process. **57**, 2365–2381 (2009)
70. Schizas, I. D., Ribeiro, A., Giannakis, G. B.: Consensus in ad hoc WSNs with noisy links - Part I: Distributed estimation of deterministic signals. IEEE Trans. Signal Process. **56**, 350–364 (2008)
71. Schölkopf, B., Smola, A.: Learning with Kernels. Support Vector Machines, Regularization, Optimization, and Beyond. MIT Press, Boston (2002)
72. Shi, W., Ling, Q., Wu, G., Yin, W.: EXTRA: An exact first-order algorithm for decentralized consensus optimization. SIAM Journal on Optimization, **25**(2), 944–966 (2015)
73. Shi, W., Ling, Q., Yuan, K., Wu, G., Yin, W.: On the linear convergence of the ADMM in decentralized consensus optimization. IEEE Transactions on Signal Processing (2014)
74. Solo, V., Kong., X.: Adaptive signal processing algorithms: Stability and performance. Prentice-Hall, Englewood Cliffs (1995)
75. Srebro, N., Shraibman, A.: Rank, trace-norm and max-norm. Proc. 44th Learning Theory. Springer, 545–560 (2005)
76. Stoica, P., Moses, R.: Spectral Analysis of Signals. Prentice Hall (2005)
77. Tsianos, K. and Rabbat, M.: Distributed dual averaging for convex optimization under communication delays. Proc. of American Control Conference (2012)
78. Wang, H., Banerjee, A.: Online alternating direction method. arXiv preprint arXiv:1306.3721 (2013)
79. Wei, E., Ozdaglar, A.: Distributed alternating direction method of multipliers. Proc. Decision and Control Conference (2012)
80. Wen, Z., Goldfarb, D., Yin W.: Alternating direction augmented Lagrangian methods for semidefinite programming. Math. Prog. Comp. **2**, 203–230 (2010)
81. Wolfe, J., Haghighi, A., Klein, D.: Fully distributed EM for very large datasets. Proc. 25th Intl. Conference on Machine Learning. Helsinki, Finland (2008)
82. Wood A. J., Wollenberg, B. F.: Power Generation, Operation, and Control. Wiley & Sons, New York, NY (1984)

83. Xiao, L., Boyd, S.: Fast linear iterations for distributed averaging. Syst. Control Lett. **53**, 65–78(2004)
84. Xiao, L., Boyd, S., Kim, S. J.: Distributed average consensus with least-mean-square deviation. Journal of Parallel and Distributed Computing. **67**, 33–46 (2007)
85. Yuan, K., Ling, Q., Yin, W.: On the convergence of decentralized gradient descent. SIAM Journal Optimization, **26**(3): 1835–1854 (2016)
86. Zhang, Y., Ge, Z., Greenberg, A., Roughan, M.: Network anomography. Proc. ACM SIGCOM Conf. on Internet Measurements. Berkeley, CA (2005)
87. Zhu, H., Giannakis, G. B.: Power System Nonlinear State Estimation using Distributed Semidefinite Programming. IEEE Journal of Special Topics in Signal Processing, **8**, 1039–1050 (2014)
88. Zhu, H., Cano, A., Giannakis, G. B.: Distributed consensus-based demodulation: algorithms and error analysis. IEEE Trans. on Wireless Comms. **9**, 2044–2054 (2010)
89. Zhu, H., Giannakis, G. B., Cano, A.: Distributed in-network channel decoding. IEEE Trans. on Signal Processing. **57**, 3970–3983 (2009)

Chapter 15
Splitting Methods for SPDEs: From Robustness to Financial Engineering, Optimal Control, and Nonlinear Filtering

Christian Bayer and Harald Oberhauser

Abstract In this survey chapter we give an overview of recent applications of the splitting method to stochastic (partial) differential equations, that is, differential equations that evolve under the influence of noise. We discuss weak and strong approximations schemes. The applications range from the management of risk, financial engineering, optimal control, and nonlinear filtering to the viscosity theory of nonlinear SPDEs.

1 Introduction

The theory of (ordinary/partial) differential equations has been very successful in modeling quantities that evolve over time. Many of these quantities can be profoundly affected by stochastic fluctuations, noise, and randomness. The theory of stochastic differential equations aims for a qualitative and quantitative understanding of the effects of such stochastic perturbations. This requires insights from pure mathematics and to deal with them in practice requires us to revisit and extend classic numerical techniques. Splitting methods turn out to be especially useful since they often allow to separate the problem into a deterministic and a stochastic part.

C. Bayer
Weierstrass Institute, Mohrenstr. 39, 10117 Berlin, Germany
e-mail: christian.bayer@wias-berlin.de

H. Oberhauser (✉)
Department of Mathematics, University of Oxford, Andrew Wiles Building, Woodstock Road, Oxford OX2 6GG, UK
e-mail: harald.oberhauser@oxford-man.ox.ac.uk

R. Glowinski et al. (eds.), *Splitting Methods in Communication, Imaging, Science, and Engineering*, Scientific Computation, DOI 10.1007/978-3-319-41589-5_15

White Noise and Brownian Motion

The arguably simplest case of such a stochastic perturbation is an ODE driven by a vector field V that is affected by noise. Let us model this perturbation by a sequence of random variables $N = (N_t)_{t\geq 0}$ which are picked up by a vector field W,

$$\frac{dy_t}{dt} = V(y_t) \underbrace{+W(y_t)N_t}_{\text{Noise}}.$$

Often a reasonable assumption is that $N = (N_t)_{t\geq 0}$ is *white noise*, that is

1. (independence) $\forall s \neq t$, N_t and N_s are independent,
2. (stationarity) $\forall t_1 \leq \cdots \leq t_n$ the law of $(N_{t_1+t}, \cdots, N_{t_n+t})$ does not depend on t,
3. (centered) $\mathbb{E}[N_t] = 0$, $\forall t \geq 0$.

Above properties imply that the trajectory $t \mapsto N_t$ cannot be continuous, and even worse if we assume that $\mathbb{E}[N_t^2] = 1$ then $(\omega, t) \mapsto N_t(\omega)$ is not even measurable (see [60, 41]). Putting mathematical rigor aside, let us rewrite the above differential equation as an integral equation, i.e., we work with $B_t = \int_0^t N_r dr$ and since integration smoothes out we expect $B = (B_t)_{t\geq 0}$ to have nicer trajectories than N. In this case the above becomes

$$\mathrm{d}y_t = V(y_t)\,\mathrm{d}t + W(y_t)\,\mathrm{d}B_t \text{ resp. } y_t = \int_0^t V(y_r)\,\mathrm{d}r + \int_0^t W(y_r)\,\mathrm{d}B_r. \tag{15.1}$$

It turns out that $B = (B_t)_{t\geq 0}$ can be rigorously defined as a stochastic process — i.e., a collection of (ω, t)-measurable random variables carried on some probability space $(\Omega, \mathscr{F}, \mathbb{P})$. This process B is the well-known *Brownian motion*[1].

Definition 1. We call a real-valued stochastic process $B = (B_t)_{t\geq 0}$ defined on a probability space $(\Omega, \mathbb{P})$ a one-dimensional Brownian motion if

1. $B_0 = 0$ and $t \mapsto B_t$ is continuous (a.s.),
2. $\forall t_1 < \cdots < t_n$ and $n \in \mathbb{N}$, $B_{t_2} - B_{t_1}, \ldots, B_{t_n} - B_{t_{n-1}}$ are independent,
3. $\forall s, t, t - s > 0$, $B_t - B_s \sim \mathscr{N}(0, t - s)$.

The trajectories $t \mapsto B_t(\omega)$ are "degenerate": they are highly oscillatory, of infinite length, (statistically) self-similar, and possess a rich fractal structure; see Figure 15.1. Developing a theory that can deal with such trajectories is what makes stochastic calculus such a fascinating and rich subject. Finally, let us note that while Brownian motion is probably the most important stochastic process, there are many other classes of noise that appear in the real-world and are not covered by the

[1] Named after the botanist Robert Brown who observed in 1827 that pollen grains suspended in water execute continuous but jittery motions. The physical explanation was given by Albert Einstein in 1905 (his "annus mirabilis": small water molecules hit the pollen) and a little earlier Marian Smoluchowski had already emphasized the importance of this process for physics. Further important contributions are due to Louis Bachelier, Andrey Kolmogorov, Paul Lévy, Joseph Doob, Norbert Wiener, and finally Kyoshi Ito

Brownian (e.g., the so-called fractional Brownian motion [56]) and many of the methods we present here are not limited to the Brownian or even semimartingale setting.

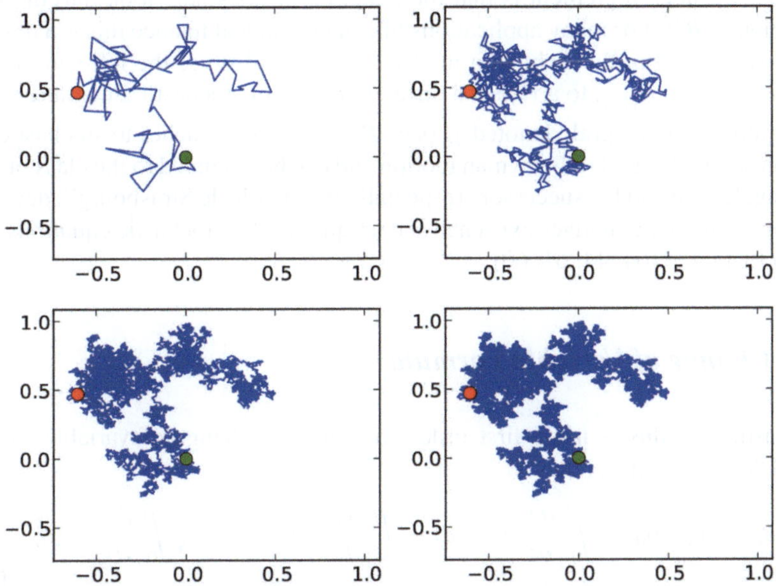

Fig. 15.1 The piecewise linear interpolation between the points of a two-dimensional Brownian motion started at $t = 0$ at $(0,0)$ (green circle), stopped at $t = 1$ (red circle) and sampled at time steps of size $\{10^{-2}, 10^{-3}, 10^{-4}, 10^{-5}\}$.

Stochastic Integrals

The Gaussianity of Brownian increments implies $B_t - B_s \sim N\sqrt{t-s}$ for $N \sim \mathcal{N}(0,1)$, hence we can expect at best a Hölder-modulus of $1/2$ and the problem of giving meaning to $\int_0^t W(y_r)\, dB_r$ appears. To see what goes wrong with Riemann–Stieltjes integrals consider integrating a one-dimensional Brownian trajectory against itself: with dyadic partitions of $[0,1]$, $t_i^n = i.2^{-n}$ a direct calculation shows that

$$\sum_i B_{\frac{t_{i+1}^n + t_i^n}{2}} (B_{t_{i+t}^n} - B_{t_i^n}) \text{ and } \sum_i B_{t_i^n} (B_{t_{i+1}^n} - B_{t_i^n}) \tag{15.2}$$

both converge (for a.e. Brownian trajectory $B(\omega)$) but to different limits. The difference of their limits equals $1/2$ times the $n \to \infty$ limit of

$$\sum_i (B_{t_{i+1}^n} - B_{t_i^n})^2 \tag{15.3}$$

and the quantity (15.3) is the so-called *quadratic variation process* $([B]_t)_{t\geq 0}$ of the Brownian motion B. Kyoshi Ito developed a powerful integration theory by generalizing the above limit construction (15.2). He gave meaning to $\int_0^t \phi_r dB_r$ for a large class of stochastic processes ϕ by taking the $L^2(\Omega)$-limit of $\sum_i \phi_{t_i^n}(B_{t_{i+1}^n} - B_{t_i^n})$ as $n \to \infty$. A crucial ingredient is that the integrand ϕ does not "look into the future evolution of B"[2]. For many applications like mathematical finance this is a desirable property. Instead of the right sum in (15.2) one can also use the left sum, i.e., the mid-points $\phi_{(t_i^n + t_{i+1}^n)/2}$, to arrive at a different notion of stochastic integration called the Stratonovich integral, denoted $\int_0^t \phi_r \circ dB_r$. Above approaches to stochastic integration are not limited to Brownian motion and can be extended to the class of semimartingales. Ito and his successors (especially the "Ecole de Strasbourg") developed a complete theory that gives existence and uniqueness for stochastic equations of the form (15.1); see [64, 44, 61, 60].

Ito's Change of Variable Formula

Stochastic calculus is not a first order calculus: the change of variable formula, called "Ito's Lemma", reads as

$$f(t,B_t) = f(0,B_0) + \int_0^t \frac{\partial f}{\partial t}(r,B_r)dr + \int_0^t \frac{\partial f}{\partial x}(r,B_r)dB_r + \frac{1}{2}\int_0^t \frac{\partial^2 f}{\partial x^2}(r,B_r)d[B]_r. \tag{15.4}$$

A big advantage of the Stratonovich integral is that it follows a first order calculus,

$$f(t,B_t) = f(0,B_0) + \int_0^t \frac{\partial f}{\partial t}(r,B_r)dr + \int_0^t \frac{\partial f}{\partial x}(r,B_r) \circ dB_r. \tag{15.5}$$

All this is only the starting point for one of the most exciting mathematical developments of the twentieth century and to make the above rigorous requires much more care — we refer the reader to the many excellent introductory texts [62, 60, 42, 64].

A Drawback: Discontinuity of the Solution Map

While stochastic calculus had tremendous impact on theory and applications it has several shortcomings; two which are relevant for this article are that it is, firstly, limited to the class of semimartingales as noise (this for example excludes fractional Brownian motion) and secondly, that a very basic object, namely the solution map associated with (15.1),

$$B \mapsto Y,$$

[2] More precisely, the relevant property of the Brownian motion here is that B is a martingale. Geometrically, this is an orthogonality relation between the increments $B_t - B_s$ and the path up to time s. Hence, the construction works in a geometric $L^2(\Omega)$ sense which allows to take advantage of this structure.

is not continuous in uniform norm (or any other reasonable norm). Over the last 20 years, Terry Lyons and collaborators [55, 53, 50, 33, 36] developed a robust and completely analytic/algebraic approach to such differential equations; this is the so-called "theory of rough paths". It is not meant to replace stochastic calculus but it complements it where it runs into trouble; especially in view of splitting results this robustness becomes very useful and gives for example continuity of the solution map.

Structure of This Chapter

In Section 2 we introduce the main topic of this chapter, namely that splitting schemes can be derived from robustness of the solution map. In Section 3 we recall some key results from the theory of rough paths which give a quantitative and qualitative understanding of the regularity of this solution map.

Splitting methods for S(P)DEs are naturally divided in strong and weak schemes. The goal of strong schemes is to approximate the solution Y of a S(P)DE (or a function of it, $f(Y)$) for a given realization of the noise. On the other hand, for many applications it is sufficient to only approximate the expected value $E[f(Y)]$. Strong approximations are discussed in Section 4 and applications to nonlinear filtering and optimal control are given in Section 5. In Section 6 we discuss weak splitting schemes for S(P)DEs and their rate of convergence; we recall a popular weak approximation scheme called "cubature on Wiener space" and show that it has a natural interpretation as a splitting scheme. In Section 7 we present three applications of splitting schemes in financial engineering: efficient implementations for popular stochastic local volatility models [2]; a calibration of the Gatheral Double Mean Reverting model to market data [3]; and finally the Heath–Jarrow–Morton interest rate model [24].

Background

This chapter is inspired by a view on stochastic differential equations that emerged over the last 15 years, namely the theory of rough paths due to Terry Lyons and collaborators; for further developments and introduction see [55, 52, 53, 36, 33, 51, 30]). This theory complements classic Ito-calculus and provides new, if not revolutionary insights, on how differential equations react to complex input signals. One of the earliest new applications was the so-called "cubature on Wiener space" of Kusuoka–Lyons–Victoir [54, 47]. Bayer, Dörsek, Teichmann among others [24, 68, 23, 5] then showed that these methods can be applied to the infinite-dimensional setting that is needed by SPDEs. More recently more applications were developed both in finite and in infinite dimensions (we survey some of these in Section 7). In a somewhat different direction, the work of Friz–Oberhauser [31] combined robustness from rough path theory with viscosity PDE methods to derive splitting schemes for strong approximations of (nonlinear) SPDEs.

Of course, splitting-up methods have appeared much earlier in stochastic calculus and we emphasize that these techniques remain highly relevant and form the basis of much of the recent developments that we present here. However, instead of giving a "horizontal" historical account we decided to give a "vertical" snapshot of what we believe are some exciting current developments in theory and applications. Unfortunately, this implies that we cannot do full justice to the existing rich literature. Nevertheless, we would like to point the reader to some classic articles as a starting point: one of the earliest motivations comes from the theory of nonlinear filtering and we mention pars pro toto the work of Bensoussan and Glowinski [6] and Bensoussan, Glowinski and Răşcanu [7, 8], Elliott and Glowinski [25], Florchinger and Le Gland [48, 28], Gyöngy and Krylov [37], Nagase [57], Sun and Glowinski [66], and Lototsky, Mikulevicius, and Rozovskii [49]. The more general field of splitting is overwhelmingly large, so that we again cannot hope to give a balanced literature review. Some general works we want to mention are Jentzen and Kloeden [43], Debussche [20], Gyöngy and Krylov [38], Răşcanu, and Tudor [63] Hausenblas [39] and, finally, Yan [69]. Let us finally stress that we consider partial differential equations driven by a temporal (possibly *also* spacial) noise, not partial differential equations with spacial noise, another very active research field in applied mathematics (see, for instance, Schwab and Gittelson [65]).

2 From Robustness to Splitting Schemes

On an abstract level, we have to understand how the output path (the solution of a differential equation) of a complex system (a differential equation) responds to an input path (e.g., time and noise). In this section we show that if a continuous dependence between output and input signal holds, then splitting results follow immediately.

A Toy Example

Let us consider the simple example of a quantity y whose evolution over time is described by the differential equation

$$\frac{\mathrm{d}y_t}{\mathrm{d}t} = V(y_t) + W(y_t), \ y_0 \in \mathbb{R}^e$$

where V, W are Lipschitz continuous vector fields on $\mathbb{R}^e$. We identify this differential equation as a special case of the integral equation

$$y_t = y_0 + \int_0^t V(y_r)\, da_r + \int_0^t W(y_r)\, db_r,\ y_0 \in \mathrm{R}^c \tag{15.6}$$

where a and b are continuous, real-valued paths that are regular enough that above integrals have meaning. While finite 1-variation as recalled in Definition 2 is sufficient for the Riemann–Stieltjes integrals, we will treat paths having much less regularity in later parts of this chapter. Equations of type (15.6) are often called *controlled (differential/integral) equations* and a, b are referred to as the *controls* or also as the *driving paths/signals*. Such equations arise naturally in the engineering sciences and have been very well studied (see the seminal work of Brockett, Sussmann, Fliess, et al. [11, 27, 67]). We henceforth use the shorthand/differential notation

$$\mathrm{d}y_t = V(y_t)\, da_t + W(y_t)\, db_t,\ y_0 \in \mathbb{R}^e \tag{15.7}$$

to denote (15.6). A basic question is the regularity of the solution map

$$(a,b) \mapsto y. \tag{15.8}$$

Obviously, the answer depends on what norms we use to measure distances between paths. What might be somewhat surprising is that the above mapping, defined on smooth paths

$$C^1\left([0,T],\mathbb{R}^2\right) \to C^1\left([0,T],\mathbb{R}^e\right),$$

is not even continuous under the usual uniform norm $|a|_\infty = \sup_{t\in[0,T]} |a_t|$; we invite the reader to find an example for this discontinuity and come back to this issue in detail in Example 1. Motivated by this, we introduce a cascade of metrics that are stronger than the uniform norm.

Definition 2. Let x be a continuous path defined on $[0,T]$ that takes values in a complete metric space (E,d). For every $p \geq 1$ the p-variation norm of x is defined as

$$|x|_{p-var} = \sup_{\substack{n\in\mathrm{N},(t_1,\dots,t_n):\\ 0\leq t_1<\dots<t_n\leq T}} \left(\sum_{i=1}^{n} d(x_{t_{i+1}}, x_{t_i})^p\right)^{1/p}$$

We denote the subset of $C([0,T],E)$ of paths finite p-variation norm by C^{p-var} $([0,T],E)$.

Standard arguments show that $\left(C^{p-var}([0,T],E), |.|_{p-var}\right)$ is a Banach space. We now see that the p-variation norm resolves the non-continuity of the uniform norm.

Theorem 1 (Robustness [33]). *Let $V, W : \mathbb{R}^e \to \mathbb{R}^e$ be Lipschitz continuous and $(a,b) \in C^{1-var}\left([0,T],\mathbb{R}^2\right)$. Then there exists a unique solution $y \in C^{1-var}\left([0,T],\mathbb{R}^e\right)$ to the controlled differential equation*

$$\mathrm{d}y_t = V(y_t)\,\mathrm{d}a_t + W(y_t)\,\mathrm{d}b_t,\ y_0 \in \mathbb{R}^d$$

and the map $(a,b) \mapsto y$ is continuous in 1-variation norm $|.|_{1-var}$.

Lie and Strang Splitting

The connection with splitting is now immediate: Fix $\Delta > 0$ and divide $[0,T]$ into intervals of size $\Delta > 0$; further, denote $t_\Delta = \lfloor \frac{t}{\Delta} \rfloor \Delta$, $t^\Delta = t_\Delta + \Delta$ and define two time-changes (real-valued increasing paths) $\phi^\Delta, \varphi^\Delta$,

$$\phi_t^\Delta = \begin{cases} t_\Delta + 2(t - t_\Delta) & , \text{if } t \in \left[t_\Delta, \frac{t_\Delta + t^\Delta}{2}\right] \\ t^\Delta & , \text{if } t \in \left[\frac{t_\Delta + t^\Delta}{2}, t^\Delta\right] \end{cases}, \quad \varphi_t^\Delta = \phi_{t+\frac{\Delta}{2}}^\Delta. \tag{15.9}$$

In other words, we approximate $t \mapsto (t,t)$ with $t \mapsto (\phi_t^\Delta, \varphi_t^\Delta)$ as $\Delta \to 0$, as depicted in Figure 15.2. Basic arguments show that this convergence holds in p-variation norm for every $p > 1$ (for $p = 1$ it is not true).

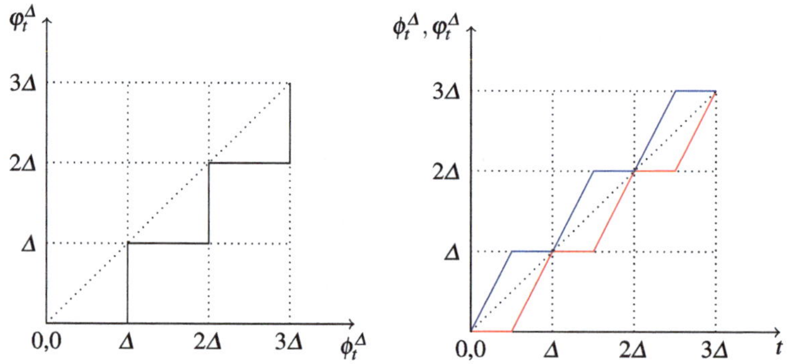

Fig. 15.2 (Lie-Splitting) The two-dimensional path $t \mapsto (\phi^\Delta, \varphi^\Delta)$ approximates the identity $t \mapsto (t,t)$ and exactly one of $\frac{d\phi_t^\Delta}{dt}, \frac{d\varphi_t^\Delta}{dt}$ is 0 for any given time $t \geq 0$. This gives rise to the so-called Lie-splitting scheme.

This particular choice of control paths immediately implies a splitting result since by composition of a (resp. b) with ϕ^Δ (resp. φ^Δ) we flow at any moment in time either along V or along W. This approach is quite different from applying the Trotter–Kato formula to semigroups but it has already been used several times in the literature; see the work of Le Gland [48], Gyöngy and Krylov [38]).

The above choice of a^Δ, b^Δ yields the classic Lie-splitting but obviously other choices are possible, for example we recover the Strang-Splitting scheme by using

$$\tilde{\phi}_t^\Delta = \phi_{t+\frac{\Delta}{4}}^\Delta \text{ and } \tilde{\varphi}_t^\Delta = \tilde{\phi}_{t+\frac{\Delta}{2}}^\Delta$$

To state it precisely, we introduce the notion of solution operators (Figure 15.3).

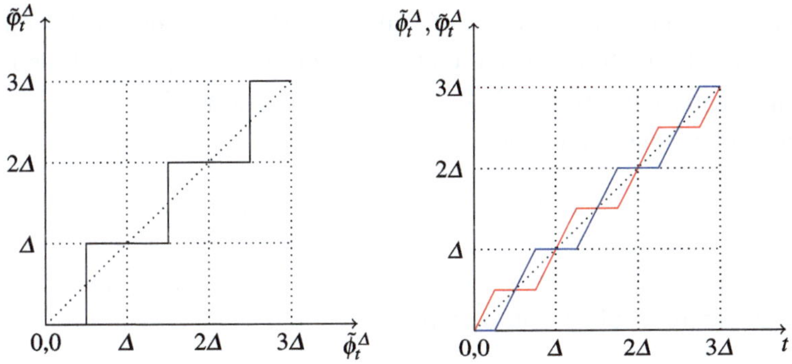

Fig. 15.3 (Strang-Splitting) The two-dimensional path $t \mapsto \left(\tilde{\phi}^\Delta, \tilde{\varphi}^\Delta\right)$ approximates the identity $t \mapsto (t,t)$ better than $t \mapsto \left(\phi_t^\Delta, \varphi_t^\Delta\right)$ as depicted in Figure 15.2. Therefore it should be not surprising that Strang-splitting leads to better rates than Lie-splitting.

Definition 3. For every $y_0 \in \mathbb{R}^e$ denote by $P_t^{\Delta,V} y_0$ the solution at time t of the controlled differential equation

$$\mathrm{d}y_t^\Delta = V\left(y_t^\Delta\right)\mathrm{d}(a \circ \phi^\Delta)_t$$

started at $y_0^\Delta = y_0$. Similarly denote by $Q_t^{\Delta,W} y_0$ the solution at time t of the controlled differential equation

$$\mathrm{d}y_t^\Delta = W\left(y_t^\Delta\right)\mathrm{d}(b \circ \varphi^\Delta)_t$$

started at $y_0^\Delta = y_0$.

Corollary 1 (Splitting). *We have* $\forall t > 0$

$$\lim_{\Delta \to 0}\left|\left(P_\Delta^{\Delta,V} Q_\Delta^{\Delta,W}\right)^{\lfloor t/\Delta \rfloor} y_0 - y_t\right| = 0$$

where y *is the solution of the differential equation (15.6) started at time* 0 *with* y_0. *Moreover, the convergence even holds uniformly in* t.

Proof. A simple calculation shows that the path $\left(a \circ \phi^\Delta, b \circ \varphi^\Delta\right)$ converges to the path $t \mapsto (t,t)$ with uniform 1-variation bounds, i.e., $\sup_{\Delta>0} |a^\Delta|_{1-var} + \sup_{\Delta>0} |b^\Delta|_{1-var} < \infty$. The claim then follows from a slight variation of Theorem 1; see [31].

Highly Oscillatory Paths and the Lie Brackets of Vector Fields

We now replace the path b in (15.6) by one with highly oscillatory trajectories (which is typical for many stochastic processes).

Example 1. Consider the sequence of paths $(a_t^n, b_t^n)_n = \left(\frac{1}{n}\cos 2\pi n^2 t, \frac{1}{n}\sin 2\pi n^2 t\right)$ and note that it converges for the uniform norm to $\left(a^0, b^0\right) = (0,0)$ as $n \to \infty$. Define the vector fields $W(y^1,y^2,y^3) := (1,0,\frac{-y^2}{2})^T$ and $V(y^1,y^2,y^3) := (0,1,\frac{y^1}{2})^T$ and denote by y^Δ the solution of

$$\mathrm{d}y_t^\Delta = V\left(y_t^\Delta\right)\mathrm{d}a_t^\Delta + W\left(y_t^\Delta\right)\mathrm{d}b_t^\Delta \text{ with } y_0^\Delta = y_0 \in \mathbb{R}^3. \tag{15.10}$$

A simple calculation then shows that y_1^Δ does not converge as $\Delta \to 0$ to y_0 (the solution of (15.10) applied with $\Delta \equiv 0$).

In the above Example 1, the highly oscillatory motions of the driving signals affect the evolution of y not directly via V or W but via their Lie bracket $[V,W] = V \cdot W - W \cdot V$ which picks up the signed area (recall the Green/Stokes formula)

$$(s,t) \mapsto \frac{1}{2}\left(\int_s^t a_r^n db_r^n - \int_s^t b_r^n da_r^n\right)$$

swept out by (a^n, b^n) during the time interval (s,t). To sum up, the highly oscillatory behavior of the driving signal leads to a subtle interplay between the iterated integrals of the driving signal and the Lie brackets of the involved vector fields that can destroy continuity of the solution map. However, was central to our derivation above of the Lie and Strang-splittings. Below we show how the theory of rough paths provides the needed continuity and gives us a very robust way to solve differential equations driven by such highly oscillatory paths.

3 Rough Path Theory

We still have to give meaning to differential equations driven by non-smooth paths and study the properties of the associated solution map. Example 1 suggests that the iterated integrals

$$\int_{s<r_1<t} \mathrm{d}x_{r_1}, \int_{s<r_1<r_2<t} \mathrm{d}x_{r_1} \otimes \mathrm{d}x_{r_2}, \dots, \int_{s<r_1<\dots<r_n<t} \mathrm{d}x_{r_1} \otimes \dots \otimes \mathrm{d}x_{r_n}.$$

of the driving signal x (resp. their linear combinations) play a special role when the path is highly oscillatory. In general, these integrals will not make sense as Riemann–Stieltjes integrals if x is of unbounded variation. However, the theory of rough paths shows that it is enough to find a sequence of tensors that "behaves algebraically" like such a sequence of iterated integrals to derive the well posedness of differential equations driven by this "iterated integrals".

The Space of Iterated Integrals

The sequence of iterated integrals has a rich algebraic structure. Let us first give the definition for the case of a bounded variation path.

Definition 4. For every u,v such that $0 \leq u \leq v \leq T$ define $\Delta^1_{u,v} = \{(s,t) : u \leq s \leq t \leq v\}$. Let $x \in C^{1-var}([0,T],\mathbb{R}^d)$, $(s,t) \in \Delta_{0,T}$ and $k \in \mathbb{N}$. We define the iterated integrals $\int_{\Delta^k_{s,t}} \mathrm{d}x \otimes \cdots \otimes \mathrm{d}x \in (\mathbb{R}^d)^{\otimes k}$ recursively as

$$\int_{\Delta^1_{s,t}} \mathrm{d}x := x(t) - x(s) \text{ and } \int_{\Delta^k_{s,t}} \underbrace{\mathrm{d}x \otimes \cdots \otimes \mathrm{d}x}_{k \text{ times}} := \int_s^t \int_{\Delta^{k-1}_{s,r}} \underbrace{\mathrm{d}x \otimes \cdots \otimes \mathrm{d}x}_{(k-1) \text{ times}} \otimes \mathrm{d}x_r.$$

Recall that the space $(\mathbb{R}^d)^{\otimes k}$ used above is the space of k-tensors which has as basis $(e_{i_1} \otimes \cdots \otimes e_j)_{i,j \in \{1,\ldots,d\}}$.

Definition 5. Let $x \in C^{1-var}([0,T],\mathbb{R}^d)$ and $(s,t) \in \Delta^1_{0,T}$. The signature of x over $[s,t]$, denoted by $S(x)_{s,t}$, is the element of $\bigoplus_{k=0}^{\infty} (\mathbb{R}^d)^{\otimes k}$ given as

$$S(x)_{s,t} = \left(1, \int_{\Delta^1_{s,t}} \mathrm{d}x, \int_{\Delta^2_{s,t}} \mathrm{d}x \otimes \mathrm{d}x, \ldots \right)$$

with the convention that $(\mathbb{R}^d)^{\otimes 0} = \{1\}$. Similarly, we define for $n \in \mathbb{N}$ the truncated signature of x over $[s,t]$, denoted $S^n(x)_{s,t}$, as the element of $\bigoplus_{k=0}^{n} (\mathbb{R}^d)^{\otimes k}$ given as

$$S^n(x)_{s,t} = \left(1, \int_{\Delta^1_{s,t}} \mathrm{d}x, \int_{\Delta^2_{s,t}} \mathrm{d}x \otimes \mathrm{d}x, \ldots, \int_{\Delta^n_{s,t}} \underbrace{\mathrm{d}x \otimes \cdots \otimes \mathrm{d}x}_{n \text{ times}} \right).$$

We call the path $t \mapsto S^n(x)_{0,t}$ the step-n lift of x.

The above definition is not efficient concerning the state space since it does not account for the recursive structure of $S^n(x)$ and we can hope to work with a much smaller subspace of $\bigoplus_{k=0}^{n} (\mathbb{R}^d)^{\otimes k}$. With slight abuse of notation denote by $\otimes : \bigoplus_{k=0}^{n} (\mathbb{R}^d)^{\otimes k} \to \bigoplus_{k=0}^{n} (\mathbb{R}^d)^{\otimes k}$ the natural extension of the tensor multiplication to the graded space $\bigoplus_{k=0}^{n} (\mathbb{R}^d)^{\otimes k}$, i.e. for

$$g = \sum_{k=0}^{n} \sum_{i_1,\ldots,i_k} g^{i_1\ldots i_k} e_{i_1} \otimes \cdots \otimes e_{i_k}, h = \sum_{k=0}^{n} \sum_{i_1,\ldots,i_k} h^{i_1\ldots i_k} e_{i_1} \otimes \cdots \otimes e_{i_k} \in \bigoplus_{k=0}^{n} \left(\mathbb{R}^d\right)^{\otimes k}$$

define

$$g \otimes h = \sum_{k=0}^{n} \sum_{l,m:l+m=k} g^{i_1\ldots i_l} h^{i_1\ldots i_m} e_{i_1} \otimes \cdots \otimes e_{i_k} \otimes e_{i_1} \otimes \cdots \otimes e_{i_m}.$$

We can now describe the algebraic structure of the subspace of $\bigoplus_{k=0}^{n}\left(\mathbb{R}^d\right)^{\otimes k}$ that contains the iterated integrals.

Theorem 2 ([33]). *For $n \geq 1$ and $d \geq 1$ define $G_{n,d} := \left\{S(x)_{0,1} : x \in C^{1-var}\left([0,T],\mathbb{R}^d\right)\right\}$. Then*

1. *$\left(G_{n,d},\otimes\right)$ is a Lie group,*
2. *$G_{n,d} = \exp \mathfrak{g}_{n,d}$ where $(\mathfrak{g}_{n,d},[\cdot,\cdot])$ is Lie algebra and*
3. *$\mathfrak{g}_{n,d} = \mathbb{R}^d \oplus \left[\mathbb{R}^d,\mathbb{R}^d\right] \oplus \cdots \oplus \left[\mathbb{R}^d,\left[\mathbb{R}^d,\left[\cdots,\left[\mathbb{R}^d,\mathbb{R}^d\right]\cdots\right]\right]\right]$*

We call $G_{n,d}$ the free step-n Lie group with d generators and $\mathfrak{g}_{n,d}$ the free step-n Lie algebra with d generators. The geodesic (so-called Carnot–Caratheodory) distance d_{CC} turns $(G_{n,d},d_{CC})$ in a metric space.

(Weak) Geometric Rough Paths

Since $(G_{n,d},d_{CC})$ is a complete metric space, Definition 2 applies and we can speak of paths of bounded p-variation — this is exactly the definition of a weak geometric p-rough path.

Definition 6. Let $p \geq 1$ and $n = \lfloor p \rfloor$. We define the space of weak geometric p rough paths as

$$C^{p-var}([0,T],G_{n,d}) := \left\{\mathbf{x} \in C([0,T],G_{n,d}) : d_{p-var}(0,\mathbf{x}) < \infty\right\}$$

(here 0 denotes constant path that takes the value of the neutral element of the group $G_{n,d}$).

Example 2 (The Brownian Rough Path). Let B be a two-dimensional Brownian motion. This gives rise to the $G_{2,2}$-valued path

$$\begin{aligned}\mathbf{B}_t &= \underbrace{\left(1, B_t, \int_0^t dB \otimes dB\right)}_{\in G_{2,2}} \\ &= \exp \underbrace{\left(B_t^1 e_1 + B_t^2 e_2 + \frac{1}{2}\left(\int_0^t B_r^1 dB_r^2 - \int_0^t B_r^2 dB_r^1\right)(e_1 \otimes e_2 - e_2 \otimes e_1)\right)}_{\in \mathfrak{g}_{2,2}}\end{aligned}$$

where the integrals are understood as (Stratonovich) stochastic integrals. One can show that $\mathbf{B} \in C^{p-var}([0,T],G_{2,2})$ for any $p > 2$, see [55, 33].

Differential Equations Driven by Rough Paths

Ito's approach to differential equations driven by highly oscillatory stochastic processes exploits the underlying probabilistic structure of the driving signal. Lyons [55, 52, 51] developed a different approach that relies only on analytic and algebraic methods; most important for us, it comes with a cascade of metrics which provide the needed continuity of the solution map.

Theorem 3 ("Universal Limit Theorem": Existence, Uniqueness and Continuity of RDEs; See [55, 33]). *Let $p \in (2,3)$, $d \geq 1$, $\mathbf{x} \in C^{p-var}([0,T], G_{2,d})$ and $V_i \in C_b^3(\mathbb{R}^e, \mathbb{R}^e)$. There exists a $y \in C^{p-var}([0,T], \mathbb{R}^e)$ such that for every sequence $(x^n)_n \subset C^{1-var}([0,T], \mathbb{R}^d)$ such that $d_{p-var}(S_2(x), \mathbf{x}) \to 0$, the solutions of the ODE*

$$dy_t^n = V(y_t^n)dx_t^n \equiv \sum_i V_i(y_t^n)d(x^n)_t^i$$

converge uniformly to y. We say that y is a solution of the RDE driven by $\mathbf{x}$ and write

$$dy_t = V(y_t)d\mathbf{x}_t.$$

The solution map is uniformly continuous on compact sets, that is for every $R > 0$ the map

$$\begin{aligned} (y_0, x) &\mapsto y \\ \mathbb{R}^e \times \left\{ d_{p-var}(x, 0) < R \right\} &\to C^{p-var}\left([0,T], G_{2,d}\right) \end{aligned}$$

is uniformly continuous in d_{p-var}-metric.

Summary

Rough path theory provides us with a machinery to solve differential equations driven by non-smooth signals (like Brownian motion, semimartingales but also many other classes of noise that are not covered by classic stochastic calculus). As opposed to the Ito-theory it not only requires the trajectory of the driving signal as input but also its "iterated integrals"; to be precise, it requires a set of tensors that "behave like" classical Riemann–Stieltjes iterated integrals. Finding efficient state spaces for these "enhanced paths" required us to work with nonlinear spaces, i.e., Lie groups. In return we get a completely analytic and algebraic approach that provides the well posedness of such differential equations, and the rough path theory comes with a cascade of metrics which makes the solution map continuous (the metric d_{p-var} for $p \geq 1$). Such a robustness is in stark contrast with Ito's theory and allows us to translate our simple splitting proof from the toy example in Section 2 to the case of S(P)DEs.

4 Strong Splitting Schemes for SPDEs

In this section we extend the splitting method to parabolic PDEs that evolve under the influence of noise. A large class of such stochastic partial differential equations (SPDEs) is of the form

$$\begin{cases} du = F\left(t,x,u,Du,D^2u\right)dt + \sum_{i=1}^d \Lambda_i\left(t,x,u,Du\right)dz_t^i & \text{on } [0,T]\times\mathbb{R}^n \\ u(0,x) = u_0(x) & \text{on } \mathbb{R}^n \end{cases} \tag{15.11}$$

where $u = u(t,x)$ is scalar-valued, F denotes a nonlinear, (possibly degenerate) elliptic differential operator, Λ is affine linear in (u,Du), and $z \in C\left([0,T],\mathbb{R}^d\right)$ is a multidimensional path with the same (or worse) regularity properties as Brownian trajectories.

Several issues appear: firstly, even if $\Lambda \equiv 0$, then the nonlinearity of F implies that we cannot hope for a smooth solution $u \in C^{1,2}([0,T]\times\mathbb{R}^n,\mathbb{R})$. Therefore we have to work with a suitable concept of generalized solutions. Secondly, the path z is not differentiable and similar to our toy example, we have to give appropriate meaning to $\Lambda(t,x,u,Du)\circ dz_t$. Put simply, we solve the first problem by working with the theory of viscosity solutions and the second problem with the theory of rough paths.

Approximating Time

As in our toy example in Section 2, we now want to look at equation (15.11) as a special case of

$$\begin{cases} du^\Delta = F\left(t,x,u^\Delta,Du^\Delta,D^2u^\Delta\right)d(a\circ\phi^\Delta)_t + \sum_{i=1}^d \Lambda\left(t,x,u^\Delta,Du^\Delta\right)\circ d(z^i\circ\varphi^\Delta)_t, \\ u^\Delta(0,x) = u_0(x). \end{cases} \tag{15.12}$$

However, the situation is more subtle.

Example 3. Consider $n = 1$, $F\left(t,x,u,Du,D^2u\right) = D^2u$ and $\Lambda \equiv 0$ in which the above reduces to the one-dimensional heat equation: $du^\Delta = D^2u^\Delta d(a\circ\phi^\Delta)_t$. Then one cannot hope for continuity of $(a,z)\mapsto u$ since this requires to give meaning to the heat equation when $\frac{d(a\circ\phi^\Delta)_t}{dt} < 0$, i.e., when time is run backwards which is in general not well posed.

We simply resolve the above issue by replacing C^{1-var} by a smaller class of paths.

Proposition 1 ([31]). *Define*

$$C_0^{1,+}([0,T],\mathbb{R}) = \left\{\xi \in C^1([0,T],\mathbb{R}) : \xi_T = T, \dot{\xi}_t > 0\ \forall t\right\}$$

and its closure $C_0^{1-var,+}([0,T],\mathbb{R}) := \overline{C_0^{1,+}([0,T],\mathbb{R})}^{|\cdot|_\infty}$ *where* $|a|_\infty \equiv \sup_{t\in[0,T]}|a_t|$. *Then*

$$C_0^{1-var,+}([0,T],\mathbb{R}) = \Big\{ \xi \in C_0([0,T],\mathbb{R}) : \xi_T = T \text{ and } \exists \xi^{cont} \in L^1([0,T],\mathbb{R}), \\ \exists \xi^{sing} \in C^{1-var}([0,T],\mathbb{R}_{\geq 0}), \dot{\xi}^{sing} = 0 \text{ a.s. and } \xi_t = \xi_t^{sing} + \int_0^t \xi_r^{cont} dr \Big\}$$

and $C_0^{1-var,+}([0,T],\mathbb{R}) \subsetneq C_0^{1-var}([0,T],\mathbb{R})$.

Viscosity Solutions of PDEs

Given a map

$$F : [0,T] \times \mathbb{R}^n \times \mathbb{R} \times \mathbb{R}^n \times \mathbb{S}^n \to \mathbb{R}$$

(with $\mathbb{S}^n$ denoting the set of symmetric $(n \times n)$-matrices) that is *proper* in the sense that

$$F(t,x,r,p,A) \leq F(t,x,r,A+B) \ \forall A \in \mathbb{S}^n \text{ and } B \geq 0 \\ r \mapsto F(t,x,r,A) \text{ is increasing,}$$

then the theory of viscosity solutions provides well posedness for parabolic PDEs of the form

$$\begin{cases} \partial_t u - F(t,x,u,Du,D^2u) = 0 & \text{on } [0,T] \times \mathbb{R}^n, \\ u(0,x) = u_0(x) & \text{on } \mathbb{R}^n. \end{cases} \tag{15.13}$$

More precisely, if $u : [0,T] \times \mathbb{R}^n \to \mathbb{R}$ is bounded and uniformly continuous then we call u a *subsolution* of the PDE (15.13) if for every $\varphi \in C^{1,2}([0,T] \times \mathbb{R}^n, \mathbb{R})$ it holds that whenever $(\hat{t},\hat{x})$ is a local maximum of

$$(t,x) \mapsto u(t,x) - \varphi(t,x)$$

then

$$\partial_t \varphi(\hat{t},\hat{x}) - F(\hat{t},\hat{x},\varphi,D\varphi,D^2\varphi) \leq 0. \tag{15.14}$$

Similarly, we define *supersolutions* and call u a solution if it is a sub- and supersolution. Viscosity theory provides *comparison results*, that is given a subsolution v and a supersolution w of (15.13) this guarantees that

$$v \leq w$$

(note that this immediately implies uniqueness of solutions), see [15, 26].

Robustness for (Nonlinear) SPDEs

We can now have an educated guess of a good solution concept for the non-linear SPDE (15.11): let us approximate $t \mapsto (t, z_t)$ by a sequence $(\xi^n, z^n)_n \subset C^1([0,T], \mathbb{R}^d)$ of smooth paths. Then for every fixed $n \in \mathbb{N}$ we can speak of a viscosity solution $u^n \in \mathrm{BUC}([0,T] \times \mathbb{R}^n)$ — the space of real-valued, bounded, and uniformly continuous functions — of

$$\begin{cases} \mathrm{d}u^n = F\left(t,x,u^n,Du^n,D^2u^n\right)\mathrm{d}\xi^n + \sum_{i=1}^d \Lambda\left(t,x,u^n,Du^n\right)\mathrm{d}z_t^{n;i} & \text{on } [0,T]\times\mathbb{R}^n, \\ u^n(0,x) = u_0(x) & \text{on } \mathbb{R}^n. \end{cases} \tag{15.15}$$

We expect that $(u^n)_n$ converges to a function $u \in \mathrm{BUC}([0,T] \times \mathbb{R}^n)$ as $n \to \infty$ and it is natural to identify this function u as the solution of the SPDE (15.11). It turns out that it is natural to define convergence of the sequence $(\xi^n, z^n)_n$ to $(t, \mathbf{z}_t)$ if

$$\begin{aligned} \sup_n \|S(z^n)\|_{p-var;[0,T]} + \sup_n |\xi^n|_{1-var} &< \infty \\ d_0(z, S(z^n)) + |\xi_t^n - t|_\infty &\to_n 0 \end{aligned} \tag{15.16}$$

holds. Here $d_0(\mathbf{x},\mathbf{y}) \equiv \sup \sum_{t_i} d_{CC}(\mathbf{x}_{t_i,t_{i+1}}, \mathbf{y}_{t_i,t_{i+1}})$ where the sup is taken over all partitions (t_i) and we use the notation $\mathbf{x}_{t_i,t_{i+1}} \equiv \mathbf{x}_{t_i}^{-1}\mathbf{x}_{t_{i+1}}$ for increments in the group. Let us take this as definition of a solution.

Definition 7. Let $z \in C_0^{0,p-var}([0,T], G_{[p],d}), \xi \in C_0^{1-var,+}([0,T], \mathbb{R})$. Let

$$(z^n, \xi^n)_n \subset C_0^{0,p-var}\left([0,T], G_{[p],d}\right) \times C_0^{1-var,+}([0,T], \mathbb{R})$$

be a sequence that converges to (t, z_t) in the sense of (15.16) and assume that there exists for every n a unique viscosity solution u^n of the PDE (15.15). We call every accumulation point (in the metric of uniform convergence on compacts) of (u^n) a solution of the RPDE

$$\begin{cases} \mathrm{d}u = F\left(t,x,u,Du,D^2u\right)\mathrm{d}\xi_t + \Lambda\left(t,x,u,Du\right)\circ\mathrm{d}z_t & \text{on } [0,T]\times\mathbb{R}^n, \\ u(0,x) = u_0(x) & \text{on } \mathbb{R}^n. \end{cases} \tag{15.17}$$

If this limit is unique and does not depend on the choice of the approximating sequence $(\xi^n, z^n)_n$ and the solution map

$$(\xi, z) \mapsto u$$

is continuous then we say that (15.17) is robust in the rough path sense.

It is clear that the above robustness in rough path sense immediately gives a splitting result when use the time changes $(\phi^{1/n}, \varphi^{1/n})$ from Section 2 to define the approximating sequence

$$\left(\xi \circ \phi^{1/n}, z \circ \varphi^{1/n}\right)_n.$$

In Section 5 below we show that large classes of SPDEs are robust in rough path sense as defined above.

5 Applications of Strong Schemes to Nonlinear Filtering and Optimal Control

Nonlinear Filtering

In many areas of science, the quantities of interest are not available for direct measurement. Fortunately, we can make reasonable inferences about them by combining mathematical models that describe their evolution with partial observations of these quantities. These partial observations are typically corrupted by noise and we need to account for this. Applications range from cryptography, tracking and guidance, the study of the global climate, to the management of risk in a economic context (see for example [17, 10, 29, 35]). Consider a Markov process (X,Y) that takes its values in $\mathbb{R}^{d_{sig}+d_{obs}}$ with its dynamics given by

$$\begin{cases} \mathrm{d}X_t = \mu(X_t)\,\mathrm{d}t + \sigma(X_t)\,\mathrm{d}B_t & \text{(signal)}, \\ \mathrm{d}Y_t = h(X_t)\,\mathrm{d}t + \mathrm{d}\tilde{B}_t & \text{(observation)}. \end{cases}$$

Here, B and $\tilde{B}$ are multidimensional Brownian motions that are defined on some probability space $(\Omega, \mathscr{F}, \mathbb{P})$. The goal is to compute for a given real-valued function f the conditional expectation

$$\pi_t f \equiv \mathbb{E}\left[f(X_t) \,|\, \mathscr{Y}_t\right],$$

i.e., to find the best estimate for $f(X_t)$ given the observation σ-algebra[3] $\mathscr{Y}_t = \sigma(\{Y_r, r \in [0,t]\}) \vee \mathscr{N}$ with $\mathscr{N}$ denoting the $\mathbb{P}$-null-sets. From basic principles it follows that there exists a measurable map $\phi_t^f : C([0,t], \mathbb{R}^{d_Y}) \to \mathbb{R}$ such that

$$\phi_t^f\left(Y|_{[0,t]}\right) = \pi_t f \;\; \mathbb{P}-\text{a.s.} \tag{15.18}$$

and our problem reduces to effectively calculate this functional ϕ_t^f.

Clark's Robustness Problem

In practice, only a finite number of observations $(Y_{t_i})_i$ of Y is available and we evaluate ϕ_t^f along some continuous interpolation of these points, $Y^{\text{interpolated}}$. Of course we expect that

$$\phi_t^f\left(Y^{\text{interpolated}}|_{[0,t]}\right) \simeq \phi_t^f\left(Y|_{[0,t]}\right)$$

but this is not guaranteed by (15.18), as the interpolation is a path of bounded variation, hence a null-set under the Wiener measure or any equivalent measure, see [16] for a detailed discussion. Clark [13] sketched a proof (a rigorous argument was given later by Clark and Crisan [14]) that if B and $\tilde{B}$ are uncorrelated, then there exists a functional $\phi_t^{f,\text{robust}}$ that is continuous in supremum norm and fulfills (15.18). In the correlated case, such a functional cannot exist but recently (see [16]), it was

[3] There are some subtle measure-theoretic issues which we gloss over but refer the reader to [1] for more details.

shown that in the correlated case there exists also a functional $\phi_t^{f,\text{robust}}$ defined on the space of rough paths such that

$$\phi_t^{f,\text{robust}}(Y) = \pi_t f \;\mathbb{P} - \text{a.s and } Y \mapsto \phi_t^{f,\text{robust}}(Y)$$

is continuous in rough path metric. This solves Clark's robustness problem (for semimartingale piecewise linear approximations converge in the appropriate rough path metric).

The Kallianpur–Striebel and Zakai Equations

Theorem 4 (Kallianpur–Striebel). *There exists a probability measure $\tilde{\mathbb{P}}$ on $(\Omega, \mathscr{F})$ such that*

1. *$\tilde{\mathbb{P}}$ is equivalent to $\mathbb{P}$,*
2. $\frac{d\tilde{\mathbb{P}}}{d\mathbb{P}}|_{\mathscr{F}_t} = \exp\left(-\int_0^t h(X_s)\cdot dB_s - \frac{1}{2}\int_0^t |h(X_s)|^2 ds\right)$,
3. *the observation process Y is a Brownian motion under $\tilde{\mathbb{P}}$,*
4. *for every $f \in B\left(\mathbb{R}^{d_{sig}}\right)$—the space of real-valued, bounded, measurable functions on $\mathbb{R}^d$—and every fixed $t > 0$*

$$\pi_t f = \frac{\tilde{\mathbb{E}}\left[f(X_t)\exp\left(\int_0^t h(X_s)\cdot dY_s - \frac{1}{2}\int_0^t |h(X_s)|^2 ds\right)|\mathscr{Y}_t\right]}{\tilde{\mathbb{E}}\left[\exp\left(\int_0^t h(X_s)\cdot dY_s - \frac{1}{2}\int_0^t |h(X_s)|^2 ds\right)\right]} \quad \mathbb{P} \text{ and } \tilde{\mathbb{P}} \text{ a.s.}$$

Proof. This can be found in every text book on nonlinear filtering; see for example [1, 17].

It turns out that it is advantageous to work with a non-normalized version of the inferred probability measure π. Indeed, if we define for every $f \in B\left(\mathbb{R}^{d_{sig}}\right)$

$$\rho_t f = \pi_t f \cdot \tilde{\mathbb{E}}\left[\exp\left(\int_0^t h(X_s)\cdot dY_s - \frac{1}{2}\int_0^t |h(X_s)|^2 ds\right)\right]$$

then obviously $\pi_t f = \frac{\rho_t f}{\rho_t 1}$. The Fokker–Planck/Kolmogorov forward equation is a PDE given by the generator of X that describes the time evolution of the density of the diffusion X via a parabolic PDE with the elliptic differential operator

$$A = \sum_{i,j}(\sigma^T\cdot\sigma)^{i,j}\frac{\partial^2}{\partial x_i \partial x_j} + \sum_i \mu^i \frac{\partial}{\partial x_i}.$$

The Zakai equation can be seen as an extension that incorporates the additional information we get from the observation process Y. Indeed, set $h \equiv 0$ in Theorem below to recover the Fokker–Planck equation.

Theorem 5 (The Zakai SDE; Uncorrelated Case). *Under standard assumptions*[4] *we have* $\tilde{\mathbb{P}}$*-a.s. for every* $t \geq 0$ *and every* $f \in B\left(\mathbb{R}^{d_{sig}}\right)$ *that*

$$(\rho_t f) = \pi_0 f + \int_0^t \rho_s (Af)\,\mathrm{d}s + \int_0^t \rho_s \left(fh^T\right)\mathrm{d}Y_s.$$

Proof. See for example [1, Chapter 3].

The above applies to the case when B and $\tilde{B}$ are uncorrelated. In the correlated case a slight variation of the above Zakai SDE holds (an extra differential operator appears in the stochastic integral against Y).

Splitting for the Zakai SPDE

It is advantageous to work with densities instead of measures. Indeed, under well-known conditions ρ has a density u and we can write

$$\rho_t (A) = \int_A u(t,x)\,\mathrm{d}x$$

for some $u \in \mathrm{BUC}\left([0,T] \times \mathbb{R}^n\right)$. In this case we can rewrite the above (infinite-dimensional) Zakai SDE from Theorem 5 for the unnormalized measure ρ as a SPDE for the density u. Since the generator of the signal X is linear, second order parabolic it is not surprising that the resulting SPDE will be linear (with linear noise). In fact, our setup is more general than needed by the nonlinear filtering application and below we treat general semi-linear PDEs (of which the SPDE for the density u is a special case).

Assumption 1. *Let*

$$L(t,x,r,p,M) = \mathrm{Tr}\left[M(x) \cdot X\right] + b(x) \cdot p + f(x,r)$$

with $M(x) = \sigma(x)\sigma^T(x)$, $\sigma : \mathbb{R}^n \to \mathbb{R}^{n\times m}$ *and* $b : \mathbb{R}^e \to \mathbb{R}^e$ *bounded, Lipschitz in* x*. Further, let* $f : \mathbb{R}^n \times \mathbb{R} \to \mathbb{R}$ *be continuous, bounded whenever* r *is bounded and with a lower Lipschitz bound, i.e.*

$$f(x,r) - f(x,s) \geq c(r-s)\ \forall r \geq s, x \in \mathbb{R}^n.$$

Assumption 2. *Let*

$$\Lambda(t,x,r,p) = p \cdot \sigma_k(t,x) + r \cdot \nu_k(t,x) + g_k(t,x)$$

where σ, ν *and* g *are* Lip^γ *for* $\gamma > p+2$.

[4] For example $E\left[\int_0^t |h(X_s)|^2 ds\right] < \infty, E\left[\int_0^t Z_s |h(X_s)|^2 ds\right] < \infty$ and $\tilde{\mathbb{P}}\left[\int_0^t [\rho_s(|h|)]^2 ds < \infty\right] = 1$ is sufficient where $Z_s = \exp\left(-\sum_i \int_0^s h^i(X_r) dB_r^i - \frac{1}{2}\int_0^s h^i(X_r)^2 dr\right)$; see [1, Chapter 3]

Theorem 6 (Well Posedness of Linear RPDEs). *Let $z \in C^{0,p-var}\left([0,T],\mathbb{R}^d\right)$ and let L and Λ fulfill assumption (1) resp. (2). Then*

$$\begin{cases} \mathrm{d}u = L\left(t,x,u,Du,D^2u\right)\mathrm{d}t + \sum_{i=1}^{d}\Lambda\left(t,x,u,Du\right)\circ \mathrm{d}z_t^i & \text{on } [0,T]\times\mathbb{R}^n, \\ u\left(0,x\right) = u_0\left(x\right) & \text{on } \mathbb{R}^n. \end{cases} \tag{15.19}$$

is robust in rough path sense.

Proof. We only sketch the idea of the proof for the case

$$\begin{cases} \mathrm{d}u = \sigma^2(t,x)D^2u\mathrm{d}t + \sum_{i=1}^{d} V_i\left(x\right)Du\circ \mathrm{d}z_t^i & \text{on } [0,T]\times\mathbb{R}^n, \\ u\left(0,x\right) = u_0\left(x\right) & \text{on } \mathbb{R}^n. \end{cases} \tag{15.20}$$

First assume that z is a smooth path and denote by ϕ the ODE flow

$$\mathrm{d}\phi(t,x) = V(\phi(t,x))\mathrm{d}z_t, \phi(0,x) = x \in \mathbb{R}^n. \tag{15.21}$$

Then (at least formally) we see that the function $v(t,x) := u(t,\phi(t,x))$ solves the standard parabolic heat equation

$$\begin{cases} \mathrm{d}v = \sigma_\phi^2(t,x)D^2v\mathrm{d}t & \text{on } [0,T]\times\mathbb{R}^n, \\ v\left(0,x\right) = u_0\left(x\right) & \text{on } \mathbb{R}^n, \end{cases} \quad \text{where } \sigma_\phi^2(t,x) := \sigma^2(t,\phi(t,x)). \tag{15.22}$$

An obvious idea for the case that z is no longer a smooth path is to approximate z by a sequence of smooth paths (z^n). For each fixed $n \in \mathbb{N}$ one can solve the ODE flow ϕ^n (the ODE (15.21) with z replaced by z^n) and subsequently the corresponding simple PDE (15.22) to arrive at the sequence of PDE solutions (v^n). Since the flow ϕ^n will be a diffeomorphism we also know that

$$u^n(t,x) = v^n(t,(\phi_t^n)^{-1}(x)) \tag{15.23}$$

where u^n denotes the solution of (15.20) where the driving signal z is replaced by z^n. Obviously we expect that $(v^n)_n$ as well as $(\phi^n)_n$ converge as $n \to \infty$: for $(v^n)_n$ this should follow from the robust approximations of operators from viscosity theory and for $(\phi^n)_n$ this should follow if we consider convergence in rough path metric—recall Sections 2 and 3 on the problems caused by highly oscillatory driving signals z. If this holds, then (15.23) implies that $(u^n)_n$ converges to some u and this function is a natural candidate for a solution. Of course, all the above was completely formal and the convergence can go wrong. However, with more care it can be made rigorous even for fully nonlinear operators; for the detailed argument see [12, 22, 32].

Corollary 2 (Splitting for the Zakai SPDE). *Denote by $\{P_t, t \geq 0\}$ the solution operator*

$$\varphi \mapsto v \text{ where } v \text{ is the viscosity solution of } \mathrm{d}v = L\left(t,x,Dv,D^2v\right)\mathrm{d}t,\ v(0,\cdot) = \varphi(\cdot)$$

and by $\{Q_{s,t}, 0 \leq s \leq t\}$ the solution operator

$$\varphi \mapsto v \text{ where } v \text{ is the SDE solution of } \mathrm{d}y = \Lambda\left(t,x,Dv\right)\circ \mathrm{d}B_t,\ y(0,\cdot) = \varphi(\cdot).$$

Then for a.e. ω

$$u^n(t,x) := \prod_{i=0}^{\lfloor t/n \rfloor -1} \left[Q_{i/n,i/n+1/n} \circ P_{1/n} \right] (u_0(x))$$

converges locally uniformly (in (t,x)) as $n \to \infty$ to the unique solution u of (15.19) given by Theorem 6 with $z_t = \mathbf{B}_t(\omega) \equiv \left(1, B_t(\omega), \left(\int_0^t B \otimes \circ \mathrm{d}B\right)(\omega)\right)$.

Pathwise Optimal Control

Consider the SDE

$$\mathrm{d}X_t = a(X_t, \alpha_t)\,\mathrm{d}t + b(X_t, \alpha_t) \circ \mathrm{d}B_t + c(X_t) \circ \mathrm{d}\tilde{B}_t$$

where $t \mapsto \alpha_t$ is a path, B and $\tilde{B}$ are multi-dimensional, independent Brownian motions and (a,b,c) are (sufficiently regular) vector fields. In applications (engineering, economics, etc.) one often faces the problem that one can influence the evolution of X by controlling the path α. The aim is then to minimize a cost function (consisting of a terminal cost g and a running cost f) of the form

$$v(t,x) = \inf_{\alpha} \mathbb{E}\left[g\left(X_T^{t,x}\right) + \int_t^T f\left(X_s^{t,x}, \alpha_s\right) \mathrm{d}s | \tilde{B} \right]. \tag{15.24}$$

It turns out that we can use the Bellman principle to describe the change in the cost function over time by a SPDE, the so-called Hamilton–Jacobi–Bellman (S)PDE. Indeed, a formal computation (see [21] for a rigorous derivation from basic principles) shows that after the time reversal $u(t,x) := v(T-t,x)$, we get a SPDE of the form

$$\begin{cases} \mathrm{d}u|_{t,x} + \inf_\alpha \left[b(x,\alpha_t) Du|_{t,x} + L_\alpha u|_{t,x} + f(x,\alpha_t) \right] \mathrm{d}t + Du|_{t,x} \cdot c(x) \circ \mathrm{d}\tilde{B}_t = 0 & \text{on } [0,T] \times \mathbb{R}^n, \\ u(0,x) = g(x) & \text{on } \mathbb{R}^n, \end{cases} \tag{15.25}$$

where L_α is the linear differential operator with (a,b). Using

Corollary 3. *Let $z \in C^{0,p-var}\left([0,T], \mathbb{R}^d\right)$. The SPDE* (15.25) *is robust in rough path sense.*

The proof is a slight modification of the proof of Theorem 6 since the usual comparison results from viscosity theory is stable under taking $\inf_\alpha$.

6 Weak Splitting Schemes for SPDEs

In the previous sections we have concentrated on strong approximation of (partial) differential equations driven by random signals, i.e., on the approximation of the solution $y_T = y_T(\omega)$ of the rough or stochastic (partial) differential equation as a random variable, ω-for-ω (resp. rough path by rough path). However, in many applications one is only interested in the law of the solution y_T of the equation. Indeed, if the quantity of interest is just the expectation of a functional of the solution, say

$$E\left[f(y_T)\right],$$

then it is sufficient to only approximate the law of y_T. This corresponds to the notion of *weak convergence* of random variables, and hence schemes for approximating the law of the solution of a stochastic (partial) differential equations are referred to as *weak schemes*. More precisely, let us consider the solution y_T of a stochastic differential equation defined on the Banach space X (which is infinite-dimensional in the case of an SPDE) and let us consider a sequence of approximations $\overline{y}_N$ taking values in X indexed by $N \in \mathbb{N}$. Fix a space of sufficiently regular test functions $f : X \to \mathbb{R}$ (classically chosen to be $C_b(X)$ in theoretical probability theory, but more flexibility is needed in numerics). Then we say that $\overline{y}_N$ converges to y_T in the weak sense if for any test function f we have

$$E\left[f(\overline{y}_N)\right] \xrightarrow{N\to\infty} E\left[f(y_T)\right].$$

In particular, note that weak schemes — unlike strong ones — do not have to operate on the same probability space as the true solution.

Of course, if the space of test functions is a subspace of the Lipschitz continuous functions, then strong convergence (i.e., convergence in $L^1(\Omega;X)$) implies weak convergence, and the rate of weak convergence is at least as good as the rate of strong convergence. However, in many cases the weak rate of convergence is, in fact, much better than the strong one.

Cubature on Wiener Space

For simplicity, let us concentrate on the finite-dimensional case first — we will come back to the infinite-dimensional (SPDE) setting at the end of this section. Consider the stochastic differential equation

$$dy_t = V_0(y_t)dt + \sum_{i=1}^{d} V_i(y_t) \circ dB_t^i, \tag{15.26}$$

with $y_0 \in \mathbb{R}^e$ fixed, B denoting a d-dimensional standard Brownian motion and "$\circ$" indicating that the stochastic integral is understood in the Stratonovich sense. We furthermore introduce the notation $B_t^0 \equiv t$ to simplify the presentation.

Assumption 3. *We assume that the vector fields $V_0, \dots, V_d : \mathbb{R}^e \to \mathbb{R}^e$ are C^∞-bounded, i.e., they are smooth and all the derivatives are bounded (but not necessarily the functions themselves). Moreover, the test function f is smooth and bounded.*

Remark 1. Of course, these assumptions can be relaxed. For instance, the boundedness requirements can be removed by working with properly weighted norms [23, 46]. Moreover, assuming a (hypo-)ellipticity condition for the vector fields, we can actually rely on the smoothing property of the diffusion equation and drop the smoothness assumption for the test function f – at the cost of possibly having to work with nonuniform grids, see [54].

In order to derive appropriate weak splitting schemes for the equation (15.26), we first recall the short time behavior of the solution using the *stochastic Taylor expansion*, see for instance [45]. By iterating the Ito-formula for the Stratonovich-SDE (15.26) m times, we obtain

$$f(y_t) = f(y_0) + \sum_{k=1}^{m} \sum_{(i_1,\dots,i_k)\in\{0,\dots,d\}^k} V_{i_1}\cdots V_{i_k} f(y_0) \int_{0<t_1<\cdots<t_k<t} \circ dB_{t_1}^{i_1}\cdots\circ dB_{t_k}^{i_k} + \mathcal{O}\left(t^{(m+1)/2}\right), \tag{15.27}$$

where we iteratively use the geometrical notion $Vf(x) \equiv \nabla f(x)\cdot V(x)$ for a function f and a vector field V. We also denote

$$B_t^I = B_t^{(i_1,\dots,i_k)} \equiv \int_{0<t_1<\cdots<t_k<t} \circ dB_{t_1}^{i_1}\cdots\circ dB_{t_k}^{i_k}, \quad I = (i_1,\dots,i_k) \in \{0,\dots,d\}^k. \tag{15.28}$$

Remark 2. We once again see that the short-time behavior of the solution y is controlled by the truncated signature.

Remark 3. As a matter of fact, sharper versions of (15.27) are possible, in so far that (15.27) ignores the different scaling of $t = B_t^0$ and $B_t^1,\dots,B_t^d$. Once again, we refer to [54].

Remark 4. Of course, analogous stochastic Taylor expansions can also be formulated in terms of the Ito integral, which would then lead to the Ito-signature. We prefer the geometrically more intuitive Stratonovich versions in this chapter.

This motivates the following methodology for constructing higher order weak approximation schemes termed the *ODE method* (originally introduced as *cubature on Wiener space* by [54] and, independently, [47]).

Theorem 7. *In the setting of Assumption 3, we are given a time-grid* $0 = t_0 < t_1 < \cdots < t_N = T$ *with corresponding increments* Δt_i, $i = 1,\dots,N$. *Let* $W_i : [0,\Delta t_i] \to \mathbb{R}^{d+1}$ *be a* $(d+1)$*-dimensional path of bounded variation satisfying*

$$\forall 0 \le k \le m,\ I \in \{0,\dots,d\}^k:\ E\left[B_{\Delta t_i}^I\right] = E\left[W_i^I(\Delta t_i)\right].$$

Moreover, let $W : [0,T] \to \mathbb{R}^{d+1}$ *be the bounded-variation process obtained by concatenating the processes* W_1, …, W_N. *Finally, let* $\overline{y}_N \equiv y_T(W)$ *be defined as the solution of the* ODE

$$\frac{dy(W)_t}{dt} = V_0(y(W)_t)\dot{W}_t^0 + \sum_{i=1}^{d} V_i(y(W)_t)\dot{W}_t^i \tag{15.29}$$

formally obtained from (15.26) by replacing B by W. Then there is a constant $C > 0$ such that

$$|E[f(y_T)] - E[f(y(W)_T)]| \leq C \left(\max_{i=1,\dots,N} \Delta t_i \right)^{(m-1)/2}.$$

Proof. We do not give a detailed proof, as the underlying argument is quite standard in numerical analysis. Indeed, by (15.27) the local error of the approximation is of order $(\Delta t_i)^{(m+1)/2}$. Thus, by summing up the local errors we obtain that the global error is of order $(\max_i \Delta t_i)^{(m-1)/2}$

Remark 5. The above theorem is somewhat imprecise, as the constant C depends on T, f, the vector fields $V_0, \dots, V_d$ and the *method* of constructing the processes $W_1, \dots, W_N$, but not on the grid. E.g., in the case of the Ninomiya-Victoir method introduced below, C will only depend on T, f, $V_0, \dots, V_d$.

6.1 The Ninomiya–Victoir Splitting

If liberally interpreted — e.g., for Euler schemes, when the path W is actually a step-function — Theorem 7 encompasses a large class of discretization schemes for the stochastic differential equation (15.26). In particular, it allows for a simple construction of *stochastic splitting schemes*, as we shall exemplify by the arguably most popular version, the *Ninomiya–Victoir scheme* [59]. In that case, the paths of the process W are *axis-paths*, i.e., the paths are continuous and piecewise-parallel to the axis in $\mathbb{R}^{d+1}$, similar to the construction used in Definition 3, see (15.9). More precisely, choose $1 \leq i \leq N$ and a sequence of independent (of all other sources of randomness) random variables Λ_i, $i = 1, \dots, N$ with $P(\Lambda_i = 1) = P(\Lambda_i = -1) = 1/2$. Construct a process W_i on $[0, \Delta t_i]$ in the following way: set $\delta_i \equiv \Delta t_i/(d+1)$ and when $\Lambda_i = +1$, set

$$\dot{W}_i(t) = \begin{cases} \Delta t_i/\delta_i e_0, & 0 \leq t < 1/2\delta_i, \\ \Delta B_i^j/\delta_i e_j, & (1/2 + (j-1))\delta_i \leq t < (1/2 + j)\delta_i,\ 1 \leq j \leq d, \\ \Delta t_i/\delta_i e_0, & \Delta t_i - 1/2\delta_i \leq t \leq \Delta t_i, \end{cases} \tag{15.30a}$$

where we recall that $\Delta B_i^j \equiv B_{t_i}^j - B_{t_{i-1}}^j$ and where we denote by $(e_0, e_1, \dots, e_d)$ the standard basis of $\mathbb{R}^{d+1}$. In the other case ($\Lambda_i = -1$), we define W_i by

$$\dot{W}_i(t) = \begin{cases} \Delta t_i/\delta_i e_0, & 0 \leq t < 1/2\delta_i, \\ \Delta B_i^{d-j+1}/\delta_i e_{d-j+1}, & (1/2 + (j-1))\delta_i \leq t < (1/2 + j)\delta_i,\ 1 \leq j \leq d, \\ \Delta t_i/\delta_i e_0, & \Delta t_i - 1/2\delta_i \leq t \leq \Delta t_i. \end{cases} \tag{15.30b}$$

As in the general construction, the independent processes $W_1, \dots, W_N$ are then concatenated to form the process W defined on $[0, T]$.

Inserting the process W just constructed into the general methodology (15.29), we see that the Ninomiya–Victoir method boils down to solving the ODEs driven by the individual vector fields $V_0,\dots,V_d$ on $\mathbb{R}^e$. Indeed, let $e^{sV_i}x$ denote the flow associated with the vector field V_i at time s, i.e., $e^{sV_i}x = z(s)$ solution to

$$\dot{z}(t) = V_i(z(t)), \quad z(0) = x \in \mathbb{R}^e,$$

then the solution $\overline{y}_l \equiv y(W)_{t_l}$, $l = 0,\dots,N$, of (15.29) for the Ninomiya–Victoir process W satisfies $\overline{y}_0 = y_0$ and

$$\overline{y}_l = \begin{cases} e^{\frac{\Delta t_l}{2}V_0} e^{\Delta B_l^d V_d} \dots e^{\Delta B_l^1 V_1} e^{\frac{\Delta t_l}{2}V_0} \overline{y}_{l-1}, & \Lambda_l = +1, \\ e^{\frac{\Delta t_l}{2}V_0} e^{\Delta B_l^1 V_1} \dots e^{\Delta B_l^d V_d} e^{\frac{\Delta t_l}{2}V_0} \overline{y}_{l-1}, & \Lambda_l = -1, \end{cases} \tag{15.31}$$

$l = 1,\dots,N$. This explains from the SDE side, why we consider the Ninomiya–Victoir scheme a stochastic splitting scheme for the SDE (15.26).

Theorem 8. *Under Assumption 3, the Ninomiya–Victoir scheme is a weak scheme of second order, i.e., there is a constant $C > 0$ (depending on $T, f, V_0,\dots,V_d$, but not on the grid) such that*

$$|E[f(y_T)] - E[f(y(W)_T]| \leq C\left(\max_{i=1,\dots,N} \Delta t_i\right)^2.$$

Proof. We show that $E\left[S_{0,t}^5(B)\right] = E\left[S_{0,t}^5(W)\right]$ for $t = \Delta t_i$ and any $W = W_i$, which implies the conclusion by Theorem 7.

Let us first consider the (Stratonovich) signature of the Brownian motion. By the construction of the Stratonovich integral in terms of the Ito integral, we have

$$\begin{aligned} B_t^{(i_1,\dots,i_k)} &= \int_{0<t_1<\dots<t_k<t} \circ dB_{t_1}^{i_1} \cdots \circ dB_{t_k}^{i_k} \\ &= \begin{cases} \int_0^t B_{t_k}^{(i_1,\dots,i_{k-1})} dB_{t_k}^{i_k} + \frac{1}{2}\int_0^t B_s^{(i_1,\dots,i_{k-2})} ds \delta_{i_k i_{k-1}}, & i_k \neq 0, \\ \int_0^t B_{t_k}^{(i_1,\dots,i_{k-1})} dB_{t_k}^{i_k}, & i_k = 0. \end{cases} \end{aligned}$$

Using the Ito isometry, the expectation of the iterated Stratonovich integral is iteratively given by

$$E\left[B_t^{(i_1,\dots,i_k)}\right] = \begin{cases} \frac{1}{2}\int_0^t E\left[B_s^{(i_1,\dots,i_{k-2})}\right] ds \delta_{i_k i_{k-1}}, & i_k \neq 0, \\ \int_0^t E\left[B_s^{(i_1,\dots,i_{k-1})}\right] ds, & i_k = 0. \end{cases}$$

As regards the Ninomiya–Victoir cubature formula defined above, we see that

$$\dot{W}_i^j(t) = \Delta B_i^j / \delta_i \mathbf{1}_{A_i^j}(t), \quad j = 0,\dots,d, \quad i = 1,\dots,N,$$

where we tacitly let $\Delta B_i^0 = \Delta t_i$ and define the set A_i^j by

$$A_i^0 = [0, 1/2\delta_i[\cup[\Delta t_i - 1/2\delta_i, \Delta t_i],$$

$$A_i^j = \begin{cases} [(j-1/2)\delta_i, (j+1/2)\delta_i], & \Lambda_i = +1, \\ [(d-j+1/2)\delta_i, (d-j+3/2)\delta_i], & \Lambda_i = -1, \end{cases} \quad j = 1, \ldots, d.$$

So we have the general formula

$$E\left[W_i^{(i_1,\ldots,i_k)}(\Delta t_i)\right] = E\left[\Delta B_i^{i_1} \cdots \Delta B_i^{i_k}\right] E\left[\int_{0<t_1<\cdots<t_k} \mathbf{1}_{A_i^{i_1}}(t_1) \cdots \mathbf{1}_{A_i^{i_k}}(t_k) dt_1 \cdots dt_k\right],$$

where the last expectation is necessary due to the random choice of intervals above, and, in fact, only involves the two alternatives $\Lambda_i = \pm 1$.

The verification of the theorem now boils down to a simple, but tedious exercise. For instance, for multi-indices of length 3, we see that the only nonzero components of the expectation of the signature restricted to multi-indices of length 3 for either B and W_i are

$$E\left[B_{\Delta t_i}^{(0,0,0)}\right] = \frac{\Delta t_i^3}{6} = E\left[W_i^{(0,0,0)}(\Delta t_i)\right],$$
$$E\left[B_{\Delta t_i}^{(0,j,j)}\right] = \frac{\Delta t_i^2}{4} = E\left[W_i^{(0,j,j)}(\Delta t_i)\right],$$
$$E\left[B_{\Delta t_i}^{(j,j,0)}\right] = \frac{\Delta t_i^2}{4} = E\left[W_i^{(j,j,0)}(\Delta t_i)\right],$$

$j = 1, \ldots, d,$

A Path-Wise Interpretation of the Ninomiya–Victoir Splitting Scheme

Interpreting the Ninomiya–Victoir scheme in the Lie/Strang splitting picture drawn in (15.9) and below, we define functions $a^{\Delta t_i}, b_1^{\Delta t_i}, \ldots, b_d^{\Delta t_i}$ on the interval $[0, \Delta t_i]$ by

$$\dot{a}^{\Delta t_i}(t) = \frac{\Delta t_i}{\delta_i} \mathbf{1}_{[0,\delta_i/2[}(t),$$
$$\dot{b}_j^{\Delta t_i}(t) = \frac{\Delta B_i^j}{\delta_i} \mathbf{1}_{[(1/2+j-1)\delta_i,(1/2+j)\delta_i[}(t), \quad j = 1, \ldots, d,$$
$$\dot{c}^{\Delta t_i}(t) = \frac{\Delta t_i}{\delta_i} \mathbf{1}_{[\Delta t_i - \delta_i/2, \Delta t_i]}(t).$$

After concatenating these paths, we could immediately construct a Lie-type splitting following Definition 3 (in fact, we would not need to split the time component in the a and c paths) or a Strang-type splitting. However, taking the scaling of Brownian motion into account, we realize that we need to take care of Lie brackets of order up to 5 in order to obtain a high order scheme. Hence, we need even more "re-orderings" than in the ordinary Strang splitting. Thus, we further define paths

$$\dot{\widetilde{b}}_j^{\Delta t_i}(t) = \frac{\Delta B_i^j}{\delta_i}\mathbf{1}_{[(1/2+d-j)\delta_i,(1/2+d-j+1)\delta_i[}(t), \quad j=1,\ldots,d.$$

The two alternatives (15.30a) and (15.30b) of $W_i(t)$ are then given by

$$W_i(t) = (a^{\Delta t_i}(t) + c^{\Delta t_i}(t))e_0 + \sum_{j=1}^{d}\left(b_j^{\Delta t_i}(t)\mathbf{1}_{\Lambda_i=+1} + \widetilde{b}_j^{\Delta t_i}(t)\mathbf{1}_{\Lambda_i=-1}\right)e_j,$$

and the corresponding splitting scheme is indeed given by (15.31), taking into account that the solutions of the ODEs driven by $b_j^{\Delta t_i}$ and $\widetilde{b}_j^{\Delta t_i}$ eventually coincide at time Δt_i.

The Ninomiya–Victoir Scheme as a Splitting Scheme for PDEs

It is well known that the function $u(t,y) \equiv E[f(y_t)]$ with $y_0 = y$ satisfies the linear Cauchy problem

$$\frac{\partial}{\partial t}u(t,y) = Lu(t,y), \quad u(0,y) = f(y), \tag{15.32}$$

for $t > 0$ and $y \in \mathbb{R}^e$, respectively. Here, the partial differential operator L is defined by

$$Lg(y) = V_0 g(y) + \frac{1}{2}\sum_{i=1}^{d} V_i^2 g(y), \tag{15.33}$$

where we recall that for any vector field $V : \mathbb{R}^e \to \mathbb{R}^e$ and any smooth function $g : \mathbb{R}^e \to \mathbb{R}$ we set $Vg(y) \equiv \nabla g(y) \cdot V(y)$. Iterating this procedure also defines $V^2 g$, with V^2 a second order differential operator. Hence, for any weak approximation $\overline{y}_N$ of y_T, we have that

$$E[f(\overline{y}_N)] \approx E[f(y_T)] = u(T, y_0), \tag{15.34}$$

and the order of the weak approximation is the order of the approximation in the solution of the PDE (15.32). In semi-group notation, we can denote the solution operator associated with L by $P_t \equiv \exp(tL)$, i.e.,

$$u(t,y) = P_t f(y).$$

Remark 6. Obviously, solving the SDE (15.26) is only one step for the solution of the PDE (15.32): in addition, one needs to approximate the expectation in (15.34).

In principle, for the Ninomiya–Victoir method this is a numerical integral in dimension $d \times N$. As this dimension is typically quite high, one usually resorts to Monte Carlo or Quasi Monte Carlo methods for computing the integral. Numerically, the computational cost of the integration step is often much higher than the computational cost of the discretization of the SDE, as the rate of convergence of the integration schemes is only $\frac{1}{2}$ (for Monte Carlo) or (at best) 1 (for Quasi Monte Carlo). Nonetheless, higher order weak approximation methods can reduce the overall computational cost considerably as compared to low order methods, partly because they actually lead to a considerable reduction of the dimension of the integration problem in the second step. The advantages of using higher order schemes have been observed in many numerical studies, for instance [59, 2, 4, 46].

Remark 7. As compared to classical numerical solvers for the Cauchy problem (15.32), the stochastic approximation scheme presented here has some very different features. On the one hand, most standard numerical methods such as finite element or finite difference schemes produce approximate solutions $u(t,y)$ for all values of t and y simultaneously – within a certain region in time and space, and up to interpolation. On the other hand, using the stochastic representation (15.34), one only obtains an approximation of $u(t,y)$ for one particular t and one particular y. Moreover, the stochastic method crucially relies on the performance of the (Q)MC approximation for the expected values, and shares its strengths and weaknesses. Hence, for low-dimensional problems classical numerical PDE solvers are typically more efficient, whereas for high dimensions $e \gg 1$, the stochastic method is competitive or superior, as it does not suffer from the curse of dimensionality.

In light of (15.32), the question arises whether the Ninomiya–Victoir scheme can be naturally associated with a (PDE) splitting scheme for (15.32). To this end let us first consider the situation when there is only one time-step. Let Q_t^i, $i = 1,\dots,d$, be the semi-groups corresponding to the second order differential operators $\frac{1}{2}V_i^2$, $i = 1,\dots,d$, respectively. Formally, we write $Q_t^i = e^{\frac{t}{2}V_i^2}$. Moreover, we denote by Q_t^0 the semi-group associated with the operator V_0, i.e.,

$$Q_t^0 f(x) = f\left(e^{tV_0}x\right).$$

While the correspondence between e^{tV_0} and Q_t^0 is obvious, we note that Q_t^i is in some sense the expectation of $e^{B_t^i V_i}$. More precisely, stochastic Taylor expansion shows that for any C^∞-bounded test function f and any initial value $y \in \mathbb{R}^e$ we have

$$Q_t^i f(y) = E\left[f\left(e^{B_t^i V_i} y\right)\right].$$

Hence, in the case with only one time-step (with $\Delta t = T$, $\Delta B = B_T$) we obtain that

$$\begin{aligned} E\left[f(\bar{y}_1)\right] &= \frac{1}{2}E\left[f\left(e^{\frac{\Delta t}{2}V_0}e^{\Delta B^d V_d}\cdots e^{\Delta B^1 V_1}e^{\frac{\Delta t}{2}V_0}y\right)\right] + \frac{1}{2}E\left[f\left(e^{\frac{\Delta t}{2}V_0}e^{\Delta B^1 V_1}\cdots e^{\Delta B^d V_d}e^{\frac{\Delta t}{2}V_0}y\right)\right] \\ &= \frac{1}{2}Q^0_{\Delta t/2}Q^d_{\Delta t}\cdots Q^1_{\Delta t}Q^0_{\Delta t/2}f(y) + \frac{1}{2}Q^0_{\Delta t/2}Q^1_{\Delta t}\cdots Q^d_{\Delta t}Q^0_{\Delta t/2}f(y) \equiv Q_{\Delta t}f(y). \end{aligned}$$

Note that the weight "$\frac{1}{2}$" comes from the probability $\frac{1}{2}$ to choose either of the two alternatives in (15.31).

Iterating this construction along a discretization of the time interval $[0,T]$ as above, we recover a well-known splitting scheme from the PDE literature, sometimes referred to as "symmetrically weighted sequential splitting" scheme, see [18]. In terms of the solution operator $P_t = \exp(tL)$, Theorem 8 thus says that

$$\left|P_T f(y) - Q_{\Delta t_N} \cdots Q_{\Delta t_1} f(y)\right| \leq C \left(\max_{i=1,\ldots,N} \Delta t_i \right)^2. \tag{15.35}$$

The Ninomiya–Victoir Stochastic Splitting for SPDEs

The stochastic splitting methodology introduced above can be directly generalized to the infinite-dimensional case, i.e., to the case of SPDEs instead of SDEs. This was first done by Bayer and Teichmann [5] for the abstract formulation of Theorem 7 under strong regularity conditions. Later on, Dörsek and Teichmann [23] have given a careful analysis of the Ninomiya–Victoir splitting and other splitting techniques for weak approximation of SPDEs under weaker assumptions. We will mainly follow their approach here.

Consider a stochastic partial differential equation of the form

$$dy_t = (Ay_t + V(y_t))dt + \sum_{i=1}^{d} V_i(y_t)dB_t^i, \quad y_0 \in X, \tag{15.36}$$

that is we assume that the stochastic fluctuation only depend on y_t, but not on derivatives of y_t. The state space X of the equation (15.36) is assumed to be a separable Hilbert space and the vector fields $V, V_1, \ldots, V_d : X \to X$ are Frechet-differentiable and Lipschitz continuous, whereas the operator $A : \mathscr{D}(A) \subset X \to X$ is the generator of a strongly continuous, pseudo-contractive semigroup on X — more regularity conditions on the coefficients are deferred until later. Then, a *mild* solution y_t of the SPDE exists. For details of the solution theory of this class of SPDEs we refer to the monograph [19].

As mild solutions to SPDEs of the form (15.36) are generally not semimartingales, we cannot rewrite (15.36) in Stratonovich form from the beginning, but have to work with the Ito formulation. Nonetheless, if we use the Ninomiya–Victoir stochastic splitting approach, all the resulting (simpler) SPDEs can, in fact, be written in Stratonovich form, hence we proceed just as above. Indeed, define

$$e^{sV_i}y = z_s, \text{ where } \dot{z}_t = V_i(z_t),\ z_0 = y \in X, \quad i = 1, \ldots, d.$$

Moreover, set $V_0(y) \equiv V(y) - \frac{1}{2}\sum_{i=1}^d DV_i(y) \cdot V_i(y)$, $y \in X$, which would precisely correspond to the Stratonovich drift if it actually were to exist, and define $e^{s(A+V_0)}y$ analogously, i.e., as solution z_s at time s of the Cauchy problem

$$\frac{\partial}{\partial t} z_t = Az_t + V_0(z_t), \quad z_0 = y \in X,$$

which may be represented in terms of the semi-group $\exp(tA)$ generated by A as

$$e^{s(A+V_0)}y = z_s = \exp(sA)y + \int_0^s \exp((s-t)A)V_0(z_t)dt.$$

Now we can define the Ninomiya–Victoir splitting essentially as in (15.31), i.e., we set $\overline{y}_0 = y_0$ and

$$\overline{y}_l = \begin{cases} e^{\frac{\Delta t_l}{2}(A+V_0)} e^{\Delta B_l^d V_d} \cdots e^{\Delta B_l^1 V_1} e^{\frac{\Delta t_l}{2}(A+V_0)} \overline{y}_{l-1}, & \text{with prob. } \frac{1}{2}, \\ e^{\frac{\Delta t_l}{2}(A+V_0)} e^{\Delta B_l^1 V_1} \cdots e^{\Delta B_l^d V_d} e^{\frac{\Delta t_l}{2}(A+V_0)} \overline{y}_{l-1}, & \text{else}, \end{cases} \tag{15.37}$$

$l = 1, \ldots, N$.

We formulate assumptions given in [23], which can, in fact, be weakened using suitably weighted spaces.

Assumption 4. *Consider the coefficients $A, V, V_1, \ldots, V_d$ of the SPDE* (15.36) *and a function $f : X \to \mathbb{R}$. We assume that*

- *$A : \mathscr{D}(A) \subset X \to X$ generates a strongly continuous, pseudo-contractive semi-group on X and has a compact resolvent.*
- *$V, V_1, \ldots, V_d \in C^6(X, X)$ and have bounded derivatives.*
- *$V, V_1, \ldots, V_d$ are Lipschitz when considered as maps $\mathscr{D}(A^l) \to \mathscr{D}(A^l)$, $l = 1, \ldots, 5$, where $\mathscr{D}(A^l)$ is equipped with the graph norm, i.e., the Hilbert norm given by $\|x\|^2_{\mathscr{D}(A^l)} \equiv \|x\|^2_X + \sum_{k=1}^l \|A^k x\|^2_X$.*
- *$f \in C_b^6(X)$.*

Theorem 9 ([23, Cor. 7.11, Th. 7.20]). *Under Assumption 4, there is a constant C such that*

$$\left| E\left[f(y_T)\right] - E\left[f(\overline{y}_N)\right] \right| \le C \left(\max_{i=1,\ldots,N} \Delta t_i \right)^2.$$

Remark 8. The theorem can also be re-formulated completely deterministically in the fashion of (15.35), i.e., as a deterministic splitting method for a PDE on an infinite-dimensional state space. This is the version actually given in [23].

7 Applications of Weak Schemes in Financial Engineering

Given a financial model of the form (15.26), where y_t could be a (one- or multi-dimensional) vector of asset (forward) prices, or a vector of asset prices and stochastic volatilities, or the individual factors of a multi-factor model, ... Disregarding financial technicalities (discounting, change to the pricing measure), we are concerned in computing a European option price with payoff function $f : \mathbb{R}^e \to \mathbb{R}$ and maturity T, i.e., our quantity of interest is

$$E\left[f(y_T)\right].$$

This simple option pricing problem is mostly relevant for *calibration* purposes, i.e., for identifying the model parameters which provide the best fit to the observed market prices. Hence, the speed of the pricing algorithm is extremely important for this application — more important than accuracy, due to the usually non-negligible model error.[5] In this section we give an overview of some successful applications of weak stochastic splitting methods in this context. We begin with two numerical studies on the performance of the Ninomiya–Victoir scheme for two popular (finite-dimensional) models often used in financial engineering — the SABR model and the CEV model, a special case of the former [2, 4]. Then we report the performance of the Ninomiya–Victoir method in an actual calibration routine for yet another related model, the double-mean-reverting model, [3]. Finally, we present the performance of the weak stochastic splitting method for SPDEs, again in the context of a calibration problem, this time for an interest rate model, the Heath–Jarrow–Morton model, see [24] and [40].

Option Pricing in High Dimensions

The SABR model is a prominent example of a stochastic interest rate model. We consider the following generalizations of the classical SABR model (cf. [2]).

$$\begin{aligned} dy_1(t) &= a y_2(t)^{\alpha} y_1(t)^{\beta} dB_t^1, \\ dy_2(t) &= \kappa(\theta - y_2(t))dt + b y_2(t) \left(\rho dB_t^1 + \sqrt{1-\rho^2} dB_t^2 \right), \end{aligned} \tag{15.38}$$

with $X_1(0) = x_1$ and $X_2(0) = x_2$. We assume that the parameters satisfy $\frac{1}{2} \le \beta \le 1$, $\theta, \kappa \ge 0$, $\alpha > 0$, $a, b > 0$, $-1 < \rho < 1$. Here, the first component $y_1(t)$ models the (discounted) price of a stock, and $y_2(t)$ can be interpreted as some kind of stochastic volatility. In fact, the dynamics of y_1 depend in a nonlinear way on y_1 (*local volatility*) and on a second stochastic process y_2 (*stochastic volatility*). Hence, models of this kind are known as stochastic local volatility models. Moreover, note that this model can be easily generalized to the multi-asset case by just adding new processes y with the same kind of dynamics, but driven by correlated Brownian motions for every new asset to be included in the model, see [2] for more details. We concentrate on the one-asset case for ease of presentation.

For the SABR model (15.38), the Stratonovich drift vector field and the diffusion vector fields are given by

$$V_0(y) = \begin{pmatrix} -\frac{1}{2} a^2 \beta y_2^{2\alpha} y_1^{2\beta-1} - \frac{1}{2} \alpha a b \rho y_2^{\alpha} y_1^{\beta} \\ \kappa\theta - \left(\kappa + \frac{1}{2} b^2\right) y_2 \end{pmatrix}, \quad V_1(y) = \begin{pmatrix} a y_2^{\alpha} y_1^{\beta} \\ b \rho y_2 \end{pmatrix}, \quad V_2(y) = \begin{pmatrix} 0 \\ b\sqrt{1-\rho^2} y_2 \end{pmatrix}. \tag{15.39}$$

[5] That is, even without any numerical error, it is generally not possible to obtain a perfect fit to market prices, due to the model limitations.

Quite typically for models in financial engineering, the Stratonovich drift vector field is considerably more complicated than the Ito drift vector field or the diffusion vector fields which can be seen as a consequence of the Stratonovich correction $V_0 = V - \frac{1}{2}\sum_{i=1}^{d} DV_i \cdot V_i$, noting that models in financial engineering are typically formulated in Ito form. As a consequence, it is not surprising that we have explicit formulas for the flow of the diffusion vector field, but not for the flow of the Stratonovich vector field V_0. In fact, we have

$$e^{sV_1}y = \begin{pmatrix} g_1(s) \\ y_2 \exp(b\rho s) \end{pmatrix}, \quad e^{sV_2}y = \begin{pmatrix} y_1 \\ y_2 \exp\left(b\sqrt{1-\rho^2}s\right) \end{pmatrix},$$

where

$$g_1(s) = \begin{cases} \left[(1-\beta)ay_2^{\alpha}\frac{e^{\alpha b\rho s}-1}{\alpha b\rho} + y_1^{1-\beta}\right]_+^{1/(1-\beta)}, & \frac{1}{2} \le \beta < 1, \\ y_1 \exp\left(ax_2^{\alpha}\frac{e^{\alpha b\rho s}-1}{\alpha b\rho}\right), & \beta = 1. \end{cases}$$

Of course, it is possible to compute $e^{sV_0}y$ numerically, but for efficiency (and often also for geometrical reasons) it is preferable to have explicit solutions whenever possible.[6] We therefore propose to slightly adjust the Ninomiya–Victoir splitting formula, taking the Stratonovich drift correction into account. That means, we replace the splitting $L = V_0 + \frac{1}{2}\sum_{i=1}^{d} V_i^2$ by the splitting

$$L = V_0^{(\gamma)} + \frac{1}{2}\sum_{i=1}^{d}\left(V_i^2 + 2\gamma_i V_i\right), \text{ where } V_0^{(\gamma)} = V_0 - \sum_{i=1}^{d}\gamma_i V_i,$$

and $\gamma \in \mathbb{R}^d$ is chosen such that the flow of $V_0^{(\gamma)}$ has an explicit solution. Note that $\frac{1}{2}V_i^2 + \gamma_i V_i$ corresponds to the stochastic equation

$$dz_t = \gamma_i V_i(z_t)dt + V_i(z_t) \circ dB_t^i = V_i(z_t) \circ d\left(B_t^i + \gamma_i t\right),$$

i.e., the stochastic weak splitting scheme actually looks exactly like the standard Ninomiya–Victoir scheme (15.31), but with ΔB_l^i replaced by $\Delta B_l^i + \gamma_i \Delta t_l$ and V_0 replaced by $V_0^{(\gamma)}$:

$$\bar{y}_l = \begin{cases} e^{\frac{\Delta t_l}{2}V_0^{(\gamma)}} e^{(\Delta B_l^d + \gamma_d \Delta t_l)V_d} \cdots e^{(\Delta B_l^1 + \gamma_1 \Delta t_l)V_1} e^{\frac{\Delta t_l}{2}V_0^{(\gamma)}} \bar{y}_{l-1}, & \text{with prob. } \frac{1}{2}, \\ e^{\frac{\Delta t_l}{2}V_0^{(\gamma)}} e^{(\Delta B_l^1 + \gamma_1 \Delta t_l)V_1} \cdots e^{(\Delta B_l^d + \gamma_d \Delta t_l)V_d} e^{\frac{\Delta t_l}{2}V_0^{(\gamma)}} \bar{y}_{l-1}, & \text{else}, \end{cases} \tag{15.40}$$

see [2]. As a matter of fact, we can easily find such a choice of γ for the generalized SABR model given by

$$\gamma_1 = -\frac{1}{2}\alpha b\rho, \quad \gamma_2 = \frac{\alpha b\rho^2 - 2\kappa/b - b}{2\sqrt{1-\rho^2}},$$

---

[6] By which we do not mean difficult-to-evaluate series expansions, Bessel functions or similar solutions. Instead, we mean formulas with comparable complexity to the vector fields themselves.

leading to

$$V_0^{(\gamma)}(y) = \begin{pmatrix} -\frac{1}{2}a^2\beta y_2^{2\alpha} y_1^{2\beta-1} \\ \kappa\theta \end{pmatrix}, \quad e^{sV_0^{(\gamma)}} y = \begin{pmatrix} g_0(s;y) \\ \kappa\theta s + y_2 \end{pmatrix},$$

with

$$g_0(s;y) = \begin{cases} \left(-a^2\beta(1-\beta)P(s;y) + y_1^{2(1-\beta)}\right)_+^{\frac{1}{2(1-\beta)}}, & \frac{1}{2} < \beta < 1, \\ y_1 \exp\left(-\frac{1}{2}a^2 P(s;y)\right), & \beta = 1, \\ -\frac{1}{4}a^2 P(s;y) + y_1, & \beta = \frac{1}{2}, \end{cases}$$

and

$$P(s;y) = \frac{1}{(2\alpha+1)\kappa\theta}\left((\kappa\theta s + y_2)^{2\alpha+1} - y_2^{2\alpha+1}\right).$$

Finally, let us present the results from one of the numerical experiments in [2] with real-world data. The parameters there were chosen to be $\beta = 1.0$, $\theta = 0.3$, $\kappa = 2.0$, $\alpha = 0.5$, $a = 1.0$, $b = 0.5$, $\rho = -0.7$, $y_1 = 1.0$, and $y_2 = 0.2$. The option has strike price $K = 1.05$ and time to maturity $T = 1.0$ years. The estimated "true result" is 0.1767505855. Note that the expectation is computed by quasi Monte Carlo based on Sobol numbers.

In Figure 15.4 the discretization error is plotted against the number of time steps for three different schemes: the standard Euler scheme, the classical Ninomiya–Victoir splitting (15.31) (where e^{sV_0} is computed by a standard second order Taylor expansion) and our adjusted Ninomiya–Victoir scheme (15.40). We clearly see the second order weak convergence of the Ninomiya–Victoir scheme in both variants as compared to the first order weak convergence of the standard Euler scheme.

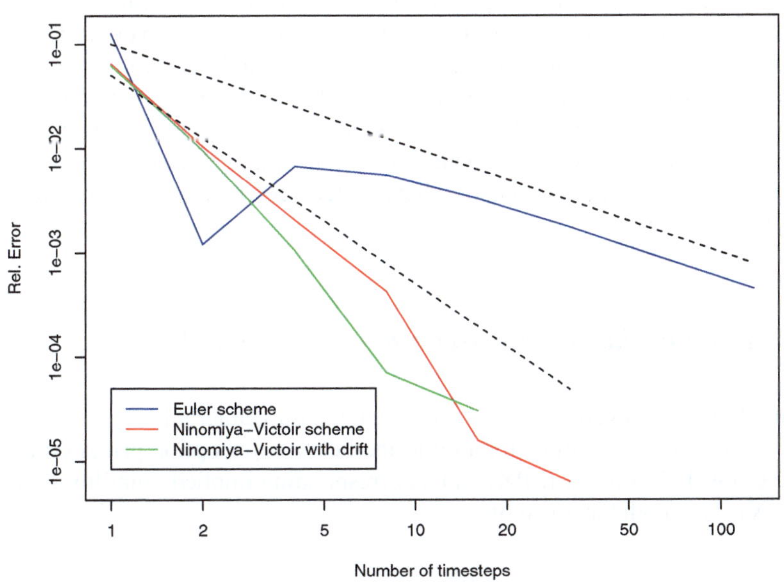

Fig. 15.4 Order of convergence for the generalized SABR model. Figure from [2].

Note that the error of the Euler scheme changes its sign at around two time steps, which explains the visible kink of the error of the Euler scheme in Figure 15.4. We consider this a numerical artifact which we disregard for the comparison.

Method	K	M	Rel. Error	Time
Euler	32	8192000	0.00174	91.94 sec
Ninomiya–Victoir	4	2048000	0.00204	13.93 sec
NV with drift	4	1024000	0.00104	2.88 sec

Table 15.1 Computational time for the generalized SABR model

In Table 15.1 the computational times are reported for the generalized SABR model. Here, the computational parameter N (the number of uniform time-steps) is chosen such that the weak error is of order 10^{-3}. We see that the computational time needed for the adjusted splitting method (15.40) is indeed considerably smaller than the time for the classical one (15.31). The other numerical parameter (the number M of samples for the quasi Monte Carlo integration, restricted to be a power of 2) was chosen such that the integration error (i.e., the error in the computation of the expected value) is of order 10^{-5}. Indeed, we focus on the discretization problem here, and we do not want our results to be overshadowed by the integration error. Note that in the case of the Euler scheme, one has to compute a 64-dimensional integral, whereas in the case of the Ninomiya–Victoir scheme (with or without drift), the integration only needs to be performed on an eight-dimensional space. This explains why the M can be chosen smaller for the Ninomiya–Victoir splitting as compared to the Euler scheme, as quasi Monte Carlo is known to work better when the dimension is smaller – despite not suffering from the curse of dimensionality.

In [2] similar results were reported for the multi-asset case. More precisely, the authors of [2] also applied it to the case of four assets, meaning an eight-dimensional model. But, in fact, the stochastic splitting method can be used in even higher dimensions. For instance, we used it in [4] in order to obtain reference solutions for options depending on up to 100 assets for a pure local volatility model, coupled with quasi Monte Carlo or Monte Carlo methods for the integration step. In that case, it is difficult if not impossible to obtain reliable reference values, but the method seems to perform very well.

Calibration of the Double Mean Reverting Model

The double mean reverting model goes back to Jim Gatheral [34]. Its main advantage is that it allows joint calibration to market data for option prices on an index like the S & P 500 index (SPX) and a corresponding implied volatility index (like the VIX). The model is given by

$$dS_t = \sqrt{v_t} S_t dW_t^1, \tag{15.41a}$$

$$dv_t = \kappa_1 (v_t' - v_t)dt + \xi_1 v_t^{\alpha_1} dW_t^2, \tag{15.41b}$$

$$dv_t' = \kappa_2 (\theta - v_t')dt + \xi_2 v_t'^{\alpha_2} dW_t^3, \tag{15.41c}$$

where the Brownian motions W_i are all correlated with $E[dW_t^i dW_t^j] = \rho_{ij} dt$. Again, it is natural to interpret v_t as the (stochastic) volatility of the (discounted) asset price process S_t – or rather, as the stochastic variance. Conforming to stylized facts about the volatility, v_t is a mean-reverting process due to the form of the drift, but unlike traditional stochastic volatility models, the long-term mean v_t' is itself a (mean-reverting) stochastic process, hence the name "double mean reverting" model.

Typically, one of the most numerically challenging tasks in financial engineering is the calibration of a model such as (15.41), i.e., the fitting of the model parameters (κ_1, κ_2, ξ_1, ξ_2, α_1, α_2, ρ_{12}, ρ_{13}, ρ_{23}, but also v_0 and v_0' which, unlike S_0, are not directly observable) to market data, in particular to market option prices. Indeed, even though the model itself assumes these parameters to be constant, in reality they typically change frequently, which means that the model has to be *re-calibrated* on a regular bases, say daily.

In the case of the double mean reverting model, practical experience seems to show that θ, κ_1, κ_2, ρ_{23} α_1 and α_2 are fairly constant in time, implying that they can be excluded from the daily re-calibration. In fact, [3] found that the data available did not suffice to successfully estimate α_2. Hence, it was *assumed* to have the same value as α_1, which was calibrated to $\alpha_1 = 0.94$. Hence, for the purpose of their numerical study, [3] assumed θ, κ_1, κ_2, ρ_{23} α_1 and α_2 to be given (by parameters which where themselves, of course, calibrated to the market data) – leaving us with the task of fast calibration of the parameters ξ_1, ξ_2, ρ_{12} and ρ_{13}. In the context of [3], "market data" mean the prices of vanilla (i.e., European put and call options) on SPX and on VIX. The general calibration procedure proposed was the following:

1. Given a time series of VIX data, linear regression allows to construct time series for the processes v_t and v_t'. Out of these, a least-square optimization is used to estimate θ, κ_1 and κ_2. Moreover, the correlation between v_t and v_t' gives ρ_{23}. A similar regression on VIX time series data gives an estimate for α_1.
2. Note that options on VIX depend only on v_t and v_t' now, but not on S_t. Hence, one can calibrate ξ_1 and ξ_2 directly to VIX options, without needing to simulate the S_t component, i.e., without adding any constraints to ρ_{12} and ρ_{13}. The calibration boils down to a least-squares minimization of misfits of VIX-option prices from the model to the quoted market prices. The minimization was done using a Levenberg-Marquardt algorithm, for the option pricing algorithm the authors of [3] tested the Euler scheme and a variant of the Ninomiya–Victoir scheme for the discretization of the SDE and both classical and quasi Monte Carlo for the computation of the expected value.
3. Having obtained ξ_1 and ξ_2 from the previous step, they used options on the SPX to calibrate the remaining parameters ρ_{12} and ρ_{13}. The procedure is very similar to the calibration of ξ_1 and ξ_2, except that now the full three-dimensional SDE needs to be solved.

In [3] the calibration was done for two particular days, namely April 3, 2007 (before the financial crisis) and September 15, 2011 (after the financial crisis). The fits to SPX options are quite good, especially for maturities which are not too small. The fit to VIX options is slightly worse, but in that time VIX options were also less liquid than today. Regarding the numerical algorithms, the Ninomiya–Victoir splitting method (with an additional splitting in the drift, not unlike the one presented in [2]) performs much better for the calculation of VIX options, where the classical Euler method would requires 500 time-steps as compared to 30 time-steps for the splitting method in order to achieve the required accuracy. Thereby, for this example, the Ninomiya–Victoir scheme reduces the calibration time for the VIX-step by a factor of around five. For the SPX options, the Euler scheme surprisingly gave sufficiently accurate results already for 30 time-steps, which implies that for this case the Euler scheme turned out to be seemingly preferable to the Ninomiya–Victoir scheme, taking costs into account. Note, however, that the error plots in Figure 15.5 reveal that this conclusion is deceptive, as the weak approximation error changes its sign around the critical number of 30 time-steps. Even though this effect was persistent in both for both market data (and the accordingly calibrated parameters), the picture might change in other regime, leading to added benefits for applying the Ninomiya–Victoir splitting scheme also for the SPX-data calibration step. In total, the authors of [3] report that their implementation can do the re-calibration to market data in about 5 seconds using the Ninomiya–Victoir splitting scheme.

Calibration of the Heath–Jarrow–Morton Model

Finally, we want to present an application of the Ninomiya–Victoir weak splitting method to a true SPDE given by Dörsek and Teichmann [24], namely the fast calibration of a general, infinite-dimensional Heath–Jarrow–Morton model for interest

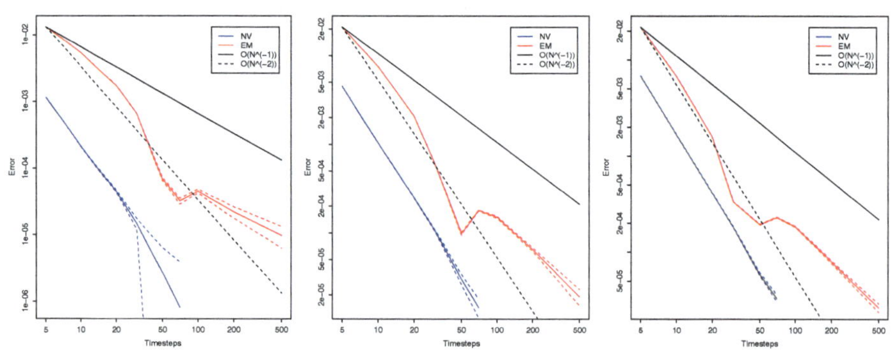

Fig. 15.5 Errors for the Euler and the Ninomiya Victoir schemes for three different SPX options. (Data from September 15, 2011, dotted lines indicate Monte Carlo confidence intervals. Figure taken from [3].)

rate dynamics, see [40]. We should also note alternative numerical treatments of the full, infinite dimensional HJM model in [9] and [58].

We start with a short description of the model. Let $P(t,T)$ denote the price of a zero-coupon bond with maturity T at time t. Of course, this entity only makes sense if $t \leq T$. The (instantaneous) *forward rate* at time t for the maturity T is defined by

$$f(t,T) = -\frac{\partial}{\partial T} \log P(t,T),$$

implying the natural inverse relation $P(t,T) = \exp\left(-\int_t^T f(t,u)du\right)$, which explains why we call f an interest rate. Unlike many other models, in which only the short rate $f(t,t)$ or only the rates $f(t,T_1),\ldots,f(t,T_n)$ for finitely many maturities are modeled, [40] propose an infinite-dimensional model for the whole forward rate curve $(f(t,T))_{T\in[t,\infty[}$. In order to avoid working with time-dependent state spaces, we introduce the parametrization $r_t(x) = f(t,t+x)$ in *time to maturity* $x = T-t \geq 0$. Then the HJM model corresponds to the SPDE

$$dr_t = \left(\frac{\partial}{\partial x} r_t + \alpha_{HJM}(r_t)\right) dt + \sum_{i=1}^{d} \sigma_i(r_t) dB_t^i. \tag{15.42}$$

Here, we restrict ourselves to a finite number d of driving Brownian motions, which can be justified empirically, but is not strictly necessary for the HJM model. Moreover, we note that there are inherent restrictions on the vector fields imposed by no-arbitrage constraints, which boil down to the relation

$$\alpha_{HJM}(h)(x) = \sum_{i=1}^{d} \sigma_i(h)(x) \int_0^x \sigma_i(h)(y)dy,$$

where $x \geq 0$ and h takes values in the state space H, a suitably weighted Sobolev space, see [24] for details on H and on further regularity requirements on σ_i.

Next, we describe the splitting. Note that the diffusion vector fields do not pose any additional complications as compared to the finite dimensional case, as they are (assumed to be) continuous vector fields on H. This is evidently not the case for the Stratonovich drift vector field $\sigma_0(h) = \frac{\partial}{\partial x} h + \alpha_{HJM}(h) - \frac{1}{2} D\sigma_i(h) \cdot \sigma_i(h)$, where the unbounded operator $\frac{\partial}{\partial x}$ appears. (Recall that the solution r_t cannot, in fact, be written in Stratonovich form.) Hence, they additionally split $\sigma_0 = \sigma_{0,1} + \sigma_{0,2}$ with $\sigma_{0,1} = \frac{\partial}{\partial x}$ and $\sigma_{0,2} = \alpha_{HJM} - \frac{1}{2} D\sigma_i \cdot \sigma_i$. Here we note that the flow of $\sigma_{0,1}$ is obviously given by the shift operator $S_t(h)(x) = h(x+t)$, so that the $e^{s\sigma_{0,1}} = S_s$ is given in closed form. Regarding the diffusion vector fields, the authors suggest to use the parametric form

$$\sigma_j(h,v)(x) = \left(\alpha_{j,0} + \alpha_{j,1} x\right) e^{-\beta x} \tanh\left(c_j e^v \int_0^{t_j} h(s)ds\right),$$

which includes a stochastic volatility component v, and $\alpha_{j,i}$, β, c_j, and t_j are parameters, which need to be estimated. Moreover, they choose $d = 3$.

The authors of [24] calibrate against 120 market prices of caplets, again using a Levenberg-Marquardt type algorithm for the optimization. In total, they report that it takes about half a second to compute these 120 option prices to the required accuracy (on a workstation with 16 cores), and the total calibration can be done in 14.5 minutes.

Acknowledgements Harald Oberhauser is grateful for the support of the ERC (grant agreement No.291244 Esig) and the Oxford-Man Institute of Quantitative finance.

References

1. Bain, A., Crişan, D.: Fundamentals of stochastic filtering. Applications of mathematics. Springer (2008)
2. Bayer, C., Friz, P., Loeffen, R.: Semi-closed form cubature and applications to financial diffusion models. Quant. Finance **13**(5), 769–782 (2013). DOI 10.1080/14697688.2012.752102. URL http://dx.doi.org/10.1080/14697688.2012.752102
3. Bayer, C., Gatheral, J., Karlsmark, M.: Fast Ninomiya-Victoir calibration of the double-mean-reverting model. Quantitative Finance **13**(11), 1813–1829 (2013)
4. Bayer, C., Laurence, P.: Asymptotics beats Monte Carlo: The case of correlated CEV baskets. Comm. Pure Appl. Math. **67**(10), 1618–1657 (2014)
5. Bayer, C., Teichmann, J.: Cubature on Wiener space in infinite dimension. Proc. R. Soc. Lond. Ser. A Math. Phys. Eng. Sci. **464**(2097), 2493–2516 (2008). DOI 10.1098/rspa.2008.0013. URL http://dx.doi.org/10.1098/rspa.2008.0013
6. Bensoussan, A., Glowinski, R.: Approximation of Zakai equation by the splitting up method. In: Stochastic systems and optimization (Warsaw, 1988), *Lecture Notes in Control and Inform. Sci.*, vol. 136, pp. 257–265. Springer, Berlin (1989)
7. Bensoussan, A., Glowinski, R., Răşcanu, A.: Approximation of the Zakai equation by the splitting up method. SIAM J. Control Optim. **28**(6), 1420–1431 (1990). DOI 10.1137/0328074. URL http://dx.doi.org/10.1137/0328074
8. Bensoussan, A., Glowinski, R., Răşcanu, A.: Approximation of some stochastic differential equations by the splitting up method. Appl. Math. Optim. **25**(1), 81–106 (1992). DOI 10.1007/BF01184157. URL http://dx.doi.org/10.1007/BF01184157
9. Björk, T., Szepessy, A., Tempone, R., Zouraris, G.E.: Monte Carlo Euler approximations of HJM term structure financial models. BIT **53**(2), 341–383 (2013)
10. Brigo, D., Hanzon, B.: On some filtering problems arising in mathematical finance. Insurance: Mathematics and Economics **22**(1), 53–64 (1998)
11. Brockett, R.W.: Volterra series and geometric control theory. Automatica **12**(2), 167–176 (1976). DOI 10.1016/0005-1098(76)90080-7. URL http://www.sciencedirect.com/science/article/pii/0005109876900807
12. Caruana, M., Friz, P., Oberhauser, H.: A (rough) pathwise approach to a class of non-linear stochastic partial differential equations. Ann. Inst. H. Poincaré Anal. Non Linéaire **28**(1), 27–46 (2011). DOI 10.1016/j.anihpc.2010.11.002. URL http://dx.doi.org/10.1016/j.anihpc.2010.11.002
13. Clark, J.: The design of robust approximations to stochastic differential equations in nonlinear filtering. In: J. Skwirzynsky (ed.) Communication Systems in Random Processes Theory. Sijthoff, Nordhoff (1978)
14. Clark, J., Crisan, D.: On a robust version of the integral representation formula of nonlinear filtering. Probability theory and related fields **133**(1), 43–56 (2005)
15. Crandall, M.G., Ishii, H., Lions, P.L.: User's guide to viscosity solutions of second order partial differential equations. Bull. Amer. Math. Soc. (N.S.) **27**(1), 1–67 (1992)

16. Crisan, D., Diehl, J., Friz, P. K., Oberhauser, H.: Robust filtering: correlated noise and multidimensional observation. The Annals of Applied Probability **23**(5), 2139–2160 (2013)
17. Crisan, D., Rozovskii, B.: The Oxford Handbook of Nonlinear Filtering. Oxford Handbooks. Oxford University Press (2011). URL http://books.google.com/books?id=XYP_3szwkIoC
18. Csomós, P., Faragó, I., Havasi, Á.: Weighted sequential splittings and their analysis. Comput. Math. Appl. **50**(7), 1017–1031 (2005). DOI 10.1016/j.camwa.2005.08.004. URL http://dx.doi.org/10.1016/j.camwa.2005.08.004
19. Da Prato, G., Zabczyk, J.: Stochastic equations in infinite dimensions, *Encyclopedia of Mathematics and its Applications*, vol. 44. Cambridge University Press, Cambridge (1992). DOI 10.1017/CBO9780511666223. URL http://dx.doi.org/10.1017/CBO9780511666223
20. Debussche, A.: Weak approximation of stochastic partial differential equations: the nonlinear case. Math. Comp. **80**(273), 89–117 (2011). DOI 10.1090/S0025-5718-2010-02395-6. URL http://dx.doi.org/10.1090/S0025-5718-2010-02395-6
21. Diehl, J., Friz, P., Gassiat, P.: Stochastic control with rough paths. P. Appl. Math. Optim. (2016). URL http://dx.doi.org/10.1007/s00245-016-9333-9
22. Diehl, J., Oberhauser, H., Riedel, S.: A Lévy area between Brownian motion and rough paths with applications to robust nonlinear filtering and rough partial differential equations. Stoch. Process. Appl. **125**(1), 161–181 (2015)
23. Dörsek, P., Teichmann, J.: A semi-group point of view on splitting schemes for stochastic (partial) differential equations (2010). Preprint
24. Dörsek, P., Teichmann, J.: Efficient simulation and calibration of general HJM models by splitting schemes. SIAM J. Financial Math. **4**(1), 575–598 (2013). DOI 10.1137/110860173. URL http://dx.doi.org/10.1137/110860173
25. Elliott, R.J., Glowinski, R.: Approximations to solutions of the Zakai filtering equation. Stochastic Analysis and Applications **7**(2), 145–168 (1989)
26. Fleming, W.H., Soner, H.M.: Controlled Markov processes and viscosity solutions, *Stochastic Modelling and Applied Probability*, vol. 25, second edn. Springer, New York (2006)
27. Fliess, M.: Fonctionnelles causales non linéaires et indéterminées non commutatives. Bull. Soc. Math. France **109**(1), 3–40 (1981)
28. Florchinger, P., Le Gland, F.: Time-discretization of the Zakai equation for diffusion processes observed in correlated noise. Stochastics Stochastics Rep. **35**(4), 233–256 (1991)
29. Frey, R., Runggaldier, W.J.: A nonlinear filtering approach to volatility estimation with a view towards high frequency data. International Journal of Theoretical and Applied Finance **4**(02), 199–210 (2001)
30. Friz, P., Hairer, M.: A Course on Rough Paths: With an Introduction to Regularity Structures. Universitext. Springer International Publishing (2014)
31. Friz, P., Oberhauser, H.: On the splitting-up method for rough (partial) differential equations. Journal of Differential Equations **251**(2), 316–338 (2011). DOI 10.1016/j.jde.2011.02.009. URL http://www.sciencedirect.com/science/article/pii/S0022039611000763
32. Friz, P., Oberhauser, H.: Rough path stability of (semi-)linear SPDEs. Probability Theory and Related Fields pp. 1–34 (2013). DOI 10.1007/s00440-013-0483-2. URL http://dx.doi.org/10.1007/s00440-013-0483-2
33. Friz, P.K., Victoir, N.B.: Multidimensional stochastic processes as rough paths: theory and applications. Cambridge Studies in Advanced Mathematics. Cambridge University Press, Cambridge (2010)
34. Gatheral, J.: Consistent modeling of SPX and VIX options (2008). Presentation at Bachelier Congress: London
35. Gençay, R., Selçuk, F., Whitcher, B.J.: An introduction to wavelets and other filtering methods in finance and economics. Access Online via Elsevier (2001)
36. Gubinelli, M.: Controlling rough paths. J. Funct. Anal. **216**(1), 86–140 (2004)
37. Gyöngy, I., Krylov, N.: On the splitting-up method and stochastic partial differential equations. Ann. Probab. **31**(2), 564–591 (2003). DOI 10.1214/aop/1048516528
38. Gyöngy, I., Krylov, N.: Accelerated numerical schemes for PDEs and SPDEs. In: Stochastic analysis 2010, pp. 131–168. Springer, Heidelberg (2011)

39. Hausenblas, E.: Approximation for semilinear stochastic evolution equations. Potential Anal. **18**(2), 141–186 (2003). DOI 10.1023/A:1020552804087. URL http://dx.doi.org/10.1023/A:1020552804087
40. Heath, D., Jarrow, R., Morton, A.: Bond pricing and the term structure of interest rates: A new methodology for contingent claims valuation. Econometrica **60**(1), 77–105 (1992)
41. Hida, T.: Brownian motion. Springer (1980)
42. Ikeda, N., Watanabe, S.: Stochastic differential equations and diffusion processes, second edn. North-Holland Publishing Co., Amsterdam (1989)
43. Jentzen, A., Kloeden, P.E.: The numerical approximation of stochastic partial differential equations. Milan J. Math. **77**, 205–244 (2009). DOI 10.1007/s00032-009-0100-0. URL http://dx.doi.org/10.1007/s00032-009-0100-0
44. Karatzas, I., Shreve, S.E.: Brownian motion and stochastic calculus, *Graduate Texts in Mathematics*, vol. 113, second edn. Springer-Verlag, New York (1991)
45. Kloeden, P.E., Platen, E.: Numerical solution of stochastic differential equations, *Applications of Mathematics (New York)*, vol. 23. Springer-Verlag, Berlin (1992)
46. Kohatsu-Higa, A., Tankov, P.: Jump-adapted discretization schemes for Lévy-driven SDEs. Stochastic Process. Appl. **120**(11), 2258–2285 (2010). DOI 10.1016/j.spa.2010.07.001. URL http://dx.doi.org/10.1016/j.spa.2010.07.001
47. Kusuoka, S.: Approximation of expectation of diffusion processes based on Lie algebra and Malliavin calculus. In: Advances in mathematical economics. Vol. 6, *Adv. Math. Econ.*, vol. 6, pp. 69–83. Springer, Tokyo (2004)
48. Le Gland, F.: Splitting-up approximation for SPDEs and SDEs with application to nonlinear filtering. In: Stochastic partial differential equations and their applications (Charlotte, NC, 1991), *Lecture Notes in Control and Inform. Sci.*, vol. 176, pp. 177–187. Springer, Berlin (1992). DOI 10.1007/BFb0007332. URL http://dx.doi.org/10.1007/BFb0007332
49. Lototsky, S., Mikulevicius, R., Rozovskii, B.L.: Nonlinear filtering revisited: a spectral approach. SIAM J. Control Optim. **35**(2), 435–461 (1997). DOI 10.1137/S0363012993248918. URL http://dx.doi.org/10.1137/S0363012993248918
50. Lyons, T.: Differential equations driven by rough signals. I. An extension of an inequality of L. C. Young. Math. Res. Lett. **1**(4), 451–464 (1994)
51. Lyons, T.: Differential equations driven by rough signals. Rev. Mat. Iberoamericana **14**(2), 215–310 (1998)
52. Lyons, T.: Systems controlled by rough paths. In: European Congress of Mathematics, pp. 269–281. Eur. Math. Soc., Zürich (2005)
53. Lyons, T., Qian, Z.: System Control and Rough Paths. Oxford University Press (2002). Oxford Mathematical Monographs
54. Lyons, T., Victoir, N.: Cubature on Wiener space. Proc. R. Soc. Lond. Ser. A Math. Phys. Eng. Sci. **460**(2041), 169–198 (2004). DOI 10.1098/rspa.2003.1239. URL http://dx.doi.org/10.1098/rspa.2003.1239
55. Lyons, T.J., Caruana, M., Lévy, T.: Differential equations driven by rough paths, *Lecture Notes in Mathematics*, vol. 1908. Springer, Berlin (2007). Lectures from the 34th Summer School on Probability Theory held in Saint-Flour, July 6–24, 2004, With an introduction concerning the Summer School by Jean Picard
56. Mandelbrot, B.B., Van Ness, J.W.: Fractional Brownian motions, fractional noises and applications. SIAM review **10**(4), 422–437 (1968)
57. Nagase, N.: Remarks on nonlinear stochastic partial differential equations: an application of the splitting-up method. SIAM J. Control Optim. **33**(6), 1716–1730 (1995). DOI 10.1137/S036301299324618X. URL http://dx.doi.org/10.1137/S036301299324618X
58. Ninomiya, M.: Application of the Kusuoka approximation with a tree-based branching algorithm to the pricing of interest-rate derivatives under the HJM model. LMS J. Comput. Math. **13**, 208–221 (2010). DOI 10.1112/S146115700800048X. URL http://dx.doi.org/10.1112/S146115700800048X
59. Ninomiya, S., Victoir, N.: Weak approximation of stochastic differential equations and application to derivative pricing. Appl. Math. Finance **15**(1-2), 107–121 (2008). DOI 10.1080/13504860701413958. URL http://dx.doi.org/10.1080/13504860701413958

60. Øksendal, B.: Stochastic differential equations. Springer (2003)
61. Protter, P.E.: Stochastic integration and differential equations, *Applications of Mathematics (New York)*, vol. 21, second edn. Springer-Verlag, Berlin (2004). Stochastic Modelling and Applied Probability
62. Protter, P.E.: Stochastic integration and differential equations, *Stochastic Modelling and Applied Probability*, vol. 21. Springer-Verlag, Berlin (2005). Second edition. Version 2.1, Corrected third printing
63. Răşcanu, A., Tudor, C.: Approximation of stochastic equations by the splitting up method. In: Qualitative problems for differential equations and control theory, pp. 277–287. World Sci. Publ., River Edge, NJ (1995)
64. Revuz, D., Yor, M.: Continuous martingales and Brownian motion, *Grundlehren der Mathematischen Wissenschaften [Fundamental Principles of Mathematical Sciences]*, vol. 293, third edn. Springer-Verlag, Berlin (1999)
65. Schwab, C., Gittelson, C.J.: Sparse tensor discretizations of high-dimensional parametric and stochastic PDEs. Acta Numer. **20**, 291–467 (2011). DOI 10.1017/S0962492911000055. URL http://dx.doi.org/10.1017/S0962492911000055
66. Sun, M., Glowinski, R.: Pathwise approximation and simulation for the Zakai filtering equation through operator splitting. Calcolo **30**(3), 219–239 (1994) (1993). DOI 10.1007/BF02575854. URL http://dx.doi.org/10.1007/BF02575854
67. Sussmann, H.: Semigroup representations, bilinear approximation of input-output maps, and generalized inputs. Mathematical systems theory **131** (1975)
68. Teichmann, J.: Another approach to some rough and stochastic partial differential equations. Stochastics and Dynamics **11**(02n03), 535–550 (2011)
69. Yan, Y.: Galerkin finite element methods for stochastic parabolic partial differential equations. SIAM J. Numer. Anal. **43**(4), 1363–1384 (electronic) (2005). DOI 10.1137/040605278. URL http://dx.doi.org/10.1137/040605278

Chapter 16
Application of Operator Splitting Methods in Finance

Karel in 't Hout and Jari Toivanen

Abstract Financial derivatives pricing aims to find the fair value of a financial contract on an underlying asset. Here we consider option pricing in the partial differential equations framework. The contemporary models lead to one-dimensional or multidimensional parabolic problems of the convection-diffusion type and generalizations thereof. An overview of various operator splitting methods is presented for the efficient numerical solution of these problems.

Splitting schemes of the Alternating Direction Implicit (ADI) type are discussed for multidimensional problems, e.g., given by stochastic volatility (SV) models. For jump models Implicit-Explicit (IMEX) methods are considered which efficiently treat the nonlocal jump operator. For American options an easy-to-implement operator splitting method is described for the resulting linear complementarity problems.

Numerical experiments are presented to illustrate the actual stability and convergence of the splitting schemes. Here European and American put options are considered under four asset price models: the classical Black–Scholes model, the Merton jump-diffusion model, the Heston SV model, and the Bates SV model with jumps.

K. in 't Hout (✉)
Department of Mathematics and Computer Science, University of Antwerp,
Middelheimlaan 1, B-2020 Antwerp, Belgium
e-mail: karel.inthout@uantwerp.be

J. Toivanen
Institute for Computational and Mathematical Engineering, Stanford University,
Stanford, CA 94305, USA

Department of Mathematical Information Technology, FI-40014 University
of Jyväskylä, Finland
e-mail: toivanen@stanford.edu; jari.toivanen@jyu.fi

R. Glowinski et al. (eds.), *Splitting Methods in Communication, Imaging, Science, and Engineering*, Scientific Computation, DOI 10.1007/978-3-319-41589-5_16

1 Introduction

In the contemporary international financial markets option products are widely traded. The average daily turnover in the global over-the-counter derivatives markets is huge. For example, in the foreign exchange market this was approximately equal to 337 billion US dollars in April 2013 [5]. In addition to standard call and put options, the so-called vanilla options, a broad range of exotic derivatives exists. One of the primary goals of financial mathematics is to determine the fair values of these derivatives as well as their sensitivities to underlying variables and parameters, which are crucial for hedging. To this purpose, advanced mathematical models are employed nowadays, yielding initial-boundary value problems for time-dependent partial differential equations (PDEs) and generalizations thereof, see, e.g., [4, 14, 59, 75, 77, 85]. These problems are in general multidimensional and of the convection-diffusion kind. In some cases analytical formulas in semi-closed form for the exact solutions have been obtained in the literature. For the majority of option valuation problems, however, such formulas are not available. In view of this, one resorts to numerical methods for their approximate solution. To banks and other financial institutions, the efficient, stable, and robust numerical approximation of option values and their sensitivities is of paramount importance.

A well-known and versatile approach to the numerical solution of time-dependent convection-diffusion equations is given by the *method of lines*. It consists of two general, consecutive steps. In the first step the PDE is discretized in the spatial variables, e.g., by finite difference, finite volume, or finite element methods. This leads to a so-called semidiscrete system of ordinary differential equations. In the second step the obtained semidiscrete system is numerically solved by applying a suitable, implicit time-discretization method. If the PDE is multidimensional, then the latter task can be computationally very intensive when standard application of classical implicit methods, such as the Crank–Nicolson scheme, is used. In the recent years, a variety of operator splitting methods have been developed that enable a highly efficient and stable numerical solution of semidiscretized multidimensional PDEs and generalizations thereof that arise in financial mathematics.

The aim of this chapter to give an overview of main classes of operator splitting methods with applications in finance. Here we have chosen to consider a variety of, increasingly sophisticated, models that are well known in the financial option valuation literature.

We deal in the following with two basic types of options, involving a given so-called strike price $K > 0$ and a given maturity time $T > 0$, where today is always denoted by time 0. A *European call (put) option* is a contract between two parties, the holder and the writer, which gives the holder the right to buy from (sell to) the writer a prescribed asset for the price K at the future date T. An *American call (put) option* is the same, except that the holder can exercise at any time between today and the maturity date. An option is a right and not an obligation. The underlying asset can be a stock, a foreign currency, a commodity, etc. For a detailed introduction to financial options we refer to [45]. Clearly, an option has value and a central question in financial mathematics is what its fair value is.

2 Models for Underlying Assets

2.1 Geometric Brownian Motion

The seminal papers by Black & Scholes [7] and Merton [63] present a key equation for the fair values of European call and put options. In these papers the dynamics of the underlying asset price is modeled by the stochastic differential equation (SDE)

$$dS(t) = \mu S(t)dt + \sigma S(t)dW(t) \quad (t \geq 0). \tag{16.1}$$

Here $W(t)$ denotes the Wiener process or standard Brownian motion, and μ, σ are given real parameters that are called the drift and the volatility, respectively. The volatility is a degree for the uncertainty of the return realized on the asset.

The SDE (16.1) describes a so-called geometric Brownian motion, which satisfies $S(t) \geq 0$ whenever $S(0) \geq 0$. Under this asset price model and several additional assumptions, Black, Scholes, and Merton derived the famous partial differential equation (PDE)

$$\frac{\partial u}{\partial t} = \frac{1}{2}\sigma^2 s^2 \frac{\partial^2 u}{\partial s^2} + rs\frac{\partial u}{\partial s} - ru \quad (s > 0,\ 0 < t \leq T). \tag{16.2}$$

Here $u(s,t)$ represents the fair value at time $T-t$ of a European vanilla option if $S(T-t) = s$. The quantity r in (16.2) is the risk-free interest rate and is given. A main consequence of the Black, Scholes, and Merton analysis is that the drift μ actually does not appear in the option pricing PDE. This observation has led to the important risk-neutral valuation theory. It is beyond the scope of the present chapter to discuss this theory in more detail, but see, e.g., [45, 75].

In formulating (16.2) we have chosen t as the time till maturity. Thus the time runs in the opposite direction compared to (16.1). Accordingly, the payoff function ϕ, which defines the value of the option contract at maturity time T, leads to an *initial condition*

$$u(s,0) = \phi(s) \quad (s \geq 0). \tag{16.3}$$

For a European vanilla option with given strike price K there holds

$$\phi(s) = \begin{cases} \max(s-K,0) & \text{for } s \geq 0 \text{ (call)}, \\ \max(K-s,0) & \text{for } s \geq 0 \text{ (put)}, \end{cases} \tag{16.4}$$

and at $s = 0$ one has the Dirichlet boundary condition

$$u(0,t) = \begin{cases} 0 & \text{for } 0 \leq t \leq T \text{ (call)}, \\ e^{-rt}K & \text{for } 0 \leq t \leq T \text{ (put)}. \end{cases} \tag{16.5}$$

Equation (16.2) is called the *Black–Scholes PDE* or *Black–Scholes–Merton PDE*. It is fully deterministic and it can be viewed as a time-dependent convection-diffusion-reaction equation. For European vanilla options, an analytical solution u in semi-closed form was derived in [7], constituting the well-known Black–Scholes formula.

The Black–Scholes PDE is generic in the sense that it is valid for a wide range of European-style options. The initial and boundary conditions are determined by the specific option. As an example, for a European up-and-out call option with given barrier $B > K$, the PDE (16.2) holds whenever $0 < s < B$, $0 < t \leq T$. In this case, the initial condition is

$$u(s,0) = \max(s-K,0) \quad \text{for } 0 \leq s < B$$

and one has the Dirichlet boundary conditions

$$u(0,t) = u(B,t) = 0 \quad \text{for } 0 \leq t \leq T.$$

The homogeneous condition at $s = B$ corresponds to the fact that, by construction, an up-and-out call option becomes worthless whenever the underlying asset price moves above the barrier.

For many types of options, including (continuous) barrier options, semi-analytical pricing formulas have been obtained in the literature in the Black–Scholes framework, see e.g. [45]. At present it is well known, however, that each of the assumptions underlying this framework are violated to a smaller or larger extent in practice. In particular, the interest rate r and the volatility σ are not constant, but vary in time. In view of this, more advanced asset pricing models have been developed and, as a consequence, more advanced option valuation PDEs are obtained. In this chapter we do not enter into the details of the mathematical connection between asset price SDEs and option valuation PDEs, but mention that a main tool is the celebrated Feynman–Kac theorem, see, e.g., [75]. In the following we discuss typical, contemporary instances of more advanced option valuation PDEs.

2.2 Stochastic Volatility and Stochastic Interest Rate Models

Heston [38] modeled the volatility itself by an SDE. The Heston stochastic volatility model is popular especially in the foreign exchange markets. The corresponding option valuation PDE is

$$\frac{\partial u}{\partial t} = \tfrac{1}{2}s^2 v \frac{\partial^2 u}{\partial s^2} + \rho\sigma s v \frac{\partial^2 u}{\partial s \partial v} + \tfrac{1}{2}\sigma^2 v \frac{\partial^2 u}{\partial v^2} + rs\frac{\partial u}{\partial s} + \kappa(\eta - v)\frac{\partial u}{\partial v} - ru \qquad (16.6)$$

for $s > 0$, $v > 0$, and $0 < t \leq T$. Here $u(s,v,t)$ represents the fair value of a European-style option if at t time units before maturity the asset price equals s and the variance equals v. We note that by definition the variance is the square of the volatility. The positive parameters κ and η are the mean-reversion rate and long-term mean, respectively, of the variance, $\sigma > 0$ is the volatility-of-variance, and $\rho \in [-1,1]$ denotes the correlation between the two underlying Brownian motions. Equation (16.6) is called the *Heston PDE*. It can be viewed as a time-dependent convection-diffusion-reaction equation on an unbounded, two-dimensional spatial domain. If the correlation ρ is nonzero, which almost always holds in practice, then the Heston PDE contains a mixed spatial derivative term.

For a European vanilla option under the Heston model, one has an initial condition as well as a boundary condition at $s = 0$ that are the same as in the Black–Scholes case discussed above. In the Heston case there is also a boundary $v = 0$. Observe that as $v \downarrow 0$, then all second-order derivative terms vanish in (16.6). It has been proved in [25] that for the fair option value function u the Heston PDE is fulfilled if $v = 0$, which constitutes the (nonstandard) boundary condition at $v = 0$.

For the Heston asset pricing model (which we did not explicitly formulate) the so-called Feller condition $2\kappa\eta \geq \sigma^2$ is often considered in the literature. This condition determines whether or not the variance process can attain the value zero (given a strictly positive initial variance): it cannot attain zero if and only if Feller holds. The situation where the Feller condition is violated is well-known to be challenging when numerically solving the Heston asset pricing model. For the Heston option valuation PDE (16.6), on the other hand, it turns out that this issue is not critical in the numerical solution.

A refinement of the Heston model is obtained by considering also a stochastic interest rate, see, e.g., [32, 33, 35, 36]. As an illustration we consider the case where the interest rate is described by the well-known Hull–White model [45, 46]. This leads to the following so-called *Heston–Hull–White (HHW) PDE* for the option value function $u = u(s, v, r, t)$:

$$\begin{aligned}\frac{\partial u}{\partial t} &= \tfrac{1}{2}s^2 v \frac{\partial^2 u}{\partial s^2} + \tfrac{1}{2}\sigma_1^2 v \frac{\partial^2 u}{\partial v^2} + \tfrac{1}{2}\sigma_2^2 \frac{\partial^2 u}{\partial r^2} + \rho_{12}\sigma_1 s v \frac{\partial^2 u}{\partial s \partial v} + \rho_{13}\sigma_2 s \sqrt{v} \frac{\partial^2 u}{\partial s \partial r} \\ &\quad + \rho_{23}\sigma_1\sigma_2\sqrt{v}\frac{\partial^2 u}{\partial v \partial r} + rs\frac{\partial u}{\partial s} + \kappa(\eta - v)\frac{\partial u}{\partial v} + a(b(T-t) - r)\frac{\partial u}{\partial r} - ru\end{aligned} \tag{16.7}$$

for $s > 0$, $v > 0$, $-\infty < r < \infty$, and $0 < t \leq T$. Here κ, η, σ_1, a, and σ_2 are given positive real constants and b denotes a given deterministic, positive function of time. Further, there are given correlations $\rho_{12}, \rho_{13}, \rho_{23} \in [-1, 1]$. Clearly, the HHW PDE is a time-dependent convection-diffusion-reaction equation on an unbounded, three-dimensional spatial domain with three mixed derivative terms. For a European vanilla option, initial and boundary conditions are the same as in the Heston case above. Note that if $v \downarrow 0$, then all second-order derivative terms, apart from the $\partial^2 u / \partial r^2$ term, vanish in (16.7).

The Heston and HHW models are two of many instances of asset pricing models that lead to multidimensional option valuation PDEs. Multidimensional PDEs are also obtained when considering other types of options, e.g., options on a basket of assets. Then, in the Black–Scholes framework, the dimension of the PDE is equal to the number of assets. In general, analytical solutions in (semi-)closed form to these PDEs are not available.

2.3 Jump Models

Sometimes the value of the underlying asset changes so rapidly that this would have very tiny probability under the above Brownian motion based models. For example, the stock price during a market crash or after a major news event can move very fast. Already in 1976, Merton proposed in [64] to add a jump component in the model of the underlying asset price. In his model, the jumps are log-normally distributed and their arrival times follow a Poisson process. After a jump the value of the asset is obtained by multiplying the value before the jump by a random variable with the probability density function (PDF)

$$f(y) = \frac{1}{y\delta\sqrt{2\pi}} \exp\left(-\frac{(\log y - \gamma)^2}{2\delta^2}\right) \tag{16.8}$$

for $y > 0$, where γ is the mean of the normal distribution and δ is its standard deviation. Kou proposed in [56] a log-double-exponential distribution defined by the PDF

$$f(y) = \begin{cases} q\alpha_2 y^{\alpha_2 - 1}, & 0 < y < 1, \\ p\alpha_1 y^{-\alpha_1 - 1}, & y \geq 1, \end{cases} \tag{16.9}$$

where $p, q, \alpha_1 > 1$, and α_2 are positive constants such that $p + q = 1$. These models have finite jump activity which is denoted by λ here. There are also many popular infinite jump activity models like the CGMY model [11]. In the following we shall consider only finite activity models.

The value $u(s,t)$ of a European option satisfies the partial integro-differential equation (PIDE)

$$\frac{\partial u}{\partial t} = \tfrac{1}{2}\sigma^2 s^2 \frac{\partial^2 u}{\partial s^2} + (r - \lambda\zeta)s\frac{\partial u}{\partial s} - (r + \lambda)u + \lambda \int_0^\infty u(sy,t)f(y)dy \tag{16.10}$$

for $s > 0$ and $0 < t \leq T$, where ζ is the mean jump size given by

$$\zeta = \int_0^\infty (y-1)f(y)dy. \tag{16.11}$$

For the Merton and Kou models the mean jumps are $\zeta = e^{\gamma + \delta^2/2} - 1$ and $\zeta = \frac{q\alpha_2}{\alpha_2+1} + \frac{p\alpha_1}{\alpha_1 - 1} - 1$, respectively.

Bates proposed to combine the Heston stochastic volatility model and the Merton jump model in [6]. Under this model the value $u(s,v,t)$ of a European option satisfies the PIDE

$$\begin{aligned} \frac{\partial u}{\partial t} &= \tfrac{1}{2}s^2 v \frac{\partial^2 u}{\partial s^2} + \rho\sigma s v \frac{\partial^2 u}{\partial s \partial v} + \tfrac{1}{2}\sigma^2 v \frac{\partial^2 u}{\partial v^2} + (r - \lambda\zeta)s\frac{\partial u}{\partial s} + \kappa(\eta - v)\frac{\partial u}{\partial v} \\ &\quad - (r+\lambda)u + \lambda \int_0^\infty u(sy,v,t)f(y)dy \end{aligned} \tag{16.12}$$

for $s > 0$, $v > 0$, and $0 < t \leq T$, where the PDF f is given by (16.8). For an extensive discussion on jump models in finance see, e.g., [16].

3 Linear Complementarity Problem for American Options

Unlike European-style options, American-style options can be exercised at any time up to the maturity date. Hence, the fair value of an American option is always greater than or equal to the instantaneous payoff,

$$u \geq \phi. \tag{16.13}$$

Due to this early exercise constraint, the P(I)DE does not hold everywhere anymore. Instead, a linear complementarity problem (LCP) or partial (integro-)differential complementarity problem is obtained in general for the fair value of an American option:

$$\begin{cases} \dfrac{\partial u}{\partial t} \geq \mathscr{A}u, \qquad u \geq \phi, \\ \left(\dfrac{\partial u}{\partial t} - \mathscr{A}u\right)(u - \phi) = 0, \end{cases} \tag{16.14}$$

where $\mathscr{A}$ stands for the pertinent spatial differential operator. For example, for the Black–Scholes model,

$$\mathscr{A}u = \frac{1}{2}\sigma^2 s^2 \frac{\partial^2 u}{\partial s^2} + rs\frac{\partial u}{\partial s} - ru.$$

The above inequalities and equation hold pointwise. The equation in (16.14) is the complementarity condition. It states that at each point one of the two inequalities has to be an equality. The paper [44] discusses the LCP formulation for American-style options under various asset price models and studies the structure and properties of the obtained fully discrete LCPs.

We note that the penalty approach is a popular alternative for LCPs. Here a penalty term is added to the P(I)DE for a European option with the aim to enforce the early exercise constraint (16.13). The resulting problems are nonlinear and their efficient numerical solution is considered in [27], for example. For several other alternative formulations and approximations for LCPs, we refer to [80].

4 Spatial Discretization

In this chapter we employ finite difference (FD) discretizations for the spatial derivatives. An alternative approach would be to use finite element discretizations; see e.g. [1, 74]. It is common practice to first truncate the infinite s-domain $[0, \infty)$ to $[0, S_{\max}]$

with a sufficiently large, real $S_{\max}$. Typically one wishes $S_{\max}$ to be such that the error caused by this truncation is a small fraction of the error due to the discretization of the differential (and integral) operators. Similarly, with multidimensional models including the variance v or the interest rate r, their corresponding infinite domains are truncated to sufficiently large bounded domains. The truncation requires additional boundary conditions to be specified. For an actual choice of these conditions for the models considered in Sections 2, 3 we refer to Section 7.

Let the grid in the s-direction be defined by the $m_1 + 1$ grid points $0 = s_0 < s_1 < \cdots < s_{m_1} = S_{\max}$. The corresponding grid sizes are denoted by $\Delta s_i = s_i - s_{i-1}$, $i = 1, 2, \ldots, m_1$. For multidimensional models, we use tensor product grids. For example, in the case of a stochastic volatility model, if a grid for the variance v is given by $0 = v_0 < v_1 < \cdots < v_{m_2} = V_{\max}$, then $(m_1 + 1) \times (m_2 + 1)$ spatial grid points are defined by (s_i, v_j) with $i = 0, 1, \ldots, m_1$ and $j = 0, 1, \ldots, m_2$. In financial applications nonuniform grids are often preferable over uniform grids. The use of suitable nonuniform grids will be illustrated in Section 7.

For discretizing the first derivative $\frac{\partial u_i}{\partial s}$ and the second derivative $\frac{\partial^2 u_i}{\partial s^2}$ at $s = s_i$, we employ in this chapter the well-known central FD schemes

$$\frac{\partial u_i}{\partial s} \approx \frac{-\Delta s_{i+1}}{\Delta s_i(\Delta s_i + \Delta s_{i+1})} u_{i-1} + \frac{\Delta s_{i+1} - \Delta s_i}{\Delta s_i \Delta s_{i+1}} u_i + \frac{\Delta s_i}{(\Delta s_i + \Delta s_{i+1})\Delta s_{i+1}} u_{i+1} \quad (16.15)$$

and

$$\frac{\partial^2 u_i}{\partial s^2} \approx \frac{2}{\Delta s_i(\Delta s_i + \Delta s_{i+1})} u_{i-1} - \frac{2}{\Delta s_i \Delta s_{i+1}} u_i + \frac{2}{(\Delta s_i + \Delta s_{i+1})\Delta s_{i+1}} u_{i+1}. \quad (16.16)$$

With multidimensional models the analogous schemes are used for the other spatial directions, thus, e.g., for $\frac{\partial u_j}{\partial v}$ and $\frac{\partial^2 u_j}{\partial v^2}$ at $v = v_j$. For the mixed derivative $\frac{\partial^2 u_{i,j}}{\partial s \partial v}$ at $(s, v) = (s_i, v_j)$ we consider the 9-point stencil obtained by successively applying the central FD schemes for the first derivative in the s- and v-directions. With sufficiently smooth varying grid sizes, the above central FDs give second-order accurate approximations for the derivatives.

We mention that in financial applications other FD schemes are employed as well, such as upwind discretization for first derivative terms or alternative discretizations for mixed derivative terms.

With the jump models the integral term needs to be discretized at grid points s_i. First the integral is divided into two parts

$$\int_0^\infty u(s_i y, t) f(y) dy = \int_0^{S_{\max}/s_i} u(s_i y, t) f(y) dy + \int_{S_{\max}/s_i}^\infty u(s_i y, t) f(y) dy,$$

which correspond to the values of u in the computational domain $[0, S_{\max}]$ and outside of it, respectively. The second part can be estimated using knowledge about u in the far field $[S_{\max}, \infty)$. For example, for put options u is usually assumed to be close to zero for $s \geq S_{\max}$ and, thus, the second integral is approximated by zero in this case. The PDFs f are smooth functions apart from the potential jump at $y = 1$

in the Kou model. Due to the smoothness of the integrand the trapezoidal rule leads to second-order accuracy with respect to the grid size. This gives the approximation

$$\int_0^{S_{\max}/s_i} u(s_i y,t)f(y)dy \approx \sum_{j=1}^{m_1} \frac{\Delta s_j}{2s_i}\left(u(s_{j-1},t)f(s_{j-1}/s_i)+u(s_j,t)f(s_j/s_i)\right).$$

For example, the papers [71] and [78] describe more accurate quadrature rules for the Merton and Kou jumps models, respectively. The discretization of the integral term leads to a dense matrix. The integral can be transformed into a convolution integral and due to this FFT can be used to compute it more efficiently; see [2, 3, 22, 77], for example. In the case of the Kou model, efficient recursion formulas can be used [12, 78].

5 Time Discretization

5.1 The θ-Method

For any P(I)DE from Section 2, the spatial discretization outlined in Section 4 leads to an initial value problem for a system of ordinary differential equations,

$$\dot{U}(t) = \mathbf{A}(t)U(t) + G(t) \quad (0 \le t \le T), \quad U(0) = U_0. \tag{16.17}$$

Here $\mathbf{A}(t)$ for $0 \le t \le T$ is a given square real matrix and $G(t)$ is a given real vector that depends on the boundary conditions. The entries of the solution vector $U(t)$ represent approximations to the exact solution of the option valuation P(I)DE at the spatial grid points, ordered in a convenient way. The vector U_0 is given by direct evaluation of the option's payoff function at these grid points.

The semidiscrete system (16.17) is stiff in general and, hence, implicit time discretization methods are natural candidates for its numerical solution. Let parameter $\theta \in (0,1]$ be given. Let time step $\Delta t = T/N$ with integer $N \ge 1$ and temporal grid points $t_n = n\Delta t$ for integers $0 \le n \le N$. The *θ-method* forms a well-known implicit time discretization method. It generates approximations U_n to $U(t_n)$ successively for $n = 1,2,\ldots,N$ by

$$U_n = U_{n-1} + (1-\theta)\Delta t\,\mathbf{A}(t_{n-1})U_{n-1} + \theta\Delta t\,\mathbf{A}(t_n)U_n + \Delta t\,G_{n-1+\theta}, \tag{16.18}$$

where $G_{n-1+\theta}$ denotes an approximation to $G(t)$ at $t = (n-1+\theta)\Delta t$. This can also be written as

$$(\mathbf{I} - \theta\Delta t\mathbf{A}(t_n))U_n = (\mathbf{I} + (1-\theta)\Delta t\,\mathbf{A}(t_{n-1}))U_{n-1} + \Delta t\,G_{n-1+\theta},$$

with $\mathbf{I}$ the identity matrix of the same size as $\mathbf{A}(t)$. For $\theta = 1$ one obtains the first-order *backward Euler method* and for $\theta = \frac{1}{2}$ the second-order *Crank–Nicolson method* or *trapezoidal rule*. For simplicity we consider in this chapter only constant time steps, but most of the presented time discretization methods can directly be extended to variable time steps.

When applying the Crank–Nicolson method, it is common practice in finance to first perform a few backward Euler steps to start the time stepping. This is often called Rannacher smoothing [67]. It helps to damp high-frequency components in the numerical solution, due to the nonsmooth initial (payoff) function, which are usually not sufficiently damped by the Crank–Nicolson method itself.

Clearly, in order to compute the vector U_n defined by (16.18), one has to solve a linear system of equations with the matrix $\mathbf{I}-\theta\Delta t\mathbf{A}(t_n)$. When the option valuation PDE is multidimensional, the size of this matrix is usually very large and it possesses a large bandwidth. For a PIDE, this matrix is dense. In these situations, the solution of the linear system can be computationally demanding when standard methods, like LU decomposition, are applied. Time discretization methods based on operator splitting can then form an attractive alternative. The key idea is to split the matrix $\mathbf{A}(t)$ into several parts, each of which is numerically handled more easily than the complete matrix itself.

5.2 Operator Splitting Methods Based on Direction

For multidimensional PDEs, splitting schemes of the Alternating Direction Implicit (ADI) type are often applied in financial practice. To illustrate the idea, the two-dimensional Heston PDE and three-dimensional HHW PDE, given in Section 2.2, are considered. For the Heston PDE the semidiscrete system (16.17) is autonomous; we split

$$\mathbf{A}=\mathbf{A}_0+\mathbf{A}_1+\mathbf{A}_2.$$

Next, for the HHW PDE,

$$\mathbf{A}(t)=\mathbf{A}_0+\mathbf{A}_1+\mathbf{A}_2+\mathbf{A}_3(t).$$

Here $\mathbf{A}_0$ is chosen as the part that represents all mixed derivative terms. It is nonzero whenever (one of) the correlation factor(s) is nonzero. The parts $\mathbf{A}_1$, $\mathbf{A}_2$, and $\mathbf{A}_3(t)$ represent all spatial derivatives in the s-, v-, and r-directions, respectively. The latter three matrices have, possibly up to permutation, all a fixed small bandwidth. The vector $G(t)$ in the semidiscrete system is split in a similar way. For notational convenience, define functions $\mathbf{F}_j$ by

$$\mathbf{F}_j(t,V)=\mathbf{A}_jV+G_j\ \ (j=0,1,2)\ \text{ and }\ \mathbf{F}_3(t,V)=\mathbf{A}_3(t)V+G_3(t)$$

for $0\le t\le T$, $V\in\mathbb{R}^m$. Set $\mathbf{F}=\sum_{j=0}^{k}\mathbf{F}_j$ with $k=2$ for Heston and $k=3$ for HHW. We discuss in this section four contemporary ADI-type splitting schemes:

Douglas (Do) scheme

$$\begin{cases} Y_0 = U_{n-1} + \Delta t\, \mathbf{F}(t_{n-1}, U_{n-1}), \\ Y_j = Y_{j-1} + \theta \Delta t\, (\mathbf{F}_j(t_n, Y_j) - \mathbf{F}_j(t_{n-1}, U_{n-1})) \quad (j = 1, 2, \ldots, k), \\ U_n = Y_k. \end{cases} \tag{16.19}$$

Craig–Sneyd (CS) scheme

$$\begin{cases} Y_0 = U_{n-1} + \Delta t\, \mathbf{F}(t_{n-1}, U_{n-1}), \\ Y_j = Y_{j-1} + \theta \Delta t\, (\mathbf{F}_j(t_n, Y_j) - \mathbf{F}_j(t_{n-1}, U_{n-1})) \quad (j = 1, 2, \ldots, k), \\ \widetilde{Y}_0 = Y_0 + \frac{1}{2}\Delta t\, (\mathbf{F}_0(t_n, Y_k) - \mathbf{F}_0(t_{n-1}, U_{n-1})), \\ \widetilde{Y}_j = \widetilde{Y}_{j-1} + \theta \Delta t\, (\mathbf{F}_j(t_n, \widetilde{Y}_j) - \mathbf{F}_j(t_{n-1}, U_{n-1})) \quad (j = 1, 2, \ldots, k), \\ U_n = \widetilde{Y}_k. \end{cases} \tag{16.20}$$

Modified Craig–Sneyd (MCS) scheme

$$\begin{cases} Y_0 = U_{n-1} + \Delta t\, \mathbf{F}(t_{n-1}, U_{n-1}), \\ Y_j = Y_{j-1} + \theta \Delta t\, (\mathbf{F}_j(t_n, Y_j) - \mathbf{F}_j(t_{n-1}, U_{n-1})) \quad (j = 1, 2, \ldots, k), \\ \widehat{Y}_0 = Y_0 + \theta \Delta t\, (\mathbf{F}_0(t_n, Y_k) - \mathbf{F}_0(t_{n-1}, U_{n-1})), \\ \widetilde{Y}_0 = \widehat{Y}_0 + (\frac{1}{2} - \theta)\Delta t\, (\mathbf{F}(t_n, Y_k) - \mathbf{F}(t_{n-1}, U_{n-1})), \\ \widetilde{Y}_j = \widetilde{Y}_{j-1} + \theta \Delta t\, (\mathbf{F}_j(t_n, \widetilde{Y}_j) - \mathbf{F}_j(t_{n-1}, U_{n-1})) \quad (j = 1, 2, \ldots, k), \\ U_n = \widetilde{Y}_k. \end{cases} \tag{16.21}$$

Hundsdorfer–Verwer (HV) scheme

$$\begin{cases} Y_0 = U_{n-1} + \Delta t\, \mathbf{F}(t_{n-1}, U_{n-1}), \\ Y_j = Y_{j-1} + \theta \Delta t\, (\mathbf{F}_j(t_n, Y_j) - \mathbf{F}_j(t_{n-1}, U_{n-1})) \quad (j = 1, 2, \ldots, k), \\ \widetilde{Y}_0 = Y_0 + \frac{1}{2}\Delta t\, (\mathbf{F}(t_n, Y_k) - \mathbf{F}(t_{n-1}, U_{n-1})), \\ \widetilde{Y}_j = \widetilde{Y}_{j-1} + \theta \Delta t\, (\mathbf{F}_j(t_n, \widetilde{Y}_j) - \mathbf{F}_j(t_n, Y_k)) \quad (j = 1, 2, \ldots, k), \\ U_n = \widetilde{Y}_k. \end{cases} \tag{16.22}$$

In the Do scheme (16.19), a forward Euler predictor step is followed by k implicit but unidirectional corrector steps that serve to stabilize the predictor step. The CS scheme (16.20), the MCS scheme (16.21), and the HV scheme (16.22) can be viewed as different extensions to the Do scheme. Indeed, their first two lines are identical to those of the Do scheme. They next all perform a second predictor step, followed by k unidirectional corrector steps. Observe that the CS and MCS schemes are equivalent if (and only if) $\theta = \frac{1}{2}$.

Clearly, in all four ADI schemes the $\mathbf{A}_0$ part, representing all mixed derivatives, is always treated in an *explicit* fashion. In the original formulation of ADI schemes mixed derivative terms were not considered. It is a common and natural use in the literature to refer to the above, extended schemes also as ADI schemes. In the special case where $\mathbf{F}_0 = 0$, the CS scheme reduces to the Do scheme, but the MCS scheme (with $\theta \neq \frac{1}{2}$) and the HV scheme do not. Following the original ADI approach, the $\mathbf{A}_1$, $\mathbf{A}_2$, $\mathbf{A}_3(t)$ parts are treated in an *implicit* fashion. In every step of each scheme, systems of linear equations need to be solved involving the matrices $(\mathbf{I} - \theta\,\Delta t\,\mathbf{A}_j)$ for $j = 1,2$ as well as $(\mathbf{I} - \theta\,\Delta t\,\mathbf{A}_3(t_n))$ if $k = 3$. Since all these matrices have a fixed, small bandwidth, this can be done very efficiently by means of LU decomposition, cf. also Section 6.1. Because for $j = 1,2$ the pertinent matrices are further independent of the step index n, their LU decompositions can be computed once, beforehand, and then used in all time steps. Accordingly, for each ADI scheme, the number of floating point operations per time step is directly proportional to the number of spatial grid points, which is a highly favorable property.

By Taylor expansion one obtains (after some elaborate calculations) the classical order of consistency[1] of each ADI scheme. For any given θ, the order of the Do scheme is just *one* whenever $\mathbf{A}_0$ is nonzero. This low order is due to the fact that the $\mathbf{A}_0$ part is treated in a simple, forward Euler fashion. The CS scheme has order *two* provided $\theta = \frac{1}{2}$. The MCS and HV schemes are of order two for any given θ. A virtue of ADI schemes, compared to other operator splitting schemes based on direction, is that the internal vectors Y_j, $\widetilde{Y}_j$ form consistent approximations to $U(t_n)$.

The Do scheme can be regarded as a generalization of the original ADI schemes for two-dimensional diffusion equations by Douglas & Rachford [23] and Peaceman & Rachford [66] to the situation where mixed derivative terms are present. This generalization was first considered by McKee & Mitchell [61] for diffusion equations and subsequently in [62] for convection-diffusion equations.

The CS scheme was developed by Craig & Sneyd [18] with the aim to obtain a stable second-order ADI scheme for diffusion equations with mixed derivative terms.

The MCS scheme was constructed by In 't Hout & Welfert [43] so as to arrive at more freedom in the choice of θ as compared to the second-order CS scheme.

The HV scheme was designed by Hundsdorfer [47] and Verwer et al. [83] for the numerical solution of convection-diffusion-reaction equations arising in atmospheric chemistry, cf. also [48]. The application of the HV scheme to equations containing mixed derivative terms was first studied in [42, 43].

[1] That is, the order for fixed nonstiff ODE systems.

The Do and CS schemes are well known for PDEs in finance, see, e.g., [4, 59]. More recently, the MCS and HV schemes have gained interest, see, e.g., [14, 20, 24, 35, 36, 39, 54].

The formulation of the ADI schemes (16.19)–(16.22) is analogous to the type of formulation used in [47]. In the literature, ADI schemes are also sometimes referred to as Stabilizing Correction schemes, and are further closely related to Approximate Matrix Factorization methods and Implicit-Explicit (IMEX) Runge–Kutta methods, cf., e.g., [48].

In [40, 41, 42, 43] comprehensive stability results in the von Neumann sense have been derived for the four schemes (16.19)–(16.22) in the application to multidimensional convection-diffusion equations with mixed derivative terms. These results concern unconditional stability, that is, without any restriction on the time step Δt. For each ADI scheme, lower bounds on θ guaranteeing unconditional stability have been obtained, depending in particular on the spatial dimension. Based on these theoretical stability results and the numerical experience in [35, 36, 39] the following values are found to be useful for $k = 2,3$:

- Do scheme with $\theta = \frac{1}{2}$ (if $k = 2$) and $\theta = \frac{2}{3}$ (if $k = 3$)
- CS scheme with $\theta = \frac{1}{2}$
- MCS scheme with $\theta = \frac{1}{3}$ (if $k = 2$) and $\theta = \max\{\frac{1}{3}, \frac{2}{13}(2\gamma+1)\}$ (if $k = 3$)
- HV scheme with $\theta = \frac{1}{2} + \frac{1}{6}\sqrt{3}$.

Here $\gamma = \max\{|\rho_{12}|, |\rho_{13}|, |\rho_{23}|\} \in [0,1]$, which is a measure for the relative size of the mixed derivative coefficients.

In addition to ADI schemes, there exists a variety of well-known alternative operator splitting schemes based on direction, called Locally One-Dimensional (LOD) methods, fractional step methods, or componentwise splitting schemes. These schemes originate in the 1960s in the work by Dyakonov, Marchuk, Samarskii, Yanenko, and others. Some of them are related to Strang splitting schemes, developed at the same time. For a general overview and analysis of such methods we refer to [48, 60]. Applications in financial mathematics of these schemes are considered in, for example, [50, 79].

5.3 Operator Splitting Methods Based on Operator Type

For the jump models considered in Section 2.3 the matrix **A** can be written in the form

$$\mathbf{A} = \mathbf{D} + \mathbf{J}, \tag{16.23}$$

where **D** and **J** correspond to the differential operator and integral operator, respectively. The matrix **D** is sparse while in general **J** is a dense matrix or has dense blocks. In view of the different nature of these two matrices it can be preferable to employ an operator splitting method based on them.

In [3], Andersen and Andreasen describe a generalized θ-method

$$(\mathbf{I}-\theta_D \Delta t \mathbf{D}-\theta_J \Delta t \mathbf{J})\, U_n = (\mathbf{I}+(1-\theta_D)\Delta t \mathbf{D}+(1-\theta_J)\Delta t \mathbf{J})\, U_{n-1} \qquad (16.24)$$

assuming here $G=0$. The standard choice $\theta_D = 1$ and $\theta_J = 0$ corresponds to the *IMEX Euler method*: it treats the stiff differential part implicitly, using the backward Euler method, and the nonstiff integral part explicitly, using the forward Euler method. This choice yields first-order consistency. The benefit is that it is not necessary to solve dense linear systems involving the matrix $\mathbf{J}$. Instead, in each time step only one multiplication with $\mathbf{J}$ is required. This approach has been considered and analyzed in [17].

In [26] an extrapolation approach is advocated based on the IMEX Euler method. Here approximations at a given fixed time are computed for a decreasing sequence of step sizes and then linearly combined so as to achieve a high order of accuracy.

In [3] second-order consistency is obtained through an alternating treatment of the $\mathbf{D}$ and $\mathbf{J}$ parts. They propose to take a $\Delta t/2$ substep with $\theta_D = 1$ and $\theta_J = 0$ followed by a $\Delta t/2$ substep with $\theta_D = 0$ and $\theta_J = 1$. Here linear systems involving the dense matrix $\mathbf{J}$ need to be solved, for which the authors employ FFT.

In [22] the original θ-method is analyzed, where the linear system in each time step is solved by applying a fixed-point iteration on the jump part following an idea in [77].

The following, second-order *IMEX midpoint scheme* has been considered in, e.g., [26, 57, 58, 72],

$$(\mathbf{I}-\Delta t \mathbf{D})U_n = (\mathbf{I}+\Delta t \mathbf{D})U_{n-2} + 2\Delta t \mathbf{J} U_{n-1} + 2\Delta t G_{n-1}. \qquad (16.25)$$

The scheme (16.25) can be viewed as obtained from the semidiscrete system (16.17) at t_{n-1} by the approximations $\mathbf{D}U_{n-1} \approx \frac{1}{2}\mathbf{D}(U_n + U_{n-2})$ and $\dot{U}_{n-1} \approx \frac{1}{2\Delta t}(U_n - U_{n-2})$. Two subsequent second-order IMEX methods are the *IMEX–CNAB scheme*

$$\left(\mathbf{I}-\tfrac{\Delta t}{2}\mathbf{D}\right) U_n = \left(\mathbf{I}+\tfrac{\Delta t}{2}\mathbf{D}\right) U_{n-1} + \tfrac{\Delta t}{2}\mathbf{J}\,(3U_{n-1}-U_{n-2}) + \Delta t G_{n-1/2} \qquad (16.26)$$

and the *IMEX–BDF2 scheme*

$$\left(\tfrac{3}{2}\mathbf{I}-\Delta t \mathbf{D}\right) U_n = 2U_{n-1} - \tfrac{1}{2}U_{n-2} + \Delta t \mathbf{J}\,(2U_{n-1}-U_{n-2}) + \Delta t G_n. \qquad (16.27)$$

These schemes have recently been applied for option pricing in [73] and can be regarded as obtained by approximating the semidiscrete system (16.17) at $t_{n-1/2} = \frac{1}{2}(t_n + t_{n-1})$ and at t_n, respectively.

The IMEX schemes (16.25), (16.26), and (16.27) were studied in a general framework, without application to option valuation, in [28]. Here it was noted that such schemes can be considered as starting with an implicit method and then replacing the nonstiff part of the implicit term by an explicit formula using extrapolation based on previous time steps. An overview of IMEX methods is given in [48].

In general, IMEX methods are only conditionally stable, that is, they are stable for a sufficiently small time step Δt. For example, the IMEX midpoint scheme

(16.25) and the IMEX–CNAB scheme (16.26) are stable whenever $\lambda \Delta t < 1$ and the λu term in (16.10) is included in $\mathbf{D}$; see [73]. Recall that λ denotes the jump activity.

The schemes discussed in this section are of the linear multistep type. For IMEX schemes of Runge–Kutta type applied to jump models we mention [10].

5.4 Operator Splitting Method for Linear Complementarity Problems

The fully discrete LCPs obtained by spatial and temporal discretization of (16.14) for American-style options are more difficult to solve than the corresponding systems of linear equations for the European-style counterparts. It is desirable to split these LCPs into simpler subproblems. Here we describe the operator splitting method considered in [49, 53] which was motivated by splitting methods for incompressible flows [13, 31]. To this purpose, we reformulate LCPs with Lagrange multipliers.

The θ-method discretization (16.18) naturally gives rise to the following, fully discrete LCP

$$\begin{cases} \mathbf{B}U_n - \mathbf{C}U_{n-1} - \Delta t G_{n-1+\theta} \geq 0, \\ U_n \geq U_0, \quad \left(\mathbf{B}U_n - \mathbf{C}U_{n-1} - \Delta t G_{n-1+\theta}\right)^T (U_n - U_0) = 0, \end{cases} \tag{16.28}$$

where $\mathbf{B} = \mathbf{I} - \theta \Delta t \mathbf{A}$, $\mathbf{C} = \mathbf{I} + (1-\theta)\Delta t \mathbf{A}$, and $\mathbf{A}$ is assumed to be constant in time. By introducing a Lagrange multiplier vector λ_n, the LCP (16.28) takes the equivalent form

$$\begin{cases} \mathbf{B}U_n - \mathbf{C}U_{n-1} - \Delta t G_{n-1+\theta} = \Delta t \lambda_n \geq 0, \\ U_n \geq U_0, \quad (\lambda_n)^T (U_n - U_0) = 0. \end{cases} \tag{16.29}$$

The basic idea of the operator splitting method proposed in [49] is to decouple in (16.29) the first line from the second line. This is accomplished by approximating the Lagrange multiplier λ_n in the first line by the previous Lagrange multiplier λ_{n-1}. This leads to the system of linear equations

$$\mathbf{B}\widetilde{U}_n = \mathbf{C}U_{n-1} + \Delta t G_{n-1+\theta} + \Delta t \lambda_{n-1}. \tag{16.30}$$

After solving this system, the intermediate solution vector $\widetilde{U}_n$ and the Lagrange multiplier λ_n are updated to satisfy the (spatially decoupled) equation and complementarity conditions

$$\begin{cases} U_n - \widetilde{U}_n = \Delta t(\lambda_n - \lambda_{n-1}), \\ \lambda_n \geq 0, \quad U_n \geq U_0, \quad (\lambda_n)^T (U_n - U_0) = 0. \end{cases} \tag{16.31}$$

Thus, this operator splitting method for American options leads to the solution of linear systems (16.30), which are essentially the same as for European options, and a simple update step (16.31). This update can be performed very fast, at each spatial grid point independently, with the formula

$$(U_{n,i}\,,\,\lambda_{n,i}) = \begin{cases} \left(\widetilde{U}_{n,i} - \Delta t \lambda_{n-1,i}\,,\,0\right), & \text{if } \widetilde{U}_{n,i} - \Delta t \lambda_{n-1,i} > U_{0,i}\,, \\ \left(U_{0,i}\,,\,\lambda_{n-1,i} + \frac{1}{\Delta t}\left(U_{0,i} - \widetilde{U}_{n,i}\right)\right), & \text{otherwise.} \end{cases} \tag{16.32}$$

The above operator splitting approach has been studied for more advanced time discretization schemes of both linear multistep and Runge–Kutta type in [49, 53]. Moreover, it has recently been effectively combined with IMEX schemes in [72] for the case of jump models and with ADI schemes in [37] for the case of the Heston model. For instance, the pertinent adaptations of the IMEX–CNAB scheme and the MCS scheme are

$$(\mathbf{I} - \tfrac{\Delta t}{2}\mathbf{D})\,\widetilde{U}_n = (\mathbf{I} + \tfrac{\Delta t}{2}\mathbf{D})\,U_{n-1} + \tfrac{\Delta t}{2}\mathbf{J}\,(3U_{n-1} - U_{n-2}) + \Delta t G_{n-1/2} + \Delta t\,\lambda_{n-1},$$

and

$$\begin{cases} Y_0 = U_{n-1} + \Delta t\,\mathbf{F}(t_{n-1}, U_{n-1}) + \Delta t\,\lambda_{n-1}, \\ Y_j = Y_{j-1} + \theta\Delta t\,(\mathbf{F}_j(t_n, Y_j) - \mathbf{F}_j(t_{n-1}, U_{n-1})) \quad (j = 1, 2, \ldots, k), \\ \widehat{Y}_0 = Y_0 + \theta\Delta t\,(\mathbf{F}_0(t_n, Y_k) - \mathbf{F}_0(t_{n-1}, U_{n-1})), \\ \widetilde{Y}_0 = \widehat{Y}_0 + (\tfrac{1}{2} - \theta)\Delta t\,(\mathbf{F}(t_n, Y_k) - \mathbf{F}(t_{n-1}, U_{n-1})), \\ \widetilde{Y}_j = \widetilde{Y}_{j-1} + \theta\Delta t\,(\mathbf{F}_j(t_n, \widetilde{Y}_j) - \mathbf{F}_j(t_{n-1}, U_{n-1})) \quad (j = 1, 2, \ldots, k), \\ \widetilde{U}_n = \widetilde{Y}_k, \end{cases}$$

respectively, followed by the update (16.32). The other three ADI schemes from Section 5.2 are adapted analogously. Note that only a $\Delta t \lambda_{n-1}$ term has been added to the first line of the MCS scheme (16.21). Accordingly, like for the θ-method, the amount of computational work per time step is essentially the same as for the corresponding European-style option.

6 Solvers for Algebraic Systems

The implicit time discretizations described in Section 5 lead, in each time step, to systems of linear equations of the form

$$\mathbf{B}U = \Psi \tag{16.33}$$

or LCPs of the form

$$\begin{cases} \mathbf{B}U \geq \Psi, & U \geq \Phi, \\ (\mathbf{B}U - \Psi)^T (U - \Phi) = 0 \end{cases} \tag{16.34}$$

with given matrix **B** and given vectors Φ, Ψ. For models without jumps, semidiscretization by finite difference, finite volume, and finite element methods yields sparse matrices **B**. For one-dimensional models, the central FDs (16.15) and (16.16) lead to tridiagonal **B**. For higher dimensional models they give rise to matrices **B** with a large bandwidth whenever classical (non-splitting) time stepping schemes are applied. On the other hand, for the operator splitting methods based on direction (cf. Section 5.2) one also acquires tridiagonal matrices (possibly after renumbering the unknowns). Wider FD stencils lead to additional nonzero diagonals. Time discretization of jump models with an implicit treatment of jumps makes **B** dense.

6.1 Direct Methods

The system of linear equations (16.33) can be solved by a direct method using LU decomposition. This method first forms a lower triangular matrix **L** and an upper triangular matrix **U** such that $\mathbf{B} = \mathbf{LU}$. After this the solution vector U is obtained by solving first $\mathbf{L}V = \Psi$ and then $\mathbf{U}U = V$.

Let m denote the dimension of the matrix **B**. For tridiagonal **B**, or more generally matrices with a fixed small bandwidth, the LU decomposition yields optimal computational cost in the sense that the number of floating point operations is of order m. Hence, it is very efficient for one-dimensional models and for higher-dimensional models when operator splitting schemes based on direction are applied.

For two-dimensional models with classical time stepping schemes, a LU decomposition can be formed by order $m^{3/2}$ floating point operations if a nested dissection method can be used and then the computational cost of the solution is of order $m \log m$, see [21, 29]. For higher-dimensional models with classical time stepping schemes, the computational cost is less favorable.

For a general matrix **B**, solving the LCP (16.34) requires iterative methods. However, in the special case that **B** is tridiagonal, the solution vector satisfies $U_i = \Phi_i$ $(1 \leq i \leq i_0)$, $U_i > \Phi_i$ $(i_0 < i \leq m)$ for certain i_0 and some additional assumptions hold, the *Brennan–Schwartz algorithm* [9] gives a direct method to solve the LCP; see also [1, 51, 55]. After inverting the numbering of the unknowns to be from right to left, represented by a permutation matrix **P**, this algorithm is equivalent to applying the LU decomposition method to the corresponding linear system with matrix **PBP** where the projection step is carried out directly after computing each component in the back substitution step with **U**. More precisely the back substitution step reads after the renumbering of unknowns:

$$\begin{cases} U_m = \max\{V_m/\mathbf{U}_{m,m}\,,\,\Phi_m\}, \\ U_i = \max\{(V_i - \mathbf{U}_{i,i+1}U_{i+1})/\mathbf{U}_{i,i}\,,\,\Phi_i\} \quad (i = m-1, m-2, \ldots, 1). \end{cases} \tag{16.35}$$

The Brennan–Schwartz algorithm is essentially as fast as the LU decomposition method for linear systems and, thus, it has optimal computational cost.

6.2 Iterative Methods

There are many iterative methods for solving systems of linear equations. The two most important method categories are the stationary iterative methods and the Krylov subspace methods. Well-known Krylov subspace methods for the, typically asymmetric, system (16.33) are the generalized minimal residual (GMRES) method [70] and the BiCGSTAB method [84]. In the following we discuss a stationary iterative method in some more detail which is familiar in finance applications. The *successive over-relaxation (SOR) method* reads

$$U_i^{(k+1)} = U_i^{(k)} + \frac{\omega}{\mathbf{B}_{i,i}} \left(\Psi_i - \sum_{j=1}^{i-1} \mathbf{B}_{i,j} U_j^{(k+1)} - \sum_{j=i}^{m} \mathbf{B}_{i,j} U_j^{(k)} \right) \tag{16.36}$$

for $i = 1, 2, \ldots, m$, $k = 0, 1, 2, \ldots$, where ω is a relaxation parameter. This method reduces to the Gauss–Seidel method in the case $\omega = 1$. The convergence rate of the iteration (16.36) can be improved significantly by an optimal choice of ω. Still the number of iterations to reach a given accuracy typically grows with m, that is, when the spatial grid is refined the convergence slows down.

The SOR iteration can be generalized for LCPs by performing a projection after each update [19]; see also [30]. This method is called the *projected SOR (PSOR) method* and it reads

$$U_i^{(k+1)} = \max\left\{ U_i^{(k)} + \frac{\omega}{\mathbf{B}_{i,i}} \left(\Psi_i - \sum_{j=1}^{i-1} \mathbf{B}_{i,j} U_j^{(k+1)} - \sum_{j=i}^{m} \mathbf{B}_{i,j} U_j^{(k)} \right),\, \Phi_i \right\} \tag{16.37}$$

($i = 1, 2, \ldots, m$, $k = 0, 1, 2, \ldots$). As can be expected, the PSOR method suffers from the same drawback as the SOR method mentioned above.

6.3 Multigrid Methods

The aim of multigrid methods for solving linear systems (16.33) is to render the number of iterations essentially independent of the problem size m. The stationary iterative methods typically reduce high frequency errors quickly, while low frequency errors are reduced much more slowly. The idea of multigrid methods is to compute efficiently corrections to these slowly varying errors on coarser spatial grids. The multigrid methods can be divided into geometrical and algebraic methods. With the geometrical methods discretizations are explicitly constructed on a sequence of grids and transfer operators between these grids are explicitly defined. Algebraic multigrid (AMG) methods [69, 76] build the coarse problems and the transfer operators automatically using the properties of the matrix **B**. The details of these methods are beyond the scope of this chapter and we refer to, e.g., [82] for details and extensive literature on this.

Several versions of multigrid methods also exist for LCPs. Brandt and Cryer introduced in [8] a projected full approximation scheme (PFAS) multigrid method for LCPs. American options under stochastic volatility were priced using the PFAS method in [15, 65]. A projected multigrid (PMG) method for LCPs introduced in [68] resembles more closely a classical multigrid method for linear problems. This method has been used to price American options in [52, 68]. Recently, an AMG method was generalized for LCPs in [81]. The resulting method is called the projected algebraic multigrid (PAMG) method and resembles the PMG method in the treatment of the complementarity conditions.

7 Numerical Illustrations

In the following we price European and American put options under a hierarchy of models: Black–Scholes, Merton, Heston, and Bates. The interest rate, the maturity time, and the strike price are always taken as

$$r = 0.03, \quad T = 0.5, \quad \text{and} \quad K = 100.$$

For the purpose of illustration, Figures 16.1 and 16.2 show fair values of European and American options, respectively, under the four considered models with the model parameters described in the following sections.

7.1 Black–Scholes Model

In the case of the Black–Scholes model, we price American put options. The volatility in the model (16.1) is taken as

$$\sigma = 0.2$$

and the following boundary conditions are employed:

$$u(0,t) = K \quad \text{for} \quad 0 < t \leq T, \tag{16.38}$$

$$u_s(S_{\max},t) = 0 \quad \text{for} \quad 0 < t \leq T. \tag{16.39}$$

The Neumann boundary condition (16.39) introduces a modeling error as it is not exactly fulfilled by the actual option price function. If $S_{\max}$ is taken sufficiently large, however, this error will be small in the region of interest.

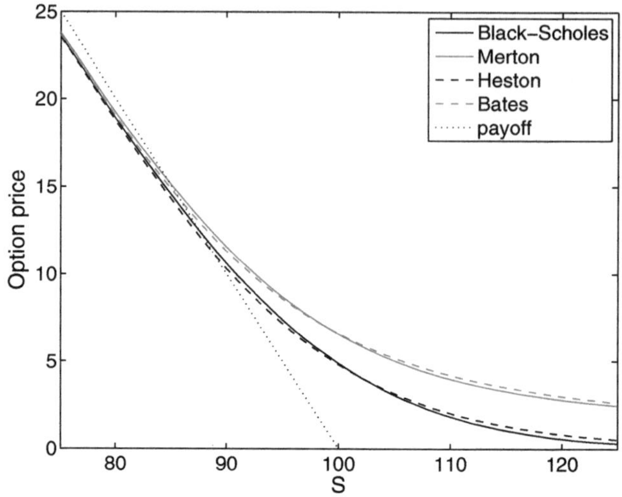

Fig. 16.1 The fair values of European put options for the asset prices $75 \leq s \leq 125$ and the volatility $\sigma = 0.2$ (the variance $v = 0.04$) under the four considered models.

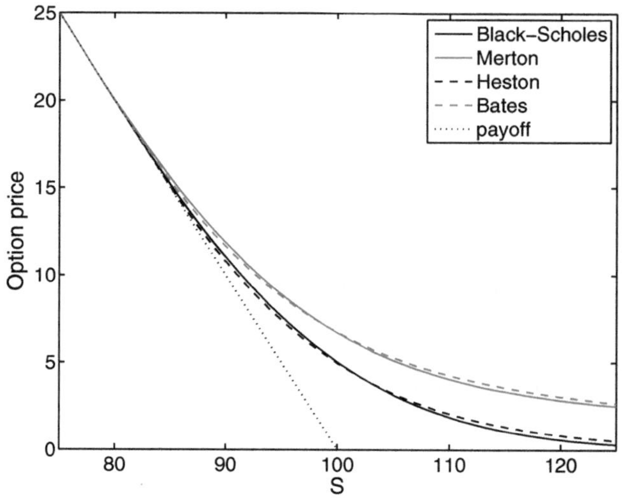

Fig. 16.2 The fair values of American put options for the asset prices $75 \leq s \leq 125$ and the volatility $\sigma = 0.2$ (the variance $v = 0.04$) under the four considered models.

For the spatial discretization of the Black–Scholes PDE (16.2), we apply FD formulas on nonuniform grids such that a large fraction of the grid points lie in the region of interest, that is, in the neighborhood of $s = K$.

For the construction of the spatial grid we adopt [36]. Let integer $m_1 \geq 1$, constant $c > 0$, and $0 < S_{\text{left}} < K < S_{\text{right}} < S_{\max}$ be given. Let equidistant points $\xi_{\min} = \xi_0 < \xi_1 < \ldots < \xi_{m_1} = \xi_{\max}$ be given with distance $\Delta\xi$ and

$$\xi_{\min} = \sinh^{-1}\left(\frac{-S_{\text{left}}}{c}\right),$$
$$\xi_{\text{int}} = \frac{S_{\text{right}} - S_{\text{left}}}{c},$$
$$\xi_{\max} = \xi_{\text{int}} + \sinh^{-1}\left(\frac{S_{\max} - S_{\text{right}}}{c}\right).$$

Then we define a nonuniform grid $0 = s_0 < s_1 < \ldots < s_{m_1} = S_{\max}$ by the transformation

$$s_i = \varphi(\xi_i) \quad (0 \le i \le m_1), \tag{16.40}$$

where

$$\varphi(\xi) = \begin{cases} S_{\text{left}} + c \cdot \sinh(\xi) & (\xi_{\min} \le \xi \le 0), \\ S_{\text{left}} + c \cdot \xi & (0 < \xi < \xi_{\text{int}}), \\ S_{\text{right}} + c \cdot \sinh(\xi - \xi_{\text{int}}) & (\xi_{\text{int}} \le \xi \le \xi_{\max}). \end{cases}$$

The grid (16.40) is uniform inside $[S_{\text{left}}, S_{\text{right}}]$ and nonuniform outside. The parameter c controls the fraction of grid points s_i that lie inside $[S_{\text{left}}, S_{\text{right}}]$. The grid is smooth in the sense that there exist real constants $C_0, C_1, C_2 > 0$ such that the grid sizes $\Delta s_i = s_i - s_{i-1}$ satisfy

$$C_0\,\Delta\xi \le \Delta s_i \le C_1\,\Delta\xi \quad \text{and} \quad |\Delta s_{i+1} - \Delta s_i| \le C_2\,(\Delta\xi)^2 \tag{16.41}$$

uniformly in i and m_1. For the parameters in the grid we make the (heuristic) choice

$$S_{\max} = 8K,\ \ c = \frac{K}{10},\ \ S_{\text{left}} = \max\left(\tfrac{1}{2}, e^{-T/10}\right)K\,,\ \ S_{\text{right}} = \min\left(\tfrac{3}{2}, e^{T/10}\right)K.$$

The semidiscretization of the initial-boundary value problem for the Black–Scholes PDE is then performed as follows. At the interior grid points each spatial derivative appearing in (16.2) is replaced by its corresponding second-order central FD formula described in Section 4. At the boundary $s = S_{\max}$ the Neumann condition (16.39) gives $\partial u/\partial s$. Next, $\partial^2 u/\partial s^2$ is approximated by the central formula with the value at the virtual point $S_{\max} + \Delta s_{m_1}$ defined by linear extrapolation using (16.39).

Concerning the initial condition, we always replace the value of the payoff function ϕ at the grid point s_i nearest to the strike K by its cell average,

$$\frac{1}{h}\int_{s_{i-1/2}}^{s_{i+1/2}} \max(K - s, 0)\,ds,$$

where

$$s_{i-1/2} = \tfrac{1}{2}(s_{i-1} + s_i),\ \ s_{i+1/2} = \tfrac{1}{2}(s_i + s_{i+1}),\ \ h = s_{i+1/2} - s_{i-1/2}.$$

This reduces the dependency of the discretization error on the location of the strike relative to the s-grid, see, e.g., [77].

The time discretization is performed by the Crank–Nicolson method with Rannacher smoothing. The time stepping is started by taking two backward Euler steps using the time step $\frac{1}{2}\Delta t$. With this choice all time steps are performed with the same coefficient matrix $\mathbf{I} - \frac{1}{2}\Delta t\mathbf{A}$. Furthermore, halving the time step with the Euler method helps to reduce the additional error caused by this method. Note that we count these two Euler steps as one time step in order to keep the notations convenient.

We define the *temporal discretization error* to be

$$\widehat{e}(m_1,N) = \max\left\{ |U_{N,i} - U_i(T)| : \tfrac{1}{2}K < s_i < \tfrac{3}{2}K \right\}, \qquad (16.42)$$

where $U_{N,i}$ denotes the component of the vector U_N associated to the grid point s_i. We study the temporal discretization errors on the grids $(m_1,N) = (160,2^k)$ for $k = 0,1,\ldots,10$. The reference price vector $U(T)$ is computed using the space-time grid $(160,5000)$. Figure 16.3 compares the temporal errors of the smoothed Crank–Nicolson method with and without the operator splitting method for LCPs described in Section 5.4. For larger time steps the Crank–Nicolson method without splitting is more accurate. In this example the convergence rate of the splitting method is slightly less than second-order and a bit higher than the convergence rate of the non-splitting method. Thus, for smaller time steps the operator splitting method is slightly more accurate.

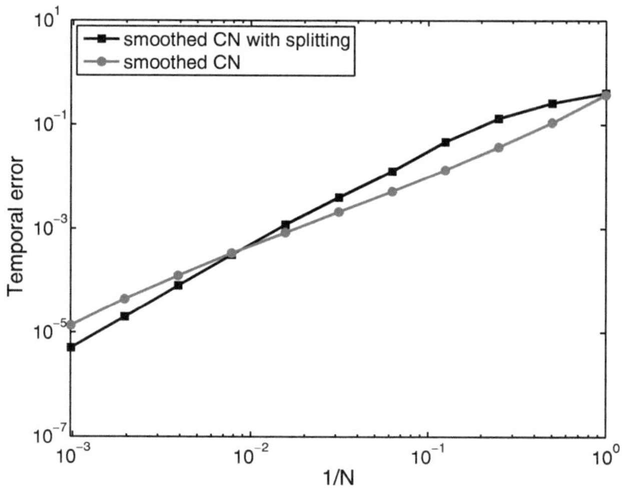

Fig. 16.3 The temporal discretization errors for the American option under the Black–Scholes model for the smoothed Crank–Nicolson method with and without the operator splitting method for LCPs.

7.2 Merton Model

Under the Merton jump diffusion model, we price European and American put options. For the jump part of the model, the jump activity, the mean of the normal distribution, and its standard deviation are taken as

$$\lambda = 0.2, \quad \delta = 0.4, \quad \text{and} \quad \gamma = -0.5, \tag{16.43}$$

respectively; see (16.8). The boundary condition at $s = 0$ is given by (16.5) for the European put option and by (16.38) for the American put option. At the truncation boundary $s = S_{\max}$, we use the Neumann boundary condition (16.39).

The same space-time grids are considered as with the Black–Scholes model in Section 7.1 and also the spatial derivatives are discretized in the same way. For the integral term, we use a linear interpolation for u between grid points and take u to be zero for $s > S_{\max}$. The formulas for the resulting matrix $\mathbf{J}$ are given in [71], for example.

For the time discretization, we apply the IMEX–CNAB scheme, which is always smoothed by two Euler steps with the time step $\frac{1}{2}\Delta t$. In these first steps the backward Euler method is used for the discretized differential part $\mathbf{D}$ and the forward Euler method is used for the discretized integral part $\mathbf{J}$. For European options, these steps are given by

$$\begin{aligned} \left(\mathbf{I} - \tfrac{\Delta t}{2}\mathbf{D}\right) U_{1/2} &= U_0 + \tfrac{\Delta t}{2}\mathbf{J}U_0 + \tfrac{\Delta t}{2}G_{1/2}, \\ \left(\mathbf{I} - \tfrac{\Delta t}{2}\mathbf{D}\right) U_1 &= U_{1/2} + \tfrac{\Delta t}{2}\mathbf{J}U_{1/2} + \tfrac{\Delta t}{2}G_1. \end{aligned}$$

In the absence of jumps, these steps reduce to the same Rannacher smoothing used with the Black–Scholes model. After these two steps the IMEX–CNAB scheme defined by (16.26) is employed.

We study the temporal discretization errors for European and American options on the same grids $(m_1, N) = (160, 2^k)$, $k = 0, 1, \ldots, 10$, and using the same error measure (16.42) as before. Figure 16.4 shows the temporal errors for the European option using the IMEX–CNAB scheme and the Crank–Nicolson method with classical Rannacher smoothing. We observe that the temporal errors for the two methods are essentially the same and they exhibit second-order convergence.

Figure 16.5 shows the same temporal errors for American options using the IMEX–CNAB scheme with operator splitting for LCPs and the Crank–Nicolson method without splitting. The convergence result for the two methods is very similar to the case of the Black–Scholes model in Section 7.1. Thus, for larger time steps the Crank–Nicolson method is more accurate while for smaller time steps the IMEX–CNAB scheme with splitting is more accurate.

In order to gauge the effectiveness of the proposed discretizations, we report the total discretization errors for the European option on the space-time refining grids $(m_1, N) = 2^k(10, 2)$, $k = 0, 1, \ldots, 6$. The *total discretization error* is defined by

$$e(m_1, N) = \max\left\{ |U_{N,i} - u(s_i, T)| : \frac{1}{2}K < s_i < \frac{3}{2}K \right\}. \tag{16.44}$$

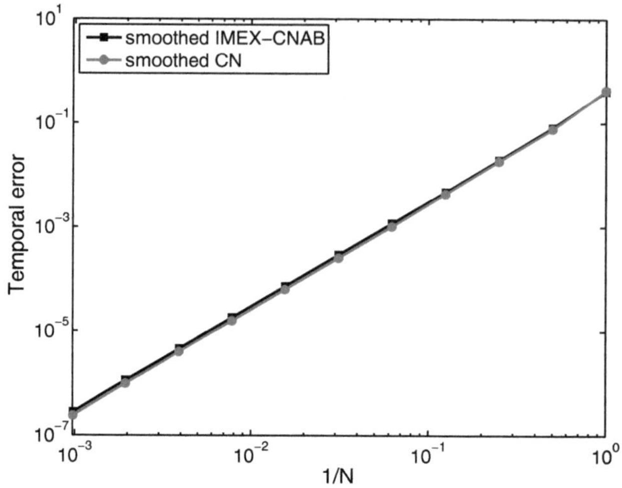

Fig. 16.4 The temporal discretization errors for the European option under the Merton model with the IMEX–CNAB scheme and the Crank–Nicolson method, both with smoothing.

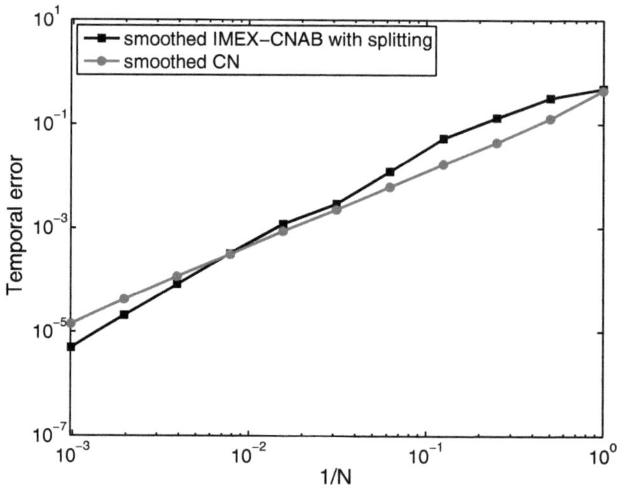

Fig. 16.5 The temporal discretization errors for the American option under the Merton model with the IMEX–CNAB scheme together with the operator splitting method for LCPs, and the Crank–Nicolson method, both with smoothing.

The reference price function u is computed on the space-time grid $(10240, 2048)$. Figure 16.6 shows the total error for the European option using the IMEX–CNAB scheme and the Crank–Nicolson method. As with the temporal errors the total errors for both methods are essentially the same and both show a second-order convergence behavior.

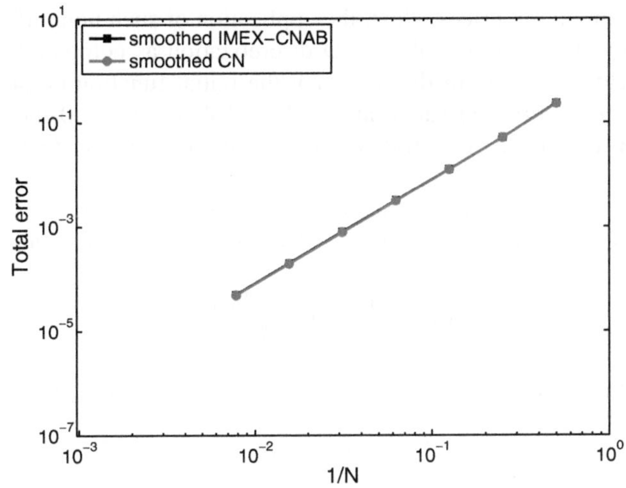

Fig. 16.6 The total discretization errors for the European option under the Merton model with the IMEX–CNAB scheme and the Crank–Nicolson method, both with smoothing.

7.3 Heston Model

Under the Heston stochastic volatility model we consider European and American put options as well. For the mean-reversion rate, the long-term mean, the volatility-of-variance and the correlation the following values are taken:

$$\kappa = 2, \quad \eta = 0.04, \quad \sigma = 0.25, \quad \text{and} \quad \rho = -0.5. \tag{16.45}$$

The spatial domain is truncated to $[0, S_{\max}] \times [0, V_{\max}]$ with $S_{\max} = 8K$ and $V_{\max} = 5$. The following boundary conditions are imposed:

$$u(0, v, t) = df \cdot K \quad \text{for} \quad 0 \le v \le V_{\max},\ 0 < t \le T, \tag{16.46}$$

$$u_s(S_{\max}, v, t) = 0 \quad \text{for} \quad 0 \le v \le V_{\max},\ 0 < t \le T, \tag{16.47}$$

$$u_v(s, V_{\max}, t) = 0 \quad \text{for} \quad 0 \le s \le S_{\max},\ 0 < t \le T, \tag{16.48}$$

where $df = e^{-rt}$ in the European case and $df = 1$ in the American case. At the degenerate boundary $v = 0$ the Heston PDE holds in the European case and it is assumed that the Heston LCP holds in the American case. The two conditions at $s = S_{\max}$ and $v = V_{\max}$ introduce a modeling error, as they are not exactly fulfilled by the actual option price function, but in our experiments this error is small on the region of interest in the (s, v)-domain.

For the spatial discretization of the Heston PDE and Heston LCP we apply FD formulas on Cartesian grids. Here nonuniform grids are used in both the s- and v-directions such that a large fraction of the grid points lie in the neighborhoods of $s = K$ and $v = 0$, respectively. This is the region in the (s, v)-domain where one wishes to obtain option prices. Next, the application of such nonuniform grids can greatly improve the accuracy of the FD discretization as compared to using uniform grids. This is related to the facts that the initial function (16.4) possesses a discontinuity in its first derivative at $s = K$ and that for $v \approx 0$ the Heston PDE is convection-dominated. The grid in the s-direction is taken identical to that in Section 7.1.

To construct the grid in the v-direction, let integer $m_2 \geq 1$ and constant $d > 0$ and let equidistant points be given by $\psi_j = j \cdot \Delta\psi$ for $j = 0, 1, \ldots, m_2$ with

$$\Delta\psi = \frac{1}{m_2} \sinh^{-1}\left(\frac{V_{\max}}{d}\right).$$

Then a smooth, nonuniform grid $0 = v_0 < v_1 < \ldots < v_{m_2} = V_{\max}$ is defined by

$$v_j = d \cdot \sinh(\psi_j) \quad (0 \leq j \leq m_2). \tag{16.49}$$

The parameter d controls the fraction of grid points v_j that lie near $v = 0$. We heuristically choose

$$d = \frac{V_{\max}}{500}.$$

The semidiscretization of the initial-boundary value problem for the Heston PDE and Heston LCP is performed as follows. In view of the Dirichlet condition (16.46), the grid in $[0, S_{\max}] \times [0, V_{\max}]$ is given by $\{(s_i, v_j) : 1 \leq i \leq m_1,\ 0 \leq j \leq m_2\}$. At this grid, each spatial derivative is replaced by its corresponding second-order central FD formula described in Section 4 with a modification for the boundaries $v = 0$, $s = S_{\max}$, and $v = V_{\max}$.

At the boundary $v = 0$ the derivative $\partial u/\partial v$ is approximated using a second-order forward formula. All other derivative terms in the v-direction vanish at $v = 0$, and therefore do not require further treatment.

At the boundary $s = S_{\max}$ the spatial derivatives in the s-direction are dealt with as in Section 7.1. Note that the Neumann condition (16.47) at $s = S_{\max}$ implies that the mixed derivative $\partial^2 u/\partial s \partial v$ vanishes there.

At the boundary $v = V_{\max}$ the spatial derivatives in the v-direction need to be considered. This is done fully analogously to those in the s-direction at $s = S_{\max}$ using now the Neumann condition (16.48).

Define the temporal discretization error by

$$\widehat{e}(m_1, m_2, N) = \max\left\{ |U_{N,l} - U_l(T)| : \tfrac{1}{2}K < s_i < \tfrac{3}{2}K,\ 0 < v_j < 1 \right\}, \tag{16.50}$$

where the index l corresponds to the grid point (s_i, v_j). The reference vector $U(T)$ is computed using $(m_1, m_2, N) = (160, 80, 5000)$. We study these errors for $(m_1, m_2, N) = (160, 80, 2^k)$ with $k = 0, 1, \ldots, 10$ and three methods: the Do scheme with $\theta = \frac{1}{2}$ and smoothing, the MCS scheme with $\theta = \frac{1}{3}$ without smoothing, and the Crank–Nicolson scheme with smoothing.

Figure 16.7 displays the obtained results for the European put option. As a first observation, for all three methods the temporal errors are bounded from above by a moderate value and decrease monotonically as N increases. The error graphs for the MCS and Crank–Nicolson schemes are almost identical and reveal a second-order convergence behavior. The Do scheme only shows first-order convergence. Clearly, the convergence orders observed for the three methods agree with their respective classical orders of consistency. Additional experiments by substantially changing (m_1,m_2) indicate that for all three methods the temporal errors are almost unaffected, which is a desirable property and suggests convergence in the so-called *stiff sense*. Whereas their results are not displayed, we mention that the CS scheme with $\theta = \frac{1}{2}$ and smoothing and the HV scheme with $\theta = \frac{1}{2} + \frac{1}{6}\sqrt{3}$ without smoothing behave similarly to the MCS scheme in this experiment, with slightly larger errors.

Figure 16.8 displays the obtained results for the American put option. Our observations are analogous to those made above in the case of the European option. It is interesting to note, however, that the Do scheme often has temporal errors that are almost the same as for the MCS and Crank–Nicolson schemes. But if N gets sufficiently large, then a first-order convergence behavior for this method indeed sets in. For the Crank–Nicolson scheme a small deviation from second-order is seen when N is large. This disappears however when other values (m_1,m_2) are considered. Additional experiments by substantially changing (m_1,m_2) indicate that for all three methods the temporal errors are at most mildly affected.

We next consider, in the European put option case, the total discretization error defined by

$$e(m_1,m_2,N) = \max\left\{ |U_{N,l} - u(s_i,v_j,T)| : \tfrac{1}{2}K < s_i < \tfrac{3}{2}K,\ 0 < v_j < 1 \right\}, \quad (16.51)$$

with index l corresponding to the grid point (s_i,v_j). Here exact solution values u are computed by a suitable implementation of Heston's semi-closed form analytical formula [38]. Note that the modeling error, which is due to the truncation of the domain of the Heston PDE to a bounded set, is also contained in $e(m_1,m_2,N)$. In our experiment, this contribution is negligible.

Figure 16.9 displays the total discretization errors for $(m_1,m_2,N) = 2^k(10,5,2)$ with $k = 0,1,\ldots,6$ and the three schemes under consideration in this section. With the MCS and Crank–Nicolson schemes the total errors are essentially the same and a second-order convergence behavior is observed. With the Do scheme, the total errors are almost same as these two schemes up to $k = 4$, but then the convergence drops to the expected first-order.

For a more extensive numerical study of ADI schemes in the (two-dimensional) Heston model we refer to [39] for European-style options and to [37] for American-style options. For three-dimensional PDEs in finance, such as the HHW PDE, the numerical convergence of ADI schemes has been investigated in [35, 36] and for a four-dimensional PDE in [34]. In these references a variety of parameter sets has been considered, including long maturity times and cases where the Feller condition is strongly violated, together with various barrier options and the approximation of hedging quantities.

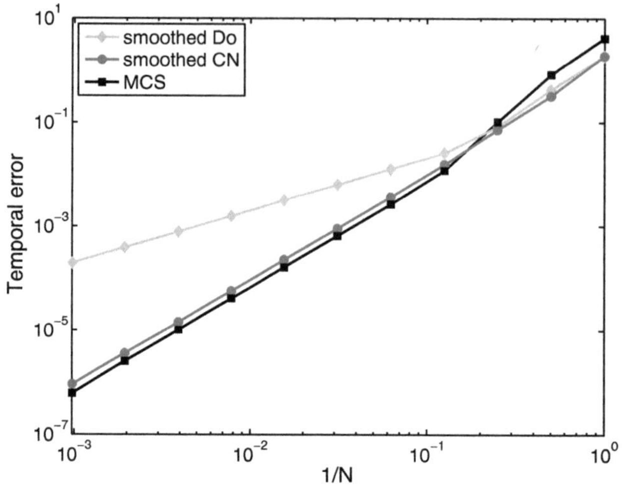

Fig. 16.7 Temporal discretization errors in the case of the European put option under the Heston model. The time discretization methods are: the Do scheme with $\theta = \frac{1}{2}$ and smoothing, the MCS scheme with $\theta = \frac{1}{3}$ without smoothing, and the Crank–Nicolson scheme with smoothing.

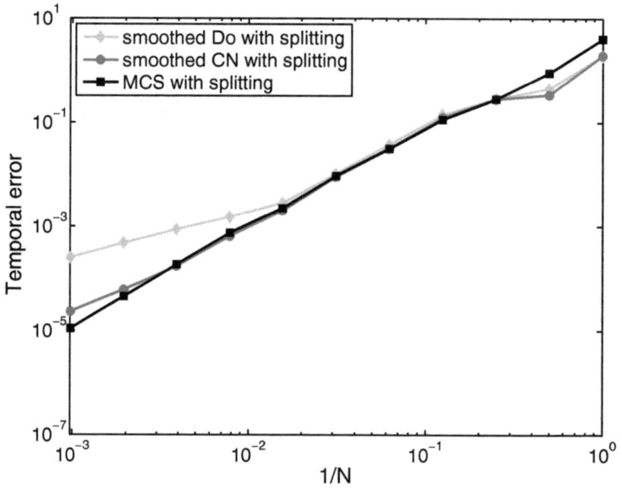

Fig. 16.8 Temporal discretization errors in the case of the American put option under the Heston model. The time discretization methods are: the Do scheme with $\theta = \frac{1}{2}$ and smoothing, the MCS scheme with $\theta = \frac{1}{3}$ without smoothing, and the Crank–Nicolson scheme with smoothing.

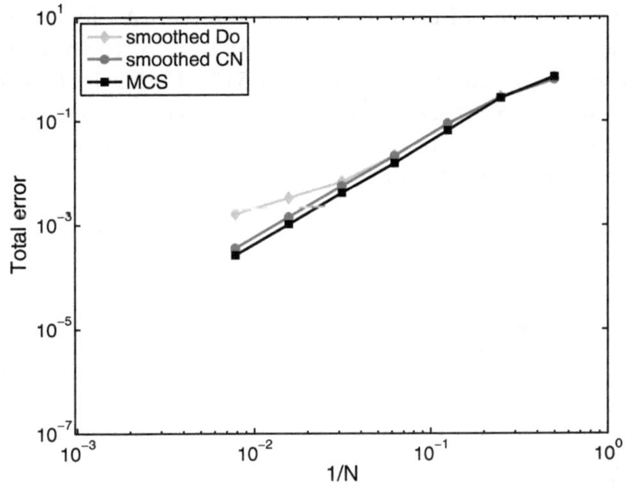

Fig. 16.9 Total discretization errors in the case of the European put option under the Heston model. The time discretization methods are: the Do scheme with $\theta = \frac{1}{2}$ and smoothing, the MCS scheme with $\theta = \frac{1}{3}$ without smoothing, and the Crank–Nicolson scheme with smoothing.

7.4 Bates Model

We price European and American put options under the Bates model. The boundary conditions are given by (16.46)–(16.48). For the stochastic volatility part of the model the parameters are taken the same as for the Heston model and they are given by (16.45). For the jump part, the parameters are the same as for the Merton model and they are given by (16.43). The discretizations are based on the same grids and the spatial derivatives are discretized in the same way as with the Heston model in Section 7.3. For the jump integral, the same discretization is used as with the Merton model in Section 7.2. We consider here the IMEX–CNAB scheme and Crank–Nicolson method both applied with smoothing as for the Merton model.

As with the Heston model, we consider the temporal discretization errors on the grids $(m_1, m_2, N) = (160, 80, 2^k)$, $k = 0, 1, \ldots, 10$. The reference price vector $U(T)$ is computed using the space-time grid $(160, 80, 5000)$. The temporal discretization errors $\widehat{e}(m_1, m_2, N)$ are shown for the European option in Figure 16.10 and for the American option in Figure 16.11. The plots show the errors for the IMEX–CNAB scheme and the Crank–Nicolson method. For the American option the operator splitting method for LCPs is used with the IMEX–CNAB scheme. As with other models, the temporal errors for the European option are very similar for both methods and they both exhibit second-order convergence. For the American option, the difference between the methods is less pronounced than with the Black–Scholes and Merton models. Still the Crank–Nicolson method is slightly more accurate than the operator splitting method for large time steps and the reverse is true for small time steps. In this example the convergence rates seem to be between 1.5 and 2.0.

We computed the total discretization errors $e(m_1,m_2,N)$ for the European option on the grids $(m_1,m_2,N) = 2^k(10,5,2)$, $k = 0,1,\ldots,6$. The reference prices are computed on the space-time grid $(2560,1280,512)$. Figure 16.12 shows the total errors for the IMEX–CNAB scheme and the Crank–Nicolson method. As with the other models, the total errors for both methods are virtually the same and both have second-order convergence of the total error.

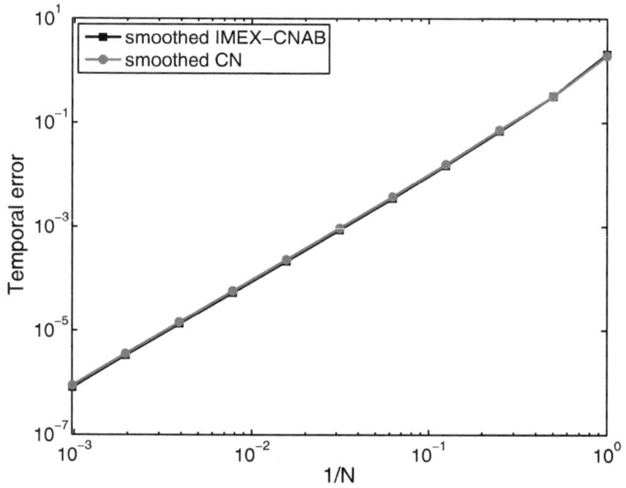

Fig. 16.10 The temporal discretization errors for the European option under the Bates model with the IMEX–CNAB scheme and the Crank–Nicolson method, both with smoothing.

8 Conclusions

We have discussed numerical solution methods for financial option valuation problems in the contemporary partial(-integro) differential equations framework. These problems are often multidimensional and can involve nonlocal integral operators due to jumps incorporated in the underlying asset price models. The early exercise feature of American-style options gives rise to linear complementarity problems, which are nonlinear. All these properties add complexity to the discrete problems obtained by classical implicit numerical methods and renders their efficient solution a challenging task. The efficient computation of option values is, however, crucial in many applications. In this chapter an overview has been given of various types of operator splitting methods for the discretization in time, which yield in each time step a sequence of discrete subproblems that can be handled much more easily and

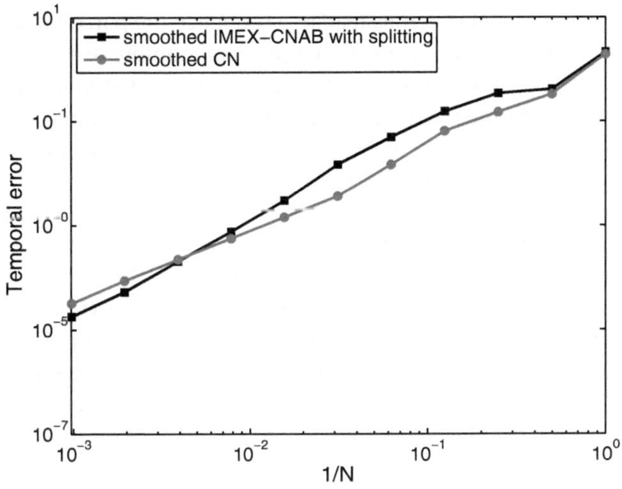

Fig. 16.11 The temporal discretization errors for the American option under the Bates model with the IMEX–CNAB scheme together with the operator splitting method for LCPs, and the Crank–Nicolson method, both with smoothing.

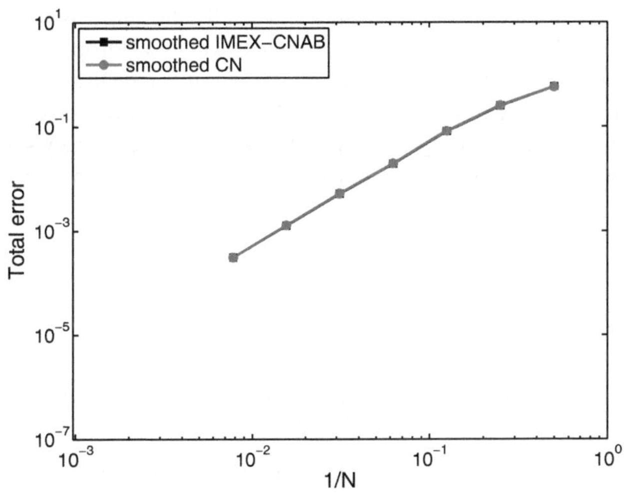

Fig. 16.12 The total discretization errors for the European option under the Bates model with the IMEX–CNAB scheme and the Crank–Nicolson method, both with smoothing.

efficiently without essentially influencing the accuracy of the underlying discretization. The following highlights the different operator splitting methods presented in this chapter.

For multidimensional models the directional splitting methods of the ADI type offer a fast, accurate, and easy-to-implement way for the numerical time stepping. They are adapted to effectively deal with mixed spatial derivative terms, which are ubiquitous in finance. ADI schemes lead to a sequence of sparse linear subproblems that can be solved by LU decomposition at optimal computational cost, that is, the number of required operations is directly proportional to the number of unknowns. The MCS and HV schemes, with a proper choice of their parameter θ, are recommended as these show stability and second-order convergence and reveal a better inherent smoothing than second-order CS.

The spatial discretization of jumps models for the underlying asset price yields dense matrices. All classical implicit time discretization schemes require solving systems with these dense matrices. By employing an IMEX method like the IMEX–CNAB scheme advocate here, with an explicit treatment of (finite activity) jumps and an implicit treatment of the remainder of the operator, each time step involves only multiplications with these dense matrices. This is computationally a much easier task and can be often performed very fast using FFT. The accuracy and stability of the IMEX–CNAB scheme are good when the jump activity is not very high, e.g., less than several jumps per year.

Iterative methods like the PSOR method for solving LCPs resulting from the pricing of American-style options often converge slowly. We discussed an operator splitting method based on a Lagrange multiplier formulation, treating in each time step the early exercise constraint and complementarity condition in separate subproblems, where the main subproblem is essentially the same as for the European-style counterpart. With this approach it is easy to adapt a European option pricer to American options. We presented such an adaptation for ADI and IMEX methods. Also, it is applicable for most models of underlying asset prices. Numerical experience with this operator splitting method indicates that the accuracy stays essentially the same as in the case of the original LCP, but there can be a major reduction in computational time.

References

1. Achdou, Y., Pironneau, O.: Computational methods for option pricing, *Frontiers in Applied Mathematics*, vol. 30. Society for Industrial and Applied Mathematics (SIAM), Philadelphia, PA (2005)
2. Almendral, A., Oosterlee, C.W.: Numerical valuation of options with jumps in the underlying. Appl. Numer. Math. **53**(1), 1–18 (2005)
3. Andersen, L., Andreasen, J.: Jump-diffusion processes: Volatility smile fitting and numerical methods for option pricing. Rev. Deriv. Res. **4**(3), 231–262 (2000)
4. Andersen, L.B.G., Piterbarg, V.V.: Interest rate modeling, volume I: foundations and vanilla models. Atlantic Financial Press (2010)
5. Bank for International Settlements: Triennial Central Bank Survey, Foreign exchange turnover in April 2013: preliminary global results (2013)
6. Bates, D.S.: Jumps and stochastic volatility: Exchange rate processes implicit in Deutsche Mark options. Review Financial Stud. **9**(1), 69–107 (1996)
7. Black, F., Scholes, M.: The pricing of options and corporate liabilities. J. Political Economy **81**, 637–654 (1973)
8. Brandt, A., Cryer, C.W.: Multigrid algorithms for the solution of linear complementarity problems arising from free boundary problems. SIAM J. Sci. Statist. Comput. **4**(4), 655–684 (1983)

9. Brennan, M.J., Schwartz, E.S.: The valuation of American put options. J. Finance **32**, 449–462 (1977)
10. Briani, M., Natalini, R., Russo, G.: Implicit-explicit numerical schemes for jump-diffusion processes. Calcolo **44**(1), 33–57 (2007)
11. Carr, P., Geman, H., Madan, D.B., Yor, M.: The fine structure of asset returns: an empirical investigation. J. Business **75**, 305–332 (2002)
12. Carr, P., Mayo, A.: On the numerical evaluation of option prices in jump diffusion processes. Eur. J. Finance **13**, 353–372 (2007)
13. Chorin, A.J.: Numerical solution of the Navier-Stokes equations. Math. Comp. **22**, 745–762 (1968)
14. Clark, I.J.: Foreign exchange option pricing. Wiley, Chichester (2011)
15. Clarke, N., Parrott, K.: Multigrid for American option pricing with stochastic volatility. Appl. Math. Finance **6**, 177–195 (1999)
16. Cont, R., Tankov, P.: Financial modelling with jump processes. Chapman & Hall/CRC, Boca Raton, FL (2004)
17. Cont, R., Voltchkova, E.: A finite difference scheme for option pricing in jump diffusion and exponential Lévy models. SIAM J. Numer. Anal. **43**(4), 1596–1626 (2005)
18. Craig, I.J.D., Sneyd, A.D.: An alternating-direction implicit scheme for parabolic equations with mixed derivatives. Comput. Math. Appl. **16**, 341–350 (1988)
19. Cryer, C.W.: The solution of a quadratic programming problem using systematic overrelaxation. SIAM J. Control **9**, 385–392 (1971)
20. Dang, D.M., Christara, C.C., Jackson, K.R., Lakhany, A.: A PDE pricing framework for cross-currency interest rate derivatives. Proc. Comp. Sc. **1**, 2371–2380 (2010)
21. Davis, T.A.: Direct methods for sparse linear systems, *Fundamentals of Algorithms*, vol. 2. Society for Industrial and Applied Mathematics (SIAM), Philadelphia, PA (2006)
22. d'Halluin, Y., Forsyth, P.A., Vetzal, K.R.: Robust numerical methods for contingent claims under jump diffusion processes. IMA J. Numer. Anal. **25**(1), 87–112 (2005)
23. Douglas, J., Rachford, H.H.: On the numerical solution of heat conduction problems in two and three space variables. Trans. Amer. Math. Soc. **82**, 421–439 (1956)
24. Egloff, D.: GPUs in financial computing part III: ADI solvers on GPUs with application to stochastic volatility. Wilmott Magazine (March), 51–53 (2011)
25. Ekström, E., Tysk, J.: The Black–Scholes equation in stochastic volatility models. J. Math. Anal. Appl. **368**, 498 –507 (2010)
26. Feng, L., Linetsky, V.: Pricing options in jump-diffusion models: an extrapolation approach. Oper. Res. **56**(2), 304–325 (2008)
27. Forsyth, P.A., Vetzal, K.R.: Quadratic convergence for valuing American options using a penalty method. SIAM J. Sci. Comput. **23**(6), 2095–2122 (2002)
28. Frank, J., Hundsdorfer, W., Verwer, J.G.: On the stability of implicit-explicit linear multistep methods. Appl. Numer. Math. **25**(2–3), 193–205 (1997)
29. George, A.: Nested dissection of a regular finite element mesh. SIAM J. Numer. Anal. **10**, 345–363 (1973). Collection of articles dedicated to the memory of George E. Forsythe
30. Glowinski, R.: Numerical methods for nonlinear variational problems. Scientific Computation. Springer-Verlag, New York (1984)
31. Glowinski, R.: Splitting methods for the numerical solution of the incompressible Navier-Stokes equations. In: Vistas in applied mathematics, Transl. Ser. Math. Engrg., pp. 57–95. Optimization Software, New York (1986)
32. Grzelak, L.A., Oosterlee, C.W.: On the Heston model with stochastic interest rates. SIAM J. Financial Math. **2**(1), 255–286 (2011)
33. Grzelak, L.A., Oosterlee, C.W., Van Weeren, S.: Extension of stochastic volatility equity models with the Hull-White interest rate process. Quant. Finance **12**(1), 89–105 (2012)
34. Haentjens, T.: ADI schemes for the efficient and stable numerical pricing of financial options via multidimensional partial differential equations. PhD thesis, University of Antwerp (2013)
35. Haentjens, T.: Efficient and stable numerical solution of the Heston–Cox–Ingersoll–Ross partial differential equation by alternating direction implicit finite difference schemes. Int. J. Comput. Math. **90**(11), 2409–2430 (2013)

36. Haentjens, T., in 't Hout, K.J.: Alternating direction implicit finite difference schemes for the Heston–Hull–White partial differential equation. J. Comput. Finance **16**(1), 83–110 (2012)
37. Haentjens, T., in 't Hout, K.J.: ADI schemes for pricing American options under the Heston model Appl. Math. Finan. **22**, 207–237 (2015)
38. Heston, S.L.: A closed-form solution for options with stochastic volatility with applications to bond and currency options. Review Financial Stud. **6**, 327–343 (1993)
39. in 't Hout, K.J., Foulon, S.: ADI finite difference schemes for option pricing in the Heston model with correlation. Int. J. Numer. Anal. Model. **7**(2), 303–320 (2010)
40. in 't Hout, K.J., Mishra, C.: Stability of the modified Craig–Sneyd scheme for two-dimensional convection-diffusion equations with mixed derivative term. Math. Comput. Simulation **81**(11), 2540–2548 (2011)
41. in 't Hout, K.J., Mishra, C.: Stability of ADI schemes for multidimensional diffusion equations with mixed derivative terms. Appl. Numer. Math. **74**, 83–94 (2013)
42. in 't Hout, K.J., Welfert, B.D.: Stability of ADI schemes applied to convection-diffusion equations with mixed derivative terms. Appl. Numer. Math. **57**(1), 19–35 (2007)
43. in 't Hout, K.J., Welfert, B.D.: Unconditional stability of second-order ADI schemes applied to multi-dimensional diffusion equations with mixed derivative terms. Appl. Numer. Math. **59**(3–4), 677–692 (2009)
44. Huang, J., Pang, J.S.: Option pricing and linear complementarity. J. Comput. Finance **2**, 31–60 (1998)
45. Hull, J.C.: Options, futures and other derivatives. Pearson Education, Harlow (2011)
46. Hull, J.C., White, A.: Pricing interest-rate-derivative securities. Review Financial Stud. **3**, 573–592 (1990)
47. Hundsdorfer, W.: Accuracy and stability of splitting with Stabilizing Corrections. Appl. Numer. Math. **42**, 213–233 (2002)
48. Hundsdorfer, W., Verwer, J.G.: Numerical solution of time-dependent advection-diffusion-reaction equations, *Computational Mathematics*, vol. 33. Springer, Berlin (2003)
49. Ikonen, S., Toivanen, J.: Operator splitting methods for American option pricing. Appl. Math. Lett. **17**(7), 809–814 (2004)
50. Ikonen, S., Toivanen, J.: Componentwise splitting methods for pricing American options under stochastic volatility. Int. J. Theor. Appl. Finance **10**(2), 331–361 (2007)
51. Ikonen, S., Toivanen, J.: Pricing American options using LU decomposition. Appl. Math. Sci. **1**(49–52), 2529–2551 (2007)
52. Ikonen, S., Toivanen, J.: Efficient numerical methods for pricing American options under stochastic volatility. Numer. Methods Partial Differential Equations **24**(1), 104–126 (2008)
53. Ikonen, S., Toivanen, J.: Operator splitting methods for pricing American options under stochastic volatility. Numer. Math. **113**(2), 299–324 (2009)
54. Itkin, A., Carr, P.: Jumps without tears: a new splitting technology for barrier options. Int. J. Numer. Anal. Model. **8**(4), 667–704 (2011)
55. Jaillet, P., Lamberton, D., Lapeyre, B.: Variational inequalities and the pricing of American options. Acta Appl. Math. **21**(3), 263–289 (1990)
56. Kou, S.G.: A jump-diffusion model for option pricing. Management Sci. **48**(8), 1086–1101 (2002)
57. Kwon, Y., Lee, Y.: A second-order finite difference method for option pricing under jump-diffusion models. SIAM J. Numer. Anal. **49**(6), 2598–2617 (2011)
58. Kwon, Y., Lee, Y.: A second-order tridiagonal method for American options under jump-diffusion models. SIAM J. Sci. Comput. **33**(4), 1860–1872 (2011)
59. Lipton, A.: Mathematical methods for foreign exchange. World Scientific, Singapore (2001)
60. Marchuk, G.: Splitting and alternating direction methods. In: P. Ciarlet, P. Lions (eds.) Handbook of Numerical Analysis, vol. 1, pp. 197–462. North-Holland, Amsterdam (1990)
61. McKee, S., Mitchell, A.R.: Alternating direction methods for parabolic equations in two space dimensions with a mixed derivative. Computer J. **13**, 81–86 (1970)
62. McKee, S., Wall, D.P., Wilson, S.K.: An alternating direction implicit scheme for parabolic equations with mixed derivative and convective terms. J. Comput. Phys. **126**, 64–76 (1996)

63. Merton, R.C.: Theory of rational option pricing. Bell J. Econom. Management Sci. **4**, 141–183 (1973)
64. Merton, R.C.: Option pricing when underlying stock returns are discontinuous. J. Financial Econ. **3**, 125–144 (1976)
65. Oosterlee, C.W.: On multigrid for linear complementarity problems with application to American-style options. Electron. Trans. Numer. Anal. **15**, 165–185 (2003)
66. Peaceman, D.W., Rachford, H.H.: The numerical solution of parabolic and elliptic differential equations. J. Soc. Ind. Appl. Math. **3**, 28–41 (1955)
67. Rannacher, R.: Finite element solution of diffusion problems with irregular data. Numer. Math. **43**(2), 309–327 (1984)
68. Reisinger, C., Wittum, G.: On multigrid for anisotropic equations and variational inequalities: pricing multi-dimensional European and American options. Comput. Vis. Sci. **7**(3–4), 189–197 (2004)
69. Ruge, J.W., Stüben, K.: Algebraic multigrid. In: Multigrid methods, *Frontiers Appl. Math.*, vol. 3, pp. 73–130. SIAM, Philadelphia, PA (1987)
70. Saad, Y., Schultz, M.H.: GMRES: a generalized minimal residual algorithm for solving nonsymmetric linear systems. SIAM J. Sci. Statist. Comput. **7**(3), 856–869 (1986)
71. Salmi, S., Toivanen, J.: An iterative method for pricing American options under jump-diffusion models. Appl. Numer. Math. **61**(7), 821–831 (2011)
72. Salmi, S., Toivanen, J.: Comparison and survey of finite difference methods for pricing American options under finite activity jump-diffusion models. Int. J. Comput. Math. **89**(9), 1112–1134 (2012)
73. Salmi, S., Toivanen, J.: IMEX schemes for option pricing under jump-diffusion models. Appl. Numer. Math. **84**, 33–45 (2014)
74. Seydel, R.U.: Tools for computational finance. Springer, London (2012)
75. Shreve, S.E.: Stochastic calculus for finance II. Springer, New York (2008)
76. Stüben, K.: Algebraic multigrid: An introduction with applications. In: Multigrid. Academic Press Inc., San Diego, CA (2001)
77. Tavella, D., Randall, C.: Pricing financial instruments: The finite difference method. John Wiley & Sons (2000)
78. Toivanen, J.: Numerical valuation of European and American options under Kou's jump-diffusion model. SIAM J. Sci. Comput. **30**(4), 1949–1970 (2008)
79. Toivanen, J.: A componentwise splitting method for pricing American options under the Bates model. In: Applied and numerical partial differential equations, *Comput. Methods Appl. Sci.*, vol. 15, pp. 213–227. Springer, New York (2010)
80. Toivanen, J.: Finite difference methods for early exercise options. In: R. Cont (ed.) Encyclopedia of Quantitative Finance. John Wiley & Sons (2010)
81. Toivanen, J., Oosterlee, C.W.: A projected algebraic multigrid method for linear complementarity problems. Numer. Math. Theory Methods Appl. **5**(1), 85–98 (2012)
82. Trottenberg, U., Oosterlee, C.W., Schüller, A.: Multigrid. Academic Press Inc., San Diego, CA (2001). With contributions by A. Brandt, P. Oswald and K. Stüben
83. Verwer, J.G., Spee, E.J., Blom, J.G., Hundsdorfer, W.: A second-order Rosenbrock method applied to photochemical dispersion problems. SIAM J. Sci. Comput. **20**, 1456–1480 (1999)
84. van der Vorst, H.A.: Bi-CGSTAB: A fast and smoothly converging variant of Bi-CG for the solution of nonsymmetric linear systems. SIAM J. Sci. Stat. Comp. **13**, 631–644 (1992)
85. Wilmott, P.: Derivatives. John Wiley & Sons Ltd., Chichester (1998)

Chapter 17
A Numerical Method to Solve Multi-Marginal Optimal Transport Problems with Coulomb Cost

Jean-David Benamou, Guillaume Carlier, and Luca Nenna

Abstract In this chapter, we present a numerical method, based on iterative Bregman projections, to solve the optimal transport problem with Coulomb cost. This problem is related to the strong interaction limit of Density Functional Theory. The first idea is to introduce an entropic regularization of the Kantorovich formulation of the Optimal Transport problem. The regularized problem then corresponds to the projection of a vector on the intersection of the constraints with respect to the Kullback-Leibler distance. Iterative Bregman projections on each marginal constraint are explicit which enables us to approximate the optimal transport plan. We validate the numerical method against analytical test cases.

1 Introduction

1.1 On Density Functional Theory

Quantum mechanics for a molecule with N electrons can be studied in terms of the many-electron Schrödinger equation for a wave function $\psi \in L^2(\mathbb{R}^{3N};\mathbb{C})$ (in this chapter, we neglect the spin variable). The practical limitation of this approach is computational: in order to predict the chemical behavior of H_2O (10 electrons)

J.-D. Benamou (✉) • L. Nenna
INRIA, MOKAPLAN, Domaine de Voluceau Le Chesnay, France

CEREMADE, Université Paris Dauphine, Paris, France
e-mail: jean-david.benamou@inria.fr; luca.nenna@inria.fr

G. Carlier
CEREMADE, Université Paris Dauphine, Paris, France

INRIA, MOKAPLAN, Domaine de Voluceau Le Chesnay, France
e-mail: carlier@ceremade.dauphine.fr

R. Glowinski et al. (eds.), *Splitting Methods in Communication, Imaging, Science, and Engineering*, Scientific Computation, DOI 10.1007/978-3-319-41589-5_17

using a 10 gridpoints discretization of $\mathbb{R}$, we need to solve the Schrödinger equation at 10^{30} gridpoints. This is why Hohenberg, Kohn, and Sham introduced, in [20] and [22], the Density Functional Theory (DFT) as an approximate computational method for solving the Schrödinger equation at a more reasonable cost.

The main idea of the DFT is to compute only the marginal density for one electron

$$\rho(x_1) = \int \gamma_N(x_1, x_2 \cdots, x_N) dx_2 \cdots dx_N,$$

where $\gamma_N = |\psi(x_1, \cdots, x_N)|^2$ is the joint probability density of electrons at positions $x_1, \cdots, x_N \in \mathbb{R}^3$, instead of the full wave function ψ. One scenario of interest for the DFT is when the repulsion between the electrons largely dominates over the kinetic energy. In this case, the problem can, at least formally, be reformulated as an Optimal Transport (OT) problem as emphasized in the pioneering works of Buttazzo, De Pascale, and Gori-Giorgi [6] and Cotar, Friesecke, and Klüppelberg [11].

1.2 Optimal Transport

Before discussing the link between DFT and OT, let us recall the standard optimal transport problem and its extension to the multi-marginal framework. Given two probability distributions μ and ν (on $\mathbb{R}^d$, say) and a transport cost c: $\mathbb{R}^d \times \mathbb{R}^d \to \mathbb{R}$, the optimal transport problem consists in finding the minimal cost to transport μ to ν for the cost c. A transport map between μ and ν is a Borel map T such that $T_\#\mu = \nu$, i.e., $\nu(A) = \mu(T^{-1}(A))$ for every Borel subset A of $\mathbb{R}^d$. The Monge problem (which dates back to 1781 when Monge [27] formulated the problem of finding the optimal way to move a pile of dirt to a hole of the same volume) then reads

$$\min_{T_\#\mu=\nu} \int_{\mathbb{R}^d} c(x, T(x))\mu(dx). \tag{17.1}$$

This is a delicate problem since the mass conservation constraint $T_\#\mu = \nu$ is highly nonlinear (and the feasible set may even be empty for instance if μ is a Dirac measure and ν is not). This is why, in 1942, Kantorovich [21] proposed a relaxed formulation of (17.1) which allows mass splitting

$$\min_{\gamma\in\Pi(\mu,\nu)} \int_{\mathbb{R}^d\times\mathbb{R}^d} c(x,y)\gamma(dx,dy) \tag{17.2}$$

where $\gamma \in \Pi(\mu, \nu)$ consists of all probability measures on $\mathbb{R}^d \times \mathbb{R}^d$ having μ and ν as marginals, that is:

$$\gamma(A \times \mathbb{R}) = \mu(A), \quad \forall A \text{ Borel subset of } \mathbb{R}^d, \tag{17.3}$$

$$\gamma(\mathbb{R} \times B) = \nu(B), \quad \forall B \text{ Borel subset of } \mathbb{R}^d. \tag{17.4}$$

Note that this is a linear programming problem and that there exists solutions under very mild assumptions (e.g., c continuous and μ and ν compactly supported). A solution γ of (17.2) is called an optimal transport plan and it gives the probability that a mass element in x be transported in y. Let us remark that if T is a transport map then it induces a transport plan $\gamma_T(x,y) := \mu(x)\delta(y-T(x))$; so if an optimal plan of (17.2) has the form γ_T (which means that no splitting of mass occurs and γ is concentrated on the graph of T) then T is actually an optimal transport map, i.e., a solution to (17.1). The linear problem (17.2) also has a convenient dual formulation

$$\max_{u,v|u(x)+v(y)\leq c(x,y)} \int_{\mathbb{R}^d} u(x)\mu(dx) + \int_{\mathbb{R}^d} v(y)\nu(dy) \tag{17.5}$$

where $u(x)$ and $v(y)$ are the so-called Kantorovich potentials. OT theory for two marginals has developed very rapidly during the 25 last years; there are well-known conditions on c, μ, and ν which guarantee that there is a unique optimal plan which is in fact induced by a map (e.g., $c = |x-y|^2$ and μ absolutely continuous, see Brenier [4]) and we refer to the textbooks of Villani [38, 39] for a detailed exposition.

Let us now consider the multi-marginal problems, i.e., OT problems involving N marginals $\mu_1, \cdots, \mu_N$ and a cost $c : \mathbb{R}^{dN} \to \mathbb{R}$, which leads to the following generalization of (17.2)

$$\min_{\gamma\in\Pi(\mu_1,\cdots,\mu_N)} \int_{\mathbb{R}^{\times\mathrm{N}}} c(x_1,\cdots,x_N)\gamma(dx_1,\cdots,dx_N) \tag{17.6}$$

where $\Pi(\mu_1,\cdots,\mu_N)$ is the set of probability measures on $(\mathbb{R}^d)^N$ having $\mu_1,\cdots,\mu_N$ as marginals. The corresponding Monge problem then becomes

$$\min_{T_{i\#}\mu_1=\mu_i,\, i=2,\cdots,N} \int_{\mathbb{R}^d} c(x_1, T_2(x_1), \cdots, T_N(x_1))\mu_1(dx_1). \tag{17.7}$$

Such multi-marginals problems first appeared in the work of Gangbo and Święch [17] who solved the quadratic cost case and proved the existence of Monge solutions. In recent years, there has been a lot of interest in such multi-marginal problems because they arise naturally in many different settings such as economics [7, 32], polar factorization of vector fields and theory of monotone maps [18] and, of course, DFT [6, 11, 9, 15, 25, 12], as recalled below. Few results are known about the structure of optimal plans for (17.7) apart from the general results of Brendan Pass [31], in particular the case of *repulsive costs* such as the Coulomb's cost from DFT is an open problem.

The chapter is structured as follows: In Section 2, we recall the link between Density Functional Theory and Optimal Transportation and we present some analytical solutions of the OT problem (e.g., optimal maps for radially symmetric marginals, for 2 electrons). In Section 3, we introduce a numerical method, based on iterative Bregman projections, and an algorithm which aims at refining the mesh where the transport plan is concentrated. In Section 4 we present some numerical results. Section 5 concludes.

2 From Density Functional Theory to Optimal Transportation

2.1 Optimal Transportation with Coulomb Cost

In Density Functional Theory [20] the ground state energy of a system (with N electrons) is obtained by minimizing the following functional w.r.t. the electron density $\rho(x)$:

$$E[\rho] = \min_{\rho \in \mathscr{R}} F_{HK}[\rho] + \int_{\mathbb{R}^3} v_{ext}(x)\rho(x)dx \tag{17.8}$$

where $\mathscr{R} = \{\rho : \mathbb{R}^3 \to \mathbb{R} | \rho \geq 0, \sqrt{\rho} \in H^1(\mathbb{R}^3), \int_{\mathbb{R}^3} \rho(x)dx = N\}$,

$v_{ext} := -\dfrac{Z}{|x-R|}$ is the electron-nuclei potential (Z and R are the charge and the position of the nucleus, respectively) and F_{HK} is the so-called Hohenberg-Kohn functional, which is defined by minimizing over all wave functions ψ which yield ρ:

$$F_{HK}[\rho] = \min_{\psi \to \rho} \hbar^2 T[\psi] + V_{ee}[\psi], \tag{17.9}$$

where $\hbar^2$ is a semiclassical constant factor,

$$T[\psi] = \frac{1}{2} \int \cdots \int \sum_{i=1}^{N} |\nabla_{x_i} \psi|^2 dx_1 \cdots dx_N$$

is the kinetic energy and

$$V_{ee} = \int \cdots \int \sum_{i=1}^{N} \sum_{j>i}^{N} \frac{1}{|x_i - x_j|} |\psi|^2 dx_1 \cdots dx_N$$

is the Coulomb repulsive energy operator.

Let us now consider the *Semiclassical* limit

$$\lim_{\hbar \to 0} \min_{\psi \to \rho} \hbar^2 T[\psi] + V_{ee}[\psi]$$

and assume that taking the minimum over ψ commutes with passing to the limit $\hbar \to 0$ (Cotar, Friesecke and Klüppelberg in [11] proved it for $N = 2$), we obtain the following functional

$$V_{ee}^{SCE}[\rho] = \min_{\psi \to \rho} \int \cdots \int \sum_{i=1}^{N} \sum_{j>i}^{N} \frac{1}{|x_i - x_j|} |\psi|^2 dx_1 \cdots dx_N \tag{17.10}$$

where V_{ee}^{SCE} is the minimal Coulomb repulsive energy whose minimizer characterizes the state of *Strictly Correlated Electrons* (SCE).

Problem (17.10) gives rise to a multi-marginal optimal transport problem as (17.6) by considering that

- according to the indistinguishability of electrons, all the marginals are equal to ρ,

- the cost function c originates from the electron-electron Coulomb repulsion, that is

$$c(x_1,\ldots,x_N)=\sum_{i=1}^{N}\sum_{j>i}^{N}\frac{1}{|x_i-x_j|}, \tag{17.11}$$

- $\gamma_N=|\psi(x_1,\cdots,x_N)|^2$ (which is the joint probability density of electrons at positions $x_1,\cdots,x_N\in\mathbb{R}^3$) is the transport plan.

The Coulomb cost function (17.11) is different from the costs usually considered in OT as it is not bounded at the origin and it decreases with distance. So it requires a generalized formal framework, but this framework is beyond the scope of this work (see [6] and [11]). Finally (17.10) can be re-formulated as a Kantorovich problem

$$V_{ee}^{SCE}[\rho]=\min_{\pi_i(\gamma_N)=\rho,i=1,\cdots,N}\int_{\mathbb{R}^{3N}}c(x_1,\cdots,x_N)\gamma_N(x_1,\cdots,x_N)dx_1\cdots dx_N \tag{17.12}$$

where

$$\pi_i(\gamma_N)=\int_{\mathbb{R}^{3(N-1)}}\gamma_N(x_1,\cdots,x_i,\cdots,x_N)dx_1,\cdots,dx_{i-1},dx_{i+1},\cdots,dx_N$$

is the i-th marginal. As mentioned in Section 1.2, if the optimal transport plan γ_N has the following form

$$\gamma_N(x_1,\cdots,x_N)=\rho(x_1)\delta(x_2-f_2^\star(x_1))\cdots\delta(x_N-f_N^\star(x_1)) \tag{17.13}$$

then the functions $f_i^\star:\mathbb{R}^3\to\mathbb{R}^3$ are the optimal transport maps (or *co-motion* functions) of the Monge problem

$$\begin{aligned}V_{ee}^{SCE}[\rho]&=\min_{\{f_i:\mathbb{R}^3\to\mathbb{R}^3\}_{i=1}^N}\int\sum_{i=1}^{N}\sum_{j>i}^{N}\frac{1}{|f_i(x)-f_j(x)|}\rho(x)dx\\ s.t.\quad & f_{i\#}\rho=\rho,\ i=2,\ldots,N,\quad f_1(x)=x.\end{aligned} \tag{17.14}$$

Remark 1. Physical Meaning of the Co-motion Function The quantity $f_i(x)$ determines the position of the i-th electron in terms of x which is the position of the "1st"electron: V_{ee}^{SCE} defines a system with the maximum possible correlation between the relative electronic positions.

In full generality, problem (17.14) is delicate and proving the existence of the co-motion functions is difficult. However, the co-motion functions can be obtained via semianalytic formulations for spherically symmetric atoms and strictly 1D systems (see [11, 37, 24, 9]) and we will give some examples in the following section.

Problem (17.12) admits a useful dual formulation in which the so-called Kantorovich potential u plays a central role

$$V_{ee}^{SCE}=\max_u\{N\int u(x)\rho(x)dx\quad s.t.\quad \sum_{i=1}^{N}u(x_i)\le c(x_1,\ldots,x_N)\}. \tag{17.15}$$

Because c is invariant under permutations, there is a single dual Kantorovich potential for all marginal constraints. Moreover, this potential $u(x)$ is related to the co-motion functions via the classical equilibrium equation (see [37])

$$\nabla u(x) = -\sum_{i=2}^{N} \frac{x - f_i(x)}{|x - f_i(x)|^3}. \tag{17.16}$$

Remark 2. (Physical Meaning of (17.16)) The gradient of the Kantorovich potential equals the total net force exerted on the electron in x by electrons in $f_2(x), \cdots, f_N(x)$.

2.2 Analytical Examples

2.2.1 The Case $N = 2$ and $d = 1$

In order to better understand the problem we have formulated in the previous section, we recall some analytical examples (see [6] for the details).

Let us consider 2 particles in one dimension and marginal densities

$$\rho_1(x) = \rho_2(x) = \begin{cases} a & if\ |x| \leq a/2 \\ 0 & otherwise. \end{cases} \tag{17.17}$$

After a few computations, we obtain the following associated co-motion function

$$f(x) = \begin{cases} x + \frac{a}{2} & if\ x \leq 0 \\ x - \frac{a}{2} & otherwise \end{cases}. \tag{17.18}$$

If we take

$$\rho_1(x) = \rho_2(x) = \frac{a - |x|}{a^2} \quad defined \quad in \quad [-a, a], \tag{17.19}$$

we get

$$f(x) = \frac{x}{|x|}(\sqrt{2a|x| - x^2} - a) \quad on \quad [-a, a] \tag{17.20}$$

Figure 17.1 shows the co-motion functions for (17.17) and (17.19).

2.2.2 The Case $N > 2$ and $d = 1$

In [9], the authors proved the existence of optimal transport maps for problem (17.14) in dimension $d = 1$ and provided an explicit construction of the optimal maps. Let ρ be the normalized electron density and $-\infty = x_0 < x_1 < \cdots < x_N = +\infty$ be such that

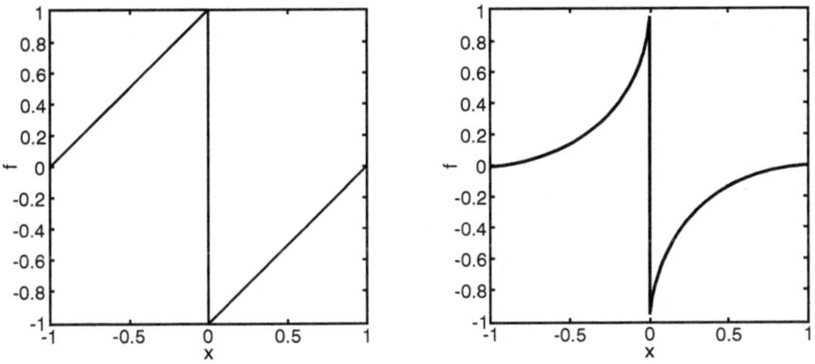

Fig. 17.1 *Right: Co-motion function for (17.17) with $a = 2$. Left: Co-motion function for (17.19) with $a = 1$.*

$$\int_{x_i}^{x_{i+1}} \rho(x)dx = 1/N \; \forall i = 0, \cdots, N-1.$$

Thus, there exists a unique increasing function $\tilde{f} : \mathbb{R} \to \mathbb{R}$ on each interval $[x_i, x_{i+1}]$ such that for every test-function φ one has

$$\int_{[x_i, x_{i+1}]} \varphi(\tilde{f}(x))\rho(x)dx = \int_{[x_{i+1}, x_{i+2}]} \varphi(x)\rho(x)dx \quad \forall i = 0, \cdots, N-2, \tag{17.21}$$

$$\int_{[x_{N-1}, x_N]} \varphi(\tilde{f}(x))\rho(x)dx = \int_{[x_0, x_1]} \varphi(x)\rho(x)dx, \tag{17.22}$$

The optimal maps are then given by

$$f_2(x) = \tilde{f}(x) \tag{17.23}$$

$$f_i(x) = f_2^{(i)}(x) \quad \forall i = 2, \cdots, N, \tag{17.24}$$

where $f_2^{(i)}$ stands for the i-th composition of f_2 with itself. Here, we present an example given in [6]. We consider the case where ρ is the Lebesgue measure on the unit interval $I = [0, 1]$, the construction above gives the following optimal co-motion functions

$$\begin{aligned} f_2(x) &= \begin{cases} x + 1/3 & if \quad x \leq 2/3 \\ x - 2/3 & if \quad x > 2/3 \end{cases}, \\ f_3(x) = f_2(f_2(x)) &= \begin{cases} x + 2/3 & if \quad x \leq 1/3 \\ x - 1/3 & if \quad x > 1/3 \end{cases}. \end{aligned} \tag{17.25}$$

Furthermore, we know that the Kantorovich potential u satisfies the relation (here we take $N = 3$)

$$u'(x) = -\sum_{i=2}^{N} \frac{x - f_i(x)}{|x - f_i(x)|^3} \tag{17.26}$$

and by substituting the co-motion functions in (17.26) (and integrating it) we get

$$u(x) = \begin{cases} \frac{45}{4}x & 0 \le x \le 1/3 \\ \frac{15}{4} & 1/3 \le x \le 2/3 \\ -\frac{45}{4}x + \frac{45}{4} & 2/3 \le x \le 1 \end{cases} \tag{17.27}$$

Figure 17.2 illustrates this example.

When $N \geq 4$ similar arguments as above can be developed and we can similarly compute the co-motion functions and the Kantorovich potential.

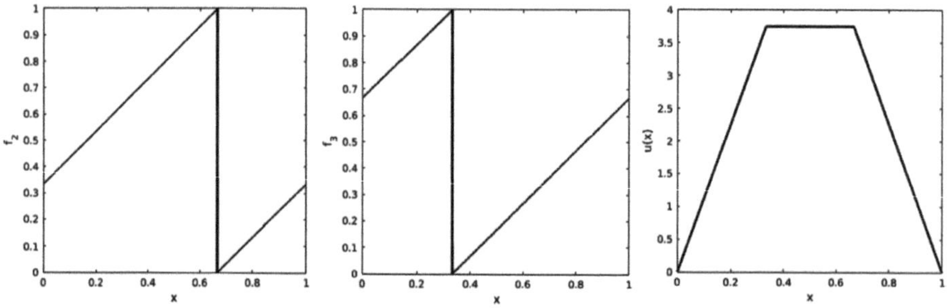

Fig. 17.2 *Right: co-motion function f_2 for (17.25). Center: co-motion function f_3 for (17.25). Left: Kantorovich Potential $u(x)$ (17.27).*

2.2.3 The Radially Symmetric Marginal Case for $N = 2, d \geq 2$

We discuss now the radial $d-$dimensional ($d \geq 2$) case for $N = 2$. We assume that the marginal ρ is radially symmetric, then we recall the following theorem from [11]:

Theorem 1. *[11] Suppose that $\rho(x) = \rho(|x|)$, then the optimal transport map is given by*

$$f^{\star}(x) = \frac{x}{|x|}g(|x|), \quad x \in \mathbb{R}^d, \tag{17.28}$$

with $g(r) = -F_2^{-1}(F_1(r))$, $F_1(t) := C(d)\int_0^t \rho(s)s^{d-1}ds$, $F_2(t) := C(d)\int_t^{\infty} \rho(s)s^{d-1}ds$ where $C(d)$ denotes the measure of S^{d-1}, the unit sphere in $\mathbb{R}^d$.

Example 1. (Spherical Coordinates System) If ρ is radially symmetric $\rho(x) = \rho(|x|)$, it is convenient to work in spherical coordinates and then to set for every radius $r > 0$

$$\lambda(r) = C(d)r^{d-1}\rho(r) \tag{17.29}$$

so that for every test-function φ we have

$$\int_{\mathbb{R}^d} \varphi(x)\rho(|x|)dx = \int_0^{+\infty} \Big(\int_{S^{d-1}} \varphi(r,\omega)\frac{d\sigma(\omega)}{C_d}\Big)\lambda(r)dr$$

with $C(d)$ the measure of S^{d-1} and σ the $d-1$ measure on S^{d-1} which in particular implies that $\lambda := |.|_{\#}\rho$ i.e.

$$\int_{\mathbb{R}^d} \varphi(|x|)\rho(|x|)dx = \int_0^{+\infty} \varphi(r)\lambda(r)dr, \; \forall \varphi \in C_c(\mathbb{R}_+). \tag{17.30}$$

The radial part of the optimal co-motion function $a(r) = -g(r)$ can be computed by solving the ordinary differential equation

$$a'(r)\lambda(a(r)) = \lambda(r)$$

which gives

$$\int_0^{a(r)} \lambda(s)ds = 2 - \int_0^r \lambda(s)ds. \tag{17.31}$$

We define $R(r) = \int_0^r \lambda(s)ds$, since $r \mapsto R(r)$ is increasing, its inverse $R^{-1}(w)$ is well defined for $w \in [0,1)$. Thus, we see that $a(r)$ has the form

$$a(r) = R^{-1}(2 - R(r)). \tag{17.32}$$

2.2.4 Reducing the Dimension Under Radial Symmetry

In the case where the marginal $\rho(x) = \rho(|x|)$ is radially symmetric, the multi-marginal problem with Coulomb cost

$$M(\rho) := \inf_{\gamma \in \Pi(\rho,\cdots,\rho)} \int_{\mathbb{R}^{dN}} c(x_1,\cdots,x_N)d\gamma(x_1,\cdots,x_N) \tag{17.33}$$

with c the Coulomb cost given by (17.11) involves plans on $\mathbb{R}^{dN}$, which is very costly to discretize. Fortunately, due to the symmetries of the problem, it can actually be solved by considering a multi-marginal problem only on $\mathbb{R}_+^N$. Let us indeed define for every $(r_1,\cdots,r_N) \in (0,+\infty)^N$:

$$\widetilde{c}(r_1,\cdots,r_N) := \inf\{c(x_1,\cdots,x_N) \; : \; |x_1| = r_1,\cdots,|x_N| = r_N\}. \tag{17.34}$$

Defining λ by (17.29) (or equivalently (17.30)) and defining $\Pi(\lambda,\cdots,\lambda)$ as the set of probability measures on $\mathbb{R}_+^N$ having each marginal equal to λ, consider

$$\widetilde{M}(\lambda) := \inf_{\widetilde{\gamma} \in \Pi(\lambda,\cdots,\lambda)} \int_{\mathbb{R}_+^N} \widetilde{c}(r_1,\cdots,r_N)d\widetilde{\gamma}(r_1,\cdots,r_N). \tag{17.35}$$

We then have

Lemma 1. $M(\rho) = \widetilde{M}(\lambda)$.

Proof. The inequality $M(\rho) \geq \widetilde{M}(\lambda)$ is easy: take $\gamma \in \Pi(\rho, \cdots, \rho)$ and define its *radial component* $\widetilde{\gamma}$ by

$$\int_{\mathbb{R}_+^N} F(r_1, \cdots, r_N) d\widetilde{\gamma}(r_1, \cdots, r_N) := \int_{\mathbb{R}^{dN}} F(|x_1|, \cdots, |x_N|) d\gamma(x_1, \cdots, x_N), \ \forall F \in C_c(\mathbb{R}_+^N), \quad (17.36)$$

it is obvious that $\widetilde{\gamma} \in \Pi(\lambda, \cdots, \lambda)$ and since $c(x_1, \cdots, x_N) \geq \widetilde{c}(|x_1|, \cdots, |x_N|)$, the inequality $M(\rho) \geq \widetilde{M}(\lambda)$ easily follows. To show the converse inequality, we use duality. Indeed, by standard convex duality, we have

$$M(\rho) = K(\rho) := \sup_u \Big\{ N \int_{\mathbb{R}^d} u(x) \rho(x) dx \ : \ \sum_{i=1}^N u(x_i) \leq c(x_1, \cdots, x_N) \Big\} \quad (17.37)$$

and similarly

$$\widetilde{M}(\lambda) = \widetilde{K}(\lambda) := \sup_v \Big\{ N \int_{\mathbb{R}_+} v(r) \lambda(r) dr \ : \ \sum_{i=1}^N v(r_i) \leq \widetilde{c}(r_1, \cdots, r_N) \Big\}. \quad (17.38)$$

Now since ρ is radially symmetric and the constraint of (17.37) is invariant by changing u by $u \circ R$ with R a rotation (see (17.11)), there is no loss of generality in restricting the maximization in (17.37) to potentials of the form $u(x_i) = w(r_i)$, but then the constraint of (17.37) implies that w satisfies the constraint of (17.38). Then we have $M(\rho) = K(\rho) \leq \widetilde{K}(\lambda) = \widetilde{M}(\lambda)$.

Note that $\gamma \in \Pi(\rho, \cdots, \rho)$ solves (17.33) if and only if its radial component $\widetilde{\gamma}$ solves (17.33) and $c(x_1, \cdots, x_N) = \widetilde{c}(|x_1|, \cdots, |x_N|)$ γ-a.e. Therefore (17.33) gives the optimal radial component, whereas the extra condition $c(x_1, \cdots, x_N) = \widetilde{c}(|x_1|, \cdots, |x_N|)$ γ-a.e. gives an information on the angular distribution of γ.

3 Iterative Bregman Projections

Numerics for multi-marginal problems have so far not been extensively developed. Discretizing the multi-marginal problem leads to the linear program (17.41) where the number of constraints grows exponentially in N, the number of marginals.

In a recent paper, Carlier, Oberman, and Oudet studied the matching for teams problem and they managed to reformulate the problem as a linear program whose number of constraints grows only linearly in N [8].

Here, we present a numerical method which is not based on linear programming techniques, but on an entropic regularization and the so-called alternate projection method. It has recently been applied to various optimal transport problems in [13] and [2].

The initial idea goes back to von Neumann [28, 29] who proved that the sequence obtained by projecting orthogonally iteratively onto two affine subspaces converges to the projection of the initial point onto the intersection of these affine subspaces. Since the seminal work of Bregman [3], it is by now well known that one can extend this idea not only to several affine subspaces (the extension to convex sets is due to Dyskstra but we won't use it in the sequel) but also by replacing the Euclidean distance by a general Bregman divergence associated with some suitable strictly and differentiable convex function f (possibly with a domain) where we recall that the Bregman divergence associated with f is given by

$$D_f(x,y) = f(x) - f(y) - \langle \nabla f(y), x-y \rangle. \tag{17.39}$$

In what follows, we shall only consider the Bregman divergence (also known as the Kullback-Leibler distance) associated with the Boltzmann/Shannon entropy $f(x) := \sum_i x_i(\log x_i - 1)$ for non-negative x_i. This Bregman divergence (restricted to probabilities, i.e., imposing the normalization $\sum_i x_i = 1$) is the Kullback-Leibler distance or relative entropy:

$$D_f(x,y) = \sum_i x_i \log\left(\frac{x_i}{y_i}\right).$$

Bregman distances are used in many other applications most notably image processing, see [19] for instance.

3.1 The Discrete Problem and Its Entropic Regularization

In this section we introduce the discrete problem solved using the iterative Bregman projections [3]. From now on, we consider the problem (17.12)

$$\min_{\gamma_N \in \mathscr{C}} \int_{(\mathbb{R}^d)^N} c(x_1, \cdots, x_N) \gamma_N(x_1, \cdots, x_N) dx_1 \cdots dx_N, \tag{17.40}$$

where N is the number of marginals (or electrons), $c(x_1, \ldots, x_N)$ is the Coulomb cost, γ_N the transport plan, is the probability distribution over $(\mathbb{R}^d)^N$ and $\mathscr{C} := \bigcap_{i=1}^N \mathscr{C}_i$ with $\mathscr{C}_i := \{\gamma_N \in Prob\{(\mathbb{R}^d)^N\} | \pi_i \gamma_N = \rho\}$ (we remind the reader that electrons are indistinguishable so the N marginals coincide with ρ).

In order to discretize (17.40), we use a discretization with M_d points of the support of the kth electron density as $\{x_{j_k}\}_{j_k=1,\cdots,M_d}$. If the densities ρ are approximated by $\sum_{j_k} \rho_{j_k} \delta_{x_{j_k}}$, we get

$$\min_{\gamma \in \mathscr{C}} \sum_{j_1, \cdots j_N} c_{j_1, \cdots, j_N} \gamma_{j_1, \cdots, j_N}, \tag{17.41}$$

where $c_{j_1,\cdots,j_N} = c(x_{j_1}, \cdots, x_{j_N})$ and the transport plan support for each coordinate is restricted to the points $\{x_{j_k}\}k = 1, \cdots, M_d$ thus becoming a $(M_d)^N$ matrix again denoted γ with elements $\gamma_{j_1,\cdots,j_N}$. The marginal constraints $\mathscr{C}_i$ becomes

$$\mathscr{C}_i := \{\gamma \in \mathbb{R}_+^{(M_d)^N} \mid \sum_{j_1,\dots,j_{i-1},j_{i+1},\dots,j_N} \gamma_{j_1,\dots,j_N} = \rho_{j_i}, \forall j_i = 1,\cdots,M_d\}. \quad (17.42)$$

Recall that the electrons are indistinguishable, meaning that they have same densities : $\rho_{j_k} = \rho_{j_{k'}}, \forall j, \forall k \neq k'$.

The discrete optimal transport problem (17.41) is a linear program problem and is dual to the discretization of (17.15)

$$\begin{aligned} &\max_{u_j} \sum_{j=1}^{M} N u_j \rho_j \\ &s.t. \sum_{i=1}^{N} u_{j_i} \leq c_{j_1 \cdots j_N} \quad \forall j_i = 1,\cdots,M_d \end{aligned} \quad (17.43)$$

where $u_j = u_{j_i} = u(x_{j_i})$. Thus the primal (17.41) has $(M_d)^N$ unknowns and $M_d \times N$ linear constraints and the dual (17.43) only M_d unknowns but still $(M_d)^N$ constraints. They are computationally not solvable with standard linear programming methods even for small cases in the multi-marginal case.

A different approach consists in computing the problem (17.41) regularized by the entropy of the joint coupling. This regularization dates to E. Schrödinger [36] and it has been recently introduced in machine learning [13] and economics [16] (we refer the reader to [2] for an overview of the entropic regularization and the iterative Bregman projections in OT). Thus, we consider the following discrete regularized problem

$$\min_{\gamma \in \mathscr{C}} \sum_{j_1,\cdots j_N} c_{j_1,\cdots,j_N} \gamma_{j_1,\cdots,j_N} + \varepsilon E(\gamma), \quad (17.44)$$

where $E(\gamma)$ is defined as follows

$$E(\gamma) = \begin{cases} \sum_{j_1,\cdots j_N} \gamma_{j_1,\cdots,j_N} \log(\gamma_{j_1,\cdots,j_N}) \text{ if } \gamma \geq 0 \\ +\infty \text{ otherwise.} \end{cases} \quad (17.45)$$

After elementary computations, we can rewrite the problem as

$$\min_{\gamma \in \mathscr{C}} KL(\gamma|\bar{\gamma}) \quad (17.46)$$

where $KL(\gamma|\bar{\gamma}) = \sum_{i_1,\dots,i_N} \gamma_{i_1,\dots,i_N} \log\left(\frac{\gamma_{i_1,\dots,i_N}}{\bar{\gamma}_{i_1,\dots,iN}}\right)$ is the Kullback-Leibler distance and

$$\bar{\gamma}_{i_1,\dots,i_N} = e^{-\frac{c_{j_1,\cdots,j_N}}{\varepsilon}}. \quad (17.47)$$

As explained in Section 1.2, when the transport plan γ is concentrated on the graph of a transport map which solves the Monge problem, after discretization of the densities, this property is lost. However we still expect the γ matrix to be sparse. The entropic regularization will spread the support and this helps to stabilize the computation: it defines a strongly convex program with a unique solution γ^ε which can be obtained through elementary operations (we detail this in Section 3.3 for both the continuous and discrete framework). The regularized solutions γ^ε then converge to $\gamma^\star$, the solution of (17.41) with minimal entropy, as $\varepsilon \to 0$ (see [10] and [23] in our case for detailed proofs of convergence). We also remark that the choice of the regularization parameter ε is quite delicate for two reasons: *(i)* some of the quantities in the proposed algorithms could become smaller than machine precision whenever ε is small; *(ii)* the convergence speed deteriorates significantly as $\varepsilon \to 0$, see Tables 17.1 and 17.2.

3.2 Alternate Projections

The main idea of the iterative Bregman projections (we call it *Bregman* as the Kullback-Leibler distance is also called Bregman distance, see [3]) is to construct a sequence γ^n (which converges to the minimizer of (17.46)) by alternately projecting on each set $\mathscr{C}_i$ with respect to the Kullback-Leibler distance. Thus, the iterative KL (or Bregman) projections can be written

$$\begin{cases} \gamma^0 & = \bar{\gamma} \\ \gamma^n & = P^{KL}_{\mathscr{C}_n}(\gamma^{n-1}) \quad \forall n > 0 \end{cases} \tag{17.48}$$

where we have extended the indexing of the set by $N-$periodicity such that $\mathscr{C}_{n+N} = \mathscr{C}_n \quad \forall n \in \mathbb{N}$ and $P^{KL}_{\mathscr{C}_n}$ denotes the KL projection on $\mathscr{C}_n$.

The convergence of γ^n to the unique solution of (17.46) is well known, it actually holds for large classes of Bregman distances and in particular the Kullback-Leibler divergence as was proved by Bauschke and Lewis [1]

$$\gamma^n \to P^{KL}_{\mathscr{C}}(\bar{\gamma}) \text{ as } n \to \infty.$$

Remark 3. If the convex sets $\mathscr{C}_i$ are not affine sub-sets (that is not our case), γ^n converges toward a point of the intersection which is not the KL projection of $\bar{\gamma}$ anymore so that a correction term is needed as provided by Dykstra's algorithm (we refer the reader to [2]).

The KL projection on $\mathscr{C}_i \quad i = 1, \ldots, N$ can be computed explicitly as detailed in the following proposition

Proposition 1. *For* $\bar{\gamma} \in (\mathbb{R}_+)^{M_d{}^N}$ *the projection* $P^{KL}_{\mathscr{C}_i}(\bar{\gamma})$ *is given by*

$$P^{KL}_{\mathscr{C}_i}(\bar{\gamma})_{j_1,\ldots,j_N} = \rho_{j_i} \frac{\bar{\gamma}_{j_1,\ldots,j_N}}{\sum_{k_1,\ldots,k_{i-1},k_{i+1},\ldots,k_N} \bar{\gamma}_{k_1,\ldots,k_N}} \quad \forall j_i = 1, \ldots, M_d. \tag{17.49}$$

Proof. Introducing Lagrange multipliers λ_{j_i} associated with the constraint $\mathscr{C}_i$

$$\sum_{j_1,\ldots,j_{i-1},j_{i+1},\ldots,j_N} \gamma_{j_1,\ldots,j_N} = \rho_{j_i} \tag{17.50}$$

the KL projection is given by the optimality condition :

$$\log\left(\frac{\gamma_{j_1,\ldots,j_N}}{\bar{\gamma}_{j_1,\ldots,j_N}}\right) - \lambda_{j_i} = 0 \tag{17.51}$$

so that

$$\gamma_{j_1,\ldots,j_N} = C_{j_i}\bar{\gamma}_{j_1,\ldots,j_N}, \tag{17.52}$$

where $C_{j_i} = e^{\lambda_{j_i}}$. If we substitute (17.52) in (17.50), we get

$$C_{j_i} = \rho_{j_i}\frac{1}{\sum_{k_1,\ldots,k_{i-1},k_{i+1},\ldots,k_N}\bar{\gamma}_{k_1,\ldots,k_N}} \tag{17.53}$$

which gives (17.49).

3.3 From the Alternate Projections to the Iterative Proportional Fitting Procedure

The alternate projection procedure (17.48) is performed on M_d^N matrices. Moreover each projection (17.49) involves computing partial sum of this matrix. The total operation cost of each Bregman iteration scales like $O(M_d^{2N-1})$.

In order to reduce the cost of the problem, we use an equivalent formulation of the Bregman algorithm known as the Iterative Proportional Fitting Procedure (IPFP). Let us consider the problem (17.46) in a continuous measure setting and, for simplicity, 2-marginals framework

$$\min_{\{\gamma|\pi_1(\gamma)=\rho,\pi_2(\gamma)=\rho\}} \int \log\left(\frac{d\gamma}{d\bar{\gamma}}\right) d\gamma, \tag{17.54}$$

where ρ, γ, and $\bar{\gamma}$ are nonnegative measures. The aim of the IPFP is to find the KL projection of $\bar{\gamma}$ on $\Pi(\rho,\rho)$ (see (17.47) for the definition of $\bar{\gamma}$ which depends on the cost function).

Under the assumption that the value of (17.54) is finite, Rüschendorf and Thomsen (see [34]) proved that a unique KL-projection γ^* exists and that it is of the form

$$\gamma^*(x,y) = a(x)b(y)\bar{\gamma}(x,y), \quad a(x) \geq 0, \quad b(y) \geq 0. \tag{17.55}$$

From now on, we consider (with a slightly abuse of notation) Borel measures with densities γ, $\bar{\gamma}$ and ρ w.r.t. the suitable Lebesgue measure. The functions a and b can be uniquely determined by the marginal condition as follows

$$\begin{aligned} a(x) &= \frac{\rho(x)}{\int \bar{\gamma}(x,y)b(y)dy}, \\ b(y) &= \frac{\rho(y)}{\int \bar{\gamma}(x,y)a(x)dx}. \end{aligned} \tag{17.56}$$

Then, IPFP is defined by the following recursion

$$b_0 = 1, \quad a_0 = \rho,$$

$$\begin{aligned} b_{n+1}(y) &= \frac{\rho(y)}{\int \bar{\gamma}(x,y)a_n(x)dx}, \\ a_{n+1}(x) &= \frac{\rho(x)}{\int \bar{\gamma}(x,y)b_{n+1}(y)dy}. \end{aligned} \tag{17.57}$$

Moreover, we can define the sequence of joint densities (and of the corresponding measures)

$$\gamma^{2n}(x,y) := a^n(x)b^n(y)\bar{\gamma}(x,y) \quad \gamma^{2n+1} := a^n(x)b^{n+1}(y)\bar{\gamma}(x,y), \quad n \geq 0. \tag{17.58}$$

Rüschendorf proved (see [33]) that γ^n converges to the KL-projection of $\bar{\gamma}$. We can, now, recast the IPFP in a discrete framework, which reads as

$$\gamma_{ij} = a_i b_j \bar{\gamma}_{ij}, \, b_j^0 = 1, \quad a_i^0 = \rho_i, \tag{17.59}$$

$$\begin{aligned} b_j^{n+1} &= \frac{\rho_j}{\sum_i \bar{\gamma}_{ij} a_i^n}, \\ a_i^{n+1} &= \frac{\rho_i}{\sum_j \bar{\gamma}_{ij} b_j^{n+1}}, \end{aligned} \tag{17.60}$$

$$\gamma_{ij}^{2n} = a_i^n \bar{\gamma}_{ij} b_j^n \quad \gamma_{ij}^{2n+1} = a_i^n \bar{\gamma}_{ij} b_j^{n+1}. \tag{17.61}$$

By definition of γ_{ij}^n, notice that

$$\bar{\gamma}_{ij} b_j^n = \frac{\gamma_{ij}^{2n-1}}{a_i^{n-1}} \text{ and } a_i^n \bar{\gamma}_{ij} = \frac{\gamma_{ij}^{2n}}{b_j^n}$$

and if (17.61) is rewritten as follows

$$\begin{aligned} \gamma_{ij}^{2n} &= \rho_i \frac{\bar{\gamma}_{ij} b_j^n}{\sum_k \bar{\gamma}_{ik} b_k^n} \\ \gamma_{ij}^{2n+1} &= \rho_j \frac{\bar{\gamma}_{ij} a_i^n}{\sum_k \bar{\gamma}_{kj} a_k^n} \end{aligned} \tag{17.62}$$

then we obtain

$$\begin{aligned} \gamma_{ij}^{2n} &= \rho_i \frac{\gamma_{ij}^{2n-1}}{\sum_k \gamma_{ik}^{2n-1}} \\ \gamma_{ij}^{2n+1} &= \rho_j \frac{\gamma_{ij}^{2n}}{\sum_k \gamma_{kj}^{2n}}. \end{aligned} \tag{17.63}$$

Thus, we exactly recover the Bregman algorithm described in the previous section, for 2 marginals.

The extension to the multi-marginal framework is straightforward but cumbersome to write. It leads to a problem set on N M_d-dimensional vectors $a_{j,i_{(\cdot)}}, \quad j = 1,\cdots,N, \quad i_{(\cdot)} = 1,\cdots,M_d$. Each update takes the form

$$a_{j,i_j}^{n+1} = \frac{\rho_{i_j}}{\sum_{i_1,i_2,\ldots i_{j-1},i_{j+1},\ldots,i_N} \bar{\gamma}_{i_1,\ldots,i_N} a_{1,i_1}^{n+1} a_{2,i_2}^{n+1} \cdots a_{j-1,i_{j-1}}^{n+1} a_{j+1,i_{j+1}}^{n} \cdots a_{N,i_N}^{n}}, \tag{17.64}$$

Where each i_k takes values in $\{1,\cdots,M_d\}$.

Note that we still need a constant M_d^N cost matrix $\bar{\gamma}$. Thanks to the symmetry and separability properties of the cost function (see (17.11) and (17.47)), it is possible to replace it by a $N(N-1)/2$ product of M_d^2 matrices. This is already a big improvement from the storage point of view. Further simplifications are under investigations but the brute force IPFP operational cost therefore scales like $O(N M_d^{N+1})$ which provides a small improvement over the Bregman iterates option.

3.4 A Heuristic Mesh Refinement Strategy

We will use a heuristic mesh refinement strategy allowing to obtain more accuracy without increasing the computational cost and memory requirements. This idea was introduced in [30] and [35] for the adaptative resolution of the pure linear programming formulation of the Optimal Transportation problem, i.e., without the entropic regularization. It can also be remotely connected to the multiscale approach in [26] which does not use the linear programming approach at all.

If the optimal transport plan is supported by a lower dimensional set, we expect the entropic regularization to be concentrated on a mollified version of this set. Its width should decrease with the entropic parameter ε if the discretization is fine enough. Working with a fixed ε, the idea is to apply coarse to fine progressive resolution and work with a sparse matrix γ. At each level, values below a threshold are filtered out (set to 0), then new positive values are interpolated on a finer grid (next level) where γ is strictly positive.

To simplify the exposition, we describe the algorithm for $2-$marginals in $1D$ and take a $\sqrt{M}$ gridpoints discretization of $I = [a,b] \in \mathbb{R}$:

1. we start with a cartesian M gridpoints mesh on $I \times I$ to approximate transport plan γ^{ε}, obtained by running the IPFP on a coarse grid.

2. we take $m_c(j) = max_i \gamma^{\varepsilon}_{ij}$ and $m_r(i) = max_j \gamma^{\varepsilon}_{ij}$ which are the maximum values over the rows and over the columns respectively, and we define

$$m = \min[\min_j(m_c(j)), \min_i(m_r(i))].$$

We will refine the grid only inside the level curve $\gamma^{\varepsilon} = \xi m$ where we expect the finer solution is supported, see Figure 17.3.

3. In order to keep approximately the same number of elements in the sparse matrix γ at each level we refine the grid as follows : Let $\mathscr{T} := \{(i,j) | \gamma^{\varepsilon}_{ij} \geq \xi m\}$ and $M_{\mathscr{T}} := \sharp \mathscr{T}$ and $r := M_{\mathscr{T}}/M$, then the size of the grid at the next level is $M^{new} = M/r$.

4. We compute the interpolation $\gamma_{M^{new}}$ of the old transport plan γ_M on the finer grid.
5. Elements of $\gamma_{M^{new}}$ below the fixed threshold ξm are filtered out, i.e., are fixed to 0 and are not used in the IPFP sum computations, see Figure 17.3.
6. Finally, a new IPFP computation is performed and it can be initialized with an interpolation of the data at the previous level ($\bar{\gamma}$ can be easily recomputed on the gridpoints where $\gamma_{M^{new}}$ is strictly positive).

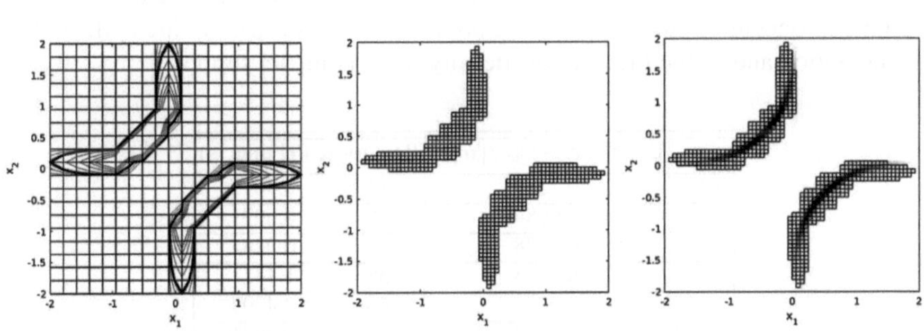

Fig. 17.3 *Left:* $\mathscr{T}$ *is the set of grid points inside the level curve* $\gamma = \xi m$ $(\xi = 0.9)$ *(the bold line curve). Center: The new grid after the refinement. Right: The transport Plan after a new IPFP computation*

4 Numerical Results

4.1 $N = 2$ *Electrons: Comparison Between Numerical and Analytical Results*

In order to validate the numerical method, we now compare some numerical results for 2 electrons in dimension $d = 1,2,3$ with the analytical results from Section 2.2. Let us first consider a uniform density (as (17.17) with $a = 2$) in $1D$. In Table 17.1, we analyze the performance of the numerical method by varying the parameter ε. We notice that the error becomes smaller by decreasing the regularizing parameter, but the drawback is that the method needs more iterations to converge. Figure 17.4 shows the Kantorovich potential, the co-motion function which can be recovered from the potential by using (17.16) and the transport plan. The simulation is performed with a discretization of (17.17) with $a = 2$, $M = 1000$ (gridpoints) and $\varepsilon = 0.004$.

As explained in Section 2.2.3, we can also compute the co-motion for a radially symmetric density. We have tested the method in $2D$ and $3D$, Figures 17.5 and 17.6 respectively, by using the normalized uniform density on the unit ball. Moreover, in the radial case we have proved that the OT problem can be reduced to a $1-$dimensional problem by computing $\widetilde{c}$ which is trivial for the 2 electrons case: let us set the problem in $2D$ in polar coordinates (r_1, θ_1) and (r_2, θ_2), for the first and the second electron respectively (without loss of generality we can set $\theta_1 = 0$), then it is easy to verify that the minimum is achieved with $\theta_2 = \pi$. Figure 17.5 shows the Kantorovich potential (the radial component $v(r)$ as defined in Section 2.2.4), the co-motion and the transport plan for the $2-$dimensional case, the simulation is performed with $M = 1000$ and $\varepsilon = 0.002$. In Figure 17.6 we present the result for th $3-$dimensional case, the simulation is performed with $M = 1000$ and $\varepsilon = 0.002$.

Remark 4. One can notice that, in the case of a uniform density, the transport plan presents a concentration of mass on the boundaries. This is a combined effect of the regularization and of the fact that the density has a compact support.

ε	Error ($\|u^{\varepsilon} - u\|_{\infty}/\|u\|_{\infty}$)	Iteration	CPU time (s)
0.256	0.1529	11	0.4017
0.128	0.0984	16	0.5977
0.064	0.0578	25	0.9282
0.032	0.0313	38	1.4411
0.016	0.0151	66	2.4297
0.008	0.0049	114	4.2674
0.004	0.0045	192	7.0638

Table 17.1 *Numerical results for uniform density in 1D. u^{ε} is the numerical Kantorovich potential and u is the analytical one.*

4.2 $N = 2$ Electrons in Dimension $d = 3$: Helium Atom

Once we have validated the method with some analytical examples, we solve the regularized problem for the Helium atom by using the electron density computed in [14]. In Figure 17.7, we present the electron density, the Kantorovich potential, and the transport plan. The simulation is performed with a discretization of $[0,4]$ with $M = 1000$ and $\varepsilon = 5\,10^{-3}$. We can notice the potential correctly fits the asymptotic behavior from [37], namely $v(r) \sim \dfrac{N-1}{|r|}$ for $r \to \infty$, where N is the number of electrons.

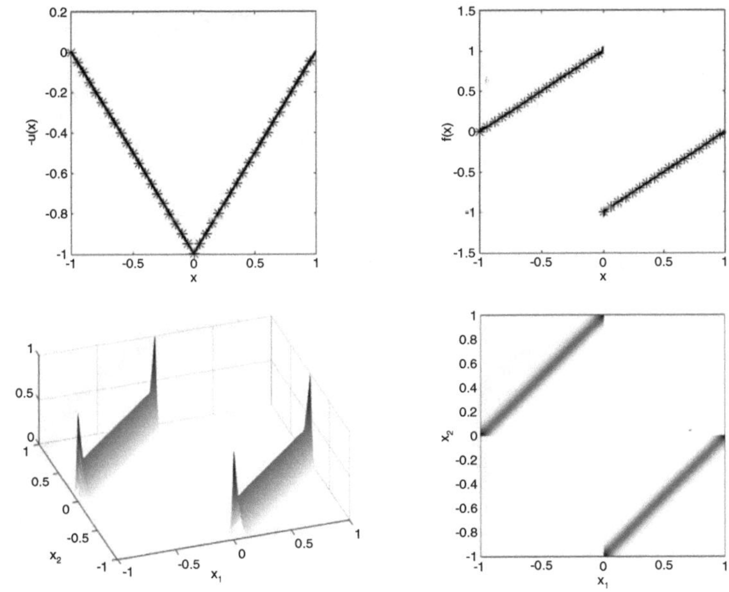

Fig. 17.4 *Top-Left: Kantorovich Potential $u(x)$. Top-Right: Numerical co-motion function (solid line) and analytical co-motion (star-solid line) . Bottom-Left: Transport plan $\tilde{\gamma}$. Bottom-Right: Support of $\tilde{\gamma}$.*

4.3 $N = 3$ *Electrons in Dimension* $d = 1$

We present now some results for the 1−dimensional multi-marginal problem with $N = 3$. They are validated against the analytical solutions given in Section 2.2.2. We recall that splitting ρ into three tertiles ρ_i with equal mass, we will have $\rho_1 \to \rho_2$, $\rho_2 \to \rho_3$ and $\rho_3 \to \rho_1$.

In Table 17.2, we present the performance of the method for a uniform density on $[0,1]$ by varying ε and, as expected, we see the same behavior as in the 2 marginals case. Figure 17.8 shows the Kantorovich potential and the projection of the transport plan onto two marginals (namely $\gamma^2 = \pi_{12}(\gamma^\varepsilon)$). The support gives the relative positions of two electrons.

The simulation is performed on a discretization of $[0,1]$ with a uniform density, $M = 1000$ and $\varepsilon = 0.02$. If we focus on the support of the projected transport plan we can notice that the numerical solution correctly reproduces the prescribed behavior. The concentration of mass is again due to the compact support of the density, which is not the case of the Gaussian as one can see in Figure 17.9. In Figure 17.9 we present the numerical results for $\rho(x) = e^{-x^2}/\sqrt{\pi}$. The simulation is performed on the discretization of $[-2.5, 2.5]$ with $M = 1000$ and $\varepsilon = 0.008$.

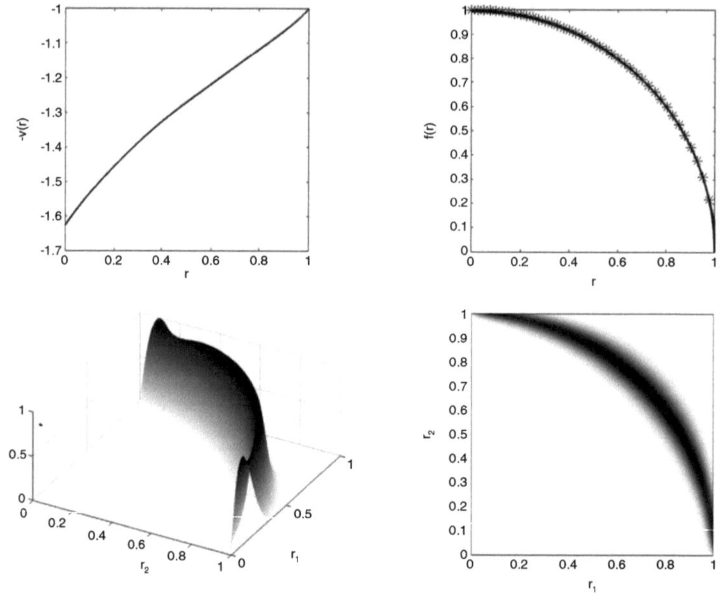

Fig. 17.5 *Top-Left: Kantorovich Potential* $v(r)$. *Top-Right: Numerical co-motion function (solid line) and analytical co-motion (star-solid line) . Bottom-Left: Transport plan* $\tilde{\gamma}$. *Bottom-Right: Support of* $\tilde{\gamma}$.

ε	Error ($\|u^{\varepsilon} - u\|_{\infty}/\|u\|_{\infty}$)	Iteration	CPU time (s)
0.32	0.0658	15	5.8372
0.16	0.0373	27	20.061
0.08	0.0198	64	55.718
0.04	0.0097	178	194.22
0.02	0.0040	374	542.63

Table 17.2 *Numerical results for uniform density in 1D and three electrons.* u^{ε} *is the numerical Kantorovich potential and* u *is the analytical one.*

4.4 $N = 3$ *Electrons in Dimension* $d = 3$ *Radial Case: Lithium Atom*

We finally perform some simulations for the radial 3−dimensional case for $N = 3$. As for the 3−dimensional case with 2 marginals we solve the reduced problem: let us consider the spherical coordinates (r_i, θ_i, ϕ_i) with $i = 1, \cdots, 3$ and we fix $\theta_1 = 0$ and $\phi_1 = \phi_2 = 0$ (the first electrons defines the z axis and the second one is on the xz plane). We then notice that $\phi_3 = 0$ as the electrons must be on the same plane of the nucleus to achieve compensation of forces (one can see it by computing the optimality conditions), so we have to minimize on θ_2 and θ_3 in order to obtain $\widetilde{c}$.

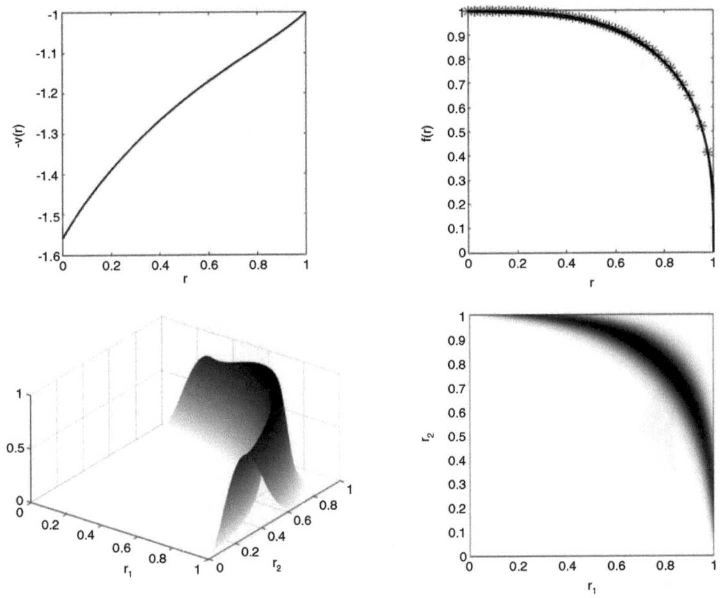

Fig. 17.6 *Top-Left: Kantorovich Potential $v(r)$. Top-Right: Numerical co-motion function (solid line) and analytical co-motion (star-solid line) . Bottom-Left: Transport plan $\tilde{\gamma}$. Bottom-Right: Support of $\tilde{\gamma}$.*

Figure 17.10 shows the electron density of the Lithium (computed in [5]), the Kantorovich Potential (and the asymptotic behavior) and the projection of the transport plan onto two marginals $\tilde{\gamma}^2 = \pi_{12}(\tilde{\gamma}^\varepsilon)$. The support gives the relative positions of two electrons.

The simulation is performed on a discretization of $[0,8]$ with $M = 300$ and $\varepsilon = 0.007$. Our results show (taking into account the regularization effect) a concentrated transport plan for this kind of density and they match analogous result obtained in [37].

5 Conclusion

We have presented a numerical scheme for solving multi-marginal OT problems arising from DFT. This is a challenging problem, not only because of the unusual features of the Coulomb cost which is singular and repulsive but also due to the high dimension of the space of transport plans.

Using an entropic regularization gives rise to a Kullback-Leibler projection problem onto the intersection of affine subsets given by the marginal constraints. Because each projection is explicit, one can use Bregman's iterative projection algorithm to approximate the solution.

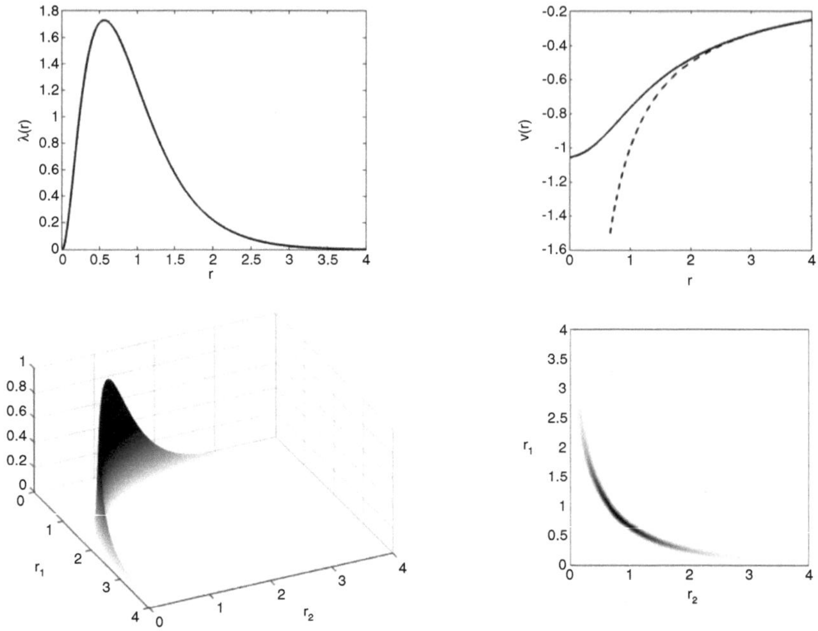

Fig. 17.7 *Top-Left: Helium density* $\lambda(r) = 4\pi r^2 \rho(r)$. *Top-Right: Kantorovich Potential* $v(r)$ *(blue) and asymptotic behavior (red)* $v(r) \sim \frac{1}{r}$ $r \to \infty$. *Bottom-Left: Transport plan* $\tilde{\gamma}$. *Bottom-Right: Support of* $\tilde{\gamma}$. *All quantities are in Hartree atomic units.*

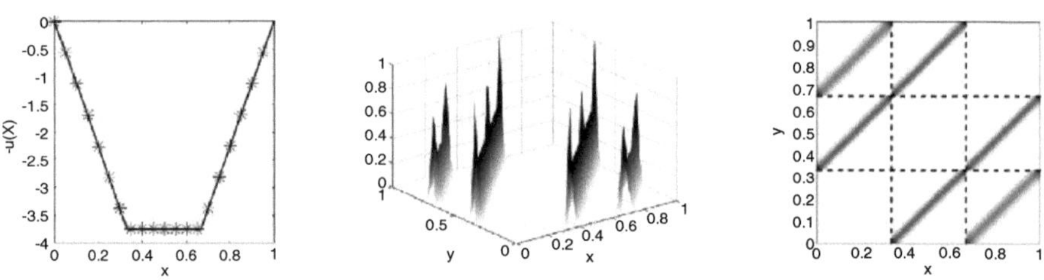

Fig. 17.8 *Left: Numerical Kantorovich potential* $u(x)$ *(solid line) and analytical potential (star-solid line). Center: Projection of the transport plan* $\pi_{12}(\gamma(x,y,z))$. *Right: Support of* $\pi_{12}(\gamma(x,y,z))$ *The dot-dashed lines delimit the intervals where* ρ_i, *with* $i = 1, \cdots, 3$, *are defined.*

The power of such an iterative projection approach was recently emphasized in [13, 2] for the entropic regularization of optimal transport problems. We showed that it is also well suited to treat the multi-marginal OT problem with Coulomb cost and leads to the same benefits in terms of convexification of the problem and simplicity of implementation. However, the general DFT problem in dimension 3 for a large number of electrons is unfeasible due to computational cost and we need to use radial symmetry simplification and also a heuristic refinement mesh strategy.

The choice of the regularization parameter ε remains to be investigated both from the analysis ([23] may be of help here) and numerical points of view when combined with the necessary refinement strategy.

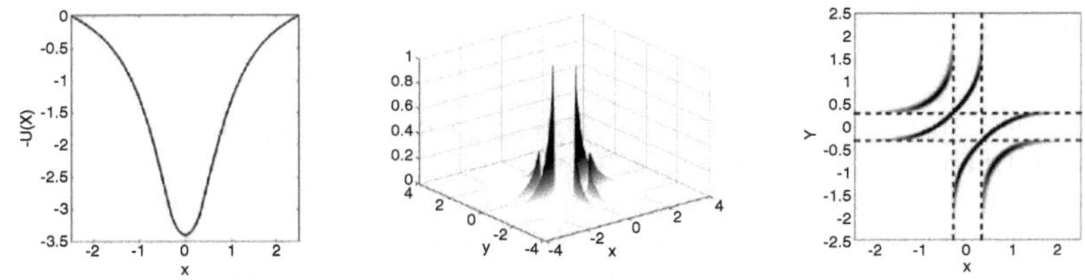

Fig. 17.9 *Left: Kantorovich potential $u(x)$. Center: Projection of the transport plan $\pi_{12}(\gamma(x,y,z))$. Right: Support of $\pi_{12}(\gamma(x,y,z))$. The dot-dashed lines delimit the intervals where ρ_i, with $i = 1, \cdots, 3$, are defined.*

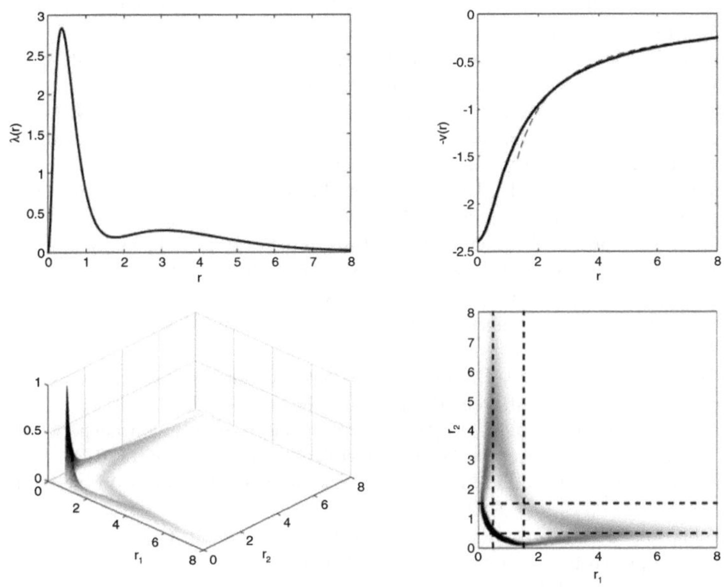

Fig. 17.10 *Top-Left: lithium density $\lambda(r) = 4\pi r^2 \rho(r)$. Top-Right: Kantorovich Potential $v(r)$ (blue) and asymptotic behavior (red) $v(r) \sim \frac{2}{r}$ $r \to \infty$. Bottom-Left: Projection of the Transport plan $\tilde{\gamma}^2 = \pi_{12}(\tilde{\gamma}^\varepsilon)$. Bottom-Right: Support of the projected transport plan $\tilde{\gamma}^2$. The dot-dashed lines delimit the three regions that the electrons must occupy, we computed them numerically following the idea in [37]. All quantities are in Hartree atomic units.*

Acknowledgements

We would like to thank Adam Oberman and Brendan Pass for many helpful and stimulating discussions as well as Paola Gori-Giorgi for sharing numerical details concerning the Helium and Lithium atom.

We gratefully acknowledge the support of the ANR, through the project ISOTACE (ANR-12-MONU-0013), and INRIA through the “action exploratoire” MOKAPLAN.

References

1. Bauschke, H.H., Lewis, A.S.: Dykstra's algorithm with Bregman projections: a convergence proof. Optimization **48**(4), 409–427 (2000)
2. Benamou, J.D., Carlier, G., Cuturi, M., Nenna, L., Peyré, G.: Iterative Bregman projections for regularized transportation problems. arXiv preprint arXiv:1412.5154 (2014)
3. Bregman, L.M.: The relaxation method of finding the common point of convex sets and its application to the solution of problems in convex programming. USSR Computational Mathematics and Mathematical Physics **7**(3), 200–217 (1967)
4. Brenier, Y.: Polar factorization and monotone rearrangement of vector-valued functions. Comm. Pure Appl. Math. **44**(4), 375–417 (1991)
5. Bunge, C.: The full CI density of the Li atom has been computed with a very large basis set with 8 s functions and up to k functions (private communication)
6. Buttazzo, G., De Pascale, L., Gori-Giorgi, P.: Optimal-transport formulation of electronic density-functional theory. Phys. Rev. A **85**, 062,502 (2012)
7. Carlier, G., Ekeland, I.: Matching for teams. Econom. Theory **42**(2), 397–418 (2010)
8. Carlier, G., Oberman, A., Oudet, E.: Numerical methods for matching for teams and Wasserstein barycenters. arXiv preprint arXiv:1411.3602 (2014)
9. Colombo, M., De Pascale, L., Di Marino, S.: Multimarginal optimal transport maps for one-dimensional repulsive costs. Canad. J. Math. **67**, 350–368 (2015)
10. Cominetti, R., Martin, J.S.: Asymptotic analysis of the exponential penalty trajectory in linear programming. Mathematical Programming **67**(1–3), 169–187 (1994)
11. Cotar, C., Friesecke, G., Klüppelberg, C.: Density functional theory and optimal transportation with Coulomb cost. Communications on Pure and Applied Mathematics **66**(4), 548–599 (2013)
12. Cotar, C., Friesecke, G., Pass, B.: Infinite-body optimal transport with Coulomb cost. Calculus of Variations and Partial Differential Equations **54**(1), 717–742 (2013)
13. Cuturi, M.: Sinkhorn distances: Lightspeed computation of optimal transport. In: Advances in Neural Information Processing Systems (NIPS) 26, pp. 2292–2300 (2013)
14. Freund, D.E., Huxtable, B.D., Morgan, J.D.: Variational calculations on the helium isoelectronic sequence. Phys. Rev. A **29**, 980–982 (1984)
15. Friesecke, G., Mendl, C.B., Pass, B., Cotar, C., Klüppelberg, C.: N-density representability and the optimal transport limit of the Hohenberg-Kohn functional. Journal of Chemical Physics **139**(16), 164,109 (2013)
16. Galichon, A., Salanié, B.: Matching with trade-offs: Revealed preferences over competing characteristics. Tech. rep., Preprint SSRN-1487307 (2010)
17. Gangbo, W., Święch, A.: Optimal maps for the multidimensional Monge-Kantorovich problem. Comm. Pure Appl. Math. **51**(1), 23–45 (1998)
18. Ghoussoub, N., Maurey, B.: Remarks on multi-marginal symmetric Monge-Kantorovich problems. Discrete Contin. Dyn. Syst. **34**(4), 1465–1480 (2014)
19. Goldstein, T., Bresson, X., Osher, S.: Geometric applications of the split Bregman method: Segmentation and surface reconstruction. Journal of Scientific Computing **45**(1–3), 272–293 (2010)
20. Hohenberg, P., Kohn, W.: Inhomogeneous electron gas. Phys. Rev. **136**, B864–B871 (1964)
21. Kantorovich, L.: On the transfer of masses (in Russian). Doklady Akademii Nauk **37**(2), 227–229 (1942)
22. Kohn, W., Sham, L.J.: Self-consistent equations including exchange and correlation effects. Phys. Rev. **140**, A1133–A1138 (1965)
23. Léonard, C.: From the Schrödinger problem to the Monge-Kantorovich problem. Journal of Functional Analysis **262**(4), 1879–1920 (2012)
24. Malet, F., Gori-Giorgi, P.: Strong correlation in Kohn-Sham density functional theory. Phys. Rev. Lett. **109**, 246,402 (2012)
25. Mendl, C.B., Lin, L.: Kantorovich dual solution for strictly correlated electrons in atoms and molecules. Physical Review B **87**(12), 125,106 (2013)

26. Mérigot, Q.: A multiscale approach to optimal transport. In: Computer Graphics Forum, vol. 30, pp. 1583–1592. Wiley Online Library (2011)
27. Monge, G.: Mémoire sur la théorie des déblais et des remblais. De l'Imprimerie Royale (1781)
28. von Neumann, J.: On rings of operators. reduction theory. Annals of Mathematics **50**(2), pp. 401–485 (1949)
29. von Neumann, J.: Functional Operators. Princeton University Press, Princeton, NJ (1950)
30. Oberman, A., Yuanlong, R.: An efficient linear programming method for optimal transportation. In preparation
31. Pass, B.: Uniqueness and Monge solutions in the multimarginal optimal transportation problem. SIAM Journal on Mathematical Analysis **43**(6), 2758–2775 (2011)
32. Pass, B.: Multi-marginal optimal transport and multi-agent matching problems: uniqueness and structure of solutions. Discrete Contin. Dyn. Syst. **34**(4), 1623–1639 (2014)
33. Ruschendorf, L.: Convergence of the iterative proportional fitting procedure. The Annals of Statistics **23**(4), 1160–1174 (1995)
34. Ruschendorf, L., Thomsen, W.: Closedness of sum spaces and the generalized Schrodinger problem. Theory of Probability and its Applications **42**(3), 483–494 (1998)
35. Schmitzer, B.: A sparse algorithm for dense optimal transport. In: Scale Space and Variational Methods in Computer Vision, pp. 629–641. Springer, Berlin Heidelberg (2015)
36. Schrodinger, E.: Uber die umkehrung der naturgesetze. Sitzungsberichte Preuss. Akad. Wiss. Berlin. Phys. Math. **144**, 144–153 (1931)
37. Seidl, M., Gori-Giorgi, P., Savin, A.: Strictly correlated electrons in density-functional theory: A general formulation with applications to spherical densities. Phys. Rev. A **75**, 042,511 (2007)
38. Villani, C.: Topics in Optimal Transportation. Graduate Studies in Mathematics Series. American Mathematical Society (2003)
39. Villani, C.: Optimal Transport: Old and New. Springer, Berlin Heidelberg (2009)

Chapter 18
Robust Split-Step Fourier Methods for Simulating the Propagation of Ultra-Short Pulses in Single- and Two-Mode Optical Communication Fibers

Ralf Deiterding and Stephen W. Poole

Abstract Extensions of the split-step Fourier method (SSFM) for Schrödinger-type pulse propagation equations for simulating femto-second pulses in single- and two-mode optical communication fibers are developed and tested for Gaussian pulses. The core idea of the proposed numerical methods is to adopt an operator splitting approach, in which the nonlinear sub-operator, consisting of Kerr nonlinearity, the self-steepening and stimulated Raman scattering terms, is reformulated using Madelung transformation into a quasilinear first-order system of signal intensity and phase. A second-order accurate upwind numerical method is derived rigorously for the resulting system in the single-mode case; a straightforward extension of this method is used to approximate the four-dimensional system resulting from the nonlinearities of the chosen two-mode model. Benchmark SSFM computations of prototypical ultra-fast communication pulses in idealized single- and two-mode fibers with homogeneous and alternating dispersion parameters and also high nonlinearity demonstrate the reliable convergence behavior and robustness of the proposed approach.

R. Deiterding (✉)
Engineering and the Environment, University of Southampton Highfield Campus, Southampton, SO45 4NZ, United Kingdom
e-mail: r.deiterding@soton.ac.uk

S.W. Poole
Oak Ridge National Laboratory, Computer Science and Mathematics Division, P.O. Box 2008, MS6164, Oak Ridge, TN 37831, USA
e-mail: spoole@ornl.gov

R. Glowinski et al. (eds.), *Splitting Methods in Communication, Imaging, Science, and Engineering*, Scientific Computation, DOI 10.1007/978-3-319-41589-5_18

1 Introduction

As computational capabilities are continuously rising, so is the demand for enhanced networking speed. One possible approach for increasing data throughput is the design of networks with transmission speeds well in the Tb/s range. While the maximal single channel communication speed in demonstrated wavelength division multiplexing systems is generally below 100 Gb/s, cf. [7], we are in here concerned with the modeling of single- and two-mode optical fibers that are suitable in particular for long-distance data transmission.

At present, computational models for investigating the propagation of light pulses in fibers have been developed primarily for pulses with a temporal half width well in the pico-second regime. Pulses with half widths $T_0 \gg 1\,\mathrm{ps}$ are sufficient for representing even on-off-key modulated bit streams with up to $100\,\mathrm{Gb/s}$ frequency. However, bit streams in the Tb/s regime can only be represented with *ultra-fast* pulses satisfying $T_0 < 100\,\mathrm{fs}$. Yet, in the ultra-fast pulse regime nonlinear pulse self-steepening and nonlinear stimulated Raman scattering are not negligible anymore and an extended version of the Schrödinger-type pulse propagation equation has to be considered.

Numerical solutions of the Schrödinger-type pulse propagation equation are primarily obtained with split-step Fourier schemes that perform spatial propagation steps considering firstly only the linearities in the equation by discrete Fourier transformation and then secondly only the nonlinear terms. While the construction of such *split-step Fourier methods* (SSFM) is very well established, cf. [1, 10], the topic of how to incorporate both self-steepening and Raman scattering reliably into the SSFM has received little attention. Here, we will describe a new class of extended SSFM that properly consider the hyperbolic nature of the nonlinear sub-operator for single- and coupled two-mode optical communication fibers.

The chapter is organized as follows: In Section 2, we recall the governing equations of pulse propagation in single-mode fibers. Section 3 first discusses the construction principles of split-step Fourier methods and then proceeds by describing our new type of single-mode SSFM for ultra-fast pulses as first- and second-order accurate numerical schemes, cf. [5]. An ultra-fast Gaussian pulse benchmark confirming robust second-order accuracy of the overall SSFM and demonstrating its application for simulating pulse propagation through an idealized dispersion-managed single-mode communication line are given. In Section 4, we describe an extended two-mode model for considering the simultaneous and fully coupled propagation of two ultra-fast pulses in a single fiber cable. The subsequent Section 5 presents a fractional step approach for effectively extending the derived single-mode nonlinear sub-operator to the corresponding system in the two-mode case. A two-mode benchmark of two interacting ultra-fast Gaussian communication pulses confirms the reliability of the method and its straightforward applicability in the dispersion-managed case is also shown. The conclusions are given in Section 6.

2 Governing Equation for Ultra-Fast Pulses in a Single-Mode Fiber

The most general equation representing single-mode pulse propagation in a one-dimensional optical fiber reads

$$\frac{\partial A}{\partial z}+\frac{\alpha}{2}A+\left(\sum_{k\geq 1}\beta_k\frac{i^{k-1}}{k!}\frac{\partial^k}{\partial t^k}\right)A=i\gamma\left(1+\frac{i}{\omega_0}\frac{\partial}{\partial t}\right)\times\left[A\int_{-\infty}^{\infty}R(t')|A(t-t')|^2dt'\right]. \tag{18.1}$$

Equation (18.1) is derived from the electric field of the Maxwell equations, cf. [1], and describes the evolution of the slowly varying field envelope $A(z,t)$ of the complex-valued signal over the propagation distance z and time t. The coefficients β_k model signal dispersion. Since the refractive index n of the fiber material is dependent on the light's circular frequency ω, different spectral components associated with a pulse travel at slightly different velocities, given by $c/n(\omega)$, with c denoting the speed of light in vacuum. This effect is mathematically modeled by expressing the mode propagation constant β in a Taylor series about the central frequency $\omega_0 = 2\pi c/\lambda_0$ as

$$\beta(\omega) = n(\omega)\frac{\omega}{c} = \sum_{k\geq 0}\frac{1}{k!}\beta_k(\omega-\omega_0)^k. \tag{18.2}$$

Here, the wavelength of the injected laser light is denoted by λ_0 and the parameters α and γ model linear signal loss and fiber nonlinearity, respectively. The function $R(t)$ represents intrapulse Raman scattering, a nonlinear effect transferring energy from higher to lower light frequencies. Using $R(t) = (1-f_R)\delta(t)+f_R h_R(t)$ with $f_R = 0.18$ [4] as Raman response function, applying a Taylor series expansion and neglecting higher order terms, Eq. (18.1) eventually becomes

$$\frac{\partial A}{\partial z}+\frac{\alpha}{2}A+\beta_1\frac{\partial A}{\partial t}+i\frac{\beta_2}{2}\frac{\partial^2 A}{\partial t^2}-\frac{\beta_3}{6}\frac{\partial^3 A}{\partial t^3}=i\gamma\left(A|A|^2+\frac{i}{\omega_0}\frac{\partial}{\partial t}\left(A|A|^2\right)-T_R A\frac{\partial|A|^2}{\partial t}\right). \tag{18.3}$$

In general, Eq. (18.4) is widely accepted as a valid model for modeling the propagation of pulses with a half width $T_0 > 10\,\text{fs}$ [1]. For $\lambda_0 = 1550\,\text{nm}$, a typical value for the Raman response parameter is $T_R = 3\,\text{fs}$. The first nonlinear term on the right-hand side of Eq. (18.3) is called the *Kerr* nonlinearity and the second represents nonlinear pulse self-steepening.

Introducing the signal group velocity v with $\beta_1 = 1/v$ and using the transformation $T \equiv t - z/v$ into *retarded* time T, Eq. (18.3) is transformed into the frame of reference of the pulse to read

$$\frac{\partial A}{\partial z}+\frac{\alpha}{2}A+i\frac{\beta_2}{2}\frac{\partial^2 A}{\partial T^2}-\frac{\beta_3}{6}\frac{\partial^3 A}{\partial T^3}=i\gamma\left(A|A|^2+iS\frac{\partial}{\partial T}\left(A|A|^2\right)-T_R A\frac{\partial|A|^2}{\partial T}\right), \tag{18.4}$$

where we have also introduced $S = \omega_0^{-1}$. For $T_0 \gg 1\,\text{ps}$, the last two terms can be neglected and Eq. (18.4) reduces to

$$\frac{\partial A}{\partial z} + \frac{\alpha}{2}A + i\frac{\beta_2}{2}\frac{\partial^2 A}{\partial T^2} - \frac{\beta_3}{6}\frac{\partial^3 A}{\partial T^3} = i\gamma A|A|^2, \tag{18.5}$$

where $\beta_3 \equiv 0$ can be employed if λ_0 is not close to the zero-dispersion wavelength.

3 Numerical Methods for Ultra-Fast Pulses in Single-Mode Fibers

3.1 Split-Step Fourier Approach

In order to develop a numerical solution method, Eq. (18.4) is commonly written in the form

$$\frac{\partial A}{\partial z} = \underbrace{\left(-\frac{\alpha}{2} - i\frac{\beta_2}{2}\frac{\partial^2}{\partial T^2} + \frac{\beta_3}{6}\frac{\partial^3}{\partial T^3}\right)}_{D} A + \underbrace{i\gamma\left(|A|^2 + iS\frac{1}{A}\frac{\partial}{\partial T}\left(A|A|^2\right) - T_R\frac{\partial|A|^2}{\partial T}\right)}_{N} A, \tag{18.6}$$

where we denote with $D(A)$ the operator of all terms linear in A and with $N(A)$ the operator of all nonlinearities. Using these definitions, we write Eq. (18.6) in short as

$$\frac{\partial A}{\partial z} = (D+N)A. \tag{18.7}$$

If one assumes D and N to be independent of z, Eq. (18.7) can be integrated exactly and the solution at $z+h$ reads

$$A(z+h,T) = \exp(h(D+N))A(z,T). \tag{18.8}$$

The last expression forms the basis of split-step numerical methods [1]. Note, however, that the operators D and N in general do not commute and that it corresponds to an $O(h)$ approximation to replace Eq. (18.8) with $\exp(hD)\exp(hN)A(z,T)$. A commonly used symmetric approximation is [24, 6]

$$A(z+h,T) = \exp\left(\frac{h}{2}D\right)\exp(hN)\exp\left(\frac{h}{2}D\right)A(z,T). \tag{18.9}$$

Utilizing the Baker-Campbell-Hausdorff formula for expanding two non-commuting operators, Eq. (18.9) can be proven to be an $O(h^2)$ approximation [19]. Comprehensive descriptions of the split-step approach for simulating pulse propagation in fibers are given for instance by Agrawal [1] and Hohage & Schmidt [10]. The efficiency of the SSFM, especially for longer propagation distances, as required for modeling optical communication lines, can be improved by taking solution adaptive steps in space as proposed by Sinkin *et al.* [21].

Alternatively, one may also construct a fractional step splitting method by solving

$$\frac{\partial A}{\partial z} = DA\,, \quad \frac{\partial A}{\partial z} = N(A)A = \bar{N}(A) \tag{18.10}$$

successively, which we approximate with the symmetric fractional step method

$$A^{*} = \exp\left(\frac{h}{2}D\right)A(z,T)\,, \tag{18.11a}$$

$$A^{**} = A^{*} + h\bar{N}(A^{*})\,, \tag{18.11b}$$

$$A(z+h,T) = \exp\left(\frac{h}{2}D\right)A^{**}. \tag{18.11c}$$

Note that step (18.11b) is written here as a simple explicit Euler method to motivate the fundamental idea but schemes described below are in fact more complicated.

3.2 Linear Sub-steps

Since the dispersion parameters β_2 and β_3 are very small, discretization of the temporal derivatives in D by finite differences and approximation in physical time is no viable option. Instead, Fourier transformation into frequency space is commonly applied. The linear operator then becomes

$$\exp\left(\frac{h}{2}D\right)A(z,T) = F^{-1}\exp\left[\frac{h}{2}\left(i\frac{\beta_2}{2}\omega^2 - i\frac{\beta_3}{6}\omega^3 - \frac{\alpha}{2}\right)\right]FA(z,T), \tag{18.12}$$

where F and F^{-1} denote Fourier and inverse Fourier transformation, respectively. In the practical implementation, discrete Fourier transformation needs to be used and for ω we employ the discrete frequency spectrum

$$\{j\Delta\omega : j \in \mathbb{Z} \wedge -N \leq j \leq N-1\} \tag{18.13}$$

with spectral width $\Delta\omega = \pi/(N\Delta T)$. Here, it is assumed that the temporal window traveling with the pulse is discretized with $2N$ points (note that discrete Fourier transformation algorithms are specially efficient if the number of points is a power of 2), ΔT denotes the temporal discretization width and the temporal window has the extensions $[-N\Delta T, (N-1)\Delta T]$.

3.3 Nonlinear Sub-steps

The nonlinear operator N of the split-step method (18.9) is discretized in physical space. Utilizing $|A|^2 = A\bar{A}$ to eliminate $1/A$, we write $N(A)$ in the form

$$N(A) = i\gamma\left(|A|^2 + iS\bar{A}\frac{\partial A}{\partial T} + [iS - T_R]\frac{\partial |A|^2}{\partial T} \right). \tag{18.14}$$

A consistent numerical method can be constructed by simply approximating the temporal derivatives in Eq. (18.14) by complex-valued first-order central differences and applying Eq. (18.9). The resulting split-step scheme would be second-order accurate in time and space. However, it is also clear that central finite differences will result in Gibbs phenomena (cf. [13]) when strong self-steepening occurs or the propagation of an initially discontinuous signal needs to be simulated.

An alternative approach for handling $N(A)$ is to apply forward and inverse Fourier transformation individually to the derivatives, cf. [16]. For instance, in (18.14) one simply replaces $\bar{A}\partial_T A$ and $\frac{\partial |A|^2}{\partial T}$ with $\bar{A}F^{-1}(i\omega F(A))$ and F^{-1} $(i\omega F(|A|^2)$, respectively, thereby neglecting the dependence of $\bar{A}$ on T. The result is class of numerical operators that would generally not be consistent in the strict mathematical sense with $N(A)$ and that are not uniquely defined, with different authors arriving at slightly different discretizations of Eq. (18.14), cf. [16] and [2]. Therefore, we have opted to pursue a different approach, which can handle self-steepening and arbitrary signal shapes without artificial numerical oscillations. This method is based on solving

$$\frac{\partial A}{\partial z} = \underbrace{\left(-\frac{\alpha}{2} - i\frac{\beta_2}{2}\frac{\partial^2}{\partial T^2} + \frac{\beta_3}{6}\frac{\partial^3}{\partial T^3} \right)}_{D} A + \underbrace{i\gamma\left(A|A|^2 + iS\frac{\partial}{\partial T}\left(A|A|^2\right) - T_R A\frac{\partial |A|^2}{\partial T} \right)}_{\bar{N}(A)} \tag{18.15}$$

within the fractional step method (18.11). Specific to our approach is that we discretize and numerically solve the complete sub-operator

$$\frac{\partial A}{\partial z} = \bar{N}(A) = i\gamma\left(A|A|^2 + iS\frac{\partial}{\partial T}\left(A|A|^2\right) - T_R A\frac{\partial |A|^2}{\partial T} \right) \tag{18.16}$$

directly. Using the Madelung transformation [17, 23] $A(z,t) = \sqrt{I(z,t)}e^{i\phi(z,t)}$, one can transform Eq. (18.16) into the equivalent system of partial differential equations

$$\frac{\partial I}{\partial z} + 3\gamma S I\frac{\partial I}{\partial T} = 0, \tag{18.17a}$$

$$\frac{\partial \phi}{\partial z} + \gamma S I\frac{\partial \phi}{\partial T} + \gamma T_R\frac{\partial I}{\partial T} = \gamma I \tag{18.17b}$$

of the real-valued quantities intensity I and phase ϕ. If we write the latter in the form

$$\frac{\partial}{\partial z}\begin{bmatrix} I \\ \phi \end{bmatrix} + \begin{bmatrix} 3\gamma SI & 0 \\ \gamma T_R & \gamma SI \end{bmatrix} \frac{\partial}{\partial T}\begin{bmatrix} I \\ \phi \end{bmatrix} = \begin{bmatrix} 0 \\ \gamma I \end{bmatrix}, \tag{18.18}$$

its structure as a hyperbolic advection problem

$$\frac{\partial \mathbf{q}}{\partial z} + \mathbf{M}(\mathbf{q})\frac{\partial \mathbf{q}}{\partial T} = \mathbf{s}(\mathbf{q}) \tag{18.19}$$

with $\mathbf{q} = (I, \phi)^T$ becomes apparent. The matrix $\mathbf{M}(\mathbf{q})$ has the eigenvalues $\lambda_1 = 3\gamma SI$, $\lambda_2 = \gamma SI$ and a unique eigendecomposition for $I \neq 0$. Here we propose a numerical method for (18.18) that considers the characteristic information, i.e., the sign of the eigenvalues of $\mathbf{M}(\mathbf{q})$ for constructing one-sided (aka "upwinded") differences for the temporal derivatives, as it is required for a reliable and robust method following the theory of hyperbolic problems (cf. [22]).

Again, we adopt an operator splitting technique and, instead of discretizing (18.19) directly, alternate between solving the homogeneous partial differential equation

$$\partial_z \mathbf{q} + \mathbf{M}(\mathbf{q})\partial_T \mathbf{q} = 0 \tag{18.20}$$

and the ordinary differential equation

$$\partial_z \mathbf{q} = \mathbf{s}(\mathbf{q}) \tag{18.21}$$

successively, using the updated data from the preceding step as initial condition. A first-order accurate upwind scheme for (18.20) can be derived easily based on the discrete update formula [15]

$$\mathbf{q}_j^{n+1} = \mathbf{q}_j^n - \frac{h}{\Delta T}\left(\hat{\mathbf{M}}^-(\mathbf{q}_{j+1}, \mathbf{q}_j)\Delta \mathbf{q}_{j+1/2}^n + \hat{\mathbf{M}}^+(\mathbf{q}_j, \mathbf{q}_{j-1})\Delta \mathbf{q}_{j-1/2}^n\right) \tag{18.22}$$

with $\Delta \mathbf{q}_{j+1/2}^n = \mathbf{q}_{j+1}^n - \mathbf{q}_j^n$, where we assume a computational grid with equidistant mesh widths ΔT in time indexed with $j \in \mathbb{Z}$, where $-N \leq j \leq N-1$, cf. Section 3.2. The spatial update steps are indexed by $n \in \mathbb{N}_0$. In general, the matrices $\hat{\mathbf{M}}^+$ and $\hat{\mathbf{M}}^-$ indicate decompositions of $\mathbf{M}$ with only positive and negative eigenvalues, respectively. However, in the case of Eq. (18.18) the eigenvalues have the same sign, which depends solely on the sign of γ (since $I \geq 0$). Based on (18.22), we construct a straightforward upwind scheme for Eq. (18.18) that reads

$$I_j^{n+1} = I_j^n - \frac{h}{\Delta T}\left[3\gamma S\tilde{I}_j^n \Delta I_j^n\right], \tag{18.23a}$$

$$\bar{\phi}_j^{n+1} = \phi_j^n - \frac{h}{\Delta T}\left[\gamma T_R \Delta I_j^n + \gamma S\tilde{I}_j^n \Delta \phi_j^n\right], \tag{18.23b}$$

$$\phi_j^{n+1} = \bar{\phi}_j^{n+1} + h\gamma I_j^{n+1} \tag{18.23c}$$

with

$$\tilde{I}_j^n = \frac{1}{2}\left(I_j^n + I_{j-1}^n\right), \quad \Delta I_j^n = I_j^n - I_{j-1}^n \qquad \text{for } \gamma > 0,$$
$$\tilde{I}_j^n = \frac{1}{2}\left(I_j^n + I_{j+1}^n\right), \quad \Delta I_j^n = I_{j+1}^n - I_j^n \qquad \text{for } \gamma < 0.$$

When computing the phase difference $\Delta\phi_j^n$, it is of crucial importance to remember that phase is given only modulo 2π. Here, we have obtained reliable and stable results by ensuring that the smallest possible difference $\Delta\phi_j^n$ modulo 2π is applied in (18.23b). Using the auxiliary variable

$$\Delta\theta = \begin{cases} \phi_j^n - \phi_{j-1}^n, \text{ for } \gamma > 0, \\ \phi_{j+1}^n - \phi_j^n, \text{ for } \gamma < 0. \end{cases} \tag{18.24}$$

and $\Delta\tau_j^n = \min\left\{|\Delta\theta_j^n|, |\Delta\theta_j^n + 2\pi|, |\Delta\theta_j^n - 2\pi|\right\}$ we evaluate $\Delta\phi_j^n$ as

$$\Delta\phi_j^n = \begin{cases} \Delta\theta_j^n, & \text{if } |\Delta\theta_j^n| = \Delta\tau_j^n, \\ \Delta\theta_j^n + 2\pi, & \text{if } |\Delta\theta_j^n + 2\pi| = \Delta\tau_j^n, \\ \Delta\theta_j^n - 2\pi, & \text{if } |\Delta\theta_j^n - 2\pi| = \Delta\tau_j^n. \end{cases} \tag{18.25}$$

The scheme (18.23) is of first-order accuracy and thereby entirely free of producing numerical oscillations in the approximation of Eq. (18.15) provided that the stability condition

$$3|\gamma| S \max_j \left\{I_j^n\right\} \frac{h}{\Delta T} \leq 1 \tag{18.26}$$

is satisfied. Our present implementation guarantees (18.26) under all circumstances by having the ability to adaptively take k steps with step size Δz with $h = k\Delta z$ within the central, nonlinear sub-step (18.11b) when required. Note, however, that for all computations presented in here the stability conditions (18.26) was always already satisfied for $k = 1$.

To complete the algorithmic description we remark that we set $I_j^0 := |A_j^*|^2$ and $\phi_j^0 := \arg(A_j^*)$ after sub-step (18.11a) and compute $A_j^{**} = \sqrt{I_j^k} e^{i\phi_j^k}$ before step (18.11c). Periodic boundary conditions could be implemented by one layer of halo points. But note that thanks to the directional dependence, inherent to (18.23) and (18.24), it suffices to update only the upstream halo point, that is the one with index $j = -N-1$ for $\gamma > 0$ and the one with $j = N$ in case $\gamma < 0$ before applying the upwind scheme.

3.4 High-Resolution Upwind Scheme

To enable overall second-order numerical accuracy of the fractional step method (18.11), in case the solution is smooth and differentiable, it is necessary to extend the homogeneous nonlinear update (18.22) to a *high-resolution scheme*. For this purpose, we have developed a special MUSCL-type slope-limiting technique of the solution vector $\mathbf{q}$. Originally proposed by van Leer for hyperbolic equations in conservation law form [14], application to quasilinear systems is not apparent. Inspired by Ketcheson & LeVeque [12], we formulate our high-resolution method as

$$\mathbf{q}_j^{n+1} = \mathbf{q}_j^n - \frac{h}{\Delta T}\left(\hat{\mathbf{M}}^{-}\Delta\mathbf{q}_{j+1/2}^{\star} + \hat{\mathbf{M}}^{+}\Delta\mathbf{q}_{j-1/2}^{\star} + \hat{\mathbf{M}}\Delta\mathbf{q}_j^{\star}\right) \tag{18.27}$$

with $\Delta\mathbf{q}_{j+1/2}^{\star} = \mathbf{q}_{j+1}^{l} - \mathbf{q}_j^{r}$, $\Delta\mathbf{q}_{j-1/2}^{\star} = \mathbf{q}_j^{l} - \mathbf{q}_{j-1}^{r}$, and $\Delta\mathbf{q}_j^{\star} = \mathbf{q}_j^{r} - \mathbf{q}_j^{l}$. Here, $\mathbf{q}_j^{l/r}$ refers to slope-limited values constructed for each component of $\mathbf{q}$ separately as

$$q_j^r = \bar{q}_j + \tfrac{1}{4}\sigma_j, \qquad q_j^l = \bar{q}_j - \tfrac{1}{4}\sigma_j \tag{18.28}$$

with reconstructed linear local slope

$$\sigma_j = \Phi\left(\frac{\Delta_{j-\frac{1}{2}}}{\Delta_{j+\frac{1}{2}}}\right)\Delta_{j+\frac{1}{2}} + \Phi\left(\frac{\Delta_{j+\frac{1}{2}}}{\Delta_{j-\frac{1}{2}}}\right)\Delta_{j-\frac{1}{2}} \tag{18.29}$$

with $\Delta_{j-1/2} = \bar{q}_j - \bar{q}_{j-1}$, $\Delta_{j+1/2} = \bar{q}_{j+1} - \bar{q}_j$. In the latter, $\Phi(\cdot)$ is a typical limiter function, where we utilize in here exclusively the *van Albada* limiter

$$\Phi(r) = \max\left(0, (r^2 + r)/(1 + r^2)\right). \tag{18.30}$$

To permit second-order accuracy overall, we do not utilize in (18.28) the discrete values from the previous step $\mathbf{q}^n$ but instead intermediate values $\bar{\mathbf{q}}$ computed as

$$\bar{\mathbf{q}}_j = \mathbf{q}_j^n - \frac{h}{2\Delta T}\left(\hat{\mathbf{M}}^{-}\Delta\mathbf{q}_{j+1/2}^{n} + \hat{\mathbf{M}}^{+}\Delta\mathbf{q}_{j-1/2}^{n}\right). \tag{18.31}$$

The consecutive application of (18.27) and (18.31) corresponds to an explicit 2-step Runge-Kutta method in the spatial update. Finally, a second-order accurate symmetric operator splitting [24, 6] is employed to integrate Eq. (18.21) before and after the high-resolution scheme. Thanks to the simplicity of $\mathbf{s}(\mathbf{q})$ using an explicit Euler method for this step is equivalent to an explicit 2-step Runge-Kutta update.

We want to point out that the first-order method (18.23) as well as the MUSCL-based second-order scheme are equally applicable for $T_R = 0$ and especially in the singular case $S = 0$, which allows deactivation of Raman scattering and/or self-steepening if desired. Note that for $S = 0$ or $\max_j \left\{I_j^n\right\} = 0$, the stability condition (18.26) is trivially satisfied.

3.5 Simulation of a Propagating Pulse

In order to demonstrate the described numerical method we simulate the propagation of a Gaussian pulse with initial shape

$$A(0,T) = \sqrt{P_0}\exp\left(-\frac{1+iC}{2}\frac{T^2}{T_0^2}\right) \tag{18.32}$$

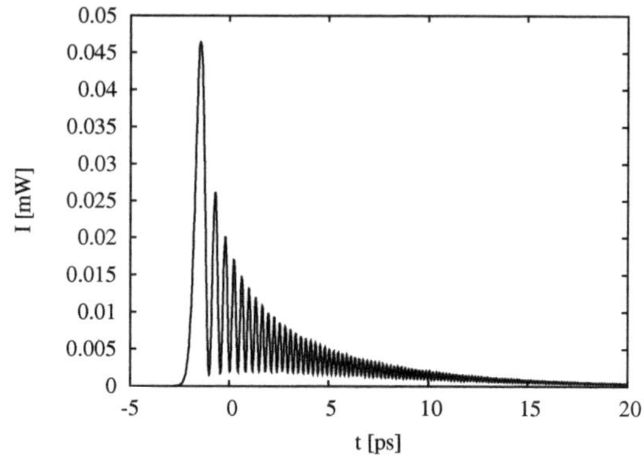

Fig. 18.1 Simulated signal at $L_{\max} = 1\,\mathrm{km}$ (temporal window enlarged) for Benchmark 1. The initially Gaussian pulse, cf. Eq. (18.32), with half width $T_0 = 80\,\mathrm{fs}$ has broadened severely because of second-order dispersion. Asymmetric high-frequency oscillations have been added by third-order dispersion effects. Maximal signal strength is reduced by a factor of ~ 13.9.

in a homogeneous fiber. The fiber is assumed to be lossless ($\alpha = 0$) for simplicity as the omitted linear weakening of the signal is unproblematic for any numerical scheme. We use the SSFM in line with Eq. (18.11) with second-order accurate upwind-based nonlinear operator, cf. Section 3.4, and Van Albada slope-limiter (18.30).

Used parameters correspond to a typical ultra-short communication pulse with $P_0 = 0.625\,\mathrm{mW}$, $T_0 = 80\,\mathrm{fs}$, and no chirp, i.e., $C = 0$. The central wavelength is set to $\lambda_0 = 1550\,\mathrm{nm}$, from which one computes the self-steepening parameter $S = \lambda_0/2\pi c$, with c denoting the speed of light in vacuum. Raman scattering is activated with $T_R = 3\,\mathrm{fs}$. Realistic fiber parameters $\beta_2 = 0.5\,\mathrm{ps}^2\,\mathrm{km}^{-1}$, $\beta_3 = 0.07\,\mathrm{ps}^3\,\mathrm{km}^{-1}$ and $\gamma = 0.1\,\mathrm{W/m}$ are used. For this configuration, the second-order dispersion length is just $L_d = T_0^2/|\beta_2| \approx 12.8\,\mathrm{m}$, the third-order dispersion length is $T_0^3/|\beta_3| \approx 7.31\,\mathrm{m}$, and the nonlinear length is $L_{nl} = (\gamma P_0)^{-1} = 16\,\mathrm{km}$. The pulse is assumed to travel a distance of just $L_{\max} = 1\,\mathrm{km}$ and the simulated temporal window moving with the pulse has the width $[-30\,\mathrm{ps}, 30\,\mathrm{ps} - \Delta T]$.

Figure 18.1 shows the computed solution using a temporal discretization of $2N$ points for $N = 2048$ and after taking $M = 100$ spatial steps of equal size of $h = 10\,\mathrm{m}$. Because of the very small second- and third-order dispersion lengths, typical for ultra-fast pulses, the final signal shape is clearly dominated by dispersion effects. Second-order dispersion has introduced severe pulse broadening, reducing the maximum in power by a factor of ~ 13.9; third-order dispersion has added high-frequency oscillations.

A detailed numerical analysis verifies the convergence and expected order of accuracy of the scheme. Starting from $N = 512$ and $h = 40\,\mathrm{m}$ ($M = 25$ steps), in each successive computation the number of Fourier modes and spatial steps is doubled. The numerical error at $L_{\max}$ is measured for the intensity of the signal in the discrete maximum norm

$$E_\infty = \max_{j \in \{-N, N-1\}} |I_j - I^{\mathrm{ref}}(j\Delta T)|, \qquad (18.33)$$

where a highly resolved result with $N = 131{,}072$ and $M = 6400$ is used as reference solution I^{ref}. Figure 18.2 visualizes the numerical error E_∞ over h and it is

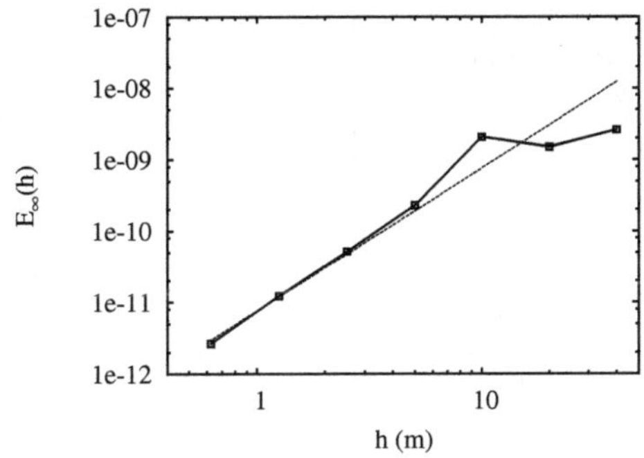

Fig. 18.2 Numerical error E_∞ over h for Benchmark 1. The dotted line corresponds to ideal second order approximation accuracy.

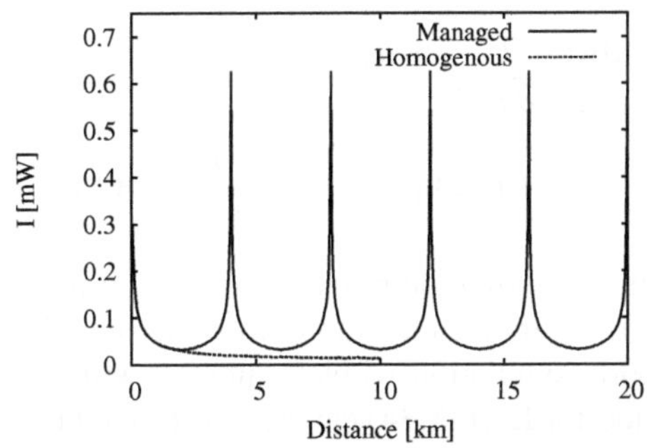

Fig. 18.3 Benchmark 2: Maximal power over distance with and without dispersion management.

eminent that the method achieves almost perfect second-order approximation accuracy and reliable, robust convergence. A more detailed numerical study of the second-order accurate upwind-based SSFM including comparisons with several alternative numerical methods can be found in [5].

3.6 Spatially Dependent Fiber Parameters

Continued propagation of the pulse of Figure 18.1 will invariably lead to a signal which has broadened to such an extent that it cannot be used for digital communication. Yet, this problem can be compensated surprisingly easily by combining fiber sections with positive and negative dispersion characteristics into a single communication line. This technique is called *dispersion management* and has been studied extensively both theoretically and numerically because of its practical significance for long-distance fiber optical communication [18, 20, 3]. Instead of Eq. (18.15), one considers the extended variant

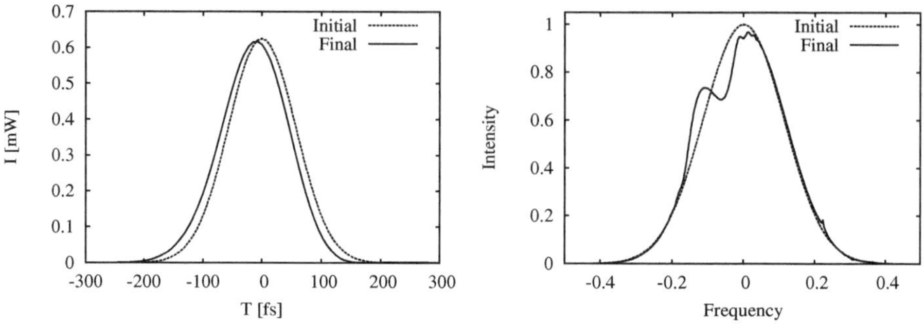

Fig. 18.4 Benchmark 2: Pulse shape and spectrum after propagating 100km or experiencing 25 soliton-like oscillations from alternating signs of dispersion parameters.

$$\frac{\partial A}{\partial z} = \left(-\frac{\alpha(z)}{2} - i\frac{\beta_2(z)}{2}\frac{\partial^2}{\partial T^2} + \frac{\beta_3(z)}{6}\frac{\partial^3}{\partial T^3}\right)A$$
$$+ i\gamma(z)\left(A|A|^2 + iS\frac{\partial}{\partial T}\left(A|A|^2\right) - T_R A\frac{\partial |A|^2}{\partial T}\right) \quad (18.34)$$

as governing equation. Adopting the practical viewpoint that the spatial numerical steps of any SSFM will be significantly larger than the spatial extension corresponding to the used temporal simulation window moving with the pulse, a straightforward numerical method for Eq. (18.34) can be constructed by simply averaging the spatially dependent parameters between discrete propagation steps, i.e., by using

$$\bar{\beta}_{\{2,3\},j} = \frac{2}{h}\int_{z_j}^{z_j+\frac{h}{2}} \beta_{\{2,3\}}(\xi)d\xi\,, \quad \bar{\alpha}_j = \frac{2}{h}\int_{z_j}^{z_j+\frac{h}{2}} \alpha(\xi)d\xi \quad (18.35)$$

in the linear numerical operator (18.12) and by using

$$\bar{\gamma}_j = \frac{1}{h}\int_{z_j}^{z_j+h} \gamma(\xi)d\xi \quad (18.36)$$

in the nonlinear operator approximating (18.16).

In practice, very sophisticated dispersion management designs might be employed (for instance, Guo & Huang [8] propose an exponential decrease of $|\beta_2|$ to accommodate better for linear loss). Here, we simply extend the example of Section 3.5 and alternate the sign of β_2 and β_3 every 2 km. All other parameters are unaltered and for an example computation we use $N = 4096$ and $h = 40$ m ($M = 2500$) to simulate a pulse propagation over a distance of 100 km. In the fiber sections with negative dispersion parameters, the pulse deterioration is effectively reversed and the pulse shape mostly recovered. The pulse is undergoing a soliton-like oscillation with a period of 4 km, which can be inferred from Figure 18.3. This graphic compares the pulse power peak over distance in the simulation with periodic dispersion management and when the computation of the previous section is continued to a length of 10 km. In Figure 18.4 are compared the shape and spectra of the initial Gaussian pulse and of the signal after propagating for 100 km. The observed slight signal delay and spectral modification is the combined effects of the nonlinearities. If $\gamma = 0$ is used, the initial signal is exactly recovered.

4 Governing Equations for Two Interacting Ultra-Fast Pulses

Data throughput can be increased significantly if multiple optical fields of different wavelengths propagate simultaneously inside the fiber. However, these fields would interact with one another through all the fiber nonlinearities. Additionally if three or more fields are initially present, even new signal fields can be induced (aka *four-wave mixing* [1]). Therefore, we consider in the following only the case of two interacting signal fields propagating through an optical fiber, for which there is already some agreement about the structure of the governing equations in the literature [11]. Extensions of the ultra-fast pulse propagation equation (18.3) to three or more interacting fields are still a topic of active research.

We assume two pulses at carrier frequencies $\omega_0^{(1)}$, $\omega_0^{(2)}$, and two nonlinear constants γ_1, γ_2. It is further assumed that the cross-phase modulation of each frequency can be expressed for all higher order nonlinear terms by positive factors B_1, B_2, the cross-phase modulation in the Kerr nonlinearity by factors C_1, C_2. Extending Eq. (18.3) accordingly, we use the model equations

$$\begin{aligned}\frac{\partial A_1}{\partial z} =& -\frac{\alpha_1}{2}A_1 - \beta_1^{(1)}\frac{\partial A_1}{\partial t} - i\frac{\beta_2^{(1)}}{2}\frac{\partial^2 A_1}{\partial t^2} + \frac{\beta_3^{(1)}}{6}\frac{\partial^3 A_1}{\partial t^3} + i\gamma_1\left(|A_1|^2 + C_1|A_2|^2\right)A_1 \\ & - \frac{\gamma_1}{\omega_0^{(1)}}\left[\frac{\partial\left(|A_1|^2 A_1\right)}{\partial t} + B_1\frac{\partial\left(|A_2|^2 A_1\right)}{\partial t}\right] - i\gamma_1 T_R\left[\frac{\partial |A_1|^2}{\partial t} + B_1\frac{\partial |A_2|^2}{\partial t}\right]A_1,\end{aligned} \tag{18.37a}$$

$$\begin{aligned}\frac{\partial A_2}{\partial z} =& -\frac{\alpha_2}{2}A_2 - \beta_1^{(2)}\frac{\partial A_2}{\partial t} - i\frac{\beta_2^{(2)}}{2}\frac{\partial^2 A_2}{\partial t^2} + \frac{\beta_3^{(2)}}{6}\frac{\partial^3 A_2}{\partial t^3} + i\gamma_2\left(|A_2|^2 + C_2|A_1|^2\right)A_2 \\ & - \frac{\gamma_2}{\omega_0^{(2)}}\left[\frac{\partial\left(|A_2|^2 A_2\right)}{\partial t} + B_2\frac{\partial\left(|A_1|^2 A_2\right)}{\partial t}\right] - i\gamma_2 T_R\left[\frac{\partial |A_2|^2}{\partial t} + B_2\frac{\partial |A_1|^2}{\partial t}\right]A_2.\end{aligned} \tag{18.37b}$$

Note that (18.37) encompasses the model actually adopted for simulation by Kalithasan *et al.* in [11]. Using $\beta_1^{(j)} = 1/v_j$ and the transformation $T \equiv t - \beta_1^{(j)} z$ into retarded time yields

$$\frac{\partial A_1}{\partial z} = -\frac{\alpha_1}{2}A_1 \qquad -i\frac{\beta_2^{(1)}}{2}\frac{\partial^2 A_1}{\partial T^2} + \frac{\beta_3^{(1)}}{6}\frac{\partial^3 A_1}{\partial T^3} + i\gamma_1\left(|A_1|^2 + C_1|A_2|^2\right)A_1$$
$$-\gamma_1 S_1\left[\frac{\partial\left(|A_1|^2 A_1\right)}{\partial T} + B_1\frac{\partial\left(|A_2|^2 A_1\right)}{\partial T}\right] - i\gamma_1 T_R\left[\frac{\partial |A_1|^2}{\partial T} + B_1\frac{\partial |A_2|^2}{\partial T}\right]A_1, \tag{18.38a}$$

$$\frac{\partial A_2}{\partial z} = -\frac{\alpha_2}{2}A_2 - \delta\frac{\partial A_2}{\partial T} - i\frac{\beta_2^{(2)}}{2}\frac{\partial^2 A_2}{\partial T^2} + \frac{\beta_3^{(2)}}{6}\frac{\partial^3 A_2}{\partial T^3} + i\gamma_2\left(|A_2|^2 + C_2|A_1|^2\right)A_2$$
$$-\gamma_2 S_2\left[\frac{\partial\left(|A_2|^2 A_2\right)}{\partial T} + B_2\frac{\partial\left(|A_1|^2 A_2\right)}{\partial T}\right] - i\gamma_2 T_R\left[\frac{\partial |A_2|^2}{\partial T} + B_2\frac{\partial |A_1|^2}{\partial T}\right]A_2, \tag{18.38b}$$

with $\delta = (v_1 - v_2)/(v_1 v_2)$ representing the group velocity mismatch between both fields. As before we use $S_k = 1/\omega_0^{(k)}$ for $k = 1,2$.

In the regime of pico-second pulses, that is for pulses with $T_0 \gg 1\,\mathrm{ps}$, two-mode extensions of Eq. (18.5) are well established. Setting $S_k = 0$, $T_R = 0$ and using $C_k = 2$ in (18.38), we obtain the frequently used [1] cross-phase modulation model

$$\frac{\partial A_1}{\partial z} = \underbrace{\left(-\frac{\alpha_1}{2} \qquad -i\frac{\beta_2^{(1)}}{2}\frac{\partial^2}{\partial T^2} + \frac{\beta_3^{(1)}}{6}\frac{\partial^3}{\partial T^3}\right)}_{D^{(1)}}A_1 + \underbrace{i\gamma_1\left(|A_1|^2 + 2|A_2|^2\right)}_{N^{(1)}}A_1, \tag{18.39a}$$

$$\frac{\partial A_2}{\partial z} = \underbrace{\left(-\frac{\alpha_2}{2} - \delta\frac{\partial}{\partial T} - i\frac{\beta_2^{(2)}}{2}\frac{\partial^2}{\partial T^2} + \frac{\beta_3^{(2)}}{6}\frac{\partial^3}{\partial T^3}\right)}_{D^{(2)}}A_2 + \underbrace{i\gamma_2\left(|A_2|^2 + 2|A_1|^2\right)}_{N^{(2)}}A_2, \tag{18.39b}$$

which we write as

$$\frac{\partial A_1}{\partial z} = \left(D^{(1)} + N^{(1)}(A_1,A_2)\right)A_1, \quad \frac{\partial A_2}{\partial z} = \left(D^{(2)} + N^{(2)}(A_1,A_2)\right)A_2. \tag{18.40}$$

5 Numerical Methods for Two Interacting Ultra-Fast Pulses

5.1 Extended Split-Step Fourier Method

Taking advantage of the fact that the linear operators $D^{(k)}$ only need to be applied to each field A_k, an SSFM for approximating solutions of system (18.39) – in line with Eq. (18.9) – is easily constructed as

$$A_1^* = \exp\left(\frac{h}{2}D^{(1)}\right)A_1, \qquad A_2^* = \exp\left(\frac{h}{2}D^{(2)}\right)A_2, \tag{18.41a}$$

$$A_1^{**} = \exp\left(hN^{(1)}(A_1^*,A_2^*)\right)A_1^*, \qquad A_2^{**} = \exp\left(hN^{(2)}(A_1^{**},A_2^*)\right)A_2^* \tag{18.41b}$$

$$A_1(z+h) = \exp\left(\frac{h}{2}D^{(1)}\right)A_1^{**}, \qquad A_2(z+h) = \exp\left(\frac{h}{2}D^{(2)}\right)A_2^{**}. \tag{18.41c}$$

Obviously, the numerical operators of (18.41b) and (18.41c) acting on each fields can be executed consecutively. The linear operator $D^{(1)}$ is identical to (18.12). For $D^{(2)}$ we have

$$\exp\left(\frac{h}{2}D^{(2)}\right)A_2 = F^{-1}\exp\left[\frac{h}{2}\left(-i\delta\omega + i\frac{\beta_2^{(2)}}{2}\omega^2 - i\frac{\beta_3^{(2)}}{6}\omega^3 - \frac{\alpha_2}{2}\right)\right]FA_2. \tag{18.42}$$

A second-order accurate scheme can be expected if (18.41b) is replaced with a symmetric splitting scheme such as

$$A_1^\star = \exp\left(\frac{h}{2}N^{(1)}(A_1^*,A_2^*)\right)A_1^*, \tag{18.43a}$$

$$A_2^{**} = \exp\left(hN^{(2)}(A_1^\star,A_2^*)\right)A_2^*, \tag{18.43b}$$

$$A_1^{**} = \exp\left(\frac{h}{2}N^{(1)}(A_1^\star,A_2^{**})\right)A_1^\star. \tag{18.43c}$$

5.2 Nonlinear Sub-steps

While the derivation of an SSFM for the simplified system (18.39) is apparently a straightforward task, formulation of a reliable numerical method for the system of propagation equations for two coupled ultra-fast pulsed signals, (18.38), is more involved. In particular, when the equations of (18.38) are written in the form $\partial_z A_k = (D^{(k)} + N^{(k)})A_k$ one quickly finds that due to the cross-phase coupling the factor $1/A_k$ of the self-steepening term cannot be eliminated from $N^{(k)}$ as it was done to obtain Eq. (18.14). This leaves a singularity in the operator for vanishing signals and neither the centered difference method nor particularly an ad hoc Fourier transformation technique, sketched both in the beginning of Section 3.3, are available anymore for numerical method construction. However, we will demonstrate subsequently how our upwind-based discretization technique of Section 3.3 can be easily extended to (18.38), yielding a reliable and robust numerical method.

We start the derivation of the method by inserting the linear operators from (18.39) into (18.38) to obtain

$$\begin{aligned}\frac{\partial A_k}{\partial z} &= D^{(k)}A_k + i\gamma_k\left(|A_k|^2 + C_k|A_l|^2\right)A_k \\ &\quad -\gamma_k S_k\left[\frac{\partial\left(|A_k|^2 A_k\right)}{\partial T} + B_k\frac{\partial\left(|A_l|^2 A_k\right)}{\partial T}\right] - i\gamma_k T_R\left[\frac{\partial|A_k|^2}{\partial T} + B_k\frac{\partial|A_l|^2}{\partial T}\right]A_k\end{aligned} \tag{18.44}$$

for $k,l \in \{1,2\}$ and $k \neq l$. In analogy to Section 3.3, we assume a fractional step approach in the spirit of Eq. (18.11) that considers the linear operators with the update steps (18.41a) and (18.41c) and approximates the nonlinear sub-operator equations

$$\begin{aligned}\frac{\partial A_k}{\partial z} &= \bar{N}^{(k)}(A_k, A_l) = i\gamma_k\left(|A_k|^2 + C_k|A_l|^2\right)A_k \\ &\quad -\gamma_k S_k\left[\frac{\partial\left(|A_k|^2 A_k\right)}{\partial T} + B_k\frac{\partial\left(|A_l|^2 A_k\right)}{\partial T}\right] - i\gamma_k T_R\left[\frac{\partial|A_k|^2}{\partial T} + B_k\frac{\partial|A_l|^2}{\partial T}\right]A_k.\end{aligned} \tag{18.45}$$

Using again Madelung transformation for each field, i.e., $A_k(z,t) = \sqrt{I_k(z,t)}e^{i\phi_k(z,t)}$, we obtain the transport equations for the intensities I_k and the phases ϕ_k instead of (18.45) as

$$\frac{\partial I_k}{\partial z} + \gamma_k S_k\left[(3I_k + B_k I_l)\frac{\partial I_k}{\partial T} + 2B_k I_k\frac{\partial I_l}{\partial T}\right] = 0, \tag{18.46a}$$

$$\frac{\partial \phi_k}{\partial z} + \gamma_k S_k(I_k + B_k I_l)\frac{\partial \phi_k}{\partial T} + \gamma_k T_R\left[\frac{\partial I_k}{\partial T} + B_k\frac{\partial I_l}{\partial T}\right] = \gamma_k\left(I_k + C_k I_l\right). \tag{18.46b}$$

The latter defines a single system of advection equations that couples the fields A_k and A_l. Using the state vector $\mathbf{u} = (I_1, \phi_1, I_2, \phi_2)^T$, this system reads

$$\frac{\partial \mathbf{u}}{\partial z} + \mathbf{B}(\mathbf{u})\frac{\partial \mathbf{u}}{\partial T} = \mathbf{r}(\mathbf{u}), \tag{18.47}$$

with matrix

$$\mathbf{B}(\mathbf{u}) = \begin{bmatrix} \gamma_1 S_1(3I_1 + B_1 I_2) & 0 & 2\gamma_1 S_1 B_1 I_1 & 0 \\ \gamma_1 T_R & \gamma_1 S_1(I_1 + B_1 I_2) & \gamma_1 T_R B_1 & 0 \\ 2\gamma_2 S_2 B_2 I_2 & 0 & \gamma_2 S_2(3I_2 + B_2 I_1) & 0 \\ \gamma_2 T_R B_2 & 0 & \gamma_2 T_R & \gamma_2 S_2(I_2 + B_2 I_1) \end{bmatrix} \tag{18.48}$$

and right-hand side

$$\mathbf{r}(\mathbf{u}) = \left(0, \gamma_1\left(I_1 + C_1 I_2\right), 0, \gamma_2\left(I_2 + C_2 I_1\right)\right)^T. \tag{18.49}$$

In order to verify the hyperbolicity of Eq. (18.47) and for constructing an upwind scheme, one would require the eigendecomposition $\mathbf{B} = \mathbf{R}\Lambda\mathbf{R}^{-1}$. However, the necessary linear algebra is very involved and is additionally complicated by the singular cases $I_k = 0$, which have to be considered separately in order to construct a generally robust numerical scheme. To simplify the latter, we have opted to use a splitting approach and update the fields A_k and A_l successively. Instead of solving the combined system (18.47) we construct an approximation to (18.46) under the assumption that I_l is independent of z. Proceeding then as in Section 3.3, we write (18.46) as the advection system

$$\frac{\partial \mathbf{q}^{(\mathbf{k})}}{\partial z} + \mathbf{M}^{(\mathbf{k})}(\mathbf{q}^{(\mathbf{k})})\frac{\partial \mathbf{q}^{(\mathbf{k})}}{\partial T} = \mathbf{s}^{(\mathbf{k})}(\mathbf{q}^{(\mathbf{k})}), \tag{18.50}$$

with vector of state $\mathbf{q}^{(k)} = (I_k, \phi_k, I_l)^T$, matrix

$$\mathbf{M}^{(\mathbf{k})}(\mathbf{q}^{(\mathbf{k})}) = \begin{bmatrix} \gamma_k S_k(3I_k + B_k I_l) & 0 & 2\gamma_k S_k B_k I_k \\ \gamma_k T_R & \gamma_k S_k(I_k + B_k I_l) & \gamma_k T_R B_k \\ 0 & 0 & 0 \end{bmatrix} \tag{18.51}$$

and source term

$$\mathbf{s}^{(\mathbf{k})}(\mathbf{q}^{(\mathbf{k})}) = \left(0, \gamma_k\left(I_k + C_k I_l\right), 0\right)^T. \tag{18.52}$$

The nonzero eigenvalues of $\mathbf{M}^{(\mathbf{k})}$ are $\gamma_k S_k(3I_k + B_k I_l)$ and $\gamma_k S_k(I_k + B_k I_l)$. Since $B_k \geq 0$ and $I_{k/l} \geq 0$ hold true, both eigenvalues have again the same sign, solely determined by the sign of γ_k. Following the upwind approach again we construct a first-order accurate method for (18.50) as

$$I_{k,j}^{n+1} = I_{k,j}^n - \frac{h}{\Delta T}\gamma_k S_k\left[(3\tilde{I}_{k,j}^n + B_k\tilde{I}_{j,l}^n)\Delta I_{k,j}^n + 2B_k\tilde{I}_{k,j}^n\Delta I_{l,j}^n\right], \tag{18.53a}$$

$$\bar{\phi}_{k,j}^{n+1} = \phi_{k,j}^n - \frac{h}{\Delta T}\gamma_k\left[T_R(\Delta I_{k,j}^n + B_k\Delta I_{l,j}^n) + S_k(\tilde{I}_{k,j}^n + B_k\tilde{I}_{l,j}^n)\Delta\phi_{k,j}^n\right], \tag{18.53b}$$

$$\phi_{k,j}^{n+1} = \bar{\phi}_{k,j}^{n+1} + h\gamma_k\left(I_{k,j}^{n+1} + C_k I_{l,j}^n\right), \tag{18.53c}$$

where

$$\tilde{I}_{k/l,j}^n = \frac{1}{2}\left(I_{k/l,j}^n + I_{k/l,j-1}^n\right), \quad \Delta I_{k/l,j}^n = I_{k/l,j}^n - I_{k/l,j-1}^n \qquad \text{for } \gamma_k > 0,$$

$$\tilde{I}_{k/l,j}^n = \frac{1}{2}\left(I_{k/l,j}^n + I_{k/l,j+1}^n\right), \quad \Delta I_{k/l,j}^n = I_{k/l,j+1}^n - I_{k/l,j}^n, \qquad \text{for } \gamma_k < 0.$$

As before, $\Delta\phi_{k,j}^n$ is evaluated modulo 2π using Eqs. (18.24) and (18.25) and the stability condition reads

$$|\gamma_k| S_k \max_j\left\{3I_{k,j} + B_k I_{l,j}\right\}\frac{h}{\Delta T} \leq 1. \tag{18.54}$$

By construction, the single-field upwind method (18.53) computes only new values for I_k and ϕ_k, while the intensity of the other field, I_l, is assumed to remain unchanged. In order to achieve an update of both fields, and thereby approximation of (18.47), we apply the single-field upwind scheme within a symmetric fractional-step splitting method, i.e.

$$A_1^{\star} = A_1^{*} + \frac{h}{2}\bar{N}^{(1)}(A_1^{*}, A_2^{*}), \tag{18.55a}$$

$$A_2^{\star\star} = A_2^{*} + h\bar{N}^{(2)}(A_1^{\star}, A_2^{*}), \tag{18.55b}$$

$$A_1^{\star\star} = A_1^{\star} + \frac{h}{2}\bar{N}^{(1)}(A_1^{\star}, A_2^{\star\star}). \tag{18.55c}$$

A symmetric SSFM is obtained by applying expressions (18.41a), (18.55), and (18.41c) after another. Finally, the high-resolution technique, described in Section 3.4, is adopted to implement a second-order accurate approximation to $\bar{N}^{(k)}$, where we presently apply slope-limited reconstruction to I_k and ϕ_k but not to I_l.

5.3 Simulation of Two Interacting Propagating Pulses

We use a configuration with very strong nonlinearity and thereby nonlinear pulse interaction to assess the reliability of the derived two-mode method. A fiber without linear loss and third-order dispersion is assumed, i.e., $\alpha_{1,2} = 0$ and $\beta_3^{(1,2)} = 0$, and Raman scattering is also deactivated by setting $T_R = 0$. To enforce a strong influence of the nonlinearities we use $\beta_2^{(1,2)} = 4 \times 10^{-5}\,\text{ps}^2\,\text{km}^{-1}$, $\gamma_1 = 1\,\text{W/m}$, and $\gamma_2 = 1.2\,\text{W/m}$. Two unchirped pulses in the range of ultra-short communication pulses with $T_0^{(1,2)} = 80\,\text{fs}$ and power levels of $P_0^{(1)} = 0.625\,\text{mW}$ and $P_0^{(2)} = 0.3125\,\text{mW}$ are used. The first central wavelength is set to $\lambda_0^{(1)} = 1550\,\text{nm}$ and the second to $\lambda_0^{(2)} = 1300\,\text{nm}$. The group velocity mismatch parameter is set to $\delta = 0.015625\,\text{fs/m}$ and the cross-phase modulation parameters read $B_{1,2} = C_{1,2} = 2$.

For this configuration, the second-order dispersion length is $L_d = 160\,\text{km}$ and the nonlinear lengths $L_{nl}^{(1)} = 1.6\,\text{km}$ and $L_{nl}^{(2)} = 2.667\,\text{km}$, respectively. The approximate optical shock distances [1], $z_s^{(1,2)} = \sqrt{e}L_{nl}^{(1,2)}\omega_0^{(1,2)}T_0/(3\sqrt{2})$ are $\sim 60.491\,\text{km}$ and $\sim 120.216\,\text{km}$, respectively. We use a propagation distance of $L_{\text{max}} = 64\,\text{km}$, yielding a temporal shift of the second pulse by exactly 1 ps, and the temporal window has the width $[-4\,\text{ps}, 4\,\text{ps} - \Delta T]$.

In Figure 18.5 is shown the computed solution using a temporal discretization of $2N$ points for $N = 2048$ and after taking $M = 3200$ spatial steps of equal size of $h = 20\,\text{m}$. Additionally are shown the solutions if each pulse travels individually. These solutions are computed by keeping all other parameters unchanged while setting $P_0^{(2)} \equiv 0$ and $P_0^{(1)} \equiv 0$, respectively. If only a single field is present, our two-mode SSFM is identical to the previously developed single-mode SSFM, which

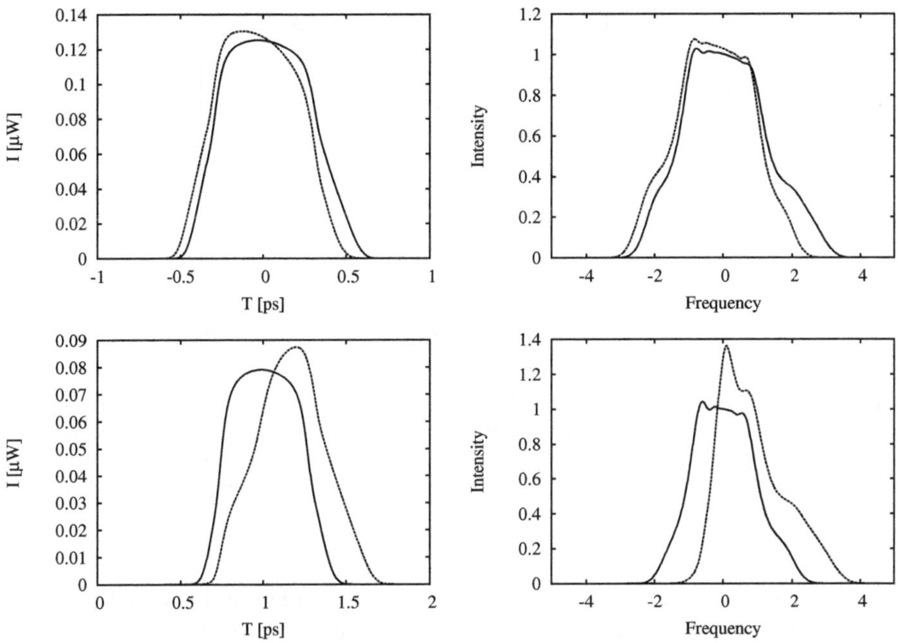

Fig. 18.5 Benchmark 3: Pulse shape and spectrum after propagating 64 km of two individual highly nonlinear ultra-short single-mode pulses (solid lines) and when the two pulses are interacting with one another in a two-mode fiber. The upper row corresponds to Pulse 1 with $P_0^{(1)} = 0.625\,\text{mW}$; the lower row to Pulse 2 with $P_0^{(2)} = 0.3125\,\text{mW}$ and $\delta = 0.015625\,\text{fs/m}$ causing the pulse to arrive 1 ps earlier.

was confirmed to be second-order accurate in Section 3.5. Note that the single-mode solution of Pulse 1 was also used as a detailed computational benchmark in [5] and is thereby available as a reference. From Figure 18.5 it can be seen that the two non-interacting single-mode pulses exhibit a very similar shape and spectrum. However, in the two-mode model particularly the faster and weaker second pulse is significantly altered. Pulse 2, visualized in the lower row of Figure 18.5, experiences considerable signal steepening from cross-phase modulation, which can be inferred especially from its spectrum.

We use the same technical approach as in Section 3.5 to quantify the numerical error and order of accuracy of the two-mode SSFM. We double the temporal resolution consecutively starting from $N = 512$ up to $N = 16{,}384$ and simultaneously divide the spatial step size by a factor of 2 respectively, starting with $h = 80\,\text{m}$ ($M = 800$ steps). The numerical error at $L_{\max}$ is measured for $I_{(1,2)}$ in the maximum norm, cf. Eq. (18.33), where results computed with $N = 32{,}768$ and $h = 1.25\,\text{m}$ ($M = 51{,}200$) are used as respective reference solutions. The computational errors of a series of fully coupled two-mode results as well as the errors of single-mode computations (cf. Figure 18.5) of both individual pulses are plotted in Figure 18.6. In general, the example confirms that the proposed two-mode SSFM converges

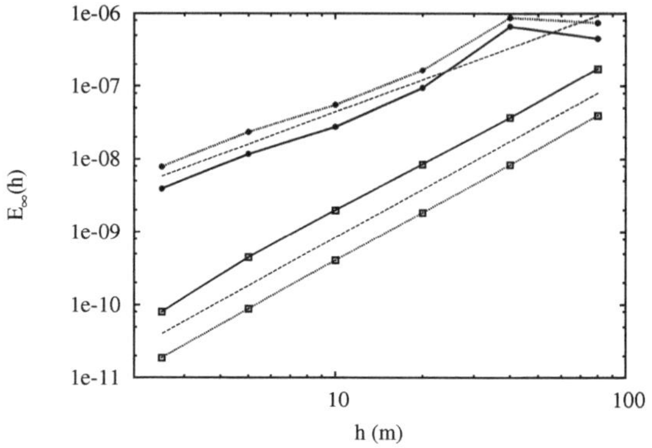

Fig. 18.6 Numerical error E_∞ over h for Benchmark 3. The respective error of Pulse 1 is marked with solid lines, the respective error of Pulse 2 with dotted lines. Single-pulse simulation results are indicated with open squares, the fully coupled simulations are marked with closed circles. The upper broken line corresponds to an order of accuracy ~ 1.47, the lower one to an order of accuracy of ~ 2.20.

reliably and robustly even for a highly nonlinear coupled problem and performs identical beside round-off errors to the single-mode method of Section 3.4 for uncoupled individual pulses. While the single-mode SSFM with limiter (18.30) actually achieves slight super-convergence in this test case (the measured order of accuracy is ~ 2.20), the two-mode SSFM of Section 5.2 with same limiter yields an approximate order of accuracy of ~ 1.47. One might attribute this behavior to the fractional splitting treatment of the nonlinear operator, (18.55). Note, however, that increasing the number of spatial steps up to a factor of 8 in order to reduce the splitting error of the nonlinear sub-operator resulted only in marginally smaller numerical errors for this case, suggesting that the reduction of the order of accuracy could also have a different origin.

5.4 Spatially Dependent Fiber Parameters

As final test case, the coupled propagation of the two Gaussian pulses of the previous benchmark through the dispersion-managed communication line of Section 3.6 is considered. Like in Section 3.6 we assume an optical communication line of 100 km length with dispersion parameters $|\beta_2^{(1,2)}| = 0.5\,\mathrm{ps^2\,km^{-1}}$ and $|\beta_3^{(1,2)}| = 0.07\,\mathrm{ps^3\,km^{-1}}$, which all change sign every 2 km, and $\alpha_{(1,2)} = 0$, $\gamma_{(1,2} = 0.1\,\mathrm{W/m}$, and $T_R = 3\,\mathrm{fs}$. As before, the parameters of the two unchirped Gaussian pulses are $P_0^{(1)} = 0.625\,\mathrm{mW}$, $P_0^{(2)} = 0.3125\,\mathrm{mW}$, and $T_0^{(1,2)} = 80\,\mathrm{fs}$. The wavelengths are again $\lambda_0^{(1)} = 1550\,\mathrm{nm}$ and $\lambda_0^{(1)} = 1300\,\mathrm{nm}$. The group velocity mismatch is $\delta = 0.015625\,\mathrm{fs/m}$ and cross-phase modulation parameters are $B_{1,2} = C_{1,2} = 2$.

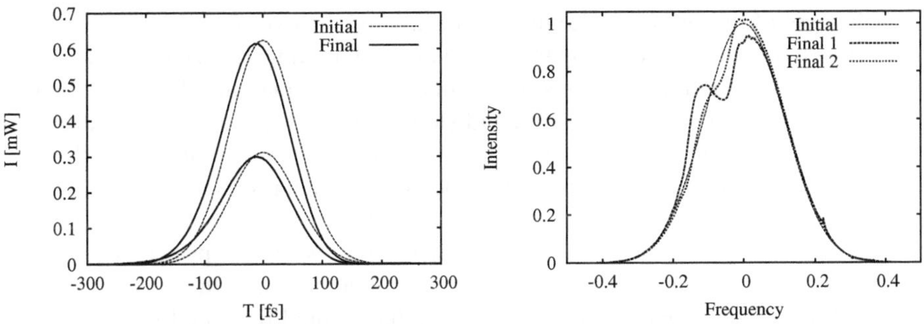

Fig. 18.7 Pulse shape and spectrum of two coupled pulses after propagating 100 km through the idealized dispersion-managed fiber of Benchmark 2.

The same computational parameters are used as in Section 3.6: The temporal window has the width $[-30\,\text{ps}, 30\,\text{ps} - \Delta T]$ and $N = 4096$, $h = 40\,\text{m}$ are applied.

During propagation both pulses are experiencing almost undisturbed soliton-like oscillations every 4 km. Figure 18.7 compares the final signal shapes and spectra with the respective initial ones, where Pulse 2 has been shifted for visualization by $-1.5625\,\text{ps}$. Both pulses are delayed by roughly 10 fs but the signal shape is quite well preserved; the spectral alteration being rather moderate in both cases. In the left graphic of Figure 18.7 Pulse 1 and 2 are easily distinguished; in the right graphic the final spectra of Pulse 1 and 2 are specially indicated.

Finally, we comment on typical run times of the proposed split-step Fourier methods. Our implementation is in FORTRAN 90 and uses the Netlib NAPACK Fast Fourier Transformation (FFT) routines, which are coded in FORTRAN 77 [9]. Compiled with usual optimizations, the two-mode computation of Figure 18.5 required ~ 48 seconds on a single Intel Xeon E5 CPU with 2.1 GHz. Dependence on the number of Fourier modes N as well as the number of spatial steps M is linear and each computation of the convergence analysis of Figure 18.6 is therefore four times more expensive than the next coarser one. On the same CPU, the dispersion-managed two-mode simulation of Figure 18.7 ran for ~ 100 seconds, its single-mode analogue of Figure 18.4 required ~ 40 seconds. These moderate run times and the given results provide evidence for the relevance of the proposed numerical methods for practical long-distance fiber optical communication line design.

6 Conclusions

Reliable extensions of the classical SSFM into the regime of ultra-fast pulses have been derived and demonstrated for typical Gaussian communication pulses in highly nonlinear and dispersion-managed long-distance optical fibers. The primary difficulty in this regime lies in the appropriate mathematical treatment of the additional nonlinear terms modeling signal self-steepening and stimulated Raman scattering.

For the case of the single-mode equation (18.3) and the two-mode system (18.37) it was shown that under Madelung transformation all nonlinearities can be effectively combined into an inhomogeneous system of advection equations of the signal intensities and phases. Following upwind and slope-limiting ideas, originally developed in the context of supersonic hydrodynamics, a robust numerical method is then derived for the single-mode nonlinear sub-operator and incorporated into a symmetric SSFM. Reliable convergence and numerical approximation accuracy of second order is demonstrated for the overall method. While it would be principally feasible to apply the exact same approach to the two-mode case and the correspondingly derived four-dimensional system (18.47), we have opted for now for a mathematically less involved fractional step approach and apply two single-field nonlinear sub-operators successively to approximate the solution of (18.47). This single-field sub-operator is derived as a straightforward extension of the slope-limited upwind method for the single-mode case. When the fractional step method for (18.47) is used in a two-mode SSFM, the overall numerical scheme converges reliably, yet, in a highly nonlinear test case only an order of accuracy of 1.5 has been measured. Future work will concentrate on developing an unsplit scheme for (18.47). It is expected that such a method should obtain an order of accuracy close to 2 while being of comparable computational expense and robustness as the two-mode SSFM proposed in here.

Acknowledgements This work was supported by the Department of Defense and used resources of the Extreme Scale Systems Center at Oak Ridge National Laboratory.

References

1. Agrawal, G.P.: Nonlinear Fiber Optics, 4th edn. Academic Press (2007)
2. Amorim, A.A., Tognetti, M.V., Oliveira, P., Silva, J.L., Bernardo, L.M., Kärtner, F.X., Crespo, H.M.: Sub-two-cycle pulses by soliton self-compression in highly-nonlinear photonic crystal fibers. Opt. Lett. **34**, 3851 (2009)
3. Atre, R., Panigrahi, P.: Controlling pulse propagation in optical fibers through nonlinearity and dispersion management. Phys. Rev. A **76**, 043,838 (2007)
4. Blow, K.J., Wood, D.: Theoretical description of transient stimulated Raman scattering in optical fibers. IEEE J. Quantum Electronics **25**(12), 2665–2673 (1989)
5. Deiterding, R., Glowinski, R., Oliver, H., Poole, S.: A reliable split-step Fourier method for the propagation equation of ultra-fast pulses in single-mode optical fibers. J. Lightwave Technology **31**, 2008–2017 (2013)
6. Glowinski, R.: Finite element methods for incompressible viscous flows. In: P.G. Ciarlet, J.L. Lions (eds.) Handbook of Numerical Analysis, vol. IX, pp. 3–1176, North-Holland, Amsterdam (2003)
7. Gnauck, A.H., Charlet, G., Tran, P., Winzer, P.J., Doerr, C.R., Centanni, J.C., Burrows, E.C., Kawanishi, T., Sakamoto, T., Higuma, K.: 25.6 Tb/s WDM transmission of polarization-multiplexed RZ-DQPSK signals. J. Lightwave Technology **26**, 79 (2008)
8. Guo, S., Huang, Z.: Densely dispersion-managed fiber transmission system with both decreasing average dispersion and decreasing local dispersion. Optical Engineering **43**, 1227 (2004)
9. Hager, W.: Applied Numerical Linear Algebra. Prentice Hall, Englewood Cliffs, NJ (1988)

10. Hohage, T., Schmidt, F.: On the numerical solution of nonlinear Schrödinger type equations in fiber optics. Tech. Rep. ZIB-Report 02–04, Konrad-Zuse-Zentrum für Informationstechnik Berlin (2002)
11. Kalithasan, B., Nakkeeran, K., Porsezian, K., Tchofo Dinda, P., Mariyappa, N.: Ultra-short pulse propagation in birefringent fibers – the projection operator method. J. Opt. A: Pure Appl. Opt. **10**, 085,102 (2008)
12. Ketcheson, D.I., LeVeque, R.J.: WENOClaw: a higher order wave propagation method. In: Hyperbolic Problems: Theory, Numerics, Applications, pp. 609–616. Springer, Berlin (2008)
13. Lax, P.D.: Gibbs phenomena. J. Scientific Comput. **28**(2/3), 445–449 (2006)
14. van Leer, B.: Towards the ultimate conservative difference scheme V. A second order sequel to Godunov's method. J. Comput. Phys. **32**, 101–136 (1979)
15. LeVeque, R.J.: Finite Volume Methods for Hyperbolic Problems. Cambridge University Press, Cambridge, New York (2002)
16. Long, V.C., Viet, H.N., Trippenback, M., Xuan, K.D.: Propagation technique for ultrashort pulses II: Numerical methods to solve the pulse propagation equation. Comp. Meth. Science Techn. **14**(1), 13–19 (2008)
17. Madelung, E.: Quantentheorie in hydrodynamischer Form. Zeitschrift für Physik **40**(3–4), 322–326 (1927)
18. Malomed, B.A.: Pulse propagation in a nonlinear optical fiber with periodically modulated dispersion: variational approach. Opt. Comm. **136**, 313–319 (1997)
19. Muslu, G.M., Erbay, H.A.: A split-step Fourier method for the complex modified Korteweg-de Vries equation. Computers and Mathematics with Applications **45**, 503–514 (2003)
20. Richardson, L.J., Forsyiak W. Blow, K.J.: Single channel 320Gbit/s short period dispersion managed transmission over 6000km. Optics Letters **36**, 2029 (2000)
21. Sinkin, O.V., Holzlöhner, R., Zweck, J., Menyuk, C.R.: Optimization of the split-step Fourier method in modeling optical-fiber communication systems. J. Lightwave Technology **21**(1), 61–68 (2003)
22. Smoller, J.: Shock Waves and Reaction-Diffusion Equations. Springer, New York (1982)
23. Spiegel, E.A.: Fluid dynamical form of the linear and nonlinear Schrödinger equations. Physica D: Nonlinear Phenomena **1**(2), 236–240 (1980)
24. Strang, G.: On the construction and comparison of difference schemes. SIAM J. Num. Anal. **5**, 506–517 (1968)

Chapter 19
Operator Splitting Methods with Error Estimator and Adaptive Time-Stepping. Application to the Simulation of Combustion Phenomena

Stéphane Descombes, Max Duarte, and Marc Massot

Abstract Operator splitting techniques were originally introduced with the main objective of saving computational costs. A multi-physics problem is thus split in subproblems of different nature with a significant reduction of the algorithmic complexity and computational requirements of the numerical solvers. Nevertheless, splitting errors are introduced in the numerical approximations due to the separate evolution of the split subproblems and can compromise a reliable simulation of the coupled dynamics. In this chapter we present a numerical technique to estimate such splitting errors on the fly and dynamically adapt the splitting time steps according to a user-defined accuracy tolerance. The method applies to the numerical solution of time-dependent stiff PDEs, illustrated here by propagating laminar flames investigated in combustion applications.

Mathematical Subject Classification (2010): 65M22, 65G20, 65Z05, 35K55

S. Descombes (✉)
Université Côte d'Azur, CNRS, Inria, LJAD, France

INRIA Sophia Antipolis-Mediterrané Research Center, Nachos Project-Team, 06902 Sophia Antipolis Cedex, France
e-mail: stephane.descombes@unice.fr

M. Duarte
CCSE, Lawrence Berkeley National Laboratory, 1 Cyclotron Rd. MS 50A-1148, Berkeley, CA 94720, USA

CD-adapco, 200 Shepherds Bush Road, London W6 7NL, UK
e-mail: max.duarte@cd-adapco.com

M. Massot
CNRS UPR 288, Laboratoire EM2C, Grande Voie des Vignes, 92295 Chatenay-Malabry Cedex, France

CentraleSupélec, Grande Voie des Vignes, 92295 Chatenay-Malabry Cedex, France

Fédération de Mathématiques de l'École Centrale Paris, CNRS FR 3487, Grande Voie des Vignes, 92295 Chatenay-Malabry Cedex, France
e-mail: marc.massot@centralesupelec.fr

R. Glowinski et al. (eds.), *Splitting Methods in Communication, Imaging, Science, and Engineering*, Scientific Computation, DOI 10.1007/978-3-319-41589-5_19

1 Context and Motivation

Let us consider a scalar reaction–diffusion equation

$$\left.\begin{array}{ll} \partial_t u - \partial_x^2 u = f(u), & x \in \mathbb{R}, t > 0, \\ u(x,0) = u_0(x), & x \in \mathbb{R}, \end{array}\right\} \tag{19.1}$$

and represent the solution $u(.,t)$ as $T^t u_0$, where T^t is the semi-flow associated with (19.1). Given v_0 and w_0, an operator splitting approach amounts to consider the following subproblems:

$$\left.\begin{array}{l} \partial_t v - \partial_x^2 v = 0, \;\; x \in \mathbb{R}, t > 0, \\ v(x,0) = v_0(x), \; x \in \mathbb{R}, \end{array}\right\} \tag{19.2}$$

and

$$\left.\begin{array}{ll} \partial_t w = f(w), & x \in \mathbb{R}, t > 0, \\ w(x,0) = w_0(x), & x \in \mathbb{R}. \end{array}\right\} \tag{19.3}$$

We denote by $X^t v_0$ and $Y^t w_0$, respectively, the solutions of (19.2) and (19.3). The Lie (or Lie–Trotter [55]) splitting approximations to the solution of problem (19.1) are thus given by

$$\mathscr{L}_1^t u_0 = X^t Y^t u_0, \qquad \mathscr{L}_2^t u_0 = Y^t X^t u_0. \tag{19.4}$$

Lie approximations are of first order in time; second order can be achieved by using symmetric Strang (or Marchuk [41]) formulas [53] to obtain

$$\mathscr{S}_1^t u_0 = X^{t/2} Y^t X^{t/2} u_0, \qquad \mathscr{S}_2^t u_0 = Y^{t/2} X^t Y^{t/2} u_0. \tag{19.5}$$

Even though higher order splitting schemes have been also developed, more sophisticated numerical implementations are required and their applicability is currently limited to specific linear or non-stiff problems (see, e.g., [17, 11, 33, 9] and discussions therein). The main advantage of such a splitting approach is that problems of different mathematical nature, in this case diffusion and reaction equations, can be solved separately with dedicated numerical methods. The latter involves a significant reduction of the algorithmic complexity of the overall method to advance the fully coupled problem with a potential reduction of computational requirements.

However, the separate time evolution of the split subproblems during a given splitting time step Δt introduces the so-called *splitting errors*. These errors have been mathematically characterized in the literature for general nonlinear problems and sufficiently small splitting time steps, relying mainly on the Baker–Campbell–Hausdorff formula on composition of exponentials together with Lie derivative calculus (see, e.g., [31] for ODEs and [36] for PDEs). In particular, Lanser & Verwer explicitly derived in [40] the splitting errors arising in the solution of reaction–diffusion–convection equations. Within this theoretical framework and considering an appropriate functional space, the following estimates can be thus derived for the scalar nonlinear reaction–diffusion equation (19.1).

Theorem 1. *Given $C_b^\infty(\mathbb{R})$, the space of functions of class C^∞ on $\mathbb{R}$ and bounded over $\mathbb{R}$, let us introduce the Schwartz space $\mathbb{S}(\mathbb{R})$ defined by*

$$\mathbb{S}(\mathbb{R}) = \{g \in C^\infty(\mathbb{R}) \mid \sup_{v\in\mathbb{R}} |v^{\alpha_1}\partial_v^{\alpha_2} g(v)| < \infty \quad \textit{for all integers } \alpha_1, \alpha_2\}$$

and define the space $\mathbb{S}_1(\mathbb{R})$ made out of functions v belonging to $C_b^\infty(\mathbb{R})$ such that v' belongs to $\mathbb{S}(\mathbb{R})$.

Assume that u_0 belongs to $\mathbb{S}_1(\mathbb{R})$ and that f belongs to $C^\infty(\mathbb{R})$. For Δt small enough, the following asymptotics hold

$$T^{\Delta t}u_0 - \mathscr{L}_2^{\Delta t}u_0 = \frac{\Delta t^2}{2} f''(u_0)(\partial_x u_0)^2 + \mathscr{O}(\Delta t^3), \tag{19.6}$$

and

$$\begin{aligned} T^{\Delta t}u_0 - \mathscr{S}_2^{\Delta t}u_0 = {} & \frac{\Delta t^3}{24}\left[f'(u_0)f''(u_0) + f(u_0)f^{(3)}(u_0)\right](\partial_x u_0)^2 \\ & -\frac{\Delta t^3}{12} f^{(4)}(u_0)(\partial_x u_0)^4 - \frac{\Delta t^3}{3} f^{(3)}(u_0)(\partial_x u_0)^2 \partial_x^2 u_0 \\ & -\frac{\Delta t^3}{6} f''(u_0)(\partial_x^2 u_0)^2 + \mathscr{O}(\Delta t^4). \end{aligned} \tag{19.7}$$

Proof. It suffices to consider the Baker–Campbell–Hausdorff formula (see [31, 36]) and compute the corresponding Lie brackets (commutators in the case of linear operators) containing the Lie derivatives associated with the nonlinear function f and the Laplace operator Δ (see [20]). □

Similar estimates can be derived for $\mathscr{L}_1^{\Delta t}u_0$ and $\mathscr{S}_1^{\Delta t}u_0$. More refined estimates that characterize the dependences with respect to the norms of the initial data and the nonlinearity can be also obtained using exact error representations. The following theorem shows, for instance, the representation of $T^{\Delta t}u_0 - \mathscr{L}_2^{\Delta t}u_0$.

Theorem 2. *Let us denote by D_2 the derivative with respect to the initial condition, under the same assumptions of Theorem 1 we have*

$$\begin{aligned} T^{\Delta t}u_0 - \mathscr{L}_2^{\Delta t}u_0 = {} & \int_0^{\Delta t}\int_0^s D_2 T^{t-s}(Y^s X^s u_0)\exp\left(\int_0^{s-r} f'(Y^{\sigma+r}X^s u_0)\,d\sigma\right) \times \\ & f''(Y^r X^s u_0)\exp\left(2\int_0^r f'(Y^\sigma X^s u_0)\,d\sigma\right)(\partial_x X^s u_0)^2\, drds. \end{aligned}$$

Similar representations can be derived for $\mathscr{L}_1^{\Delta t}u_0$, $\mathscr{S}_1^{\Delta t}u_0$, and $\mathscr{S}_2^{\Delta t}u_0$. These results are due to a long series of papers and especially those of Michelle Schatzman, a great contributor to operator splitting methods. Originally introduced in [49] for linear operators, these exact representations of the local errors have been extended in a more general functional setting in [24, 22] and in the nonlinear case in [23, 19].

Even though theoretical estimates of splitting errors can be formally established for general problems like (19.1), computing them in practice in multi-dimensional configurations or for more complex models may rapidly become cumbersome. Developing a splitting error estimator based on these theoretical estimates can hence be inappropriate except for particular configurations like, for example, linear problems as proposed in [2]. On the other hand, splitting solvers that do not account for splitting errors may yield numerical approximations that poorly reproduce the coupled physical dynamics of the problem under investigation. The latter is even more relevant if one takes into account that practical considerations often suggest the use of relatively large splitting time steps in order to ease heavy computational costs related to the numerical simulation of complex applications. In what follows we present a numerical strategy to estimate splitting errors *on the fly* and hence adapt splitting time steps to guarantee numerical approximations within a user-defined accuracy tolerance. The scheme was originally introduced in [20] along with its corresponding mathematical analysis. Throughout this chapter we consider the scalar nonlinear reaction–diffusion equation (19.1), although the same ideas are easily extended to multi-dimensional or more complex configurations, as well as to other time-dependent stiff PDEs.

2 Splitting Error Estimator and Adaptive Time-stepping

In general an adaptive time-stepping technique relies on a dynamic numerical estimate of local errors; time steps are consequently set according to a predefined accuracy tolerance. In our case estimating the splitting errors, for instance, (19.6) and (19.7), constitutes the key issue since the physics of the problem may be substantially altered by the splitting procedure. Inspired by ODE solvers one way to compute such an estimate considers a lower order scheme, embedded if possible in the main numerical integration solver (see, e.g., [32]). This is a standard approach, for instance, for Runge–Kutta methods.

Based on the $\mathscr{S}_2$-scheme in (19.5), let us consider the shifted Strang formula introduced in [20],

$$\mathscr{S}_{2,\varepsilon}^{\Delta t} u_0 = Y^{(1/2-\varepsilon)\Delta t} X^{\Delta t} Y^{(1/2+\varepsilon)\Delta t} u_0, \tag{19.8}$$

for ε in $(-1/2, 0) \cup (0, 1/2)$. To simplify the notations, we will denote $\mathscr{S}_2$ by $\mathscr{S}$ and $\mathscr{S}_{2,\varepsilon}$ by $\mathscr{S}_\varepsilon$. Similar to Theorem 1 the following one was demonstrated in [20].

Theorem 3. *Assume that u_0 belongs to $\mathbb{S}_1(\mathbb{R})$ and that f belongs to $\mathrm{C}^\infty(\mathbb{R})$. For Δt and ε small enough, the following asymptotic holds*

$$
\begin{aligned}
T^{\Delta t} u_0 - \mathscr{S}_\varepsilon^{\Delta t} u_0 = & -\varepsilon \Delta t^2 f''(u_0)(\partial_x u_0)^2 \\
& \frac{\Delta t^3}{24} \left[f'(u_0) f''(u_0) + f(u_0) f^{(3)}(u_0) \right] (\partial_x u_0)^2 \\
& -\frac{\Delta t^3}{12} f^{(4)}(u_0)(\partial_x u_0)^4 - \frac{\Delta t^3}{3} f^{(3)}(u_0)(\partial_x u_0)^2 \partial_x^2 u_0 \\
& -\frac{\Delta t^3}{6} f''(u_0)(\partial_x^2 u_0)^2 + \mathcal{O}(\varepsilon \Delta t^3) + \mathcal{O}(\Delta t^4). \qquad (19.9)
\end{aligned}
$$

Proof. See [20] Theorem 3.2. □

Therefore, just like Lie schemes, the shifted Strang formula (19.8) yields first order approximations. For ε equal to $-1/2$ (resp., 0) estimate (19.9) becomes (19.6) (resp., (19.7)). In particular, from (19.7) and (19.9), we have that

$$
\mathscr{S}^{\Delta t} u_0 - \mathscr{S}_\varepsilon^{\Delta t} u_0 = \varepsilon \Delta t^2 f''(u_0)(\partial_x u_0)^2 + \mathcal{O}(\varepsilon \Delta t^3).
$$

Given u_0, we can thus compute two splitting approximations,

$$
\begin{pmatrix} \mathscr{S}^{\Delta t} u_0 \\ \mathscr{S}_\varepsilon^{\Delta t} u_0 \end{pmatrix} = \begin{pmatrix} Y^{\Delta t/2} X^{\Delta t} Y^{\Delta t/2} u_0 \\ Y^{(1/2-\varepsilon)\Delta t} X^{\Delta t} Y^{(1/2+\varepsilon)\Delta t} u_0 \end{pmatrix}, \qquad (19.10)
$$

where the $\mathscr{S}_\varepsilon$-scheme is a lower order, embedded method with respect to the standard $\mathscr{S}$-scheme. Embedding is accomplished as long as ε is different from $-1/2$, i.e., $\mathscr{S}_{2,\varepsilon}$ is not $\mathscr{L}_2$. Taking into account that

$$
\begin{aligned}
\mathscr{S}^{\Delta t} u_0 - \mathscr{S}_\varepsilon^{\Delta t} u_0 &= \mathscr{S}^{\Delta t} u_0 - T^{\Delta t} u_0 + T^{\Delta t} u_0 - \mathscr{S}_\varepsilon^{\Delta t} u_0 \\
&= \mathcal{O}(\Delta t^3) + \mathcal{O}(\Delta t^2) \approx \mathcal{O}(\Delta t^2), \qquad (19.11)
\end{aligned}
$$

we define a splitting error estimator, *err*, and for a given accuracy tolerance, η, the following must be verified

$$
err = \left\| \mathscr{S}^{\Delta t} u_0 - \mathscr{S}_\varepsilon^{\Delta t} u_0 \right\| \le \eta,
$$

to assure local splitting errors bounded by η. Supposing that $err \approx C\Delta t^2$ following (19.11), we define a new splitting time step, Δt_{new}, such that $\eta \approx C\Delta t_{\text{new}}^2$. Therefore, the adaptive time-stepping is defined by

$$
\Delta t_{\text{new}} = \upsilon \Delta t \sqrt{\frac{\eta}{err}}, \qquad (19.12)
$$

with a security factor $\upsilon > 0$, close to 1.

To summarize, given the numerical approximation u_n at a given time and the splitting time step Δt_n, the time-stepping technique for a given η is performed as follows:

1. Compute both splitting approximations: $\mathscr{S}^{\Delta t_n} u_n$ and $\mathscr{S}_\varepsilon^{\Delta t_n} u_n$, following (19.10).

2. Compute the splitting error estimator: $err = \|\mathscr{S}^{\Delta t_n} u_n - \mathscr{S}_{\varepsilon}^{\Delta t_n} u_n\|$, and the new splitting time step, Δt_{new}, with (19.12).
3. If $err \leq \eta$, the solution $\mathscr{S}^{\Delta t_n} u_n$ is accepted: $u_{n+1} = \mathscr{S}^{\Delta t_n} u_n$ and $\Delta t_{n+1} = \Delta t_{\text{new}}$. Otherwise, the solution is rejected and the time-stepping technique is restarted with $\Delta t_n = \Delta t_{\text{new}}$.

Notice that whenever a lower order, embedded scheme is used to estimate local errors, the error estimator is actually measuring the numerical errors associated with the low order approximation, computed here by the $\mathscr{S}_{\varepsilon}$-scheme. The splitting error in the numerical solution is therefore over-estimated since the $\mathscr{S}$-scheme is formally of higher order. However, this is a safety choice to guarantee numerical approximations within the user-defined accuracy tolerance. In particular a complementary numerical procedure was developed in [20] in order to dynamically choose ε such that the estimator *err* yields closer estimates to the actual splitting errors $\|T^{\Delta t_n} u_n - \mathscr{S}^{\Delta t_n} u_n\|$, even for relatively large splitting time steps.

The shifted Strang formula (19.8) could also become the $\mathscr{L}_1$-scheme, if we let ε be equal to $1/2$. In this case the $\mathscr{L}_1$-scheme in (19.4) acts as the lower order, embedded method for the $\mathscr{S}_2$-scheme, as proposed in [38] for non-stiff problems. Nevertheless, it is well known in the context of stiff PDEs that order reductions may arise, due to short-life transients associated with the fastest variables, when one considers splitting time steps larger than the fastest scales. In particular it has been mathematically proved in [52, 21, 39] that better performances are expected when the splitting scheme ends with the stiffest operator. Having the $\mathscr{S}_{\varepsilon}$-scheme as the embedded scheme, built analogously to the standard Strang formula with the same stiffest operator as the ending step, involves similar behaviors in terms of order reduction and overall numerical performance for relatively large splitting time steps. Finally, the mathematical analysis was conducted in [20] in the case of the scalar nonlinear reaction–diffusion equation (19.1); however, the asymptotic estimate (19.9) would in general remain of $\mathscr{O}(\varepsilon \Delta t^2)$ when considering other PDEs.

3 Dedicated Splitting Solver for Stiff PDEs

An operator splitting approach allows one to use appropriate solvers for each split subproblems. In particular for splitting schemes resulting from composition methods like (19.4)–(19.5), the numerical stability of the splitting scheme is assured depending on the stability properties of these solvers. That is, if the numerical solvers used to advance each split subproblems are stable during a given splitting time step, then the splitting approximation will remain stable. The latter involves that relatively large splitting time steps can be considered without having any numerical issue; however, the validity of results may be undermined by the splitting errors. The technique summarized in Section 2 aims at tracking these errors, but it relies in practice on a splitting solver that must be capable to cope with stiff PDEs.

Let us make the following observations on splitting schemes for stiff PDEs:

1. The exact solutions of the split subproblems are considered in the classical analysis of splitting schemes, given here by the semi-flows $X^t\cdot$ and $Y^t\cdot$, associated with equations (19.2) and (19.3), respectively. This is also the case for the $\mathscr{S}_\varepsilon$-scheme introduced in [20].
2. For sufficiently large splitting time steps, the intermediate splitting approximations may drift away from the coupled dynamics. The latter may introduce potentially fast transients or boundary layers immediately after a split operator has been applied [58, 46].
3. As previously said, for relatively large splitting time steps, better performances are expected with splitting schemes that end with the stiffest operator [52, 21, 39].

We consider here the dedicated splitting solver developed in [28] to address the latter three remarks. This solver was originally conceived for stiff reaction–diffusion models, but the same ideas can be extended to other configurations.

In terms of the numerical methods used to solve the split subproblems, we consider one-step and high order schemes with appropriate stability properties and time-stepping features, based on the following precepts [28]:

- One-step schemes are preferred over multi-step ones because of the initial procedure needed by the latter to start the algorithm. For splitting schemes this starting procedure would be performed at least once at every splitting time step and might coincide with the fast transient regimes, hence involving more computational effort (see, e.g., the study conducted in [56]).
- Integration schemes of approximation order higher than the splitting method are chosen such that the corresponding integration errors remain lower than the splitting ones. In this way integration errors do not interfere with the splitting ones and the global accuracy of the method is set by the splitting scheme as considered in the corresponding mathematical analysis.
- Methods with time-stepping features based on stability or a user-defined accuracy tolerance are preferred, computing as many time steps as necessary within a given splitting time step. The latter is particularly relevant to adapt time steps during short non-physical transients that may arise within splitting time steps. The splitting time step can be therefore defined independently of the numerical integration of the split subproblems avoiding stability constraints associated with mesh size or stiff source time scales. When only one integration time step is needed and it is therefore equal to the splitting one, the higher order solver yields numerical errors potentially lower than the splitting ones.
- When implicit methods are used to cope with the stiffness of a given subproblem, L-stable schemes should be considered to rapidly damp out potentially fast transients. As before one-step schemes are favored as multi-step methods cannot be L-stable with an order higher than 2 [14].

Further discussions on these aspects can be found in [25].

As an illustration, the splitting technique proposed in [28] to solve stiff reaction–diffusion equations like (19.1) considers the following solvers: Radau5 [32] for the reaction subproblem (19.3) and the ROCK4 method [1] for the diffusion one (19.2). Radau5 is a fifth order implicit Runge–Kutta method exhibiting A- and L-stability properties to efficiently solve stiff systems of ODEs. However, the high performance of implicit Runge–Kutta methods for stiff ODEs is adversely affected when applied to large systems of nonlinear equations arising in the numerical integration of semi-discrete stiff PDEs. Significant effort is actually required to achieve numerical implementations that solve the corresponding algebraic problems at reasonable computational expenses. A splitting approach offers a much simpler alternative since the split subproblem (19.3) becomes a system of ODEs which can be separately solved point-wise over the computational domain. On the other hand, ROCK4 is a fourth order stabilized explicit Runge–Kutta method with extended stability domain along the negative real axis, well suited to numerically treat mildly stiff elliptic operators. The diffusion equation (19.2) is thus solved over the entire domain with an explicit scheme and therefore with a limited memory requirement with respect to an implicit one. Both methods implement adaptive time-stepping techniques to guarantee computations within a prescribed accuracy tolerance.

Within this framework one can prescribe relatively fine tolerances for the numerical solvers, Radau5 and ROCK4, and the only input parameter of the splitting solver is the splitting time step. In [28] a constant splitting time step was considered to simulate propagating reaction waves. In this case the wavefront velocity is retained as the key physical parameter to define a splitting time step that guarantees a sufficiently accurate resolution of the coupled dynamics. The numerical solvers are thus in charge of solving the split subproblems during the given splitting time step, assuring the numerical stability of the computations and coping with the stiffness of the equations. The performance of this strategy was assessed in the context of multi-dimensional chemical waves [28] and for a model of human ischemic stroke with several variables and complex mechanisms in the stiff source terms [27, 29]. In both cases the $\mathscr{S}_2$-scheme that ends with the reaction operator was used according to the theoretical insights derived in [21] for stiff nonlinear reaction–diffusion equations. Notice that localized reacting fronts are simulated in the case of propagating waves; therefore, intense computation of the stiff reaction problem (19.3) is required only along the fronts where important reactive activity is present. Consequently, Radau5 adapts its time steps only where it is necessary with important computational enhancements, yielding local time steps that may substantially differ according to the vicinity of the fronts. The latter cannot be done in the same simple way without splitting the original model.

While preliminary studies were required to determine an appropriate constant time step for propagating waves (as shown in [28, 27]), they are no longer necessary if one considers the adaptive splitting technique with error estimator introduced in [20] and previously described in Section 2. These methods complement each other since the adaptive strategy also requires a dedicated splitting solver to guarantee the theoretical framework of the analysis and to efficiently handle in practice stiffness and stability issues. As a result we end up with an adaptive, dedicated splitting solver for stiff PDEs. The only input parameter now is the splitting tolerance accuracy η, noting that the accuracy parameters of the numerical solvers must be set lower than η. Notice that lower order schemes for the split subproblems could be also used with the adaptive splitting technique as long as these solvers implement adaptive time-stepping with error control and their accuracy tolerances are set lower than the splitting one. However, for a given prescribed accuracy, higher order solvers as proposed in [28] yield solutions with potentially larger time steps.

4 Operator Splitting for Combustion Problems

Operator splitting methods have been used in the literature for decades, and were widely implemented and exploited for combustion problems to overcome classical restrictions on computational resources (see, e.g., [30, 15, 59, 44, 48, 50, 51, 47, 13, 45]). A good example is given by the numerical strategy developed in [37, 43] for reactive flows in a low Mach number formulation with detailed chemical kinetics and transport parameters. The splitting scheme introduced by these authors combines the implicit multi-step VODE solver [10] for stiff ODEs for the chemical reaction terms with a second order explicit RKC scheme [35, 57] for the diffusion problem. In this way important gains of computational efficiency are achieved with a splitting time step not limited by the stiff reactive scales and set according to the extended stability domain of the RKC solver (when convective stability constraints are less restrictive). Moreover, another efficient low Mach solver was introduced in [16] with an operator splitting method coupled with an AMR (Adaptive Mesh Refinement) technique [3, 8]. The problem is thus solved level-wise throughout a set of grids with different resolution, with splitting time steps set by the corresponding CFL condition associated with each grid size. Reaction terms are locally solved with VODE. With these bases, further developments in terms of algorithm implementation and parallel computing techniques led to the effective simulation of three-dimensional turbulent premixed flames with detailed chemistry (see, e.g., [5, 4]), a remarkable achievement for laboratory-scale turbulent flames (see, e.g., [7, 6]).

Considering the state of art and these recent advances, one may note as previously remarked that splitting schemes favor the use of dedicated numerical solvers of different nature as well as straightforward coupling with other techniques, with important gains in computational efficiency. Nevertheless, there are some open issues related to the construction of splitting schemes and the interaction of splitting errors with those originating from the inner implicit–explicit solvers (the influence of the latter ones on the global integration error was numerically illustrated, for instance, in [37, 43]). In particular a critical matter underlined in the literature is the lack of precise criteria to properly choose the splitting time steps according to the physical decoupling capabilities of the problem and for a given accuracy. Another question is the extension of these strategies to highly dynamic problems for which neither a constant nor a stability-based variable splitting time step is adequate, taking into account that the explicit schemes are intended to handle the slow, non-stiff part of the equations.

To address these problems the adaptive operator splitting scheme, briefly recalled in Section 2, was considered in [26] combined with a dedicated splitting solver for stiff time dependent PDEs built in practice under the precepts described in Section 3. Additionally, this splitting solver was coupled in [28] with a dynamic mesh refinement technique based on multiresolution (MR) analysis [34, 12, 42]. For a given semi-discretized problem, the MR mathematical background allows a better monitoring of numerical errors introduced by the compressed spatial representations with respect to the original uniform grid discretization. The resulting time–space adaptive solution scheme was thus described and analyzed in [26] and constitutes a fundamental building block for solvers used in combustion simulations. It provides an efficient algorithm in terms of both memory storage and computational performance, which allows multi-dimensional simulations assuring a given error tolerance, fixed in advance by the user.

The study conducted in [26] considered multi-dimensional laminar flames interacting with vortex structures, including the propagation of flame fronts and self-ignition processes of reactive mixtures. In what follows we recall some interesting aspects resulting from the numerical simulation of the ignition process and generation of a diffusion flame, investigated in [26].

5 Numerical Illustration

Let us consider the mathematical model derived in [54] to investigate the ignition dynamics of a diffusion flame, formed while a reactive layer is being rolled–up in a vortex. The hydrodynamics is decoupled from species and energy transport equations by adopting the standard thermo-diffusive approximation, leading to a reaction–diffusion–convection model. A two-dimensional computational domain is considered where pure and fresh hydrogen at temperature $T_{\mathrm{F},0}$ initially occupies the upper half part, while the remaining lower part of the domain is occupied by hot air at $T_{\mathrm{O},0}$. By defining a Schvab–Zeldo'vich variable Z and a reduced temperature θ given by

$$\theta = \frac{T - T_{\mathrm{O},0}}{T_{\mathrm{F},0} - T_{\mathrm{O},0}},$$

the mathematical model is given by a system of equations of the form [54]:

$$\left.\begin{aligned} &\partial_t Z + v_x \partial_x Z + v_y \partial_y Z - \left(\partial_x^2 Z + \partial_y^2 Z\right) = 0, \\ &\partial_t \theta + v_x \partial_x \theta + v_y \partial_y \theta - \left(\partial_x^2 \theta + \partial_y^2 \theta\right) = F(Z,\theta), \end{aligned}\right\}$$

where $F(Z,\theta)$ is a highly nonlinear function. The velocity field (v_x, v_y) is given by a single vortex centered on the planar interface between the two media, which varies strongly in time and space.

The physics of the phenomenon can be briefly described as follows. A rotating vortex is introduced immediately at $t = 0$. The resulting forced convection superposes to the diffusive mechanisms and accelerates the mixture of the gases. As a consequence, a diffusion flame ignites along the contact surface of both media, taking into account the important difference of temperatures in those regions. If the velocity field is sufficiently strong, it will entrain fresh gases into the vortex core which will react with an intensity set by the mixing temperature of gases of about $(T_{F,0}+T_{O,0})/2$. These locally lower temperatures result in a delayed ignition of the core unless air of sufficiently high temperature is initially considered. Once the flame is completely ignited, it propagates outwards from the center of the computational domain. The complete phenomenon encompasses thus very different physical regimes like mixing, ignition, propagation, which can be characterized depending on the initial reactants configuration and on the imposed velocity field, as studied in detail in [54].

As an illustration we recall a configuration investigated in [26] with fresh fuel initially at $T_{F,0}$ equal to 300 K and hot air at a temperature $T_{O,0}$ of 1000 K, with a strongly varying velocity field. Figure 19.1 shows three different stages of the ignition phenomenon in terms of the heat release rate $F(Z,\theta)$. The adaptive splitting scheme of Section 2 with a predefined accuracy tolerance of $\eta = 10^{-3}$, yields splitting time steps as shown in Figure 19.1. An initial splitting time step of $\Delta t = 10^{-8}$ is chosen to properly handle the inclusion of the vortex and the fast variation of the velocity field. The splitting step increases until $t \approx 6.5 \times 10^{-5}$ ($\Delta t \approx 2 \times 10^{-5}$) during the mixing phase, and one then finds a series of rejected steps. The splitting time step is thus reduced down to the time scale needed to guarantee the prescribed accuracy: $\Delta t \approx 10^{-7}$. This behavior naturally coincides with the sudden ignition of the flame and the subsequent fast propagation along the contact surface, once a certain temperature is locally reached after the initial mixing of reactants. The last part shown in Figure 19.1 corresponds to the beginning of the propagation stage with $\Delta t \approx 10^{-5}$, where the core has not ignited yet.

A dynamic adaptation of the splitting time step is hence mandatory to identify these changes in the physical behavior of the phenomenon and to suitably describe the entire process. In particular it takes approximately 207.5 minutes to solve this problem with an adaptive splitting technique on a uniform grid, compared to 674.7 minutes with a constant splitting time step of 10^{-7}, of the order of the convective time steps. Greater gains are observed for longer periods of time integration as the impact of the initial transients and hence small splitting time steps is less important on the overall computational performance. Using dynamic grid adaptation combined with the adaptive splitting solver further reduces the CPU time to 8.9 minutes, taking into account the highly localized flame fronts in this particular problem. Most importantly, with this time–space adaptive technique time integration errors can be tracked and controlled as well as those originating from the compressed spatial representations.

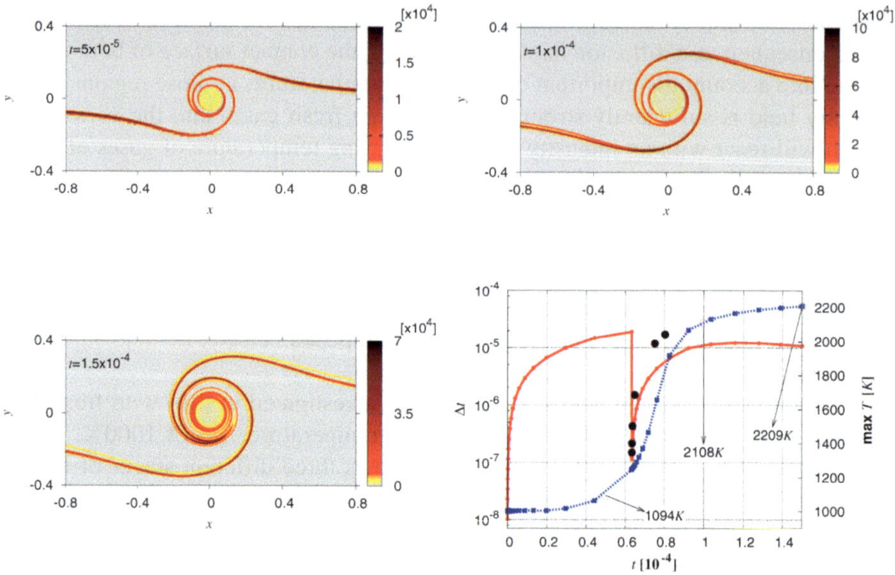

Fig. 19.1 Instantaneous heat release rate F at $t = 5 \times 10^{-5}$ (top left), 10^{-4} (top right), and 1.5×10^{-4} (bottom left). Bottom right: time evolution of splitting time steps and maximum temperature T, deduced from θ. Rejected time steps are indicated with black bullets (•), while maximum temperatures for previous snapshots are marked with (∘).

6 Conclusion

In this chapter we have introduced operator splitting methods with error estimators and a time adaptive technique. The latter was further coupled with a space adaptive, finite volume multiresolution method. Numerical results obtained with this time-space adaptive technique support the conclusions that different multi-scale physical configurations can be successfully simulated and that numerical errors can be effectively controlled. The time and space adaptive techniques are clearly independent allowing the compatibility with any space discretization scheme such as finite volumes, finite elements, or discontinuous Galerkin methods. The method allows us also to exploit the current computational resources and to obtain high efficiency in terms of load balancing on parallel architectures as it is shown in [18], where a task-based parallelism is used on multi-core architectures in conjunction with a work stealing approach.

References

1. Abdulle, A.: Fourth order Chebyshev methods with recurrence relation. SIAM J. Sci. Comput. **23**(6), 2041–2054 (2002)
2. Auzinger, W., Koch, O., Thalhammer, M.: Defect-based local error estimators for splitting methods, with application to Schrödinger equations, Part I: The linear case. J. Comput. Appl. Math. **236**(10), 2643–2659 (2012)

3. Bell, J., Berger, M., Saltzman, J., Welcome, M.: Three-dimensional adaptive mesh refinement for hyperbolic conservation laws. SIAM J. Sci. Comput. **15**, 127–138 (1994)
4. Bell, J., Day, M., Almgren, A., Lijewski, M., Rendleman, C., Cheng, R., Shepherd, I.: Simulation of lean premixed turbulent combustion. J. Phys. Conf. Ser. **46**, 1–15 (2006)
5. Bell, J., Day, M., Grcar, J.: Numerical simulation of premixed turbulent methane combustion. Proc. Combust. Inst. **29**(2), 1987–1993 (2002)
6. Bell, J., Day, M., Grcar, J., Lijewski, M., Driscoll, J., Filatyev, S.: Numerical simulation of a laboratory-scale turbulent slot flame. Proc. Combust. Inst. **31**(1), 1299–1307 (2007)
7. Bell, J., Day, M., Shepherd, I., Johnson, M., Cheng, R., Grcar, J., Beckner, V., Lijewski, M.: Numerical simulation of a laboratory-scale turbulent V-flame. Proc. Nat. Acad. Sci. **1021**, 10,006–10,011 (2005)
8. Berger, M., Oliger, J.: Adaptive mesh refinement for hyperbolic partial differential equations. J. Comput. Phys. **53**, 484–512 (1984)
9. Blanes, S., Casas, F., Chartier, P., Murua, A.: Optimized high-order splitting methods for some classes of parabolic equations. Math. Comp. **82**(283), 1559–1576 (2013)
10. Brown, P.N., Byrne, G., Hindmarsh, A.: VODE: A variable-coefficient ODE solver. SIAM J. Sci. Stat. Comput. **10**, 1038–1051 (1989)
11. Castella, F., Chartier, P., Descombes, S., Vilmart, G.: Splitting methods with complex times for parabolic equations. BIT Numer. Math. **49**, 487–508 (2009)
12. Cohen, A., Kaber, S., Müller, S., Postel, M.: Fully adaptive multiresolution finite volume schemes for conservation laws. Math. Comp. **72**, 183–225 (2003)
13. Cuoci, A., Frassoldati, A., Faravelli, T., Ranzi, E.: Numerical modeling of laminar flames with detailed kinetics based on the operator-splitting method. Energy & Fuels **27**(12), 7730–7753 (2013)
14. Dahlquist, G.: A special stability problem for linear multistep methods. Nordisk Tidskr. Informations-Behandling **3**, 27–43 (1963)
15. D'Angelo, Y., Larrouturou, B.: Comparison and analysis of some numerical schemes for stiff complex chemistry problems. RAIRO Modél. Math. Anal. Numér. **29**(3), 259–301 (1995)
16. Day, M., Bell, J.: Numerical simulation of laminar reacting flows with complex chemistry. Combust. Theory Model. **4**, 535–556 (2000)
17. Descombes, S.: Convergence of a splitting method of high order for reaction-diffusion systems. Math. Comp. **70**(236), 1481–1501 (2001)
18. Descombes, S., Duarte, M., Dumont, T., Guillet, T., Louvet, V., Massot, M.: Task-based adaptive multiresolution for time-space multi-scale reaction-diffusion systems on multi-core architectures. arXiv preprint arXiv:1506.04651 p. 24 (2015)
19. Descombes, S., Duarte, M., Dumont, T., Laurent, F., Louvet, V., Massot, M.: Analysis of operator splitting in the nonasymptotic regime for nonlinear reaction-diffusion equations. Application to the dynamics of premixed flames. SIAM J. Numer. Anal. **52**(3), 1311–1334 (2014)
20. Descombes, S., Duarte, M., Dumont, T., Louvet, V., Massot, M.: Adaptive time splitting method for multi-scale evolutionary partial differential equations. Confluentes Math. **3**(3), 413–443 (2011)
21. Descombes, S., Massot, M.: Operator splitting for nonlinear reaction-diffusion systems with an entropic structure: Singular perturbation and order reduction. Numer. Math. **97**(4), 667–698 (2004)
22. Descombes, S., Schatzman, M.: Strang's formula for holomorphic semi-groups. J. Math. Pures Appl. (9) **81**(1), 93–114 (2002)
23. Descombes, S., Thalhammer, M.: The Lie-Trotter splitting for nonlinear evolutionary problems with critical parameters: A compact local error representation and application to nonlinear Schrödinger equations in the semiclassical regime. IMA J. Numer. Anal. **33**(2), 722–745 (2013)
24. Dia, B.O., Schatzman, M.: Commutateurs de certains semi-groupes holomorphes et applications aux directions alternées. RAIRO Modél. Math. Anal. Numér. **30**(3), 343–383 (1996)
25. Duarte, M.: Méthodes numériques adaptatives pour la simulation de la dynamique de fronts de réaction multi-échelles en temps et en espace. Ph.D. thesis, Ecole Centrale Paris, France (2011)

26. Duarte, M., Descombes, S., Tenaud, C., Candel, S., Massot, M.: Time-space adaptive numerical methods for the simulation of combustion fronts. Combust. Flame **160**, 1083–1101 (2013)
27. Duarte, M., Massot, M., Descombes, S., Tenaud, C., Dumont, T., Louvet, V., Laurent, F.: New resolution strategy for multi-scale reaction waves using time operator splitting and space adaptive multiresolution: Application to human ischemic stroke. ESAIM: Proc. **34**, 277–290 (2011)
28. Duarte, M., Massot, M., Descombes, S., Tenaud, C., Dumont, T., Louvet, V., Laurent, F.: New resolution strategy for multiscale reaction waves using time operator splitting, space adaptive multiresolution and dedicated high order implicit/explicit time integrators. SIAM J. Sci. Comput. **34**(1), A76–A104 (2012)
29. Dumont, T., Duarte, M., Descombes, S., Dronne, M.A., Massot, M., Louvet, V.: Simulation of human ischemic stroke in realistic 3D geometry. Commun. Nonlinear Sci. Numer. Simul. **18**(6), 1539–1557 (2013)
30. Goyal, G., Paul, P., Mukunda, H., Deshpande, S.: Time dependent operator-split and unsplit schemes for one dimensional premixed flames. Combust. Sci. Technol. **60**, 167–189 (1988)
31. Hairer, E., Lubich, C., Wanner, G.: Geometric Numerical Integration. Structure-Preserving Algorithms for Ordinary Differential Equations, 2nd edn. Springer–Verlag, Berlin (2006)
32. Hairer, E., Wanner, G.: Solving Ordinary Differential Equations II. Stiff and Differential-Algebraic Problems, 2nd edn. Springer–Verlag, Berlin (1996)
33. Hansen, E., Ostermann, A.: High order splitting methods for analytic semigroups exist. BIT Numer. Math. **49**, 527–542 (2009)
34. Harten, A.: Multiresolution algorithms for the numerical solution of hyperbolic conservation laws. Comm. Pure Appl. Math. **48**, 1305–1342 (1995)
35. van der Houwen, P., Sommeijer, B.: On the internal stability of explicit, m-stage Runge-Kutta methods for large m-values. Z. Angew. Math. Mech. **60**(10), 479–485 (1980)
36. Hundsdorfer, W., Verwer, J.: Numerical Solution of Time-Dependent Advection-Diffusion-Reaction Equations. Springer–Verlag, Berlin (2003)
37. Knio, O., Najm, H., Wyckoff, P.: A semi-implicit numerical scheme for reacting flow. II. Stiff, operator-split formulation. J. Comput. Phys. **154**(2), 428–467 (1999)
38. Koch, O., Neuhauser, C., Thalhammer, M.: Embedded exponential operator splitting methods for the time integration of nonlinear evolution equations. Appl. Numer. Math. **63**, 14–24 (2013)
39. Kozlov, R., Kværnø, A., Owren, B.: The behaviour of the local error in splitting methods applied to stiff problems. J. Comput. Phys. **195**(2), 576–593 (2004)
40. Lanser, D., Verwer, J.: Analysis of operator splitting for advection-diffusion-reaction problems from air pollution modelling. J. Comput. Appl. Math. **111**(1–2), 201–216 (1999)
41. Marchuk, G.: Some application of splitting-up methods to the solution of mathematical physics problems. Appl. Math. **13**(2), 103–132 (1968)
42. Müller, S.: Adaptive Multiscale Schemes for Conservation Laws, *Lect. Notes Comput. Sci. Eng.*, vol. 27. Springer-Verlag (2003)
43. Najm, H., Knio, O.: Modeling low Mach number reacting flow with detailed chemistry and transport. J. Sci. Comput. **25**(1–2), 263–287 (2005)
44. Oran, E., Boris, J.: Numerical Simulation of Reacting Flows, 2nd edn. Cambridge University Press (2001)
45. Ren, Z., Xu, C., Lu, T., Singer, M.A.: Dynamic adaptive chemistry with operator splitting schemes for reactive flow simulations. J. Comput. Phys. **263**(0), 19–36 (2014)
46. Ropp, D., Shadid, J.: Stability of operator splitting methods for systems with indefinite operators: Reaction-diffusion systems. J. Comput. Phys. **203**(2), 449–466 (2005)
47. Safta, C., Ray, J., Najm, H.: A high-order low-Mach number AMR construction for chemically reacting flows. J. Comput. Phys. **229**(24), 9299–9322 (2010)
48. Schwer, D., Lu, P., Green, W., Semião, V.: A consistent-splitting approach to computing stiff steady-state reacting flows with adaptive chemistry. Combust. Theory Model. **7**(2), 383–399 (2003)
49. Sheng, Q.: Global error estimates for exponential splitting. IMA J. Numer. Anal. **14**(1), 27–56 (1994)

50. Singer, M., Pope, S.: Exploiting ISAT to solve the reaction-diffusion equation. Combust. Theory Model. **8**(2), 361–383 (2004)
51. Singer, M., Pope, S., Najm, H.: Modeling unsteady reacting flow with operator splitting and ISAT. Combust. Flame **147**(1–2), 150–162 (2006)
52. Sportisse, B.: An analysis of operator splitting techniques in the stiff case. J. Comput. Phys. **161**(1), 140–168 (2000)
53. Strang, G.: On the construction and comparison of difference schemes. SIAM J. Numer. Anal. **5**, 506–517 (1968)
54. Thévenin, D., Candel, S.: Ignition dynamics of a diffusion flame rolled up in a vortex. Phys. Fluids **7**(2), 434–445 (1995)
55. Trotter, H.: On the product of semi-groups of operators. Proc. Am. Math. Soc. **10**, 545–551 (1959)
56. Valorani, M., Goussis, D.: Explicit time-scale splitting algorithm for stiff problems: Auto-ignition of gaseous mixtures behind a steady shock. J. Comput. Phys. **169**(1), 44–79 (2001)
57. Verwer, J.: Explicit Runge-Kutta methods for parabolic partial differential equations. Appl. Numer. Math. **22**(1–3), 359–379 (1996)
58. Verwer, J.G., Spee, E.J., Blom, J.G., Hundsdorfer, W.: A second-order Rosenbrock method applied to photochemical dispersion problems. SIAM J. Sci. Comput. **20**(4), 1456–1480 (1999)
59. Yang, B., Pope, S.: An investigation of the accuracy of manifold methods and splitting schemes in the computational implementation of combustion chemistry. Combust. Flame **112**(1–2), 16–32 (1998)

Chapter 20
Splitting Methods for Some Nonlinear Wave Problems

Annalisa Quaini and Roland Glowinski

Abstract The main goal of this chapter is to discuss the numerical solution of nonlinear wave equations associated with the first of the celebrated Painlevé transcendent ordinary differential equations and the Bratu problem nonlinearity. In order to solve numerically the above equations, whose solutions blow up in finite time in most cases, we advocate a numerical methodology based on the Strang's symmetrized operator-splitting scheme. With this approach, we can decouple nonlinearity and differential operators. The resulting schemes, combined with a finite element space discretization and adaptive time-stepping to monitor possible blow-up of the solution, provide a robust and accurate solution methodology, as shown by the results of the numerical experiments reported here.

1 Introduction

In this chapter, we discuss the numerical solution of two *nonlinear wave equations*. The first equation under consideration is the following one:

$$\frac{\partial^2 u}{\partial t^2} - c^2 \nabla^2 u = 6u^2 + t \quad \text{in } \Omega \times (0, T_{max}). \tag{20.1}$$

A. Quaini (✉)
Department of Mathematics, University of Houston, 4800 Calhoun Rd, Houston, TX 77204, USA
e-mail: quaini@math.uh.edu

R. Glowinski
Department of Mathematics, University of Houston, Houston, TX 77204, USA
e-mail: roland@math.uh.edu

R. Glowinski et al. (eds.), *Splitting Methods in Communication, Imaging, Science, and Engineering*, Scientific Computation, DOI 10.1007/978-3-319-41589-5_20

Eq. (20.1) is a classical wave equation with forcing term given by the first Painlevé equation, that is

$$\frac{d^2y}{dt^2} = 6y^2 + t. \tag{20.2}$$

Although discovered from purely mathematical considerations, the *six Painlevé 'transcendent' ordinary differential equations* arise in a variety of important physical applications (from plasma physics to quantum gravity). There is an abundant literature concerning the Painlevé equations (see, for example, [14, 23, 5] and the references therein). Surprisingly very few of the related publications are of numerical nature, notable exceptions being [6] and [5], which contain also additional references on the numerical solution of the Painlevé equations. Actually, we are going to consider the numerical solution of two *initial/boundary value problems* associated with (20.1), namely we supplement (20.1) with *initial conditions* and *pure homogeneous Dirichlet boundary conditions* (resp., *mixed Dirichlet-Sommerfeld boundary conditions*), that is

$$\begin{cases} u = 0 \text{ on } \partial\Omega \times (0, T_{max}), \\ u(0) = u_0, \dfrac{\partial u}{\partial t}(0) = u_1, \end{cases} \tag{20.3}$$

(resp.,

$$\begin{cases} u = 0 \text{ on } \Gamma_0 \times (0, T_{max}), \dfrac{1}{c}\dfrac{\partial u}{\partial t} + \dfrac{\partial u}{\partial n} = 0 \text{ on } \Gamma_1 \times (0, T_{max}), \\ u(0) = u_0, \dfrac{\partial u}{\partial t}(0) = u_1). \end{cases} \tag{20.4}$$

In (20.1)–(20.4):

- c (> 0) is the speed of propagation of the linear waves solutions of the equation

$$\frac{\partial^2 u}{\partial t^2} - c^2\nabla^2 u = 0.$$

- Ω is a bounded domain of $\mathbb{R}^d$, $\partial\Omega$ being its boundary.
- Γ_0 and Γ_1 are two disjoint non-empty subsets of $\partial\Omega$ verifying $\Gamma_0 \cup \Gamma_1 = \partial\Omega$.
- $\phi(t)$ denotes the function $x \to \phi(x,t)$.
- $(0, T_{max})$ is the solution existence interval.

Problems (20.1), (20.3) and (20.1), (20.4) are of multi-physics (reaction-propagation type) and multi-time scales natures. Thus, it makes sense to apply an *operator-splitting method* for their numerical solution, in order to decouple nonlinearity and differential operators and to treat the resulting sub-initial value problems with appropriate (and necessarily variable) time discretization sub-steps. Among the available operator-splitting methods, we have chosen the *Strang's symmetrized operator-splitting scheme* (introduced in [20]), because it provides a good compromise between accuracy and robustness as shown in, e.g., [9, 3, 13, 11] and other chapters of this book (see also the references therein).

In the second part of the chapter, we extend the methodology adopted for problems (20.1), (20.3) and (20.1), (20.4) to achieve the numerical solution of a slightly more challenging wave equation, namely:

$$\frac{\partial^2 u}{\partial t^2} + \frac{k}{\varepsilon + t}\frac{\partial u}{\partial t} - c^2\nabla^2 u = \lambda e^u \quad \text{in } \Omega \times (0, T_{\max}), \tag{20.5}$$

where $k, \varepsilon, \lambda > 0$. Eq. (20.5) is of the *Euler-Poisson-Darboux* type and the forcing term is given by the celebrated *Bratu problem* nonlinearity (see, e.g., [2] for applications to solid combustion). The differences between eq. (20.5) and eq. (20.1) are an extra damping term in eq. (20.5) and the nonlinear forcing term, which in eq. (20.5) does not depend on t explicitly. Eq. (20.5) is completed with initial conditions and homogeneous Dirichlet conditions (20.3). The extension to mixed Dirichlet-Sommerfeld (20.4) conditions is straightforward. The existence of solutions to nonlinear wave equations very close to (20.5) has been investigated by J.B. Keller in [15], assuming that $u_1 = 0$ in (20.3). In order to solve (20.5), we advocate a five-stage operator splitting scheme of the Strang symmetric type, which is an extension of the three-stage scheme used for (20.1).

This chapter is structured as follows: In Section 2, we discuss the time discretization of problems (20.1), (20.3) and (20.1), (20.4) by the Strang's symmetrized scheme. In Sections 3 and 4, we discuss the solution of the initial value subproblems originating from the splitting; the discussion includes the finite element approximation of the linear wave steps and the adaptive in time solution of the nonlinear ODE steps. In Section 5 and 6, we discuss the time discretization of problem (20.5), (20.3) by the Strang's symmetrized scheme and its realization. In Section 7, we present the results of numerical experiments validating the numerical methodologies discussed in the previous sections.

Remark 1. Strictly speaking, it is the *solution* of the Painlevé equations which is *transcendent*, not the equations themselves.

Remark 2. The numerical methodology discussed here would apply more or less easily to other nonlinear wave equations of the following type

$$\frac{\partial^2 u}{\partial t^2} + \frac{k}{\varepsilon + t}\frac{\partial u}{\partial t} - c^2\nabla^2 u = f\left(u, \frac{\partial u}{\partial t}, x, t\right).$$

Remark 3. The analysis of quasilinear parabolic equations with blow-up has motivated a substantial number of publications (see, e.g., [19] and references therein). Similarly, much literature has been devoted to the analysis and numerical analysis of nonlinear Schrödinger equations with blow-up (see, e.g., [16] and references therein). Concerning the Euler-Poisson-Darboux problem, J.B. Keller has proved blow-up in finite time properties and has provided an estimate of the blow-up time (see [15] for details). Albeit bearing the name of some of the most famous mathematicians of all times, the Euler-Poisson-Darboux problem has not attracted much attention from a numerical standpoint, a notable exception being [7] which focuses on linear cases. We are not aware of any publication addressing the numerical solution of the nonlinear Euler-Poisson-Darboux problem.

2 Application of the Strang's Symmetrized Operator-Splitting Scheme to the Solution of Problems (20.1), (20.3) and (20.1), (20.4)

2.1 A Brief Discussion on the Strang's Operator-Splitting Scheme

Although the Strang's symmetrized scheme is quite well known, it may be useful to present briefly this scheme before applying it to the solution of problems (20.1), (20.3) and (20.1), (20.4). Our presentation closely follows the ones in [9] (Chapter 6) and [10]. See also Chapters 1, 2, and 3 of this volume.

Let us consider the following *non-autonomous abstract initial value problem* (taking place in a Banach space, for example):

$$\begin{cases} \dfrac{d\phi}{dt} + A(\phi,t) + B(\phi,t) = 0 \text{ on } (0, T_{max}), \\ \phi(0) = \phi_0, \end{cases} \tag{20.6}$$

where in (20.6) the operators A and B can be *nonlinear* and even *multivalued* (in which case one has to replace $= 0$ by $\ni 0$ in (20.6)). Let Δt be a time-step (fixed, for simplicity) and let us denote $(n+\alpha)\Delta t$ by $t^{n+\alpha}$. When applied to the time discretization of (20.6), the basic Strang's symmetrized scheme reads as follows:

- Step 0: Set

$$\phi^0 = \phi_0. \tag{20.7}$$

For $n \geq 0$, ϕ^n being known, compute ϕ^{n+1} as follows:

- Step 1: Set $\phi^{n+1/2} = \phi(t^{n+1/2})$, ϕ being the solution of

$$\begin{cases} \dfrac{d\phi}{dt} + A(\phi,t) = 0 \text{ on } (t^n, t^{n+1/2}), \\ \phi(t^n) = \phi^n. \end{cases} \tag{20.8}$$

- Step 2: Set $\hat{\phi}^{n+1/2} = \phi(\Delta t)$, ϕ being the solution of

$$\begin{cases} \dfrac{d\phi}{dt} + B(\phi, t^{n+1/2}) = 0 \text{ on } (0, \Delta t), \\ \phi(0) = \phi^{n+1/2}. \end{cases} \tag{20.9}$$

- Step 3: Set $\phi^{n+1} = \phi(t^{n+1})$, ϕ being the solution of

$$\begin{cases} \dfrac{d\phi}{dt} + A(\phi,t) = 0 \text{ on } (t^{n+1/2}, t^{n+1}), \\ \phi(t^{n+1/2}) = \hat{\phi}^{n+1/2}. \end{cases} \tag{20.10}$$

If the operators A and B are "smooth" functions of their arguments, the above scheme is second order accurate. In addition to [20, 9, 3, 13, 10], useful information about the operator-splitting solution of partial differential equations can be found in [4, 1, 17, 18, 21] (and references therein) and in various chapters of this book.

Remark 4. The generalization to decompositions involving more than two operators is not difficult in principle. Focusing on the three-operator situation, that is

$$\begin{cases} \dfrac{d\phi}{dt} + A(\phi,t) + B(\phi,t) + C(\phi,t) = 0 \text{ on } (0, T_{max}), \\ \phi(0) = \phi_0, \end{cases}$$

we return immediately to the two-operator situation by observing that, for example,

$$A + B + C = A + (B + C) \quad \text{or} \quad A + B + C = (A + B) + C. \tag{20.11}$$

The first decomposition in (20.11) leads to a five-stage scheme, namely:

- Step 0: Set

$$\phi^0 = \phi_0.$$

For $n \geq 0$, ϕ^n being known, compute ϕ^{n+1} as follows:

- Step 1: Set $\phi^{n+1/5} = \phi(t^{n+1/2})$, ϕ being the solution of

$$\begin{cases} \dfrac{d\phi}{dt} + A(\phi,t) = 0 \text{ on } (t^n, t^{n+1/2}), \\ \phi(t^n) = \phi^n. \end{cases}$$

- Step 2: Set $\phi^{n+2/5} = \phi\left(\dfrac{\Delta t}{2}\right)$, ϕ being the solution of

$$\begin{cases} \dfrac{d\phi}{dt} + B(\phi, t^{n+1/2}) = 0 \text{ on } \left(0, \dfrac{\Delta t}{2}\right), \\ \phi(0) = \phi^{n+1/5}. \end{cases}$$

- Step 3: Set $\phi^{n+3/5} = \phi(\Delta t)$, ϕ being the solution of

$$\begin{cases} \dfrac{d\phi}{dt} + C(\phi, t^{n+1/2}) = 0 \text{ on } (0, \Delta t), \\ \phi(0) = \phi^{n+2/5}. \end{cases}$$

- Step 4: Set $\phi^{n+4/5} = \phi\,(\Delta t)$, ϕ being the solution of

$$\begin{cases} \dfrac{d\phi}{dt} + B(\phi, t^{n+1/2}) = 0 \text{ on } \left(\dfrac{\Delta t}{2}, \Delta t\right), \\ \phi\left(\dfrac{\Delta t}{2}\right) = \phi^{n+3/5}. \end{cases}$$

- Step 5: Set $\phi^{n+1} = \phi(t^{n+1})$, ϕ being the solution of

$$\begin{cases} \dfrac{d\phi}{dt} + A(\phi, t) = 0 \text{ on } (t^{n+1/2}, t^{n+1}), \\ \phi(t^{n+1/2}) = \phi^{n+4/5}. \end{cases}$$

Using the second decomposition in (20.11) would lead to a seven-stage scheme.

2.2 *Application to the Solution of the Nonlinear Wave Problem (20.1), (20.3)*

In order to apply the symmetrized scheme to the solution of (20.1), (20.3) we reformulate the above problem as a *first order* in time system by introducing the function $p = \dfrac{\partial u}{\partial t}$. We obtain:

$$\begin{cases} \dfrac{\partial u}{\partial t} - p = 0 \text{ on } \Omega \times (0, T_{max}), \\ \dfrac{\partial p}{\partial t} - c^2 \nabla^2 u = 6u^2 + t \quad \text{in } \Omega \times (0, T_{max}), \end{cases} \tag{20.12}$$

with boundary and initial conditions

$$\begin{cases} u = 0 \text{ on } \partial\Omega \times (0, T_{max}), \\ u(0) = u_0, p(0) = u_1. \end{cases} \tag{20.13}$$

Clearly, formulation (20.12), (20.13) is equivalent to (20.1), (20.3).

With Δt as in Section 2.1, we introduce $\alpha, \beta \in (0,1)$ such that $\alpha + \beta = 1$. Applying scheme (20.7)–(20.10) to the solution of (20.12), (20.13), we obtain

- Step 0: Set

$$u^0 = u_0,\ p^0 = u_1. \tag{20.14}$$

For $n \geq 0$, $\{u^n, p^n\}$ being known, compute $\{u^{n+1}, p^{n+1}\}$ as follows:

- Step 1: Set $u^{n+1/2} = u(t^{n+1/2})$, $p^{n+1/2} = p(t^{n+1/2})$, $\{u, p\}$ being the solution of

$$\begin{cases} \dfrac{\partial u}{\partial t} - \alpha p = 0 \text{ in } \Omega \times (t^n, t^{n+1/2}), \\ \dfrac{\partial p}{\partial t} = 6u^2 + t \text{ in } \Omega \times (t^n, t^{n+1/2}), \\ u(t^n) = u^n, p(t^n) = p^n. \end{cases} \tag{20.15}$$

- Step 2: Set $\hat{u}^{n+1/2} = u(\Delta t)$, $\hat{p}^{n+1/2} = p(\Delta t)$, $\{u, p\}$ being the solution of

$$\begin{cases} \dfrac{\partial u}{\partial t} - \beta p = 0 \text{ in } \Omega \times (0, \Delta t), \\ \dfrac{\partial p}{\partial t} - c^2 \nabla^2 u = 0 \text{ in } \Omega \times (0, \Delta t), \\ u = 0 \text{ on } \partial\Omega \times (0, \Delta t), \\ u(0) = u^{n+1/2}, p(0) = p^{n+1/2}. \end{cases} \tag{20.16}$$

- Step 3: Set $u^{n+1} = u(t^{n+1})$, $p^{n+1} = p(t^{n+1})$, $\{u, p\}$ being the solution of

$$\begin{cases} \dfrac{\partial u}{\partial t} - \alpha p = 0 \text{ in } \Omega \times (t^{n+1/2}, t^{n+1}), \\ \dfrac{\partial p}{\partial t} = 6u^2 + t \text{ in } \Omega \times (t^{n+1/2}, t^{n+1}), \\ u(t^{n+1/2}) = \hat{u}^{n+1/2}, p(t^{n+1/2}) = \hat{p}^{n+1/2}. \end{cases} \tag{20.17}$$

By partial elimination of p, (20.14)–(20.17) reduce to:

- Step 0 as in (20.14).

For $n \geq 0$, $\{u^n, p^n\}$ being known, compute $\{u^{n+1}, p^{n+1}\}$ as follows:

- Step 1: Set $u^{n+1/2} = u(t^{n+1/2})$, $p^{n+1/2} = \frac{1}{\alpha}\frac{\partial u}{\partial t}(t^{n+1/2})$, u being the solution of

$$\begin{cases} \frac{\partial^2 u}{\partial t^2} = \alpha(6u^2 + t) \text{ in } \Omega \times (t^n, t^{n+1/2}), \\ u(t^n) = u^n, \frac{\partial u}{\partial t}(t^n) = \alpha p^n. \end{cases} \tag{20.18}$$

- Step 2: Set $\hat{u}^{n+1/2} = u(\Delta t)$, $\hat{p}^{n+1/2} = \frac{1}{\beta}\frac{\partial u}{\partial t}(\Delta t)$, u being the solution of

$$\begin{cases} \frac{\partial^2 u}{\partial t^2} - \beta c^2 \nabla^2 u = 0 \text{ in } \Omega \times (0, \Delta t), \\ u = 0 \text{ on } \partial\Omega \times (0, \Delta t), \\ u(0) = u^{n+1/2}, \frac{\partial u}{\partial t}(0) = \beta p^{n+1/2}. \end{cases} \tag{20.19}$$

- Step 3: Set $u^{n+1} = u(t^{n+1})$, $p^{n+1} = \frac{1}{\alpha}\frac{\partial u}{\partial t}(t^{n+1})$, u being the solution of

$$\begin{cases} \frac{\partial^2 u}{\partial t^2} = \alpha(6u^2 + t) \text{ in } \Omega \times (t^{n+1/2}, t^{n+1}), \\ u(t^{n+1/2}) = \hat{u}^{n+1/2}, \frac{\partial u}{\partial t}(t^{n+1/2}) = \alpha \hat{p}^{n+1/2}. \end{cases} \tag{20.20}$$

2.3 *Application to the Solution of the Nonlinear Wave Problem (20.1), (20.4)*

Proceeding as in Section 2.2, we introduce $p = \frac{\partial u}{\partial t}$ in order to reformulate (20.1), (20.4) as a first order in time system. We obtain system (20.12) supplemented with the following boundary and initial conditions

$$\begin{cases} u(0) = 0 \text{ on } \Gamma_0 \times (0, T_{max}), \frac{p}{c} + \frac{\partial u}{\partial n} = 0 \text{ on } \Gamma_1 \times (0, T_{max}), \\ u(0) = u_0, p(0) = u_1. \end{cases} \tag{20.21}$$

Applying scheme (20.7)–(20.10) for the solution of the equivalent problem (20.12), (20.21), we obtain

- Step 0: as in (20.14).

For $n \geq 0$, $\{u^n, p^n\}$ being known, compute $\{u^{n+1}, p^{n+1}\}$ as follows:

- Step 1: as in (20.18).
- Step 2: Set $\hat{u}^{n+1/2} = u(\Delta t)$, $\hat{p}^{n+1/2} = \dfrac{1}{\beta}\dfrac{\partial u}{\partial t}(\Delta t)$, u being the solution of

$$\begin{cases} \dfrac{\partial^2 u}{\partial t^2} - \beta c^2 \nabla^2 u = 0 \text{ in } \Omega \times (0, \Delta t), \\ u = 0 \text{ on } \Gamma_0 \times (0, \Delta t), \dfrac{1}{\beta c}\dfrac{\partial u}{\partial t} + \dfrac{\partial u}{\partial n} = 0 \text{ on } \Gamma_1 \times (0, \Delta t), \\ u(0) = u^{n+1/2}, \dfrac{\partial u}{\partial t}(0) = \beta p^{n+1/2}. \end{cases} \tag{20.22}$$

- Step 3: as in (20.20).

3 On the Numerical Solution of the Sub-initial Value Problems (20.19) and (20.22)

Since problem (20.19) is the particular case of (20.22) corresponding to $\Gamma_1 = \emptyset$, we are going to consider the second problem only. The linear wave problem is itself a particular case of

$$\begin{cases} \dfrac{\partial^2 \phi}{\partial t^2} - \beta c^2 \nabla^2 \phi = 0 \text{ in } \Omega \times (t_0, t_f), \\ \phi = 0 \text{ on } \Gamma_0 \times (t_0, t_f), \dfrac{1}{\beta c}\dfrac{\partial \phi}{\partial t} + \dfrac{\partial \phi}{\partial n} = 0 \text{ on } \Gamma_1 \times (t_0, t_f), \\ \phi(t_0) = \phi_0, \dfrac{\partial \phi}{\partial t}(t_0) = \phi_1. \end{cases} \tag{20.23}$$

Assuming that ϕ_0 and ϕ_1 have enough regularity, a variational (weak) formulation of problem (20.23) is given by: Find $\phi(t) \in V_0$, a.e. on (t_0, t_f), such that

$$\begin{cases} \left\langle \dfrac{\partial^2 \phi}{\partial t^2}, \theta \right\rangle + \beta c^2 \displaystyle\int_\Omega \nabla \phi \cdot \nabla \theta dx + c \int_{\Gamma_1} \dfrac{\partial \phi}{\partial t} \theta d\Gamma = 0, \quad \forall \theta \in V_0, \\ \phi(t_0) = \phi_0, \dfrac{\partial \phi}{\partial t}(t_0) = \phi_1, \end{cases} \tag{20.24}$$

where in (20.24):

- V_0 is the *Sobolev space* defined by

$$V_0 = \{\theta | \theta \in H^1(\Omega),\ \theta = 0 \text{ on } \Gamma_0\}. \tag{20.25}$$

- $\langle \cdot, \cdot \rangle$ is the duality pairing between V_0' (the dual of V_0) and V_0, coinciding with the canonical inner product of $L^2(\Omega)$ if the first argument is smooth enough.
- $dx = dx_1 \dots dx_d$.

3.1 A Finite Element Method for the Space Discretization of the Linear Wave Problem (20.23)

From now on, we are going to assume that Ω is a bounded *polygonal* domain of $\mathbb{R}^2$. Let $\mathscr{T}_h$ be a classical finite element *triangulation* of Ω, as considered in, e.g., [8] (Appendix 1) and related references therein. We approximate the space V_0 in (20.25) by

$$V_{0h} = \{\theta | \theta \in C^0(\overline{\Omega}),\ \theta|_{\Gamma_0} = 0,\ \theta|_K \in \mathbb{P}_1,\ \forall K \in \mathscr{T}_h\}, \tag{20.26}$$

where $\mathbb{P}_1$ is the space of the polynomials of two variables of degree ≤ 1. If $\Gamma_1 \neq \emptyset$, the points at the interface of Γ_0 and Γ_1 have to be (for consistency reasons) vertices of $\mathscr{T}_h$ at which any element of V_{0h} has to vanish. It is natural to approximate the wave problem (20.24) as follows: Find $\phi_h(t) \in V_{0h}$, a.e. on $(t_0, t_f]$, such that

$$\begin{cases} \displaystyle\int_\Omega \frac{\partial^2 \phi_h}{\partial t^2} \theta dx + \beta c^2 \int_\Omega \nabla \phi_h \cdot \nabla \theta dx + c \int_{\Gamma_1} \frac{\partial \phi_h}{\partial t} \theta d\Gamma = 0, \quad \forall \theta \in V_{0h}, \\ \phi_h(t_0) = \phi_{0h}, \dfrac{\partial \phi_h}{\partial t}(t_0) = \phi_{1h}, \end{cases} \tag{20.27}$$

where ϕ_{0h} and ϕ_{1h} belong to V_{0h} and approximate ϕ_0 and ϕ_1, respectively.

In order to formulate (20.27) as a second order in time system of linear ordinary differential equations, we introduce first the set $\Sigma_{0h} = \{P_j\}_{j=1}^{N_{0h}}$ of the vertices of $\mathscr{T}_h$ which do not belong to $\overline{\Gamma}_0$ and associate with it the following basis of V_{0h}:

$$\mathscr{B}_{0h} = \{w_j\}_{j=1}^{N_{0h}},$$

where the basis function w_j is defined by

$$w_j \in V_{0h},\ w_j(P_j) = 1, w_j(P_k) = 0,\ \forall k \in \{1, \dots, N_{0h}\},\ k \neq j.$$

Expanding the solution ϕ_h of (20.27) over the above basis, we obtain:

$$\phi_h(t) = \sum_{j=1}^{N_{0h}} \phi_h(P_j,t)w_j.$$

Denoting $\phi_h(P_j,t)$ by $\phi_j(t)$ and the N_{0h}-dimensional vector $\{\phi_j(t)\}_{j=1}^{N_{0h}}$ by $\boldsymbol{\Phi}_h(t)$, we can easily show that the approximated problem (20.27) is equivalent to the following ordinary differential system

$$\begin{cases} \mathbf{M}_h\ddot{\boldsymbol{\Phi}}_h + \beta c^2\mathbf{A}_h\boldsymbol{\Phi}_h + c\mathbf{C}_h\dot{\boldsymbol{\Phi}}_h = 0 \text{ on } (t_0,t_f), \\ \boldsymbol{\Phi}_h(t_0) = \boldsymbol{\Phi}_{0h}(= (\phi_{0h}(P_j))_{j=1}^{N_{0h}}), \dot{\boldsymbol{\Phi}}_h(t_0) = \boldsymbol{\Phi}_{1h}(= (\phi_{1h}(P_j))_{j=1}^{N_{0h}}), \end{cases} \tag{20.28}$$

where the *mass matrix* $\mathbf{M}_h$, the *stiffness matrix* $\mathbf{A}_h$, and the *damping matrix* $\mathbf{C}_h$ are defined by

$$\mathbf{M}_h = (m_{ij})_{1\le i,j\le N_{0h}}, \text{ with } m_{ij} = \int_\Omega w_i w_j dx,$$
$$\mathbf{A}_h = (a_{ij})_{1\le i,j\le N_{0h}}, \text{ with } a_{ij} = \int_\Omega \nabla w_i \cdot \nabla w_j dx,$$
$$\mathbf{C}_h = (c_{ij})_{1\le i,j\le N_{0h}}, \text{ with } c_{ij} = \int_{\Gamma_1} w_i w_j d\Gamma,$$

respectively.

The matrices $\mathbf{M}_h$ and $\mathbf{A}_h$ are *sparse* and *positive definite*, while matrix $\mathbf{C}_h$ is *'very' sparse* and *positive semi-definite*. Indeed, if P_i and P_j are not neighbors, i.e., they are not vertices of a same triangle of $\mathscr{T}_h$, we have $m_{ij} = 0$, $a_{ij} = 0$, and $c_{ij} = 0$. All these matrix coefficients can be computed exactly, using, for example, the *two-dimensional Simpson's rule* for the m_{ij} and the *one-dimensional Simpson's rule* for the c_{ij}; since ∇w_i and ∇w_j are piecewise constant, computing a_{ij} is (relatively) easy. See, e.g., [9] (Chapter 5) for more details on these calculations.

Remark 5. Using the *trapezoidal rule*, instead of the Simpson's one, to compute the m_{ij} and c_{ij} brings simplification: the resulting $\mathbf{M}_h$ and $\mathbf{C}_h$ will be diagonal matrices, retaining the positivity properties of their Simpson's counterparts. The drawback is some accuracy loss associated with this simplification.

3.2 A Centered Second Order Finite Difference Scheme for the Time Discretization of the Initial Value Problem (20.28)

Let Q be a positive integer (≥ 3, in practice). We associate with Q a time discretization step $\tau = (t_f - t_0)/Q$. After dropping the subscript h, a classical time discretization for problem (20.28) reads as: Set

$$\Phi^0 = \Phi_0,\ \Phi^1 - \Phi^{-1} = 2\tau\Phi_1, \tag{20.29}$$

then for $q = 0,\ldots,Q$, compute Φ^{q+1} from Φ^{q-1} and Φ^q via

$$\mathbf{M}(\Phi^{q+1} + \Phi^{q-1} - 2\Phi^q) + \beta c^2\tau^2 \mathbf{A}\Phi^q + c\frac{\tau}{2}\mathbf{C}(\Phi^{q+1} - \Phi^{q-1}) = 0. \tag{20.30}$$

It follows from, e.g., [9] (Chapter 6), that the above second order accurate scheme is *stable* if the following condition holds:

$$\tau < \frac{2}{c\sqrt{\beta\lambda_{\max}}}, \tag{20.31}$$

where $\lambda_{\max}$ is the *largest eigenvalue* of $\mathbf{M}^{-1}\mathbf{A}$.

Remark 6. To obtain Φ^{q+1} from (20.30), one has to solve a linear system associated with the *symmetric positive definite* matrix

$$\mathbf{M} + \frac{\tau}{2}c\mathbf{C}. \tag{20.32}$$

If the above matrix is *diagonal* from the use of the trapezoidal rule (see remark 5), computing Φ^{q+1} is particularly easy and the time discretization scheme (20.30) is fully explicit. Otherwise, scheme (20.30) is not explicit, strictly speaking. However, matrix (20.32) being well conditioned, a *conjugate gradient algorithm with diagonal preconditioning* will have a very fast convergence, particularly if one uses Φ^q to initialize the computation of Φ^{q+1}.

Remark 7. In order to initialize the discrete analogue of the initial value problem (20.20), we will use

$$\Phi^Q \quad \text{and} \quad \frac{\alpha}{\beta}\frac{\Phi^{Q+1} - \Phi^{Q-1}}{2\tau}. \tag{20.33}$$

Remark 8. As the solution of the nonlinear wave problem under consideration gets closer to blow-up, the norms of the corresponding initial data in (20.29) will go to infinity. In order to off-set (partly, at least) the effect of round-off errors we suggest the following *normalization strategy*:

1. Denote by $||\phi_{0h}||_{0h}$ and $||\phi_{1h}||_{0h}$ the respective approximations of
$$\left(\int_\Omega |\phi_{0h}|^2\,dx\right)^{1/2} \quad \text{and} \quad \left(\int_\Omega |\phi_{1h}|^2\,dx\right)^{1/2}$$
obtained by the trapezoidal rule.
2. Divide by $\max[1, \sqrt{||\phi_{0h}||_{0h}^2 + ||\phi_{1h}||_{0h}^2}]$ the initial data Φ_0 and Φ_1 in (20.29).
3. Apply scheme (20.30) with normalized initial data to compute Φ^{Q-1}, Φ^Q, and Φ^{Q+1}.

4. Prepare the initial data for the following nonlinear sub-step by multiplying (20.33) by the normalization factor $\max[1, \sqrt{||\phi_{0h}||_{0h}^2 + ||\phi_{1h}||_{0h}^2}]$.

4 On the Numerical Solution of the Sub-initial Value Problems (20.18) and (20.20)

From $n = 0$ until blow-up, we have to solve the initial value sub-problems (20.18) and (20.20) for almost every point of Ω. Following what we discussed in Section 3 (whose notation we keep) for the solution of the linear wave equation subproblems, we will consider only those nonlinear initial value sub-problems associated with the N_{0h} vertices of $\mathscr{T}_h$ not located on $\overline{\Gamma}_0$. Each of these sub-problem is of the following type:

$$\begin{cases} \dfrac{d^2\phi}{dt^2} = \alpha(6\phi^2 + t) \text{ on } (t_0, t_f), \\ \\ \phi(t_0) = \phi_0, \dfrac{d\phi}{dt}(t_0) = \phi_1, \end{cases} \tag{20.34}$$

with initial data in (20.34) as in algorithm (20.14), (20.18), (20.22), (20.20), after space discretization. A time discretization scheme of (20.34), with automatic adjustment of the time step will be discussed in the following section.

4.1 A Centered Scheme for the Time Discretization of Problem (20.34)

Let M be a positive integer (> 2 in practice). With M, we associate a time discretization step $\sigma = (t_f - t_0)/M$. For the time discretization of the initial value problem (20.34) we suggest the following nonlinear variant of (20.30): Set

$$\phi^0 = \phi_0, \; \phi^1 - \phi^{-1} = 2\sigma\phi_1,$$

then for $m = 0, \ldots, M$, compute ϕ^{m+1} from ϕ^{m-1} and ϕ^m via

$$\phi^{m+1} + \phi^{m-1} - 2\phi^m = \alpha\sigma^2(6|\phi^m|^2 + t^m), \tag{20.35}$$

with $t^m = t^0 + m\sigma$.

Considering the blowing-up properties of the solutions of the nonlinear wave problems (20.1), (20.3) and (20.1), (20.4), we expect that at one point in time, the solution of problem (20.34) will start growing very fast before becoming infinite. In order to track such a behavior we have to decrease σ in (20.35), until the solution reaches some threshold at which one decides to stop computing . A practical method for the adaptation of the time step σ is described below.

4.2 On the Dynamical Adaptation of the Time Step σ

The starting point of our adaptive strategy will be the following observation: if ϕ is the solution of (20.34), at a time t before blow-up and for σ sufficiently small we have (*Taylor's expansion*):

$$\begin{aligned}\phi(t+\sigma) &= \phi(t)+\sigma\dot{\phi}(t)+\frac{\sigma^2}{2}\ddot{\phi}(t)+\frac{\sigma^3}{6}\dddot{\phi}(t+\theta\sigma)\\ &= \phi(t)+\sigma\dot{\phi}(t)+\frac{\sigma^2}{2}\alpha(6|\phi(t)|^2+t)+\sigma^3\alpha\left(2\phi(t+\theta\sigma)\dot{\phi}(t+\theta\sigma)+\frac{1}{6}\right),\end{aligned} \tag{20.36}$$

with $0 < \theta < 1$. Suppose that we drop the σ^3-term in the above expansion and that we approximate by finite differences the resulted truncated expansion at $t = t^m$; we obtain then

$$\phi^{m+1} = \phi^m + \sigma\frac{\phi^{m+1} - \phi^{m-1}}{2\sigma} + \frac{\sigma^2}{2}\alpha(6|\phi^m|^2 + t^m),$$

that is the explicit scheme (20.35). Moreover, from expansion (20.36) we can derive the following estimator of the relative error at $t = t^{m+1}$:

$$E^{m+1} = \sigma^3\alpha\frac{\left|(\phi^{m+1}+\phi^m)\dfrac{(\phi^{m+1}-\phi^m)}{\sigma}\right| + \dfrac{1}{6}}{\max[1, |\phi^{m+1}|]}. \tag{20.37}$$

Another possible estimator would be

$$\sigma^3\alpha\frac{\left|(\phi^{m+1}+\phi^m)\dfrac{(\phi^{m+1}-\phi^m)}{\sigma}\right| + \dfrac{1}{6}}{\max\left[1, \dfrac{1}{2}|\phi^m + \phi^{m+1}|\right]}.$$

In order to adapt σ using E^{m+1}, we may proceed as follows: If ϕ^{m+1} obtained from scheme (20.35) verifies

$$E^{m+1} \leq tol, \tag{20.38}$$

keep integrating with σ as time discretization step. If criterion (20.38) is not verified, we have two possible situations, one for $m = 0$ and one for $m \geq 1$. If $m = 0$:

- Divide σ by 2 as many times as necessary to have

$$E^1 \leq \frac{tol}{5}. \tag{20.39}$$

  Each time σ is divided by 2, double M accordingly.
- Still calling σ the first time step for which (20.39) holds after successive divisions by 2, apply scheme (20.35) to the solution of (20.34), with the new σ and the associated M.

If $m \geq 1$:

- Go to $t = t^{m-1/2} = t_0 + (m-1/2)\sigma$.
- $t^{m-1/2} \to t_0$, $\dfrac{\phi^{m-1} + \phi^m}{2} \to \phi_0$, $\dfrac{\phi^m - \phi^{m-1}}{\sigma} \to \phi_1$.
- $\sigma \to \sigma/2$.
- $2(M-m)+1 \to M$.
- Apply scheme (20.35) on the new interval (t_0, t_f). If criterion (20.38) is not verified, then proceed as above, according to the value of m (that is, $m = 0$ or $m \geq 1$).

For the numerical results reported in Section 5, we used $tol = 10^{-4}$.

Remark 9. In order to initialize the discrete analogues of the initial value problems (20.18) and (20.19), we will use

$$\phi^M \quad \text{and} \quad \frac{\phi^{M+1} - \phi^{M-1}}{2\sigma},$$

and

$$\phi^M \quad \text{and} \quad \frac{\beta}{\alpha} \frac{\phi^{M+1} - \phi^{M-1}}{2\sigma},$$

respectively.

5 Application of the Strang's Symmetrized Operator-Splitting Scheme to the Solution of Problem (20.5), (20.3)

In this section, we extend the methodology discussed in Section 2.2 for problem (20.1), (20.3) to problem (20.5), (20.3). We will use the same notation as in Section 2.2.

First, we introduce $p = \dfrac{\partial u}{\partial t}$ to reformulate the above problem as a *first order* in time system on which we will apply the Strang's symmetrized scheme repeatedly. This first order system reads as:

$$\begin{cases} \dfrac{\partial u}{\partial t} - p = 0 \quad \text{on } \Omega \times (0, T_{\max}), \\ \dfrac{\partial p}{\partial t} + \dfrac{k}{\varepsilon + t} p - c^2 \nabla^2 u = \lambda e^u \quad \text{in } \Omega \times (0, T_{max}), \\ u = 0 \quad \text{on } \partial\Omega \times (0, T_{max}), \\ u(0) = u_0, p(0) = u_1. \end{cases} \tag{20.40}$$

Inspired by the three-operator situation discussed in Section 2.2, we suggest the following five-stage operator splitting scheme for the time-discretization of problem (20.40):

- Step 0: Set

$$u^0 = u_0, \ p^0 = u_1. \tag{20.41}$$

For $n \geq 0$, $\{u^n, p^n\}$ being known, compute $\{u^{n+1}, p^{n+1}\}$ as follows:

- Step 1: Set $u^{n+1/5} = u(t^{n+1/2})$, $p^{n+1/5} = p(t^{n+1/2})$, $\{u, p\}$ being the solution of

$$\begin{cases} \dfrac{\partial u}{\partial t} - \alpha p = 0 \text{ in } \Omega \times (t^n, t^{n+1/2}), \\[2ex] \dfrac{\partial p}{\partial t} = \lambda e^u \text{ in } \Omega \times (t^n, t^{n+1/2}), \\[2ex] u(t^n) = u^n, p(t^n) = p^n. \end{cases} \tag{20.42}$$

- Step 2: Set $u^{n+2/5} = u\left(\dfrac{\Delta t}{2}\right)$, $p^{n+2/5} = p\left(\dfrac{\Delta t}{2}\right)$, $\{u, p\}$ being the solution of

$$\begin{cases} \dfrac{\partial u}{\partial t} = 0 \text{ in } \Omega \times \left(0, \dfrac{\Delta t}{2}\right), \\[2ex] \dfrac{\partial p}{\partial t} + \dfrac{k}{\varepsilon + t^{n+1/2}} p = 0 \text{ in } \Omega \times \left(0, \dfrac{\Delta t}{2}\right), \\[2ex] u(0) = u^{n+1/5}, p(0) = p^{n+1/5}. \end{cases} \tag{20.43}$$

- Step 3: Set $u^{n+3/5} = u(\Delta t)$, $p^{n+3/5} = p(\Delta t)$, $\{u, p\}$ being the solution of

$$\begin{cases} \dfrac{\partial u}{\partial t} - \beta p = 0 \text{ in } \Omega \times (0, \Delta t), \\[2ex] \dfrac{\partial p}{\partial t} - c^2 \nabla^2 u = 0 \text{ in } \Omega \times (0, \Delta t), \\[2ex] u = 0 \text{ on } \partial\Omega \times (0, \Delta t), \\[2ex] u(0) = u^{n+2/5}, p(0) = p^{n+2/5}. \end{cases} \tag{20.44}$$

- Step 4: Set $u^{n+4/5} = u(\Delta t)$, $p^{n+4/5} = p(\Delta t)$, $\{u,p\}$ being the solution of

$$\begin{cases} \dfrac{\partial u}{\partial t} = 0 \text{ in } \Omega \times \left(\dfrac{\Delta t}{2}, \Delta t\right), \\ \dfrac{\partial p}{\partial t} + \dfrac{k}{\varepsilon + t^{n+1/2}} p = 0 \text{ in } \Omega \times \left(\dfrac{\Delta t}{2}, \Delta t\right), \\ u\left(\dfrac{\Delta t}{2}\right) = u^{n+3/5}, p\left(\dfrac{\Delta t}{2}\right) = p^{n+3/5}. \end{cases} \tag{20.45}$$

- Step 5: Set $u^{n+1} = u(t^{n+1})$, $p^{n+1} = p(t^{n+1})$, $\{u,p\}$ being the solution of

$$\begin{cases} \dfrac{\partial u}{\partial t} - \alpha p = 0 \text{ in } \Omega \times (t^{n+1/2}, t^{n+1}), \\ \dfrac{\partial p}{\partial t} = \lambda e^u \text{ in } \Omega \times (t^{n+1/2}, t^{n+1}), \\ u(t^{n+1/2}) = u^{n+4/5}, p(t^{n+1/2}) = p^{n+4/5}. \end{cases} \tag{20.46}$$

By partial elimination of p, (20.41)–(20.46) reduce to:

- Step 0 as in (20.41).

For $n \geq 0$, $\{u^n, p^n\}$ being known, compute $\{u^{n+1}, p^{n+1}\}$ as follows:

- Step 1: Set $u^{n+1/5} = u(t^{n+1/2})$, $p^{n+1/5} = \dfrac{1}{\alpha}\dfrac{\partial u}{\partial t}(t^{n+1/2})$, u being the solution of

$$\begin{cases} \dfrac{\partial^2 u}{\partial t^2} = \alpha \lambda e^u \text{ in } \Omega \times (t^n, t^{n+1/2}), \\ u(t^n) = u^n, \dfrac{\partial u}{\partial t}(t^n) = \alpha p^n. \end{cases} \tag{20.47}$$

- Step 2: Set $u^{n+2/5} = u^{n+1/5}$, $p^{n+2/5} = p\left(\dfrac{\Delta t}{2}\right)$, p being the solution of

$$\begin{cases} \dfrac{\partial p}{\partial t} + \dfrac{k}{\varepsilon + t^{n+1/2}} p = 0 \text{ in } \Omega \times \left(0, \dfrac{\Delta t}{2}\right), \\ p(0) = p^{n+1/5}. \end{cases} \tag{20.48}$$

- Step 3: Set $u^{n+3/5} = u(\Delta t)$, $p^{n+3/5} = \dfrac{1}{\beta}\dfrac{\partial u}{\partial t}(\Delta t)$, u being the solution of

$$\begin{cases} \dfrac{\partial^2 u}{\partial t^2} - \beta c^2 \nabla^2 u = 0 \text{ in } \Omega \times (0, \Delta t), \\ u = 0 \text{ on } \partial\Omega \times (0, \Delta t), \\ u(0) = u^{n+2/5}, \dfrac{\partial u}{\partial t}(0) = \beta p^{n+2/5}. \end{cases} \tag{20.49}$$

- Step 4: Set $u^{n+4/5} = u^{n+3/5}$, $p^{n+4/5} = p(\Delta t)$, p being the solution of

$$\begin{cases} \dfrac{\partial p}{\partial t} + \dfrac{k}{\varepsilon + t^{n+1/2}} p = 0 \text{ in } \Omega \times \left(\dfrac{\Delta t}{2}, \Delta t\right), \\ p\left(\dfrac{\Delta t}{2}\right) = p^{n+3/5}. \end{cases} \tag{20.50}$$

- Step 5: Set $u^{n+1} = u(t^{n+1})$, $p^{n+1} = \dfrac{1}{\alpha}\dfrac{\partial u}{\partial t}(t^{n+1})$, u being the solution of

$$\begin{cases} \dfrac{\partial^2 u}{\partial t^2} = \alpha \lambda e^u \text{ in } \Omega \times (t^{n+1/2}, t^{n+1}), \\ u(t^{n+1/2}) = u^{n+4/5}, \dfrac{\partial u}{\partial t}(t^{n+1/2}) = \alpha p^{n+4/5}. \end{cases} \tag{20.51}$$

Sub-problem (20.49) is the particular case of (20.23) corresponding to $\Gamma_1 = \emptyset$. Thus, we refer to Section 3 for a discussion of the numerical solution of sub-problem (20.49). The numerical solution of the initial value sub-problems (20.47) and (20.51) will be discussed in the next section, generalizing what we already presented in Section 4. The sub-problems (20.48) and (20.50) are new in the context of these nonlinear wave problems. Fortunately, they have closed-form solutions given by:

$$p^{n+2/5} = e^{-\frac{k\Delta t}{2(\varepsilon + t^{n+1/2})}} p^{n+1/5}, \quad p^{n+4/5} = e^{-\frac{k\Delta t}{2(\varepsilon + t^{n+1/2})}} p^{n+3/5},$$

respectively.

6 On the Numerical Solution of the Sub-initial Value Problems (20.47) and (20.51)

From $n = 0$ until blow-up, we have to solve the initial value sub-problems (20.47) and (20.51) for almost every point of Ω. Each of these sub-problem is of the following type:

$$\begin{cases} \dfrac{d^2\psi}{dt^2} = \alpha\lambda e^{\psi} \text{ on } (t_0, t_f), \\ \psi(t_0) = \psi_0, \dfrac{d\psi}{dt}(t_0) = \psi_1. \end{cases} \tag{20.52}$$

For the time discretization of (20.52), we adopt the same centered scheme with automatic adjustment of the time-step discussed in Section 4.1. Keeping the same notation as in Section 4.1, the time discrete problem reads: Set

$$\psi^0 = \psi_0, \ \psi^1 - \psi^{-1} = 2\sigma\psi_1,$$

then for $m = 0, \ldots, M$, compute ψ^{m+1} from ψ^{m-1} and ψ^m via

$$\psi^{m+1} + \psi^{m-1} - 2\psi^m = \alpha\sigma^2\lambda e^{\psi^m},$$

with $t^m = t^0 + m\sigma$.

For the adaptation of the time-step σ, we will use the method described in Section 4.2, where the estimator of the relative error is given by:

$$E^{m+1} = \frac{\sigma^3}{6}\alpha\lambda \frac{e^{\frac{\psi^{m+1}+\psi^m}{2}} \left| \dfrac{\psi^{m+1} - \psi^m}{\sigma} \right|}{\max[1, |\psi^{m+1}|]},$$

instead of eq. (20.37).

7 Numerical Experiments

In this section, we are going to report on the results of numerical experiments concerning the solution of the nonlinear wave problems (20.1), (20.3) and (20.1), (20.4), and (20.5), (20.3). The role of these experiments is twofold: (i) validate the numerical methodologies discussed in Sections 2 to 6, and (ii) investigate how c influences the solutions. We will also check how the boundary conditions influence the solution of eq. (20.1) and how the value of k impacts the solution of problem (20.5), (20.3).

For all the problems, we took $\Omega = (0,1)^2$. For problem (20.1), (20.4), we took $\Gamma_1 = \{\{x_1, x_2\}, x_1 = 1, 0 < x_2 < 1\}$. The simplicity of the geometry suggests the use of finite differences for the space discretization. Actually, the finite difference schemes we employ can be obtained via the finite element approximation discussed in Section 3, combined with the trapezoidal rule to compute the mass matrix $\mathbf{M}_h$ and the damping matrix $\mathbf{C}_h$; this supposes that the triangulations we employ are uniform like the one depicted in Figure 20.1.

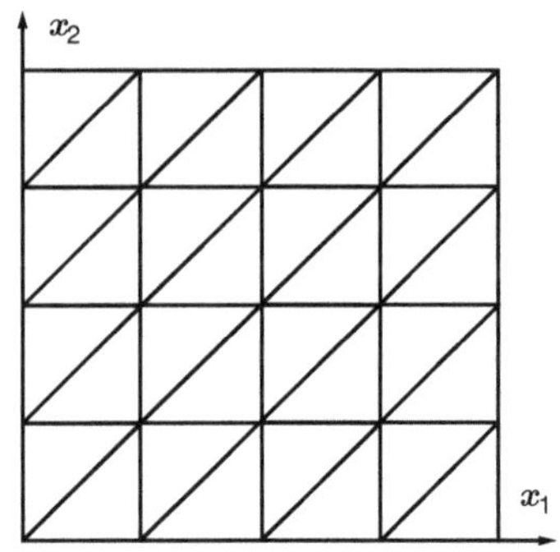

Fig. 20.1 A uniform triangulation of Ω.

7.1 Numerical Experiments for the Nonlinear Wave Problem (20.1), (20.3)

Using well-known notation, let us assume that the directional space discretization steps Δx_1 and Δx_2 are equal and we denote by h their common value. We also assume that $h = 1/(I+1)$, I being a positive integer. For $0 \leq i, j \leq I+1$, we denote by M_{ij} the point $\{ih, jh\}$ and $u_{ij}(t) \simeq u(M_{ij}, t)$. Using finite differences, we obtain the following continuous in time, discrete in space analogue of problem (20.1), (20.3):

$$\begin{cases} u_{ij}(0) = u_0(M_{ij}),\ 0 \leq i, j \leq I+1, \text{ and } \dot{u}_{ij}(0) = u_1(M_{ij}),\ 1 \leq i, j \leq I, \\ \\ \ddot{u}_{ij}(t) + \left(\dfrac{c}{h}\right)^2 (4u_{ij} - u_{i+1j} - u_{i-1j} - u_{ij+1} - u_{ij-1})(t) = 6|u_{ij}(t)|^2 + t \\ \text{on } (0, T_{\max}),\ 1 \leq i, j \leq I, \\ \\ u_{kl}(t) = 0 \text{ on } (0, T_{\max}) \text{ if } M_{kl} \in \partial\Omega. \end{cases} \tag{20.53}$$

In (20.53), we assume that u_0 (resp., u_1) belongs to $C^0(\overline{\Omega}) \cap H_0^1(\Omega)$ (resp., $C^0(\overline{\Omega})$).

The application of the discrete analogue of the operator-splitting scheme (20.14), (20.18)–(20.20) to problem (20.53) leads to the solution at each time step of:

- a *discrete linear wave problem* of the following type

$$\begin{cases} \phi_{ij}(t_0) = \phi_0(M_{ij}),\ 0 \leq i, j \leq I+1, \text{ and } \dot{\phi}_{ij}(t_0) = \phi_1(M_{ij}),\ 1 \leq i, j \leq I, \\ \\ \ddot{\phi}_{ij}(t) + \beta \left(\dfrac{c}{h}\right)^2 (4\phi_{ij} - \phi_{i+1j} - \phi_{i-1j} - \phi_{ij+1} - \phi_{ij-1})(t) = 0 \\ \text{on } (t_0, t_f),\ 1 \leq i, j \leq I, \\ \\ \phi_{kl}(t) = 0 \text{ on } (t_0, t_f) \text{ if } M_{kl} \in \partial\Omega. \end{cases} \tag{20.54}$$

- $2I^2$ nonlinear initial value problems (2 for each interior grid point M_{ij}) like (20.34).

The numerical solution of problem (20.34) has been addressed in Sections 4.1 and 4.2. Concerning problem (20.54), it follows from Section 3 that its time discrete analogue reads as follows: Set

$$\phi_{ij}^0 = \phi_0(M_{ij}),\ 0 \le i,j \le I+1 \text{ and } \phi_{ij}^1 - \phi_{ij}^{-1} = 2\tau\phi_1(M_{ij}),\ 1 \le i,j \le I,$$

then, for $q = 0,\dots,Q$, $1 \le i,j \le I$, we have

$$\begin{cases} \phi_{ij}^{q+1} + \phi_{ij}^{q-1} - 2\phi_{ij}^q + \beta\left(\frac{\tau}{h}c\right)^2 (4\phi_{ij}^q - \phi_{i+1j}^q - \phi_{i-1j}^q - \phi_{ij+1}^q - \phi_{ij-1}^q) = 0, \\ \phi_{kl}^{q+1} = 0 \text{ if } M_{kl} \in \partial\Omega, \end{cases} \tag{20.55}$$

with $\tau = (t_f - t_0)/Q$. In the particular case of scheme (20.55), the stability condition (20.31) takes the following form:

$$\tau < \frac{h}{c\sqrt{2\beta}}, \tag{20.56}$$

For the numerical results presented below, we took:

- $u_0 = 0$ and $u_1 = 0$.
- c ranging from 0 to 1.5.
- $\alpha = \beta = 1/2$.
- $Q = 3$.
- For $h = 1/100$: $\Delta t = 10^{-2}$ for $c \in [0, 0.6]$, $\Delta t = 8 \times 10^{-3}$ for $c = 0.7, 0.8$, $\Delta t = 5 \times 10^{-3}$ for $c = 0.9, 1, 1.25$, $\Delta t = 10^{-3}$ for $c = 1.5$.
- For $h = 1/150$: $\Delta t = 6 \times 10^{-3}$ for $c \in [0, 0.6]$, $\Delta t = 4 \times 10^{-3}$ for $c = 0.7, 0.8$, $\Delta t = 3 \times 10^{-3}$ for $c = 0.9, 1, 1.25$, $\Delta t = 6 \times 10^{-4}$ for $c = 1.5$.

We initialized with $M = 3$ (see Section 4.1) and then adapted M following the procedure described in Section 4.2.

We considered that the blow-up time was reached as soon as the maximum value of the discrete solution reached 10^4. Let us remark that the numerical results obtained with $h = 1/100$ and $h = 1/150$ (and the respective associated values of Δt) are essentially identical.

In Figure 20.2, we reported the results obtained by our methodology when $c = 0$. They compare quite well with the results reported by Wikipedia [22].

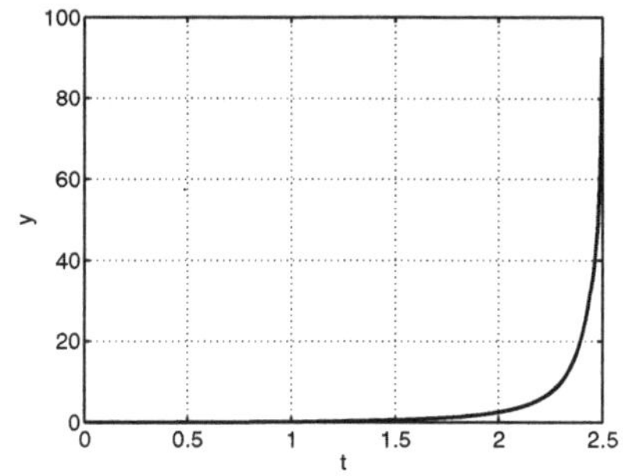

Fig. 20.2 Case $c = 0$: results obtained by our methodology.

In Figure 20.3, we visualized for $c = 0.8$ and $t \in [0, 14.4]$ (the blow-up time being close to $T_{\max} \simeq 15.512$) the evolution of the computed approximations of the functions

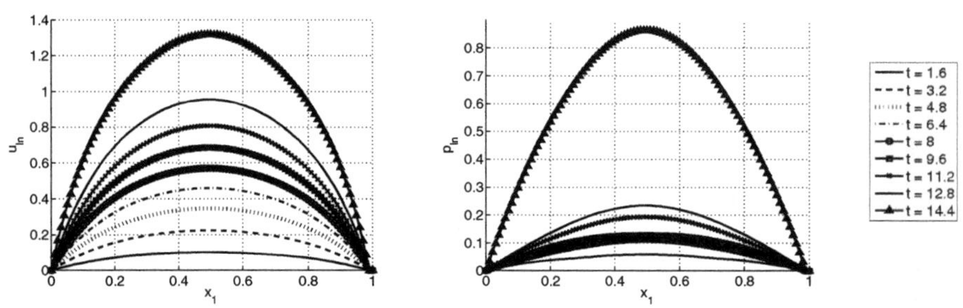

(a) evolution of $u_{\ln}$ (b) evolution of $p_{\ln}$ (c) caption

Fig. 20.3 Case $c = 0.8$, pure Dirichlet boundary conditions: Evolution of quantities (a) $u_{\ln}$ and (b) $p_{\ln}$. The caption in (c) is common to (a) and (b).

$$u_{\ln} = \operatorname{sgn}(u)\ln(1 + |u|) \quad \text{and} \quad p_{\ln} = \operatorname{sgn}(p)\ln(1 + |p|), \tag{20.57}$$

restricted to the segment $\{\{x_1, x_2\}, 0 \leq x_1 \leq 1, x_2 = 1/2\}$. The *oscillatory behavior* of the solution appears clearly in Figure 20.3(b). In Figure 20.4, we reported the graph of the computed approximations of u and p for $c = 0.8$ at $t = 15.512$, very close to the blow-up time.

In Figure 20.5, we showed for $c = 1$ the approximated evolution for $t \in [0, 35.03]$ of the function

$$t \to \max_{\{x_{,1}, x_2\} \in \Omega} u(x_1, x_2, t) \tag{20.58}$$

The computed maximum value is always achieved at $\{0.5, 0.5\}$. The explosive nature of the solution is obvious from this figure.

In order to better understand the evolution of the function (20.58), we analyzed its restriction to the time interval $[0, 28]$ in both the time and frequency domains (see Figure 20.6). Actually, concerning the frequency domain we spectrally analyzed

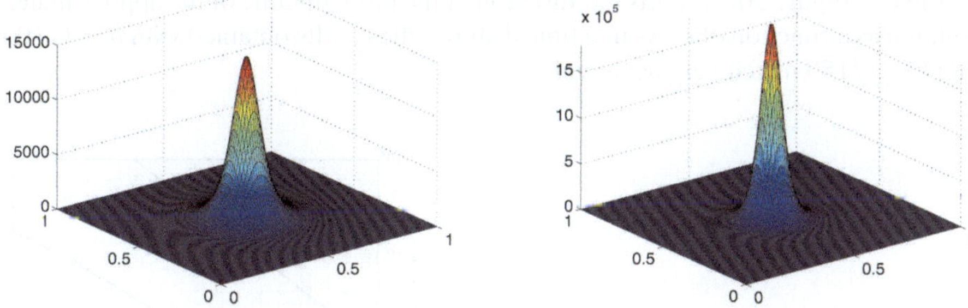

(a) u (b) p

Fig. 20.4 Case $c = 0.8$, pure Dirichlet boundary conditions: Computed approximations for (a) u and (b) p at $t = 15.512$.

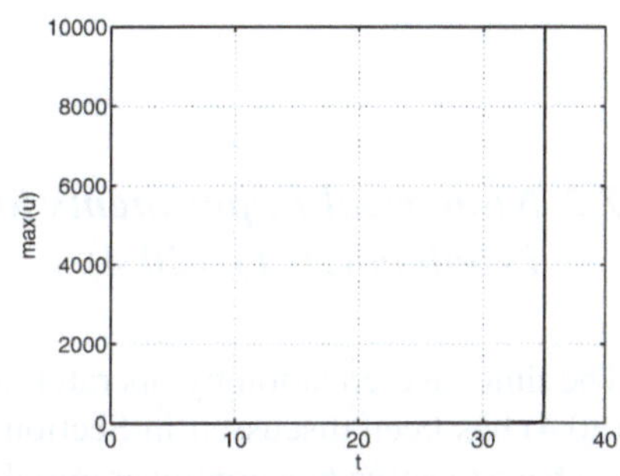

Fig. 20.5 Case $c = 1$, pure Dirichlet boundary conditions: Evolution of the computed approximation of the function (20.58) for $t \in [0, 35.03]$.

the modulation of the above function, that is the signal obtained after subtracting from the function (20.58) its convex component. Figure 20.6(b) suggests that the modulation observed in Figure 20.6(a) is quasi-monochromatic, with $f \simeq 0.9$ Hz.

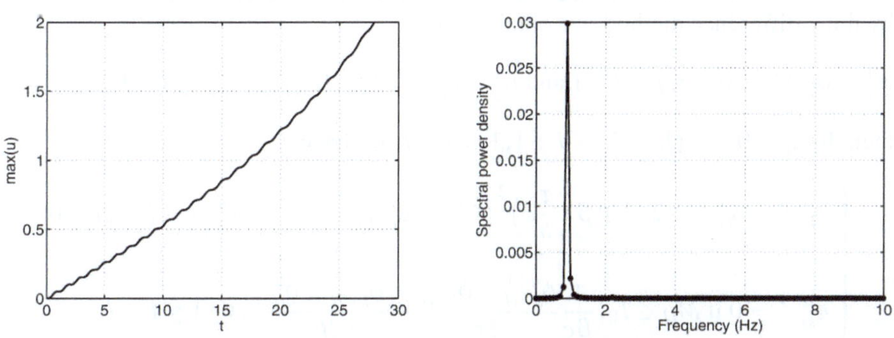

(a) Zoom of Figure 5 (b) Spectrul power density for (a)

Fig. 20.6 Case $c = 1$, pure Dirichlet boundary conditions: (a) Evolution of the computed approximation of the function (20.58) for $t \in [0, 28]$ and (b) spectrum of the modulation.

Finally, Figure 20.7 reports the variation of the blow-up time of the approximated solution as a function of c. As mentioned above, the results obtained with $h = 1/100$ and $h = 1/150$ match very accurately.

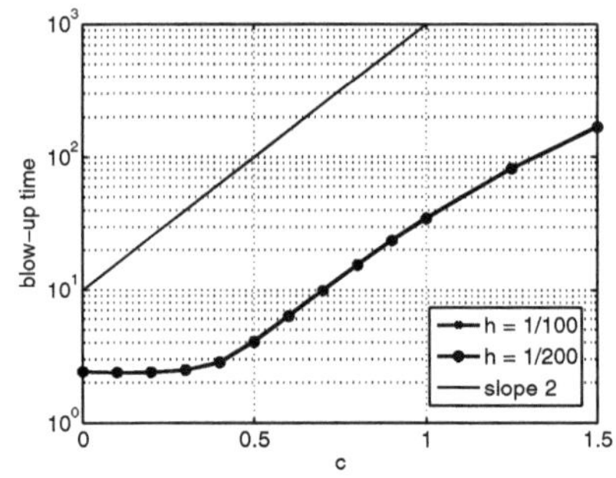

Fig. 20.7 The blow-up time as a function of c (semi-log scale).

7.2 Numerical Experiments for the Nonlinear Wave Problem (20.1), (20.4)

The time discretization by operator-splitting of the nonlinear wave problem (20.1), (20.4) has been discussed in Section 2.3, where we showed that at each time step we have to solve *two* nonlinear initial value problems such as (20.34) and *one* linear wave problem such as (20.23).

The simplicity of the geometry of this test problem (see Figure 20.1) suggests the use of *finite differences* for the space discretization. Using the notation of Section 7.1, at each time level we will have to solve $2I(I+1)$ initial value problems such as (20.34): two for each grid point M_{ij}, with $1 \leq i \leq I+1$, $1 \leq j \leq I$. The solution method discussed in Section 4 still applies. By discretizing problem (20.23) by finite difference method, we obtain: Set

$$\phi_{ij}^0 = \phi_0(M_{ij}),\ 0 \leq i,j \leq I+1 \text{ and } \phi_{ij}^1 - \phi_{ij}^{-1} = 2\tau\phi_1(M_{ij}),\ 1 \leq i \leq I+1, 1 \leq j \leq I,$$

then, for $q = 0, \ldots, Q$, $1 \leq i \leq I+1, 1 \leq j \leq I$, we have

$$\begin{cases} \phi_{ij}^{q+1} + \phi_{ij}^{q-1} - 2\phi_{ij}^q + \beta\left(\frac{\tau}{h}c\right)^2 (4\phi_{ij}^q - \phi_{i+1j}^q - \phi_{i-1j}^q - \phi_{ij+1}^q - \phi_{ij-1}^q) = 0, \\ \phi_{kl}^{q+1} = 0 \text{ if } M_{kl} \in \Gamma_0, \frac{1}{\beta c}\frac{\phi_{I+1l}^{q+1} - \phi_{I+1l}^{q-1}}{2\tau} + \frac{\phi_{I+2l}^q - \phi_{Il}^q}{2h} = 0,\ 1 \leq l \leq I, \end{cases} \tag{20.59}$$

where $\tau = (t_f - t_0)/Q$ and the "ghost" value ϕ^q_{I+2l} has been introduced to impose the Sommerfeld condition at the discrete level. Upon elimination of ϕ^q_{I+2l}, we can derive a more practical formulation of the fully discrete problem, namely for $q = 0, \ldots, Q$, $1 \leq i \leq I, 1 \leq j \leq I$, instead of (20.59) we have

$$\begin{cases} \phi^{q+1}_{ij} + \phi^{q-1}_{ij} - 2\phi^q_{ij} + \beta \left(\frac{\tau}{h} c\right)^2 (4\phi^q_{ij} - \phi^q_{i+1j} - \phi^q_{i-1j} - \phi^q_{ij+1} - \phi^q_{ij-1}) = 0, \\ \phi^{q+1}_{kl} = 0 \text{ if } M_{kl} \in \Gamma_0, \end{cases} \tag{20.60}$$

and for $q = 0, \ldots, Q$, $i = I+1$, $1 \leq j \leq I$, we have

$$\left(1 + \frac{\tau}{h} c\right) \phi^{q+1}_{I+1j} + \left(1 - \frac{\tau}{h} c\right) \phi^{q-1}_{I+1j} - 2\phi^q_{I+1j} + \beta \left(\frac{\tau}{h} c\right)^2 (4\phi^q_{I+1j} - 2\phi^q_{Ij} - \phi^q_{I+1j+1} - \phi^q_{I+1j-1}) = 0. \tag{20.61}$$

Via (20.61), the discrete Sommerfeld boundary condition has been included in the discrete wave equation.

We chose the same values for u_0, u_1, c, α, β, Q, h, and Δt as in Section 7.1. Once again, the results obtained with $h = 1/100$ and $h = 1/150$ match very accurately.

In Figure 20.8, we visualized for $c = 0.8$ and $t \in [0, 6.4]$ (the blow-up time being close to $T_{\max} \simeq 7.432$) the evolution of the computed approximations of the quantities in (20.57) restricted to the segment $\{\{x_1, x_2\}, 0 \leq x_1 \leq 1, x_2 = 1/2\}$. These results (and the ones below) show that the blow-up occurs sooner that in the pure Dirichlet boundary condition case. In Figure 20.9, we reported the graph of the computed approximations of u and p for $c = 0.8$ at $t = 7.432$, very close to the blow-up time.

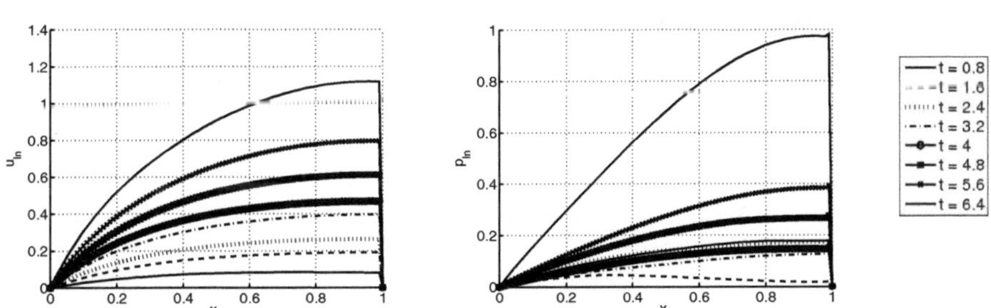

(a) evolution of u_{in} (b) evolution of p_{in} (c) caption

Fig. 20.8 Case $c = 0.8$, mixed Dirichlet-Sommerfeld boundary conditions: Evolution of quantities (a) u_{In} and (b) p_{In}. The caption in (c) is common to (a) and (b).

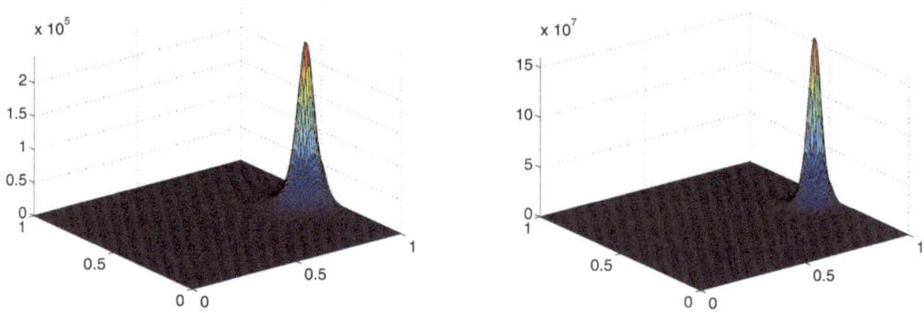

(a) u (b) p

Fig. 20.9 Case $c = 0.8$, mixed Dirichlet-Sommerfeld boundary conditions: Computed approximations for (a) u and (b) p at $t = 7.432$.

Figure 20.10 reports the graph of the computed approximations of u and p for $c = 0.3$ at $t = 2.44$, very close to the blow-up time. Figures 20.9 and 20.10 show that for c sufficiently small (resp., large), the blow-up takes place inside Ω (resp., on Γ_1).

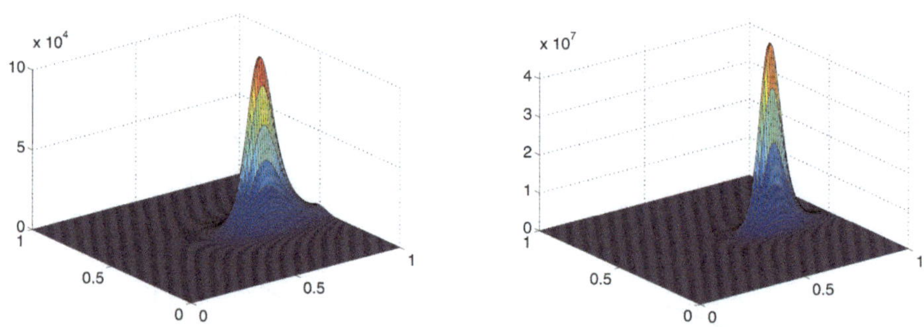

(a) u (b) p

Fig. 20.10 Case $c = 0.3$, mixed Dirichlet-Sommerfeld boundary conditions: Computed approximations for (a) u and (b) p at $t = 2.44$.

In Figure 20.11(a), we reported for $c = 1$ the approximated evolution of the function (20.58) for $t \in [0, 15.135]$. In order to have a better view of the expected modulation of the above function, we reported in Figure 20.11(b) its evolution for $t \in [0, 13.5]$. These figures show the dramatic growth of the solution as t nears $T_{\max}$.

Finally, we reported in Figure 20.12 the variation versus c of the blow-up time for both the pure Dirichlet and the mixed Dirichlet-Sommerfeld boundary conditions. It is interesting to observe how the presence of a boundary condition with (rather) good transparency properties decreases significantly the blow-up time, everything

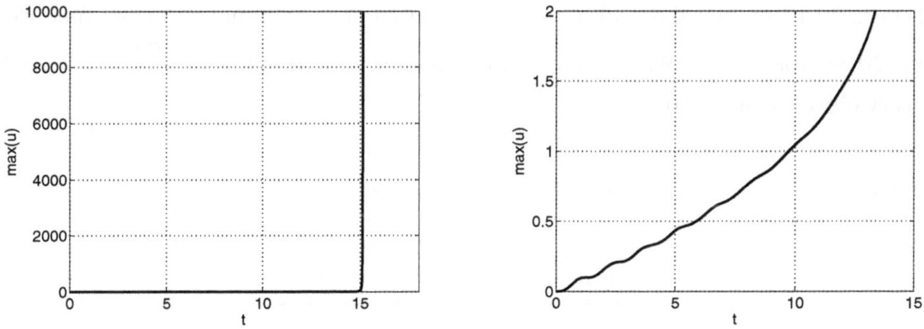

(a) function (58) (a) zoom of (a)

Fig. 20.11 Case $c = 1$, mixed Dirichlet-Sommerfeld boundary conditions: (a) Evolution of the computed approximation of the function (20.58) for $t \in [0, 15.135]$ and (b) zoomed view for $t \in [0, 13.5]$.

else being the same. Also, the above figure provides a strong evidence of the very good matching of the approximate solutions obtained for $h = 1/100$ and $h = 1/150$ (and the related time discretization steps).

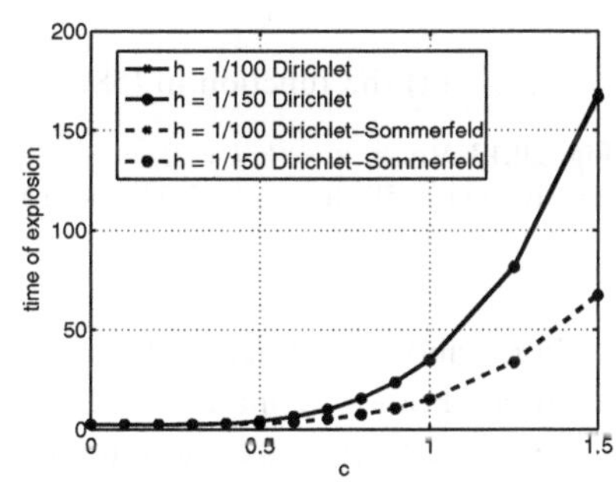

Fig. 20.12 The blow-up time as a function of c for both the pure Dirichlet and the mixed Dirichlet-Sommerfeld boundary conditions.

7.3 Numerical Experiments for the Nonlinear Wave Problem (20.5), (20.3)

We consider now problem (20.5), (20.3). We take the same values for u_0, u_1, α, β, and h as in Section 7.1. Moreover, let us start by setting $k = 0$ and $\lambda = 1$ in (20.5). For c small enough, we observe the same explosive nature of the solutions that we have seen in Sections 7.1 and 7.2; see Table 20.1 for the blow-up times. For $c = 0.32$,

Figure 20.13 reports the evolution of the computed approximation of the function in (20.58) for $t \in [0, 2.94]$ and the computed approximation for $p = \dfrac{\partial u}{\partial t}$ at $t = 2.94$ (very close to blow-up).

Table 20.1 Bratu, $k = 0$: Blow-up times for different values of c.

c	0.1	0.2	0.3	0.31	0.32	0.33
blow-up time	1.99	1.99	2.48	2.66	2.94	3.72

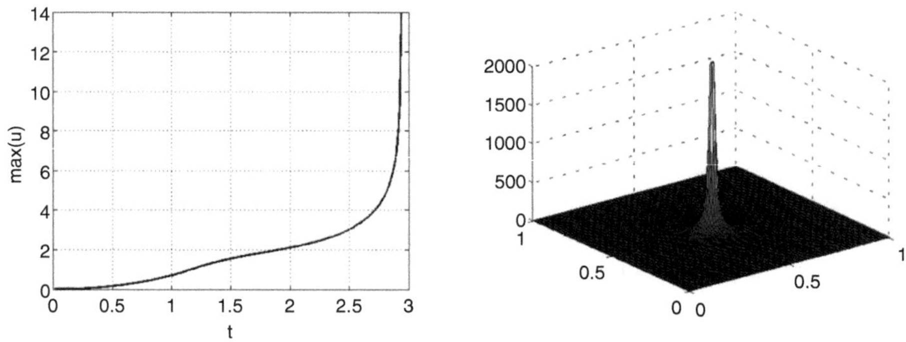

(a) the function in (58) (b) p at t = 2.94

Fig. 20.13 Bratu, $\lambda = 1$, $k = 0$, $c = 0.32$: (a) Evolution of the computed approximation of the function in (20.58) for $t \in [0, 2.94]$ and (b) Computed approximation of p at $t = 2.94$.

For c above a critical value c_{cr}, the solution to (20.5), (20.3) does not blow-up anymore. For $k = 0$ and $\lambda = 1$, the numerical results suggest that $0.33 < c_{cr} < 0.34$. In Figure 20.14(a), we show the evolution of the computed approximation of the function in (20.58) for $c = 0.34$. Figure 20.14(b) suggests that the modulation for $c = 0.34$ observed in Figure 20.14(a) is a quasi-monochromatic signal, with $f \simeq 0.22$ Hz. It is possible to get a good estimate of the critical value c_{cr} by considering the static version of the equation under consideration, that is:

$$\begin{cases} -\nabla^2 u = \dfrac{\lambda}{c^2} e^u & \text{in } \Omega, \\ u = 0 & \text{on } \partial\Omega. \end{cases}$$

It is well known that the above problem possesses a *turning point* for $\dfrac{\lambda}{c^2} \simeq 6.81$ (see, e.g., [9], Chapter 3). This means that c_{cr} can be estimated by $c_{cr} \simeq \sqrt{\lambda/6.81}$. So, for $\lambda = 1$ we get $c_{cr} \simeq 0.38$, which is not too far from the value suggested by the numerical experiments.

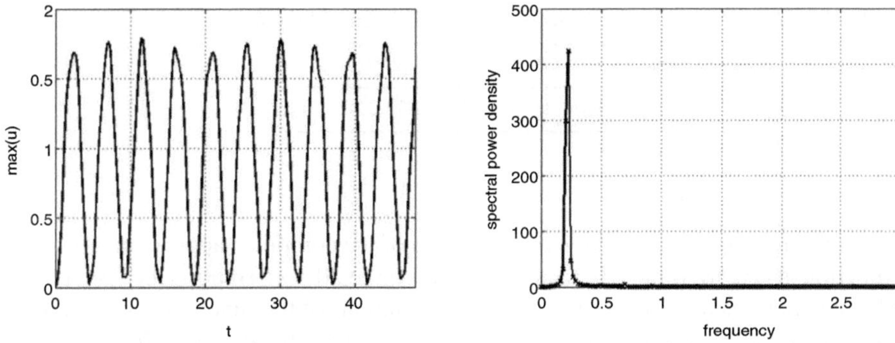

(a) the function in (58) (b) spectral power density for (a)

Fig. 20.14 Bratu, $\lambda = 1$, $k = 0$, $c = 0.34$: (a) Evolution of the computed approximation of the function in (20.58) and (b) Spectrum of the modulation.

If we solve problem (20.5), (20.3) with $k = 0$, $\lambda = 1$, and $c = 0.4$, the oscillations of the function in (20.58) have smaller amplitude and higher frequency than for $c = 0.34$, which is closer to c_{cr} (see Figure 20.15). Again, the spectral power density (in Figure 20.15(b)) suggests that the modulation of function (20.58) in Figure 20.15(a) is a quasi-monochromatic signal, with $f \simeq 0.32$ Hz in this case.

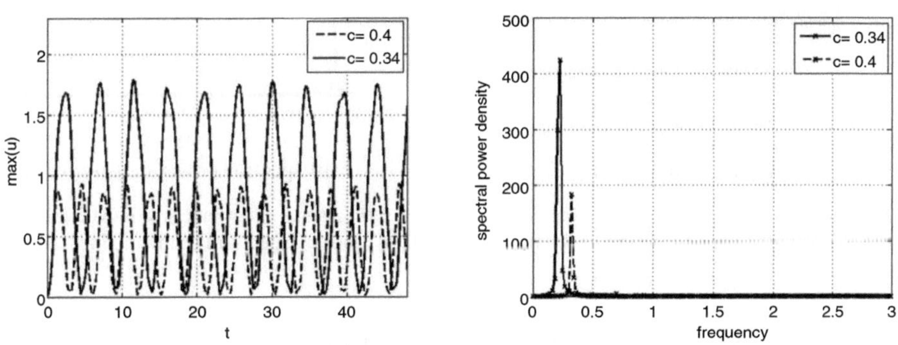

Fig. 20.15 Bratu, $\lambda = 1, k = 0$: Comparison between $c = 0.34$ and $c = 0.4$ in terms of (a) Evolution of the computed approximation of the function in (20.58) and (b) Spectrum of the modulation.

Next, let us set $k = 1$ and $\varepsilon = 0.1$ (it was shown in [12] that the value of ε has little impact on the solution), while keeping $\lambda = 1$. As for $k = 0$, for small values of c the solution to (20.5), (20.3) displays an explosive nature. See Table 20.2 for the blow-up times, which are higher than in the non-damped case (compare with the times in Table 20.1). Also, a higher k has the effect of reducing the value of c_{cr}: the numerical results suggest that for $k = 1$ we have $0.31 < c_{cr} < 0.32$.

Table 20.2 Bratu, $\lambda = 1$, $k = 1$: Blow-up times for different values of c.

c	0.1	0.2	0.3	0.31
blow-up time	2.52	2.57	3.82	4.63

In Figure 20.16(a), we show the evolution of the computed approximation of the function in (20.58) for $c = 0.32$ and $c = 0.4$. Figure 20.16(b) reports the spectrum of the modulations in 20.16(a): the damped oscillations of the function in (20.58) have frequency $f \simeq 0.18$ Hz for $c = 0.32$ and $f \simeq 0.33$ Hz for $c = 0.4$. Notice that the effective damping coefficient in (20.5) is $k/(\varepsilon + t)$, so as $t \to \infty$ it approaches zero. We let the simulations whose results are reported in Figure 20.16 run till $t = 1000$. For $c = 0.32$ (resp., 0.4) the amplitude of the oscillations is 0.12 (resp., 0.035) at $t = 200$ and 0.055 (resp., 0.016) at $t = 1000$.

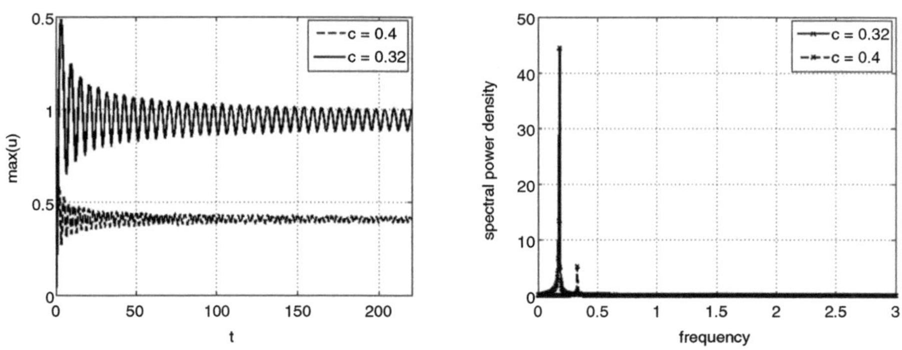

(a) the function in (58)

(b) spectral power density for (a)

Fig. 20.16 Bratu, $\lambda = 1$, $k = 1$: Comparison between $c = 0.32$ and $c = 0.4$ in terms of (a) Evolution of the computed approximation of the function in (20.58) and (b) Spectrum of the modulation.

Next, we set $k = 10$, while keeping $\lambda = 1$. The value of c_{cr} is further reduced: the numerical results suggest that we have $0.3 < c_{cr} < 0.31$. See Table 20.3 for the blow-up times for $c = 0.1, 0.2, 0.3$. In Figure 20.17(a), we show the evolution of the computed approximation of the function in (20.58) for $c = 0.31$ and $c = 0.4$. Unlike the cases $k = 0$ and $k = 1$, the function in (20.58) does not display an oscillatory behavior. The computed solution u approaches a steady state (which is clearly the solution of the associated steady Bratu's problem) after the initial transitory phase. Figure 20.17(b) shows the computed approximation of u at $t = 70$ (close to the steady state) for $c = 0.4$.

Table 20.3 Bratu, $\lambda = 1$, $k = 10$: Blow-up times for different values of c.

c	0.1	0.2	0.3
blow-up time	4.85	5.22	10.02

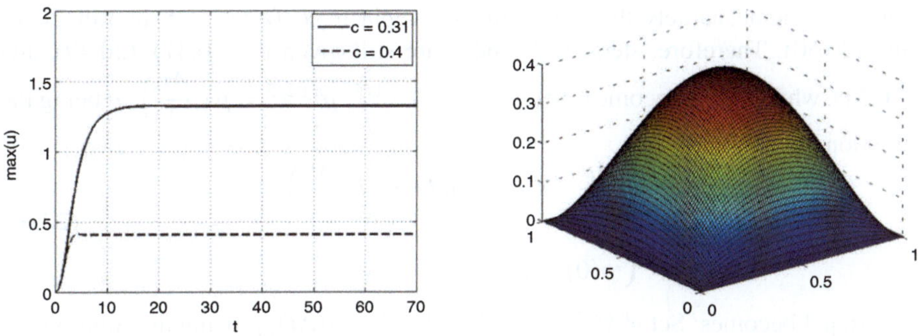

(a) the function in (58) (b) u at t = 70 for c = 0.4

Fig. 20.17 Bratu, $\lambda = 1$, $k = 10$: (a) Comparison between $c = 0.31$ and $c = 0.4$ in terms of evolution of the computed approximation of the function in (20.58) and (b) Computed approximation of u at $t = 70$ for $c = 0.4$.

Figure 20.16(a) and 20.17(a) suggests that there is a critical value of k between 1 and 10 at which the oscillatory behavior of the function in (20.58) disappears. To estimate such a value, we fixed $\lambda = 1$, $c = 0.4$, and progressively increased the value of k. For $k = 2$, the function in (20.58) is still oscillating at $t = 1000$ with amplitude $5 \cdot 10^{-4}$. For $k = 3$, amplitude $5 \cdot 10^{-4}$ is already reached at $t = 100$, while at $t = 1000$ the amplitude is $2 \cdot 10^{-5}$. For $k = 8$, at $t = 50$ the amplitude of the oscillations is 10^{-6}, whereas for $k = 9$ there are no oscillations. Thus, the critical value of k is between 8 and 9.

Figure 20.18 is a zoom of Figure 20.15(a), 20.16(a), and 20.17(a), that is it shows the evolution of the computed approximation of the function in (20.58) for $k = 0, 1, 10$ over the interval $[0, 2]$, when the damping coefficient $k/(\varepsilon + t)$ is large.

Fig. 20.18 Bratu, $\lambda = 1$, $c = 0.4$: evolution of the computed approximation of the function in (20.58) for $k = 0, 1, 10$ over the interval $[0, 2]$. This figure is a zoom of the corresponding curves in Figure 20.15(a), 20.16(a), and 20.17(a).

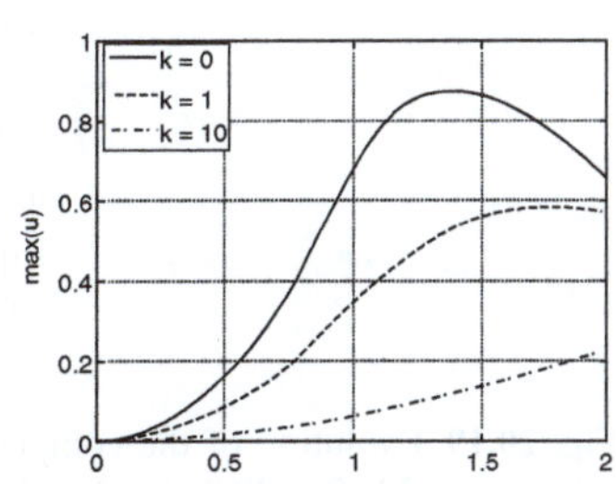

Finally, we replaced the time dependent damping coefficient $k/(\varepsilon+t)$ by k to check how the solution u varies. So, instead of (20.5), we now consider

$$\frac{\partial^2 u}{\partial t^2}+k\frac{\partial u}{\partial t}-c^2\nabla^2 u=\lambda e^u \quad \text{in } \Omega \in (0,T_{\max}). \tag{20.62}$$

Problem (20.62) can be easily solved using a three-stage operator-splitting scheme of the Strang's type; however, for commonality we still used a five-stage operator-splitting scheme, namely, the one obtained by replacing $k/(\varepsilon+t^{n+1/2})$ by k in (20.48) and (20.50). Therefore, steps 1, 3, and 5 are still given by (20.47), (20.49), and (20.51), while step 2 becomes: Set $u^{n+2/5}=u^{n+1/5}$, $p^{n+2/5}=p\left(\frac{\Delta t}{2}\right)$, p being the solution of

$$\begin{cases} \dfrac{\partial p}{\partial t}+kp=0 \text{ in } \Omega\times\left(0,\dfrac{\Delta t}{2}\right), \\ p(0)=p^{n+1/5}; \end{cases} \tag{20.63}$$

and step 4 becomes: Set $u^{n+4/5}=u^{n+3/5}$, $p^{n+4/5}=p(\Delta t)$, p being the solution of

$$\begin{cases} \dfrac{\partial p}{\partial t}+kp=0 \text{ in } \Omega\times\left(\dfrac{\Delta t}{2},\Delta t\right), \\ p\left(\dfrac{\Delta t}{2}\right)=p^{n+3/5}. \end{cases} \tag{20.64}$$

The solutions to sub-problems (20.63) and (20.64) are given by:

$$p^{n+2/5}=e^{-\frac{k\Delta t}{2}}p^{n+1/5}, \quad p^{n+4/5}=e^{-\frac{k\Delta t}{2}}p^{n+3/5}, \tag{20.65}$$

respectively.

In Figure 20.19, we show the evolution of the computed approximation of the function in (20.58) for $c=0.4$ and $k=1$. Since the damping coefficient is constant, the oscillations are damped out quickly (compare Figure 20.19 with the dashed line in Figure 20.16(a)).

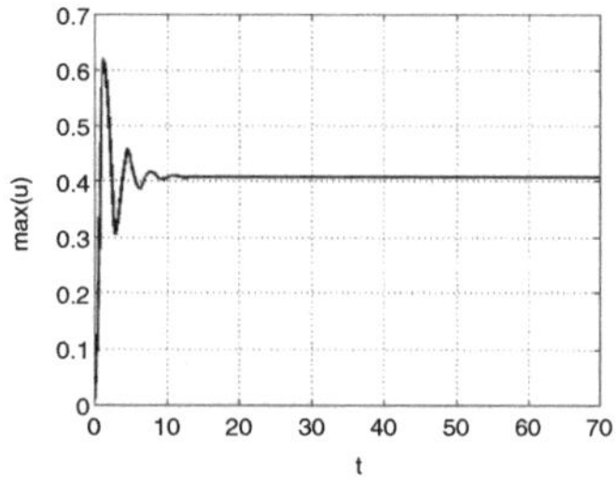

Fig. 20.19 Evolution of the computed approximation of the function in (20.58) where u is the solution to (20.62)–(20.3) for $\lambda=1$, $c=0.4$, and $k=1$.

8 Conclusions

We have investigated the numerical solution of two nonlinear wave equations: a classical wave equation with a forcing term given by the first Painlevé equation and a wave equation of the Euler-Poisson-Darboux type with a forcing term given by the Bratu problem nonlinearity. Depending on the respective values of the various parameters in the model and on the nonlinearity, solutions can blow-up in finite time or evolve to a limit cycle. The key ingredient to capture the solutions has been a three-stage (for the classical wave equation) or five-stage (for the Euler-Poisson-Darboux equation) symmetrized splitting scheme for the time discretization coupled to a well-chosen time-step adaptation technique for treating the fractional steps associated with the nonlinear forcing term of the equation.

The methods discussed in this chapter can be generalized to the coupling of the linear wave equation with nonlinear equations.

References

1. Beale, J.T., and Majda, A., Rates of convergence for viscous splitting of the Navier-Stokes equations, *Math. Comp.*, **37**(156), 1981, 243–259.
2. Bebernes, J., and Eberly, D., Mathematical Problems from Combustion Theory, Springer-Verlag, New York, NY, 1989.
3. Bokil, V.A., and Glowinski, R., An operator-splitting scheme with a distributed Lagrange multiplier based fictitious domain method for wave propagation problems, *J. Comput. Phys.*, **205**(1), 2005, 242–268.
4. Chorin, A.J., Hughes, T.J.R., McCracken, M.F., and Marsden, J.E., Product formulas and numerical algorithms, *Comm. Pure Appl. Math.*, **31**, 1978, 205–256.
5. Clarkson, P.A., Painlevé transcendents, In *NIST Handbook of Mathematical Functions*, Olver, F.W.J., Lozier, D.W., Boisvert, R.F., and Clark, C.W., eds., Cambridge University Press, Cambridge, UK, 2010, 723–740.
6. Fornberg, B. and Weideman, J.A.C., A numerical methodology for the Painlevé equations, *J. Comp. Phys.*, **230**(15), 2011, 5957–5973.
7. Genis, A.M., On finite element methods for the Euler-Poisson-Darboux equation, *SIAM J. Numer. Anal.*, **12**(6), 1984, 1080–1106.
8. Glowinski, R., Numerical Methods for Nonlinear Variational Problems, Springer, New York, NY, 1984. (second printing: 2008)
9. Glowinski, R., Finite element methods for incompressible viscous flow, In *Handbook of Numerical Analysis*, **Vol. IX**, Ciarlet, P.G. & Lions, J.L., eds., North-Holland, Amsterdam, 2003, 3–1176.
10. Glowinski, R., Dean, E.J., Guidoboni, G., Juarez, L.H., and Pan, T.W., Application of operator-splitting methods to the direct numerical simulation of particulate and free-surface flows and to the numerical solution of the two-dimensional elliptic Monge-Ampère equation, *Japan J. Indust. Appl. Math.*, **25**(1), 2008, 1–63.
11. Glowinski, R., and Quaini, A., On the numerical solution of a nonlinear wave equation associated with the first Painlevé equation: An operator-splitting approach, *Chin. Ann. of Math., Series B*, 34(2), 2013, 237–254.
12. Glowinski, R., and Quaini, A., When Euler-Poisson-Darboux meets Painlevé and Bratu: On the numerical solution of nonlinear wave equations., *Methods and Applications of Analysis*, 20(4), 2013, 405–424.

13. Glowinski, R., Shiau, L., and Sheppard, M., Numerical methods for a class of nonlinear integro-differential equations, *Calcolo*, 2012, DOI: 10.1007/s10092-012-0056-2.
14. Jimbo, M., Monodromy problem and the boundary condition for some Painlevé equations, *Publ. Res. Inst. Sci.*, **18**(3), 1982, 1137–1161.
15. Keller, J.B., On solutions of nonlinear wave equations, *Comm. Pure Appl. Math.*, **10**(4), 1957, 523–530.
16. Landman, M.J., Papanicolaou, G.C., Sulem, C., Sulem, P.L., Wang, X.P., Stability of isotropic singularities for the nonlinear Schrödinger equation, *Physica D*, **47**, 1991, 393–415.
17. Leveque, R.J., and Oliger, J., Numerical methods based on additive splittings for hyperbolic partial differential equations, *Math. Comp.*, **40**(162), 1983, 469–497.
18. Marchuk, G.I., Splitting and alternating direction method, In *Handbook of Numerical Analysis*, **Vol. I**, Ciarlet, P.G. & Lions, J.L., eds., North-Holland, Amsterdam, 1990, 197–462.
19. Samarski, A.A., Galaktionov, V.A., Kurdyumov, S.P., and Mikhailov, A.P., Blow-up in quasilinear parabolic equations, de Gruyter Expositions in Mathematics, vol. 19, de Gruyter, Berlin and Hawthorne, NY, 1995.
20. Strang, G., On the construction and comparison of difference schemes, *SIAM J. Numer. Anal.*, **5**(3), 1968, 506–517.
21. Temam, R., Navier-Stokes Equations: Theory and Numerical Analysis, AMS, Providence, RI, 2001.
22. `http://en.wikipedia.org/wiki/Painlevé_transcendents`
23. Wong, R. and Zhang, H.Y., On the connection formulas of the fourth Painlevé transcendent, *Analysis and Applications*, **7**(4), 2009, 419–448.

Chapter 21
Operator Splitting Algorithms for Free Surface Flows: Application to Extrusion Processes

Andrea Bonito, Alexandre Caboussat, and Marco Picasso

Abstract We investigate the benefits of operator splitting methods in the context of computational fluid dynamics. In particular, we exploit their capacity at handling free surface flows and a large variety of physical phenomena in a flexible way. A mathematical and computational framework is presented for the numerical simulation of free surface flows, where the operator splitting strategy allows to separate inertial effects from the other effects. The method of characteristics on a fine structured grid is put forward to accurately approximate the inertial effects while continuous piecewise polynomial finite element associated with a coarser subdivision made of simplices is advocated for the other effects. In addition, the splitting strategy also allows modularity, and in a straightforward manner rheological model change for the fluid. We will emphasize this flexibility by treating Newtonian flows, visco-elastic flows, multi-phase, and multi-density immiscible incompressible Newtonian flows. The numerical framework is thoroughly presented; the test case of the filling of a cylindrical tube with potential die swell in an extrusion process is taken as the main illustration of the advantages of operator splitting.

A. Bonito (✉)
Department of Mathematics, Texas A&M University, College Station, TX 77843-3368, USA
e-mail: bonito@math.tamu.edu

A. Caboussat
Haute Ecole de Gestion de Genève, University of Applied Sciences Western Switzerland (HES-SO), Rue de la Tambourine 17, 1227 Carouge, Switzerland
e-mail: alexandre.caboussat@hesge.ch

M. Picasso
MATHICSE, Ecole Polytechnique Fédérale de Lausanne, 1015 Lausanne, Switzerland
e-mail: marco.picasso@epfl.ch

R. Glowinski et al. (eds.), *Splitting Methods in Communication, Imaging, Science, and Engineering*, Scientific Computation, DOI 10.1007/978-3-319-41589-5_21

1 Introduction

Complex free surface phenomena involving multi-phase Newtonian and/or Non-Newtonian flows are nowadays a topic of active research in many fields of physics, engineering, and bioengineering. Numerous mathematical models and associated numerical approximations for complex liquid-gas free surfaces problems are also available.

The purpose of this chapter is to present a comprehensive review of a computational methodology developed in the group of Jacques Rappaz at *Ecole polytechnique fédérale de Lausanne (EPFL)*, called *cfsFlow* and commercialized by a spin-off company of EPFL named Ycoor Systems S.A. [40]. Originally proposed for two-dimensional cases by Maronnier, Picasso, and Rappaz [25], it evolved to handle three-dimensional flows [26], account for surrounding compressible gas [11, 12] and surface tension [8], allow complex rheology [6], include space adaptive interface tracking [9], and recently integrate multi-phase fluids [19]. Besides the typical fluid flows applications, it is worth noting that these methods have been also applied successfully to predict the evolution of glaciers [20, 21, 33].

Many algorithms are available to approximate free boundary problems, see for instance [2, 29, 31, 37, 38]. The novelty in *cfsFlow* is to use a time splitting approach [15] and a two-grids method to decouple advection and diffusion regimes. This allows the use of well-suited numerical techniques for each of the two regimes separately. In particular, the advection phenomenon describing the evolution of each liquid phases is approximated on structured grids by the forward method of characteristics [34] on the volume-of-fluid representation of each phase. On the other hand, finite element approximations on simplices determined as liquid are implemented to handle diffusion-like phenomena.

We start by discussing in Section 2 the basic model for Newtonian fluids with free surface. The type of operator splitting strategies considered and their applications to free boundary problems are presented in Section 3, the associated numerical algorithms being presented in Section 4. Fluids verifying more complex rheology are discussed in Section 5, where the upper convected Maxwell constitutive relation for the extra stress tensor is chosen as our model problem. Multi-phase fluids are considered in Section 6 and perspectives on emulsion processes are put forward in Section 7.

The filling of a cylindrical tube with potential die swell in an extrusion process is taken as the main illustration of the advantages of the presented numerical algorithm and is used throughout this chapter to evaluate the effect of each component in the final model. We note in passing that the numerical simulations of extrusion is of great importance for instance in industrial processes involving pasta dough [22] or textile products [1].

Acknowledgements

All the numerical simulations have been performed using the software `cfsFlow` developed by EPFL and Ycoor Systems S.A. The authors would like to thank A. Masserey and G. Steiner (Ycoor Systems S.A.) for implementation support. They are also pleased to acknowledge the valuable contributions of S. Boyaval (EDF & ENPC, Paris) and N. James (Université de Poitiers) on the multiphase model, and of P. Clausen (formerly at EPFL) on the implementation of efficient numerical algorithms and adaptive techniques for multiphase flows and surface tension effects. This work is partially funded by the Commission for technology and innovation (CTI grant number 14359.1 PFES-ES), the Swiss national science foundation (grant number 200021_143470), and the American National Science Foundation (NSF grant number DMS-1254618).

2 Mathematical Modeling of Newtonian Fluids with Free Surfaces

We present in this section the mathematical model used to describe the evolution of an incompressible Newtonian fluid with a free surface, neglecting the effect of the ambient fluid. A simple model for the treatment of the ambient fluid has been proposed in [12] and the addition of surface tension effects has been described in [8, 9].

The computational domain is denoted by $\Lambda \subset \mathbb{R}^d$, $d = 2,3$, and $T > 0$ stands for the final time. We describe in Section 2.1 the Navier-Stokes equations for fluids subject to free boundaries and detail in Section 2.2 the Eulerian approach used to track the liquid domain evolution.

2.1 Navier-Stokes System

We denote by $\Omega(t) \subset \Lambda$ the domain occupied by the liquid at time $t \in [0,T]$ and by

$$Q := \{(\mathbf{x},t) \in \Lambda \times (0,T] \mid \mathbf{x} \in \Omega(t)\},$$

the space-time liquid domain. The fluid is assumed to be incompressible and Newtonian so that its velocity $\mathbf{u} : Q \to \mathbb{R}^d$ and pressure $p : Q \to \mathbb{R}$ are the solutions to the Navier-Stokes equations:

$$\begin{cases} \rho\left(\dfrac{\partial}{\partial t}\mathbf{u} + (\mathbf{u}\cdot\nabla)\mathbf{u}\right) - 2\nabla\cdot(\mu\mathbf{D}(\mathbf{u})) + \nabla p = \mathbf{f} & \text{in} \quad Q, \\ \nabla\cdot\mathbf{u} = 0 & \text{in} \quad Q, \end{cases} \tag{21.1}$$

where $\mathbf{f} : Q \to \mathbb{R}^d$ is a given external force (typically $\mathbf{f} := \rho\mathbf{g}$, where $\mathbf{g}$ is the gravitational acceleration), $\mathbf{D}(\mathbf{u}) := \frac{1}{2}(\nabla\mathbf{u} + \nabla\mathbf{u}^t)$ is the symmetric part of the velocity gradient and $\rho > 0$, $\mu > 0$ are respectively the fluid density and viscosity. Notice that the Navier-Stokes equations are only defined in the liquid domain Q, the effect of the outside fluid being neglected. Hence, the velocity and pressure outside Q are not defined.

We now discuss the boundary/interface conditions associated with the above system and refer to Figure 21.1 for an illustration in the die swell context. We separate the boundary of the computational domain in two disjoint open sets $\Gamma_{\mathscr{D}}$ and $\Gamma_{\mathscr{N}}$ such that $\partial\Lambda = \overline{\Gamma_{\mathscr{D}}} \cup \overline{\Gamma_{\mathscr{N}}}$. The velocity is prescribed on $\Gamma_{\mathscr{D}}$ (Dirichlet boundary condition), i.e. for a given $\mathbf{g}_{\mathscr{D}} : \Gamma_{\mathscr{D}} \times [0,T] \to \mathbb{R}^d$, we have

$$\mathbf{u} = \mathbf{g}_{\mathscr{D}} \quad \text{on} \quad \partial Q_{\mathscr{D}} := \{(\mathbf{x},t) \in \Gamma_{\mathscr{D}} \times [0,T] \mid \mathbf{x} \in \partial\Omega(t)\}. \tag{21.2}$$

On the other hand, a force is applied on $\Gamma_{\mathscr{N}}$ (Neumann boundary condition), i.e. for a given $\mathbf{g}_{\mathscr{N}} : \Gamma_{\mathscr{N}} \times [0,T] \to \mathbb{R}^d$, we have

$$(2\mu\mathbf{D}(\mathbf{u}) - p\mathbf{I})\mathbf{n} = \mathbf{g}_{\mathscr{N}} \quad \text{on} \quad \partial Q_{\mathscr{N}} := \{(\mathbf{x},t) \in \Gamma_{\mathscr{N}} \times [0,T] \mid \mathbf{x} \in \partial\Omega(t)\}, \tag{21.3}$$

where $\mathbf{n}(.,t)$ is the outward pointing unit vector normal to $\partial\Omega(t)$ and $\mathbf{I}$ is the identity tensor. More general boundary conditions such as slip boundary conditions could be considered in a similar way but are not included here to keep the presentation as simple as possible.

Regarding the free interface condition, we assume that no force is exerted to the liquid, that is

$$(2\mu\mathbf{D}(\mathbf{u}) - p\mathbf{I})\mathbf{n} = \mathbf{0} \quad \text{on} \quad \mathscr{F} := \{(\mathbf{x},t) \in \Lambda \times (0,T] \mid \mathbf{x} \in \partial\Omega(t) \setminus \partial\Lambda\}, \tag{21.4}$$

and, since the interface evolves with the fluid velocity, that the interface velocity $\mathbf{u}_{\mathscr{F}}$ satisfies

$$\mathbf{u}_{\mathscr{F}} = \mathbf{u} \quad \text{on} \quad \mathscr{F}. \tag{21.5}$$

Finally, an initial condition $\mathbf{u}_0 : \Omega(0) \to \mathbb{R}^d$ is provided for the velocity

$$\mathbf{u}(.,0) = \mathbf{u}_0 \quad \text{on} \quad \Omega(0). \tag{21.6}$$

2.2 Implicit Representation of the Liquid Domain

The liquid domain $\Omega(t)$ is mathematically represented during the evolution via its characteristic function $\phi : \Lambda \times [0,T] \to \{0,1\}$, implying that

$$\Omega(t) = \{\mathbf{x} \in \Lambda \mid \phi(\mathbf{x},t) = 1\}, \qquad t \in [0,T]. \tag{21.7}$$

In view of the interface velocity condition (21.5), we interpret the evolution of $\Omega(t)$ as the transport of its characteristic function with the fluid velocity:

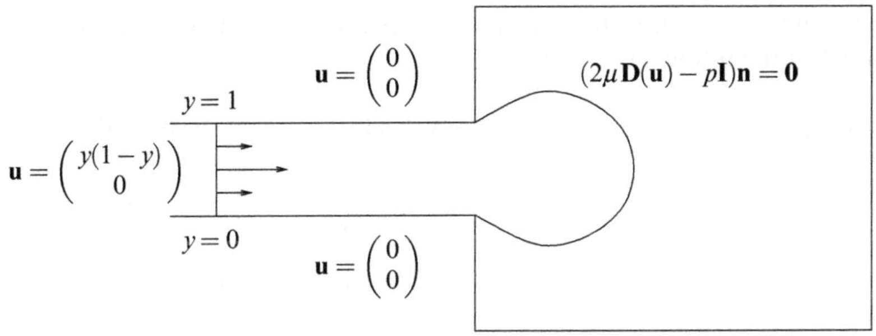

Fig. 21.1 Boundary conditions in the context of die swell. The liquid enters the cavity with an initial velocity **u**, the horizontal component is a parabolic profile and the vertical component vanishes. The velocity is imposed to vanish on the rest of the boundary. Another setup would be to enforce slip boundary conditions on the lateral walls of the extruder, implying a constant, instead of parabolic, profile of velocities in the tube.

$$\frac{\partial}{\partial t}\phi + \mathbf{u} \cdot \nabla \phi = 0 \quad \text{in} \quad Q, \qquad \phi = 0 \quad \text{in} \quad \Lambda \setminus Q, \tag{21.8}$$

where **u** is the fluid velocity only defined on the space-time fluid domain Q as noted in Section 2.1. An illustration is provided in Figure 21.2.

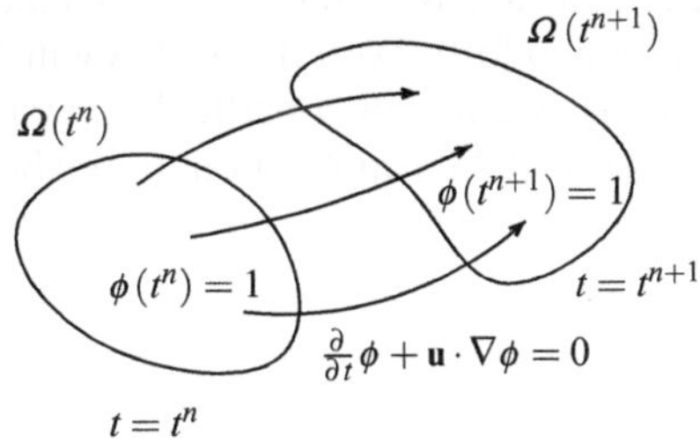

Fig. 21.2 Deformation of the liquid domain $\Omega(t)$ for $t \in [t^n, t^{n+1}]$ deduced from the transport of the characteristic function ϕ with the liquid velocity **u** according to (21.8).

Remark 1 (Ambient Fluid and Computational Cost). As the effect of the outside fluid is neglected, the Navier-Stokes relations (21.1) for the velocity-pressure pair are only considered in the space-time liquid domain Q. As a consequence, the velocity is only defined on Q and so is the transport equation for the characteristic function in (21.8). A possible equivalent alternative would consist in finding an extension of the velocity field to $\Lambda \times (0,T]$, thereby extending the transport relation to the entire space-time computational domain $\Lambda \times (0,T]$. However, the numerical scheme described in Section 4 takes full advantage of the representation (21.8) in order to reduce the overall computational cost.

We supplement the transport equation in (21.8) by the value of the characteristic function ϕ at the inflow boundary

$$\partial Q_{\text{inflow}} := \{(\mathbf{x},t) \in \partial\Lambda \times (0,T] \mid \mathbf{x} \in \partial\Omega(t) \quad \text{and} \quad \mathbf{u}\cdot\mathbf{n} < 0\}, \tag{21.9}$$

namely,

$$\phi = 1 \quad \text{on} \quad \partial Q_{\text{inflow}}. \tag{21.10}$$

The initial value of the characteristic equation is chosen to match the initial given domain $\Omega(0) := \Omega$,

$$\phi(.,0) = 1 \qquad \text{on} \quad \Omega(0) \qquad \text{and} \qquad \phi(.,0) = 0 \qquad \text{elsewhere.} \tag{21.11}$$

3 Operator Splitting Algorithm

We take advantage of an operator splitting scheme to separate the numerical issues inherent to the approximation of the diffusion and advection operators in the approximation of the system of equations (21.1) and (21.8). In this context, it allows to treat separately Stokes systems on given non-moving liquid domains and transport equations for the velocity and the liquid characteristic function.

Several operator splitting algorithms are available in the literature, starting from the early works of Peaceman and Rachford [32], Douglas and Rachford [14], Marchuk [23, 24], and Yanenko [39]. We refer to Glowinski [15] for a survey of these methods. In Section 3.1, we review a particular version of the so-called *Lie scheme* and we detail in Section 3.2 its application to free boundary problems.

3.1 The Lie Scheme

We advocate in this work a particular version of the *Lie scheme* and follow the description provided in [15, 16] (see also Chapters 1 and 2 in this book). Assume that we are interested in the solution of the Cauchy problem

$$\begin{cases} \dfrac{d}{dt}\mathrm{v} + A(\mathrm{v},t) = 0,\ t \in (0,T], \\ \mathrm{v}(0) = \mathrm{v}_0, \end{cases}$$

where the operator A can be decomposed as the sum of q operators

$$A = \sum_{i=1}^{q} A_i. \tag{21.12}$$

The scheme starts with a subdivision $0 =: t^0 < t^1 < \ldots < t^N := T$ of the time interval $[0,T]$. Over each sub-interval $I^{n+1} := (t^n, t^{n+1}]$ the approximation of $\mathrm{v}(t^{n+1})$ (an approximation of $\mathrm{v}(t^n)$ being given) is obtained in q steps (corresponding to q alternating "directions"):

set $\mathrm{v}^0 = \mathrm{v}_0$ and $\mathrm{w}^0(t) \equiv \mathrm{v}^0$ for $t \in [0,t^1]$;
for $n = 0,\ldots,N-1$
  for $i = 1,\ldots,q$
    find $\mathrm{w}^{n+i/q}(t)$ as the solution of
    $\frac{d}{dt}\mathrm{v} + A_i(\mathrm{v},t) = 0$ on $(t^n, t^{n+1}]$
    and satisfying the initial condition $\mathrm{v}(t^n) = \mathrm{w}^{n+(i-1)/q}(t^{n+1})$;
  end for
  set $\mathrm{v}^{n+1} := \mathrm{w}(t^{n+1})$
end for

It turns out that if the operators A_i are *linear*, *time independent*, and they *commute*, then $\mathrm{v}^n = \mathrm{v}(t^n)$ for $n = 0,\ldots,N$. However, in generic situations, the above scheme is, at most, first order accurate [15]. Nevertheless, this motivates the introduction of first order discretizations in time and space for each sub-step of the algorithm, as described in Sections 4.1 and 4.2.

3.2 Application to Free Surface Flows

In the context of fluid flows with free boundaries, we set up a splitting of type (21.12) using two alternating "directions" ($q = 2$). We call these two steps the *prediction* and *correction* steps which are now described on each time subinterval $I^{n+1} := (t^n, t^{n+1}]$. They consist in separating the hyperbolic regime from the parabolic regime in order to apply numerical methods well suited to each situation; see Section 4.

We assume that an approximation of the liquid characteristic function ϕ^n is given, and therefore so is an approximation of the liquid domain Ω^n via the relation

$$\Omega^n := \{\mathbf{x} \in \Lambda \mid \phi^n(\mathbf{x}) = 1\}.$$

This relation corresponds to (21.7) after approximating the liquid characteristic function at time $t = t^n$. We also assume to be given a velocity approximation $\mathbf{u}^n(\mathbf{x})$ of $\mathbf{u}(\mathbf{x},t^n)$. The prediction step determines an approximation of the liquid domain at time t^{n+1} together with a prediction of the velocity on the new domain. The correction step provides an update of the velocity and pressure on the liquid domain that remains unchanged. Figure 21.3 provides an illustration of the process for the die swelling example.

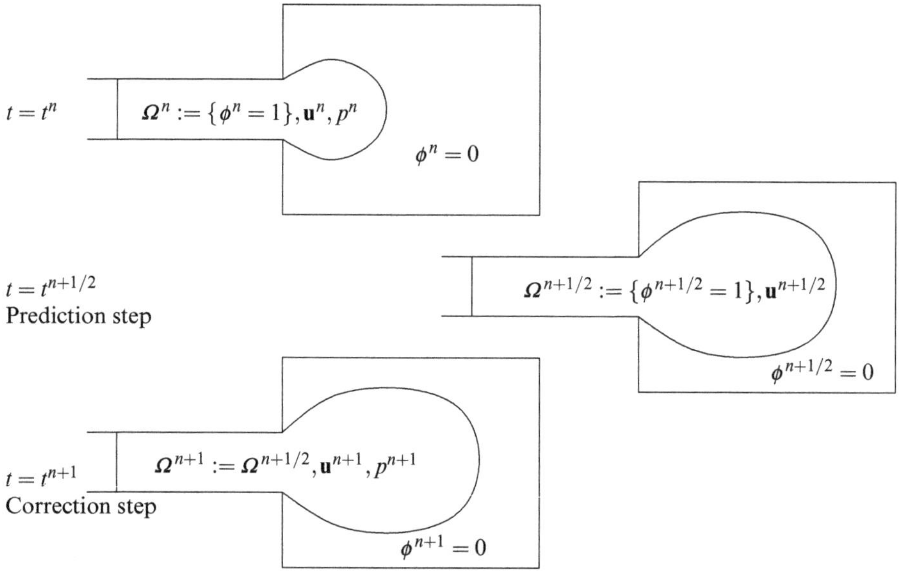

Fig. 21.3 Alternating direction splitting applied to free boundary problems. Given approximations ϕ^n and $\mathbf{u}^n$ of the liquid domain characteristic function ϕ and velocity $\mathbf{u}$ at time $t = t^n$, the first step consists in finding updated approximations of the characteristic function ϕ (and thus of the liquid domain Ω) as well as of the fluid velocity $\mathbf{u}$. On the new liquid domain, the second step determines a velocity correction together with its associated pressure. In particular, the liquid domain does not change during the correction step.

3.2.1 The Prediction Step

The *prediction step* encompasses the advection components of (21.1) and (21.8). It consists in simultaneously finding approximations of the characteristic function ϕ and the velocity field $\mathbf{u}$ satisfying

$$\frac{\partial}{\partial t}\phi + \mathbf{u}\cdot\nabla\phi = 0 \quad \text{and} \quad \frac{\partial}{\partial t}\mathbf{u} + (\mathbf{u}\cdot\nabla)\mathbf{u} = \mathbf{0} \quad \text{in} \quad Q^{n+1} := Q \cap \left(\Lambda \times I^{n+1}\right). \tag{21.13}$$

The numerical scheme proposed here relies on the so-called method of characteristics and is detailed now. For any point $\overline{\mathbf{x}} \in \Omega^n$ in the liquid domain, we define the characteristic trajectory $\mathbf{y}(.;\overline{\mathbf{x}})$ starting at $\overline{\mathbf{x}}$ by

$$\frac{d}{dt}\mathbf{y}(t;\overline{\mathbf{x}}) = \mathbf{u}(\mathbf{y}(t;\overline{\mathbf{x}}),t), \quad \text{for} \quad t \in I^{n+1}, \quad \text{and} \quad \mathbf{y}(t^n;\overline{\mathbf{x}}) = \overline{\mathbf{x}}. \tag{21.14}$$

Along this characteristic trajectory, the transport relations in (21.8) read

$$\frac{d}{dt}\phi(\mathbf{y}(t;\overline{\mathbf{x}}),t) = 0 \qquad \text{and} \qquad \frac{d}{dt}\mathbf{u}(\mathbf{y}(t;\overline{\mathbf{x}}),t) = \mathbf{0}.$$

Hence, from the initial conditions $\phi(t^n) = \phi^n$ and $\mathbf{u}(t^n) = \mathbf{u}^n$ on Ω^n, we obtain

$$\phi(\mathbf{y}(t;\overline{\mathbf{x}}),t) = \phi^n(\overline{\mathbf{x}}) = 1 \qquad \text{and} \qquad \mathbf{u}(\mathbf{y}(t;\overline{\mathbf{x}}),t) = \mathbf{u}^n(\overline{\mathbf{x}}) \tag{21.15}$$

as long as $\mathbf{y}(t;\overline{\mathbf{x}}) \in \Lambda$. We set $\phi(\mathbf{x},t) = 0$ whenever $\mathbf{x} \in \Lambda \setminus \{\mathbf{y}(t;\overline{\mathbf{x}}) \mid \overline{\mathbf{x}} \in \Omega^n\}$ so that these relations defines ϕ on $\Lambda \times I^{n+1}$ (and the associated liquid domain) as well as the velocity $\mathbf{u}$ on the liquid domain. As pointed out earlier, the algorithm does not need the velocity $\mathbf{u}(\mathbf{x},t)$ whenever $\mathbf{x} \in \Lambda \setminus \{\mathbf{y}(t;\overline{\mathbf{x}}) \mid \overline{\mathbf{x}} \in \Omega^n\}$. The *prediction step* ends upon setting

$$\phi^{n+\frac{1}{2}} := \phi(t^{n+1}) \qquad \text{in} \quad \Lambda,$$

and consequently

$$\Omega^{n+\frac{1}{2}} := \left\{\mathbf{x} \in \Lambda \mid \phi^{n+\frac{1}{2}}(\mathbf{x}) = 1\right\} := \{\mathbf{y}(t^{n+1},\overline{\mathbf{x}}) \mid \overline{\mathbf{x}} \in \Omega^n\} \cap \Lambda \tag{21.16}$$

as well as

$$\mathbf{u}^{n+\frac{1}{2}} := \mathbf{u}(t^{n+1}) \qquad \text{in} \quad \Omega^{n+\frac{1}{2}}.$$

3.2.2 The Correction Step

After the prediction step, the approximation of the liquid domain remains unchanged. In the framework of the splitting scheme described in Section 3.1, the "corrected" characteristic function satisfies

$$\frac{\partial}{\partial t}\phi = 0 \quad \text{in} \quad \Omega^{n+\frac{1}{2}} \times I^{n+1} \qquad \text{with} \qquad \phi(t^n) = \phi^{n+\frac{1}{2}} \quad \text{in} \quad \Omega^{n+\frac{1}{2}}.$$

As a consequence, we set $\phi^{n+1} := \phi^{n+\frac{1}{2}}$, $\Omega^{n+1} := \Omega^{n+\frac{1}{2}}$ and we note that the predicted velocity is now defined over Ω^{n+1}, i.e., $\mathbf{u}^{n+\frac{1}{2}} : \Omega^{n+1} \to \mathbb{R}^d$.

Then, the updated velocity $\mathbf{u} : \Omega^{n+1} \times I^{n+1} \to \mathbb{R}^d$ as well as the associated pressure $p : \Omega^{n+1} \times I^{n+1} \to \mathbb{R}$ are defined as the solution to the following Stokes system on a given non-moving domain:

$$\begin{cases} \rho\dfrac{\partial}{\partial t}\mathbf{u} - 2\nabla\cdot(\mu\mathbf{D}(\mathbf{u})) + \nabla p = \mathbf{f} \\ \nabla\cdot\mathbf{u} = 0 \end{cases} \quad \text{in} \quad \Omega^{n+1} \times I^{n+1}, \tag{21.17}$$

supplemented by the boundary conditions

$$\mathbf{u} = \mathbf{g}_D \quad \text{on} \quad \partial\Omega^{n+1} \cap \Gamma_{\mathscr{D}}, \quad (2\mu\mathbf{D}(\mathbf{u}) - p\mathbf{I})\mathbf{n}^{n+1} = \mathbf{g}_N \quad \text{on} \quad \partial\Omega^{n+1} \cap \Gamma_{\mathscr{N}},$$

and the free interface condition

$$(2\mu\mathbf{D}(\mathbf{u}) - p\mathbf{I})\mathbf{n}^{n+1} = \mathbf{0} \quad \text{on} \quad \partial\Omega^{n+1} \setminus \partial\Lambda,$$

where $\mathbf{n}^{n+1}$ is the outer pointing unit vector normal to $\partial\Omega^{n+1}$. Finally, we define the corrected velocity approximation $\mathbf{u}^{n+1} : \Omega^{n+1} \to \mathbb{R}^d$ by $\mathbf{u}^{n+1} := \mathbf{u}(t^{n+1})$ and the associated pressure by $p^{n+1} : \Omega^{n+1} \to \mathbb{R}$ by $p^{n+1} := p(t^{n+1})$.

4 Numerical Approximation of Free Surface Flows

We are now in a position to describe the numerical algorithm for the approximation of the solution to the free boundary problem (21.1) and (21.8). It takes full advantage of the splitting into *prediction* and *correction* steps discussed in Section 3.2. The time and space discretizations are presented in Sections 4.1 and 4.2 respectively. This section ends with Section 4.3, where numerical illustrations are given, in particular, in the context of die swell.

4.1 Time Discretization

We recall that the time interval $[0,T]$ is decomposed in N subintervals $I^n := (t^n, t^{n+1}]$, $n = 0, \ldots, N-1$ and we denote the associated time steps by $\delta t^{n+1} := t^{n+1} - t^n$. In what follows, we discuss the algorithm over the time interval I^n.

4.1.1 Prediction Step

An *explicit Euler* approximation $\mathbf{Y}^{n+1}$ of the characteristic curve $\mathbf{y}(t^{n+1};\overline{\mathbf{x}})$ in (21.14) is advocated for the prediction step. For all $\overline{\mathbf{x}} \in \Omega^n$, we set

$$\mathbf{Y}^{n+1}(\overline{\mathbf{x}}) := \overline{\mathbf{x}} + \delta t^{n+1}\mathbf{u}^n(\overline{\mathbf{x}}). \tag{21.18}$$

In view of (21.16), the approximation of the liquid domain $\Omega^{n+\frac{1}{2}}$, denoted $\Omega_N^{n+\frac{1}{2}}$, is defined as

$$\Omega_N^{n+\frac{1}{2}} := \left\{\mathbf{Y}^{n+1}(\overline{\mathbf{x}}) \mid \overline{\mathbf{x}} \in \Omega^n\right\} \cap \Lambda.$$

The characteristic curves $\mathbf{Y}^{n+1}$ determine also the approximations $\Phi^{n+\frac{1}{2}}$ and $\mathbf{U}^{n+\frac{1}{2}}$ of $\phi^{n+\frac{1}{2}}$ and $\mathbf{u}^{n+\frac{1}{2}}$ according to the relations

$$\Phi^{n+\frac{1}{2}}(\mathbf{Y}^{n+1}(\overline{\mathbf{x}})) = \phi^n(\overline{\mathbf{x}}) := 1, \qquad \mathbf{U}^{n+\frac{1}{2}}(\mathbf{Y}^{n+1}(\overline{\mathbf{x}})) = \mathbf{u}^n(\overline{\mathbf{x}}), \tag{21.19}$$

whenever $\mathbf{Y}^{n+1}(\overline{\mathbf{x}}) \in \Lambda$. In addition, we set $\Phi^{n+\frac{1}{2}}(\mathbf{x}) = 0$ for $\mathbf{x} \in \Lambda \setminus \Omega_N^{n+\frac{1}{2}}$.

4.1.2 Correction Step

The approximation of the liquid domain characteristic function is not modified in this step, i.e.,

$$\Phi^{n+1} := \Phi^{n+\frac{1}{2}} \qquad \text{and} \qquad \Omega_N^{n+1} := \Omega_N^{n+\frac{1}{2}}.$$

An *implicit Euler* method is advocated for the solution of the Stokes system (21.17). This consists in seeking $\mathbf{U}^{n+1} : \Omega_N^{n+1} \to \mathbb{R}^d$ and $P^{n+1} : \Omega_N^{n+1} \to \mathbb{R}$ satisfying

$$\begin{cases} \rho \dfrac{\mathbf{U}^{n+1} - \mathbf{U}^{n+\frac{1}{2}}}{\delta t^{n+1}} - 2\nabla \cdot (\mu \mathbf{D}(\mathbf{U}^{n+1})) + \nabla P^{n+1} = \mathbf{f}(.,t^{n+1}), \\ \nabla \cdot \mathbf{U}^{n+1} = 0, \end{cases} \tag{21.20}$$

in Ω_N^{n+1}, subject to the boundary conditions

$$\mathbf{U}^{n+1} = \mathbf{g}_D(.,t^{n+1}) \quad \text{on} \quad \partial\Omega_N^{n+1} \cap \Gamma_{\mathscr{D}},$$
$$(2\mu \mathbf{D}(\mathbf{U}^{n+1}) - P^{n+1}\mathbf{I})\mathbf{n}_N^{n+1} = \mathbf{g}_N(.,t^{n+1}) \quad \text{on} \quad \partial\Omega_N^{n+1} \cap \Gamma_{\mathscr{N}},$$

and to the free interface condition

$$(2\mu \mathbf{D}(\mathbf{U}^{n+1}) - P^{n+1}\mathbf{I})\mathbf{n}_N^{n+1} = \mathbf{0} \quad \text{on} \quad \partial\Omega_N^{n+1} \setminus \partial\Lambda.$$

Here $\mathbf{n}_N^{n+1}$ is the outer pointing unit vector normal to $\partial\Omega_N^{n+1}$.

4.2 Two-Grid Spatial Discretization

The space discretization takes also full advantage of the alternating splitting described above. The prediction and correction steps are approximated using different subdivisions and numerical techniques. On the one hand, a subdivision made of structured cells is advocated for the characteristic relation (21.18) coupled with a Simple Linear Interface Calculation (SLIC) [28] procedure in order to limit the numerical diffusion when approximating the volume fraction of liquid in (21.19). On the other hand, a standard stabilized finite element method is proposed for the approximation of the solution of the Stokes system (21.20). We start with the description of the two subdivisions and define the associated discrete approximation spaces. Then we detail the numerical techniques tailored to each discrete spaces.

4.2.1 Two Subdivisions and Associated Discrete Spaces

The prediction and correction steps rely on two different subdivisions. A volume-of-fluid type method on a structured grid is advocated for the prediction step consisting of two transport equations (21.19). The computational domain Λ is bounded and therefore can be included into a structured grid of cells C_i, $i = 1,\dots,M$. We denote by $\mathscr{T}^S := \{C_i,\ i = 1,\dots,M\}$ the collection of all those structured cells and by $h := \max_{C\in\mathscr{T}^S} \text{diameter}(C)$ the typical size of the elements. An example of such mesh is shown in Figure 21.4.

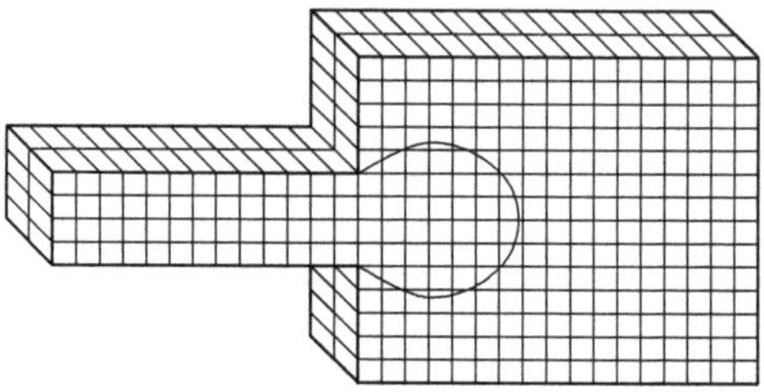

Fig. 21.4 Structured subdivision used for the space discretization during the prediction step.

We denote by $\mathbb{V}^S$ the approximation space which consists of all piecewise constant functions associated with the partition $\mathscr{T}^S$:

$$\mathbb{V}^S := \left\{ v : \Lambda \to \mathbb{R} \ \middle|\ v|_C \text{ is constant } \forall C \in \mathscr{T}^S \right\}.$$

Note that $\mathbb{V}^S$ will be used as the approximation space for the liquid characteristic function $\boldsymbol{\Phi}^n$ and the predicted velocity $\mathbf{U}^{n+\frac{1}{2}}$. In particular, the approximation of the former does not necessarily take values in $\{0,1\}$ but in $\mathbb{R}$.

The second discretization considered is a typical conforming finite element subdivision made of triangles when $d = 2$ or tetrahedra when $d = 3$. The collection of these elements is denoted $\mathscr{T}^{FEM}$ and we denote by $H := \max_{T\in\mathscr{T}^{FEM}} \text{diameter}(T)$ the typical size of the elements. An example of such a discretization is shown in Figure 21.5.

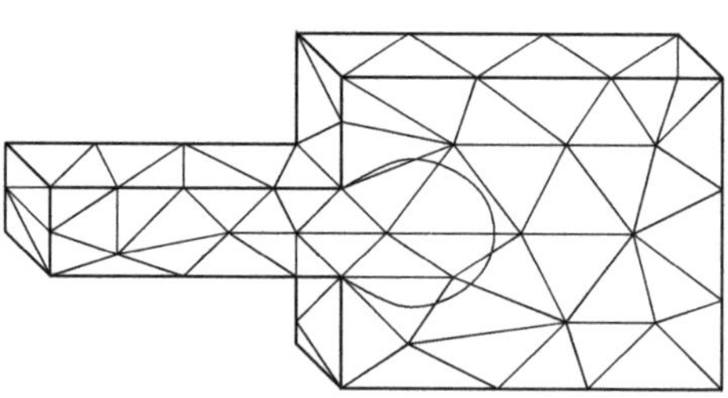

Fig. 21.5 Finite element subdivision used for the space discretization during the correction step.

For any subset $\tau \subset \mathscr{T}^{FEM}$, we denote by $\mathscr{V}(\tau)$ the collection of vertices in τ and by $\mathbb{V}^{FEM}(\tau)$ the space of globally continuous, piecewise polynomials of degree ≤ 1 associated with the subdivision τ:

$$\mathbb{V}^{FEM}(\tau) := \left\{ V : \bigcup_{T \in \tau} T \to \mathbb{R} \mid V \text{ continuous, } V|_T \text{ is a polynomial of degree } 1, \forall T \in \tau \right\}$$

and

$$\mathbb{V}_0^{FEM}(\tau) := \left\{ V \in \mathbb{V}^{FEM}(\tau) \mid V|_{\Gamma_{\mathcal{D}} \cap \partial T} = 0 \quad \forall T \in \tau \right\}.$$

In the sequel, the subset τ will represent the "liquid" elements, i.e., an approximate subdivision of $\Omega(t)$ at a given time t.

In order to fully exploit the potentialities of this two-grid method, we consider a structured grid $\mathcal{T}^S$ that is finer than the finite element mesh $\mathcal{T}^{FEM}$. This allows us to improve the accuracy on the approximation of the transport equations (21.18), without having a computationally prohibitive approximation of the diffusion problem (21.20). As it turns out, this allows choices of relatively large CFL, see Section 4.3. Typically the value of H is between $5h$ and $10h$, namely the structured grid is five to ten times finer than the finite element mesh. Further comments about the choice of the sizes of both discretizations can be found in [12, 26].

To alternate between the prediction and correction steps, we need projection operators to map functions in $\mathbb{V}^S$ into functions in $\mathbb{V}^{FEM}(\tau)$ and vice versa. We start with the projection $\pi_{S \to FEM} : \mathbb{V}^S \to \mathbb{V}^{FEM}(\tau)$ mapping the structured grid into the finite element mesh. Note that a function in $\mathbb{V}^{FEM}(\tau)$ is uniquely determined by its values on the vertex set $\mathcal{V}(\tau)$ and it is thus sufficient for the projection operator to set these values. Hence, for any $V \in \mathbb{V}^S$ and any $\mathbf{v} \in \mathcal{V}(\tau)$, we define

$$(\pi_{S \to FEM} V)(\mathbf{v}) := \frac{\sum_{T \in \tau \,:\, \mathbf{v} \in \mathcal{V}(T)} \sum_{C \in \mathcal{T}^S,\, C \subset T} \phi_{\mathbf{v}}(\text{center}(C)) V(\text{center}(C))}{\sum_{T \in \tau \,:\, \mathbf{v} \in \mathcal{V}(T)} \sum_{C \in \mathcal{T}^S,\, C \subset T} \phi_{\mathbf{v}}(\text{center}(C))}, \tag{21.21}$$

where center(C) denotes the (barycentric) center of the cell C and $\phi_{\mathbf{v}}$ denotes the Lagrange piecewise linear basis function associated with the vertex $\mathbf{v}$. The notation $C \subset T$ indicates that center(C) $\in T$. We denote identically the projection of a scalar-valued function or of a vector-valued function for which the projection is applied component-wise. In Figure 21.6, we have depicted a sketch in two dimensions of the set of cells of $\mathcal{T}^S$ appearing in the above summation for the calculation of the value at a vertex of $\mathcal{V}(\tau)$ in a typical structured subdivision.

The projection from the finite element subdivision to the structured mesh is defined for any $V \in \mathbb{V}^{FEM}(\tau)$ as follows. First, we extend the function V to the entire computational domain Λ by 0. Then, for each cell $C \in \mathcal{T}^S$, we denote by $T(C) \in \mathcal{T}^{FEM}$ the element in the finite element subdivision containing the center of the cell C and define

$$(\pi_{FEM \to S} V)|_C := \sum_{\mathbf{v} \in \mathcal{V}(T(C))} \phi_{\mathbf{v}}(\text{center}(C)) V(\mathbf{v}). \tag{21.22}$$

If the cell center is exactly at the boundary of several elements of the finite element mesh, then one arbitrary (but fixed) element is chosen among the possible elements. Again, we denote identically the projection of a scalar-valued function or of a vector-valued function (computed component-wise).

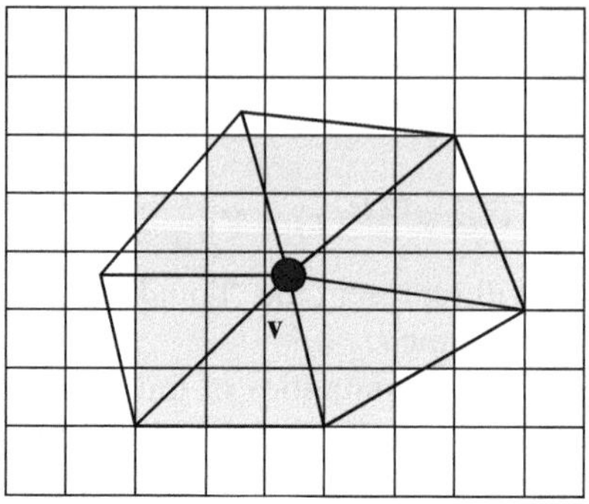

Fig. 21.6 The shaded cells correspond to all the cells appearing in the average projection (21.21) in order to determine the value of the function in $\mathbb{V}^{FEM}$ at the node $\mathbf{v}$. The Lagrange basis function $\phi_{\mathbf{v}}$ is the piecewise linear function (subordinate to the finite triangular subdivision) with value 1 at $\mathbf{v}$ and 0 at the other vertices.

Remark 2 (Implementation). The two projection operators (21.21) and (21.22) require a data structure mapping each cell of the structured mesh to an element of the finite element subdivision (the element containing the cell center). This array of indices is computed once and for all at the beginning of the simulation. However, when allowing for mesh adaptations an updated array is required after each mesh modification.

4.2.2 Prediction Step

The prediction steps start with given approximations $\Phi_M^n \in \mathbb{V}^S$ and $\mathbf{U}_M^n \in (\mathbb{V}^S)^d$ of the liquid fraction and velocity respectively, on the structured grid of cells (recall that M denotes the number of structured cells in the subdivision). As noted earlier, although $\phi(\mathbf{x},t) \in \{0,1\}$, its approximation takes values in $\mathbb{R}$. However, the resulting numerical diffusion is counter-balanced by the SLIC and decompression algorithms described below.

We define the approximation $\mathbf{Y}_M^{n+1} \in (\mathbb{V}^S)^d$ of the characteristic trajectories $\mathbf{Y}^{n+1}$ as follows. As $\mathbf{Y}_M^{n+1}$ is constant over each cell, it suffices to determine its values at the centers $\mathbf{x}_i$ of each cell C_i, $i = 1, \ldots, M$ and we set

$$\mathbf{Y}_M^{n+1}(\mathbf{x}_i) := \mathbf{x}_i + \delta t^{n+1} \mathbf{U}_M^n(\mathbf{x}_i). \tag{21.23}$$

The image via $\mathbf{Y}_M^{n+1}$ of each cell C_i is denoted $\widetilde{C}_i$, i.e., $\widetilde{C}_i := \mathbf{Y}_M^{n+1}(C_i)$, so that

$$\left\{ C \in \mathscr{T}^S \ : \ C_i \cap \widetilde{C} \neq \emptyset \right\}$$

corresponds to all the cells (at least partially) transported to the cell C_i. As a consequence, the approximation $\Phi_M^{n+\frac{1}{2}}(\mathbf{x}_i)$ of the liquid characteristic function $\Phi^{n+\frac{1}{2}}(\mathbf{x}_i)$

defined by (21.15) is obtained by adding the (weighted) contribution from cells transported to C_i, that is

$$\Phi_M^{n+\frac{1}{2}}(\mathbf{x}_i) := \sum_{C \in \mathcal{T}^S} \Phi_M^n(\text{center}(C))|C_i \cap \widetilde{C}|, \tag{21.24}$$

where $|C_i \cap \widetilde{C}|$ denotes the measure of $C_i \cap \widetilde{C}$. Due to the Cartesian properties of the structured grid $\mathcal{T}^S$, this measure is straightforward to compute as $C_i \cap \widetilde{C}$ are parallelepiped rectangles. Figure 21.7 illustrates the transport of one (two-dimensional) cell C into $\widetilde{C}$, which overlaps four other cells.

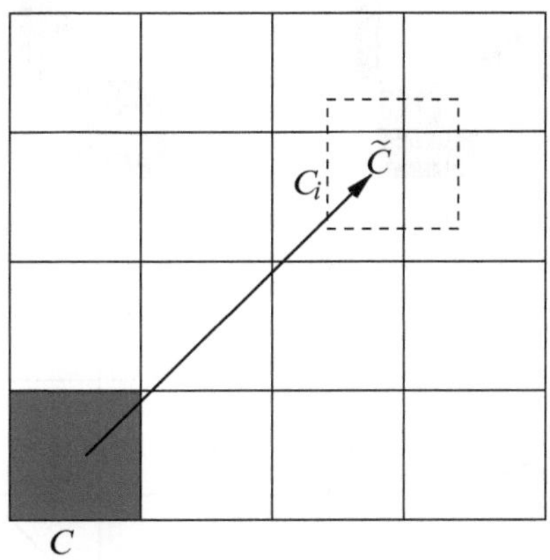

Fig. 21.7 Approximation of $\Phi^{n+\frac{1}{2}}$ using the method of characteristics. The cell C is transported to $\widetilde{C}$ and the quantity $\Phi_M^n(\mathbf{x}_j)$ is distributed among the intersecting cells. The contribution to $\Phi_M^{n+\frac{1}{2}}(\mathbf{x}_i)$ from $\Phi_M^n(\text{center}(C))$ is $\Phi_M^n(\text{center}(C))|C_i \cap \widetilde{C}|$ according to relation (21.24). The elements $C_i \cap \widetilde{C}$ are rectangles, making the computation of $|C_i \cap \widetilde{C}|$ straightforward.

We emphasize again that in view of relation (21.24) the liquid domain characteristic approximation $\Phi_M^{n+\frac{1}{2}}$ values are thus not necessarily 0 or 1 but could be any positive real number. In fact, this numerical diffusion (values strictly between 0 and 1) and numerical compression (values strictly larger than 1) are the two drawbacks of the projection formula (21.7), and are addressed now.

Numerical diffusion manifests itself when cells are partially filled, i.e., $0 < \Phi_M^{n+\frac{1}{2}}(\text{center}(C)) < 1$ for some $C \in \mathcal{T}^S$. Since the exact volume fraction of liquid ϕ is a step function and discontinuous at the free surface, numerical diffusion around the interfaces has to be controlled by the numerical scheme. It is reduced by the so-called SLIC algorithm [28], where the contribution to $\Phi_M^{n+\frac{1}{2}}(\text{center}(C_i))$ of partially filled cell C_i is still proportional to $|\widetilde{C} \cap C_i|$ but depends in addition on the values of the Φ_M^n on the neighboring cells of C. More precisely, before being transported along the characteristics, the quantities $\Phi_M^n(\text{center}(C))$ are concentrated near the boundary of the cell C instead of being spread out in the entire cell. This procedure is illustrated in Figure 21.8 (bottom), and allows to reduce the error due to the projection of the transported quantity in $\tilde{C}$ across several cells C_i. The way the quantity $\Phi_M^n(\text{center}(C_i))$ is pushed towards the boundary of the cell depends on the neighboring values of the volume fraction. Examples in two dimensions of space are illustrated in Figure 21.8 (top). We refer to [25, 26] for a more detailed description of the algorithm.

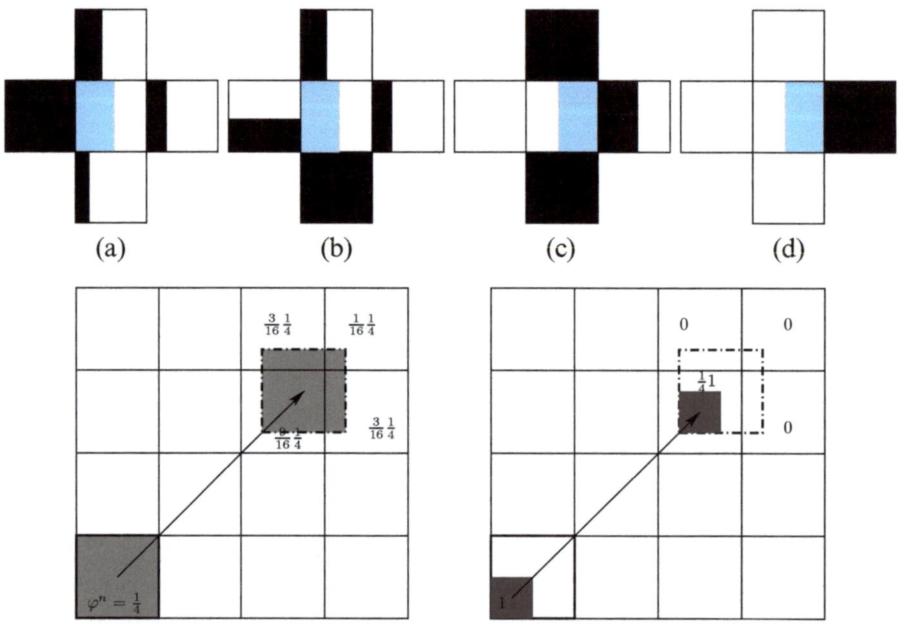

Fig. 21.8 (Top) Effect of the two dimensional SLIC algorithm on the cell center for four possible interfaces. The quantity $\Phi_M^n(\mathbf{x}_j)$, in blue, is pushed back to the sides of C depending on the values of Φ_M^n in the neighboring cells, in black. (Bottom) An example of two dimensional advection and projection when the volume fraction of liquid in the cell is $\Phi_M^n(\mathbf{x}_j) = \frac{1}{4}$. Left: without SLIC, the volume fraction of liquid is advected and projected on four cells, with contributions (from the top left cell to the bottom right cell) $\frac{3}{16}\frac{1}{4}$, $\frac{1}{16}\frac{1}{4}$, $\frac{9}{16}\frac{1}{4}$, $\frac{3}{16}\frac{1}{4}$. Right: with SLIC, the volume fraction of liquid is first pushed at one corner, then it is advected and projected on one cell only, with contribution $\frac{1}{4}$.

Let us discuss the compression case, i.e., $\Phi_M^{n+\frac{1}{2}}(C) > 1$ for some cell C. In that case, the excess $\Phi_M^{n+\frac{1}{2}}(C) - 1$ is stored in a buffer and redistributed into partially filled cells in order to *decompress* the field $\Phi_M^{n+\frac{1}{2}}$. Our algorithm redistributes the excess of liquid in a global way into cells that are in a neighborhood of the interface first (cells that are partially filled). Therefore it allows to conserve the mass in a global sense, in a way that is similar to *global repair* algorithms [35].

More precisely, we proceed in two steps: first, we compute the excess of liquid $\Phi_M^{n+\frac{1}{2}}(C) - 1$ in each cell C after advection and projection onto $\mathscr{T}^S$; second, we redistribute these amounts into partially filled cells, starting with cells that are nearly full. The detailed algorithm can be found in [25, 26] and is illustrated in Figure 21.9 when $\mathscr{T}^S$ is a single layer of cells. Although this figure represents a one-dimensional, over-simplified situation, it illustrates the rebalancing principle that allows to conserve the mass at each time step.

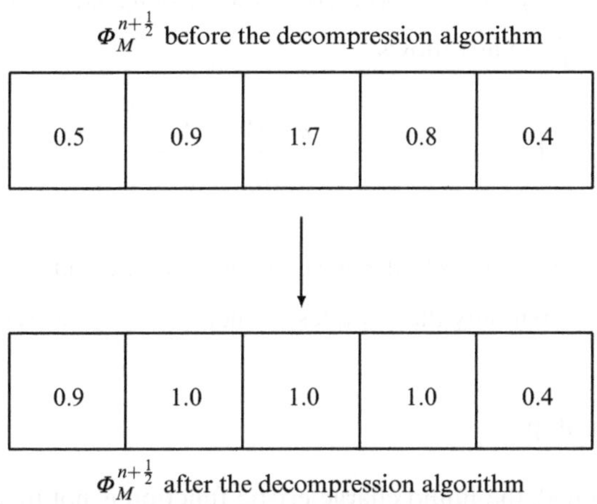

Fig. 21.9 Decompression algorithm. The volume fraction in excess in some cells is redistributed into the partially filled cells. Here the excess of 0.7 in the middle cell is redistributed in the partially filled cells, starting with the ones that are nearly full (0.9, 0.8 and 0.5 in order).

Similarly to (21.24), the velocity approximation $\mathbf{U}_M^{n+\frac{1}{2}}$ is given by the formula

$$\mathbf{U}_M^{n+\frac{1}{2}}(\mathbf{x}_i) := \sum_{C\in\mathscr{T}^S} \mathbf{U}_M^n(\text{center}(C))|C_i \cap \widetilde{C}|; \tag{21.25}$$

but the SLIC and decompression algorithm are not applied to the approximation of the velocity.

At the end of the prediction step, the projection onto the finite element space $\Phi_K^{n+\frac{1}{2}} \in \mathbb{V}^{FEM}$ of $\Phi_M^{n+\frac{1}{2}}$ is computed using the operators defined in Section 4.2.1

$$\Phi_K^{n+\frac{1}{2}} := \pi_{S\to FEM}\Phi_M^{n+\frac{1}{2}} \in \mathbb{V}^{FEM}. \tag{21.26}$$

The latter is used to define the liquid domain: An element $T \in \mathscr{T}^{FEM}$ is said to be liquid if $\max_{\mathbf{x}\in T} \Phi_K^{n+1/2}(\mathbf{x}) \geq 1/2$, the set of all liquid elements is then denoted by $\tau_K^{n+\frac{1}{2}}$, and the liquid domain $\Omega_K^{n+\frac{1}{2}}$ is the union of all liquid elements. The choice of the value $1/2$ for the threshold is arbitrary. It has been empirically discussed in

[26], but results have shown little sensitivity with respect to the value of this parameter. However, this definition of the liquid domain $\Omega_K^{n+\frac{1}{2}}$ implies an approximation error of the order $O(H)$ on the approximation of the free surface. Mesh refinement techniques have been designed to address this drawback [9], but are not developed further here.

The velocity is not directly projected onto the finite element space as the projection would depend on the values of the velocity outside $\Omega_K^{n+\frac{1}{2}}$ (which do not exist). Instead, we project $\Phi_M^{n+\frac{1}{2}}\mathbf{U}_M^{n+\frac{1}{2}}$ and recover the velocity $\mathbf{U}_K^{n+\frac{1}{2}} \in \mathbb{V}^{FEM}(\tau_K^{n+\frac{1}{2}})^d$ at each vertex $\mathbf{v} \in \mathscr{V}(\tau_K^{n+\frac{1}{2}})$ as follows:

$$\mathbf{U}_K^{n+\frac{1}{2}}(\mathbf{v}) := \frac{(\pi_{S\to FEM}(\Phi_M^{n+\frac{1}{2}}\mathbf{U}_M^{n+\frac{1}{2}}))(\mathbf{v})}{\Phi_K^{n+\frac{1}{2}}(\mathbf{v})} \tag{21.27}$$

if $\mathbf{v} \notin \Gamma_{\mathscr{D}}$ and $\mathbf{U}_K^{n+\frac{1}{2}}(\mathbf{v}) = \mathbf{g}_{\mathscr{D}}(\mathbf{v})$ otherwise. Notice that the above expression defines $\mathbf{U}_K^{n+\frac{1}{2}}$ only on $\Omega_K^{n+\frac{1}{2}}$ but only these values are needed in the correction step.

4.2.3 Correction Step

As already mentioned, the liquid characteristic function is not modified during this step; so we set

$$\Phi_K^{n+1}(\mathbf{v}) := \Phi_K^{n+\frac{1}{2}}(\mathbf{v})$$

for all $\mathbf{v} \in \mathscr{V}(\tau_K^{n+\frac{1}{2}})$, and

$$\Omega_K^{n+1} := \Omega_K^{n+\frac{1}{2}} \qquad \text{and} \qquad \tau_K^{n+1} := \tau_K^{n+\frac{1}{2}}.$$

Then, the Stokes system (21.20) on the *fixed* liquid domain Ω_K^{n+1} and with vanishing Dirichlet boundary condition $\mathbf{g}_{\mathscr{D}} \equiv 0$[1] reads as follows.

Seek $\mathbf{U}_K^{n+1} \in \mathbb{V}_0^{FEM}(\tau_K^{n+1})^d$ and $P_K^{n+1} \in \mathbb{V}^{FEM}(\tau_K^{n+1})$ satisfying

$$B^{n+1}((\mathbf{U}_K^{n+1},P_K^{n+1}),(\mathbf{V},R)) + S^{n+1}((\mathbf{U}_K^{n+1},P_K^{n+1}),(\mathbf{V},R)) = L^{n+1}(\mathbf{V}) \tag{21.28}$$

for any $(\mathbf{V},R) \in \mathbb{V}_0^{FEM}(\tau_K^{n+1})^d \times \mathbb{V}^{FEM}(\tau_K^{n+1})$. The bilinear functional $B^{n+1} : (\mathbb{V}_0^{FEM}(\tau_K^{n+1})^d \times \mathbb{V}^{FEM}(\tau_K^{n+1})) \times (\mathbb{V}_0^{FEM}(\tau_K^{n+1})^d \times \mathbb{V}^{FEM}(\tau_K^{n+1})) \to \mathbb{R}$ is defined as

$$\begin{aligned} B^{n+1}((\mathbf{U},P),(\mathbf{V},R)) :=& \frac{\rho}{\delta t^{n+1}} \int_{\Omega_K^{n+1}} \mathbf{U}\cdot\mathbf{V}\,d\mathbf{x} + 2\mu \int_{\Omega_K^{n+1}} \mathbf{D}(\mathbf{U}) :: \mathbf{D}(\mathbf{V})\,d\mathbf{x} \\ &- \int_{\Omega_K^{n+1}} R\,\nabla\cdot\mathbf{V}\,d\mathbf{x} + \int_{\Omega_K^{n+1}} P\,\nabla\cdot\mathbf{V}\,d\mathbf{x}, \end{aligned}$$

[1] The case of non-vanishing Dirichlet boundary conditions reads similarly upon defining a lifting of the boundary conditions.

where $A :: B := \sum_{i,j=1}^{d} A_{ij}B_{ij}$, for $A, B \in \mathbb{R}^{d\times d}$. The right-hand side $L^{n+1} : \mathbb{V}_0^{FEM}(\tau_K^{n+1})^d \to \mathbb{R}$ is given by

$$L^{n+1}(\mathbf{V}) := \frac{\rho}{\delta t^{n+1}} \int_{\Omega_K^{n+1}} \mathbf{U}^n \cdot \mathbf{V}\, d\mathbf{x} + \int_{\Omega_K^{n+1}} \mathbf{f}(t^{n+1}) \cdot \mathbf{V}\, d\mathbf{x}.$$

The functionals S^{n+1} in (21.28) are the Galerkin Least-Square stabilization terms to cope with the fact that the choice of the finite element spaces is not inf-sup stable. They are given by:

$$S^{n+1}((\mathbf{U},P),(\mathbf{V},R)) := \sum_{T\subset\Omega_K^{n+1}} \alpha_T \int_T \left(\rho \frac{\mathbf{U}-\mathbf{U}^n}{\delta t^{n+1}} + \nabla P - \mathbf{f}(t^{n+1})\right) \cdot \nabla R\, d\mathbf{x},$$

where α_T is the stabilization coefficient locally defined as:

$$\alpha_T := \begin{cases} C_{SUPG}\dfrac{\text{diam}(T)^2}{12\mu} & \text{if } 0 \le Re_T \le 3 \\[2ex] C_{SUPG}\dfrac{\text{diam}(T)^2}{4Re_T\mu} & \text{if } 3 \le Re_T \end{cases}$$

where the local Reynolds number is defined by $Re_T := \frac{\rho\,\text{diam}(T)\max_{\mathbf{x}\in T}|\mathbf{U}^n|}{2\mu}$ and C_{SUPG} is an dimensionless constant typically set to 1.0.

At the end of the correction step, the velocity is projected onto the structured grid

$$\mathbf{U}_M^{n+1} := \pi_{FEM\to S}\mathbf{U}_K^{n+1} \in (\mathbb{V}^S)^d,$$

while the volume fraction of liquid remains unchanged

$$\Phi_M^{n+1} := \Phi_M^{n+\frac{1}{2}} \in \mathbb{V}^S.$$

4.3 Numerical Results for Newtonian Flows

The example that serves as a guideline in this work is the experiment of the die swell with contraction in an extrusion process. This benchmark not only illustrates the advantages of the splitting approach presented in this work, but it is also worth noting that numerical simulation of extrusion is of great importance in industrial processes, for instance for pasta dough in food engineering [22].

We consider an axisymmetric capillary die with a contraction at the entrance. The fluid is injected into the die and then expands at the exit. The behavior of the fluid depends strongly on the fluid rheology. In this section, we consider Newtonian fluids and refer to Sections 5 and 6 for more complex fluids.

4.3.1 Extrusion with Initial Contraction: Computational Domain

We describe the computational domain used for the subsequent extrusion experiments with initial contraction. The computational domain Λ is depicted in Figures 21.10 and 21.11, together with the finite element mesh used for (most of) the simulations presented in this work. It consists of three cylinder as depicted in Figure 21.10. The first cylinder of diameter 0.010 m and length 0.005 m is where the liquid is injected. Then, the liquid enters the die, a second cylinder of diameter 0.001 m (contraction) and length 0.010 m. When, exiting the die, the liquid enters the third cylinder of diameter 0.010 m and length 0.015 m. The total length of the domain is therefore 0.030 m. The size of the finite elements in the die is characterized by $H = 0.0001$ m. The structured grid consists of a subdivision made of cubic cells of length 0.000025 m.

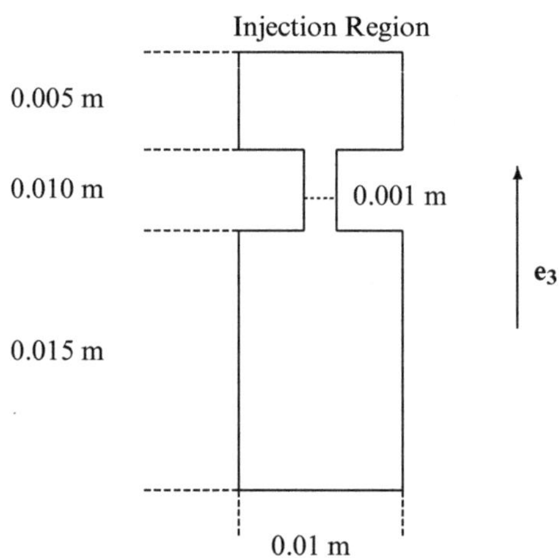

Fig. 21.10 Extrusion with Initial Contraction: Dimensions of the computational domain.

4.3.2 Slip Boundary Conditions

We consider a Newtonian fluid with density $\rho = 1300$ kg m^{-3}, and viscosity $\mu = 10$ kg(ms)$^{-1}$. The fluid is injected with a constant speed of 0.00023 ms^{-1}, such that the speed in the die is approximately 0.05 ms^{-1}. No-slip boundary conditions are imposed at the bottom of the domain and slip boundary conditions are imposed on the other parts of $\partial\Lambda$. We postpone to Section 4.3.3 for a discussion on the effect of different types of boundary conditions. Gravity forces, $\mathbf{g} = -9.81\mathbf{e_3}$ ms^{-2} are oriented along the die (see Figure 21.10). The time step is constant and equal to $\delta t = 0.005$ s, which implies a CFL number of about 10 during the prediction step and of about 2.5 during the correction step. At time $t = 0.9$ s approximately, the

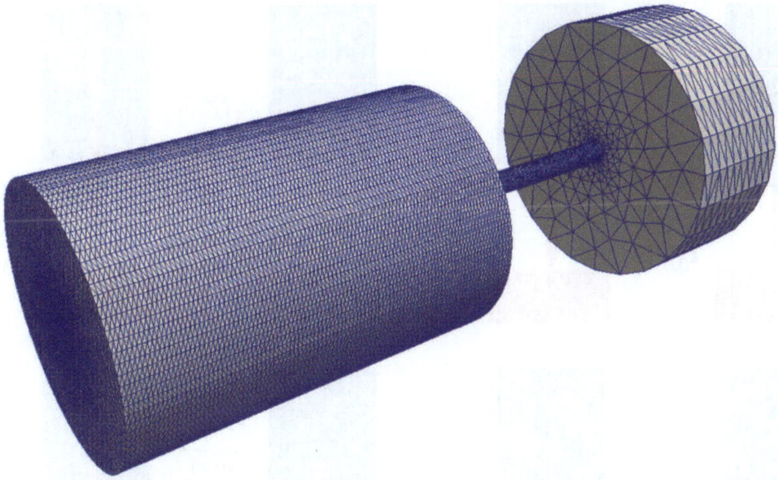

Fig. 21.11 Extrusion with Initial Contraction: Computational domain and finite element mesh.

jet hits the bottom of the computational domain and the fluid buckles. Figure 21.12 visualizes, in a medium plane inside the tube, snapshots of the volume fraction of liquid Φ and of the corresponding velocity field $\mathbf{U}$. Figure 21.13 visualizes the buckling of the jet of Newtonian fluid once it hits the bottom of the computational domain.

In this case, we observe that the operator splitting scheme does not introduce any additional error as long as the flow is laminar and does not touch the bottom of the domain. This allows to consider large time steps if needed, without any CFL condition. Little oscillations in the jet are observed due to the spatial discretization and the unstructured finite element mesh. The buckling effect when the flow touches the boundary requires smaller time steps to retain accuracy.

4.3.3 No-Slip Boundary Conditions

When enforcing slip boundary conditions on the lateral side as in the previous test case, the liquid has a constant velocity (until it hits the bottom) so that the operator splitting produces the exact solution for any value of the time step; see Figure 21.12. For this simulation, we impose no-slip boundary condition on $\partial\Lambda$ except at the inflow where we keep the constant velocity profile of magnitude 0.00023 ms^{-1}. A Poiseuille profile is observed for the velocity in the die with a slight perturbation due to the contraction. Figure 21.14 shows the results obtained with the setup described in Sections 4.3.1 and 4.3.2 but with no-slip boundary conditions in the cavity before the die and in the die; compare with Figure 21.12.

The effect of boundary conditions is amplified for liquids with larger viscosities. Figure 21.15 provides a similar simulation when the viscosity is 10 times larger ($\mu = 100$ kg(ms)^{-1}). It demonstrates the effect of no-slip boundary conditions on the shape of the liquid front and on the free surface front velocity, which decreases as the viscosity increases.

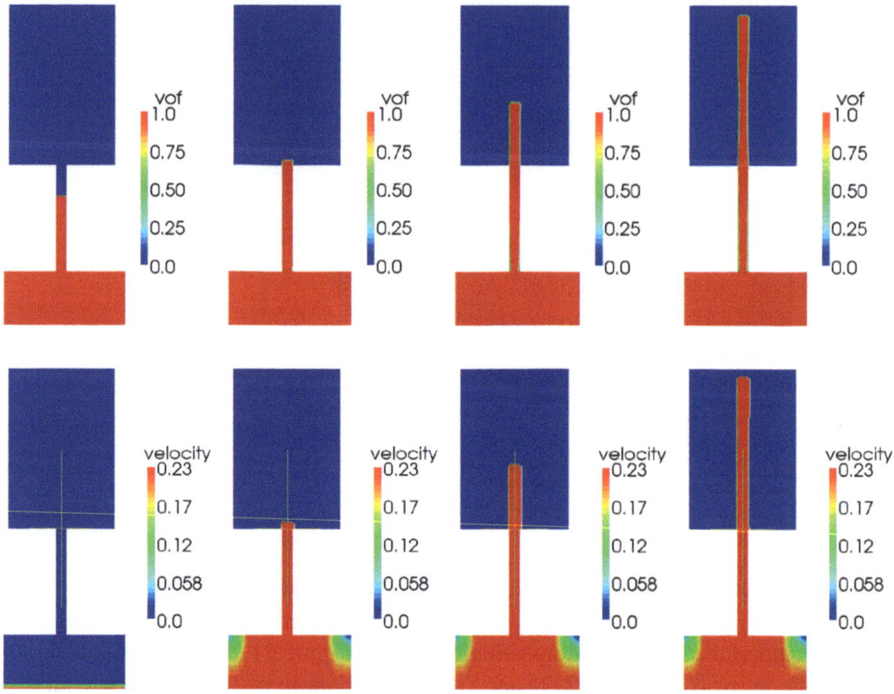

Fig. 21.12 Die swell with extrusion of a Newtonian fluid. Snapshots of the solution at times $t = 0$, 0.3, 0.6, and 0.9 s on a plane located in the middle of the tubes. Top: representation of volume fraction of liquid Φ; bottom: speed $|\mathbf{U}|$.

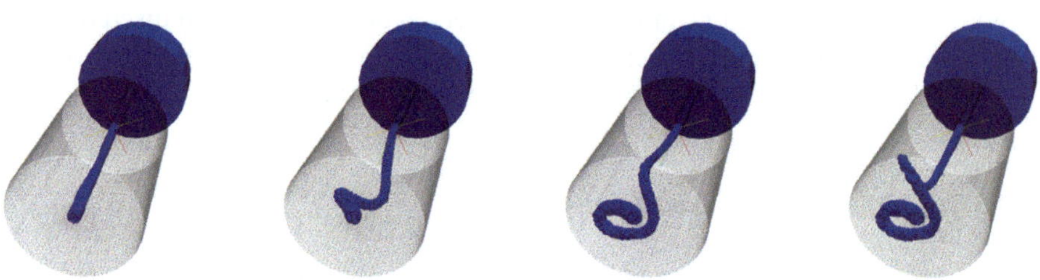

Fig. 21.13 Die swell with extrusion of a Newtonian fluid. Snapshots of the buckling of the jet at times $t = 1.0, 1.2, 1.4$, and 1.6 s (left to right).

Figures 21.16 and 21.17 show snapshots of the buckling effects for $\mu = 10$ kg(ms)$^{-1}$ and $\mu = 100$ kg(ms)$^{-1}$ respectively, and no-slip boundary conditions. The boundary conditions change drastically the shape of the liquid during the buckling. In addition, larger viscosities slow the liquid front propagation and reduce the buckling effect as it was already noted in the simple cavity setting [6, 36].

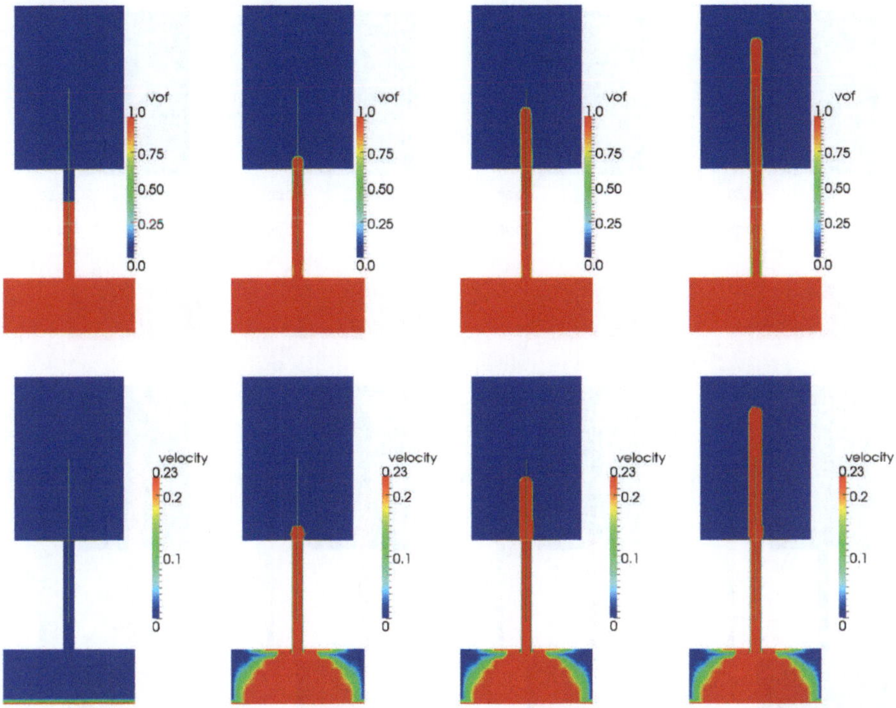

Fig. 21.14 Die swell with extrusion of a Newtonian fluid, with no-slip boundary conditions in the die ($\mu = 10$ kg(ms)$^{-1}$). Snapshots of the solution at times $t = 0, 0.3, 0.6$, and 0.9 s on a plane located in the middle of the tubes. Top: representation of volume fraction of liquid Φ; bottom: speed $|\mathbf{U}|$.

5 Visco-Elastic Flows with Free Surfaces

We now discuss an extension to liquids with more complex rheology and in particular the modification of the Navier-Stokes system (21.1) to account for visco-elastic effects. The upper-convected Maxwell model is chosen to describe the complex rheology but the algorithm presented here is not restricted to specific models.

5.1 Mathematical Modeling of Visco-Elastic Flows with Free Surfaces

Visco-elastic fluids are characterized by the presence of an extra-stress tensor denoted by $\sigma \in \mathbb{R}^{\frac{d(d+1)}{2}} \simeq \mathbb{R}^{d\times d}_{sym}$, the space of $d \times d$ symmetric tensors, supplementing the Cauchy stress tensor $2\mu\mathbf{D}(\mathbf{u}) - p\mathbf{I}$ in (21.1). Hence, the velocity $\mathbf{u}$, pressure p and visco-elastic stress σ satisfy in Q:

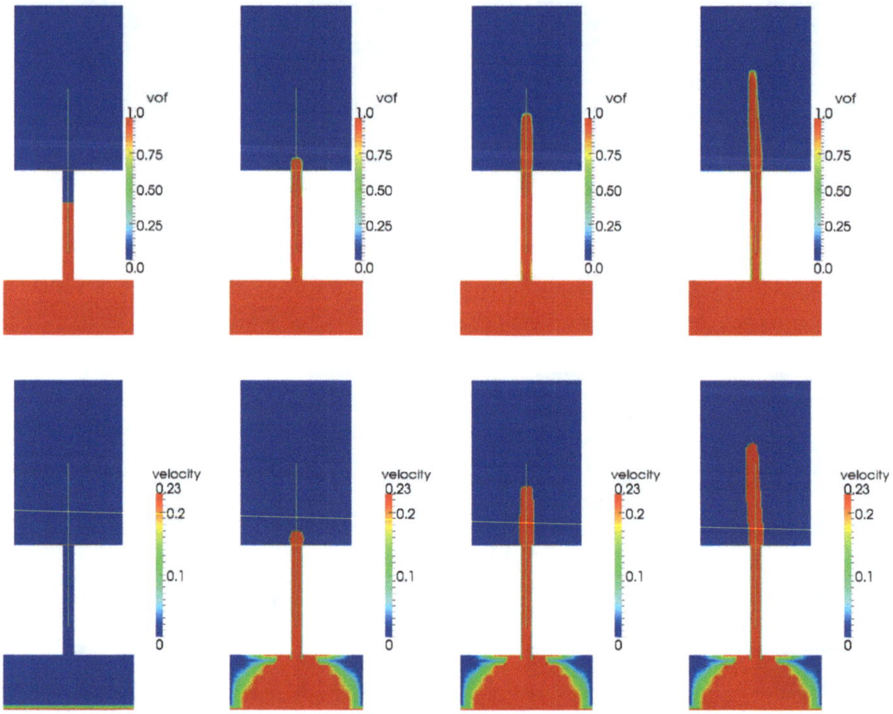

Fig. 21.15 Die swell with extrusion of a Newtonian fluid, with no-slip boundary conditions in the die ($\mu = 100$ kg(ms)$^{-1}$). Snapshots of the solution at times $t = 0, 0.3, 0.6$, and 0.9 s on a plane located in the middle of the tubes. Top: representation of volume fraction of liquid Φ; bottom: speed $|\mathbf{U}|$. Compare with Figure 21.14 representing the same setting but with a fluid of smaller viscosity.

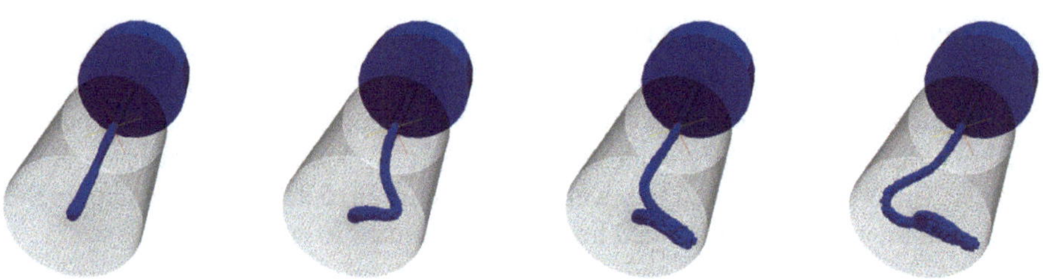

Fig. 21.16 Die swell with extrusion of a Newtonian fluid, with no-slip boundary conditions in the die ($\mu = 10$ kg(ms)$^{-1}$). Snapshots of the buckling of the jet at times $t = 1.0, 1.2, 1.4$, and 1.6 s (left to right).

$$\begin{cases} \rho\left(\dfrac{\partial}{\partial t}\mathbf{u} + (\mathbf{u}\cdot\nabla)\mathbf{u}\right) - \nabla\cdot(2\mu\mathbf{D}(\mathbf{u})) + \nabla p - \nabla\cdot\sigma = \mathbf{f}, \\ \nabla\cdot\mathbf{u} = 0. \end{cases} \tag{21.29}$$

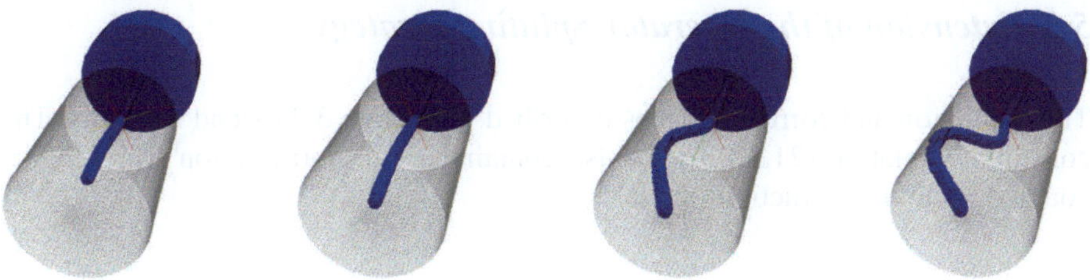

Fig. 21.17 Die swell with extrusion of a Newtonian fluid, with no-slip boundary conditions in the die ($\mu = 100$ kg(ms)$^{-1}$). Snapshots of the buckling of the jet at times $t = 1.0, 1.2, 1.4$, and 1.6 s (left to right).

The Dirichlet condition (21.2) on the velocity field remains unchanged but the Neumann (21.3) as well as the interface (21.4) conditions are modified to account for the presence of the visco-elastic stress

$$(2\mu\mathbf{D}(\mathbf{u}) - p\mathbf{I} + \sigma)\mathbf{n} = \mathbf{g}_{\mathcal{N}} \quad \text{on} \quad \partial Q_{\mathcal{N}}, \tag{21.30}$$

$$(2\mu\mathbf{D}(\mathbf{u}) - p\mathbf{I} + \sigma)\mathbf{n} = \mathbf{0} \quad \text{on} \quad \mathcal{F}. \tag{21.31}$$

As model problem, we consider the upper-convected Maxwell model to provide the constitutive relation for σ, namely the extra-stress σ satisfies:

$$\sigma + \lambda\left(\frac{\partial}{\partial t}\sigma + (\mathbf{u}\cdot\nabla)\sigma - \nabla\mathbf{u}\,\sigma - \sigma\nabla\mathbf{u}^t\right) = 2\mu_p\mathbf{D}(\mathbf{u}) \qquad \text{in} \quad Q, \tag{21.32}$$

where λ is the fluid relaxation time, μ_p is the so-called polymer viscosity [4, 5, 30]. The problem is thus coupled via the introduction of the extra-stress σ in the Navier-Stokes equations, and reciprocally, the velocity $\mathbf{u}$ in the constitutive equation for σ. The values of the stress tensor are set to a given tensor $\mathbf{G} : \partial Q_{inflow} \to \mathbb{R}^{\frac{d(d+1)}{2}}$ on the inflow boundary of the domain:

$$\sigma = \mathbf{G}, \qquad \text{on} \quad \partial Q_{inflow}.$$

Similar to the initial conditions (21.6) for the velocity field, the initial viscoelastic stress is set to be

$$\sigma(0) = \sigma_0 \qquad \text{on} \quad \Omega(0)$$

for a given $\sigma_0 : \Omega(0) \to \mathbb{R}^{\frac{d(d+1)}{2}}$.

5.2 Extension of the Operator Splitting Strategy

The prediction and correction steps described in Section 3.2 extend naturally. The constitutive relation (21.32) for σ also contains a transport relation which is accounted for in the prediction step.

5.2.1 The Prediction Step

The prediction step (21.13) becomes: find the characteristic function ϕ, the velocity field $\mathbf{u}$, and the extra-stress σ satisfying

$$\begin{aligned} &\frac{\partial}{\partial t}\phi + \mathbf{u}\cdot\nabla\phi = 0 \\ &\frac{\partial}{\partial t}\mathbf{u} + (\mathbf{u}\cdot\nabla)\mathbf{u} = 0 \qquad \text{in} \quad Q^{n+1} := Q \cap \left(\Lambda \times I^{n+1}\right). \\ &\frac{\partial}{\partial t}\sigma + (\mathbf{u}\cdot\nabla)\sigma = 0 \end{aligned} \tag{21.33}$$

It ends upon setting

$$\sigma^{n+\frac{1}{2}} := \sigma(t^{n+1}) \qquad \text{in} \quad \Omega^{n+\frac{1}{2}},$$

in addition to the values for $\phi^{n+\frac{1}{2}}$ and $\mathbf{u}^{n+\frac{1}{2}}$. As for the velocity, the method of characteristics transports each component of the symmetric tensor (namely six fields when $d = 3$ and three fields when $d = 2$). Their values are obtained from the characteristics lines as in (21.15).

After space discretization, and using the notations introduced in Section 4, the prediction $\Sigma_M^{n+\frac{1}{2}} \in (\mathbb{V}^S)^{\frac{d(d+1)}{2}}$ of $\sigma(t^{n+1})$ is given at each cell center $\mathbf{x}_i$ by

$$\Sigma_M^{n+\frac{1}{2}}(\mathbf{x}_i) := \sum_{C \in \mathscr{T}^S} \Sigma_M^n(\text{center}(C))|C_i \cap \widetilde{C}|;$$

compare with (21.24).

At the end of the prediction step, the projection of the tensor $\Sigma_M^{n+\frac{1}{2}}$ into the finite element space into the finite element space $\mathbb{V}^{FEM}(\tau_K^{n+1})^{\frac{d(d+1)}{2}}$ is computed according to a formula similar to (21.27): i.e., for every $\mathbf{v} \in \mathscr{V}(\tau_K^{n+\frac{1}{2}})$

$$\Sigma_K^{n+\frac{1}{2}}(\mathbf{v}) := \frac{(\pi_{S\to FEM}(\Phi_M^{n+\frac{1}{2}}\Sigma_M^{n+\frac{1}{2}}))(\mathbf{v})}{\Phi_K^{n+\frac{1}{2}}(\mathbf{v})} \tag{21.34}$$

if $\mathbf{v}$ is not a vertex at the inflow boundary, and $\Sigma_K^{n+\frac{1}{2}}(\mathbf{v}) = \mathbf{G}(\mathbf{v})$ otherwise.

5.2.2 The Correction Step

After incorporation of the extra-stress related terms, the correction step reads as follows:

$$\begin{cases} \rho \frac{\partial}{\partial t}\mathbf{u} - \nabla\cdot(2\mu\mathbf{D}(\mathbf{u})) + \nabla p - \nabla\cdot\sigma = \mathbf{f} \\ \nabla\cdot\mathbf{u} = 0 \\ \sigma + \lambda\left(\frac{\partial}{\partial t}\sigma - \nabla\mathbf{u}\,\sigma - \sigma\nabla\mathbf{u}^t\right) = 2\mu_p\mathbf{D}(\mathbf{u}) \end{cases} \quad \text{in} \quad \Omega^{n+1}\times I^{n+1}, \tag{21.35}$$

supplemented by the appropriate boundary conditions and free interface conditions.

As in Section 4, the volume fraction and liquid domain remain unchanged during the correction step. Problem (21.35) allows to obtain a correction of the velocity $\mathbf{u}$, the extra-stress tensor σ, and the pressure p. This correction step consists in two sub-steps decoupling the velocity-pressure corrections and the extra-stress correction. The first sub-step consists of solving a Stokes problem of the (21.28) type with a modified functional $L^{n+1}(.)$ accounting for the extra-stress tensor term:

$$L^{n+1}(\mathbf{V}) := \frac{\rho}{\delta t^{n+1}}\int_{\Omega_K^{n+1}} \mathbf{U}_K^n\cdot\mathbf{V}\,d\mathbf{x} + \int_{\Omega_K^{n+1}} \mathbf{f}(t^{n+1})\cdot\mathbf{V}\,d\mathbf{x} - \int_{\Omega_K^{n+1}} \Sigma_K^{n+\frac{1}{2}} :: \mathbf{D}(\mathbf{V})\,d\mathbf{x}.$$

This corresponds to an explicit treatment of the visco-elastic effect $\Sigma_K^{n+\frac{1}{2}}$ in the first equation in (21.35). We then solve the third relation of (21.35) to update the extra-stress tensor $\Sigma_K^{n+\frac{1}{2}}$. The time discretization considered consists of an explicit treatment of the nonlinear terms, while continuous piecewise linear finite elements are used for the space discretization: Seek $\Sigma_K^{n+1} \in (\mathbb{V}^{FEM}(\tau_K^{n+1}))^{\frac{d(d+1)}{2}}$, the subspace of $(\mathbb{V}^{FEM}(\tau_K^{n+1}))^{d\times d}$ consisting in those symmetric matrices, satisfying

$$\begin{aligned} &\int_{\Omega_K^{n+1}} (\delta t^{n+1}\Sigma_K^{n+1} + \lambda\Sigma_K^{n+1}) :: \Theta\,d\mathbf{x} \\ &\quad = \int_{\Omega_K^{n+1}} \Big(\Sigma_K^{n+\frac{1}{2}} + \delta t^{n+1}\Sigma_K^{n+\frac{1}{2}} \\ &\qquad + \delta t^{n+1}\nabla\mathbf{U}_K^{n+1}\,\Sigma_K^{n+\frac{1}{2}} + \delta t^{n+1}\Sigma_K^{n+\frac{1}{2}}(\nabla\mathbf{U}_K^{n+1})^t\Big) :: \Theta\,d\mathbf{x} \\ &\qquad + 2\mu_p\delta t^{n+1}\int_{\Omega_K^{n+1}} \mathbf{D}(\mathbf{U}_K^{n+1}) :: \Theta\,d\mathbf{x}, \qquad \forall\Theta \in \mathbb{V}_M^{FEM}(\tau_K^{n+1})^{\frac{d(d+1)}{2}}. \end{aligned}$$

In addition, the Elastic Viscous Stress Splitting (EVSS) stabilization procedure could be activated to compensate for possible small viscosities μ. Details can be found in [6].

5.3 Numerical Results for Visco-Elastic Flows

We first consider again the extrusion with initial contraction experiment presented in Section 4.3.1. The goal of this section is to discuss the visco-elastic influence, via the presence of the extra-stress σ_{33}.

5.3.1 Extrusion with Die Swell and Contraction

Let us consider a visco-elastic fluid, which has density $\rho = 1300$ kg m^{-3}, and viscosity $\mu = 0$. Its relaxation time is $\lambda = 0.1$ s and the polymer viscosity is $\mu_p = 10$ kg(ms)$^{-1}$. The boundary condition at the inflow boundary is a Poiseuille flow with velocity given by $\mathbf{u} = (0, 0, u_z)$ and $u_z(r) = -100(r^2 - 0.01^2)$ ms^{-1} (where r is the radial distance to the central axis of the die). Slip boundary conditions are imposed on the lateral sides of $\partial\Lambda$ and no-slip boundary conditions are applied on the bottom plate. Gravity forces, $\mathbf{g} = -9.81\mathbf{e_3}$ ms^{-2} are oriented along the die. The time step is constant and equal to $\delta t = 0.005$ s.

Figure 21.18 provides the volume fraction Φ, extra-stress σ_{33}, and speed $|\mathbf{U}|$ fields in a median cut in the middle of the domain at various times. Figure 21.19 visualizes the buckling effect. Since slip boundary conditions are applied along the die, no die swell occurs after exiting the die. However, when the jet hits the wall we observe a different buckling behavior compared to the Newtonian case; compare Figures 21.13 and 21.19.

5.3.2 Influence of the Polymer Viscosity and Relaxation Time

The influence of the polymer viscosity μ_p and relaxation time λ is now investigated, keeping slip boundary conditions along the die. Figures 21.20 and 21.21 represent the volume fraction Φ, extra-stress σ_{33}, and speed $|\mathbf{U}|$ fields in a median cut in the middle of the domain at various times, as well as the buckling effect of the liquid domain, for a relaxation time $\lambda = 1$ s. and a polymer viscosity $\mu_p = 10$ kg(ms)$^{-1}$ (larger relaxation time compared to the simulations in Section 5.3.1). Figures 21.20 and 21.21 illustrate the same quantities, for a relaxation time $\lambda = 0.1$ s. and a polymer viscosity $\mu_p = 100$ kg(ms)$^{-1}$ (larger viscosity compared to the simulations in Section 5.3.1). Clearly, both polymer viscosity μ_p and the relaxation time λ have a significant influence on the jet shape during buckling.

5.3.3 Influence of Boundary Conditions

In this section, we discuss the influence of the boundary conditions (typically slip vs. no-slip boundary conditions) on the die boundary, on the buckling phenomena, and on the visco-elastic fluid behavior. The polymer viscosity and relaxation time are kept as in Figures 21.20–21.23, $\mu_p = 100$ kg(ms)$^{-1}$, $\lambda = 0.1$ s, whereas no-slip

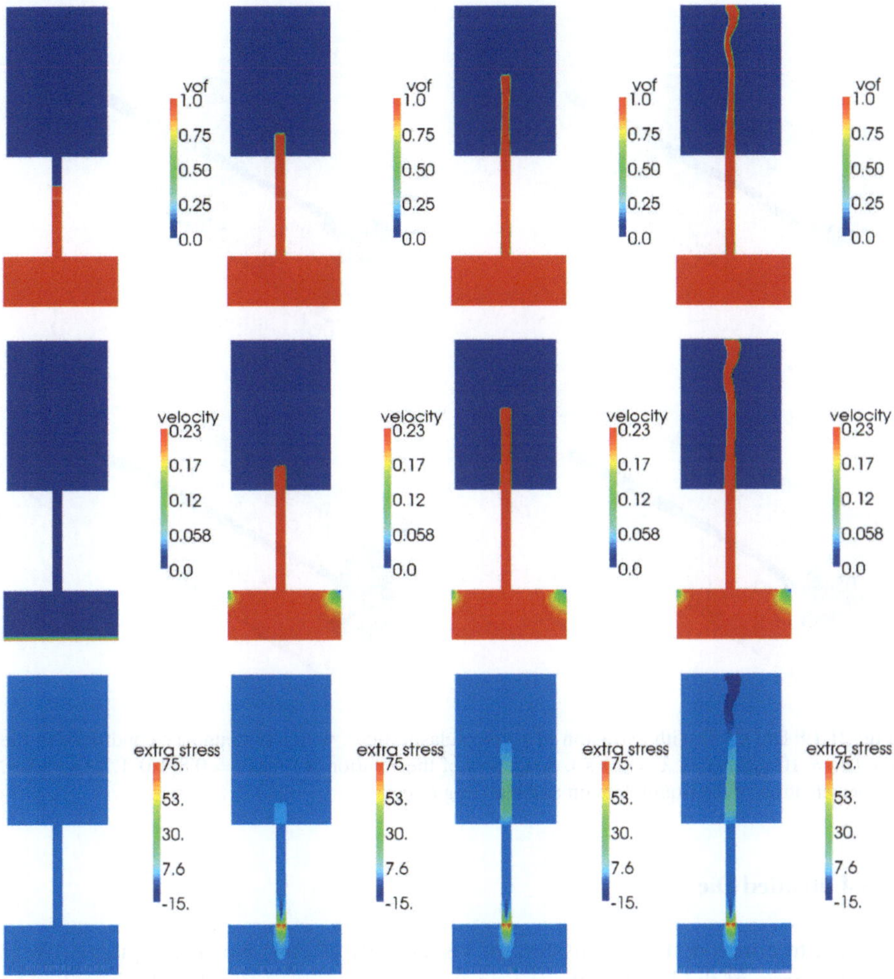

Fig. 21.18 Die swell with extrusion of a visco-elastic fluid, with slip boundary conditions in the die ($\mu_p = 10$ kg(ms)$^{-1}$, $\lambda = 0.1$ s.). Snapshots of the solution at times $t = 0, 0.4, 0.6$, and 0.8 s on a plane located in the middle of the tubes. Top: representation of volume fraction of liquid Φ; middle: speed $|\mathbf{U}|$; bottom: representation of extra-stress σ_{33}.

boundary conditions now apply along the die. The shape of the jet is significantly different as shown in Figures 21.24 and 21.25. The die swell is significant, therefore we decrease the value of the relaxation time to $\lambda = 0.005$ s, still keeping the same polymer viscosity $\mu_p = 100$ kg(ms)$^{-1}$. The swelling of the die is now much smaller as shown in Figures 21.26 and 21.27. Figures 21.28 and 21.29 illustrate the same quantities, for a smaller relaxation time, $\lambda = 0.002$ s, still keeping the same polymer viscosity $\mu_p = 100$ kg(ms)$^{-1}$. We therefore conclude that the type of boundary conditions applied along the die has a significant impact on the extrusion process.

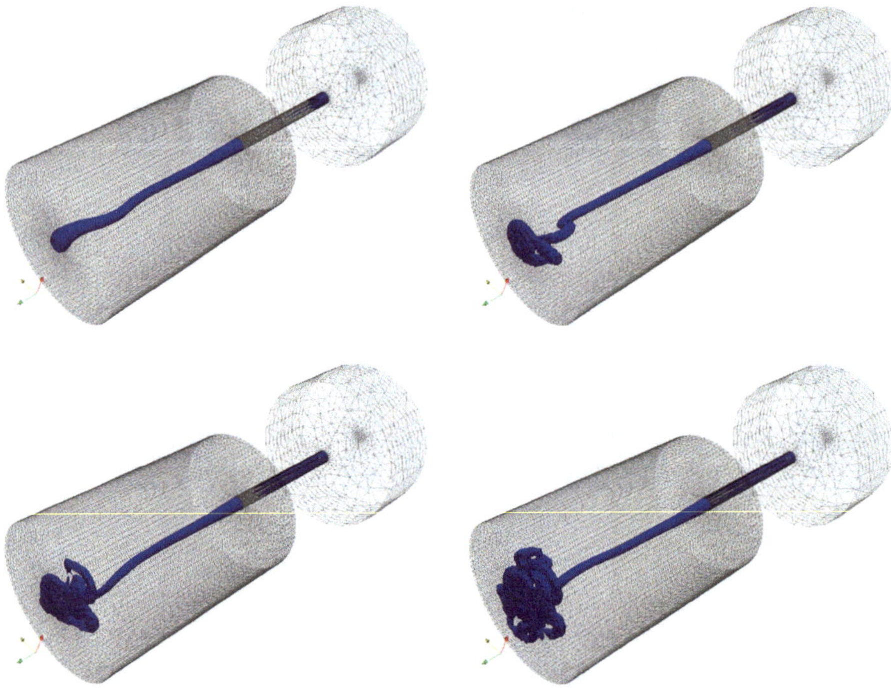

Fig. 21.19 Die swell with extrusion of a visco-elastic fluid, with slip boundary conditions in the die ($\mu_p = 10$ kg(ms)$^{-1}$, $\lambda = 0.1$ s.). Snapshots of the solution at times $t = 0.8, 1.0, 1.2,$ and 1.6 s. Representation of the liquid domain and buckling effect.

5.3.4 Bended Die

Finally, to conclude the discussion on viscoelastic effects, we study quantitatively the influence of the bend of the die on the extrusion. In particular, the distribution of the extra-stress and the differences of amplitude are fundamental in industrial processes, as they induce a different behavior of the visco-elastic material at the exit of the die, and thus a different final production.

We consider a geometry that is similar to the one before, with the addition of a ninety degree angle bend in the die. All the other geometrical dimensions remain the same. Figure 21.30 illustrates the geometry together with the associated finite element mesh.

As before, the visco-elastic fluid has density $\rho = 1300$ kg m^{-3}, and viscosity $\mu = 0$. Its relaxation time is $\lambda = 0.1$ s and the polymer viscosity is $\mu_p = 10$ kg(ms)$^{-1}$. The boundary condition at the inflow boundary is a Poiseuille flow, slip boundary conditions are imposed along the die, as well as at the exit of the domain. Gravity forces with amplitude $|\mathbf{g}| = 9.81$ ms^{-2} are oriented along the inflow. The time step is constant and equal to $\delta t = 0.005$ s.

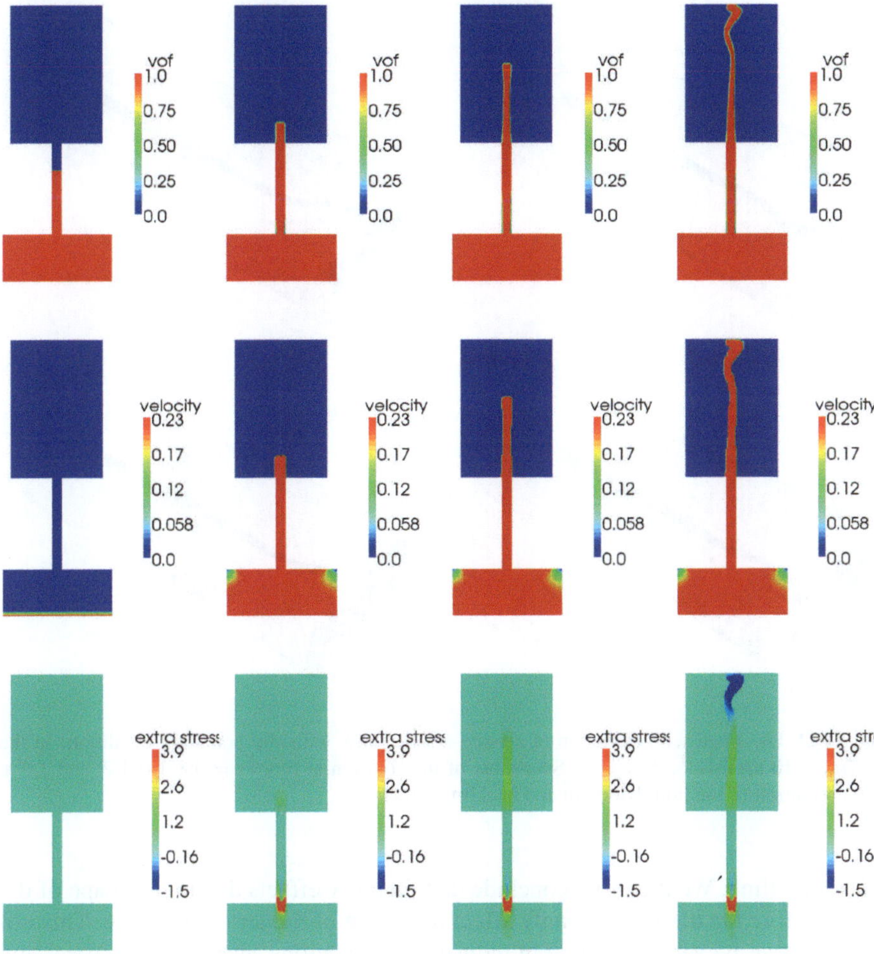

Fig. 21.20 Die swell with extrusion of a visco-elastic fluid, with slip boundary conditions in the die ($\mu_p = 10$ kg(ms)$^{-1}$, $\lambda = 1$ s.). Snapshots of the solution at times $t = 0, 0.4, 0.6$, and 0.8 s on a plane located in the middle of the tubes. Top: representation of volume fraction of liquid Φ; middle: speed $|\mathbf{U}|$; bottom: representation of extra-stress σ_{33}.

Figure 21.31 illustrates representations of the volume fraction Φ, speed $|\mathbf{U}|$, and extra-stress σ_{33} fields in a median cut in the middle of the domain at various instants of time. Figure 21.32 illustrates snapshots of the liquid domain. These results should be compared with those of Figures 21.18 and 21.19 which correspond to a straight die. One can observe a significant buckling effect as the liquid is switching directions. This behavior is caused by the variation of the extra stress inside the curved die. Figure 21.33 illustrates snapshots of the liquid domain in the same situation but for a relaxation time $\lambda = 0.02$ s. One can observe larger effects due to the shorter

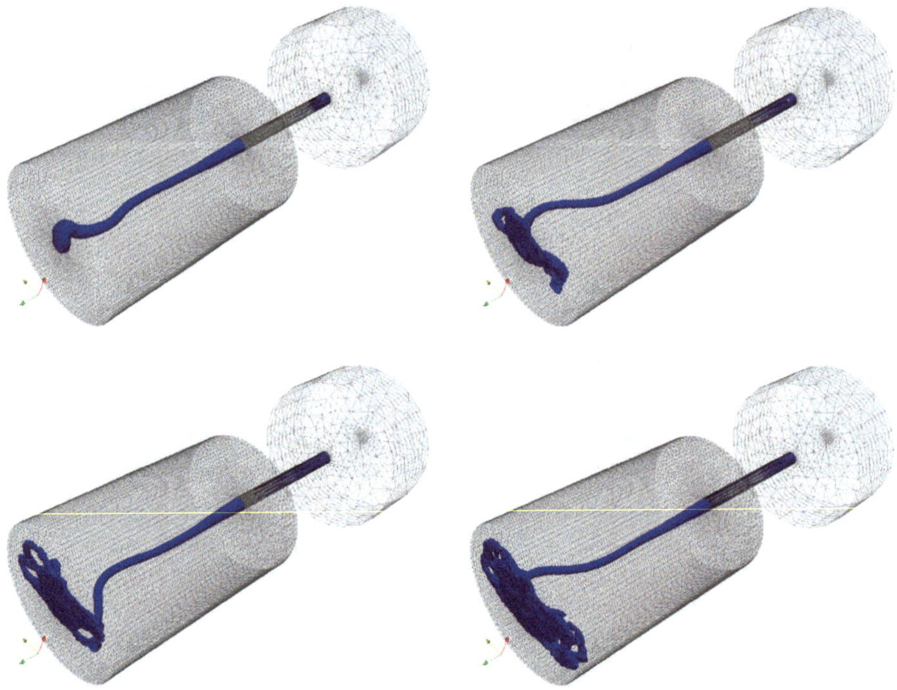

Fig. 21.21 Die swell with extrusion of a visco-elastic fluid, with slip boundary conditions in the die ($\mu_p = 10$ kg(ms)$^{-1}$, $\lambda = 1$ s.). Snapshots of the solution at times $t = 0.8, 1.0, 1.2$, and 1.6 s. Representation of the liquid domain and buckling effect.

relaxation time. We therefore conclude that memory effects due to the shape of the cavity before the die may strongly affect the shape of the jet after the die. This observation can be very important in industrial applications such as pasta processing for instance.

6 Multiphase Flows with Free Surfaces

6.1 Mathematical Modeling of Multiphase Flows with Free Surfaces

We extend here the previous model to the case of multiple liquid phases with a free surface. More precisely, we consider P liquid phases, and the ambient gas is the phase numbered $P + 1$. We assume that the liquid phases are incompressible and immiscible, and thus rely on the *density-dependent Navier-Stokes equations* for the

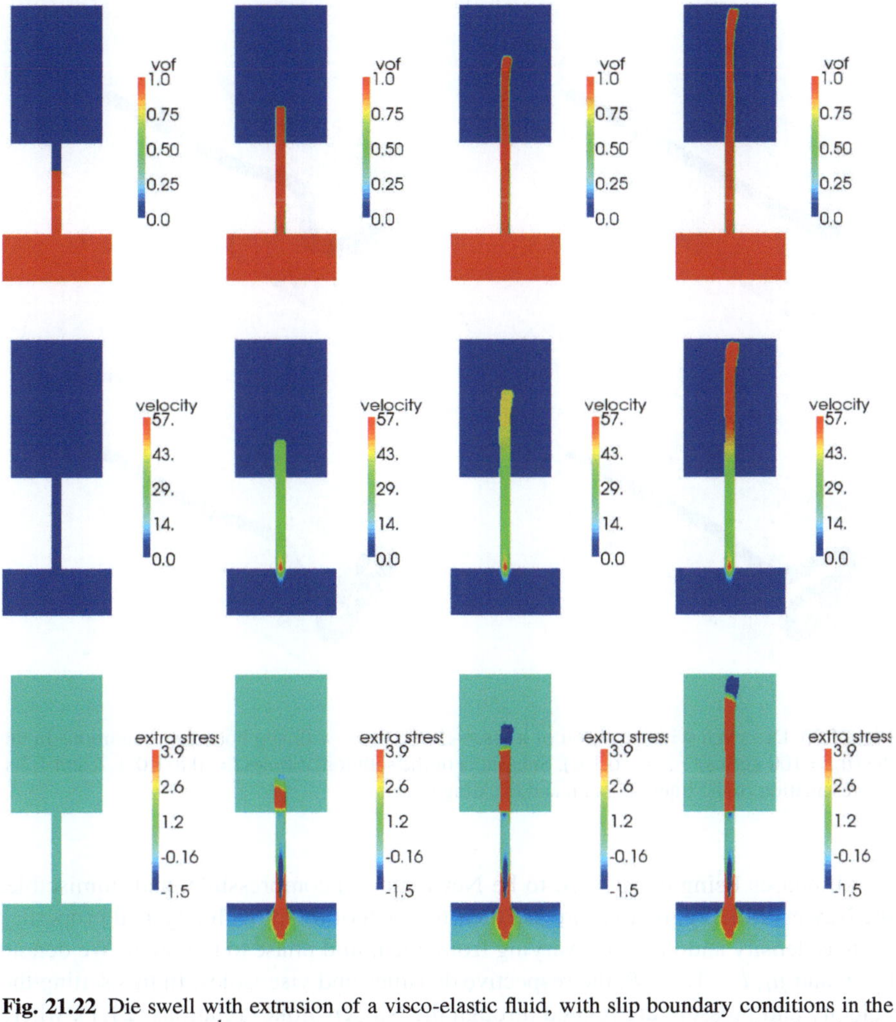

Fig. 21.22 Die swell with extrusion of a visco-elastic fluid, with slip boundary conditions in the die ($\mu_p = 100$ kg(ms)$^{-1}$, $\lambda = 0.1$ s.). Snapshots of the solution at times $t = 0, 0.4, 0.6$, and 0.8 s on a plane located in the middle of the tubes. Top: representation of volume fraction of liquid Φ; middle: speed $|\mathbf{U}|$; bottom: representation of extra-stress σ_{33}.

modeling of the global set of liquid phases. This model is based on [10, 19]. In [19], the emphasis has been put on the simulation of landslide-generated impulse waves. Here we show that the applications are numerous and that our algorithm can apply at different time and space scales.

We denote by $\Omega_\ell(t) \subset \Lambda$, $\ell = 1, \ldots, P$, the domain occupied by the ℓth liquid phase at time $t \in [0, T]$ and by $\Omega(t) = \bigcup_{\ell=1}^{P} \Omega_\ell(t)$ the global liquid domain. The subdomain $\Omega_\ell(t)$ is defined by its characteristic function $\phi_\ell : \Lambda \times [0, T] \to \{0, 1\}$:

$$\Omega_\ell(t) = \{\mathbf{x} \in \Lambda \mid \phi_\ell(\mathbf{x}, t) = 1\}, \quad \ell = 1, \ldots, P. \tag{21.36}$$

As a consequence, $\phi := \sum_{\ell=1}^{P} \phi_\ell$ is the characteristic function of $\Omega(t)$.

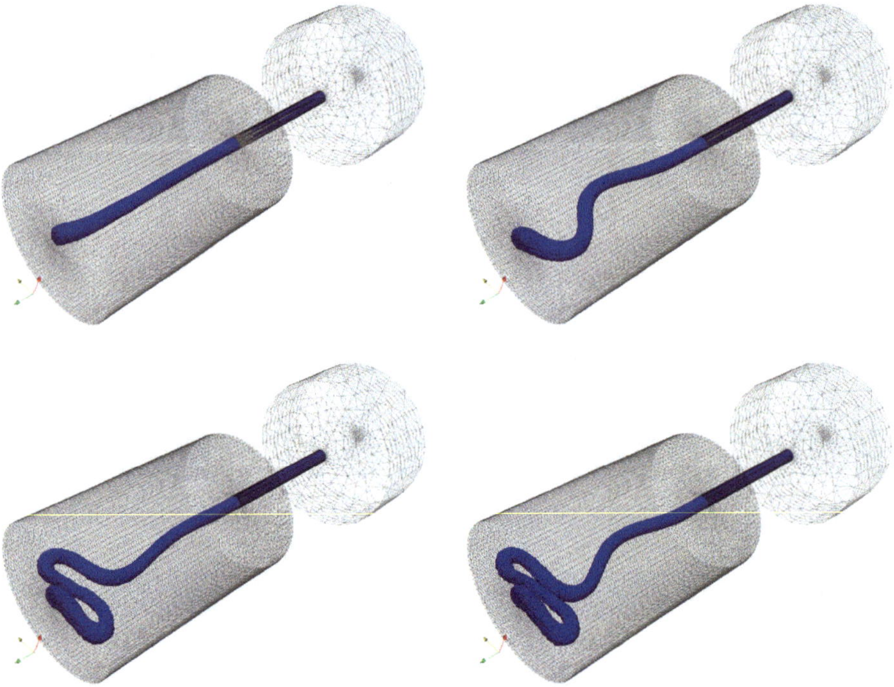

Fig. 21.23 Die swell with extrusion of a visco-elastic fluid, with slip boundary conditions in the die ($\mu_p = 100$ kg(ms)$^{-1}$, $\lambda = 0.1$ s.). Snapshots of the solution at times $t = 0.8, 1.0, 1.2$, and 1.3 s. Representation of the liquid domain and buckling effect.

All phases being considered to be Newtonian, incompressible, and immiscible, the Navier-Stokes equations are satisfied for each of them, with physical properties such as density and viscosity varying from one liquid phase to the other. We denote by ρ_l and μ_l, $l = 1,\ldots,P$, the respective densities and viscosities. In this setting the velocity and pressure are related through the Navier-Stokes relations (21.1), where the mass density is recovered as $\rho := \sum_{\ell=1}^{P} \phi_\ell \, \rho_\ell$, and similarly for the viscosity $\mu := \sum_{\ell=1}^{P} \phi_\ell \, \mu_\ell$.

The boundary conditions, interface conditions, and initial conditions are imposed in a similar way as in the single liquid phase case. At the interfaces between liquid phases, natural continuity conditions are imposed so that no forces are applied. Figure 21.34 illustrates a 2D sketch of multiple liquid phases in the case of die swell extrusion.

The evolution of each domain $\Omega_\ell(t)$ is governed by the transport of its characteristic function with the fluid velocity, that is:

$$\frac{\partial}{\partial t}\phi_\ell + \mathbf{u}\cdot\nabla\phi_\ell = 0 \quad \text{in} \quad Q_\ell, \qquad \phi_\ell = 0 \quad \text{in} \quad \Lambda \setminus Q_\ell, \tag{21.37}$$

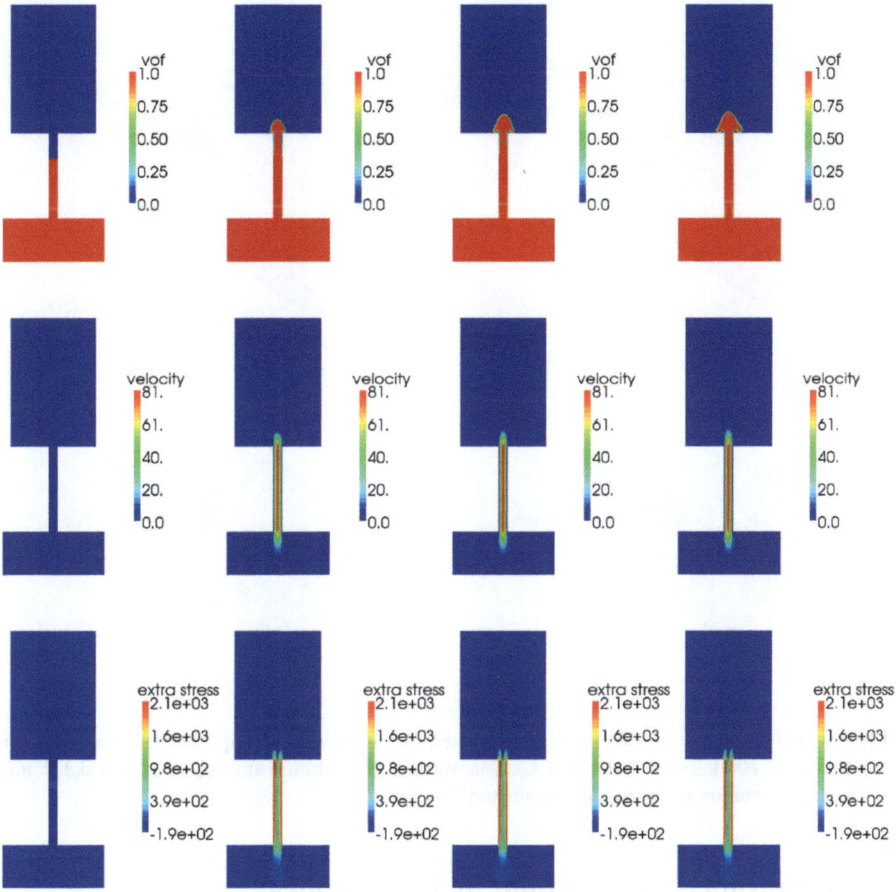

Fig. 21.24 Die swell with extrusion of a visco-elastic fluid, with no-slip boundary conditions in the die ($\mu_p = 100$ kg(ms)$^{-1}$, $\lambda = 0.1$ s.). Snapshots of the solution at times $t = 0, 0.4, 0.6$, and 0.8 s on a plane located in the middle of the tubes. Top: representation of volume fraction of liquid Φ; middle: speed $|\mathbf{U}|$; bottom: representation of extra-stress σ_{33}.

where

$$Q_\ell := \{(\mathbf{x},t) \in \Lambda \times (0,T] \mid \mathbf{x} \in \Omega_\ell(t)\}$$

and where $\mathbf{u}$ is the fluid velocity only defined on the space-time fluid domain Q.

The inflow boundary conditions supplementing the equations (21.37) have to be imposed for each liquid phase on the boundary of Λ, the same way it is imposed for one liquid phase. The initial value of the characteristic functions ϕ_ℓ are chosen to match the initial given domains $\Omega_\ell(0)$,

$$\phi_\ell(.,0) = 1 \quad \text{on} \quad \Omega_\ell(0) \quad \text{and} \quad \phi_\ell(.,0) = 0 \quad \text{otherwise.} \tag{21.38}$$

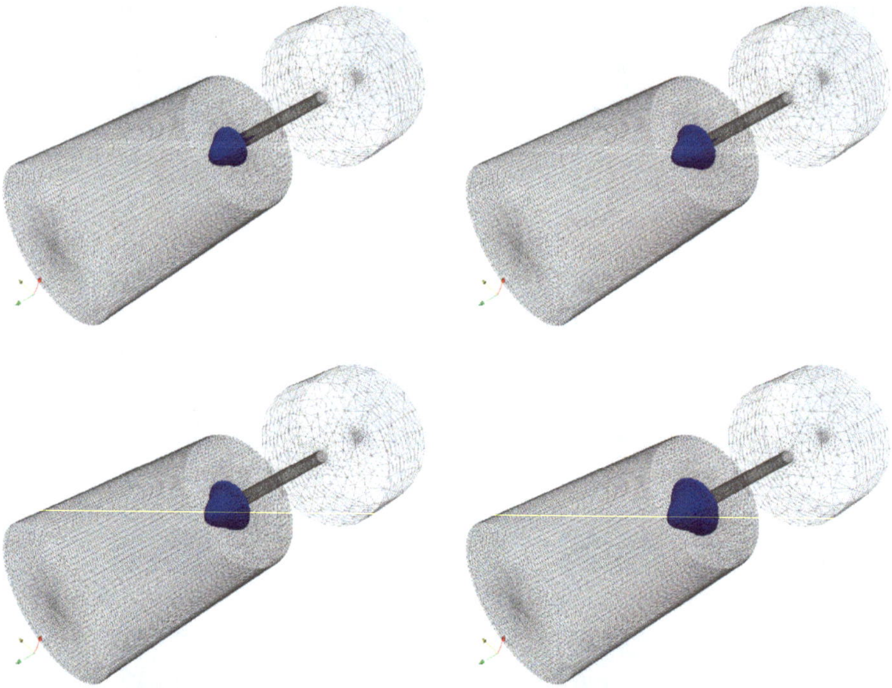

Fig. 21.25 Die swell with extrusion of a visco-elastic fluid, with no-slip boundary conditions in the die ($\mu_p = 100$ kg(ms)$^{-1}$, $\lambda = 0.1$ s.). Snapshots of the solution at times $t = 0.8, 1.0, 1.2$, and 1.6 s. Representation of the liquid domain and buckling effect.

6.2 Extension of the Operator Splitting Strategy

The extension of the operator splitting method to multiphase flows includes mainly the transport of multiple volume fractions, which is a natural extension of the transport of a single phase. However a significant step of the algorithm involves the reconstruction of the interfaces and the numerical methods to avoid artificial diffusion and compression, which have to be re-designed in the context of multiphase flows.

The operator splitting algorithm to approximate the system of equations (21.1) and (21.37) again decouples the approximation of the diffusion and advection operators. In this case, the diffusion operators correspond to a Stokes problem on a stationary domain with piecewise constant density and viscosity fields. The advection operator includes the transport equations for the Navier part of the incompressible fluid, as well as for the transport of the P characteristic functions.

The time splitting scheme reads as follows. We assume to be given an approximation of the liquid domain characteristic functions ϕ_ℓ^n , $\ell = 1, \ldots, P$ at time t^n. This entails an approximation of the liquid domains Ω_ℓ^n and of the global liquid domain Ω^n via the relations:

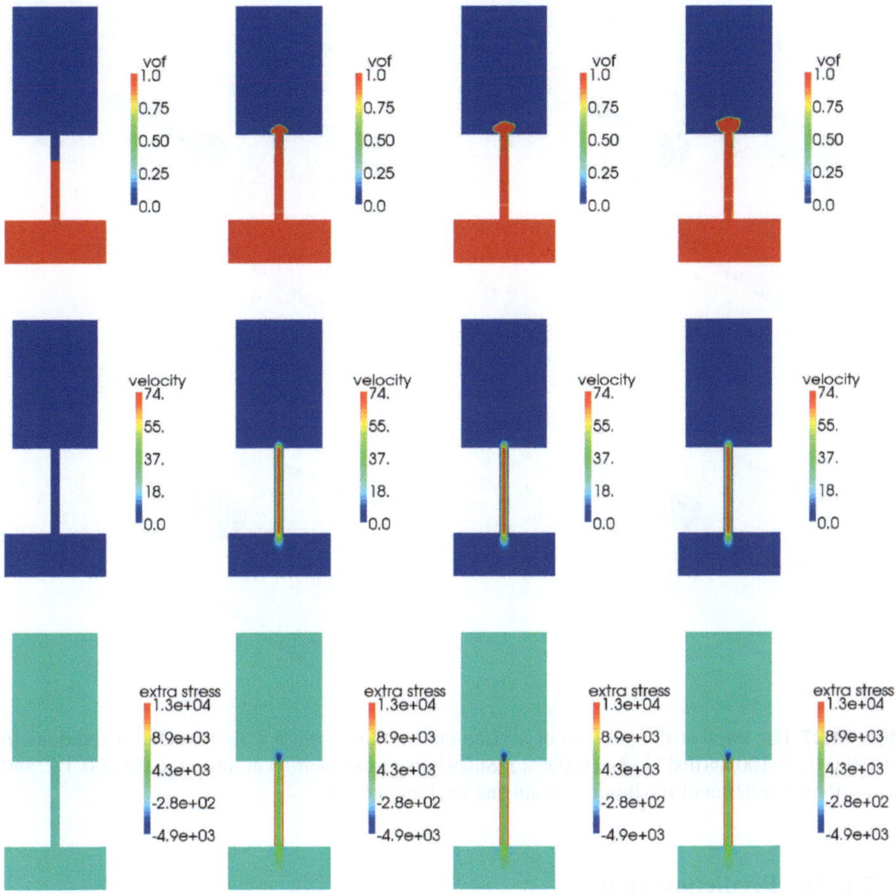

Fig. 21.26 Die swell with extrusion of a visco-elastic fluid, with no-slip boundary conditions in the die ($\mu_p = 100$ kg(ms)$^{-1}$, $\lambda = 0.005$ s). Snapshots of the solution at times $t = 0, 0.4, 0.6$, and 0.8 s on a plane located in the middle of the tubes. Top: representation of volume fraction of liquid Φ; middle: speed $|\mathbf{U}|$; bottom: representation of extra-stress σ_{33}.

$$\Omega_\ell^n := \{\mathbf{x} \in \Lambda \mid \phi_\ell^n(\mathbf{x}) = 1\}, \quad \Omega^n := \bigcup_{\ell=1}^{P} \Omega_\ell^n.$$

We also assume to be given a velocity approximation $\mathbf{u}^n(\mathbf{x})$ of $\mathbf{u}(\mathbf{x}, t^n)$. The prediction step determines the new approximation of the liquid domain at time t^{n+1}, together with a prediction of the velocity on the new domain. The correction step provides an update of the velocity and pressure while the liquid domain remains unchanged.

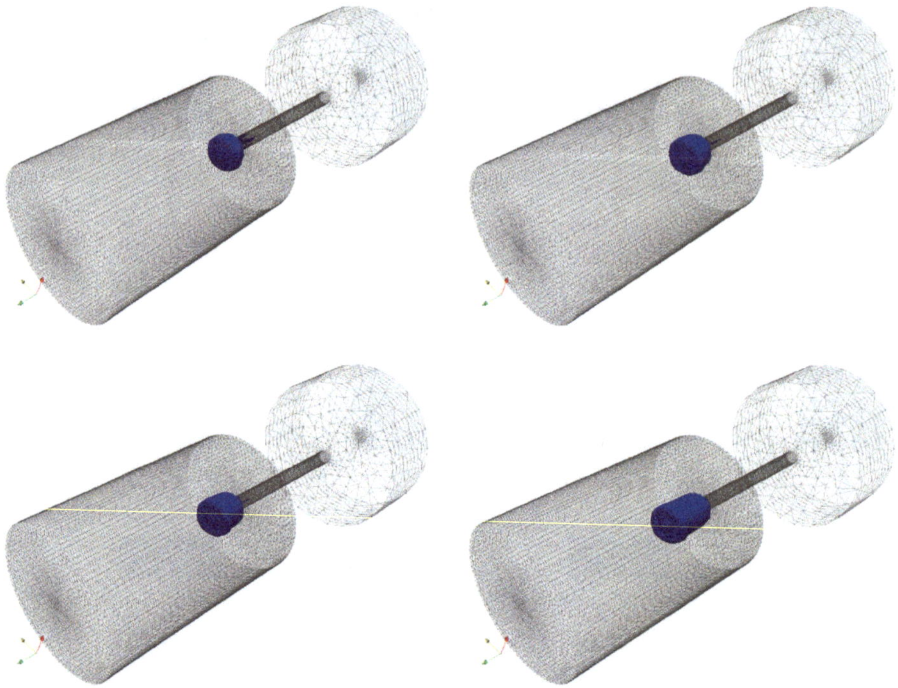

Fig. 21.27 Die swell with extrusion of a visco-elastic fluid, with no-slip boundary conditions in the die ($\mu_p = 100\ \mathrm{kg(ms)}^{-1}$, $\lambda = 0.005$ s.). Snapshots of the solution at times $t = 0.8, 1.0, 1.2,$ and 1.6 s. Representation of the liquid domain and buckling effect.

6.2.1 The Prediction Step

The *projection step* encompasses the advection components of (21.1) and (21.37). It consists in solving the $P+1$ transport equations:

$$\begin{aligned} &\frac{\partial}{\partial t}\phi_\ell + \mathbf{u}\cdot\nabla\phi_\ell = 0, \quad \ell = 1,\ldots,P, \\ &\frac{\partial}{\partial t}\mathbf{u} + (\mathbf{u}\cdot\nabla)\mathbf{u} = \mathbf{0} \end{aligned} \tag{21.39}$$

in $Q^{n+1} := Q \cap \left(\Lambda \times I^{n+1}\right)$. Outside the liquid domain Ω_ℓ^n, we set $\phi_\ell(\mathbf{x},t) = 0$ whenever $\mathbf{x} \in \Lambda \setminus \left\{\mathbf{y}(t;\overline{\mathbf{x}}) \mid \overline{\mathbf{x}} \in \Omega_\ell^n\right\}$ and $\mathbf{u}$ is not required outside Ω^n. Eventually, we end up setting $\phi_\ell^{n+1} := \phi_\ell(t^{n+1})$ in Λ, and consequently

$$\Omega_\ell^{n+1} := \left\{\mathbf{x} \in \Lambda \mid \phi_\ell^{n+1}(\mathbf{x}) = 1\right\}, \quad \Omega^{n+1} = \bigcup_{\ell=1}^{P} \Omega_\ell^{n+1}, \tag{21.40}$$

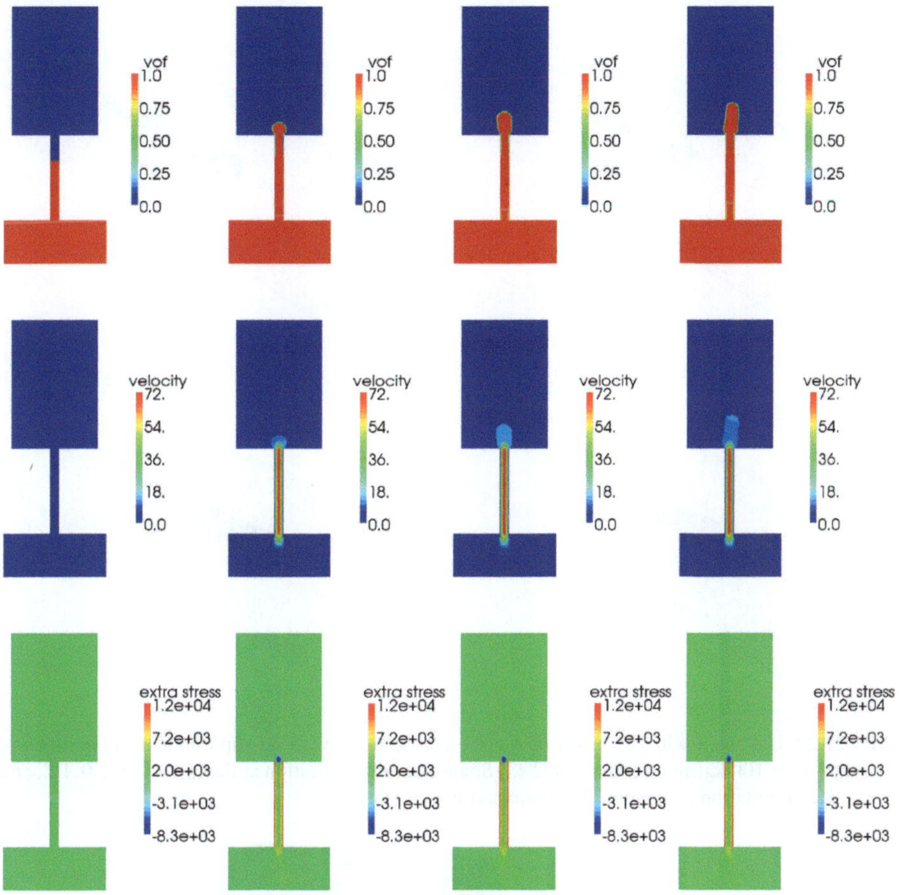

Fig. 21.28 Die swell with extrusion of a visco-elastic fluid, with no-slip boundary conditions in the die ($\mu_p = 100$ kg(ms)$^{-1}$, $\lambda = 0.002$ s.). Snapshots of the solution at times $t = 0, 0.4, 0.6$, and 0.8 s on a plane located in the middle of the tubes. Top: representation of volume fraction of liquid Φ; middle: speed $|\mathbf{U}|$; bottom: representation of extra-stress σ_{33}.

as well as $\mathbf{u}^{n+\frac{1}{2}} := \mathbf{u}(t^{n+1})$ in Ω^{n+1}. At the continuous level, these problems are highly similar to those encountered for one single phase. However, after space discretization, the complexity is quite different.

Indeed, the prediction steps starts with given approximations $\Phi^n_{\ell,M} \in \mathbb{V}^S$, $\ell = 1,\ldots,P$, and $\mathbf{U}^n_M \in (\mathbb{V}^S)^d$ of the liquid fractions and velocity respectively. The predictions $\Phi^{n+\frac{1}{2}}_{\ell,M}$ and $\mathbf{U}^{n+\frac{1}{2}}_M$, in $\mathbb{V}^S$ and $(\mathbb{V}^S)^d$ respectively, are computed as in (21.24) by transport of the quantities in each cell C and projection on the grid $\mathscr{T}^S$. Notice that each of the transport equations for $\Phi^n_{\ell,M}$ are solved in parallel for each liquid phase, and the redistribution is achieved sequentially. Details can be found in [19].

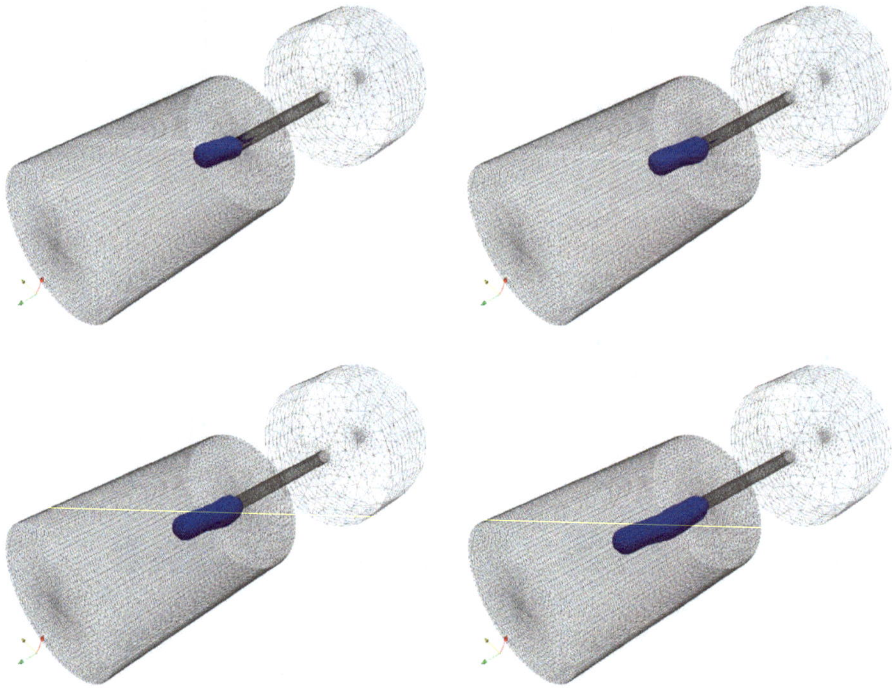

Fig. 21.29 Die swell with extrusion of a visco-elastic fluid, with no-slip boundary conditions in the die ($\mu_p = 100$ kg(ms)$^{-1}$, $\lambda = 0.002$ s.). Snapshots of the solution at times $t = 0.8, 1.0, 1.2$, and 1.6 s. Representation of the liquid domain and buckling effect.

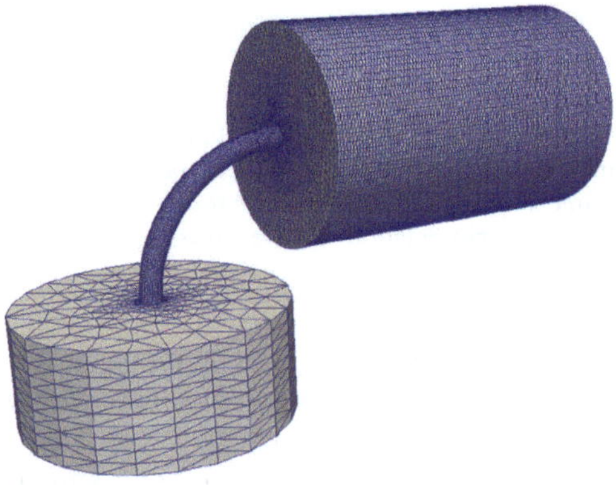

Fig. 21.30 Die swell with a bend, for the extrusion of a visco-elastic fluid. Visualization of the geometrical domain and finite element mesh.

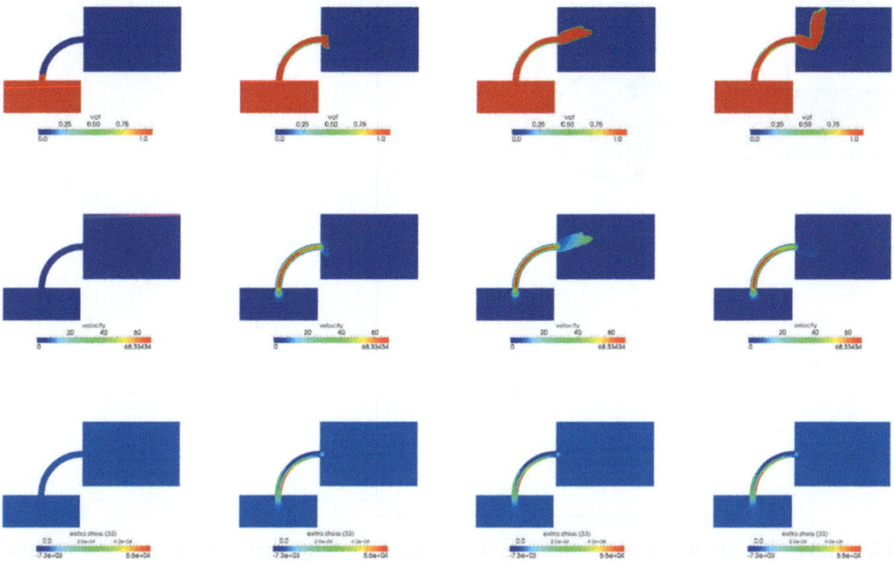

Fig. 21.31 Die swell of a visco-elastic fluid for a die with a 90 degrees bend. Snapshots of the solution at times $t = 0, 0.4, 0.8$, and 1.2 s (left to right). Top: volume fraction of liquid Φ in a median cut in the middle of the domain; middle: speed $|\mathbf{U}|$ in a median cut in the middle of the domain; bottom: Extra-stress field σ_{33} in a median cut in the middle of the domain.

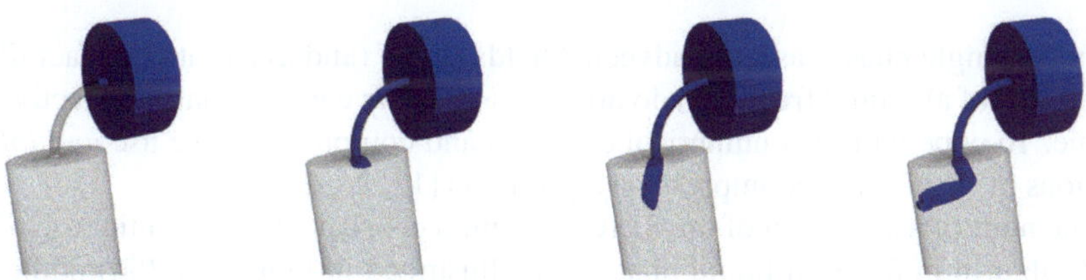

Fig. 21.32 Die swell of a visco-elastic fluid for a die with a 90 degrees bend. Snapshots of the solution at times $t = 0, 0.4, 0.8$, and 1.2 s (left to right).

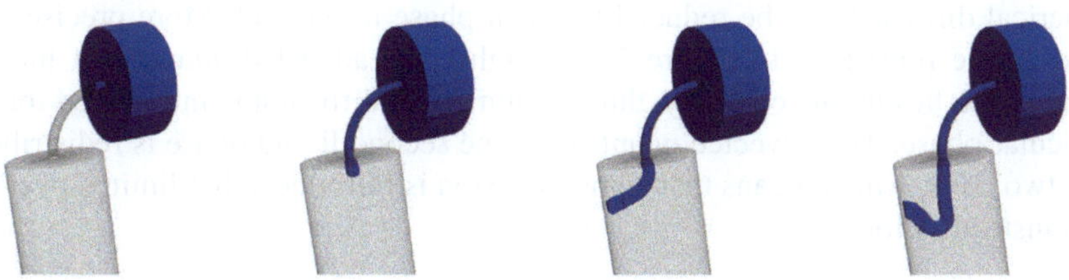

Fig. 21.33 Die swell of a visco-elastic fluid for a die with a 90 degrees bend. Snapshots of the solution at times $t = 0, 0.4, 0.8$, and 1.2 s (left to right) for $\lambda = 0.02$ s.

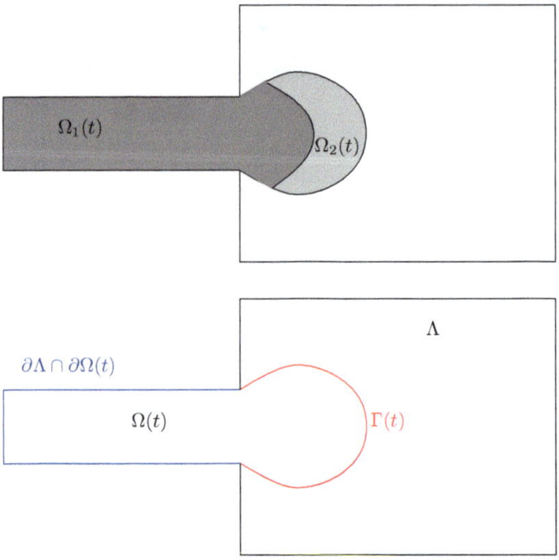

Fig. 21.34 Die swell with two liquid phases (phase 1 pushing phase 2). Geometrical notation for the VOF formulation for two liquid phases with free surface included into the computational domain Λ.

6.2.2 Numerical Diffusion vs Numerical Compression

As in the single phase case, the advected fields $\Phi_{\ell,M}^{n+\frac{1}{2}}$ (and as a matter of fact $\Phi_M^{n+\frac{1}{2}}$ as the sum of all liquid fractions) do not necessarily have values that are exactly zero or one. To cope with this numerical diffusion and compression, we use *multiphase* versions of SLIC and decompression algorithms [12, 26, 28].

The multiphase version of the SLIC algorithm consists of a sequential use of the SLIC algorithm for each liquid phase. It is illustrated in Figure 21.35 (middle and right, for one liquid phase or two liquid phases). Each of the liquid phases is pushed against the sides/corners of the cell to be transported. The transport and projection of the cell are then made for each phase independently and sequentially. Thus the numerical diffusion can be reduced for each phase in parallel. More precisely, on the example illustrated in Figure 21.35 (right), the advected quantity of the first liquid phase lies in one cell only, thus no numerical diffusion is introduced for that particular phase. The advected quantity for the second liquid phase is redistributed over two cells, which means that some diffusion is introduced but limited over two cells instead of four.

Remark 3 (SLIC vs PLIC). The SLIC procedure has been preferred for instance over the higher order PLIC procedure for its handling simplicity within the two-grid framework. The rationale behind this approach is to use a low order interface reconstruction technique, like SLIC, *on a very fine mesh.* The mesh size guarantees the accuracy of the algorithm and compensates for the low order of the reconstruction technique. Replacing the SLIC algorithm with a PLIC algorithm on the structured

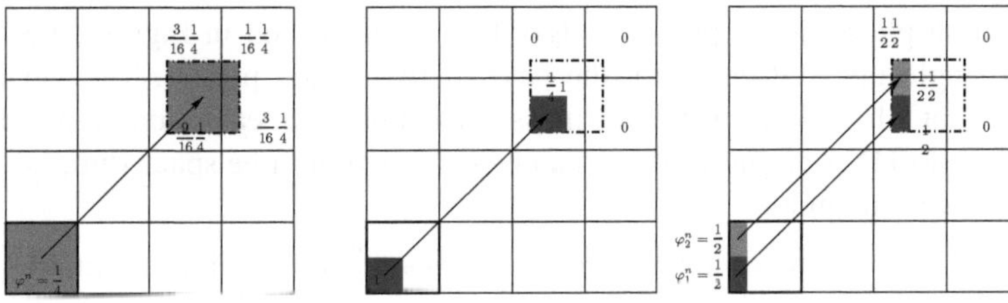

Fig. 21.35 An example of two dimensional advection and projection when the volume fraction of liquid in the cell is $\Phi_M^n = \frac{1}{4}$. Left: without SLIC and with one liquid phase, the volume fraction of liquid is advected and projected on four cells, with contributions (from the top left cell to the bottom right cell) $\frac{3}{16}\frac{1}{4}$, $\frac{1}{16}\frac{1}{4}$, $\frac{9}{16}\frac{1}{4}$, $\frac{3}{16}\frac{1}{4}$. Middle: with SLIC and with one liquid phase, the volume fraction of liquid is first pushed at one corner, then it is advected and projected on one cell only, with contribution $\frac{1}{4}$. Right: with SLIC and with two liquid phases, the volume fractions of liquid are first pushed along one side of the cell, then they are advected. The first liquid phase (corresponding to a volume of $\frac{1}{8}$) is projected on one cell only, with contribution $1\frac{1}{8}$; the second liquid phase (corresponding also to a volume of $\frac{1}{8}$) is projected on two cells, with contribution $1\frac{1}{16}$ and $1\frac{1}{16}$.

grid of small cells is not a fundamental problem, but a technical difficulty. Moreover, the PLIC procedure applied before transport of a cell, and coupled with a projection operator on the finite element mesh would be of little benefit, and expensive from the computational viewpoint.

Note that the sequential treatment of liquid phases implicitly requires them to be sorted; the arbitrary phase ordering influences the reconstruction of the interfaces, as already stated in [13] for three phases. However, numerical experiments show that the effect of the ordering of phases is not a crucial factor for the final results, especially at the limit when the mesh size tends to zero.

After the interface reconstruction and advection steps, it may happen that some cell C_i in the grid $\mathscr{T}^S$ is over-filled, i.e., $\Phi_M^{n+\frac{1}{2}} = \sum_\ell \Phi_{\ell,M}^{n+\frac{1}{2}} > 1$. Such physically non-admissible values can indeed occur even if $\Phi_{\ell,M}^n \in [0,1]$ since the transport-and-project algorithm is not a divergence-free process.

We thus need to *decompress* the fields $\Phi_{\ell,M}^{n+\frac{1}{2}}$ and $\Phi_M^{n+\frac{1}{2}}$ with a numerical technique that allows to conserve the mass in a global sense [19, 26]. This algorithm is applied after the solution of the transport equations, but before the solution of the diffusion equations. It proceeds in two steps: first, we compute the excess of each liquid phase in each cell after advection and projection onto $\mathscr{T}^S$; second, we redistribute these amounts *proportionally to the amount already included in the cell* in a given arbitrary order, in a way that is similar to *global repair* algorithms [35]. This method is robust, but requires to order the liquid phases (arbitrarily) to know which phase is redistributed first into the other cells. Numerical experiments have shown in our case that, besides guaranteeing the mass conservation globally, the error due to this decompression algorithm is reduced as the time step decreases. This heuristic algorithm can be found in [19] and is not detailed more extensively here.

This multiphase decompression algorithm is illustrated in Figure 21.36, when $\mathscr{T}^S$ is a single layer of cells, for the case of two liquid phases. The rebalancing principle that allows to conserve the mass in each phase at each time step is detailed in this pseudo 1D configuration, but can be extended in three space dimensions in a straightforward manner.

After the decompression, the approximations $\Phi_{\ell,M}^{n+\frac{1}{2}}$, $\ell = 1,\dots,P$ and $\mathbf{U}_M^{n+\frac{1}{2}}$ are projected into the finite element spaces

$$\Phi_{\ell,K}^{n+1} := \pi_{S\to FEM}\Phi_{\ell,M}^{n+\frac{1}{2}} \in \mathbb{V}^{FEM}, \qquad \ell = 1,\dots,P,$$

and $\mathbf{U}_K^{n+\frac{1}{2}} \in \mathscr{V}(\tau_K^{n+1})$ is defined according to formula (21.27) and where τ_K^{n+1} is the collection of liquid tetrahedra, see Section 4.2.

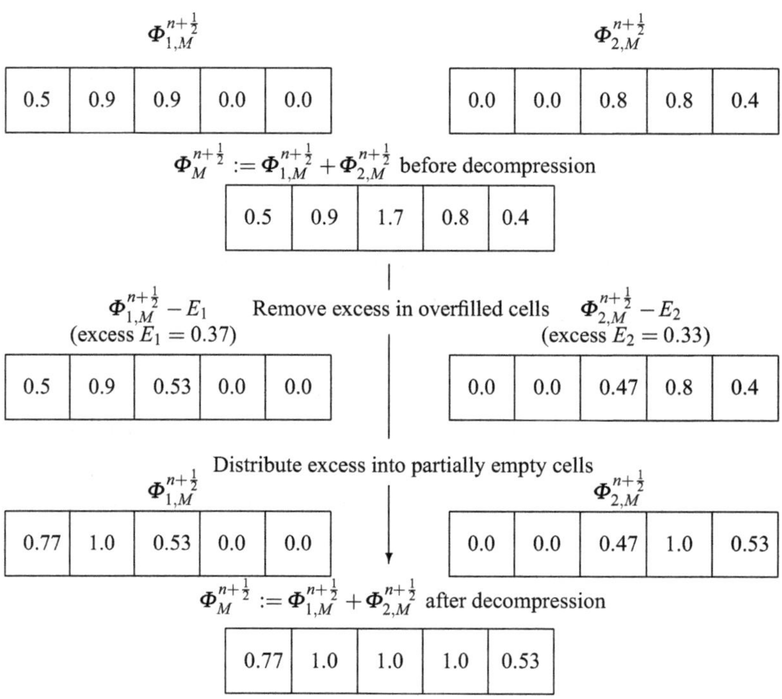

Fig. 21.36 Decompression algorithm in the case of two liquid phases. The volume fractions in excess in some cells are redistributed into the under filled cells, proportionally to the contribution of each phase. The total liquid volume fraction is given by $\Phi_M^{n+\frac{1}{2}} = \Phi_{1,M}^{n+\frac{1}{2}} + \Phi_{2,M}^{n+\frac{1}{2}}$. The excesses are first removed in overfilled cells proportionally to the contribution of each phase $(0.37 = 0.7\cdot(0.9/1.7)$ and $0.33 = 0.7\cdot(0.8/1.7))$. The excesses are then redistributed into each phase independently before recalculating the total liquid volume fraction $\Phi_M^{n+\frac{1}{2}} = \Phi_{1,M}^{n+\frac{1}{2}} + \Phi_{2,M}^{n+\frac{1}{2}}$.

6.2.3 The Correction Step

After the prediction step, the approximations of the liquid domains Ω_K^{n+1} and the set of liquid elements τ_K^{n+1} are defined as in the single phase case. Note that the approximation of the liquid domains $\Omega_{\ell,K}^{n+1}$ can be defined similarly, but they are not used explicitly in the correction step. Indeed the global Stokes system is defined and solved on the global liquid domain, and the interfaces between phases are implicitly taken into account in a diffuse modeling via the density and viscosity fields. In fact, the velocity and pressure correction $\mathbf{U}_K^{n+1} \in (\mathbb{V}^{FEM}(\tau_K^{n+1}))^d$, $P_K^{n+1} \in \mathbb{V}^{FEM}(\tau_K^{n+1})$ are defined as the solution to (21.28) upon redefining on each tetrahedral element $T \in \tau_K^{n+1}$ the density and viscosity as

$$\rho|_T := \rho^{n+1}|_T := \frac{1}{d+1}\sum_{i=1}^{d+1} \frac{\sum_{\ell=1}^{P} \Phi_{\ell,K}^{n+1}(\mathbf{v}_i^T)\rho_\ell}{\sum_{\ell=1}^{P} \Phi_{\ell,K}^{n+1}(\mathbf{v}_i^T)},$$

$$\mu| := \mu^{n+1}|_T := \frac{1}{d+1}\sum_{i=1}^{d+1} \frac{\sum_{\ell=1}^{P} \Phi_{\ell,K}^{n+1}(\mathbf{v}_i^T)\mu_\ell}{\sum_{\ell=1}^{P} \Phi_{\ell,K}^{n+1}(\mathbf{v}_i^T)},$$

where $\{\mathbf{v}_i^T, i = 1,\ldots,d+1\}$ denotes the vertices of T.

6.3 Numerical Results for Multiphase Flows

We consider again the extrusion with initial contraction described in Section 4.3.1. The computational domain is still the one reported in Figure 21.11.

6.3.1 Successive Phases

We consider three incompressible and immiscible liquid phases, each of them a Newtonian fluid, with equal densities $\rho_1 = \rho_2 = \rho_3 = 1300$ kg m^{-3}, and corresponding (equal) viscosities $\mu_1 = \mu_2 = \mu_3 = 10$ kg(ms)$^{-1}$. As the goal of this example is to study the accuracy of the splitting algorithm, the choice of the three phases is artificial. The liquids are initially located in a successive sequence such that the liquid 1 is pushing the liquid 2 and then the liquid 3. The boundary conditions at the inflow boundary are $\mathbf{u} = 0.00023$ ms^{-1}, such that the order of magnitude of the velocity in the die is approximately 0.05 ms^{-1}. There is only a liquid from phase 1 that flows inside the computational domain. Slip boundary conditions are imposed on $\partial\Lambda$, except at the bottom of the computational domain where no-slip boundary conditions are enforced. Gravity forces, with amplitude $|\mathbf{g}| = 9.81$ ms^{-2} are oriented along the die. The time step is constant and equal to $\delta t = 0.005$ s.

Figure 21.37 illustrates, in a medium plane inside the tube, snapshots of the volume fractions of liquid Φ_ℓ and of the magnitude of the corresponding speed $|\mathbf{U}|$. We observe that the operator splitting algorithm does not introduce any additional error as long as the flow is laminar (i.e., does not hit the boundary of the computational domain and starts to buckle). The liquid 1 is perfectly pushing the liquids 2 and 3. The velocity is perfectly aligned with the direction of the die, even when using different values of the time step if needed, without the drawback of a CFL condition. The mass in each phase is conserved.

The previous results have been obtained when the three phases have the same densities and viscosities. We now provide a short sensitivity analysis with respect to the value of the viscosities, all the other physical quantities remaining the same. More precisely, we consider again three successive liquid phases. The initial configuration, denoted (a), with $\mu_1 = \mu_2 = \mu_3 = 10$, is compared with two cases, namely (b) $\mu_1 = 10$, $\mu_2 = 1$, $\mu_3 = 0.1$, and (c) $\mu_1 = 10$, $\mu_2 = \mu_3 = 0.1$. Figure 21.38 illustrates in a medium plane inside the tube, snapshots of the volume fractions of liquid

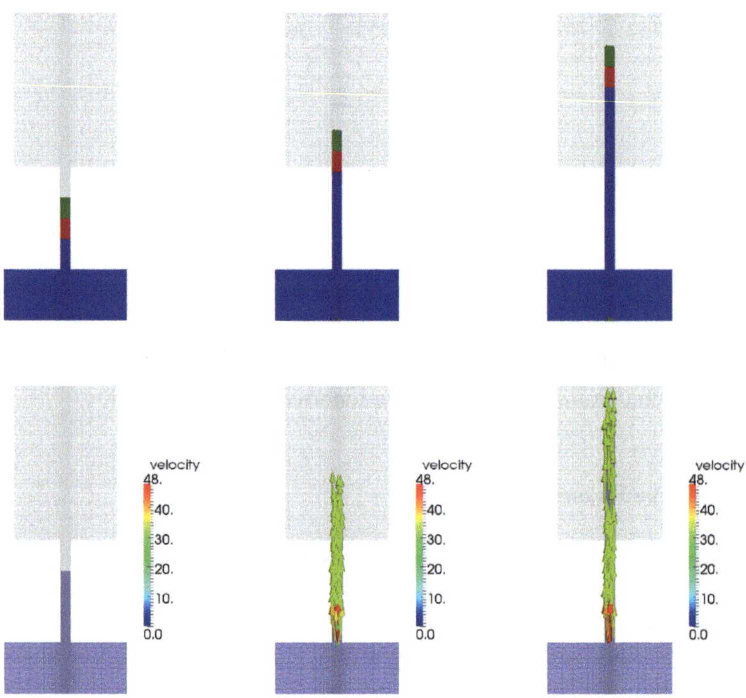

Fig. 21.37 Die swell with extrusion of a Newtonian fluid with three phases (liquid 1 in blue, liquid 2 in red, liquid 3 in green). Snapshots of the solution at times $t = 0, 0.5$ and 1.0 s on a plane located in the middle of the tubes. Top: representation of volume fractions of liquid Φ_ℓ; bottom: velocity field $\mathbf{U}$.

Φ_ℓ for the three configurations. One can observe that, when the two liquid phases at the front of the jet have a smaller viscosity, they are crashed by the more viscous phases when the jet hits the boundary of the domain; in that particular case, the less viscous liquid phases do not contribute to the buckling effect. The difference between configurations (b) and (c) is not remarkable. Before touching the boundary of the domain, the laminar behavior of the three liquid phases is identical to that illustrated in Figure 21.37.

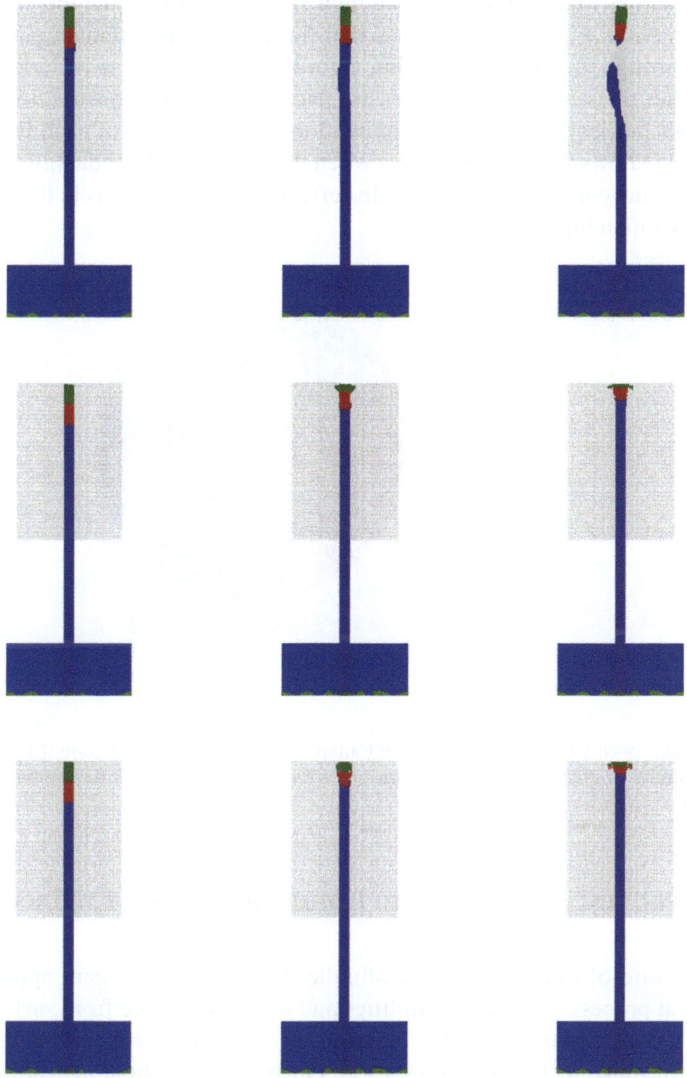

Fig. 21.38 Die swell with extrusion of a Newtonian fluid with three phases (liquid 1 in blue, liquid 2 in red, liquid 3 in green). Snapshots of the solution at times $t = 1.2, 1.3$, and 1.4 s on a plane located in the middle of the tubes. First row: (a) $\mu_1 = \mu_2 = \mu_3 = 10$; Second row: (b) $\mu_1 = 10$, $\mu_2 = 1$, $\mu_3 = 0.1$; Third row: (c) $\mu_1 = 10$, $\mu_2 = \mu_3 = 0.1$.

6.3.2 Parallel Phases

Finally, let us consider the configuration where the three liquid phases (with equal viscosities) are next to each other. The initial configuration is illustrated in Figure 21.39 and shows that each phase is contained in one-third of the total angle along the die direction. Figure 21.40 shows that the three phases remain parallel when advected through the operator splitting algorithm (the red phase is 'hidden' behind the two other phases!). The velocities are parallel and the reconstruction of the interface does not jeopardize the approximation of the location of each liquid phase. Figure 21.41 shows the buckling effect when the jet hits the boundary of the domain; as all phases have the same viscosity, this effect is quite similar as in the case of one single Newtonian liquid.

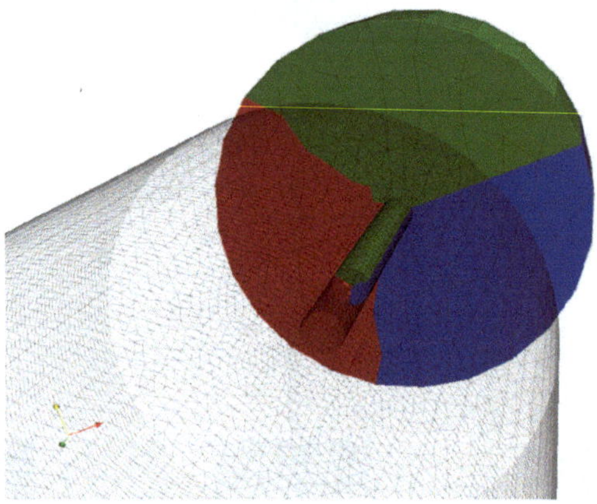

Fig. 21.39 Die swell with extrusion of a Newtonian fluid with three phases (liquid 1 in blue, liquid 2 in red, liquid 3 in green). Initial configuration of the three phases, each having one-third of the total volume.

7 Perspectives: Application to Emulsion in Food Engineering

The simulation of emulsion in microfluidic devices is a stringent application, as the physical process involves instabilities and strong influence from surface tension effects. Thus the numerical method requires an accurate approximation of the interfaces and of those surface tension effects. Further details about microfluidic emulsions can be found in [3, 7, 27] and references therein. Applications of interest exist in food engineering when producing types of mayonnaise for instance [17, 18]. Furthermore, from the numerical viewpoint, adaptive mesh refinement techniques help tremendously to increase the accuracy of the method and sharpen the approximation of the interfaces. Details about an adaptive method making a first attempt into this direction can be found in [9] when discussing the mesh refinement between one liquid phase and a vacuum. The same type of techniques have been extended in

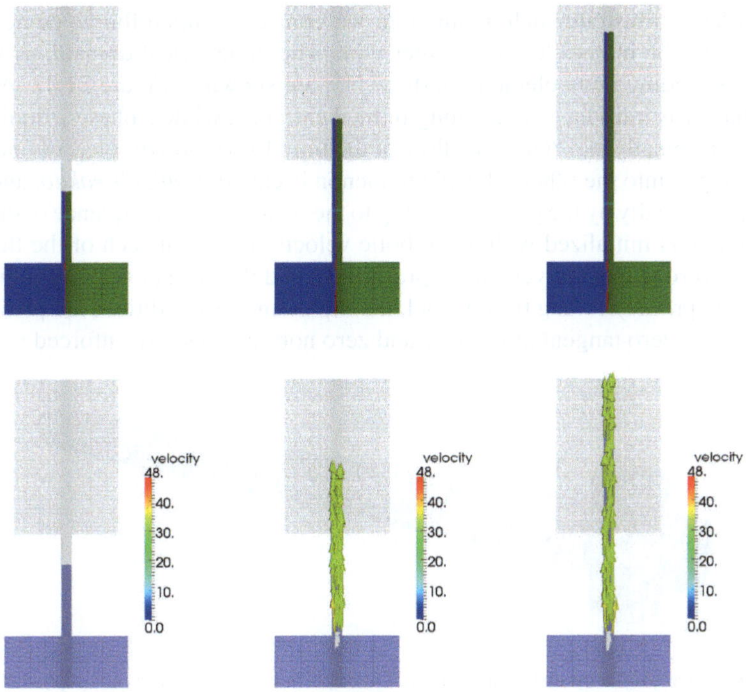

Fig. 21.40 Die swell with extrusion of a Newtonian fluid with three phases (liquid 1 in blue, liquid 2 in red, liquid 3 in green). Snapshots of the solution at times $t = 0, 0.5$, and 1.0 s on a plane located in the middle of the tubes. Top: representation of volume fraction of liquid Φ_ℓ; bottom: velocity field $\mathbf{U}$.

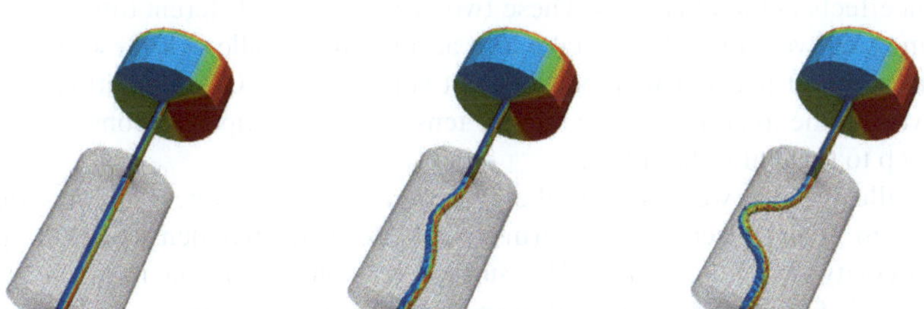

Fig. 21.41 Die swell with extrusion of a Newtonian fluid with three phases (liquid 1 in blue, liquid 2 in red, liquid 3 in green). Snapshots of the approximation of the liquid domain at times $t = 1.2, 1.4$, and 1.6 s.

subsequent work to mesh refinement around interfaces between two liquid phases. The results presented in this section have been obtained by P. Clausen while staying at EPFL on a postdoctoral position. Details of the method will be presented in a forthcoming paper.

In order to illustrate such a situation, we consider a microfluidic device composed by a tube intersected by another tube. The geometrical domain, as well as the corresponding finite element mesh $\mathcal{T}^{FEM}$, are shown in Figure 21.42. A liquid from phase 1 is introduced at the longitudinal entrance, while a phase 2 liquid is injected transversally and "cuts" the flow of the liquid 1 to form droplets of one liquid phase trapped into the other. This phenomenon is called *droplet breakup*, and is repeated periodically by the process leading to the formation of a sequence of droplets. The velocity is initialized with a parabolic velocity profile at each of the three entrances (a zero tangential velocity is prescribed, and the normal velocity is given by a parabolic profile). Along the channel, no-slip boundary conditions are prescribed. At the outlet, zero tangential velocity and zero normal stress are enforced.

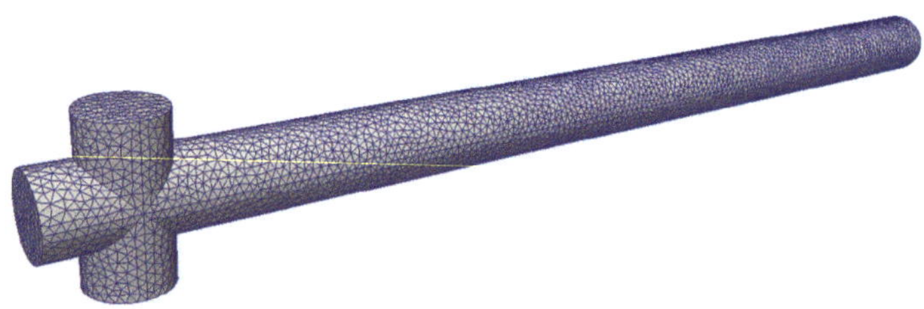

Fig. 21.42 Microfluidic emulsion simulation. Description of the geometry and representation of the finite element mesh $\mathcal{T}^{FEM}$

In an emulsion, two motions are interacting: first the movement initiated by the flow induced by the inlet velocities; second the displacements induced by the surface tension effects at the interfaces. These two effects are on different time scales, and the time step we choose has to take into account the smallest of these two scales. Despite the fact that our method does not suffer from a CFL condition, here we observe that the treatment of the surface tension effects impose a constraint on the time step to prevent instabilities.

For illustration, we consider the injection of oil in water. Oil has density 1000 $\mathrm{kg\,m^{-3}}$ and viscosity 0.5 $\mathrm{kg(ms)^{-1}}$, while water has density 1000 $\mathrm{kg\,m^{-3}}$ and viscosity 0.001 $\mathrm{kg(ms)^{-1}}$. The surface tension coefficient is given by $\gamma = 0.02\ \mathrm{Nm^{-1}}$. Oil is introduced with a maximum velocity of $\mathbf{u}_{max,oil} = 0.01\ \mathrm{ms^{-1}}$, while water is injected with a maximum velocity of $\mathbf{u}_{max,water} = 0.02\ \mathrm{ms^{-1}}$.

Figure 21.43 illustrates, in a medium plane inside the tube, snapshots of the two liquid phases and the corresponding velocity field.

The results of these numerical experiments show the difficulty of producing a regular succession of droplet breakings, which is controlled by the balance between surface energy and viscosity effects. Numerical difficulties include droplets generated with different sizes and volume losses when the stabilization terms are large.

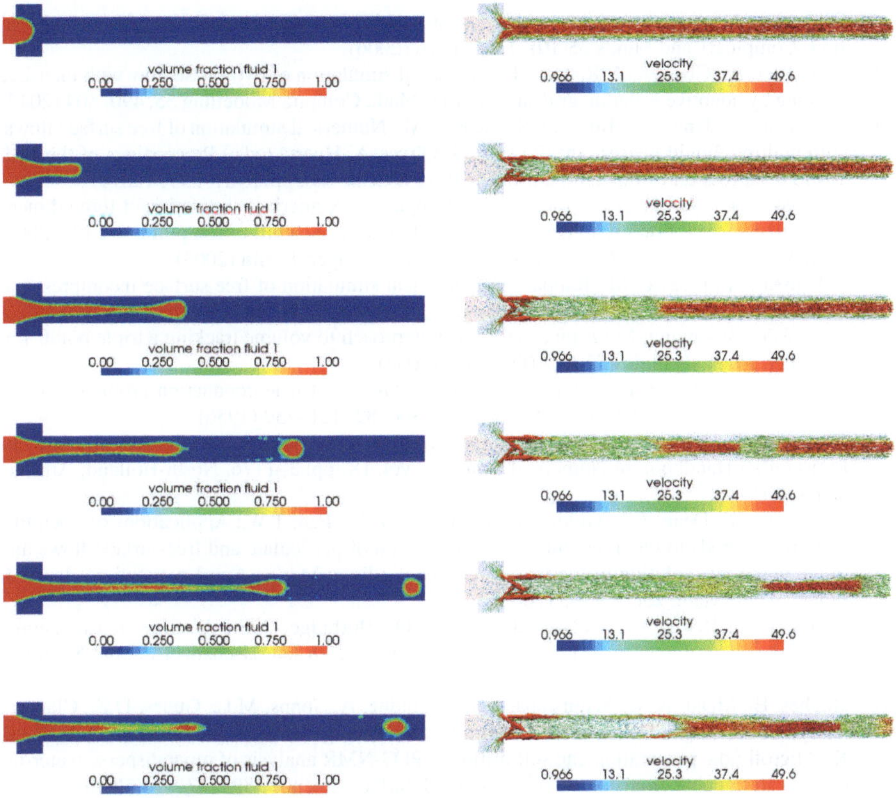

Fig. 21.43 Microfluidic emulsion simulation. Snapshots of the solution at times $t = 0.2$, 0.5,0.8,1.1,1.4, and 1.7 s on a plane located in the middle of the longitudinal tube. Left: representation of volume fraction of liquid ; right velocity field.

References

1. Antonietti, P.F., Fadel, N.A., Verani, M.: Modelling and numerical simulation of the polymeric extrusion process in textile products. Commun. Appl. Ind. Math. **1**(2), 1–13 (2010)
2. Bänsch, E.: Finite element discretization of the Navier-Stokes equations with a free capillary surface. Numer. Math. **88**(2), 203–235 (2001)
3. Baroud, C.N., Gallaire, F., Dangla, R.: Dynamics of microfluidic droplets. Lab on a Chip **10**(16), 2032 (2010)
4. Bird, R., Curtiss, C., Armstrong, R., Hassager, O.: Dynamics of Polymeric Liquids, vol. 1 and 2. John Wiley & Sons, New-York (1987)
5. Bonito, A., Clément, P., Picasso, M.: Mathematical analysis of a simplified Hookean dumbbells model arising from viscoelastic flows. J. Evol. Equ. **6**(3), 381–398 (2006)
6. Bonito, A., Picasso, M., Laso, M.: Numerical simulation of 3D viscoelastic flows with free surfaces. J. Comput. Phys. **215**(2), 691–716 (2006)
7. Brun, P.T., Nagel, M., Gallaire, F.: Generic path for droplet relaxation in microfluidic channels. Physical Review E **88**(4) (2013)

8. Caboussat, A.: A numerical method for the simulation of free surface flows with surface tension. Computers and Fluids **35**(10), 1205–1216 (2006)
9. Caboussat, A., Clausen, P., Rappaz, J.: Numerical simulation of two-phase flow with interface tracking by adaptive Eulerian grid subdivision. Math. Comput. Modelling **55**, 490–504 (2012)
10. Caboussat, A., James, N., Boyaval, S., Picasso, M.: Numerical simulation of free surface flows, with multiple liquid phases. In: E. Onate, J. Oliver, A. Huerta (eds.) Proceedings of the 11th World Congress on Computational Mechanics (WCCM XI), pp. 5381–5391 (2014)
11. Caboussat, A., Maronnier, V., Picasso, M., Rappaz, J.: Numerical simulation of three dimensional free surface flows with bubbles. In: Challenges in Scientific Computing—CISC 2002, *Lect. Notes Comput. Sci. Eng.*, vol. 35, pp. 69–86. Springer, Berlin (2003)
12. Caboussat, A., Picasso, M., Rappaz, J.: Numerical simulation of free surface incompressible liquid flows surrounded by compressible gas. J. Comput. Phys. **203**(2), 626–649 (2005)
13. Choi, B.Y., Bussmann, M.: A piecewise linear approach to volume tracking a triple point. Int. J. Numer. Methods Fluids **53**(6), 1005–1018 (2007)
14. Douglas, J., Rachford, H.H.: On the numerical solution of heat conduction problems in two and three space variables. Trans. Amer. Math. Soc. **82**, 421–439 (1956)
15. Glowinski, R.: Finite element methods for incompressible viscous flow. In: P.G. Ciarlet, J.L. Lions (eds.) Handbook of Numerical Analysis, vol. IX, pp. 3–1176. North-Holland, Amsterdam (2003)
16. Glowinski, R., Dean, E.J., Guidoboni, G., Juárez, L.H., Pan, T.W.: Applications of operator-splitting methods to the direct numerical simulation of particulate and free-surface flows and to the numerical solution of the two-dimensional elliptic Monge-Ampère equation. Japan J. Indust. Appl. Math. **25**(1), 1–63 (2008)
17. Gunes, D.Z., Bercy, M., Watske, B., Breton, O., Burbidge, A.S.: A study of extensional flow induced coalescence in microfluidic geometries with lateral channels. Soft Matter **9**, 7526–7537 (2013)
18. Hughes, E., Maan, A.A., Acquistapace, S., Burbidge, A., Johns, M.L., Gunes, D.Z., Clausen, P., Syrbe, A., Hugo, J., Schroen, K., Miralles, V., Atkins, T., Gray, R., Homewood, P., Zick, K.: Microfluidic preparation and self diffusion PFG-NMR analysis of monodisperse water-in-oil-in-water double emulsions. J. Colloid and Interface Science **389**, 147–156 (2013)
19. James, N., Boyaval, S., Caboussat, A., Picasso, M.: Numerical simulation of 3D free surface flows, with multiple incompressible immiscible phases. Applications to impulse waves. Int. J. Numer. Meth. Fluids. **76**(12), 1004–1024 (2014)
20. Jouvet, G., Huss, M., Blatter, H., Picasso, M., Rappaz, J.: Numerical simulation of Rhonegletscher from 1874 to 2100. J. Comput. Phys. **228**(17), 6426–6439 (2009)
21. Jouvet, G., Picasso, M., Rappaz, J., Huss, M., Funk, M.: Modelling and numerical simulation of the dynamics of glaciers including local damage effects. Math. Model. Nat. Phenom. **6**(5), 263–280 (2011)
22. Kratzer, A., Handschin, S., Lehmann, V., Gross, D., Escher, F., Conde-Petit, B.: Hydration dynamics of durum wheat endosperm as studied by magnetic resonance imaging and soaking experiments. Cereal Chemistry **85**(5), 660–666 (2008)
23. Marchuk, G.I.: Methods of Numerical Mathematics. Springer-Verlag, New York (1975)
24. Marchuk, G.I.: Splitting and alternating direction methods. In: P.G. Ciarlet, J.L. Lions (eds.) Handbook of Numerical Analysis, vol. I, pp. 197–462. North-Holland, Amsterdam (1990)
25. Maronnier, V., Picasso, M., Rappaz, J.: Numerical simulation of free surface flows. J. Comput. Phys. **155**(2), 439–455 (1999)
26. Maronnier, V., Picasso, M., Rappaz, J.: Numerical simulation of three-dimensional free surface flows. Int. J. Numer. Meth. Fluids **42**(7), 697–716 (2003)
27. Nagel, M., Brun, P.T., Gallaire, F.: A numerical study of droplet trapping in microfluidic devices. Physics of Fluids **26**(3), 032,002 (2014)
28. Noh, W.F., Woodward, P.: SLIC (Simple Line Interface Calculation). Proceedings of the Fifth International Conference on Numerical Methods in Fluid Dynamics **59**, 330–340 (1976)
29. Osher, S., Sethian, J.A.: Fronts propagating with curvature-dependent speed: algorithms based on Hamilton-Jacobi formulations. J. Comput. Phys. **79**(1), 12–49 (1988)

30. Öttinger, H.C.: Stochastic Processes in Polymeric Fluids. Springer-Verlag, Berlin (1996)
31. Parolini, N., Burman, E.: A finite element level set method for viscous free-surface flows. In: Applied and industrial mathematics in Italy, *Ser. Adv. Math. Appl. Sci.*, vol. 69, pp. 416–427. World Sci. Publ., Hackensack, NJ (2005)
32. Peaceman, D.W., Rachford, H.H.: The numerical solution of parabolic and elliptic differential equations. J. Soc. Indust. Appl. Math. **3**, 28–41 (1955)
33. Picasso, M., Rappaz, J., Reist, A.: Numerical simulation of the motion of a three-dimensional glacier. Ann. Math. Blaise Pascal **15**(1), 1–28 (2008)
34. Pironneau, O.: On the transport-diffusion algorithm and its applications to the Navier-Stokes equations. Numer. Math. **38**(3), 309–332 (1981/82)
35. Shashkov, M., Wendroff, B.: The repair paradigm and application to conservation laws. J. Comput. Phys. **198**(1), 265–277 (2004)
36. Tome, M., McKee, S.: Numerical simulation of viscous flow: Buckling of planar jets. Int. J. Numer. Meth. Fluids **29**(6), 705–718 (1999)
37. Tryggvason, G., Scardovelli, R., Zaleski, S.: Direct Numerical Simulations of Gas-Liquid Multiphase Flows. Cambridge University Press, Cambridge (2011)
38. Turek, S.: Efficient Solvers for Incompressible Flow Problems, *Lecture Notes in Computational Science and Engineering*, vol. 6. Springer-Verlag, Berlin (1999)
39. Yanenko, N.N.: The Method of Fractional Steps. The Solution of Problems of Mathematical Physics in Several Variables. Springer-Verlag, New York (1971)
40. Ycoor Systems S.A.: cfsflow. http://www.ycoorsystems.com/. Online; accessed September 2014

Chapter 22
An Operator Splitting Approach to the Solution of Fluid-Structure Interaction Problems in Hemodynamics

Martina Bukač, Sunčica Čanić, Boris Muha, and Roland Glowinski

Abstract We present a loosely coupled partitioned method for the numerical simulation of a class of fluid-structure interaction problems in hemodynamics. This method is based on a time discretization by an operator-splitting scheme of the Lie's type. The structure is assumed to be thin and modeled by the Koiter shell or membrane equations, while the fluid is modeled by the 3D Navier-Stokes equations for an incompressible viscous fluid. The fluid and structure are coupled via a full two-way coupling taking place at the moving fluid-structure interface, thus giving rise to a nonlinear moving-boundary problem. The Lie splitting decouples the fluid and structure sub-problems and is designed in such a way that the resulting partitioned scheme is unconditionally stable, without the need for any sub-iterations at every time step. Unconditional stability of the scheme is discussed using energy estimates, and several numerical examples are presented, showing that the scheme is first-order

---

M. Bukač (✉)
Department of Applied and Computational Mathematics and Statistics,
University of Notre Dame, 153 Hurley Hall, Notre Dame, IN 46556, USA
e-mail: mbukac@nd.edu

S. Čanić
Department of Mathematics, University of Houston, 4800 Calhoun Rd., Houston, TX 77025, USA
e-mail: canic@math.uh.edu

B. Muha
Faculty of Science, Department of Mathematics, University of Zagreb, Bijenicka 30, 10000 Zagreb, Croatia
e-mail: borism@math.hr

R. Glowinski
Department of Mathematics, University of Houston, Houston, TX 77204, USA
e-mail: roland@math.uh.edu

R. Glowinski et al. (eds.), *Splitting Methods in Communication, Imaging, Science, and Engineering*, Scientific Computation, DOI 10.1007/978-3-319-41589-5_22

accurate in time. Implementation simplicity, computational efficiency, modularity, and unconditional stability make this scheme particularly appealing for solving FSI in hemodynamics.

1 Introduction

We consider the flow of an incompressible, viscous fluid in a 3D domain, see Figure 22.1, with compliant (elastic/viscoelastic) walls, which are assumed to be thin. The fluid flow is modeled by the 3D Navier-Stokes equations, while the elastodynamics of the structure, i.e., the elastic walls, is modeled by the Koiter shell, or membrane equations. The fluid and structure are coupled via a two-way coupling:

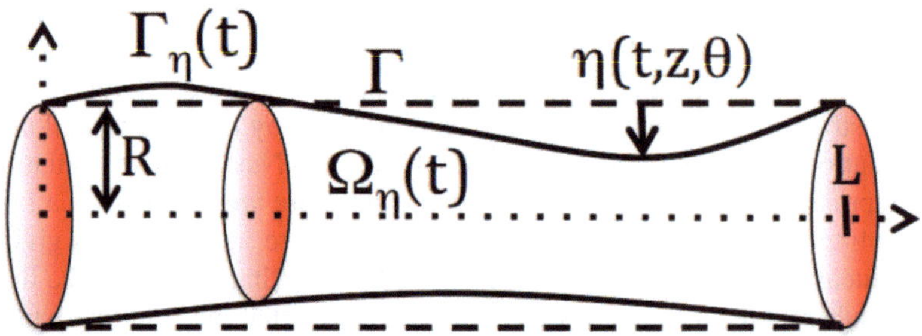

Fig. 22.1 Domain sketch and notation.

the fluid influences the motion of the structure via the normal fluid stress, while the structure influences the motion of the fluid through the motion of the fluid domain boundary. This coupling is assumed through two coupling conditions: the kinematic coupling condition stating the continuity of velocity at the fluid-structure interface (the no-slip condition), and the dynamic coupling condition stating the second Newton's law of motion describing the elastodynamics of the thin structure loaded by the normal fluid stress. The resulting fluid-structure interaction (FSI) problem is a nonlinear moving-boundary problem.

This is a classical problem in hemodynamics describing the interaction between blood flow and elastic/viscoelastic arterial walls. The main difficulty in studying this problem stems from the fact that the fluid and structure have comparable densities, which is associated with the well-known added mass effect. The structure moves within the fluid as if an additional mass was added to it due to the presence of the surrounding fluid. Mathematically, this gives rise to a highly nonlinear moving-boundary problem, where the geometric nonlinearity due to the motion of the relatively light structure driven by the fluid of comparable density, needs to be resolved carefully. It is now well known that this is the main reason for the instabilities in Dirichlet-Neumann loosely coupled schemes that are based on numerically solving this FSI problem by iterating once between the fluid and structure sub-problems

[14], employing the Dirichlet boundary condition in the fluid sub-problem. The added mass effect, the associated geometric nonlinearities, and the multi-physics nature of the problem incorporating different physical effects (wave propagation vs. diffusion) taking place at disparate time scales, are the main reasons why this class of FSI problems remains to be challenging, both from the computational and theoretical points of view.

The development of numerical solvers for fluid-structure interaction problems has become particularly active since the 1980s [67, 68, 26, 34, 55, 59, 40, 39, 42, 41, 23, 47, 46, 53, 54, 71, 69, 3, 74, 25, 27, 51, 52, 20, 33].

Until recently, only monolithic algorithms seemed applicable to blood flow simulations [33, 36, 66, 76, 6, 7]. These algorithms are based on solving the entire nonlinear coupled problem as one monolithic system. They are, however, generally quite expensive in terms of computational time, programming time, and memory requirements, since they require solving a sequence of strongly coupled problems using, e.g., fixed point and Newton's methods [57, 66, 22, 31, 46, 72].

The multi-physics nature of the blood flow problem strongly suggests to employ partitioned (or staggered) numerical algorithms, where the coupled fluid-structure interaction problem is separated into a fluid and a structure sub-problem. The fluid and structure sub-problems are integrated in time in an alternating way, and the coupling conditions are enforced asynchronously. When the density of the structure is much larger than the density of the fluid, as is the case in aeroelasticity, it is sufficient to solve, at every time step, just one fluid sub-problem and one structure sub-problem to obtain a solution. The classical loosely coupled partitioned schemes of this kind typically use the structure velocity in the fluid sub-problem as Dirichlet data for the fluid velocity (enforcing the no-slip boundary condition at the fluid-structure interface), while in the structure sub-problem the structure is loaded by the fluid normal stress calculated in the fluid sub-problem. These Dirichlet-Neumann loosely coupled partitioned schemes work well for problems in which the structure is much heavier than the fluid. Unfortunately, when fluid and structure have comparable densities, which is the case in blood flow applications, the simple strategy of separating the fluid from the structure suffers from severe stability issues [14, 58] associated with the added mass effect. The added mass effect reflects itself in Dirichlet-Neumann loosely coupled partitioned schemes by causing poor approximation of the total energy of the coupled problem at every time step of the scheme. A partial solution to this problem is to iterate several times between the fluid and structure sub-solvers at every time step until the energy of the continuous problem is well approximated. These strongly coupled partitioned schemes, however, are computationally expensive and may suffer from convergence issues for certain parameter values [14].

To get around these difficulties, and to retain the main advantages of loosely coupled partitioned schemes such as modularity, implementation simplicity, and low computational costs, several new loosely coupled algorithms have been proposed recently. In general, they behave quite well for FSI problems containing a thin fluid-structure interface with mass [4, 10, 8, 43, 66, 28, 32, 29, 30, 1, 2, 5, 69, 64, 22, 21].

Recently, a novel loosely coupled partitioned scheme, called the Kinematically Coupled β-Scheme, was introduced by Bukač, Čanić et al. in [10, 8], and applied to 2D FSI problems with thin elastic and viscoelastic structures, modeled by the membrane or shell equations. This method was then extended to thick structure problems modeled by the equations of 2D elasticity [9], to 2D FSI problems with composite structures composed of multiple structural layers [63, 11], to 2D FSI problems with multiple poroelastic layers [12], FSI problems involving endovascular stents [60], and to an FSI problem with non-Newtonian fluids [56, 48]. This scheme deals successfully with the stability issues associated with the added mass effect in a way different from those reported above. Stability is achieved by combining the structure inertia with the fluid sub-problem to mimic the energy balance of the continuous, coupled problem. It was shown in [13] by considering a simplified problem, first used in [14] to study stability of loosely coupled schemes, that our scheme is unconditionally stable for all $0 \leq \beta \leq 1$, even for the parameters associated with blood flow applications. Additionally, Muha and Čanić showed that a version of this scheme with $\beta = 0$ converges to a weak solution of the fully nonlinear FSI problem [61]. The case $\beta = 0$ considered in [61] corresponds to the classical kinematically coupled scheme, first introduced in [43]. Parameter β was introduced in [10] to increase the accuracy of the scheme. A different approach to increasing the accuracy of the classical kinematically coupled scheme was recently proposed by Fernández et al. [28, 32, 29]. Their modified kinematically coupled scheme called "the incremental displacement-correction scheme" treats the structure displacement explicitly in the fluid sub-step and then corrects it in the structure sub-step. Fernández et al. showed that the accuracy of the incremental displacement-correction scheme is first-order in time. The results were obtained for a FSI problem involving a thin elastic structure.

These recent results indicate that the kinematically coupled scheme and its modifications provide an appealing way to study multi-physics problems involving FSI.

While all the results so far related to the kinematically coupled β-scheme have been presented in 2D, here we show that this scheme, in combination with the Arbitrary Lagrangian-Eulerian approach, can successfully be extended to three space dimensions, and to problems without axial symmetry. We consider a FSI problem which consists of the 3D Navier-Stokes equations for an incompressible, viscous fluid, coupled with the linearly elastic Koiter membrane/shell equations. We show an energy estimate for the fully coupled nonlinear problem with $\beta = 0$, which, together with the convergence result of Muha and Čanić in [62], implies unconditional stability of the scheme. Using FreeFem++ [44, 45] we implemented the scheme for a few examples in 3D geometries: a 3D straight tube, a 3D curved tube, and a complex stenotic geometry which is not axially symmetric. We tested our solver against a monolithic solver on a 2D benchmark problem in blood flow [35], showing excellent agreement. Based on numerical results we show that the scheme has at least 1st-order accuracy in time both in 2D and 3D.

2 Model Description

We consider the flow of an incompressible, viscous fluid in a three-dimensional cylindrical domain which is not necessarily axially symmetric. See Figure 22.1. We will be assuming that the lateral boundary of the cylinder is deformable and that its location is not known *a priori*. The motion of the lateral boundary is fully coupled via a two-way coupling to the flow of the incompressible, viscous fluid occupying the fluid domain. Furthermore, it will be assumed that the lateral boundary is a thin, isotropic, homogeneous structure, whose displacement depends on both the axial variable z and on the azimuthal angle θ, thereby accounting for both axially symmetric and non-axially symmetric displacements. Additionally, for simplicity, we will be assuming that only the radial component of displacement is non-negligible. The radial displacement from the reference configuration will be denoted by $\eta(t,z,\theta)$. See Figure 22.1. This is a common assumption in blood flow modeling [71]. *Neither the fluid flow nor the displacement of the lateral boundary of the fluid domain will be required to satisfy the conditions of axial symmetry.*

Remark on notation: We will be using (z,x,y) to denote the Cartesian coordinates of points in $\mathbb{R}^3$, and (z,r,θ) to denote the corresponding cylindrical coordinates. We will be working with the fluid flow equations written in Cartesian coordinates, while the structure equations will be given in cylindrical coordinates. A function f given in Cartesian coordinates defines a function

$$\tilde{f}(z,r,\theta) = f(z,x,y)$$

defined in cylindrical coordinates. Since no confusion is possible, to simplify notation we will omit the superscript $\tilde{}$ and both functions, f and $\tilde{f}$, will be denoted by f.

The structural problem: Consider a clamped cylindrical shell of thickness h, length L, and reference radius of the middle surface equal to R. See Figure 22.1. This reference configuration, which we denote by Γ, can be defined via the parameterization

$$\varphi : \omega \to \mathbb{R}^3, \quad \varphi(z,\theta) = (R\cos\theta, R\sin\theta, z)^t,$$

where $\omega = (0,L) \times (0,2\pi)$ and $R > 0$. Therefore, the reference configuration is

$$\Gamma = \{\mathbf{x} = (R\cos\theta, R\sin\theta, z) \in \mathbb{R}^3 \ : \ \theta \in (0,2\pi), z \in (0,L)\}. \tag{22.1}$$

The associated covariant A_c and contravariant A^c metric tensors of this (undeformed) cylinder are given by:

$$\mathbf{A}_c = \begin{pmatrix} 1 & 0 \\ 0 & R^2 \end{pmatrix}, \ \mathbf{A}^c = \begin{pmatrix} 1 & 0 \\ 0 & \frac{1}{R^2} \end{pmatrix},$$

and the area element along cylinder Γ is $dS = \sqrt{a}dy := \sqrt{\det A_c}dy = Rdy$. The corresponding curvature tensor in covariant components is given by

$$\mathbf{B}_c = \begin{pmatrix} 0 & 0 \\ 0 & R \end{pmatrix}.$$

Under the action of force, the Koiter shell is deformed. The displacement from the reference configuration Γ of the deformed shell will be denoted by $\boldsymbol{\eta} = \boldsymbol{\eta}(t,z,\theta) = (\eta_z, \eta_\theta, \eta_r)$. We will be assuming that only the radial component of the displacement is different from zero, and will be denoting that component of the displacement by $\eta(t,z,\theta) := \eta_r(t,z,\theta)$, so that $\boldsymbol{\eta} = \eta \mathbf{e}_r$, where $\mathbf{e}_r = \mathbf{e}_r(\theta) = (\cos\theta, \sin\theta, 0)^t$ is the unit vector in the radial direction.

The cylindrical Koiter shell is assumed to be clamped at the end points, giving rise to the following boundary conditions:

$$\eta = \frac{\partial \eta}{\partial n} = 0 \text{ on } \partial\omega.$$

Deformation of a given Koiter shell depends on its elastic properties. The elastic properties of our cylindrical Koiter shell are defined by the following elasticity tensor $\mathscr{A}$:

$$\mathscr{A}\mathbf{E} = \frac{4\lambda\mu}{\lambda + 2\mu}(\mathbf{A}^c \cdot \mathbf{E})\mathbf{A}^c + 4\mu \mathbf{A}^c \mathbf{E} \mathbf{A}^c, \quad \mathbf{E} \in \mathrm{Sym}(\mathscr{M}_2), \tag{22.2}$$

where μ and λ are the Lamé coefficients. Using the following relationships between the Lamé constants and the Young's modulus of elasticity E and Poisson ratio σ:

$$\frac{2\mu\lambda}{\lambda+2\mu} + 2\mu = 4\mu\frac{\lambda+\mu}{\lambda+2\mu} = \frac{E}{1-\sigma^2}, \quad \frac{2\mu\lambda}{\lambda+2\mu} = 4\mu\frac{\lambda+\mu}{\lambda+2\mu}\frac{1}{2}\frac{\lambda}{\lambda+\mu} = \frac{E}{1-\sigma^2}\sigma, \tag{22.3}$$

the elasticity tensor $\mathscr{A}$ can also be written as:

$$\mathscr{A}\mathbf{E} = \frac{2E\sigma}{1-\sigma^2}(\mathbf{A}^c \cdot \mathbf{E})\mathbf{A}^c + \frac{2E}{1+\sigma}\mathbf{A}^c \mathbf{E} \mathbf{A}^c, \quad \mathbf{E} \in \mathrm{Sym}\,(\mathscr{M}_2).$$

A Koiter shell can undergo stretching of the middle surface, and flexure (bending). Namely, the Koiter shell model accounts for both the membrane effects (stretching) and shell effects (flexure). Stretching of the middle surface is measured by the change of metric tensor, while flexure is measured by the change of curvature tensor. By assuming only the radial component of displacement $\eta = \eta(t,r,\theta)$ to be different from zero, the linearized change of metric tensor $\boldsymbol{\gamma}$, and the linearized change of curvature tensor $\boldsymbol{\rho}$, are given by the following:

$$\boldsymbol{\gamma}(\eta) = \begin{pmatrix} 0 & 0 \\ 0 & R\eta \end{pmatrix}, \quad \boldsymbol{\rho}(\eta) = \begin{pmatrix} -\partial_z^2 \eta & -\partial_{z\theta}^2 \eta \\ -\partial_{z\theta}^2 \eta & -\partial_\theta^2 \eta + \eta \end{pmatrix}. \tag{22.4}$$

With the corresponding change of metric and change of curvature tensors we can now formally write the corresponding elastic energy of the deformed shell [17, 18, 16, 50]:

$$E_{el}(\eta) = \frac{h_s}{4}\int_\omega \mathscr{A}\boldsymbol{\gamma}(\eta):\boldsymbol{\gamma}(\eta)Rdzd\theta + \frac{h^3}{48}\int_\omega \mathscr{A}\boldsymbol{\rho}(\eta):\boldsymbol{\rho}(\eta)Rdzd\theta, \tag{22.5}$$

where h_s is the thickness of the shell, and : denotes the Frobenius inner product

$$\mathbf{A}:\mathbf{B} := \mathrm{Tr}\left(\mathbf{A}\mathbf{B}^T\right)\ \mathbf{A},\mathbf{B}\in \mathbf{M}_2(\mathbb{R}) \cong \mathbb{R}^4. \tag{22.6}$$

Given a force $\mathbf{f} = f\mathbf{e}_r$, with surface density f (the radial component), the loaded shell deforms under the applied force, and the corresponding displacement η is a solution to the following elastodynamics problem for the cylindrical linearly elastic Koiter shell, written in weak form: Find $\eta \in H_0^2(\omega)$ such that $\forall\psi \in H_0^2(\omega)$:

$$\rho_K h_s \int_\omega \partial_t^2\eta\,\psi Rdzd\theta + \frac{h_s}{2}\int_\omega \mathscr{A}\boldsymbol{\gamma}(\eta):\boldsymbol{\gamma}(\psi)Rdzd\theta + \frac{h_s^3}{24}\int_\omega \mathscr{A}\boldsymbol{\rho}(\eta):\boldsymbol{\rho}(\psi)Rdzd\theta$$
$$= \int_\omega f\psi Rdzd\theta. \tag{22.7}$$

The operator accounting for the elastic membrane and shell effects in the above equation will be denoted by $\mathscr{L}$:

$$\int_\omega \mathscr{L}\eta\,\psi Rdzd\theta := \frac{h_s}{2}\int_\omega \mathscr{A}\boldsymbol{\gamma}(\eta):\boldsymbol{\gamma}(\psi)Rdzd\theta + \frac{h_s^3}{24}\int_\omega \mathscr{A}\boldsymbol{\rho}(\eta):\boldsymbol{\rho}(\psi)Rdzd\theta, \tag{22.8}$$

for all $\psi \in H_0^2(\omega)$, so that the above weak formulation can be written as

$$\rho_K h_s \int_\omega \partial_t^2\eta\,\psi Rdzd\theta + \int_\omega \mathscr{L}\eta\,\psi Rdzd\theta = \int_\omega f\psi Rdzd\theta,\ \forall\psi\in H_0^2(\omega). \tag{22.9}$$

A calculation shows that the operator $\mathscr{L}$, written in differential form, reads:

$$\begin{aligned}\mathscr{L}\eta &= \frac{h_s^3\mu}{3R^3(\lambda+2\mu)}\Big((\lambda+\mu)\partial_\theta^4\eta + R^4(\lambda+\mu)\partial_z^4\eta + 2R^2(\lambda+\mu)\partial_z^2\partial_\theta^2\eta\\ &\quad R^2\lambda\partial_z^2\eta - 2(\lambda+\mu)\partial_\theta^2\eta + (\lambda+\mu)\eta\Big) + \frac{4h_s}{R}\frac{(\lambda+\mu)\mu}{\lambda+2\mu}\eta.\end{aligned} \tag{22.10}$$

In terms of the Young's modulus of elasticity, and the Poisson ratio, operator $\mathscr{L}$ can be written as:

$$\begin{aligned}\mathscr{L}\eta &= \frac{h_s^3 E}{12R^4(1-\sigma^2)}\left(\partial_\theta^4\eta + R^4\partial_z^4\eta + 2R^2\partial_z^2\partial_\theta^2\eta - 2\partial_\theta^2\eta + \eta\right)\\ &\quad + \frac{h_s^3 E\sigma}{6R^2(1-\sigma^2)}\partial_z^2\eta + \frac{h_s E}{R^2(1-\sigma^2)}\eta.\end{aligned} \tag{22.11}$$

Thus, the elastodynamics of the cylindrical Koiter shell with only radial displacement different from zero, and without the assumption of axial symmetry, is modeled by

$$\rho_K h_s \frac{\partial^2\eta}{\partial t^2} + \mathscr{L}\eta = f, \tag{22.12}$$

where $\mathscr{L}$ is defined by (22.11), and η and f are functions of t, z, and θ, where η denotes the radial component of displacement.

If only the membrane effects are taken into account, the resulting cylindrical Koiter membrane model is given by:

$$\rho_K h_s \frac{\partial^2 \eta}{\partial t^2} + \frac{hE}{R^2(1-\sigma^2)}\eta = f. \tag{22.13}$$

It was stated in [66, 19] that the general Koiter membrane model in *Cartesian coordinates*, with only normal displacement different from zero, takes the form:

$$\rho_K h_s \frac{\partial^2 \eta}{\partial t^2} + C\eta = f, \tag{22.14}$$

where η here is the normal component of displacement in Cartesian coordinates, and the coefficient C is given by

$$C := \frac{h_s E}{1-\sigma^2}(4\kappa_1^2 - 2(1-\sigma)\kappa_2), \tag{22.15}$$

where κ_1 and κ_2 are the mean and Gaussian curvature, respectively.

We mention one more reduced (thin-structure) model which has been used in modeling fluid-structure interaction in hemodynamics. The model was introduced in [33] by integrating the equations of linear elasticity defined on a cylindrical domain in 3D, with respect to the radial direction, after assuming that the material is homogeneous, isotropic, and that all the physical quantities, including the radial stress, are constant in the radial direction. In [33] this model was included in the fluid solver and solved using the so-called *coupled momentum method.* The model was also studied in [19, 75]. It was shown in [75] that this model is well approximated by the following simplified membrane shell model:

$$\rho_K h_s \frac{\partial^2 \eta}{\partial t^2} + C\eta - \frac{Eh_s}{2(1+\sigma)}\frac{\partial^2 \eta}{\partial z^2} = f, \tag{22.16}$$

where C is given by (22.15), and η denoted the normal component of displacement in Cartesian coordinates. The model captures the membrane effects in Cartesian coordinates by the "spring term" $C\eta$, as well as wave propagation modeled by the second-order derivative term.

While the membrane models (22.13), (22.14) do not allow any boundary conditions to be imposed on the displacement at the "inlet" or "outlet" boundaries of the tube, model (22.16) requires two boundary conditions. This model will be considered in Section 5 where we impose zero displacement $\eta = 0$, both at the inlet and outlet of the tube.

The fluid problem: The fluid domain, which depends on time and is not known *a priori*, will be denoted by

$$\Omega_\eta(t) = \{(z,x,y) \in \mathbb{R}^3 : \sqrt{x^2+y^2} < R + \eta(t,z,\theta),\ z \in (0,L)\},$$

and the corresponding lateral boundary by

$$\Gamma_\eta(t) = \{(z,x,y) \in \mathbb{R}^3 : \sqrt{x^2+y^2} = R + \eta(t,z,\theta),\ z \in (0,L)\}.$$

The corresponding reference cylinder is

$$\Omega = \{(z,x,y) \in \mathbb{R}^3 : \sqrt{x^2+y^2} < R,\ z \in (0,L)\}.$$

The lateral boundary of this cylinder, Γ, is defined in (22.1). The inlet and outlet sections of the fluid domain boundary will be denoted by $\Gamma_{in} = \{0\} \times (0,R)$, $\Gamma_{out} = \{L\} \times (0,R)$. See Figure 22.1.

The flow of an incompressible, viscous fluid in $\Omega_\eta(t)$ is modeled by the Navier-Stokes equations, which read, in Cartesian coordinates, as follows:

$$\left.\begin{array}{r}\rho_f(\partial_t \mathbf{u} + \mathbf{u}\cdot\nabla\mathbf{u}) = \nabla\cdot\sigma, \\ \nabla\cdot\mathbf{u} = 0,\end{array}\right\} \text{ in } \Omega_\eta(t),\ t \in (0,T), \tag{22.17}$$

where ρ_f denotes the fluid density, $\mathbf{u}$ the fluid velocity, p the fluid pressure,

$$\sigma = -p\mathbf{I} + 2\mu_F \mathbf{D}(\mathbf{u})$$

is the fluid Cauchy stress tensor, μ_F is the kinematic viscosity coefficient, and $\mathbf{D}(\mathbf{u}) = \frac{1}{2}(\nabla\mathbf{u} + \nabla^t\mathbf{u})$ is the symmetrized gradient of $\mathbf{u}$.

At the inlet and outlet boundary we prescribe the normal stress via:

$$\boldsymbol{\sigma n}_{in} = -p_{in}(t)\boldsymbol{n}_{in} \qquad \text{on } \Gamma_{in} \times (0,T), \tag{22.18}$$

$$\boldsymbol{\sigma n}_{out} = -p_{out}(t)\boldsymbol{n}_{out} \qquad \text{on } \Gamma_{out} \times (0,T), \tag{22.19}$$

where $\boldsymbol{n}_{in}$ and $\boldsymbol{n}_{out}$ are the outward normals to the inlet and outlet fluid boundaries, respectively. Even though not physiologically optimal, these boundary conditions are common in blood flow modeling [4, 65].

Another set of boundary conditions, often helpful in the analysis of this FSI problem, is the dynamic pressure data with zero tangential velocity:

$$\left.\begin{array}{r}p + \dfrac{\rho_f}{2}|\mathbf{u}|^2 = P_{in/out}(t), \\ \mathbf{u}\times\mathbf{e}_z = 0,\end{array}\right\} \quad \text{on } \Gamma_{in/out}, \tag{22.20}$$

where $P_{in/out} \in L^2_{loc}(0,\infty)$ are given. It was shown in [62] that the FSI problem we study in this chapter, with the dynamics pressure data given by (22.20), has a weak solution.

Remark on the inlet and outlet data: In this chapter we will be using the normal stress inlet and outlet data in all the numerical examples, while the analysis of the stability of the scheme will be performed with the dynamic pressure inlet and outlet data.

The **coupling** between the fluid and structure is defined by two sets of boundary conditions satisfied at the lateral boundary $\Gamma_\eta(t)$. They are the kinematic and dynamic lateral boundary conditions describing continuity of velocity (the no-slip condition), and balance of contact forces (i.e., the Second Newton's Law of motion). Written in the Lagrangian framework, with $(z,\theta) \in \omega$, and $t \in (0,T)$, they read:

- **The kinematic condition:**

$$\partial_t \eta(t,z,\theta)\mathbf{e}_r(\theta) = \mathbf{u}(t,z,R+\eta(t,z,\theta),\theta), \tag{22.21}$$

  where $\mathbf{e}_r(\theta) = (\cos\theta, \sin\theta, 0)^t$ is the unit vector in the radial direction.
- **The dynamic condition:**

$$\rho_K h_s \partial_t^2 \eta + \mathscr{L}\eta = -J(t,z,\theta)(\sigma\mathbf{n})|_{(t,z,R+\eta(t,z,\theta))} \cdot \mathbf{e}_r(\theta), \tag{22.22}$$

  where $\mathscr{L}$ is defined by (22.10), or equivalently by (22.11), and

$$J(t,z,\theta) = \sqrt{(1+\partial_z\eta(t,z,\theta)^2)(R+\eta(t,z,\theta))^2 + \partial_\theta\eta(t,z,\theta)^2}$$

  denotes the Jacobian of the composition of the transformation from Eulerian to Lagrangian coordinates and the transformation from cylindrical to Cartesian coordinates.

System (22.17)–(22.22) is supplemented with the following **initial conditions**:

$$\mathbf{u}(0,.) = \mathbf{u}_0, \ \eta(0,.) = \eta_0, \ \partial_t\eta(0,.) = v_0. \tag{22.23}$$

For regularity purposes, used in the existence proof presented in [62], we will be assuming that the initial data satisfies the following compatibility conditions:

$$\begin{gathered}\mathbf{u}_0(z,R+\eta_0(z),\theta)\cdot\mathbf{n}(z,\theta) = v_0(z,\theta)\mathbf{e}_r(\theta)\cdot\mathbf{n}(z,\theta), \quad z\in(0,L),\ \theta\in(0,2\pi),\\ \eta_0 = 0, \quad \text{on } \partial\omega,\\ R+\eta_0(z,\theta) > 0, \quad z\in[0,L],\ \theta\in(0,2\pi).\end{gathered} \tag{22.24}$$

Notice that the last condition requires that the initial displacement is such that the fluid domain has radius strictly greater than zero (i.e., the lateral boundary never collapses).

In summary, we study the following fluid-structure interaction problem:

Problem 1. Find $\mathbf{u} = (u_z(t,z,x,y), u_x(t,z,x,y), u_y(t,z,x,y)), p(t,z,x,y)$, and $\eta(t,z,\theta)$ such that

$$\left.\begin{aligned}\rho_f(\partial_t\mathbf{u}+(\mathbf{u}\cdot\nabla)\mathbf{u}) &= \nabla\cdot\sigma\\ \nabla\cdot\mathbf{u} &= 0\end{aligned}\right\} \text{ in } \Omega_\eta(t),\ t\in(0,T), \tag{22.25}$$

$$\left.\begin{aligned}\mathbf{u} &= \partial_t\eta\mathbf{e}_r,\\ \rho_K h_s\partial_t^2\eta + \mathscr{L}\eta &= -J\sigma\mathbf{n}\cdot\mathbf{e}_r,\end{aligned}\right\} \text{ on } \Gamma,\ t\in(0,T), \tag{22.26}$$

$$\left.\begin{aligned}\sigma\boldsymbol{n}_{in} &= -p_{in}(t)\boldsymbol{n}_{in},\\ \sigma\boldsymbol{n}_{out} &= -p_{out}(t)\boldsymbol{n}_{out},\end{aligned}\right\} \text{on } \Gamma_{in/out},\ t\in(0,T), \tag{22.27}$$

$$\left.\begin{aligned}\mathbf{u}(0,.) &= \mathbf{u}_0,\\ \eta(0,.) &= \eta_0,\\ \partial_t\eta(0,.) &= v_0.\end{aligned}\right\}\text{at } t=0. \tag{22.28}$$

This is a nonlinear, moving-boundary problem in 3D, which captures the full, two-way fluid-structure interaction coupling. The nonlinearity in the problem is represented by the quadratic term in the fluid equations, and by the nonlinear coupling between fluid and structure defined at the lateral boundary $\Gamma_\eta(t)$, which is one of the unknowns in the problem.

2.1 Energy Inequality

To simplify notation, we introduce the following energy norms defined by the membrane and flexural effects of the linearly elastic Koiter shell:

$$\|f\|_\gamma := \int_\omega \mathscr{A}\boldsymbol{\gamma}(f):\boldsymbol{\gamma}(f)R dz d\theta, \quad \|f\|_\sigma := \int_\omega \mathscr{A}\boldsymbol{\sigma}(f):\boldsymbol{\sigma}(f)R dz d\theta. \tag{22.29}$$

Notice that norm $\|.\|_\gamma$ is equivalent to the standard $L^2(\omega)$ norm, and that norm $\|.\|_\sigma$ is equivalent to the standard $H_0^2(\omega)$ norm. Assuming sufficient regularity, and the inlet and outlet data given by a prescribed dynamic pressure, see (22.20), the following energy inequality holds:

Proposition 1. *Assuming sufficient regularity, and the inlet and outlet data given by a prescribed dynamic pressure, the solutions of* (22.25), (22.26), *and* (22.28) *satisfy the following energy estimate:*

$$\frac{d}{dt}(E_{kin}(t)+E_{el}(t))+D(t)\leq C(P_{in}(t),P_{out}(t)), \tag{22.30}$$

where

$$\begin{aligned}E_{kin}(t) &:= \frac{1}{2}\left(\rho_f\|\mathbf{u}\|^2_{L^2(\Omega_\eta(t))}+\rho_K h_s\|\partial_t\eta\|^2_{L^2(\Gamma)}\right),\\ E_{el}(t) &:= \frac{h_s}{4}\|\eta\|_\gamma+\frac{h_s^3}{48}\|\eta\|_\sigma,\end{aligned} \tag{22.31}$$

denote the kinetic and elastic energy of the coupled problem, respectively, and the term $D(t)$ captures viscous dissipation in the fluid:

$$D(t) := \mu_F\|\mathbf{D}(\mathbf{u})\|^2_{L^2(\Omega_\eta(t))}. \tag{22.32}$$

The constant $C(P_{in}(t),P_{out}(t))$ depends only on the inlet and outlet pressure data, which are both functions of time.

The proof of inequality (22.30) is standard (see, e.g., [61]), so we omit it here. This says that if a smooth solution to the coupled fluid-structure interaction problem (22.25) - (22.28) exists, then it satisfies the energy inequality (22.30). This inequality states that the rate of change of the kinetic energy of the fluid, and the elastic energy of the structure, plus the viscous dissipation of the fluid, is balanced by the work done by the inlet and outlet data.

2.2 ALE Formulation

Since the fluid-structure coupling studied here is preformed along the moving fluid-structure interface, the fluid domain $\Omega(t)$ is not fixed. This is a problem from many points of view. In particular, defining the time discretization of the time derivative $\partial\mathbf{u}/\partial t$, for example $\partial\mathbf{u}/\partial t \approx (\mathbf{u}(t^{n+1},.) - \mathbf{u}(t^n,.))/(t^{n+1}-t^n)$, is not well-defined since $\mathbf{u}(t^{n+1},.)$ and $\mathbf{u}(t^n,.)$ are not defined on the same domain at two different time-steps. To resolve this difficulty, a classical approach is to map the fluid domain $\Omega_\eta(t)$ onto a fixed, reference domain Ω via a smooth, invertible ALE mapping [23]:

$$A_\eta : \Omega \rightarrow \Omega_\eta(t).$$

An example of such a mapping is the harmonic extension of the boundary $\partial\Omega_\eta(t)$ onto the fluid domain. This will be used in our numerical simulations. By using the chain rule, one can see that the time derivative of the transformed fluid velocity will contain an additional advection term with its coefficient given by the domain velocity $\mathbf{w}^\eta := (A_\eta)_t \circ (A_\eta)^{-1}$, where $(A_\eta)_t$ denotes the time derivative of A_η.

Another example is an ALE mapping A_η defined by:

$$A_\eta(t) : \Omega \rightarrow \Omega_\eta(t), \quad A_\eta(t)(z,r,\theta) := \begin{pmatrix} z \\ (R+\eta(t,z,\theta))r \\ \theta \end{pmatrix}, \quad (z,r,\theta) \in \Omega, \tag{22.33}$$

where (z,r,θ) denote the cylindrical coordinates in the reference domain Ω. We will be using this explicit formula for ALE mapping in the energy estimate associated with the stability of our splitting scheme, proved in Section 3.2. Since we work with the Navier-Stokes equations written in Cartesian coordinates, it is useful to write an explicit form of the ALE mapping A_η in Cartesian coordinates as well:

$$A_\eta(t)(z,x,y) := \begin{pmatrix} z \\ (R+\eta(t,z,\theta))x \\ (R+\eta(t,z,\theta))y \end{pmatrix}, \quad (z,x,y) \in \Omega. \tag{22.34}$$

Mapping $A_\eta(t)$ is a bijection, and its Jacobian is given by

$$|\det\nabla A_\eta(t)| = (R+\eta(t,z,\theta))^2. \tag{22.35}$$

Composite functions with the ALE mapping will be denoted by

$$\mathbf{u}^\eta(t,.) = \mathbf{u}(t,.) \circ A_\eta(t) \quad \text{and} \quad p^\eta(t,.) = p(t,.) \circ A_\eta(t). \tag{22.36}$$

The derivatives of composite functions satisfy:

$$\nabla \mathbf{u} = \nabla \mathbf{u}^\eta (\nabla A_\eta)^{-1} =: \nabla^\eta \mathbf{u}^\eta, \quad \partial_t \mathbf{u} = \partial_t \mathbf{u}^\eta - (\mathbf{w}^\eta \cdot \nabla^\eta)\mathbf{u}^\eta,$$

where the ALE domain velocity, $\mathbf{w}^\eta$, is given by:

$$\mathbf{w}^\eta = \partial_t \eta \begin{pmatrix} 0 \\ x \\ y \end{pmatrix}. \tag{22.37}$$

The following notation will also be useful:

$$\sigma^\eta = -p^\eta \mathbf{I} + 2\mu \mathbf{D}^\eta(\mathbf{u}^\eta), \quad \mathbf{D}^\eta(\mathbf{u}^\eta) = \frac{1}{2}(\nabla^\eta \mathbf{u}^\eta + (\nabla^\eta)^\tau \mathbf{u}^\eta).$$

Finally, the mapped fluid equations in Ω_η read:

$$\left.\begin{aligned} \rho_F\left(\partial_t \mathbf{u} + ((\mathbf{u} - \mathbf{w}^\eta)\cdot \nabla^\eta)\mathbf{u}\right) &= \nabla^\eta \cdot \boldsymbol{\sigma}^\eta \\ \nabla^\eta \cdot \mathbf{u} &= 0 \end{aligned}\right\} \text{ in } \Omega_\eta(t) \times (0,T). \tag{22.38}$$

Here, the notation $\boldsymbol{\sigma}^\eta$ reflects the dependence of $\mathbf{D}^\eta(\mathbf{u}) = \frac{1}{2}(\nabla^\eta \mathbf{u} + \nabla^{\eta T} \mathbf{u})$ on η. Existence of a weak solution for problem (22.38), (22.26), (22.20), (22.28), was shown in [62]. In this chapter we focus on the design of a computational scheme for this problem. The computational scheme will follow the main steps in the proof, presented in [62], which is based on the Lie operator splitting approach.

The actual numerical simulations at each time step are typically performed on the current (fixed) domain $\Omega_\eta(t^n)$, at a given fixed time t^n, with only the time-derivative calculated on Ω, thereby avoiding the need to calculate the transformed gradients ∇^η. The corresponding continuous problem in ALE form can be written as follows:

Problem 2. Find $\mathbf{u}$, p, and η such that:

$$\left.\begin{aligned} \rho_F\left(\partial_t \mathbf{u}|_\Omega + ((\mathbf{u} - \mathbf{w}^\eta)\cdot \nabla)\mathbf{u}\right) &= \nabla \cdot \boldsymbol{\sigma} \\ \nabla \cdot \mathbf{u} &= 0 \end{aligned}\right\} \text{ in } \Omega_\eta(t) \times (0,T), \tag{22.39}$$

$$\left.\begin{aligned} \mathbf{u} &= \partial_t \eta \mathbf{e}_r, \\ \rho_K h_s \partial_t^2 \eta + \mathscr{L}\eta &= -J\sigma \mathbf{n} \cdot \mathbf{e}_r, \end{aligned}\right\} \text{ on } \Gamma,\ t \in (0,T), \tag{22.40}$$

$$\left.\begin{aligned} \boldsymbol{\sigma n}_{in} &= -p_{in}(t)\boldsymbol{n}_{in}, \\ \boldsymbol{\sigma n}_{out} &= -p_{out}(t)\boldsymbol{n}_{out}, \end{aligned}\right\} \text{ on } \Gamma_{in/out},\ t \in (0,T), \tag{22.41}$$

$$\left.\begin{aligned} \mathbf{u}(0,.) &= \mathbf{u}_0, \\ \eta(0,.) &= \eta_0, \\ \partial_t \eta(0,.) &= v_0. \end{aligned}\right\} \text{ at } t = 0. \tag{22.42}$$

Here, $\partial_t \mathbf{u}|_\Omega$ denotes the time derivative calculated on a reference domain Ω.

3 The Splitting Scheme

3.1 Description of the Splitting Scheme

To solve problem (22.39)–(22.42), we use the Lie or Marchuk-Yanenko splitting strategy. The Lie splitting is particularly useful for multi-physics problems like the one we are studying here. The coupled problem is split so that the different physics in the problem can be solved separately. The main difficulty is to design the Lie splitting strategy so that the resulting numerical scheme is stable and sufficiently accurate. We present here a splitting which leads to an unconditionally stable loosely coupled partitioned scheme. This splitting was first designed in [10] where a 2D benchmark problem was solved. In this chapter we extend this scheme to 3D problems, which, additionally, do not have to satisfy the property of axial symmetry.

It follows from [37] Chapter 6 that the Lie splitting scheme can be described as follows, the differential problem being written as a first-order system in time, namely:

$$\frac{\partial \phi}{\partial t} + F(\phi) = 0 \text{ in } (0,T), \tag{22.43}$$

$$\phi(0) = \phi_0, \tag{22.44}$$

where F is an operator from a Hilbert space into itself. Operator F is then split, in a nontrivial decomposition as

$$F = \sum_{i=1}^{I} F_i. \tag{22.45}$$

The problem is discretized in time by choosing the time step $\triangle t > 0$ and denoting $t^n = n\triangle t$, and $\phi^n = \phi(t^n)$. The initial approximation is given by the initial data $\phi^0 = \phi_0$. For $n \geq 0$, ϕ^{n+1} is computed by solving

$$\frac{\partial \phi_i}{\partial t} + F_i(\phi_i) = 0 \text{ in } (t^n, t^{n+1}), \tag{22.46}$$

$$\phi_i(t^n) = \phi^{n+(i-1)/I}, \tag{22.47}$$

then set $\phi^{n+i/I} = \phi_i(t^{n+1})$, for $i = 1, \ldots .I$. Thus, the value at $t = t^{n+1}$ of the solution of the i-th problem is taken as the initial data for the $(i+1)$-st problem on (t^n, t^{n+1}).

This method is first-order accurate in time. More precisely, if (22.43) is defined on a finite-dimensional space, and if operators F_i are smooth enough, then $\|\phi(t^n) - \phi^n\| = O(\Delta t)$ [37].

To solve the FSI problem (22.39)–(22.42), we split the problem into two subproblems as follows:

1. An elastodynamics problem for the structure, and
2. A fluid problem with suitable boundary conditions involving structure velocity and fluid stress at the boundary.

The structure and the fluid sub-problems are defined in such a way that the energy of the discretized problem approximates well the energy of the continuous problem. To achieve this goal, a key role is played by the kinematic coupling condition, which will be enforced implicitly in both steps of the splitting scheme, keeping the two sub-problems tightly coupled at all times. Indeed, we show below an energy estimate of the semi-discretized problem which is associated with unconditional stability of the scheme, and shows that the energy of the discretized problem mimics well the energy of the continuous problem.

More precisely, we begin by rewriting our coupled problem in first-order form with respect to time. For this purpose we introduce v to denote the trace of the fluid velocity at the moving interface $\Gamma(t)$:

$$v\mathbf{e_r} := \mathbf{u}|_{\Gamma(t)}.$$

The kinematic coupling condition (no-slip) then reads $\partial_t \eta = v$. The system in ALE form is now rewritten by using the above-mentioned notation, and by employing the *kinematic coupling condition* in the thin structure model. This way the kinematic coupling condition will be enforced implicitly everywhere, in all the steps of the splitting scheme. The resulting coupled problem in first-order ALE form is given by the following:

Problem 3. Find $\mathbf{u}$, p, η, and v such that:

$$\left.\begin{aligned} \rho_F\left(\partial_t \mathbf{u}|_{\Omega} + ((\mathbf{u}-\mathbf{w}^{\eta})\cdot\nabla)\mathbf{u}\right) &= \nabla\cdot\boldsymbol{\sigma}, \\ \nabla\cdot\mathbf{u} &= 0, \end{aligned}\right\} \text{on } \Omega_\eta(t),\ t\in(0,T), \tag{22.48}$$

$$\left.\begin{aligned} \mathbf{u} &= v\mathbf{e}_r, \\ v &= \partial_t \eta, \\ \rho_K h_s \partial_t v + \mathscr{L}\eta &= -J\boldsymbol{\sigma}\mathbf{n}\cdot\mathbf{e}_r, \end{aligned}\right\} \text{on } \Gamma,\ t\in(0,T), \tag{22.49}$$

$$\left.\begin{aligned} \boldsymbol{\sigma}\boldsymbol{n}_{in} &= -p_{in}(t)\boldsymbol{n}_{in}, \\ \boldsymbol{\sigma}\boldsymbol{n}_{out} &= -p_{out}(t)\boldsymbol{n}_{out}, \end{aligned}\right\} \text{on } \Gamma_{in/out},\ t\in(0,T), \tag{22.50}$$

$$\mathbf{u}^{\eta}(0,.) = \mathbf{u}_0, \eta(0,.) = \eta_0, v(0,.) = v_0, \quad \text{at} \quad t=0. \tag{22.51}$$

We are now ready to split the problem. For this purpose, observe that the portion $\rho_K h_s \partial_t v = -J\boldsymbol{\sigma}\mathbf{n}\cdot\mathbf{e}_r$ of the dynamic coupling condition is formulated in terms of the trace v of the fluid velocity on Γ (recall that $\boldsymbol{\sigma}$ depends on v); we can, therefore, use this as the lateral boundary condition for the fluid sub-problem. This observation is crucial because keeping the structure inertia term $\rho_K h_s \partial_t v$ together with the inertia of the fluid in the fluid sub-problem is of paramount importance for designing a stable and convergent scheme. This mimics the added mass effect associated with the coupled physical problem, in which the coupled FSI solution dynamics corresponds to structure having combined fluid and structure inertia.

To achieve higher accuracy, we apply the following strategy: the normal fluid stress is split into two parts:

$$\sigma \mathbf{n} = \underbrace{\sigma \mathbf{n} + \beta p \mathbf{n}}_{(I)} \underbrace{-\beta p \mathbf{n}}_{(II)},$$

where $\beta \in [0,1]$, and part (I) is used in the fluid sub-problem, while part (II) in the structure sub-problem. The higher accuracy for $\beta > 0$ is achieved because the new splitting enhances the communication between the fluid and structure by loading the structure with a β portion of the normal fluid stress, which is not present for $\beta = 0$. For $\beta = 0$ we recover the classical kinematically coupled scheme, first introduced in [43]. In this chapter, $\beta = 1$ is used for the numerical simulations since it provides the highest accuracy. The choice of β does not influence the stability of the scheme [10].

The operators F_1 and F_2 in the operator splitting scheme are defined by the following two differential sub-problems:

Problem F1 : STRUCTURE

$$\left.\begin{aligned} \partial_t \eta &= v, \\ \rho_K h_s \partial_t v + \mathscr{L}\eta &= \beta \hat{p}\mathbf{n}, \end{aligned}\right\} \text{ on } \Gamma,$$

Problem F2 : FLUID

$$\left.\begin{aligned} \partial_t \mathbf{u}|_\Omega + ((\hat{\mathbf{u}} - \mathbf{w}^\eta)\cdot\nabla)\mathbf{u} &= \nabla\cdot\sigma, \\ \nabla\cdot\mathbf{u} &= 0, \end{aligned}\right\} \text{ in } \Omega_\eta(t),$$

$$\left.\begin{aligned} \mathbf{u}|_{R+\eta} &= v\mathbf{e}_r, \\ \rho_K h_s \partial_t v + J\sigma\mathbf{n}|_{R+\eta} &= -\beta \hat{p}\mathbf{n}|_{R+\eta}. \end{aligned}\right\} \text{ on } \Gamma.$$

Here $\hat{\mathbf{u}}$ is the value of $\mathbf{u}$ from the previous time step, $\hat{p}$ is the value of p from the previous time step, and $\mathbf{w}^\eta$, which is the domain velocity (the time derivative of the ALE mapping), is obtained from the just calculated Problem F1. The initial data for $\mathbf{u}$ in the fluid domain is given by the solution from the previous time step, while the initial data for the trace v of the fluid velocity on Γ in Problem $F2$ is given by the just calculated velocity of the thin structure $\partial_t \eta$ in Problem $F1$. The corresponding operator splitting scheme is given by the block diagram shown in Figure 22.2.

This is different from the classical loosely coupled schemes. In classical Dirichlet-Neumann loosely coupled scheme, the boundary condition for the fluid subproblem is the Dirichlet condition for the fluid velocity v on Γ given in terms of the structure velocity $\partial\eta/\partial t$, namely $v = \partial\eta/\partial t$, where $\partial\eta/\partial t$ is *calculated at the previous time step*! This inclusion of the structure inertia from the previous time step (explicitly) makes the fluid subproblem unstable for certain parameters values [14]. The main reason for this is that the kinetic energy at this time step includes only the fluid kinetic energy from the current time step, and not the structure kinetic energy, since the thin structure velocity enters in an explicit way.

Therefore, our above-mentioned splitting strategy, that is to keep the thin structure inertia together with the fluid inertia in the fluid sub-step, respects the physical property of added mass effect in FSI problem where the fluid and structure have

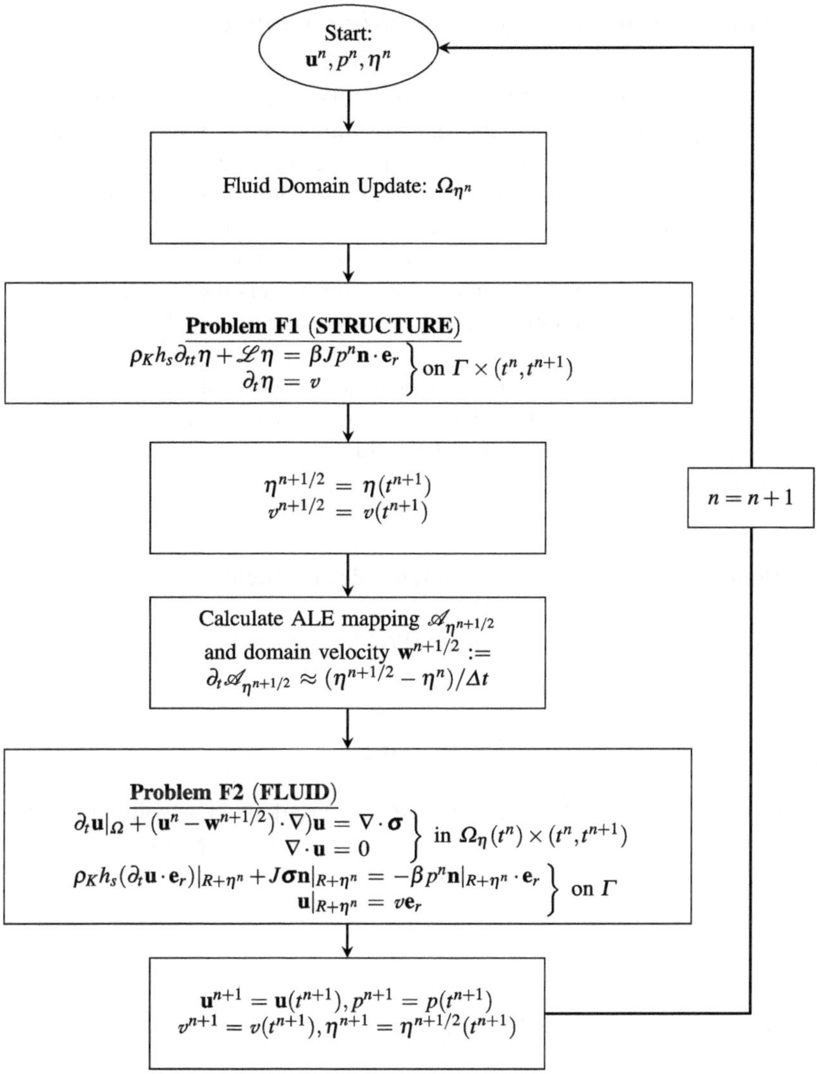

Fig. 22.2 A block diagram showing the main steps of the Kinematically Coupled β-Scheme.

comparable densities, and will give rise to the kinetic energy of the discretized problem which approximates well the kinetic energy of the continuous problem, as we will show next.

3.2 Unconditional Stability of the Splitting Scheme

We will show that the nonlinear FSI problem (22.39)–(22.42), semi-discretized via the Lie operator splitting described above, and summarized in the block diagram, shown in Figure 22.2, satisfies an energy estimate associated with unconditional stability of the operator splitting scheme. Combined with the compactness argument obtained in [62], which shows that the approximating (sub-)sequences of this splitting algorithm converge to a weak solution of problem (22.39)–(22.42), this estimate provides unconditional stability of the splitting scheme. This stability estimate is obtained for the problem containing the dynamic inlet and outlet pressure data (22.20).

To do this, we map the entire problem onto a fixed domain Ω via the ALE mapping (22.34), and perform the operator splitting, described above. The resulting structure elastodynamics problem and the fluid dynamics problem, written in weak form, are given by the following.

3.2.1 Problem F1: The Structure Elastodynamics Problem

The weak form of a semi-discrete version of Problem F1 reads as follows:

- In this problem $\mathbf{u}$ does not change, and so

$$\mathbf{u}^{n+\frac{1}{2}} = \mathbf{u}^n;$$

- The functions $(v^{n+\frac{1}{2}}, \eta^{n+\frac{1}{2}}) \in H_0^2(\omega) \times H_0^2(\omega)$ are defined as solutions of the following problem, written in weak form, where we denote by $d\omega$ the measure $d\omega = R dz d\theta$:

$$\begin{aligned}
&\int_\omega \frac{\eta^{n+\frac{1}{2}} - \eta^n}{\Delta t} \phi \, d\omega = \int_\omega v^{n+\frac{1}{2}} \phi \, d\omega, \\
&\rho_K h_s \int_\omega \frac{v^{n+\frac{1}{2}} - v^n}{\Delta t} \psi d\omega + \frac{h_s}{2} \int_\omega \mathscr{A} \boldsymbol{\gamma}(\eta^{n+\frac{1}{2}}) : \boldsymbol{\gamma}(\psi) d\omega \\
&\qquad + \frac{h_s^3}{24} \int_\omega \mathscr{A} \boldsymbol{\rho}(\eta^{n+\frac{1}{2}}) : \boldsymbol{\rho}(\psi) d\omega = 0,
\end{aligned} \tag{22.52}$$

  for all $(\phi, \psi) \in L^2(\omega) \times H_0^2(\omega)$. The first equation is a weak form of the semi-discretized kinematic coupling condition, while the second equation corresponds to a weak form of the semi-discretized elastodynamics equation.

We will assume that the Lamé coefficients are such that operator $\mathscr{A}$ is coercive, e.g. $\lambda, \mu > 0$. It was shown in [62] that the following existence result and energy estimate hold for this problem.

Proposition 2. *For each fixed $\Delta t > 0$, problem (22.52) with $\lambda, \mu > 0$ has a unique solution $(v^{n+\frac{1}{2}}, \eta^{n+\frac{1}{2}}) \in H_0^2(\omega) \times H_0^2(\omega)$. Moreover, the solution of problem (22.52) satisfies the following discrete energy equality:*

$$E^{n+\frac{1}{2}} + \frac{1}{2}\Big(\rho_K h_s \|v^{n+\frac{1}{2}} - v^n\|^2 + \frac{h_s}{2}\|\eta^{n+\frac{1}{2}} - \eta^n\|_\gamma^2 + \frac{h_s^3}{24}\|\eta^{n+\frac{1}{2}} - \eta^n\|_\sigma^2\Big) = E^n, \tag{22.53}$$

where E^n denotes the kinetic energy of the fluid and structure, and the elastic energy of the Koiter shell of the n-th approximate solution:

$$E^n = \frac{1}{2}\Big(\rho_f \int_\Omega (R+\eta^n)^2|\mathbf{u}^n|^2 dx + \rho_K h_s \|v^n\|^2_{L^2(\omega)} + \frac{h_s}{2}\|\eta^n\|_\gamma^2 + \frac{h_s^3}{24}\|\eta^n\|_\sigma^2\Big), \tag{22.54}$$

while $E^{n+1/2}$ is defined by:

$$E^{n+\frac{1}{2}} = \frac{1}{2}\Big(\rho_f \int_\Omega (R+\eta^n)^2|\mathbf{u}^{n+\frac{1}{2}}|^2 dx + \rho_K h_s \|v^{n+\frac{1}{2}}\|^2_{L^2(\omega)} + \frac{h_s}{2}\|\eta^{n+\frac{1}{2}}\|_\gamma^2 + \frac{h_s^3}{24}\|\eta^{n+\frac{1}{2}}\|_\sigma^2\Big). \tag{22.55}$$

Notice how the three terms in (22.53) that are not included in the expressions E^n and $E^{n+1/2}$ account for the kinetic and elastic energy due to the motion of the fluid domain.

3.2.2 Problem F2: The Fluid Problem

We start by defining the solution space for the fluid velocity on the moving domain $\Omega_\eta(t)$ [15]:

$$\begin{aligned} \mathscr{V}_F(t) = \{&\mathbf{u} = (u_z, u_x, u_y) \in H^1(\Omega_\eta(t))^3 : \nabla\cdot\mathbf{u} = 0, \\ &\mathbf{u}\times\mathbf{e}_r = 0 \text{ on } \Gamma(t),\ \mathbf{u}\times\mathbf{e}_z = 0 \text{ on } \Gamma_{in/out}\}, \end{aligned} \tag{22.56}$$

and then define the solution space for the fluid velocity defined on the mapped, fixed domain Ω by the following:

$$\mathscr{V}_F^\eta = \{\mathbf{u}^\eta(t,\cdot) = \mathbf{u}(t,\cdot)\circ\mathscr{A}_\eta(t) : \mathbf{u}\in\mathscr{V}_F(t)\}.$$

It was shown in [62] that $\mathscr{V}_F^\eta$ is a Hilbert space with the scalar product:

$$\begin{aligned} (\mathbf{u}^\eta, \mathbf{v}^\eta)_{\mathscr{V}_F^\eta} &= \int_\Omega (R+\eta)^2(\mathbf{u}^\eta\cdot\mathbf{v}^\eta + \nabla^\eta\mathbf{u}^\eta : \nabla^\eta\mathbf{v}^\eta)dx \\ &= \int_{\Omega_\eta(t)} (\mathbf{u}\cdot\mathbf{v} + \nabla\mathbf{u} : \nabla\mathbf{v})dx = (\mathbf{u},\mathbf{v})_{H^1(\Omega_\eta(t))}. \end{aligned}$$

The weak form of a semi-discrete version of Problem F2 reads as follows:

- In this problem η does not change, and so

$$\eta^{n+1} = \eta^{n+\frac{1}{2}};$$

- The function $(\mathbf{u}^{n+1}, v^{n+1}) \in \mathscr{V}_F^{\eta^n} \times L^2(\omega)$ is defined as a solution of the fluid sub-problem, written in weak form:

$$
\begin{gathered}
\rho_f \int_\Omega (R+\eta^n)^2 \left(\frac{\mathbf{u}^{n+1} - \mathbf{u}^{n+\frac{1}{2}}}{\Delta t} \cdot \mathbf{q} + \frac{1}{2} \left[(\mathbf{u}^n - \mathbf{w}^{n+\frac{1}{2}}) \cdot \nabla^{\eta^n} \right] \mathbf{u}^{n+1} \cdot \mathbf{q} \right. \\
\left. -\frac{1}{2} \left[(\mathbf{u}^n - \mathbf{w}^{n+\frac{1}{2}}) \cdot \nabla^{\eta^n} \right] \mathbf{q} \cdot \mathbf{u}^{n+1} \right) dx + \rho_f \int_\Omega (R + \frac{\eta^n + \eta^{n+1}}{2}) v^{n+\frac{1}{2}} \mathbf{u}^{n+1} \cdot \mathbf{q} \, dx \\
+2\mu \int_\Omega (R+\eta^n)^2 \mathbf{D}^{\eta^n}(\mathbf{u}^{n+1}) : \mathbf{D}^{\eta^n}(\mathbf{q}) \, dx + R\rho_K h_s \int_\omega \frac{v^{n+1} - v^{n+\frac{1}{2}}}{\Delta t} \psi dz d\theta \\
= P_{in}^n \int_{\Gamma_{in}} (q_z)_{|z=0} dx\, dy - P_{out}^n \int_{\Gamma_{out}} (q_z)_{|z=L} dx\, dy, \\
\text{with } \nabla^{\eta^n} \cdot \mathbf{u}^{n+1} = 0, \quad \mathbf{u}^{n+1}_{|\Gamma} = v^{n+1} \mathbf{e}_r,
\end{gathered}
\tag{22.57}
$$

for all $(\mathbf{q}, \psi) \in \mathscr{V}_F^{\eta^n} \times L^2(\omega)$ such that $\mathbf{q}_{|\Gamma} = \psi \mathbf{e}_r$.

Here $P_{in/out}^n = \frac{1}{\Delta t} \int_{n\Delta t}^{(n+1)\Delta t} P_{in/out}(t) dt$ and $\mathbf{w}^{n+\frac{1}{2}}$, which is the domain velocity defined via the ALE mapping (22.37), is given by

$$
\mathbf{w}^{n+\frac{1}{2}} = v^{n+\frac{1}{2}} \begin{pmatrix} 0 \\ x \\ y \end{pmatrix}.
$$

It was shown in [62] that the following existence result and energy estimate hold for this sub-problem:

Proposition 3. *Let $\Delta t > 0$, and assume that η^ns are such that $R + \eta^n \geq R_{\min} > 0, n = 0, \ldots, N$. Then, the fluid sub-problem defined by (22.57) has a unique weak solution $(\mathbf{u}^{n+1}, v^{n+1}) \in \mathscr{V}_F^{\eta^n} \times L^2(\omega)$. Moreover, the solution of (22.57) satisfies the following energy estimate:*

$$
\begin{gathered}
E^{n+1} + \frac{\rho_f}{2} \int_\Omega (R+\eta^n)^2 |\mathbf{u}^{n+1} - \mathbf{u}^n|^2 dx + \frac{\rho_K h_s}{2} \| v^{n+1} - v^{n+\frac{1}{2}} \|^2_{L^2(\omega)} \\
+ D^{n+1} \leq E^{n+\frac{1}{2}} + C\Delta t ((P_{in}^n)^2 + (P_{out}^n)^2),
\end{gathered}
\tag{22.58}
$$

where P_{in}^n and P_{out}^n are the average inlet and outlet dynamic pressure data, given over the time interval (t^n, t^{n+1}): $P_{in/out} = \frac{1}{\Delta t} \int_{n\Delta t}^{(n+1)\Delta t} P_{in/out}(t) dt$, E^n is the kinetic and elastic energy defined in (22.54), and D^n, the contribution from fluid dissipation is defined by

$$
D^{n+1} = \Delta t \mu_F \int_\Omega (R+\eta^n)^2 |D^{\eta^n}(\mathbf{u}^{n+1})|^2 dx, \; n = 0, \ldots, N-1. \tag{22.59}
$$

The constant C depends only on the parameters in the problem, and not on Δt.

By combining these two results we obtain an energy estimate for the semi-discretized problem in the following way. We begin by bounding the kinetic energy and the elastic energy at time step t^{n+1}:

$$
\begin{aligned}
E^{n+1} &+ \frac{\rho_f}{2}\int_\Omega (R+\eta^n)^2|\mathbf{u}^{n+1}-\mathbf{u}^n|^2 dx + \frac{\rho_K h_s}{2}\|v^{n+1}-v^n\|^2_{L^2(\omega)} \\
&+\frac{h_s}{4}\|\eta^{n+1}-\eta^n\|^2_\gamma + \frac{h_s^3}{48}\|\eta^{n+1}-\eta^n\|^2_\sigma + D^{n+1} \\
\le E^{n+1} &+ \frac{\rho_f}{2}\int_\Omega (R+\eta^n)^2|\mathbf{u}^{n+1}-\mathbf{u}^n|^2 dx + \frac{\rho_K h_s}{2}\|v^{n+1}-v^{n+\frac{1}{2}}\|^2_{L^2(\omega)} \\
+\frac{\rho_K h_s}{2}&\|v^{n+\frac{1}{2}}-v^n\|^2 + \frac{h_s}{4}\|\eta^{n+1}-\eta^n\|^2_\gamma + \frac{h_s^3}{48}\|\eta^{n+1}-\eta^n\|^2_\sigma + D^{n+1}.
\end{aligned}
$$

We use the fact that $\eta^{n+1} = \eta^{n+\frac{1}{2}}$ in the last line to obtain that the above expression equals:

$$
\begin{aligned}
= E^{n+1} &+ \frac{\rho_f}{2}\int_\Omega (R+\eta^n)^2|\mathbf{u}^{n+1}-\mathbf{u}^n|^2 dx + \frac{\rho_K h_s}{2}\|v^{n+1}-v^{n+\frac{1}{2}}\|^2_{L^2(\omega)} \\
+\frac{\rho_K h_s}{2}&\|v^{n+\frac{1}{2}}-v^n\|^2 + \frac{h_s}{4}\|\eta^{n+\frac{1}{2}}-\eta^n\|^2_\gamma + \frac{h_s^3}{48}\|\eta^{n+\frac{1}{2}}-\eta^n\|^2_\sigma + D^{n+1}.
\end{aligned}
$$

From the energy inequality (22.58) we can estimate the first line in the above expression by

$$
\begin{aligned}
E^{n+\frac{1}{2}} + \frac{\rho_K h_s}{2}\|v^{n+\frac{1}{2}}-v^n\|^2 + \frac{h_s}{4}\|\eta^{n+\frac{1}{2}}-\eta^n\|^2_\gamma + \frac{h_s^3}{48}\|\eta^{n+\frac{1}{2}}-\eta^n\|^2_\sigma \\
+C\Delta t((P_{in}^n)^2+(P_{out}^n)^2),
\end{aligned}
$$

and by the energy equality (22.53), the above expression is equal to

$$
= E^n + C\Delta t((P_{in}^n)^2+(P_{out}^n)^2). \tag{22.60}
$$

Therefore, we have just shown that the split, semi-discretized problem satisfies the following energy estimate:

$$
\begin{aligned}
E^{n+1} &+ \frac{\rho_f}{2}\int_\Omega (R+\eta^n)^2|\mathbf{u}^{n+1}-\mathbf{u}^n|^2 dx + \frac{\rho_K h_s}{2}\|v^{n+1}-v^n\|^2_{L^2(\omega)} \\
&+\frac{h_s}{4}\|\eta^{n+1}-\eta^n\|^2_\gamma + \frac{h_s^3}{48}\|\eta^{n+1}-\eta^n\|^2_\sigma + D^{n+1} \\
&\le E^n + C\Delta t((P_{in}^n)^2+(P_{out}^n)^2).
\end{aligned} \tag{22.61}
$$

By using this estimate to further bound the right-hand side from the time level n all the way down to 0, and by recalling that P_{in}^n and P_{out}^n are the average inlet and outlet data over the time interval $(n\Delta t,(n+1)\Delta t)$, one obtains

$$
\begin{aligned}
E^{n+1} &+ \frac{\rho_f}{2}\int_\Omega (R+\eta^n)^2|\mathbf{u}^{n+1}-\mathbf{u}^n|^2 dx + \frac{\rho_K h_s}{2}\|v^{n+1}-v^n\|^2_{L^2(\omega)} \\
&+\frac{h_s}{4}\|\eta^{n+1}-\eta^n\|^2_\gamma + \frac{h_s^3}{48}\|\eta^{n+1}-\eta^n\|^2_\sigma + D^{n+1} \\
\leq E_0 + C&\left\{\Delta t\sum_{n=0}^{N-1}\left(\frac{1}{\Delta t}\int_{n\Delta t}^{(n+1)\Delta t} P_{in}(t)dt\right)^2 + \Delta t\sum_{n=0}^{N-1}\left(\frac{1}{\Delta t}\int_{n\Delta t}^{(n+1)\Delta t} P_{in}(t)dt\right)^2\right\} \\
&\leq E^0 + C\|P_{in}\|^2_{L^2(0,T)} + \|P_{out}\|^2_{L^2(0,T)}.
\end{aligned}
\tag{22.62}
$$

We have just shown an energy estimate associated with the unconditional stability of the splitting scheme for the semi-discretized nonlinear FSI problem. Namely, the following theorem holds:

Theorem 1. *Under the assumption that the diameter of the fluid domain $\Omega_\eta(t)$ is greater than zero, the solutions of the semi-discrete splitting algorithm summarized in the block diagram of Figure 22.2 satisfy the following energy estimate:*

$$
\begin{aligned}
E^{n+1} &+ \frac{\rho_f}{2}\int_\Omega (R+\eta^n)^2|\mathbf{u}^{n+1}-\mathbf{u}^n|^2 dx + \frac{\rho_K h_s}{2}\|v^{n+1}-v^n\|^2_{L^2(\omega)} \\
&+\frac{h_s}{4}\|\eta^{n+1}-\eta^n\|^2_\gamma + \frac{h_s^3}{48}\|\eta^{n+1}-\eta^n\|^2_\sigma + D^{n+1} \\
&\leq E^0 + C\|P_{in}\|^2_{L^2(0,T)} + \|P_{out}\|^2_{L^2(0,T)},
\end{aligned}
\tag{22.63}
$$

where the constant $C > 0$ depends only on the parameters of the problem, E^0 is the kinetic and elastic energy of the initial data, and E^{n+1} denotes the kinetic and elastic energy of the semi-discretized solution at $t^{n+1} = (n+1)\Delta t$, defined by (22.54).

Combined with the compactness arguments in [62], which show that the approximating sequence of the Lie splitting scheme converges strongly to a weak solution of the nonlinear FSI, the energy estimate (22.63) provides unconditional stability of the splitting scheme studied in this chapter.

4 The Numerical Implementation of the Scheme

In this section we present the details of the numerical scheme. As mentioned in Section 2 (**Remark on the inlet and outlet data**), in this section we use the normal stress inlet and outlet data (22.18), (22.19), to drive the problem.

4.1 The Structure Sub-problem

The structure problem is discretized using the Backward Euler scheme, giving rise to the weak formulation of the structure sub-problem which is similar to the one presented in (22.52), except that (22.52) is presented for $\beta = 0$ for which unconditional stability is proved, and here we present this scheme for a general $\beta \in [0,1]$. More precisely, the structure sub-problem reads:

- In this sub-problem the fluid velocity in $\Omega(t^n)$ does not change, and so

$$\mathbf{u}^{n+\frac{1}{2}} = \mathbf{u}^n.$$

- Using the notation introduced in (22.8), the weak formulation for the cylindrical Koiter shell can be written as: Find $(v^{n+\frac{1}{2}}, \eta^{n+\frac{1}{2}}) \in L^2(\omega) \times H_0^2(\omega)$ such that $\forall(\phi, \psi) \in L^2(\omega) \times H_0^2(\omega)$:

$$\int_\omega \frac{\eta^{n+\frac{1}{2}} - \eta^n}{\Delta t} \phi R dz d\theta = \int_\omega v^{n+\frac{1}{2}} \phi R dz d\theta,$$

$$\rho_K h_s \int_\omega \frac{v^{n+\frac{1}{2}} - v^n}{\Delta t} \psi R dz d\theta + \int_\omega \mathscr{L} \eta^{n+\frac{1}{2}} \psi R dr dz = \int_\omega \beta \tilde{p}^n J^n \psi R dz d\theta, \tag{22.64}$$

with $v^n = \mathbf{u}^n|_{\Gamma^n}$,

where ω is the reference domain for the structure, L is defined in (22.8), and J^n is the Jacobian of the transformation from Eulerian to Lagrangian coordinates. Here, by $\tilde{p}^n$ we denoted the trace of the fluid pressure, calculated at time t^n, defined on the *reference configuration* ω via the ALE mapping $A^n : \Omega \to \Omega_{\eta^n}(t^n)$ as follows:

$$\tilde{p}^n = p^n \circ A^n. \tag{22.65}$$

In the numerical implementation of the scheme, however, to avoid calculating the Jacobian J^n, the integral on the right-hand side can be calculated along the current configuration of the structure $\Gamma^n = \Gamma(t^n)$, so that

$$\int_\omega \beta \tilde{p}^n J^n \psi R dz d\theta = \int_{\Gamma^n} \beta p^n \psi dS^n, \tag{22.66}$$

where dS^n is the surface element of Γ^n, and the functions p and $\tilde{p}$, are related through the ALE mapping A^n via (22.65). The same holds for the test functions: the ψ on the left-hand side is defined on ω, while the test function ψ on the right-hand side is defined on Γ^n.

In the case when the Koiter shell equations are reduced to the membrane equation, all the terms multiplying $h_s^3/24$ are considered negligible, and the only term that survives is the non-differentiated term $C\eta$, so that the weak formulation reads: Find $(\eta^{n+\frac{1}{2}}, v^{n+\frac{1}{2}}) \in L^2(\omega) \times L^2(\omega)$ such that $\forall(\phi, \psi) \in L^2(\omega) \times L^2(\omega)$:

$$\int_\omega \frac{\eta^{n+\frac{1}{2}} - \eta^n}{\Delta t} \phi R dr dz = \int_\omega v^{n+\frac{1}{2}} \phi R dr dz,$$

$$\rho_K h_s \int_\omega \frac{v^{n+\frac{1}{2}} - v^n}{\Delta t} \psi R dr dz + \int_\omega C \eta^{n+\frac{1}{2}} \psi R dr dz = \int_\omega \beta \tilde{p}^n J^n \psi R dr dz, \tag{22.67}$$

with $v^n = \mathbf{u}^n|_{\Gamma^n}$.

where ω is the reference domain for the structure, and $\mathbf{u}|_{\Gamma^n}$ is the trace of the fluid velocity on the fluid-structure interface calculated in the previous time-step. For the cylindrical Koiter membrane, the coefficient C is given by

$$C = \frac{h_s E}{R^2(1-\sigma^2)}.$$

For a smooth enough domain which is **not necessarily cylindrical**, the weak form in Cartesian coordinates reads: Find $(\eta^{n+\frac{1}{2}}, v^{n+\frac{1}{2}}) \in L^2(\Gamma) \times L^2(\Gamma)$

$$\begin{aligned}
&\int_\Gamma \frac{\eta^{n+\frac{1}{2}} - \eta^n}{\Delta t} \phi dS = \int_\Gamma v^{n+\frac{1}{2}} \phi dS, \ \forall \phi \in L_0^2(\Gamma), \\
&\rho_K h_s \int_\Gamma \frac{v^{n+\frac{1}{2}} - v^n}{\Delta t} dS + \int_\Gamma C \eta^{n+\frac{1}{2}} \psi dS = \int_\Gamma \beta \tilde{p}^n J^n \psi dS, \ \forall \psi \in L_0^2(\Gamma), \\
&\text{with } v^n = \mathbf{u}^n|_{\Gamma^n},
\end{aligned} \tag{22.68}$$

where Γ is the reference configuration of the structure in Cartesian coordinates, and $\mathbf{u}|_{\Gamma^n}$ is the trace of the fluid velocity on the fluid-structure interface calculated in the previous time-step. The coefficient C is given by (see [66, 19]):

$$C := \frac{h_s E}{1-\sigma^2}(4\kappa_1^2 - 2(1-\sigma)\kappa_2), \tag{22.69}$$

with κ_1 and κ_2 being the mean and Gaussian curvature, respectively. Function η here is the normal component of displacement written in Cartesian coordinates. As before, to avoid calculating the Jacobian J^n, the right-hand side of equation (22.68) can be calculated by converting everything to the current domain so that

$$\int_\Gamma \beta \tilde{p}^n J^n \psi dS = \int_{\Gamma^n} \beta p^n \psi dS^n. \tag{22.70}$$

In the examples that follow, we will be using the membrane models, first in cylindrical coordinates, and then in Cartesian coordinates for a stenotic geometry which is not axially symmetric.

Since the structure displacement does not change in the fluid sub-problem, we define:

$$\eta^{n+1} = \eta^{n+\frac{1}{2}}.$$

4.2 Calculation of the ALE Mapping and ALE Velocity $\mathbf{w}^{n+1}$

Using the just-calculated new position of the thin structure we calculate the ALE mapping A^{n+1} associated with the new structure position as a harmonic extension of the boundary to the entire fluid domain:

$$\nabla^2 A^{n+1} = 0 \quad \text{in } \Omega,$$
$$A^{n+1}|_{\Gamma} = \eta^{n+1},$$
$$A^{n+1}|_{\partial\Omega^f \setminus \Gamma} = 0.$$

Using this ALE mapping we calculate the new ALE velocity $\mathbf{w}$ via

$$\mathbf{w}^{n+1} = \frac{\partial A^{n+1}}{\partial t} = \frac{\partial \boldsymbol{x}}{\partial t} \approx \frac{\boldsymbol{x}^{n+1} - \boldsymbol{x}^n}{\Delta t},$$

which remains unchanged in the fluid sub-problem, below.

4.3 The Fluid Sub-problem

We discretize the fluid problem using the Backward Euler scheme, giving rise to the following weak formulation: Find $(\mathbf{u}^{n+1}, p^{n+1}) \in V^f(t^n) \times Q(t^n)$ and $v^{n+1} \in L^2(\Gamma)$ such that for all $(\varphi, q) \in V^f(t^n) \times Q(t^n)$ and $\psi \in L^2(\Gamma)$ satisfying $\varphi|_{\Gamma^n} = [\varphi \circ (A^n)^{-1}]|_{\Gamma} = \psi \mathbf{n}^f$, the following holds:

$$\rho_f \int_{\Omega^f(t^n)} \frac{\mathbf{u}^{n+1} - \mathbf{u}^{n+\frac{1}{2}}}{\Delta t} \cdot \varphi dx + \rho_f \int_{\Omega^f(t^n)} ((\mathbf{u}^{n+\frac{1}{2}} - \mathbf{w}^{n+1}) \cdot \nabla) \mathbf{u}^{n+1} \cdot \varphi dx$$
$$+ 2\mu_f \int_{\Omega^f(t^n)} \boldsymbol{D}(\mathbf{u}^{n+1}) : \boldsymbol{D}(\varphi) dx - \int_{\Omega^f(t^n)} p^{n+1} \nabla \cdot \varphi dx + \int_{\Omega^f(t^n)} q \nabla \cdot \mathbf{u}^{n+1} dx$$
$$+ \rho_s h_s \int_{\Gamma} \frac{v^{n+1} - v^{n+\frac{1}{2}}}{\Delta t} \cdot \psi dS = - \int_{\Gamma} J^n \beta \tilde{p}^n \psi dS$$
$$+ \int_{\Gamma_{in}} p_{in}(t^{n+1}) \varphi|_{z=0} \cdot \boldsymbol{n}^f \, dx \, dy - \int_{\Gamma_{out}} p_{out}(t^{n+1}) \varphi|_{z=L} \cdot \boldsymbol{n}^f \, dx \, dy. \tag{22.71}$$

Here, again, we can use (22.70) to simplify the calculation of the pressure integral over Γ in terms of the integral over Γ^n without the Jacobian J^n:

$$\int_{\Gamma} \beta \tilde{p}^n J^n \psi dS = \int_{\Gamma^n} \beta p^n \psi dS^n.$$

We employed FreeFem++ [44, 45] to solve this problem in 3D, using a finite element approach. Finite dimensional spaces of globally continuous piecewise affine functions (P_1) were used for the space approximation of the structure sub-problem (written in terms of velocity). Concerning the space approximation of the fluid sub-problems (the fluid advection and a quasi-Stokes problem), we proceeded as follows:

(i) Let us denote by $\mathscr{T}_h$ the finite element mesh used to approximate the fluid sub-problem (since we are in 3D, $\mathscr{T}_h$ consists of tetrahedra).

(ii) We divided each element of T_h into four tetrahedra by joining its center of mass to each of its four vertices, the resulting mesh being denoted by $\mathscr{T}_{h/4}$.

(iii) To approximate the pressure (resp. the velocity) we used globally continuous functions, piecewise affine over the elements of $\mathscr{T}_h$ (resp., $\mathscr{T}_{h/4}$).

The resulting approximation of the Stokes problem is known as the P_1 + bubble/P_1, and does not require stabilization (a detailed discussion of the P_1 + bubble/P_1 approximation, for 2D incompressible viscous flow, can be found in e.g., [37]; see also the references therein). For our simulations, the number of elements of T_h was of the order of $8,000$.

However, for the first example presented below, which is a 2D benchmark problem, we used our custom-made code. For this 2D problem, P_1 elements based approximations were used for the structure sub-problem, while the Bercovier-Pironneau method (also known as P_1-iso-P_2/P_1) was used to approximate the fluid sub-problem; again, no stabilization is needed with this approach where each triangle of the pressure mesh T_h is divided into four sub-triangles (by joining the edge mid-points) to define the twice finer mesh $T_{h/2}$ used to approximate the velocity (see Chapter 5 of [37] for more details).

5 Numerical Examples

We begin by presenting a benchmark problem in hemodynamics. Our solver will be validated on this benchmark problem against a monolithic scheme, and the classical kinematically coupled scheme ($\beta = 0$). We show that the accuracy of our operator splitting scheme with $\beta = 1$ is comparable to the accuracy of the monolithic scheme, and has higher accuracy than the classical kinematically coupled scheme ($\beta = 0$). This benchmark problem is in 2D. The remaining examples presented here will be in 3D.

5.1 Example 1: A 2D Benchmark Problem

We consider a classical test problem proposed by Formaggia et al. in [35]. This problem has been used in several works as a benchmark problem for testing the results of fluid-structure interaction algorithms in hemodynamics [4, 65, 5, 70, 43, 10]. The structure model for this benchmark problem is of the form

$$\rho_s h_s \frac{\partial^2 \eta_r}{\partial t^2} - kGh_s \frac{\partial^2 \eta_r}{\partial z^2} + \frac{Eh_s}{1-\sigma^2}\frac{\eta_r}{R^2} - \gamma \frac{\partial^3 \eta_r}{\partial z^2 \partial t} = f, \qquad (22.72)$$

with absorbing boundary conditions at the inlet and outlet boundaries:

$$\frac{\partial \eta_r}{\partial t} - \sqrt{\frac{kG}{\rho_s}} \frac{\partial \eta_r}{\partial z} = 0 \quad \text{at } z = 0 \tag{22.73}$$

$$\frac{\partial \eta_r}{\partial t} + \sqrt{\frac{kG}{\rho_s}} \frac{\partial \eta_r}{\partial z} = 0 \quad \text{at } z = L. \tag{22.74}$$

Here $G = \frac{E}{2(1+\sigma)}$ is the *shear modulus* and k is the *Timoshenko shear correction factor*. The flow is driven by the time-dependent pressure data:

$$p_{in}(t) = \begin{cases} \frac{p_{max}}{2}\left[1-\cos\left(\frac{2\pi t}{t_{max}}\right)\right] & \text{if } t \le t_{max} \\ 0 & \text{if } t > t_{max} \end{cases}, \quad p_{out}(t) = 0 \; \forall t \in (0,T), \tag{22.75}$$

where $p_{max} = 2 \times 10^4$ (dynes/cm^2) and $t_{max} = 0.005$ (s). The values of all the parameters in this model are given in Table 22.1. The problem was solved over the time interval $[0, 0.012]$ s, which is the time it takes the inlet pressure wave to reach the end of the tube.

Parameters	**Values**	**Parameters**	**Values**
Radius R (cm)	0.5	**Length** L (cm)	6
Fluid density ρ_f (g/cm^3)	1	**Dyn. viscosity** μ (poise)	0.035
Wall density ρ_s(g/cm^3)	1.1	**Wall thickness** h_s (cm)	0.1
Young's mod. E(dynes/cm^2)	0.75×10^6	**Poisson's ratio** σ	0.5
Shear mod. G(dynes/cm^2)	0.25×10^6	**Viscoelasticity** γ (poise cm)	0.01
Timoshenko factor k	1		

Table 22.1 Geometry, fluid, and structure parameters for Example 5.1.

Propagation of the corresponding pressure pulse in 2D is shown in Figure 22.6.

The numerical results obtained using the kinematically coupled β scheme with $\beta = 1$ were compared with the numerical results obtained using the classical kinematically coupled scheme (i.e., $\beta = 0$) proposed in [43], and the monolithic scheme proposed in [70]. Figures 22.3, 22.4, and 22.5 show the comparison between tube diameter, flow rate, and mean pressure, respectively, at six different times.

These results were obtained with the same mesh as the one used for a monolithic scheme in [70], containing 31×11 $\mathbb{P}_1$ fluid velocity vertices. More precisely, we used an iso-parametric version (thoroughly discussed in [37] Chapter 5; see also [38]) of the Bercovier-Pironneau element spaces, also known as $\mathbb{P}_1$-iso-$\mathbb{P}_2/\mathbb{P}_1$ approximation of the Stokes problem in which a coarse mesh (mesh size h_p) is used to approximate the pressure, and a twice finer mesh (mesh size $h_v = h_p/2$) is used for the velocity.

The time step used was $\Delta t = 10^{-4}$ which is the same as the time step used for the monolithic scheme, while the time step used for the kinematically coupled scheme in [43] was $\Delta t = 5 \times 10^{-5}$. It is well known that splitting schemes require smaller time step due to the splitting error. However, the splitting studied in this

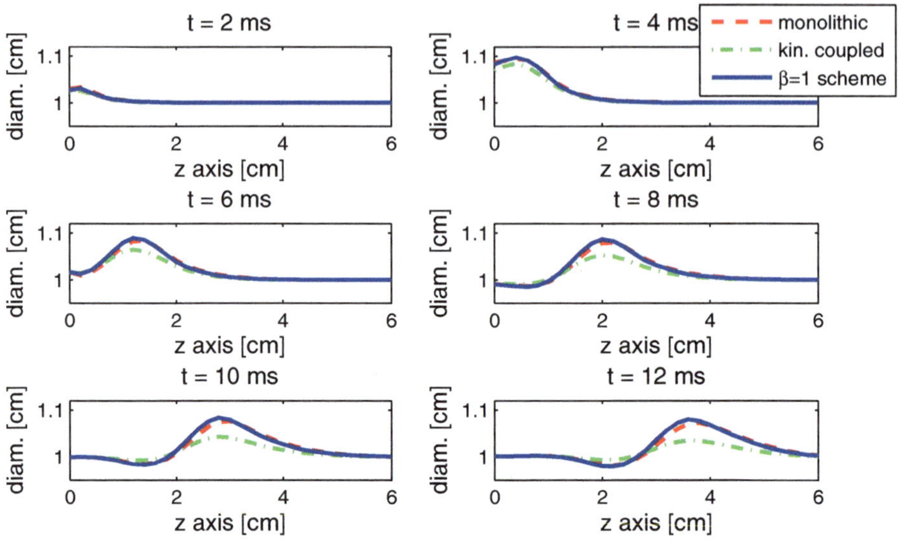

Fig. 22.3 Example 1: Diameter of the tube computed with the kinematically coupled scheme ($\beta = 0$) with time step $\triangle t = 5 \times 10^{-5}$ (dash-dot line), implicit scheme used by Quaini in [70] with the time step $\triangle t = 10^{-4}$ (dashed line), and the kinematically coupled β-scheme ($\beta = 1$) with the time step $\triangle t = 10^{-4}$ (solid line).

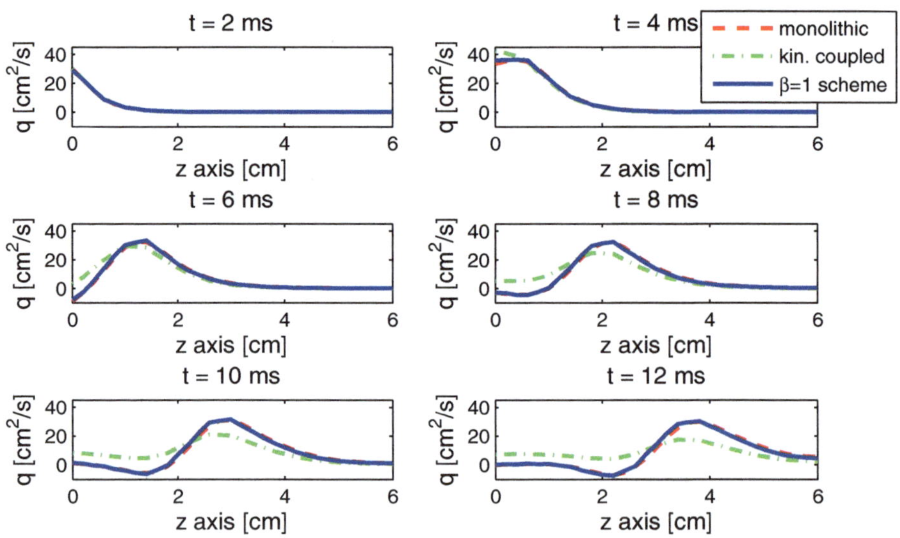

Fig. 22.4 Example 1: Flow rate computed with the kinematically coupled scheme ($\beta = 0$) with time step $\triangle t = 5 \times 10^{-5}$ (dash-dot line), the implicit scheme used by Quaini in [70] with the time step $\triangle t = 10^{-4}$ (dashed line), and our kinematically coupled β-scheme ($\beta = 1$) with the time step $\triangle t = 10^{-4}$ (solid line).

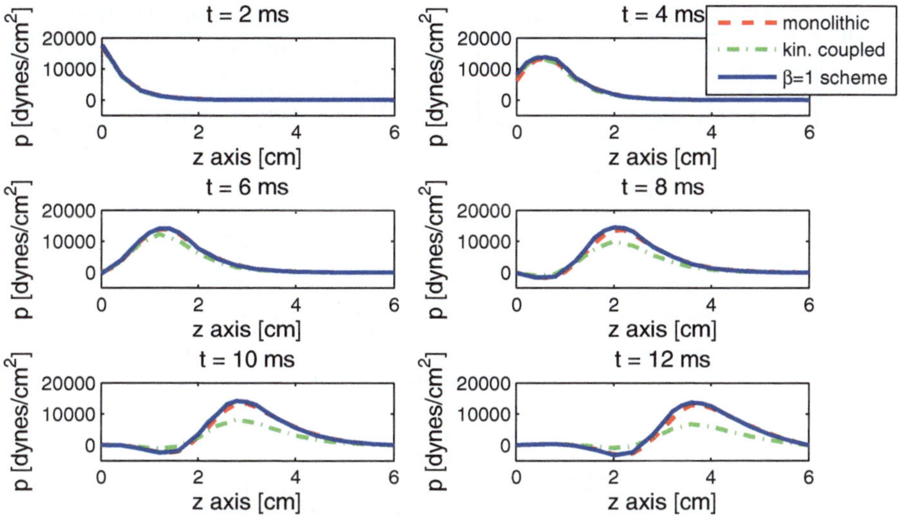

Fig. 22.5 Example 1: Mean pressure computed with the kinematically coupled scheme with time step $\triangle t = 5 \times 10^{-5}$ (dash-dot line), implicit scheme used by Quaini in [70] with the time step $\triangle t = 10^{-4}$ (dashed line) and our scheme with the time step $\triangle t = 10^{-4}$ (solid line).

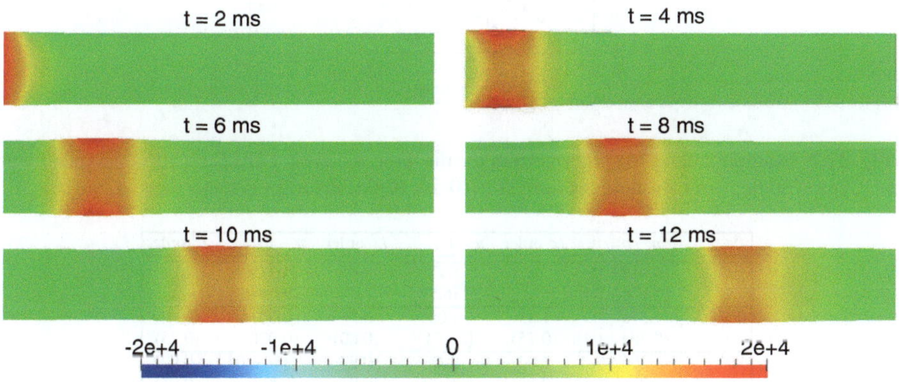

Fig. 22.6 Example 1: Propagation of the pressure wave.

chapter allows us to use the same time step as in the monolithic method, obtaining comparable accuracy, as it will be shown next. This is exciting since we obtain the same accuracy while retaining the main benefits of the partitioned schemes, such as modularity, implementation simplicity, and low computational cost.

Figure 22.7 shows a comparison between the time convergence of the kinematically coupled β-scheme (with $\beta = 1$), the classical kinematically coupled scheme (i.e., $\beta = 0$), and the monolithic scheme used in [70]. The reference solution was defined to be the one obtained with $\triangle t = 10^{-6}$. We calculated the absolute L^2 error for the velocity, pressure, and displacement between the reference solution and the solutions obtained using $\triangle t = 5 \times 10^{-6}, 10^{-5}$, 5×10^{-5} and 10^{-4}. Figure 22.7 and

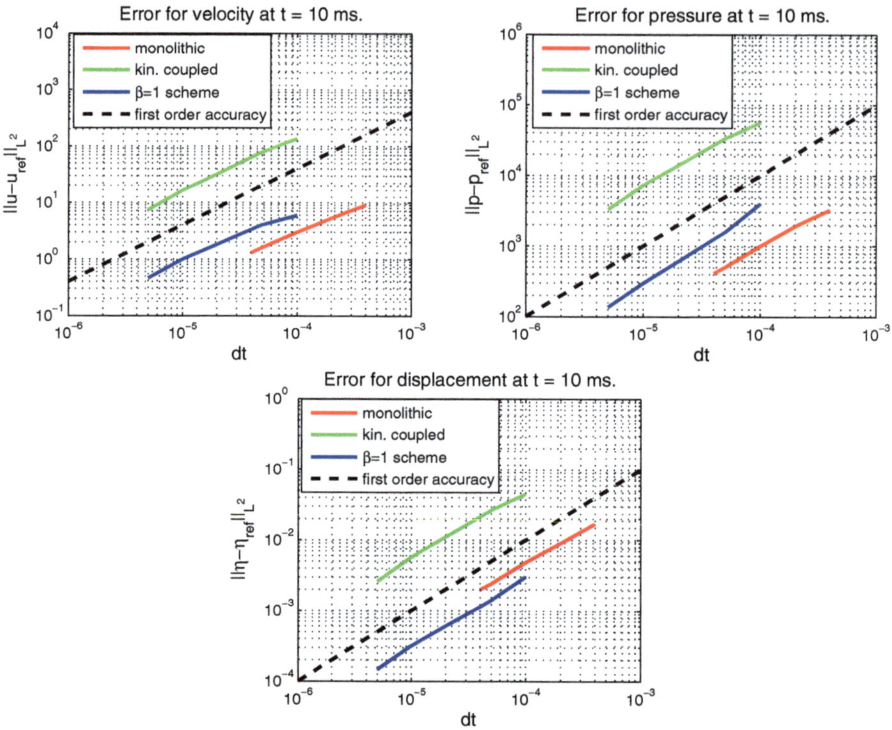

Fig. 22.7 Example 1: Log-log plot of errors for the three schemes. Left: Error for fluid velocity at t=10 ms. Middle: Error for fluid pressure at t=10 ms. Right: Error for displacement at t=10 ms.

$\triangle t$	$\|\|p-p_{ref}\|\|_{L^2}$	L^2 order	$\|\|\boldsymbol{u}-\boldsymbol{u}_{ref}\|\|_{L^2}$	L^2 order	$\|\|\boldsymbol{\eta}-\boldsymbol{\eta}_{ref}\|\|_{L^2}$	L^2 order
10^{-4}	4.01e+03	-	5.97	-	0.003	-
	(5.65e+04)	-	(136.32)	-	(0.0446)	-
5×10^{-5}	1.57e+03	1.35	4.05	0.56	0.0014	1.1
	(3.36e+04)	(0.75)	(77.91)	(0.80)	(0.0264)	(0.75)
10^{-5}	296.36	1.04	1.0	0.87	3.17e−04	0.92
	(7.27e+03)	(0.95)	(16.27)	(0.97)	(0.00576)	(0.95)
5×10^{-6}	134.33	1.14	0.46	1.12	1.45e−04	1.13
	(3.3e+03)	(1.14)	(7.36)	(1.14)	(0.0026)	(1.14)

Table 22.2 Example 1: Convergence in time calculated at $t = 10$ ms. The numbers in the parenthesis show the convergence rate for the kinematically coupled scheme ($\beta = 0$) presented in [43].

Table 22.2 show first-order in time convergence for the velocity, pressure, and displacement obtained by the kinematically coupled scheme, monolithic scheme, and our scheme. Notice how the error of our method is comparable to the error obtained by the monolithic scheme on this 2D benchmark problem.

5.2 Example 2: A 3D Straight Tube Test Case

Here we study the flow in a straight, compliant 3D tube, whose elastodynamics is modeled by the cylindrical membrane shell equation (22.16). Notice that, in relation to the previous example, since the reference configuration is a straight cylinder, this model can be written as

$$\rho_s h_s \frac{\partial^2 \eta}{\partial t^2} - G h_s \frac{\partial^2 \eta}{\partial z^2} + \frac{E h_s}{1-\sigma^2}\frac{\eta}{R^2} = f, \tag{22.76}$$

where $G = \frac{E}{2(1+\sigma)}$ is the shear modulus, as in the previous example, and η denotes the radial component of displacement. We impose the zero-displacement boundary conditions $\eta = 0$ at the "inlet" and "outlet" boundary of the cylinder.

The flow is driven by the time-dependent pressure (normal stress) data:

$$p_{in}(t) = \begin{cases} \frac{p_{max}}{2}\left[1 - \cos\left(\frac{2\pi t}{t_{max}}\right)\right] & \text{if } t \le t_{max} \\ 0 & \text{if } t > t_{max} \end{cases}, \quad p_{out}(t) = 0 \; \forall t \in (0,T), \tag{22.77}$$

where $p_{max} = 1.3333 \times 10^4$ (dyne/cm^2) and $t_{max} = 0.003$ (s). The values of all the parameters in this model are given in Table 22.3.

Fluid Parameters	Values	Structure Parameters	Values
Tube length L(cm)	5	**Thickness** h_s(cm)	0.1
Tube radius R(cm)	0.5	**Density** ρ(g/cm^3)	1.1
Fluid density ρ(g/cm^3)	1	**Young's modulus** E(dyns/cm^2)	10^6
Fluid viscosity μ(poise)	0.035	**Poisson ratio** σ	0.5

Table 22.3 Example 1: The structure parameters for Example 1.

The value of the time step is $\Delta t = 10^{-4}$, and the finite element approximation contains 8571 degrees of freedom.

In contrast with the previous example, the cylindrical membrane model does not contain the bending rigidity term(s), described by the second-order spatial derivative term in (22.72), which is associated with wave propagation phenomena, making equation (22.72) of hyperbolic type (assuming $\gamma = 0$). As a result, the pressure wave and displacement look slightly different in this example when compared with the previous example, as shown in Figures 22.8–22.11.

In particular, Figure 22.8 shows the 3D tube with the corresponding pressure wave propagation at four different times within the time interval from $t = 0$ until $t = 14$ milliseconds, which is the time it takes the pressure wave to reach the outlet boundary. The corresponding values of the pressure along the symmetry axis of the tube are shown in Figure 22.9.

Similarly, Figure 22.10 shows the magnitude of displacement along the 3D tube, at the same four time snap-shots as used in Figures 22.8 and 22.9. The corresponding values of the displacement along the symmetry axis of the tube are shown in Figure 22.11. One can see how the energy dissipates very quickly in this case, and the amplitude of displacement decreases along the tube. The results in Figures 22.8–22.11 are very similar to the results reported in [66], where a pressure wave propagation was shown in a semicircular tube, modeled by the membrane model (22.14), (22.15).

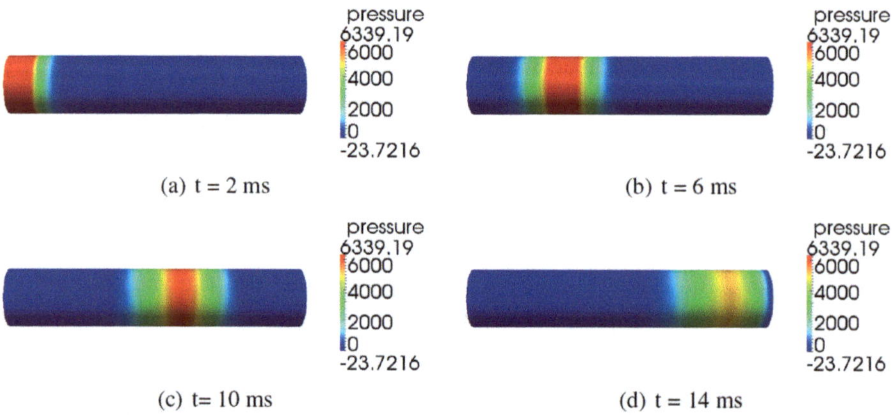

Fig. 22.8 Example 2: Pressure wave propagation in a 3D cylindrical tube, modeled by the cylindrical membrane equation (22.72).

We studied the time-convergence of the scheme solving this 3D problem by refining the time step from $\Delta t = 10^{-4}, 5 \times 10^{-5}, 10^{-5}$, with the reference solution corresponding to the one obtained with $\Delta t = 5 \times 10^{-6}$. Figure 22.12 shows the log-log plot of the error for the fluid velocity versus the time step. A table with the corresponding numbers, showing an "almost" second order convergence, is given in Table 22.4.

Δt	$\|\|\boldsymbol{u} - \boldsymbol{u}_{ref}\|\|_{L^2}$	Conv. Order
10^{-4}	0.71614	–
5×10^{-5}	0.201347	1.83
10^{-5}	0.0122303	1.74

Table 22.4 Example 2: A table showing an "almost" second order convergence.

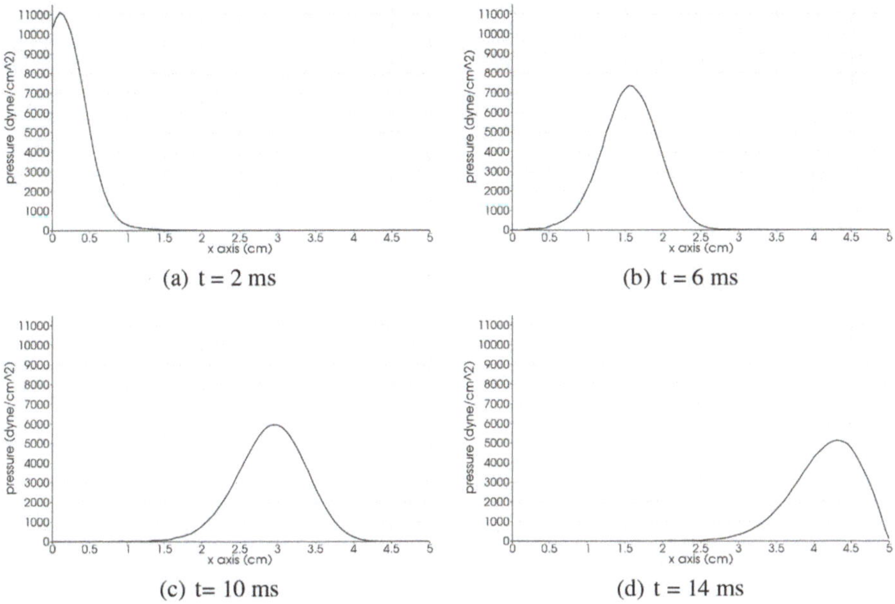

Fig. 22.9 Example 2: Pressure along the axis of symmetry of the tube corresponding to Figure 22.8.

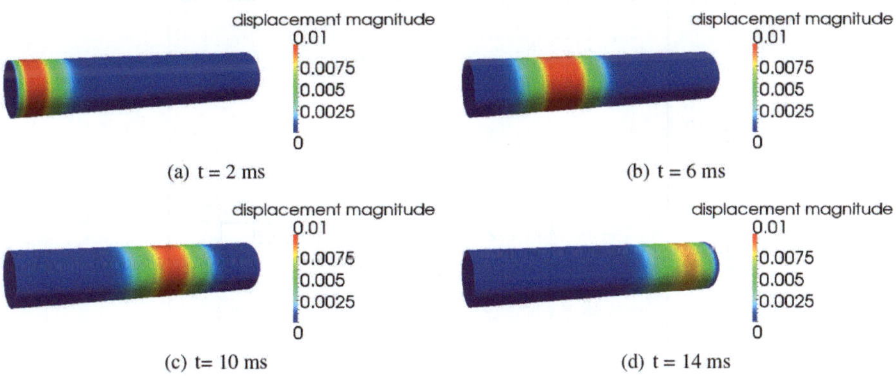

Fig. 22.10 Example 2: Displacement of the 3D cylindrical elastic tube from Figure 22.8.

5.3 Example 3: A 3D Curved Cylinder

Here we consider the structure model (22.16) with C given by (22.15), where η denotes the normal component of displacement. For completeness, we state the model here:

$$\rho_K h_s \frac{\partial^2 \eta}{\partial t^2} - \frac{E h_s}{2(1+\sigma)} \frac{\partial^2 \eta}{\partial z^2} + \frac{h_s E}{1-\sigma^2}(4\kappa_1^2 - 2(1-\sigma)\kappa_2)\eta = f, \tag{22.78}$$

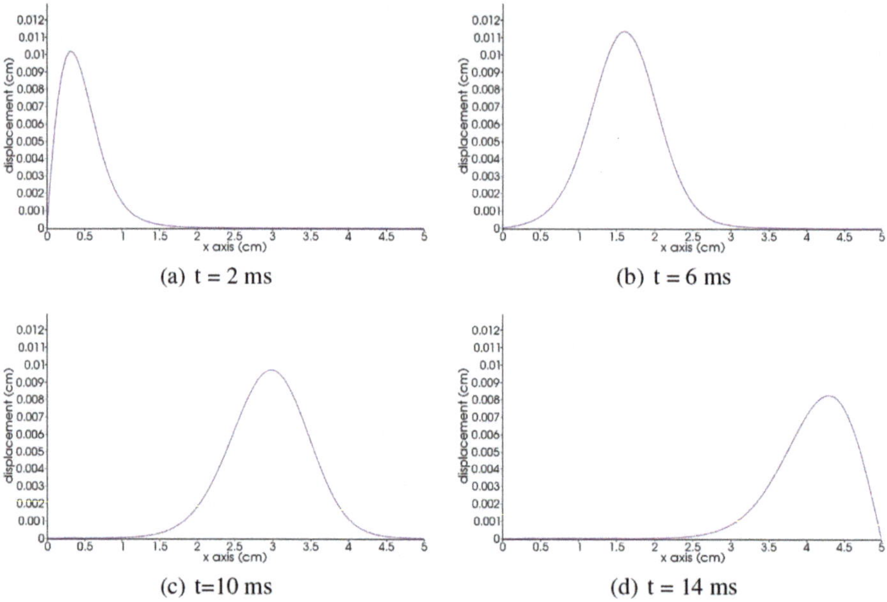

Fig. 22.11 Example 2: Displacement along the tube axis corresponding to Figure 22.10.

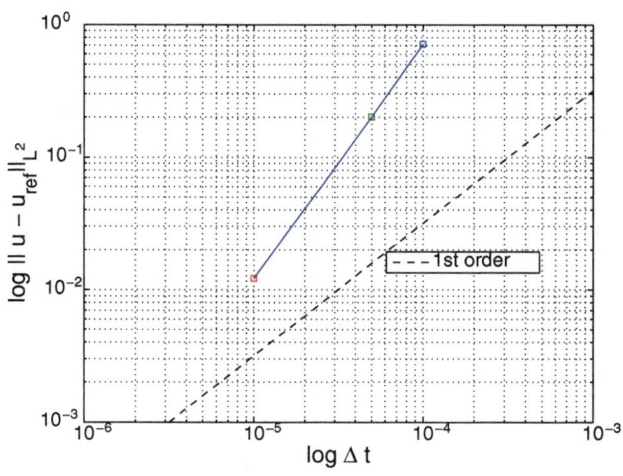

Fig. 22.12 Example 2: The time-convergence test showing the accuracy of order larger than 1. The dashed line in the figure shows the slope corresponding to 1st-order accuracy.

where κ_1 and κ_2 are the mean and Gaussian curvature, respectively. The reference domain is now a semicircular tube, approximating an idealized geometry of the ascending/descending aorta.

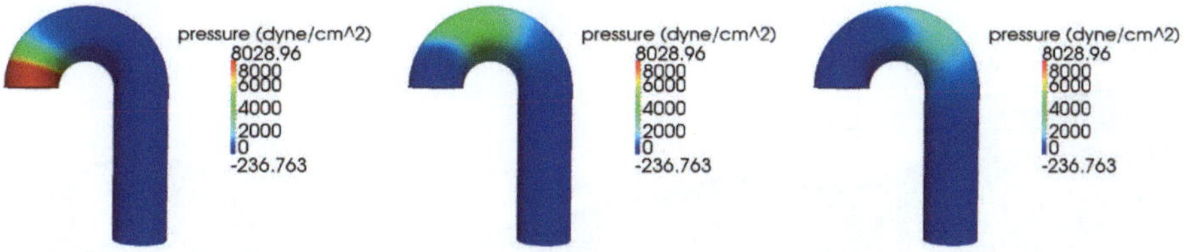

(a) t = 2 ms (b) t = 4 ms (c) t = 7 ms

Fig. 22.13 Example 3: Pressure wave propagation along the axis of symmetry of the curved tube.

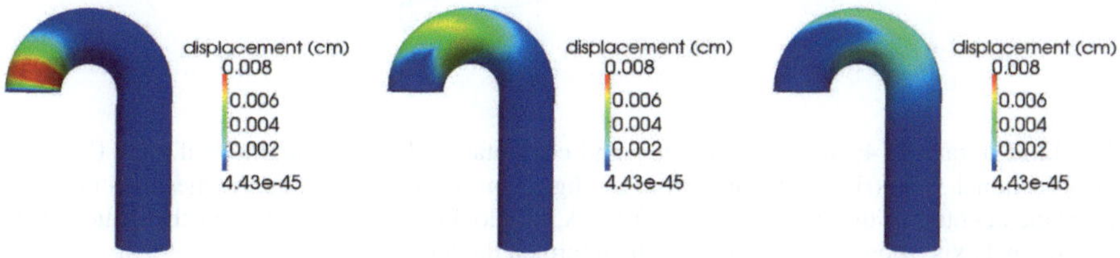

(a) t = 2 ms (b) t = 4 ms (c) t = 7 ms

Fig. 22.14 Example 3: Displacement of the curved tube.

The diameter of the cylinder is constant and equal to $R = 0.5$cm, while the two principal curvatures are given by $4\cos(\theta)/(2+\cos(\theta))$ for $\theta \in [0, 2\pi)$ (Gaussian curvature), and $2(1+\cos(\theta))/(2+\cos(\theta)), \theta \in [0, 2\pi)$ (mean curvature) for the portion of the domain that corresponds to a torus [49]. The other parameters are the same as in the above example and are given in Table 22.3. Figures 22.13 and 22.14 show the pressure and displacement in the curved cylinder. They are very similar to the results obtained by Nobile and Vergara in [66] using the membrane model as a Robin boundary condition in the fluid problem.

5.4 Example 4: Stenosis

In this example we consider a stenotic geometry which is not axially symmetric. Figure 22.15 shows two views of the geometry: the axial view and the cross-sectional view, cut by plane A, shown in Figure 22.15. The corresponding computational mesh is also shown in this figure. The cross-section in Figure 22.15(b) shows around 50% stenosis of the vessel lumen.

The structure elastodynamics is modeled by equation (22.79), where the coefficients now depend on the spatial variable $\mathbf{x}$, since the radius and curvature of the reference configuration are not constant:

$$\rho_K h_s \frac{\partial^2 \eta}{\partial t^2} - \frac{E h_s}{2(1+\sigma)} \frac{\partial^2 \eta}{\partial z^2} + \frac{h_s E}{1-\sigma^2}(4\kappa_1(\boldsymbol{x})^2 - 2(1-\sigma)\kappa_2(\boldsymbol{x}))\eta = f, \quad (22.79)$$

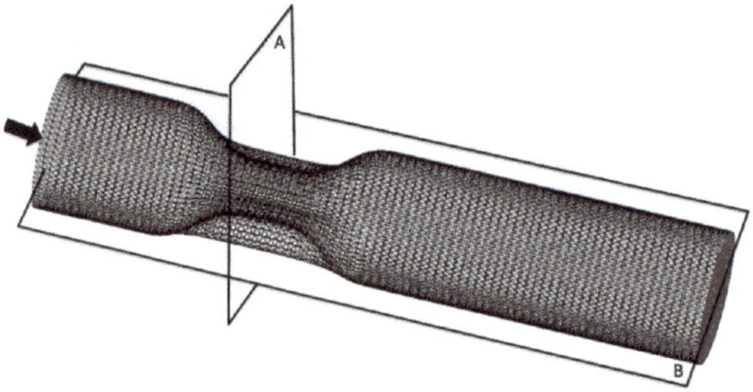

(a) Axial view of stenotic domain.

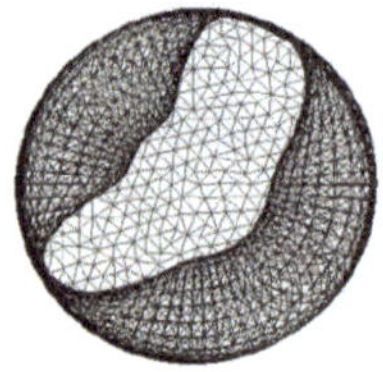

(b) Cross-section A.

Fig. 22.15 Example 4: Stenotic geometry and computational mesh: longitudinal view (left) and cross-sectional view (right) obtained from the figure on the left by cutting the mesh geometry by the plane denoted in the figure on the left by A, and looking at the mesh from the center of the longitudinal axis, shown by the arrow in the figure on the left.

Thus, the structure model and the coupling conditions have to be modified accordingly, as studied in [73]. The remaining values of the fluid and structure parameters are the same as in the previous example, and are shown in Table 22.3. The time step for the simulation is $\Delta t = 10^{-4}$.

Figures 22.16, 22.17, 22.18, and 22.19 show the numerical solution for the velocity, pressure, and displacement, at different times. In particular, Figure 22.16 shows 2D velocity snap shots taken at 4 different times. The 2D velocity snap shots are taken at the cross-section of the 3D domain by the plane denoted by B in Fig-

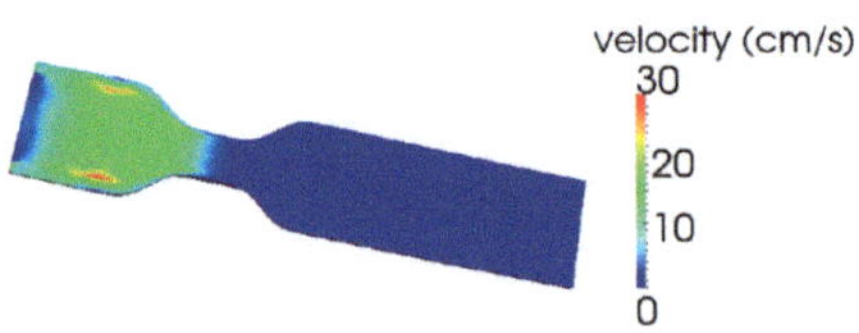

(a) t = 4 ms

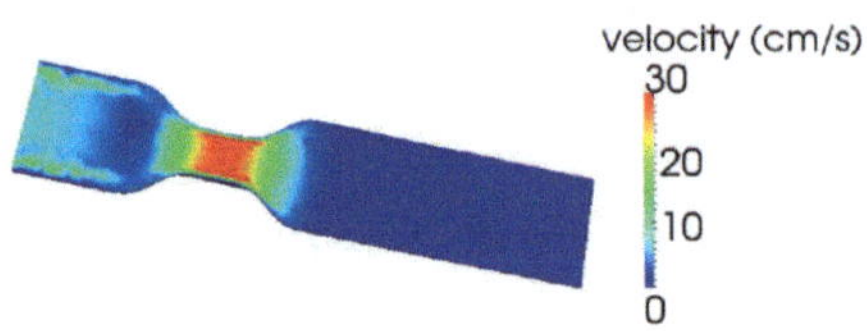

(b) t = 7 ms

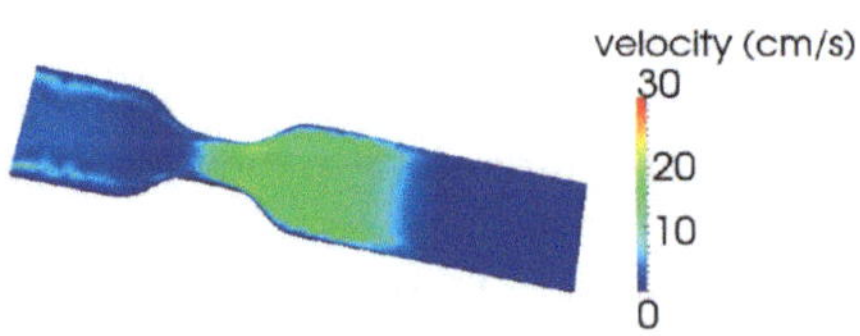

(c) t = 10 ms

(d) t = 13 ms

Fig. 22.16 A 2D cut of the 3D velocity through an asymmetric compliant stenotic region at four different times. The 2D cut plane is denoted in Figure 22.15 by B. The corresponding pressure plots are shown in Figure 22.18.

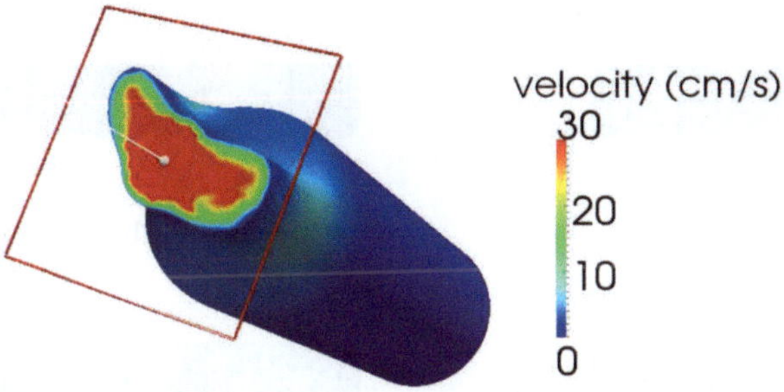

Fig. 22.17 A 2D cut of the 3D velocity at the stenotic throat at $t = 7$ ms. The 2D cut plane is denoted in Figure 22.15 by A.

ure 22.15. Figure 22.17 shows the velocity at the throat, taken at $t = 7$ ms when the velocity reaches its maximum at the throat. The 2D velocity cross-section is taken in the plane denoted by A in Figure 22.15. Figure 22.16 shows the beginning rush of fluid into the tube ($t = 4$ ms), the acceleration of the fluid at the proximal throat location ($t = 7$ ms), the high velocity region at the distal location of the stenotic throat ($t = 10$ ms), and the velocity ahead of the pressure wave exiting the tube ($t = 13$ ms). The corresponding pressure wave propagation is shown in Figure 22.18.

Finally, Figure 22.19 shows the displacement of the structure at four different times. Notice how due to the high pressure in the proximal region to stenosis, the highest displacement can be observed exactly in that region. Within the stenotic region, the smallest displacement is observed at the most narrow part of the channel in the stenotic throat (visible at the bottom part of the stenotic throat in Figure 22.19), where the velocity is highest. Notice also that the overall displacement at the distal site to stenosis if much smaller compared to that at the proximal region. The high

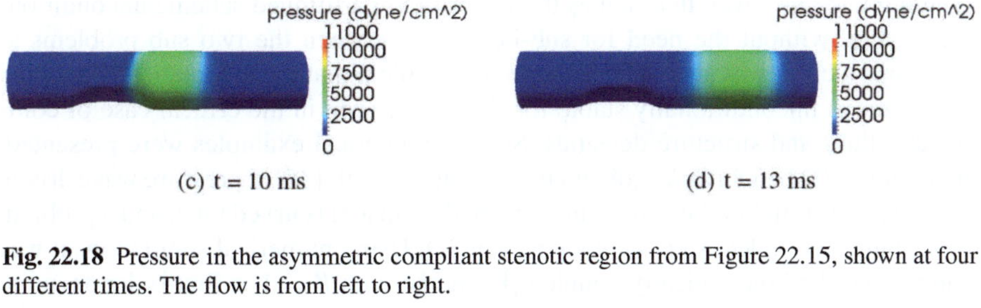

Fig. 22.18 Pressure in the asymmetric compliant stenotic region from Figure 22.15, shown at four different times. The flow is from left to right.

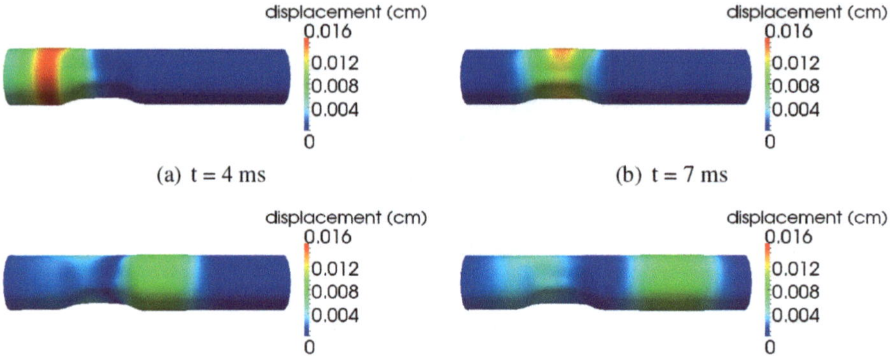

(a) t = 4 ms (b) t = 7 ms

(c) t = 10 ms (d) t = 13 ms

Fig. 22.19 Displacement in the asymmetric compliant stenotic region from Figure 22.15, shown at four different times. The flow is from left to right.

pressure and high displacement in the region proximal to stenosis is an important piece of information from the clinical point of view. Namely, it has been reported in the medical literature (see, e.g., [24]) that the region most prone to the vulnerable plaque rupture is exactly the region proximal to the most stenotic region in a coronary artery.

6 Conclusions

In this chapter we presented a review of the kinematically coupled β-scheme as it applies to 3D fluid-structure interaction (FSI) problems between an incompressible, viscous, Newtonian fluid, and a thin, elastic structure modeled by the Koiter shell or membrane equations. This class of problems arises in computational hemodynamics modeling blood flow in compliant arteries. The proposed scheme is a loosely coupled partitioned scheme, which is based on the Lie operator splitting approach (or Marchuk-Yanenko scheme). Using this operator splitting approach, the multi-physics FSI problem is partitioned into a fluid and a structure sub-problem, which communicate in a way that makes the underlying partitioned scheme unconditionally stable, without the need for sub-iterations between the two sub-problems at each time step. It was shown on a simplified problem that the kinematically coupled β-scheme is unconditionally stable for $\beta \in [0,1]$, even in the critical case of comparable fluid and structure densities. Several numerical examples were presented, including a 2D benchmark problem by Formaggia et al. [35], a pressure wave driven flow in a 3D straight tube, a pressure-driven flow in a 3D curved tube, and a problem describing a complex, stenotic geometry in 3D. Using numerical simulations it was shown that the kinematically coupled β-scheme with $\beta = 1$ is first-order accurate in time. Modularity, low computational cost, and implementation simplicity make this scheme particularly appealing for the use in biofluidic FSI problems. Future developments include extensions of this scheme to study FSI with heart valves, FSI involving endovascular stents, and FSI involving composite structures.

Acknowledgements The authors would like to acknowledge Olivier Pironneau and Frédéric Hecht for a discussion regarding FreeFem++, and Simone Deparis for a suggestion regarding the numerical implementation of the coupling conditions. Many thanks to the National Science Foundation for partial research support under grants: DMS-1318763 (Bukač), DMS-1263572, DMS-1318763, DMS-1311709, DMS-1262385 and DMS-1109189 (Čanić), DMS-1311709 (Muha), and DMS-0914788 (Glowinski).

References

1. Astorino, M., Chouly, F., Fernández, M.A.: An added-mass free semi-implicit coupling scheme for fluid–structure interaction. Comptes Rendus Mathematique **347**(1–2), 99–104 (2009)
2. Astorino, M., Chouly, F., Fernández, M.A.: Robin based semi-implicit coupling in fluid-structure interaction: Stability analysis and numerics. SIAM Journal on Scientific Computing **31**(6), 4041–4065 (2010)
3. Baaijens, F.P.T.: A fictitious domain/mortar element method for fluid-structure interaction. International Journal for Numerical Methods in Fluids **35**(7), 743–761 (2001)
4. Badia, S., Nobile, F., Vergara, C.: Fluid–structure partitioned procedures based on Robin transmission conditions. Journal of Computational Physics **227**(14), 7027–7051 (2008)
5. Badia, S., Quaini, A., Quarteroni, A.: Splitting methods based on algebraic factorization for fluid-structure interaction. SIAM Journal on Scientific Computing **30**(4), 1778–1805 (2008)
6. Bazilevs, Y., Calo, V.M., Hughes, T.J.R., Zhang, Y.: Isogeometric fluid-structure interaction: theory, algorithms, and computations. Computational Mechanics **43**(1), 3–37 (2008)
7. Bazilevs, Y., Calo, V.M., Zhang, Y., Hughes, T.J.R.: Isogeometric fluid–structure interaction analysis with applications to arterial blood flow. Computational Mechanics **38**(4–5), 310–322 (2006)
8. Bukač, M., Čanić, S.: Longitudinal displacement in viscoelastic arteries: A novel fluid-structure interaction computational model, and experimental validation. Mathematical Biosciences and Engineering **10**(2), 295–318 (2013)
9. Bukač, M., Čanić, S., Glowinski, R., Muha, B., Quaini, A.: A modular, operator-splitting scheme for fluid-structure interaction problems with thick structures. International Journal for Numerical Methods in Fluids **74**(8), 577–604 (2014)
10. Bukač, M., Čanić, S., Glowinski, R., Tambača, J., Quaini, A.: Fluid–structure interaction in blood flow capturing non-zero longitudinal structure displacement. Journal of Computational Physics **235**, 515–541 (2013)
11. Bukač, M., Čanić, S., Muha, B.: A partitioned scheme for fluid–composite structure interaction problems. Journal of Computational Physics **281**, 493–517 (2015)
12. Bukač, M., Zunino, P., Yotov, I.: Explicit partitioning strategies for interaction of the fluid with a multilayered poroelastic structure: An operator-splitting approach. Computer Methods in Applied Mechanics and Engineering **292**, 138–170 (2015)
13. Čanić, S., Muha, B., Bukač, M.: Stability of the kinematically coupled β-scheme for fluid-structure interaction problems in hemodynamics. International Journal of Numerical Analysis and Modeling **12**(1), 54–80 (2015)
14. Causin, P., Gerbeau, J., Nobile, F.: Added-mass effect in the design of partitioned algorithms for fluid–structure problems. Computer Methods in Applied Mechanics and Engineering **194**(42–44), 4506–4527 (2005)

15. Chambolle, A., Desjardins, B., Esteban, M.J., Grandmont, C.: Existence of weak solutions for the unsteady interaction of a viscous fluid with an elastic plate. Journal of Mathematical Fluid Mechanics **7**(3), 368–404 (2005)
16. Ciarlet, C., Roquefort, A.: Justification of a two-dimensional shell model of Koiter type. CR Acad. Sci. Paris, Ser I Math **331**(5), 411–416 (2000)
17. Ciarlet, P.G.: A two-dimensional non-linear shell model of Koiter's type. In: M. de Gosson (ed.) Jean Leray '99 Conference Proceedings, pp. 437–449. Springer Netherlands, Dordrecht (2003)
18. Ciarlet, P.G., Coutand, D.: An existence theorem for nonlinearly elastic 'flexural' shells. Journal of Elasticity **50**(3), 261–277 (1998)
19. Colciago, C., Deparis, S., Quarteroni, A.: Comparisons between reduced order models and full 3D models for fluid–structure interaction problems in haemodynamics. Journal of Computational and Applied Mathematics **265**, 120–138 (2014)
20. Cottet, G.H., Maitre, E., Milcent, T.: Eulerian formulation and level set models for incompressible fluid-structure interaction. ESAIM: Mathematical Modelling and Numerical Analysis **42**(3), 471–492 (2008)
21. Deparis, S., Discacciati, M., Fourestey, G., Quarteroni, A.: Fluid–structure algorithms based on Steklov–Poincaré operators. Computer Methods in Applied Mechanics and Engineering **195**(41–43), 5797–5812 (2006)
22. Deparis, S., Fernández, M.A., Formaggia, L.: Acceleration of a fixed point algorithm for fluid-structure interaction using transpiration conditions. ESAIM: Mathematical Modelling and Numerical Analysis **37**(4), 601–616 (2003)
23. Donea, J.: Arbitrary Lagrangian Eulerian finite element methods. In: T. Belytschko, T.J.R. Hughes (eds.) Computer Methods for Transient Analysis, pp. 473–516. North-Holland, Amsterdam (1983)
24. Falk, E., Shah, P.K., Fuster, V.: Coronary plaque disruption. Circulation **92**(3), 657–671 (1995)
25. Fang, H., Wang, Z., Lin, Z., Liu, M.: Lattice Boltzmann method for simulating the viscous flow in large distensible blood vessels. Physical Review E **65**(5) (2002)
26. Fauci, L.J., Dillon, R.: Biofluidmechanics of reproduction. Annual Review of Fluid Mechanics **38**(1), 371–394 (2006)
27. Feng, Z.G., Michaelides, E.E.: The immersed boundary-lattice Boltzmann method for solving fluid–particles interaction problems. Journal of Computational Physics **195**(2), 602–628 (2004)
28. Fernández, M.A.: Incremental displacement-correction schemes for the explicit coupling of a thin structure with an incompressible fluid. Comptes Rendus Mathematique **349**(7–8), 473–477 (2011)
29. Fernández, M.A.: Incremental displacement-correction schemes for incompressible fluid-structure interaction: Stability and convergence analysis. Numerische Mathematik **123**(1), 21–65 (2013)
30. Fernández, M.A., Gerbeau, J.F., Grandmont, C.: A projection algorithm for fluid–structure interaction problems with strong added-mass effect. Comptes Rendus Mathematique **342**(4), 279–284 (2006)
31. Fernández, M.Á., Moubachir, M.: A Newton method using exact jacobians for solving fluid–structure coupling. Computers and Structures **83**(2–3), 127–142 (2005)
32. Fernández, M.A., Mullaert, J.: Displacement-velocity correction schemes for incompressible fluid–structure interaction. Comptes Rendus Mathematique **349**(17–18), 1011–1015 (2011)
33. Figueroa, C.A., Vignon-Clementel, I.E., Jansen, K.E., Hughes, T.J., Taylor, C.A.: A coupled momentum method for modeling blood flow in three-dimensional deformable arteries. Computer Methods in Applied Mechanics and Engineering **195**(41–43), 5685–5706 (2006)
34. Fogelson, A.L.: Platelet-wall interactions in continuum models of platelet thrombosis: formulation and numerical solution. Mathematical Medicine and Biology **21**(4), 293–334 (2004)
35. Formaggia, L., Gerbeau, J., Nobile, F., Quarteroni, A.: On the coupling of 3D and 1D Navier–Stokes equations for flow problems in compliant vessels. Computer Methods in Applied Mechanics and Engineering **191**(6–7), 561–582 (2001)

36. Gerbeau, J.F., Vidrascu, M.: A quasi-Newton algorithm based on a reduced model for fluid-structure interaction problems in blood flows. ESAIM: Mathematical Modelling and Numerical Analysis **37**(4), 631–647 (2003)
37. Glowinski, R.: Finite element methods for incompressible viscous flow. In: P.G. Ciarlet, P.L. Lions (eds.) Handbook of Numerical Analysis, vol. 9, pp. 3–1176. Elsevier (2003)
38. Glowinski, R., Guidoboni, G., Pan, T.W.: Wall-driven incompressible viscous flow in a two-dimensional semi-circular cavity. Journal of Computational Physics **216**(1), 76–91 (2006)
39. Griffith, B.E.: Immersed boundary model of aortic heart valve dynamics with physiological driving and loading conditions. International Journal for Numerical Methods in Biomedical Engineering **28**(3), 317–345 (2012)
40. Griffith, B.E.: On the volume conservation of the immersed boundary method. Communications in Computational Physics **12**(2), 401–432 (2012)
41. Griffith, B.E., Hornung, R.D., McQueen, D.M., Peskin, C.S.: An adaptive, formally second order accurate version of the immersed boundary method. Journal of Computational Physics **223**(1), 10–49 (2007)
42. Griffith, B.E., Luo, X., McQueen, D.M., Peskin, C.S.: Simulating the fluid dynamics of natural and prosthetic heart valves using the immersed boundary method. International Journal of Applied Mechanics **01**(01), 137–177 (2009)
43. Guidoboni, G., Glowinski, R., Cavallini, N., Čanić, S.: Stable loosely-coupled-type algorithm for fluid–structure interaction in blood flow. Journal of Computational Physics **228**(18), 6916–6937 (2009)
44. Hecht, F.: Freefem++. http://www.freefem.org/ff++/
45. Hecht, F.: New development in freefem++. Journal of Numerical Mathematics **20**(3–4), 251–266 (2013)
46. Heil, M.: An efficient solver for the fully coupled solution of large-displacement fluid–structure interaction problems. Computer Methods in Applied Mechanics and Engineering **193**(1–2), 1–23 (2004)
47. Hughes, T.J.R., Liu, W.K., Zimmermann, T.K.: Lagrangian-Eulerian finite element formulation for incompressible viscous flows. Computer Methods in Applied Mechanics and Engineering **29**(3), 329–349 (1981)
48. Hundertmark-Zaušková, A., Lukácová-Medvid'ová, M., Rusnáková, G.: Fluid-structure interaction for shear-dependent non-newtonian fluids. Topics in Mathematical Modeling and Analysis **7**, 109–158 (2012)
49. Irons, M.L.: The curvature and geodesics of the torus. http://www.rdrop.com/~half/math/torus/torus.geodesics.pdf
50. Johnson, M.W., Reissner, E.: On the foundations of the theory of thin elastic shells. Journal of Mathematics and Physics **37**(1–4), 371–392 (1958)
51. Krafczyk, M., Cerrolaza, M., Schulz, M., Rank, E.: Analysis of 3D transient blood flow passing through an artificial aortic valve by Lattice–Boltzmann methods. Journal of Biomechanics **31**(5), 453–462 (1998)
52. Krafczyk, M., Tölke, J., Rank, E., Schulz, M.: Two-dimensional simulation of fluid–structure interaction using lattice-Boltzmann methods. Computers and Structures **79**(22–25), 2031–2037 (2001)
53. Le Tallec, P., Mouro, J.: Fluid structure interaction with large structural displacements. Computer Methods in Applied Mechanics and Engineering **190**(24–25), 3039–3067 (2001)
54. Leuprecht, A., Perktold, K., Prosi, M., Berk, T., Trubel, W., Schima, H.: Numerical study of hemodynamics and wall mechanics in distal end-to-side anastomoses of bypass grafts. Journal of Biomechanics **35**(2), 225–236 (2002)
55. Lim, S., Peskin, C.: Simulations of the whirling instability by the immersed boundary method. SIAM Journal on Scientific Computing **25**(6), 2066–2083 (2004)
56. Lukáčová-Medvid'ová, M., Rusnáková, G., Hundertmark-Zaušková, A.: Kinematic splitting algorithm for fluid–structure interaction in hemodynamics. Computer Methods in Applied Mechanics and Engineering **265**, 83–106 (2013)

57. M. Cervera, R. Codina, M. Galindo: On the computational efficiency and implementation of block-iterative algorithms for nonlinear coupled problems. Engineering Computations **13**(6), 4–30 (1996)
58. Michler, C., Hulshoff, S.J., van Brummelen, E.H., de Borst, R.: A monolithic approach to fluid–structure interaction. Computers and Fluids **33**(5–6), 839–848 (2004)
59. Miller, L.A., Peskin, C.S.: A computational fluid dynamics of 'clap and fling' in the smallest insects. Journal of Experimental Biology **208**(2), 195–212 (2005)
60. Muha, B., Čanić, S.: Existence of a weak solution to a fluid-structure interaction problem motivated by blood-artery-stent interaction. In preparation
61. Muha, B., Čanić, S.: Existence of a weak solution to a nonlinear fluid–structure interaction problem modeling the flow of an incompressible, viscous fluid in a cylinder with deformable walls. Archive for Rational Mechanics and Analysis **207**(3), 919–968 (2012)
62. Muha, B., Čanić, S.: A nonlinear, {3D} fluid-structure interaction problem driven by the time-dependent dynamic pressure data: a constructive existence proof. Communications in Information and Systems **13**(3), 357–397 (2013)
63. Muha, B., Čanić, S.: Existence of a solution to a fluid–multi-layered-structure interaction problem. Journal of Differential Equations **256**(2), 658–706 (2014)
64. Murea, C.M., Sy, S.: A fast method for solving fluid–structure interaction problems numerically. International Journal for Numerical Methods in Fluids **60**(10), 1149–1172 (2009)
65. Nobile, F.: Numerical Approximation of Fluid-Structure Interaction Problems with Application to Haemodynamics. Phd thesis, Federal Institute of Technology, Department of Mathematics, Lausanne, Switzerland (2001)
66. Nobile, F., Vergara, C.: An effective fluid-structure interaction formulation for vascular dynamics by generalized Robin conditions. SIAM Journal on Scientific Computing **30**(2), 731–763 (2008)
67. Peskin, C.S.: Numerical analysis of blood flow in the heart. Journal of Computational Physics **25**(3), 220–252 (1977)
68. Peskin, C.S., McQueen, D.M.: Modeling prosthetic heart valves for numerical analysis of blood flow in the heart. Journal of Computational Physics **37**(1), 113–132 (1980)
69. Quaini, A., Quarteroni, A.: A semi-implicit approach for fluid-structure interaction based on an algebraic fractional step method. Mathematical Models and Methods in Applied Sciences **17**(06), 957–983 (2007)
70. Quaini, Annalisa: Algorithms for fluid-structure interaction problems arising in hemodynamics. Phd thesis, Federal Institute of Technology, Department of Mathematics, Lausanne, Switzerland (2009)
71. Quarteroni, A., Tuveri, M., Veneziani, A.: Computational vascular fluid dynamics: problems, models and methods. Computing and Visualization in Science **2**(4), 163–197 (2014)
72. Steindorf, J., Matthies, H.G.: Numerical efficiency of different partitioned methods for fluid-structure interaction. Journal of Applied Mathematics and Mechanics **80**(S2), 557–558 (2000)
73. Tambača, J., Čanić, S., Mikelić, A.: Effective model of the fluid flow through elastic tube with variable radius. Grazer mathematische Berichte **348**, 91–112 (2005)
74. van Loon, R., Anderson, P.D., de Hart, J., Baaijens, F.P.T.: A combined fictitious domain/adaptive meshing method for fluid–structure interaction in heart valves. International Journal for Numerical Methods in Fluids **46**(5), 533–544 (2004)
75. Velčić, I.: Private communication
76. Zhao, S.Z., Xu, X.Y., Collins, M.W.: The numerical analysis of fluid-solid interactions for blood flow in arterial structures. Part 2: development of coupled fluid-solid algorithms. Proceedings of the Institution of Mechanical Engineers, Part H: Journal of Engineering in Medicine **212**(4), 241–252 (1998)

Chapter 23
On Circular Cluster Formation in a Rotating Suspension of Non-Brownian Settling Particles in a Fully Filled Circular Cylinder: An Operator Splitting Approach to the Numerical Simulation

Suchung Hou and Tsorng-Whay Pan

Abstract In this chapter, we investigate, via direct numerical simulation, the circular cluster formation taking place in a circular cylinder rotating around its axis and fully filled with fluid–rigid particle mixtures. The phenomenon is modeled by the Navier–Stokes equations coupled to the Euler–Newton equations describing the rigid solid motion of non-neutrally buoyant particles. The formation of circular clusters studied in this chapter is mainly caused by the interaction between the particles themselves. Within a circular cluster, the part of the cluster formed by the particles moving from the front to the back through the upper portion of the cylinder becomes more compact due to the particle interaction strengthened by the speedup of the particle speeds, first by rotation and later by rotation and gravity. The part of a cluster formed by the particles moving from the back to the front through the lower portion of the cylinder is always loosening up and spreading out due to the slowdown of the particle motion, first by rotation and then by rotation and the counter effect of gravity. To have a compact circular cluster, particles have to interact among themselves continuously through the entire circular cluster at an angular speed such that the separation of particles can be balanced by their aggregation.

S. Hou
Department of Mathematics, National Cheng Kung University, Tainan 701, Taiwan, Republic of China
e-mail: schou@mail.ncku.edu.tw

T.-W. Pan (✉)
Department of Mathematics, University of Houston, Houston, TX 77204, USA
e-mail: pan@math.uh.edu

R. Glowinski et al. (eds.), *Splitting Methods in Communication, Imaging, Science, and Engineering*, Scientific Computation, DOI 10.1007/978-3-319-41589-5_23

1 Introduction

Non-equilibrium systems often self-organize into interesting spatio-temporal structures or patterns. Examples include patterns in pure fluid flow systems, such as the Taylor–Couette flow between two concentric rotating cylinders and well-defined periodic clusters of particles in a partially or fully filled horizontally rotating cylinder. Particulate flows exhibiting circular clusters in a partially filled horizontal rotating cylinder are in part attributed to the presence of the free surface caused by the partial filling of the cylinder (e.g., see [12, 25, 26]). In a fully filled horizontally rotating cylinder, cluster and other pattern formations were also found in the suspensions of non-Brownian settling particles in [2, 15, 16, 17, 18, 19, 23]. For probably the most complete overview in the literature on the pattern formation and segregation in rotating-drum flows, see the recent extensive review article by Seiden and Thomas [22]. Lee and Ladd [13, 14] addressed the experimental observations made by Matson *et al.* [17, 18, 19] in creeping flow regime. The ratio of the particle diameter and the inner cylinder diameter in [13, 14, 17, 18, 19] is about 1%. In [13, 14], numerical simulations within the Stokes-flow approximation have been used to investigate the mechanism underlying circular cluster formation. The numerical results show that the formation of circular clusters is correlated with an inhomogeneous particle distribution in the radial plane, which is itself driven by the competition between gravity and the viscous drag. The circular cluster structure develops during the transition between a low-frequency segregated phase and a high-frequency dispersed phase. In this chapter, we have focused on the understanding of cluster formations similar to those observed in [16, 23]; however the values of the Reynolds number, Re= $2aU/\nu$, and Ekman number, E= $\nu/\Omega R^2$, for the cases considered here are in a different regime (here a is the ball radius, $U = \Omega R$ is the characteristic velocity; Ω being the cylinder angular speed, R the cylinder radius, and ν the fluid kinematic viscosity). Thus the numerical simulations discussed in this publication are, strictly speaking, not comparable with the experiments reported in [16, 23]. The fluid–particle mixtures considered here are not in the creeping flow regime as considered computationally by Lee and Ladd in [13, 14]. In [15], Lipson used a horizontal rotating cylinder filled with an over-saturated solution to grow crystal without any interaction with a substrate and found that crystals accumulate in well-defined periodic clusters, normal to the axis of rotation. Lipson and Seiden [16] just suggested, with no further discussion, that this could be due to the interaction between particles and fluid in the cylinder. In [23], Seiden *et al.* did an experimental investigation of the dependence of the formation of clusters on particle characteristics, tube diameter and length, and fluid viscosity. They suggested that the segregation of particles occurs as a result of mutual interaction between the particles and inertial waves excited in the bounded fluid. In [24] Seiden *et al.* believed according to their general dimensionless analysis that the axial pressure gradient associated with an inertial-mode excitation within a bounded fluid is responsible for the formation of clusters. A single ball motion was discussed by solving the equation of motion for the ball with a one-way coupling in a filled and horizontally rotating cylinder;

a stability analysis and a phase diagram based on one ball motion are addressed; but [24] did not consider the effect of the ball on the fluid and the interaction between particles themselves.

Using a distributed Lagrange multiplier/fictitious domain method combined with time discretization by operator splitting, we have observed via direct numerical simulations that the formation of circular clusters is mainly caused by the interaction between particles themselves. In our simulations, the particles form a layer inside a horizontally rotating cylinder similar to the one in Figure 7 in [23]. These particles are partially coated on the inner wall of the rotating cylinder under the influence of a strong centrifugal force. Within a circular cluster, the part of the cluster formed by the particles moving from the front to the back through the upper portion of the cylinder becomes more compact due to the particle interaction strengthened by the speedup of the particle speeds first by rotation and then later by rotation and gravity. The part of a cluster formed by the particles moving from the back to the front through the lower portion of the cylinder is always loosening up and spreading out due to the slowdown of the particle motion first by rotation and then by rotation and the counter effect of gravity. To have a compact circular cluster, particles have to interact among themselves continuously through the entire circular cluster at a angular speed such that the separation of these particles can be balanced by their aggregation. Hence the balance of gravity, angular speed, fluid flow inertia, and the number of particles is important for the formation of circular clusters in a fully filled cylinder.

The content of this chapter is as follows: We discuss the models and numerical methods in Section 2. In Section 3, we study the effect of the particle number, the angular speed, and the initial gap size on the formation of circular clusters and then present the flow field development under the influence of the particle interaction. The conclusions are summarized in Section 4.

2 Governing Equations

To perform the direct numerical simulation of the interaction between rigid bodies and fluid, we have developed a methodology which combines a distributed volume Lagrange multiplier based fictitious domain method with operator splitting and finite element methods (e.g., see [6, 7, 8, 20, 21, 27]). For a ball B moving in a Newtonian viscous incompressible fluid, of viscosity μ and density ρ_f, contained in a truncated cylinder C under the effect of gravity, as depicted in Figure 23.1, the flow is modeled by the Navier–Stokes equations, namely,

$$\rho_f\left[\frac{\partial \mathbf{u}}{\partial t}+(\mathbf{u}\cdot\nabla)\mathbf{u}\right]-\mu\Delta\mathbf{u}+\nabla p=\rho_f\mathbf{g} \text{ in } \{(\mathbf{x},t)|\mathbf{x}\in C\setminus\overline{B(t)},\ t\in(0,T)\}, \tag{23.1}$$

$$\nabla\cdot\mathbf{u}(t)=0 \text{ in } \{(\mathbf{x},t)|\mathbf{x}\in C\setminus\overline{B(t)},\ t\in(0,T)\}, \tag{23.2}$$

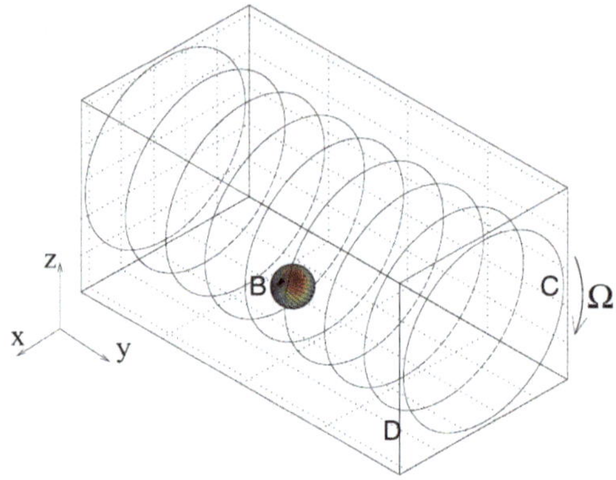

Fig. 23.1 Flow region with a ball B in a truncated cylinder C.

$$\mathbf{u}(0) = \mathbf{u}_0(\mathbf{x}),\ (with\ \nabla \cdot \mathbf{u}_0 = 0), \tag{23.3}$$

$$\mathbf{u} = \mathbf{g}_0\ on\ \Gamma_0 \times (0,T),\ (with \int_{\Gamma_0} \mathbf{g}_0 \cdot \mathbf{n}\, d\Gamma = 0), \tag{23.4}$$

where Γ_0 is the entire surface of cylinder C, $\mathbf{g}$ denotes gravity, $\mathbf{g}_0$ is the given velocity field, $\mathbf{u}_0$ is the initial condition of the flow field, and $\mathbf{n}$ is the unit normal vector pointing outward to the flow region. We assume a *no-slip condition* on $\gamma(=\partial B)$. The motion of the rigid body B satisfies the Euler–Newton's equations, namely

$$\mathbf{v}(\mathbf{x},t) = \mathbf{V}(t) + \boldsymbol{\omega}(t) \times \mathbf{G}(t)\mathbf{x},\ \forall \mathbf{x} \in \overline{B(t)},\ \forall t \in (0,T), \tag{23.5}$$

$$\frac{d\mathbf{G}}{dt} = \mathbf{V}, \tag{23.6}$$

$$M_p \frac{d\mathbf{V}}{dt} = M_p \mathbf{g} + \mathbf{F}_H, \tag{23.7}$$

$$\mathbf{I}_p \frac{d\boldsymbol{\omega}}{dt} = \mathbf{T}_H, \tag{23.8}$$

with the resultant and torque of the hydrodynamical forces given by, respectively,

$$\mathbf{F}_H = -\int_\gamma \boldsymbol{\sigma} \mathbf{n}\, d\gamma,\ \ \mathbf{T}_H = -\int_\gamma \mathbf{Gx} \times \boldsymbol{\sigma} \mathbf{n}\, d\gamma, \tag{23.9}$$

with $\boldsymbol{\sigma} = \mu(\nabla \mathbf{u} + \nabla \mathbf{u}^t) - p\mathbf{I}$. Equations (23.1)–(23.9) are completed by the following initial conditions

$$\mathbf{G}(0) = \mathbf{G}_0,\ \mathbf{V}(0) = \mathbf{V}_0,\ \boldsymbol{\omega}(0) = \boldsymbol{\omega}_0, B(0) = B_0. \tag{23.10}$$

Above, M_p, $\mathbf{I}_p$, $\mathbf{G}$, $\mathbf{V}$, and $\boldsymbol{\omega}$ are the mass, inertia, center of mass, velocity of the center of mass, and angular velocity of the rigid body B, respectively. The gravity is pointed downward in the direction of z.

In order to take a full advantage of the fictitious domain approach we will embed the truncated cylinder C in a rectangular parallelepiped (denoted by D) with a square cross section whose edge length is slightly larger than the diameter of the cylinder C as shown in Figure 23.1. The region outside C is denoted by $A = D\backslash\overline{C}$ and the boundary of D is denoted by Γ. Also we assume that $\mathbf{g}_0$ defined on Γ_0 is nothing but the velocity field on the surface of a horizontal rotating cylinder; hence we can easily extend it on $\overline{A}$ according to the angular velocity of the cylinder. For extended value on Γ, we still use $\mathbf{g}_0$ in the following. To solve numerically the coupled problem (23.1)–(23.10), we have first applied a distributed Lagrange multiplier-based fictitious domain method (see, [7] and [8] for details) and obtain then an equivalent formulation of (23.1)–(23.10) defined on the whole domain D, namely

For a.e. $t > 0$, *find* $\mathbf{u}(t) \in \mathbf{W}_{\mathbf{g}_0}(t)$, $p(t) \in L_0^2(D)$, $\mathbf{V}(t) \in I\!R^3$, $\mathbf{G}(t) \in I\!R^3$, $\boldsymbol{\omega}(t) \in I\!R^3$, $\boldsymbol{\lambda}(t) \in \Lambda(t)$, $\boldsymbol{\lambda}_A \in \Lambda_A$ *such that*

$$\begin{cases} \rho_f \int_D \left[\frac{\partial \mathbf{u}}{\partial t} + (\mathbf{u}\cdot\boldsymbol{\nabla})\mathbf{u}\right]\cdot\mathbf{v}d\mathbf{x} - \int_D p\boldsymbol{\nabla}\cdot\mathbf{v}d\mathbf{x} \\ \quad +\mu_f \int_D \boldsymbol{\nabla}\mathbf{u} : \boldsymbol{\nabla}\mathbf{v}d\mathbf{x} - <\boldsymbol{\lambda}, \mathbf{v} - \mathbf{Y} - \boldsymbol{\theta}\times\mathbf{G}\mathbf{x}>_{\Lambda(t)} - <\boldsymbol{\lambda}_A, \mathbf{v}>_{\Lambda_A} \\ \quad +(1-\frac{\rho_f}{\rho_s})[M_p\frac{d\mathbf{V}}{dt}\cdot\mathbf{Y} + \mathbf{I}_p\frac{d\boldsymbol{\omega}}{dt}\cdot\boldsymbol{\theta}] - \mathbf{F}^r\cdot\mathbf{Y} \\ = (1-\frac{\rho_f}{\rho_s})M_p\,\mathbf{g}\cdot\mathbf{Y} + \rho_f\int_D \mathbf{g}\cdot\mathbf{v}d\mathbf{x},\ \forall\mathbf{v}\in\mathbf{W}_0,\ \forall\mathbf{Y}\in\mathbb{R}^3,\ \forall\boldsymbol{\theta}\in\mathbb{R}^3, \end{cases} \tag{23.11}$$

$$\int_D q\boldsymbol{\nabla}\cdot\mathbf{u}(t)d\mathbf{x} = 0,\ \forall q\in L^2(D), \tag{23.12}$$

$$<\mu, \mathbf{u}(t) - \mathbf{V}(t) - \boldsymbol{\omega}(t)\times\mathbf{G}(t)\mathbf{x}>_{\Lambda(t)} = 0,\ \forall\mu\in\Lambda(t), \tag{23.13}$$

$$<\mu_A, \mathbf{u}(t) - \mathbf{g}_0(t)>_{\Lambda_A} = 0,\ \forall\mu_A\in\Lambda_A, \tag{23.14}$$

$$\frac{d\mathbf{G}}{dt} = \mathbf{V}, \tag{23.15}$$

$$\mathbf{V}(0) = \mathbf{V}_0,\ \boldsymbol{\omega}(0) = \boldsymbol{\omega}_0,\ \mathbf{G}(0) = \mathbf{G}_0, B(0) = B_0, \tag{23.16}$$

$$\mathbf{u}(\mathbf{x},0) = \tilde{\mathbf{u}}_0(\mathbf{x}) = \begin{cases} \mathbf{u}_0(\mathbf{x}),\ \forall\mathbf{x}\in C\backslash B(0), \\ \mathbf{V}_0 + \boldsymbol{\omega}_0\times\mathbf{G}_0\mathbf{x},\ \forall\mathbf{x}\in\overline{B(0)}, \\ \mathbf{g}_0(0),\ \forall\mathbf{x}\in\overline{A}, \end{cases} \tag{23.17}$$

with the following functional spaces

$$\mathbf{W} = (H^1(D))^3,\ \mathbf{W}_0 = (H_0^1(D))^3,\ \mathbf{W}_{\mathbf{g}_0}(t) = \{\mathbf{v}|\mathbf{v}\in\mathbf{W},\ \mathbf{v} = \mathbf{g}_0(t)\ on\ \Gamma\},$$
$$L_0^2(D) = \{q|q\in L^2(D),\ \int_D q\,d\mathbf{x} = 0\},$$
$$\Lambda(t) = (H^1(B(t)))^3,\ \Lambda_A = \{\mu|\mu\in(H^1(A))^3\}.$$

In (23.11), (23.13), and (23.14), $<\cdot,\cdot>_{\Lambda(t)}$ and $<\cdot,\cdot>_{\Lambda_A}$ are inner product on $\Lambda(t)$ and Λ_A, respectively. Various examples are given in [6] (Chapter 8) and [8]. The velocity field inside A is enforced in (23.11) and (23.14) via the Lagrange multiplier $\boldsymbol{\lambda}_A$ supported by $\overline{A}$. The second gravity term in the right-hand-side of the (23.11) can be combined with the pressure. Hence in the following, we will not use this term anymore. The numerical solution of other examples of particulate flow problems is discussed in Section 4 of Chapter 2 of this volume.

3 Time and Space Discretization

3.1 A First Order Operator-Splitting Scheme: Lie's Scheme

Many operator-splitting schemes can be applied to problem (23.11)–(23.17). One of the advantages of operator-splitting schemes is that we can decouple difficulties such as (i) the incompressibility condition, (ii) the nonlinear advection term, (iii) the diffusion and prescribed flow condition outside the cylinder, (iv) the particle motion and collision, and (v) the rigid body motion, so that each one of them can be handled separately, and in principle optimally. Let $\triangle t$ be a time discretization step and $t^{n+s} = (n+s)\triangle t$. The Lie's scheme is a *first order* operator-splitting scheme [3] (see also Chapters 1 and 2 of this volume), which, when applied to problem (23.11)–(23.17), yields:

$$\mathbf{u}^0 = \tilde{\mathbf{u}}_0,\ \mathbf{G}^0 = \mathbf{G}_0,\ \mathbf{V}^0 = \mathbf{V}_0,\ \boldsymbol{\omega}^0 = \boldsymbol{\omega}_0 \ \textit{given}; \tag{23.18}$$

for $n \geq 0, \mathbf{u}^n (\simeq \mathbf{u}(t^n))$, $\mathbf{G}^n$, $\mathbf{V}^n$ and $\boldsymbol{\omega}^n$ *being known, we first compute* $\mathbf{u}^{n+\frac{1}{6}}$, $p^{n+\frac{1}{6}}$ *via the solution of*

$$\begin{cases} \rho_f \int_D \dfrac{\partial \mathbf{u}}{\partial t} \cdot \mathbf{v} d\mathbf{x} - \int_D p \boldsymbol{\nabla} \cdot \mathbf{v} d\mathbf{x} = 0,\ \forall \mathbf{v} \in \mathbf{W}_0,\ a.e.\ on\ (t^n,\ t^{n+1}), \\ \int_D q \boldsymbol{\nabla} \cdot \mathbf{u} d\mathbf{x} = 0,\ \forall q \in L^2(D), \\ \mathbf{u}(t^n) = \mathbf{u}^n, \\ \mathbf{u}(t) \in \mathbf{W},\ \mathbf{u}(t) = \mathbf{g}_0(t^{n+1})\ on\ \Gamma \times (t^n,\ t^{n+1}),\ p(t) \in L_0^2(D), \end{cases} \tag{23.19}$$

and set $\mathbf{u}^{n+\frac{1}{6}} = \mathbf{u}(t^{n+1}),\ \ p^{n+\frac{1}{6}} = p(t^{n+1})$.

Next, compute $\mathbf{u}^{n+\frac{2}{6}}$ *via the solution of*

$$\begin{cases} \int_D \dfrac{\partial \mathbf{u}}{\partial t} \cdot \mathbf{v} d\mathbf{x} + \int_D (\mathbf{u}^{n+\frac{1}{6}} \cdot \boldsymbol{\nabla}) \mathbf{u} \cdot \mathbf{v} d\mathbf{x} = 0,\ \forall \mathbf{v} \in \mathbf{W}_0^{n+1,-},\ a.e.\ on\ (t^n,\ t^{n+1}), \\ \mathbf{u}(t^n) = \mathbf{u}^{n+\frac{1}{6}}, \\ \mathbf{u}(t) \in \mathbf{W},\ \mathbf{u}(t) = \mathbf{g}_0(t^{n+1})\ on\ \Gamma_-^{n+1} \times (t^n,\ t^{n+1}), \end{cases} \tag{23.20}$$

and set $\mathbf{u}^{n+\frac{2}{6}} = \mathbf{u}(t^{n+1})$.

Then, compute $\mathbf{u}^{n+\frac{3}{6}}$ *via the solution of*

$$\begin{cases} \rho_f \int_D \frac{\partial \mathbf{u}}{\partial t} \cdot \mathbf{v} d\mathbf{x} + \alpha \mu_f \int_D \nabla \mathbf{u} : \nabla \mathbf{v} d\mathbf{x} \\ \quad = < \boldsymbol{\lambda}_A, \mathbf{v} >_{\Lambda_A}, \forall \mathbf{v} \in \mathbf{W}_0, \ a.e. \ on \ (t^n, \ t^{n+1}), \\ < \mu_A, \mathbf{u} - \mathbf{g}_0(t^{n+1}) >_{\Lambda_A} = 0, \ \forall \mu_A \in \Lambda_A, \\ \mathbf{u}(t^n) = \mathbf{u}^{n+\frac{2}{6}}, \ \ \mathbf{u}(t) \in \mathbf{W}, \end{cases} \tag{23.21}$$

and set $\mathbf{u}^{n+\frac{3}{6}} = \mathbf{u}(t^{n+1})$.

Now predict the motion of the center of mass of the particle via

$$\frac{d\mathbf{G}}{dt} = \mathbf{V}(t)/2, \tag{23.22}$$

$$(1 - \frac{\rho_f}{\rho_s}) M_p \frac{d\mathbf{V}}{dt} = \mathbf{F}_r/2, \tag{23.23}$$

$$\mathbf{G}(t^n) = \mathbf{G}^n, \ \mathbf{V}(t^n) = \mathbf{V}^n, \tag{23.24}$$

for $t^n < t < t^{n+1}$. *Then set* $\mathbf{G}^{n+\frac{4}{6}} = \mathbf{G}(t^{n+1})$ *and* $\mathbf{V}^{n+\frac{4}{6}} = \mathbf{V}(t^{n+1})$.

Using $\mathbf{G}^{n+\frac{4}{6}}$ *obtained in the above step, we enforce the rigid body motion in the region occupied by* $B^{n+\frac{4}{6}}$:

$$\begin{cases} \rho_f \int_D \frac{\partial \mathbf{u}}{\partial t} \cdot \mathbf{v} d\mathbf{x} + \beta \mu_f \int_D \nabla \mathbf{u} : \nabla \mathbf{v} d\mathbf{x} \ + (1 - \frac{\rho_f}{\rho_s}) M_p \frac{d\mathbf{V}}{dt} \cdot \mathbf{Y} \\ \quad + (1 - \frac{\rho_f}{\rho_s}) \mathbf{I}_p \frac{d\boldsymbol{\omega}}{dt} \cdot \boldsymbol{\theta} = (1 - \frac{\rho_f}{\rho_s}) M_p \mathbf{g} \cdot \mathbf{Y} \\ \quad + < \boldsymbol{\lambda}, \ \mathbf{v} - \mathbf{Y} - \boldsymbol{\theta} \times \overrightarrow{\mathbf{G}^{n+\frac{4}{6}} \mathbf{x}} >_{\Lambda^{n+\frac{4}{6}}}, \\ \forall \mathbf{v} \in \mathbf{W}_0, \ \mathbf{Y} \in \mathbb{R}^3, \ \boldsymbol{\theta} \in \mathbb{R}^3, \ a.e. \ on \ (t^n, \ t^{n+1}), \\ \mathbf{u}(t^n) = \mathbf{u}^{n+\frac{3}{6}}, \ \mathbf{V}(t^n) = \mathbf{V}^{n+\frac{4}{6}}, \ \boldsymbol{\omega}(t^n) = \boldsymbol{\omega}^n, \\ \mathbf{u} \in \mathbf{W}, \ \mathbf{u}(t) = \mathbf{g}_0(t^{n+1}) \ on \ \Gamma \times (t^n, \ t^{n+1}), \\ \boldsymbol{\lambda} \in \Lambda^{n+\frac{4}{6}}, \ \mathbf{V} \in \mathbb{R}^3, \ \boldsymbol{\omega} \in \mathbb{R}^3, \end{cases} \tag{23.25}$$

$$< \mu, \ \mathbf{u} - \mathbf{V} - \boldsymbol{\omega} \times \overrightarrow{\mathbf{G}^{n+\frac{4}{6}} \mathbf{x}} >_{\Lambda^{n+\frac{4}{6}}} = 0, \ \forall \mu \in \Lambda^{n+\frac{4}{6}}, \tag{23.26}$$

and set $\mathbf{u}^{n+1} = \mathbf{u}(t^{n+1})$, $\mathbf{V}^{n+\frac{5}{6}} = \mathbf{V}(t^{n+1})$, $\boldsymbol{\omega}^{n+1} = \boldsymbol{\omega}(t^{n+1})$.

Correct the motion of the center of mass of the particle via

$$\frac{d\mathbf{G}}{dt} = \mathbf{V}(t)/2, \tag{23.27}$$

$$(1 - \frac{\rho_f}{\rho_s}) M_p \frac{d\mathbf{V}}{dt} = \mathbf{F}_r/2, \tag{23.28}$$

$$\mathbf{G}(t^n) = \mathbf{G}^{n+\frac{4}{6}}, \ \mathbf{V}(t^n) = \mathbf{V}^{n+\frac{5}{6}}, \tag{23.29}$$

for $t^n < t < t^{n+1}$. *Then set* $\mathbf{G}^{n+1} = \mathbf{G}(t^{n+1})$ *and* $\mathbf{V}^{n+1} = \mathbf{V}(t^{n+1})$.

In (23.18)–(23.29), $\Gamma_-^{n+1} = \{\mathbf{x}|\mathbf{x} \in \Gamma,\ \mathbf{g}_0(t^{n+1})(\mathbf{x})\cdot\mathbf{n}(\mathbf{x}) < 0\}$ and $\mathbf{W}_0^{n+1,-} = \{\mathbf{v}|\mathbf{v} \in \mathbf{W},\ \mathbf{v} = \mathbf{0}\ on\ \Gamma_-^{n+1}\}$, $\Lambda^{n+\frac{4}{6}} = (H^1(B^{n+\frac{4}{6}}))^3$, $B^{n+\frac{4}{6}}$ is the region occupied by the ball B according to $\mathbf{G}^{n+\frac{4}{6}}$, and $\alpha+\beta = 1$. In the numerical simulation, we usually choose $\alpha = 1$ and $\beta = 0$ to lower the computational cost when solving (23.25) and (23.26).

3.2 Space Discretization

We assume that $D \subset \mathbb{R}^3$ and is a rectangular parallelepiped. Concerning the *finite element approximation* of problems (23.11)–(23.17), we have

$$\mathbf{W}_h = \{\mathbf{v}_h|\mathbf{v}_h \in (C^0(\overline{D}))^3, \mathbf{v}_h|_T \in (P_1)^3, \forall T \in \mathscr{T}_h,\}, \tag{23.30}$$

$$\mathbf{W}_{0h} = \{\mathbf{v}_h|\mathbf{v}_h \in \mathbf{W}_h,\ \mathbf{v}_h = \mathbf{0}\ on\ \Gamma\}, \tag{23.31}$$

$$L_h^2 = \{q_h|q_h \in C^0(\overline{D}),\ q_h|_T \in P_1,\ \forall T \in \mathscr{T}_{2h}\}, \tag{23.32}$$

$$L_{0h}^2 = \{q_h|q_h \in L_h^2,\ \int_D q_h\, d\mathbf{x} = 0\} \tag{23.33}$$

where $\mathscr{T}_h$ is a tetrahedral partition of D, $\mathscr{T}_{2h}$ is twice coarser than $\mathscr{T}_h$, and P_1 is the space of the polynomials in three variables of degree ≤ 1. A finite dimensional space approximating $\Lambda(t)$ is as follows: let $\{\boldsymbol{\xi}_i\}_{i=1}^N$ be a set of points from $\overline{B(t)}$ which cover $\overline{B(t)}$ (uniformly, for example); we define then

$$\Lambda_h(t) = \{\boldsymbol{\mu}_h|\boldsymbol{\mu}_h = \sum_{i=1}^N \boldsymbol{\mu}_i\delta(\mathbf{x}-\boldsymbol{\xi}_i),\ \boldsymbol{\mu}_i \in \mathbb{R}^3,\ \forall i = 1,\ldots,N\}, \tag{23.34}$$

where $\delta(\cdot)$ is the Dirac measure at $\mathbf{x} = \mathbf{0}$. Then we shall use $< \cdot,\cdot >_{\Lambda_h(t)}$ defined by

$$< \boldsymbol{\mu}_h, \mathbf{v}_h >_{\Lambda_h(t)} = \sum_{i=1}^N \boldsymbol{\mu}_i \cdot \mathbf{v}_h(\boldsymbol{\xi}_i),\ \forall \boldsymbol{\mu}_h \in \Lambda_h(t),\ \mathbf{v}_h \in \mathbf{W}_h. \tag{23.35}$$

A typical choice of points for defining (23.34) is to take the grid points of the velocity mesh internal to the particle B and whose distance to the boundary of B is greater than, e.g., $h/2$ (used in the simulation), and to complete with selected points from the boundary of $B(t)$. As we did for $\Lambda_h(t)$ and $< \cdot,\cdot >_{\Lambda_h(t)}$, we define the finite dimensional space Λ_{A_h} and the inner product $< \cdot,\cdot >_{\Lambda_{A_h}}$ via a set of points of the velocity mesh internal to the region $\overline{A}$ and whose distance to the surface of the cylinder C is greater than, e.g., h, and a set of the points chosen from the surface of the surface of the cylinder C. In practice, we have chosen D so that its square cross section is slightly larger than the cross section of the cylinder in order to have collocation points between the surface of the cylinder C and Γ so that the enforcement of the constraint over $\overline{A}$ can be done much more easily.

Remark 1. The inner product like bracket $< \cdot,\cdot >_{\Lambda_h(t)}$ in (23.35) makes little sense for the continuous problem, but it is meaningful for the discrete problem; it amounts to forcing the rigid body motion of $B(t)$ via a *collocation method.* A similar technique has been used to enforce Dirichlet boundary conditions by F. Bertrand *et al.* in [1]. □

Using the above finite dimensional spaces and the backward Euler's method for most of the subproblems in schemes (23.18)–(23.29), we obtain the following scheme after dropping some of the subscripts h (similar ones are discussed in [8]):

$$\mathbf{u}^0 = \tilde{\mathbf{u}}_0,\ \mathbf{G}^0 = \mathbf{G}_0,\ \mathbf{V}^0 = \mathbf{V}_0,\ \boldsymbol{\omega}^0 = \boldsymbol{\omega}_0 \ \textit{given}; \tag{23.36}$$

for $n \geq 0, \mathbf{u}^n (\simeq \mathbf{u}(t^n))$, $\mathbf{G}^n$, $\mathbf{V}^n$, and $\boldsymbol{\omega}^n$ *being known, we compute* $\mathbf{u}^{n+\frac{1}{6}}$, $p^{n+\frac{1}{6}}$ *via the solution of*

$$\begin{cases} \rho_f \int_D \dfrac{\mathbf{u}^{n+\frac{1}{6}} - \mathbf{u}^n}{\triangle t} \cdot \mathbf{v} d\mathbf{x} - \int_D p^{n+\frac{1}{6}} \boldsymbol{\nabla} \cdot \mathbf{v} d\mathbf{x} = 0,\ \forall \mathbf{v} \in \mathbf{W}_{0h}, \\ \int_D q \boldsymbol{\nabla} \cdot \mathbf{u}^{n+\frac{1}{6}} d\mathbf{x} = 0,\ \forall q \in L_h^2, \\ \mathbf{u}^{n+\frac{1}{6}} \in \mathbf{W}_h,\ \mathbf{u}^{n+\frac{1}{6}} = \mathbf{g}_{0h}^{n+1} \ \textit{on}\ \Gamma,\ p^{n+\frac{1}{6}} \in L_{0h}^2. \end{cases} \tag{23.37}$$

Next, compute $\mathbf{u}^{n+\frac{2}{6}}$ *via the solution of*

$$\begin{cases} \int_D \dfrac{\partial \mathbf{u}}{\partial t} \cdot \mathbf{v} d\mathbf{x} + \int_D (\mathbf{u}^{n+\frac{1}{6}} \cdot \boldsymbol{\nabla}) \mathbf{u} \cdot \mathbf{v} d\mathbf{x} = 0, \forall \mathbf{v} \in \mathbf{W}_{0h}^{n+1,-},\ \textit{a.e. on}\ (t^n,\ t^{n+1}), \\ \mathbf{u}(t^n) = \mathbf{u}^{n+\frac{1}{6}}, \\ \mathbf{u}(t) \in \mathbf{W}_h,\ \mathbf{u}(t) = \mathbf{g}_{0h}^{n+1} \ \textit{on}\ \Gamma_-^{n+1} \times (t^n,\ t^{n+1}), \end{cases} \tag{23.38}$$

and set $\mathbf{u}^{n+\frac{2}{6}} = \mathbf{u}(t^{n+1})$.

Then, compute $\mathbf{u}^{n+\frac{3}{6}}$ *and* $\boldsymbol{\lambda}_{A_h}^{n+\frac{3}{6}}$ *via the solution of*

$$\begin{cases} \rho_f \int_D \dfrac{\mathbf{u}^{n+\frac{3}{6}} - \mathbf{u}^{n+\frac{2}{6}}}{\triangle t} \cdot \mathbf{v} d\mathbf{x} + \alpha \mu_f \int_D \boldsymbol{\nabla} \mathbf{u}^{n+\frac{3}{6}} : \boldsymbol{\nabla} \mathbf{v} d\mathbf{x} \\ \quad = < \boldsymbol{\lambda}_{A_h}^{n+\frac{3}{6}}, \mathbf{v} >_{\Lambda_{A_h}}, \forall \mathbf{v} \in \mathbf{W}_{0h}, \\ < \boldsymbol{\mu}_A, \mathbf{u}^{n+\frac{3}{6}} - \mathbf{g}_{0h}^{n+1} >_{\Lambda_{A_h}} = 0,\ \forall \boldsymbol{\mu}_A \in \boldsymbol{\Lambda}_{A_h}; \\ \mathbf{u}^{n+\frac{3}{6}} \in \mathbf{W}_h, \boldsymbol{\lambda}_{A_h}^{n+\frac{3}{6}} \in \boldsymbol{\Lambda}_{A_h}. \end{cases} \tag{23.39}$$

Now predict the motion of the center of mass of the particle via

$$\frac{d\mathbf{G}}{dt} = \mathbf{V}(t)/2, \tag{23.40}$$

$$(1 - \frac{\rho_f}{\rho_s}) M_p \frac{d\mathbf{V}}{dt} = \mathbf{F}_r/2, \tag{23.41}$$

$$\mathbf{G}(t^n) = \mathbf{G}^n,\ \mathbf{V}(t^n) = \mathbf{V}^n, \tag{23.42}$$

for $t^n < t < t^{n+1}$. *Then set* $\mathbf{G}^{n+\frac{4}{6}} = \mathbf{G}(t^{n+1})$ *and* $\mathbf{V}^{n+\frac{4}{6}} = \mathbf{V}(t^{n+1})$.

With the center $\mathbf{G}^{n+\frac{4}{6}}$ *obtained at the above step, we enforce the rigid body motion in the region* $B^{n+\frac{4}{6}}$ *occupied by the particle*

$$\begin{cases} \rho_f \int_D \frac{\mathbf{u}^{n+1} - \mathbf{u}^{n+\frac{4}{6}}}{\triangle t} \cdot \mathbf{v}\, d\mathbf{x} + \beta \mu_f \int_D \nabla \mathbf{u}^{n+1} : \nabla \mathbf{v}\, d\mathbf{x} \\ +(1 - \frac{\rho_f}{\rho_s}) M_p \frac{\mathbf{V}^{n+\frac{5}{6}} - \mathbf{V}^{n+\frac{4}{6}}}{\triangle t} \cdot \mathbf{Y} + (1 - \frac{\rho_f}{\rho_s}) \mathbf{I}_p \frac{\boldsymbol{\omega}^{n+1} - \boldsymbol{\omega}^n}{\triangle t} \cdot \boldsymbol{\theta} \\ = (1 - \frac{\rho_f}{\rho_s}) M_p \mathbf{g} \cdot \mathbf{Y} + < \boldsymbol{\lambda}^{n+\frac{4}{6}}, \mathbf{v} - \mathbf{Y} - \boldsymbol{\theta} \times \overrightarrow{\mathbf{G}^{n+\frac{4}{6}}\mathbf{x}} >_{\Lambda_h^{n+\frac{4}{6}}}, \\ \forall \mathbf{v} \in \mathbf{W}_{0h},\ \mathbf{Y} \in \mathbb{R}^3,\ \boldsymbol{\theta} \in \mathbb{R}^3; \\ \mathbf{u}^{n+1} \in \mathbf{W}_h,\ \mathbf{u}^{n+1} = \mathbf{g}_{0h}^{n+1}\ on\ \Gamma,\ \boldsymbol{\lambda}^{n+\frac{4}{6}} \in \Lambda_h^{n+\frac{4}{6}},\ \mathbf{V}^{n+\frac{5}{6}} \in \mathbb{R}^3,\ \boldsymbol{\omega}^{n+1} \in \mathbb{R}^3, \end{cases} \tag{23.43}$$

$$< \mu, \mathbf{u}^{n+1} - \mathbf{V}^{n+\frac{5}{6}} - \boldsymbol{\omega}^{n+1} \times \overrightarrow{\mathbf{G}_j^{n+\frac{4}{6}}\mathbf{x}} >_{\Lambda_h^{n+\frac{4}{6}}} = 0,\ \forall \mu \in \Lambda_h^{n+\frac{4}{6}}. \tag{23.44}$$

Correct the motion of the center of mass of the particle via

$$\frac{d\mathbf{G}}{dt} = \mathbf{V}(t)/2, \tag{23.45}$$

$$(1 - \frac{\rho_f}{\rho_s}) M_p \frac{d\mathbf{V}}{dt} = \mathbf{F}_r/2, \tag{23.46}$$

$$\mathbf{G}(t^n) = \mathbf{G}^{n+\frac{4}{6}},\ \mathbf{V}(t^n) = \mathbf{V}^{n+\frac{5}{6}}, \tag{23.47}$$

for $t^n < t < t^{n+1}$. *Then set* $\mathbf{G}^{n+1} = \mathbf{G}(t^{n+1})$ *and* $\mathbf{V}^{n+1} = \mathbf{V}(t^{n+1})$.

In (23.36)–(23.47), $\Gamma_-^{n+1} = \{\mathbf{x} | \mathbf{x} \in \Gamma,\ \mathbf{g}_{0h}^{n+1}(\mathbf{x}) \cdot \mathbf{n}(\mathbf{x}) < 0\}$ and $\mathbf{W}_{0h}^{n+1,-} = \{\mathbf{v} | \mathbf{v} \in \mathbf{W}_h,\ \mathbf{v} = \mathbf{0}\ on\ \Gamma_-^{n+1}\}$, $\Lambda_h^{n+s} = \Lambda_h(t^{n+s})$, $\mathbf{g}_{0h}^{n+1}$ is an approximation of $\mathbf{g}_0^{n+1}$ belonging to

$$\gamma \mathbf{W}_h = \{\mathbf{z}_h | \mathbf{z}_h \in (C^0(\Gamma))^3,\ \mathbf{z}_h = \tilde{\mathbf{z}}_h|_\Gamma\ with\ \tilde{\mathbf{z}}_h \in \mathbf{W}_h\}$$

and verifying $\int_\Gamma \mathbf{g}_{0h}^{n+1} \cdot \mathbf{n}\, d\Gamma = 0$.

3.3 On the Solution of Subproblems (23.37), (23.38), (23.39), (23.40)–(23.42), and (23.43)–(23.44)

The degenerated quasi-Stokes problem (23.37) is solved by an Uzawa/preconditioned conjugate gradient algorithm as in [6]. The advection problem (23.38) for the velocity field is solved by a wave-like equation method as in [4, 6].

Systems (23.40)–(23.42) and (23.45)–(23.47) are systems of ordinary differential equations, thanks to operator splitting. For their solution one can use a time-marching scheme with a time step smaller than $\triangle t$ (i.e., we can divide $\triangle t$ into smaller steps) to compute the translation velocity of the center of mass and the position of the center of mass and then the regions occupied by each particle so that the repulsion forces can be effective to prevent particle–particle and particle–wall overlapping; see, e.g., [8].

In (23.43)–(23.44), the hydrodynamical forces and gravity acting on the particles are also taken into account in order to update the translation and angular velocities of the particles. At the same time the rigid body motion is enforced in $B^{n+4/6}$, via equation (23.44). To solve (23.43)–(23.44), we use a conjugate gradient algorithm as discussed in [7]. Since we take $\beta = 0$ in (23.43) for the simulation, we actually do not need to solve any nontrivial linear systems for the velocity field; this saves a lot of computing time.

Problem (23.39) is a classical *saddle-point problem*, which is a particular case of

$$\begin{cases} \alpha \int_D \mathbf{u} \cdot \mathbf{v} d\mathbf{x} + \mu \int_D \nabla \mathbf{u} : \nabla \mathbf{v} d\mathbf{x} = \int_D \mathbf{f} \cdot \mathbf{v} d\mathbf{x} + < \boldsymbol{\lambda}, \mathbf{v} >, \ \forall \mathbf{v} \in \mathbf{W}_{0h}, \\ < \mu, \mathbf{u} - \mathbf{g} > = 0, \ \forall \mu \in \Lambda, \\ \mathbf{u} \in \mathbf{W}_h, \ \mathbf{u} = \mathbf{g} \ on \ \Gamma, \ \boldsymbol{\lambda} \in \Lambda. \end{cases} \tag{23.48}$$

To solve problem (23.48), we have applied the following conjugate gradient method:

$$\boldsymbol{\lambda}^0 \in \Lambda \ is \ given, \tag{23.49}$$

solve

$$\begin{cases} \alpha \int_D \mathbf{u}^0 \cdot \mathbf{v} d\mathbf{x} + \mu \int_D \nabla \mathbf{u}^0 : \nabla \mathbf{v} d\mathbf{x} = \int_D \mathbf{f} \cdot \mathbf{v} d\mathbf{x} + < \boldsymbol{\lambda}^0, \mathbf{v} >, \\ \forall \mathbf{v} \in \mathbf{W}_{0h}; \ \mathbf{u}^0 \in \mathbf{W}_h, \ \mathbf{u} = \mathbf{g} \ on \ \Gamma, \end{cases} \tag{23.50}$$

then solve

$$< \mu, \mathbf{g}^0 > = < \mu, \mathbf{u}^0 - \mathbf{g} >, \ \forall \mu \in \Lambda; \ \mathbf{g}^0 \in \Lambda, \tag{23.51}$$

and set

$$\mathbf{w}^0 = \mathbf{g}^0. \tag{23.52}$$

For $m \geq 0$, assuming that $\boldsymbol{\lambda}^m, \mathbf{u}^m, \mathbf{w}^m, \mathbf{g}^m$ are known, compute $\boldsymbol{\lambda}^{m+1}$, $\mathbf{u}^{m+1}$, $\mathbf{w}^{m+1}$, $\mathbf{g}^{m+1}$ as follows:

Solve

$$\begin{cases} \alpha \int_D \overline{\mathbf{u}}^m \cdot \mathbf{v} d\mathbf{x} + \mu \int_D \nabla \overline{\mathbf{u}}^m : \nabla \mathbf{v} d\mathbf{x} = < \mathbf{w}^m, \mathbf{v} >, \\ \forall \mathbf{v} \in \mathbf{W}_{0h}; \ \overline{\mathbf{u}}^m \in \mathbf{W}_{0h}, \end{cases} \tag{23.53}$$

and set

$$< \mu, \overline{\mathbf{g}}^m > = < \mu, \overline{\mathbf{u}}^m >, \forall \mu \in \Lambda; \ \overline{\mathbf{g}}^m \in \Lambda. \tag{23.54}$$

Then compute

$$\rho_m = < \mathbf{g}^m, \mathbf{g}^m > / < \overline{\mathbf{g}}^m, \mathbf{w}^m >, \tag{23.55}$$

and set

$$\boldsymbol{\lambda}^{m+1} = \boldsymbol{\lambda}^m - \rho_m \mathbf{w}^m, \ \ \mathbf{u}^{m+1} = \mathbf{u}^m - \rho_m \overline{\mathbf{u}}^m, \ \ \mathbf{g}^{m+1} = \mathbf{g}^m - \rho_m \overline{\mathbf{g}}^m. \tag{23.56}$$

If $< \mathbf{g}^{m+1}, \mathbf{g}^{m+1} > / < \mathbf{g}^0, \mathbf{g}^0 > \leq \varepsilon$, then take $\mathbf{u} = \mathbf{u}^{m+1}$. If not, compute

$$\gamma_m = < \mathbf{g}^{m+1}, \mathbf{g}^{m+1} > / < \mathbf{g}^m, \mathbf{g}^m >, \tag{23.57}$$

and set

$$\mathbf{w}^{m+1} = \mathbf{g}^{m+1} + \gamma_m \mathbf{w}^m. \tag{23.58}$$

Do $m = m+1$ *and go back to* (23.53).

Remark 2. The above conjugate gradient algorithm is similar to the one discussed in [9, 10]; here a distributed Lagrange multiplier has been used instead of the boundary Lagrange multiplier used in [9, 10]. Those distributed volume multipliers are also considered in [6].

4 Numerical Experiments and Discussion

To investigate and reproduce the cluster formations observed in experiments for a settling suspension of uniform non-Brownian particles in a *fully* filled horizontal rotating cylinder, we have applied the methodology discussed in the above sections to simulate the motion of up to 128 balls in a truncated cylinder rotating around its central axis. We have first considered the cases of two balls to find out the interaction between two balls in a *fully* filled horizontal rotating cylinder and then we have studied (i) the effect of the angular speed and of the number of particles on the formation of the clusters by placing the balls in positions which cannot be achieved by laboratory experiments and (ii) the effect of ball clusters in the fluid flow field.

4.1 The Interaction of Two Balls Side by Side Initially

In this test case, we consider the simulation of the motion of two balls in a truncated cylinder rotating around its central axis. The computational domain is $D = (0, 1.0625) \times (0,4) \times (0, 1.0625)$. The diameter of the rotating cylinder is 1 and its length is 4. The central axis of the cylinder is located at $x = 0.53125$ and $z = 0.53125$. The fluid density is $\rho_f = 1$ and the fluid viscosity is $\mu_f = 0.15$. The density of the balls is $\rho_s = 1.25$. The diameters of the balls are 0.15. The gravity force is pointed downward (in the negative z direction). The initial positions of the two ball mass centers are $(0.13125, 2.109375, 0.53125)^t$ and $(0.13125, 1.890625, 0.53125)^t$, respectively. Thus they are initially side by side at the front of the cylinder. The flow field initial condition is $\mathbf{u}_0(x,y,z) = (-\Omega(z-0.53125), 0, \Omega(x-0.53125))^t$ with $\Omega = 8$ and 12, respectively. Hence $\mathbf{u}_0$ is also the extension of the boundary condition to the region outside the cylinder used in the simulation. To check the convergence, we have chosen the following three pairs of mesh size and time step $\{h_v, \triangle t\}$=$\{1/64, 0.001\}$, $\{1/96, 0.001\}$, and $\{1/128, 0.001\}$. The mesh size of the pressure grid is always $h_p = 2h_v$. The histories of the mass centers obtained from three different pairs of mesh size and time step are shown in Figure 23.2. We have obtained a good agreement between the corresponding computational results.

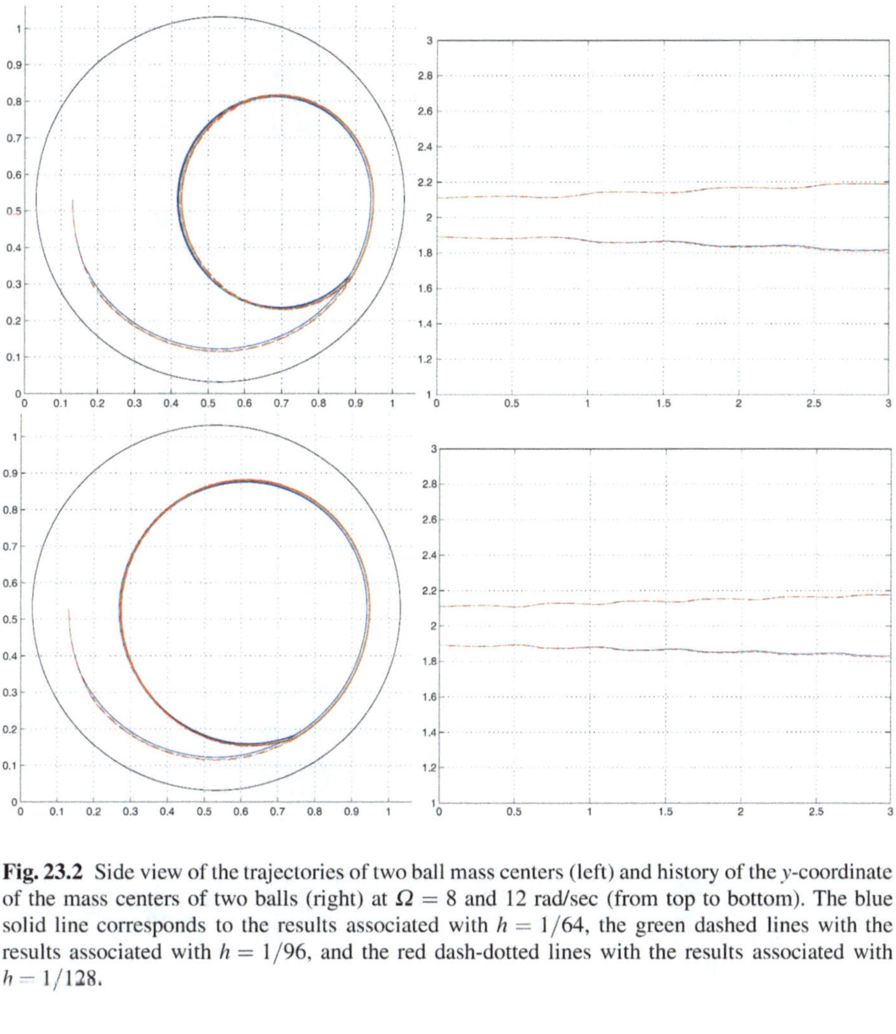

Fig. 23.2 Side view of the trajectories of two ball mass centers (left) and history of the y-coordinate of the mass centers of two balls (right) at $\Omega = 8$ and 12 rad/sec (from top to bottom). The blue solid line corresponds to the results associated with $h = 1/64$, the green dashed lines with the results associated with $h = 1/96$, and the red dash-dotted lines with the results associated with $h = 1/128$.

For the case of $\Omega = 8$, the two balls rotate side by side at about the same speed with respect to the central axis of the cylinder (see Figure 23.2); but in the y direction they attract and expel each other in a way that the distance between the two balls decreases when they move from the upper-front region of cylinder to the back of cylinder and the distance between the two balls increases when they move from the back of cylinder to the upper-front region of cylinder (see Figure 23.2). We also obtained similar results for the case of $\Omega = 12$, but the distance is smaller when the rotation speed becomes higher (see Figure 23.2). The averaged particle Reynolds numbers of the two balls are 2.24 and 4.09, respectively, for Ω=8 and 12 based on the averaged velocity for $2.5 \leq t \leq 3$.

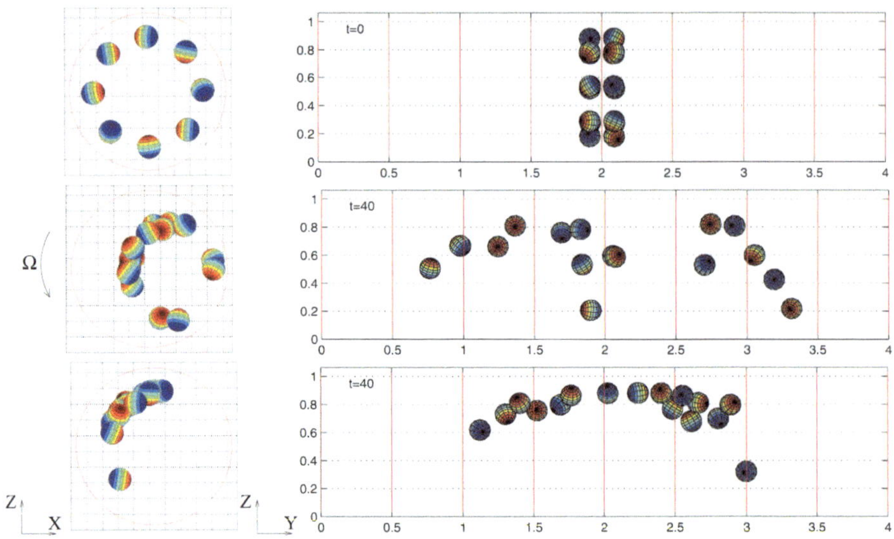

Fig. 23.3 Side view (left) and front view (right) of the initial position of 16 balls (top) and position obtained at the angular speeds $\Omega = 8$ (middle) and 12 (bottom) rad/sec at $t = 40$ seconds.

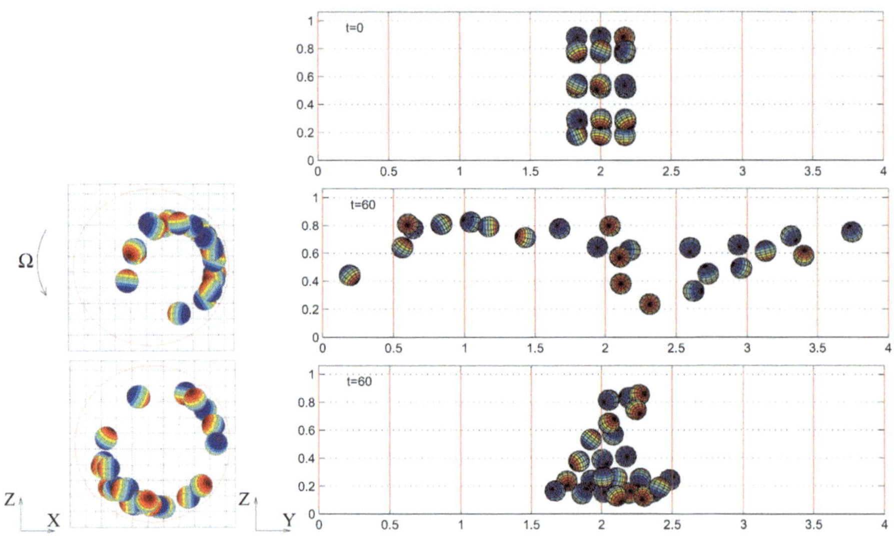

Fig. 23.4 Side view (left) and front view (right) of the initial position of 24 balls (top) and position obtained at the angular speeds $\Omega = 8$ (middle) and 12 (bottom) rad/sec at $t = 60$ seconds.

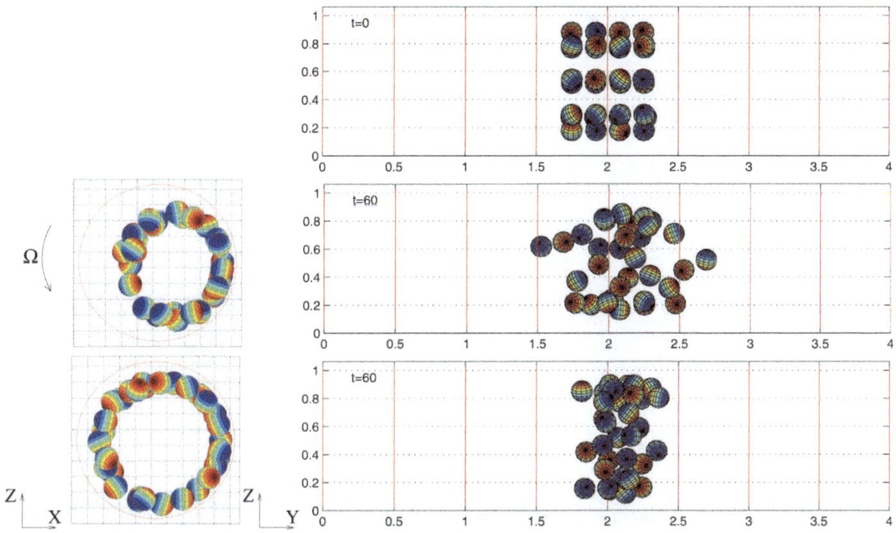

Fig. 23.5 Side view (left) and front view (right) of the initial position of 32 balls (top) and position obtained at the angular speeds $\Omega = 8$ (middle) and 12 (bottom) rad/sec at $t = 60$ seconds.

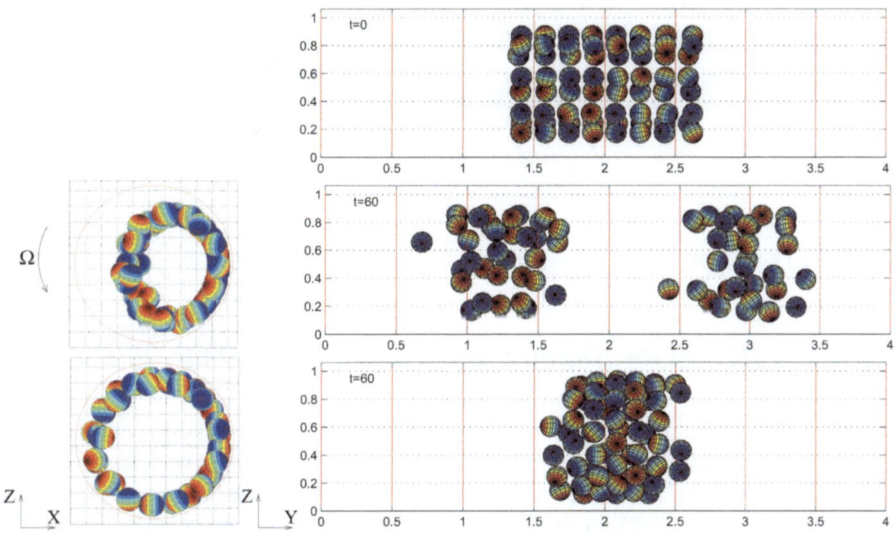

Fig. 23.6 Side view (left) and front view (right) of the initial position of 64 balls (top) and position obtained at the angular speeds $\Omega = 8$ (middle) and 12 (bottom) rad/sec at $t = 60$ seconds.

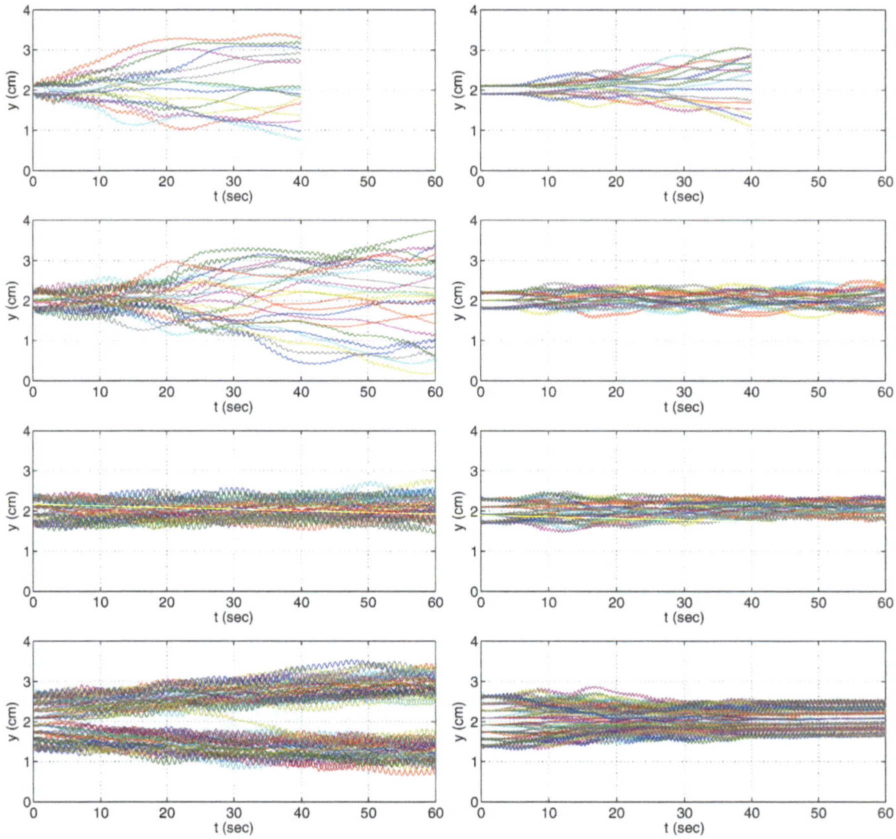

Fig. 23.7 Histories of the y-coordinate of the mass centers of 16, 24, 32, and 64 balls (from top to bottom) at $\Omega = 8$ (left) and 12 (right) rad/sec.

4.2 The Effect of the Angular Speed and of the Number of Particles

To investigate the circular cluster formation for the suspensions of particles in a fully filled horizontally rotating cylinder, we have studied first the formation of a single circular cluster in the cases of 16, 24, 32, and 64 balls of radius a =0.075 cm and density $\rho_p = 1.25$ g/cm^3 in a truncated cylinder of diameter $2R = 1$ cm and length 4 cm filled with a fluid of density 1 g/cm^3 and kinematic viscosity $\nu = 0.15$ cm^2/sec. The solid fractions are 0.9%, 1.35%, 1.8%, and 2.7%, respectively, for the cases of 16, 24, 32, and 64 balls. The initial positions of the ball mass centers are on the circles of radius 0.35 cm centered at the cylinder central axis with eight balls in each circle (see Figures 23.3–23.6). We have perturbed each mass center randomly in the direction of the cylinder axis to break the symmetry of the initial pattern. The distance between two neighboring circles is about $2.25a$ hence the initial gap size d_g between balls in the cylinder axis direction is about $a/4$. In the simulations, the

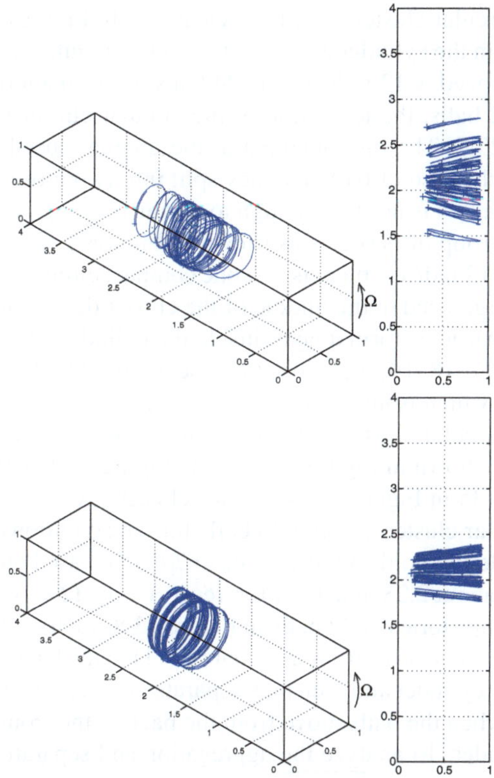

Fig. 23.8 Trajectories and top view of the trajectories of 32 balls at the angular speeds $\Omega = 8$ (left two) and 12 (right two) rad/sec for $59 \leq t \leq 60$ seconds. The length unit is cm.

cylinder rotates around the cylinder axis parallel to the y–axis in a clockwise direction with angular speed Ω of either 8 or 12 rad/sec (see Figure 23.1). The Reynolds numbers, Re= $2aU/\nu$, with the characteristic velocity $U = \Omega R$ are 4 and 6, respectively, for Ω= 8 and 12 rad/sec. The Reynolds numbers of the cases considered here are about two orders less than those considered in [23]. The Ekman numbers, E= $\nu/\Omega R^2$, are 0.075 and 0.05, respectively, for $\Omega = 8$ and 12 rad/sec and both are an order larger than those considered in [23]. Both numbers for the cases considered here are in a different regime, thus the numerical simulation is, strictly speaking, not comparable with the experiments reported in [16, 23] even though the circular cluster formation is similar to those observed in the two above publications. Since the thickness of the Ekman boundary layer is the order of $E^{1/2}$, our meshes can resolve the Ekman boundary layer for the cases studied in this chapter.

The histories of the y–coordinate of the particle mass centers and the positions of 16 balls at $t = 40$ seconds obtained with the angular speed $\Omega = 8$ and 12 rad/sec in Figures 23.3 and 23.7 clearly show that the 16 balls spread out in the cylinder axis direction and do not form a circular cluster at all. For the cases of 24 balls, the formation of the circular cluster is still not clear yet. In Figures 23.4 and 23.7, the 24 balls spread out in the cylinder axis direction at the angular speed $\Omega = 8$ rad/sec. When the angular speed is 12 rad/sec, the 24 balls do form a loose circular cluster. For the cases of 32 balls, the formation of the circular cluster is clearly shown in Figures 23.5 and 23.7. The one obtained at the angular speed $\Omega = 12$ rad/sec is very compact. For the case of 64 balls, they split into two loose circular clusters at $\Omega = 8$ rad/sec since this is not fast enough to produce strong particle interaction to sustain the whole group of particles as shown in Figures 23.6 and 23.7. But at the angular speed $\Omega = 12$ rad/sec, there is just one compact circular cluster in which the particles are well organized in the middle of the cluster due to the pushing from the outside balls. The particles form a layer inside the cylinder which is different from those observed in [17, 18, 19], but close to those in [16, 23]. These results give us a simple observation which is that there is a need to have enough particles so that the particles within a circular cluster can continuously interact among themselves. For the case of 64 balls shown in Figure 23.6 (resp., Figure 23.14), there are 33 and 31 balls (resp., 29 and 35 in Figure 23.14) in two clusters, respectively. The threshold for forming a circular cluster is about 30 balls for the conditions considered in this chapter. The particle Reynolds numbers $\mathrm{Re}_p = 2aU_p/\nu$ based on the average speed U_p of particles are about 2.28 and 4.28, for, respectively, $\Omega = 8$ and 12 rad/sec.

Observing the trajectories of 32 balls in Figure 23.8, we have found that the balls aggregate when the balls move from the front ($x = 1$) to the back ($x = 0$) through the upper portion of the cylinder and then they separate (spreading out in the direction of the cylinder axis) when the balls move from the back to the front through the lower portion of the cylinder. To analyze the aggregation and separation of the particles, we define the speed as $V_r = \sqrt{V_1^2 + V_3^2}$ in the xz-plane from the particle translation velocity $\mathbf{V} = (V_1, V_2, V_3)$. The speed V_r tell us how fast each particle moves in the plane perpendicular to the cylinder axis directions, especially how it moves within a cluster when it is part of such a cluster. When each particle moves up from the front of the cylinder to the top of the cylinder, the speed in the x-direction, $|V_1|$, is increasing by the rotation and the one in the z-direction, $|V_3|$, is suppressed by the rotation and the gravity. Once it passes the top position and moves into the back portion of the cylinder, the speed in the z-direction is increasing dramatically since the rotation and the gravity work together even the one in the x-direction is decreasing to zero. This explains when the balls of a cluster move through the upper portion of the cylinder, their speeds V_r are increasing as shown in Figure 23.9 (the left ones). When one ball enters the wake of another ball which is speeding up, it experiences reduced drag and drafts closer to the leading ball (e.g., see [5, 11] for the drafting, kissing, and tumbling between two balls). Thus the group of balls with increasing speeds can aggregate due to the hydrodynamical interaction between balls. For the part of a circular cluster formed by the balls moving from the back to the front through the lower portion of the cylinder, the balls separate and spread out due to the slowdown of the speed V_r (see Figure 23.9). The slowdown is caused by the rotation when each ball moves from the back to the bottom of the cluster since the gravity cannot compete with the rotation. Once the ball starts moving up from the

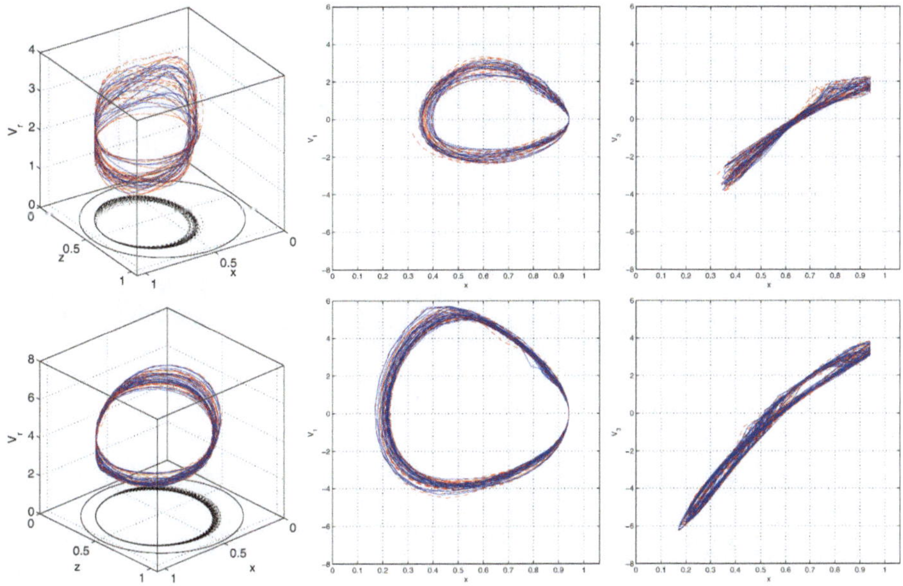

Fig. 23.9 Speed $V_r = \sqrt{V_1^2 + V_3^2}$ (left) of particles in the xz–plane versus the particle's x and z–coordinates, V_1 (middle) and V_3 (right) for the cases of $\Omega = 8$ (top) and 12 (bottom) rad/sec for $59 \leq t \leq 60$ seconds. The blue solid (resp., red dashed) lines are associated with the particles whose average mass centers are located to the right (resp., left) of the average mass center of all particles in the cylinder axis direction. The black dotted lines (in the left figure) are the projected particle trajectories in the xz–plane and the black line in the xz–plane is the boundary of the cylinder.

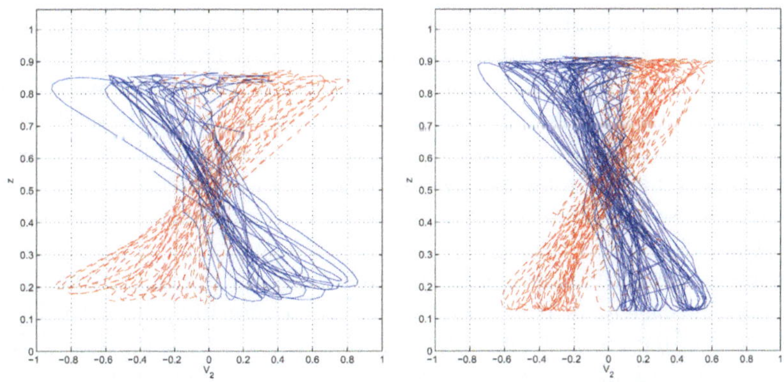

Fig. 23.10 Particle speed V_2 in the cylinder central axis direction versus the particle z–coordinate for the cases of $\Omega = 8$ (left) and 12 (right) rad/sec for $59 \leq t \leq 60$ seconds. The blue solid (resp., red dashed) lines are associated with the particles whose average mass centers are located to the right (resp., left) of the average mass center of all particles in the cylinder axis direction.

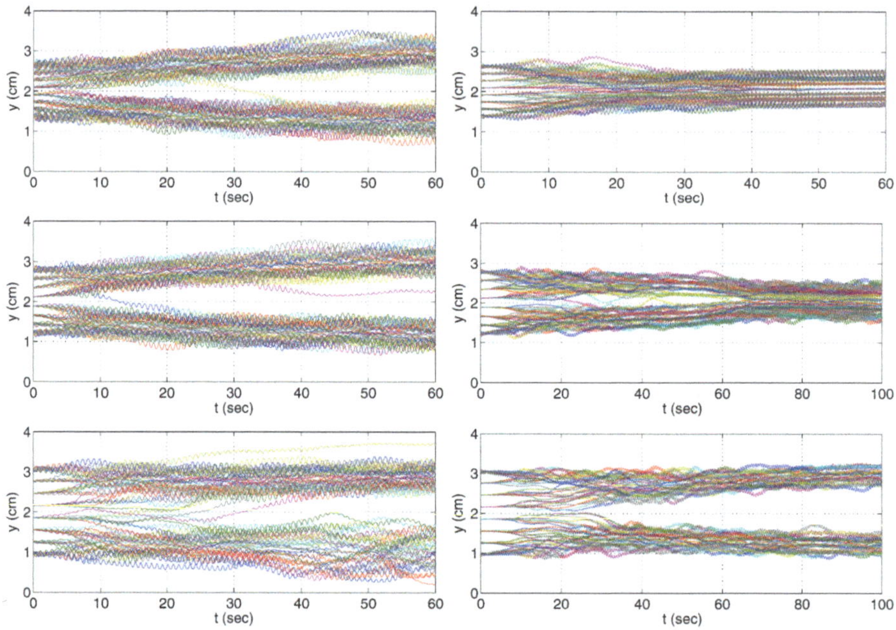

Fig. 23.11 Histories of the y-coordinate of the mass centers of 64 balls with initial gap sizes $a/4$, a, and $2a$ (from top to bottom): $\Omega = 8$ (left) and 12 (right) rad/sec.

bottom to the front, the speed is suppressed further by the rotation and the counter effect of the gravity as in Figure 23.9. Due to these effects of the speedup and slowdown, the particle speed V_2 in the cylinder central axis direction does have different sign as shown in Figure 23.10. Consider those particles whose average mass centers are located to the right of the average mass center of all particles in the cylinder central axis direction: when they move from the front to the back through the upper (resp., lower) portion of the cylinder, the speed V_2 is negative (resp., positive). For those located to the left of the average mass center of all particles, the speeds V_2 are opposite to those located to the right. Hence the balls aggregate during the speedup of V_r when the balls move from the front to the back through the upper portion of the cylinder and they separate because of the slowdown of the speed V_r when the balls move from the back to the front through the lower portion of the cylinder. Therefore the histories of the y–coordinate of the particle mass centers in Figures 23.7, 23.11 and 23.15 show oscillations in the y-direction. To have a stabilized and compact circular cluster, a sufficiently large number of balls and a fast enough angular speed are needed in order to balance both effects, e.g., the results of the 32 ball cases in Figures 23.5 and 23.8 show that at the angular speed $\Omega = 8$ rad/sec, the particle speeds are just fast enough to have the aggregation which can overcome the separation. But at the angular speed $\Omega = 12$ rad/sec, the particle interaction is stronger so that a compact circular cluster is formed. Similarly for the 64 ball case at lower angular speed $\Omega = 8$ rad/sec in Figures 23.6 and 23.7, the balls spread out a little bit and segregate into two loose circular clusters; but at $\Omega = 12$ rad/sec the particle interaction can pull all 64 balls together in one compact circular cluster. Thus the particle segregation also depends on the relative motion between the particles and the rotating flow field. Actually the distance between particles does matter concerning the formation of the circular clusters. In Figure 23.11, the histories of the y-coordinate of the particle mass centers are shown for different values of the initial gap size d_g. The particle interaction at Ω =8 rad/sec cannot pull all 64 balls into one circular cluster. The balls split into two groups for all three initial gap sizes and form circular clusters except for one group of balls for the case $d_g = 2a$. At the angular speed Ω =12 rad/sec, the threshold of the initial gap size for forming a circular cluster is $d_g = a$. There are two circular clusters formed for 64 balls with $d_g = 2a$, but 64 balls with the initial gap size $d_g = a$ interact and finally come together to form a circular cluster at $t = 60$ seconds as in Figure 23.11. The formation of a circular cluster of 64 balls with the initial gap size $d_g = a$ is much slower than the one with $d_g = a/4$ at Ω =12 rad/sec. These results show that the particle interaction has short range effect on the formation of circular clusters.

4.3 The Cluster Effect on the Fluid Flow Field

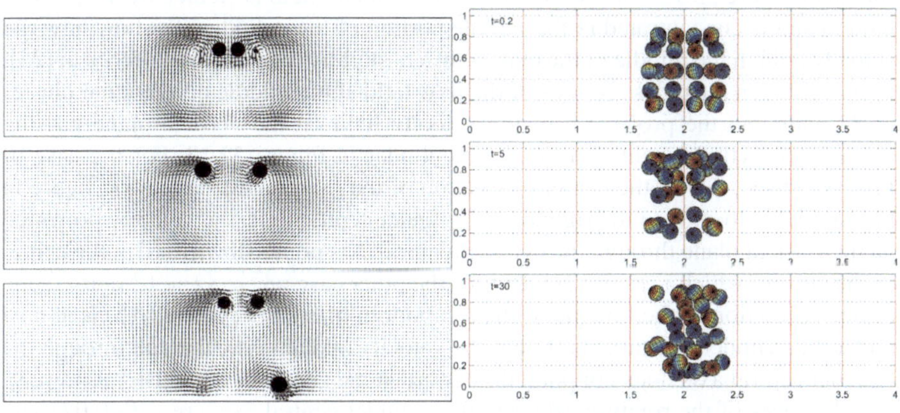

Fig. 23.12 Projection of the velocity field on the vertical plane passing through the central axis of the cylinder for the case of 32 balls (left) and front view of the position of 32 balls (right) at t = 0.2, 5, and 30 seconds (from top to bottom) with $\Omega = 12$ rad/sec.

Despite that fact that in [23] the Reynolds numbers and Ekman numbers are in regimes different from those in this chapter, we have obtained circular clusters like those reported in the above reference. In the experiments, it is not easy to set up the initial positions of the particles like those chosen in direct numerical simulations; but those initial positions help us to understand the formation of circular clusters and the

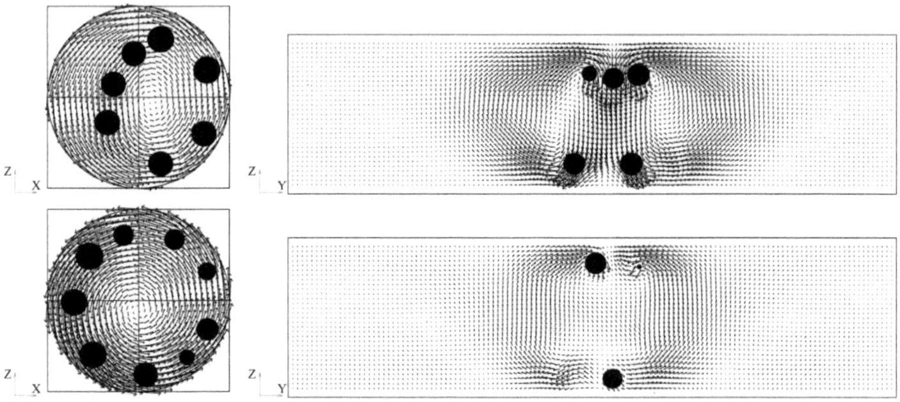

Fig. 23.13 Projection of the velocity field on the vertical plane at the middle of the circular cluster (left) and on the vertical plane passing through the central axis of the cylinder (right) for the case of 32 balls at t =60 seconds: $\Omega = 8$ (top) and 12 (bottom) rad/sec.

development of the flow field inside the cylinder. For the case of 32 balls at $\Omega = 12$ rad/sec studied in the previous subsection, the projections of the velocity field on the vertical plane passing through the central axis of the cylinder at different time are shown in Figure 23.12. The circulation of the velocity field is created by the particle motion and is concentrated in the middle portion of the cylinder. To show how the velocity field in the middle of the cluster differs slightly from those observed in [23], we have shown the cross sections of the flow field at the middle of a circular cluster of 32 balls and the projection of the velocity field on the vertical plane passing through the cylinder central axis of the cylinder in Figure 23.13. For the case of $\Omega = 8$ rad/sec, due to the center of rotation of the flow field (the left top figure in Figure 23.13) located to the right of the cylinder central axis, the velocity projected on the vertical plane through the cylinder central axis points downward at the center of the cluster, which is as in [23]. But for the other case where $\Omega = 12$ rad/sec, the center of rotation of the flow field (the left bottom figure in Figure 23.13) is almost under the cylinder central axis so that the velocity on the vertical plane through the cylinder central axis does not point downward at the center of the cluster as in [23]. The distances of the rotating center to the cylinder central axis are $\triangle R$=0.073 and 0.058 cm for Ω =8 and 12 rad/sec, respectively. Then the Rossby numbers Ro= $U/\Omega R$, where $U = \Omega\triangle R$ is the relative velocity of the secondary flow associated with the clusters as considered in [18], are 0.146 and 0.116. For both angular speeds, the Rossby numbers are not small and the inertial effect cannot be ignored as in [23].

The evolution of the flow field related to two circular clusters in the case of 64 balls with initial gap size $d_g = 2a$ and the angular speed $\Omega = 12$ rad/sec are shown in Figures 23.14 and 23.15. We have observed no specific pattern concerning the flow field from the beginning as in Figure 23.14. The particles break into two circular clusters between $t = 16$ and 19 seconds and then the two clusters move away from each other as in Figure 23.15. The projected velocity fields at $t = 19$

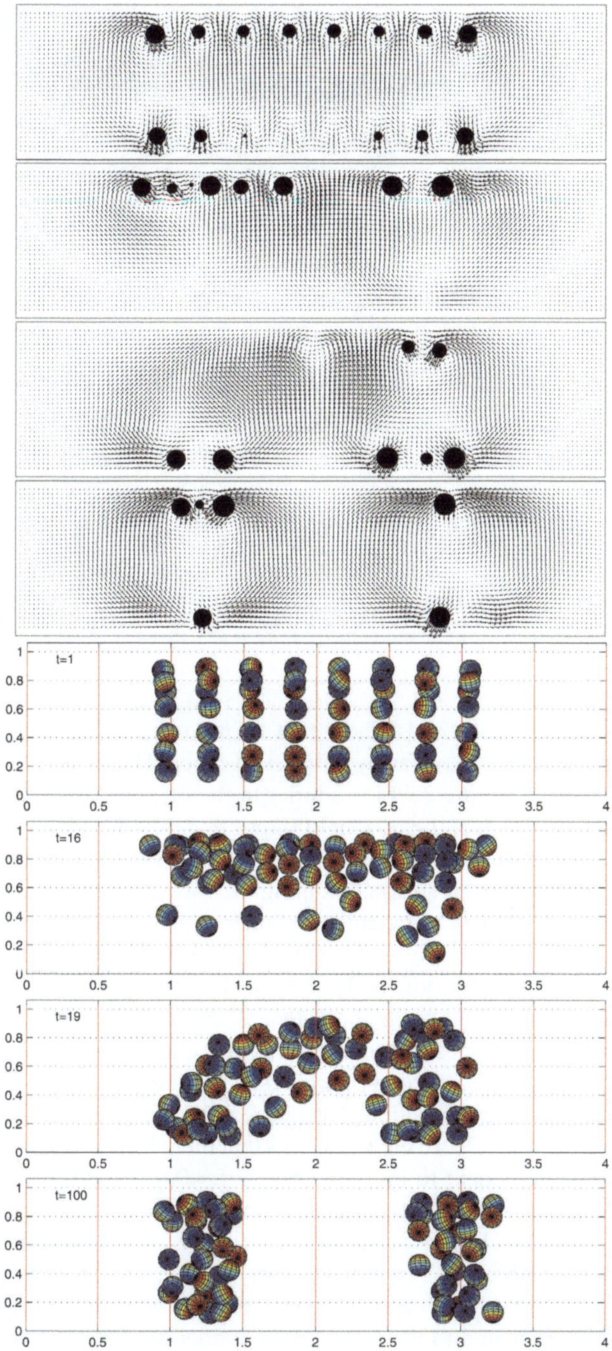

Fig. 23.14 Projection of the velocity field on the vertical plane passing through the central axis of the cylinder for the case of 64 balls (top four pictures) and front view of the positions of 64 balls (lower four pictures) at $t = 1, 16, 19$, and 100 seconds (from top to bottom) with $\Omega = 12$ rad/sec; the initial gap size $d_g = 2a$.

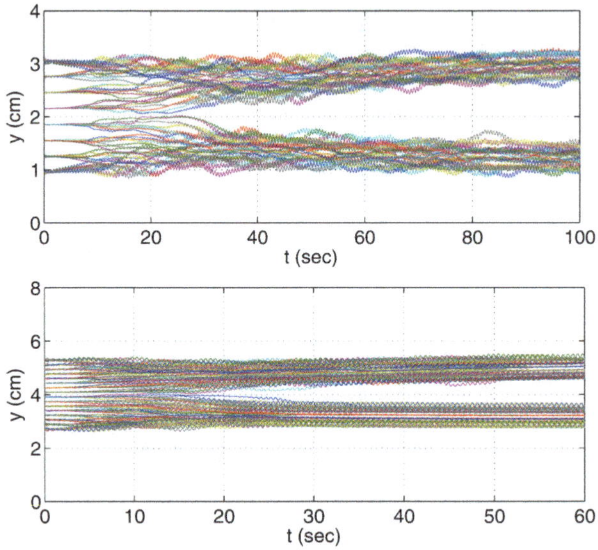

Fig. 23.15 Histories of the y-coordinate of the mass centers of the 64 balls with the initial gap size $d_g = 2a$ (top) and 128 balls with the initial gap size $d_g = a/4$ (bottom).

and 100 seconds in Figure 23.14 show that the two circulations move apart since the two circular clusters move away from each other. The projected velocity field at $t = 100$ seconds in Figure 23.14 is similar to the one obtained experimentally in [23], but the development of the flow field shows that the circulation of the flow field is caused by the motion of the particles in the two circular particle clusters and there are no secondary flows occurring and helping the formation of the circular clusters. For the case of 128 balls in a truncated cylinder of length L =8 cm at the angular speed $\Omega = 12$ rad/sec with the initial gap size $d_g = a/4$, the particles are initially placed on 16 circles in the middle of the cylinder as in the previous subsection. Later they break into two compact circular clusters as shown in Figures 23.15 and 23.16. There are 63 and 65 particles in these two circular clusters, respectively, which are consistent with the results of the 64 particles at the angular speed $\Omega = 12$ rad/sec discussed in the previous subsection. The figure of the circulation of the flow field at $t = 0.4$ seconds in Figure 23.16 clearly shows that there is only one large circulation. Two small circulations next to the large one at t =0.4 seconds are created by the strong advection due to the particle motion and stay there all the time even when the balls split into two clusters as the one at t =40 seconds. These secondary flows are very weak for both cases. For both clusters in Figure 23.16, the cross section of the flow field at the middle of each circular cluster shows that the center of rotation of the flow field is located to the left of the cylinder central axis, and the velocity projected on the vertical plan through the cylinder central axis points upward at the center of the cluster as shown in the middle pictures of Figure 23.16.

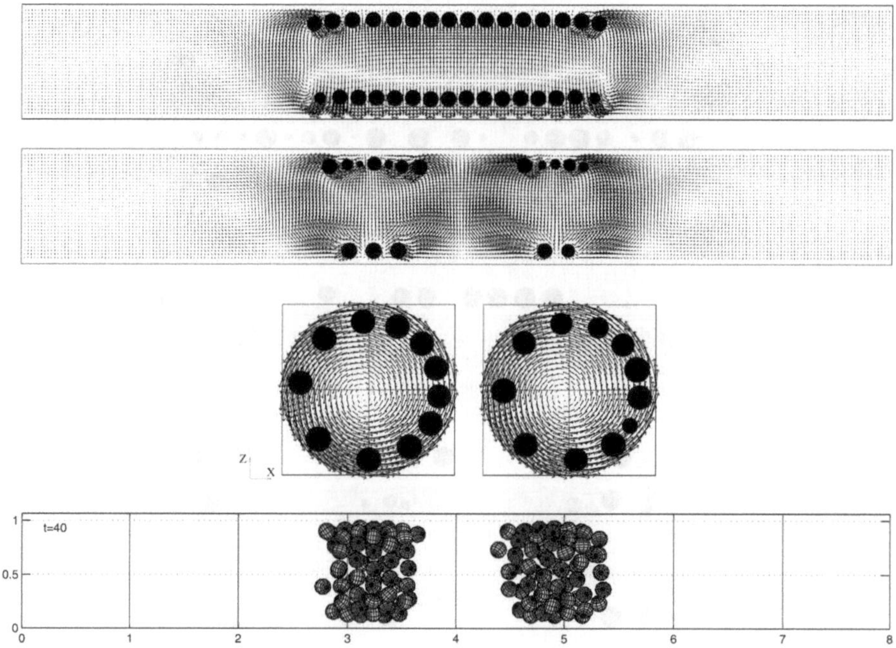

Fig. 23.16 Projection of the velocity field on the vertical plane passing through the central axis of the cylinder for the case of 128 balls at $t = 0.4$ (top picture) and 40 seconds (second picture from top), projection of the velocity field on two cluster middle planes, $y = 3.21875$ and $y = 5$, at $t = 40$ seconds (third pictures from top), and front view of the position of 128 balls (bottom picture) at $t = 40$ seconds.

To produce circular clusters like those observed in [16, 23], we have considered the case of 128 balls in a truncated cylinder of length L =4 cm. We have first placed 128 balls on 16 circles in the middle of the cylinder with the initial gap size $d_g = 0.25a$ as in the previous subsection and then let them settle at zero angular speed. The balls settle down at the bottom of the cylinder after 2 seconds as shown in Figures 23.17 and 23.18. Then we rotate the cylinder at the angular speed Ω =12 rad/sec. First the 128 balls move up and down inside the rotating cylinder and interact with the fluid. At $t = 5$ seconds, there is no specific flow field pattern in the cylinder. About $t = 20$ seconds, two outer clusters next to the two ends of the cylinder start forming. Gradually three circular clusters, which are similar to the one obtained experimentally in [16, 23], are formed as shown in Figures 23.17 and 23.18. The wavelength between two left clusters at $t = 150$ seconds is $3.25R$ and the distance from the leftmost cluster to the left end of the cylinder is also about half of the above wavelength. The right circular cluster has been pushed to the right end of the cylinder with no room to move. The wavelength is in a good agreement with the wavelengths obtained in [23], which are between $3.2R$ and $3.3R$, for the case of $L/R = 8$. The cross sections of the flow field at the middle of each circular clusters (y =0.875, 2.5, and 3.75 cm) are shown in Figure 23.19. We observe that

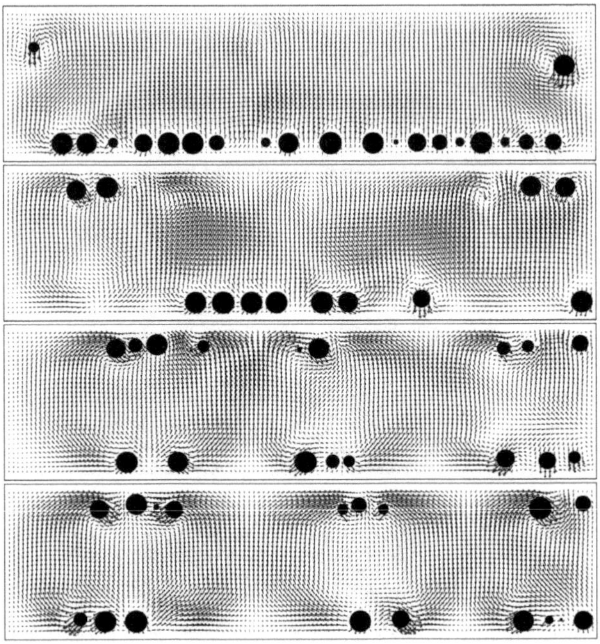

Fig. 23.17 Projection of the velocity field on the vertical plane passing through the central axis of the cylinder for the case of 128 balls at t = 5, 20, 44, and 150 seconds (from top to bottom) with $\Omega = 12$ rad/sec.

the centers of rotation of the flow field of each cross section are located either to the left of the cylinder central axis or right under the cylinder central axis. Thus the velocity projected on the vertical plan through the cylinder central axis points either upward or really hard to tell at the center of the cluster as shown in the lower left one in Figures 23.17 and 23.18. The distances of the rotating center to the cylinder central axis are $\triangle R$=0.0572, 0.0473, and 0.0481 cm, respectively, for y =0.875, 2.5, and 3.75 cm. The Rossby numbers Ro= $U/\Omega R$ are 0.1144, 0.0946, and 0.0962, respectively, for y =0.875, 2.5, and 3.75 cm. The Ekman number for this case is 0.05 as discussed at the beginning of Section 3.1. Since both Ekman number and Rossby number are not too small, the inertial effect and diffusion cannot be ignored for the perturbation analysis in [23].

5 Conclusion

In this chapter we have applied a distributed Lagrange multiplier fictitious domain method combined with finite element and operator splitting methods to simulate rotating suspension of particles and then to study the interaction between balls and fluid in a fully filled and horizontally rotating cylinder. The formation of circular clusters studied in this chapter is mainly caused by the interaction between particles

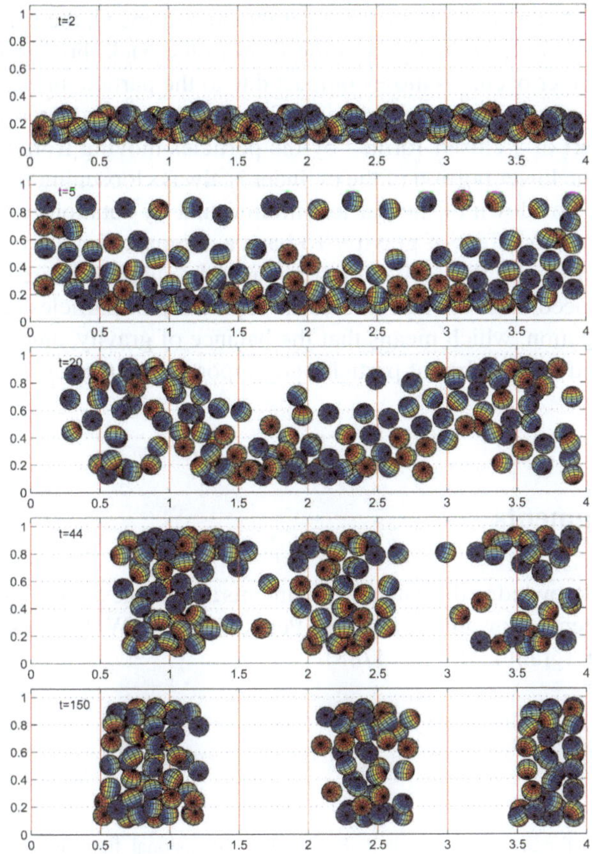

Fig. 23.18 Front view of the position of 128 balls (right) at $t = 2, 5, 20, 44$, and 150 seconds (from top to bottom) with $\Omega = 12$ rad/sec.

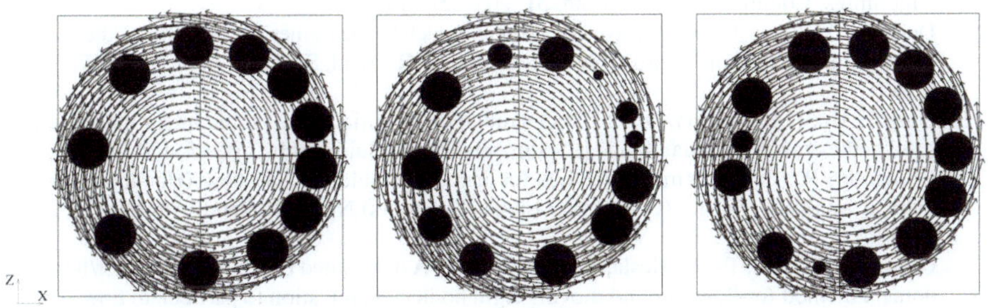

Fig. 23.19 Projection of the velocity field on the vertical plane at the middle of each circular clusters for the case of 128 balls at $t = 150$ seconds: $y = 0.875$, 2.5, and 3.75 cm (from left to right).

themselves. Within a circular cluster, at larger enough speed, the part of the cluster formed by the particles moving from the front to the back through the upper portion of the cylinder becomes more compact due to the particle interaction strengthened by the speedup of the particle speeds first by rotation and later by rotation and gravity. The part of a cluster formed by the particles moving from the back to the front through the lower portion of the cylinder is always loosening up and spreading out due to the slowdown of the particle motion first by rotation and later by rotation and the counter effect of gravity. To have a compact circular cluster, particles have to interact among themselves continuously through the entire circular cluster at an angular speed large enough so that the separation of particles can be balanced by their aggregation, which means that the balance of gravity, angular speed, fluid flow inertia, and the number of particles are important on the formation of circular clusters.

Acknowledgments

The authors acknowledge the valuable comments and suggestions of Roland Glowinski, James J. Feng, Howard H. Hu, and Penger Tong. T.W. Pan acknowledges the support of NSF (grant DMS–0914788).

References

1. Bertrand, T., Tanguy, P.A., Thibault, F.: A three-dimensional fictitious domain method for incompressible fluid flow problems. Int. J. Num. Meth. Fluids **25**, 719–736 (1997).
2. Breu, A.P.J., Kruelle, C.A., Rehberg, I.: Pattern formation in a rotating aqueous suspension. Europhys. Lett. **62**, 491–497 (2003).
3. Chorin, A.J., Hughes, T.J.R., Marsden, J.E., McCracken, M.: Product Formulas and Numerical Algorithms. Comm. Pure Appl. Math. **31**, 205–256 (1978).
4. Dean, E.J., Glowinski, R.: A wave equation approach to the numerical solution of the Navier-Stokes equations for incompressible viscous flow. C.R. Acad. Sc. Paris, Série 1, **325**, 783–791 (1997).
5. Feng, J., Hu, H.H., Joseph, D.D.: Direct simulation of initial value problems for the motion of solid bodies in a Newtonian fluid. 1. Sedimentation, J. Fluid Mech. **261**, 95–134 (1994).
6. Glowinski, R.: Finite element methods for incompressible viscous flow (Handbook of Numerical Analysis, Vol. IX (Ciarlet PG, Lions JL, editors) North–Holland, Amsterdam, 2003, 3–1176).
7. Glowinski, R., Pan, T.–W., Hesla, T., Joseph, D.D.: A distributed Lagrange multiplier/fictitious domain method for flows around moving rigid bodies: Application to particulate flows. Int. J. Multiphase Flow **25**, 755–794 (1999).
8. Glowinski, R., Pan, T.–W., Hesla, T., Joseph, D.D., Periaux, J.: A fictitious domain approach to the direct numerical simulation of incompressible viscous flow past moving rigid bodies: Application to particulate flow. J. Comput. Phys. **169**, 363–426 (2001).
9. Glowinski., R., Pan, T.–W., Périaux, J.: A fictitious domain method for Dirichlet problem and applications. Comp. Meth. Appl. Mech. Eng., **111**, 283–303 (1994).

10. Glowinski., R., Pan, T.–W., Périaux, J.: A fictitious domain method for external incompressible viscous flow modeled by Navier-Stokes equations. Comp. Meth. Appl. Mech. Eng., **112**, 133–148 (1994).
11. Joseph, D.D.: Interrogations of Direct Numerical Simulation of Solid–Liquid Flows, www.efluids.com (2002).
12. Joseph, D.D., Wang, J., Bai, R., Yang, B.H., Hu, H.H.: Particle motion in a liquid film rimming the inside of a partially filled rotating cylinder. J. Fluid Mech. **496**, 139–163 (2003).
13. Lee, J.H., Ladd, A.J.C.: Axial segregation of a settling suspension in a rotating cylinder. Phys. Rev. Lett. **95**, 048001 (2005).
14. Lee, J.H., Ladd, A.J.C.: Particle dymanics and pattern formation in a rotating suspension. J. Fluid Mech. **577**, 183–209 (2007).
15. Lipson, S.G.: Periodic banding in crystallization from rotating supersaturated solutions. J. Phys.: Condens. Matters **13**, 5001–5008 (2001).
16. Lipson, S.G., Seiden, G.: Particles banding in rotating fluids: a new pattern–forming system. Physica A **314**, 272–277 (2002).
17. Matson, W.R., Ackerson, B.J., Tong, P.: Pattern formation in a rotating suspension of non-Brownian settling particles. Phys. Rev. E **67**, 050301 (2003).
18. Matson, W.R., Ackerson, B.J., Tong, P.: Measured scaling properties of the transition boundaries in a rotating suspension of non-Brownian settling particles. J. Fluid Mech. **597**, 233–259 (2008).
19. Matson, W.R., Kalyankar, M., Ackerson, B.J., Tong, P.: Concentration and velocity patterns in a horizontal rotating suspension of non-Brownian settling particles. Phys. Rev. E **71**, 031401 (2005).
20. Pan, T.–W., Glowinski, R., Hou, S.: Direct numerical simulation of pattern formation in a rotating suspension of non-Brownian settling particles in a fully filled cylinder. Computers and Structures **85**, 955–969 (2007).
21. Pan, T.–W., Joseph, D.D., Bai, R., Glowinski, R., Sarin, V.: Fluidization of 1204 spheres: simulation and experiment. J. Fluid Mech. **451**, 169–191 (2002).
22. Seiden, G., Thomas, P.J.: Complexity, segretation, and pattern formation in rotating–drum flows. Rev. Mod. Phys. **83**, 1323–1365 (2011).
23. Seiden, G., Ungarish, M., Lipson, S.G.: Banding of suspended particles in a rotating fluid–filled horizontal cylinder. Phys. Rev. E **72**, 021407 (2005).
24. Seiden, G., Ungarish, M., Lipson, S.G.: Formation and stability of band patterns in a rotating suspension–filled cylinder. Phys. Rev. E **76**, 026221 (2007).
25. Tirumkudulu, M., Tripathi, A., Acrivos, A.: Particle segregation in monodisperse sheared suspensions. Phys. Fluids **11**, 507–509 (1999).
26. Tirumkudulu, M., A. Mileo, A., Acrivos, A.: Particle segregation in monodisperse sheared suspensions in a partially filled rotating horizontal cylinder. Phys. Fluids **12**, 1615–1618 (2000).
27. Yang, B.H., Wang, J., Joseph, D.D., Hu, H.H., T.–W. Pan, Glowinski, R.: Numerical Study of Particle Migration in Tube and Plane Poiseuille Flows. J. Fluid Mech. **540**, 109–131 (2005).

Index

R. Glowinski et al. (eds.), *Splitting Methods in Communication, Imaging, Science, and Engineering*, Scientific Computation, DOI 10.1007/978-3-319-41589-5

G

H

I

MIX
Papier aus verantwortungsvollen Quellen
Paper from responsible sources
FSC® C105338

If you have any concerns about our products,
you can contact us on
ProductSafety@springernature.com

In case Publisher is established outside the EU,
the EU authorized representative is:
Springer Nature Customer Service Center GmbH
Europaplatz 3, 69115 Heidelberg, Germany

Printed by Libri Plureos GmbH
in Hamburg, Germany